T0189708

# Lecture Notes in Computer Science 10031

*Commenced Publication in 1973*
Founding and Former Series Editors:
Gerhard Goos, Juris Hartmanis, and Jan van Leeuwen

More information about this series at http://www.springer.com/series/7410

Jung Hee Cheon · Tsuyoshi Takagi (Eds.)

# Advances in Cryptology – ASIACRYPT 2016

22nd International Conference on the Theory
and Application of Cryptology and Information Security
Hanoi, Vietnam, December 4–8, 2016
Proceedings, Part I

 Springer

*Editors*
Jung Hee Cheon
Seoul National University
Seoul
Korea (Republic of)

Tsuyoshi Takagi
Kyushu University
Fukuoka
Japan

ISSN 0302-9743            ISSN 1611-3349   (electronic)
Lecture Notes in Computer Science
ISBN 978-3-662-53886-9       ISBN 978-3-662-53887-6   (eBook)
DOI 10.1007/978-3-662-53887-6

Library of Congress Control Number: 2016956613

LNCS Sublibrary: SL4 – Security and Cryptology

Printed on acid-free paper

This Springer imprint is published by Springer Nature
The registered company is Springer-Verlag GmbH Germany
The registered company address is: Heidelberger Platz 3, 14197 Berlin, Germany

# Preface

ASIACRYPT 2016, the 22nd Annual International Conference on Theory and Application of Cryptology and Information Security, was held at InterContinental Hanoi Westlake Hotel in Hanoi, Vietnam, during December 4–8, 2016. The conference focused on all technical aspects of cryptology, and was sponsored by the International Association for Cryptologic Research (IACR).

Asiacrypt 2016 received a total of 240 submissions from all over the world. The Program Committee selected 67 papers from these submissions for publication in the proceedings of this conference. The review process was made via the usual double-blind pier review by the Program Committee comprising 43 leading experts in the field. Each submission was reviewed by at least three reviewers and five reviewers were assigned to submissions co-authored by Program Committee members. This year, the conference operated a two-round review system with a rebuttal phase. In the first-round review the Program Committee selected the 140 submissions that were considered of value for proceeding to the second round. In the second-round review the Program Committee further reviewed the submissions by taking into account their rebuttal letter from the authors. The selection process was assisted by a total of 309 external reviewers. These two-volume proceedings contain the revised versions of the papers that were selected. The revised versions were not reviewed again and the authors are responsible for their contents.

The program of Asiacrypt 2016 featured three excellent invited talks. Nadia Heninger gave a talk on "The Reality of Cryptographic Deployments on the Internet," Hoeteck Wee spoke on "Advances in Functional Encryption," and Neal Koblitz gave a non-technical lecture on "Cryptography in Vietnam in the French and American Wars." The conference also featured a traditional rump session that contained short presentations on the latest research results of the field. The Program Committee selected the work "Faster Fully Homomorphic Encryption: Bootstrapping in Less Than 0.1 Seconds" by Ilaria Chillotti, Nicolas Gama, Mariya Georgieva, and Malika Izabachène for the Best Paper Award of Asiacrypt 2016. Two more papers, "Nonlinear Invariant Attack—Practical Attack on Full SCREAM, iSCREAM, and Midori64" by Yosuke Todo, Gregor Leander, Yu Sasaki and "Cliptography: Clipping the Power of Kleptographic Attacks" by Alexander Russell, Qiang Tang, Moti Yung, Hong-Sheng Zhou were solicited to submit full versions to the *Journal of Cryptology*.

Many people contributed to the success of Asiacrypt 2016. We would like to thank the authors for submitting their research results to the conference. We are very grateful to all of the Program Committee members as well as the external reviewers for their fruitful comments and discussions on their areas of expertise. We are greatly indebted to Ngo Bao Chau and Phan Duong Hieu, the general co-chairs for their efforts and overall organization. We would also like to thank Nguyen Huu Du, Nguyen Quoc Khanh, Nguyen Duy Lan, Duong Ngoc Thai, Nguyen Ta Toan Khoa, Nguyen Ngoc Tuan,

Le Thi Lan Anh, and the local Organizing Committee for their continuous supports. We thank Steven Galbraith for expertly organizing and chairing the rump session.

Finally we thank Shai Halevi for letting us use his nice software for supporting the paper submission and review process. We also thank Alfred Hofmann, Anna Kramer, and their colleagues at Springer for handling the editorial process of the proceedings. We would like to express our gratitude to our partners and sponsors: XLIM, Microsoft Research, CISCO, Intel, Google.

December 2016                                               Jung Hee Cheon
                                                          Tsuyoshi Takagi

# ASIACRYPT 2016

## The 22nd Annual International Conference on Theory and Application of Cryptology and Information Security

Sponsored by the International Association for Cryptologic Research (IACR)

December 4–8, 2016, Hanoi, Vietnam

## General Co-chairs

| | |
|---|---|
| Ngo Bao Chau | VIASM, Vietnam and University of Chicago, USA |
| Phan Duong Hieu | XLIM, University of Limoges, France |

## Program Co-chairs

| | |
|---|---|
| Jung Hee Cheon | Seoul National University, Korea |
| Tsuyoshi Takagi | Kyushu University, Japan |

## Program Committee

| | |
|---|---|
| Elena Andreeva | KU Leuven, Belgium |
| Xavier Boyen | Queensland University of Technology, Australia |
| Anne Canteaut | Inria, France |
| Chen-Mou Cheng | National Taiwan University, Taiwan |
| Sherman S.M. Chow | Chinese University of Hong Kong, Hong Kong, SAR China |
| Nico Döttling | University of California, Berkeley, USA |
| Thomas Eisenbarth | Worcester Polytechnic Institute, USA |
| Georg Fuchsbauer | École Normale Supérieure, France |
| Steven Galbraith | Auckland University, New Zealand |
| Sanjam Garg | University of California, Berkeley, USA |
| Vipul Goyal | Microsoft Research, India |
| Jens Groth | University College London, UK |
| Sylvain Guilley | Secure-IC S.A.S., France |
| Alejandro Hevia | Universidad de Chile, Chile |
| Antoine Joux | Foundation UPMC and LIP6, France |
| Xuejia Lai | Shanghai Jiaotong University, China |
| Hyung Tae Lee | Nanyang Technological University, Singapore |
| Kwangsu Lee | Sejong University, Korea |
| Dongdai Lin | Chinese Academy of Sciences, China |
| Feng-Hao Liu | Florida Atlantic University, USA |
| Takahiro Matsuda | AIST, Japan |
| Alexander May | Ruhr University Bochum, Germany |

| | |
|---|---|
| Florian Mendel | Graz University of Technology, Austria |
| Amir Moradi | Ruhr University Bochum, Germany |
| Svetla Nikova | KU Leuven, Belgium |
| Tatsuaki Okamoto | NTT, Japan |
| Elisabeth Oswald | University of Bristol, UK |
| Thomas Peyrin | Nanyang Technological University, Singapore |
| Rei Safavi-Naini | University of Calgary, Canada |
| Peter Schwabe | Radboud University, The Netherlands |
| Jae Hong Seo | Myongji University, Korea |
| Damien Stehlé | ENS de Lyon, France |
| Ron Steinfeld | Monash University, Australia |
| Rainer Steinwandt | Florida Atlantic University, USA |
| Daisuke Suzuki | Mitsubishi Electric, Japan |
| Mehdi Tibouchi | NTT, Japan |
| Yosuke Todo | NTT, Japan |
| Hoang Viet Tung | University of California Santa Barbara, USA |
| Dominique Unruh | University of Tartu, Estonia |
| Ivan Visconti | University of Salerno, Italy |
| Huaxiong Wang | Nanyang Technological University, Singapore |
| Meiqin Wang | Shandong University, China |
| Aaram Yun | UNIST, Korea |

## External Reviewers

Dung Hoang Duong
Maria Eichlseder
Martianus Frederic
   Ezerman
Xiong Fan
Pooya Farshim
Serge Fehr
Max Fillinger
Dario Fiore
Victor Fischer
Marc Fischlin
Thomas Fuhr
Jake Longo Galea
David Galindo
Peter Gazi
Essam Ghadafi
Mohona Ghosh
Zheng Gong
Rishab Goyal
Hannes Gross
Vincent Grosso
Berk Gulmezoglu
Chun Guo
Jian Guo
Qian Guo
Divya Gupta
Iftach Haitner
Dong-Guk Han
Kyoohyung Han
Shuai Han
Goichiro Hanaoka
Christian Hanser
Mitsuhiro Hattori
Gottfried Herold
Felix Heuer
Takato Hirano
Shoichi Hirose
Wei-Chih Hong
Yuan-Che Hsu
Geshi Huang
Guifang Huang
Jialin Huang
Xinyi Huang
Pavel Hubacek
Ilia Iliashenko
Mehmet Sinan Inci

Vincenzo Iovino
Gorka Irazoqui
Ai Ishida
Takanori Isobe
Tetsu Iwata
Aayush Jain
Sune Jakobsen
Yin Jia
Shaoquan Jiang
Chethan Kamath
Sabyasachi Karati
Sayasachi Karati
Yutaka Kawai
Carmen Kempka
HeeSeok Kim
Hyoseung Kim
Jinsu Kim
Myungsun Kim
Taechan Kim
Paul Kirchner
Elena Kirshanova
Fuyuki Kitagawa
Susumu Kiyoshima
Jessica Koch
Markulf Kohlweiss
Vladimir Kolesnikov
Thomas Korak
Yoshihiro Koseki
Ashutosh Kumar
Ranjit Kumaresan
Po-Chun Kuo
Robert Kübler
Thijs Laarhoven
Ching-Yi Lai
Russell W.F. Lai
Virginie Lallemand
Adeline Langlois
Sebastian Lauer
Su Le
Gregor Leander
Kwangsu Lee
Gaëtan Leurent
Anthony Leverrier
Jingwei Li
Ming Li
Wen-Ding Li

Benoit Libert
Fuchun Lin
Tingting Lin
Meicheng Liu
Yunwen Liu
Zhen Liu
Zidong Lu
Yiyuan Luo
Atul Luykx
Vadim Lyubashevsky
Bernardo Magri
Mary Maller
Alex Malozemoff
Antonio Marcedone
Benjamin Martin
Daniel Martin
Marco Martinoli
Daniel Masny
Maike Massierer
Mitsuru Matsui
Willi Meier
Bart Mennink
Peihan Miao
Kazuhiko Minematsu
Nicky Mouha
Pratyay Mukherjee
Sean Murphy
Jörn Müller-Quade
Valérie Nachef
Michael Naehrig
Matthias Nagel
Yusuke Naito
Mridul Nandi
María Naya-Plasencia
Kartik Nayak
Khoa Nguyen
Ivica Nikolic
Ventzislav Nikov
Ryo Nishimaki
Anca Nitulescu
Koji Nuida
Maciej Obremski
Toshihiro Ohigashi
Miyako Ohkubo
Sumit Kumar Pandey
Jong Hwan Park

Seunghwan Park
Alain Passelègue
Christopher Patton
Bo-Yuan Peng
Rachel Player
Antigoni Polychroniadou
Bertram Pöttering
Sebastian Ramacher
Vanishree Rao
Shuqin Ren
Reza Reyhanitabar
Bastian Richter
Thomas Ristenpart
Mike Rosulek
Hansol Ryu
Akshayaram Srinivasan
Yusuke Sakai
Kochi Sakumoto
Amin Sakzad
Simona Samardjiska
Yu Sasaki
Pascal Sasdrich
Falk Schellenberg
Benedikt Schmidt
Tobias Schneider
Jacob Schuldt
Okan Seker
Nicolas Sendrier
Jae Hong Seo
Minhye Seo
Yannick Seurin
Masoumeh Shafienejad
Barak Shani
Danilo Sijacic
Alice Silverberg
Siang Meng Sim
Dave Singelee

Luisa Siniscalchi
Daniel Slamanig
Nigel Smart
Raphael Spreitzer
Douglas Stebila
Christoph Striecks
Takeshi Sugawara
Yao Sun
Berk Sunar
Koutarou Suzuki
Alan Szepieniec
Mostafa Taha
Somayeh Taheri
Junko Takahashi
Katsuyuki Takashima
Benjamin Tan
Jean-Pierre Tillich
Junichi Tomida
Yiannis Tselekounis
Himanshu Tyagi
Thomas Unterluggauer
Damien Vergnaud
Gilles Villard
Vanessa Vitse
Damian Vizar
Michael Walter
Han Wang
Hao Wang
Qiungju Wang
Wei Wang
Yuyu Wang
Yohei Watanabe
Hoeteck Wee
Wei Wei
Mor Weiss
Mario Werner
Bas Westerbaan

Carolyn Whitnall
Alexander Wild
Baofeng Wu
Keita Xagawa
Zejun Xiang
Hong Xu
Weijia Xue
Shota Yamada
Takashi Yamakawa
Hailun Yan
Jun Yan
Bo-Yin Yang
Bohan Yang
Guomin Yang
Mohan Yang
Shang-Yi Yang
Kan Yasuda
Xin Ye
Wentan Yi
Scott Yilek
Kazuki Yoneyama
Rina Zeitoun
Fan Zhang
Guoyan Zhang
Liang Feng Zhang
Liangfeng Zhang
Tao Zhang
Wentao Zhang
Yusi Zhang
Zongyang Zhang
Jingyuan Zhao
Yongjun Zhao
Yixin Zhong
Hong-Sheng Zhou
Xiao Zhou
Jincheng Zhuang

## Local Organizing Committee

### Co-chairs

Ngo Bao Chau        VIASM, Vietnam and University of Chicago, USA
Phan Duong Hieu      XLIM, University of Limoges, France

**Members**

| | |
|---|---|
| Nguyen Huu Du | VIASM, Vietnam |
| Nguyen Quoc Khanh | Vietcombank, Vietnam |
| Nguyen Duy Lan | Microsoft Research, USA |
| Duong Ngoc Thai | Google, USA |
| Nguyen Ta Toan Khoa | NTU, Singapore |
| Nguyen Ngoc Tuan | VIASM, Vietnam |
| Le Thi Lan Anh | VIASM, Vietnam |

# Sponsors

XLIM
Microsoft Research
CISCO
Intel
Google

# Invited Talks

# Advances in Functional Encryption

Hoeteck Wee

ENS, Paris, France
wee@di.ens.fr

**Abstract.** Functional encryption is a novel paradigm for public-key encryption that enables both fine-grained access control and selective computation on encrypted data, as is necessary to protect big, complex data in the cloud. In this talk, I will provide a brief introduction to functional encryption and an overview of the state of the art, with a focus on constructions based on lattices.

CNRS, INRIA and Columbia University. Supported in part by ERC Project aSCEND (H2020 639554) and NSF Award CNS-1445424.

# The Reality of Cryptographic Deployments on the Internet

Nadia Heninger

University of Pennsylvania, Philadelphia, USA

**Abstract.** Security proofs for cryptographic primitives and protocols rely on a number of (often implicit) assumptions about the world in which these components live. They assume that implementations are correct, that specifications are followed, that systems make sensible choices about error conditions, and that reliable sources of random numbers are present. However, a number of real world studies examining cryptographic deployments have shown that these assumptions are often not true on a large scale, with catastrophic effects for security. In addition to simple programming errors, many real-world cryptographic vulnerabilities can be traced back to more complex underlying causes, such as backwards compatibility, legacy protocols and software, hard-coded resource limits, and political interference in design choices.

Many of these issues appear on the surface to be at an entirely different level of abstraction from the cryptographic primitives used in their construction. However, by taking advantage of the structure of many cryptographic primitives when used at Internet scale, it is possible to uncover fundamental vulnerabilities in implementations. I will discuss the interplay between mathematical crypt-analysis techniques and the thorny implementation issues that lead to vulnerable cryptographic deployments in the real world.

# Contents – Part I

**Block Cipher II**

**Mathematical Analysis II**

**SCA and Leakage Resilience II**

# Contents – Part II

## Digital Signature

## Functional and Homomorphic Cryptography

## ABE and IBE

**Foundation**

**Cryptographic Protocol**

**Multi-party Computation**

# Asiacrypt 2016 Best Paper

# Faster Fully Homomorphic Encryption: Bootstrapping in Less Than 0.1 Seconds

Ilaria Chillotti[1]([✉]), Nicolas Gama[2,1], Mariya Georgieva[3]([✉]),
and Malika Izabachène[4]([✉])

[1] Laboratoire de Mathématiques de Versailles, UVSQ, CNRS,
Université Paris-Saclay, 78035 Versailles, France
ilaria.chillotti@uvsq.fr
[2] Inpher, Lausanne, Switzerland
nicolas.gama@gmail.com
[3] Gemalto, 6 rue de la Verrerie, 92190 Meudon, France
mariya.georgieva@gemalto.com
[4] CEA LIST, Point Courrier 172, 91191 Gif-sur-Yvette Cedex, France
malika.izabachene@cea.fr

**Abstract.** In this paper, we revisit fully homomorphic encryption (FHE) based on GSW and its ring variants. We notice that the internal product of GSW can be replaced by a simpler external product between a GSW and an LWE ciphertext.

We show that the bootstrapping scheme FHEW of Ducas and Micciancio [11] can be expressed only in terms of this external product. As a result, we obtain a speed up from less than 1 s to less than 0.1 s. We also reduce the 1 GB bootstrapping key size to 24 MB, preserving the same security levels, and we improve the noise propagation overhead by replacing exact decomposition algorithms with approximate ones.

Moreover, our external product allows to explain the unique asymmetry in the noise propagation of GSW samples and makes it possible to evaluate deterministic automata homomorphically as in [13] in an efficient way with a noise overhead only linear in the length of the tested word.

Finally, we provide an alternative practical analysis of LWE based scheme, which directly relates the security parameter to the error rate of LWE and the entropy of the LWE secret key.

**Keywords:** Fully homomorphic encryption · Bootstrapping · Lattices · LWE · GSW

## 1 Introduction

Fully homomorphic encryption (FHE) allows to perform computations over encrypted data without decrypting them. This concept has long been regarded as an open problem until the breakthrough paper of Gentry in 2009 [15] which demonstrates the feasibility of computing any function on encrypted data. Since

© International Association for Cryptologic Research 2016
J.H. Cheon and T. Takagi (Eds.): ASIACRYPT 2016, Part I, LNCS 10031, pp. 3–33, 2016.
DOI: 10.1007/978-3-662-53887-6_1

then, many constructions have appeared involving new mathematical and algorithmic concepts and improving efficiency.

In homomorphic encryption, messages are encrypted with a noise that grows at each homomorphic evaluation of an elementary operation. In a somewhat encryption scheme, the number of homomorphic operations is limited, but can be made asymptotically large using bootstrapping [15]. This technical trick introduced by Gentry allows to evaluate arbitrary circuits by essentially evaluating the decryption function on encrypted secret keys. This step has remained very costly until the recent paper of Ducas and Micciancio [11], which presented a very fast bootstrapping procedure running in around 0.69 s, making an important step towards practical FHE for arbitrary NAND circuits. In this paper, we further improve the bootstrapping procedure.

We first provide an intuitive formalization of LWE/RingLWE on numbers or polynomials over the real torus, obtained by combining the Scale-Invariant-LWE problem of [9] or the LWE normal form of [10] with the General-LWE problem of Brakerski-Gentry-Vaikutanathan [5]. We call TLWE this unified representation of LWE ciphertexts, which encode polynomials over the Torus. Its security relies either on the hardness of general or ideal lattice reduction, depending on the choice of dimensions. Using the same formalism, we extend the GSW/RingGSW ciphertexts to TGSW, which is the combined analogue of Gentry-Sahai-Water's ciphertexts from [3,16], and which can also instantiate the ring version used in Ducas-Micciancio scheme [11] in the FHEW cryptosystem. Similarly, a TGSW ciphertext encodes an integer polynomial message, and depending on the choice of dimensions, its security is also based on (worst-case) generic or ideal lattice reduction algorithms. TLWE and TGSW are basically dual to each other, and the main idea of our efficiency result comes from the fact that these two schemes can directly be combined together to map the external product of their two messages into a TLWE sample. Since a TGSW sample is essentially a matrix whose individual rows are TLWE samples, our external product TGSW times TLWE is much quicker than the usual internal product TGSW times TGSW used in previous work. This could mostly be understood as comparing the speed of the computation of a matrix-vector product to a matrix-matrix product. As a result, we obtain a significant improvement (12 times faster) of the most efficient bootstrapping procedure [11]; it now runs in less than 0.052 s.

We also analyze the case of leveled encryption. Using an external product means that we lose some composability properties in the design of homomorphic circuits. This corresponds to circuits where boolean gates have different kinds of wires that cannot be freely interconnected. Still, we show that we maintain the expressiveness of the whole binary decision diagram and automata-based logic, which was introduced in [13] with the GSW-GSW internal product, and we tighten the analysis. Indeed, while it was impractical (10 transitions per second in the ring case, and impractical in the non-ring case), we show that the TGSW-TLWE external product enables to evaluate up to 5000 transitions per second, in a leveled homomorphic manner. We also refine the mapping between automata and homomorphic gates, and reduce the number of homomorphic operations to test a word with a deterministic automata. This allows to compile and evaluate constant-time algorithms (i.e. with data-independent control flow) in a leveled homomorphic manner, with only sub-linear noise overhead in the running time.

We also propose a new security analysis where the security parameter is directly expressed as a function of the entropy of the secret and the error rate. For the parameters that we propose in our implementation, we predict 188-bits of security for both the bootstrapping key and the keyswitching key.

**Roadmap.** In Sect. 2, we give mathematical definitions and a quick overview of the classical version of LWE-based schemes. In Sect. 3, we generalize LWE and GSW schemes using a *torus representation* of the samples. We also review the arithmetic operations over the torus and introduce our main theorem characterizing the new morphism between TLWE and TGSW. As a proof of concept, we present two main applications in Sect. 4 where we explain our fast bootstrapping procedure, and in Sect. 5, we present efficient leveled evaluation of deterministic automata, and apply it on a constant-time algorithm with logarithmic memory. Finally, we provide a practical security analysis in Sect. 6.

## 2  Background

**Notation.** In the rest of the paper we will use the following notations. The security parameter will be denoted as $\lambda$. The set $\{0, 1\}$ (without any structure) will be written $\mathbb{B}$. The real Torus $\mathbb{R}/\mathbb{Z}$, called $\mathbb{T}$ set of real numbers modulo 1. $\mathfrak{R}$ denotes the ring of polynomials $\mathbb{Z}[X]/(X^N + 1)$. $\mathbb{T}_N[X]$ denotes $\mathbb{R}[X]/(X^N + 1)$ mod 1. Finally, we note by $\mathcal{M}_{p,q}(E)$ the set of matrices $p \times q$ with entries in $E$.

This section combines some algebra theory, namely abelian groups, commutative rings, $R$-modules, and on some metrics of the continuous field $\mathbb{R}$.

**Definition 2.1 ($R$-module).** *Let $(R, +, \times)$ be a commutative ring. We say that a set $M$ is a $R$-module when $(M, +)$ is an abelian group, and when there exists an external operation $\cdot$ which is bi-distributive and homogeneous. Namely, $\forall r, s \in R$ and $x, y \in M$, $1_R \cdot x = x$, $(r + s) \cdot x = r \cdot x + s \cdot x$, $r \cdot (x + y) = r \cdot x + r \cdot y$, and $(r \times s) \cdot x = r \cdot (s \cdot x)$.*

Any abelian group is by construction a $\mathbb{Z}$-module for the iteration (or exponentiation) of its own law. In this paper, one of the most important abelian group we use is the real torus $\mathbb{T}$, composed of all reals modulo 1 ($\mathbb{R}$ mod 1). The torus is not a ring, since the real internal product is not compatible with the modulo 1 projection (expressions like $0 \times \frac{1}{2}$ are undefined). But as an additive group, it is a $\mathbb{Z}$-module, and the external product $\cdot$ from $\mathbb{Z} \times \mathbb{T}$ to $\mathbb{T}$, like in $0 \cdot \frac{1}{2} = 0$, is well defined. More importantly, we recall that for all positive integers $N$ and $k$, $(\mathbb{T}_N[X]^k, +, \cdot)$ is a $\mathfrak{R}$-module.

A $R$-module $M$ shares many arithmetic operations and constructions with vector spaces: vectors $M^n$ or matrices $\mathcal{M}_{n,m}(M)$ are also $R$-modules, and their left dot product with a vector in $R^n$ or left matrix product in $\mathcal{M}_{k,n}(R)$ are both well defined.

**Gaussian Distributions.** Let $\sigma \in \mathbb{R}^+$ be a parameter and $k \geq 1$ the dimension. For all $\boldsymbol{x}, \boldsymbol{c} \in \mathbb{R}^k$, we note $\rho_{\sigma,\boldsymbol{c}}(\boldsymbol{x}) = \exp(-\pi \|\boldsymbol{x} - \boldsymbol{c}\|^2 / \sigma^2)$. If $\boldsymbol{c}$ is omitted, then it is implicitly 0. Let $S$ be a subset of $\mathbb{R}^k$, $\rho_{\sigma,\boldsymbol{c}}(S)$ denotes $\sum_{x \in S} \rho_{\sigma,\boldsymbol{c}}(\boldsymbol{x})$ or $\int_{x \in S} \rho_{\sigma,\boldsymbol{c}}(\boldsymbol{x}).d\boldsymbol{x}$. For all closed (continuous or discrete) additive subgroup $M \subseteq \mathbb{R}^k$, then $\rho_{\sigma,\boldsymbol{c}}(M)$ is finite, and defines a *(restricted) Gaussian Distribution* of parameter $\sigma$, standard deviation $\sqrt{2/\pi}\sigma$ and center $\boldsymbol{c}$ over $M$, with the density function $\mathcal{D}_{M,\sigma,\boldsymbol{c}}(\boldsymbol{x}) = \rho_{\sigma,\boldsymbol{c}}(\boldsymbol{x})/\rho_{\sigma,\boldsymbol{c}}(M)$. Let $L$ be a discrete subgroup of $M$, then the *Modular Gaussian distribution* over $M/L$ exists and is defined by the density $\mathcal{D}_{M/L,\sigma,\boldsymbol{c}}(\boldsymbol{x}) = \mathcal{D}_{M,\sigma,\boldsymbol{c}}(\boldsymbol{x} + L)$. Furthermore, when $\mathrm{span}(M) = \mathrm{span}(L)$, then $M/L$ admits a uniform distribution of constant density $\mathcal{U}_{M/L}$. In this case, the *smoothing parameter* $\eta_{M,\varepsilon}(L)$ of $L$ in $M$ is defined as the smallest $\sigma \in \mathbb{R}$ such that $\sup_{x \in M} |\mathcal{D}_{M/L,\sigma,\boldsymbol{c}}(\boldsymbol{x}) - \mathcal{U}_{M/L}| \leq \varepsilon \cdot \mathcal{U}_{M/L}$. If $M$ is omitted, it implicitly means $\mathbb{R}^k$.

**Subgaussian Distributions.** A distribution $X$ over $\mathbb{R}$ is $\sigma$-subgaussian iff it satisfies the Laplace-transformation bound: $\forall t \in \mathbb{R}, \mathbb{E}(\exp(tX)) \leq \exp(\sigma^2 t^2/2)$. By Markov's inequality, this implies that the tails of $X$ are bounded by the Gaussian function of standard deviation $\sigma$: $\forall x > 0, \mathbb{P}(|X| \geq x) \leq 2\exp(-x^2/2\sigma^2)$. As an example, the Gaussian distribution of standard deviation $\sigma$ (i.e. parameter $\sqrt{\pi/2}\sigma$), the equi-distribution on $\{-\sigma, \sigma\}$, and the uniform distribution over $[-\sqrt{3}\sigma, \sqrt{3}\sigma]$, which all have standard deviation $\sigma$, are $\sigma$-subgaussian[1]. If $X$ and $X'$ are two independent $\sigma$ and $\sigma'$-subgaussian variables, then for all $\alpha, \beta \in \mathbb{R}$, $\alpha X + \beta X'$ is $\sqrt{\alpha^2\sigma^2 + \beta^2\sigma'^2}$-subgaussian.

**Distance and Norms.** We use the standard $\|\cdot\|_p$ and $\|\cdot\|_\infty$ norms for scalars and vectors over the real field or over the integers. By extension, the norm $\|P(X)\|_p$ of a real or integer polynomial $P \in \mathbb{R}[X]$ is the norm of its coefficient vector. If the polynomial is modulo $X^N + 1$, we take the norm of its unique representative of degree $\leq N - 1$.

By abuse of notation, we write $\|\boldsymbol{x}\|_p = \min_{u \in \boldsymbol{x} + \mathbb{Z}^k}(\|u\|_p)$ for all $\boldsymbol{x} \in \mathbb{T}^k$. It is the $p$-norm of the representative of $\boldsymbol{x}$ with all coefficients in $]-\frac{1}{2}, \frac{1}{2}]$. Although it satisfies the separation and the triangular inequalities, this notation is not a norm, because it lacks homogeneity[2], and $\mathbb{T}^k$ is not a vector space either. But we have $\forall m \in \mathbb{Z}, \|m \cdot \boldsymbol{x}\|_p \leq |m| \|\boldsymbol{x}\|_p$. By extension, we define $\|a\|_p$ for a polynomial $a \in \mathbb{T}_N[X]$ as the $p$- norm of its unique representative in $\mathbb{R}[X]$ of degree $\leq N - 1$ and with coefficients in $]-\frac{1}{2}, \frac{1}{2}]$.

---

[1] For the first two distributions, it is tight, but the uniform distribution over $[-\sqrt{3}\sigma, \sqrt{3}\sigma]$ is even $0.78\sigma$-subgaussian.

[2] Mathematically speaking, a more accurate notion would be $\mathrm{dist}_p(\boldsymbol{x}, \boldsymbol{y}) = \|\boldsymbol{x} - \boldsymbol{y}\|_p$, which is a distance. However, the norm symbol is clearer for almost all practical purposes.

**Definition 2.2 (Infinity norm over $\mathcal{M}_{p,q}(\mathbb{T}_N[X])$).** *Let $A \in \mathcal{M}_{p,q}(\mathbb{T}_N[X])$. We define the infinity norm of $A$ as*

$$\|A\|_\infty = \max_{\substack{i \in [\![1,p]\!] \\ j \in [\![1,q]\!]}} \|a_{i,j}\|_\infty .$$

**Concentrated Distribution on the Torus, Expectation and Variance**
A distribution $\mathcal{X}$ on the torus is *concentrated* iff. its support is included in a ball of radius $\frac{1}{4}$ of $\mathbb{T}$, except for negligible probability. In this case, we define the *variance* $\mathsf{Var}(\mathcal{X})$ and the expectation $\mathbb{E}(\mathcal{X})$ of $\mathcal{X}$ as respectively $\mathsf{Var}(\mathcal{X}) = \min_{\bar{x} \in \mathbb{T}} \sum p(x)|x - \bar{x}|^2$ and $\mathbb{E}(\mathcal{X})$ as the position $\bar{x} \in \mathbb{T}$ which minimizes this expression. By extension, we say that a distribution $\mathcal{X}'$ over $\mathbb{T}^n$ or $\mathbb{T}_N[X]^k$ is concentrated iff. each coefficient has an independent concentrated distribution on the torus. Then the expectation $\mathbb{E}(\mathcal{X}')$ is the vector of expectations of each coefficient, and $\mathsf{Var}(\mathcal{X}')$ denotes the maximum of each coefficient's Variance.

These expectation and variance over $\mathbb{T}$ follow the same linearity rules than their classical equivalent over the reals.

**Fact 2.3.** *Let $\mathcal{X}_1, \mathcal{X}_2$ be two independent concentrated distributions on either $\mathbb{T}, \mathbb{T}^n$ or $\mathbb{T}_N[X]^k$, and $e_1, e_2 \in \mathbb{Z}$ such that $\mathcal{X} = e_1 \cdot \mathcal{X}_1 + e_2 \cdot \mathcal{X}_2$ remains concentrated, then $\mathbb{E}(\mathcal{X}) = e_1 \cdot \mathbb{E}(\mathcal{X}_1) + e_2 \cdot \mathbb{E}(\mathcal{X}_2)$ and $\mathsf{Var}(\mathcal{X}) \leq e_1^2 \cdot \mathsf{Var}(\mathcal{X}_1) + e_2^2 \cdot \mathsf{Var}(\mathcal{X}_2)$.*

Also, subgaussian distributions with small enough parameters are necessarily concentrated:

**Fact 2.4.** *Every distribution $\mathcal{X}$ on either $\mathbb{T}, \mathbb{T}^n$ or $\mathbb{T}_N[X]^k$ where each coefficient is $\sigma$-subgaussian where $\sigma \leq 1/\sqrt{32 \log(2)(\lambda + 1)}$ is a concentrated distribution: a fraction $1 - 2^{-\lambda}$ of its mass is in the interval $[-\frac{1}{4}, \frac{1}{4}]$.*

## 2.1 Learning with Error Problem

The Learning With Errors (LWE) problem was introduced by Regev in 2005 [21]. The Ring variant, called RingLWE, was introduced by Lyubashevsky, Peikert and Regev in 2010 [19]. Both variants are nowadays extensively used for the construction of lattice-based Homomorphic Encryption schemes. In the original definition [21], a LWE sample has its right member on the torus and is defined using continuous Gaussian distributions. Here, we will work entirely on the real torus, employing the same formalism as the Scale Invariant LWE (SILWE) scheme in [9], or LWE scale-invariant normal form in [10]. Without loss of generality, we refer to it as LWE.

**Definition 2.5 ((Homogeneous) LWE).** *Let $n \geq 1$ be an integer, $\alpha \in \mathbb{R}^+$ be a noise parameter and $s$ be a uniformly distributed secret in some bounded set $\mathcal{S} \in \mathbb{Z}^n$. Denote by $\mathcal{D}^{\mathsf{LWE}}_{s,\alpha}$ the distribution over $\mathbb{T}^n \times \mathbb{T}$ obtained by sampling a couple $(a, b)$, where the left member $a \in \mathbb{T}^n$ is chosen uniformly random and the right member $b = a \cdot s + e$. The error $e$ is a sample from a gaussian distribution with parameter $\alpha$.*

- *Search problem: given access to polynomially many* LWE *samples, find* $s \in \mathcal{S}$.
- *Decision problem: distinguish between* LWE *samples and uniformly random samples from* $\mathbb{T}^n \times \mathbb{T}$.

Both the LWE search or decision problems are reducible to each other, and their average case is asymptotically as hard as worst-case lattice problems. In practice, both problems are also intractable, and their hardness increases with the the entropy of the key set $\mathcal{S}$ (*i.e.* $n$ if keys are binary) and $\alpha \in ]0, \eta_\varepsilon(\mathbb{Z})[$.

Regev's encryption scheme [21] is the following: Given a discrete message space $\mathcal{M} \in \mathbb{T}$, for instance $\{0, \frac{1}{2}\}$, a message $\mu \in \mathcal{M}$ is encrypted by summing up the *trivial* LWE sample $(\mathbf{0}, \mu)$ of $\mu$ to a Homogeneous LWE sample $(\mathbf{a}, b) \in \mathbb{T}^{n+1}$ with respect to a secret key $s \in \mathbb{B}^n$ and a noise parameter $\alpha \in \mathbb{R}^+$. The semantic security of the scheme is equivalent to the LWE decisional problem. The decryption of a sample $c = (\mathbf{a}, b)$ consists in computing this quantity $\varphi_s(\mathbf{a}, b) = b - s \cdot \mathbf{a}$, which we call the *phase* of $c$, and to round it to the nearest element in $\mathcal{M}$. Decryption is correct with overwhelming probability $1 - 2^{-p}$ provided that the parameter $\alpha$ is $O(R/\sqrt{p})$ where $R$ is the packing radius of $\mathcal{M}$.

## 3 Generalization

In this section we extend this presentation to rings, following the generalization of [5], and also to GSW [16].

### 3.1 TLWE

We first define TLWE samples, together with the search and decision problems. In the following, ciphertexts are viewed as normal samples.

**Definition 3.1 (TLWE samples).** *Let* $k \geq 1$ *be an integer,* $N$ *a power of 2, and* $\alpha \geq 0$ *be a noise parameter. A TLWE secret key* $s \in \mathbb{B}_N[X]^k$ *is a vector of* $k$ *polynomials* $\in \mathfrak{R} = \mathbb{Z}[X]/X^N + 1$ *with binary coefficients. For security purposes, we assume that private keys are uniformly chosen, and that they actually contain* $n \approx Nk$ *bits of entropy. The message space of* TLWE *samples is* $\mathbb{T}_N[X]$. *A fresh* TLWE *sample of a message* $\mu \in \mathbb{T}_N[X]$ *with noise parameter* $\alpha$ *under the key* $s$ *is an element* $(\mathbf{a}, b) \in \mathbb{T}_N[X]^k \times \mathbb{T}_N[X]$, $b \in \mathbb{T}_N[X]$ *has Gaussian distribution* $\mathcal{D}_{\mathbb{T}_N[X],\alpha,s\cdot a+\mu}$ *around* $\mu + s \cdot a$. *The sample is* random *iff its left member* $\mathbf{a}$ *(also called mask) is uniformly random* $\in \mathbb{T}_N[X]^k$ *(or a sufficiently dense submodule[3]),* trivial *if* $\mathbf{a}$ *is fixed to* $\mathbf{0}$, noiseless *if* $\alpha = 0$, *and* homogeneous *iff its message* $\mu$ *is* 0.

---

[3] A submodule $G$ is sufficiently dense if there exists an intermediate submodule $H$ such that $G \subseteq H \subseteq \mathbb{T}^n$, the relative smoothing parameter $\eta_{H,\varepsilon}(G)$ is $\leq \alpha$, and $H$ is the orthogonal in $\mathbb{T}^n$ of at most $n-1$ vectors of $\mathbb{Z}^n$. This definition allows to convert any (Ring)-LWE with non-binary secret to a TLWE instance via binary decomposition.

- *Search problem: given access to polynomially many fresh random homogeneous TLWE samples, find their key $s \in \mathbb{B}_N[X]^k$.*
- *Decision problem: distinguish between fresh random homogeneous TLWE samples from uniformly random samples from $\mathbb{T}_N[X]^{k+1}$.*

This definition is the analogue on the torus of the General-LWE problem of [5]. It allows to consider both LWE and RingLWE as a single problem. Choosing $N$ large and $k = 1$ corresponds to the classical (bin)RingLWE (over cyclotomic rings, and up to a scaling factor $q$). When $N = 1$ and $k$ large, then $\mathfrak{R}$ and $\mathbb{T}_N[X]$ respectively collapses to $\mathbb{Z}$ and $\mathbb{T}$, and TLWE is simply bin-LWE (up to the same scaling factor $q$). Other choices of $N, k$ give some continuum between the two extremes, with a security that varies between worst-case ideal lattices to worst-case regular lattices.

Thanks to the underlying $\mathfrak{R}$-module structure, we can sum TLWE samples, or we can make integer linear or polynomial combinations of samples with coefficients in $\mathfrak{R}$. However, each of these combinations increases the noise inside the samples. They are therefore limited to small coefficients.

We additionally define a function called the phase of a TLWE sample, that will be used many times. The phase computation is the first step of the classical decryption algorithm, and uses the secret key.

**Definition 3.2 (Phase).** *Let $c = (a, b) \in \mathbb{T}_N[X]^k \times \mathbb{T}_N[X]$ and $s \in \mathbb{B}_N[X]^k$, we define the phase of the sample as $\varphi_s(c) = b - s \cdot a$.*

*The phase is linear over $\mathbb{T}_N[X]^{k+1}$ and is $(kN + 1)$-lipschitzian for the $\ell_\infty$ distance: $\forall x, y \in \mathbb{T}_N[X]^{k+1}, \|\varphi_s(x) - \varphi_s(y)\|_\infty \leq (kN + 1) \|x - y\|_\infty$.*

Note that a TLWE sample contains noise, that its semantic is only function of its phase, and that the phase has the nice property to be lipschitzian. Together, these properties have many interesting implications. In particular, we can always work with approximations, since two samples at a short distance on $\mathbb{T}_N[X]^{k+1}$ share the same properties: they encode the same message, and they can in general be swapped. This fact explains why we can work and describe our algorithms on the infinite Torus.

Given a finite message space $\mathcal{M} \subseteq \mathbb{T}_N[X]$, the (classical) decryption algorithm computes the phase $\varphi_s(c)$ of the sample, and returns the closest $\mu \in \mathcal{M}$. It is easy to see that if $c$ is a fresh TLWE sample of $\mu \in \mathcal{M}$ with gaussian noise parameter $\alpha$, the decryption of $c$ over $\mathcal{M}$ is equal to $\mu$ as soon as $\alpha$ is $\Theta(\sqrt{\lambda})$ times smaller than the packing radius of $\mathcal{M}$. However decryption is harder to define for non-fresh samples. In this case, correctness of the decryption procedure involves a recurrence formula between the decryption of the sum and the sum of the decryption of the inputs conditioned by the noise parameters. In addition, message spaces of the input samples can be in different subgroups of $\mathbb{T}$. To raise the limitations of the decryption function, we will instead use a mathematical definition of message and error by reasoning directly on the following $\Omega$-probability space.

**Definition 3.3 (The $\Omega$-probability space).** *Since samples are either independent (random, noiseless, or trivial) fresh $c \leftarrow TLWE_{s,\alpha}(\mu)$, or linear combination $\tilde{c} = \sum_{i=1}^{p} e_i \cdot c_i$ of other samples, the probability space $\Omega$ is the product of the probability spaces of each individual fresh samples $c$ with the TLWE distributions defined in Definition 3.1, and of the probability spaces of all the coefficients $(e_1, \ldots, e_p) \in \mathfrak{R}^p$ or $\mathbb{Z}^p$ that are obtained with randomized algorithm.*

In other words, instead of viewing a TLWE sample as a fixed value which is the result of one particular event in $\Omega$, we will consider all the possible values at once, and make statistics on them.

We now define functions on TLWE samples: message, error, noise variance, and noise norm. These functions are well defined mathematically, and can be used in the analysis of various algorithms. However, they cannot be directly computed or approximated in practice.

**Definition 3.4.** *Let $c$ be a random variable $\in \mathbb{T}_N[X]^{k+1}$, which we'll interpret as a TLWE sample. All probabilities are on the $\Omega$-space. We say that $c$ is a valid TLWE sample iff there exists a key $s \in \mathbb{B}_N[X]^k$ such that the distribution of the phase $\varphi_s(c)$ is concentrated. If $c$ is trivial, all keys $s$ are equivalent, else the mask of $c$ is uniformly random, so $s$ is unique. We then define:*

- *the message of $c$, denoted as $\mathsf{msg}(c) \in \mathbb{T}_N[X]$ is the expectation of $\varphi_s(c)$;*
- *the error, denoted $\mathsf{Err}(c)$, is equal to $\varphi_s(c) - \mathsf{msg}(c)$;*
- *$\mathsf{Var}(\mathsf{Err}(c))$ denotes the variance of $\mathsf{Err}(c)$, which is by definition also equal to the variance of $\varphi_s(c)$;*
- *finally, $\|\mathsf{Err}(c)\|_\infty$ denotes the maximum amplitude of $\mathsf{Err}(c)$ (possibly with overwhelming probability).*

Unlike the classical decryption algorithm, the message function can be viewed as an ideal black box decryption function, which works with infinite precision even if the message space is continuous. Provided that the noise amplitude remains smaller than $\frac{1}{4}$, the message function is perfectly linear. Using these intuitive and intrinsic functions will considerably ease the analysis of all algorithms in this paper. In particular, we have:

**Fact 3.5.** Given $p$ valid and independent TLWE samples $c_1, \ldots, c_p$ under the same key $s$, and $p$ integer polynomials $e_1, \ldots, e_p \in \mathfrak{R}$, if the linear combination $c = \sum_{i=1}^{p} e_i \cdot c_i$ is a valid TLWE sample, it satisfies: $\mathsf{msg}(c) = \sum_{i=1}^{p} e_i \cdot \mathsf{msg}(c_i)$, with variance $\mathsf{Var}(\mathsf{Err}(c)) \leq \sum_{i=1}^{p} \|e_i\|_2^2 \cdot \mathsf{Var}(\mathsf{Err}(c_i))$ and noise amplitude $\|\mathsf{Err}(c)\|_\infty \leq \sum_{i=1}^{p} \|e_i\|_1 \cdot \|\mathsf{Err}(c_i)\|_\infty$. If the last bound is $< \frac{1}{4}$, then $c$ is necessarily a valid TLWE sample (under the same key $s$).

In order to characterize the average case behaviour of our homomorphic operations, we shall rely on the heuristic assumption of independence below. This heuristic will only be used for practical average-case bounds. Our worst-case theorems and lemma based on the infinite norm do not use it at all.

**Assumption 3.6 (Independence Heuristic).** All the coefficients of the error of TLWE or TGSW samples that occur in all the linear combinations we consider are independent and concentrated. More precisely, they are $\sigma$-subgaussian where $\sigma$ is the square-root of their variance.

This assumption allows us to bound the variance of the noise instead of its norm, and to provide realistic average-case bounds which often correspond to the square root of the worst-case ones. The error can easily be proved subgaussian, since each coefficients are always obtained by convolving Gaussians or zero-centered bounded uniform distributions. But the independence assumption between all the coefficients remains heuristic. Dependencies between coefficients may affect the variance of their combinations in both directions. The independence of coefficients can be obtained by adding enough entropy in all our decomposition algorithms and by increasing some parameters accordingly, but as noticed in [11], this work-around seems more as a proof artefact, and is experimentally not needed. Since average case corollaries should reflect practical results, we leave the independence of subgaussian samples as a heuristic assumption.

## 3.2   TGSW

In this section we present a generalized scale invariant version of the FHE scheme GSW [16], that we call TGSW. GSW was proposed Gentry, Sahai and Waters in 2013 [16], and improved in [3] and its security is based on the LWE problem. The scheme relies on a gadget decomposition function, which we also extend to polynomials, but most importantly, the novelty is that our function is an approximate decomposition, up to some precision parameter. This allows to improve running time and memory requirements for a small amount of additional noise.

**Definition 3.7 (Approximate Gadget Decomposition).** *Let $h \in \mathcal{M}_{p,k+1}$ $(\mathbb{T}_N[X])$ as in (1). We say that $Dec_{h,\beta,\epsilon}(v)$ is a decomposition algorithm on the gadget $h$ with quality $\beta$ and precision $\epsilon$ if and only if for any TLWE sample $v \in \mathbb{T}_N[X]^{k+1}$, it efficiently and publicly outputs a small vector $u \in \mathfrak{R}^{(k+1)\ell}$ such that $\|u\|_\infty \le \beta$ and $\|u \cdot h - v\|_\infty \le \epsilon$. Furthermore, the expectation of $u \cdot h - v$ must to be 0 when $v$ is uniformly distributed in $\mathbb{T}_N[X]^{k+1}$*

Definition 3.7 is generic, but in the rest of the paper, we will only use this fixed gadget:

$$
h = \begin{pmatrix} 1/B_g & \cdots & 0 \\ \vdots & \ddots & \vdots \\ 1/B_g^\ell & \cdots & 0 \\ \vdots & \ddots & \vdots \\ 0 & \cdots & 1/B_g \\ \vdots & \ddots & \vdots \\ 0 & \cdots & 1/B_g^\ell \end{pmatrix} \in \mathcal{M}_{p,k+1}(\mathbb{T}_N[X]). \tag{1}
$$

The matrix $h$ consists in a diagonal of columns, each containing a super-increasing sequence of constant polynomials in $\mathbb{T}$. Algorithm 1 represents an efficient decomposition of TLWE samples on $h$, and the following lemma proves its correctness. In theory, decomposition algorithms should be randomized to guarantee that the distribution of all error coefficients remain independent. In practice, we already rely on Heuristic 3.6. We just need that the expectation of the small errors induced by the approximations remains null, so that the message is not changed.

**Lemma 3.8** *Let $\ell \in \mathbb{N}$ and $B_g \in \mathbb{N}$. Then for $\beta = B_g/2$ and $\epsilon = 1/2B_g^\ell$, Algorithm 1 is a valid $Dec_{h,\beta,\epsilon}$.*

---

**Algorithm 1.** Gadget Decomposition of a TLWE sample

---

**Input:** A TLWE sample $(a, b) = (a_1, \ldots, a_k, b = a_{k+1}) \in \mathbb{T}_N[X]^k \times \mathbb{T}_N[X]$
**Output:** A combination $[e_{1,1}, \ldots, e_{k+1,\ell}] \in \mathfrak{R}^{(k+1)\ell}$
1: For each $a_i$ choose the unique representative $\sum_{j=0}^{N-1} a_{i,j} X^j$, with $a_{i,j} \in \mathbb{T}$, and set $\bar{a}_{i,j}$ the closest multiple of $\frac{1}{B_g^\ell}$ to $a_{i,j}$
2: Decompose each $\bar{a}_{i,j}$ uniquely as $\sum_{p=1}^{\ell} \bar{a}_{i,j,p} \frac{1}{B_g^p}$ where each $\bar{a}_{i,j,p} \in [\![ -B_g/2, B_g/2 [\![$
3: **for** $i = 1$ to $k + 1$
4:      **for** $p = 1$ to $\ell$
5:         $e_{i,p} = \sum_{j=0}^{N-1} \bar{a}_{i,j,p} X^j \in \mathfrak{R}$
6: **Return**    $(e_{i,p})_{i,p}$

---

*Proof.* Let $v = (a, b) = (a_1, \ldots, a_k, b = a_{k+1}) \in \mathbb{T}_N[X]^{k+1}$ be a TLWE sample, given as input to Algorithm 1. Let $u = [e_{1,1}, \ldots, e_{k+1,\ell}] \in \mathfrak{R}^{(k+1)\ell}$ be the corresponding output by construction $\|u\|_\infty \leq B_g/2 = \beta$.

Let $\epsilon_{dec} = u \cdot h - v$. For all $i \in [\![1, k+1]\!]$ and $j \in [\![1, \ell]\!]$, we have by construction $\epsilon_{dec_{i,j}} = \sum_{p=0}^{\ell} e_{i,p} \cdot \frac{1}{B_g^p} - a_{i,j} = \bar{a}_{i,j} - a_{i,j}$. Since $\bar{a}_{i,j}$ is defined as the nearest multiple of $\frac{1}{B_g^\ell}$ on the torus, we have $|\bar{a}_{i,j} - a_{i,j}| \leq 1/2B_g^\ell = \epsilon$. $\epsilon_{dec}$ has therefore a concentrated distribution when $v$ is uniform. We now verify that it is zero-centered. Finally, if we call $f$ the function from $\mathbb{T}$ to $\mathbb{T}$ which rounds an element $x$ to its closest multiple of $\frac{1}{B_g^\ell}$ and the function $g$ the symmetry defined by $g(x) = 2f(x) - x$ on the torus; we easily verify that the $\mathbb{E}(\epsilon_{dec_{i,j}})$ is equal to $\mathbb{E}(a_{i,j} - f(a_{i,j}))$ when $a_{i,j}$ has uniform distribution, which is equal to $\mathbb{E}(g(a_{i,j}) - f(g(a_{i,j})))$ when $g(a_{i,j})$ has uniform distribution also equal to $\mathbb{E}(f(a_{i,j}) - a_{i,j}) = -\mathbb{E}(\epsilon_{dec_{i,j}})$. Thus, the expectation of $\epsilon_{dec}$ is 0. $\quad\square$

We are now ready to define TGSW samples, and to extend the notions of phase of valid sample, message and error of the samples.

**Definition 3.9 (TGSW samples).** *Let $\ell$ and $k \geq 1$ be two integers, $\alpha \geq 0$ be a noise parameter and $h$ the gadget defined in Eq. (1). Let $s \in \mathbb{B}_N[X]^k$*

be a RingLWE *key, we say that* $C \in \mathcal{M}_{(k+1)\ell,k+1}(\mathbb{T}_N[X])$ *is a fresh* TGSW *sample of* $\mu \in \mathfrak{R}/h^\perp$ *with noise parameter* $\alpha$ *iff* $C = Z + \mu \cdot h$ *where each row of* $Z \in \mathcal{M}_{(k+1)\ell,k+1}(\mathbb{T}_N[X])$ *is an Homogeneous TLWE sample (of 0) with Gaussian noise parameter* $\alpha$. *Reciprocally, we say that an element* $C \in \mathcal{M}_{(k+1)\ell,k+1}(\mathbb{T}_N[X])$ *is a valid TGSW sample iff there exists a unique polynomial* $\mu \in \mathfrak{R}/h^\perp$ *and a unique key* $s$ *such that each row of* $C - \mu \cdot h$ *is a valid TLWE sample of 0 for the key* $s$. *We call the polynomial* $\mu$ *the message of* $C$, *and we denote it by* msg$(C)$.

**Definition 3.10 (Phase, Error).** *Let* $A = \in \mathcal{M}_{(k+1)\ell,k+1}(\mathbb{T}_N[X])$ *be a* TGSW *sample for a secret key* $s \in \mathbb{B}_N[X]^k$ *and noise parameter* $\alpha \geq 0$.

*We define the phase of* $A$, *denoted as* $\varphi_s(A) \in (\mathbb{T}_N[X])^{(k+1)\ell}$, *as the list of the* $(k+1)\ell$ *TLWE phases of each line of* $A$. *In the same way, we define the error of* $A$, *denoted* Err$(A)$, *as the list of the* $(k+1)\ell$ *TLWE errors of each line of* $A$.

Since TGSW samples are essentially vectors of TLWE samples, they are naturally compatible with linear operations. And both phase and message functions remain linear.

**Fact 3.11.** Given $p$ valid TGSW samples $C_1, \ldots, C_p$ of messages $\mu_1, \ldots, \mu_p$ under the same key, and with independent error coefficients, and given $p$ integer polynomials $e_1, \ldots, e_p$, the linear combination $C = \sum_{i=1}^{p} e_i \cdot C_i$ is a sample of $\mu = \sum_{i=1}^{p} e_i \cdot \mu_i$, with variance $\mathsf{Var}(C) = \left(\sum_{i=1}^{p} \|e_i\|_2^2 \cdot \mathsf{Var}(C_i)\right)^{1/2}$ and noise infinity norm $\|\mathsf{Err}(C)\|_\infty = \sum_{i=1}^{p} \|e_i\|_1 \cdot \|\mathsf{Err}(C)\|_\infty$.

Also, the phase remains $1 + kN$ lipschitzian for the infinity norm.

**Fact 3.12.** For all $A \in \mathcal{M}_{p,k+1}(\mathbb{T}_N[X])$, $\|\varphi_s(A)\|_\infty \leq (Nk+1) \|A\|_\infty$.

We finally define the homomorphic product between TGSW and TLWE samples, whose corresponding message is simply the product of the two messages of the initial samples. Since the left member encodes an integer polynomial, and the right one a torus polynomial, this operator performs a homomorphic evaluation of their external product. Theorem 3.14 (resp. Corollary 3.15) analyzes the worst-case (resp. average-case) noise propagation of this product. Then, Corollary 3.16 relates this new morphism to the classical internal product between TGSW samples.

**Definition 3.13 (External product).** *We define the product* $\boxdot$ *as*

$$\boxdot: \text{TGSW} \times \text{TLWE} \longrightarrow \text{TLWE}$$
$$(A, b) \longmapsto A \boxdot b = Dec_{h,\beta,\epsilon}(b) \cdot A.$$

The formula is almost identical to the classical product defined in the original GSW scheme in [16], except that only one vector needs to be decomposed. For this reason, we get almost the same noise propagation formula, with an additional term that comes from the approximations in the decomposition.

**Theorem 3.14 (Worst-case External Product).** *Let $A$ be a valid TGSW sample of message $\mu_A$ and let $b$ be a valid TLWE sample of message $\mu_b$. Then $A \boxdot b$ is a TLWE sample of message $\mu_A \cdot \mu_b$ and $\|Err(A \boxdot b)\|_\infty \leq (k+1)\ell N\beta \|Err(A)\|_\infty + \|\mu_A\|_1 (1 + kN)\epsilon + \|\mu_A\|_1 \|Err(b)\|_\infty$ (worst case), where $\beta$ and $\epsilon$ are the parameters used in the decomposition $Dec_{h,\beta,\epsilon}(b)$. If $\|Err(A \boxdot b)\|_\infty \leq 1/4$ we are guaranteed that $A \boxdot b$ is a valid TLWE sample.*

*Proof.* As $A = \mathrm{TGSW}(\mu_A)$, then by definition it is equal to $A = Z_A + \mu_A \cdot h$, where $Z_A$ is a TGSW encryption of 0 and $h$ is the gadget matrix. In the same way, as $b = \mathrm{TLWE}(\mu_b)$, then by definition it is equal to $b = z_b + (0, \mu_b)$, where $z_b$ is a TLWE encryption of 0. Let

$$\begin{cases} \|Err(A)\|_\infty = \|\varphi_s(Z_A)\|_\infty = \eta_A \\ \|Err(b)\|_\infty = \|\varphi_s(z_b)\|_\infty = \eta_b. \end{cases}$$

Let $u = Dec_{h,\beta,\epsilon}(b) \in \Re^{(k+1)\ell}$. By definition $A \boxdot b$ is equal to

$$A \boxdot b = u \cdot A$$
$$= u \cdot Z_A + \mu_A \cdot (u \cdot h).$$

From Definition 3.7, we have that $u \cdot h = b + \epsilon_{dec}$, where $\|\epsilon_{dec}\|_\infty = \|u \cdot h - b\|_\infty \leq \epsilon$. So

$$A \boxdot b = u \cdot Z_A + \mu_A \cdot (b + \epsilon_{dec})$$
$$= u \cdot Z_A + \mu_A \cdot \epsilon_{dec} + \mu_A \cdot z_b + (0, \mu_A \cdot \mu_b).$$

Then the phase (linear function) of $A \boxdot b$ is

$$\varphi_s(A \boxdot b) = u \cdot Err(A) + \mu_A \cdot \varphi_s(\epsilon_{dec}) + \mu_A \cdot Err(b) + \mu_A\mu_b.$$

Taking the expectation, we get that $msg(A \boxdot b) = 0 + 0 + 0 + \mu_A\mu_b$, and so $Err(A \boxdot b) = \varphi_s(A \boxdot b) - \mu_A\mu_b$. Then thanks to Fact 3.12, we have

$$\|Err(A \boxdot b)\|_\infty \leq \|u \cdot Err(A)\|_\infty + \|\mu_A \cdot \varphi(\epsilon_{dec})\|_\infty + \|\mu_A \cdot Err(b)\|_\infty$$
$$\leq (k+1)\ell N\beta\eta_A + \|\mu_A\|_1 (1 + kN) \|\epsilon_{dec}\|_\infty + \|\mu_A\|_1 \eta_b.$$

The result follows. □

We similarly obtain the more realistic average-case noise propagation, based on the independence heuristic, by bounding the Gaussian variance instead of the amplitude.

**Corollary 3.15 (Average-case External Product).** *Under the same conditions of Theorem 3.14 and by assuming the Heuristic 3.6, we have that $Var(Err(A \boxdot b)) \leq (k+1)\ell N\beta^2 Var(Err(A)) + (1+kN)\|\mu_A\|_2^2 \epsilon^2 + \|\mu_A\|_2^2 Var(Err(b))$.*

*Proof.* Let $\vartheta_A = \mathsf{Var}(\mathsf{Err}(A)) = \mathsf{Var}(\varphi_s(Z_A))$ and $\vartheta_b = \mathsf{Var}(\mathsf{Err}(b)) = \mathsf{Var}(\varphi_s(z_b))$. By using the same notations as in the proof of Theorem 3.14 we have that the error of $A \boxdot b$ is $\mathsf{Err}(A \boxdot b) = u \cdot \mathsf{Err}(A) + \mu_A \cdot \varphi_s(\epsilon_{dec}) + \mu_A \cdot \mathsf{Err}(b)$ and thanks to Assumption 3.6 and Fact 3.12, we have:

$$\mathsf{Var}(\mathsf{Err}(A \boxdot b)) \le \mathsf{Var}(u \cdot \mathsf{Err}(A))) + \mathsf{Var}(\mu_A \cdot \varphi(\epsilon_{dec})) + \mathsf{Var}(\mu_A \cdot \mathsf{Err}(b))$$
$$\le (k+1)\ell N \beta^2 \vartheta_A + (1+kN)\|\mu_A\|_2^2 \epsilon^2 + \|\mu_A\|_2^2 \vartheta_b.$$

$\square$

The last corollary describes exactly the classical internal product between two TGSW samples, already presented in [3,11,13,16] with adapted notations. As we mentioned before, it is much slower to evaluate, because it consists in $(k+1)\ell$ independent computations of the $\boxdot$ product, which we illustrate now.

**Corollary 3.16 (Internal Product).** *Let the product*

$$\boxtimes: \mathsf{TGSW} \times \mathsf{TGSW} \longrightarrow \mathsf{TGSW}$$

$$(A, B) \longmapsto A \boxtimes B = \begin{bmatrix} A \boxdot b_1 \\ \vdots \\ A \boxdot b_{(k+1)\ell} \end{bmatrix} = \begin{bmatrix} Dec_{h,\beta,\epsilon}(b_1) \cdot A \\ \vdots \\ Dec_{h,\beta,\epsilon}(b_{(k+1)\ell}) \cdot A \end{bmatrix},$$

*with $A$ and $B$ two valid TGSW samples of messages $\mu_A$ and $\mu_B$ respectively and $b_i$ corresponding to the $i$-th line of $B$. Then $A \boxtimes B$ is a TGSW sample of message $\mu_A \cdot \mu_B$ and $\|\mathsf{Err}(A \boxdot B)\|_\infty \le (k+1)\ell N \beta \|\mathsf{Err}(A)\|_\infty + \|\mu_A\|_1 (1+kN)\epsilon + \|\mu_A\|_1 \|\mathsf{Err}(B)\|_\infty$ (worst case). If $\|\mathsf{Err}(A \boxdot B)\|_\infty \le 1/4$ we are guaranteed that $A \boxdot B$ is a valid TGSW sample.*

*Furthermore, by assuming the Heuristic 3.6, we have that $\mathsf{Var}(\mathsf{Err}(A \boxdot B)) \le (k+1)\ell N \beta^2 \mathsf{Var}(\mathsf{Err}(A)) + (1+kN)(\mu_A \epsilon)^2 + \mu_A^2 \mathsf{Var}(\mathsf{Err}(b))$ (average case).*

*Proof.* Let $A$ and $B$ be two TGSW samples, and $\mu_A$ and $\mu_B$ their message. By definition, the $i$-th row of $B$ encodes $\mu_B \cdot h_i$, so the $i$-th row of $A \boxtimes B$ encodes $(\mu_A \mu_B) \cdot h_i$. This proves that $A \boxtimes B$ encodes $\mu_A \mu_B$. Since the internal product $A \boxtimes B$ consists in $(k+1)\ell$ independent runs of the external products $A \boxdot b_i$, the noise propagation formula directly follows from Theorem 3.14 and Corollary 3.15. $\square$

In the next section, we show that all internal products in the bootstrapping procedure can be replaced with the external one. Consequently, we expect a speed-up of a factor at least $(k+1)\ell$.

## 4   Application: Single Gate Bootstrapping in Less Than 0.1 Seconds

In this section, we show how to use Theorem 3.14 to speed-up the bootstrapping presented in [11]. With additional optimizations, we drastically reduce the bootstrapping key size, and also reduce a bit the noise overhead. To bootstrap a LWE sample $(a, b) \in \mathbb{T}^{n+1}$, which is rescaled as $(\bar{a}, \bar{b}) \mod 2N$, using relevant encryptions of its secret key $s \in \mathbb{B}^n$, the overall idea is the following. We start from a fixed polynomial testv $\in \mathbb{T}_N[X]$, which is our phase detector: its $i$-th coefficient is set to the value that the bootstrapping should return if $\varphi_s(a, b) = i/2N$. testv is first encoded in a trivial LWE sample. Then, we iteratively rotate its coefficients, using external multiplications with TGSW encryptions of the hidden monomials $X^{-s_i \bar{a}_i}$. By doing so, the original testv gets rotated by the (hidden) phase of $(a, b)$, and in the end, we simply extract the constant term as a LWE sample.

## 4.1   TLWE to LWE Extraction

Like in previous work, extracting a LWE sample from a TLWE sample simply means rewriting polynomials into their list of coefficients, and discarding the $N - 1$ last coefficients of $b$. This yields a LWE encryption of the constant term of the initial polynomial message.

**Definition 4.1 (TLWE Extraction).** *Let $(a'', b'')$ be a TLWE$_{s''}(\mu)$ sample with key $s'' \in \mathfrak{R}^k$, We call KeyExtract$(s'')$ the integer vector $s' = (\mathsf{coefs}(s_1''(X)), \ldots, \mathsf{coefs}(s_k''(X)) \in \mathbb{Z}^{kN}$ and SampleExtract$(a'', b'')$ the LWE sample $(a', b') \in \mathbb{T}^{kN+1}$ where $a' = (\mathsf{coefs}(a_1''(1/X)), \ldots, \mathsf{coefs}(a_k''(1/X))$ and $b' = b_0''$ the constant term of $b''$. Then $\varphi_{s'}(a', b')$ (resp. $\mathsf{msg}(a', b')$) is equal to the constant term of $\varphi_{s''}(a'', b'')$ (resp. to the constant term of $\mu = \mathsf{msg}(a'', b''))$. And $\|\mathsf{Err}(a', b')\|_\infty \leq \|\mathsf{Err}(a'', b'')\|_\infty$ and $\mathsf{Var}(\mathsf{Err}(a', b')) \leq \mathsf{Var}(\mathsf{Err}(a'', b''))$.*

## 4.2   LWE to LWE Key-Switching Procedure

Given a LWE$_{s'}$ sample of a message $\mu \in \mathbb{T}$, the key switching procedure initially proposed in [5, 7] outputs a LWE$_s$ sample of the same $\mu$ without increasing the noise too much. Contrary to previous exact keyswitch procedures, here we tolerate approximations.

**Definition 4.2.** *Let $s' \in \{0, 1\}^{n'}$, $s \in \{0, 1\}^n$, a noise parameter $\gamma \in \mathbb{R}$ and a precision parameter $t \in \mathbb{N}$, we call key switching secret KS$_{s' \to s, \gamma, t}$ a sequence of fresh LWE samples KS$_{i,j} \in$ LWE$_{s, \gamma}(s_i' \cdot 2^{-j})$ for $i \in [1, n']$ and $j \in [1, t]$.*

**Lemma 4.3 (Key switching).** *Given $(a', b') \in$ LWE$_{s'}(\mu)$ where $s' \in \{0, 1\}^{n'}$ with noise $\eta' = \|\mathsf{Err}(a', b')\|_\infty$ and a keyswitching key KS$_{s' \to s, \gamma, t}$, where $s \in \{0, 1\}^n$, the key switching procedure outputs a LWE sample $(a, b) \in$ LWE$_{s_n}(\mu)$ where $\|\mathsf{Err}(a, b)\|_\infty \leq \eta' + n't\gamma + n'2^{-(t+1)}$.*

---

**Algorithm 2.** KeySwitch procedure

**Input:** A LWE sample $(a' = (a'_1, \ldots, a'_{n'}), b') \in \mathsf{LWE}_{s'}(\mu)$, a switching key $\mathsf{KS}_{s' \to s}$
   where $s' \in \{0,1\}^{n'}$, $s \in \{0,1\}^n$ and $t \in \mathbb{N}$ a precision parameter
**Output:** A LWE sample $\mathsf{LWE}_s(\mu)$
1: Let $\bar{a}'_i$ be the closest multiple of $\frac{1}{2^t}$ to $a'_i$, thus $|\bar{a}'_i - a'_i| < 2^{-(t+1)}$
2: Binary decompose each $\bar{a}'_i = \sum_{j=1}^t a_{i,j} \cdot 2^{-j}$ where $a_{i,j} \in \{0,1\}$

3: Return $\quad (0, b') - \sum_{i=1}^{n'} \sum_{j=1}^{t} a_{i,j} \cdot \mathsf{KS}_{i,j}$

---

*Proof.* We have

$$\varphi_s(a, b) = \varphi_s(0, b') - \sum_{i=1}^{n'} \sum_{j=1}^{t} a_{i,j} \varphi_s(\mathsf{KS}_{i,j})$$

$$= b' - \sum_{i=1}^{n'} \sum_{j=1}^{t} a_{i,j} \left( 2^{-j} s'_i + \mathsf{Err}(\mathsf{KS}_{i,j}) \right)$$

$$= b' - \sum_{i=1}^{n'} \bar{a}'_i s'_i - \sum_{i=1}^{n'} \sum_{j=1}^{t} a_{i,j} \mathsf{Err}(\mathsf{KS}_{i,j})$$

$$= b' - \sum_{i=1}^{n'} a'_i s'_i - \sum_{i=1}^{n'} \sum_{j=1}^{t} a_{i,j} \mathsf{Err}(\mathsf{KS}_{i,j}) + \sum_{i=1}^{n'} (a'_i - \bar{a}'_i) s'_i$$

$$= \varphi_{s'}(a', b') - \sum_{i=1}^{n'} \sum_{j=1}^{t} a_{i,j} \mathsf{Err}(\mathsf{KS}_{i,j}) + \sum_{i=1}^{n'} (a'_i - \bar{a}'_i) s'_i.$$

The expectation of the left side of the equality is equal to $\mathsf{msg}(a, b)$. For the right side, each $a_{i,j}$ is uniformly distributed in $\{0,1\}$ and $(a'_i - \bar{a}'_i)$ is a 0-centered variable so the expectation of the sum is 0. Thus, $\mathsf{msg}(a, b) = \mathsf{msg}(a', b')$. We obtain $\|\varphi_s(a, b) - \mathsf{msg}(a, b)\|_\infty \leq \eta' + n' \cdot t \cdot \gamma + n' 2^{-(t+1)}$. $\qquad \square$

**Corollary 4.4.** *Let $t$ be an integer parameter. Under Assumption 3.6 Given $(a', b') \in \mathsf{LWE}_{s'}(\mu)$ with noise variance $\eta' = \mathsf{Var}(\mathsf{Err}(a', b'))$ and a key switching key $\mathsf{KS}_{s' \to s, \gamma, \ell}$, the key switching procedure outputs an LWE sample $(a', b') \in \mathsf{LWE}_s(\mu)$ where $\mathsf{Var}(\mathsf{Err}(a, b)) \leq \eta' + n' \cdot t \cdot \gamma^2 + n' 2^{-2(t+1)}$.*

### 4.3 Bootstrapping Procedure

Given a LWE sample $\mathsf{LWE}_s(\mu) = (a, b)$, the bootstrapping procedure constructs an encryption of $\mu$ under the same key $s$ but with a fixed amount of noise. As in [11], we will use TLWE as an intermediate encryption scheme to perform a homomorphic evaluation of the phase but here we will use its external product from Theorem 3.14 with a TGSW encryption of the key $s$.

**Definition 4.5.** *Let $s \in \mathbb{B}^n$, $s'' \in \mathbb{B}_N[X]^k$ and $\alpha$ be a noise parameter. We define the bootstrapping key $BK_{s \to s'', \alpha}$ as the sequence of $n$ TGSW samples where $BK_i \in \mathrm{TGSW}_{s'', \alpha}(s_i)$.*

---

**Algorithm 3.** Bootstrapping procedure

---

**Input:** A LWE sample $(\boldsymbol{a}, b) \in \mathsf{LWE}_{\boldsymbol{s}, \eta}(\mu)$, a bootstrapping key $\mathsf{BK}_{\boldsymbol{s} \to \boldsymbol{s}'', \alpha}$, a keyswitch
    key $\mathsf{KS}_{\boldsymbol{s}' \to \boldsymbol{s}, \gamma}$ where $\boldsymbol{s}' = \mathsf{KeyExtract}(\boldsymbol{s}'')$, two fixed messages $\mu_0, \mu_1 \in \mathbb{T}$

**Output:** A LWE sample $\mathsf{LWE}_{\boldsymbol{s}} \left( \mu_0 \text{ if } \varphi_s(\boldsymbol{a}, b) \in \left] -\frac{1}{4}, \frac{1}{4} \right[ ; \mu_1 \text{ else} \right)$

1: Let $\bar{\mu} = \frac{\mu_1 + \mu_0}{2}$ and $\bar{\mu}' = \mu_0 - \bar{\mu}$
2: Let $\bar{b} = \lfloor 2Nb \rceil$ and $\bar{a}_i = \lfloor 2Na_i \rceil$ **for each** $i \in [1, n]$
3: Let $\mathrm{testv} := (1 + X + \ldots + X^{N-1}) \times X^{-\frac{2N}{4}} \cdot \bar{\mu}' \in \mathbb{T}_N[X]$
4: $\mathsf{ACC} \leftarrow \left( X^{\bar{b}} \cdot (0, \mathrm{testv}) \right) \in \mathbb{T}_N[X]^{k+1}$
5: **for** $i = 1$ **to** $n$
6:     $\mathsf{ACC} \leftarrow \left[ \boldsymbol{h} + (X^{-\bar{a}_i} - 1) \cdot \mathsf{BK}_i \right] \boxdot \mathsf{ACC}$
7: Let $\boldsymbol{u} := (\mathbf{0}, \bar{\mu}) + \mathsf{SampleExtract}(\mathsf{ACC})$
8: **Return** $\mathsf{KeySwitch}_{\mathsf{KS}}(\boldsymbol{u})$

---

We first provide a comparison between the bootstrapping of Algorithm 3 and [11, Algorithms 1 and 2] proposal.

- Like [11], we rescale the computation of the phase of the input LWE sample so that it is modulo $2N$ (line 2) and we map all the corresponding operations in the multiplicative cyclic group $\{1, X, \ldots, X^{2N-1}\}$. Since our LWE samples are described over the real torus, the rescaling is done explicitly in line 2. This rescaling may induce a cumulated rounding error of amplitude at most $\delta \approx \sqrt{n}/4N$ in the average case and $\delta \leq (n+1)/4N$ in the worst case. In the best case, this amplitude can decrease to zero ($\delta = 0$) if in the actual representation of LWE samples, all the coefficients are restricted to multiple of $\frac{1}{2N}$, which would be the analogue of [11]'s setting.
- As in [11], messages are encoded as roots of unity in $\mathcal{R}$. Our accumulator is a TLWE sample instead of a TGSW sample in [11]. Also accumulator operations use the external product from Theorem 3.14 instead of the slower classical internal product. The test vector $(1 + X + \ldots + X^{N-1})$ is embedded in the accumulator from the very start, when the accumulator is still noiseless while in [11], it is added at the very end. This removes a factor $\sqrt{N}$ to the final noise overhead.
- All the TGSW ciphertexts of $X^{-\bar{a}_i s_i}$ required to update the accumulator internal value are computed dynamically as a very small polynomial combination of $BK_i$ in the for loop (line 5). This completely removes the need to decompose each $\bar{a}_i$ on an additional base $B_r$, and to precompute all possibilities in the bootstrapping key. In other words, this makes our bootstrapping key 46 times smaller than in [11], for the exact same noise overhead. Besides, due to this squashing technique, two accumulator operations were performed per iteration instead of one in our case. This gives us an additional 2X speed up.

**Theorem 4.6 (Bootstrapping Theorem).** *Let $\boldsymbol{h} \in \mathcal{M}_{\ell(k+1), k+1}(\mathbb{T}_N[X])$ be the gadget defined in Eq. 1 and let $Dec_{\boldsymbol{h}, \epsilon, \beta}$ be the associated vector gadget decomposition function.*

Let $s \in \mathbb{B}^n$, $s'' \in \mathbb{B}_N[X]^k$ and $\alpha, \gamma$ be noise amplitudes. Let $BK = BK_{s \to s'', \alpha}$ be a bootstrapping key, let $s' = \mathsf{KeyExtract}(s'') \in \mathbb{B}^{kN}$ and $\mathsf{KS} = \mathsf{KS}_{s' \to s, \gamma, t}$ be a keyswitching secret.

Given $(a, b) \in \mathsf{LWE}_s(\mu)$ for $\mu \in \mathbb{T}$, two fixed messages $\mu_0, \mu_1$, Algorithm 3 outputs a sample in $\mathsf{LWE}_s(\mu')$ s.t. $\mu' = \mu_0$ if $|\varphi_s(a, b)| < -1/4 - \delta$ and $\mu' = \mu_1$ if $|\varphi_s(a, b)| > 1/4 + \delta$ where $\delta$ is the cumulated rounding error equal to $\frac{n+1}{4N}$ in the worst case and $\delta = 0$ if the all coefficients of $(a, b)$ are multiple of $\frac{1}{2N}$. Let $v$ be the output of Algorithm 3. Then $\|\mathsf{Err}(v)\|_\infty \leq 2n(k+1)\ell\beta N\alpha + kNt\gamma + n(1 + kN)\epsilon + kN2^{-(t+1)}$.

*Proof.* Line 1: the division by two over torus gives two possible values for $(\bar{\mu}, \bar{\mu}')$. In both cases, $\bar{\mu} + \bar{\mu}' = \mu_0$ and $\bar{\mu} - \bar{\mu}' = \mu_1$.

Line 2: let $\bar{\varphi} \overset{def}{=} \bar{b} - \sum_{i=1}^n \bar{a}_i s_i \mod 2N$. We have

$$\left| \varphi - \frac{\bar{\varphi}}{2N} \right| = b - \frac{\lfloor 2Nb \rfloor}{2N} + \sum_{i=1}^n \left( a_i - \frac{\lfloor 2Na_i \rfloor}{2N} \right) s_i \leq \frac{1}{4N} + \sum_{i=1}^n \frac{1}{4N} \leq \frac{n+1}{4N}. \quad (2)$$

And if the coefficients $(a, b) \in \frac{1}{2N}\mathbb{Z}/\mathbb{Z}$, then $\varphi = \frac{\bar{\varphi}}{2N}$. In all cases, $|\varphi - \frac{\bar{\varphi}}{2N}| < \delta$.

At line 3, the test vector $\mathsf{testv} := (1 + X + \ldots + X^{N-1}) \cdot X^{-\frac{2N}{4}} \cdot \bar{\mu}'$ is defined such that for all $p \in [0, 2N]$, the constant term of $X^p \cdot \mathsf{testv}$ is either $\bar{\mu}'$ if $p \in ] -\frac{N}{2}, \frac{N}{2} [$ and $-\bar{\mu}'$ else.

In the loop for (from line 5 to 6), we will prove the following invariant: At the beginning of iteration $i+1 \in [1, n+1]$ (i.e. at the end of iteration $i$), $\mathsf{msg}(ACC_i) = X^{b - \sum_{j=1}^i \bar{a}_j s_j} \cdot \mathsf{testv}$ and $\|\mathsf{Err}(ACC_i)\|_\infty \leq \sum_{j=1}^i \left( 2(k+1)\ell N\beta \|\mathsf{Err}(BK_j)\|_\infty + (1 + kN)\epsilon \right)$.

At the beginning of iteration $i = 1$, the accumulator contains a trivial ciphertext $\mathsf{msg}(ACC_1) = \left( X^{\bar{b}} \cdot \mathsf{testv} \right)$, so $\|\mathsf{Err}(ACC_1)\|_\infty = 0$.

During iteration $i$, $A_i = h + (X^{-\bar{a}_i} - 1) \cdot BK_i$ is a TGSW sample of message $X^{-\bar{a}_i s_i}$ (this can be seen by replacing $s_i$ with its two possible values 0 and 1) and of noise $\|\mathsf{Err}(A_i)\|_\infty \leq 2 \|\mathsf{Err}(BK_i)\|_\infty$. This inequality holds from Fact 3.11. Then, we have:

$$\mathsf{msg}(ACC_i) = \mathsf{msg}\left( A_i \boxdot ACC_{i-1} \right)$$
$$= \mathsf{msg}\left( A_i \right) \cdot \mathsf{msg}(ACC_{i-1}) \qquad \text{(from Theorem 3.14)}$$
$$= X^{-\bar{a}_i s_i} \cdot \left( X^{b - \sum_{j=1}^{i-1} \bar{a}_j s_j} \cdot \mathsf{testv} \right)$$

and from the norm inequality of Theorem 3.14,

$$\|\mathsf{Err}(ACC_i)\|_\infty \leq (k+1)\ell N\beta \|\mathsf{Err}(A_i)\|_\infty + \|\mathsf{msg}(A_i)\|_1 (1 + kN)\epsilon +$$
$$+ \|\mathsf{msg}(A_i)\|_1 \|\mathsf{Err}(ACC_{i-1})\|_\infty$$
$$\leq (k+1)\ell N\beta 2 \|\mathsf{Err}(BK_i)\|_\infty + (1 + kN)\epsilon + \|\mathsf{Err}(ACC_{i-1})\|_\infty.$$

This proves the invariant by induction on $i$.

After SampleExtract (line 7), the message of $u$ is equal to the constant term of the message of $ACC_n$, i.e. $X^{\bar{\varphi}} \cdot testv$ where $\bar{\varphi} = \bar{b} - \sum_{i=1}^{n} \bar{a}_i s_i$. If $\bar{\varphi} \in [\![-N/2, N/2[\![$, the constant term is equal to $\bar{\mu}'$ and $-\bar{\mu}'$ otherwise.

In other words, $|\varphi_s(a, b)| < 1/4 - \delta$, then $\varphi_s(a, b) < 1/4 - \delta$ and $\varphi_s(a, b) \geq -1/4 + \delta$ and thus using Eq. (2), we obtain that $\bar{\varphi} \in ]\!] - \frac{N}{2}, \frac{N}{2}[\![$ and thus, the message of $u$ is equal to $\bar{\mu}'$. And if $|\varphi_s(a, b)| > 1/4 + \delta$ then $\varphi_s(a, b) > 1/4 + \delta$ or $\varphi_s(a, b) < -1/4 - \delta$ and using Eq. (2), we obtain the message of $u$ is equal to $-\bar{\mu}'$.

Since SampleExtract does not add extra noise, $\|\mathsf{Err}(u)\|_{\infty} \leq \|\mathsf{Err}(ACC_n)\|$. Since the KeySwitch procedure preserves the message, the message of $v = \mathsf{KeySwitch}_{\mathsf{KS}}(u)$ is equal to the message of $u$. And $\|\mathsf{Err}(v)\|_{\infty} \leq \|\mathsf{Err}(u)\|_{\infty} + kNt\gamma + kN2^{-(t+1)}$. □

**Corollary 4.7.** *Let $\vartheta_{BK} = \mathsf{Var}(\mathsf{Err}(BK_i)) = 2/\pi \cdot \alpha^2$ and $V_{\mathsf{KS}} = \mathsf{Var}(\mathsf{Err}(\mathsf{KS}_i)) = 2/\pi \cdot \gamma^2$. Under the same conditions of Theorem 4.6, and assuming Assumption 3.6, then the Variance of the output $v$ of Algorithm 3 satisfies $\mathsf{Var}(\mathsf{Err}(v)) \leq 2Nn(k+1)\ell\beta^2\vartheta_{BK} + kNtV_{\mathsf{KS}} + n(1+kN)\epsilon^2 + kN2^{-2(t+1)}$.*

*Proof.* The proof is the same as for the proof of the bound on $\|\mathsf{Err}(v)\|_{\infty}$ replacing all $\|\|_{\infty}$ inequalities by $\mathsf{Var}()$ inequalities. □

### 4.4 Application to Circuits

In [11], the homomorphic evaluation of a NAND gate between LWE samples is achieved with 2 additions (one with a noiseless trivial sample) and a bootstrapping. Let $BK = BK_{s \to s'', \alpha}$ be a bootstrapping key and $KS = KS_{s' \to s, \gamma, t}$ be a keyswitching secret defined as in Theorem 4.6 such that $2n(k+1)\ell\beta N\alpha + kNt\gamma + n(1+kN)\epsilon + kN2^{-(t+1)} < \frac{1}{16}$, We denote as Bootstrap $(c)$ the output of the bootstrapping procedure described in Algorithm 3 applied to $c$ with $\mu_0 = 0$ and $\mu_1 = \frac{1}{4}$. Let consider two LWE samples $c_1$ and $c_2$, with message space $\{0, 1/4\}$ and $\|\mathsf{Err}(c_1)\|_{\infty}, \|\mathsf{Err}(c_2)\|_{\infty} \leq \frac{1}{16}$. The result is obtained by computing $\tilde{c} = (\mathbf{0}, \frac{5}{8}) - c_1 - c_2$, plus a bootstrapping. Indeed the possible values for the messages of $\tilde{c}$ are $\frac{5}{8}, \frac{3}{8}$ if either $c_1$ or $c_2$ encode 0, and $\frac{1}{8}$ if both encode $\frac{1}{4}$. Since the noise amplitude $\|\mathsf{Err}(\tilde{c})\|_{\infty}$ is $< \frac{1}{8}$, then $|\varphi_s(\tilde{c})| > \frac{1}{4}$ iff. $\mathsf{NAND}(\mathsf{msg}(c_1), \mathsf{msg}(c_2)) = 1$. This explains why it suffices to bootstrap $\tilde{c}$ with parameters $(\mu_1, \mu_0) = (\frac{1}{4}, 0)$ to get the answer. By using a similar approach, it is possible to directly evaluate with a single bootstrapping all the basic gates:

- HomNOT$(c) = (\mathbf{0}, \frac{1}{4}) - c$ (no bootstrapping is needed);
- HomAND$(c_1, c_2) = $ Bootstrap $((\mathbf{0}, -\frac{1}{8}) + c_1 + c_2)$;
- HomNAND$(c_1, c_2) = $ Bootstrap $((\mathbf{0}, \frac{5}{8}) - c_1 - c_2)$;
- HomOR$(c_1, c_2) = $ Bootstrap $((\mathbf{0}, \frac{1}{8}) + c_1 + c_2)$;
- HomXOR$(c_1, c_2) = $ Bootstrap $(2 \cdot (c_1 - c_2))$.

The HomXOR$(c_1, c_2)$ gate can be achieved also by performing Bootstrap $(2 \cdot (c_1 + c_2))$.

## 4.5  Parameters Implementation and Timings

In this section, we review our implementation parameters and provide a comparison with previous works.

*Samples.* From a theoretical point of view, our scale invariant scheme is defined over the real torus $\mathbb{T}$, where all the operations are modulo 1. In practice, since we can work with approximations, we chose to rescale the elements over $\mathbb{T}$ by a factor $2^{32}$, and to map them to 32-bit integers. Thus, we take advantage of the native and automatic mod $2^{32}$ operations, including for the external multiplication with integers. Except for some FFT operations, this seems more stable and efficient than working with floating point numbers and reducing modulo 1 regularly. Polynomials mod $X^N + 1$ are either represented as the classical list of the $N$ coefficients, either using the Lagrange half-complex representation, which consists in the complex $(2 \cdot 64\text{bits})$ evaluations of the polynomial over the roots of unity $\exp(i(2j+1)\pi/N)$ for $j \in [\![0, \frac{N}{2}[\![$. Indeed, the $\frac{N}{2}$ other evaluations are the conjugates of the first ones, and do not need to be stored. The conversion between both representations is done via Fast Fourier Transform (FFT) (using the library *FFTW* [12], also used by [11]). Note that the direct FFT transform is $\sqrt{2N}$ lipschitzian, so the lagrange half-complex representation tolerates approximations, and 53 bits of precision is indeed more than enough, provided that the real representative remains small. However, the modulo 1 that can reduce the coefficients of Torus polynomials cannot be applied from the Lagrange representation: we need to perform regular transformations to and from the classical representation. Luckily, it does not represent an overhead, since these conversions are needed anyway, at each iteration of the bootstrapping in order to decompose the accumulator in base $h$.

*Parameters.* We take the same or even stronger security parameters as [11], but we adapt them to our notations. We used $n = 500$, $N = 1024$, $k = 1$.

- LWE samples: $32 \cdot (n + 1)$ bits $\approx 2$ KBytes.
  The mask of all LWE samples (initial and KeySwitch) are clamped to multiples of $\frac{1}{2048}$. Therefore, the phase computation in the bootstrapping is exact ($\delta = 0$).
- TLWE samples: $(k + 1) \cdot N \cdot 32$ bits $\approx 8$ KBytes.
- TGSW samples: $(k + 1) \cdot \ell$ TLWE samples $\approx 48$ KBytes.
  To define $h$ and $\text{Dec}_{h,\beta,\epsilon}$, we used $\ell = 3$, $B_g = 1024$, so $\beta = 512$ and $\epsilon = 2^{-31}$.
- Bootstrapping Key: $n$ TGSW samples $\approx 23.4$ MBytes.
  We used $\alpha = 9.0 \cdot 10^{-9}$. Since we have a lower noise overhead, our parameter is higher than the parameter $\approx 3.25 \cdot 10^{-10}$ of [11], (i.e. ours is more secure), but in counterpart, our TLWE key is binary. See Sect. 6 for more details on the security analysis.
- Key Switching Key: $k \cdot N \cdot t$ LWE samples $\approx 29.2$ MBytes.
  we used $\gamma = 3.05 \cdot 10^{-5}$, $t = 15$ (The decomposition in the key switching has an precision $2^{-16}$).
- Correctness: The final error variance after bootstrapping is $9.24.10^{-6}$, by Corollary 4.7. It corresponds to a standard deviation of $\sigma = 0.00961$.

In [11], the final standard deviation is larger $0.01076$. In other words, the noise amplitude after our bootstrapping is $< \frac{1}{16}$ with very high probability $\mathsf{erf}(1/16\sqrt{2}\sigma) \geq 1 - 2^{-33.56}$ (this is comparable to probability $\geq 1 - 2^{-32}$ in [11]).

Note that the size of the key switching key can be reduced by a factor $n + 1 = 501$ if all the masks are the output of a pseudo random function; we may for instance just give the seed. The same technique can be applied to the bootstrapping key, on which the size is only reduced by a factor $k + 1 = 2$.

*Implementation Tools and Source Code.* The source code of our implementation is available on github https://github.com/tfhe/tfhe. We implemented the FHE scheme in C/C++, and run the bootstrapping algorithm on a 64-bit single core (i7-4930MX) at $3.00$ GHz. This seems to correspond to the machine used in [11]. We implemented a version with classical representation for polynomials, and a version in Lagrange half-complex representation. The following table compares the number of multiplications or FFT that are required to complete one external product and the full bootstrapping.

|  | #(Classical products) | #(FFT + Lagrange repr.) |
|---|---|---|
| External product | 12 | 8 |
| Bootstrapping | 6000 | 4006 |
| Bootstrapping in [11] | (72000) | 48000 |

In practice, we obtained a running time of 52ms per bootstrapping using the Lagrange half-complex representation. It is coherent with the 12x speed-up predicted by the table. Profiling the execution shows that the FFTs and complex multiplications are still taking more than $90\%$ of the total time. Other operations like keyswitch have a negligible running time compared to the main loop of the bootstrapping.

# 5   Leveled Homomorphic Encryption

In the previous section, we showed how to accelerate the bootstrapping computation in FHE. In this section, we focus on the improvement of Leveled Homomorphic encryption schemes. We present an efficient way to evaluate any deterministic automata homomorphically.

## 5.1   Boolean Circuits Interpretation

In order to express our external product in a circuit, we consider two kinds of wires: *control wires* which encode either a small integer or a small integer polynomial. They will be represented by a TGSW sample; and *data wires* which encode either a sample in $\mathbb{T}$ or in $\mathbb{T}_N[X]$. They will be represented by a TLWE sample. The gates we present contain three kinds of slots: *control input*, *data input* and *data output*. In this following section, the rule to build valid circuits is that all control wires are freshly generated by the user, and the data input ports of our gates can be either freshly generated or connected to a data output or to another gate.

We now give an interpretation of our leveled scheme, to simulate boolean circuits only. In this case, the message space of the input TLWE samples will be restricted to $\{0, \frac{1}{2}\}$, and the message space of control gates to $\{0, 1\}$.

- The constant source $\mathtt{Cst}(\mu)$ for $\mu \in \{0, \frac{1}{2}\}$ is defined with a single data output equal to $(\mathbf{0}, \mu)$.
- The negation gate $\mathtt{Not}(\boldsymbol{d})$ takes a single data input $\boldsymbol{d}$ and outputs $(\mathbf{0}, \frac{1}{2}) - \boldsymbol{d}$.
- The controlled And gate $\mathtt{CAnd}(C, \boldsymbol{d})$ takes one control input $C$ and one data input $\boldsymbol{d}$, and outputs $C \boxdot \boldsymbol{d}$.
- The controlled Mux gate $\mathtt{CMux}(C, \boldsymbol{d_1}, \boldsymbol{d_0})$ takes one control input $C$ and two data inputs $\boldsymbol{d_1}, \boldsymbol{d_0}$ and returns $C \boxdot (\boldsymbol{d_1} - \boldsymbol{d_0}) + \boldsymbol{d_0}$.

Unlike classical circuits, these gates have to be composed with each other depending on the type of inputs/outputs. In our applications, the TGSW encryptions are always fresh ciphertexts.

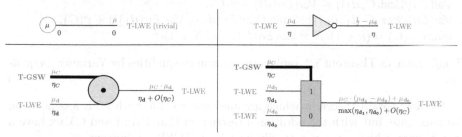

**Theorem 5.1 (Correctness).** *Let* $\mu \in \{0, \frac{1}{2}\}$, $\boldsymbol{d}, \boldsymbol{d_1}, \boldsymbol{d_0} \in \mathrm{TLWE}_{\boldsymbol{s}}(\{0, \frac{1}{2}\})$ *and* $C \in \mathrm{TGSW}_{\boldsymbol{s}}(\{0, 1\})$.

- $msg(\mathtt{Cst}(\mu)) = \mu$
- $msg(\mathtt{Not}(\boldsymbol{d})) = \frac{1}{2} - \mu = not\ \mu$
- $msg(\mathtt{CAnd}(C, \boldsymbol{d})) = msg(C) \cdot msg(\boldsymbol{d})$
- $msg(\mathtt{CMux}(C, \boldsymbol{d_1}, \boldsymbol{d_0})) = msg(C)\,?\,msg(\boldsymbol{d_1}):msg(\boldsymbol{d_0})$

**Theorem 5.2 (Worst-case noise).** *In the conditions of Theorem 5.1, we have*

- $\|Err(\mathtt{Cst}(\mu))\|_\infty = 0$
- $\|Err(\mathtt{Not}(\boldsymbol{d}))\|_\infty = \|Err(\boldsymbol{d})\|_\infty$
- $\|Err(\mathtt{CAnd}(C, \boldsymbol{d}))\|_\infty \leq \|Err(\boldsymbol{d})\|_\infty + \eta(C)$
- $\|Err(\mathtt{CMux}(C, \boldsymbol{d_1}, \boldsymbol{d_0}))\|_\infty \leq \max(\|Err(\boldsymbol{d_0})\|_\infty, \|Err(\boldsymbol{d_1})\|_\infty) + \eta(C)$,
  *where* $\eta(C) = (k+1)\ell N \beta \|Err(C)\|_\infty + (kN+1)\epsilon$.

*Proof.* The noise is indeed null for constant gates, and negated for the Not gate, which preserves the norm. The noise bound for the CAnd gate is exactly the one from Theorem 3.14, however, we need to explain why there is a max in the CMux formula instead of the sum we would obtain by blindly applying Theorem 3.14. Let $d = d_1 - d_0$, recall that in the proof of Theorem 3.14, the expression of $C \boxdot d$ is $\mathrm{Dec}_{h,\beta,\epsilon}(d) \cdot z_C + \mu_C \epsilon_{\mathrm{dec}} + \mu_C z_d + (0, \mu_C \cdot \mu_d)$, where $C = z_C + \mu_C \cdot h$ and $d = z_d + \mu_d$, $z_C$ and $z_d$ are respectively TGSW and TLWE samples of 0, and $\|\epsilon_{\mathrm{dec}}\|_\infty \leq \epsilon$. Thus, $\mathtt{CMux}(C, d_1, d_0)$ is the sum of four terms:

- $\mathrm{Dec}_{h,\beta,\epsilon}(d) \cdot z_C$ of norm $\leq (k+1)\ell N \beta \eta_C$;
- $\mu_C \epsilon_{\mathrm{dec}}$ of norm $\leq (kN+1)\epsilon$;
- $z_{d_0} + \mu_C(z_{d_1} - z_{d_0})$, which is either $z_{d_1}$ or $z_{d_0}$, depending on the value of $\mu_C$;
- $\mu_{d_0} + \mu_C \cdot (\mu_{d_1} - \mu_{d_0})$, which is the output message $\mu_C?\mu_{d_1}:\mu_{d_0}$, and is not part of the noise.

Thus, summing the three terms concludes the proof. $\qquad\square$

**Corollary 5.3 (Average noise of boolean gates).** *In the conditions of Theorem 5.1, and in the conditions of Assumption 3.6, we have:*

- $\mathsf{Var}(\mathsf{Err}(\mathtt{Cst}(\mu))) = 0;$
- $\mathsf{Var}(\mathsf{Err}(\mathtt{Not}(d))) = \mathsf{Var}(\mathsf{Err}(d));$
- $\mathsf{Var}(\mathsf{Err}(\mathtt{CAnd}(C, d))) \leq \mathsf{Var}(\mathsf{Err}(d)) + \vartheta(C);$
- $\mathsf{Var}(\mathsf{Err}(\mathtt{CMux}(C, d_1, d_0))) \leq \max(\mathsf{Var}(\mathsf{Err}(d_0)), \mathsf{Var}(\mathsf{Err}(d_1))) + \vartheta(C),$
  *where* $\vartheta(C) = (k+1)\ell N \beta^2 \mathsf{Var}(\mathsf{Err}(C)) + (kN+1)\epsilon^2.$

*Proof.* Same as Theorem 5.2, replacing all norm inequalities by Variance inequalities. $\qquad\square$

We now obtain theorems which are analogue to [13], with a bit less noise on the mux gate, but with the additional restriction that CAnd and CMux have a control wire, which must necessarily be a fresh TGSW ciphertext.

The next step is to understand the meaning of this additional restriction in terms of expressiveness of the resulting homomorphic circuits.

It is clear that we cannot build a random boolean circuit, and just apply the noise recurrence formula from Theorem 5.2 or Corollary 5.3 to get the output noise level. Indeed, it is not allowed to connect a data wire to an control input.

In the following section, we will show that we can still obtain the two most important circuits of [13], namely the deterministic automata circuits, which can evaluate any permutation of regular languages with noise propagation sublinear in the word length and the lookup table, which evaluates arbitrary functions with sublinear noise propagation.

## 5.2   Deterministic Automata

It is folklore that every deterministic program which reads its input bit-by-bit in a pre-determined order, uses less than $B$ bits of memory, and produces a boolean answer, is equivalent to a deterministic automata of at most $2^B$ states (independently of the time complexity). This is in particular the case for every boolean function of $p$ variables, that can be trivially executed with $p - 1$ bits of internal memory by reading and storing its input bit-by-bit before returning the final answer. It is of particular interest for most arithmetic functions, like addition, multiplication, or CRT operations, whose naive evaluation only requires $O(\log(p))$ bits of internal memory.

Let $\mathcal{A} = (Q, i, T_0, T_1, F)$ be a deterministic automata (over the alphabet $\{0,1\}$, where $Q$ is the set of states, $i \in Q$ denotes the initial state, $T_0, T_1$ are the two transitions (deterministic) functions from $Q$ to $Q$ and $F \subset Q$ is the set of final states. Such automata is used to evaluate (rational) boolean functions on words where the image of $(w_1, \ldots, w_p) \in \mathbb{B}^p$ is equal to 1 iff. $T_{w_p}(T_{w_{p-1}}(\ldots(T_{w_1}(i)))) \in F$, and 0 otherwise.

Following the construction of [13], we show that we are able to evaluate any deterministic automata homomorphically using only constant and CMux gates efficiently. The noise propagation remains linear in the length of the word $w$, but compared to [13, Theorem 7.11], we reduce the number of evaluated CMux gates by a factor $|w|$ for a specific class of acyclic automata that are linked to fixed-time algorithms.

**Theorem 5.4 (Evaluating Deterministic Automata).** *Let $\mathcal{A} = (Q, i, T_0, T_1, F)$ be a deterministic automata. Given $p$ valid TGSW samples $C_1, \ldots, C_p$ encrypting the bits of a word $w \in \mathbb{B}^p$, with noise amplitude $\eta = \max_i \|\mathsf{Err}(C_i)\|_\infty$ and $\vartheta = \max_i \mathsf{Var}(\mathsf{Err}(C_i))$, by evaluating at most $\leq p\#Q$ Cmux gates, one can produce a TLWE sample $d$ which encrypts $\frac{1}{2}$ iff $\mathcal{A}$ accepts $w$, and 0 otherwise such that $\|\mathsf{Err}(d)\|_\infty \leq p \cdot ((k+1)\ell N \beta \eta + (kN+1)\epsilon)$. Assuming Heuristic 3.6, $\mathsf{Var}(\mathsf{Err}(d)) \leq p \cdot ((k+1)\ell N \beta^2 \vartheta + (kN+1)\epsilon^2)$. Furthermore, the number of evaluated CMux can be decreased to $\leq \#Q$. if $\mathcal{A}$ satisfies either one of the conditions:*

*(i) for all $q \in Q$ (except KO states), all the words that connect $i$ to $q$ have the same length;*

*(ii) $\mathcal{A}$ only accepts words of the same length.*

*Proof.* We initialize $\#Q$ noiseless ciphertexts $d_{q,p}$ for $q \in Q$ with $d_{q,p} = (\mathbf{0}, \frac{1}{2}) = \mathsf{Cst}(\frac{1}{2})$ if $q \in F$ and $d_{q,p} = (\mathbf{0}, 0) = \mathsf{Cst}(0)$ otherwise. Then for each letter of $w$, we map the transitions as follow for all $q \in Q$ an $j \in [\![0, p-1]\!]$: $d_{q,j-1} = \mathsf{CMux}(C_j, d_{T_1(q),j}, d_{T_0(q),j})$. And we finally output $d_{i,0}$.

Indeed, with this construction, we have

$$\mathsf{msg}(d_{i,0}) = \mathsf{msg}(d_{T_{w_1}(i),1}) = \ldots = \mathsf{msg}(d_{T_{w_p}(T_{w_{p-1}}\ldots(T_{w_1}(i))\ldots),p}),$$

which encrypts $\frac{1}{2}$ iff $T_{w_p}(T_{w_{p-1}} \ldots (T_{w_1}(i)) \ldots) \in F$, i.e. iff $w_1 \ldots w_p$ is accepted by $\mathcal{A}$. This proves correctness.

For the complexity, each $d_{q,j}$ for all $q \in Q$ an $j \in [\![0, p-1]\!]$ is computed with a single CMux. By applying the noise propagation inequalities of Theorem 5.2 and Corollary 5.3, it follows by an immediate induction on $j$ from $p$ down to 0, that for all $j \in [\![0, p]\!]$, $\|\mathsf{Err}(d_{q,j})\|_\infty \leq (p-j) \cdot ((k+1)\ell N \beta \eta + (kN+1)\epsilon)$ and $\mathsf{Var}(\mathsf{Err}(d_{q,j})) \leq (p-j) \cdot ((k+1)\ell N \beta^2 \vartheta + (kN+1)\epsilon^2)$.

Note that it is sufficient to evaluate only the $d_{q,j}$ when $q$ is accessible by at least one word of length $j$. Thus, if the $\mathcal{A}$ satisfies the additional condition (i), then for each $q \in Q$, we only need to evaluate $d_{q,j}$ for at most one position $j$. Thus, we evaluate less than $\#Q$ CMux gates in total.

Finally, if $\mathcal{A}$ satisfies (ii), then we first compute the minimal deterministic automata of the same language (and removing the KO state if it is present), then with an immediate proof by contradiction, this minimal automata satisfies (i), and has less than $\#Q$ states. $\qquad\square$

For sake of completeness, since every boolean function with $p$ variables can be evaluated by an Automata (that accepting only words of length $p$), we obtain the evaluation of arbitrary boolean function as an immediate corollary, which is the leveled variant of [13, Corollary 7.9].

**Lemma 5.5 (Arbitrary Functions).** *Let $f$ be any boolean function with $p$ inputs, and $c_1, \ldots, c_p$ be $p$ $\mathrm{TGSW}_s(\{0,1\})$ ciphertexts of $x_1, \ldots, x_p \in \{0,1\}$, with noise $\|\mathsf{Err}(c_i)\|_\infty \leq \eta$ for all $i \in [1,p]$. Then the CMux-based Reduced Binary Decision Diagram of $f$ computes a $\mathrm{TLWE}_s$ ciphertext $d$ of $\frac{1}{2}f(x_1, \ldots, x_p)$ with noise $\|\mathsf{Err}(d)\|_\infty \leq p((k+1)\ell N \beta \eta + (kN+1)\epsilon)$ by evaluating $\mathcal{N}(f) \leq 2^p$ CMux gates where $\mathcal{N}(f)$ is the number of distinct partial functions $(x_l, \ldots, x_p) \to f(x_1, \ldots, x_p)$ for all $l \in [1, p+1]$, $(x_1, \ldots, x_{l-1}) \in \mathbb{B}^{l-1}$.*

*Proof (sketch).* A trivial automata which evaluates $f$ consists in its full binary decision tree, with the initial state $i = q_{0,0}$ as the root, each state $q_{l,j}$ depth $l \in [0, p-1]$ and $j \in [0, 2^l-1]$ is connected with $T_0(q_{l,j}) = q_{l+1,2j}$ and $T_1(q_{l,j}) = q_{l+1,2j+1}$, and at depth $p$, $q_{p,j} \in F$ iff $f(x_1, \ldots, x_p) = 1$ where $j = \sum_{l=1}^p x_l 2^{p-l}$. The minimal version of this automaton has at most $\mathcal{N}(f)$ states, the rest follows from Theorem 5.4. $\qquad\square$

**Application: Compilation for Leveled Homomorphic Circuits.** We now give an example of how we can map a problem to an automata in order to perform a leveled homomorphic evaluation. We will illustrate this concept on the computation of the $p$-th bit of an integer product $a \times b$ where $a$ and $b$ are given in base 2. We do not claim that the automata approach is the fastest way to solve the problem, arithmetic circuits based on bitDecomp/recomposition are likely to be faster. But the goal is to clarify the generality and simplicity of the process. All we need is a fixed-time algorithm that solves the problem using the least possible memory. Among all algorithms that compute a product, the most naive ones are in general the best: here, we choose the elementary-school multiplication algorithm that computes the product bit-by-bit, starting from the LSB, and counting the current carry with the fingers. The pseudocode of this algorithm is recalled in Algorithm 4. The pseudo-code is almost given as a deterministic automata, since each step reads a single input bit, and uses it to update its internal state $(x, y)$, that can be stored in only $M = \log_2(4p)$ bits of memory. More precisely, the states $Q$ of the corresponding automata $\mathcal{A}$ would be all $(j, (x, y))$ where $j \in [0, j_{\max}]$ is the step number (*i.e.* number of reads from the beginning) and $(x, y) \in \mathbb{B} \times [0, 2p[$ are the $4p$ possible values of the internal memory. The initial state is $(0, 0, 0)$, the total number of reads $j_{max}$ is $\leq p^2$, and the final states are all $(j_{max}, x, y)$ where $y$ is odd. This automata satisfies condition (i), since a state $(j, x, y)$ can only be reached after reading $j$ inputs, so by Theorem 5.4, the output can be homomorphically computed by evaluating less than $\#Q \leq 4p^3$ CMux gates, with some $O(p)$ noise overhead. The number of Mux can decrease by a factor 8 by minimizing the automata. Using the same parameters as the bootstrapping key, for $p = 32$, evaluating one Mux gate takes about 0.0002 s, so the whole program (16384 Cmux) would be homomorphically evaluated in 3.2 s.

We mapped a problem from its high-level description to an algorithm using very few bits of memory. Since low memory programs are in general more naive, it should be easier to find them than obtaining a circuit with low multiplicative depth that would be required for other schemes such as BGV, FHE over integers. Once a suitable program is found, as in the previous example, compiling it to a net-list of CMux gates is straightforward by our Theorem 5.4.

---

**Algorithm 4.** Elementary fixed time algorithm that computes the $p$-th bit of the product of $a$ and $b$

**Input:** $a$ and $b$ as little endian bits
**Output:** $p$-th bit of $ab$
1: Internal memory: $x \in \{0, 1\}, y \in [0, 2p[$
2: **initialize** $x = 0, y = 0$
3: **for** $k = 0$ to $p - 1$ **do**
4:　　**for** $i = 0$ to $k - 1$ **do**
5:　　　　**read** $a_i$; $x = a_i$
6:　　　　**read** $b_{k-i}$; $y = y + xb_{k-i}$
7:　　**end for**
8:　　**read** $a_k$; $x = a_k$
9:　　**read** $b_0$; $y = \lfloor (y + xb_0)/2 \rfloor$
10: **end for**
11: **for** $i = 0$ to $p$ **do**
12:　　**read** $a_i$; $x = a_i$
13:　　**read** $b_{p-i}$; $y = y + xb_{p-i}$
14: **end for**
15: **accept if** $y == 1 \mod 2$

---

# 6    Practical Security Parameters

For an asymptotical security analysis, since the phase is lipschitzian, TLWE samples can be equivalently mapped to their closest binLWE (or bin-RingLWE), which in turn can be reduced to standard LWE/ringLWE with full secret using the modulus-dimension reduction [6] or group-switching techniques [13]. It can

then be reduced to worst case BDD instances. It is also easy to write a direct
and tighter search-to-decision reductions for TLWE, or a direct worst-case to
average-case reductions from TLWE to Gap-SVP or BDD.

In this section, we will rather focus on the practical hardness of LWE, and
express after all the security parameter $\lambda$ directly as a function of the entropy
of the secret $n$ and the error rate $\alpha$.

Our analysis is based on the work described in [2]. This paper studies many
attacks against LWE, ranging from a direct BDD approach with standard lattice reduction, sieving, or with a variant of BKW [4], resolution via man in
the middle attacks. Unfortunately, they found out that there is no single-best
attack. According to their results table [2, Sect. 8, Tables 7 and 8] for the range
of dimensions and noise used for FHE, it seems that the SIS-distinguisher attack
is often the best candidate (related to the Lindner-Peikert [17] model, and also
used in the parameter estimation of [11]). However, since $q$ is not a parameter in
our definition of TLWE, we need to adapt their results. This section relies on the
following heuristics concerning the experimental behaviour of lattice reduction
algorithms. They have been extensively verified and used in practice.

1. The fastest lattice reduction algorithms in practice are blockwise lattice algorithms (like BKZ-2.0 [8], D-BKZ [20], or the slide reduction with large blocksize [14, 20]).
2. Practical blockwise lattice reduction algorithms have an intrinsic quality $\delta >$
   $1$ (which depends on the blocksize), and given a $m$-dimensional real basis $B$
   of volume $V$, they compute short vectors of norm $\delta^m V^{1/m}$.
3. The running time of BKZ-2.0 (expressed in bit operations) as a function of
   the quality parameter is: $\log_2(t_{\text{BKZ}})(\delta) = \frac{0.009}{\log_2(\delta)^2} - 27$ (According to the
   extrapolation by Albrecht et al. [1] of Liu-Nguyen datasets [18]).
4. The coordinates of vectors produced by lattice reduction algorithms are bal-
   anced. Namely, if the algorithm produces vectors of norm $\|v\|_2$, each coeffi-
   cient has a marginal Gaussian distribution of standard deviation $\|v\|_2 / \sqrt{n}$.
   Provided that the geometry of the lattice is not too skewed in particular directions, this fact can sometimes be proved, especially if the reduction algorithm
   samples vectors with Gaussian distribution over the input lattice. This simple
   fact is at the heart of many attacks based on Coppersmith techniques with
   lattices.
5. For mid-range dimensions and polynomially small noise, the SIS-distinguisher
   plus lattice reduction algorithms combined with the search-to-decision is the
   best attack against LWE; (but this point is less clear, according to the analysis
   of [1], at least, this attack model tends to over-estimate the power of the
   attacker, so it should produce more conservative parameters).
6. Except for small polynomial speedups in the dimension, we don't know better
   algorithms to find short vectors in random anti-circulant lattices than generic
   algorithms. This folklore assumption seems still up-to date at the time of
   writing.

If one finds a small integer combination that cancels the mask of homogeneous
LWE samples, one may use it to distinguish them from uniformly chosen random

samples. If this distinguisher has small advantage $\varepsilon$, we repeat it about $1/\varepsilon^2$ times. Then, thanks to the search to decision reduction (which is particularly tight with our TLWE formulation), each successful answer of the distinguisher reveals one secret key bit. To handle the continuous torus, and since $q$ is not a parameter of TLWE either, we show how to extend the analysis of [2] to our scheme.

Let $(\boldsymbol{a_1}, b_1), \ldots, (\boldsymbol{a_m}, b_m)$ be either $m$ LWE samples of parameter $\alpha$ or $m$ uniformly random samples of $\mathbb{T}^{n+1}$, we need to find a small combination $v_1, \ldots, v_m$ of samples such that $\sum v_i \boldsymbol{a_i}$ is small. This condition differs from most previous models, were working on a discrete group, and required an exact solution. By allowing approximations, we may find solutions for much smaller $m$ than the usual bound $n \log q$, even $m < n$ can be valid. Now, consider the $(m+n)$-dimensional lattice, generated by the rows of the following basis $B \in \mathcal{M}_{n+m, n+m}(\mathbb{R})$:

$$
B = \begin{bmatrix}
1 & & & 0 & & & \\
 & \ddots & & & & 0 & \\
0 & & 1 & & & & \\
a_{1,1} & \cdots & a_{1,n} & 1 & & 0 & \\
\vdots & \ddots & \vdots & & \ddots & & \\
a_{m,1} & \cdots & a_{m,n} & 0 & & 1 &
\end{bmatrix}.
$$

Our target is to find a short vector $\boldsymbol{w} = [x_1, \ldots, x_n, v_1, \ldots, v_m]$ in the lattice of $B$, whose first $n$ coordinates $(x_1, \ldots, x_n) = \sum_{i=1}^{m} v_i \boldsymbol{a_i} \mod 1$ are shorter than the second part $(v_1, \ldots, v_m)$. To take this skewness into account, we choose a real parameter $q > 1$ (that will be optimized later), and apply the unitary transformation $f_q$ to the lattice, which multiplies the first $n$ coordinates by $q$ and the last $m$ coordinates by $1/q^{n/m}$. Although this matrix looks like a classical LWE matrix instance, the variable $q$ is a real parameter, and it doesn't need to be an integer. It then suffices to find a regular short vector with balanced coordinates in the transformed lattice, defined by this basis:

$$
f_q(B) = \begin{bmatrix}
q & & & 0 & & & \\
 & \ddots & & & & 0 & \\
0 & & q & & & & \\
qa_{1,1} & \cdots & qa_{1,n} & \frac{1}{q^{n/m}} & & 0 & \\
\vdots & \ddots & \vdots & & \ddots & & \\
qa_{m,1} & \cdots & qa_{m,n} & 0 & & \frac{1}{q^{n/m}} &
\end{bmatrix}, \text{ with } q \in \mathbb{R} > 1.
$$

The direct approach is to apply the fastest algorithm (BKZ-2.0 or slide reduction) directly to $f_q(B)$, which outputs a vector $f_q(\boldsymbol{w})$ of standard deviation $\delta^{n+m}/\sqrt{n+m}$ where $\delta \in ]1, 1.1]$ is the quality of the reduction.

Once we have a vector $\boldsymbol{w}$, all we need is to analyse the term $\sum_{i=1}^{m} v_i b_i = \sum_{i=1}^{m} v_i(\boldsymbol{a_i} s + e_i) = \boldsymbol{s} \cdot \sum_{i=1}^{m}(v_i \boldsymbol{a_i}) + \sum_{i=1}^{m} v_i e_i = \boldsymbol{s} \cdot \boldsymbol{x} + \boldsymbol{v} \cdot \boldsymbol{e}$.

It has Gaussian distribution of square parameter $\sigma^2 = \frac{\delta^{2(m+n)}\pi}{2q^2} \cdot \frac{nS^2}{m+n} +$ $\frac{q^{2n/m}\delta^{2(m+n)}\alpha^2 m}{m+n} = \delta^{2(m+n)}\left(\frac{\pi S^2}{2q^2} \cdot \frac{n}{m+n} + q^{2n/m}\alpha^2 \frac{m}{m+n}\right)$. Here $S = \frac{\|s\|}{\sqrt{n}} \approx \frac{1}{\sqrt{2}}$. By definition of the smoothing parameter, it may be distinguished from the uniform distribution with advantage $\varepsilon$ as long as $\sigma^2 \geq \eta_\varepsilon^2(\mathbb{Z})$. To summarize, the security parameter of LWE is (bounded by) the solution of the following system of equations

$$\lambda(n, \alpha) = \log_2(t_{attack}) = \min_{0 < \varepsilon < 1} \log_2\left(\frac{n}{\varepsilon^2}t_{BKZ}(n, \alpha, \varepsilon)\right) \tag{3}$$

$$\log_2(t_{BKZ})(n, \alpha, \varepsilon) = \frac{0.009}{\log_2(\delta)^2} - 27 \tag{4}$$

$$\ln(\delta)(n, \alpha, \varepsilon) = \max_{\substack{m > 1 \\ q > 1}} \frac{1}{2(m+n)}\left(\ln(\eta_\varepsilon^2(\mathbb{Z})) - \ln\left(\frac{\pi S^2}{2q^2}\frac{n}{m+n} + q^{\frac{2n}{m}}\alpha^2\frac{m}{m+n}\right)\right) \tag{5}$$

$$\eta_\varepsilon(\mathbb{Z}) \approx \sqrt{\frac{1}{\pi}\ln(\frac{1}{\varepsilon})}. \tag{6}$$

Here, Eq. (3) means that we need to run the distinguisher $\frac{1}{\varepsilon^2}$ times per unknown key bit (by Chernoff's bound), and we need to optimize the advantage $\varepsilon$ accordingly. Equation (4) is the heuristic prediction of the running time of lattice reduction. In Eq. (5) $q$ and $m$ need to be chosen in order to maximize the targeted approximation factor of the lattice reduction step.

Differentiating Eq. (5) in $q$, we find that its maximal value is

$$q_{best} = \left(\frac{\pi S^2}{2\alpha^2}\right)^{\frac{m}{2(m+n)}}.$$

Replacing this value and setting $t = \frac{n}{m+n}$, Eq. (5) becomes:

$$\ln(\delta)(n, \alpha, \varepsilon) = \max_{t > 0} \frac{1}{2n}\left(t^2\ell_2 + t(1-t)\ell_1\right) \text{ where } \begin{cases} \ell_1 = \ln\left(\frac{\eta_\varepsilon^2(\mathbb{Z})}{\alpha^2}\right) \\ \ell_2 = \ln\left(\frac{2\eta_\varepsilon^2(\mathbb{Z})}{\pi S^2}\right). \end{cases}$$

Finally, by differentiating this new expression in $t$, the maximum of $\delta$ is reached for $t_{best} = \frac{\ell_1}{2(\ell_1 - \ell_2)}$, because $\ell_1 > \ell_2$, which gives the best choices of $m$ and $q$ and $\delta$. Finally, we optimize $\varepsilon$ numerically in Eq. (3).

All previous results are summarized in Fig. 1, which displays the security parameter $\lambda$ as a function of $n, \log_2(\alpha)$.

In particular, in the following table we precise the values for the keyswitching key and the bootstrapping key (for our implementation and for the one in [11]).

|  | $n$ | $\alpha$ | $\lambda$ | $\varepsilon_{best}$ | $m_{best}$ | $q_{best}$ | $\delta_{best}$ |
|---|---|---|---|---|---|---|---|
| Switch key | 500 | $2^{-15}$ | 136 | $2^{-12}$ | 444 | 125.7 | 1.0058 |
| Boot. key | 1024 | $9.0 \cdot 10^{-9}$ | 194 | $2^{-10}$ | 968 | 7664 | 1.0048 |
| Boot.key [11] | 1024 | $3.25 \cdot 10^{-10}$ | 141 | $2^{-7}$ | 993 | 44096 | 1.0055 |

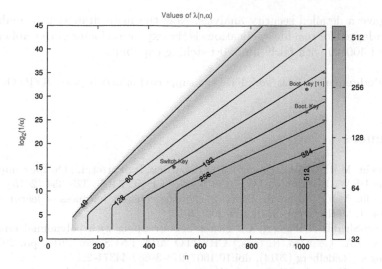

**Fig. 1.** Security parameter $\lambda$ as a function of $n$ and $\alpha$ for LWE samples. *This curve shows the security parameter levels $\lambda$ (black levels) as a function of $n = kN$ (along the x-axis) and $\log_2(1/\alpha)$ (along the y-axis) for TLWE (also holds for bin-LWE), considering both the attack of this section and the collision attack in time $2^{n/2}$.*

The table shows that the strength of the lattice reduction is compatible with the values announced in [11]. Our model predicts that the lattice reduction phase is harder ($\delta = 1.0055$ in our analysis and $\delta = 1.0064$ in [11]), but the value of $\varepsilon$ is bigger in our case. Overall, the security of their parameters-set is evaluated by our model to 136-bits of security, which is larger than the $\geq$ 100-bits of security announced in [11]. The main reason is that we take into account the number of times we need to run the SIS-distinguisher to obtain a non negligible advantage. Since our scheme has a smaller noise propagation overhead, we were able to raise the input noise levels in order to strengthen the system, so with the parameters we chose in our implementation, our model predicts 194-bits of security for the bootstrapping key and 136-bits for the keyswitching key (which remains the bottleneck).

## 7 Conclusion

In this paper, we presented a generalization of the LWE and GSW homomorphic encryption schemes. We improved the execution timing of the bootstrapping procedure and we reduced the size of the keys by keeping at least the same security as in previous fast implementations. This result has been obtained by simplifying the multiplication morphism, which is the main operation used in the scheme we described. As a proof of concept we implemented the scheme itself and we gave concrete parameters and timings. Furthermore, we extend the applicability of the external product to leveled homomorphic encryption. We finally gave a detailed security analysis. Now the main drawback to make our scheme adapted for real life applications is the expansion factor of the ciphertexts of around 400000 with fairly limited batching capabilities.

**Acknowledgements.** This work has been supported in part by the CRYPTOCOMP project.

# References

1. Albrecht, M.R., Cid, C., Faugère, J., Fitzpatrick, R., Perret, L.: On the complexity of the BKW algorithm on LWE. Des. Codes Crypt. **74**(2), 325–354 (2015)
2. Albrecht, M.R., Player, R., Scott, S.: On the concrete hardness of learning with errors. J. Math. Crypt. **9**(3), 169–203 (2015)
3. Alperin-Sheriff, J., Peikert, C.: Faster bootstrapping with polynomial error. In: Garay, J.A., Gennaro, R. (eds.) CRYPTO 2014. LNCS, vol. 8616, pp. 297–314. Springer, Heidelberg (2014). doi:10.1007/978-3-662-44371-2_17
4. Blum, A., Kalai, A., Wasserman, H.: Noise-tolerant learning, the parity problem, and the statistical query model. J. ACM **50**(4), 506–519 (2003)
5. Brakerski, Z., Gentry, C., Vaikuntanathan, V.: (Leveled) fully homomorphic encryption without bootstrapping. In: ITCS, pp. 309–325 (2012)
6. Brakerski, Z., Langlois, A., Peikert, C., Regev, O., Stehlé, D.: Classical hardness of learning with errors. In: Proceedings of 45th STOC, pp. 575–584. ACM (2013)
7. Brakerski, Z., Vaikuntanathan, V.: Efficient fully homomorphic encryption from (standard) LWE. In: FOCS, pp. 97–106 (2011)
8. Chen, Y., Nguyen, P.Q.: BKZ 2.0: better lattice security estimates. In: Lee, D.H., Wang, X. (eds.) ASIACRYPT 2011. LNCS, vol. 7073, pp. 1–20. Springer, Heidelberg (2011). doi:10.1007/978-3-642-25385-0_1
9. Cheon, J.H., Stehlé, D.: Fully homomophic encryption over the integers revisited. In: Oswald, E., Fischlin, M. (eds.) EUROCRYPT 2015. LNCS, vol. 9056, pp. 513–536. Springer, Heidelberg (2015). doi:10.1007/978-3-662-46800-5_20
10. Chillotti, I., Gama, N., Georgieva, M., Izabachène, M.: A homomorphic LWE based e-voting scheme. In: Takagi, T. (ed.) PQCrypto 2016. LNCS, vol. 9606, pp. 245–265. Springer, Heidelberg (2016). doi:10.1007/978-3-319-29360-8_16
11. Ducas, L., Micciancio, D.: FHEW: bootstrapping homomorphic encryption in less than a second. In: Oswald, E., Fischlin, M. (eds.) EUROCRYPT 2015. LNCS, vol. 9056, pp. 617–640. Springer, Heidelberg (2015). doi:10.1007/978-3-662-46800-5_24
12. Frigo, M., Johnson, S.G.: The design, implementation of FFTW3. In: Proceedings of the IEEE, vol. 93, no. 2, pp. 216–231 (2005). Special issue on "Program Generation, Optimization, and Platform Adaptation"
13. Gama, N., Izabachène, M., Nguyen, P.Q., Xie, X.: Structural lattice reduction: generalized worst-case to average-case reductions. IACR Crypt. ePrint Arch. **2014**, 48 (2014)
14. Gama, N., Nguyen, P.Q.: Predicting lattice reduction. In: Smart, N. (ed.) EUROCRYPT 2008. LNCS, vol. 4965, pp. 31–51. Springer, Heidelberg (2008). doi:10.1007/978-3-540-78967-3_3
15. Gentry, C.: Fully homomorphic encryption using ideal lattices. In: 41st ACM STOC, pp. 169–178 (2009)

16. Gentry, C., Sahai, A., Waters, B.: Homomorphic encryption from learning with errors: conceptually-simpler, asymptotically-faster, attribute-based. In: Canetti, R., Garay, J.A. (eds.) CRYPTO 2013. LNCS, vol. 8042, pp. 75–92. Springer, Heidelberg (2013). doi:10.1007/978-3-642-40041-4_5

17. Lindner, R., Peikert, C.: Better key sizes (and Attacks) for LWE-based encryption. In: Kiayias, A. (ed.) CT-RSA 2011. LNCS, vol. 6558, pp. 319–339. Springer, Heidelberg (2011). doi:10.1007/978-3-642-19074-2_21

18. Liu, M., Nguyen, P.Q.: Solving BDD by enumeration: an update. In: Dawson, E. (ed.) CT-RSA 2013. LNCS, vol. 7779, pp. 293–309. Springer, Heidelberg (2013). doi:10.1007/978-3-642-36095-4_19

19. Lyubashevsky, V., Peikert, C., Regev, O.: On ideal lattices and learning with errors over rings. In: Gilbert, H. (ed.) EUROCRYPT 2010. LNCS, vol. 6110, pp. 1–23. Springer, Heidelberg (2010). doi:10.1007/978-3-642-13190-5_1

20. Micciancio, D., Walter, M.: Practical, predictable lattice basis reduction. In: Fischlin, M., Coron, J.-S. (eds.) EUROCRYPT 2016. LNCS, vol. 9665, pp. 820–849. Springer, Heidelberg (2016). doi:10.1007/978-3-662-49890-3_31

21. Regev, O.: On lattices, learning with errors, random linear codes, and cryptography. In: STOC, pp. 84–93 (2005)

# Mathematical Analysis I

# A General Polynomial Selection Method and New Asymptotic Complexities for the Tower Number Field Sieve Algorithm

Palash Sarkar and Shashank Singh[(✉)]

Applied Statistics Unit, Indian Statistical Institute, Kolkata, India
palash@isical.ac.in, sha2nk.singh@gmail.com

**Abstract.** In a recent work, Kim and Barbulescu had extended the tower number field sieve algorithm to obtain improved asymptotic complexities in the medium prime case for the discrete logarithm problem on $\mathbb{F}_{p^n}$ where $n$ is not a prime power. Their method does not work when $n$ is a composite prime power. For this case, we obtain new asymptotic complexities, e.g., $L_{p^n}(1/3, (64/9)^{1/3})$ (resp. $L_{p^n}(1/3, 1.88)$ for the multiple number field variation) when $n$ is composite and a power of 2; the previously best known complexity for this case is $L_{p^n}(1/3, (96/9)^{1/3})$ (resp. $L_{p^n}(1/3, 2.12)$). These complexities may have consequences to the selection of key sizes for pairing based cryptography. The new complexities are achieved through a general polynomial selection method. This method, which we call Algorithm-$\mathcal{C}$, extends a previous polynomial selection method proposed at Eurocrypt 2016 to the tower number field case. As special cases, it is possible to obtain the generalised Joux-Lercier and the Conjugation method of polynomial selection proposed at Eurocrypt 2015 and the extension of these methods to the tower number field scenario by Kim and Barbulescu. A thorough analysis of the new algorithm is carried out in both concrete and asymptotic terms.

## 1 Introduction

The discrete logarithm problem (DLP) over the multiplicative group of a finite field is a basic problem in cryptography. Two general approaches are known for tackling the DLP on such groups. These are the function field sieve (FFS) [1,2, 12,14] algorithm and the number field sieve (NFS) [8,13,15] algorithm.

Let $p$ be a prime, $n \geq 1$ be an integer and $Q = p^n$. Suppose that $p = L_Q(a, c_p)$ where

$$L_Q(a, c_p) = \exp\left((c_p + o(1))(\ln Q)^a (\ln \ln Q)^{1-a}\right).$$

Depending on the value of $a$, fields $\mathbb{F}_Q$ are classified into the following types: small characteristic, if $a \leq 1/3$; medium characteristic, if $1/3 < a < 2/3$; boundary, if $a = 2/3$; and large characteristic, if $a > 2/3$.

For fields of small characteristic, there has been tremendous progress in the FFS algorithm leading to a quasi-polynomial time algorithm [4]. Based on the FFS algorithms given in [4,11], a record computation of discrete log in the binary

© International Association for Cryptologic Research 2016
J.H. Cheon and T. Takagi (Eds.): ASIACRYPT 2016, Part I, LNCS 10031, pp. 37–62, 2016.
DOI: 10.1007/978-3-662-53887-6_2

extension field $\mathbb{F}_{2^{9234}}$ was reported by Granger et al. [9]. Applications of the FFS algorithm to the medium prime case have been reported in [10, 14, 19].

For medium to large characteristic finite fields, the NFS algorithm is generally considered to be the state-of-the-art. NFS was initially proposed for solving the factoring problem. Its application to DLP was first proposed by Gordon [8] for prime order fields. Application to composite order fields was shown by Schirokauer [21]. Important improvements to the NFS for prime order fields was given by Joux and Lercier [13].

A major step in the application of NFS was by Joux, Lercier, Smart and Vercauteren [15] who showed that the NFS algorithm is applicable to all finite fields. When the prime $p$ is of a special form, Joux and Pierrot [16] showed the application of the special number field sieve algorithm to obtain improved complexity.

The NFS algorithm proceeds by constructing two polynomials $f(x)$ and $g(x)$ over the integers which have a common factor $\varphi(x)$ of degree $n$ modulo $p$. The polynomial $\varphi(x)$ defines the field $\mathbb{F}_{p^n}$ while the polynomials $f(x)$ and $g(x)$ define two number fields. The efficiency of the NFS algorithm is crucially dependent on the properties of the polynomials $f(x)$ and $g(x)$ used to construct the number fields. Consequently, polynomial selection is an important step in the NFS algorithm and is an active area of research.

There has been a recent spurt of interest in the study of the NFS algorithm for DLP in finite fields. The work [3] by Barbulescu et al. extends a previous method [13] for polynomial selection and also presents a new method. The extension of [13] is called the generalised Joux-Lercier (GJL) method while the new method proposed in [3] is called the Conjugation method. The paper also provides a comprehensive comparison of the trade-offs in the complexity of the NFS algorithm offered by the various polynomial selection methods.

The NFS based algorithm has been extended to multiple number field sieve algorithm (MNFS). The work [6] showed the application of the MNFS to medium to high characteristic finite fields. More recently, Pierrot [18] proposed MNFS variants of the GJL and the Conjugation methods. Sarkar and Singh proposed [20] a new polynomial selection method which subsumes both the GJL and the Conjugation methods. Using this method, the asymptotic complexity of both the NFS and the MNFS were worked out in [20].

The minimum asymptotic complexities using the NFS algorithm of Barbulescu et al. [3] can be written as $L_Q(1/3, (c/9)^{1/3})$ where $c = 96$ for the medium characteristic case; $c = 48$ for the boundary case and $c = 64$ for the large characteristic case. The multiple number field sieve algorithm [18] improves these complexities. Further, the minimum complexities are achievable for a certain value of $c_p$. The analysis in [20] improves the asymptotic complexity of the boundary case for a range of values of $c_p$.

When the extension degree $n$ is composite, the finite field $\mathbb{F}_{p^n}$ can be represented as a tower of fields. The idea of using this in the context of DLP is due to Schirokauer [21]. This variant is called the tower number field sieve (TNFS) algorithm.

At Asiacrypt 2015, Barbulescu et al., [5] presented a detailed analysis of the tower number field sieve (TNFS) variant. In a recent paper, Kim and Barbulescu [17] extended the TNFS algorithm and applied previous polynomial selection methods to the TNFS, the multiple TNFS (MTNFS) and the special TNFS variants. These were respectively called the exTNFS, MexTNFS and the SexTNFS algorithms. The polynomial selection methods considered in [17] include the methods from Joux-Lercier-Smart-Vercauteren [15], the GJL and the Conjugation methods from [3] and the polynomial selection method from [20].

**Consequences to the Medium Prime Case.** An important achievement of the work by Kim and Barbulescu [17] is to improve the asymptotic complexity of the medium prime case when $n$ is not a prime power. In this case, they show that the complexity $L_Q(1/3, (48/9)^{1/3})$ is achievable. Further, if $p$ is of a special form, then the complexity of $L_Q(1/3, (32/9)^{1/3})$ is achievable. The condition $n$ is not a prime power is equivalent to saying that $n$ can be written as $\eta\kappa$ with $\gcd(\eta, \kappa) = 1$. How restrictive is the condition $\gcd(\eta, \kappa) = 1$?

One way of removing this restriction is to embed $\mathbb{F}_{p^n}$ into $\mathbb{F}_{p^{nm}}$ with $\gcd(n, m) = 1$ and compute discrete logarithms in $\mathbb{F}_{p^{nm}}$. Let $Q = p^n$ and $Q' = p^{nm}$. The complexity of the NFS algorithm in $\mathbb{F}_{Q'}$ can be written as $L_{Q'}(1/3, \mu)$ where $\mu$ is a constant. Note that $L_{Q'}(1/3, \mu)$ is $L_Q(1/3, \mu m^{1/3})$ (ignoring small terms). The best complexity obtained by Kim and Barbulescu is $\mu = (48/9)^{1/3}$. So, the best complexity achieved for solving DLP in $\mathbb{F}_{p^n}$ by embedding into $\mathbb{F}_{p^{nm}}$ is $L_Q(1/3, \nu)$ where $\nu = (48m/9)^{1/3}$.

Since $m \geq 2$, $\nu \geq (96/9)^{1/3}$. For $p = L_Q(a, c_p)$ with $1/3 < a < 2/3$, the complexity of NFS for directly solving DLP in $\mathbb{F}_{p^n}$ is $L_Q(1/3, (96/9)^{1/3})$. So, we see that trying to solve DLP in $\mathbb{F}_{p^n}$ by embedding into a larger field increases the complexity. This motivates the problem of finding a variant of NFS for fields $\mathbb{F}_{p^n}$ where $n$ is a composite prime-power with complexity $L_Q(1/3, \nu)$ with $\nu < (96/9)^{1/3}$.

## Our Contributions

This paper makes two contributions.

The first contribution is to present a general polynomial selection method which we call Algorithm-$\mathcal{C}$. The polynomial selection method of [20] can be obtained as a special case and so, in turn, the GJL and the Conjugation methods are also obtained as special cases. Further, the exTNFS variants of the GJL and the Conjugation methods are also obtained as special cases of Algorithm-$\mathcal{C}$.

One important feature of Algorithm-$\mathcal{C}$ is that both prime-power and non prime-power $n$ can be covered. For the medium prime case, we have the following consequences.

1. For non prime-power $n$, the minimum complexity achievable is that obtained by Kim and Barbulescu [17]. The analysis, however, reveals improvement over the complexities achieved by Kim and Barbulescu in certain ranges of the relevant parameters.

2. For composite prime-power $n$, the complexities achieved by the new polynomial selection method are currently the best known. For some small values of $n$, the minimum achievable complexities using the exTNFS and the MexTNFS algorithms are shown in Table 1. For $n = 4, 8, 9$ and 16 the new complexities may have consequences to choosing the key sizes for pairing based cryptography.

**Table 1.** Improved minimum complexities $L_Q(1/3, c)$ for some composite prime-power $n$. The entries in the table are the various values of $c$ in different cases.

| | NFS | | | MNFS | |
|---|---|---|---|---|---|
| $n$ | new | [3] | | new | [18] |
| $2^i, i \geq 2$ | $(64/9)^{1/3} \approx 1.92$ | $(96/9)^{1/3} \approx 2.2$ | | 1.88 | 2.12 |
| 9 | $(112/15)^{1/3} \approx 1.95$ | $(96/9)^{1/3} \approx 2.2$ | | 1.92 | 2.12 |
| 25 | $(880/117)^{1/3} \approx 1.96$ | $(96/9)^{1/3} \approx 2.2$ | | 1.94 | 2.12 |

# 2    The Set-Up of the Tower Number Field Sieve Algorithm

The target is to compute discrete logarithm in the field $\mathbb{F}_{p^n}$ where $n$ is composite. Suppose that $n = \eta\kappa$ is a non-trivial factorisation of $n$. We do not necessarily require $\gcd(\eta, \kappa) = 1$.

Let $h(z)$ be a monic polynomial of degree $\eta$ which is irreducible over both $\mathbb{Z}$ and $\mathbb{F}_p$. Let $R = \mathbb{Z}[z]/(h(z))$. Also, note that $\mathbb{F}_{p^\eta} = \mathbb{F}_p[z]/(h(z))$.

Let $f(x)$ and $g(x)$ be polynomials in $R[x]$ whose leading coefficients are from $\mathbb{Z}$. The other coefficients of $f$ and $g$ are polynomials in $z$ of degrees at most $\eta - 1$. In particular, $f$ and $g$ can be viewed as bi-variate polynomials in $x$ and $z$ with coefficients in $\mathbb{Z}$. The following properties are required.

1. Both $f(x)$ and $g(x)$ are irreducible over $R$.
2. Over $\mathbb{F}_{p^\eta}$, $f(x)$ and $g(x)$ have a common factor $\varphi(x)$ of degree $\kappa$.

The field $\mathbb{F}_{p^n}$ is realised as $\mathbb{F}_{p^\eta}[x]/(\varphi(x)) = (R/pR)[x]/(\varphi(x))$.

Let $K_f$ and $K_g$ be the number fields associated with the polynomials $f$ and $g$ respectively. The above set-up provides two different decompositions of a homomorphism from $R[x]$ to $\mathbb{F}_{p^n}$. One of these goes through $R[x]/(f(x))$ and the other goes through $R[x]/(g(x))$.

With this set-up, it is possible to set up a factor base and perform the three main steps (relation collection, linear algebra and descent) of the NFS algorithm. For details we refer to [5,17]. In this work, we will need only the following facts.

1. The factor base consists of $B$ elements for some value $B$ which determines the overall complexity of the algorithm.
2. A polynomial $\phi(x) \in R[x]$ generates a relation if both the norms $N(\phi, f)$ and $N(\phi, g)$ are $B$-smooth, where

$$N(\phi, f) := \mathrm{Res}_z(\mathrm{Res}_x(\phi(x), f(x)), h(z));$$
$$N(\phi, g) := \mathrm{Res}_z(\mathrm{Res}_x(\phi(x), g(x)), h(z)).$$

In this work, we describe a method to choose $h(z), f(x), g(x)$ and $\varphi(x)$ such that the above norms are suitably bounded. Consequences to the complexity of the NFS algorithm are analysed.

## 2.1 Bounds on Resultants

Let $f(z, x)$ be a bivariate polynomial with integer coefficients where $f_{i,j}$ is the coefficient of $x^i z^j$. Then

$$\|f\|_\infty = \max |f_{i,j}|.$$

We summarise bounds on resultants of univariate and bivariate polynomials given in [7].

**Univariate Polynomials:** Let $a(u)$ and $b(u)$ be two polynomials with integer coefficients. From [7], we have

$$|\mathrm{Res}_u(a(u), b(u))|$$
$$\leq (\deg(a) + 1)^{\deg(b)/2}(\deg(b) + 1)^{\deg(a)/2}\|a\|_\infty^{\deg(b)} \times \|b\|_\infty^{\deg(a)}. \quad (1)$$

**Bivariate Polynomials:** Let $a(u, v)$ and $b(u, v)$ be two polynomials with integer coefficients. Let $c(u) = \mathrm{Res}_v(a(u, v), b(u, v))$. Then

$$\|c\|_\infty$$
$$\leq (\deg_v(a) + \deg_v(b))!(\max(\deg_u(a), \deg_u(b)) + 1)^{\deg_v(a)+\deg_v(b)+1}$$
$$\times \|a\|_\infty^{\deg_v(b)} \times \|b\|_\infty^{\deg_v(a)}. \quad (2)$$

The bounds given by (1) and (2) combine to provide bounds on $N(\phi, f)$.

Let $\phi(x, z)$ and $f(x, z)$ be two polynomials and

$$\rho(z) = \mathrm{Res}_x(\phi(x, z), f(x, z)).$$

Further, suppose $\deg_x \phi \leq t - 1$ and $\deg_z \phi \leq \eta - 1$. For $\|\phi\|_\infty = E^{2/(t\eta)}$, the number of possible $\phi(x, z)$'s is $E^2$. Assuming that $t, \eta, \deg_x f$ and $\deg_z f$ are small in comparison to $E$, using (2) we have

$$\|\rho\|_\infty = O\left(E^{2\deg_x(f)/(t\eta)} \cdot \|f\|_\infty^{t-1}\right).$$

Suppose $h(z)$ is a polynomial of degree $\eta$ with $\|h\|_\infty = H$. Let

$$\Gamma = \mathrm{Res}_z(\mathrm{Res}_x(\phi(x), f(x)), h(z)).$$

Assuming that $H = O(\log Q)$, using (1) we have

$$|\Gamma| = O\left(\|\rho\|_\infty^\eta \cdot \|h\|_\infty^{\deg(\rho)}\right)$$
$$= \left(E^{2\deg_x f/t} \cdot \|f\|_\infty^{\eta(t-1)}\right)^{1+o(1)}.$$

Note that in the TNFS set-up described above $N(\phi, f) = \Gamma$.

**Sieving Polynomials:** Sieving is done using polynomials $\phi(x) \in R[x]$ of degrees at most $t-1$ with $\|\phi\|_\infty = E^{2/\eta t}$. Then the number of sieving polynomials is $E^2$.

# 3   Using the LLL Algorithm for Polynomial Selection

The work [3] provides two methods for selecting polynomials for the classical
NFS algorithm. These are called the generalised Joux-Lercier (GJL) and the
Conjugation method. The GJL method is based on an earlier method due to
Joux and Lercier [13] and uses the LLL algorithm to select polynomials.

**The GJL matrix:** Given a vector $\mathbf{a} = [a_0, \ldots, a_{n-1}] \in \mathbb{F}_p^n$ and $r \geq n$, define an
$(r+1) \times (r+1)$ matrix in the following manner.

$$
\begin{bmatrix}
p & & & & & & & \\
 & \ddots & & & & & & \\
 & & \ddots & & & & & \\
 & & & p & & & & \\
a_0 & a_1 & \cdots & a_{n-1} & 1 & & & \\
 & \ddots & \ddots & & & \ddots & & \\
 & & a_0 & a_1 & \cdots & a_{n-1} & 1 &
\end{bmatrix}
\tag{3}
$$

We extend the idea of the GJL to work for tower fields. In the TNFS set-up,
$Q = p^n$ where $n = \eta\kappa$. Recall that $h(z)$ is a monic irreducible polynomial of
degree $\eta$ over the integers and $R = \mathbb{Z}[z]/(h(z))$.

Let $\varphi(x) \in R[x]$ be a monic polynomial of degree $k$. We can write

$$\varphi(x) = x^k + \varphi_{k-1}(z)x^{k-1} + \cdots + \varphi_1(z)x + \varphi_0(z),$$

where each

$$\varphi_i(z) = \varphi_{i,0} + \varphi_{i,1}z + \cdots + \varphi_{i,\eta-1}z^{\eta-1}$$

is a polynomial of degree less than $\eta$ with the coefficients $\varphi_{i,j}$ in $\mathbb{Z}$.

Let $\lambda$ be an integer such $\deg(\varphi_i) \leq \lambda - 1$ for $i = 0, \ldots, k$. The possible values
of $\lambda$ are $1, \ldots, \eta$. The quantity $\lambda$ will be a parameter of the polynomial selection
algorithm and the asymptotic complexity. Though in theory $\lambda$ can take any
value in the range $1, \ldots, \eta$, in practice the values of $\lambda$ which can be achieved are
$1$ and $\eta$. Later we will consider these values of $\lambda$ in more details. Note that the
condition $\eta = 1$ reduces to the classical NFS and in this case $\lambda$ is necessarily $1$.

The polynomial $\varphi_i(z)$ can be uniquely encoded by the vector
$\boldsymbol{\varphi}_i = (\varphi_{i,0}, \ldots, \varphi_{i,\lambda-1})$ and the polynomial $\varphi(x)$ is uniquely encoded by the
vector

$$\boldsymbol{\varphi} = (\varphi_{0,0}, \ldots, \varphi_{0,\lambda-1}, \ldots, \varphi_{k-1,0}, \ldots, \varphi_{k-1,\lambda-1}) \tag{4}$$

which is the concatenation of the vectors $\boldsymbol{\varphi}_0, \ldots, \boldsymbol{\varphi}_{k-1}$.

We introduce some matrix notation.

1. $\mathrm{diag}_i(p)$: the $i \times i$ diagonal matrix having all the diagonal entries to be $p$.
2. $\mathbf{0}_{i,j}$: the $i \times j$ matrix all of whose entries are $0$.

3. For a vector $\mathbf{a}$, let $\text{shift}_i(\mathbf{a})$ be the vector $(\underbrace{0, \ldots, 0}_{i}, \mathbf{a})$.

Given the polynomial $\varphi(x)$ and an integer $r \geq k$, we define a lower triangular matrix $M_{\varphi,r}$ as follows:

$$M_{\varphi,r} = \begin{bmatrix} \text{diag}_{\lambda k}(p) \\ \varphi \quad 1 \\ \mathbf{0}_{\lambda-1,1+\lambda k} \ \text{diag}_{\lambda-1}(p) \\ \quad \text{shift}_\lambda(\varphi) \qquad 1 \\ \qquad \mathbf{0}_{\lambda-1,1+\lambda(k+1)} \qquad \text{diag}_{\lambda-1}(p) \\ \qquad\quad \text{shift}_{2\lambda}(\varphi) \qquad\qquad 1 \\ \qquad\qquad \ddots \qquad\qquad\qquad\qquad \ddots \\ \qquad\quad \mathbf{0}_{\lambda-1,1+\lambda(r-1)} \qquad\qquad\qquad \text{diag}_{\lambda-1}(p) \\ \qquad\qquad \text{shift}_{(r-k)\lambda}(\varphi) \qquad\qquad\qquad\qquad 1 \end{bmatrix}_{(r\lambda+1)\times(r\lambda+1)} \tag{5}$$

Note that for $\lambda = 1$, the matrix given by (5) becomes identical to the matrix given by (3).

Apply the LLL algorithm to $M_{\varphi,r}$ and let the first row of the resulting LLL-reduced matrix be written as

$$[\psi_{0,0}, \ldots, \psi_{0,\lambda-1}, \psi_{1,0}, \ldots, \psi_{1,\lambda-1}, \ldots, \psi_{r-1,0}, \ldots, \psi_{r-1,\lambda-1}, \psi_r].$$

This vector is taken to represent a polynomial $\psi(x) \in R[x]$ of degree $r$ where

$$\psi(x) = \psi_0(z) + \psi_1(z)x + \cdots + \psi_{r-1}(z)x^{r-1} + \psi_r x^r;$$
$$\psi_i(z) = \psi_{i,0} + \psi_{i,1}z + \cdots + \psi_{i,\lambda-1}z^{\lambda-1}.$$

We denote $\psi(x)$ as

$$\psi(x) = \text{LLL}(M_{\varphi,r}). \tag{6}$$

The number of rows of $M_{\varphi,r}$ which are constructed from $\varphi$ is $r - k + 1$. Each of these rows contribute 1 as the diagonal entry. All the other rows contribute $p$ as the diagonal entry and there are $r\lambda + 1 - (r - k + 1) = r(\lambda - 1) + k$ such rows. Since $M_{\varphi,r}$ is a lower triangular matrix, its determinant is the product of its diagonal entries which is equal to $p^{r(\lambda-1)+k}$. Since the matrix has $r\lambda+1$ rows, each entry of the first row of the matrix formed by applying LLL to $M_{\varphi,r}$ is at most

$$p^{\frac{r(\lambda-1)+k}{r\lambda+1}}.$$

So, each $\psi_{i,j}$ and also $\psi_r$ is at most this value. Consequently,

$$\|\psi\|_\infty = p^{\frac{r(\lambda-1)+k}{r\lambda+1}} = Q^{\frac{1}{n} \cdot \frac{r(\lambda-1)+k}{r\lambda+1}} = Q^{\varepsilon/n} \tag{7}$$

where

$$\varepsilon = \frac{r(\lambda - 1) + k}{r\lambda + 1}. \tag{8}$$

Note that for $k \leq r$, $\varepsilon < 1$. The quantity $\varepsilon$ will be another parameter in the asymptotic analysis.

## 4    A New Polynomial Selection Method for TNFS

Algorithm $\mathcal{C}$ describes the polynomial selection method for TNFS. It extends Algorithm-$\mathcal{A}$ in [20] to the setting of tower fields.

---

**Algorithm.** $\mathcal{C}$: Polynomial selection for TNFS.

**Input:** $p$, $n = \eta\kappa$, $d$ (a factor of $\kappa$), $r \geq \kappa/d$ and $\lambda \in \{1, \eta\}$.
**Output:** $f(x)$, $g(x)$ and $\varphi(x)$.

Let $k = \kappa/d$;
Let $R = \mathbb{Z}[z]/(h(z))$;
Let $\mathbb{F}_{p^\eta} = \mathbb{F}_p[z]/(h(z))$;
**repeat**

    Randomly choose a monic polynomial $A_1(x) \in R[x]$ having the following properties:
      $\deg A_1(x) = r + 1$;
      $A_1(x)$ is irreducible over $\mathbb{Q}[z]/(h(z))$ and hence over $R$;
      $A_1(x)$ has coefficient polynomials of size $O(\ln(p))$;
      over $\mathbb{F}_{p^\eta}$, $A_1(x)$ has an irreducible factor $A_2(x)$ of degree $k$ such that all the coefficient polynomials of $A_2(x)$ have degrees at most $\lambda - 1$.

    Randomly choose monic polynomials $C_0(x)$ and $C_1(x)$ with small integer coefficients such that $\deg C_0(x) = d$ and $\deg C_1(x) < d$.
    Define

$$f(x) = \mathrm{Res}_y\left(A_1(y), C_0(x) + y\, C_1(x)\right);$$
$$\varphi(x) = \mathrm{Res}_y\left(A_2(y), C_0(x) + y\, C_1(x)\right) \bmod p;$$
$$\psi(x) = \mathrm{LLL}(M_{A_2, r});$$
$$g(x) = \mathrm{Res}_y\left(\psi(y), C_0(x) + y\, C_1(x)\right).$$

**until** $f(x)$ and $g(x)$ are irreducible over $\mathbb{Q}[z]/(h(z))$ (and hence over $R$) and $\varphi(x)$ is irreducible over $\mathbb{F}_{p^\eta} = F_p[z]/(h(z))$.

**return** $f(x)$, $g(x)$ and $\varphi(x)$.

---

In Algorithm-$\mathcal{C}$, there is only one loop. It is possible to rewrite the algorithm with a nested loop structure. Such a description will have an outer loop which will construct suitable $A_1(x)$, $A_2(x)$ and $\psi(x)$. For each such $(A_1(x), A_2(x), \psi(x))$ the inner loop will try to find suitable $C_0(x)$ and $C_1(x)$ such that the required conditions on $f(x)$, $g(x)$ and $\varphi(x)$ are satisfied. This approach would have been necessary if it had been difficult to find the required polynomials. As things stand, however, the current description of Algorithm-$\mathcal{C}$ finds the required polynomials within a few trials. So, we did not implement the more complex nested version.

The following result states the basic properties of Algorithm $\mathcal{C}$.

**Proposition 1.** *The outputs $f(x)$, $g(x)$ and $\varphi(x)$ of Algorithm $\mathcal{C}$ satisfy the following.*

1. $\deg(f) = d(r + 1)$; $\deg(g) = rd$ and $\deg(\varphi) = \kappa$;
2. over $\mathbb{F}_{p^n}$, both $f(x)$ and $g(x)$ have $\varphi(x)$ as a factor;
3. $\|f\|_\infty = O(\ln(p))$ and $\|g\|_\infty = O(Q^{\varepsilon/n})$.

*Consequently, if $\phi$ is a sieving polynomial, then*

$$N(\phi, f) = E^{2d(r+1)/t} \times L_Q(2/3, o(1)); \tag{9}$$

$$N(\phi, g) = E^{2dr/t} \times Q^{(t-1)\varepsilon/\kappa} \times L_Q(2/3, o(1)); \tag{10}$$

$$N(\phi, f) \times N(\phi, g) = E^{(2d(2r+1))/t} \times Q^{(t-1)\varepsilon/\kappa} L_Q(2/3, o(1)). \tag{11}$$

We note the following points.

1. If $\eta = 1$, then $\lambda$ must be 1 and we obtain Algorithm-$\mathcal{A}$ of [20]. As has been noted in [20], Algorithm-$\mathcal{A}$ generalises and also subsumes the GJL and the Conjugation methods for polynomial selection for the classical NFS given in [3].
2. If $\eta > 1$ and $\lambda = 1$, then $\varphi(x)$ produced by Algorithm-$\mathcal{C}$ has coefficients in $\mathbb{F}_p$ and is of degree $\kappa$. For such a $\varphi(x)$ to be irreducible over $\mathbb{F}_{p^n}$ it is required that $\gcd(\eta, \kappa) = 1$.
3. TNFS variants of the GJL and the Conjugation methods were described in [17]. These can be seen as special cases of Algorithm-$\mathcal{C}$: Suppose $\eta > 1$ and $\lambda = 1$; if $k = \kappa$, then we obtain the TNFS variant of the GJL algorithm; and if $r = k = 1$, then we obtain the TNFS variant of the Conjugation method.
4. The case $\lambda = \eta > 1$ has not been considered earlier. For this case, Algorithm-$\mathcal{C}$ allows $\varphi(x)$ to have coefficients in $\mathbb{F}_{p^n}$. As a result, for irreducibility of $\varphi(x)$, the condition $\gcd(\eta, \kappa) = 1$ is no longer required. Later we show that this case leads to new asymptotic complexity when $n$ is a composite prime-power.
5. Algorithm-$\mathcal{C}$ has the condition $\lambda \in \{1, \eta\}$. It is possible to generalise the condition to $\lambda \in \{1, \ldots, \eta\}$. However, as mentioned earlier, the case $1 < \lambda < \eta$ is difficult to achieve in practice and so we do not consider this case.

## 5    Non-asymptotic Analysis and Examples

In Table 2, we compare the expressions for norm bounds for the various algorithms. As has already been mentioned in [20], the NFS-GJL and the NFS-Conj methods can be seen as special cases of NFS-$\mathcal{A}$: for the former choose $d = 1$ while for the latter, choose $d = n$ and $r = k = 1$. We explain that NFS-$\mathcal{A}$, exTNFS-GJL and exTNFS-Conj can be seen as special cases of exTNFS-$\mathcal{C}$.

1. Choose $\eta = \lambda = 1$ in exTNFS-$\mathcal{C}$ to obtain NFS-$\mathcal{A}$.
2. Choose $\eta > 1$, $\lambda = 1$ and $d = 1$ in exTNFS-$\mathcal{C}$ to obtain exNFS-GJL.
3. Choose $\eta > 1$, $\lambda = 1$, $d = \kappa$ and $r = k$ in exTNFS-$\mathcal{C}$ to obtain exNFS-Conj. Choosing $\eta > 1$, $\lambda = 1$, $d = \kappa$ and $r > k$ in exTNFS-$\mathcal{C}$ provides a generalisation of exNFS-Conj.

We note that NFS-JLSV1 cannot be derived as a special case of NFS-$\mathcal{A}$ and similarly, exTNFS-JLSV1 cannot be derived as a special case of exTNFS-$\mathcal{C}$.

The exTNFS-JLSV1, exTNFS-GJL and exTNFS-Conj algorithms are applicable only for non-prime power $n$. These algorithms cannot be applied when $n$ is a composite prime-power. In Table 3, we compare concrete norm bounds for $n = 4, 8$ and 9 for NFS-JLSV1, NFS-GJL, NFS-Conj, NFS-$\mathcal{A}$ with exTNFS-$\mathcal{C}$. This shows that new trade-offs are achievable with exTNFS-$\mathcal{C}$. In Table 4, we compare concrete norm bounds for $n = 6$ and 12. This shows that exTNFS-GJL and exTNFS-Conj can be seen as special cases of exTNFS-$\mathcal{C}$; also, by choosing $r > k$, new trade-offs are achievable.

## 5.1  Plots of Norm Bounds

In Fig. 1, we provide plots of norm bound for various finite fields of composite prime power extension degree. It is clear from the plots that for composite prime power extension degree, Algorithm-$\mathcal{C}$ provides the lowest norm bound. Note that we have used the estimates of $Q$-$E$ pairs given in the Table 2 of the paper [3] for plotting the norm bounds.

Plots of norm bound for extension degrees 12 and 24 are given in the Fig. 2. Note that for these extension degrees, two types of towers are possible; one for which $\gcd(\eta, \kappa) = 1$, and the other for which $\gcd(\eta, \kappa) \neq 1$. Let us denote by Algorithm-$\mathcal{B}$, the special case of Algorithm-$\mathcal{C}$ where $\lambda = 1$ and so $\gcd(\eta, \kappa) = 1$. Plots for Algorithm-$\mathcal{B}$ are shown separately in Fig. 2. It is interesting to note that, in the certain range of finite fields, the minimum norm bound achieved by Algorithm-$\mathcal{C}$ is lower than the minimum norm bound achieved by Algorithm-$\mathcal{B}$, i.e., it is not necessarily the best to choose $\gcd(\eta, \kappa) = 1$. While this appears in the concrete comparison, it is not captured by the asymptotic analysis.

**Table 2.** Parameterised efficiency estimates for NFS obtained from the different polynomial selection methods.

| Method | Norms product | Conditions |
|---|---|---|
| NFS-JLSV1 [15] | $E^{\frac{4n}{t}} Q^{\frac{t-1}{n}}$ | |
| NFS-GJL [3] | $E^{\frac{2(2r+1)}{t}} Q^{\frac{t-1}{r+1}}$ | $r \geq n$ |
| NFS-Conj [3] | $E^{\frac{6n}{t}} Q^{\frac{t-1}{2n}}$ | |
| NFS-$\mathcal{A}$ [20] | $E^{\frac{2d(2r+1)}{t}} Q^{\frac{t-1}{d(r+1)}}$ | $d\|n, r \geq n/d$ |
| exTNFS-JLSV1 [17] | $E^{\frac{4\kappa}{t}} Q^{\frac{t-1}{\kappa}}$ | $n = \eta\kappa, \gcd(\eta, \kappa) = 1, \eta$ small |
| exTNFS-GJL [17] | $E^{\frac{2(2r+1)}{t}} Q^{\frac{t-1}{r+1}}$ | $n = \eta\kappa, \gcd(\eta, \kappa) = 1, \eta$ small, $r \geq \kappa$ |
| exTNFS-Conj [17] | $E^{\frac{6\kappa}{t}} Q^{\frac{t-1}{2\kappa}}$ | $n = \eta\kappa, \gcd(\eta, \kappa) = 1, \eta$ small |
| exTNFS-$\mathcal{C}$ | $E^{\frac{2d(2r+1)}{t}} Q^{\frac{(t-1)(r(\lambda-1)+k)}{\kappa(r\lambda+1)}}$ | $n = \eta\kappa, k = \kappa/d, r \geq k$; NFS: $\eta = \lambda = 1$; exTNFS $(\gcd(\eta, \kappa) = 1)$: $\eta > 1, \lambda = 1$; exTNFS: $\eta = \lambda$ |

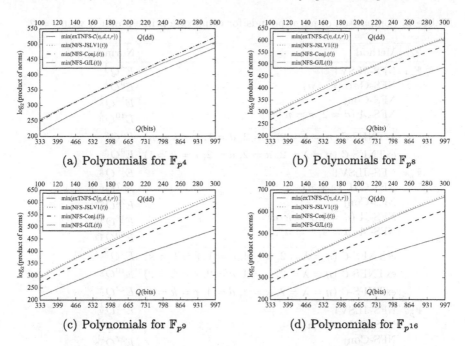

**Fig. 1.** Product of norms for various polynomial selection methods

## 5.2 Examples for Non Prime-Power $n$

We provide concrete examples for the following settings.

- $n = 6$ with $(\eta, \kappa) = (2, 3)$ or $(3, 2)$: Example 1 below.
- $n = 12$ with $(\eta, \kappa) = (3, 4)$: Example 2 below.

In both cases, $\eta > 1$ and so the obtained examples cannot be generated by Algorithm-$\mathcal{A}$. Since $\gcd(\eta, \kappa) = 1$, we have taken $\lambda = 1$ and we provide both examples which can and cannot be generated by the TNFS variant of the Conjugation method from [17].

*Example 1.* Let $p$ be the 201-bit prime given below

$$p = 1606938044258990275541962092341162602522202993782792835301611 \quad (12)$$

and $n = 6$.

*Case 1:* Let $(\eta, \kappa) = (2, 3)$ so we can take $\lambda = 1$. Choose $d = \kappa$, and so $k = \kappa/d = 1$. Taking $r = k$, we get the following polynomials.

$$h(z) = z^2 + 14\,z + 20$$
$$f(x) = x^6 + 5\,x^5 + 6\,x^4 + 18\,x^3 + 73\,x^2 + 52\,x + 20$$
$$g(x) = 5163787857847060995607487014014\,x^3 + 1874354673374387667869084608560\,x^2$$
$$\qquad + 4592761622761020079997668111670\,x + 1683194203609950937495174411516$$

**Table 3.** Comparison of norm bounds for composite prime-power $n$ with $t = 2$.

| $\mathbb{F}_Q$ | Method | Norm bound |
|---|---|---|
| $\mathbb{F}_{p^4}$ | NFS-JLSV1 | $E^8 Q^{\frac{1}{4}}$ |
| | NFS-GJL $(r = n)$ | $E^9 Q^{\frac{1}{5}}$ |
| | NFS-Conj | $E^{12} Q^{\frac{1}{8}}$ |
| | NFS-$\mathcal{A}$ $(d = 2, r = n/d)$ | $E^{10} Q^{\frac{1}{6}}$ |
| | exTNFS-$\mathcal{C}$ $(\eta = \lambda = 2, \kappa = 2, d = 1, r = k = \kappa)$ | $E^5 Q^{\frac{2}{5}}$ |
| | exTNFS-$\mathcal{C}$ $(\eta = \lambda = 2, \kappa = 2, d = 2, r = k = 1)$ | $E^6 Q^{\frac{1}{3}}$ |
| $\mathbb{F}_{p^8}$ | NFS-JLSV1 | $E^{16} Q^{\frac{1}{8}}$ |
| | NFS-GJL $(r = n)$ | $E^{17} Q^{\frac{1}{9}}$ |
| | NFS-Conj | $E^{24} Q^{\frac{1}{16}}$ |
| | NFS-$\mathcal{A}$ $(d = 2, r = n/d)$ | $E^{18} Q^{\frac{1}{10}}$ |
| | NFS-$\mathcal{A}$ $(d = 4, r = n/d)$ | $E^{20} Q^{\frac{1}{12}}$ |
| | exTNFS-$\mathcal{C}$ $(\eta = \lambda = 2, \kappa = 4, d = 1, r = k = \kappa)$ | $E^9 Q^{\frac{2}{9}}$ |
| | exTNFS-$\mathcal{C}$ $(\eta = \lambda = 2, \kappa = 4, d = 2, r = k = 2)$ | $E^{10} Q^{\frac{1}{5}}$ |
| | exTNFS-$\mathcal{C}$ $(\eta = \lambda = 2, \kappa = 4, d = 4, r = k = 4)$ | $E^{12} Q^{\frac{1}{6}}$ |
| $\mathbb{F}_{p^9}$ | NFS-JLSV1 | $E^{18} Q^{\frac{1}{9}}$ |
| | NFS-GJL $(r = n)$ | $E^{19} Q^{\frac{1}{10}}$ |
| | NFS-Conj | $E^{27} Q^{\frac{1}{18}}$ |
| | NFS-$\mathcal{A}$ $(d = 3, r = n/d)$ | $E^{21} Q^{\frac{1}{12}}$ |
| | exTNFS-$\mathcal{C}$ $(\eta = \lambda = 3, \kappa = 3, d = 1, r = k = \kappa)$ | $E^7 Q^{\frac{3}{10}}$ |
| | exTNFS-$\mathcal{C}$ $(\eta = \lambda = 3, \kappa = 3, d = 3, r = k = 1)$ | $E^9 Q^{\frac{1}{4}}$ |

$$\phi(x) = x^3 + {\scriptstyle 4370464675316262929768958368698673612607491294431378655895}\, x^2$$
$$+ {\scriptstyle 1311139402594878878930687510609602083782247388329413596 7675}\, x$$
$$+ {\scriptstyle 874092935063252585953791673739734722521498258886275731 1786}$$

Clearly, the above polynomials represents the polynomials generated by Conjugation method and we have $\|g\|_\infty \approx 2^{101}$.

If we choose $r = k + 1$ i.e., $r = 2$, we get the following polynomials.

$$h(z) = z^2 + z + {\scriptstyle 20}$$
$$f(x) = x^9 + 14\,x^8 + 74\,x^7 + 183\,x^6 + 200\,x^5 - 32\,x^4 - 375\,x^3 - 232\,x^2 - 48\,x - 1$$

$$g(x) = {\scriptstyle 46647198736133019425}\, x^6 + {\scriptstyle 530869201059776791498}\, x^5 + {\scriptstyle 2094297655062561189093}\, x^4$$
$$+ {\scriptstyle 3465328474724235168588}\, x^3 + {\scriptstyle 2717008192279799547052}\, x^2$$
$$+ {\scriptstyle 1322043132032704860464}\, x + {\scriptstyle 290748395825577445032}$$

$$\phi(x) = x^3 + {\scriptstyle 315444052193803149917391335705534526435873425227915090402562}\, x^2$$
$$+ {\scriptstyle 126177620877521259966956534282213810574349370091166036161 0232}\, x$$
$$+ {\scriptstyle 315444052193803149917391335705534526435873425227915090402559}$$

We note that $\|g\|_\infty \approx 2^{72}$. Thus taking $r > k$, gives us the polynomials which are not obtained by Conjugation method.

**Table 4.** Comparison of norm bounds for non prime-power $n$ with $t = 2$.

| $\mathbb{F}_Q$ | Method | Norm bound |
|---|---|---|
| $\mathbb{F}_{p^6}$ | NFS-JLSV1 | $E^{12}Q^{\frac{1}{6}}$ |
| | NFS-GJL $(r = n)$ | $E^{13}Q^{\frac{1}{7}}$ |
| | NFS-Conj | $E^{18}Q^{\frac{1}{12}}$ |
| | NFS-$\mathcal{A}$ $(d = 2, r = n/d)$ | $E^{14}Q^{\frac{1}{8}}$ |
| | exTNFS-JLSV1 $(\eta = 2, \kappa = 3)$ | $E^{6}Q^{\frac{1}{3}}$ |
| | exTNFS-GJL $(\eta = 2, r = \kappa = 3)$ | $E^{7}Q^{\frac{1}{4}}$ |
| | exTNFS-Conj $(\eta = 2, \kappa = 3)$ | $E^{9}Q^{\frac{1}{6}}$ |
| | exTNFS-$\mathcal{C}$ $(\eta = 2, \lambda = 1, d = 1, r = k = \kappa = 3)$ | $E^{7}Q^{\frac{1}{4}}$ |
| | exTNFS-$\mathcal{C}$ $(\eta = 2, \lambda = 1, d = 3, \kappa = 3, r = k = 1)$ | $E^{9}Q^{\frac{1}{6}}$ |
| | exTNFS-$\mathcal{C}$ $(\eta = 2, \lambda = 1, d = 3, \kappa = 3, k = 1, r = 2)$ | $E^{15}Q^{\frac{1}{9}}$ |
| $\mathbb{F}_{p^{12}}$ | NFS-JLSV1 | $E^{24}Q^{\frac{1}{12}}$ |
| | NFS-GJL $(r = n)$ | $E^{25}Q^{\frac{1}{13}}$ |
| | NFS-Conj | $E^{36}Q^{\frac{1}{24}}$ |
| | NFS-$\mathcal{A}$ $(d = 2, r = n/d)$ | $E^{26}Q^{\frac{1}{14}}$ |
| | exTNFS-JLSV1 $(\eta = 3, \kappa = 4)$ | $E^{8}Q^{\frac{1}{4}}$ |
| | exTNFS-GJL $(\eta = 3, r = \kappa = 4)$ | $E^{9}Q^{\frac{1}{5}}$ |
| | exTNFS-Conj $(\eta = 3, \kappa = 4)$ | $E^{12}Q^{\frac{1}{8}}$ |
| | exTNFS-$\mathcal{C}$ $(\eta = 3, \lambda = 1, d = 1, r = k = \kappa = 4)$ | $E^{9}Q^{\frac{1}{5}}$ |
| | exTNFS-$\mathcal{C}$ $(\eta = 3, \lambda = 1, d = 4, \kappa = 4, r = k = 1)$ | $E^{12}Q^{\frac{1}{8}}$ |
| | exTNFS-$\mathcal{C}$ $(\eta = 3, \lambda = 1, d = 4, \kappa = 4, k = 1, r = 2)$ | $E^{20}Q^{\frac{1}{12}}$ |

*Case 2:* Let $(\eta, \kappa) = (3, 2)$. Taking $d = \kappa$ and $r = 1$, we get the following polynomials.

$$h(z) = z^3 + z^2 + 15\,z + 7$$
$$f(x) = x^4 - x^3 - 2\,x^2 - 7\,x - 3$$
$$g(x) = 7171755614869845772782428430019\,x^2 + 21894353131977750564429465543188\,x$$
$$+ 29066108746847596337211893862{07}$$
$$\phi(x) = x^2 + 1313968758518166109156841236000601376540005423373691304025{54}\,x$$
$$+ 1313968758518166109156841236000601376540005423373691304025{55}$$

Note that $\|g\|_\infty \approx 2^{102}$. If we take $d = \kappa$ and $r = 2$, we get the following set of polynomials where $\|g\|_\infty \approx 2^{69}$.

$$h(z) = z^3 + z^2 + 15\,z + 7$$
$$f(x) = x^6 - 4\,x^5 - 53\,x^4 - 147\,x^3 - 188\,x^2 - 157\,x - 92$$
$$g(x) = 15087279002722300985\,x^4 + 124616743720753879934\,x^3 + 451785460058994237397\,x^2$$
$$+ 749764394939964245000\,x + 567202989572349792620$$

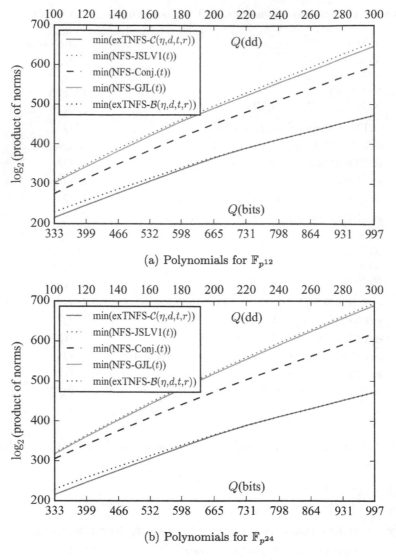

(a) Polynomials for $\mathbb{F}_{p^{12}}$

(b) Polynomials for $\mathbb{F}_{p^{24}}$

**Fig. 2.** Product of norms for various polynomial selection methods. Note that algorithm-$\mathcal{B}$ is the algorithm-$\mathcal{C}$ with $\gcd(\eta, \kappa) = 1$.

$$\phi(x) = x^2 + \text{\small 4597432113076247879730918301514182563567790998604530481656 28 } x$$
$$+ \text{\small 137922963392287436391927549045425476907033729958135914449 6879}$$

*Example 2.* Consider $p$ given by (12) and $n = 12$. Take $\eta = 3$, so we have $\kappa = 4$. Since $\gcd(\eta, \kappa) = 1$, we can take $\lambda = 1$. For $d = 4$ and $r = 1$, we get the following set of polynomials.

$$h(z) = z^3 + 4\,z^2 + z + 10$$

$$f(x) = x^8 - 76\,x^7 - 2425\,x^6 - 18502x^5 - 29145x^4 - 27738x^3 - 19029x^2 - 5470x - 899$$

$$g(x) = 6716755184000388685097611858847\,x^4 + 922925477134968745315513948219 3\,x^3$$
$$+\ 2644321248368967746217849131611 1\,x^2 + 1037326889529552052877683744140 9\,x$$
$$+\ 12363161023892249178889813706137$$

$$\phi(x) = x^4 + 646864792711457069399567439420493376414881652645022449494547\,x^3$$
$$+\ 1021416876665447593884133690181094257197328168116350116415450x^2$$
$$+\ 667312667750761865313480451840635053444883928304549026363993\,x$$
$$+\ 408957500786095918278260248402833540600045513190531537389 07$$

Note that $\|g\|_\infty \approx 2^{105}$. If we take $d = 2$ and $r = 2$, we get the following set of polynomials.

$$h(z) = z^3 + 9\,z^2 + 16\,z + 6$$

$$f(x) = x^6 - 31\,x^5 - 1368\,x^4 - 12769\,x^3 - 25114\,x^2 + 80676\,x + 46152$$

$$g(x) = -3110542872966491216142377505541399497324\,x^4$$
$$-5426446159044675843838047008564401026162 8\,x^3$$
$$-3147851405357699755698076582420151735725 25\,x^2$$
$$-4943164354795189719934785414688038890322 52\,x$$
$$+1282345843739963030376594369830360797777 868$$

$$\phi(x) = x^4 + 111638887953462514640700077445807615721516795538687963473069 38\,x^3$$
$$+\ 244260557761228308164124096832151544783881431251339247716776\,x^2$$
$$+\ 143158516928131538002656218627939244574673300192006062636096 0\,x$$
$$+\ 320118736190538295384063259633892825807420618998673420293 65$$

Note that $\|g\|_\infty \approx 2^{140}$.

## 5.3   Examples for Composite Prime-Power $n$

*Example 3.* Consider again the prime $p$ given by the Eq. (12). Let $n = 4$. Take $\eta = 2$, so we have $\kappa = 2$ and $\gcd(\eta, \kappa) \neq 1$. For $d = 2$ and $r = 1$, we get the following set of polynomials.

$$h(z) = z^2 + 3\,z + 9$$

$$f(x) = x^4 - 63\,x^3 + (z-2252)\,x^2 + (26\,z - 16788)\,x + 169\,z - 4547$$

$$g(x) = 1383414878882125995926103183619409643753\,x^2 + \big(-12055618797162796264$$
$$473546996019291321934\,z + 1401267231113193698909077587836051312914 5\big)\,x$$
$$-1567230443631163514381561109482507871851 42\,z - 64083108396303246416$$
$$666280265568245909149$$

$$\phi(x) = x^2 + \big({\scriptstyle 79841662233750009138191057515828855406255518668802407863546}6\, z$$
$$+{\scriptstyle 78583049089685779542962824510444458915079587893508069025034}5\big)\, x$$
$$+{\scriptstyle 73778782483355953471306492301077558767999946424755601045139}2\, z$$
$$+{\scriptstyle 5741681161052096873333946323108040438271284634592919614446}41$$

Note that $\|g\|_\infty \approx 2^{137}$.

*Example 4.* Consider $n = 8$ and $p$ as given by the Eq. (12). Take $\eta = 2$, so we have $\kappa = 4$ and $\gcd(\eta, \kappa) \neq 1$. For $d = 2$ and $r = 2$, we get the following set of polynomials.

$$h(z) = z^2 + 5\, z + 1$$
$$f(x) = x^6 - 12\, x^5 - 34\, x^4 + (-z+555)\, x^3 + (-21\, z+2768)\, x^2 + (-147\, z+9405)\, x$$
$$+(-343\, z+23477)$$

$$g(x) = -{\scriptstyle 854222881267358737695287657076641386620058484405}\, x^4$$
$$+\big({\scriptstyle 4674143459783379952663678415322804544750219136 34}\, z$$
$$-{\scriptstyle 18784759735246185343960111938850905267931156589 79}\big)\, x^3$$
$$+\big({\scriptstyle 3248230356563661967038490496701146071676382792940}\, z$$
$$-{\scriptstyle 1935512919110472958110118658980468260409847426884 7}\big)\, x^2$$
$$+\big({\scriptstyle 1070862123949157988233922008249401828342277305930}\, z$$
$$-{\scriptstyle 2076186916416931940139138564374407993594773947083 3}\big)\, x$$
$$+{\scriptstyle 8655868066594601909115589364961851171185700663912}\, z$$
$$-{\scriptstyle 1000785457280073411552855562025875027867253289268 40}$$

$$\phi(x) = x^4 + \big({\scriptstyle 21911219752524993948902114872339619506052675724045230559991}1\, z$$
$$+{\scriptstyle 23219545975627720766829074418966645813990243339055134165652}8\big)\, x^3$$
$$+\big({\scriptstyle 5949160735000408237882940244016247461008039729961137950811 68}\, z$$
$$+{\scriptstyle 14917331880357671382697148636511995301975657172021282812247 36}\big)\, x^2$$
$$+\big({\scriptstyle 3686706181737494859148039516294087348388376343649667815576 95}\, z$$
$$+{\scriptstyle 43263399991339548283034948543964250199739186251021131507767 1}\big)\, x$$
$$+{\scriptstyle 3771491491052670142526117076162654150270566837365472736477 76}\, z$$
$$+{\scriptstyle 154186426074930945711033287451850266367654257871233943721298 5}$$

Note that $\|g\|_\infty \approx 2^{167}$. If we take $d = 4$ and $r = 1$, we get the following polynomials.

$$h(z) = z^2 + 12\, z + 7$$
$$f(x) = x^8 - 33\, x^7 + (z-732)\, x^6 + (14\, z-3424)\, x^5 + (57\, z-2627)\, x^4 + (68\, z-5218)\, x^3$$
$$+(100\, z-3524)\, x^2 + (48\, z-2940)\, x + (36\, z-1764)$$

$$g(x) = -8459635622214131881453154771645357881453\,x^4 + \big(2792953719200328914185612401694261426806\,z - 755890305610455751417352305473498741035 23\big)\,x^3 + \big(1955067603440230239929928681185982 9987642\,z - 4625805350042849342051017989436930074575\big)\,x^2 + \big(11171814876801315656742449606777045707224\,z - 4856705357775834412334627904003875997 0502\big)\,x + 16757722315201973485113674410165568560836\,z - 474716734999951205406599542451220663113 94$$

$$\phi(x) = x^4 + \big(5654752046099492711523076367081283120169586847337989073 53217\,z + 11847567849244634596347447136982244860033582290932911315840 98\big)\,x^3 + \big(744450343751664346982229272274572979074304805571006680869 297\,z + 258607273176292839733402534181758389412492634739073744580 569\big)\,x^2 + \big(654962774180806809067268454491350645545631745152402794111257\,z + 1525151051179873287455054670110572738969026928807578855 733140\big)\,x + 17897513914171507582992163556644466705734612083720 7773516080\,z + 680788532510819655640619912824696505931337399428575448298096$$

Note that $\|g\|_\infty \approx 2^{136}$.

*Example 5.* Consider $n = 9$ and $p$ as given by the Eq. (12). Take $\eta = 3$, so we have $\kappa = 3$. For $d = 1$ and $r = 3$, we get the following set of polynomials.

$$h(z) = z^3 + z^2 + 18\,z + 15$$
$$f(x) = x^4 - 6\,x^3 - 211\,x^2 - 1187\,x + z - 2034$$
$$g(x) = 2698140291270948534773782584704649727969933070 5965517\,x^3$$
$$+\big(-14528721852302226470323283323743148459708080792182 6676\,z^2$$
$$+5041139398333626569424296143995787688546468585878 2189\,z$$
$$+20688147989640478752125253465002089729398216768 9590409\big)\,x^2$$
$$+\big(-562799080702299135013029013687164984984908940586 595961\,z^2$$
$$+349365561960939643979968647853372949952345188547 313416\,z$$
$$-2387723443262139639657867923812539404807336156225 8711\big)\,x$$
$$+1285779122778936362366127131791594871482581595497 499229\,z^2$$
$$+68771362075856705638794695798484555912964355806087 4123\,z$$
$$-83012938176333676194723672703681662866114646956995 5030$$
$$\phi(x) = x^3 + \big(669339476643413131528050298510860109656533927528 567342123649\,z^2$$
$$+155266446751696420973178819178735779443468114072 3383971203939\,z$$
$$+943932691068840507491372697519702885049901068217 108999449340\big)\,x^2$$
$$+\big(119185392336022577784894438687795788351626109687 7478017413866\,z^2$$
$$+421341580961908534729044227924897299449513889433 708503340901\,z$$
$$+235622039392351511019273446915854970293312748291 080468943554\big)\,x$$
$$+209551211497370380856126797068962682591000324511 877123851147\,z^2$$

$+1000724369592593057730299648522737075992111849892300346453870\,z$

$+10599747836799489598174239488437943742027432070177936374203 70$

Note that $\|g\|_\infty \approx 2^{180}$.

# 6  Asymptotic Complexity Analysis for the Medium Prime Case

For $1/3 < a \le 2/3$, write

$$p = L_Q(a, c_p), \text{ where } c_p = \frac{1}{n}\left(\frac{\ln Q}{\ln \ln Q}\right)^{1-a} \text{ and so } n = \frac{1}{c_p}\left(\frac{\ln Q}{\ln \ln Q}\right)^{1-a} \quad (13)$$

For each $c_p$, the runtime of the NFS algorithm is the same for the family of finite fields $\mathbb{F}_{p^n}$ where $p$ is given by (13).

Recall that $n = \eta\kappa$, $k = \kappa/d$, $r \ge k$ and $\varepsilon$ is given by (8). Suppose that $\eta$ can be written as

$$\eta = c_\eta \left(\frac{\ln Q}{\ln \ln Q}\right)^{2/3-a}. \quad (14)$$

The boundary case arises when $a = 2/3$ and in this case $\eta = c_\eta$. If further, we have $\eta = 1$, then $c_\eta$ is also 1.

From $n = \eta\kappa$, we get

$$\kappa = \frac{1}{c_\theta}\left(\frac{\ln Q}{\ln \ln Q}\right)^{1/3} \text{ where} \quad (15)$$
$$c_\theta = c_p c_\eta.$$

So, given $Q$ and $\kappa$, the value of $c_\theta$ is fixed. We recall the following.

1. The number of polynomials to be considered for sieving is $E^2$, so the cost of relation collection step is $O(E^2)$.
2. The factor base is of size $B$ and hence cost of linear algebra step is $O(B^2)$.

Let

$$B = L_Q(1/3, c_b). \quad (16)$$

Set

$$E = B \quad (17)$$

so that asymptotically, the cost of relation collection step is same as the cost of linear algebra step.

Let $\pi = \Psi(\Gamma, B)$ be the probability that a random positive integer which is at most $\Gamma$ is $B$-smooth. Let $\Gamma = L_Q(z, \zeta)$ and $B = L_Q(b, c_b)$. Using the L-notation version of the Canfield-Erdős-Pomerance theorem,

$$(\Psi(\Gamma, B))^{-1} = L_Q\left(z - b, (z - b)\frac{\zeta}{c_b}\right). \tag{18}$$

Following the usual convention, we assume that the same smoothness probability $\pi$ holds for the event that a random sieving polynomial $\phi(x)$ is smooth over the factor base.

Since the total number of polynomials considered for sieving is $E^2$, the number of relations obtained after sieving is $E^2\pi$. For the linear algebra step to be successful, we need $E^2\pi = B$ and so

$$\pi^{-1} = B. \tag{19}$$

Obtaining $\pi^{-1}$ from (18) and setting it to be equal to $B$ allows solving for $c_b$. Balancing the costs of the sieving and the linear algebra phases leads to the runtime of the NFS algorithm to be $B^2 = L_Q(b, 2c_b)$. So, to determine the runtime, we need to determine $c_b$.

**Lemma 1.** *Let $n = \eta\kappa$ and $\kappa = kd$ for positive integers $\eta, k$ and $d$. For a fixed value of $t$, using the expressions for $p$ and $E$ ($= B$) given by (13) and (16) and $\eta = c_\eta(\ln Q/\ln\ln Q)^{2/3-a}$, we obtain the following.*

$$\left.\begin{array}{r}E^{\frac{2}{t}d(2r+1)} = L_Q\left(2/3, \frac{2c_b(2r+1)}{c_\theta kt}\right); \\ Q^{\frac{(t-1)\varepsilon}{\kappa}} = L_Q\left(2/3, (t-1)c_\theta\varepsilon\right);\end{array}\right\} \tag{20}$$

*where $\varepsilon$ is given by the Eq. (8).*

**Theorem 1.** *Let $n = \eta\kappa$; $\kappa = kd$; $r \geq k$; $t \geq 2$; $p = L_Q(a, c_p)$ with $1/3 < a \leq 2/3$; and $\eta = c_\eta(\ln Q/\ln\ln Q)^{2/3-a}$. It is possible to ensure that the runtime of the NFS algorithm with polynomials chosen by Algorithm $\mathcal{B}$ is $L_Q(1/3, 2c_b)$ where*

$$c_b = \frac{2(2r+1)}{6c_\theta kt} + \sqrt{\left(\frac{2r+1}{3c_\theta kt}\right)^2 + \frac{(t-1)c_\theta\varepsilon}{3}}. \tag{21}$$

*Proof.* The product of the norms given by (20) is

$$\Gamma = L_Q\left(\frac{2}{3}, \frac{2c_b(2r+1)}{c_\theta kt} + (t-1)c_\theta\varepsilon\right).$$

Then $\pi^{-1}$ given by (18) is

$$L_Q\left(\frac{1}{3}, \frac{1}{3}\left(\frac{2(2r+1)}{c_\theta kt} + \frac{(t-1)c_\theta\varepsilon}{c_b}\right)\right).$$

From the condition $\pi^{-1} = B$, we get

$$c_b = \frac{1}{3}\left(\frac{2(2r+1)}{c_\theta kt} + \frac{(t-1)c_\theta \varepsilon}{c_b}\right). \tag{22}$$

Solving the quadratic for $c_b$ and choosing the positive root gives

$$c_b = \frac{2(2r+1)}{6c_\theta kt} + \sqrt{\left(\frac{2r+1}{3c_\theta kt}\right)^2 + \frac{(t-1)c_\theta \varepsilon}{3}}.$$

$\square$

We wish to minimise the value of $c_b$ with respect to $c_\theta$. To do this, we differentiate (21) with respect to $c_\theta$ and set to 0 to obtain the following equation which has to be solved for $c_\theta$.

$$0 = \frac{-2(2r+1)}{6ktc_\theta^2} + \frac{1}{2}\left(\left(\frac{2r+1}{3c_\theta kt}\right)^2 + \frac{(t-1)c_\theta \varepsilon}{3}\right)^{-1/2}\left(\frac{-2(2r+1)^2}{9k^2t^2c_\theta^3} + \frac{(t-1)\varepsilon}{3}\right)$$

This can be seen as a quadratic in $c_\theta^3$ which can be solved using standard algebraic manipulations to obtain

$$c_\theta^3 = 8\left(\frac{2r+1}{3kt}\right)^2 \cdot \frac{3}{(t-1)\varepsilon}.$$

Taking cube roots on both sides gives the value of $c_\theta$. Substituting this value of $c_\theta$ in (21) we obtain

$$2c_b = \left(\frac{64(2r+1)(t-1)\varepsilon}{9kt}\right)^{1/3} = \left(\frac{64(2r+1)(t-1)}{9kt} \cdot \frac{r(\lambda-1)+k}{r\lambda+1}\right)^{1/3} \tag{23}$$

The expression on the right hand side of (23) clearly increases as $t$ increases. So, to minimise $2c_b$, we should choose the minimum value of $t$ which is $t = 2$. With $t = 2$, the right hand side of (23) becomes

$$\left(\frac{32(2r+1)}{9k} \cdot \frac{r(\lambda-1)+k}{r\lambda+1}\right)^{1/3} \tag{24}$$

We consider several cases:

**Case $\lambda = 1$:** The right hand side of (24) becomes

$$\left(\frac{32(2r+1)}{9(r+1)}\right)$$

which takes the minimum value of $(48/9)^{1/3}$ for $r = 1$. This can arise in the following ways.

1. $\eta = 1$, $a = 2/3$: This corresponds to the boundary case and the minimum complexity of $(48/9)^{1/3}$ has already been reported in [3].

2. $\eta > 1$, $1/3 < a < 2/3$: Again, the minimum complexity of $(48/9)^{1/3}$ for this case has already been reported in [17]. Note that since $\lambda = 1$ and $\eta > 1$, this case requires $\gcd(\eta, \kappa) = 1$ and hence applies to non prime-power values of $n$.

In both the above cases, the minimum complexity is not achievable for all values of $c_\theta$. The minimum achievable values of $2c_b$ as $c_\theta$ varies depends on the values of $r, k$ and $t$. This is shown in Fig. 3 by the plot of $2c_b$ against $c_\theta$ where $c_b$ is given by (21). This plot extends a similar plot provided in [20] for the case $\eta = 1$.

**Case $\lambda = \eta > 1$:** For a fixed $k$, increasing $r$ leads to increase in the value of (24) which shows that this expression is minimised for the minimum value of $r$ which is $r = k$. Setting $r = k$, and using $\lambda = \eta$, (24) becomes

$$\left( \frac{32(2k+1)}{9} \cdot \frac{\eta}{k\eta + 1} \right)^{1/3}. \tag{25}$$

The expression given by (25) decreases as $k$ increases and so the minimum is achieved for the maximum value of $k$ which is $k = \kappa$ implying that $d = 1$. Using $k = \kappa$ in (25) we obtain the minimum possible value of $2c_b$ in this case to be

$$\left( \frac{32(2\kappa+1)}{9} \cdot \frac{\eta}{\kappa\eta + 1} \right)^{1/3} = \left( \frac{32(2n+\eta)}{9(n+1)} \right)^{1/3}. \tag{26}$$

We consider composite prime-power values of $n$. Suppose that $n$ can be written as $n = \eta^i$ for some prime $\eta$ and some $i > 1$.

1. If $\eta = 2$, then the minimum possible value of $2c_b$ for the case $\lambda = \eta = 2$ is $(64/9)^{1/3} \approx 1.92$ for all $n = 2^i$. In particular, this case covers $n = 4, 8, 16$.
2. If $\eta = 3$ and $n = 9$, then the minimum possible value of $2c_b$ for the case $\lambda = \eta = 3$ is $(112/15)^{1/3} \approx 1.95$.
3. If $\eta = 5$ and $n = 25$, then the minimum possible value of $2c_b$ for the case $\lambda = \eta = 5$ is $(880/117)^{1/3} \approx 1.96$.

The above covers the small composite prime-power values of $n$ and the minimum value of $2c_b$ that can be achieved in each case. Note that similar to the case of $\lambda = 1$, this minimum is achieved at a particular value of $c_\theta$. The more general picture of the variation in complexity is given by $2c_b$ where the expression for $c_b$ is given by (21). Figure 3 shows the plots of $2c_b$ (minimised over $t$, $k$ and $r$) against $c_\theta$ for different values of $\lambda$.

# 7    Multiple Number Field Sieve Variant

In the multiple number field sieve (MNFS) algorithm, several number fields are considered. These number fields are generated by the irreducible polynomials in $R[x]$, having a common irreducible factor over $\mathbb{F}_{p^n}$. There are two variants of MNFS algorithm. We discuss the second variant of MNFS only where the image of $\phi(x)$ needs to be smooth in the first number field and at least one of the other $V$ number fields.

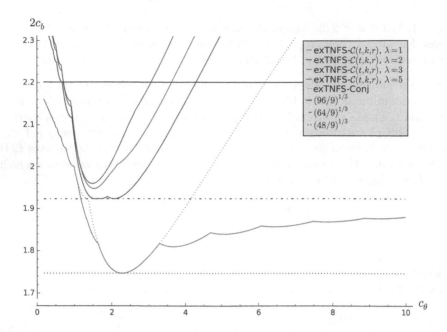

**Fig. 3.** Complexity plots for the medium prime case using the exTNFS algorithm.

Methods for obtaining the collection of number fields for MNFS algorithm have been mentioned in [18]. We adapt one of these methods to our setting. Note that the Algorithm $\mathcal{C}$ produces two polynomials $f(x)$ and $g(x)$ of degrees $d(r + 1)$ and $dr$ respectively. The polynomial $g(x)$ is defined as $\mathrm{Res}_y(\psi(y), C_0(x) + yC_1(x))$ where $\psi(x) = \mathrm{LLL}(M_{A_2,r})$, i.e., $\psi(x)$ is defined from the first row of the matrix obtained after applying the LLL-algorithm to $M_{A_2,r}$. We use $f(x)$ for constructing the first number field. Let $g_1(x) = g(x)$ and $g_2(x) = \mathrm{Res}_y(\psi_2(y), C_0(x) + yC_1(x))$, where $\psi_2(x)$ is the polynomial defined from the second row of the matrix $M_{A_2,r}$. For $i = 3, \ldots, V$, we consider $g_i(x) = s_i g_1(x) + t_i g_2(x)$ where the coefficients $s_i$ and $t_i$ are of the size of $V^{1/(2\eta)}$. These $g_i(x)$ are used for constructing the other $V$ number fields.

Clearly the $g_i$'s have degree $dr$. Asymptotically, we have $\|\psi_2\|_\infty = \|\psi_1\|_\infty = Q^{1/(d(r+1))}$. If we choose $V = L_Q(1/3)$, all the $g_i$'s have their infinity norms given by Proposition 1.

Let $B$ and $B'$ be the bounds on the norms of the ideals which are in the factor basis defined by $f$ and each of $g_i$'s respectively. So, the size of the entire factor basis is $B + VB'$. We further use the following condition to balances the factor basis.

$$B = VB'. \tag{27}$$

With this condition, the size of the factor basis is $B^{1+o(1)}$ (see [17] for the role of ECM based smoothness testing in this setting) and so asymptotically, the linear algebra step takes time $B^2$. Similar to the analysis of NFS variant, the number of sieving polynomials is $E^2$ and the coefficient polynomials of $\phi(x)$ can take $E^{2/t}$ distinct values. Since we require that the cost of relation collection should be same as the cost of linear algebra, we have $E^2 = B^2$ i.e., $E = B$.

As before, let $\pi$ be the probability that a random sieving polynomial $\phi(x)$ gives rise to a relation. Let $\pi_1$ be the probability that $\phi(x)$ is smooth over the factor basis for the first number field and $\pi_2$ be the probability that $\phi(x)$ is smooth over *at least* one of the other $V$ factor bases. Further, let $\Gamma_1 = \text{Res}_x(f(x), \phi(x))$ be the bound on the norm corresponding to the first number field and $\Gamma_2 = \text{Res}_x(g_i(x), \phi(x))$ be the bound on the norm for any of the other number fields. Recall that $\Gamma_2$ is determined only by the degree and the $L_\infty$-norm of $g_i(x)$ and hence is the same for all $g_i(x)$'s. Heuristically, we have

$$
\begin{aligned}
\pi_1 &= \Psi(\Gamma_1, B); \\
\pi_2 &= V\Psi(\Gamma_2, B'); \\
\pi &= \pi_1 \times \pi_2.
\end{aligned}
\tag{28}
$$

One relation is obtained in about $\pi^{-1}$ trials and so total number of relations obtained after sieving would be $E^2\pi$ and this should be equal to $B$ for linear algebra step to go through. Hence we have, as before, $B = E = \pi^{-1}$.

The following choices of $B$ and $V$ are made.

$$
\begin{aligned}
E = B &= L_Q\left(\tfrac{1}{3}, c_b\right); \\
V &= L_Q\left(\tfrac{1}{3}, c_v\right); \text{ and so} \\
B' = B/V &= L_Q\left(\tfrac{1}{3}, c_b - c_v\right).
\end{aligned}
\tag{29}
$$

**Theorem 2.** *Let* $n = \eta\kappa$; $p = L_Q(a, c_p)$ *with* $1/3 < a < 2/3$; *and* $\eta = c_\eta(\ln Q/\ln\ln Q)^{2/3-a}$. *It is possible to ensure that the runtime of the MNFS algorithm is* $L_Q(1/3, 2c_b)$ *where*

$$
c_b = \frac{2r+1}{3c_\theta kt} + \sqrt{\frac{r(3r+2)}{9c_\theta^2 k^2 t^2} + \frac{(t-1)c_\theta\varepsilon}{3}}.
\tag{30}
$$

*Proof.* For a sieving polynomial $\phi$,

$$
\begin{aligned}
\Gamma_1 = N(\phi, f) &= E^{2d(r+1)/t} L_Q(2/3, o(1)) \\
&= L_Q(2/3, (2c_b(r+1))/(c_\theta kt)); \\
\pi_1^{-1} &= L_Q(1/3, 2(r+1)/(3c_\theta kt)); \\
\Gamma_2 = N(\phi, g) &= E^{2dr/t} \times Q^{(t-1)\varepsilon/\kappa} L_Q(2/3, o(1)) \\
&= L_Q(2/3, 2c_b r/(c_\theta kt) + (t-1)c_\theta\varepsilon); \\
\pi_2^{-1} &= L_Q\left(\frac{1}{3}, -c_v + \frac{1}{3(c_b - c_v)}\left(\frac{2c_b r}{c_\theta kt} + (t-1)c_\theta\varepsilon\right)\right); \\
\pi^{-1} &= L_Q\left(\frac{1}{3}, \frac{2(r+1)}{3c_\theta kt} - c_v + \frac{1}{3(c_b - c_v)}\left(\frac{2c_b r}{c_\theta kt} + (t-1)c_\theta\varepsilon\right)\right);
\end{aligned}
$$

From the condition $\pi^{-1} = B$, we obtain the following equation.

$$c_b = \frac{2(r+1)}{3c_\theta kt} - c_v + \frac{1}{3(c_b - c_v)}\left(\frac{2c_b r}{c_\theta kt} + (t-1)c_\theta \varepsilon\right). \tag{31}$$

Simplifying, we obtain

$$3c_\theta kt(c_b^2 - c_v^2) = 2(2r+1)c_b - 2(r+1)c_v + (t-1)c_\theta^2 \varepsilon kt. \tag{32}$$

We wish to find $c_v$ such that $c_b$ is minimised subject to the constraint (32). Using the method of Lagrange multipliers, the partial derivative of (32) with respect to $c_v$ gives

$$c_v = \frac{(r+1)}{3c_\theta kt}.$$

Using this value of $c_v$ in (32) provides the following quadratic in $c_b$.

$$(3c_\theta kt)^2 c_b^2 - (6(2r+1)c_\theta kt)c_b + (r+1)^2 - 3(t-1)c_\theta^3 k^2 t^2 \varepsilon = 0.$$

Solving this and taking the positive square root, we obtain the expression for $c_b$ given by (30).    □

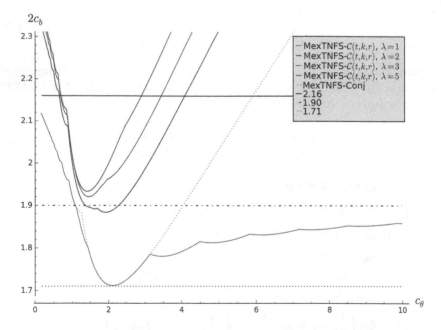

**Fig. 4.** Complexity plots for the medium prime case using the MexTNFS algorithm.

To find the absolute minimum complexity, we need to minimise the expression for $c_b$ given by (30) with respect to $c_\theta$. The standard way of doing this is to differentiate with respect to $c_\theta$ and set to 0 to find the value of $c_\theta$ for which the minimum value of $c_b$ is attained. Differentiating the right hand side of (30) with respect to $c_\theta$ and setting to 0 yields (after some simplifications) a quadratic in $c_\theta^3$ which can be solved to obtain:

$$c_\theta^3 = \frac{2}{3\varepsilon k^2 t^2(t-1)} \cdot \left(4r^2 + 9r + 1 + \sqrt{7r^2 + 16r + 1}\right). \tag{33}$$

Substituting the value of $c_\theta$ in (30) provides the expression for the corresponding value of $2c_b$ in terms of $t, r, k$ and $\lambda$. For each value of $\lambda$, we wish to obtain the minimum possible value of $2c_b$. This is achieved with $t = 2$ and $r = k$. The actual value of $r$ depends on the value of $\lambda$: for $\lambda = 1$ the minimum value of $2c_b$ is $\approx 1.71$ and is achieved for $r = 1$; for $\lambda = 2$ the minimum value of $2c_b$ is $\approx 1.88$ and is achieved for $r = 1$; for $\lambda = 3$ the minimum value of $2c_b$ is $\approx 1.92$ and is achieved for $r = 4$; for $\lambda = 5$ the minimum value of $2c_b$ is $\approx 1.94$ and is achieved for $r = 4$.

The variation of $2c_b$ with $c_\theta$ is more complex. Figure 4 shows these plots for various values of $\lambda$. From [17], the complexities for the medium characteristic case, the large characteristic case and the best complexity for the boundary case are respectively $L_Q(1/3, 2.12)$, $L_Q(1/3, 1.90)$ and $L_Q(1/3, 1.71)$. For composite prime-power $n$, these are the previously known best known complexities for these cases. To make the comparison of the new complexities easier, Fig. 4 shows the lines for 2.12, 1.90 and 1.71.

## 8   Conclusion

In this paper, we have presented a new polynomial selection method for exTNFS algorithm. The new polynomial selection method subsumes GJL, Conjugation and Sarkar-Singh polynomial selection methods. The exTNFS algorithm combined with new polynomial selection method provides new asymptotic complexities for the extension fields with composite prime power extension degrees.

## References

1. Adleman, L.M.: The function field sieve. In: Adleman, L.M., Huang, M.-D. (eds.) ANTS 1994. LNCS, vol. 877, pp. 108–121. Springer, Heidelberg (1994). doi:10. 1007/3-540-58691-1_48
2. Adleman, L.M., Huang, M.-D.A.: Function field sieve method for discrete logarithms over finite fields. Inf. Comput. 151(1–2), 5–16 (1999)
3. Barbulescu, R., Gaudry, P., Guillevic, A., Morain, F.: Improving NFS for the discrete logarithm problem in non-prime finite fields. In: Oswald, E., Fischlin, M. (eds.) EUROCRYPT 2015. LNCS, vol. 9056, pp. 129–155. Springer, Heidelberg (2015). doi:10.1007/978-3-662-46800-5_6
4. Barbulescu, R., Gaudry, P., Joux, A., Thomé, E.: A heuristic quasi-polynomial algorithm for discrete logarithm in finite fields of small characteristic. In: Nguyen, P.Q., Oswald, E. (eds.) EUROCRYPT 2014. LNCS, vol. 8441, pp. 1–16. Springer, Heidelberg (2014). doi:10.1007/978-3-642-55220-5_1

5. Barbulescu, R., Gaudry, P., Kleinjung, T.: The tower number field sieve. In: Iwata, T., Cheon, J.H. (eds.) ASIACRYPT 2015. LNCS, vol. 9453, pp. 31–55. Springer, Heidelberg (2015). doi:10.1007/978-3-662-48800-3_2
6. Barbulescu, R., Pierrot, C.: The multiple number field sieve for medium and high characteristic finite fields. LMS J. Comput. Math. **17**, 230–246 (2014)
7. Bistritz, Y., Lifshitz, A.: Bounds for resultants of univariate and bivariate polynomials. Linear Algebra Appl. **432**(8), 1995–2005 (2010). Special issue devoted to the 15th ILAS Conference at Cancun, Mexico, 16–20 June 2008
8. Gordon, D.M.: Discrete logarithms in GF(p) using the number field sieve. SIAM J. Discrete Math. **6**(1), 124–138 (1993)
9. Granger, R., Kleinjung, T., Zumbrägel, J.: Discrete logarithms in $GF(2^{9234})$. NMBRTHRY list, January 2014
10. Joux, A.: Faster index calculus for the medium prime case application to 1175-bit and 1425-bit finite fields. In: Johansson, T., Nguyen, P.Q. (eds.) EUROCRYPT 2013. LNCS, vol. 7881, pp. 177–193. Springer, Heidelberg (2013). doi:10.1007/978-3-642-38348-9_11
11. Joux, A.: A new index calculus algorithm with complexity $L(1/4 + o(1))$ in small characteristic. In: Lange, T., Lauter, K., Lisoněk, P. (eds.) SAC 2013. LNCS, vol. 8282, pp. 355–379. Springer, Heidelberg (2014). doi:10.1007/978-3-662-43414-7_18
12. Joux, A., Lercier, R.: The function field sieve is quite special. In: Fieker, C., Kohel, D.R. (eds.) ANTS 2002. LNCS, vol. 2369, pp. 431–445. Springer, Heidelberg (2002). doi:10.1007/3-540-45455-1_34
13. Joux, A., Lercier, R.: Improvements to the general number field sieve for discrete logarithms in prime fields. A comparison with the gaussian integer method. Math. Comput. **72**(242), 953–967 (2003)
14. Joux, A., Lercier, R.: The function field sieve in the medium prime case. In: Vaudenay, S. (ed.) EUROCRYPT 2006. LNCS, vol. 4004, pp. 254–270. Springer, Heidelberg (2006). doi:10.1007/11761679_16
15. Joux, A., Lercier, R., Smart, N.P., Vercauteren, F.: The number field sieve in the medium prime case. In: Dwork, C. (ed.) CRYPTO 2006. LNCS, vol. 4117, pp. 326–344. Springer, Heidelberg (2006). doi:10.1007/11818175_19
16. Joux, A., Pierrot, C.: The special number field sieve in $\mathbb{F}_{p^n}$ - application to pairing-friendly constructions. In: Cao, Z., Zhang, F. (eds.) Pairing 2013. LNCS, vol. 8365, pp. 45–61. Springer, Heidelberg (2014). doi:10.1007/978-3-319-04873-4_3
17. Kim, T., Barbulescu, R.: Extended tower number field sieve: a new complexity for the medium prime case. In: Robshaw, M., Katz, J. (eds.) CRYPTO 2016. LNCS, vol. 9814, pp. 543–571. Springer, Heidelberg (2016). doi:10.1007/978-3-662-53018-4_20
18. Pierrot, C.: The multiple number field sieve with conjugation and generalized Joux-Lercier methods. In: Oswald, E., Fischlin, M. (eds.) EUROCRYPT 2015. LNCS, vol. 9056, pp. 156–170. Springer, Heidelberg (2015). doi:10.1007/978-3-662-46800-5_7
19. Sarkar, P., Singh, S.: Fine tuning the function field sieve algorithm for the medium prime case. IEEE Trans. Inf. Theory **62**(4), 2233–2253 (2016). http://ieeexplore.ieee.org/xpl/articleDetails.jsp?reload=true&arnumber=7405328
20. Sarkar, P., Singh, S.: New complexity trade-offs for the (multiple) number field sieve algorithm in non-prime fields. In: Fischlin, M., Coron, J.-S. (eds.) EUROCRYPT 2016. LNCS, vol. 9665, pp. 429–458. Springer, Heidelberg (2016). doi:10.1007/978-3-662-49890-3_17
21. Schirokauer, O.: Using number fields to compute logarithms in finite fields. Math. Comput. **69**(231), 1267–1283 (2000)

# On the Security of Supersingular Isogeny Cryptosystems

Steven D. Galbraith[1]([✉]), Christophe Petit[2], Barak Shani[1], and Yan Bo Ti[1]

[1] Mathematics Department, University of Auckland, Auckland, New Zealand
{s.galbraith,barak.shani}@auckland.ac.nz, yanbo.ti@gmail.com
[2] Mathematical Institute, Oxford University, Oxford OX2 6GG, UK
christophe.petit@maths.ox.ac.uk

**Abstract.** We study cryptosystems based on supersingular isogenies. This is an active area of research in post-quantum cryptography. Our first contribution is to give a very powerful active attack on the supersingular isogeny encryption scheme. This attack can only be prevented by using a (relatively expensive) countermeasure. Our second contribution is to show that the security of all schemes of this type depends on the difficulty of computing the endomorphism ring of a supersingular elliptic curve. This result gives significant insight into the difficulty of the isogeny problem that underlies the security of these schemes. Our third contribution is to give a reduction that uses partial knowledge of shared keys to determine an entire shared key. This can be used to retrieve the secret key, given information leaked from a side-channel attack on the key exchange protocol. A corollary of this work is the first bit security result for the supersingular isogeny key exchange: Computing any component of the $j$-invariant is as hard as computing the whole $j$-invariant.

Our paper therefore provides an improved understanding of the security of these cryptosystems. We stress that our work does not imply that these systems are insecure, or that they should not be used. However, it highlights that implementations of these schemes will need to take account of the risks associated with various active and side-channel attacks.

**Keywords:** Isogenies · Supersingular elliptic curves

## 1 Introduction

In 2011, Jao and De Feo [17] introduced the supersingular isogeny Diffie–Hellman key exchange protocol as a candidate for a post-quantum key exchange. The security of this scheme is based on so-called supersingular isogeny problems. Similar problems had appeared in a previous hash function construction by Charles–Lauter–Goren [6], and were subsequently used to build other cryptographic functions such as public-key encryption, undeniable signatures and designated verifier signatures [13,18,34]. As with classical Diffie–Hellman, the basic

© International Association for Cryptologic Research 2016
J.H. Cheon and T. Takagi (Eds.): ASIACRYPT 2016, Part I, LNCS 10031, pp. 63–91, 2016.
DOI: 10.1007/978-3-662-53887-6_3

version of the key exchange protocol uses ephemeral elements, but the encryption scheme and some of the more sophisticated applications use static values for at least one element.

The idea behind the supersingular isogeny key exchange protocol is largely based on the isogeny protocol for ordinary elliptic curves proposed in [29]. However, there is a (subexponential) quantum algorithm [7] to break the system in the ordinary case (in part since the ordinary case is based on commutative ring theory). In contrast, the case of supersingular curves is non-commutative and seems to be a promising candidate for a post-quantum-secure system [2].

One particular feature of Jao and De Feo's protocols compared to other schemes based on isogeny problems is the publication of auxiliary points, which are used to get around the difficulties of non-commutativity. These auxiliary points open the door to active attacks on the encryption scheme (or key exchange where one party uses a static key). To be precise, one could try to perform some kind of "small subgroup" or "invalid curve" attacks such as have been proposed for DLP cryptosystems in the past [8,23]. The possibility of active attacks has been mentioned by Kirkwood, Lackey, McVey, Motley, Solinas and Tuller [20] and Costello, Longa and Naehrig [9]. Both papers discuss "validation" techniques that are designed to prevent such attacks, but neither paper demonstrates all the details of the attacks. Some of the validation methods discussed in [9] use pairings, but we observe a stronger property of pairings that makes detecting such attacks easier. Note that [9] is only concerned with ephemeral Diffie–Hellman key exchange, and so their scheme is not subject to attacks on static keys.

The first contribution of our paper (Sect. 3) is to describe a general active attack against the static-key variant of the protocol. Our attack allows to recover the whole static key with the minimum number of queries and negligible computation. Our attack is not prevented by any of the validation techniques introduced in [9], nor by our stronger validation technique using pairings. Our attack is prevented by the method in [20] (see Sect. 2.5), but this adds significant cost to the running time of the system.

The second contribution of our paper (Sect. 4) is to explore the security of the schemes assuming there is an efficient algorithm to compute the endomorphism ring of a supersingular elliptic curve. It is known that computing endomorphism rings of supersingular curves is equivalent to computing isogenies between supersingular elliptic curves, and it is believed that both these problems are hard [6,17]. But previous techniques were not sufficient to break the Jao–De Feo cryptosystems if the endomorphism ring was known (the resulting isogeny would have too high degree). We present a new method to find an isogeny of the correct degree in the special case of the isogeny problem arising in these cryptosystems. This shows that the hardness of computing endomorphism rings is necessary for the security of any cryptosystem based on the Jao and De Feo concept (it is not restricted to ElGamal or key exchange, and requires no interaction with a user). We give heuristic and experimental evidence that our algorithm is practical.

Our third contribution (Sect. 5) is to define and analyse an isogeny analogue of the hidden number problem. Our main result is an algorithm to compute the $j$-invariant of a "hidden" elliptic curve given partial information of the $j$-invariants of "nearby" curves. We believe that, as with the original hidden number problem in finite fields, this result will have applications of two flavours. On the one hand, our theorem shows how to mount a type of side-channel attack on the key exchange protocol: An attacker can compute the shared secret with high probability if they can get partial information of the shared key during "correlated" executions of the key exchange protocol. On the other hand, the result gives the first bit security result for the supersingular isogeny key exchange: Computing one component of the finite field representation of the $j$-invariant is as hard as computing the whole $j$-invariant. A consequence of this result is that it is secure for an implementation to use only one component of the $j$-invariant of the shared key.

The paper is organised as follows. Section 2 quickly reviews the Jao–De Feo cryptosystem and other preliminaries. Our results and discussions are given in Sects. 3, 4 and 5. In Sect. 6 we present our conclusions.

## 2  Preliminaries

### 2.1  Supersingular Elliptic Curves and Isogenies

Fix a prime $p$ and a prime power $q = p^k$ and let $E_1$ and $E_2$ be elliptic curves defined over $\mathbb{F}_q$. An isogeny between $E_1$ and $E_2$ is a non-constant morphism defined over $\mathbb{F}_q$ that sends the identity in $E_1$ to the identity in $E_2$. Then $\phi$ is a group homomorphism from $E_1(\overline{\mathbb{F}}_q)$ to $E_2(\overline{\mathbb{F}}_q)$ [30, III.4.8]. The degree of $\phi$ as an isogeny is equal to the degree of $\phi$ as a morphism. In addition, if $\phi$ is separable, then $\deg \phi = \# \ker \phi$ [30, III.4.10]. In this case, we say that $E_1$ and $E_2$ are isogeneous.

The isogeny is defined by its kernel in the sense that for every finite subgroup $G \subset E_1$, there is a unique $E_2$ (up to isomorphism) and a separable isogeny $\phi : E_1 \rightarrow E_2$ such that $\ker \phi = G$ [30, III.4.12]. We sometimes write $E_1/G$ for $E_2$. Vélu [32] gave an algorithm to construct an isogeny given a finite subgroup. Notice that the total number of distinct isogenies with degree $\ell$, which we now call $\ell$-isogenies, is equal to the number of distinct subgroups of $E_1$ of order $\ell$. For every prime $\ell$ not dividing $p$, there are $\ell + 1$ isogenies of degree $\ell$ since the group of $\ell$-torsion points form a subgroup $E[\ell] = \mathbb{Z}/\ell\mathbb{Z} \oplus \mathbb{Z}/\ell\mathbb{Z}$ [30, III.6.4].

If $G = \langle P \rangle \subset E_1$ is a cyclic group of order $\ell^n$ then the isogeny with kernel $G$ factors as a chain of isogenies

$$E_1 \rightarrow E_2 \rightarrow \cdots \rightarrow E_{\ell+1}$$

such that each $\phi_i : E_i \rightarrow E_{i+1}$ is an isogeny of degree $\ell$ with kernel in $E_i[\ell]$. We will use the following notation

$$G_1 = G, \quad G_{i+1} = \phi_i(G_i),$$
$$P_1 = P, \quad P_{i+1} = \phi_i(P_i).$$

Now, note that $\phi_i(G_i) = \langle \phi_i(P_i) \rangle \subseteq E_{i+1}[\ell^{n-i}]$. The kernel of $\phi_1$ is $\langle [\ell^{n-1}]P \rangle$ and for $i > 1$ the kernel of $\phi_i$ is $\langle [\ell^{n-i}]\phi_{i-1}(P_{i-1}) \rangle$.

For every $\phi : E_1 \rightarrow E_2$, there exists an isogeny $\hat{\phi} : E_2 \rightarrow E_1$ such that

$$\phi \circ \hat{\phi} = [\deg \phi] = \hat{\phi} \circ \phi.$$

We call $\hat{\phi}$ the dual isogeny of $\phi$. This allows us to define an equivalence relation on elliptic curves that are isogenous.

If we have a pair of isogenies $\phi : E_1 \rightarrow E_2$ and $\psi : E_2 \rightarrow E_1$ such that $\phi \circ \psi$ and $\psi \circ \phi$ are the identity, then we say that $\phi$ and $\psi$ are isomorphisms. We also then say that $E_1$ and $E_2$ are isomorphic curves. This naturally defines an equivalence relation and the isomorphism classes can be represented by the $j$-invariants [30, III.1.4(b)].

Isogenies that have the same domain and range are known as endomorphisms. For an elliptic curve $E$, we write $\text{End}(E)$ for the set of all endomorphisms $\phi : E \rightarrow E$ together with the zero morphism. In fact, we can define addition and multiplication on endomorphisms by setting $(\phi + \psi)(P) = \phi(P) + \psi(P)$ and $(\phi \cdot \psi)(P) = \phi(\psi(P))$ for all $\phi, \psi \in \text{End}(E)$ and $P \in E$. This gives it a ring structure. The multiplication-by-$n$ maps are examples of endomorphisms and so $\mathbb{Z} \hookrightarrow \text{End}(E)$. In fact, over a finite field, $\text{End}(E)$ is isomorphic to either a maximal order in a quaternion algebra or to an order in an imaginary quadratic field [30, III.9.3]. In the former case, we say that $E$ is supersingular, otherwise, we say that it is ordinary.

An elliptic curve $E/\mathbb{F}_{p^k}$ is supersingular if and only if $|E(\mathbb{F}_{p^k})| \equiv 1 \pmod{p}$. It is known that there are approximately $p/12$ isomorphism classes of supersingular elliptic curves $E$ over $\overline{\mathbb{F}}_p$ [30, V.4.1]. It is also known that every supersingular curve is isomorphic to one defined over $\mathbb{F}_{p^2}$ [30, V.3.1(a)(iii)]. A theorem of Tate states that $E_1$ and $E_2$ are isogenous over $\mathbb{F}_{p^k}$ if and only if $|E_1(\mathbb{F}_{p^k})| = |E_2(\mathbb{F}_{p^k})|$ [31, Sect. 3].

## 2.2  Hard Problem Candidates Related to Isogenies

Starting from the work of Charles–Lauter–Goren [6] and later Jao–De Feo [17], several recent cryptosystems have been based on the computational hardness of computing isogenies between supersingular elliptic curves. The main problem in this area can be described as follows:

**Definition 1 (Supersingular Isogeny Problem).** *Given a finite field $K$ and two supersingular elliptic curves $E_1, E_2$ defined over $K$ such that $|E_1| = |E_2|$, compute an isogeny $\varphi : E_1 \rightarrow E_2$.*

We stress that this isogeny is not unique (in fact there are infinitely many of them without additional restrictions). Further, the most natural representations of an isogeny are either as a pair of rational maps or as a kernel, and both these representations generally require exponential space. However, one can also represent an isogeny of smooth degree as a composition of low degree isogenies, and this can be done in polynomial space. Hence the computational problem makes sense.

This problem has been studied in a number of previous works. The cryptanalysis of Charles–Lauter–Goren's hash function requires to compute isogenies of degree $\ell^e$ for some small, fixed prime $\ell$. Similarly, the Jao–De Feo schemes involve isogenies of the same form with an additional condition on $e$.

Another important problem in this area is the problem of computing the endomorphism ring of a given elliptic curve.

**Definition 2 (Endomorphism Ring Computation).** *Given an elliptic curve $E$ defined over a finite field $K$, compute its endomorphism ring.*

This problem was studied by Kohel [21]. In the supersingular case Kohel described a probabilistic algorithm running in time $\tilde{O}(p)$, where $p$ is the characteristic of the field. This was later improved to $\tilde{O}(\sqrt{p})$ by Galbraith [15] using birthday paradox arguments. We remark that for some supersingular elliptic curves the problem is easy (for example when $j = 0$), but the problem is believed to be hard on average.

Heuristically, one can turn an algorithm that computes isogenies into an algorithm that computes the full endomorphism ring of an elliptic curve; the reduction actually underlies Kohel's algorithm.

It turns out that the converse is also true, at least heuristically. There is an equivalence of categories between the set of supersingular curves and the set of maximal orders of a quaternion algebra (see [12,21,22]). Given the endomorphism rings of the two elliptic curves, one can identify the corresponding maximal orders in the quaternion algebra, and then use techniques developed in [22] to compute paths between them in the quaternion algebra and translate these paths into isogeny paths.

The algorithm in [22] solves the quaternion algebra analog of the supersingular isogeny problem, which requires to compute an ideal with a smooth norm connecting two given maximal orders. However, the degree of the ideal returned by this algorithm is about $p^7$ in general and $p^{7/2}$ if one of the orders is special (a $p$-extremal order, as defined in [22]), whereas a degree about $p$ is expected to suffice in general, and a degree about $p^{1/2}$ would be needed to break the Jao–De Feo cryptosystems. Here $p$ is the characteristic of the field.

### 2.3 Jao–De Feo scheme

**Key Exchange Protocol.** There are three steps in the key exchange protocol: The set-up, the key exchange and the key derivation.

In the set-up, a prime of the form $p = 2^n \cdot 3^m \cdot f - 1$ is generated where $f$ is small and $2^n \approx 3^m$ (more generally $p = \ell_A^n \ell_B^m f \pm 1$ where $\ell_A, \ell_B$ are small primes). A supersingular elliptic curve $E$ over $\mathbb{F}_{p^2}$ is constructed, and linearly independent points $P_A, Q_A \in E[2^n]$ and $P_B, Q_B \in E[3^m]$ are chosen. Here "linearly independent" means that the group $\langle P_A, Q_A \rangle$ generated by $P_A$ and $Q_A$ has order $2^{2n}$, and similarly, $|\langle P_B, Q_B \rangle| = 3^{2m}$.

In the key exchange, Alice picks random integers $0 \le a_1, a_2 < 2^n$ (not both divisible by 2) and Bob picks random integers $0 \le b_1, b_2 < 3^m$ (not both divisible by 3). Alice and Bob compute

$$G_A = \langle [a_1]P_A + [a_2]Q_A \rangle, \quad G_B = \langle [b_1]P_B + [b_2]Q_B \rangle$$

respectively. Using Vélu's formulas [32], they will then be able to compute the isogenies $\phi_A$ and $\phi_B$ with respective kernels $G_A$ and $G_B$. They then compute $E_A = \phi_A(E) = E/G_A$, $\phi_A(P_B)$, $\phi_A(Q_B)$ and $E_B = \phi_B(E) = E/G_B$, $\phi_B(P_A)$, $\phi_B(Q_A)$ respectively. Their respective messages in the protocol will be

$$(E_A, \phi_A(P_B), \phi_A(Q_B)), \quad (E_B, \phi_B(P_A), \phi_B(Q_A)).$$

Upon receipt of Bob's message, to derive the shared key, Alice would compute

$$\langle [a_1]\phi_B(P_A) + [a_2]\phi_B(Q_A) \rangle = \langle \phi_B([a_1]P_A + [a_2]Q_A) \rangle = \phi_B(G_A).$$

Alice then computes the isogeny from $E_B$, with kernel equal to this subgroup. Bob will perform a similar computation and the resulting isogeny will be generated by $G_A$ and $G_B$ (since the subgroups have a trivial intersection). The shared secret will be

$$E_{AB} := E/\langle G_A, G_B \rangle = E_A/\langle \phi_A(G_B) \rangle = E_B/\langle \phi_B(G_A) \rangle.$$

This can be summarised in the following diagram, where we use the notation from above.

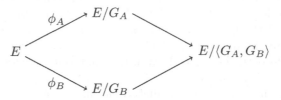

The Jao–De Feo key exchange scheme originates from a similar scheme for ordinary elliptic curves proposed by Rostovtsev and Stolbunov [29]. The ordinary case is based on a commutative mathematical structure, however this structure enables a subexponential-time quantum algorithm [7] to break the system. On the other hand, the supersingular curves variant is based on a non-commutative structure and so it seems to be a promising candidate for a post-quantum-secure system. The auxiliary points included in the protocol messages allow Jao and De Feo to get around the difficulties of non-commutativity.

We stress that the isogeny problem involved here differs from a general one in several ways. On the one hand, the special primes used and the auxiliary points given to an attacker may make the supersingular isogeny problem easier than the general isogeny problem. On the other hand there is a very strong constraint imposed on the degree of the isogeny, and this might a priori make the problem harder; we discuss this issue in more detail in Sect. 4. We remark that our first and third results use the auxiliary points in essential ways. However the result of Sect. 4 does not use the auxiliary points and only uses the fact that the required isogeny has a strongly constrained degree.

**Encryption Protocol.** The public-key encryption scheme is constructed from the key exchange scheme with a few adaptations [13]. Namely, the shared secret would be used as a key for a symmetric encryption scheme (below we use the one-time pad) to encrypt the message. We will use the same notation as above and assume that Bob wants to send a message to Alice. There are four steps to the encryption protocol: The set-up, key generation, encryption and decryption.

The set-up is almost identical to the key exchange protocol, where the two parties Alice and Bob agree on a prime of the form $p = 2^n \cdot 3^m \cdot f - 1$, a supersingular elliptic curve over $\mathbb{F}_{p^2}$, and linearly independent points $P_A, Q_A \in E[2^n]$ and $P_B, Q_B \in E[3^m]$. In addition, they agree on a keyed hash function $H_k$ that sends $\mathbb{F}_{p^2}$ to the set $\{0,1\}^w$ of $w$-bit strings.

In the key generation phase, Alice picks random integers $0 \le a_1, a_2 < 2^n$ (not both divisible by 2) and computes

$$E_A, \phi_A(P_B), \phi_A(Q_B)$$

as above. She also chooses a random ephemeral key, $k$, for the hash and publishes the tuple

$$(E_A, \phi_A(P_B), \phi_A(P_B), k)$$

as her public key. She retains $(a_1, a_2)$ as her private key.

Upon the receipt of Alice's public keys, Bob selects a $w$-bit message $m \in \{0,1\}^w$ and chooses random integers $0 \le b_1, b_2 < 3^m$ (not both divisible by 3) and computes

$$E_B, \phi_B(P_A), \phi_B(Q_A).$$

Using his randomly generated keys $b_1$ and $b_2$, he can also compute $E_{AB}$ as in the key-exchange protocol. He then computes

$$c = m \oplus H_k(j(E_{AB}))$$

and sends the tuple

$$(E_B, \phi_B(P_A), \phi_B(Q_A), c)$$

to Alice.

To decrypt Bob's message, Alice computes $E_{AB}$ using $E_B, \phi_B(P_A), \phi_B(Q_A)$ and $a_1, a_2$ and recovers the message $m$ by computing

$$m = c \oplus H_k(j(E_{AB})).$$

We stress that encryption is just one possible application where a static key may be used for at least one element in the protocol. We anticipate that as the subject develops further there will be more protocols of this type.

**Equivalent Keys and Normalisation.** The Vélu formulas tell us that the isogeny is determined solely by its kernel. In Alice's case, there are $3 \cdot 2^{n-1}$ choices of kernels, and the total number of choices for $(a_1, a_2)$ is about $2^{2n}$, so there will be private keys that correspond to the same public keys.

We define an equivalence relation on the private keys, by saying $(a_1, a_2) \sim$ $(a_1', a_2')$ if the two keys lead to the same subgroup for all possible input points. The relation is satisfied by $(a_1', a_2') = (\theta a_1, \theta a_2)$ for any $\theta \in \mathbb{Z}_{2^n}^*$, and so the equivalence class is a point in projective space over a ring. We may define a unique equivalence class representative by "normalising" as explained in the following lemma (this fact is also used by [9]).

**Lemma 1.** *Let* $P, Q \in E[2^n]$ *be linearly independent generators of* $E[2^n]$. *Then for some* $(a_1, a_2) \in \mathbb{Z}^2$ *(not simultaneously even), we have that* $(a_1, a_2) \sim (1, \alpha)$ *or* $(a_1, a_2) \sim (\alpha, 1)$ *for some* $\alpha \in \mathbb{Z}$ *(using the equivalence relation defined above).*

*Proof.* If $a_1$ is odd, then it is invertible modulo the order of the group, so let $\theta \equiv a_1^{-1} \pmod{2^n}$, then $\theta$ must be odd, hence

$$\langle [a_1]P_A + [a_2]Q_A \rangle = \langle [\theta a_1]P_A + [\theta a_2]Q_A \rangle = \langle P_A + [\alpha]Q_A \rangle,$$

where the first equality stems from the fact that $\theta$ is co-prime to the order of the generator, and the last equality is obtained by setting $\alpha = \theta a_2$.

If $a_1$ is even, then $a_2$ must be odd, and repeating the procedure gives $(\alpha, 1)$. $\square$

This result tells us that there is no loss of generality for Alice to restrict her secret key to be $(1, \alpha)$ or $(\alpha, 1)$. This was noted by [9]. However, even if Alice does not employ such a simplification, the result also tells us that there is no loss of generality for an attacker to assume the secret key is of one of these two forms. This observation is used repeatedly in the adaptive attack presented in Sect. 3.

## 2.4 Active Attacks and Validation Methods

Active attacks are a standard type of attack on cryptosystems that use a static private key. These first arose in the setting of protocols based on the discrete logarithm problem, where a user can be treated as an oracle that takes as input a group element $g$ and returns $g^a$ for some long-term secret value $a$. A first kind of attack is the "small subgroup" attack of Lim and Lee [23]. Here a group element $g$ of small order $\ell$ is sent, so that on receipt of the value $g^a$ one can do a search and learn $a \pmod{\ell}$. Similar ideas have been used based on "invalid curve" attacks, which involve providing a point that lies in a different group altogether (see Ciet and Joye [8]).

In the context of the isogeny cryptosystem, if Alice has a fixed key $(a_1, a_2)$ then a dishonest Bob can send her $(E, P, Q)$ and then Alice will compute an isogeny $\phi : E \to E'$ with kernel $\langle [a_1]P + [a_2]Q \rangle$. The idea is to try to learn something about Alice's secret key $(a_1, a_2)$ using knowledge of $E'$. The possibility of such attacks is mentioned in [9,20], but neither paper presented full details of them.

The concept of "validation" is intended to prevent active attacks. In the case of protocols based on the DLP, the typical countermeasures check that $g$ does lie in the correct group, and that the order of $g$ is the correct value. In the context of supersingular isogeny cryptosystems the validation of $(E, P, Q)$ should test that $E$ really is a supersingular elliptic curve, that $P$ and $Q$ lie on the curve and have the correct order, and that $P$ and $Q$ are independent. Methods to do this are given in [9].

In particular, Sect. 9 of [9] presented some explicit validation steps. Their two requirements are: The points in the public key have full order and they are independent. They use the Weil pairing of the two points to check independence. We remark that it is not necessary to use the Weil pairing: Since the DLP is easy in a group of order $2^n$ one can just try to solve the DLP of $Q$ to the base $P$, and if the algorithm fails then the points are independent. In particular, to show that $\langle P, Q \rangle = E[2^n]$ it suffices to compute $[2^{n-1}]P$ and $[2^{n-1}]Q$ and verify that these points are both different, and neither is the identity.

*Remark 1.* We now observe that the Weil pairing can be used to check a lot more than just independence. A standard fact is that if $\phi : E \to E'$ is an isogeny and if $P, Q \in E[N]$ then

$$e_N(\phi(P), \phi(Q)) = e_N(P, Q)^{\deg(\phi)}$$

where the first Weil pairing is computed on $E'$ and the second on $E$ (for details see [30, III.8.2] or [4, IX.9]). This allows to validate not only that the points are independent but also that they are consistent with being the image of the correct points under an isogeny of the correct degree. Hence, a natural validation step for Alice to run in the Jao–De Feo scheme is to check

$$e_{2^n}(\phi_B(P_A), \phi_B(Q_A)) = e_{2^n}(P_A, Q_A)^{3^m}.$$

This will give her some assurance that the points $\phi_B(P_A), \phi_B(Q_A)$ provided by Bob are consistent with being the images of the correct points under an isogeny of the correct degree. However, as we will show, this validation step is not sufficient to prevent all adaptive attacks. It will be necessary to use a much stronger protection, which we describe in the next section.

## 2.5   The Kirkwood et al. Validation Method

The Fujisaki-Okamoto transform [14] leads to a general method to secure any key exchange protocol of a certain type. This is explained in Sect. 5.2 of Peikert [28] and, in the context of the isogeny cryptosystem, it is discussed by Kirkwood et al. [20].

The idea is to complete the key exchange protocol and then for each party to encrypt to the other party the randomness used in the protocol so that they can check that the protocol has been performed correctly. Note that [20] does not contain a formal analysis of the security of the resulting protocol.

We now briefly describe the key exchange protocol that arises when this transform is applied to the Jao–De Feo protocol. In the following description, we show what Bob should do and how Alice can verify that Bob has followed the protocol correctly (this is suited for the case where Alice is using a static key and where Bob is a potential adversary).

(1) Bob obtains Alice's static public key $(E_A, \phi_A(P_B), \phi_A(Q_B))$.
(2) Bob chooses a random seed $r_B$ and derives his private key using a pseudo-random function PRF (Kirkwood et al. call this a key derivation function).

$$(b_1, b_2) = \text{PRF}(r_B).$$

He then computes his message $(E_B, \phi_B(P_A), \phi_B(Q_A))$ where $\phi_B$ is defined to have kernel $\langle [b_1]P_B + [b_2]Q_B \rangle$.
(3) Bob derives the shared secret value $E_{AB}$ from $(E_A, \phi_A(P_B), \phi_A(Q_B))$ and $(b_1, b_2)$ and computes a session key $(SK)$ and validation key $(VK)$ via a key derivation function (KDF)

$$SK \mid VK = \text{KDF}(j(E_{AB})).$$

(4) Bob then sends $(E_B, \phi_B(P_A), \phi_B(Q_A))$ and $c_B = \text{Enc}_{VK}(r_B \oplus SK)$ to Alice.
(5) From $(a_1, a_2)$ and $(E_B, \phi_B(P_A), \phi_B(Q_A))$, Alice derives $E'_{AB}$, then $SK'$ and $VK'$.
(6) Alice computes

$$r'_B = \text{Dec}_{VK'}(c_B) \oplus SK'.$$

She then computes $\text{PKDF}(r'_B)$ and recomputes Bob's operations. If the resulting message is equal to the value $(E_B, \phi_B(P_A), \phi_B(Q_A))$ originally sent by Bob then Alice terminates the protocol correctly and uses $SK' = SK$ for future communicate with Bob. If not, the protocol terminates in a non-accepting state.

Notice that this protocol requires that Bob reveals his secret key to Alice, so it compels him to change his secret key after each verification. This validation method can be used for both the key-exchange and the encryption protocols.

# 3   Adaptive Attack

In this section, we will assume that Alice is using a static key $(a_1, a_2)$, and that a dishonest user is playing the role of Bob and trying to learn her key. Our discussion is entirely about Alice's key and points in $E[2^n]$, but it should be clear that the same methods would work for points in $E[\ell^m]$ for any small prime $\ell$ (see Remark 2 for further discussion).

There are two attack models that can be defined in terms of access to an oracle $O$:

1. $O(E, R, S) = E/\langle [a_1]R + [a_2]S \rangle$. This corresponds to Alice taking Bob's protocol message, completing her side of the protocol, and outputting the shared key.
2. $O(E, R, S, E')$ which returns 1 if $j(E') = j(E/\langle [a_1]R + [a_2]S \rangle)$ and 0 otherwise. This corresponds to Alice taking Bob's protocol message, completing her side of the protocol, and then performing some operations using the shared key that return an error message if the shared key is not the same as the $j$-invariant provided (e.g., the protocol involves verifying a MAC corresponding to a key derived from the session key).

Our attacks can be mounted in both models. To emphasise their power we explain them in the context of the second, weaker, model.

## 3.1 First Step of the Attack

From Lemma 1, we may assume that the private key is normalised. In the following exposition, we will assume that the normalisation is $(1, \alpha)$. The case where we have $(\alpha', 1)$ where $\alpha'$ is even is performed in exactly the same way with some tweaks. Note that if $\alpha'$ is odd then it can be converted to the $(1, \alpha)$ case, so we may assume $\alpha'$ is even in the second case.

To differentiate between $(1, \alpha)$ and $(\alpha', 1)$ an attacker honestly generates Bob's ephemeral values $(E_B, R = \phi_B(P_A), S = \phi_B(Q_A))$ and follows the protocol to compute the resulting key $E_{AB}$. Then the attacker sends $(E_B, R, S + [2^{n-1}]R)$ to Alice and tests the resulting $j$-invariant. Expressing this in terms of the oracle access: The attacker queries an oracle of the second type on $(E_B, R, S + [2^{n-1}]R, E_{AB})$. If the oracle returns 1 then the curve $E_B/\langle [a_1]R + [a_2](S + [2^{n-1}]R)\rangle$ is isomorphic to $E_{AB}$ and so $\langle [a_1]R + [a_2](S + [2^{n-1}]R)\rangle = \langle [a_1]R + [a_2]S\rangle$. Hence, by the following Lemma, $a_2$ is even and we are in the first case. If the oracle returns 0 then $a_2$ is odd.

**Lemma 2.** *Let $R, S \in E[2^n]$ be linearly independent points of order $2^n$ and let $a_1, a_2 \in \mathbb{Z}$. Then*

$$\langle [a_1]R + [a_2](S + [2^{n-1}]R)\rangle = \langle [a_1]R + [a_2]S\rangle$$

*if and only if $a_2$ is even.*

*Proof.* If $a_2$ is even then $[a_2][2^{n-1}]R = 0$ and so the result follows. Conversely, if the two groups are equal then there is some $\lambda \in \mathbb{Z}_{2^n}^*$ such that

$$\lambda([a_1]R + [a_2](S + [2^{n-1}]R)) = [a_1]R + [a_2]S.$$

Since the points are independent we have $\lambda a_2 = a_2$ and so $\lambda = 1$. Hence, since $S$ has order $2^n$, we have $a_2 2^{n-1} \equiv 0 \pmod{2^n}$ and $a_2$ is even. $\qquad\square$

Note that the Weil pairing

$$e_{2^n}(R, S + [2^{n-1}]R) = e_{2^n}(R, S) = e_{2^n}(P_A, Q_A)^{3^m}$$

and so the attack is not detectable using pairings.

Similarly one can call the oracle on $(E_B, R + [2^{n-1}]S, S, E_{AB})$. The oracle returns 1 if and only if $a_1$ is even. Hence, we can determine which of the two cases we are in and determine if $\alpha$ is even or odd. Having recovered a single bit of $\alpha$, we will now explain how to use similar ideas to recover the rest of the bits of $\alpha$.

## 3.2   Continuing the Attack

We now assume that Alice's static key is of the form $(1, \alpha)$ and we write

$$\alpha = \alpha_0 + 2^1 \alpha_1 + 2^2 \alpha_2 + \cdots + 2^{n-1} \alpha_{n-1}.$$

The attacker will learn one bit of $\alpha$ for each query of the oracle. Algorithm 1 gives pseudo-code for the attack.

We now give some explanation and present the derivation of the algorithm. Suppose an attacker has recovered the first $i$ bits of $\alpha$, so that

$$\alpha = K_i + 2^i \alpha_i + 2^{i+1} \alpha',$$

where $K_i$ is known but $\alpha_i \in \{0, 1\}$ and $\alpha' \in \mathbb{Z}$ are not known.

The attacker generates $E_B, R = \phi_B(P_A), S = \phi_B(Q_A)$ and $E_{AB}$ as in the protocol. To recover $\alpha_i$, the attacker will choose suitable integers $a, b, c, d$ and query the oracle on

$$(E_B, [a]R + [b]S, \ [c]R + [d]S, E_{AB}).$$

The integers $a$, $b$, $c$, and $d$ will be chosen to satisfy the following conditions:

1. If $\alpha_i = 0$, then $\langle [a + \alpha c]R + [b + \alpha d]S \rangle = \langle R + [\alpha]S \rangle$.
2. If $\alpha_i = 1$, then $\langle [a + \alpha c]R + [b + \alpha d]S \rangle \neq \langle R + [\alpha]S \rangle$.
3. $[a]R + [b]S$ and $[c]R + [d]S$ both have order $2^n$.
4. The Weil pairing $e_{2^n}([a]R + [b]S, [c]R + [d]S)$ must be equal to

$$e_{2^n}(\phi_B(P_A), \phi_B(Q_A)) = e_{2^n}(P_A, Q_A)^{\deg \phi_B} = e_{2^n}(P_A, Q_A)^{3^m}.$$

The first two conditions help us distinguish the bit $\alpha_i$ and the latter two prevent the attack from being detected via order checking and Weil pairing validation checks respectively.

Consider the following integers:

$$a_i = 1, \quad b_i = -2^{n-i-1} K_i,$$
$$c_i = 0, \quad d_i = 1 + 2^{n-i-1}.$$

One can verify that they satisfy the third condition. To satisfy the fourth condition we need to use a scaling by $\theta$ that we will discuss later.

To show that the first two conditions are satisfied, note that $\langle [a]R + [b]S + [\alpha]([c]R + [d]S) \rangle$ is equal to

$$\langle R - [2^{n-i-1} K_i]S + [\alpha][1 + 2^{n-i-1}]S \rangle$$
$$= \langle R + [\alpha]S + [-2^{n-i-1} K_i + 2^{n-i-1}(K_i + 2^i \alpha_i + 2^{i+1} \alpha')]S \rangle$$
$$= \langle R + [\alpha]S + [\alpha_i 2^{n-1}]S \rangle$$
$$= \begin{cases} \langle R + [\alpha]S \rangle & \text{if } \alpha_i = 0, \\ \langle R + [\alpha]S + [2^{n-1}]S \rangle & \text{if } \alpha_i = 1. \end{cases}$$

By the following Lemma, these two subgroups are different. Hence the response of the oracle tells us $\alpha_i$.

**Lemma 3.** *Let $R$ and $S$ be linearly independent elements of the group $E[2^n]$ with full order, then the subgroups*

$$\langle R + [\alpha]S + [2^{n-1}]S \rangle \quad and \quad \langle R + [\alpha]S \rangle$$

*are different.*

*Proof.* The proof is very similar to the proof of Lemma 2. The subgroups have order $2^n$, since $R$ has order $2^n$, and $R$ and $S$ are linearly independent. Then if the subgroups are the same, we must have some $\lambda$ such that

$$[\lambda]R + [\lambda\alpha]S = R + [\alpha]S + [2^{n-1}]S.$$

By the linear independence of $R$ and $S$, we can compare coefficients and conclude that $\lambda = 1$, and that $[2^{n-1}]S = \mathcal{O}$, which implies that $S$ has order a factor of $2^{n-1}$, which is a contradiction. □

---

**Algorithm 1.** Adaptive attack using oracle $O(E, R, S, E')$.

**Data:** $n$, $E$, $P_A$, $Q_A$, $P_B$, $Q_B$, $E_A$, $\phi_A(P_B)$, $\phi_A(Q_B)$
**Result:** $\alpha$

1   Set $K_0 \leftarrow 0$;
2   **for** $i \leftarrow 0$ **to** $n-3$ **do**
3      Set $\alpha_i \leftarrow 0$;
4      Choose random $(b_1, b_2)$;
5      Set $G_B \leftarrow \langle [b_1]P_B + [b_2]Q_B \rangle$;
6      Set $E_B \leftarrow E/G_B$ and let $\phi_B : E \rightarrow E_B$ be the isogeny with kernel $G_B$;
7      Set $(R, S) \leftarrow (\phi_B(P_A), \phi_B(Q_A))$;
8      Set $E_{AB} \leftarrow E_A/\langle [b_1]\phi_A(P_B) + [b_2]\phi_A(Q_B) \rangle$;
9      Set $\theta \leftarrow \sqrt{(1 + 2^{n-i-1})^{-1}} \pmod{2^n}$;
10     Query the oracle on $\left( E_B, [\theta](R - [2^{n-i-1}K_i]S), [\theta][1 + 2^{n-i-1}]S, E_{AB} \right)$ ;
11     **if** *Response is false* **then** $\alpha_i = 1$;
12     Set $K_{i+1} \leftarrow K_i + 2^i\alpha_i$;
13 **end**
14 Brute force $\alpha_{n-2}, \alpha_{n-1}$ using $E$ and $E_A$ and $K_{n-2} = \alpha \pmod{2^{n-2}}$ to find $\alpha$ (this requires no oracle calls);
15 Return $\alpha$;

---

Finally, we address the fourth condition. We need that

$$e_{2^n}([a]R + [b]S, [c]R + [d]S) = e_{2^n}(R, S)^{ad-bc} = e_{2^n}(P_A, Q_A)^{3^m}.$$

The idea is that we can mask the points chosen from the attack above to satisfy the fourth condition. Recall that the points we wish to send to Alice are

$$(R', S') = (R - [2^{n-i-1}K_i]S, [1 + 2^{n-i-1}]S).$$

Computing the Weil pairing of the two points, we have

$$
\begin{aligned}
& e_{2^n}(R', S') \\
=& e_{2^n}(R - [K_i 2^{n-i-1}]S, [1 + 2^{n-i-1}]S) \\
=& e_{2^n}(R, [1 + 2^{n-i-1}]S) \cdot e_{2^n}(-[K_i 2^{n-i-1}]S, [1 + 2^{n-i-1}]S) \\
=& e_{2^n}(R, S)^{1+2^{n-i-1}},
\end{aligned}
$$

which is not the correct value. So we choose $\theta$ such that

$$
e_{2^n}(\theta R', \theta S') = e_{2^n}(R, S)^{\theta^2 (1+2^{n-i-1})} = e_{2^n}(P_A, Q_A)^{3^m} = e_{2^n}(R, S).
$$

Note that $\langle [\theta] R' + [\alpha][\theta] S' \rangle = \langle [\theta](R' + [\alpha]S') \rangle = \langle R' + [\alpha]S' \rangle$ as long as $\theta$ is coprime to the order $2^n$. Hence we need $\theta$ to be the square root of $1 + 2^{n-i-1}$ modulo $2^n$. The following lemma shows that such a square root exists as long as $n - i - 1 \geq 3$. Note that $\theta$ will be odd, as required.

**Lemma 4.** *If $a$ is an odd number and $m = 8$, $16$, or some higher power of $2$, then $a$ is a quadratic residue modulo $m$ if and only if $a \equiv 1 \pmod 8$.*

The condition $n - i - 1 \geq 3$ means we may not be able to launch the attack in an undetected way for the last two bits. This is why we use a brute force method to determine these bits.

The attack in the case $(\alpha', 1)$ follows by swapping the roles of $R$ and $S$.

## 3.3  Analysis and Complexity of the Attack

The attack requires fewer than $n \approx \frac{1}{2}\log_2(p)$ interactions with Alice. This seems close to optimal for attack model 2, where the attacker only gets one bit of information at each query. We can reduce the number of queries by doing more computation (increasing the range of the brute-force search).

We now consider the attack in the context of [9,20]. Due to our third and fourth conditions, the attack passes the validation steps in [9], and even the stronger check of taking the degree of the isogeny into account as mentioned in Remark 1.

The approach in [20] would be able to detect the attack. This is because the auxiliary points sent to Alice in the attack are not the correct values generated in an honest protocol run.

*Remark 2.* We now say a few words about attacking odd prime power isogenies. Let $\ell$ be an odd prime such that $\ell^n \mid (p+1)$ and $E[\ell^n] \subset E(\mathbb{F}_{p^2})$. Let $P_A$, $Q_A$ be generators of $E[\ell^n]$. Alice would compute an $\ell^n$-isogeny with kernel $\langle [a_1]P_A + [a_2]Q_A \rangle$ and a dishonest user Bob is trying to learn her key $a_1$, $a_2$, where $a_1$ and $a_2$ are not simultaneously divisible by $\ell$. As above, we take Alice's secret key to be $(1, \alpha)$.

The obvious generalisation for this attack is to set $R = \phi_B(P_A)$ and $S = \phi_B(Q_A)$ and to send Alice points

$$
(R - [x\ell^{n-i-1}]S, [1 + \ell^{n-i-1}]S).
$$

In her computation for the subgroup, Alice would compute

$$\langle R + [\alpha]S + [\ell^{n-i-1}][\alpha - x]S \rangle.$$

Since we want to compare this subgroup against $\langle R + [\alpha]S \rangle$, we need

$$(\ell^{n-i-1})(\alpha - x) \equiv 0 \pmod{\ell^n}$$

to ensure the subgroups computed are the same. Hence for each coefficient of a power of $\ell$ in the $\ell$-expansion of $\alpha$, we will need at most $\ell - 1$ queries to recover it.

For $\ell = 3$ this is as good as one would expect (at most two queries), but for primes $\ell \geq 5$ this seems not optimal since one would hope that given an oracle that returns one bit of information one could learn the value with only $\lceil \log_2(\ell) \rceil$ queries. In Appendix B we specify a simple attack, that is easily detectable and uses a stronger oracle, but can be used to efficiently handle the case $\ell > 3$.

## 4 Solving the Isogeny Problem When the Endomorphism Ring Is Known

Let $p = \ell_A^n \ell_B^m f - 1$ as in the Jao–De Feo cryptosystems, and let $E$ and $E_A$ be two supersingular elliptic curves such that there exists an isogeny $\phi_A : E \to E_A$ of degree $\ell_A^n$ between them. In this section we additionally suppose that we know (or can compute) the endomorphism rings $\mathrm{End}(E)$ and $\mathrm{End}(E_A)$, and we provide an efficient algorithm to recover $\phi_A$ assuming a certain natural heuristic holds. A formal statement of our reduction is below and we will prove this in Sect. 4.2.

**Theorem 1.** *Let $E$ and $E_A$ be supersingular elliptic curves over $\mathbb{F}_{p^2}$ such that $E[\ell_A^n] \subseteq E(\mathbb{F}_{p^2})$ and there is an isogeny $\phi_A : E \to E_A$ of degree $\ell_A^n$ from $E$ to $E_A$. Suppose there is no isogeny $\phi : E \to E_A$ of degree $< \ell_A^n$. Then, given an explicit description of $\mathrm{End}(E)$ and $\mathrm{End}(E_A)$, there is an efficient algorithm to compute $\phi_A$.*

As recalled in Sect. 2.2, computing the endomorphism ring of a supersingular elliptic curve is a problem essentially equivalent to computing an arbitrary isogeny between two supersingular elliptic curves. However, the the algorithm of [22] does not produce an isogeny that satisfies the additional constraint that it must be of small degree, as is required in the Jao–De Feo cryptosystems ($\ell_A^n \approx p^{1/2}$). Hence the current state of knowledge does not give a reduction of the form we require. The aim of this section is to present an alternative method to [22] in this context. We use the notation of [22].

### 4.1 The Importance of the Correct Isogeny

We first explain that to break the Jao–De Feo protocol it is not sufficient to compute *any* isogeny from $E$ to $E_A$. There are infinitely many such isogenies, but to break the Jao and De Feo cryptosystems it is necessary to find the right sort of isogeny, as we now explain.

Suppose there are curves $E$ and isogenies $\phi_A : E \to E_A$, $\phi_B : E \to E_B$ with $\ker(\phi_A) = G_A, \ker(\phi_B) = G_B$ satisfying the usual isogeny diagram from Sect. 2.3:

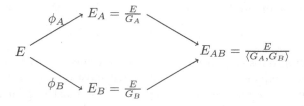

The correctness of the protocol follows from the fact that $E/\langle G_A, G_B\rangle = E_A/\langle\phi_A(G_B)\rangle = E_B/\langle\phi_B(G_A)\rangle$ and that $\phi_A(G_B)$ and $\phi_B(G_A)$ can be computed by the honest parties.

Suppose an attacker given $E, E_A, E_B$ can compute an isogeny $\phi' : E \to E_A$. So the picture now looks like:

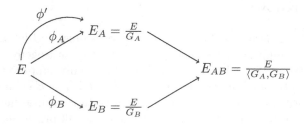

The natural approach for an attacker to try to compute $E_{AB}$ is to compute $\phi_B(\ker(\phi'))$ and hence an isogeny from $E_B$ with this kernel. However, the attacker only has the points $\phi_B(P_A), \phi_B(Q_A)$ to work with, and so can only compute $\phi_B(\ker(\phi'))$ if $\ker(\phi') \subseteq \langle P_A, Q_A\rangle$ (in which case $\phi'$ is an isogeny of degree dividing $2^n$). A random isogeny $\phi'$ is unlikely to have this property. Indeed, $\phi_A$ is likely to be the only isogeny from $E$ to $E_A$ with kernel in $\langle P_A, Q_A\rangle$ (apart from composing with an automorphism, which is of no consequence).

This is the crux of the difficulty in giving a reduction from computing endomorphism rings to computing the secret key in the Jao–De Feo cryptosystem: Known algorithms to compute an isogeny from $E$ to $E_A$, given $\mathrm{End}(E)$ and $\mathrm{End}(E_A)$, are not likely to give an isogeny of the correct degree. However, as we now explain, the particularly small degree of the secret key gives the reduction an advantage that does not arise in the general case.

## 4.2   Reduction of Problem to Computation of Endomorphism Ring

We show how the existence of a small degree isogeny actually *helps* the cryptanalysis of Jao–De Feo's cryptosystems, assuming we know (or we are able to compute) the endomorphism rings of the curves in play.

We write $B_{p,\infty}$ for the quaternion algebra ramified at $p$ and $\infty$ and use the standard notions of reduced trace and reduced norm (see Vigneras [33] for background). One extends the reduced norm to ideals in $B_{p,\infty}$.

Given two maximal orders $\mathcal{O}$ and $\mathcal{O}_A$, one can compute in polynomial time an ideal $I$ that connects them (see [22, Lemma 8]). Computing an isogeny of the correct degree corresponds to computing an equivalent ideal of the correct norm. In order to find such an equivalent ideal we use the following lemma.

**Lemma 5.** *[22, Lemma 5] Let $I$ be a left $\mathcal{O}$-ideal of reduced norm $N$ and $\alpha$ an element in $I$. Then $I\gamma$, where $\gamma = \bar{\alpha}/N$, is a left $\mathcal{O}$-ideal of norm $n(\alpha)$.*

We observe that in the context of Jao–De Feo cryptosystems, there exists by construction an element $\alpha$ of small norm $N\ell_A^n$ in $I$, corresponding via this lemma to an ideal of norm $\ell_A^n$. Moreover as Minkowski bases can be computed in polynomial time for lattices of dimension up to 4 [27], this element $\alpha$ can be efficiently recovered as long as it is in fact the smallest element in $I$. These observations lead to the following first simple algorithm:

---

**Algorithm 2.** Computing small degree isogenies in Jao–De Feo cryptosystems given an algorithm to compute the endomorphism ring of a random supersingular elliptic curve.

---

**Data**: $\ell_A$, $n$, $E$, $E_A$, $\mathcal{O} = \mathrm{End}(E)$, $\mathcal{O}_A = \mathrm{End}(E_A)$ such that $E$ and $E_A$ are connected by an isogeny of degree $\ell_A^n$

**Result**: Isogeny $\varphi_A : E \to E_A$ of small degree $\ell_A^n$, or *failure*

1 Compute an ideal $I$ connecting $\mathcal{O}$ and $\mathcal{O}_A$ as in [22, Lemma 8];
2 Compute a Minkowski-reduced basis of $I$;
3 Let $\alpha$ be the non-zero element in $I$ of minimal norm;
4 **if** $n(\alpha) \neq n(I)\ell_A^n$ **then** return *failure*;
5 Compute an ideal $I' = I\bar{\alpha}/n(I)$ ;
6 Compute the isogeny $\varphi_A$ that corresponds to $I'$ using Vélu's formulae;
7 Return $\varphi_A$;

---

All the steps in this algorithm can be performed in polynomial time. The above discussion forms the proof of Theorem 1.

*Proof.* (Theorem 1). Given an explicit representation of the endomorphism rings, we can translate the endomorphism rings into maximal orders of quaternion algebras. One can then find, in polynomial time, an ideal $I$ connecting them by [22, Lemma 8].

By Lemma 5, it is sufficient to find an element of $I$ of the correct norm. But given that the norm we seek is the smallest norm in the ideal, we can use lattice reduction methods to recover the smallest norm in polynomial time. Then using methods in [22], we can recover the isogeny we seek. □

In the remainder of this section, we study the success probability of this algorithm on average, and show how to use it to achieve a very large success probability.

Heuristically, we can approximate the probability that $E$ and $E_A$ are connected by an isogeny of degree $\ell$ by estimating the probability that two randomly chosen supersingular elliptic curves are connected by an isogeny of the same degree.[1]

Random pairs of elliptic curves over $\mathbb{F}_{p^2}$ are unlikely to be connected by isogenies of degrees significantly smaller than $\sqrt{p}$. Indeed, when $\ell = \prod_i p_i^{e_i}$, there are exactly

$$a(\ell) := \prod_i (p_i + 1)p_i^{e_i - 1}$$

isogenies of degree $\ell$ from any curve $E$, hence any curve $E$ is connected to at most $\sum_{\ell \leq D} a(\ell)$ curves $E_A$ by an isogeny of degree at most $D$. A calculation given in Appendix A shows that this sum converges to

$$\frac{15}{2\pi^2}D^2$$

as $D$ tends to infinity. As there are roughly $p/12$ supersingular invariants over $\mathbb{F}_{p^2}$ we can evaluate the success probability of the above algorithm as

$$SR \approx \max\left(0,\ 1 - \frac{90}{\pi^2}\frac{\ell_A^{2n}}{p}\right).$$

For the parameters used in Jao–De Feo's cryptosystems we expect this basic attack to succeed with a probability larger than $50\%$ as soon as $f > \frac{180}{\pi^2} \approx 18.23$, where $f$ is the cofactor in $p = \ell_A^n \ell_B^m f \pm 1$.

The success rate of our attack can be easily improved in two ways. First, we can apply the algorithm separately on all curves that are at distance $\ell_A^e$ of $E_A$ for some small constant $e$, until it succeeds for one of them. Clearly one of these curves will be connected to $E$ by an isogeny of degree $\ell_A^{n-e}$, and as a result the success rate will increase to

$$SR \approx \max\left(0,\ 1 - \frac{90}{\pi^2}\frac{\ell_A^{2(n-e)}}{p}\right).$$

With $\ell_A = 2$ and $e = 10$ this method will lead to a success rate above $99\%$, even when $f = 1$. Second, we can try to use the Minkowski-reduced basis computed in Step 3 of the algorithm to find an element $\alpha$ of the appropriate norm, even when it is not the smallest element. We explore two heuristic methods in that direction in our experiments below.

---

[1] The argument is not totally accurate as $E$ and $E_A$ are slightly closer in the $\ell_A$-isogeny graph than random pair of curves would be. This may a priori impact the probabilities, however a significant distortion of these probabilities would reveal some unexpected properties of the graph, such as the existence of more or fewer loops of certain degrees than expected.

## 4.3 Experimental Results

We tested our algorithm in Magma with $\ell_A = 2$ and with a $\lambda$-bit prime $p$, a randomly selected maximal order, another random maximal order connected to the first by a path of length $\lceil \log_{\ell_A}(p)/2 \rceil + \delta$, with $\delta \in \{-5, \ldots, 5\}$. One can traverse from the first order to the second via $\lceil \log_{\ell_A}(p)/2 \rceil + \delta$ steps in the $\ell_A$-isogeny tree.

The first three columns of Table 1 ("First basis element") correspond to the attack described in the previous section. The next three columns ("All basis elements") correspond to a variant where instead of considering only the smallest element in Step 4 of the algorithm, we try all elements in the Minkowski-reduced basis. Finally, the last three columns ("Linear combinations") correspond to a variant where we search for $\alpha$ of the right norm amongst all elements of the form $\sum_{i=1}^{4} c_i \beta_i$, where $c_i \in \{-4, \ldots, 4\}$ and $\beta_i$ are the Minkowski-reduced basis elements. Each percentage in the table corresponds to a success rate over 100 experiments.

**Table 1.** Experimental results for $\delta$ values. $\ell = 2$.

|   |   | First basis element | | | All basis elements | | | Linear combinations | | |
|---|---|---|---|---|---|---|---|---|---|---|
|   |   | $\lambda$ | | | $\lambda$ | | | $\lambda$ | | |
|   |   | 100 | 150 | 200 | 100 | 150 | 200 | 100 | 150 | 200 |
|   | $-5$ | 100 % | 99 % | 99 % | 100 % | 100 % | 99 % | 100 % | 100 % | 100 % |
|   | $-4$ | 93 % | 99 % | 94 % | 98 % | 99 % | 100 % | 100 % | 100 % | 100 % |
|   | $-3$ | 83 % | 84 % | 88 % | 92 % | 95 % | 99 % | 100 % | 100 % | 100 % |
|   | $-2$ | 40 % | 43 % | 45 % | 81 % | 74 % | 76 % | 100 % | 100 % | 100 % |
|   | $-1$ | 0 % | 2 % | 0 % | 35 % | 42 % | 35 % | 100 % | 100 % | 99 % |
| $\delta$ | 0 | 0 % | 0 % | 0 % | 3 % | 4 % | 3 % | 100 % | 100 % | 100 % |
|   | 1 | 0 % | 0 % | 0 % | 1 % | 0 % | 0 % | 97 % | 99 % | 98 % |
|   | 2 | 0 % | 0 % | 0 % | 0 % | 0 % | 0 % | 95 % | 94 % | 91 % |
|   | 3 | 0 % | 0 % | 0 % | 0 % | 0 % | 0 % | 57 % | 68 % | 70 % |
|   | 4 | 0 % | 0 % | 0 % | 0 % | 0 % | 0 % | 25 % | 28 % | 18 % |
|   | 5 | 0 % | 0 % | 0 % | 0 % | 0 % | 0 % | 0 % | 3 % | 1 % |

The experimental results are entirely convincing, so we leave better strategies to identify $\alpha$ from the Minkowski-reduced basis to further work.

## 5 Isogeny Hidden Number Problem

In this section we present an algorithm that takes partial information about the shared $j$-invariant $j(E_{AB})$ of Alice and Bob, and recovers the entire $j$-invariant, i.e. their shared key. This algorithm can therefore be used as a tool to obtain the shared key from a side-channel attack and to prove a bit security result.

Influenced by work on Diffie–Hellman key exchange in $\mathbb{Z}_p^*$, we propose the *isogeny hidden number problem* as a useful abstraction for analysing different cases where partial information is provided.

Hidden number problems have been used in other research. For example, [5] proved that some bits are hardcore for Diffie–Hellman shared keys in $\mathbb{Z}_p^*$, [16,25, 26] studied partial leakage of nonces in DSA and EC-DSA signatures, and [1,24] discussed side-channel attacks in the context of signatures.

**Definition 3. (Isogeny hidden number problem).** *Let $E_s$ be an unknown supersingular elliptic curve over $\mathbb{F}_{p^2}$. The isogeny hidden number problem is to compute the $j$-invariant $j(E_s)$ given an oracle $O$ such that $O(r)$ outputs partial information on $j(E')$ for some curve $E'$ which is $r$-isogenous to $E_s$.*

We now explain how the oracle $O$ in this problem can be realized in the context of the supersingular isogeny Diffie–Hellman key exchange. We use the same notation as earlier in the paper, so that $P_A, Q_A, P_B, Q_B \in E$ are known, and so are Alice and Bob's session values: $E_A, E_B, \phi_A(P_B), \phi_A(Q_B), \phi_B(P_A), \phi_B(Q_A)$. We set $E_s := E_{AB}$ to be the unknown elliptic curve. We suppose we have another oracle $O'$ that takes these values and produces some partial information on $j(E_{AB})$, which we interpret as the oracle query $O(1)$.

As a second stage, the adversary chooses a small integer $r$ (coprime to Alice's prime $\ell$) and a point $R \in E_B[r]$ of full order. Let $\phi_{BC} : E_B \to E_C$ be an isogeny of degree $r$ with kernel $\langle R \rangle$, that is $E_C = E_B / \langle R \rangle$. Note that there is a curve $E' := E_{AC}$ and an $r$-isogeny $E_{AB} \to E_{AC}$ corresponding to the image of $R$ under the isogeny from $E_B$ to $E_{AB}$. We also have that $E_{AC} = E_C / \phi_C(G_A)$ where $G_A$ is the kernel of $\phi_A$ and $\phi_C = \phi_{BC} \circ \phi_B$. This situation is pictured below.

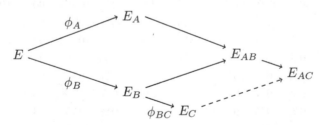

The curves $E_A$, $E_C$ and the corresponding values $\phi_A(P_B)$, $\phi_A(Q_B)$, $\phi_C(P_A) = \phi_{BC}(\phi_B(P_A))$, $\phi_C(Q_A) = \phi_{BC}(\phi_B(Q_A))$ can be used to perform a key exchange, which will constitute the curve $E_{AC}$ (this is the dotted arrow in the figure).

Querying the oracle $O'$ on these values results in some partial information on $j(E_{AC})$. We interpret this as the oracle query $O(r)$.

We give a full solution to the isogeny hidden number problem in the case where the oracle outputs an entire component of the $j$-invariant, and propose an attack where the oracle outputs some most significant bits of both components. This leads to a bit security result and to an active attack, which can be realized by a side-channel attack, when Alice uses a static key.

## 5.1    Algorithms for the Isogeny Hidden Number Problem

We recall that each $j$-invariant is an element in $\mathbb{F}_{p^2}$. Let $\mathbb{F}_{p^2} = \mathbb{F}_p(\theta)$, where $\theta^2 + A\theta + B = 0$, with $A, B \in \mathbb{F}_p$. We write $j = j_1 + j_2\theta$. For simplicity we only consider two cases of partial knowledge:

1. Oracle returns an entire component $j_i$ of each $j$-invariant.
2. Oracle returns the most significant bits of both components.

Other models of partial information could be considered as well.

We first remark that, since there are only around $p/12$ supersingular $j$-invariants, one might expect that knowledge of one component $j_i$ uniquely determines the entire $j$-invariant. This is not true in general, and it seems to be the case that there is no bound independent of $p$ on the number of supersingular $j$-invariants in $\mathbb{F}_{p^2}$ with a fixed value for $j_i$ (one exception is the rare class of $j$-invariants that actually lie in $\mathbb{F}_p$ and so are uniquely determined by their first component; the number of such $j$-invariants grows proportional to $\sqrt{p}$). Furthermore, there seems to be no known efficient algorithm that computes the other component $j_{3-i}$ given the value $j_i$ together with the fact that the curve is supersingular. Hence, even the first case is not trivial.

Our result is based on the modular polynomials $\Phi_r(x, y)$, which have the property that there is an isogeny $\phi : E \to E'$ of degree $r$ with cyclic kernel if and only if $\Phi_r(j(E), j(E')) = 0$. We refer to [10, Sect. 11.C], [3, Sect. 3.8] for background. These polynomials give a way to relate the known information on the different $j$-invariants. The degree of $\Phi_r(x, y)$, as well as their number of monomials, grow with $r$. Since the degree of these polynomials influences the complexity of the computation, it is desirable to work with the smallest possible $r$ (in practice we can take either $r = 2$ or $r = 3$). For $r = 2$ we have

$$\Phi_2(x, y) = x^3 + y^3 - x^2y^2 + 1488x^2y + 1488xy^2$$
$$- 162000x^2 - 162000y^2 + 40773375xy$$
$$+ 8748000000x + 8748000000y - 157464000000000.$$

The framework is the following. Let $x = x_1 + x_2\theta, y = y_1 + y_2\theta$. We call $x_1$ a "coefficient of 1" and $x_2$ a "coefficient of $\theta$". Then $\Phi_2(x, y) = F_1(x_1, x_2, y_1, y_2) + F_2(x_1, x_2, y_1, y_2)\theta$ for $F_1, F_2 \in \mathbb{F}_p[x_1, x_2, y_1, y_2]$, of total degree 4. Let $j = j(E) = j_1 + j_2\theta$ and $j' = j(E') = j_1' + j_2'\theta$, then if $\Phi_2(j, j') \equiv 0 \pmod p$ it holds that $F_1(j_1, j_2, j_1', j_2') = F_2(j_1, j_2, j_1', j_2') \equiv 0 \pmod p$.

Given some most significant bits of $x$, a common approach is to write

$$h := \mathrm{MSB}_k(x) = x - e, \quad \text{for } |e| < \frac{p}{2^{k+1}},$$

so $e$ is a relatively small integer. If all the bits are given, then $e = 0$. Substituting the known values that the oracle provides into each $F_i$, one constructs new polynomials $G_i$ whose roots can be used to fully recover the $j$-invariant $j(E)$. The problem reduces to the problem of recovering desired roots of $G_i$.

**Complete Component.** In this case we assume the attacker has a whole component for each $j$-invariant. We show that two samples are sufficient to recover the secret $j$-invariant $j(E_s)$. That is, we need one component of $j(E_s)$ and one component of another $j(E')$. Moreover, we can work with any pair of components (the components do not have to be in the same position).

**Theorem 2.** *Let the oracle $O$ in the isogeny hidden number problem output one component of the finite field representation of $j(E') \in \mathbb{F}_{p^2}$. Then there is an algorithm to solve the isogeny hidden number problem that makes two queries to $O$ and succeeds with probability at least $1/18$ if both components are coefficients of 1, with probability at least $1/12$ if both components are coefficients of $\theta$, and with probability at least $1/15$ otherwise.*

*Proof.* Let $E_s$ be the desired elliptic curve. The query $O(1)$ gives one component of $j(E_s)$ and the query $O(2)$ gives one component of $j(E')$ where $\Phi_2(j(E_s), j(E')) = 0$.

Writing $j(E_s) = j_1 + j_2\theta$ and $j(E') = j_1' + j_2'\theta$ then, as explained, $\Phi_2(j, j') = 0$ can be expressed as $F_1(j_1, j_2, j_1', j_2') = F_2(j_1, j_2, j_1', j_2') = 0$ for two polynomials $F_1, F_2$.

The oracle queries provide values $x_{3-k} = j_{3-k}, y_{3-l} = j_{3-l}'$ for $k, l \in \{1, 2\}$. Plugging these values into the polynomials $F_i$, we construct two bivariate polynomials $G_i$ in variables $x_k, y_l$ where the highest degree of each variable is at most 3. By taking the resultant of these polynomials with respect to $y_l$ we get a univariate polynomial in $x_k$ of degree at most 18. We show in Appendix C that the resultant is not the constant zero. One can then factor this polynomial to get at most 18 roots over $\mathbb{F}_p$, where one of the roots is $j_k$. As we have $j_k$ and $j_{3-k}$, we can construct $j(E_s)$. Hence, taking one of these solutions at random, we have determined the unknown $j$-invariant of $E_s$ with probability at least $1/18$.

Note that if the oracle queries yield $j_2, j_2'$, then $G_2$ is of degree 2, and so the resultant is of degree at most 12 (see Appendix C). Therefore, there are at most 12 possibilities of $\mathbb{F}_p$-solutions to the remaining unknown, which bound the success probability by $1/12$. Similarly, if only one of the components is a coefficient of $\theta$, then the degree of the variable associated to this component in $G_2$ is 2, and so the resultant is of degree at most 15. □

*Remark 3.* The solution given in Theorem 2 applies directly to any degree $r$. Note that the degree of $\Phi_r(x, y)$ increases with $r$, so we get more candidates for $j_k$. The proof holds with non-negligible probability for any low degree $r$. Notice that one can run the algorithm for several different degrees $r$ and test if there is only one root which is common to all lists of candidates, this will be $j_k$.

This solution assumes the oracle always gives the correct answer. An oracle that gives correct answers with some probability can be treated using the ideas in the next partial information model.

Theorem 2 provides the following bit security result for the supersingular isogeny key-exchange in a manner analogous to how the hidden number problem is used to give bit security results for Diffie–Hellman key exchange in $\mathbb{Z}_p^*$ [5].

**Theorem 3.** *Computing any component of the shared j-invariant $j(E_{AB})$ in the supersingular isogeny key exchange is as hard as computing the entire j-invariant $j(E_{AB})$.*

Indeed, the isogeny hidden number problem in this case can be derived from the oracle $O'$ described above, that takes the public parameters as well as the values $E_A$, $E_C$, $\phi_A(P_B)$, $\phi_A(Q_B)$, $\phi_C(P_A)$, $\phi_C(Q_A)$ and outputs a component of $j(E_{AC})$ (if Alice's prime $\ell$ is 2, one can take $r = 3$ or work with Bob's values and $E_{BC}$). We have just shown that, given an algorithm that computes a component of the shared $j$-invariant from the public keys, there is an algorithm that computes the entire $j$-invariant.

**Partial Components.** In this case we assume the attacker has most significant bits of both component for each $j$-invariant. Therefore, we write $j_i = h_i + e_i$ and $j'_i = h'_i + e'_i$ for $i = 1, 2$ and for a pair of $j$-invariants $j, j'$. Substituting these values to the equations of $F_i$, we construct two new polynomials $G_1, G_2 \in \mathbb{F}_p[u_1, u_2, v_1, v_2]$ of degree 4, such that

$$G_1(e_1, e_2, e'_1, e'_2) = G_2(e_1, e_2, e'_1, e'_2) \equiv 0 \pmod{p}.$$

The problem of computing the hidden $j$-invariant can therefore be expressed in terms of finding a small solution to a system of multivariate polynomial equations modulo $p$. One can then solve the problem by applying the well-known lattice-based techniques due to Coppersmith and Howgrave-Graham. We refer to [19] for a survey of these methods, where multivariate polynomials are considered.

These lattice methods require several relations, so we expect to need more than the six relations that are coming from the three 2-isogenous curves to $E_s$. To get more relations one can take isogenies of higher degrees, but we suggest working with degree 2 to get a stronger attack. That is, instead of fixing $E_s$ and taking several $r$-isogenous curves $E'$ for increasing $r$, we suggest following a (short) path in the 2-isogeny graph rooted at $E_s$. This ensures that the only polynomial being used is $\Phi_2$, which has minimal degree and the minimal number of monomials.

The main idea is to consider a part of the 2-isogeny graph close to $E_s$ (typically it will be a tree rooted at $E_s$). For every edge in the graph we obtain partial information on a $j$-invariant, which gives rise to two polynomials, namely $G_1, G_2$, which are satisfied by a simultaneous "small" solution.

Once enough polynomials are gathered, one can apply the techniques mentioned above to get a solution to the entire system where some of the roots are small (coming from the coordinates of a short vector in a corresponding lattice). Given these roots, one can recover the $j$-invariant for a curve $E_d$ in this path. Using the modular polynomials, we can "travel back" to find the $j$-invariant of the root $E_s$. Indeed, suppose our path is $E_0 = E_s, E_1, \ldots, E_k$. Then as we know $j(E_d)$ for some $d \le k$, we can use $\Phi_2$ to compute $j(E_{d-1})$ by solving $\Phi_2(j(E_d), y) \equiv 0 \pmod{p}$. We get at most 3 candidates for $j(E_{d-1})$, and we proceed recursively to find candidates for $j(E_{d-2}), \ldots, j(E_0)$. Since the distance from $E_d$ to the root $E_s$ is short, this results in a small list of candidates for $j(E_s)$.

We remark that in practice the polynomials $G_1, G_2$ consist of many monomials, and therefore this approach would require knowledge of many bits. However, Coopersmith's method shows how to generate more relations, which help to reduce the number of bits, and as an attack one can also rely on lattice algorithms working better in practice than theoretically guaranteed.

## 5.2   Active Attack When Alice Uses a Static Key

We assume that Alice uses a static key for encryption or key exchange. A legitimate key exchange protocol takes place between Alice and Bob, and an adversary Eve who sees the protocol messages wishes to obtain the resulting shared $j$-invariant $j_{AB}$. Hence Eve knows $(E, E_A, E_B)$ and the corresponding points.

We further assume that Eve can (adaptively) engage in protocol sessions with Alice (who always uses the same static secret key) and that, through some side-channel or other means, Eve is able to obtain partial information on the shared key computed by Alice on each protocol session.

Here, Alice acts as the oracle $O$ that provides the partial information. Eve first observes a protocol exchange between Alice and Bob, and so sees $(E_B, \phi_B(P_A), \phi_B(Q_A))$. She learns some partial information on $j(E_{AB})$.

Eve then chooses a small integer $r$ coprime to Alice's prime $\ell$, and as described above computes an isogeny $\phi_C$, the curve $E_C$ and the corresponding points $\phi_C(P_A)$, $\phi_C(Q_A)$. She sends $(E_C, \phi_C(P_A), \phi_C(Q_A))$ to Alice as part of a key exchange session. Alice then computes $E_{AC} = E_C/\phi_C(G_A)$ and some partial information about this $j$-invariant $j(E_{AC})$ is leaked. This leads to the scenario described in the isogeny hidden number problem, and using one of the solutions to this problem yields the desired $j$-invariant $j(E_{AB})$.

Note that this attack can be detected by the countermeasure of Kirkwood et al. [20], since the query on $E_C$ is not on a correct execution of the protocol. However, the protocol still requires Alice to compute $E_{AC}$ and so in the context of a side-channel attack, an attacker might already have received enough information to determine the desired secret key $j(E_{AB})$.

## 6   Conclusion

We have given several results on the security of cryptosystems based on the Jao–De Feo concept. Our main conclusion is that it seems very hard to prevent all active attacks using simple methods. Our first active attack seems to be undetectable using pairings or any other tools, as the curves and points appear to be indistinguishable from correct executions of the protocol. Similarly, our side-channel attack based on leakage of partial knowledge of the key seems to be hard to detect (without storing all previous sessions and each user checking that all curves $E_C$ sent to her are not related to previous curves $E_B$ by an isogeny of small degree). However, both these active attacks are detected by the heavy-duty countermeasure of Kirkwood et al. [20]. The latter attack comes from a reduction that gives the first bit security result for the supersingular isogeny key exchange.

Our paper therefore suggests that there is no way to avoid the use of such general countermeasures. It also shows that there is a risk of side-channel and fault attacks on these protocols, and these topics will no doubt generate a small following of literature in the coming years.

We have also discussed the connection between the problem of computing endomorphism rings and computing isogenies. In general, knowledge of $\text{End}(E_A)$ does not immediately lead to a 2-power isogeny of low degree from $E$ to $E_A$. But in the setting of the Jao and De Feo scheme such an isogeny can be efficiently computed when $\text{End}(E)$ and $\text{End}(E_A)$ are known. This demonstrates that the isogenies considered in these cryptosystems are special, which is natural to suspect since they are too short to provide good mixing in the expander graph.

**Acknowledgement.** We thank the anonymous reviewers for their comments. We would like to thank Roger Heath-Brown for his help with the calculation in Appendix A. The idea to study bit security of the isogeny scheme, which led to our third result, was suggested to us by Katsuyuki Takashima. We thank David Jao for comments on the Kirkwood et al. validation. The second author is supported by a GCHQ grant on post-quantum cryptography.

## A    Number of Isogenies of Degree Smaller Than $D$

To the sum $\sum_{n=2}^{D} a(n)$ with $a(n) = \prod_{p^e \mid n} (p+1)p^{e-1}$ we can associate a Dirichlet series $d(s) = \sum_{n \geq 1} \frac{a(n)}{n^s}$. This Dirichlet series is in fact equal to $d(s) = \frac{\zeta(s)\zeta(s-1)}{\zeta(2s)}$ by applying Euler's product formula. The function has a pole at $s = 2$ with residue equal to $\zeta(2)/\zeta(4)$. Using Perron's formula and Cauchy's Residue theorem, we arrive at

$$\sum_{n \leq D} a(n) \sim c \cdot D^2$$

where

$$c = \text{Res}(s = 2) = \frac{1}{2}\frac{\zeta(2)}{\zeta(4)} = \frac{15}{2\pi^2}.$$

## B    Low Order Adaptive Attack

In this appendix, we will discuss an adaptive attack that is easily detected but can be more powerful than the attack in Sect. 3. This adaptive attack uses points of small order; in particular, the attacker uses points $(R, [\ell^k]S)$, where $R, S \in E[\ell^n]$. We will illustrate the attack using the first oracle model and when $\ell > 3$.

As with the attack presented in Sect. 3, we will assume that Alice is using a static key $(1, \alpha)$, and that a dishonest user is playing the role of Bob to learn her key. It will be immediately clear that the attack will not stand up to the validations proposed by [9].

Let Alice be working in $E[\ell^n] \subset E(\mathbb{F}_{p^2})$, where $\ell^n \mid (p+1)$ and $\ell > 3$. Suppose that an attacker has recovered the first $i$ bits of $\alpha$, so that

$$\alpha = K_i + \ell^i \alpha_i + \ell^{i+1} \alpha'$$

where $K_i$ is known but $\alpha_i \in \{0, 1, \ldots, \ell - 1\}$ and $\alpha'$ are not known.

The attacker computes $E_B$, $R = \phi_B(P_A)$, $S = \phi_B(Q_A)$ and queries the oracle on $(E_B, R, [\ell^{n-i-1}]S)$. The resulting elliptic curve that the oracle computes is

$$E_B/\langle R + [\alpha][\ell^{n-i-1}]S \rangle = E_B/\langle R + [\ell^{n-i-1}][K_i + \ell^i \alpha_i + \ell^{i+1} \alpha']S \rangle$$
$$= E_B/\langle R + [\ell^{n-i-1}][K_i]S + [\ell^{n-1}\alpha_i]S \rangle.$$

Since the component $R + [\ell^{n-i-1}][K_i]S$ is known, the attacker can recover $\alpha_i$ if he knows the $j$-invariant by trying all of the $\ell$ different values of $\alpha_i$. For each $\ell$-ary bit, we only need one oracle interaction. This therefore solves the problem mentioned in Remark 2. The pseudo-code for this attack is presented in Algorithm 3.

Notice that with the second oracle model the attacker would need to make at most $\ell$ queries to the $O(E, R, S, E')$ oracle to recover $\alpha_i$.

---

**Algorithm 3.** Low order adaptive attack using oracle $O(E, R, S)$.

**Data**: $n$, $E$, $P_A$, $Q_A$, $P_B$, $Q_B$, $E_A$, $\phi_A(P_B)$, $\phi_A(Q_B)$
**Result**: $\alpha$

1  Set $K_0 \leftarrow 0$;
2  **for** $i \leftarrow 0$ **to** $n - 1$ **do**
3       Choose random $(b_1, b_2)$;
4       Set $G_B \leftarrow \langle [b_1]P_B + [b_2]Q_B \rangle$;
5       Set $E_B \leftarrow E/G_B$ and let $\phi_B : E \to E_B$ be the isogeny with kernel $G_B$;
6       Set $(R, S) \leftarrow (\phi_B(P_A), \phi_B(Q_A))$;
7       Set $j_i \leftarrow$ Query the oracle on $(E_B, R, [\ell^{n-i-1}]S)$ ;
8       **for** $x \leftarrow 0$ **to** $\ell - 1$ **do**
9           Set $j_{att} \leftarrow j(E_B/\langle R + [K_i]S + [x]S \rangle)$ ;
10          **if** $j_{att} = j_i$ **then** $\alpha_i \leftarrow x$;
11      **end**
12      Set $K_{i+1} \leftarrow K_i + \alpha_i \ell^i$;
13 **end**
14 Return $K_n$;

---

## C    The Resultant of $G_1(x_k, y_l)$ and $G_2(x_k, y_l)$

Let $p, q \in k[x, y]$ be two polynomials, and $k$ some field. The *resultant* of $p$ and $q$ with respect to $y$, denoted $\text{Res}(p, q, y)$, is given by the determinant of the Sylvester matrix of $p$ and $q$ as univariate polynomials in $y$, that is, we consider

$p, q \in k(x)[y]$. The resultant $\mathrm{Res}(p, q, y)$ is a univariate polynomial in $x$, so belongs to $k[x]$. For background on the resultant we refer to Sects. 5 and 6 of Chap. 3 in [11].

We show that the resultant $\mathrm{Res}(G_1, G_2, y_l)$, considered in Sect. 5.1, is not identically zero. We will use the fact that the modular polynomial $\Phi_r(X, Y) \in \mathbb{F}_p[X, Y]$ is absolutely irreducible (irreducible over the algebraic closure). We therefore consider $\Phi_r$, as well as $G_1, G_2$, in $\overline{\mathbb{F}}_p[X, Y]$. Recall that there are four cases depending on the values of $(k, l)$. For example when $(k, l) = (1, 2)$ we have $G_1(x_1, y_2) + G_2(x_1, y_2)\theta = \Phi_2(x_1 + j_2\theta, j_1' + y_2\theta)$.

Assume for contradiction that $\mathrm{Res}(G_1, G_2, y_l) \equiv 0$. By Proposition 1(ii) in [11, Chap.3, §6], $\mathrm{Res}(G_1, G_2, y_l) \equiv 0$ if and only if there exists a polynomial $h \in \overline{\mathbb{F}}_p[x_k, y_l]$ with positive degree in $y_l$ such that $h \mid G_1$ and $h \mid G_2$.

Consider the following linear substitution of variables:

- If $k = 1$ then set $x_1 = X - j_2\theta$ and if $k = 2$ then set $x_2 = (X - j_1)\theta^{-1}$.
- If $l = 1$ then set $y_1 = Y - j_2'\theta$ and if $l = 2$ then set $y_2 = (Y - j_1')\theta^{-1}$.

One can check that these substitutions give

$$G_1(x_k, y_l) + G_2(x_k, y_l)\theta = \Phi_r(X, Y).$$

Hence, letting $\bar{h}(X, Y)$ be the polynomial obtained by evaluating $h(x_k, y_l)$ with these substitutions we have

$$\bar{h}(X, Y) \mid \Phi_r(X, Y).$$

From the facts that the degree of $\bar{h}$ is equal to the degree of $h$, and that $\Phi_r$ is irreducible, it follows that (since we assumed $h$ is non-constant) that $h$ is a constant multiple of both $G_1$ and $G_2$. But by comparing the monomials in $G_1, G_2$, it is easy to see that they are not constant multiples of each other. Hence we have a contradiction and the resultant is non-zero.

We now explain the degrees arising in the proof of Theorem 2. Given the components $j_{3-k}, j_{3-l}'$, consider $\Phi_2(x, y)$ and the corresponding polynomials $G_1(x_k, y_l)$, $G_2(x_k, y_l)$. We have

$$\deg_{x_k} \mathrm{Res}(G_1, G_2, y_l) = \begin{cases} 12 & \text{if } k = l = 1, \\ 18 & \text{if } k = l = 2, \\ 15 & \text{otherwise.} \end{cases}$$

It follows from the following lemma, since $\deg_{x_1} F_1 = \deg_{y_1} F_1 = 3$, $\deg_{x_2} F_1 = \deg_{y_2} F_1 \leq 3$, $\deg_{x_1} F_2 = \deg_{y_1} F_2 \leq 2$ and $\deg_{x_2} F_2 = \deg_{y_2} F_2 \leq 3$.

**Lemma 6.** *Let $p, q \in k[x, y]$ be two polynomials with*

$$\deg_x p = n_x, \ \deg_y p = n_y,$$
$$\deg_x q = m_x, \ \deg_y q = m_y.$$

*Then* $\deg_x \mathrm{Res}(p, q, y) \leq m_y n_x + n_y m_x.$

*Proof.* The Sylvester matrix of $p$ and $q$ with respect to $y$ is a $(m_y+n_y) \times (m_y+n_y)$ matrix. The first $m_y$ rows, coming from the coefficients of $p$, contain polynomials in $x$ of degree at most $n_x$. Similarly, the last $n_y$ rows contain polynomials in $x$ of degree at most $m_x$. The resultant $\mathrm{Res}(p,q,y)$ is given by the determinant of this matrix, which is formed by summing products of an entry from each row. The first $m_y$ rows contribute at most $m_y n_x$ to the degree of $x$, and the last $n_y$ rows contribute at most $n_y m_x$.                                    □

# References

1. Aranha, D.F., Fouque, P.-A., Gérard, B., Kammerer, J.-G., Tibouchi, M., Zapalow-icz, J.-C.: GLV/GLS decomposition, power analysis, and attacks on ECDSA signatures with single-bit nonce bias. In: Sarkar, P., Iwata, T. (eds.) ASIACRYPT 2014. LNCS, vol. 8873, pp. 262–281. Springer, Heidelberg (2014). doi:10.1007/978-3-662-45611-8_14
2. Biasse, J.-F., Jao, D., Sankar, A.: A quantum algorithm for computing isogenies between supersingular elliptic curves. In: Meier, W., Mukhopadhyay, D. (eds.) INDOCRYPT 2014. LNCS, vol. 8885, pp. 428–442. Springer, Heidelberg (2014). doi:10.1007/978-3-319-13039-2_25
3. Blake, I.F., Seroussi, G., Smart, N.P.: Elliptic Curves in Cryptography. Cambridge University Press, Cambridge (1999)
4. Blake, I.F., Seroussi, G., Smart, N.P.: Advances in Elliptic Curve Cryptography. Cambridge University Press, Cambridge (2005)
5. Boneh, D., Venkatesan, R.: Hardness of computing the most significant bits of secret keys in diffie-hellman and related schemes. In: Koblitz, N. (ed.) CRYPTO 1996. LNCS, vol. 1109, pp. 129–142. Springer, Heidelberg (1996). doi:10.1007/3-540-68697-5_11
6. Charles, D.X., Lauter, K.E., Goren, E.Z.: Cryptographic hash functions from expander graphs. J. Cryptol. **22**(1), 93–113 (2009)
7. Childs, A.M., Jao, D., Soukharev, V.: Constructing elliptic curve isogenies in quantum subexponential time. J. Math. Cryptol. **8**(1), 1–29 (2014)
8. Ciet, M., Joye, M.: Elliptic curve cryptosystems in the presence of permanent and transient faults. Des. Codes Crypt. **36**(1), 33–43 (2005)
9. Costello, C., Longa, P., Naehrig, M.: Efficient algorithms for supersingular isogeny Diffie-Hellman. In: Robshaw, M., Katz, J. (eds.) CRYPTO 2016. LNCS, vol. 9814, pp. 572–601. Springer, Heidelberg (2016). doi:10.1007/978-3-662-53018-4_21
10. Cox, D.A.: Primes of the Form $x^2+ny^2$. John Wiley & Sons Inc, New York (1989)
11. Cox, D.A., Little, J., O'Shea, D.: Ideals, Varieties, and Algorithms: An Introduction to Computational Algebraic Geometry and Commutative Algebra. Undergraduate Texts in Mathematics, 3rd edn. Springer, Secaucus (2007)
12. Deuring, M.: Die typen der multiplikatoren ringe elliptischer funktionenkörper. Abh. Math. Sem. Hansischen Univ. **14**, 197–272 (1941)
13. De Feo, L., Jao, D., Plût, J.: Towards quantum-resistant cryptosystems from supersingular elliptic curve isogenies. J. Math. Cryptol. **8**(3), 209–247 (2014)
14. Fujisaki, E., Okamoto, T.: Secure integration of asymmetric and symmetric encryption schemes. In: Wiener, M. (ed.) CRYPTO 1999. LNCS, vol. 1666, pp. 537–554. Springer, Heidelberg (1999). doi:10.1007/3-540-48405-1_34
15. Galbraith, S.D.: Constructing isogenies between elliptic curves over finite fields. LMS J. Comput. Math. **2**, 118–138 (1999)

16. Howgrave-Graham, N.A., Smart, N.P.: Lattice attacks on digital signature schemes. Des. Codes Crypt. **23**(3), 283–290 (2001)
17. Jao, D., Feo, L.: Towards quantum-resistant cryptosystems from supersingular elliptic curve isogenies. In: Yang, B.-Y. (ed.) PQCrypto 2011. LNCS, vol. 7071, pp. 19–34. Springer, Heidelberg (2011). doi:10.1007/978-3-642-25405-5_2
18. Jao, D., Soukharev, V.: Isogeny-based quantum-resistant undeniable signatures. In: Mosca, M. (ed.) PQCrypto 2014. LNCS, vol. 8772, pp. 160–179. Springer, Heidelberg (2014). doi:10.1007/978-3-319-11659-4_10
19. Jochemsz, E., May, A.: A strategy for finding roots of multivariate polynomials with new applications in attacking RSA variants. In: Lai, X., Chen, K. (eds.) ASIACRYPT 2006. LNCS, vol. 4284, pp. 267–282. Springer, Heidelberg (2006). doi:10.1007/11935230_18
20. Kirkwood, D., Lackey, B.C., McVey, J., Motley, M., Solinas, J.A., Tuller, D.: Failure is not an option: standardization issues for post-quantum key agreement. In: Workshop on Cybersecurity in a Post-Quantum World (2015)
21. Kohel, D.: Endomorphism rings of elliptic curves over finite fields. Ph.D. thesis, University of California, Berkeley (1996)
22. Kohel, D., Lauter, K., Petit, C., Tignol, J.-P.: On the quaternion $\ell$-isogeny path problem. LMS J. Comput. Math. **17**(Special issue A), 418–432 (2014)
23. Lim, C.H., Lee, P.J.: A key recovery attack on discrete log-based schemes using a prime order subgroup. In: Kaliski, B.S. (ed.) CRYPTO 1997. LNCS, vol. 1294, pp. 249–263. Springer, Heidelberg (1997). doi:10.1007/BFb0052240
24. De Mulder, E., Hutter, M., Marson, M.E., Pearson, P.: Using Bleichenbacher's solution to the hidden number problem to attack nonce leaks in 384-bit ECDSA: extended version. J. Crypt. Eng. **4**(1), 33–45 (2014)
25. Nguyen, P.Q., Shparlinski, I.E.: The insecurity of the digital signature algorithm with partially known nonces. J. Crypt. **15**(3), 151–176 (2002)
26. Nguyen, P.Q., Shparlinski, I.E.: The insecurity of the elliptic curve digital signature algorithm with partially known nonces. Des. Codes Crypt. **30**(2), 201–217 (2003)
27. Nguyen, P.Q., Stehlé, D.: Low-dimensional lattice basis reduction revisited. In: Buell, D. (ed.) ANTS 2004. LNCS, vol. 3076, pp. 338–357. Springer, Heidelberg (2004). doi:10.1007/978-3-540-24847-7_26
28. Peikert, C.: Lattice cryptography for the internet. In: Mosca, M. (ed.) PQCrypto 2014. LNCS, vol. 8772, pp. 197–219. Springer, Heidelberg (2014). doi:10.1007/978-3-319-11659-4_12
29. Rostovtsev, A., Stolbunov, A.: Public-key cryptosystem based on isogenies. Cryptology ePrint Archive, Report 2006/145 (2006). http://eprint.iacr.org/
30. Silverman, J.H.: The Arithmetic of Elliptic Curves. Graduate Texts in Mathematics, vol. 106, 2nd edn. Springer, New York (2009)
31. Tate, J.: Endomorphisms of abelian varieties over finite fields. Inventiones mathematicae **2**(2), 134–144 (1966)
32. Vélu, J.: Isogénies entre courbes elliptiques. C.R. Acad. Sci. Paris Sér. A. **273**, 238–241 (1971)
33. Vignéras, M.-F.: Arithmétique des Algèbres de Quaternions. Lecture Notes in Mathematics, vol. 800. Springer, New York (1980)
34. Xi, S., Tian, H., Wang, Y.: Toward quantum-resistant strong designated verifier signature from isogenies. Int. J. Grid Util. Comput. **5**(2), 292–296 (2012)

# AES and White-Box

# Simpira v2: A Family of Efficient Permutations Using the AES Round Function

Shay Gueron[1,2] and Nicky Mouha[3,4,5,6](✉)

[1] Department of Mathematics, University of Haifa, Haifa, Israel
shay@math.haifa.ac.il
[2] Israel Development Center, Intel Corporation, Haifa, Israel
[3] Department of Electrical Engineering-ESAT/COSIC, KU Leuven, Leuven, Belgium
nicky@mouha.be
[4] iMinds, Ghent, Belgium
[5] Project-team SECRET, Inria, Paris, France
[6] National Institute of Standards and Technology, Gaithersburg, MD, USA

**Abstract.** This paper introduces Simpira, a family of cryptographic permutations that supports inputs of $128 \times b$ bits, where $b$ is a positive integer. Its design goal is to achieve high throughput on virtually all modern 64-bit processors, that nowadays already have native instructions for AES. To achieve this goal, Simpira uses only one building block: the AES round function. For $b = 1$, Simpira corresponds to 12-round AES with fixed round keys, whereas for $b \geq 2$, Simpira is a Generalized Feistel Structure (GFS) with an $F$-function that consists of two rounds of AES. We claim that there are no structural distinguishers for Simpira with a complexity below $2^{128}$, and analyze its security against a variety of attacks in this setting. The throughput of Simpira is close to the theoretical optimum, namely, the number of AES rounds in the construction. For example, on the Intel Skylake processor, Simpira has throughput below 1 cycle per byte for $b \leq 4$ and $b = 6$. For larger permutations, where moving data in memory has a more pronounced effect, Simpira with $b = 32$ (512 byte inputs) evaluates 732 AES rounds, and performs at 824 cycles (1.61 cycles per byte), which is less than 13 % off the theoretical optimum. If the data is stored in interleaved buffers, this overhead is reduced to less than 1 %. The Simpira family offers an efficient solution when processing wide blocks, larger than 128 bits, is desired.

**Keywords:** Cryptographic permutation · AES-NI · Generalized Feistel structure (GFS) · Beyond birthday-bound (BBB) security · Hash function · Lamport signature · Wide-block encryption · Even-Mansour

## 1 Introduction

The introduction of AES instructions by Intel (subsequently by AMD, and recently ARM) has changed the playing field for symmetric-key cryptography on modern processors, which lead to a significant reduction of the encryption

© International Association for Cryptologic Research 2016
J.H. Cheon and T. Takagi (Eds.): ASIACRYPT 2016, Part I, LNCS 10031, pp. 95–125, 2016.
DOI: 10.1007/978-3-662-53887-6_4

overheads. The performance of these instructions has been steadily improving in every new generation of processors. By now, on the latest Intel Architecture Codename Skylake, the AESENC instruction that computes one round of AES has latency of 4 cycles and throughput of 1 cycle. The improved AES performance trend can be expected to continue, with the increasing demand for fast encryption of more and more data.

To understand the impact of the AES instructions in practice, consider for example the way that Google Chrome browser connects to https://google.com. In this situation, Google is in a privileged position, as it controls both the client and the server side. To speed up connections, Chrome (the client) is configured to identify the processor's capabilities. If AES-NI are available, it would offer (to the server) to use AES-128-GCM for performing authenticated encryption during the TLS handshake. The high-end server would accept the proposed cipher suite, due to the high performance of AES-GCM on its side. This would capture any recent 64-bit PC, tablet, desktop, or even smartphone. On older processors, or architectures without AES instructions, Chrome resorts to proposing the ChaCha20-Poly1305 algorithm during the secure handshake negotiation.

An advantage of AES-GCM is that the message blocks can be processed independently for encryption. This allows pipelining of the AES round instructions, so that the observed performance is dominated by their throughput, and not by their latency [43,44]. We note that even if a browser negotiates to use an inherently sequential mode such as CBC encryption, the web server can process multiple independent data buffers in parallel to achieve high throughput (see [43,44]), and this technique is already used in the recent OpenSSL version 1.0.2. This performance gain by collecting multiple independent encryption tasks and pipelining their execution, is important for the design rationale of Simpira.

**Setting.** This paper should be understood in the following setting. We focus only on processors with AES instructions. Assuming that several independent data sources are available, we explore several symmetric-key cryptographic constructions with the goal of achieving a high throughput. Our reported benchmarks are performed on the latest Intel processor, namely Architecture Codename Skylake, but we expect to achieve similar performance on any processor that has AES instructions with throughput 1.

In particular, we focus here on applications where the 128-bit block size of AES is not sufficient, and support for a wider range of block sizes is desired. This includes various use cases such as permutation-based hashing and wide-block encryption, or just to easily achieve security beyond $2^{64}$ input blocks without resorting to (often inefficient) modes of operation with "beyond birthday-bound" security. For several concrete suggestions of applications, we refer to Sect. 7.

Admittedly, our decision to focus on only throughput may result in unoptimized performance in certain scenarios where the latency is critical. However, we point out that this is not only a property of Simpira, but also of AES itself, when it is implemented on common architectures with AES instructions. To achieve optimal performance on such architectures, AES needs to be used in a parallelizable mode of operation, or in a protocol that supports processing independent inputs. Similarly, this is the case for Simpira as well. In fact, for 128-bit inputs, Simpira is the same as 12-round AES with fixed round keys.

**Origin of the Name.** Simpira is named after a mythical animal of the Peruvian Amazon. According to the legend, one of its front legs has the form of a spiral that can be extended to cover the entire surface of the earth [26]. In a similar spirit, the Simpira family of permutations extends itself to a very wide range of input sizes. Alternatively, Simpira can be seen as an acronym for "SIMple Permutations based on the Instruction for a Round of AES."

**Update.** This paper proposes Simpira v2. Compared to Simpira v1, the Type-1.x GFS by Yanagihara and Iwata was found to have a problem (see Sect. 8), and is replaced by a new construction that performs the same number of AESENCs. We also updated the round constants (see Sect. 4). Although no attack is currently known on Simpira v2 with the old rotation constants, the new constants seem to strengthen Simpira without affecting its performance in our benchmarks. Unless otherwise specified, Simpira in this document is assumed to refer to Simpira v2.

## 2    Related Work

Block ciphers that support wide input blocks have been around for a long time. Some of the earliest designs are Bear and Lion [2], and Beast [62]. They are higher-level constructions, in the sense that they use hash functions and stream ciphers as underlying components.

Perhaps the first wide-block block cipher that is not a higher-level construction is the Hasty Pudding Cipher [75], which supports block sizes of any positive number of bits. Another early design is the Mercy block cipher that operates on 4096-bit blocks [27]. More recently, low-level constructions that can be scaled up to large input sizes are the SPONGENT [17,18] permutations and the LowMC [1] block ciphers.

Our decision to use only the AES round function as a building block for Simpira means that some alternative constructions are not considered in this paper. Of particular interest are the EGFNs [7] used in Lilliput [6], the AESQ permutation of PAEQ [13], and Haraka[1] [56]. The security claims and benchmark targets of these designs are very different from those of Simpira. We only claim security up to $2^{128}$ blocks of input. However unlike Haraka, we consider all distinguishing attacks up to this bound. Also, we focus only on throughput, and not on latency. An interesting topic for future work is to design variants of these constructions with similar security claims, and to compare their security and implementation properties with Simpira.

---

[1] The first version of Haraka was vulnerable to an attack by Jérémy Jean [52] due to a bad choice of round constants. We therefore refer to the second version of Haraka, which prevents the attack.

# 3   Design Rationale of Simpira

AES [31] is a block cipher that operates on 128-bit blocks. It iterates the AES round function 10, 12 or 14 times, using round keys that are derived from a key of 128, 192 or 256 bits, respectively. On Intel (and AMD) processors, the AES round function is implemented by the AESENC instruction. It takes a 128-bit state and a 128-bit round key as inputs, and returns a 128-bit output that is the result of applying the SubBytes, ShiftRows, MixColumns and AddRoundKey operations. An algorithmic description of AESENC is given in Algorithm 1 of Sect. 4, where we give the full specification of Simpira.

A cryptographic permutation can be obtained by setting the AES round keys to fixed, publicly-known values. It is a bad idea to set all round keys to zero. Such a permutation can easily be distinguished from random: if all input bytes are equal to each other, the AES rounds preserve this property. Such problems are avoided when round constants are introduced: this breaks the symmetry inside every round, as well as the symmetry between rounds. Several ciphers are vulnerable to attacks resulting from this property, such as the CAESAR candidate PAES [53,54] and the first version of Haraka [52]. The aforementioned design criterion, already present in Simpira v1, excludes the round constants of these designs.

We decided to use two rounds of AES in Simpira as the basic building block. As the AESENC instruction includes an XOR with a round key, this can be used to introduce a round constant in one AES round, and to do a "free XOR" in the other AES round. An added advantage is that two rounds of AES achieve *full bit diffusion*: every output bit depends on every input bit, and every input bit depends on every output bit.

Another design choice that we made, is to use only AES round functions in our construction, and no other operations. Our hope is that this design would maximize the contribution of every instruction to the security of the cryptographic permutation. It also simplifies the analysis and the implementation. From the performance viewpoint, the theoretically optimal software implementation would be able to dispatch a new AESENC instruction in every CPU clock cycle. A straightforward way to realize this design strategy is to use a (Generalized) Feistel Structure (GFS) for $b \geq 2$ that operates on $b$ input subblocks of 128 bits each, as shown in Fig. 1.

As with any design, our goal is to obtain a good trade-off between security and efficiency. In order to explore a large design space, we use simple metrics to quickly estimate whether a given design reaches a sufficient level of security, and to determine its efficiency. In subsequent sections, we will formally introduce the designs, and study them in detail to verify the accuracy of our estimates.

## 3.1   Design Criteria

Our design criteria are as follows. The significance of both criteria against cryptanalysis attacks will be explained in Sect. 6.

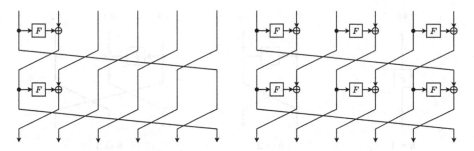

**Fig. 1.** Two common classes of Generalized Feistel Structures (GFSs) are the Type-1 GFS (left) and the Type-2 GFS (right). For each example, two rounds are shown of a GFS that operates on $b = 6$ subblocks. We will initially consider these GFSs in this paper, as well as other GFSs with a different number of $F$-functions per round, and other *subblock shuffles* at the end of every round. At a later stage, we will consider more advanced constructions as well.

- **Security**: We calculate the number of Feistel rounds to achieve either *full bit diffusion*, as well as the number of Feistel rounds to achieve at least 25 (linearly or differentially) active S-boxes. To ensure a sufficient security margin against known attacks, we require that the number of rounds is three times the largest of these two numbers.
- **Efficiency**: As explained in Sect. 1, we will only focus on throughput. Given that we use no other operations besides the AES round function, we will use the number of AES round functions as an estimate for the total number of cycles.

Suzaki and Minematsu [76] formally defined DRmax to calculate how many Feistel rounds are needed for an input subblock to affect all the output subblocks. We will say that *full subblock diffusion* is achieved after DRmax rounds of the permutation or its inverse, whichever is greater. To achieve the strictly stronger criterion of *full bit diffusion*, one or two additional Feistel rounds may be required.

To obtain a lower bound for the minimum number of active S-boxes, we use a simplified representation that assigns one bit to every pair of bytes, to indicate whether or not they contain a non-zero difference (or linear mask). This allows us to use the Mixed-Integer Linear Programming (MILP) technique introduced by Mouha et al. [70] to quickly find a lower bound for the minimum number of active S-boxes.

## 3.2   Design Space Exploration

For each input size of the permutation, we explore a range of designs, and choose the one that maximizes the design criteria. If the search returns several alternatives, it does not really matter which one we choose. In that case, we arbitrarily choose the "simplest" design. The resulting Simpira design is shown in Figs. 2 and 3.

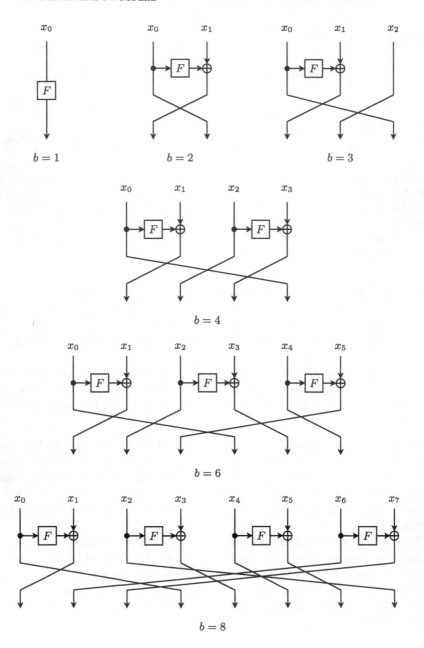

**Fig. 2.** One round of the Simpira construction for $b \in \{1, 2, 3, 4, 6, 8\}$. The total number of rounds is 6 for $b = 1$, 15 for $b = 2$, $b = 4$ and $b = 6$, 21 for $b = 3$, and 18 for $b = 8$. $F$ is shorthand for $F_{c,b}$, where $c$ is a counter that is initialized by one, and incremented after every evaluation of $F_{c,b}$. Every $F_{c,b}$ consists of two AES round evaluations, where the round constants that are derived from $(c, b)$. The last round is special: the MixColumns is omitted when $b = 1$, and the final subblocks may be output in a different order. See Sect. 4 for a full specification.

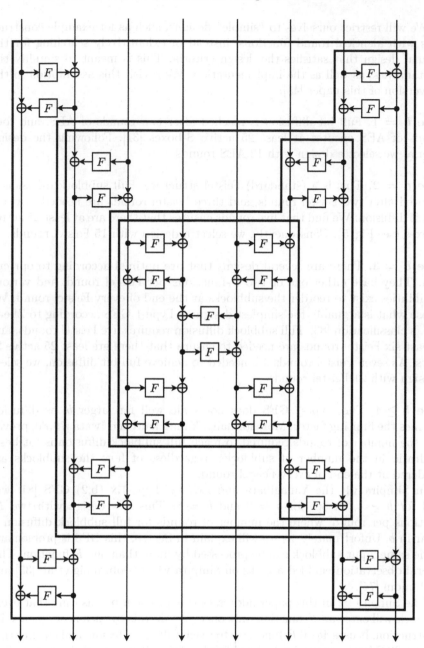

**Fig. 3.** The Simpira construction for $b \notin \{1, 2, 3, 4, 6, 8\}$. $F$ is shorthand for $F_{c,b}$, which consists of two rounds of AES as specified in Algorithm 2. A generic construction is shown for all $b \geq 4$, however for $b \in \{4, 6, 8\}$ we will use the construction of Fig. 2. By convention, the leftmost $F$-function is from left to right; when this is not the case in the diagram, the direction of every $F$-function should be inverted. The full-round Simpira iterates the construction in this diagram three times. See Sect. 4 for a full specification.

We will restrict ourselves to "simple" designs, such as for example constructions with identical round functions, instead of exhaustively searching for the
optimal design that satisfies the design criteria. This is meant to simplify the
cryptanalysis, as well as the implementation. We revisit this assumption in the
full version of this paper [45].

**Case $b = 1$.** Full bit diffusion is reached after two rounds of AES, and four
rounds of AES ensures at least 25 active S-boxes [31]. Following the design
criteria, we select a design with 12 AES rounds.

**Case $b = 2$.** This is a (standard) Feistel structure. Full subblock diffusion is
achieved after two Feistel rounds, and three Feistel rounds are needed to reach
full bit diffusion. We find that five rounds ensures that there are at least 25 active
S-boxes (see Fig. 5). Consequently, we select a design with 15 Feistel rounds.

**Case $b = 3$.** There are several designs that are optimal according to our criteria. They have either one or two $F$-functions per Feistel round, and various
possibilities exist to reorder the subblocks at the end of every Feistel round. We
choose what is arguably the simplest design: a Type-1 GFS according to Zheng
et al.'s classification [83]. Full subblock diffusion requires five Feistel rounds, and
at least six Feistel rounds are needed to ensure that there are least 25 active S-
boxes. As seven Feistel rounds are needed to achieve full bit diffusion, we select
a design with 21 Feistel rounds.

**Case $b \geq 4$.** The Type-1 GFS does not scale well for larger $b$, as diffusion
becomes the limiting factor. More formally, Yanagihara and Iwata [79,80] proved
that the number of rounds required to reach full subblock diffusion is (at best)
quadratic in the number of subblocks, regardless of how the subblocks are
reordered at the end of every Feistel round.

In Simpira v1, the Yanagihara and Iwata's Type-1.x $(b,2)$ GFS [81] was
used for $b \geq 4$, except for $b = 6$ and $b = 8$. This is a design with two $F$-
functions per round, where the number of rounds for full subblock diffusion is
linear in $b$. Unfortunately, as we will explain in Sect. 8, this GFS is problematic
as the same input subblock can be processed by more than one $F$-function. This
general observation enabled attacks on Simpira v1 by Dobraunig et al. [35] and
by Rønjom [74].

The Simpira v2 in this paper addresses this problem by ensuring that every
subblock will enter an $F$-function only once. We do this by means of a new GFS
construction. It uses $4b-6$ $F$-functions to reach full bit diffusion, and ensures that
at least 30 S-boxes are active (as explained in the full version of this paper [45]).
This construction is iterated three times, resulting in a design with $12b - 18$
$F$-functions, which is the same number of $F$-function as in Simpira v1.

We could also have used this construction for $b = 4$. However, we instead
chose to go for a Type-2 GFS with 15 rounds. This not only results in a simpler

construction, but also has the advantage ensuring at least 40 active S-boxes (instead of only 30) after five rounds.

But even if we had considered Yanagihara and Iwata's Type-1.x (b,2) GFS, we should also consider GFSs with more than two $F$-functions per Feistel round, which reach full subblock diffusion even quicker. However, this seems to come at the cost of using more $F$-functions in total. Looking only at the tabulated values of DRmax$(\pi)$ and DRmax$(\pi^{-1})$ in literature [76,79–81], we can immediately rule out almost all alternative designs. Nevertheless, two improved Type-2 GFS designs by Suzaki and Minematsu [76] turned out be superior. Instead of a cyclic left shift, they reorder the subblocks in a different way at the end of every Feistel round. We now explore these in detail.

**Case $b = 6$.** Let the *subblock shuffle* at the end of every Feistel round be presented by a list of indices that indicates which input subblock is mapped to which output subblock, e.g. $\{b - 1, 0, 1, 2, \ldots, b - 2\}$ denotes a cyclic left shift. Suzaki and Minematsu's improved Type-2 GFS with subblock shuffle $\{3, 0, 1, 4, 5, 2\}$ reaches full subblock diffusion and full bit diffusion after five Feistel rounds. At least 25 active S-boxes (in fact at least 30) are reached after four Feistel rounds. Following the design criteria, we end up with a design with 15 Feistel rounds. As this design has three $F$-functions in every Feistel round, it evaluates $3 \cdot 15 = 45$ $F$-functions. This is less than the general $b \geq 4$ case that requires $6b - 9$ Feistel rounds with 2 $F$-functions per round, which corresponds to $(6 \cdot 6 - 9) \cdot 2 = 54$ $F$-functions.

**Case $b = 8$.** Suzaki and Minematsu's improved Type-2 GFS with subblock shuffle $\{3, 0, 7, 4, 5, 6, 1, 2\}$ ensures both full subblock diffusion and full bit diffusion after six rounds. After four Feistel rounds, there are at least 25 active S-boxes (in fact at least 30). According to the design criteria, we end up with a design with 18 Feistel rounds, or $18 \cdot 4 = 72$ $F$-functions in total. The general $b \geq 4$ design would have required $(6b - 9) \cdot 2$ $F$-functions, which for $b = 8$ corresponds to $(6 \cdot 8 - 9) \cdot 2 = 78$ $F$-functions.

### 3.3 Design Alternatives

Until now, the only designs we discussed were GFS constructions where the $F$-function consists of two rounds of AES. We now take a step back, and briefly discuss alternative design choices.

As explained earlier, it is convenient to use two rounds of AES as a building block. It not only means that we reach full bit diffusion, but also that a "free XOR" is available to add a round constant on Intel and AMD architectures.

It is nevertheless possible to consider GFS designs with an $F$-function that consists of only one AES round. A consequence of this design choice is that extra XOR instructions will be needed to introduce round constants, which could increase the cycle count. But this design choice also complicates the analysis. For example when $b = 2$, we find that 25 Feistel rounds are then needed to ensure at least 25 linearly active S-boxes. As shown in Fig. 4, this is because the tool can only ensure one active S-box for every Feistel round. Using two rounds of AES avoids this problem (see Fig. 5), and also significantly speeds up the tool: it makes bounding the minimum number of active S-boxes rather easy, instead of becoming increasingly complicated for a reasonably large value of $b$.

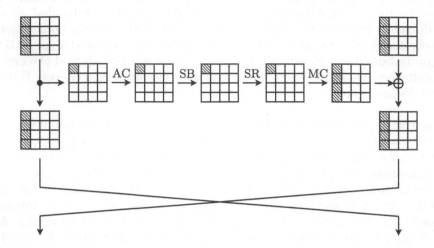

**Fig. 4.** A linear characteristic for an AES-based Feistel that uses only one round of AES inside its $F$-function. Crosshatches represent bytes with non-zero linear masks. The AES round consists of the AddConstant (AC), SubBytes (AC), ShiftRows (SR), and MixColumns (MC) operations. This round has only one active S-box. Therefore, 25 rounds are needed to ensure that there are least 25 linearly active S-boxes.

Likewise, we could also consider designs with more than two AES rounds per $F$-function. In our experiments, we have not found any cases where this results in a design where the total number of AES rounds is smaller. The intuition is as follows: the number of Feistel rounds to reach full subblock diffusion is independent of the $F$-function, therefore adding more AES rounds to every $F$ function is not expected to result in a better trade-off.

If we take another step back, we might consider to use other instructions besides AESENC. Clearly, AESDEC can be used as an alternative, and the security properties and the benchmarks will remain the same. In fact, we use AESDEC when $b = 1$, to implement the inverse permutation. We do not use the AESENCLAST and AESDECLAST instructions, as they omit the MixColumns (resp. InvMixColumns) operation that is crucial to the *wide trail design strategy* [30] of AES. We do, however, use only one AESENCLAST for the very last round of the $b = 1$ permutation, as this makes an efficient implementation of the inverse permutation possible on Intel architectures. This is equivalent to applying a linear transformation to the output of the $b = 1$ permutation, therefore it does not reduce its cryptographic properties.

Of course, it is possible to use non-AES instructions, possibly in combination with AES instructions. Actually, we do not need to be restricted to (generalized)

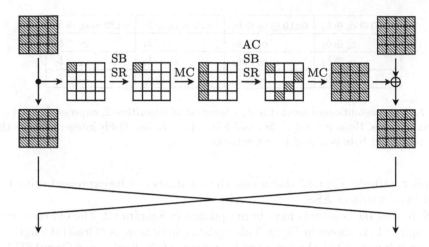

**Fig. 5.** A linear characteristic for one round of Simpira with $b = 2$ with 5 active S-boxes. Crosshatches represent bytes with non-zero linear masks. As Simpira uses two AES rounds per $F$-function, it can reach 25 active S-boxes in only 5 Feistel rounds, corresponding to 10 AES rounds in total.

Feistel designs for $b \geq 2$. However, such considerations are outside of the scope of this paper.

## 4    Specification of Simpira

An algorithmic specification of the Simpira design of Figs. 2 and 3 is given in Figs. 9, 10 and 11. It uses one round of AES as a building block, which corresponds to the `AESENC` instruction on Intel processors (see Algorithm 1). Its input is a 128-bit xmm register, which stores the AES $4 \times 4$ matrix of bytes as shown in Fig. 6. For additional details, we refer to [44].

| $s_0$ | $s_4$ | $s_8$ | $s_{12}$ |
|---|---|---|---|
| $s_1$ | $s_5$ | $s_9$ | $s_{13}$ |
| $s_2$ | $s_6$ | $s_{10}$ | $s_{14}$ |
| $s_3$ | $s_7$ | $s_{11}$ | $s_{15}$ |

**Fig. 6.** The internal state of AES can be represented by a $4 \times 4$ matrix of bytes, or as a 128-bit xmm register value $s = s_{15} \| \ldots \| s_0$, where $s_0$ is the least significant byte.

The $F$-function is specified in Algorithm 2. It is parameterized by a counter $c$ and by the number of subblocks $b$. Here, `SETR_EPI32` converts four 32-bit values into a 128-bit value, using the same byte ordering as the `_mm_setr_epi32()`

| 0x00 $\oplus c_0 \oplus b_0$ | 0x10 $\oplus c_0 \oplus b_0$ | 0x20 $\oplus c_0 \oplus b_0$ | 0x30 $\oplus c_0 \oplus b_0$ |
|---|---|---|---|
| $c_1 \oplus b_1$ | $c_1 \oplus b_1$ | $c_1 \oplus b_1$ | $c_1 \oplus b_1$ |
| $c_2 \oplus b_2$ | $c_2 \oplus b_2$ | $c_2 \oplus b_2$ | $c_2 \oplus b_2$ |
| $c_3 \oplus b_3$ | $c_3 \oplus b_3$ | $c_3 \oplus b_3$ | $c_3 \oplus b_3$ |

**Fig. 7.** The constants used inside the $F_{c,b}$ function of Algorithm 2, expressed as a $4 \times 4$ matrix of bytes. Here, $c = c_4 \| \ldots \| c_0$ and $b = b_4 \| \ldots \| b_0$ are 32-bit integers, where the least significant byte is $c_0$ and $b_0$ respectively.

compiler intrinsic. Figure 7 shows how the constants can be expressed using the $4 \times 4$ byte matrix of AES.

Note that the constants have been updated in Simpira v2. The old constants of Simpira v1 are shown in Fig. 8. This update can be seen as "Grøstl strengthening," as it is inspired by the new round constants of the final-round Grøstl SHA-3 candidate [41]. No attack is currently known on Simpira v2 with the old rotation constants. Nevertheless, this change seems to strengthen Simpira without affecting its performance in our benchmarks.

| $c_0$ | $b_0$ | 0 | 0 |
|---|---|---|---|
| $c_1$ | $b_1$ | 0 | 0 |
| $c_2$ | $b_2$ | 0 | 0 |
| $c_3$ | $b_3$ | 0 | 0 |

**Fig. 8.** The old Simpira v1 constants used inside the $F_{c,b}$ function of Algorithm 2, expressed as a $4 \times 4$ matrix of bytes. Again, $c = c_4 \| \ldots \| c_0$ and $b = b_4 \| \ldots \| b_0$ are 32-bit integers, where the least significant byte is $x_0$ and $c_0$ respectively.

Both the input and output of Simpira consist of $b$ subblocks of 128 bits. The arrays use zero-based numbering, and array subscripts should be taken modulo the number of elements of the array. The subblock shuffle is done implicitly: we do not reorder the subblocks at the end of a Feistel round, but instead we apply the $F$-functions to other subblock inputs in the subsequent round. It is rather straightforward to implement the cyclic left shift in this way. For $b = 6$ and $b = 8$, the implementation of the subblock shuffle uses a decomposition into disjoint cycles.

As a result of this implementation choice, for $b \in \{2, 3, 4, 6, 8\}$, Simpira and its reduced-round variants are not always equivalent to a (generalized) Feistel with identical rounds. For example, for $b = 2$ the $F$-function is alternatingly applied from left to right and from right to left. When the number of rounds is odd, this is not equivalent to a Feistel with identical rounds: the two output subblocks will be swapped.

When $b = 1$, an extra `InvMixColumns` operation is applied to the output. This is equivalent to omitting the `MixColumns` operation in the last round, and is required to efficiently implement the inverse Simpira permutation using Intel's AES instructions. For details on how to efficiently implement both Simpira and Simpira$^{-1}$ when $b = 1$, we refer to the full version of this paper [45].

The design strategy of Simpira is intended to be very conservative. Because we think that the security of Simpira with very large $b$ may not yet be well-understood, we recommend to use Simpira with $b \leq 65536$, corresponding to inputs of at most one megabyte. However, the external cryptanalysis of Simpira for any value of $b$ is highly encouraged.

## 5 Benchmarks

We measured the performance of Simpira on the latest Intel processor, Architecture Codename Skylake. On this platform, the latency of AESENC is 4 cycles, and its throughput is 1 cycle. It follows that the software can be written in a way that fills the pipeline, by operating on four independent inputs. To obtain maximum throughput for all permutation sizes, we wrote functions that compute Simpira on four independent inputs. All Simpira permutations are benchmarked in the same setting, to make the results comparable.

Note that when $b = 4$, Simpira uses two independent $F$-functions, which means that maximum throughput could already be reached with only two independent inputs. For $b = 8$, where Simpira has four independent $F$-functions, even a single-stream Simpira implementation would fill the pipeline.

The measurements are performed as follows. We benchmark a function that evaluates Simpira for four independent inputs, and computed the number of cycles to carry out 256 calls to this function, as a "unit." This provides us with the throughput of Simpira. The results were obtained by using the RDTSCP instruction, 250 repetitions as a "warmup" phase, averaging the measurement on subsequent 1000 runs. Finally, this experiment was repeated 30 times, and the best result was selected. The platform was set up with Hyperthreading and Turbo Boost disabled.

The four data inputs can be stored sequentially at different pointers, or in an interleaved way (i.e. A[0]B[0]C[0]D[0]A[1]B[1]C[1]D[1]...). We benchmarked both settings. The results are shown in Table 1. We present only benchmarks for the forward Simpira permutation; the benchmarks for Simpira$^{-1}$ turned out to be very similar.

We refer to the full version of this paper [45] for a comparison with other constructions.

## 6 Cryptanalysis

The design criteria of Sect. 3 are not meant to be sufficient to guarantee security. In fact, it is not difficult to come up with trivially insecure constructions that satisfy (most of) the criteria. Rather, the design criteria are meant to assist us in identifying interesting constructions, which must then pass the scrutiny of cryptanalysis. Actually, during the design process of Simpira, we stumbled

**Algorithm 1.** AESENC (see [44])

1: **procedure** AESENC(state, key)
2:     state ← SubBytes(state)
3:     state ← ShiftRows(state)
4:     state ← MixColumns(state)
5:     state ← state ⊕ key
6:     **return** state
7: **end procedure**

**Algorithm 2.** $F_{c,b}(x)$

1: **procedure** $F_{c,b}(x)$
2:     $C$ ← SETR_EPI32($0x00 \oplus c \oplus b$,
3:                 $0x10 \oplus c \oplus b$,
4:                 $0x20 \oplus c \oplus b$,
5:                 $0x30 \oplus c \oplus b$)
6:     **return** AESENC(AESENC$(x, C), 0$)
7: **end procedure**

**Algorithm 3.** Simpira $(b = 1)$

1: **procedure** SIMPIRA$(x_0)$
2:     $R \leftarrow 6$
3:     **for** $c = 1, \ldots, R$ **do**
4:         $x_0 \leftarrow F_{c,b}(x_0)$
5:     **end for**
6:     InvMixColumns$(x_0)$
7:     **return** $x_0$
8: **end procedure**

**Algorithm 4.** Simpira$^{-1}$ $(b = 1)$

1: **procedure** SIMPIRA$(x_0)$
2:     $R \leftarrow 6$
3:     MixColumns$(x_0)$
4:     **for** $c = R, \ldots, 1$ **do**
5:         $x_0 \leftarrow F_{c,b}^{-1}(x_0)$
6:     **end for**
7:     **return** $x_0$
8: **end procedure**

**Algorithm 5.** Simpira $(b \in \{2, 3, 4\})$

1: **procedure** SIMPIRA$(x_0, \ldots, x_{b-1})$
2:     **if** $(b = 2) \vee (b = 3)$ **then**
3:         $R \leftarrow 6b + 3$
4:     **else**
5:         $R \leftarrow 15$
6:     **end if**
7:     $c \leftarrow 1$
8:
9:     **for** $r = 0, \ldots, R - 1$ **do**
10:         $x_{r+1} \leftarrow x_{r+1} \oplus F_{c,b}(x_r)$
11:         $c \leftarrow c + 1$
12:         **if** $b = 4$ **then**
13:             $x_{r+3} \leftarrow x_{r+3} \oplus F_{c,b}(x_{r+2})$
14:             $c \leftarrow c + 1$
15:         **end if**
16:     **end for**
17:     **return** $(x_0, x_1, \ldots, x_{b-1})$
18: **end procedure**

**Algorithm 6.** Simpira$^{-1}$ $(b \in \{2, 3, 4\})$

1: **procedure** SIMPIRA$^{-1}(x_0, \ldots, x_{b-1})$
2:     **if** $(b = 2) \vee (b = 3)$ **then**
3:         $R \leftarrow 6b + 3$
4:         $c \leftarrow R$
5:     **else**
6:         $R \leftarrow 15$
7:         $c \leftarrow 2R$
8:     **end if**
9:     **for** $r = R - 1, \ldots, 0$ **do**
10:         **if** $b = 4$ **then**
11:             $x_{r+3} \leftarrow x_{r+3} \oplus F_{c,b}(x_{r+2})$
12:             $c \leftarrow c - 1$
13:         **end if**
14:         $x_{r+1} \leftarrow x_{r+1} \oplus F_{c,b}(x_r)$
15:         $c \leftarrow c - 1$
16:     **end for**
17:     **return** $(x_0, x_1, \ldots, x_{b-1})$
18: **end procedure**

**Fig. 9.** Algorithm 2 specifies $F_{c,b}$ using the AESENC operation that is defined in Algorithm 1. Algorithms 3–6 specify Simpira and its inverse for $b \leq 4$, where the input and output consist of $b$ subblocks of 128 bits. Note that all arrays use zero-based numbering, and array subscripts should be taken modulo the number of elements of the array.

---

**Algorithm 7.** Simpira $(b = 6)$

1: **procedure** SIMPIRA$(x_0, \ldots, x_5)$
2:     $R \leftarrow 15$
3:     $c \leftarrow 1$
4:     $s \leftarrow (0, 1, 2, 5, 4, 3)$
5:     **for** $r = 0, \ldots, R - 1$ **do**
6:         $x_{s_{r+1}} \leftarrow x_{s_{r+1}} \oplus F_{c,b}(x_{s_r})$
7:         $c \leftarrow c + 1$
8:         $x_{s_{r+5}} \leftarrow x_{s_{r+5}} \oplus F_{c,b}(x_{s_{r+2}})$
9:         $c \leftarrow c + 1$
10:        $x_{s_{r+3}} \leftarrow x_{s_{r+3}} \oplus F_{c,b}(x_{s_{r+4}})$
11:        $c \leftarrow c + 1$
12:    **end for**
13:    **return** $(x_0, x_1, \ldots, x_5)$
14: **end procedure**

---

**Algorithm 8.** Simpira$^{-1}$ $(b = 6)$

1: **procedure** SIMPIRA$^{-1}(x_0, \ldots, x_5)$
2:     $R \leftarrow 15$
3:     $c \leftarrow 45$
4:     $s \leftarrow (0, 1, 2, 5, 4, 3)$
5:     **for** $r = R - 1, \ldots, 0$ **do**
6:         $x_{s_{r+3}} \leftarrow x_{s_{r+3}} \oplus F_{c,b}(x_{s_{r+4}})$
7:         $c \leftarrow c - 1$
8:         $x_{s_{r+5}} \leftarrow x_{s_{r+5}} \oplus F_{c,b}(x_{s_{r+2}})$
9:         $c \leftarrow c - 1$
10:        $x_{s_{r+1}} \leftarrow x_{s_{r+1}} \oplus F_{c,b}(x_{s_r})$
11:        $c \leftarrow c - 1$
12:    **end for**
13:    **return** $(x_0, x_1, \ldots, x_5)$
14: **end procedure**

---

**Algorithm 9.** Simpira $(b = 8)$

1: **procedure** SIMPIRA$(x_0, \ldots, x_7)$
2:     $R \leftarrow 18$
3:     $c \leftarrow 1$
4:     $s \leftarrow (0, 1, 6, 5, 4, 3)$
5:     $t \leftarrow (2, 7)$
6:     **for** $r = 0, \ldots, R - 1$ **do**
7:         $x_{s_{r+1}} \leftarrow x_{s_{r+1}} \oplus F_{c,b}(x_{s_r})$
8:         $c \leftarrow c + 1$
9:         $x_{s_{r+5}} \leftarrow x_{s_{r+5}} \oplus F_{c,b}(x_{t_r})$
10:        $c \leftarrow c + 1$
11:        $x_{s_{r+3}} \leftarrow x_{s_{r+3}} \oplus F_{c,b}(x_{s_{r+4}})$
12:        $c \leftarrow c + 1$
13:        $x_{t_{r+1}} \leftarrow x_{t_{r+1}} \oplus F_{c,b}(x_{s_{r+2}})$
14:        $c \leftarrow c + 1$
15:    **end for**
16:    **return** $(x_0, x_1, \ldots, x_7)$
17: **end procedure**

---

**Algorithm 10.** Simpira$^{-1}$ $(b = 8)$

1: **procedure** SIMPIRA$^{-1}(x_0, \ldots, x_7)$
2:     $R \leftarrow 18$
3:     $c \leftarrow 72$
4:     $s \leftarrow (0, 1, 6, 5, 4, 3)$
5:     $t \leftarrow (2, 7)$
6:     **for** $r = R - 1, \ldots, 0$ **do**
7:         $x_{t_{r+1}} \leftarrow x_{t_{r+1}} \oplus F_{c,b}(x_{s_{r+2}})$
8:         $c \leftarrow c - 1$
9:         $x_{s_{r+3}} \leftarrow x_{s_{r+3}} \oplus F_{c,b}(x_{s_{r+4}})$
10:        $c \leftarrow c - 1$
11:        $x_{s_{r+5}} \leftarrow x_{s_{r+5}} \oplus F_{c,b}(x_{t_r})$
12:        $c \leftarrow c - 1$
13:        $x_{s_{r+1}} \leftarrow x_{s_{r+1}} \oplus F_{c,b}(x_{s_r})$
14:        $c \leftarrow c - 1$
15:    **end for**
16:    **return** $(x_0, x_1, \ldots, x_7)$
17: **end procedure**

---

**Fig. 10.** Algorithms 7–10 specify Simpira and its inverse for $b = 6$ and $b = 8$, using the $F_{c,b}$-function that is specified in Algorithm 2. The input and the output consist of $b$ subblocks of 128 bits. Note that all arrays use zero-based numbering, and array subscripts should be taken modulo the number of elements of the array.

upon designs that were either insecure, or for which the security analysis was not so straightforward. When this happened, we adjusted the design criteria and repeated the search for constructions.

As such, we will not directly use the design criteria to argue the security of Simpira. Instead, we will use the fact that Simpira uses (generalized) Feistel

**Algorithm 11.** Simpira
$(b \notin \{1, 2, 3, 4, 6, 8\})$

```
1: procedure SIMPIRA(x₀, ..., x_{b-1})
2:     k ← 0
3:     d ← 2 · ⌊b/2⌋
4:     for j = 1, ..., 3 do
5:         if d ≠ b then
6:             TwoF(b − 2, k)
7:             k ← k + 1
8:         end if
9:         for r = 0, ..., d − 2 do
10:            TwoF(r, k)
11:            k ← k + 1
12:            if r ≠ d − r − 2 then
13:                TwoF(d − r − 2, k)
14:                k ← k + 1
15:            end if
16:        end for
17:        if d ≠ b then
18:            TwoF(b − 2, k)
19:            k ← k + 1
20:        end if
21:    end for
22:    return (x₀, x₁, ..., x_{b-1})
23: end procedure

24: procedure TwoF(r, k)
25:     if r mod 2 = 0 then
26:         x_{r+1} ← x_{r+1} ⊕ F_{2k+1,b}(x_r)
27:         x_r ← x_r ⊕ F_{2k+2,b}(x_{r+1})
28:     else
29:         x_r ← x_r ⊕ F_{2k+1,b}(x_{r+1})
30:         x_{r+1} ← x_{r+1} ⊕ F_{2k+2,b}(x_r)
31:     end if
32: end procedure
```

**Algorithm 12.** Simpira⁻¹
$(b \notin \{1, 2, 3, 4, 6, 8\})$

```
1: procedure SIMPIRA⁻¹(x₀, ..., x_{b-1})
2:     k ← 6b − 10
3:     d ← 2 · ⌊b/2⌋
4:     for j = 1, ..., 3 do
5:         if d ≠ b then
6:             InvTwoF(b − 2, k)
7:             k ← k − 1
8:         end if
9:         for r = d − 2, ..., 0 do
10:            if r ≠ d − r − 2 then
11:                InvTwoF(d − r − 2, k)
12:                k ← k − 1
13:            end if
14:            InvTwoF(r, k)
15:            k ← k − 1
16:        end for
17:        if d ≠ b then
18:            InvTwoF(b − 2, k)
19:            k ← k − 1
20:        end if
21:    end for
22:    return (x₀, x₁, ..., x_{b-1})
23: end procedure

24: procedure InvTwoF(r, k)
25:     if r mod 2 = 0 then
26:         x_r ← x_r ⊕ F_{2k+2,b}(x_{r+1})
27:         x_{r+1} ← x_{r+1} ⊕ F_{2k+1,b}(x_r)
28:     else
29:         x_{r+1} ← x_{r+1} ⊕ F_{2k+2,b}(x_r)
30:         x_r ← x_r ⊕ F_{2k+1,b}(x_{r+1})
31:     end if
32: end procedure
```

**Fig. 11.** Algorithms 11–12 specify Simpira and its inverse for $b \notin \{1, 2, 3, 4, 6, 8\}$, using the $F_{c,b}$-function that is specified in Algorithm 2. Both the input and the output consist of $b$ subblocks of 128 bits.

structures and the AES round function, both of which have been extensively studied in literature. This allows us to focus our cryptanalysis efforts on the most promising attacks for this type of construction. We have tried to make this section easy to understand, which will hopefully convince the reader that Simpira should have a very comfortable security margin against all currently-known attacks.

**Table 1.** Benchmarking results for the throughput of the Simpira permutations. For every $b$, we benchmark a function that applies the $128b$-bit permutation to four independent inputs. The data is either stored sequentially at different pointers, or in interleaved buffers. We give the number of cycles to process the four inputs, as well as the overhead compared the theoretical optimum of performing only `AESENC` instructions.

| $b$ | Bits | # `AESENC` | Non-interleaved | | Interleaved | |
|---|---|---|---|---|---|---|
| | | | Cycles (4×) | Overhead | Cycles (4×) | Overhead |
| 1 | 128 | 12 | 50 | 3 % | 50 | 3 % |
| 2 | 256 | 30 | 122 | 1 % | 122 | 1 % |
| 3 | 384 | 42 | 171 | 2 % | 171 | 2 % |
| 4 | 512 | 60 | 241 | 1 % | 241 | 1 % |
| 6 | 768 | 90 | 362 | 1 % | 362 | 1 % |
| 8 | 1024 | 144 | 594 | 3 % | 594 | 3 % |
| 16 | 2048 | 348 | 1586 | 14 % | 1400 | 1 % |
| 32 | 4096 | 732 | 3295 | 13 % | 2946 | 1 % |
| 64 | 8192 | 1500 | 6791 | 13 % | 6040 | 1 % |
| 128 | 16384 | 3036 | 13942 | 15 % | 12220 | 1 % |
| 256 | 32768 | 6108 | 31444 | 29 % | 24799 | 2 % |

**Security Claim.** In what follows, we will only consider structural distinguishers [8] with a complexity up to $2^{128}$. Simpira can be used in constructions that require a random permutation, however no statements can be made for adversaries that exceed $2^{128}$ queries. This type of security argument was first made by the SHA-3 [38] design team in response to high-complexity distinguishing attacks on the underlying permutation [19–21], and has since been reused for other permutation-based designs.

**Symmetry Attacks.** As explained in Sect. 3, the round constants are meant to avoid symmetry inside a Simpira round, as well as symmetry between rounds. The round constants also depend on $b$, which means that Simpira permutations of different widths should be indistinguishable from each other. The round constants are generated by a simple counter: this not only makes the design easy to understand and to implement, but also avoids any concerns that the constants may contain a backdoor. Every $F$-function has a different round constant: this does not seem to affect performance on recent Intel platforms, but greatly reduces the probability that a symmetry property can be maintained over several rounds.

**Invariant Subspace Attacks.** In its basic form, an invariant subspace attack [59] implies that there exists a coset of a vector space, so any number of iterations of the cryptographic round function maps to cosets of the same subspace. Rønjom [74] describes such an attack on Simpira v1 with $b = 4$, which is fixed in the current version. As explained in Sect. 9, no invariant subspace attacks were found for Simpira v2.

**State Collisions.** For most block-cipher-based modes of operation, it is possible to define a "state," which is typically 128 bits long. This can be the chaining value for CBC mode, the counter for CTR mode, or the checksum in OCB. When a collision is found in this state, which is expected to happen around $2^{64}$ queries, the mode becomes insecure. For the Feistel-based Simpira ($b \geq 2$), there is no such concept of a "state." In fact: all subblocks receive roughly an equal amount of "processing." This allows Simpira to reach security beyond $2^{64}$ queries after a sufficient amount of Feistel rounds.

**Linear and Differential Cryptanalysis.** Simpira's security argument against linear [12] and differential [63] cryptanalysis (up to attacks with complexity $2^{128}$) is the same as the argument for AES, which is based on counting the number of active S-boxes. As explained in [31], four rounds of AES have at least 25 (linearly or differentially) active S-boxes. Then any four-round differential characteristic holds with a probability less than $2^{-6.25} = 2^{-150}$, and any four-round linear characteristic holds with a correlation less than $2^{-3.25} = 2^{-75}$.

Here, $2^{-6}$ refers to the maximum difference propagation probability, and $2^{-3}$ is the maximum correlation amplitude of the S-box used in AES. The aforementioned reasoning makes the common assumptions that the probabilities of every round of a characteristic can be multiplied, and that this leads to a good estimate for the probability of the characteristic, and also of the corresponding differential.

The number of rounds typically needs to be slightly higher to account for partial key guesses (for keyed constructions), and to have a reasonable security margin. For any of the Simpira designs, we have at least three times the number of rounds required to reach 25 active S-boxes. This should give a sizable security margin against linear and differential cryptanalysis, and even against more advanced variants such as saturation and integral cryptanalysis [29]. In the case of integral cryptanalysis, of particular interest are the recently proposed integral distinguishers on Feistel and Generalized Feistel Networks by Todo [77] and by Zhang and Wenling [82].

**Boomerang and Differential-Linear Cryptanalysis.** Instead of using one long characteristic, boomerang [78] and differential-linear [11,58] cryptanalysis combine two shorter characteristics. But even combined with partial key guesses, the fact that Simpira has at least three times the number of rounds that result in 25 active S-boxes, should be more than sufficient to protect against this type of attacks.

**Truncated and Impossible Differential Cryptanalysis.** When full bit diffusion is not reached, it is easy to construct a truncated differential [55] characteristic with probability one. A common way to construct an impossible differential [9,10] is the *miss in the middle* approach. It combines two probability-one truncated differentials, whose conditions cannot be met together.

However, every Simpira variant has at least three times the number of rounds to reach full bit diffusion. This should not only prevent truncated and impossible differential attacks, but result in a satisfactory security margin against such attacks.

**Meet-in-the-middle and Rebound Attacks.** Meet-in-the-middle-attacks [34] separate the equations that describe a symmetric-key primitive into two or three groups. This is done in such a way that some variables do not appear into at least one of these groups. A typical rebound attack [65] also splits a cipher into three parts: an inner part that is satisfied by meet-in-the-middle techniques (in the inbound phase), and two outer parts that are fulfilled in a probabilistic way (in the outbound phase).

With Simpira, splitting the construction in three parts will always result in one part that either has at least 25 active S-boxes, or that reaches full bit diffusion. This should not only prevent meet-in-the-middle and rebound attacks, but also provide a large security margin against these attacks.

On Simpira with $b = 1$ (corresponding to 12-round AES with fixed round keys), the best known distinguisher is a rebound attack by Gilbert and Peyrin [42] that attacks 8 rounds out of 12.

**Generic Attacks.** A substantial amount of literature exists on generic attacks of Feistel structures. In particular, we are interested in attacks in Maurer et al.'s indifferentiability setting [64], which is an extension of the indistinguishability notion for constructions that use publicly available oracles. In Simpira, the $F$-functions contain no secret key, and are therefore assumed to be publicly available.

Coron et al. [25] showed that five rounds of Feistel are not indifferentiable from a random permutation, and presented a indifferentiability proof for six rounds. Holenstein et al. [51] later showed that their proof is flawed, and provided a new indifferentiability proof for fourteen rounds. In very recent work, Dai and Steinberger [32] and independently Dachman-Soled et al. [28] announced an indifferentiability proof for the 10-round Feistel, which Dai and Steinberger subsequently improved to a proof for 8 rounds [33].

A problem with the aforementioned indifferentiability proofs is that they are rather weak: if the $F$-function is 128 bits wide, security is only proven up to about $2^{16}$ queries. The indistinguishability setting is better understood, where many proofs are available for not only Feistel, but also various generalized Feistel structures. But even in this setting, most proofs do not go beyond $2^{64}$ queries, and proving security with close to $2^{128}$ queries requires a very large number of rounds [50].

So although several of Simpira's Feistel-based permutations were proven to be indistinguishable from random permutations using [66,83], it is an open problem to prove stronger security bounds for Simpira and other generalized Feistel structures. Nevertheless, no generic attacks are known for Simpira, even when up to $2^{128}$ are made.

Note that strictly speaking, there is an exception to the previous sentence for Simpira with $b = 1$. It is guaranteed to be an even permutation [22, Theorem 4.8], and therefore $2^{128} - 1$ queries can distinguish it from a random permutation with advantage 0.5. We only mention this for completeness; actually all of Simpira's permutations can be shown to be even, but this is typically not considered to be more than just a mathematical curiosity.

**Other Attacks.** We do not consider brute-force-like attacks [71], such as the biclique attacks on AES [16]: they perform exhaustive search on a smaller number of rounds, and therefore do not threaten the practical security of the cipher. However, it will be interesting to investigate such attacks in future work, as they give an indication of the security of the cipher in the absence of other attacks. We also do not look into algebraic attacks, as AES seems to very resistant against such attacks.

# 7   Applications

Simpira can be used in various scenarios where AES does not permit an efficient construction with security up to $2^{128}$ evaluations of the permutation. We present a brief overview possible applications.

**A Block Cipher Without Round Keys.** The (single-key) Even-Mansour construction [37, 39, 40] uses a secret key $K$ to turn a plaintext $P$ into a ciphertext $C$ as follows:

$$C = E_K(P) = \pi(P \oplus K) \oplus K \ , \tag{1}$$

where $\pi$ is an $n$-bit permutation. As argued by Dunkelman et al. [37], the construction is minimal, in the sense that simplifying it, for example by removing one of its components, will render it completely insecure. Mouha and Luykx [68] showed that the Even-Mansour is in some sense optimal in the multi-key setting, where several keys are independently and uniformly drawn from the key space.

When $D$ plaintext-ciphertexts are available, the secret key $K$ of the Even-Mansour construction can be recovered in $2^n/D$ (off-line) evaluations of the permutation $\pi$ [37]. This may be acceptable in lightweight authentication algorithms which rekey regularly, but may not be sufficient for encryption purposes [67,68]. In order to achieve security up to about $2^{128}$ queries against all attacks in the multi-key setting, the Even-Mansour construction requires a permutation of at least 256 bits.

An important advantage of the Even-Mansour construction is that it avoids the need to precalculate round keys (and store them securely!) or to calculate them on the fly. Moreover, it also allows the easy construction of a tweakable block cipher. For a given tweak $T$, one can turn the Even-Mansour construction into a tweakable block cipher [60, 61]:

$$C = E_K(P) = \pi(P \oplus K \cdot T) \oplus K \cdot T \ , \tag{2}$$

that can be proven to be secure up to $2^{n/2}$ queries in the multi-key setting using the proof of [68,69]. For concreteness, we use the multiplication $K \cdot T$ in $GF(2^n)$, which restricts the tweaks to $T \neq 0$. However, any $\epsilon$-AXU hash function can be used instead of this multiplication [23].

If the cipher is computed in a parallelizable mode of operation, independent blocks can be pipelined, and the performance would be dominated by Simpira with the relevant value of $b$, plus the overhead of the key addition.

**Permutation-Based Hashing.** Achieving 128-bit collision resistance with a 128-bit permutation has been shown to be impossible [72]. Typically, a large permutation size is used to achieve a high throughput, for example 1600 bits in the sponge construction of SHA-3 [38]. The downside of using a large permutation is that performance is significantly reduced when many short messages need to be hashed, for example to compute a Lamport signature [57]. Simpira overcomes these problems by providing a family of efficient permutations with different input sizes.

In particular for hashing short messages, one may consider to use Simpira with a Davies-Meyer feed-forward: $\pi(x) \oplus x$. This construction has been shown to be optimally preimage and collision-resistant [14,15], and even preimage aware [36], but not indifferentiable from a random oracle [24] as it is easy to find a fixed point: $\pi^{-1}(0)$. To match the intended application, padding of the input and/or truncation of the output of Simpira may be required.

**Wide-Block Encryption and Robust Authenticated Encryption.** Wide-block encryption can be used to provide security against chosen ciphertext attacks when short (or even zero-length) authentication tags are used. In the context of full-disk encryption, there is usually no space to store an authentication tag. In an attempt to reduce the risk that ciphertext changes result in meaningful plaintext, a possibility is to use a wide block cipher to encrypt an entire disk sector, which typically has a size of 512 to 4096 bytes.

The same concern also exists when short authentication tags are used, and can be addressed by an encode-then-encipher approach [5]: add some bits of redundancy, and then encrypt with an arbitrary-input-length block cipher. Note that this technique achieves robust authenticated encryption [49].

Typical solutions for wide-block encryption such as the VIL [4], CMC [47] and EME [46,48] modes of operation have the disadvantage that they are patented, and do not provide security beyond $2^{64}$ blocks of input. We are unaware of any patents related to Simpira.

When used in an Even-Mansour construction, Simpira with $b \geq 2$ can provide a wide block cipher that provides security up to $2^{128}$ blocks. When the block size exceeds the key size, the Even-Mansour construction can be generalized as follows:

$$C = E_K(P) = \pi(P \oplus (K \cdot T)\|0^*) \oplus (K \cdot T)\|0^*) , \qquad (3)$$

where we set $T = 1$ if no tweak is provided. Note that this straightforward extension of the Even-Mansour construction appears in the proof for various sponge constructions. The first proof of security of this construction in the multi-key setting was given by Andreeva et al. [3].

## 8    A Problem with Yanagihara and Iwata's GFS

For $b \geq 4$ (except $b = 6$ and $b = 8$), Simpira v1 used Yanagihara and Iwata's Type-1.x (b,2) GFS [81]. This is a GFS with two $F$-functions per round, shown in Fig. 12. Strictly speaking, we consider a variant of Yanagihara and Iwata's construction, that is identical up to a reordering of the input and output sub-blocks.

This construction has a problem. As can be seen from Fig. 12, the same value $x_0$ will eventually be processed by two $F$-functions. This clearly results in a redundant calculation, as the same $F$-function is evaluated twice on the same input.

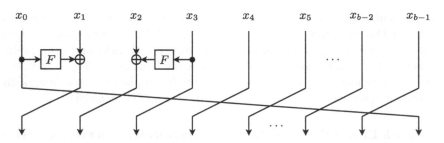

**Fig. 12.** Yanagihara and Iwata's Type-1.x (b,2) GFS [81], which was used in Simpira v1. Note that regardless of $b$, the same $x_0$ will eventually enter an $F$-function twice after a sufficient number of rounds.

In Simpira v1, the $F$-functions are not entirely identical due to the round constants. However, it can be seen that the problem in Yanagihara and Iwata's Type-1.x GFS also results in an attack on Simpira v1. In particular, the bounds on the number of active S-boxes were not correct, as the exact same S-box transitions were counted more than once. Dobraunig et al. [35] exploited the fact that the actual number of active S-boxes is much lower than expected, and constructed a series of attacks on the full 15-round Simpira v1 with $b = 4$, including a collision attack with complexity $2^{82.62}$ on Simpira when it is used in a truncated Davies-Meyer hash construction.

The problem with Yanagihara and Iwata's construction was confirmed to us by its designers. It was pointed out to us that their Type-1.x GFS was implicitly assumed to use independent round keys, but that this assumption was unfortunately not mentioned in their paper [81].

When this assumption does not hold, the counts of active S-boxes can be incorrect. This occurs when various simple key schedules are used, such as for example the Even-Mansour construction. We avoid this problem in Simpira v2 by ensuring that the same input is never processed by more than one $F$-function. This can be seen to avoid attacks on GFS in block-cipher-based constructions, when used with a uniformly random key.

But Simpira is designed to be a family of cryptographic permutations, and should therefore also be secure in unkeyed settings. In the next section, we show how the unkeyed setting leads to invariant subspace attacks on Simpira v1 for $b = 4$.

## 9    Invariant Subspace Attacks

Leander et al. [59] introduced the term *invariant subspace attack*, which applies when there exists a (large) subspace, so that any coset of this subspace is mapped to itself when the round function is applied. We now explain such an attack applies to Yanagihara and Iwata Type-1.x (4,2) GFS. Again, strictly speaking Yanagihara and Iwata defined a variant of this construction, that is however identical up to a reordering of the input and output blocks. As illustrated in Fig. 13, we find that if the second and the last subblock of the input are identical, this property is preserved after any multiple of two rounds.

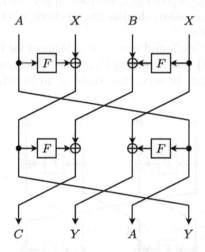

**Fig. 13.** Yanagihara and Iwata Type-1.x (4,2) GFS [81], which was used in Simpira v1 for $b = 4$. We assume that all $F$-functions are identical. Here $A$, $B$ and $X$ can be any value. The leftmost input subblock enters the same $F$-function twice, and therefore guarantees that the output value $Y$ will appear twice as well.

A similar observation also holds for Simpira v1 with $b = 4$, where only the round constants slightly destroy the symmetry property of the input. This is a consequence of the sparse round constants in Simpira v1, and the reuse of values into several $F$-functions, as explained in Sect. 8. In particular, for any even multiple of rounds up to 126, Simpira v1 round constants (see Fig. 8) only differ in the zeroth byte of the AES state. This means that if the second and the last subblock of the input are identical, this property will be preserved, except for the first column of corresponding AES states.

Rønjom [74] described an invariant subspace attack on Simpira v1 with $b = 4$. In particular, Rønjom identified a large subspace such that any coset of this space is invariant under two rounds of Simpira v2. This leads to a plaintext invariant over infinitely many even rounds. It can be seen as a generalization of the attack on Yanagihara-Iwata's Type-1.x GFS that is described in this section.

The property does not hold for an odd number of rounds, so it does not apply directly for Simpira v1 with $b = 4$, which consists of 15 Feistel rounds. For this reason, we did not detect any nonrandomess in our test vectors, although it included the all-zero input that is an element of the coset of the invariant subspace. However, simply applying the permutation twice means that the total number of rounds is even, so that the distinguisher applies.

Do such invariant subspaces attacks also exist for Simpira v2? In an attempt to find such attacks, we first look for invariant subspaces when all $F$-functions are identical. This should give a good starting point to find invariant subspaces when the real (non-identical) $F$-functions of Simpira are used. More specifically, we select a random $F$-function, and consider four values for every input subblock: $0$, $F(0)$, $F(F(0))$ and $F(F(0)) \oplus F(0)$. We then apply the Feistel round function several times, and use Gaussian elimination to check whether we stay within a particular linear subspace.

Using this technique, we found invariant subspaces for the GFS used in Simpira v2 when $b \in \{4, 6, 8\}$ (i.e. assuming identical $F$-functions), but not for other values of $b$. In fact, it can be seen that there is an invariant subspace for any

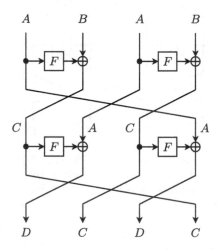

**Fig. 14.** The Type-2 GFS with $b = 4$, used in Simpira v2. We assume that all $F$-functions are identical. Here, $A$ and $B$ can be any value. If the odd-numbered input subblocks are equal, and the even-numbered input subblock are equal, then this property is preserved for any number of rounds.

Type-2 GFS with an "even-odd shuffle [76]," that is, where even-numbered input subblocks are mapped to odd-numbered output subblocks and vice versa. For $b = 4$, such an invariant subspace is shown in Fig. 14. With the introduction of appropriate round constants, however, these invariant subspace attacks are avoided.

We chose to retain Type-2 GFS in Simpira v2 for $b \in \{4, 6, 8\}$, instead of replacing them by Generalized Feistel structures that "inherently" avoid invariant subspace attacks. This is because Type-2 GFS constructions are efficient and well-analyzed, and invariant subspaces can be avoided by using round constants.

We searched for invariant subspaces in all Simpira v2 variants, but were unable to find any. A similar search was also performed by Rønjom [73], who also could not identify invariant subspaces in the updated Simpira design. Unfortunately, currently no provable arguments against invariant subspace attacks are known. This is an interesting topic for future work.

# 10   Conclusion

We introduced Simpira, which is a family of cryptographic permutations that processes inputs of $128 \times b$ bits. It is intended to be a very conservative design that achieves high throughput on processors with AES instructions. We decided to use two rounds of AES as a building block, with the goal of simplifying the design space exploration, and making the cryptanalysis and implementation straightforward.

With this building block, we explored a large number of generalized Feistel structures, and calculated how many rounds are required to reach either full bit diffusion, or 25 linearly or differentially active S-boxes, whichever is greater. To ensure a large security margin, we multiplied this number of rounds by three. Of all designs that we considered, we selected the ones with the lowest amount of $F$-functions in total.

Following these design criteria, Simpira resulted in seven different designs. For $b = 1$, we have AES with fixed round keys. Simpira uses a Feistel structure for $b = 2$, a Type-1 GFS for $b = 3$, and a Type-2 GFS for $b = 4$. The $b \geq 5$ design is a dedicated construction that we introduce in this paper. For $b = 6$ and $b = 8$, we use Suzaki and Minematsu's improved Type-2 GFS, as it has fewer $F$-functions than general construction for $b \geq 5$.

Our benchmarks on Intel Skylake showed that Simpira is close to the theoretical optimum of only executing `AESENC` instructions. For $b \leq 4$, Simpira is less than 3 % away from this optimum. For $b \leq 32$, corresponding to inputs of up to 512 bytes, Simpira is less than 13 % away from this optimum for a non-interleaved implementation, and less than 1 % away for an interleaved implementation.

It is unfortunate that many methods to encrypt wide input blocks, such as VIL, CMC, and EME, have not seen widespread adoption. The main obstacle appears to be that they are patented. We hope that Simpira can provide an interesting alternative: it is not only free from patent concerns, but offers security way beyond the $2^{64}$ limit for typical AES-based modes.

**Acknowledgments.** We thank the organizers and participants of Dagstuhl Seminar 16021, where an early version of this work was presented. The detailed comments and suggestions of the seminar participants helped to improve this manuscript significantly. Thanks to Christoph Dobraunig, Maria Eichlseder, Florian Mendel and Sondre Rønjom their attacks on Simpira v1, which lead to the updated Simpira v2 that is presented in this document. We also thank Eik List for pointing out some notation issues in an earlier version of this text, and Sébastien Duval, Brice Minaud, Kazuhiko Minematsu, and Tetsu Iwata for their insights into Feistel structures. This work was supported in part by the Research Council KU Leuven: GOA TENSE (GOA/11/007), by Research Fund KU Leuven, OT/13/071, by the PQCRYPTO project, which was partially funded by the European Commission Horizon 2020 research Programme, grant #645622, by the ISRAEL SCIENCE FOUNDATION (grant No. 1018/16), and by the French Agence Nationale de la Recherche through the BLOC project under Contract ANR-11-INS-011, and the BRUTUS project under Contract ANR-14-CE28-0015. Nicky Mouha is supported by a Postdoctoral Fellowship from the Flemish Research Foundation (FWO-Vlaanderen), and by FWO travel grant 12F9714N. Certain algorithms and commercial products are identified in this paper to foster understanding. Such identification does not imply recommendation or endorsement by NIST, nor does it imply that the algorithms or products identified are necessarily the best available for the purpose.

# References

1. Albrecht, M.R., Rechberger, C., Schneider, T., Tiessen, T., Zohner, M.: Ciphers for MPC and FHE. In: Oswald, E., Fischlin, M. (eds.) EUROCRYPT 2015. LNCS, vol. 9056, pp. 430–454. Springer, Heidelberg (2015). doi:10.1007/978-3-662-46800-5_17
2. Anderson, R.J., Biham, E.: Two practical and provably secure block ciphers: BEAR and LION. In: Gollmann, D. (ed.) FSE 1996. LNCS, vol. 1039, pp. 113–120. Springer, Heidelberg (1996). doi:10.1007/3-540-60865-6_48
3. Andreeva, E., Daemen, J., Mennink, B., Van Assche, G.: Security of keyed sponge constructions using a modular proof approach. In: Leander, G. (ed.) FSE 2015. LNCS, vol. 9054, pp. 364–384. Springer, Heidelberg (2015). doi:10.1007/978-3-662-48116-5_18
4. Bellare, M., Rogaway, P.: On the construction of variable-input-length ciphers. In: Knudsen, L. (ed.) FSE 1999. LNCS, vol. 1636, pp. 231–244. Springer, Heidelberg (1999). doi:10.1007/3-540-48519-8_17
5. Bellare, M., Rogaway, P.: Encode-then-encipher encryption: how to exploit nonces or redundancy in plaintexts for efficient cryptography. In: Okamoto, T. (ed.) ASIACRYPT 2000. LNCS, vol. 1976, pp. 317–330. Springer, Heidelberg (2000). doi:10.1007/3-540-44448-3_24
6. Berger, T.P., Francq, J., Minier, M., Thomas, G.: Extended generalized Feistel networks using matrix representation to propose a new lightweight block cipher: Lilliput. IEEE Trans. Comput. **65**(7), 2074–2089 (2016)
7. Berger, T.P., Minier, M., Thomas, G.: Extended generalized Feistel networks using matrix representation. In: Lange, T., Lauter, K., Lisoněk, P. (eds.) SAC 2013. LNCS, vol. 8282, pp. 289–305. Springer, Heidelberg (2014). doi:10.1007/978-3-662-43414-7_15
8. Bertoni, G., Daemen, J., Peeters, M., Van Assche, G.: Cryptographic sponge functions. http://sponge.noekeon.org/CSF-0.1.pdf

9. Biham, E., Biryukov, A., Shamir, A.: Cryptanalysis of skipjack reduced to 31 rounds using impossible differentials. In: Stern, J. (ed.) EUROCRYPT 1999. LNCS, vol. 1592, pp. 12–23. Springer, Heidelberg (1999). doi:10.1007/3-540-48910-X_2

10. Biham, E., Biryukov, A., Shamir, A.: Cryptanalysis of skipjack reduced to 31 rounds using impossible differentials. J. Cryptology 18(4), 291–311 (2005)

11. Biham, E., Dunkelman, O., Keller, N.: Enhancing differential-linear cryptanalysis. In: Zheng, Y. (ed.) ASIACRYPT 2002. LNCS, vol. 2501, pp. 254–266. Springer, Heidelberg (2002). doi:10.1007/3-540-36178-2_16

12. Biham, E., Shamir, A.: Differential cryptanalysis of DES-like cryptosystems. J. Cryptology 4(1), 3–72 (1991)

13. Biryukov, A., Khovratovich, D.: PAEQ: parallelizable permutation-based authenticated encryption. In: Chow, S.S.M., Camenisch, J., Hui, L.C.K., Yiu, S.M. (eds.) ISC 2014. LNCS, vol. 8783, pp. 72–89. Springer, Heidelberg (2014). doi:10.1007/978-3-319-13257-0_5

14. Black, J., Rogaway, P., Shrimpton, T.: Black-box analysis of the block-cipher-based hash-function constructions from PGV. In: Yung, M. (ed.) CRYPTO 2002. LNCS, vol. 2442, pp. 320–335. Springer, Heidelberg (2002). doi:10.1007/3-540-45708-9_21

15. Black, J., Rogaway, P., Shrimpton, T., Stam, M.: An analysis of the blockcipher-based hash functions from PGV. J. Cryptology 23(4), 519–545 (2010)

16. Bogdanov, A., Khovratovich, D., Rechberger, C.: Biclique cryptanalysis of the full AES. In: Lee, D.H., Wang, X. (eds.) ASIACRYPT 2011. LNCS, vol. 7073, pp. 344–371. Springer, Heidelberg (2011). doi:10.1007/978-3-642-25385-0_19

17. Bogdanov, A., Knežević, M., Leander, G., Toz, D., Varıcı, K., Verbauwhede, I.: SPONGENT: a lightweight hash function. In: Preneel, B., Takagi, T. (eds.) CHES 2011. LNCS, vol. 6917, pp. 312–325. Springer, Heidelberg (2011). doi:10.1007/978-3-642-23951-9_21

18. Bogdanov, A., Knežević, M., Leander, G., Toz, D., Varıcı, K., Verbauwhede, I.: SPONGENT: the design space of lightweight cryptographic hashing. IEEE Trans. Comput. 62(10), 2041–2053 (2013)

19. Boura, C., Canteaut, A.: A zero-sum property for the KECCAK-f permutation with 18 rounds. In: ISIT 2010. pp. 2488–2492. IEEE (2010)

20. Boura, C., Canteaut, A.: Zero-sum distinguishers for iterated permutations and application to KECCAK-f and Hamsi-256. In: Biryukov, A., Gong, G., Stinson, D.R. (eds.) SAC 2010. LNCS, vol. 6544, pp. 1–17. Springer, Heidelberg (2011). doi:10.1007/978-3-642-19574-7_1

21. Boura, C., Canteaut, A., De Cannière, C.: Higher-order differential properties of Keccak and Luffa. Cryptology ePrint Archive, Report 2010/589 (2010)

22. Cid, C., Murphy, S., Robshaw, M.J.B.: Algebraic Aspects of the Advanced Encryption Standard. Springer, Heidelberg (2006)

23. Cogliati, B., Lampe, R., Seurin, Y.: Tweaking Even-Mansour ciphers. In: Gennaro, R., Robshaw, M. (eds.) CRYPTO 2015. LNCS, vol. 9215, pp. 189–208. Springer, Heidelberg (2015). doi:10.1007/978-3-662-47989-6_9

24. Coron, J.-S., Dodis, Y., Malinaud, C., Puniya, P.: Merkle-Damgård revisited: how to construct a hash function. In: Shoup, V. (ed.) CRYPTO 2005. LNCS, vol. 3621, pp. 430–448. Springer, Heidelberg (2005). doi:10.1007/11535218_26

25. Coron, J.-S., Patarin, J., Seurin, Y.: The random oracle model and the ideal cipher model are equivalent. In: Wagner, D. (ed.) CRYPTO 2008. LNCS, vol. 5157, pp. 1–20. Springer, Heidelberg (2008). doi:10.1007/978-3-540-85174-5_1

26. Cossíos, D.: Breve Bestiario Peruano. Editorial Casatomada, 2nd edn. (2008)

27. Crowley, P.: Mercy: a fast large block cipher for disk sector encryption. In: Goos, G., Hartmanis, J., Leeuwen, J., Schneier, B. (eds.) FSE 2000. LNCS, vol. 1978, pp. 49–63. Springer, Heidelberg (2001). doi:10.1007/3-540-44706-7_4

28. Dachman-Soled, D., Katz, J., Thiruvengadam, A.: 10-round Feistel is indifferentiable from an ideal cipher. In: Fischlin, M., Coron, J.-S. (eds.) EUROCRYPT 2016. LNCS, vol. 9666, pp. 649–678. Springer, Heidelberg (2016). doi:10.1007/978-3-662-49896-5_23

29. Daemen, J., Knudsen, L., Rijmen, V.: The block cipher square. In: Biham, E. (ed.) FSE 1997. LNCS, vol. 1267, pp. 149–165. Springer, Heidelberg (1997). doi:10.1007/BFb0052343

30. Daemen, J., Rijmen, V.: The wide trail design strategy. In: Honary, B. (ed.) Cryptography and Coding 2001. LNCS, vol. 2260, pp. 222–238. Springer, Heidelberg (2001). doi:10.1007/3-540-45325-3_20

31. Daemen, J., Rijmen, V.: The Design of Rijndael: AES - The Advanced Encryption Standard. Springer, Heidelberg (2002)

32. Dai, Y., Steinberger, J.: Indifferentiability of 10-round Feistel networks. Cryptology ePrint Archive, Report 2015/874 (2015)

33. Dai, Y., Steinberger, J.: Indifferentiability of 8-round Feistel networks. In: Robshaw, M., Katz, J. (eds.) CRYPTO 2016. LNCS, vol. 9814, pp. 95–120. Springer, Heidelberg (2016). doi:10.1007/978-3-662-53018-4_4

34. Diffie, W., Hellman, M.E.: Exhaustive cryptanalysis of the NBS data encryption standard. Computer 10(6), 74–84 (1977)

35. Dobraunig, C., Eichlseder, M., Mendel, F.: Cryptanalysis of Simpira. Cryptology ePrint Archive, Report 2016/244 (2016)

36. Dodis, Y., Ristenpart, T., Shrimpton, T.: Salvaging Merkle-Damgård for practical applications. In: Joux, A. (ed.) EUROCRYPT 2009. LNCS, vol. 5479, pp. 371–388. Springer, Heidelberg (2009). doi:10.1007/978-3-642-01001-9_22

37. Dunkelman, O., Keller, N., Shamir, A.: Minimalism in cryptography: the Even-Mansour scheme revisited. In: Pointcheval, D., Johansson, T. (eds.) EUROCRYPT 2012. LNCS, vol. 7237, pp. 336–354. Springer, Heidelberg (2012). doi:10.1007/978-3-642-29011-4_21

38. Dworkin, M.J.: SHA-3 standard: permutation-based hash and extendable-output functions. Federal Inf. Process. Stds. (NIST FIPS) - 202, August 2015

39. Even, S., Mansour, Y.: A construction of a cipher from a single pseudorandom permutation. In: Imai, H., Rivest, R.L., Matsumoto, T. (eds.) ASIACRYPT 1991. LNCS, vol. 739, pp. 210–224. Springer, Heidelberg (1993). doi:10.1007/3-540-57332-1_17

40. Even, S., Mansour, Y.: A construction of a cipher from a single pseudorandom permutation. J. Cryptology 10(3), 151–162 (1997)

41. Gauravaram, P., Knudsen, L.R., Matusiewicz, K., Mendel, F., Rechberger, C., Schläffer, M., Thomsen, S.S.: Grøstl - a SHA-3 candidate. Submission to the NIST SHA-3 Competition (Round 3) (2011). http://www.groestl.info/Groestl.pdf

42. Gilbert, H., Peyrin, T.: Super-Sbox cryptanalysis: improved attacks for AES-like permutations. In: Hong, S., Iwata, T. (eds.) FSE 2010. LNCS, vol. 6147, pp. 365–383. Springer, Heidelberg (2010). doi:10.1007/978-3-642-13858-4_21

43. Gueron, S.: Intel's new AES instructions for enhanced performance and security. In: Dunkelman, O. (ed.) FSE 2009. LNCS, vol. 5665, pp. 51–66. Springer, Heidelberg (2009). doi:10.1007/978-3-642-03317-9_4

44. Gueron, S.: Intel® Advanced Encryption Standard (AES) new instructions set, September 2012. https://software.intel.com/en-us/articles/intel-advanced-encryption-standard-aes-instructions-set, Revision 3.01

45. Gueron, S., Mouha, N.: Simpira v2: a family of efficient permutations using the AES round function. Cryptology ePrint Archive, Report 2016/122 (2016). Full version of this paper
46. Halevi, S.: EME*: extending EME to handle arbitrary-length messages with associated data. In: Canteaut, A., Viswanathan, K. (eds.) INDOCRYPT 2004. LNCS, vol. 3348, pp. 315–327. Springer, Heidelberg (2004). doi:10.1007/978-3-540-30556-9_25
47. Halevi, S., Rogaway, P.: A tweakable enciphering mode. In: Boneh, D. (ed.) CRYPTO 2003. LNCS, vol. 2729, pp. 482–499. Springer, Heidelberg (2003). doi:10.1007/978-3-540-45146-4_28
48. Halevi, S., Rogaway, P.: A parallelizable enciphering mode. In: Okamoto, T. (ed.) CT-RSA 2004. LNCS, vol. 2964, pp. 292–304. Springer, Heidelberg (2004). doi:10.1007/978-3-540-24660-2_23
49. Hoang, V.T., Krovetz, T., Rogaway, P.: Robust authenticated-encryption AEZ and the problem that it solves. In: Oswald, E., Fischlin, M. (eds.) EUROCRYPT 2015. LNCS, vol. 9056, pp. 15–44. Springer, Heidelberg (2015). doi:10.1007/978-3-662-46800-5_2
50. Hoang, V.T., Rogaway, P.: On generalized Feistel networks. In: Rabin, T. (ed.) CRYPTO 2010. LNCS, vol. 6223, pp. 613–630. Springer, Heidelberg (2010). doi:10.1007/978-3-642-14623-7_33
51. Holenstein, T., Künzler, R., Tessaro, S.: The equivalence of the random oracle model and the ideal cipher model, revisited. In: STOC 2011, pp. 89–98. ACM (2011)
52. Jean, J.: Cryptanalysis of Haraka. Cryptology ePrint Archive, Report 2016/396 (2016)
53. Jean, J., Nikolić, I., Sasaki, Y., Wang, L.: Practical cryptanalysis of PAES. In: Joux, A., Youssef, A. (eds.) SAC 2014. LNCS, vol. 8781, pp. 228–242. Springer, Heidelberg (2014). doi:10.1007/978-3-319-13051-4_14
54. Jean, J., Nikolić, I., Sasaki, Y., Wang, L.: Practical forgeries and distinguishers against PAES. IEICE Trans. **99–A**(1), 39–48 (2016)
55. Knudsen, L.R.: Truncated and higher order differentials. In: Preneel, B. (ed.) FSE 1994. LNCS, vol. 1008, pp. 196–211. Springer, Heidelberg (1995). doi:10.1007/3-540-60590-8_16
56. Kölbl, S., Lauridsen, M.M., Mendel, F., Rechberger, C.: Haraka - efficient short-input hashing for post-quantum applications. Cryptology ePrint Archive, Report 2016/098 (2016)
57. Lamport, L.: Constructing digital signatures from a one way function. Technical report. SRI-CSL-98, SRI International Computer Science Laboratory, October 1979
58. Langford, S.K., Hellman, M.E.: Differential-linear cryptanalysis. In: Desmedt, Y.G. (ed.) CRYPTO 1994. LNCS, vol. 839, pp. 17–25. Springer, Heidelberg (1994). doi:10.1007/3-540-48658-5_3
59. Leander, G., Abdelraheem, M.A., AlKhzaimi, H., Zenner, E.: A cryptanalysis of PRINTCIPHER: the invariant subspace attack. In: Rogaway, P. (ed.) CRYPTO 2011. LNCS, vol. 6841, pp. 206–221. Springer, Heidelberg (2011). doi:10.1007/978-3-642-22792-9_12
60. Liskov, M., Rivest, R.L., Wagner, D.: Tweakable block ciphers. In: Yung, M. (ed.) CRYPTO 2002. LNCS, vol. 2442, pp. 31–46. Springer, Heidelberg (2002). doi:10.1007/3-540-45708-9_3
61. Liskov, M., Rivest, R.L., Wagner, D.: Tweakable block ciphers. J. Cryptology **24**(3), 588–613 (2011)

62. Lucks, S.: BEAST: a fast block cipher for arbitrary blocksizes. In: Horster, P. (ed.) CMS 1996. IFIP Conference Proceedings, vol. 70, pp. 144–153. Chapman & Hall, New York (1996)

63. Matsui, M.: Linear cryptanalysis method for DES cipher. In: Helleseth, T. (ed.) EUROCRYPT 1993. LNCS, vol. 765, pp. 386–397. Springer, Heidelberg (1994). doi:10.1007/3-540-48285-7_33

64. Maurer, U., Renner, R., Holenstein, C.: Indifferentiability, impossibility results on reductions, and applications to the random oracle methodology. In: Naor, M. (ed.) TCC 2004. LNCS, vol. 2951, pp. 21–39. Springer, Heidelberg (2004). doi:10.1007/978-3-540-24638-1_2

65. Mendel, F., Rechberger, C., Schläffer, M., Thomsen, S.S.: The rebound attack: cryptanalysis of reduced Whirlpool and Grøstl. In: Dunkelman, O. (ed.) FSE 2009. LNCS, vol. 5665, pp. 260–276. Springer, Heidelberg (2009). doi:10.1007/978-3-642-03317-9_16

66. Moriai, S., Vaudenay, S.: On the pseudorandomness of top-level schemes of block ciphers. In: Okamoto, T. (ed.) ASIACRYPT 2000. LNCS, vol. 1976, pp. 289–302. Springer, Heidelberg (2000). doi:10.1007/3-540-44448-3_22

67. Mouha, N.: The design space of lightweight cryptography. Cryptology ePrint Archive, Report 2015/303 (2015)

68. Mouha, N., Luykx, A.: Multi-key security: the Even-Mansour construction revisited. In: Gennaro, R., Robshaw, M. (eds.) CRYPTO 2015. LNCS, vol. 9215, pp. 209–223. Springer, Heidelberg (2015). doi:10.1007/978-3-662-47989-6_10

69. Mouha, N., Mennink, B., Herrewege, A.V., Watanabe, D., Preneel, B., Verbauwhede, I.: Chaskey: an efficient MAC algorithm for 32-bit microcontrollers. In: Joux, A., Youssef, A. (eds.) SAC 2014. LNCS, vol. 8781, pp. 306–323. Springer, Heidelberg (2014). doi:10.1007/978-3-319-13051-4_19

70. Mouha, N., Wang, Q., Gu, D., Preneel, B.: Differential and linear cryptanalysis using mixed-integer linear programming. In: Wu, C.-K., Yung, M., Lin, D. (eds.) Inscrypt 2011. LNCS, vol. 7537, pp. 57–76. Springer, Heidelberg (2012). doi:10.1007/978-3-642-34704-7_5

71. Rechberger, C.: On bruteforce-like cryptanalysis: new meet-in-the-middle attacks in symmetric cryptanalysis. In: Kwon, T., Lee, M.-K., Kwon, D. (eds.) ICISC 2012. LNCS, vol. 7839, pp. 33–36. Springer, Heidelberg (2013). doi:10.1007/978-3-642-37682-5_3

72. Rogaway, P., Steinberger, J.: Security/efficiency tradeoffs for permutation-based hashing. In: Smart, N. (ed.) EUROCRYPT 2008. LNCS, vol. 4965, pp. 220–236. Springer, Heidelberg (2008). doi:10.1007/978-3-540-78967-3_13

73. Rønjom, S.: Personal Communication, March 2016

74. Rønjom, S.: Invariant subspaces in Simpira. Cryptology ePrint Archive, Report 2016/248 (2016)

75. Schroeppel, R.: The hasty pudding cipher - a tasty morsel, submission to the NIST AES competition (1998)

76. Suzaki, T., Minematsu, K.: Improving the generalized Feistel. In: Hong, S., Iwata, T. (eds.) FSE 2010. LNCS, vol. 6147, pp. 19–39. Springer, Heidelberg (2010). doi:10.1007/978-3-642-13858-4_2

77. Todo, Y.: Structural evaluation by generalized integral property. In: Oswald, E., Fischlin, M. (eds.) EUROCRYPT 2015. LNCS, vol. 9056, pp. 287–314. Springer, Heidelberg (2015). doi:10.1007/978-3-662-46800-5_12

78. Wagner, D.: The boomerang attack. In: Knudsen, L. (ed.) FSE 1999. LNCS, vol. 1636, pp. 156–170. Springer, Heidelberg (1999). doi:10.1007/3-540-48519-8_12

79. Yanagihara, S., Iwata, T.: On permutation layer of type 1, source-heavy, and target-heavy generalized Feistel structures. In: Lin, D., Tsudik, G., Wang, X. (eds.) CANS 2011. LNCS, vol. 7092, pp. 98–117. Springer, Heidelberg (2011). doi:10.1007/978-3-642-25513-7_8

80. Yanagihara, S., Iwata, T.: Improving the permutation layer of type 1, type 3, source-heavy, and target-heavy generalized Feistel structures. IEICE Trans. **96–A**(1), 2–14 (2013)

81. Yanagihara, S., Iwata, T.: Type 1.x generalized Feistel structures. IEICE Trans. **97A**(4), 952–963 (2014)

82. Zhang, H., Wu, W.: Structural evaluation for generalized Feistel structures and applications to LBlock and TWINE. In: Biryukov, A., Goyal, V. (eds.) INDOCRYPT 2015. LNCS, vol. 9462, pp. 218–237. Springer, Heidelberg (2015). doi:10.1007/978-3-319-26617-6_12

83. Zheng, Y., Matsumoto, T., Imai, H.: On the construction of block ciphers provably secure and not relying on any unproved hypotheses. In: Brassard, G. (ed.) CRYPTO 1989. LNCS, vol. 435, pp. 461–480. Springer, Heidelberg (1990). doi:10.1007/0-387-34805-0_42

# Towards Practical Whitebox Cryptography: Optimizing Efficiency and Space Hardness

Andrey Bogdanov[1]([⊠]), Takanori Isobe[2], and Elmar Tischhauser[1]

[1] Technical University of Denmark, Kongens Lyngby, Denmark
{anbog,ewti}@dtu.dk
[2] Sony Global Manufacturing & Operations Corporation, Tokyo , Japan
Takanori.Isobe@jp.sony.com

**Abstract.** Whitebox cryptography aims to provide security for cryptographic algorithms in an untrusted environment where the adversary has full access to their implementation. Typical security goals for whitebox cryptography include *key extraction security* and *decomposition security*: Indeed, it should be infeasible to recover the secret key from the implementation and it should be hard to decompose the implementation by finding a more compact representation without recovering the secret key, which mitigates code lifting.

Whereas all published whitebox implementations for standard cryptographic algorithms such as DES or AES are prone to practical key extraction attacks, there have been two dedicated design approaches for whitebox block ciphers: ASASA by Birykov et al. at ASIACRYPT'14 and SPACE by Bogdanov and Isobe at CCS'15. While ASASA suffers from decomposition attacks, SPACE reduces the security against key extraction and decomposition attacks in the white box to the security of a standard block cipher such as AES in the standard blackbox setting. However, due to the security-prioritized design strategy, SPACE imposes a sometimes prohibitive performance overhead in the real world as it needs many AES calls to encrypt a single block.

In this paper, we address the issue by designing a family of dedicated whitebox block ciphers SPNbox and a family of underlying small block ciphers with software efficiency and constant-time execution in mind. While still relying on the standard blackbox block cipher security for the resistance against key extraction and decomposition, SPNbox attains speed-ups of up to 6.5 times in the black box and up to 18 times in the white box on Intel Skylake and ARMv8 CPUs, compared to SPACE. The designs allow for constant-time implementations in the blackbox setting and meet the practical requirements to whitebox cryptography in real-world applications such as DRM or mobile payments. Moreover, we formalize resistance towards decomposition in form of *weak* and *strong* *space hardness* at various security levels. We obtain bounds on space hardness in all those adversarial models.

Thus, for the first time, SPNbox provides a practical whitebox block cipher that features well-understood key extraction security, rigorous

J.H. Cheon and T. Takagi (Eds.): ASIACRYPT 2016, Part I, LNCS 10031, pp. 126–158, 2016.
DOI: 10.1007/978-3-662-53887-6_5

analysis towards decomposition security, demonstrated real-world efficiency on various platforms and constant-time implementations. This paves the way to enhancing susceptible real-world applications with whitebox cryptography.

**Keywords:** White-box cryptography · Space hardness · Code lifting · Decomposition · Key extraction · Mass surveillance · Trojans · Malware

# 1 Introduction

## 1.1 Black Box vs White Box

Whitebox cryptography was introduced by Chow et al. in 2002 [14] as a technique to secure software implementations of block ciphers when the adversary has full access to the execution environment. This setup is called the *whitebox setting*, which is opposed to the standard *blackbox setting* where the attacker can neither observe nor influence the internals of the block cipher. The functionality of the cipher shall be the same when implemented in the black-box and white-boxe settings. However, the *whitebox implementation* in the untrusted environment (as e.g. in the mobile client software) and *blackbox implementation* in the secure environment (as e.g. on the backend server) can vary significantly to meet distinct security demands arising from two different threat models:

- **In the black box:** The adversary is able to access inputs and outputs of the cipher with known, chosen or adaptively chosen plaintexts/ciphertexts. Given the blackbox implementation, the attacker aims to recover the secret key (*key recovery*) or to distinguish the block cipher from a randomly drawn permutation (*distinguishing*).
- **In the white box:** The attacker has full access to the execution environment of the cipher. Given the whitebox implementation of the cipher, the adversary's goal is then to extract the secret key (*key extraction*) or to decompose the implementation to find a more compact representation that can be used as an effective key to replicate the functionality (*decomposition*, or *code lifting*).

## 1.2 Whitebox Cryptography in the Wild

The seminal papers [14,15] in whitebox cryptography had the goal to provide security in digital rights management (DRM) applications where encrypted contents (e.g. a music or movie file) are decrypted on the user's device. A malicious end user may attempt to extract the key from its software and then illegally distribute it outside the DRM system.

15 years have passed since those papers were published, and the context of whitebox cryptography has drastically changed. With the rapidly increasing demand for software-only security solutions in embedded devices, laptop PCs, mobile and server systems as well as the ever growing field of cloud-based services, the target for whitebox cryptography is no longer limited to the software

128    A. Bogdanov et al.

**Fig. 1.** Cloud-based content distribution: Cloud server encrypts contents in the black box and distributes them to user devices. User devices decrypt the contents in the white box.

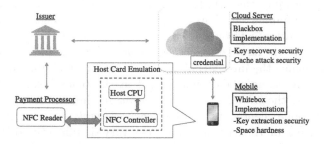

**Fig. 2.** Cloud-based mobile payments with HCE: Cloud server sends tokenized payment credentials provided by the issuer, to the mobile. Mobile phone transfers payment data with tokenized payment credentials to the payment processor via HCE. Payment processor sends it to the issuer for authorization.

implementation in the user-controlled device only. Such a device is now merely a part of a larger system, as e.g. in cloud computing or cloud-based payment. In addition, as whitebox cryptography inherently addresses resistance to malware, Trojans and zero-day vulnerabilities, it will find more and more applications in banking and other security-critical settings as well.

For illustrative purposes, we mention three application scenarios for whitebox cryptography, see also Figs. 1 and 2.

**DRM in the Cloud.** DRM-based services have moved to cloud-based contents distribution systems such as Adobe Primetime Cloud DRM [1] and Akamai's Secure Cloud-Based Workflows for Premium Content [2]. State-of-the-art contents distribution services often utilize IaaS (Infrastructure as a Service), for instance, Google cloud platform, IBM, Amazon AWS and Microsoft Azure, in order to optimize costs and to scale infrastructure. This application is illustrated in Fig. 1.

On the user device that plays the contents, whitebox implementation shall protect the contents key against key extraction and decomposition attacks [6, 28,39] and recent side-channel attacks [11,35]. A useful security property in this context is space hardness, which aims to mitigate code lifting, and discourages the adversary from illegally distributing the code due to its large size [9].

On the cloud server that distributes the contents, a blackbox implementation is used to deal with a large number of user keys simultaneously, since running whitebox implementations for all users would require a huge amount of memory. Though usually much better protected than the player devices, cloud computing infrastructures do pose additional threats to the application. Namely, they are based on co-residency and multi-tenancy, i.e. the user runs multiple virtual machines (VMs) in the hardware resources of the same physical machine. Therefore, VM isolation raises a new security concern: *cache timing attacks* which exploit the fact that cache memory access times are data dependent. This may allow one to extract the secret key, given shared cache across co-located VMs. With the rapidly increasing deployment of cloud services, cache timing attacks have lately received a lot of attention [18,22,23,34]. Thus, cloud service providers have to deal with countermeasures. Indeed, having seen the novel cache timing attacks of [23,41], VMware made memory deduplication an opt-in feature, and Amazon disabled deduplication on its EC2 cloud servers. However, Irazouqui et al. show that attacks exploiting the L3 shared cache are still applicable even if such system-level countermeasures are deployed [22]. Thus, this threat has to be addressed at the cipher implementation level as well.

All in all, for DRM applications in the cloud, the blackbox cipher implementation should be secure against cache timing attacks on the cloud server, whereas the whitebox implementation should provide key extraction security and space hardness on the consumer device.

**Host Card Emulation in Cloud-based Mobile Payments.** NFC (Near Field Communication) is extensively used in applications such as payment systems. A standard NFC payment implementation employs a mobile phone with credentials stored inside a hardware-based secure element. HCE (Host Card Emulation) is a technology that enables NFC transactions in a pure software environment without secure elements — here anyone can create a mobile application without depending on the secure element. This allows one to launch new payment services in a more flexible way with a much less complex ecosystem. Thus, HCE is expected to become a game changer for mobile payments. Google provides the HCE architecture from Android 4.4 Kitkat on, by which anyone can emulate an NFC smart card for a payment service. Moreover, Visa and MasterCard also support the cloud-based HCE payments. In the HCE, instead of expensive secure hardware, credentials are stored in alternative media such as cloud. Figure 2 provides an overview of cloud-based payment systems with HCE.

In cloud-based payments, resilient whitebox cryptography on the mobile phone is central to the overall security. More precisely [29,37], a whitebox implementation shall replace the secure element in two ways. First, it should protect sensitive data such as tokens, payment information and card data from malware and spyware possibly running on the same CPU. Second, it should ensure that legitimate devices and users are accessing their payment credentials in the cloud by means of secure authentication between the cloud and the device.

From the implementation viewpoint, a mobile phone may not have rich resources, and available memory can be restrictive. Thus, the deployed white-box cipher shall support variable sizes of its whitebox tables to meet a variety of implementation demands. In the cloud, which manages credentials, the corresponding blackbox implementation should prevent cross-VM cache timing attacks [18,22,23,34] similar to the previous application.

**Memory-Leakage Resilient Software.** Leakage of memory by vulnerabilities such as buffer overflows, cold boot attacks [20], bus monitoring attacks, Trojans and malware, or heartbleed-type vulnerabilities is a major problem in today's software. The notion of space hardness has been used to restrict the effect of memory leakage in applications where the leakage channel from the implementation environment to the adversary's backend is of limited capacity [9]. In particular, the use of space-hard whitebox cryptography can mitigate the damage of a memory-leakage vulnerability in security-critical systems. Indeed, those are typically insulated from the Internet, making it infeasible for Trojans to use low-capacity covert and side channels for the transmission of necessary key material if space-hard ciphers are employed.

Thus, for a memory-leakage resilient software implementation, the space hardness is necessary. It can be considered as a class of leakage resilient cryptography in bounded retrieval model where malware has complete control over the computer but can only send out a bounded amount of information.

## 1.3 Existing Whitebox Constructions

In order to meet some of the demands arising from applications, several whitebox constructions have been proposed.

**Whitebox Implementations of DES and AES.** Whitebox implementations of DES and AES were first proposed by Chow et al. in [14,15]. Their approach is to find a representation of the algorithm as a network of look-ups in randomized and key-dependent tables. In the wake of these seminal papers, several further variants of whitebox implementations for DES and AES were proposed [12,24, 27,40]. However, all published whitebox solutions for DES and AES to date have been *practically* broken by key extraction and table-decomposition attacks [6,26, 30,31,39].

**ASASA.** Dedicated whitebox block ciphers were proposed by Biryukov et al. in [7] at ASIACRYPT'14. They are based on the ASASA structure that consists of two secret nonlinear layers (S) and three secret affine layers (A), with affine and nonlinear layers interleaved. The security of ASASA against the key extraction in the whitebox setting relies on the hardness of the decomposition problem for ASASA. Unfortunately, efficient decomposition attacks on ASASA have been proposed [28]. The security of constructions based on multiple secret nonlinear and linear layers is still to be explored and seems hard to evaluate, despite several cryptanalytic efforts [8,10,38]. Moreover, generic ASASA-type constructions are difficult to implement in the constant-time fashion in the black box, which makes them potentially susceptible to side channel leakage.

**SPACE.** At CCS '15, Bogdanov and Isobe proposed a family of whitebox-secure block ciphers SPACE [9]. The design of SPACE is such that the security against key extraction and decomposition attacks in the whitebox setting reduces to the well-studied problem of key recovery for block ciphers in the standard blackbox setting. Their approach is to construct the whitebox table from a well-understood standard block cipher (AES in their example) by constraining the plaintext and truncating the ciphertext. Furthermore, to mitigate code lifting, they proposed the new security notion of *space hardness* which is a generalization of the weak whitebox security notion of [7]. Space hardness quantifies security against code lifting by the amount of code that needs to be extracted from an implementation by a whitebox adversary to maintain its functionality with a certain probability.

However, in order to strongly guarantee security against key extraction and space hardness in the whitebox setting, SPACE employs a very conservative design strategy. Namely, a target-heavy Feistel construction is deployed that does not allow for parallel or even pipelined implementations. Moreover, the internal F-function of SPACE requires one full 10-round AES-128 call. As estimated in [9], at least 128 full-round AES-128 calls are necessary to perform a single block encryption. That appears rather unacceptable in real-world applications. However, it's possible to derive a constant-time implementation of SPACE in the black box.

Thus, all existing designs have important practical limitations. This paper aims to bridge this gap by a novel design that addresses the key extraction security, the decomposition security (space hardness), constant-time blackbox implementation requirement as well as efficiency issues simultaneously.

## 1.4   Our Contributions

The contributions of this paper are as follows.

**Design of SPNbox: New Efficient Whitebox Block Cipher.** We propose SPNbox, a new family of space-hard block ciphers, which significantly improves upon the SPACE ciphers proposed at CCS 2015 [9]. While SPACE is based on a target-heavy Feistel construction, SPNbox is an SPN-type design with small block ciphers as the key-dependent S-boxes. In order to efficiently utilize the parallelism offered by both standard SIMD and the AES-NI instructions on contemporary microprocessors, the small block ciphers are based on the AES round transformation. The resulting parallelization opportunities allow for significantly faster implementations both in the black box and in the white box. At the same time, similarly to SPACE, SPNbox still offers all important whitebox security properties of quantifiable space hardness as well as reduction of key extraction security to the blackbox key-recovery security of the underlying block cipher. See Sect. 2.

**Security Analysis of SPNbox in the Black Box.** Our constructions come with security analysis as block ciphers. As the overall design as well as the design of underlying small block cipher follows the principles of substitution-permutation networks, we use the well-established tools of symmetric-key cryptanalysis. See Sect. 3. In addition, we stress that our ciphers are secure against new types of attacks such as differential computational and differential fault attacks [11, 35] in the white box as well as cross-VM cache timing attacks for cloud in the black box [18, 22, 23, 34].

**Refined Compression Attack Settings.** Resistance to decomposition attacks is formalized by the notions of weak whitebox security and incompressibility [7], $(M, Z)$-space hardness and strong $(M, Z)$-space hardness [9] as well as by a related notion of $(\lambda, \delta)$ compressibility [16]. As opposed to previous studies of space hardness [9] that did not go beyond a weak whitebox adversary, this paper considers various levels of space hardness for table-based whitebox implementations, which are classified with respect to the adversary's abilities such as types of table accesses, knowledge about the execution environment or reverse engineering capabilities. This covers a very wide class of real-world adversaries that are thinkable in applications. In particular, we introduce known-space, chosen-space and adaptively-chosen-space attacks on space hardness. See Sect. 4.

**Provable Bounds on Space Hardness.** Moreover, we obtain bounds on space hardness in all those adversarial models under the assumption that the underlying tables are secure against decomposition, which is in turn guaranteed by the security of the underlying small block ciphers in the standard blackbox setting. This enables us to obtain rigorous upper bounds on the success probability, given a space of size $M$, in each adversarial model. These are the *first* security bounds on space hardness for table-based whitebox implementations, while previous results only roughly evaluate the security by an attack-based approach [9]. Furthermore, we apply our bounds to SPNbox and SPACE ciphers. As a result, we update the evaluations of space hardnesses of SPACE ciphers, and show that SPNbox offers a conservative level of space hardness in each adversary model.

**Efficient Optimized Software Implementations of SPNbox and SPACE.** We implement both SPNbox and SPACE families of whitebox block ciphers on Intel Skylake and ARMv8. Our implementations use SIMD/AVX, AES-NI and NEON extensions whenever possible to optimize performance. As a result, we report that instances of SPNbox achieve speed-ups of up to 6.5 times in the black box and up to 18 times compared to SPACE in the whitebox setting. See Sect. 5.

## 2    SPNbox: Efficient Space-Hard Block Ciphers

### 2.1    Design Choices

*From Feistel to nested SPN.* The SPACE family of space-hard block ciphers employs a very conservative design strategy which involves using the full 10-round AES-128 transformation, even for 8-bit inputs. Furthermore, its Feistel structure prevents the exploitation of any parallel execution or pipelining possibilities. At the same time, it seems likely that the security margin offered by the proposed SPACE instances can be reduced without ill effects.

The requirement of parallelism immediately points to an SPN-type design. For the desired level of space hardness, key-dependent S-boxes of varying size can be employed. This can then be combined with a public linear MDS diffusion layer operating on the entire state, allowing rigorous security arguments for standard blackbox security.

Within this design framework, it remains to construct key-dependent S-boxes of different sizes (for instance 8, 16, 24 and 32 bits as in SPACE). This is accomplished by using smaller internal block ciphers, which are themselves SPNs, yielding a *nested SPN structure* [3]. For the reasons of efficiency, security and side-channel protection, it is desirable to base these internal SPNs on the AES round transformation, especially given the availability of the AES-NI instructions [19]. The efficiency requirements also dictate that little or no truncation should take place, and ideally, the AES round transformation should be used to compute some of the larger S-boxes in parallel.

In order to also have an efficient implementation for the inverse cipher, the design should employ involutory MDS matrices wherever possible. Since we mainly target high-performance software implementations, our selection criteria for efficient MDS matrices differs somewhat from the widely studied area of lightweight hardware implementations as in [36]: In software, arbitrary bit permutations are costly, which means that a matrix with smaller coefficients but higher theoretical XOR count can result in a more efficient SIMD implementation.

*Efficient Constant-Time Small Block Ciphers.* We note that these small SPN-type block ciphers used to construct the key-dependent S-boxes are of potential independent interest: Block ciphers of sizes smaller than 32 bit are virtually unstudied, and an AES-NI based implementation further allows an efficient constant-time implementation, which avoids the pitfalls of key-dependent table lookups (which is the usual way of implementing small nonlinear functions due to efficiency reasons though bit-sliced implementations may be possible as well). In addition, in order to prevent the differential computational attacks [11], this small SPN-type block cipher depends on 128 bits of key information.

## 2.2  Specification

We now define the SPNbox family of block ciphers and their concrete instantiations SPNbox-8, SPNbox-16, SPNbox-24, and SPNbox-32. SPNbox-$n_{in}$ is a substitution-permutation network (SPN) with a block length of $n$ bits, a $k$-bit secret key, and based on $n_{in}$-bit substitution boxes. For SPNbox-8, SPNbox-16 and SPNbox-32, the block length is $n = 128$ bits, whereas SPNbox-24 has $n = 120$. While SPNbox can support a wide range of key sizes, we use $k = 128$ for concreteness in the following.

*Representation of Finite Fields.* We will in the sequel sometimes view the set $\{0,1\}^m$ of bit strings as the finite field $\mathrm{GF}(2^m)$. For this, we identify $\mathrm{GF}(2^m)$ with the quotient ring $\mathrm{GF}(2)[\mathrm{x}]/(p)$ for a suitable irreducible polynomial $p \in$

GF(2)[x]. An $m$-bit string $a_{m-1}a_{m-2}\cdots a_1a_0 \in \{0,1\}^m$ then corresponds to the polynomial $a_{m-1}\mathsf{x}^{m-1} + a_{m-2}\mathsf{x}^{m-2} + \cdots + a_1\mathsf{x} + a_0 \in \mathrm{GF}(2^m)$. We write such an element in a hexadecimal representation of its bit string, e.g. $4_x$ for 100.

For $\mathrm{GF}(2^8)$, we use the same irreducible polynomial as the AES, namely $p(\mathsf{x}) = \mathsf{x}^8 + \mathsf{x}^4 + \mathsf{x}^3 + \mathsf{x} + 1$. Similarly, we use $p(\mathsf{x}) = \mathsf{x}^{16} + \mathsf{x}^5 + \mathsf{x}^3 + \mathsf{x} + 1$ for $\mathrm{GF}(2^{16})$, $p(\mathsf{x}) = \mathsf{x}^{24} + \mathsf{x}^4 + \mathsf{x}^3 + \mathsf{x} + 1$ for $\mathrm{GF}(2^{24})$ and $p(\mathsf{x}) = \mathsf{x}^{32} + \mathsf{x}^7 + \mathsf{x}^3 + \mathsf{x}^2 + 1$ for $\mathrm{GF}(2^{32})$, respectively.

**State.** The state of SPNbox-$n_{in}$ is organised as a vector of $t \stackrel{\text{def}}{=} n/n_{in}$ elements of $n_{in}$ bits each:

$$X = \{X_0, \ldots, X_{t-1}\}.$$

Each of the $n_{in}$-bit elements $X_i$ can in turn be represented by a vector of $\ell \stackrel{\text{def}}{=} n_{in}/8$ bytes: $X_i = \{X_{i,\ell-1}, \ldots, X_{i,0}\}$.

**Key Schedule.** The $k$-bit master key is expanded to $(R_{n_{in}} + 1)$ round keys $k_0, \ldots, k_{R_{n_{in}}}$ of $n_{in}$ bits using any generic key derivation function (KDF) [32]:

$$(k_0, \ldots, k_{R_{n_{in}}}) = \mathrm{KDF}(k, n_{in} \cdot (R_{n_{in}} + 1)).$$

For example, one can use the SHAKE extendable output function which is based on the SHA-3 hash [33].

**Round Transformation.** The encryption of a plaintext $X^0$ to a ciphertext $X^R$ is accomplished by applying $R$ rounds of the following round transformation to the plaintext:

$$X^R = \left( \bigcirc_{r=1}^{R} (\sigma^r \circ \theta \circ \gamma) \right) (X^0).$$

For all concrete proposals SPNbox-8, SPNbox-16, SPNbox-24 and SPNbox-32, we set the number of rounds to $R = 10$. We now define in turn each of the components $\gamma, \theta$ and $\sigma^r$. An overview of the round transformation is given in Fig. 3

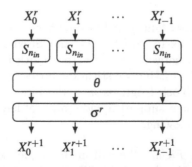

Fig. 3. Round transformation of SPNbox.

*The Nonlinear Layer* $\gamma$. $\gamma$ is a nonlinear substitution layer, in which $t$ key-dependent identical bijective $n_{in}$-bit S-boxes are applied to the state:

$$\gamma : \mathrm{GF}(2^{n_{in}})^t \to \mathrm{GF}(2^{n_{in}})^t$$
$$(X_0, \ldots, X_{t-1}) \mapsto (S_{n_{in}}(X_0), \ldots, S_{n_{in}}(X_{t-1})).$$

In SPNbox-$n_{in}$, the substitution $S_{n_{in}}$ is realised by an internal small block cipher of block length $n_{in}$, which will be defined in the next subsection.

*The Linear Layer* $\theta$. $\theta$ is a linear diffusion layer that applies a $t \times t$ MDS matrix to the state:

$$\theta : \mathrm{GF}(2^{n_{in}})^t \to \mathrm{GF}(2^{n_{in}})^t$$
$$(X_0, \ldots, X_{t-1}) \mapsto (X_0, \ldots, X_{t-1}) \cdot M_{n_{in}}.$$

We denote by $\mathrm{cir}\,(a_0, \ldots, a_{t-1})$ the $t \times t$ circulant matrix $A$ with the coefficients $a_0, \ldots, a_{t-1}$ in the first row; and by $\mathrm{had}\,(a_0, \ldots, a_{t-1})$ the $t \times t$ Hadamard matrix $A$ with coefficients $A_{i,j} = a_{i \oplus j}$, with $t$ a power of two.

For the concrete proposals SPNbox-$n_{in}$ with $n_{in} = 32, 24, 16, 8$, the matrix $M_{n_{in}}$ is then respectively defined as follows:

$$\begin{aligned}
M_{32} &= \mathrm{cir}\,(1_x, 2_x, 4_x, 6_x) & \text{for } n_{in} = 32, \\
M_{24} &= \mathrm{cir}\,(1_x, 2_x, 5_x, 3_x, 4_x) & \text{for } n_{in} = 24, \\
M_{16} &= \mathrm{had}\,(1_x, 3_x, 4_x, 5_x, 6_x, 8_x, \mathrm{b}_x, 7_x) & \text{for } n_{in} = 16,
\end{aligned}$$

and

$$M_8 = \mathrm{had}\,(08_x, 16_x, 8\mathrm{a}_x, 01_x, 70_x, 8\mathrm{d}_x, 24_x, 76_x,$$
$$\mathrm{a}8_x, 91_x, \mathrm{ad}_x, 48_x, 05_x, \mathrm{b}5_x, \mathrm{af}_x, \mathrm{f}8_x)$$
$$\text{for } n_{in} = 8.$$

Note that $M_{32}, M_{16}$ and $M_8$ are involutions. $M_{32}$ and $M_{16}$ are the matrices used in the block ciphers Anubis [4] and Khazad [5], respectively. $M_8$ is an optimised involutory Hadamard-Cauchy matrix proposed at FSE 2015 [36].

*The Affine Layer* $\sigma^r$. $\sigma^r$ is an affine layer that adds round-dependent constants to the state:

$$\sigma^r : \mathrm{GF}(2^{n_{in}})^t \to \mathrm{GF}(2^{n_{in}})^t$$
$$(X_0, \ldots, X_{t-1}) \mapsto \left(X_0 \oplus C_0^r, \ldots, X_{t-1} \oplus C_{t-1}^r\right),$$

with $C_i^r \overset{\text{def}}{=} (r-1) \cdot t + i + 1$ for $0 \leq i \leq t-1$.

**The Underlying Small Block Ciphers.** The key-dependent $n_{in}$-bit bijective S-boxes $S_{n_{in}}$ in the nonlinear layer $\gamma$ are small SPN-type block ciphers themselves. They are based on the round transformation of the AES and consist of $R_{n_{in}}$ rounds operating on a state $x = \{x_0, \ldots, x_{\ell-1}\}$ of $\ell \stackrel{\text{def}}{=} n_{in}/8$ bytes:

$$S_{n_{in}} : \mathrm{GF}(2^8)^\ell \to \mathrm{GF}(2^8)^\ell$$

$$x \mapsto \left( \bigcirc_{i=1}^{R_{n_{in}}} \left( \mathsf{AK}^i \circ \mathsf{MC}_{n_{in}} \circ \mathsf{SB} \right) \right) (\mathsf{AK}^0(x)).$$

Here, $\mathsf{SB}$ denotes the application of the AES S-box to each byte of the state. For $0 \leq i \leq R_{n_{in}}$, $\mathsf{AK}^i$ is defined as the addition of the round key $k_i$ (as expanded by the key schedule) by XOR. $\mathsf{MC}_{n_{in}}$ implements an MDS diffusion layer on all $\ell$ bytes of the state. It is based on the AES MixColumns operation. For the concrete proposals of $n_{in} = 32, 24, 16$, it is defined as the multiplication of $x$ with the matrices

$$A_{32} = \mathrm{cir}\left(2_x, 1_x, 1_x, 3_x\right) \qquad \text{for } n_{in} = 32,$$

$$A_{24} = \begin{pmatrix} 2_x & 1_x & 1_x \\ 3_x & 2_x & 1_x \\ 1_x & 3_x & 2_x \end{pmatrix} \qquad \text{for } n_{in} = 24,$$

$$A_{16} = \begin{pmatrix} 2_x & 1_x \\ 3_x & 2_x \end{pmatrix} \qquad \text{for } n_{in} = 16,$$

respectively. For $n_{in} = 8$, $\mathsf{MC}_{n_{in}}$ is the identity mapping. Note that $A_{32}$ is the AES MixColumns matrix (adjusted for Intel's byte order), while $A_{24}$ and $A_{16}$ are obtained from $A_{32}$ as $(x, y, z, 0) \times A_{32}$ and $(x, y, 0, 0) \times A_{32}$, respectively. As square submatrices of $A_{32}$, all derived matrices are also $\ell \times \ell$ MDS matrices over $\mathrm{GF}(2^8)$. An overview of the round transformation is given in Fig. 4.

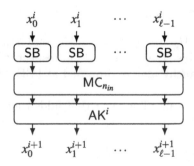

**Fig. 4.** Round transformation of the underlying block ciphers $S_{n_{in}}$.

The number of rounds for each concrete proposal is defined as $R_{32} = 16$, $R_{24} = 20$, $R_{16} = 32$ and $R_8 = 64$.

## 2.3  SPNbox vs ASASA

Both SPNbox and the ASASA construction are based on the classical substitution-permutation network structure, consisting however of secret key-dependent S-box and public linear layers. The main constructive difference is how to construct the secret key-dependent S-box. However, this is the discrepancy having far-reaching practical consequences both in terms of security arguments and implementation.

In the ASASA construction, as its name suggests, tables are based on the ASASA structure that consists of two secret nonlinear layers (S) and three secret affine layers (A), with affine and nonlinear layers interleaved. On the other hand, SPNbox is based on the SPN-type small block cipher that consists of the *public* nonlinear and linear layer, and secret key XOR layers.

Regarding the security of the whitebox implementation, the difficulty of the key extraction and the decomposition problem for ASASA relies on the hardness of the decomposition problem for ASASA, which is still to be explored and seems hard to evaluate, despite several cryptanalytic efforts [8, 10, 38]. Actually, efficient decomposition attacks on ASASA have been proposed [28]. On the other hand, SPNbox relies on well analyzed problem of the key recovery attack of the block cipher in the standard blackbox setting.

In the blackbox implementation, assuming the random choice of secret S-boxes, the substitution layer of ASASA is realized by the table based implementation due to the secrecy of underlying component, and is impossible to optimize the performance by AES-NI. The table-based blackbox implementation of the ASASA is not secure against cache timing attacks similar to the table-based blackbox AES implementation [18, 22, 23, 34].

# 3  Security in the Black Box: Analysis as a Block Cipher

We evaluate the general construction of SPNbox-8, -16, -24 and -32, modeling the underlying small block cipher as pseudorandom permutation. We furthermore analyze the security of the underlying small block ciphers $S_{n_{in}}$ against cryptanalytic attacks. Finally, we evaluate the security against cross-VM cache timing attacks for cloud application.

## 3.1  General Construction

First, we evaluate the security of the general construction of SPNbox-8, -16, -24 and -32, assuming an underlying small block cipher, i.e. the key-dependent $n_{in}$-bit bijective S-boxes $S_{n_{in}}$, is a pseudo random permutation. The generic construction of all variants is a 10-round SPN-type construction.

**Differential Cryptanalysis.** Here we analyze the differential properties of an $n_{in}$-bit permutation $S_{n_{in}} : \{0,1\}^{n_{in}} \rightarrow \{0,1\}^{n_{in}}$. Given input difference $a$ and output difference $b$, the differential probability of function $f$ is defined as

$$DP(a, b) = \#\{(v, u) | u \oplus v = a \text{ and } f(v) \oplus f(u) = b\}$$

for $u, v \in \{0, 1\}^{n_{in}}$. The bound of the maximum differential probability $MDP$ is proved as follows [21].

$$Pr\left(\frac{n \ln 2}{2^{n-1} \ln n} \leq MDP < \frac{n}{2^{n-1}}\right) \approx 1$$

Suppose that the maximum differential probability of $S_{n_{in}}$ of SPNbox-8, -16, -24 and -32 to be $2^{-4}$ ($= 8/2^7$), $2^{-11}$ ($= 16/2^{15}$), $2^{-18.42}$ ($= 24/2^{23}$) and $2^{-26}$ ($= 32/2^{31}$), respectively. Due to properties of MDS diffusion matrices. SPNbox-8, -16, -24 and -32 have at least 34, 18, 12 and 5 active $S_{n_{in}}$ after 4, 4, 4 and 2 rounds.

**Linear Cryptanalysis.** Now we analyze the linear properties of an $n_{in}$-bit permutation $S_{n_{in}} : \{0, 1\}^{n_{in}} \to \{0, 1\}^{n_{in}}$.

Given an input mask $\alpha$ and an output mask $\beta, \alpha, \beta \in \{0, 1\}^{n_{in}}$, the correlation of a linear approximation $(\alpha, \beta)$ for a function $f : \{0, 1\}^{n_{in}} \to \{0, 1\}^{n_{in}}$ is defined as

$$Cor = 2^{-n_{in}} [\#\{x \in \{0, 1\}^{n_{in}} | \alpha \cdot x \oplus \beta \cdot f(x) = 0\} -$$
$$\#\{x \in \{0, 1\}^{n_{in}} | \alpha \cdot x \oplus \beta \cdot f(x) = 1\}.$$

The linear probability $LP$ of $(\alpha, \beta)$ is defined as $Cor^2$. For a fixed-key block cipher, the maximum linear probability $MLP$ is normally distributed in mean $\approx (1.38 \cdot 2n - \ln(1.38 \cdot 2n) + 1) \cdot 2^{-n}$ and standard deviation $\approx 2.6 \times 2^{-n}$ [21].

Suppose that the maximum linear probability of $S_{n_{in}}$ of SPNbox-8, -16, -24 and -32 to be $2^{-3.67}$ ($= 19.99 \cdot 2^{-8}$), $2^{-10.62}$ ($= 41.37 \cdot 2^{-16}$), $2^{-18.02}$ ($= 63 \cdot 2^{-24}$) and $2^{-25.61}$ ($= 84 \cdot 2^{-32}$), respectively. SPNbox-8, -16, -24 and -32 have at least 51, 18, 12 and 5 active $F_i^{(j)}(x)$ after 6, 4, 4 and 2 rounds.

**Other Cryptanalysis.** Any input difference nonlinearly affects all states after one round due to the MDS matrix. Following the miss-in-the-middle approach, after 3 rounds, we have not found any useful impossible differentials for the respective variants. Also, a 2.5-round generic integral distinguisher against the SPN-type construction is proposed [8]. We also consider other-types of attacks including a higher order differential, a truncated differential, a slide, and an algebraic attack. Consequently, we expect that none of them work better than brute force attacks.

## 3.2   The Underlying Small Block Ciphers

We evaluate the security of underlying small block ciphers $S_{n_{in}}$. These are based on well-analyzed AES components such as the inversion base 8-bit S-box and the MDS circulant matrix on $GF(2^8)$.

**Differential/Linear Cryptanalysis.** The differential/linear probability of 8-bit S-box is $2^{-6}$. $S_8$, $S_{16}$, $S_{24}$, and $S_{32}$ have at least 2, 3, 4 and 10 differentially/linearly active S-boxes after 2, 2, 2 and 4 rounds, respectively. We therefore expect all $S_{n_{in}}$, for $n_{in} = 8, 16, 24, 32$, to not have any differential or linear trails with probabilities exceeding the bound $2^{-n_{in}}$ after 2, 2, 2 and 4 rounds, respectively. Since they are proposed with much higher numbers of rounds, they offer ample security margin.

**Meet-in-the-Middle and Other Cryptanalysis.** In each cipher, four times 128-bit key information is involved, and one round already achieves full diffusion. Thus, we believe that the small block ciphers are secure against MitM attacks. We developed MitM attacks on each variant using splice and cut, biclique and partial matching techniques. However, we did not find full round attacks.

Considering further attacks, the byte-oriented structure combined with full diffusion after 1 round means that for impossible (truncated) differential attacks, and integral and higher order differential attacks, we can at most construct cryptanalytic properties spanning 3 and 4 rounds, respectively. All small block ciphers are proposed with much significantly higher numbers of rounds. Finally, the use of distinct round constants in the key schedule precludes slide attacks.

### 3.3   Cache Timing Attack

There are several techniques exploiting cache information over VM isolations in the cloud: the Prime+Probe attack [22] and Flush+Reload attacks [18,23,41]. All attacks make use of timing differences between cache hits and misses. Our key-dependent small block ciphers are designed to be executed in constant time by using AES-NI, and there are no cache accesses during key-dependent operations. Thus, it is impossible to mount cache timing attacks against the blackbox implementation of SPNbox.

## 4   Security in the White Box: Analysis of Space Hardness

In this section, we first evaluate the security against key-extraction and decomposition attacks in the whitebox model. Second, we evaluate the difficulty of code lifting attacks by notions of *weak and strong space hardness* [9]. We generalize the adversarial models of space hardness to capture a wide class of adversaries: from adversaries with limited control (greybox) to stronger ones with more knowledge of the computational platform and reverse engineering abilities (whitebox). Then, we show bounds for weak and strong space hardness for table-based whitebox cryptography under the assumption that *tables are secure against key extraction and table decomposition attacks*, i.e. it is computationally infeasible to compress the tables in the whitebox models[1]. By contrast, the

---

[1] Whitebox AES implementations [12,14,24,40] and the ASASA construction [7] do not satisfy the assumption due to practical decomposition attacks [6,26,28,30,31].

authors of [9] evaluate the space hardness of their proposals only by attack-based approaches, called compression attack. Finally, we evaluate the security against recent advanced side-channel attacks [11, 35].

## 4.1   Key Extraction and Table Decomposition Attacks

As the tables are constructed from small block ciphers, the security of key-extraction and decomposition attacks in the whitebox model reduces to the key recovery problem for these small block ciphers in the blackbox model (which is evaluated in Sect. 3). The advantage of key extraction in the whitebox model for SPNbox, $\mathrm{Adv_{KE\text{-}WB}}$, is upper-bounded by the advantage of the key recovery for the underlying block cipher in the blackbox model, $\mathrm{Adv_{KR\text{-}BB}}$: $\mathrm{Adv_{KE\text{-}WB}} \leq \mathrm{Adv_{KR\text{-}BB}}$.

## 4.2   Existing Notions of Space Hardness

The difficulty of a decomposition attack is measured by space hardness that is summarized here. The whitebox implementation of a cipher should resists decomposition: Instead of a secret key, the adversary can directly use the implementation itself as a larger effective key. In particular, he can isolate the program code where the key is embedded in order to copy the functionality of encryption/decryption routines and to utilize it in a stand-alone manner. We refer to decomposition attacks as *code lifting attack*. If a code lifting attack succeeds, the adversary gets the advantage which is almost the same as key extraction, i.e. he can encrypt/decrypt any plaintext/ciphertext.

To formalize the difficulty of code lifting, the notions of weak white-box security and incompressibility have been introduced in [7]. To capture the resistance towards compression attacks in a more fine-grained fashion, two further security notions were introduced in [9]: $(M, Z)$-*space hardness* and *strong* $(M, Z)$-*space hardness*. Space hardness measures the difficulty of compressing the whitebox implementation of a cipher, and quantifies security against code lifting by the amount of code that needs to be extracted from the implementation by a white-box adversary to maintain its functionality. Moreover, Delerablee et al. propose a related notion of $(\lambda, \delta)$ compressibility [16]. However, the latter aims to evaluate the difficulty of code compression, given the full code. Space hardness [9] assesses the difficulty of isolating code from execution environments, namely, code lifting, by the amount of the data. Thus, it covers a wide class of adversaries: from the one with limited control all the way to the stronger ones with full code and complete access to the environments. For the sake of clarity, the paper at hand refers to $(M, Z)$-space hardness of [9] as *weak* $(M, Z)$-*space hardness*:

**Definition 1 (Weak** $(M, Z)$**-space hardness** [9]**).**   *An implementation of a block cipher* $E_K$ *is weakly* $(M, Z)$-*space hard if it is infeasible to encrypt (decrypt) any randomly drawn plaintext (ciphertext) with probability of more than* $2^{-Z}$ *given any code (table) of size less than* $M$ *bits.*

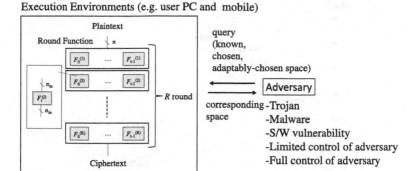

**Fig. 5.** Target block cipher construction for the white box and its adversarial models

Weak $(M, Z)$-space hardness estimates the code (table) size $M$ that needs to be isolated from the whitebox environment to be able to encrypt (decrypt) any plaintext (ciphertext) with a success probability larger than $2^{-Z}$.

**Definition 2 (Strong $(M, Z)$-space hardness [9]).** *An implementation of a block cipher $E_K$ is strongly $(M, Z)$-space hard if it is infeasible to obtain a valid plaintext and ciphertext pair with probability higher than $2^{-Z}$ given the code (table) of size less than $M$ bits.*

Strong $(M, Z)$-space hardness assumes an adversary who tries to find any valid input/output pair. It is relevant to message authentication codes in the context of forgeries.

### 4.3 Target Construction

To simplify our evaluation of space hardness in the sequel, we define a target construction: an $n$-bit block cipher that is encrypted/decrypted by *key dependent table-based* implementations in the whitebox environment as shown in Fig. 5. Let the input and output sizes of each table be $n_{in}$ and $n_{out}$, respectively, and the number of rounds be $R$, where the each round consists of $t$ tables. We denote $j$-th table in round $r$ as a function $F_j^{(r)} : \{0,1\}^{n_{in}} \to \{0,1\}^{n_{out}}$ for $j \in \{0,1,\ldots,t-1\}$ and $r \in \{1,2,\ldots,R\}$. In the cases of SPNbox and SPACE [9], all tables are identical, and the total table sizes $T$ is estimated as $T = (2^{n_{in}} \times n_{out})$.

### 4.4 Adversary Models of Space Hardness

We consider three adversary models that are classified with respect to the adversary's ability, while previous works [9] do not specify the adversary model. In particular, we simulate the action of the adversary against the execution environments by access to the table (space) functions $F_j^{(r)}$ (see Fig. 5).

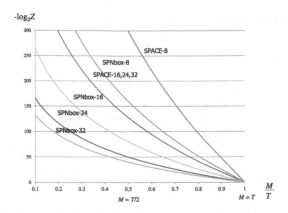

**Fig. 6.** Weak $(M, Z)$-space hardness of SPACE-8, 16, 24 and 32 and SPNbox-8, 16, 24 and 32 in known- and chosen-space attack

**Definition 3 (Known-Space (KS) Attack).** *The adversary obtains $q$ pairs of inputs and the corresponding outputs of tables $(x_i, F_j^{(r)}(x_i))$, $i \in \{0, 1, \ldots, q-1\}$, $j \in \{0, 1, \ldots, t-1\}$ and $r \in \{1, 2, \ldots, R\}$.*

**Definition 4 (Chosen-Space (CS) Attack).** *The adversary obtains $q$ pairs of inputs and the corresponding outputs of tables $(x_i, F_j^{(r)}(x_i))$ for a series of a priori chosen inputs $x_i$, $i \in \{0, 1, \ldots, q - 1\}$, $j \in \{0, 1, \ldots, t - 1\}$ and $r \in \{1, 2, \ldots, R\}$.*

**Definition 5 (Adaptively-Chosen-Space (ACS) Attack).** *The adversary obtains $q$ pairs of inputs and the corresponding outputs of tables $(x_i, F_j^{(r)}(x_i))$ for a series of adaptively chosen inputs $x_i$, $i \in \{0, 1, \ldots, q - 1\}$, $j \in \{0, 1, \ldots, t - 1\}$ and $r \in \{1, 2, \ldots, R\}$, namely he can choose $x_a$ after obtaining $(x_{a-1}, F_j^{(r)}(x_{a-1}))$.*

The known-space attack models the limited control of the adversary over the platform, where the adversary passively gets a part of space from the environments, e.g. with the aid of a trojan, malware, or a memory-leakage software vulnerability. The model is applicable to memory-leakage resilient cryptography where malware has complete control over the computer but can only send out a bounded amount of information [17].

The chosen-space attack captures the stronger adversary who has the ability to isolate any part of tables (space) with the knowledge of the memory layout, but the amount of data and the timing of access to the implementation are restricted due to the limited capacity of the communication channel and access controlled environments.

Finally, the adaptively-chosen-space attack assumes an adversary who has full access to the execution environment at *any time* by decompiler and debugger tools, e.g. IDA Pro and IL DASM, which is corresponding to the original whitebox adversary defined in [14] and the assumption of $(\lambda, \delta)$ compressibility [16].

Previous weak and strong $(M, Z)$-space hardness are evaluated by compression attacks [9]. The assumption of these attacks is classified as the known-table attack, i.e. weak KS-$(M, Z)$-space hardness and KS-$(M, Z)$-space hardness, respectively. Thus, previous evaluation of space hardness in [9] can capture only the weaker adversary than the standard whitebox adversary who has full access to the execution environment.

## 4.5   Weak Space Hardness

We show bounds for the weak $(M, Z)$-space hardness of the target construction in known-, chosen- and adaptively-chosen space attacks. Our evaluation assumes that the table decomposition is computationally infeasible as evaluated in Sect. 4.1, and input values of each table in the cipher are uniformly distributed, which is a reasonable assumption for block ciphers. The evaluation of the weak space hardness in the case where the adversary has a partial knowledge of a plaintext is provided in SubSect. 4.6.

**Known-Space Attack.** First, we introduce the following lemma.

**Lemma 1 (Inequality of Arithmetic and Geometric Means).** *For arbitary $n$ positive positive numbers $x_0, x_1, \ldots, x_{n-1}$, the inequality*

$$\sqrt[n]{x_0 \cdot x_1 \cdots x_{n-1}} \le \frac{x_0 + x_1 + \ldots, x_{n-1}}{n}$$

*holds, with equality if and only if $x_0 = x_1, \ldots, = x_{n-1}$.*

There are various proofs in the literature, and for example we refer to [13]. For known-space attacks, we have the following theorem:

**Theorem 1.** *Given known space of size $M$, the probability that a randomly-drawn plaintext can be computed is upper bounded by $(M/T)^{tR}$.*

*Proof.* Let the number of known entries of each table $F_j^{(r)}$ be $\#F_j^{(r)}$ for $j \in \{0, 1, \ldots, t-1\}$ and $r \in \{1, 2, \ldots, r\}$. The probability that an input of a tables $F_j^{(r)}$ matches with known ones is estimated as $(\#F_j^{(r)}/2^{n_{in}})$. Hence, a randomly-drawn plaintext can be computed with the probability of

$$\prod_{j=0}^{t-1} \prod_{r=1}^{R} \frac{\#F_j^{(r)}}{2^{n_{in}}} = \left(\frac{1}{2^{n_{in}}}\right)^{tR} \prod_{j=0}^{t-1} \prod_{r=1}^{R} \#F_j^{(r)}.$$

Here the sum of the numbers of known inputs is expressed as $\sum_{j=0}^{t-1} \sum_{r=1}^{R} \#F_j^{(r)}$. According to Lemma 1, we have

$$\prod_{j=0}^{t-1} \prod_{r=1}^{R} (\#F_j^{(r)}) \le \left(\frac{\sum_{j=0}^{t-1} \sum_{r=1}^{Rt} \#F_j^{(r)}}{tR}\right)^{tR}.$$

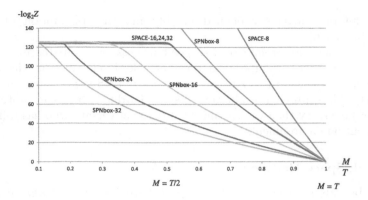

**Fig. 7.** Weak $(M, Z)$-space hardness of SPACE-8, 16, 24 and 32 and SPNbox-8, 16, 24 and 32 in adaptively-chosen attack.

Only if $\#F_0^{(1)} = \#F_1^{(1)} = \ldots = \#F_{t-1}^{(R)}$, the equation holds. Here, $M$ is estimated as $M = (\sum_{j=0}^{t-1} \sum_{r=1}^{R} \#F_j^{(r)} \cdot n_{out})/tR$ bits.

Thus, we have

$$\left(\frac{1}{2^{n_{in}}}\right)^{tR} \prod_{j=0}^{t-1} \prod_{r=1}^{R} \#F_j^{(r)} \leq \left(\frac{1}{2^{n_{in}}}\right)^{tR} \left(\frac{\sum_{j=0}^{t-1} \sum_{r=1}^{Rt} \#F_j^{(r)}}{tR}\right)^{tR}$$

$$\leq \left(\frac{1}{2^{n_{in}}}\right)^{tR} \left(\frac{M \cdot 2^{n_{in}}}{T}\right)^{tR} \square$$

From Theorem 1, we obtain weak KS-$(M, -\log_2((M/T)^{tR}))$-space hardness, i.e. given any known space of size $M$, it is infeasible to encrypt a randomly-drawn plaintext with the probability larger than $(M/T)^{tR}$. Figure 6 shows the relation between $M$ and $Z$ in terms of weak KS-$(M, Z)$ space hardness of SPACE-8, 16, 24 and 32 and SPNbox-8, 16, 24 and 32. For example, in SPNbox-16, given space of size $M = T/4$, the success probability is upper bounded by $2^{-160}$ $(= (2^{-2})^{8 \cdot 10})$ (Fig. 6).

**Chosen-Space Attack.** Due to the randomly-drawn plaintext, inputs of tables are unpredictable in advance even in the chosen-table attack. Thus, the chosen-space attack has no advantage over the known-space attack. We obtain weak CS-$(M, -\log_2((M/T)^{tR}))$-space hardness from Theorem 1.

**Adaptive-Space Attack.** The adversary is able to encrypt any plaintext by adaptively accessing the tables and computing round functions one by one. Thus he can prepare a set of pairs of plaintexts and the corresponding ciphertexts before a target plaintext is given. If the target plaintext is included in the set of prepared pairs, the corresponding ciphertext is obtained with the probability one.

Let us estimate how large space is necessary to compute $N$ plaintexts in advance. In the encryptions of $N$ plaintexts, it requires $N \cdot t \cdot R$ table accesses, and each table function $F_j^{(r)}$ has $N$ accesses. We provide the following Lemma.

**Lemma 2.** *For $q$ table accesses, the expected value of the number of used entries in the table is estimated as $(1 - ((2^{n_{in}} - 1)/2^{n_{in}})^q) \cdot 2^{n_{in}}$.*

*Proof.* An $i$-th entry of the table is used during $q$ table accesses with the probability of $(1 - ((2^{n_{in}} - 1)/2^{n_{in}})^q)$. There are $2^{n_{in}}$ entries in the table. □

Here, we define $(2^{n_{in}} - 1)/2^{n_{in}}$ as $e_{in}$. Using Lemma 2, we obtain Theorem 2.

**Theorem 2.** *Given adaptively-chosen space of size $M$, the probability that a randomly-drawn plaintext can be computed is upper bounded by $N \cdot 2^{-128} + (1 - N \cdot 2^{-128})(M/T)^{tR}$, where $N = \lceil log_{e_{in}}(1 - M/T)/tR \rceil$.*

*Proof.* According to Lemma 2, in order to compute $N$ pairs of plaintexts and the corresponding ciphertexts, it requires $(1 - (e_{in})^{N \cdot tR}) \cdot 2^{n_{in}} \cdot n_{out} = (1 - (e_{in})^{N \cdot tR}) \cdot T$-bit space. In the other words, adaptively-chosen space of size $M$ enables to compute $N(= \lceil log_{e_{in}}(1 - M/T)/tR \rceil)$ pairs of plaintexts and the corresponding ciphertexts. Then, the randomly-drawn plaintext is included in a set of the prepared pairs with probability of $2^{-128+N}$. Otherwise, given space of size $M$, the probability that the randomly-drawn plaintext can be computed is upper-bounded by $(M/T)^{tR}$ from Theorem 1. □

From Theorem 2, we obtain weak ACS-$(M, -\log_2(N \cdot 2^{-128} + (1 - N \cdot 2^{-128})(M/T)^{tR})$-space hardness. For example, in SPNbox-16, given $M = 0.46 \cdot T$ space, the success probability is upper bounded by $2^{-88.4} (= 9 \cdot 2^{-128} + (1 - 9 \cdot 2^{-128}) \cdot (0.465)^{8 \cdot 10})$ (Fig. 8).

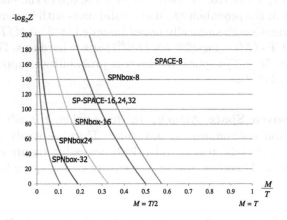

**Fig. 8.** Strong KS-$(M, Z)$-space hardness of SPACE-8, 16, 24 and 32 and SPNbox-8, 16, 24 and 32 in known/chosen space attacks

## 4.6    On (Partial) Target Plaintext for Weak Space Hardness

So far we assume that a plaintext is randomly drawn. However, the adversary might have the (partial) knowledge of a plaintext, e.g. the header of a file and the format-fixed encryption cases.

Let us estimate the security when the adversary has $z$-bit ($z \leq n$) information about the given plaintext in advance. In the known-space attack, since the adversary is not able to choose the entries of tables, the advantage in this setting is same as that in the randomly-drawn plaintext setting. In the chosen-space setting, the adversary is able to know inputs of some tables in advance. If the inputs of tables in first $y$ rounds is known, it is weak CS-$(X, -\log_2((M/T)^{t(R-y)}))$-space hardness, where $y$ depends on the $z$ and constructions. In the adaptively-chosen space setting, since the plaintext space is reduced to $2^{128-z}$, we have ACS-$(M, -\log_2(N \cdot 2^{-128+z} + (1 - N \cdot 2^{-128+z})(M/T)^{tR})$-space hardness.

## 4.7    Strong Space Hardness

Next, we show bounds for the strong $(M, Z)$-space hardness in known-, chosen- and adaptively-chosen space attacks.

**Known- and Chosen-Space Attack.** To begin with, we give the following lemma.

**Lemma 3.** *Given any space of size $M$, the expected number of the computable pairs is $2^n \cdot (M/T)^{tR}$.*

*Proof.* According to Theorem 1, given space of size $M$, a randomly-drawn plaintext can be computed with the probability $(M/T)^{tR}$ or less. It holds in any set of known/chosen-space of size $M$. Here, the entire space of the plaintext is $2^n$. □

From Lemma 3, the probability to find a valid pairs with known/chosen space of size $M$ is information-theoretically upper bounded by $2^n \cdot (M/T)^{tR}$. We prove strong KP- and CP- $(M, -\log_2(2^n \cdot (M/T)^{tR}))$-space hardness. For example, in SPNbox-16, given $M = T/4$ space, the success probability is upper bounded by $2^{-32}$ $(= 2^{128} \cdot (1/4)^{8 \cdot 10})$.

**Adaptively-Chosen Space Attack.** In this setting, the adversary has full access to execution environment at any time. Thus, he easily obtain a valid pair of plaintext and ciphertext by adaptively accessing inputs and outputs of each table $tR$ times. Therefore, we can not ensure strong space hardness in this setting.

## 4.8    Tradeoffs Between Strong Space Hardness and Time Complexity

In the previous subsection, we have obtained the upper bound of the probability to find a valid pair of plaintexts and ciphertexts given known- and chosen-space

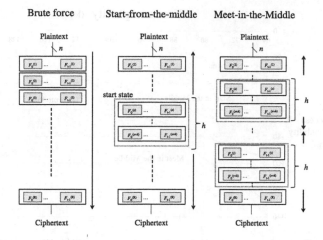

**Fig. 9.** Three types of attacks for strong space hardness

of size $M$. Here, we try to figure out how much time complexity is necessary to find the pair and reveal the tradeoff between the success probability and time complexity. In the multi-table setting, we assume $\#F_0^{(1)} = \#F_1^{(1)} = \ldots = \#F_{t-1}^{(R)}$ which is the optimal case with respect to the success probability. We consider following three types of attacks as shown in Fig. 9.

**Brute Force Attack.** The adversary simply tries to encrypt $2^b$ plaintexts with the given space of size $M$. The time complexity is estimated as $2^b$ for $b \leq n$ and the success probability is $2^b \cdot (M/T)^{tR}$. If $b = n$, the probability becomes the upper bounded value of Lemma 3.

**Start-from-the-Middle Attack.** Assume that if input values of all tables in consecutive $h$ rounds are chosen, then the $n$-bit internal state are determined. We call such states in $r$ rounds start states. We prepare a start state, and then check whether a pair of the plaintext and the ciphertext is computed from the start state through the remaining $(R - h)$ rounds with the given space of size $M$. The number of possible state states is estimated $(\#F)^{th} = 2^n \cdot (\#F/2^{n_{in}})^{th} = 2^n \cdot (M/T)^{th}$. The time complexity is estimated as $2^b$ ($\leq 2^n \cdot (M/T)^{th}$) and the success probability is $2^b \cdot (M/T)^{t(R-h)}$.

**Meet-in-the-Middle Attack.** We start with two start states in the different locations, and mount the meet-in-the-middle approach. In particular, we check whether two states match in the middle rounds, and a pair of the plaintext and the ciphertext is computed from the start states through the $(R-2h)$ rounds. The number of possible start states is estimated $2 \cdot (\#F)^{th} = 2^{(n+1)} \cdot (\#F/2^{n_{in}})^{th} = 2^{n+1} \cdot (M/T)^{th}$. The time complexity is estimated as $2^b$ ($\leq 2^n \cdot (M/T)^{th}$) and the success probability is $2^{2b-n} \cdot (M/T)^{t(R-2h)}$.

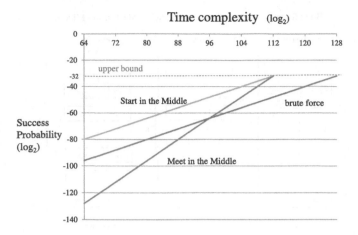

**Fig. 10.** Tradeoffs between time complexity and strong $(T/4, Z)$-space hardness of SPNbox-16 in known/chosen space attacks

**Table 1.** Summary of bounds for weak/strong space hardness against known-, chosen- and adaptively-chosen space attacks

| Known/Chosen-space attack | |
| --- | --- |
| Weak space hardness | $(M, -\log_2((M/T)^{tR}))$ |
| Strong space hardness | $(M, -\log_2(2^n \cdot (M/T)^{tR}))$ |
| Adaptively-chosen space attack | |
| Weak space hardness | $(M, -\log_2(2^{-128+N} + (M/T)^{tR})$ |
| $N = \lceil log_{e_{in}}(1 - M/T)/tR \rceil$, where $e_{in} = (2^{n_{in}} - 1)/2^{n_{in}}$ | |

**Evaluation.** Fig. 10 shows the trade off between time complexity and strong KS-$(T/4, Z)$-space hardness of SPNbox-16. As mentioned in Sect. 4.7, given $T/4$ space, the success probability is upper bounded by $2^{-32}$ $(= 2^{128} \cdot (1/4)^{8 \cdot 10})$. In our evaluations, in order to achieve it, it requires at most time complexity of $2^{112} (= 2^{128} \cdot (1/4)^{8 \cdot 1})$ by the meet-in-the-middle approach and the start-from-the-middle approach. If adversary's time complexity is restricted, the success probability decreases depending on the time complexity. If time complexity is $2^{80}$ or $2^{64}$, the probability is estimated as $2^{-64}$ $(=2^{80} \cdot 2^{-144})$ or $2^{-80}$ $(=2^{64} \cdot 2^{-144})$ by the start-from-the-middle attack.

### 4.9 Summary of Space Hardness

Table 1 provides a summary of weak and strong space hardness of the target construction in known-, chosen- and adaptively-chosen space attacks. It shows the upper bounds of the success probability against each attack, given space of size $M$; or, in other words, lower bounds for the required space with respect to the success probability of $2^{-Z}$.

Table 2 shows the lower bounds of the required space with respect to success probabilities of $2^{-64}$ and $2^{-128}$ of SPACE-8,-16,-24 and -32 and SPNbox-8,-16,-24 and -32. These results update the evaluations of SPACE-8,-16,-24 and -32 as weak KP-$(T/2^{0.44}, 128)$, $(T/2, 128)$, $(T/2, 128)$ and $(T/2, 128)$-space hardness, while previous results claim weak KP-$(T/4, 128)$-space hardness [9]. All variants SPNbox-8,-16,-24 and -32 achieve weak $(T/4, 64)$-space hardness in known, chosen and adaptively-chosen space attacks, which is a reasonable security level for practical applications. Also, all variants achieve strong $(T/2.3, 64)$ to $(T/32, 64)$-space hardness in known/chosen space attacks.

## 4.10  Advanced Side Channel Attacks

**Differential Computation Analysis.** Bos et al. proposed a new class of side channel attacks called differential computation analysis [11]. This attack exploits memory access patterns during the software execution of whitebox AES [15, 24,40] with the aid of a binary instrumentation framework such as PIN and Valgrind. Since the software execution traces contain time demarcated physical addresses of memory locations being read/written into, they essentially leak the values of the inputs to the various tables accessed, and can be used as side-channel information to extract the key.

This attack basically utilizes the fact that each table depends on only a fraction of the key, e.g. 8 and 16 bits of key [15,24,40]. A small part of the key is efficiently extracted using side-channel leakages. On the other hand, any table of SPNbox contains full 128-bit key information. Thus, even if the adversary can fully monitor the memory access patterns for the target key-dependent table, there are $2^{128}$ possible candidates of corresponding memory access patterns for

**Table 2.** Comparison of SPACE, SPNbox: Lower bounds of the required space with respected to the success probability $2^{-64}$ and $2^{-128}$

| cipher | $T$ | Weak Space hardness | | | | Strong Space hardness | |
|---|---|---|---|---|---|---|---|
| | | $Z = 64$ | | $Z = 128$ | | $Z = 64$ | $Z = 128$ |
| | | KS/CS | ACS | KS/CS | ACS | KS/CS | KS/CS |
| SPACE-8 [9] | 3.84 KB | $T/2^{0.22}$ | $T/2^{0.22}$ | $T/2^{0.44}$ | $T/2^{0.44}$ | $T/2^{0.64}$ | $T/2^{0.86}$ |
| SPACE-16 [9] | 918 KB | $T/2^{0.5}$ | $T/2^{0.5}$ | $T/2$ | - | $T/2^{1.5}$ | $T/2^2$ |
| SPACE-24 [9] | 218 MB | $T/2^{0.5}$ | $T/2^{0.5}$ | $T/2$ | - | $T/2^{1.5}$ | $T/2^2$ |
| SPACE-32 [9] | 51.5 GB | $T/2^{0.5}$ | $T/2^{0.5}$ | $T/2$ | - | $T/2^{1.5}$ | $T/2^2$ |
| SPNbox-8 | 256 B | $T/2^{0.40}$ | $T/2^{0.40}$ | $T/2^{0.80}$ | $T/2^{0.80}$ | $T/2^{1.20}$ | $T/2^{1.60}$ |
| SPNbox-16 | 132 KB | $T/2^{0.81}$ | $T/2^{0.81}$ | $T/2^{1.61}$ | - | $T/2^{2.40}$ | $T/2^{3.20}$ |
| SPNbox-24 | 50.3 MB | $T/2^{1.28}$ | $T/2^{1.28}$ | $T/2^{2.57}$ | - | $T/2^{3.68}$ | $T/2^{4.96}$ |
| SPNbox-32 | 17.2 GB | $T/2^{1.60}$ | $T/2^{1.60}$ | $T/2^{3.20}$ | - | $T/2^{4.80}$ | $T/2^{6.40}$ |

KS: Known-space attack
CP : Chosen-space attack
ACS: Adaptive-chosen-space Attack

each key value. Therefore, a differential computational attack on SPNbox is computationally infeasible.

**Differential Fault Attacks.** Sanfelix et al. propose a differential fault attack on whitebox AES and DES [35]. This attack modifies the specific byte position of internal states by injecting a fault. In the case of the AES, the fault injection targets the MixColumn operation in the 9-th round.

The tables of SPNbox compose of small block ciphers, and the internals of the small block ciphers are inaccessible in whitebox setting. Thus, any fault injection attack reduces to a differential attack on a small block cipher in the blackbox setting. Since the underlying cipher is secure against a differential attack in the blackbox setting as estimated in Sect. 3.2, SPNbox is secure against differential fault attacks.

# 5   Efficient Software Implementations

## 5.1   Setting

In this section, we discuss implementation characteristics of the SPNbox family of block ciphers. We also present experimental measurements based on our optimised high-performance software implementations and compare them to equivalent instances of the SPACE family of whitebox ciphers proposed at CCS 2015 [9]. Altogether, this provides a comprehensive implementation study of all proposed variants both in the blackbox and the whitebox setting. As target platforms for the server-side, we chose the recent Skylake generation of Intel microprocessors which support the AES-NI instruction set [19] and SSE instructions up to AVX2. As a mobile platform, we use the ARMv8 (AArch64) microarchitecture with NEON instructions.

For the blackbox implementations, we specifically focus on constant-time implementations without key-dependent table lookups on recent Intel platforms. Whenever possible, we realise the small block ciphers with AES-NI instructions.

For the whitebox implementations, both on Intel and ARM, the small block ciphers are implemented as table lookups, while the linear mixing of the table lookups is implemented using AVX2 (Intel) and NEON (ARMv8) instructions.

## 5.2   Implementation Characteristics of SPNbox

The SPNbox ciphers can efficiently utilize the parallelism offered by both standard SIMD and the AES instructions on contemporary microprocessors. With block sizes of $n = 128$ or $n = 120$ bit, one block fits naturally in the 128/256-bit SSE/AVX registers on Intel, or the 128-bit NEON registers on ARMv8. Additionally, the parallel and independent application of the S-boxes $S_{n_{in}}$, realised by the small internal block ciphers, offers opportunities for exploiting parallelism, both inside one block and across blocks of a longer message.

**In the Black Box.** In the blackbox setting on Intel platforms, the small block ciphers are implemented in a round-based fashion using the AES-NI instructions for the individual transformations. The composition of $MC_{n_{in}} \circ SB$ can be realised by first using the `pshufb` instruction reordering the bytes of the state equivalent to inverse ShiftRows, followed by an `aesenc` instruction for one full AES round. For $n_{in} = 32$, this is already sufficient. For $n_{in} = 24, 16$, we note that by construction, the matrices $A_{24}$ and $A_{16}$ are submatrices of $A_{32}$ such that their multiplication with the state corresponds to $(x, y, z, 0) \times A_{32}$ and $(x, y, 0, 0) \times A_{32}$, respectively (the last 8 resp. 16 bits are ignored). We can therefore realize the round function of the small block ciphers by XOR-ing the values $(0, 0, 0, 52_x)$ or $(0, 0, 52_x, 52_x)$ before applying inverse ShiftRows and the AES round, with $52_x$ being the inverse of 0 through the AES S-box. This allows the efficient re-use of Intel's AES-NI instructions also for smaller block sizes. For $n_{in} = 8$, the linear mixing step is the identity mapping, so can be omitted.

For the implementation of the linear layer $\theta$ in the outer rounds, it is beneficial to re-organise the internal state such that the $i$-th S-boxes of multiple message blocks are collected in one 128-bit register. This allows an efficient parallel execution of the finite field arithmetic, which vastly outweighs the overhead imposed by the input and output conversion to and from this format.

Additionally, on the Skylake platform, the AES round function has a latency of 4 cycles with a throughput of 1. Altogether, this implies that in order to both fully utilize the AES-NI instruction pipeline and fill the SSE/AVX registers for SIMD operations, our implementations for $n_{in} = 32, 24, 16, 8$ process $8/4/8/16$ consecutive blocks at a time, respectively (which is possible in any parallelizable mode, in particular ECB or CTR). By reordering the round keys accordingly, the implementation of the internal block ciphers can remain unchanged.

*Efficient and Constant-Time Parallel Finite Field Arithmetic.* Since we explicitly aim for constant-time implementations in the black box, the conditional polynomial reduction has to be carried out without branching. For this, we employ an optimized variant of the technique introduced in [25], which allows a simultaneous doubling of 4 elements of $GF(2^{32})$ and $GF(2^{24})$, or 8 elements of $GF(2^{16})$ or 16 elements of $GF(2^8)$ with just four instructions with a latency of 3 and a throughput of 1.

The in-place multiplication by two of register `%xmm0` can be implemented in constant-time as follows:

```
vpcmpgtd    MSB4_M, %xmm0, %xmm1
vpslld      $1, %xmm0, %xmm0
vpand       REDPOLY4_M, %xmm1, %xmm1
vpxor       %xmm0, %xmm1, %xmm0
```

with `MSB4_M` containing four 32-bit copies of the value $7fffffff_x$, and `REDPOLY4_M` containing four 32-bit copies of the reduction polynomial, i.e. $8d_x$.

**In the White Box.** In the whitebox setting, the small block ciphers $S_{n_{in}}$ are implemented as lookup tables of size $n_{in} \cdot 2^{n_{in}}$ bytes. The linear layer $\theta$ of SPNbox is then implemented on top of these table lookups using AVX (Intel) or NEON instructions (ARMv8).

Again, we found it beneficial to re-organize the state to collect the $i$-th S-boxes of consecutive blocks in one SSE/NEON register for a SIMD execution of the finite field arithmetic. Compared to SPACE, we have 4,5,8 and 16 parallel independent table lookups in SPNbox-32,24,16 and 8, respectively. Since memory-XMM register transfers have a throughput of 0.5 on Skylake, two of these independent table lookups can be scheduled per cycle on Intel platforms. This has to be contrasted to the situation in the serial round function of SPACE, where no simultaneous table lookups were possible.

On ARM, the smaller caches and slower memory interface imply that lookups in larger tables tend to be relatively more expensive than on Intel platforms.

## 5.3   Performance Measurements

We provide performance measurements for SPNbox and SPACE in both the black-box and the whitebox setting for the encryption of messages of length 2048 bytes. For the Intel platform, all measurements were taken on a single core of an Intel Core i7-6700 CPU at 3400 MHz with Turbo Boost and hyperthreading disabled, and averaged over 100000 repetitions, processing one message at a time. For the ARMv8 platform, a single Cortex-A57 core at 2100 MHz of a Samsung Exynos 7420 CPU as shipped in a Samsung Galaxy S6 mobile phone was used.

Our findings are summarised in Table 3 and Fig. 11 for the blackbox setting; and Table 4 for the whitebox setting. The whitebox performance is further illustrated in Fig. 12 (grouped by table size) and Fig. 13 (grouped by platform). All performance figures are given in cycles per byte (cpb).

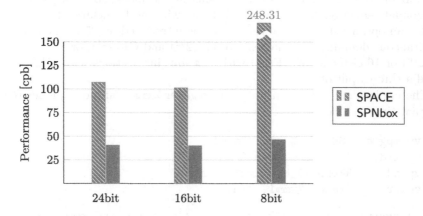

**Fig. 11.** Constant-time blackbox performance of SPACE and SPNbox on Intel Skylake platform for various table sizes in cycles per byte (lower is better).

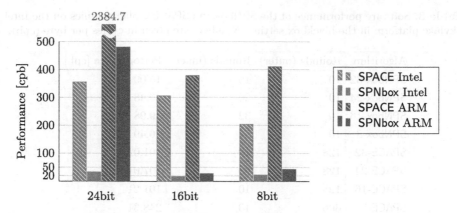

**Fig. 12.** Whitebox performance of SPACE and SPNbox on Intel Skylake and ARMv8 platforms for various table sizes in cycles per byte (lower is better).

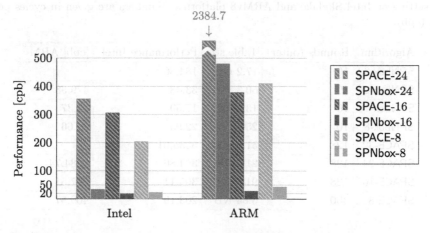

**Fig. 13.** Whitebox performance of SPACE and SPNbox on Intel Skylake and ARMv8 platforms for various table sizes in cycles per byte (lower is better).

**Discussion.** The blackbox constant-time implementation results in Table 3 indicate that for each variant with comparable space hardness, the SPNbox ciphers offer significantly increased performance compared to SPACE. Somewhat interestingly, the largest improvement (factor 4.5 speed-up) is obtained for the 32-bit variant offering the highest level of space hardness. This is due to inherent construction differences: While SPACE always uses the full AES transform and its performance is only affected by the number of Feistel rounds, SPNbox needs more and more rounds in its internal block ciphers to ensure sufficient key mixing when the block sizes becomes smaller. Additionally, the use of the AES round transformation implies increasing overhead with smaller block sizes, since increasing parts of the state are unused. For $n_{in} = 8$, the decrease in performance is caused by the heavy $16 \times 16$ MDS diffusion layer over $GF(2^8)$.

154     A. Bogdanov et al.

**Table 3.** Software performance of the SPNbox and SPACE cipher families on the Intel Skylake platform in the blackbox setting. Numbers are given in cycles per byte (cpb).

| Algorithm | Rounds (outer) | Rounds (inner) | Performance [cpb] |
|---|---|---|---|
| SPNbox-32 | 10 | 16 | 15.09 |
| SPNbox-24 | 10 | 20 | 40.48 |
| SPNbox-16 | 10 | 32 | 39.98 |
| SPNbox-8 | 10 | 64 | 46.49 |
| SPACE-32 | 128 | 10 | 101.02 |
| SPACE-24 | 128 | 10 | 107.01 |
| SPACE-16 | 128 | 10 | 101.21 |
| SPACE-8 | 300 | 10 | 248.31 |

**Table 4.** Software performance of the SPNbox and SPACE cipher families in the whitebox setting on Intel Skylake and ARMv8 platforms. Numbers are given in cycles per byte (cpb).

| Algorithm | Rounds (outer) | Table size | Performance Intel | [cpb] ARM |
|---|---|---|---|---|
| SPNbox-32 | 10 | 17.2 GB | 184.56 | — |
| SPNbox-24 | 10 | 50.3 MB | 33.48 | 479.38 |
| SPNbox-16 | 10 | 132 KB | 17.59 | 27.37 |
| SPNbox-8 | 10 | 256 B | 22.93 | 42.66 |
| SPACE-32 | 128 | 51.5 GB | 5535.01 | — |
| SPACE-24 | 128 | 218 MB | 354.86 | 2384.74 |
| SPACE-16 | 128 | 918 KB | 305.11 | 377.51 |
| SPACE-8 | 300 | 3.84 KB | 203.19 | 409.57 |

Regarding the performance of SPACE, our results largely confirm the estimation of $R$ cpb for $R$ rounds on an AES-NI platform provided in [9].

Also in the whitebox setting, SPNbox significantly outperforms SPACE for all variants, on both Intel and ARM platforms. One observes that any increases in pure lookup performance due to smaller table size is increasingly compensated for by the heavier linear MDS layers. The surprisingly good performance of SPNbox-32 can to some extent be attributed to the fact that our test platform had 16 GB of memory available.

Comparing the blackbox to the whitebox performances of each variant of SPNbox, it becomes apparent that from $n_{in} = 24$ and smaller, table-based implementations outperform round-based implementations. The latter, however, offer constant timing behaviour. Further optimizations of the constant-time implementations also remain possible.

Summarising, the constant-time blackbox performance of the proposed SPNbox ciphers outperforms the SPACE variants by factors of 2.5 to 6.5. In the whitebox setting, the new SPNbox ciphers offer performance improvements by factors of 8 to 18 (on Intel) and 5 to 13 (on ARM) over SPACE, as illustrated in Figs. 12 and 13.

# 6  Conclusion and Outlook

In this paper, we proposed SPNbox, a new family of space-hard block ciphers, which significantly improves upon the SPACE ciphers. Employing an SPN-type design with efficient constant-time small block ciphers, the resulting parallelization opportunities allow significantly faster implementations both in the black box and in the white box. Instances of SPNbox achieve speed-ups of up to 6.5 times in the black box and up to 18 times in the whitebox setting, while offering comparable space hardness. Moreover, we formalized the security models of space hardness which are classified with respect to the adversary's abilities. We proved security bounds of space hardness in all adversarial models. We then applied this analysis to SPNbox, showing that SPNbox offers sufficiently high levels of space hardness in each adversary model.

Our work also raises a couple of open research questions and directions. Concerning the design of the small internal block ciphers, there seems to be an efficiency bottleneck regarding the key mixing: The smaller the block size, the more rounds are needed to avoid meet-in-the-middle attacks, which limits their efficiency. This raises the question of how to build more efficient block ciphers with very small block lengths and a relatively large key. Especially, fast key mixing and efficient key scheduling functions for small block ciphers are essentially unknown.

A possible solution for this efficiency problem is to use table lookups for secret S-boxes. This however introduces side-channel issues with key-dependent lookups, motivating further research into how to construct secret S-boxes of various sizes with efficient constant-time implementations.

# References

1. Adobe Systems Incorporated. Adobe Primetime Technical Primer for Operators (2014)
2. Akamai Technologies. Securing Cloud-Based Workflows for Premium Content (2014)
3. Benadjila, R., Billet, O., Gilbert, H., Macario-Rat, G., Peyrin, T., Robshaw, M., Seurin, Y.: SHA-3 Proposal: ECHO. Submission to NIST (2009)
4. Barreto, P., Rijmen, V.: The Anubis Block Cipher. Submission to the NESSIE Project (2000)
5. Barreto, P., Rijmen, V.: The Khazad Legacy-level Block Cipher. Submission to the NESSIE Project (2000)
6. Billet, O., Gilbert, H., Ech-Chatbi, C.: Cryptanalysis of a white box AES implementation. In: Handschuh, H., Hasan, M.A. (eds.) SAC 2004. LNCS, vol. 3357, pp. 227–240. Springer, Heidelberg (2004). doi:10.1007/978-3-540-30564-4_16

7. Biryukov, A., Bouillaguet, C., Khovratovich, D.: Cryptographic schemes based on the **ASASA** structure: black-box, white-box, and public-key (extended abstract). In: Sarkar, P., Iwata, T. (eds.) ASIACRYPT 2014. LNCS, vol. 8873, pp. 63–84. Springer, Heidelberg (2014). doi:10.1007/978-3-662-45611-8_4
8. Biryukov, A., Shamir, A.: Structural cryptanalysis of SASAS. J. Cryptology **23**(4), 505–518 (2010)
9. Bogdanov, A., Isobe, T.: White-box cryptography revisited: space-hard ciphers. In: Proceedings of the 22nd ACM SIGSAC Conference on Computer and Communications Security, pp. 1058–1069. ACM (2015)
10. Borghoff, J., Knudsen, L.R., Leander, G., Thomsen, S.S.: Slender-set differential cryptanalysis. J. Cryptology **26**(1), 11–38 (2013)
11. Bos, J.W., Hubain, C., Michiels, W., Teuwen, P.: Differential computation analysis: hiding your white-box designs is not enough. In: Gierlichs, B., Poschmann, A.Y. (eds.) CHES 2016. LNCS, vol. 9813, pp. 215–236. Springer, Heidelberg (2016). doi:10.1007/978-3-662-53140-2_11
12. Bringer, J., Chabanne, H., Dottax, E.: White box cryptography: another attempt. IACR Cryptology ePrint Archive 2006:468 (2006)
13. Chong, K.-M.: The arithmetic mean-geometric mean inequality: a new proof. Math. Mag. **49**(2), 87–88 (1976)
14. Chow, S., Eisen, P., Johnson, H., Oorschot, P.C.: A white-box DES implementation for DRM applications. In: Feigenbaum, J. (ed.) DRM 2002. LNCS, vol. 2696, pp. 1–15. Springer, Heidelberg (2003). doi:10.1007/978-3-540-44993-5_1
15. Chow, S., Eisen, P., Johnson, H., Oorschot, P.C.: White-box cryptography and an AES implementation. In: Nyberg, K., Heys, H. (eds.) SAC 2002. LNCS, vol. 2595, pp. 250–270. Springer, Heidelberg (2003). doi:10.1007/3-540-36492-7_17
16. Delerablée, C., Lepoint, T., Paillier, P., Rivain, M.: White-box security notions for symmetric encryption schemes. In: Lange, T., Lauter, K., Lisoněk, P. (eds.) SAC 2013. LNCS, vol. 8282, pp. 247–264. Springer, Heidelberg (2014). doi:10.1007/978-3-662-43414-7_13
17. Dziembowski, S.: Intrusion-resilience via the bounded-storage model. In: Halevi, S., Rabin, T. (eds.) TCC 2006. LNCS, vol. 3876, pp. 207–224. Springer, Heidelberg (2006). doi:10.1007/11681878_11
18. Gruss, D., Spreitzer, R., Mangard, S.: Cache template attacks: automating attacks on inclusive last-level caches. In: 24th USENIX Security Symposium, USENIX Security 15, pp. 897–912. USENIX Association (2015)
19. Gueron, S.: Intel Advanced Encryption Standard (AES) Instructions Set. Intel white paper, September 2012
20. Halderman, J.A., Schoen, S.D., Heninger, N., Clarkson, W., Paul, W., Calandrino, J.A., Feldman, A.J., Appelbaum, J., Felten, E.W.: Lest we remember: cold boot attacks on encryption keys. In: Proceedings of the 17th USENIX Security Symposium, pp. 45–60. USENIX Association (2008)
21. Hawkes, P., O'Connor, L.: XOR and Non-XOR differential probabilities. In: Stern, J. (ed.) EUROCRYPT 1999. LNCS, vol. 1592, pp. 272–285. Springer, Heidelberg (1999). doi:10.1007/3-540-48910-X_19
22. Irazoqui, G., Eisenbarth, T., Sunar, B.: S$a: A shared cache attack that works across cores and defies VM sandboxing - and its application to AES. In: 2015 IEEE Symposium on Security and Privacy, SP 2015, pp. 591–604. IEEE Computer Society (2015)

23. Irazoqui, G., Inci, M.S., Eisenbarth, T., Sunar, B.: Wait a minute! A fast, cross-VM attack on AES. In: Stavrou, A., Bos, H., Portokalidis, G. (eds.) RAID 2014. LNCS, vol. 8688, pp. 299–319. Springer, Heidelberg (2014). doi:10.1007/978-3-319-11379-1_15

24. Karroumi, M.: Protecting white-box AES with dual ciphers. In: Rhee, K.-H., Nyang, D.H. (eds.) ICISC 2010. LNCS, vol. 6829, pp. 278–291. Springer, Heidelberg (2011). doi:10.1007/978-3-642-24209-0_19

25. Käsper, E., Schwabe, P.: Faster and timing-attack resistant AES-GCM. In: Clavier, C., Gaj, K. (eds.) CHES 2009. LNCS, vol. 5747, pp. 1–17. Springer, Heidelberg (2009). doi:10.1007/978-3-642-04138-9_1

26. Lepoint, T., Rivain, M., Mulder, Y., Roelse, P., Preneel, B.: Two attacks on a white-box AES implementation. In: Lange, T., Lauter, K., Lisoněk, P. (eds.) SAC 2013. LNCS, vol. 8282, pp. 265–285. Springer, Heidelberg (2014). doi:10.1007/978-3-662-43414-7_14

27. Link, H.E., Neumann, W.D.: Clarifying obfuscation: improving the security of white-box DES. In: International Symposium on Information Technology: Coding and Computing (ITCC 2005), vol. 1, pp. 679–684 (2005)

28. Minaud, B., Derbez, P., Fouque, P.-A., Karpman, P.: Key-recovery attacks on ASASA. In: Iwata, T., Cheon, J.H. (eds.) ASIACRYPT 2015. LNCS, vol. 9453, pp. 3–27. Springer, Heidelberg (2015). doi:10.1007/978-3-662-48800-3_1

29. Workgroup Mobey, H.C.E., Forum. The Host Card Emulation in Payments: Options for Financial Institutions (2014)

30. Mulder, Y., Roelse, P., Preneel, B.: Cryptanalysis of the xiao – lai white-box AES implementation. In: Knudsen, L.R., Wu, H. (eds.) SAC 2012. LNCS, vol. 7707, pp. 34–49. Springer, Heidelberg (2013). doi:10.1007/978-3-642-35999-6_3

31. De Mulder, Y., Wyseur, B., Preneel, B.: Cryptanalysis of a perturbated white-box AES implementation. In: Gong, G., Gupta, K.C. (eds.) INDOCRYPT 2010. LNCS, vol. 6498, pp. 292–310. Springer, Heidelberg (2010). doi:10.1007/978-3-642-17401-8_21

32. National Institute of Standards and Technology. Recommendation for Key Derivation Using Pseudorandom Functions. NIST Special Publication (SP) 800–108 (2009)

33. National Institute of Standards and Technology: SHA-3 Standard: Permutation-Based Hash and Extendable-Output Functions. Federal Information Processing Standards Publication 202 (2015)

34. Ristenpart, T., Tromer, E., Shacham, H., Savage, S.: Hey, you, get off of my cloud: exploring information leakage in third-party compute clouds. In: Proceedings of the 2009 ACM Conference on Computer and Communications Security, CCS 2009, pp. 199–212. ACM (2009)

35. Sanfelix, E., Mune, C., de Haas, J.: Unboxing the white-box practical attacks against obfuscated ciphers. In: Black Hat Europe 2015 (2015)

36. Sim, S.M., Khoo, K., Oggier, F., Peyrin, T.: Lightweight MDS involution matrices. In: Leander, G. (ed.) FSE 2015. LNCS, vol. 9054, pp. 471–493. Springer, Heidelberg (2015). doi:10.1007/978-3-662-48116-5_23

37. Alliance, S.C., Paper, W.: Host Card Emulation (HCE) 101 (2014)

38. Tiessen, T., Knudsen, L.R., Kölbl, S., Lauridsen, M.M.: Security of the AES with a secret S-Box. In: Leander, G. (ed.) Fast Software Encryption. LNCS, vol. 9054, pp. 175–189. Springer, Heidelberg (2015). doi:10.1007/978-3-662-48116-5_9

39. Wyseur, B., Michiels, W., Gorissen, P., Preneel, B.: Cryptanalysis of white-box DES implementations with arbitrary external encodings. In: Adams, C., Miri, A., Wiener, M. (eds.) SAC 2007. LNCS, vol. 4876, pp. 264–277. Springer, Heidelberg (2007). doi:10.1007/978-3-540-77360-3_17
40. Xiao, Y., Lai, X.: A secure implementation of white-box AES. In: 2nd International Conference on Computer Science and its Applications (CSA2009) (2009)
41. Yarom, Y., Falkner, K.: FLUSH+RELOAD: A high resolution, low noise, L3 cache side-channel attack. In: Proceedings of the 23rd USENIX Security Symposium, pp. 719–732. USENIX Association (2014)

# Efficient and Provable White-Box Primitives

Pierre-Alain Fouque[1,2]([✉]), Pierre Karpman[1,3,4,5], Paul Kirchner[6],
and Brice Minaud[1,7]

[1] Université de Rennes 1, Rennes, France
pierre-alain.fouque@ens.fr
[2] Institut Universitaire de France, Paris, France
[3] Inria, Rennes, France
[4] École Polytechnique, Paris, France
[5] Nanyang Technological University, Singapore, Singapore
pierre.karpman@inria.fr
[6] École Normale Supérieure, Paris, France
pkirchne@clipper.ens.fr
[7] Royal Holloway University of London, Egham, UK
brice.minaud@gmail.com

**Abstract.** In recent years there have been several attempts to build white-box block ciphers whose implementations aim to be incompressible. This includes the *weak white-box* ASASA construction by Bouillaguet, Biryukov and Khovratovich from ASIACRYPT 2014, and the recent *space-hard* construction by Bogdanov and Isobe from CCS 2015. In this article we propose the first constructions aiming at the same goal while offering provable security guarantees. Moreover we propose concrete instantiations of our constructions, which prove to be quite efficient and competitive with prior work. Thus provable security comes with a surprisingly low overhead.

**Keywords:** White-box cryptography · Provable security

## 1 Introduction

### 1.1 White-Box Cryptography

The notion of white-box cryptography was originally introduced by Chow et al. in the early 2000s [CEJO02a, CEJO02b]. The basic goal of white-box cryptography is to provide implementations of cryptographic primitives that offer cryptographic guarantees even in the presence of an adversary having direct access to the implementation. The exact content of these security guarantees varies, and different models have been proposed.

P.-A. Fouque, P. Karpman, P. Kirchner, B. Minaud—Partially supported by the French ANR project *BRUTUS*, ANR-14-CE28-0015.
P. Karpman—Partially supported by the Direction Générale de l'Armement and by the Singapore National Research Foundation Fellowship 2012 (NRF-NRFF2012-06).

© International Association for Cryptologic Research 2016
J.H. Cheon and T. Takagi (Eds.): ASIACRYPT 2016, Part I, LNCS 10031, pp. 159–188, 2016.
DOI: 10.1007/978-3-662-53887-6_6

Ideally, white-box cryptography can be thought of as trying to achieve security guarantees similar to a Trusted Execution Environment [ARM09] or trusted enclaves [CD16], purely through implementation means—in so far as this is feasible. Of course this line of research finds applications in many situations where code containing secret information is deployed in non-trusted environments, such as software protection (DRM) [Wys09, Gil16].

Concretely, the initial goal in [CEJO02a, CEJO02b] was to offer implementations of the DES and AES block ciphers, such that an adversary having full access to the implementation would not be able to extract the secret keys. Unfortunately both the initial constructions and later variants aiming at the same goal (such as [XL09]) were broken [BGEC04, GMQ07, WMGP07, DMRP12]: to this day no secure white-box implementation of DES or AES is known.

Beside cryptanalytic weaknesses, defining white-box security as the impossibility to extract the secret key has some drawbacks. Namely, it leaves the door open to code lifting attacks, where an attacker simply extracts the encryption function as a whole and achieves the same functionality as if she had extracted the secret key: conceptually, the encryption function can be thought of as an equivalent secret key[1].

This has led research on white-box cryptography into two related directions. One is to find new, sound and hopefully achievable definitions of white-box cryptography. The other is to propose new constructions fulfilling these definitions.

In the definitional line of work, various security goals have been proposed for white-box constructions. On the more theoretical end of the spectrum, the most demanding property one could hope to attain for a white-box construction would be that of virtual black-box obfuscation [BGI+01]. That is, an adversary having access to the implementation of a cipher would learn no more than they could from interacting with the cipher in a black-box way (i.e. having access to an oracle computing the output of the cipher). Tremendous progress has been made in recent years in the domain of general program obfuscation, starting with [GGH+13]. However the current state of the art is still far from practical use, both in terms of concrete security (see e.g. [Hal15]) and performance (see e.g. an obfuscation of AES in [Zim15]).

A less ambitious goal, proposed in [DLPR13, BBK14] is that an adversary having access to the implementation of an encryption scheme may be able to encrypt (at least via code lifting), but should remain unable to decrypt. This notion is called *strong white-box* in [BBK14] and *one-wayness* in [DLPR13]. Such a goal is clearly very similar to that of a trapdoor permutation, and indeed known constructions rely on public-key primitives. As a consequence they are no faster than public key encryption. An interesting way to partially circumvent this issue, proposed in [BBK14], is to use multivariate cryptography, where knowledge of the secret information allows encryption and decryption at a speed comparable to standard symmetric ciphers (although public key operations are quite slow). However multivariate cryptography lacks security reductions to well-established hard problems (although they are similar in flavor to MQ),

---

[1] This can be partially mitigated by the use of external encodings [CEJO02a].

and numerous instantiations have been broken, including those of [BBK14]: see [GPT15, DDKL15, MDFK15].

Finally, on the more modest but efficiently achievable end of the spectrum, one can ask that an adversary having access to the white-box implementation cannot produce a functionally equivalent program of significantly smaller size. This notion has been called *incompressibility* in [DLPR13], *weak white-box* in [BBK14] and *space-hardness* in [BI15][2]. This definition implies in particular that it is difficult for an adversary to extract a short master key, which captures the goal of the original white-box constructions by Chow *et al.* In addition, the intent behind this approach is that large, incompressible code can more easily be made resistant to code lifting when combined with engineering obfuscation techniques [BBK14, BI15, Gil16]; and make code distribution more cumbersome for a potential attacker.

As mentioned earlier, there is no known implementation of AES or DES that successfully hides the encryption key. A fortiori there is no known way to achieve incompressibility for AES, DES or indeed any pre-existing cipher. However recent constructions have proposed new, ad-hoc, and quite efficient ciphers specifically designed to meet the incompressibility criterion [BBK14, BI15]. These constructions aim for incompressibility by relying on a large pseudo-random table hard-coded into the implementation of the cipher, with repeated calls to the table being made during the course of encryption. The idea is that, without knowledge of all or most of the table, most plaintexts cannot be encrypted. This enforces incompressibility.

In [BBK14], the table is used as an S-box in a custom block cipher design. This requires building the table as a permutation, which is achieved using an ASASA construction, alternating secret affine and non-linear layers. Unfortunately this construction was broken [DDKL15, MDFK15]. This type of attack is completely avoided in the new SPACE construction [BI15], where the table is built by truncating calls to AES. This makes it impossible for an adversary to recover the secret key used to generate the table, based solely on the security of AES. However this also implies that the table is no longer a permutation and cannot be used as an S-box. Accordingly, in SPACE, the table is used as a round function in a generalized Feistel network. While an adversary seeking to extract the key is defeated by the use of AES, there is no provable resistance against an adversary trying to compress the cipher.

We also remark that the standard formalization of white-box cryptography is very close to other models. For example, the bounded-storage model considers the problem of communicating securely given a long public random string which the adversary is unable to store. Indeed, up to renaming, it is essentially the same as the incompressibility of the key, and one of our design is inspired by a solution proposed to this problem [Vad04]. Another model, even stronger than incompressibility, is intrusion-resilience [Dzi06]. The goal is to communicate securely, even when a virus may output any data to the adversary during

---

[2] Here, we lump together very similar definitions, although they are technically distinct. More details are provided in Sect. 2.1.

the computations of both parties, as long as the total data leaked is somewhat smaller than the key size. The disadvantage of this model is that it requires rounds of communication (e.g. 9 rounds in [CDD+07]), while white-box solutions need only add some computations.

## 1.2  Our Contribution

Both of the previously mentioned constructions in [BBK14, BI15] use ad-hoc designs. They are quite efficient, but cannot hope to achieve provable security. Our goal is to offer provable constructions, while retaining similar efficiency.

First, we introduce new formal definitions of incompressibility, namely weak and strong incompressibility. *Weak* incompressibility is very close to incompressibility definitions in previous work [BBK14, BI15], and can be regarded as a formalization of the space-hardness definition of [BI15]. *Strong incompressibility* on the other hand is a very demanding notion; in particular it is strictly stronger than the incompressibility definition of [DLPR13].

Our main contribution is to introduce two provably secure white-box constructions, named WhiteKey and WhiteBlock. We prove both constructions in the weak model. The bounds we obtain are close to a generic attack, and yield quite efficient parameters. Moreover we also prove WhiteKey in the strong model.

Previous work has concentrated on building white-box block ciphers. This was of course unavoidable when attempting to provide white-box implementations of AES or DES. However, it was already observed in the seminal work of Chow *et al.* that the use of white-box components could be limited to key encapsulation mechanisms [CEJO02a]. That is, the white-box component is used to encrypt and decrypt a symmetric key, which is then used to encrypt or decrypt the rest of the message. This is of course the same technique as hybrid encryption, and beneficial for the same reason: white-box component are typically slower than standard symmetric ciphers (albeit to a lesser extent than public-key schemes).

In this context, the white-box component need not be a block cipher, and our WhiteKey construction is in fact a key generator. That is, it takes a random string as input and outputs a key, which can then be used with any standard block cipher. Its main feature is that it is provably strongly incompressible. Roughly speaking, this implies it is unfeasible for an adversary, given full access to a white-box implementation of WhiteKey, to produce a significantly smaller implementation that is functionally equivalent on most inputs. In fact, an efficient adversary knowing this smaller implementation cannot even use it to distinguish, with noticeable probability, outputs of the original WhiteKey instance from random.

However, WhiteKey is not invertible, and in particular it is not a block cipher, unlike prior work. Nevertheless we also propose a white-box block cipher named WhiteBlock. WhiteBlock can be used in place of any 128-bit block cipher, and is not restricted to key generation. However this comes at some cost: WhiteBlock has a more complex design, and is slightly less efficient than WhiteKey. Furthermore, it is proved only in the weak incompressibility model (essentially the same model as that of SPACE [BI15]), using a heuristic assumption. Thus WhiteKey is a cleaner and more efficient solution, if the key generation functionality suffices (which is likely in most situations where a custom white-box design can be used).

Regarding the proof of WhiteKey in the strong incompressibility model, the key insight is that what we are trying to build is essentially an entropy extractor. Indeed, roughly speaking, the table can be regarded as a large entropy pool. If an adversary tries to produce an implementation significantly smaller than the table, then the table still has high (min-)entropy conditioned on the knowledge of the compressed implementation. Thus if the key generator functions as a good entropy extractor, then the output of the key generator looks uniform to an (efficient) adversary knowing only the compressed implementation.

Furthermore, for efficiency reasons, we want our extractor to be local, i.e. we want our white-box key generator to make as few calls to the table as possible. Hence a local extractor does precisely what we require, and as a result our proof relies directly on previous work on local extractors [Vad04]. Meanwhile our proofs in the weak incompressibility model use dedicated combinatorial arguments.

Finally, we provide concrete instantiations of WhiteKey and WhiteBlock, named PUPPYCIPHER and COUREURDESBOIS respectively. Our implementations show that these instances are quite efficient, yielding performance comparable to previous ad-hoc designs such as SPACE. Like in previous work, our instances also offer various choices in terms of the desired size of the white-box implementation.

## 1.3   Related Work

We are aware of three prior incompressible white-box schemes [DLPR13, BBK14, BI15]. In the first of these papers, incompressibility is formally defined [DLPR13]. A public-key scheme is proven in the incompressible model: in a nutshell, the scheme consists in a standard RSA encryption, except for the fact that the public key is inflated by adding an arbitrary multiple of the group order. This provably results in an incompressible scheme, which is also one-way due to its public-key nature. However it is orders of magnitude slower than a symmetric scheme (note that it is also slower than standard RSA due to the size of the exponent).

On the other hand, the authors of [BBK14, BI15] propose symmetric encryption schemes aiming at incompressibility alone. These constructions naturally achieve higher performance. The white-box construction of [BBK14] was broken in [MDFK15, DDKL15]. The construction in [BI15] provides provable guarantees against an adversary attempting to recover the secret key used to generate the table. However no proof is given against an adversary merely attempting to compress the implementation. In fact the construction relies on symmetric building blocks, and any such proof seems out of reach.

An independent work by Bellare, Kane and Rogaway was recently published at CRYPTO 2016 [BKR16]; its underlying goal and techniques are similar to our strong incompressibility model, and the WhiteKey construction in particular. Although the setting of [BKR16] is different and no mention is made of white-box cryptography, the design objective is similar. The setting considered in [BKR16] is that of the bounded-retrieval model [ADW09], and the aim is to foil key exfiltration attempts by using a large encryption key. The point is that encryption should remain secure in the presence of an adversary having access to a bounded exfiltration of the big key. The exfiltrated data is modeled as the output of an adversarially-defined function of the key with bounded output.

The compressed implementation plays the same role in our definition of strong incompressibility: interestingly, our strong model almost matches big-key security in that sense (contrary to prior work on incompressible white-box cryptography, which is closer to our weak model). Relatively minor differences include the fact that we require a bound on the min-entropy of the table/big key relative to the output of the adversarially-defined function, rather than specifically the number of bits; and we can dispense with a random oracle at the output because we do not assume that the adversary is able to see generated keys directly, after the compression phase. A notable difference is how authenticity is treated: we require that the adversary is unable to encrypt most plaintexts, given the compressed implementation; whereas the authors of [BKR16] only enforce authenticity when there is no leakage. A word-based generalization of the main result in [BKR16], as mentioned in the discussion of that paper, would be very interesting from our perspective, likely allowing better bounds for WhiteKey in the strong incompressibility model. Proofs of weak incompressibility, the WhiteBlock construction, as well as the concrete design of the WhiteKey instance using a variant of the extractor from [CMNT11], are unrelated.

As mentioned earlier in the introduction, the design of local extractors is also directly related to our proof in the strong incompressibility model, most notably [Vad04].

## 2   Models

### 2.1   Context

As noted in the introduction, the term *white-box cryptography* encompasses a variety of models, aiming to achieve related, but distinct security goals. Here we are interested in the *incompressibility* model. The basic goal is to prevent an attacker who has access to the full implementation of a cipher to produce a more compact implementation.

Incompressibility has been defined under different names and with slight variations in prior work. It is formally defined as $(\lambda, \delta)$-*Incompressibility* in [DLPR13]. A very similar notion is called *weak white-box* in [BBK14], and *space-hardness* in [BI15]. In [BBK14], the *weak white-box* model asks that an efficient adversary, given full access to the cipher implementation, is unable to produce a new implementation of the same cipher of size less than some security parameter $T$. In [BI15], this notion is refined by allowing the adversary-produced implementation to be correct up to a negligible proportion $2^{-Z}$ of the input space. Thus a scheme is considered $(T, Z)$-*space-hard* iff an efficient adversary is unable to produce an implementation of the cipher of size less than $T$, that is correct on all but a proportion $2^{-Z}$ of inputs. This is essentially equivalent to the $(\lambda, \delta)$-*incompressibility* definition of [DLPR13], where $\lambda$ and $\delta$ play the respective roles of $T$ and $2^{-Z}$.

In this work, we introduce and use two main notions of incompressibility, which we call *weak* and *strong* incompressibility. Weak incompressibility may be regarded as a formalization of space-hardness from [BI15]. As the names suggest, strong incompressibility implies weak incompressibility (see the full version of this paper [FKKM16]). The point of strong incompressibility is that it provides stronger guarantees, and is a natural fit for the WhiteKey construction.

## 2.2 Preliminary Groundwork

To our knowledge, all prior work that has attempted to achieve white-box incompressibility using symmetric means[3] has followed a similar framework. The general idea is as follows. The white-box implementation of the cipher is actually a symmetric cipher that uses a large table as a component. The table is hardcoded into the implementation. To an adversary looking at the implementation, the table looks uniformly random. An adversary attempting to compress the implementation would be forced to retain only part of the table in the compressed implementation. Because repeated pseudo-random calls to the table are made in the course of each encryption and decryption, any implementation that ignores a significant part of the table would be unable to encrypt or decrypt accurately most messages. This enforces incompressibility.

To a legitimate user in possession of the shared secret however, the table is not uniformly random. It is in fact generated using a short secret key. Of course this short master key should be hard to recover from the table, otherwise the scheme could be dramatically compressed.

Thus a white-box encryption scheme is made up of two components: an *encryption scheme*, which takes as input a short master secret key and uses it to encrypt data, and a white-box *implementation*, which is functionally equivalent, but does not use the short master secret key directly. Instead, it uses a large table (which can be thought of as an equivalent key) that has been derived from the master key. This situation is generally formalized by defining a white-box scheme as an encryption scheme together with a *white-box compiler*, which produces the white-box implementation of the scheme.

**Definition 1 (Encryption Scheme).** *An encryption scheme is a mapping $E : \mathcal{K} \times \mathcal{R} \times \mathcal{P} \rightarrow \mathcal{C}$, taking as input a key $K \in \mathcal{K}$, possibly some randomness $r \in \mathcal{R}$, and a plaintext $P \in \mathcal{P}$. It outputs a ciphertext $C \in \mathcal{C}$. Furthermore it is required that the encryption scheme be invertible, in the sense that there exists a decryption function $D : \mathcal{K} \times \mathcal{C} \rightarrow \mathcal{P}$ such that $\forall K, R, P, D(K, E(K, R, P)) = P$.*

---

[3] This excludes the incompressible construction from [DLPR13], which is based on a modified RSA.

**Definition 2 (White-box Encryption Scheme).** *A* white-box encryption scheme *is defined by a pair of two encryption schemes:*

$$E_1 : \mathcal{K} \times \mathcal{R} \times \mathcal{P} \to \mathcal{C}$$
$$E_2 : \mathcal{T} \times \mathcal{R} \times \mathcal{P} \to \mathcal{C}$$

*together with a* white-box compiler $C : \mathcal{K} \to \mathcal{T}$, *such that for all* $K \in \mathcal{K}$, $E_1(K, \cdot, \cdot)$ *is functionally equivalent to* $E_2(C(K), \cdot, \cdot)$.

In the definition above, $E_1$ can be thought of as a standard encryption scheme relying on a short (say, 128-bit) master key $K$, while $E_2$ is its white-box implementation, relying on a large table $T$ derived from $K$. To distinguish between $E_1$ and $E_2$, we will sometimes call the first scheme the *cipher*, and the second the (white-box) *implementation*.

## 2.3    Splitting the Adversaries

A white-box scheme is faced with two distinct adversaries:

- The *black-box* adversary only has black-box access to the scheme. She attempts to attack the cipher with respect to some standard black-box security notion.
- The *white-box* adversary has full access to the white-box implementation. She attempts to break incompressibility by producing a smaller implementation of the scheme.

The black-box adversary can be evaluated with respect to standard security notions such as IND-CCA. The specificity of white-box schemes is of course the second adversary, on which we now focus. The white-box adversary itself can be decomposed into two distinct adversaries:

- The *compiler* adversary attempts to recover the master key $K$ of $E_1$ given the implementation $E_2$. This is the adversary that succeeds in the cryptanalyses of many previous schemes, *e.g.* [BGEC04, GMQ07, DDKL15, MDFK15]. More generally this adversary attempts to distinguish $C(K)$ for $K \xleftarrow{\$} \mathcal{K}$ from a uniform element of $\mathcal{T}$.
- Finally, the *implementation* adversary does not attempt to distinguish $T$, and instead regards $T$ as uniformly random. She focuses purely on the white-box implementation $E_2$. She attempts to produce a functionally equivalent (up to some error rate specified by the security parameters), but smaller implementation of $E_2$.

Nicely enough, the three black-box, compiler and implementation adversaries target respectively the $E_1$, $C$, and $E_2$ components of the white-box scheme (hence their name). Of course the two white-box adversaries (targeting the compiler and implementation) break incompressibility, so they can be captured by the same security definition (as in [DLPR13]). However it is helpful to think of the two as separate adversaries, especially because they can be thwarted by separate mechanisms. Moreover it is clear that resistance to both adversaries implies incompressibility (the dichotomy being whether the table can be efficiently distinguished from random).

The authors of [BI15] introduce a new general method to make sure that the compiler adversary fails, *i.e.* $C(T)$ is indistinguishable from uniform. Namely, they propose to generate the table $T$ by truncating the output of successive calls to AES (or some other fixed block cipher). In this scenario the master key $K$ of $E_1$ is the AES key. Assuming AES cannot be distinguished from a uniformly random permutation, and the truncated output is (say) at most half of the original cipher, then the table $T$ is indistinguishable from a random function.

## 2.4  Weak Incompressibility

As noted in the previous section, using the technique from [BI15], defeating the compiler adversary is quite easy, and relies directly and provably on the security of a standard cipher. As a result, our security definition (and indeed, our constructions) focus on the *implementation* adversary.

The weak incompressibility notion we define below is very close to the space-hardness notion of [BI15], indeed it is essentially a formalization of it. Like in [BBK14,BI15], the definition is specific to the case where the table $T$ is actually a table (rather than an arbitrary binary string) which implements a function (or permutation) $T : \mathcal{I} \to \mathcal{O}$, and can be queried on inputs $i \in \mathcal{I}$.

We write weak incompressibility as ENC-TCOM: *ENC* reflects the fact that the adversary's ultimate goal is to *encrypt* a plaintext. *TCOM* stands for *table-compressed*, as the adversary is given access to a compressed form of the table. This is of course weaker than being given access to a compressed implementation defined in an arbitrary adversarially-defined way, as will be the case in the next section.

In the following definition, the encryption scheme should be thought of as the white-box implementation $E_2$ from the previous sections. In particular the "key" is a large table.

**Definition 3 (Weak Incompressibility, ENC-TCOM).** *Let* $E : \mathcal{T} \times \mathcal{R} \times \mathcal{P}$ *denote an encryption scheme. Let* $s, \lambda$ *denote security parameters. Let us further assume that the key* $T \in \mathcal{T}$ *is a function* $T : \mathcal{I} \to \mathcal{O}$ *for some input and output sets* $\mathcal{I}$ *and* $\mathcal{O}$. *The encryption scheme is said to be* $\tau$-secure *for* $(s, \lambda, \delta)$-weak *incompressibility iff, with probability at least* $1 - 2^{-\lambda}$ *over the random choice of* $T \in \mathcal{T}$ *(performed in the initial step of the game), the probability of success of an adversary running in time* $\tau$ *and playing the following game is upper-bounded by* $\delta$.

1. *The challenger* $\mathcal{B}$ *picks* $T \in \mathcal{T}$ *uniformly at random.*
2. *For* $0 \leq i < s$, *the adversary chooses* $q_i \in \mathcal{I}$, *and receives* $T(q_i)$ *from the challenger. Note that the queries are adaptive.*
   *At this point the adversary is tasked with trying to encrypt a random message:*
3. *The challenger chooses* $P \in \mathcal{P}$ *uniformly at random, and sends* $P$ *to the adversary.*
4. *The adversary chooses* $C \in \mathcal{C}$. *The adversary wins iff* $C$ *decrypts to* $P$ *(for key* $T$*).*

In other words, a scheme is $(s, \lambda, \delta)$-weakly incompressible iff any adversary allowed to adaptively query up to $s$ entries of the table $T$ can only correctly encrypt up to a proportion $\delta$ of plaintexts (except with negligible probability $2^{-\lambda}$ over the choice of $T$). Note that $(s, \lambda, \delta)$-weak incompressibility matches exactly with $(s, -\log(\delta))$-space-hardness in [BI15]. The only difference is that our definition is more formal, as is necessary since we wish to provide a security proof. In particular we specify that the adversary's queries are adaptive.

It should also be noted that the adversary's goal could be swapped for *e.g.* indistinguishability in the definition above. The reason we choose a weaker goal here is that it matches with prior white-box definitions, namely space-hardness [BI15] and weak white-box [BBK14]. Moreover it makes sense in white-box contexts such as DRM, where an attacker is attempting to create a rogue encryption or decryption algorithm: the point is that such an algorithm should fail on most inputs, unless the adversary has succeeded in extracting the whole table (or close to it), and the algorithm includes it.

It is noteworthy that in our definitions, "incompressibility" is captured as a power given to the adversary. The adversary's goal, be it encryption or indistinguishability, can be set independently of the specific form of compressed implementation she is allowed to ask for. This makes the definition conveniently modular, in the spirit of standard security notions such as IND-CCA.

## 2.5   Strong Incompressibility

We now introduce a stronger notion of incompressibility. This definition is stronger in two significant ways.

1. First, there is no more restriction on how the adversary can choose to compress the implementation. In the case of weak incompressibility, the adversary was only allowed to "compress" by learning a portion of the table. With strong incompressibility, she is allowed to compress the implementation in an arbitrary way, as long as the table $T$ retains enough randomness from the point of view of the adversary (*i.e.* she does not learn the whole secret).
2. Second, the adversary's goal is to distinguish the output of the encryption function from random, rather than being able to encrypt. This requirement may be deemed too demanding for some applications, but can be thought of as the best form of incompressibility one can ask for.

We denote strong incompressibility by IND-COM because the ultimate goal of the adversary is to break an indistinguishability game (IND), given a compressed (or compact) implementation of their choice (COM). We actually give more power to the adversary than this would seem to imply, as the adversary is also given the power to query plaintexts of her choice after receiving the compressed implementation.

Note that in the following definitions, $f$ is not computationally bounded, so generating the tables via a pseudorandom function is not possible.

**Definition 4 (Strong Incompressibility, IND-COM).** *Let* $E : T \times \mathcal{R} \times \mathcal{P}$
*denote an encryption scheme. Let* $\mu$ *denote a security parameter. Let us further
assume that the key* $T \in T$ *is chosen according to some distribution* $D$ *(typically
uniform). The scheme* $E$ *is said to be* $(\tau, \epsilon)$*-secure for* $\mu$*-strong incompressibility
iff the advantage of an adversary* $\mathcal{A}$ *running in time* $\tau$ *and playing the following
game is upper-bounded by* $\epsilon$.

1. *The adversary chooses a set* $\mathcal{S}$ *and a function* $f : T \to \mathcal{S}$*, subject only to the
   condition that for all* $s \in \mathcal{S}$*, the min-entropy of the variable* $T$ *conditioned on
   $f(T) = s$ is at least $\mu$. The function $f$ should be thought of as a compression
   algorithm chosen by the adversary.*
2. *Meanwhile the challenger* $\mathcal{B}$ *picks* $T \in T$ *according to the distribution* $D$ *(thus
   fixing an instance of the encryption scheme).*
3. *The adversary receives* $f(T)$*. At this point the adversary is tasked with breaking a standard IND-CPA game, namely:*
4. *The adversary may repeatedly choose any plaintext* $P \in \mathcal{P}$*, and learns
   $E(T, R, P)$.*
5. *The adversary chooses two plaintext messages* $P_0, P_1 \in \mathcal{P}$*, and sends* $(P_0, P_1)$
   *to* $\mathcal{B}$.
6. *The challenger chooses a uniform bit* $b \in \{0,1\}$*, randomness* $R \in \mathcal{R}$*, and
   sends $E(T, R, P_b)$ to the adversary.*
7. *The adversary computes* $b' \in \{0,1\}$ *and wins iff* $b' = b$.

It may be tempting, in the previous definition, to allow the adversary to first
query $E$, and choose $f$ based on the answers. However it is not necessary to
add such interactions to the definition: indeed, such interactions can be folded
into the function $f$, which can be regarded as an arbitrary algorithm or protocol
between the adversary and the challenger having access to $T$. The only limitation
is that the min-entropy of $T$ should remain above $\mu$ from the point of view of
the adversary. It is clear that a limitation of this sort is necessary, otherwise the
adversary could simply learn $T$.

Furthermore, while a definition based on min-entropy may seem rather
impractical, it encompasses as a special case the simpler space-hard notion of
[BI15]. In that case the table $T$ is a uniform function, and $f$ outputs a fixed
proportion $1/4$ of the table. The min-entropy $\mu$ is then simply the number of
unknown output bits of the table (namely $3/4$ of its output).

The WhiteKey construction that we define later on is actually a key generator. That is, it takes as input a uniformly random string and outputs a key.
The strong incompressibility definition expects an encryption scheme. In order
for the WhiteKey key generator to fulfill strong incompressibility, it needs to be
converted into an encryption scheme. This is achieved generically by using the
generated key (the output of WhiteKey) with a conventional symmetric encryption scheme, as in a standard hybrid cryptosystem. For instance, the plaintext
can be XORed with the output of a pseudorandom generator whose input is the
generated key. Strictly speaking, when we say that WhiteKey satisfies strong
incompressibility, we mean that this is the case when WhiteKey is thus used

as a key generator in combination with any conventional symmetric encryption process.

Note that this does not enforce authenticity. For instance, if the generated key is used as an input to a stream cipher, forgeries are trivial. More generally it is not possible to prevent existential forgeries, as the adversarially compressed implementation could include any fixed arbitrary valid ciphertext. However universal forgeries can be prevented. This is naturally expressed by the following model. The model actually captures the required goal in previous definitions of incompressibility, in fact the model as a whole is essentially equivalent to incompressibility in the sense of [DLPR13].

**Definition 5 (Encryption Incompressibility, ENC-COM).** *Let $E : \mathcal{T} \times \mathcal{R} \times \mathcal{P}$ denote an encryption scheme. Let $\mu$ denote a security parameter. Let us further assume that the key $T \in \mathcal{T}$ is chosen according to some distribution $D$ (typically uniform). The scheme $E$ is said to be $(\tau, \epsilon)$-secure for $\mu$-strong incompressibility iff the advantage of an adversary $\mathcal{A}$ running in time $\tau$ and playing the following game is upper-bounded by $\epsilon$.*

1. *The adversary chooses a distribution $\mathcal{D}$ with min-entropy at least $\mu$ on $\mathcal{P}$.*
2. *The adversary chooses a set $\mathcal{S}$ and a function $f : \mathcal{T} \to \mathcal{S}$, subject only to the condition that for all $s \in \mathcal{S}$, the min-entropy of the variable $T$ conditioned on $f(T) = s$ is at least $\mu$. The function $f$ should be thought of as a compression algorithm chosen by the adversary.*
3. *Meanwhile the challenger $\mathcal{B}$ picks $T \in \mathcal{T}$ according to the distribution $D$ (thus fixing an instance of the encryption scheme).*
4. *The adversary receives $f(T)$.*
   *At this point the adversary is tasked with forging a message, namely:*
5. *The adversary samples a plaintext $M \in \mathcal{P}$ from the distribution $\mathcal{D}$.*
6. *The adversary may repeatedly choose any plaintext $P \in \mathcal{P}$, and learns $E(T, R, P)$.*
7. *The adversary wins iff she can compute a $C \in \mathcal{C}$ such that $D(T, C) = M$.*

This model can also be fulfilled by the WhiteKey scheme, if we derive the required randomness from $H(P) + r$ where $H$ is a random oracle, $P$ is the plaintext, and $r$ is a uniform value of $\mu$ bits added to the encryption. The decryption starts by recovering the key, and then checks if the randomness used came from $H(P', r)$ where $P'$ is the decrypted plaintext. This naturally makes any encryption scheme derived from a key generator resistant to universal forgeries.

Remark that it is necessary in the model to have the forged message generated independently of $f(\mathcal{T})$, otherwise one can simply put an encryption of the message in $f(\mathcal{T})$.

Finally, observe that ENC-COM is stronger than ENC-TCOM, as ENC-TCOM it is the special case of ENC-COM where the adversary's chosen function $f$ does nothing more than querying $T$ on some adaptively chosen inputs, and returning the outputs.

# 3 Constructions

In this section, we present two constructions that are provably secure in the weak white-box model ENC-TCOM of Sect. 2 (cf. Definition 3): the WhiteBlock block cipher, and the WhiteKey key generator. WhiteKey is also provable in the strong model. We also propose PUPPYCIPHER and COUREURDESBOIS as concrete instantiations of each construction, using the AES as underlying primitive.

## 3.1 The WhiteBlock Block Cipher

The general idea of WhiteBlock is to build a Feistel network whose round function uses calls to a large table $T$. An adversary who does not extract and store a large part of this table should be unable to encrypt most plaintexts. For that purpose, it is important that the inputs of table calls be pseudo-random, or at least not overly structured. Otherwise the adversary could attempt to store a structured subset of the table that exploits this lack of randomness. In WhiteBlock, the pseudo-randomness of table calls is enforced by interleaving calls to a block cipher between each Feistel round.

Concretely, WhiteBlock defines a family of block ciphers with blocks of size $b = 128$ bits, and a key of size $\kappa = 128$ bits[4]. The family is parameterized with a *size* parameter which corresponds to the targeted size of a white-box implementation. In principle, this size can be anything from a few dozen bytes up to $\approx 2^{64}$ bytes, but we will mostly restrict this description to the smallest case considered in this article, which has an implementation of size $2^{21}$ bytes.

Formally, we define one round of WhiteBlock (with tables of input size 16 bits) as follows. Let $\mathcal{A}_k$ denote a call to the block cipher $\mathcal{A}$ with key $k$, and $\mathcal{T}_i : \{0,1\}^{16} \rightarrow \{0,1\}^{64}$ denote the $i$-th table. The Feistel round function is defined by:

$$\mathcal{F} : \{0,1\}^{64} \rightarrow \{0,1\}^{64},$$
$$x_{63} \ldots x_0 \mapsto \mathcal{T}_3(x_{63} \ldots x_{48}) \oplus \mathcal{T}_2(x_{47} \ldots x_{32}) \oplus \mathcal{T}_1(x_{31} \ldots x_{16}) \oplus \mathcal{T}_0(x_{15} \ldots x_0)$$

and one round of WhiteBlock with key $k$ is defined as:

$$\mathcal{R}_k : \{0,1\}^{128} \rightarrow \{0,1\}^{128}$$
$$x_{127} \ldots x_0 \mapsto \mathcal{A}_k \left( ((x_{127} \ldots x_{64}) \oplus \mathcal{F}(x_{63} \ldots x_0)) \| x_{63} \ldots x_0 \right).$$

A full instance of WhiteBlock is then simply the composition of a certain number of independently-keyed round functions, with the addition of one initial top call to $\mathcal{A}$: $\text{WhiteBlock}_{k_0,\ldots k_r} : \{0,1\}^{128} \rightarrow \{0,1\}^{128}, x \mapsto \mathcal{A}_{k_r} \circ \mathcal{R}_{k_{r-1}} \circ \cdots \circ \mathcal{R}_{k_0}(x)$. We give an illustration of this construction (omitting the outer sandwiching calls to $\mathcal{A}$) in Fig. 1.

---

[4] This generalizes well to other sizes.

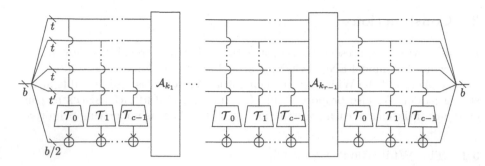

**Fig. 1.** The WhiteBlock construction, with tables on $t$ bits, without the outer calls to $\mathcal{A}$. We have $t' = (b/2) \mod s$, $c = \lfloor (b/2)/t \rfloor$.

**Constructing the Tables.** For WhiteBlock instances with small tables, the most efficient way to implement the cipher is simply to use the white-box implementation, *i.e.* use a table-based implementation of $\mathcal{F}$ (this will be clear from the results of Sect. 5). In that case, it is easy to generate the tables "perfectly" by drawing each entry uniformly at random, either by using a suitable source of randomness (in that case, no one may be able to compress the tables) or by using the output of a cryptographically-strong PRG seeded with a secret key. In the latter case, the owner of the secret knows how to compactly represent the tables, but this knowledge seems to be hard to exploit in a concrete implementation.

For larger instances, it is not true anymore that the fastest implementation is table-based, and it may be useful in some contexts to be able to compute the output of a table more efficiently than by querying it. Surely, if one knows how to compactly represent a table, it is desirable that he would be able to do so, at least for large tables. In that respect, drawing the tables at random would not be satisfactory anymore.

Consequently, the tables used in WhiteBlock are generated as follows. Let again $\mathcal{T}_i : \{0,1\}^{16} \to \{0,1\}^{64}$ be such a table (in the 16-bit case), then an instance of it is defined with two parameters $k \in \{0,1\}^{128}$, $c \in \{0,1\}^{128-16}$ as $\mathcal{T}(x) \triangleq \lfloor \mathcal{A}_k(c||x) \rfloor_{64}$, with $\lfloor \cdot \rfloor_{64}$ denoting the truncation to the 64 lowest bits.

An instance of WhiteBlock can thus always be described and implemented compactly when knowing $k$ and $c$. Of course this knowledge is not directly accessible in a white-box implementation, where a user would only be given the tables as a whole.

**Concrete Parameters for Various Instances of WhiteBlock.** We need to address two more points before finishing this high-level description of WhiteBlock: (1) given the size of the tables, how many rounds $r$ are necessary to obtain a secure white-box construction; (2) how to generate the multiple round keys $k_0, \ldots k_r$. The answer to (1) is provided by the analysis of the construction done in the full paper, specifically [FKKM16, Theorem 3]. By instantiating the formula from the theorem with concrete parameters, we obtain the results given in Table 1. As for (2), we simply suggest to use independent keys (as both their generation process and the cost of storing the precomputed subkeys are negligible w.r.t. the generation and storage of the tables). There is some flexibility in the framework and one can for instance consider using a tweakable block cipher instead, as we do in our actual instantiation of WhiteBlock presented next.

**Table 1.** Number of rounds for WhiteBlock instances with tables of selected input sizes from $t = 16$ to $32$ bits, at a white-box security level of $128 - t$ bits for a compression factor of 4. Black-box security is 128 bits in all cases.

| Instance | WB size | # Tables/round | WB security | #rounds |
|---|---|---|---|---|
| WhiteBlock 16 | $2^{21}$ B | 4 | 112 bits @ 1/4 | 18 |
| WhiteBlock 20 | $2^{24.6}$ B | 3 | 108 bits @ 1/4 | 23 |
| WhiteBlock 24 | $2^{28}$ B | 2 | 104 bits @ 1/4 | 34 |
| WhiteBlock 28 | $2^{32}$ B | 2 | 100 bits @ 1/4 | 34 |
| WhiteBlock 32 | $2^{36}$ B | 2 | 96 bits @ 1/4 | 34 |

**PuppyCipher: WhiteBlock in Practice.** So far WhiteBlock has been described from an abstract point of view, where all components are derived from a block cipher $\mathcal{A}$. In practice, we need to specify a concrete cipher; we thus define the PUPPYCIPHER family as an instantiation of WhiteBlock using AES128 [DR02] for the underlying block cipher. Furthermore, though relying on a secure block cipher is an important argument in the proof of the construction, one can wish for a less expensive round function in practice. Hence we also define the lighter, more aggressive alternative "HOUND" which trades provable security for speed. The only differences between PUPPYCIPHER and HOUND are:

1. The calls to the full AES128 are traded for calls to AES128 reduced to five rounds (this excludes the calls in the table generation, which still use the full AES).
2. The round keys $k_r \ldots k_0$ used as input to $\mathcal{A}$ are simply derived from a unique key $K$ as $k_i \triangleq K \oplus i$. Note that using a tweakable cipher such as KIASU-BC [JNP14] would also be possible here.

In Sect. 5, we discuss the efficiency of PUPPYCIPHER and HOUND implemented with the AES instructions, for tables of 16, 20, and 24-bit inputs.

## 3.2 The WhiteKey Key Generator

In WhiteBlock, we generated pseudo-random calls to a large table by interleaving a block cipher between table calls. If we are not restricted by the state size of a block cipher, generating pseudo-random inputs for the table is much easier: we can simply use a pseudo-random generator. From a single input, we are then able to generate a large number of pseudo-random values to be used as inputs for table calls. It then remains to combine the outputs of these table calls into a single output value of appropriate size. For this purpose, we use an entropy extractor. More details on our choice of extractor are provided in the design rationale below.

We now describe the WhiteKey function family, which can in some way be seen as an unrolled and parallel version of WhiteBlock, with some adjustments. As with WhiteBlock, we describe the main components of WhiteKey for use with a 128-bit block cipher and tables of 16-bit inputs, but this generalizes easily to other sizes.

Thus WhiteKey uses a table $T : \{0,1\}^{16} \rightarrow \{0,1\}^{128}$. Let $n$ denote the number of table calls (which will be determined later on by security proofs), $t \triangleq \lceil n/8 \rceil$ and $d \triangleq \lceil \sqrt{n} \rceil$. At a high level, the construction of WhiteKey can be described by the following process: (1) from a random seed, generate $t$ 128-bit values using a block cipher $\mathcal{A}$ with key $k$ in counter mode; (2) divide each such value into eight 16-bit words; (3) use these words as $n$ inputs to the table $T$ (possibly ignoring from one to seven of the last generated values), resulting in $n$ 128-bit values $Q_{i,j}, 0 \leq i,j \leq d = \lceil \sqrt{n} \rceil$ (if $n$ is not a square, the remaining values $Q_{i,j}$ are set to zero); (4) from a random seed, generate $d$ 128-bit values $a_i$ and $d$ 128-bit values $b_i$ using $\mathcal{A}$ with key $k'$ in counter mode; (5) the output of WhiteKey is $\sum_{i,j} Q_{i,j} \cdot a_i \cdot b_j$, the operations being computed in $\mathbb{F}_{2^{128}}$.

Let us now define this more formally. We write $\mathcal{A}_k^t(s)$ for the $t$ first 128-bit output blocks of $\mathcal{A}$ in counter mode with key $k$ and initial value $s$. We write $\mathcal{C}_n$ for the parallel application of $n \leq 8 \times t$ tables $T : \{0,1\}^{16} \rightarrow \{0,1\}^{128}$ (written here in the case $n = 8 \times t$ for the sake of simplicity):

$$\mathcal{C}_n : \{0,1\}^{t \times 128} \rightarrow \{0,1\}^{n \times 128}$$

$$x_{t128-1}x_{t128-2}\cdots x_0 \mapsto T(x_{t128-1}\cdots x_{t128-16}) || \cdots || T(x_{15}\cdots x_0)$$

We write $\mathcal{S}_n$ for the "matrixification" mapping; taking $d \triangleq \lceil \sqrt{n} \rceil$ (here with $n = 57$, for a not too complex general case):

$$\mathcal{S}_n : \{0,1\}^{n \times 128} \rightarrow \mathcal{M}_d(\mathbb{F}_{2^{128}})$$

$$x_{n128-1}x_{n128-2}\cdots x_0 \mapsto \begin{pmatrix} x_{127}\cdots x_0 & x_{255}\cdots x_0 & \cdots & x_{1023}\cdots x_{896} \\ \vdots & \vdots & \ddots & \vdots \\ x_{n128-1}\cdots x_{n128-128} & 0 & \cdots & 0 \end{pmatrix}.$$

Finally, we write $\mathcal{E}$ the "product" mapping:

$$\mathcal{E} : \mathbb{F}_{2^{128}}^d \times \mathbb{F}_{2^{128}}^d \times \mathcal{M}_d(\mathbb{F}_{2^{128}}) \rightarrow \mathbb{F}_{2^{128}}$$

$$a, b, Q \mapsto \sum_{i,j} Q_{i,j} \cdot a_i \cdot b_j$$

We can then describe an instance of WhiteKey parametered by $(k_1, s_1, k_2, s_2)$ over $t$ and $n$ values as $\text{WhiteKey}_{k_1,s_1,k_2,s_2}^{t,n} \triangleq \mathcal{E} \circ \mathcal{A}_{k_2}^d(s_2) \circ \mathcal{A}_{k_2}^d(s_2 + d) \circ \mathcal{S}_n \circ \mathcal{C}_n \circ \mathcal{A}^t(k_1, s_1)$ (using a Curried version of $\mathcal{E}$ for simplicity of notations).

**Constructing the Tables.** The table used in an instance of WhiteKey is built in the same way as for WhiteBlock. The only difference is that the output of $\mathcal{A}$ is not truncated and the full 128 bits are used.

**Design Rationale of WhiteKey.** The first part of the scheme consists in selecting a fraction of the secret that needs to be accessed, which is a necessary step. The fastest way to implement this part is to access the secret in parallel at locations that are thus determined independently.

The second part is to derive a short key from the table outputs, which are of high min-entropy. The standard way to build a key derivation function is to use a hash function [Kra10]. However it is slow, since even a fast hash function like BLAKE2b takes 3 cycles per byte on modern processors [ANWOW13]. Instead, we decided to use an extractor, which has also the advantage to be unconditionally secure for a uniform seed. The extractor literature focused primarily on reducing the number of seed bits and maximizing the number of extracted bits, because of their importance in theoretical computer science; see [Sha11] for a survey. In our case, we want to extract only a few bits and speed is the principal concern. The approach recommended by [BDK+11] is to generate pseudo-random elements in a large field using a standard pseudorandom generator (say, AES-CTR) and to compute a dot product with the input. The main problem of this extractor is that it uses a seed which is very large, and it takes about as much time to generate it (with AES-NI) as to use it. Hence, we decided to use the extractor introduced in [CMNT11], which has a seed length about the square root of the length of the input. Since we can evaluate $\sum_{i,j} Q_{i,j} a_i b_j$ with about one multiplication and one addition in the field per input value, the computation of the extractor takes essentially the same time. Indeed, the complexity of the extractor is similar to GHASH.

Another possibility for the extractor is to increase the degree, for instance use $\sum_{i,j,k} Q_{i,j,k} a_i b_j c_k$. While this approach, proposed by [CNT12], is indeed sound and allows to reduce the seed further, the best bound we know on the statistical distance of the output is about $q^{-1/2}$ when working over $\mathbb{F}_q$. The main problem is that the tensor decomposition of $Q_{i,j,k}$ does not have the needed properties, so that Coron et al. use a generic bound on the number of zeroes, which must account for elliptic curves and therefore a deviation of $q^{-1/2}$ is required. The specific case of $\sum_{k=0}^{1} \sum_{i,j} Q_{i,j,k} a_i b_j c_k$ can probably be tackled using linear matrix pencil theory, at the cost of a much more difficult proof.

**Concrete Parameters for Various Instances of WhiteKey.** Once the size of an instance of WhiteKey has been chosen (*i.e.* the output size of the table $\mathcal{T}$), the only parameter that needs to be determined is the number of calls to the tables $n$, and thus the number of output blocks $t$ of $\mathcal{A}$. This is obtained by instantiating the formula of [FKKM16, Theorem 2] for a given white-box security. We give the parameters for instances of various sizes in Table 2. The tables used in these instances have the same input size as the ones of the WhiteBlock instances of Table 1, but they are twice as large because of their larger output size, which impacts the size of a white-box implementation similarly. On the other hand, a single table is used in WhiteKey, whereas up to four (for input sizes of 16 bits and more) are necessary in WhiteBlock.

**Table 2.** Number of table calls for WhiteKey instances with tables of selected input sizes from 16 to 32 bits, at a white-box security level of 96 to 112 bits for a compression factor of 4. Black-box security is 128 bits in all cases.

| Instance | WB size | # Table/block | WB security | #Table calls (#blocks) |
|---|---|---|---|---|
| WhiteKey 16 | $2^{20}$ B | 8 | 112 bits @ 1/4 | 57 (8) |
| WhiteKey 20 | $2^{24}$ B | 6 | 108 bits @ 1/4 | 55 (10) |
| WhiteKey 24 | $2^{28}$ B | 5 | 104 bits @ 1/4 | 53 (11) |
| WhiteKey 28 | $2^{32}$ B | 4 | 100 bits @ 1/4 | 51 (13) |
| WhiteKey 32 | $2^{36}$ B | 4 | 96 bits @ 1/4 | 49 (13) |

**CoureurDesBois: WhiteKey in Practice.** Similarly to WhiteBlock and PUPPYCIPHER, we define the COUREURDESBOIS family as a concrete instantiation of WhiteKey. It simply consists in using AES128 for $\mathcal{A}$ and a specific representation for $\mathbb{F}_{2^{128}}$, e.g. $\mathbb{F}_2[x]/x^{128} + x^7 + x^2 + x + 1$ (the "GCM" field).

Unlike PUPPYCIPHER, the components of COUREURDESBOIS are not cascaded multiple times; hence we cannot hope for a similar tradeoff of provable security against speed. However, the main advantage of COUREURDESBOIS compared to PUPPYCIPHER is that it lends itself extremely well to parallelization. This allows to optimally hide the latency of the executions of AES and of the queries to the table in memory.

We further discuss the matter in Sect. 5, where we evaluate implementations of COUREURDESBOIS with AES instructions for tables of 16 to 24-bit inputs.

## 4  Security Proofs

For both the WhiteBlock and WhiteKey constructions, we provide proofs in the weak incompressibility model. These proofs provide concrete bounds, on which we base our implementations. This allows direct comparison to previous work [BBK14, BI15]. Moreover in the case of WhiteKey, we provide a proof in the strong incompressibility model. This proof shows the soundness of the general construction in a very demanding model. However we do not use it to derive the parameters of our constructions.

Recall that weak incompressibility (Definition 3) depends on three parameters $s$, $\lambda$, $\delta$: essentially if the number of outputs of the table known to the adversary is $s$, then $(s, \lambda, \delta)$-incompressibility says that with probability at least $1 - 2^{-\lambda}$, the adversary is unable to encrypt more than a ratio $\delta$ of plaintexts, no matter which $s$ table outputs she chooses to learn. If inputs to the table are $t$-bit long, then $\alpha = s2^{-t}$ is the fraction of the table known to the adversary. We can fix $\alpha = 1/4$ as in [BI15], hence $s = \alpha 2^t$. In that case weak incompressibility essentially matches $(s, -\log(\delta))$-space hardness from [BI15], and $-\log(\delta)$ can be thought of as the number of bits of white-box security.

However we do not claim security for $\delta = 2^{-128}$, which would express 128 bits of white-box security. Instead, we claim security for $\delta = 2^{-128+t}$. Thus for larger table of size $\approx 2^{28}$, white-box security drops to around $2^{100}$. We believe this is quite acceptable.

The reason we claim only $128 - t$ bits of white-box security rather than 128 is a result of our security proofs, as we shall see. This should be compared with the fact that an adversary allowed to store $s$ table inputs could use the same space to store $s$ outputs of the whole scheme (within a small constant factor $\lambda/t$ in the case of WhiteBlock). Such an adversary would naturally be able to encrypt a proportion $s2^{-\lambda}$ of inputs. Since $s = 2^t/4$, with a small constant factor $1/4$, this yields the $128 - t$ bits of white-box security achieved by our proofs.

Our security claims are summarized in Tables 1 and 3.2.

## 4.1  Proofs of Weak Incompressibility

We provide proofs of both WhiteKey and WhiteBlock in the weak incompressibility model. In the case of WhiteKey, a proof is also available in the strong incompressibility model. However the proof of WhiteKey for weak incompressibility is fairly straightforward, yields better bounds (as one would expect), and also serves as a warm-up for the combinatorially more involved proof of WhiteBlock. The bulk of the proofs for WhiteKey and WhiteBlock are given in the full paper [FKKM16]. In this section, we provide some context and a brief outline.

**Weak Incompressibility of WhiteKey.** First note that if the AES in counter mode used in the initial layer of WhiteKey is modeled as a pseudo-random generator (PRG), the proof is quite straightforward. Indeed, we are then free to regard the inputs of table calls as uniformly random (after paying the PRG advantage of an adversary against counter mode AES). It follows that the adversary has probability $\alpha$ of knowing the output of each individual table call, where $\alpha$ is the proportion of the table she has queried, regardless of which particular inputs she chose to query. Since the extractor in the last layer of the scheme is linear, as soon as the adversary is missing one table output, the global output of the scheme is uniformly random from her point of view.

However we focus on a different route for the proof, where the initial layer of the scheme is modeled as a pseudo-random function (PRF) rather than a PRG. The main reason we do this is that the resulting proof will be much closer to the proof of WhiteBlock, and serve to prepare it.

We thus view the initial layer of WhiteKey as being comprised of a PRF generating the inputs of the table calls. Using standard arguments, this pseudo-random function can be replaced by a random function; the effect this has on the weak incompressibility adversary is upper-bounded by the distinguishing advantage of a real-or-random adversary against the PRF.

In the weak incompressibility game, the adversary learns the output of the table on some adaptively chosen inputs. By nature of white-box security, any keying material present in the PRF is known to the adversary (formally, in our definition of white-box encryption scheme this keying material would have to be appended to the table $T$ of the white-box implementation, and could be recovered with a single or few queries). Hence the adversary can choose which table inputs she queries based on full knowledge of the initial PRF.

On the other hand, for a given PRF input, as soon as the adversary does not know a single output of the table, due to the linearity of the final layer of the construction, the output has full 128-bit entropy from the point of view of the adversary.

Thus the core of the proof, is to show that, with high probability over the random choice of the PRF, for the best possible choice of $s$ table inputs the adversary chooses to query[5], most PRF outputs still include at least one table input that is unknown to the adversary. We explicitly compute this upper bound in the complete proof.

More precisely, [FKKM16, Theorem 2] shows:

$$\log\left(\Pr\left[\mu(s) \geq k\right]\right) \leq 2^t - k\log\left(\frac{k}{\rho}\right) - (n-k)\log\left(\frac{n-k}{n-\rho}\right) \tag{1}$$

where:

- $n = 2^\lambda$ is the size of the input space of WhiteKey;
- $t$ is the number of bits at the input of a table;
- $s$ is the number of table entries stored by the adversary;
- $\rho = 2^\lambda(s/2^t)^m$, with $m$ the number of table calls in the construction;
- $k$ is the maximal number of inputs the adversary may be able to encrypt;
- and $\mu(s)$ is the maximal number of WhiteKey inputs that can be encrypted with storage size $s$; it is a random variable over the uniform choice of the initial PRF ($\mathcal{A}$ in counter mode, in the previous description).

We want this bound to be below $-\lambda$. We are now interested in what this implies, in terms of number of table calls $m$ necessary to achieve a given security level. As noted earlier, the bound imposes $k \approx 2^t$. For simplicity we let $k = 2^t$, which means we achieve $\lambda - t$ bits of white-box security (i.e. $\delta = 2^{t-\lambda}$ in the sense of Definition 3). We can also fix $s/2^t = 1/4$ for the purpose of being comparable to [BI15].

The term $(n-k)\ln\left((n-k)/(n-\rho)\right)$ is equivalent to $\rho - k$ as $k/n$ tends to zero[6]. Since we are looking for an upper bound we can approximate it by $k$. This yields a probability:

$$2^t\left(1 - k2^{-t}\left(\log\left(\frac{k}{\rho}\right) - 1\right)\right) = 2^t\left(1 - k2^{-t}\left(\log(k) - \lambda + 2m - 1\right)\right)$$

$$= -2^t\left(\log(k) - \lambda + 2m\right)$$

---

[5] In this respect, the adversary we consider is computationally unbounded.

[6] In fact, simple functional analysis shows that we can bound the right-hand term by $4(\rho - k)$ provided $\alpha^m < 1/2$ and $k < 4n$, which will always be the case.

In the end, we get that $m$ only needs to be slightly larger than $\frac{\lambda - \log(k)}{2}$. Indeed, as long as this is the case, the $2^t$ factor will ensure that the bound is (much) lower than $-128$.

This actually matches a generic attack. If the adversary just stores $s = 2^t/4$ random outputs of the table, then on average she is able to encrypt a ratio $2^{-2m}$ of inputs. This imposes $2^{-2m} < k2^{-\lambda}$, so $m > (\lambda - \log(k))/2$. When testing our parameter choices against Eq. 1, we find that it is enough to add a single table call beyond what the generic attack requires: in essence, [FKKM16, Theorem 2] implies that no strategy is significantly better than random choices.

**Weak Incompressibility of WhiteBlock.** The general approach of the proof is the same as above. However the combinatorial arguments are much trickier, essentially because table calls are no longer independent (they depend on table outputs in the previous round.). Nevertheless an explicit bound is proven in the full paper.

However, what we prove is only that w.h.p., for most inputs to WhiteBlock, during the computation of the output, at least two table calls at different rounds are unknown to the adversary. Since table outputs cover half a block, this implies that at two separate rounds during the course of the computation, 64 bits are unknown and uniform from the point of view of the adversary. At this point we heuristically assume that for an efficient adversary, this implies the output cannot be computed with probability significantly higher then $2^{-128}$. In practice the bottleneck in the bound provided by the proof comes from other phenomena, namely we prove $128 - t$ bits of security for $t$-bit tables. Nevertheless this means our proof is heuristic.

More precisely, [FKKM16, Theorem 3] shows:

$$\log\left(\Pr\left[\mu(s) \geq k\right]\right) \leq 2^t + k\left(\lambda + m\left(1 - \frac{1}{k} - \frac{1}{r}\right)\log\left(\frac{s}{2^t}\right)\right)$$

where:

- $\lambda$ is the input size of WhiteBlock;
- $t$ is the number of bits at the input of a table;
- $r$ is the number of rounds;
- $m$ is the total number of table calls in the construction ($m \triangleq \lfloor (\lambda/2)/t \rfloor \cdot r$);
- $s$ is the number of table entries stored by the adversary;
- $k$ is the maximal number of inputs the adversary may be able to encrypt;

and $\mu(s)$ is the maximal number of WhiteBlock inputs that can be encrypted with storage size $s$; it is a random variable over the uniform choice of the round permutations $\mathcal{A}_{k_i}$.

We are now interested in what this bound implies, in terms of number of rounds $r$ to achieve a given security level. Observe that the bound requires $k \approx 2^t$. For simplicity we let $k = 2^t$, which means we achieve $\lambda - t$ bits of white-box security (*i.e.* $\delta = 2^{t-\lambda}$ in the sense of Definition 3). We can also fix

$s/2^t = 1/4$ for the purpose of being comparable to [BI15]. Observe that $1/k$ is negligible compared to $1/r$. Let $c = \lfloor (\lambda/2)/t \rfloor$ be the number of table calls per round. Then our bound asks:

$$\lambda - 2m\left(1 - \frac{1}{r}\right) = \lambda - 2c(r-1) < 0$$

Indeed, as long as this value is negative, the preceding $k = 2^t$ factor will ensure that the bound is (much) lower than $-128$. We get:

$$r > \frac{\lambda}{2c} + 1$$

We can compare this bound with the previous generic attack, where the adversary stores table outputs at random. As we have seen, this attack implies $m > (\lambda - \log(k))/2$, so $r > (\lambda - \log(k))/(2c)$. Instead our proof requires $r > \frac{\lambda}{2c} + 1$. Thus the extra number of rounds required by our security proof, compared to the lower bound coming from the generic attack, is less than $\log(k)/(2c) + 1$: it is only a few extra rounds (and not, for instance, a multiplicative factor).

### 4.2 Proof of Strong Incompressibility

We first prove that $\sum_{i,j} Q_{i,j} a_i b_j \in \mathbb{F}_q$ is a strong extractor. This extractor comes mostly from Coron *et al.* [CMNT11, Sect. 4.2] but we tighten the proof.

**Definition 6.** *A family $\mathcal{H}$ of hash functions $h : X \mapsto Y$ is $\epsilon$-pairwise indepen-dent if*

$$\sum_{x \neq x'} \left( \Pr_{h \leftarrow \mathcal{H}}[h(x) = h(x')] - \frac{1}{Y} \right) \leq \frac{\epsilon |X|^2}{Y}.$$

The next lemma is a variant of the leftover hash lemma, proven in [Sti02, Theorem 8.1].

**Lemma 1.** *Let $h \in \mathcal{H}$ be uniformly sampled, and $x \in X$ be an independent ran-dom variable with min-entropy at least $k$. Then, the statistical distance between $(h(x), h)$ and the uniform distribution is at most*

$$\sqrt{|Y|2^{-k} + \epsilon}.$$

We now prove that our function is indeed pairwise independent.

**Lemma 2.** *Let $\mathcal{H} = \mathbb{F}_q^{2n}$, $X = M_n(\mathbb{F}_q)$ and $Y = \mathbb{F}_q$. Then, the function $h_{a,b}(Q) = \sum_{i,j} Q_{i,j} a_i b_j = a^t Q b$ is $11q^{-n}$-pairwise independent.*

*Proof.* We first count the number of $a, b$ such that $\sum_{i,j} Q_{i,j} a_i b_j = a^t Q b = 0$. Let $Q$ be a matrix of rank $r$. Then, there exist $r$ vectors $u, v$ such that $Q = \sum_{k=0}^{r-1} u_i v_i^t$ and the $u_i$ as well as the $v_i$ are linearly independent. Thus,

$$a^t Q b = \sum_{k=0}^{r-1} a^t u_i v_i^t b.$$

and therefore, by a change of basis, this form has the same number of zeros as

$$\sum_{k=0}^{r-1} a_i b_i$$

which is $q^{2n-1} + q^{2n-r} - q^{2n-r-1}$.

Now, there are $\prod_{k=1}^{r-1} \frac{(q^n - q^k)^2}{q^r - q^k}$ matrices of rank $r$. We deduce:

$$\sum_{x \neq x'} \left( \Pr_{h \leftarrow \mathcal{H}}[h(x) = h(x')] - \frac{1}{Y} \right) = \sum_{r=1}^{n} \left( (q^{-r} - q^{-r-1})q^{-n^2} \prod_{k=0}^{r-1} \frac{(q^n - q^k)^2}{q^r - q^k} \right)$$

$$\leq \sum_{r=1}^{n} q^{-r} q^{-n^2} q^{2nr-r^2} \prod_{k=1}^{\infty} \frac{1}{1 - 1/q^k}$$

$$\leq \frac{2 - 1/q}{1 - 1/q} q^{-n} \prod_{k=1}^{\infty} \frac{1}{1 - 1/q^k}$$

$$\leq 11 q^{-n} \qquad \qquad \square$$

Hence, if the input of our extractor has at least $2\mu$ bits of entropy, the generated key will be essentially uniform. The proof for the security of sampling the seed from a pseudorandom generator (from which we cannot build a public-key primitive) is in [BDK+11]. We now prove that the input has indeed a lot of entropy.

**Lemma 3.** *Let $f : [n] \mapsto [0;1]$ be of average $\mu$. Then, the average of the image $k$ uniform elements is at least $\mu - \delta$, except with probability*

$$\exp\left(-\frac{k^2\delta^2/2}{k/4 + \delta\mu/3}\right).$$

*Proof.* This is the result of Bernstein's inequality (see [BLB04, Theorem 3]), since the variance of all terms is at most $1/4$ and they are all positive. $\qquad \square$

We now use a lemma of Vadhan [Vad04, Lemma 9]:

**Lemma 4.** *Let $S$ be a random variable over $[n]^t$ with distinct coordinates and $\mu, \delta, \epsilon > 0$, such that for any function $f : [n] \mapsto [0;1]$ of average $(\delta - 2\tau)/\log(1/\tau)$, we have that the probability that the average of the image of the $t$ positions given by $S$ is smaller than $(\delta - 3\tau)/\log(1/\tau)$ is at most $\epsilon$.*

*Then, for every $X$ of min-entropy $\delta n$ over $\{0,1\}^n$, the variable $(S, X_S)$ where $X_S$ is the subset of bits given by $S$ is $\epsilon + 2^{-\Omega(\tau n)}$ close to $(A, B)$ where $B$ conditioned on $A = a$ has a min-entropy $(\delta - \tau)t$.*

Finally, it is clear that if a sampling done with a pseudorandom generator instead of a uniform function leads to a low min-entropy key, we have a distinguisher on the pseudorandom generator.

# 5    Implementation

In this section, we evaluate the efficiency of PuppyCipher {16,20,24}, Hound {16,20,24} and CoureurDesBois {16,20,24}, when implemented with the AES and PCLMULQDQ instructions (the latter being only used for the finite field arithmetic of CoureurDesBois) on a recent *Haswell* CPU. For each algorithm, we tested table-based white-box implementations and "secret" implementations where one has the knowledge of the key used to generate the tables.

The number of rounds we choose was directly deduced from proofs in the weak model (cf. Sects. 3 and 4). Since this model essentially matches that of previous work [BBK14, BI15], this allows for a direct comparison.

The processor on our test machine was an Intel Xeon E5-1603v3, which has a maximal clock frequency of 2.8 GHz and a 10 MB cache (which is thus larger than the implementation sizes of the '16 instances). The machine has 32 GB of memory, in four sticks of 8 GB all clocked at 2133 MHz. All measurements were done on an idle system, without Turbo Boost activated[7]. As a reference, we first measured the performance of AES128 implemented with the AES instructions, given in Table 3. We give the average (Avg.) number of clock cycles and the standard deviation (Std. Dev.) for one execution, both in the transient and steady regime (in practice, when performing series of independent runs, the transient regime only corresponds to the first run of the series). The average and standard deviation are computed from 25 series of 11 runs. The figures obtained from this test are coherent with the theoretical performance of the AES instruction set (even if slightly better): on a Haswell architecture, the `aesenc` and `aesenclast` instructions are both given for a latency of 7 cycles, and the cost of a single full AES128 is dominated by the $10 \times 7$ calls to perform the 10 rounds of encryption.

**Table 3.** Performance of a single call to AES128 with AES instructions on a Xeon E5-1603v3. All numbers are in clock cycles.

|        | Transient Avg. | Transient Std. Dev. | Steady Avg. | Steady Std. Dev. |
|--------|----------------|---------------------|-------------|------------------|
| AES128 | 79             | 3.6                 | 68          | 2.4              |

## 5.1    PuppyCipher

Writing a simple implementation of PuppyCipher is quite straightforward. The main potential for instruction-level parallelism (ILP) are the calls to the tables (or the analogous on-the-fly function calls); the rest of the cipher is chiefly sequential, especially the many intermediate calls to the (potentially reduced) AES. This parallelism is however somewhat limited, especially starting from PuppyCipher 24 where only two parallel calls to the tables can be made.

---

[7] As a matter of fact, this CPU does not have Turbo Boost support.

In all implementations, we precompute the sub-keys for the calls to AES (including calls potentially made to emulate the tables). Not doing so would only add a negligible overhead.

The performance measurements were done in a setting similar to the reference test on AES128 from above. We give the results for PUPPYCIPHER {16,20,24} in Table 4 and for HOUND {16,20,24} in Table 5. In both tables, we also express the performance in the steady regime as the number of equivalent AES128 calls (Eq. A) with AES instructions on the same platform (taken to be 68 cycles, as per Table 3) as it is a block cipher with similar expected (black-box) security, and as the number of equivalent ephemeral Diffie-Hellman key exchanges with the FourQ elliptic curve (Eq. F), one of the fastest current implementation of ECDHE [CL15] (measured at 92000 cycles on the Haswell architecture), as there is some overlap in what white-box and public-key cryptography try to achieve.

**Discussion.** As it was mentioned in Sect. 3, for a small white-box implementation such as the one of PUPPYCIPHER 16, table-based implementations may be the most efficient way of implementing the cipher, especially as the entire tables can usually fit in the cache. However, from a certain size on, the random RAM accesses inherent to such implementations cost more than recomputing the necessary outputs of the tables (when the secret is known).

It is quite easy to estimate how much time is spent in RAM accesses compared to the time spent in calls to the (potentially reduced) AES. Indeed, knowing the number of rounds and the cost of one AES execution, one can subtract this contribution to the total. For instance, based on the cycle counts in the steady and transient regimes, for PUPPYCIPHER 24, at least $2380 = 35 \times 68$ and at most $2765 = 35 \times 79$ cycles are expected to be spent in AES instructions; the real figure in this case is about 2690 cycles, for an average cost per AES call of 77 cycles. All in all, this means that in steady regime, close to 90 % of the time is spent in RAM accesses. This is understandingly slightly more for the HOUND 24 variant, where RAM accesses represent about 93 % of the execution time.

**Table 4.** Performance of a single call to PUPPYCIPHER {16,20,24} ("PC") on a Xeon E5-1603v3. All numbers are in clock cycles, rounded to the nearest ten. The "white-box" instances are table-based, and the "secret" instances uses on-the-fly computations of the tables on their queried values. All calls to AES use the AES instructions.

|  | Tr. Avg. | Tr. Std. Dev. | St. Avg. | St. Std. Dev. | Eq. A | Eq. F |
|---|---|---|---|---|---|---|
| PC 16 (white-box) | 2960 | 130 | 2800 | 70 | 41 | 0.030 |
| PC 16 (secret) | 4140 | 60 | 3940 | 10 | 58 | 0.043 |
| PC 20 (white-box) | 13660 | 1000 | 11500 | 1190 | 169 | 0.125 |
| PC 20 (secret) | 4810 | 60 | 4540 | 100 | 67 | 0.049 |
| PC 24 (white-box) | 27570 | 1410 | 23390 | 1340 | 344 | 0.25 |
| PC 24 (secret) | 6760 | 120 | 6600 | 60 | 97 | 0.072 |

**Table 5.** Performance of a single call to HOUND {16,20,24} ("HD") on a Xeon E5-1603v3. All numbers are in clock cycles, rounded to the nearest ten. The "white-box" instances are table-based, and the "secret" instances uses on-the-fly computations of the tables on their queried values. All calls to AES use the AES instructions.

|  | Tr. Avg. | Tr. Std. Dev. | St. Avg. | St. Std. Dev. | Eq. A | Eq. F |
|---|---|---|---|---|---|---|
| HD 16 (white-box) | 2300 | 180 | 2190 | 130 | 32 | 0.024 |
| HD 16 (secret) | 3520 | 80 | 3280 | 2 | 48 | 0.036 |
| HD 20 (white-box) | 11870 | 980 | 9940 | 1030 | 146 | 0.11 |
| HD 20 (secret) | 4000 | 230 | 3700 | 65 | 54 | 0.040 |
| HD 24 (white-box) | 26540 | 1450 | 21740 | 1230 | 320 | 0.24 |
| HD 24 (secret) | 5490 | 60 | 5360 | 60 | 79 | 0.058 |

It is also interesting to look at how many RAM accesses can effectively be done in parallel. As two to four table calls are independent every round, we may hope to partially hide the latency of some of these. For PUPPYCIPHER 24, removing one of the two table accesses decreases the cycle count to 19400 on average. This means that the second table call only adds less than 4000 cycles. Put another way, using a single table per round, one table access takes 490 cycles on average, but this goes down to an amortized 300 cycles when two tables are accessed per round. In the end, the 68 table access of PUPPYCIPHER 24 only cost an equivalent 42 purely sequential accesses. A similar analysis can be performed for PUPPYCIPHER 20 and PUPPYCIPHER 16, where the 69 and 72 parallel accesses cost 31 and 23 equivalent accesses respectively.

COMPARISON WITH SPACE. We can compare the performance of PUPPYCIPHER with the one of SPACE-(16,128) and SPACE-(24,128), which offer similar white-box implementation sizes as PUPPYCIPHER 16 and PUPPYCIPHER 24 respectively [BI15]. As the authors of SPACE do not provide cycle counts for their ciphers but only the number of necessary cache or RAM accesses, a few assumptions are needed for a brief comparison. Both SPACE instances need 128 table accesses, which is much more than the 72 of PUPPYCIPHER 16 and 68 of PUPPYCIPHER 24. However, there is an extra cost in PUPPYCIPHER due to the many AES calls, which need to be taken into account. On the other hand, the table accesses in SPACE are necessarily sequential, which is not the case for PUPPYCIPHER, and we have just seen that parallel accesses can bring a considerable gain. It is thus easiest to use our average sequential access times as a unit. In that respect, PUPPYCIPHER 24 and HOUND 24 cost on average $48 = 23390/490$ and $44 = 21790/490$ table accesses, which is significantly less than the 128 of SPACE-(24,128). Similarly, we measured one sequential table access for PUPPYCIPHER 16 to take 59 cycles on average, and we thus have a cost of $47 = 2800/59$ and $37 = 2190/59$ for table accesses for PUPPYCIPHER 16 and HOUND 16.

The performance gap reduces slightly when one considers the case of "secret" implementations. As the tables of SPACE use the AES as a building block, the cost of a secret SPACE (24–128) implementation should correspond to approximately 128 sequential calls to AES; the corresponding PUPPYCIPHER and HOUND implementations cost an equivalent 97 and 79 AES respectively.

## 5.2 COUREURDESBOIS

The main advantage of COUREURDESBOIS compared to PUPPYCIPHER (as far as efficiency is concerned) is the higher degree of parallelism that it offers. Unlike PUPPYCIPHER, the calls to AES can be made in parallel, and there is no limit either in the potential parallelism of table accesses. Because the output of the tables are of a bigger size, there is also fewer accesses to be made. Consequently, we expect COUREURDESBOIS to be quite more efficient than PUPPYCIPHER.

A consequence of the higher parallelism of COUREURDESBOIS is that there are more potential implementation tradeoffs than for PUPPYCIPHER. In our implementations, we chose to parallelize the AES calls up to four calls at a time, and the table accesses (or equivalent secret computations) at the level of one block (*i.e.* from eight parallel accesses for COUREURDESBOIS 16 to five for COUREUR-DESBOIS 24). The final step of COUREURDESBOIS also offers some parallelism; we have similarly regrouped the calls to AES used for randomness generation by four, and the finite field multiplications are regrouped by rows of eight.

The results for COUREURDESBOIS {16,20,24} are given in Table 6.

**Table 6.** Performance of a single call to COUREURDESBOIS {16,20,24} ("CDB") on a Xeon E5-1603v3. All numbers are in clock cycles, rounded to the nearest ten. The "white-box" instances are table-based, and the "secret" instances uses on-the-fly computations of the tables on their queried values. All calls to AES use the AES instructions.

|                     | Tr. Avg. | Tr. Std. Dev. | St. Avg. | St. Std. Dev. | Eq. A | Eq. F |
|---------------------|----------|---------------|----------|---------------|-------|-------|
| CDB 16 (white-box)  | 3190     | 460           | 2020     | 20            | 29.7  | 0.022 |
| CDB 16 (secret)     | 3100     | 380           | 2150     | 30            | 31.6  | 0.023 |
| CDB 20 (white-box)  | 7880     | 880           | 4700     | 600           | 69.1  | 0.051 |
| CDB 20 (secret)     | 4060     | 460           | 2900     | 20            | 42.6  | 0.032 |
| CDB 24 (white-box)  | 17360    | 980           | 11900    | 610           | 175   | 0.13  |
| CDB 24 (secret)     | 4470     | 560           | 3050     | 30            | 44.9  | 0.033 |

**Discussion.** We can notice a few things from these results. First, COUREUR-DESBOIS is indeed more efficient than PUPPYCIPHER; for instance, COUREUR-DESBOIS 24 is about twice as fast as HOUND 24. Second, the performance gap between secret and white-box implementations is somewhat smaller for the smaller instances of COUREURDESBOIS; on the other hand, the gap between transient and steady regime performance is slightly bigger than for PUPPYCI-PHER.

As pointed out above, more tradeoffs are possible in implementing COUREUR-DESBOIS than for PUPPYCIPHER. As a result, it would be interesting to evaluate alternatives in practice.

Implementations of our schemes will be made available at http://whitebox4. gforge.inria.fr/.

**Acknowledgments.** The authors would like to thank Florent Tardif for letting us use his test machine.

# References

[ADW09] Alwen, J., Dodis, Y., Wichs, D.: Survey: leakage resilience and the bounded retrieval model. In: Kurosawa, K. (ed.) ICITS 2009. LNCS, vol. 5973, pp. 1–18. Springer, Heidelberg (2010). doi:10.1007/978-3-642-14496-7_1

[ANWOW13] Aumasson, J.-P., Neves, S., Wilcox-O'Hearn, Z., Winnerlein, C.: BLAKE2: simpler, smaller, fast as MD5. In: Jacobson, M., Locasto, M., Mohassel, P., Safavi-Naini, R. (eds.) ACNS 2013. LNCS, vol. 7954, pp. 119–135. Springer, Heidelberg (2013). doi:10.1007/978-3-642-38980-1_8

[ARM09] ARM: Security Technology Building a Secure System Using TrustZone Technology. White paper (2009). http://infocenter.arm.com/help/topic/com.arm.doc.prd29-genc-009492c/

[BBK14] Biryukov, A., Bouillaguet, C., Khovratovich, D.: Cryptographic schemes based on the ASASA structure: black-box, white-box, and public-key (Extended Abstract). In: Sarkar, P., Iwata, T. (eds.) ASIACRYPT 2014. LNCS, vol. 8873, pp. 63–84. Springer, Heidelberg (2014). doi:10.1007/978-3-662-45611-8_4

[BDK+11] Barak, B., Dodis, Y., Krawczyk, H., Pereira, O., Pietrzak, K., Standaert, F.-X., Yu, Y.: Leftover hash lemma, revisited. In: Rogaway, P. (ed.) CRYPTO 2011. LNCS, vol. 6841, pp. 1–20. Springer, Heidelberg (2011). doi:10.1007/978-3-642-22792-9_1

[BGEC04] Billet, O., Gilbert, H., Ech-Chatbi, C.: Cryptanalysis of a white box AES implementation. In: Handschuh, H., Hasan, M.A. (eds.) SAC 2004. LNCS, vol. 3357, pp. 227–240. Springer, Heidelberg (2004). doi:10.1007/978-3-540-30564-4_16

[BGI+01] Barak, B., Goldreich, O., Impagliazzo, R., Rudich, S., Sahai, A., Vadhan, S., Yang, K.: On the (im)possibility of obfuscating programs. In: Kilian, J. (ed.) CRYPTO 2001. LNCS, vol. 2139, pp. 1–18. Springer, Heidelberg (2001). doi:10.1007/3-540-44647-8_1

[BI15] Bogdanov, A., Isobe, T.: Revisited, white-box cryptography: space-hard ciphers. In: CCM 2015, pp. 1058–1069. ACM (2015)

[BKR16] Bellare, M., Kane, D., Rogaway, P.: Big-key symmetric encryption: resisting key exfiltration. In: Robshaw, M., Katz, J. (eds.) CRYPTO 2016. LNCS, vol. 9814, pp. 373–402. Springer, Heidelberg (2016). doi:10.1007/978-3-662-53018-4_14

[BLB04] Boucheron, S., Lugosi, G., Bousquet, O.: Concentration inequalities. In: Bousquet, O., Luxburg, U., Rätsch, G. (eds.) ML -2003. LNCS (LNAI), vol. 3176, pp. 208–240. Springer, Heidelberg (2004). doi:10.1007/978-3-540-28650-9_9

[CD16]     Costan, V., Devadas, S.: Intel SGX Explained. IACR Cryptology ePrint Archive 2016:86 (2016)

[CDD+07]   Cash, D., Ding, Y.Z., Dodis, Y., Lee, W., Lipton, R., Walfish, S.: Intrusion-resilient key exchange in the bounded retrieval model. In: Vadhan, S.P. (ed.) TCC 2007. LNCS, vol. 4392, pp. 479–498. Springer, Heidelberg (2007). doi:10.1007/978-3-540-70936-7_26

[CEJO02a]  Chow, S., Eisen, P., Johnson, H., Oorschot, P.C.: White-box cryptography and an AES implementation. In: Nyberg, K., Heys, H. (eds.) SAC 2002. LNCS, vol. 2595, pp. 250–270. Springer, Heidelberg (2003). doi:10.1007/3-540-36492-7_17

[CEJO02b]  Chow, S., Eisen, P., Johnson, H., Oorschot, P.C.: A white-box DES implementation for DRM applications. In: Feigenbaum, J. (ed.) DRM 2002. LNCS, vol. 2696, pp. 1–15. Springer, Heidelberg (2003). doi:10.1007/978-3-540-44993-5_1

[CL15]     Costello, C., Longa, P.: FourQ: four-dimensional decompositions on a Q-curve over the mersenne prime. In: Iwata, T., Cheon, J.H. (eds.) ASIACRYPT 2015. LNCS, vol. 9452, pp. 214–235. Springer, Heidelberg (2015). doi:10.1007/978-3-662-48797-6_10

[CMNT11]   Coron, J.-S., Mandal, A., Naccache, D., Tibouchi, M.: Fully homomorphic encryption over the integers with shorter public keys. In: Rogaway, P. (ed.) CRYPTO 2011. LNCS, vol. 6841, pp. 487–504. Springer, Heidelberg (2011). doi:10.1007/978-3-642-22792-9_28

[CNT12]    Coron, J.-S., Naccache, D., Tibouchi, M.: Public key compression and modulus switching for fully homomorphic encryption over the integers. In: Pointcheval, D., Johansson, T. (eds.) EUROCRYPT 2012. LNCS, vol. 7237, pp. 446–464. Springer, Heidelberg (2012). doi:10.1007/978-3-642-29011-4_27

[DDKL15]   Dinur, I., Dunkelman, O., Kranz, T., Leander, G.: Decomposing the ASASA block cipher construction. IACR Cryptology ePrint Archive 2015:507 (2015)

[DLPR13]   Delerablée, C., Lepoint, T., Paillier, P., Rivain, M.: White-box security notions for symmetric encryption schemes. In: Lange, T., Lauter, K., Lisoněk, P. (eds.) SAC 2013. LNCS, vol. 8282, pp. 247–264. Springer, Heidelberg (2014). doi:10.1007/978-3-662-43414-7_13

[DMRP12]   Mulder, Y., Roelse, P., Preneel, B.: Cryptanalysis of the xiao – lai white-box AES implementation. In: Knudsen, L.R., Wu, H. (eds.) SAC 2012. LNCS, vol. 7707, pp. 34–49. Springer, Heidelberg (2013). doi:10.1007/978-3-642-35999-6_3

[DR02]     Daemen, J., Rijmen, V.: The Design of Rijndael: AES - The Advanced Encryption Standard. Information Security and Cryptography. Springer, Heidelberg (2002)

[Dzi06]    Dziembowski, S.: Intrusion-resilience via the bounded-storage model. In: Halevi, S., Rabin, T. (eds.) TCC 2006. LNCS, vol. 3876, pp. 207–224. Springer, Heidelberg (2006). doi:10.1007/11681878_11

[FKKM16]   Fouque, P.-A., Karpman, P., Kirchner, P., Minaud, B.: Efficient and Provable White-Box Primitives. IACR Cryptology ePrint Archive 2016:642 (2016)

[GGH+13]   Garg, S., Gentry, C., Halevi, S., Raykova, M., Sahai, A., Waters, B.: Candidate indistinguishability obfuscation and functional encryption for all circuits. In: FOCS 2013, pp. 40–49. IEEE (2013)

[Gil16]    Gilbert, H.: On White-Box Cryptography. invited talk, Fast Software Encryption 2016 (2016). slides https://fse.rub.de/slides/wbc_fse2016_hg_2pp.pdf

[GMQ07]    Goubin, L., Masereel, J.-M., Quisquater, M.: Cryptanalysis of white box DES implementations. In: Adams, C., Miri, A., Wiener, M. (eds.) SAC 2007. LNCS, vol. 4876, pp. 278–295. Springer, Heidelberg (2007). doi:10.1007/978-3-540-77360-3_18

[GPT15]    Gilbert, H., Plût, J., Treger, J.: Key-recovery attack on the ASASA cryptosystem with expanding S-boxes. In: Gennaro, R., Robshaw, M. (eds.) CRYPTO 2015. LNCS, vol. 9215, pp. 475–490. Springer, Heidelberg (2015). doi:10.1007/978-3-662-47989-6_23

[Hal15]    Halevi, S.: Graded Encoding, Variations on a Scheme. IACR Cryptology ePrint Archive, 2015:866 (2015)

[JNP14]    Jean, J., Nikolić, I., Peyrin, T.: Tweaks and keys for block ciphers: the TWEAKEY framework. In: Sarkar, P., Iwata, T. (eds.) ASIACRYPT 2014. LNCS, vol. 8874, pp. 274–288. Springer, Heidelberg (2014). doi:10.1007/978-3-662-45608-8_15

[Kra10]    Krawczyk, H.: Cryptographic extraction and key derivation: the HKDF scheme. In: Rabin, T. (ed.) CRYPTO 2010. LNCS, vol. 6223, pp. 631–648. Springer, Heidelberg (2010). doi:10.1007/978-3-642-14623-7_34

[MDFK15]   Minaud, B., Derbez, P., Fouque, P.-A., Karpman, P.: Key-recovery attacks on ASASA. In: Iwata, T., Cheon, J.H. (eds.) ASIACRYPT 2015. LNCS, vol. 9453, pp. 3–27. Springer, Heidelberg (2015). doi:10.1007/978-3-662-48800-3_1

[Sha11]    Shaltiel, R.: An introduction to randomness extractors. In: Aceto, L., Henzinger, M., Sgall, J. (eds.) ICALP 2011. LNCS, vol. 6756, pp. 21–41. Springer, Heidelberg (2011). doi:10.1007/978-3-642-22012-8_2

[Sti02]    Stinson, D.R.: Universal hash families and the leftover hash lemma, and applications to cryptography and computing. J. Comb. Math. Comb. Comput. **42**, 3–32 (2002)

[Vad04]    Vadhan, S.P.: Constructing locally computable extractors and cryptosystems in the bounded-storage model. J. Cryptology **17**(1), 43–77 (2004)

[WMGP07]   Wyseur, B., Michiels, W., Gorissen, P., Preneel, B.: Cryptanalysis of white-box DES implementations with arbitrary external encodings. In: Adams, C., Miri, A., Wiener, M. (eds.) SAC 2007. LNCS, vol. 4876, pp. 264–277. Springer, Heidelberg (2007). doi:10.1007/978-3-540-77360-3_17

[Wys09]    Wyseur, B.: White-box cryptography. Ph.D. thesis, KU Leuven (2009)

[XL09]     Xiao, Y., Lai, X.: A secure implementation of white-box AES. In: CSA 2009, pp. 1–6. IEEE (2009)

[Zim15]    Zimmerman, J.: How to obfuscate programs directly. In: Oswald, E., Fischlin, M. (eds.) EUROCRYPT 2015. LNCS, vol. 9057, pp. 439–467. Springer, Heidelberg (2015). doi:10.1007/978-3-662-46803-6_15

# Hash Function

# MiMC: Efficient Encryption and Cryptographic Hashing with Minimal Multiplicative Complexity

Martin Albrecht[1]([✉]), Lorenzo Grassi[3], Christian Rechberger[2,3], Arnab Roy[2], and Tyge Tiessen[2]

[1] Royal Holloway, University of London, London, UK
martinralbrecht@googlemail.com
[2] DTU Compute, Technical University of Denmark, Kongens Lyngby, Denmark
{crec,arroy,tyti}@dtu.dk
[3] IAIK, Graz University of Technology, Graz, Austria
{lorenzo.grassi,christian.rechberger}@iaik.tugraz.at

**Abstract.** We explore cryptographic primitives with low multiplicative complexity. This is motivated by recent progress in practical applications of secure multi-party computation (MPC), fully homomorphic encryption (FHE), and zero-knowledge proofs (ZK) where primitives from symmetric cryptography are needed and where linear computations are, compared to non-linear operations, essentially "free". Starting with the cipher design strategy "LowMC" from Eurocrypt 2015, a number of bit-oriented proposals have been put forward, focusing on applications where the multiplicative depth of the circuit describing the cipher is the most important optimization goal.

Surprisingly, albeit many MPC/FHE/ZK-protocols natively support operations in GF($p$) for large $p$, very few primitives, even considering all of symmetric cryptography, natively work in such fields. To that end, our proposal for both block ciphers and cryptographic hash functions is to reconsider and simplify the round function of the Knudsen-Nyberg cipher from 1995. The mapping $F(x) := x^3$ is used as the main component there and is also the main component of our family of proposals called "MiMC". We study various attack vectors for this construction and give a new attack vector that outperforms others in relevant settings.

Due to its very low number of multiplications, the design lends itself well to a large class of applications, especially when the depth does not matter but the total number of multiplications in the circuit dominates all aspects of the implementation. With a number of rounds which we deem secure based on our security analysis, we report on significant performance improvements in a representative use-case involving SNARKs.

**Keywords:** Distributed cryptography · Cryptanalysis · Block ciphers · Hash functions · Zero knowledge

© International Association for Cryptologic Research 2016
J.H. Cheon and T. Takagi (Eds.): ASIACRYPT 2016, Part I, LNCS 10031, pp. 191–219, 2016.
DOI: 10.1007/978-3-662-53887-6_7

# 1  Introduction

Modern cryptography developed many techniques that go well beyond solving traditional confidentiality and authenticity problems in two-party communication. Secure multi-party computation (MPC), zero-knowledge proofs (ZK), and fully homomorphic encryption (FHE) are some of the most striking examples. In various applications of these three technologies, part of the circuit or function that is being evaluated is in turn a cryptographic primitive such as a PRF, a symmetric encryption scheme, or a collision resistant function.

In this work, we focus on a large class of such applications where the total number of field multiplications in the underlying cryptographic primitive poses the largest performance bottleneck. Examples include MPC protocols based on Yao's garbled circuit and all ZK-proof system that we are aware of, including recent developments around SNARKs [BSCG+13] which found practical applications, e.g., in Zerocash [BCG+14]. This motivates the following question addressed in this work: *How does a construction for a secure block cipher or a secure cryptographic hash functions look like that minimizes the number of field multiplications?*

Earlier  work  on  specialized  designs  for  such  applications,  like LowMC [ARS+15], Kreyvium [CCF+16], or the very recent FLIP [MJSC16] all consider the case of *Boolean* multiplications and mostly focus on the depth of the resulting circuit.

Surprisingly, albeit many MPC/FHE/ZK-protocols natively support operations in $GF(p)$ for large $p$, very few candidates, even considering all of symmetric cryptography, exist which natively work in such fields. Our focus in this paper is hence on *multiplications in the larger fields* $GF(2^m)$ and $GF(p)$ which is motivated as follows: As many protocols support multiplications in larger fields natively, encoding of a description in $GF(2)$ is cumbersome and inefficient. Whilst it is possible to do bit operations over $\mathbb{F}_p$ using standard tricks (which turn XOR into a non-linear operation), such a conversion is expensive. Consider AES as an example: it allows for an efficient description in a variety of field sizes. This is also the reason why the bit-cased LowMC which has a lower number of AND gates can often barely, if at all, outperform AES in actual implementations of the GMW MPC protocols, despite being much better than AES in terms of $GF(2)$ metrics. See [ARS+16a, Table 6] for details of the most striking example. This is also partly due to the *very* high number of XORs computed in LowMC, resulting them to be no longer negligible.

**Contributions and Related Work.** The design we propose is extremely simple: A function $F(x) := x^3$ is iterated with subkey additions. This is described in detail in Sect. 2. In fact, our design is a simplified variant of a design by Nyberg and Knudsen [KN95] from the 1990s, which was aimed to demonstrate ways to achieve provable security against the then emerging differential and linear attacks, using a small number of rounds (smaller than, say, DES). However, not much later, [JK97] showed very efficient, even practical interpolation attacks on such proposals. Indeed, our proposal resembles $\mathcal{PURE}$, a design introduced in [JK97] in order to present their attack. We pick up this work from almost 20 years ago and study in earnest if a much higher number of rounds can make this design secure in Sect. 4. It turns out, perhaps surprisingly, that the required much higher number of rounds (in the order of 100 s instead of 10 or less) is *very competitive* when it comes to the new application areas of symmetric cryptography that motivate this work.

We propose several variants of our design called MiMC: variants for $GF(p)$ and $GF(2^n)$ as well as variants that use the cube mapping directly or in a Feistel structure. MiMC can be used for encryption as well as for collision-resistant cryptographic hashing. See Sect. 2 for the basic variant in $GF(2^n)$ and Sect. 5 for a discussion on the other variants. MiMC is distinguished from any of the many constructions that have been proposed in this field recently to the that it contradicts popular belief: A recent standard textbook [KR11, Sect. 8.4] explicitly considers such constructions as "not serious, for various reasons".

**Metrics.** Given the wide variety of applications and protocols, no simple metric will be able to reliably predict application level performance. Issues of conversion between various field types (as the conversion between $GF(2)$ and $GF(p)$ mentioned above, which can be quite costly) add to the complication. Nevertheless, in order to give at least some hint towards expected performance, we will use the minimal number of multiplication to compute an output (minMULs), and the average number of multiplications needed per input bit (MULs/bit) on various designs. For the important special case of $GF(2)$ we will use minANDs and ANDs/bit, respectively.

A discussion of various constructions in $GF(p)$ and $GF(2)$ can be found in Sect. 3. In the benchmarking part in Sect. 6.1, we will also come across the case of an extremely imbalanced LowMC-variant where this simple metric clearly fails to predict actual performance. The application performance is not independent of the size of the multiplier, but for the sizes relevant for MiMC this dependence is fairly weak. The experimental result supporting this is provided in the full version of this paper [AGR+16].

**Implementation Results.** The hashing mode for $GF(p)$ may prove to be particularly useful as it is the first of its kind, despite various applications in verifiable computing [CFH+15] and applications of SNARKS like Zerocash [BCG+14] requiring such a function. Due to a lack of an alternative, authors implemented and optimized SHA-256, which leads to a bottleneck in efficiency. We demonstrate that MiMC compares very favorably in such an application. Based on our experiments and implementations, we report a factor 10 improvement in Sect. 6.1. We briefly mention more direct implementations in Sect. 6.2 and discuss the suitability of the design for cheap (generic) protection against higher-order side-channel attacks in Sect. 6.3.

In follow-up to this work [GRR+16], it was found that MiMC is also a very competitive candidate as an MPC-friendly PRF. Compared to AES, benchmark results showed that MiMC has a more than 10 times higher throughput in the online phase, and still about six times faster in the offline/precomputation phase in the LAN setting. Even the latency, which one could expect to be relatively high for MiMC due to its serial nature and the relatively high number of rounds, is better than the latency of AES. Note that for the AES case, this does not include conversion losses due to the application not using the AES field GF($2^8$), and hence the difference in real-world application settings will likely be larger.

## 2    The MiMC Primitives

In the following, we describe a block cipher, a permutation, and a permutation-based cryptographic hash function with a low number of multiplications in a finite field $\mathbb{F}_q$ (alternatively GF($q$)) where $q$ is either a prime $p$ or a power of 2.

### 2.1    The Block Cipher

In order to achieve an efficient implementation over a field $\mathbb{F}_q$ (with $q$ either prime or a power of 2), i.e., to minimize computationally expensive multiplications in the field, our design operates entirely over $\mathbb{F}_q$, thereby avoiding S-boxes completely. More precisely, we use a permutation polynomial over $\mathbb{F}_q$ as round function. In the following, we restrict ourselves to $\mathbb{F}_{2^n}$ and we denote by MiMC-$b/\kappa$ a keyed permutation with block size $b$ and key size $\kappa$. The concept however equally applies to $\mathbb{F}_p$, which we will discuss briefly in Sect. 5.

**MiMC-$n/n$.** Our block cipher is constructed by iterating a round function $r$ times where each round consists of a key addition with the key $k$, the addition of a round constant $c_i \in \mathbb{F}_{2^n}$, and the application of a non-linear function defined as $F(x) := x^3$ for $x \in \mathbb{F}_{2^n}$. For a discussion of this particular choice of polynomial and alternatives, we refer to Sect. 5.3. The ciphertext is finally produced by adding the key $k$ again to the output of the last round. Hence, the round function is described as $F_i(x) = F(x \oplus k \oplus c_i)$ where $c_0 = c_r = 0$ and the encryption process is defined as

$$E_k(x) = (F_{r-1} \circ F_{r-2} \circ \ldots F_0)(x) \oplus k.$$

We choose $n$ to be odd and the number of rounds as $r = \left\lceil \frac{n}{\log_2 3} \right\rceil$. The $r - 1$ round constants are chosen as random elements from $\mathbb{F}_{2^n}$.

Note that the random constants $c_i$ do not need to be generated for every evaluation of MiMC. Instead the constants are fixed once and can be hard-coded into the implementation on either side. No extra communication is thus needed, just as with round constants in LowMC, AES, or in fact any other cipher.

Decryption for MiMC-$n/n$ can be realized analogously to encryption by reversing the order of the round constants and using $F^{-1}(x) := x^s$ with $s = (2^{n+1} - 1)/3$ instead of $F(x) := x^3$ (the complete derivation of $s$ is given in Sect. 4, Lemma 1). Hence, encryption and decryption need to be implemented

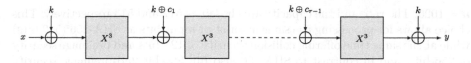

**Fig. 1.** $r$ rounds of MiMC-$n/n$

separately. Furthermore, decryption is much more expensive than encryption. Using modes where the inverse is not needed is thus advisable. We note that for our targeted applications, such as PRFs or cryptographic hash functions, computing the inverse is usually not required. We therefore provide benchmark results only for the encryption function. The fact that the inverse has a more complex algebraic description also has a beneficial effect on security as it limits cryptanalytic approaches that try to combine the encryption and decryption direction, such as inside-out approaches (Fig. 1).

**MiMC-$2n/n$ (Feistel).** By using the same non-linear permutation in a Feistel network, we can process larger blocks at the cost of increasing the number of rounds by a factor of two. The round function of MiMC-$2n/n$ is defined as following

$$x_L \| x_R \longleftarrow x_R \oplus (x_L \oplus k \oplus c_i)^3 \| x_L.$$

The round constants $c_i$ are again random elements of $\mathbb{F}_{2^n}$ except for the first and last round constants which are equal to 0. In the last round, the swap operation is not applied. The number of rounds for the Feistel version is $r' = 2 \cdot r = 2 \cdot \left\lceil \frac{n}{\log_2 3} \right\rceil$, where $r$ is the number of rounds of MiMC-$n/n$.

Decryption for MiMC-$2n/n$ can easily be realized by using the encryption function with reversed order of round constants, as usual for Feistel networks.

## 2.2 The Permutation

To construct the permutation MiMC$^P$ from the cipher MiMC as described above, we simply set the key to the all-0 string.

## 2.3 The Hash Function

For the hash function MiMChash, we propose to instantiate the permutation MiMC$^P$ in the sponge framework [BDPA08]. Given a permutation of size $n$, and a desired security level $s$, we can hash $r = n - 2s$ bits per call to the permutation. The MiMC permutation can be realized either in the SPN mode or Feistel mode by setting the key to $0^\kappa$ where $\kappa$ is the size of the key in bits. MiMCHash-$\ell$ denotes the hash function with $\ell$ bit output.

As usual, the message is first padded according to the sponge specification so that the number of message blocks is a multiple of $r$ where $r$ is the rate in sponge mode. For MiMCHash-$t$ we use MiMC-$n/n$ permutation where $n = 4 \cdot t + 1$ and $s = 2 \cdot t$. For MiMCHash-256 we thus use a MiMC-$n/n$ permutation with

$n = 1025$. The rate and the capacity are chosen as 512 and 513 respectively. This choice allows for processing the same amount of input bits as SHA-256 (512 bits) while at the same time offering collision security of 128-bits and preimage security of 256-bits, and in contrast to SHA-256 also full 256-bit 2nd-preimage security independent of the message length. We also propose MiMCHash-256b, which also offers collision resistance of 128 bits but only 128-bit security against preimage-style attacks, similar to SHAKE-256 as specified in the new SHA-3 standard. This construction makes use of a MiMC-$n/n$ permutation where $n = 769$. The rate and the capacity are chosen as 512 and 257 respectively. More generally for MiMCHash-$t$b, we use the MiMC-$n/n$ permutation where $n = 3 \cdot t + 1$ and $s = t + 1$.

# 3    Related Designs and Comparison

In this section, we give an overview of related designs, i.e. symmetric primitives which are based on arithmetic operations in some ring.

## 3.1    Knudsen-Nyberg Cipher

As discussed above, our design can be seen as a resurrection of a design due to Knudsen and Nyberg in [KN95], who proposed a DES-like cipher using a similar idea for non-linear mappings in a finite field. The Feistel round function of the 64-bit KN-cipher uses an affine mapping $e : \mathbb{F}_{2^{32}} \rightarrow \mathbb{F}_{2^{37}}$ to first transform the 32-bit input into a 37-bit value. After addition with a 37-bit round key, the resulting 37-bit value is then input to the non-linear permutation $g : x \rightarrow x^3$ in $\mathbb{F}_{2^{37}}$. Five bits of the output of $g$ are then discarded to reduce the final output again to 32 bits. In summary, one application of the round function is given as

$$x_L || x_R \rightarrow x_R || x_L \oplus f(e(x_R) \oplus k_i)$$

where $f$ consists of application $g$ followed by discarding one bit. The KN cipher is a six-round Feistel design with six 37-bit independent round keys and is provably secure against differential attacks. However, it is vulnerable to an interpolation attack (see below) because of the low algebraic degree of the polynomial corresponding to the encryption function. The Feistel variant of our design — MiMC-$2n/n$ — can be easily recognized as a variant of the KN cipher, except for that we do not discard any bits (and hence always stay in the same field), add independent round constants and have a higher number of rounds. Indeed, our design more closely resembles $\mathcal{PURE}$, the cipher used in [JK97] to demonstrate the vulnerability of the KN cipher to interpolation attacks, except for the higher number of rounds in our design. The performance of both designs essentially differs linearly in by how much we extend the number of rounds. We note that our GCD attack in Sect. 4.2 also extends to $\mathcal{PURE}$ and allows to reduce the number of plaintext-ciphertext pairs required for a successful cryptanalysis.

## 3.2   The Pohlig-Hellman Cipher

The Pohlig-Hellman cipher was described in [PH78]. Choose a prime $p$. Pick $1 \leq k \leq p - 2$ with $\gcd(k, p - 1) = 1$ and $1 \leq d \leq p - 2$ with $d = k^{-1} \bmod p - 1$, with $p$ public and $k$ and $d$ private. To encrypt the message $1 \leq m \leq p - 1$ compute $c = m^k \bmod p$. To decrypt compute $m = c^d \bmod p$. Encryption and decryption take between $\log_2 p$ and $2 \log_2 p$ multiplications depending on the Hamming weights of $k$ and $d$. A key recovery attack solves the discrete logarithm problem in $\mathbb{F}_p$. The General Number Field Sieve solves this problem in complexity $\exp\left(\left(\sqrt[3]{\frac{64}{9}} + o(1)\right)(\ln p)^{\frac{1}{3}}(\ln \ln p)^{\frac{2}{3}}\right) = L_p\left[\frac{1}{3}, \sqrt[3]{\frac{64}{9}}\right]$. Thus for $n$-bit security, the number of multiplications required grows faster than $O(n)$.

## 3.3   Naor-Reingold PRF

The Naor-Reingold PRF [NR97] is a pseudorandom function whose security can be reduced to the decisional Diffie-Hellman problem. For a given $n \in \mathbb{N}$, primes $p$ and $q$ with $q$ dividing $p - 1$, an element $g \in \mathbb{F}_p^*$ of order $q$, and $n + 1$ elements $a_0, \ldots, a_n \in \mathbb{Z}_q$, and an $n$-bit input $x_1, \ldots, x_n \in \mathbb{F}_2$ define

$$f_{p,q,g,a}(x_1, \ldots, x_n) := g^{a_0 \prod_{x_i=1} a_i}$$

where $(g, a)$ is the secret key. Evaluation of the function corresponds to one exponentiation in $\mathbb{F}_p$ and $n$ multiplications in $\mathbb{Z}_q$. Thus it takes between $p$ and $2p$ multiplications in $\mathbb{F}_p$. As the security of this primitive can be reduced to the decisional Diffie-Hellman problem, just as with the Pohlig-Hellman cipher, for $n$ bit security the number of multiplications grows faster than $O(n)$.

## 3.4   Ajtai, SWIFFT, SWIFFTX

SWIFFT [LMPR08] is a hash function family related to hard problems in lattices. In can be seen in the tradition of the work of Ajtai [Ajt96] and was used as a building block for the SWIFFTX SHA-3 submission [ADL+08]. The hash function consists of an application of the Number Theoretic Transform (NTT) over $\mathbb{Z}_{257}$ and in dimension 64 to $m = 16$ blocks of $n = 64$ bits. Each such transform costs $\frac{1}{2} n \log_2 n = 3 \cdot n = 192$ multiplications by a constant per 64 bits. The output of the NTT is then pointwise multiplied with 64 random fixed elements in $\mathbb{Z}_{257}$, costing another 64 multiplications. For $m \cdot n$ bits of input the algorithm scales linearly in $m$, so require $mn(1 + \frac{1}{2} \log_2 n)$ operations for $m \cdot n$ bits of input. On modern microprocessors most of these multiplications can be avoided by using precomputed lookup tables and some specifically chosen constants. However, it is not clear that these techniques translate to our setting. Furthermore, we note that multiplication by small constants can be more efficient than general multiplications in, e.g. homomorphic encryption schemes. On the other hand, the constants in an NTT are not small a priori. Still, our analysis might be somewhat pessimistic. We note that SWIFFT itself does not fulfil standard requirements for general purpose hash functions and that SWIFFTX addresses these issues by running four SWIFFT instances (increasing the number of multiplications accordingly) and by introducing an S-box.

### 3.5 SPRING

SPRING [BBL+15] is a PRF proposal with security related to the Learning with Errors (LWE) problem. Similarly, to SWIFFT this construction employs an NTT over $\mathbb{Z}_{257}$, but at dimension $n = 128$. This costs $\frac{1}{2} n \log_2 n = 448$ multiplications in $\mathbb{Z}_{257}$. Additional, $k$ multiplications in $\mathbb{Z}_{257}$ are required in a post-processing step for $k \in \{64, 128\}$ being the bit size of the input to the PRF. Hence, for $k = 128$ we expect 576 multiplications in $\mathbb{Z}_{257}$. We note that these multiplications can be realized efficiently on modern CPUs, but not necessarily in the scenarios targeted in this work.

### 3.6 Comparison

In Table 1 we compare MiMC with various block cipher and PRF designs. In Table 2 we compare MiMC with various cryptographic hash function proposals. In both cases, we notice a big difference between MiMC instantiations, and other designs for the two metrics that interest us: (1) the minimal number of multiplications needed to encrypt a block or at least $n$ bits (minMULs), and (2) the number of multiplications per encrypted bit. For the GF($p$) version of MiMC, the number of multiplications has to be multiplied by 2.

## 4    Design Rationale and Analysis of MiMC

In this section we explain the design rationale of the keyed permutation and argue its security. The monomial $x^3$ serves as the non-linear layer of the block cipher. Note that we can use $x^3$ to construct the cipher iff it is a permutation monomial in the field $\mathbb{F}_{2^n}$. The following well known result governs the choice of the monomial and size of the field in the design of MiMC.

**Proposition 1.** *Any monomial $x^d$ is a permutation in the field $\mathbb{F}_{2^n}$ iff $\gcd(d, 2^n - 1) = 1$.*

Hence, $x \to x^3$ is not a permutation in $\mathbb{F}_{2^n}$ when $n$ is even but only when $n$ is odd. In particular, choosing thus $n = 2^t + 1$ ensures that $x^3$ is a permutation in $\mathbb{F}_{2^n}$.

Moreover, using the previous proposition, we can compute the inverse of the non-linear permutation $x^3$ in $\mathbb{F}_{2^n}$.

**Lemma 1.** *Let $n$ an odd integer. The inverse of the non-linear function $x^3$ in $\mathbb{F}_{2^n}$ is given by $x^s$ with $s := (2^{n+1} - 1)/3$.*

**Table 1.** Comparison of ciphers in encryption mode (excluding key schedule). We list the size-optimized variants. Note that in most cases multiplication refers to the field GF(2) (minANDs and ANDs/bit) whereas in MiMC and others multiplication is in a larger field(minMULs and MULs/bit). For stream ciphers we give the minANDs needed to generate $n$ bits of output.

| Name | Security | minANDs | ANDs/bit | Remarks and Reference |
|------|----------|---------|----------|----------------------|
| AES-128 | 128 | 5120 | 40 | GF(2) rep. [BP12] ([BMP13]) |
| Simon | 128 | 4352 | 34 | [BSS+13] |
| Noekeon | 128 | 2048 | 16 | [DPVAR00] |
| Robin | 128 | 3072 | 24 | [GLSV14] |
| Fantomas | 128 | 2112 | 16.5 | [GLSV14] |
| LowMC | 128 | 1132 | 8.85 | [ARS+15] |
| Grain-128a | 128 | $4864 + 19 \cdot n$ | 19 | [ÅHJM11] |
| Trivium | 80 | $1152 + 3 \cdot n$ | 3 | [CP08] |
| Kreyvium | 128 | $1152 + 3 \cdot n$ | 3 | [CCF+16] |
| | | minMULs | MULs/bit | |
| AES-128 | 128 | 800 | 6.25 | GF($2^4$) rep. [CGP+12] |
| SPRING | 128 | 576 | 4.5 | [BBL+15] |
| Pohlig-Hellman | 128 | 3072 | $\approx 1.5$ | [PH78, ENI13] |
| MiMC-129/129 | 129 | **82** | **0.64** | This paper |
| MiMC-258/129 | 129 | 164 | **0.64** | This paper |

*Proof.* Given $y = x^3$, we are looking for an $s$ such that $x = y^s$ in $GF(2^n)$, that is $x^{3 \cdot s} = x$. By Fermat's little theorem, this is equivalent to look for an $s$ such that $3 \cdot s = 1 \pmod{2^n - 1}$. That is, there exists an integer $t$ such that $3 \cdot s = 1 + t \cdot 2^n - 1$. By Proposition 1, we have that $\gcd(3, 2^x - 1) = 1$ if and only if $x$ is odd (i.e. $\gcd(3, 2^x - 1) = 3$ if and only if $x$ is even). For $t = 1$, we obtain $3 \cdot s = 2^n$ which is a contradiction. If $t$ is equal to 2, then $3 \cdot s = 2^{n+1} - 1$. Since $n + 1$ is even (by hypothesis), then 3 divides $2^{n+1} - 1$. Finally, since $x^3$ is a permutation in $GF(2^n)$ for $n$ odd (by previous proposition), then the inverse is unique and is given by $s := (2^{n+1} - 1)/3$. ☐

## 4.1   Computation Cost Model

In most models of computation field multiplication is considered to be more computationally expensive than addition. However, note that squaring is a linear operation in a binary field $\mathbb{F}_{2^n}$. Hence, if we consider the number of non-linear multiplications in a binary field then the number required to compute $x^3$ is one. In the SNARK setting, each witness variable (and possibly each constraint) is generated from a field operation more specifically from a field multiplication. As a consequence, computing $x^3$ generates two equations $x \cdot x = y$ and $y \cdot x = x^3$. Hence, in this setting we do not benefit from the linearity of squaring over the

200    M. Albrecht et al.

**Table 2.** Comparison of hash functions. We list the size-optimized variants. Note that in most cases multiplication refers to the field GF(2) (minANDs and ANDs/bit) whereas in MiMC multiplication is in a larger field (minMULs and MULs/bit).

| Name | Coll. Resist | minANDs | ANDs/bit | Remarks and Reference |
|---|---|---|---|---|
| SHA-256 | 128 | 29000 | 56.64 | [BCG+14]) |
| SHA3-256 | 128 | 38400 | 35.29 | [NIS14] |
| SHAKE128 | 128 | 38400 | 28.57 | [NIS14] |
| | | minMULs | MULs/bit | |
| SWIFFTX | 112–256 | 16384 | 8.0 | [ADL+08] |
| MiMCHash-256 | 129 | **1293** | **2.52** | This paper |
| MiMCHash-256b | 129 | **971** | **1.89** | This paper |

fields $\mathbb{F}_{2^n}$ and computing $x^3$ costs two multiplications. However, the cost of additions in these fields is still negligible compared to that of multiplication. Note that we can also disregard the cost of multiplication by a constant. Details on the form of equations involved in SNARK is given in Sect. 6.

We stress that although the cost of an addition is considered negligible compared to a multiplication, very large number of additions can reduce the efficiency of a design.

## 4.2 Security Analysis

Our designs resist a variety of cryptanalysis techniques. The algebraic design principle of MiMC causes a natural concern about the security of the keyed permutation against algebraic cryptanalytic techniques. We describe several possible algebraic attacks (incl. a new "GCD" attack) against the design and analyze the resistance of the block cipher against these attacks. We also consider statistical attacks.

To summarize the following results, the number of rounds for the case of MiMC-$n/n$ is derived from an interpolation attack, while the number of rounds for the case of MiMC-$2n/n$ is deduced from a Meet-in-the-Middle GCD attack.

We discuss the case in which some restrictions on the memory that the attacker can use to implement the attack hold in the full version of this paper [AGR+16]. We show that in this case it is possible to reduce the total number of rounds. We have also analysed the security when the adversary has a restriction on the number of plaintexts available in [AGR+16].

**Interpolation Attack.** Interpolation attacks, introduced by Jakobsen and Knudsen [JK97], construct a polynomial corresponding to the encryption function without knowledge of the secret key. If an adversary can construct such a polynomial then for any given plaintext the corresponding cipher-text can be produced without knowledge of the secret key.

Let $E_k : \mathbb{F}_{2^n} \to \mathbb{F}_{2^n}$ be an encryption function. For a randomly fixed key $k$, the polynomial $P(x)$ representing $E_k(x)$ can be constructed using Lagrange's theorem, where $x$ is the indeterminate corresponding to the plaintext. If the polynomial has degree $d$ then we can find it using Lagrange's formula

$$P(x) = \sum_{i=1}^{d} y_i \prod_{1 \leq j \leq d, i \neq j} \frac{x - x_j}{x_i - x_j}$$

where $E_k(x_i) = y_i$ for $i = 1, 2, \ldots d$.

This method can be extended to a key recover attack. The attack proceeds by simply guessing the key of the final round, decrypting the cipher-texts and constructing the polynomial for $r - 1$ rounds. With one extra p/c pair, the attacker checks whether the polynomial is correct.

Observe that the number of unknown coefficients of the interpolation polynomial is $d + 1$ and that the complexity of constructing a Lagrangian interpolation polynomial is $\mathcal{O}(d \log d)$ [Sto85]. Hence, setting $d = 3^r$ with $r = r_{max} \approx n / \log_2(3)$ thwarts this attack. Note that no function mapping from $\text{GF}(2^n)$ to $\text{GF}(2^n)$ has degree $\geq 2^n$, since $T^{2^n - 1} \equiv 1$ for each $T \in \mathbb{F}_{2^n}$ and the degree of the interpolation polynomial does not increase for $r > r_{max}$.

By the same argument, a similar result holds for the case of the Feistel network MiMC-$2n/n$. Indeed, at each round the left/right hand part of the state can be described as a polynomial of the left and of the right hand part of the plaintext, with at most $3^{2r-1} + 3^r + 3^{r-1} + 1$ unknown coefficients (observe that at round $r$, the degree of the polynomial is at most $3^r$ in the left part of the plaintext and $3^{r-1}$ in the right part). Thus, the complexity of constructing this Lagrangian interpolation polynomial is approximately $\mathcal{O}(r \cdot 3^{2r})$, where a function mapping from $\text{GF}(2^n)^2$ to $\text{GF}(2^n)$ has degree at most $2^{2n}$.

Note that in the chosen-plaintext scenario and in the case of MiMC-$2n/n$, an attacker can reduce the degree of the interpolation polynomial. For example, for chosen plaintexts of the form $x||x^3$ the degree of the interpolation polynomial after $r$ rounds is at most $2 \cdot 3^{r-1}$ in the left part of the plaintext and $2 \cdot 3^{r-2}$ in the right part, while for chosen plaintexts of the form $0||x$ the degree of the interpolation polynomial is at most $3^{r-1}$ in the left part of the plaintext and $3^{r-2}$ in the right part. Thus, for this second case, the interpolation polynomial of the right part of the text depends only by the right part of the plaintexts and has degree $3^{r-2}$. In order to avoid the reduced degree of the polynomial, it is sufficient to add (at least) two rounds more to the number of rounds calculated for MiMC-$n/n$.

A meet-in-the-middle variant of the interpolation attack was also proposed in [JK97], constructing a polynomials $g(x) = h(y)$ instead of one polynomial $y = f(x)$. For MiMC-$n/n$, this approach does not produce an improvement due to the prohibitive degree of the inverse operation. In contrast, for MiMC-$2n/n$ we have that $g$ and $h$ may have degree $3^{r/2}$ in the left part of the plaintext and $3^{r/2-1}$ in the right part only instead of degree $3^r$ and $3^{r-1}$ respectively. However, this lower degree comes at the price of increases computational cost. Indeed, constructing $g$ and $h$ requires solving a system of equation in

$n = 2 \cdot (3^{r/2} + 1) \cdot (3^{r/2-1} + 1)$ unknowns costing $\mathcal{O}(n^\omega) = \mathcal{O}(3^r)$ operations, where the hidden constant is $\geq 1$ and we conservatively set the linear algebra constant $\omega = 2$. The chosen plaintext variant of this attack is quite similar. As before, the idea is to choose plaintexts in which the left part is fixed. In this way, one of the two interpolation polynomial depends only on one variable, the right part of the plaintext. Thus, constructing $g$ and $h$ requires solving a system of equation in $n = (3^{r/2-2} + 1) + (3^{r/2} + 1) \cdot (3^{r/2-1} + 1)$ unknowns costing $\mathcal{O}(n^2) = \mathcal{O}(3^{r-1})$ operations where the hidden constant is $\geq 1$.

We note that the complexity of an interpolation attack may decrease if the polynomial $P(x)$ is sparse for a chosen key. However, because we are adding random round constants in each round and $x^3$ is a permutation in $\mathbb{F}_{2^n}$ by construction, our $P(x)$ is not expected to be sparse[1].

**Computing GCDs.** From the description of MiMC, it is clear that factoring univariate polynomials recovers the key. However, if we are given more than one known plaintext-cipher-text pair, we can reduce the complexity further by computing a GCD of them. Denote by $E(k, x)$ the encryption of $x$ under key $k$. For a pair $(x, y) \in \mathbb{F}_q^2$, $E(K, x) - y$ denotes a univariate polynomial in $\mathbb{F}_q[K]$ corresponding to $(x, y)$. Note that in general, given plaintext/cipher text pair $(x, y)$, it should be hard for a generic encryption scheme to compute the univariate polynomial $E(K, x) - y$ explicitly in the variable $K$ (i.e. the secret key). However, this is not the case of MiMC, for which the polynomial $E(K, x) - y$ can be always computed explicitly, and it simply corresponds to the definition of encryption process (that is, the iterative application of the cubic function). Moreover, note that this attack may also be applied to $\mathcal{PURE}$, the cipher used in [JK97] to demonstrate the vulnerability of the KN cipher to interpolation attacks, assuming round keys are not independent but linearly derived from $k$.

Consider now two such polynomials $E(K, x_1) - y_1$ and $E(K, x_2) - y_2$, with $y_1 = E(k, x_1)$ and $y_2 = E(k, x_2)$ for the fixed but unknown key $k$. It is clear that these polynomials share $(K - k)$ as a factor. Indeed, with high probability the greatest common divisor will be $(K - k)$. Thus, by computing the GCD of the two polynomials, we can find the value of $k$.

---

[1] This claim is supported by our experiments. In particular, for a field $\mathbb{F}_{2^n}$ and using $x^3$ as permutation, we observed:

- after 1 round, all terms appear (percentage: 100 %);
- after 2 round, 8 terms appear instead of 10 (percentage: 80 %);
- after 3 round, 19 terms appear instead of 28 (percentage: 67.86 %);
- after 4 round, 54 terms appear instead of 82 (percentage: 65.85 %);
- after 5 round, 161 terms appear instead of 244 (percentage: 66 %);
- after 6 round, 531 terms appear instead of 730 (percentage: 72.74 %);

and so on, where the percentage of the non-null terms continues to grow for the next rounds. For example, for the particular field $GF(2^{17})$, after 10 rounds almost all the terms are non-zero.

MiMC-$n/n$ for a known plain text $x$ corresponds to a polynomial having degree $3^r$, where the leading monomial always has non-zero coefficient. Hence, we can recover $k$ with a GCD computation of two polynomials at degree $3^r$ (indeed, considering differences of two polynomials $G(K, x_i) - y_i$ reduces this degree to $3^r - 1$ by canceling the leading term). It is well-known that the complexity for finding the GCD of two polynomials of degree $d$ is $\mathcal{O}(d \log^2 d)$. Hence, the complexity of this attack is $\mathcal{O}(r^2 \cdot 3^r)$. For MiMC-$n/n$ the time complexity of this attack is higher than that of the interpolation attack.

More care must be taken for MiMC-$2n/n$, since in this case the meet-in-the-middle variant of this attack can be performed. That is, instead of constructing polynomials expressing ciphertexts as polynomials in the plaintext and the key, we can construct two polynomials $G'(K, x_i)$ and $G''(K, y_i)$ expressing the state in round $r/2$ as a polynomial in the key and the plaintext or ciphertext respectively. Then, considering $G'(K, x_1) - G''(K, y_1)$ and $G'(K, x_2) - G''(K, y_2)$ we can apply a GCD attack on polynomials of degree $3^{r/2}$, reducing the complexity to $\mathcal{O}(r^2 \cdot 3^{r/2})$. Hence, to thwart this attack we must increase the number of rounds to $r = 2 \cdot r_{max} \approx 2 \cdot n / \log_2(3)$.

**Invariant Subfields.** The algebraic structure of MiMC allows to mount an invariant subfield attack on the block cipher under a poor choice of round constants. That is, if all the round constants $c_i$ and the key $k$ are in subfield $\mathbb{F}_{2^m}$ of $\mathbb{F}_{2^n}$ then by choosing a plaintext $x \in \mathbb{F}_{2^m}$ an adversary can ensure that $E_k(x) \in \mathbb{F}_{2^m}$. This attack is thwarted by picking $n$ to be prime. The only subfield is then $\mathbb{F}_2$ such that picking constants $\neq 1$ will be enough to avoid the attack.

**Differential Attacks.** Differential cryptanalysis is one of the most widely used technique in symmetric-key cryptanalysis. The different types of cryptanalysis methods based on this technique depend on the propagation of an input difference through a given number of rounds of an iterative block cipher to yield a known output difference with high probability. The probability of the propagation often determines how many rounds can be attacked using this technique.

Given an input difference $\delta$ and an output difference $\delta'$, the differential probability of the round function is given as

$$\Pr(\delta \to \delta') = |\{x \in \mathbb{F}_{2^n} : F(x + \delta) + F(x) = \delta'\}|/2^n \qquad (1)$$

In our case the number of $x$ satisfying $F(x + \delta) + F(x) = \delta'$ is determined by the non-linear function $x^3$. Hence it is enough to determine the size of the set

$$D = \{x \in \mathbb{F}_{2^n} : (x + \delta)^3 + x^3 = \delta', \delta \neq 0\}.$$

As this is a quadratic equation in $x$ for any, there are at most two solutions to the equation. This implies $\Pr(\delta \to \delta') \leq \frac{2}{2^n}$. This is sufficient to give any differential trail of at least two rounds a probability too low to be useful in an attack. A detailed analysis of the differential property of monomials of the form $x^{2^t+1}$ in $\mathbb{F}_{2^n}$ can be found in [Nyb94, Can97].

**Linear Attacks.** Similar to differential attacks, linear attacks pose no threat to MiMC. Indeed, the cubic function is an *almost bent* or an *almost perfect nonlinear* (APN) function, i.e., differential 2-uniform, where an APN permutation provides the best resistance against linear and differential cryptanalysis. Thus, since its maximum square correlation is limited to $2^{-n+1}$ (cf. for example [AÅBL12] for details), any linear trail of the cubing function will have negligible potential after a few rounds.

**Algebraic Degree and Higher-Order Differentials.** As discussed above, the large number of rounds ensures that the algebraic degree of MiMC in its native field will be maximal or almost maximal. This naturally thwarts higher-order differential attacks when considering the difference as defined in the field (i.e., using the inverse of the field addition). But what happens to the degree when viewing the rounds as vectorial Boolean functions? As squaring is a linear operation in $\mathbb{F}_{2^n}$, it is also linear when viewed as vectorial function over $\mathbb{F}_2$. Cubing on the other hand introduces an additional multiplication which gives the round function an algebraic degree of 2 in every component when viewed as a vectorial Boolean function. Again, the large number of rounds should cause the degree to rise quickly and reach the limit of $2^n$ which is sufficient to thwart any higher-order differential attacks also when viewing the round function as a vectorial Boolean function.

**Hash-Specific Security Considerations.** For usage in the MiMC permutation in the sponge mode as described in Sect. 2.3 we require the permutation to not show non-trivial non-random behavior for up to $2^s$ input/output pairs. As specified in Sect. 2 the size of the permutation $n$ determines the number of rounds (based on the GCD attack described above). As $2s < n$ for both MiMCHash-256 and MiMCHash-256b, this choices leaves us with an additional security margin, even if an hypothetical inside-out approach could double the number of rounds in an attack.

## 5 Variants

In this section, we discuss two variants of MiMC. One for instantiating MiMC over prime fields and one for extending the key size to increase security.

### 5.1 MiMC over Prime Fields

The above descriptions of MiMC can also be used to operate over prime fields i.e. a field $\mathbb{F}_p$ where $p$ is prime. In that case, it needs to be assured that the cubing in the round function creates a permutation. For this, it is sufficient to require $\gcd(3, p - 1) = 1$.

Following the notation as above, we can consider MiMC-$p/p$ where the permutation monomial $x^3$ is defined over $\mathbb{F}_p$. The number of rounds for constructing the keyed permutation is $r = \left\lceil \frac{\log p}{\log_2 3} \right\rceil$. In the Feistel mode, we define MiMC-$2p/p$ where the round function is defined over $\mathbb{F}_p$ and where the number of rounds is double with respect to MiMC-$p/p$. In both the constructions the $r$ round constants are chosen as random elements in $\mathbb{F}_p$.

Our cryptanalysis from Sect. 4 transfers to this case except for the subfield attack which does not apply here.

## 5.2   Larger Keys

Instead of considering our simple iterative construction where we add the same key in each round, we may also consider the case where we have a key which is $\kappa$-times bigger than the block size $n$. In this case, we may consider an instance where we are cyclically adding $\kappa$ independent keys to our rounds. Our $i$-th round function then becomes:

$$F_i(x) = (x \oplus k_{i \bmod \kappa} \oplus c_i)^3$$

It is clear that differential and linear cryptanalysis are not affected by this modification if we model MiMC as a Markov cipher. However, considering a larger key size does affect algebraic attacks. In particular, a simple GCD attack is not sufficient any more to recover the keys $k_0, k_1, \ldots, k_{\kappa-1}$. Instead, we may consider Resultants or Gröbner bases.

We consider the case where $\kappa = 2$. It is well-known [BKW93] that the maximum degree reached during a Gröbner basis computation of a bivariate system of equations is $\le 2 \cdot \mathrm{maxdeg}(P) + 1$, where $\mathrm{maxdeg}(P)$ is the maximum degree of our input system (i.e. $3^r$ in our case). Hence, from e.g. [BFS14], the complexity of solving such a system of equations is

$$\mathcal{O}\left(2 \cdot 3^r \cdot \binom{2 \cdot 3^r + 3}{2 \cdot 3^r + 1}\right).$$

Applying resultants, from [LMS13] we expect a complexity of

$$\tilde{\mathcal{O}}\left(d^{4.69}\right) = \tilde{\mathcal{O}}\left(3^{4.69\,r}\right).$$

Conservatively, we may anticipate a meet-in-the-middle attack which would reduce the cost of either of these attacks to a square root of the above estimates.

## 5.3   Different Round Functions

Considering the case $\mathrm{GF}(2^n)$, we may consider a round function of the form

$$F(x) = (x \oplus k \oplus c)^d$$

for generic exponents $d$. In particular, we have decided to limit our analysis to exponents of the form $2^t + 1$ and $2^t - 1$, for positive integer $t$ (note that 3 is the only number that can be written in both ways). Remember that for MiMC-$n/n$, $d$ has to satisfy the condition $\gcd(d, 2^n - 1) = 1$ in order to be a permutation, while in the case of MiMC-$2n/n$ (that is, for Feistel Networks) this condition is not necessary.

For further analysis, we recall the Lucas's Theorem:

**Theorem 1.** *For non-negative integers $m$ and $n$ and a prime $p$, the following congruence relation holds:*

$$\binom{m}{n} \equiv \prod_{i=0}^{k} \binom{m_i}{n_i} \pmod{p},$$

*where $m = m_k p^k + m_{k-1} p^{k-1} + ... + m_1 p + m_0$ and $n = n_k p^k + n_{k-1} p^{k-1} + ... + n_1 p + n_0$ are the base $p$ expansions of $m$ and $n$ respectively, using the convention that $\binom{m}{n} = 0$ if $m < n$.*

Exponents of the form $2^t + 1$ (with $t > 1$) have the nice property that the cost to compute $x^{2^t+1}$ does not depend on $t$, i.e. it requires only one multiplication (in some applications). Moreover, the degree of the resulting $r$-round interpolation polynomial is $(2^t + 1)^r$, which is significantly higher than $3^r$ even for "small" $t$. The major problem of this kind of exponents is that the corresponding interpolation polynomials are in general sparse. For example, using Lucas's Theorem, it is very easy to note that just after one round the polynomial has only 4 terms instead of $2^t + 2$:

$$(x \oplus k)^{2^t+1} \equiv_2 (x \oplus k)^{2^t} \cdot (x \oplus k) \equiv_2$$
$$\equiv_2 (x^{2^t} \oplus k^{2^t}) \cdot (x \oplus k) \equiv_2 x^{2^t+1} \oplus k \cdot x^{2^t} \oplus k^{2^t} \cdot x \oplus k^{2^t+1}.$$

Using the same technique, after $r$ rounds, the number of terms of the polynomial is upper bounded by $3^r + 1$, which is (much) smaller than $(2^t + 1)^r + 1$. Note that $3^r + 1$ is exact the same upper bounded obtained for the exponent 3 (which corresponds to $t = 1$). Thus, the number of rounds to guarantee the security against the algebraic attacks doesn't change choosing exponent of the form $2^t + 1$ for $t > 1$. That is, both from the security point of view and from the implementation one, there is no advantage to choose exponents of the form $2^t + 1$ greater than 3.

Similar considerations can be done also for exponents of the form $2^t + 2^s = 2^s \cdot (2^{t-s} + 1)$, where $s < t$.

For this reason, coefficients of the form $2^t - 1$ are more interesting. Indeed, in this case it is very easy to prove that the interpolation polynomial is not sparse:

$$(x \oplus k)^{2^t-1} \equiv_2 \bigoplus_{i=0}^{2^t-1} x^i \cdot k^{2^t-1-i},$$

since

$$\binom{2^t - 1}{i} \equiv_2 1 \qquad \forall i \in \{0, 1, \ldots, 2^t - 1\}.$$

On the other hand, in order to compute $x^{2^t-1}$, we need more multiplications and square operations. Thus, a natural question is if it is possible to minimize the total number of multiplications necessary to compute the ciphertext choosing an exponent of the form $2^t - 1$ different from 3.

There are different ways to compute $g^e$ where $g \in \mathbb{F}_{2^n}$ and $e = 2^t - 1$ for some $t \geq 2$, the classical algorithm being the square-and-multiply algorithm, cf. [MVO96, Sect. 14.6]. For this algorithm, the number of multiplications requested for this exponent is equal to the number of squares $t - 1$. In Algorithm 1, we give a slight variation of the original algorithm.

---

**Data:** $g \in \mathbb{F}_{2^n}$ and $e = 2^t - 1$ for some $t \geq 2$
**Result:** $g^e$
$g_0 \leftarrow g$;
$g_1 \leftarrow g^2 \cdot g$;
$A \leftarrow 1$;
**for** $i$ from 0 to $\lfloor t/2 \rfloor$ **do**
$\quad | \quad A \leftarrow (A^2)^2$;
$\quad | \quad A \leftarrow A \cdot g_1$;
**end**
**if** $t \bmod 2 \neq 0$ **then**
$\quad | \quad A \leftarrow A^2$;
$\quad | \quad A \leftarrow A \cdot g_0$;
**end**
**return** $A$.

**Algorithm 1.** Modular exponentiation with cache

---

By simple computation, the number of multiplications for the previous algorithm is $\lceil t/2 \rceil$, while the number of squares is $t - 1$. Observe that with respect to the original algorithm, it requires precomputation and to store the quantity $g^2 \cdot g$. Thus, for our purpose, this algorithm is better than the original one (for the case $e = 2^t - 1$). This algorithm can be improved[2], but for our purpose it suffices.

Thus, using the previous analysis about the number of rounds, the total number of multiplications $m$ and of squares $s$ for MiMC-$n/n$ (analogous for MiMC-$2n/n$) is

$$m = \left\lceil \frac{t}{2} \right\rceil \cdot \left\lceil \frac{n}{\log_2(2^t - 1)} \right\rceil \qquad s = (t-1) \cdot \left\lceil \frac{n}{\log_2(2^t - 1)} \right\rceil.$$

---

[2] For example, suppose that $t \geq 8$. The idea is to precompute $g_0, g_1$ (defined as before) and also $g_2 := (g_1)^4 \cdot g_1$. Thus, in the *for* loop $0 \leq i \leq \lfloor t/4 \rfloor$ and $A \leftarrow A^8 \cdot g_2$. Finally, after the *for* loop and before the *if*-statement, one has to take care of the case $t \bmod 4 \neq 0$.

For example, for $n = 129$, the best result is obtained for $t = 4$ (that is for the exponent 15)[3], for which the total number of multiplications is 66 (instead of 82 for the exponent 3), while the number of squares is 99 (instead of 82 for the exponent 3).

Note that the sum of the total number of multiplications $m$ and of the total number of squares $s$ is almost constant for each choice of $t$.

Finally, only for completeness, it is also possible to extend the previous analysis to the case $GF(p)$. In this case, since the square operation is not linear, it counts as a multiplication. Thus, if we consider an exponent of the form $2^t - 1$, the total number of multiplications $m$ for MiMC-$p/p$ is

$$m = \left( \left\lceil \frac{t}{2} \right\rceil + t - 1 \right) \cdot \frac{\log(p-1)}{\log(2^t - 1)}.$$

To conclude, if the cost of a square operation is negligible with respect to the cost of a multiplication (that is, if the square operation is linear), then it is possible to minimize the total number of multiplications choosing an exponent of the form $2^t - 1$ different from 3. Instead, when the number of square operations can not be ignored (as for example in the case of SNARK settings or in the $GF(p)$ case), the choice of an exponent of the form $2^t - 1$ different from 3 does not offer any advantage due to the fact that the number $m + s$ is almost constant.

## 6    Application and Implementation

We implemented the MiMC block cipher and hash function in C++ using NTL [Sho]. Note that we put no restriction on the irreducible polynomial to represent the finite field $\mathbb{F}_{2^n}$ in our proposal.

### 6.1    Verifiable Computation and SNARK

Recently, several techniques have been proposed to achieve practical or nearly practical verifiable computation through constructions such as Pinocchio [PHGR16] and zk-SNARK. A special kind of *Succinct Non-interactive Argument of Knowledge* or SNARK was proposed in 2014 to build Zerocash [BCG+14] — a digital currency similar to Bitcoin but achieving anonymity. In [BSCG+13] an implementation of a publicly verifiable non-interactive argument system is given.

The main idea of the SNARK is to provide a circuit whose satisfiability enables a verifier to check correctness of an underlying computation. In this concrete implementation, we focus on the (zk)SNARK for arithmetic circuit satisfiability. The main target of our design proposals is to improve the efficiency of (zk)SNARK when they are used as cryptographic primitives in a SNARK setting.

---

[3] Actually, the best result is obtained for $t = 6$, that is for the exponent 63. But since $\gcd(63, 2^{129} - 1) = 7$, the round function defined using the exponent 63 is not a permutation.

An $\mathbb{F}$-arithmetic circuit takes input from the field $\mathbb{F}$ and its gates produce output in $\mathbb{F}$. Also the circuits considered here consist of bilinear gates only. Arithmetic circuit satisfiability (ACS) is defined as follows:

**Definition 1.** *The ACS problem of an $\mathbb{F}$-arithmetic circuit $\mathcal{C} : \mathbb{F}^n \times \mathbb{F}^h \to \mathbb{F}^l$ is depicted by the relation $\mathcal{R} = \{(x,a) \in \mathbb{F}^n \times \mathbb{F}^h : \mathcal{C}(x,a) = 0^l\}$ such that its language is $L = \{x \in \mathbb{F}^n : \exists a \in \mathbb{F}^h \text{ s.t } \mathcal{C}(x,a) = 0^l\}$.*

Since the circuit consists of bilinear gates only, we aim to minimize the number of NLM or field multiplications in our design. The addition in the field, which is the same as bitwise XOR, is a comparatively less expensive operation. The SNARK algorithm generates the proof for satisfiability of a system of *rank-1 quadratic constraints* over a finite field. This system of constraints is defined as below.

**Definition 2.** *A system of rank-1 quadratic equations over a field $\mathbb{F}$ is a sequence of tuples $((A_i, B_i, C_i), n)$ for $i = 1, \ldots, N_c$ and $A_i, B_i, C_i \in \mathbb{F}^{1+N'}$ such that $n \le N'$. This system is satisfiable with an input $x \in \mathbb{F}^n$ if there is a witness $w \in \mathbb{F}^{N'}$ such that*

$$\langle A_i, w \rangle \cdot \langle B_i, w \rangle = \langle C_i, w \rangle \quad \forall i = 1, \ldots, N_c$$

*Here $N_c$ is the number of constraints and $N'$ is the number of variables.*

The number of such constraints contributes to the efficiency of the SNARK algorithm. From the above definition it is also clear that in a SNARK setting over $\mathbb{F}_{2^m}$ we can not ignore the squaring as linear operation.

**MiMC in the SNARK Setting.** In MiMC, each round can be expressed with the following equations

$$X + \underbrace{k_i + C_i}_{\alpha} + U = 0 \tag{2}$$

$$U \cdot U = Y \tag{3}$$

$$Y \cdot U = Z \tag{4}$$

where $k_i, C_i$ are the round key and constants respectively. Note that the above 3 equations can be combined to form one rank-1 quadratic constraint (as in Definition 2)

$$(X + \alpha)(X + \alpha + Y) = Y + Z \tag{5}$$

For the MiMCHash the round key is fixed to a constant hence $\alpha$ can be treated as a constant in this equation. Note that the number of witness per round of MiMC is 2. Therefore the total number of witness for the fixed key permutation is $2 \cdot R$, where $R \approx \frac{n}{\log 3}$ is the number of rounds and $n$ is the block size. The witness generation requires one constant addition (XOR) and two multiplications in the corresponding field. The complexity of the prover algorithm of SNARK (Appendix E in [BSCG+13]) is dominated by $O(N_c \log N_c)$ where $N_c$ is the number of rank-1 constraints.

**LowMC in the SNARK Setting.** In LowMC, each round consists of Sbox (3-bit), matrix multiplication (over $\mathbb{F}_2$), round key and constant addition (XOR). Each 3-bit Sbox application can be written as

$$b \cdot c = a + z_1 \tag{6}$$
$$a \cdot (c + 1) = b + z_2 \tag{7}$$
$$a \cdot (b + 1) = b + c + z_3 \tag{8}$$

The above three equations can be combined to form 2 rank-1 constraints as following

$$b \cdot c = a + z_1 \tag{9}$$
$$a \cdot (b + c) = c + z_2 + z_3 \tag{10}$$

The witness generation for each Sbox requires 3 multiplications and 6 additions (out of which 2 are constant additions) over $\mathbb{F}_2$. In each round there are $m$ Sboxes. Hence per round the witness generation process will require $3\,\mathrm{m}$ multiplications and $6\,\mathrm{m}$ ($2\,\mathrm{m}$ of them are constant addition) additions per round. Suppose $N_b$ is the block size of the permutation. Then there will be approximately $(l-1) \cdot N_b$ additions over $\mathbb{F}_2$ due to linear layer of LowMC in each round, where $l$ is the average number of non-zero entries in each row of the random matrix of the linear layer. Also there will be $N_b$ constant additions over $\mathbb{F}_2$ which is due to round constant and key addition. The total number of rank-1 constraints for $R$ rounds of LowMC will be $R \cdot 2\,\mathrm{m}$. Note that the number of additions is much higher in comparison with the number of multiplication over $\mathbb{F}_2$.

*Remark 1.* For the MiMC permutation, the operations are performed over a larger field e.g. $\mathbb{F}_{2^{1025}}$. Indeed the cost of a single multiplication is higher in the larger field compared to a multiplication over $\mathbb{F}_2$. Moreover, the number of additions are significantly more than the number of multiplications (see Table 3). Although in the cost model the cost of addition is much less than the cost of multiplication, very large number of additions over $\mathbb{F}_2$ brings down the efficiency of LowMC in SNARK setting in comparison to MiMC. On the other hand, in MiMC the number of additions per round is one.

**Experimental Results.** Following the `libsnark` [Lab] implementation we have implemented a prototype of SNARK for generating the circuit and witness for MiMC permutation for different block sizes and MiMCHash-256. One important target application of MiMC is SNARK or SNARK like algorithms. We have measured the time taken by MiMCHash for processing a single block and compared it with the time taken by SHA-256 using the `libsnark` implementation.

For processing a single block i.e. for hashing a single block message our MiMC implementation in the SNARK setting requires $\approx 7.8$ milliseconds to generate the arithmetic circuit and witness while SHA-256 takes $\approx 73$ milliseconds.

Since LowMC was designed for MPC/ZK applications we have also implemented it in the SNARK setting. A comparison of LowMC with MiMC is given in Table 3.

**Table 3.** Comparison of LowMC and MiMC with block size 1025 and the corresponding parameters for LowMC and Keccak permutation with specified parameters. For all implementations we have used the -O3 optimization option of the gcc compiler. For LowMC, the number of rounds and the number of Sboxes per round are denoted as $R$ and $m$ respectively.

| | MiMC | LowMC | | Keccak-[1600, 24] |
| | | $R = 16$ $m = 196$ | $R = 55$ $m = 20$ | |
|---|---|---|---|---|
| Total time | 7.8 ms | 90.3 ms | 271.2 ms | 75.8 ms |
| Constraint generation | 6.3 ms | 13.5 ms | 9.2 ms | 65.2 ms |
| Witness generation | 1.5 ms | 76.8 ms | 262.0 ms | 10.6 ms |
| # addition | 646 | 8420888 | 28894643 | 422400 |
| # multiplication | 1293 | 9408 | 3300 | 38400 |
| # rank-1 constraint | 646 | 4704 | 2200 | 38400 |

If we intend to use the LowMC permutation to construct a hash function using Sponge mode then the block size of LowMC should be 1025 bit for achieving the same security level as SHA-256 or MiMCHash-256. We have implemented LowMC with the updated parameter-set v2 from [ARS+16b] with this block size and two possible choices for the parameters $(R, m)$, where $R$ and $m$ are number of rounds and number of Sbox per round respectively. One is minimizing the number of rounds for the given block size and security requirements, the other one is minimizing the number of ANDs/bit. Both are derived from the round formula given in [ARS+16b]. LowMC is a block cipher designed for MPC/FHE applications and the original proposal did not provide any suggestion to construct a secure hash function using the permutation. However if used in the sponge mode then the performance of the resulting hash function can be approximated by the performance of the LowMC permutation in SNARK setting.

We have also compared the performance of the Keccak-[1600, 24] [NIS14] permutation when used for the SHA-3 and SHAKE hash function in our SNARK setting. Note that the truncation after a Keccak permutation can be expressed as equality constraints. In fact the performance for the SHAKE128 or SHA3 are almost same as the Keccak-[1600, 24]. The performance comparison in the Table 3 shows that MiMC is significantly more efficient than LowMC and SHA-3 in SNARK setting.

All field operations are implemented using the NTL together with the gf2x library. All computations were carried out on an Intel Core i7 2.10 GHz processor with 16 GB memory and we took the average over $\approx 2000$ repetitions. As a design with an unusual imbalance between ANDs and XORs, the comparison with LowMC variants is interesting as it gives an example where the number multiplications alone can no longer be used as a hint for the eventual performance. Where the round-minimized LowMC variant is more than 10 times slower with about 8 times more multiplications, reducing the number of ANDs in the other

LowMC variant at the expense of many more rounds does not have the expected effect: The runtime grows again. The reason is the huge amount of XOR computations whose cost is clearly are no longer negligible. This shows the limits of a simplified metric that focuses on AND gates (or multiplication gates) also.

All implementations in C++ can be found on https://github.com/byt3bit/mimc_snark.git.

## 6.2   Direct Implementation

For the sake of completeness we provide a brief discussion of the complexity for the direct implementation MiMC, but stress that it has limited impact on the performance on our target platforms. Each round of MiMC-$n/n$ performs one multiplication in the field $\mathbb{F}_{2^n}$. For the considered values of $n$ this computation of $x^3$ becomes computationally expensive, since it is not feasible to use the efficient lookup table method even for $n = 32, 64$.

The evaluation of $x^3$ can be reduced to field multiplication. Since the problem is frequently encountered in many public-key cryptographic algorithms and protocols, efficient field multiplication is a well studied area in the literature. One strategy for efficient field multiplication is to use lookup tables. Indeed, several algorithms [GP97, DWBV+96, HMV93] are proposed in the literature which use precomputed lookup tables to improve the efficiency of finite field multiplication. We briefly describe the complexity for evaluating the monomial using several algorithms from the literature (Table 4).

**Table 4.** Complexities of different algorithms for implementing field multiplications

|          | Number of instructions | | Look-up table | |
|----------|------------------------|---------------------------------------------|-------------------|----------------------------------------|
|          | XOR                    | ADD, SUB, SHIFT, AND                        | Bit size          | No. of access                          |
| [HMV93]  | $2g^2$                 | $g^2\left(\frac{3}{2} - \frac{1}{2(2^b-1)}\right)$ | $2b2^b$     | $3g^2$                                 |
| [GP97]   | $6g^{\log 3} - 8 \cdot g + 2$ | $g^{\log 3}$                         | $2b2^b$           | $3g^{\log 3}$                          |
| [KA98]   | $4g^2$                 | —                                           | $(2b-1)2^{2b}$    | $2g^2 + g$                             |
| [Has00]  | $\left(\frac{1}{2}(g+1)(b+3) - 4\right)\lceil \frac{n}{w} \rceil$ | $(g-1)\lceil \frac{n}{w} \rceil + 4g - 2$ | $(b+d)2^b$ | $(g-1)\lceil \frac{b+d}{w} \rceil$ |

In all lookup-table based multiplication algorithms above, $b$ is the size of the internal data path of the processor. Any element in $\mathbb{F}_{2^n}$ is partitioned as a collection into $g$ groups each having $b$ bits. If $n$ is not a multiple of $b$ then the most significant group will contain $n \pmod{b}$ bits. Note that the algorithm in [HMV93] requires $n$ to be multiple of $b$. Furthermore, $d$ denotes the degree of the second highest monomial (with non-zero coefficient) in the irreducible polynomial that defines the field $\mathbb{F}_{2^n}$ and $w$ denotes the word size of processor. The resources of a processor are optimally utilized when $b = w$. For example in a 32 bit processor two polynomials can be added using $\lceil \frac{n}{32} \rceil$ XOR instructions. However choosing $b = w$ in this case increases the size of the lookup table to $2^5$

GB for the algorithms from [HMV93, GP97]. On the other hand choosing $b < w$ may imply lower utilization of processor's resources. The algorithm described in [Has00] proposes a better utilization of resources when a small value of $b$ is chosen to keep the size of the lookup table sufficiently small. Also, this algorithm does not require $n$ to be multiple of $b$.

### 6.3  Generic Masking Against Side-Channel Attack

Side-channel attacks exploit different types of physical leakage of information e.g. power consumption or EM emanations during the execution of cryptographic algorithms on a device for recovering sensitive variables (e.g. secret key). Masking is a well known technique to prevent implementations of cryptographic algorithms from such attacks. Most of the masking schemes usually protect an implementation against first-order attacks. Over the past years several higher-order side-channel attacks were proposed and demonstrated successfully against many well-known cryptographic algorithms. Higher order masking schemes are useful to protect a cryptographic algorithm against such attacks.

In a higher order masking scheme a sensitive variable (e.g. variables involving secret keys) is split into $t+1$ shares where $t$ is known as the order of masking. It has been shown that the complexity of side-channel attacks increases exponentially with the masking order.

In FSE 2012 a generic higher order masking scheme [CGP+12] was proposed by Carlet, Goubin, Prouff, Quisquater and Rivain. For masking an S-box using CGPQR scheme we need to consider the polynomial corresponding to the S-box, which can be easily computed from the S-box table using Lagrange's theorem in a field $\mathbb{F}_{2^n}$. In CGPQR masking scheme evaluation of this polynomial is protected against higher order attacks. For example, let $x$ be a secret variable for which we evaluate a function $f(x)$. Let $x_0, x_1, \ldots, x_t$ are the $t+1$ shares corresponding to this variable such that $x = \bigoplus_{i=0}^{t} x_i$. Any linear function $\ell(x)$ is easy to mask since $\ell(x) = \ell(x_0) \oplus \ldots \oplus \ell(x_t)$. However masking a non-linear function is not as easy as linear or affine functions.

The operations necessary for evaluating a polynomial in $\mathbb{F}_{2^n}$ are addition, multiplication by a scalar, squaring and regular multiplication. For $t$th order masking any affine and linear operation in $\mathbb{F}_{2^n}$ requires $\mathcal{O}(t)$ logical operations, whereas regular multiplication requires $\mathcal{O}(t^2)$ logical operations. Hence regular multiplication is significant operation in CGPQR masking scheme and its efficiency can be increased by minimizing the number of regular multiplications in a field for a cryptographic algorithm.

MiMC is constructed using a monomial $x^3$ in $\mathbb{F}_{2^n}$. Evaluation of this monomial in each round requires only one multiplication and hence is optimized for CGPQR higher order masking scheme.

## 7  Conclusions

We have reconsidered a 20-year old cipher design idea, given a thorough security analysis, and demonstrated that it can be very competitive in emerging new applications of symmetric cryptography: SNARKs. It might seem that the use-fulness of the design is limited to this setting, as the number of rounds is high compared to other more "traditional" designs for symmetric primitives. However there is evidence that the opposite is true, which was recently discovered in a follow-up work [GRR+16]. Due to its very simple design and despite the high number of rounds, it also turned out to be very competitive in a very different application setting: The currently fastest known MPC protocols with security against active adversaries. This clearly shows that there is a good use-case for designs which work natively in GF(p), and we hope that MiMC can inspire more design and cryptanalysis in this direction.

**Acknowledgements.** We thank Alessandro Chiesa, Eran Tromer and Madars Virza for helpful discussions on SNARKs. The work in this paper has been partially supported by the Austrian Science Fund (project P26494-N15) and by the EU H2020 project Prismacloud (grant agreement nr. 644962). Albrecht was supported by EPSRC grant EP/L018543/1 "Multilinear Maps in Cryptography".

# A    SNARK Prover Algorithm

Here we give a brief description of the parameters chosen to implement the prover algorithm for MiMCHash-256 using the MiMC-1025/1025 permutation with a fixed key. We also briefly describe a part the prover algorithm for MiMC in a SNARK setting. For a more detailed description of the SNARK algorithm we refer the readers to [BSCG+13].

## A.1    Complexity of the Prover Algorithm

Let $S$ be the system of rank-1 quadratic constraints as described in Definition 2 of the article with the tuples $(A_i, B_i, C_i) \in \mathbb{F}^{N'+1}$ for $i \in [N_c]$. Fix an arbitrary subset $\mathcal{X} = \{\alpha_1, \alpha_2, \ldots, \alpha_N\}$ of $\mathbb{F}$ such that $\alpha_i = \omega^{i-1}$ for $i \in [N_c]$ and $\omega$ is the $N_c$ th root of unity. Given an input $x \in \mathbb{F}^m$ and witness $w \in \mathbb{F}^{N'}$ such that $(x, w) \in \mathcal{R}$. The prover algorithm performs the following steps :

1. Choose $\delta_1, \delta_2, \delta_3$ independently at random from the field $\mathbb{F}$
2. Construct the polynomial

$$Q(z) := \frac{F(z)G(z) - H(z)}{U(z)}$$

where $U(z) := z^{N_c} - 1$ and $F, G, H$ are univariate polynomials of degree $N$ defined as

$$F(z) = \underbrace{F_0(z) + \sum_{i=1}^{N'} w_i F_i(z)}_{F'} + \delta_1 U(z), \quad G(z) = G_0(z) + \sum_{i=1}^{N'} w_i G_i(z) + \delta_2 U(z)$$

$$H(z) = H_0(z) + \sum_{i=1}^{N'} w_i H_i(z) + \delta_3 U(z)$$

Here $F_i, G_i, H_i : \mathcal{X} \to \mathbb{F}$ are the Lagrange basis functions for the corresponding polynomials satisfying the following conditions

$$F_i(\alpha_j) = A_j(i), G_i(\alpha_j) = B_j(i), H_i(\alpha_j) = C_j(i)$$

for each $i \in \{0, 1, \ldots, N'\}$ and $j \in [N]$. Note that for any input $x$ and witness $w$ if $(x, w) \in \mathcal{R}$ then $U(z)$ divides $F(z)G(z) - H(z)$.
3. Output the vector $(1, \delta_1, \delta_2, \delta_3, w, q)$ such that $q = (q_0, q_1, \ldots q_N)$ represents the polynomial $Q$.

Note that each of the polynomials $F', G', H'$ (hence $F, G, H$) can be computed using an inverse FFT which has a complexity $O(N_c \log N_c)$. Next a multiplicative coset $\mathcal{Y} := \gamma \mathcal{X}$ of $\mathcal{X} = \{\alpha_1, \ldots \alpha_{N_c}\}$ is chosen such that $\gamma \in \mathbb{F} - \mathcal{X}$. The polynomial $Q(z)$ is computed in two steps

- Evaluate $Q'(z) := \frac{F'(z)G'(z) - H'(z)}{U(z)}$ on $\mathcal{Y}$ point-by-point using the evaluations of $F', G', H', U$ on $\mathcal{Y}$
- Compute $Q'(z)$ using inverse FFT and compute $Q(z) := Q'(z) + \delta_2 F'(z) + \delta_2 G'(z) + \delta_1 \delta_2 U(z) - \delta_3$.

The first step out of the above two takes $O(N_c)$ field operations and the inverse FFT has the complexity $O(N_c \log N_c)$.

## A.2    Parameters for MiMCHash-256

**Over $\mathbb{F}_{2^n}$** We describe the parameter choices for $n = 1025$. The hash function constructed over this particular field promises the same level of security as SHA-256. For processing a single block we use the MiMC-1025/1025 over $\mathbb{F}_{2^{1025}}$. The two constraints in each round of MiMC permutation can be combined to obtain a single rank one quadratic constraint. Hence we get approximately $1025/\log(3) \approx 646$ constraints from the permutation together plus an additional constraint for compression function making the total number of constraints 647. Note that each round introduces two variables in the constraints hence the number of witness is 1293 where $w_1 = x \in \mathbb{F}_{2^{1025}}$ is the input to the hash function and $w \in (\mathbb{F}_{2^{1025}})^{1293}$.

In the prover algorithm the number of constraints $N$ should be such that the principal $N$-th root exists in $\mathbb{F}_{2^{1025}}$. To satisfy this condition we choose $N = 1801$ (since 1801 divides $|\mathbb{F}_{2^{1025}}^*|$). This is the smallest number which divides the order of the multiplicative group corresponding to the finite field and also greater than 647. We add 1154 dummy constraints of the form $0.X_i = 0$ to make the total number of constraint 1801. Note that although the complexity of the prover algorithm depends on the number of constraints (or number of multiplications) for a specific algorithm the number of constraints may not be feasible choice for the FFT algorithm. In such case the complexity actually depends on the best possible choice of the multiplicative subgroup of $\mathbb{F}_{2^n}$.

This is not only applicable to MiMC or MiMCHash but a feature of the SNARK algorithm. In [BSCG+13] a finite field $\mathbb{F}_p$ is chosen in such way that $p - 1$ is of the form $2^t \cdot q$.

**Over** $\mathbb{F}_p$ When we use MiMC-$p/p$ over $\mathbb{F}_p$ for some prime $p$ (with 1025 or more bits) to construct the hash function we have the option of choosing $p$ such that $p - 1 = 2^l \cdot q$. However this yields a very large prime number $p$. For $\approx 1025$ bit security of the keyed permutation it is enough to have $1025/\log(3) \approx 646$ rounds. Hence the number of witness will be 1293 in this case for processing a single block. Instead of choosing such large prime we can choose $p$ such that $p-1$ has a prime factor closed to and greater than 1293.

# References

[AÅBL12] Abdelraheem, M.A., Ågren, M., Beelen, P., Leander, G.: On the distribution of linear biases: three instructive examples. In: Safavi-Naini, R., Canetti, R. (eds.) CRYPTO 2012. LNCS, vol. 7417, pp. 50–67. Springer, Heidelberg (2012). doi:10.1007/978-3-642-32009-5_4

[ADL+08] Arbitman, Y., Dogon, G., Lyubashevsky, V., Micciancio, D., Peikert, C., Rosen, A.: Swifftx: a proposal for the SHA-3 standard. Submission to NIST (2008)

[AGR+16] Albrecht, M., Grassi, L., Rechberger, C., Roy, A., Tiessen, T.: MiMC: efficient encryption and cryptographic hashing with minimal multiplicative complexity. Cryptology ePrint Archive, Report 2016/492 (2016). http://eprint.iacr.org/2016/492

[ÅHJM11] Ågren, M., Hell, M., Johansson, T., Meier, W.: Grain-128a: a new version of grain-128 with optional authentication. IJWMC 5(1), 48–59 (2011)

[Ajt96] Ajtai, M.: Generating hard instances of lattice problems (extended abstract). In: 28th ACM STOC, May 1996, pp. 99–108. ACM Press (1996)

[ARS+15] Albrecht, M.R., Rechberger, C., Schneider, T., Tiessen, T., Zohner, M.: Ciphers for MPC and FHE. In: Oswald, E., Fischlin, M. (eds.) EUROCRYPT 2015. LNCS, vol. 9056, pp. 430–454. Springer, Heidelberg (2015). doi:10.1007/978-3-662-46800-5_17

[ARS+16a] Albrecht, M., Rechberger, C., Schneider, T., Tiessen, T., Zohner, M.: Ciphers for MPC and FHE. Cryptology ePrint Archive, Report 2016/687 (2016). http://eprint.iacr.org/2016/687

[ARS+16b] Albrecht, M.R., Rechberger, C., Schneider, T., Tiessen, T., Zohner, M.: Ciphers for MPC and FHE. Cryptology ePrint Archive, Report 2016 (2016). http://eprint.iacr.org/

[BBL+15] Banerjee, A., Brenner, H., Leurent, G., Peikert, C., Rosen, A.: SPRING: fast pseudorandom functions from rounded ring products. In: Cid, C., Rechberger, C. (eds.) FSE 2014. LNCS, vol. 8540, pp. 38–57. Springer, Heidelberg (2015). doi:10.1007/978-3-662-46706-0_3

[BCG+14] Ben-Sasson, E., Chiesa, A., Garman, C., Green, M., Miers, I., Tromer, E., Virza, M.: Zerocash: decentralized anonymous payments from bitcoin. In: 2014 IEEE Symposium on Security and Privacy, SP 2014, Berkeley, CA, USA, 18–21 May 2014, pp. 459–474. IEEE Computer Society (2014)

[BDPA08] Bertoni, G., Daemen, J., Peeters, M., Assche, G.: On the indifferentiability of the sponge construction. In: Smart, N. (ed.) EUROCRYPT 2008. LNCS, vol. 4965, pp. 181–197. Springer, Heidelberg (2008). doi:10.1007/978-3-540-78967-3_11

[BFS14]   Bardet, M., Faugère, J.-C., Salvy, B.: On the complexity of the F5 Gröbner basis Algorithm. J. Symb. Comput. **70**, 49–70 (2014)

[BKW93]   Becker, T., Kredel, H., Weispfenning, V.: Gröbner Bases: A Computational Approach to Commutative Algebra. Springer, New York (1993)

[BMP13]   Boyar, J., Matthews, P., Peralta, R.: Logic minimization techniques with applications to cryptology. J. Cryptology **26**(2), 280–312 (2013)

[BP12]   Boyar, J., Peralta, R.: A small depth-16 circuit for the AES S-box. In: Gritzalis, D., Furnell, S., Theoharidou, M. (eds.) Information Security and Privacy Conference (SEC). IFIP Advances in Information and Communication Technology, vol. 376, pp. 287–298. Springer, Heidelberg (2012)

[BSCG+13]   Ben-Sasson, E., Chiesa, A., Genkin, D., Tromer, E., Virza, M.: SNARKs for C: verifying program executions succinctly and in zero knowledge. In: Canetti, R., Garay, J.A. (eds.) CRYPTO 2013. LNCS, vol. 8043, pp. 90–108. Springer, Heidelberg (2013). doi:10.1007/978-3-642-40084-1_6

[BSS+13]   Beaulieu, R., Shors, D., Smith, J., Treatman-Clark, S., Weeks, B., Wingers, L.: The SIMON and SPECK families of lightweight block ciphers. Cryptology ePrint Archive, Report 2013/404 (2013). http://eprint.iacr.org/2013/404

[Can97]   Canteaut, A.: Differential cryptanalysis of feistel ciphers and differentially $\delta$-uniform mappings. In: Workshop on Selected Areas in Cryptography, SAC 1997, Workshop Record, pp. 172–184 (1997)

[CCF+16]   Canteaut, A., Carpov, S., Fontaine, C., Lepoint, T., Naya-Plasencia, M., Paillier, P., Sirdey, R.: Stream ciphers: a practical solution for efficient homomorphic-ciphertext compression. To appear in Proceedings of FSE 2016, available on Cryptology ePrint Archive, Report 2015/113 (2016). http://eprint.iacr.org/

[CFH+15]   Costello, C., Fournet, C., Howell, J., Kohlweiss, M., Kreuter, B., Naehrig, M., Parno, B., Zahur, S.: Geppetto: versatile verifiable computation. In: 2015 IEEE Symposium on Security and Privacy, SP 2015, pp. 253–270. IEEE Computer Society (2015)

[CGP+12]   Carlet, C., Goubin, L., Prouff, E., Quisquater, M., Rivain, M.: Higher-order masking schemes for S-boxes. In: Canteaut, A. (ed.) FSE 2012. LNCS, vol. 7549, pp. 366–384. Springer, Heidelberg (2012). doi:10.1007/978-3-642-34047-5_21

[CP08]   Cannière, C., Preneel, B.: TRIVIUM. In: Robshaw, M., Billet, O. (eds.) New Stream Cipher Designs. LNCS, vol. 4986, pp. 244–266. Springer, Heidelberg (2008). doi:10.1007/978-3-540-68351-3_18

[DPVAR00]   Daemen, J., Peeters, M., Van Assche, G., Rijmen, V.: Nessie proposal: Noekeon. In: First Open NESSIE Workshop (2000)

[DWBV+96]   De Win, E., Bosselaers, A., Vandenberghe, S., De Gersem, P., Vandewalle, J.: A fast software implementation for arithmetic operations in GF(2n). In: Kim, K., Matsumoto, T. (eds.) Advances in Cryptology – ASIACRYPT '96. Lecture Notes in Computer Science, vol. 1163, pp. 65–76. Springer, Berlin Heidelberg (1996)

[ENI13]   ENISA. Algorithms, key sizes and parameters report – 2013 recommendations. Technical report, European Union Agency for Network and Information Security, October 2013

[GLSV14] Grosso, V., Leurent, G., Standaert, F.-X., Varıcı, K.: LS-designs: bitslice encryption for efficient masked software implementations. In: Cid, C., Rechberger, C. (eds.) FSE 2014. LNCS, vol. 8540, pp. 18–37. Springer, Heidelberg (2015). doi:10.1007/978-3-662-46706-0_2

[GP97] Guajardo, J., Paar, C.: Efficient algorithms for elliptic curve cryptosystems. In: Kaliski, B.S. (ed.) CRYPTO 1997. LNCS, vol. 1294, pp. 342–356. Springer, Heidelberg (1997). doi:10.1007/BFb0052247

[GRR+16] Grassi, L., Rechberger, C., Rotaru, D., Scholl, P., Smart, N.: MPC-friendly symmetric key primitives. Cryptology ePrint Archive, Report 2016 (2016). http://eprint.iacr.org/

[Has00] Hasan, M.A.: Look-up table-based large finite field multiplication in memory constrained cryptosystems. IEEE Trans. Comput. 49(7), 749–758 (2000)

[HMV93] Harper, G., Menezes, A., Vanstone, S.: Public-key cryptosystems with very small key lengths. In: Rueppel, R.A. (ed.) EUROCRYPT 1992. LNCS, vol. 658, pp. 163–173. Springer, Heidelberg (1993). doi:10.1007/3-540-47555-9_14

[JK97] Jakobsen, T., Knudsen, L.R.: The interpolation attack on block ciphers. In: Biham, E. (ed.) FSE 1997. LNCS, vol. 1267, pp. 28–40. Springer, Heidelberg (1997). doi:10.1007/BFb0052332

[KA98] Koc, C.K., Acar, T.: Montgomery multiplication in GF(2k). Des. Codes Crypt. 14(1), 57–69 (1998)

[KN95] Knudsen, L.R., Nyberg, K.: Provable security against a differential attack. J. Crypt. 8(1), 27–37 (1995)

[KR11] Knudsen, L.R., Robshaw, M.: The Block Cipher Companion. Information Security and Cryptography. Springer, Heidelberg (2011)

[Lab] SCIPR lab. libsnark. https://github.com/scipr-lab/libsnark

[LMPR08] Lyubashevsky, V., Micciancio, D., Peikert, C., Rosen, A.: SWIFFT: a modest proposal for FFT hashing. In: Nyberg, K. (ed.) FSE 2008. LNCS, vol. 5086, pp. 54–72. Springer, Heidelberg (2008). doi:10.1007/978-3-540-71039-4_4

[LMS13] Lebreton, R., Mehrabi, E., Schost, É.: On the complexity of solving bivariate systems: the case of non-singular solutions. In: Kauers, M. (ed.) International Symposium on Symbolic and Algebraic Computation, ISSAC'13, Boston, MA, USA, 26–29 June 2013, pp. 251–258. ACM (2013)

[MJSC16] Méaux, P., Journault, A., Standaert, F.-X., Carlet, C.: Towards stream ciphers for efficient FHE with low-noise ciphertexts. In: Fischlin, M., Coron, J.-S. (eds.) EUROCRYPT 2016. LNCS, vol. 9665, pp. 311–343. Springer, Heidelberg (2016). doi:10.1007/978-3-662-49890-3_13

[MVO96] Menezes, A.J., Vanstone, S.A., Van Oorschot, P.C.: Handbook of Applied Cryptography, 1st edn. CRC Press Inc., Boca Raton (1996)

[NIS14] NIST. DRAFT FIPS PUB 202, SHA-3 standard: permutation-based hash and extendable-output functions (2014)

[NR97] Naor, M., Reingold, O.: Number-theoretic constructions of efficient pseudo-random functions. In: 38th Annual Symposium on Foundations of Computer Science, FOCS 1997, pp. 458–467. IEEE Computer Society (1997)

[Nyb94] Nyberg, K.: Differentially uniform mappings for cryptography. In: Helleseth, T. (ed.) EUROCRYPT 1993. LNCS, vol. 765, pp. 55–64. Springer, Heidelberg (1994). doi:10.1007/3-540-48285-7_6

[PH78]  Pohlig, S.C., Hellman, M.E.: An improved algorithm for computing logarithms over GF(p) and its cryptographic significance (corresp.). IEEE Trans. Inf. Theory **24**(1), 106–110 (1978)

[PHGR16]  Parno, B., Howell, J., Gentry, C., Raykova, M.: Pinocchio: nearly practical verifiable computation. Commun. ACM **59**(2), 103–112 (2016)

[Sho]  Shoup, V.: Number theory library 5.5.2 (NTL) for C++. http://www.shoup.net/ntl/

[Sto85]  Stoss, H.-J.: The complexity of evaluating interpolation polynomials. Theor. Comput. Sci. **41**, 319–323 (1985)

# Balloon Hashing: A Memory-Hard Function Providing Provable Protection Against Sequential Attacks

Dan Boneh[1], Henry Corrigan-Gibbs[1(✉)], and Stuart Schechter[2]

[1] Stanford University, Stanford, CA 94305, USA
{dabo,henrycg}@cs.stanford.edu
[2] Microsoft Research, Redmond, WA 98052, USA

**Abstract.** We present the Balloon password-hashing algorithm. This is the first practical cryptographic hash function that: (i) has proven memory-hardness properties in the random-oracle model, (ii) uses a password-independent access pattern, and (iii) meets—and often exceeds—the performance of the best heuristically secure password-hashing algorithms. Memory-hard functions require a large amount of working space to evaluate efficiently and, when used for password hashing, they dramatically increase the cost of offline dictionary attacks. In this work, we leverage a previously unstudied property of a certain class of graphs ("random sandwich graphs") to analyze the memory-hardness of the Balloon algorithm. The techniques we develop are general: we also use them to give a proof of security of the scrypt and Argon2i password-hashing functions, in the random-oracle model. Our security analysis uses a *sequential* model of computation, which essentially captures attacks that run on single-core machines. Recent work shows how to use massively parallel special-purpose machines (e.g., with hundreds of cores) to attack memory-hard functions, including Balloon. We discuss these important attacks, which are outside of our adversary model, and propose practical defenses against them. To motivate the need for security proofs in the area of password hashing, we demonstrate and implement a practical attack against Argon2i that successfully evaluates the function with less space than was previously claimed possible. Finally, we use experimental results to compare the performance of the Balloon hashing algorithm to other memory-hard functions.

**Keywords:** Memory-hard functions · Password hashing · Pebbling arguments · Time-space trade-offs · Sandwich graph · Argon2 · Scrypt

## 1 Introduction

The staggering number of password-file breaches in recent months demonstrates the importance of cryptographic protection for stored passwords. In 2015 alone,

---

The full version of this paper is available online at https://eprint.iacr.org/2016/027.

© International Association for Cryptologic Research 2016
J.H. Cheon and T. Takagi (Eds.): ASIACRYPT 2016, Part I, LNCS 10031, pp. 220–248, 2016.
DOI: 10.1007/978-3-662-53887-6_8

attackers stole files containing users' login names, password hashes, and contact information from many large and well-resourced organizations, including Last-Pass [79], Harvard [47], E*Trade [62], ICANN [45], Costco [41], T-Mobile [76], the University of Virginia [74], and a large number of others [65]. In this environment, systems administrators must operate under the assumption that attackers will eventually gain access to sensitive authentication information, such as password hashes and salts, stored on their computer systems. After a compromise, the secrecy of user passwords rests on the cost to an attacker of mounting an offline dictionary attack against the stolen file of hashed passwords.

An ideal password-hashing function has the property that it costs as much for an attacker to compute the function as it does for the legitimate authentication server to compute it. Standard cryptographic hashes completely fail in this regard: it takes $100\,000\times$ more energy to compute a SHA-256 hash on a general-purpose x86 CPU (as an authentication server would use) than it does to compute SHA-256 on special-purpose hardware (such as the ASICs that an attacker would use) [21]. Iterating a standard cryptographic hash function, as is done in bcrypt [66] and PBKDF2 [43], increases the absolute cost to the attacker and defender, but the attacker's $100\,000\times$ relative cost advantage remains.

*Memory-hard functions* help close the efficiency gap between the attacker and defender in the setting of password hashing [8,18,37,56,60]. Memory-hard functions exploit the observation that on-chip memory is just as costly to power on special-purpose hardware as it is on a general-purpose CPU. If evaluating the password-hashing function requires large amounts of memory, then an attacker using special-purpose hardware has little cost advantage over the legitimate authentication server (using a standard x86 machine, for example) at running the password-hashing computation. Memory consumes a large amount of on-chip area, so the high memory requirement ensures that a special-purpose chip can only contain a small number of hashing engines.

An optimal memory-hard function, with security parameter $n$, has a space-time product that satisfies $S \cdot T \in \Omega(n^2)$, irrespective of the strategy used to compute the function [60]. The challenge is to construct a function that *provably* satisfies this bound with the largest possible constant multiple on the $n^2$ term.

In this paper, we introduce the Balloon memory-hard function for password hashing. This is the first practical password-hashing function to simultaneously satisfy three important design goals [56]:

- *Proven memory-hard.* We prove, in the random-oracle model [13], that computing the Balloon function with space $S$ and time $T$ requires $S \cdot T \geq n^2/8$ (approximately). As the adversary's space usage decreases, we prove even sharper time-space lower bounds.

  To motivate our interest in memory-hardness proofs, we demonstrate in Sect. 4 an attack against the Argon2i password hashing function [18], winner of a recent password-hashing design competition [56]. The attack evaluates the function with far less space than claimed without changing the time required to compute the function. We also give a proof of security for Argon2i in the random-oracle model, which demonstrates that significantly more powerful attacks against Argon2i are impossible under our adversary model.

– *Password-independent memory-access pattern.* The memory-access pattern of the Balloon algorithm is *independent* of the password being hashed. Password-hashing functions that lack this property are vulnerable to a crippling attack in the face of an adversary who learns the memory-access patterns of the hashing computation, e.g., via cache side-channels [23,54,77]. The attack, which we describe in the full version of this paper, makes it possible to run a dictionary attack with very little memory. A hashing function with a password-independent memory-access pattern eliminates this threat.

– *Performant.* The Balloon algorithm is easy to implement and it matches or exceeds the performance of the fastest comparable password-hashing algorithms, Argon2i [18] and Catena [37], when instantiated with standard cryptographic primitives (Sect. 6).

We analyze the memory-hardness properties of the Balloon function using pebble games, which are arguments about the structure of the data-dependency graph of the underlying computation [48,57,59,72,75]. Our analysis uses the framework of Dwork, Naor, and Wee [32]—later applied in a number of cryptographic works [6,8,33,34,37]—to relate the hardness of pebble games to the hardness of certain computations in the random-oracle model [13].

The crux of our analysis is a new observation about the properties of "random sandwich graphs," a class of graphs studied in prior work on pebbling [6,8]. To show that our techniques are broadly applicable, we apply them in the full version of this paper to give simple proofs of memory-hardness, in the random-oracle model, for the Argon2i and scrypt functions. We prove stronger memory-hardness results about the Balloon algorithm, but these auxiliary results about Argon2i and scrypt may be of independent interest to the community.

The performance of the Balloon hashing algorithm is surprisingly good, given that our algorithm offers stronger proven security properties than other practical memory-hard functions with a password-independent memory access patterns. For example, if we configure Balloon to use Blake2b as the underlying hash function [10], run the construction for five "rounds" of hashing, and set the space parameter to require the attacker to use 1 MiB of working space to compute the function, then we can compute Balloon Hashes at the rate of 13 hashes per second on a modern server, compared with 12.8 for Argon2i, and 2.1 for Catena DBG (when Argon2i and Catena DBG are instantiated with Blake2b as the underlying cryptographic hash function).[1]

*Caveat: Parallel Attacks.* The definition of memory-hardness we use puts a lower-bound on the time-space product of computing a *single* instance of the Balloon function on a sequential (single-core) computer. In reality, an adversary mounting a dictionary attack would want to compute *billions* of instances of the Balloon function, perhaps using many processors running in parallel. Alwen and

---

[1] The relatively poor performance of Argon2i here is due to the attack we present in Sect. 4. It allows an attacker to save space in computing Argon2i with no increase in computation time.

Serbinenko [8], formalizing earlier work by Percival [60], introduce a new computational model— the parallel random-oracle model (pROM)—and a memory-hardness criterion that addresses the shortcomings of the traditional model. In recent work, Alwen and Blocki prove the surprising result that *no function* that uses a password-independent memory access pattern can be optimally memory-hard in the pROM [3]. In addition, they give a special-purpose pROM algorithm for computing Argon2i, Balloon, and other practical (sequential) memory-hard functions with some space savings. We discuss this important class of attacks and the relevant related work in Sect. 5.1.

**Contributions.** In this paper, we

- introduce and analyze the Balloon hashing function, which has stronger provable security guarantees than prior practical memory-hard functions (Sect. 3),
- present a practical memory-saving attack against the Argon2i password-hashing algorithm (Sect. 4), and
- explain how to ameliorate the danger of massively parallel attacks against memory-hard functions with a password-independent access pattern (Sect. 5.1)
- prove the first known time-space lower bounds for Argon2i and an idealized variant of scrypt, in the random-oracle model. (See the full version of this paper for these results.)

With the Balloon algorithm, we demonstrate that it is possible to provide provable protection against a wide class of attacks without sacrificing performance.

*Notation.* Throughout this paper, Greek symbols ($\alpha$, $\beta$, $\gamma$, $\lambda$, etc.) typically denote constants greater than one. We use $\log_2(\cdot)$ to denote a base-two logarithm and $\log(\cdot)$ to denote a logarithm when the base is not important. For a finite set $S$, the notation $x \xleftarrow{R} S$ indicates sampling an element of $S$ uniformly at random and assigning it to the variable $x$.

# 2   Security Definitions

This section summarizes the high-level security and functionality goals of a password hashing function in general and the Balloon hashing algorithm in particular. We draw these aims from prior work on password hashing [60,66] and also from the requirements of the recent Password Hashing Competition [56].

## 2.1   Syntax

The Balloon password hashing algorithm takes four inputs: a password, salt, time parameter, and space parameter. The output is a bitstring of fixed length (e.g., 256 or 512 bits). The password and salt are standard [52], but we elaborate on the role of the latter parameters below.

*Space Parameter (Buffer Size).* The space parameter, which we denote as "$n$" throughout, indicates how many fixed-size blocks of working space the hash function will require during the course of its computation, as in scrypt [60]. At a high level, a memory-hard function should be "easy" to compute with $n$ blocks of working space and should be "hard" to compute with much less space than that. We make this notion precise later on.

*Time Parameter (Number of Rounds).* The Balloon function takes as input a parameter $r$ that determines the number of "rounds" of computation it performs. As in bcrypt [66], the larger the time parameter, the longer the hash computation will take. On memory-limited platforms, a system administrator can increase the number of rounds of hashing to increase the cost of computing the function without increasing the algorithm's memory requirement. The choice of $r$ has an effect on the memory-hardness properties of the scheme: the larger $r$ is, the longer it takes to compute the function in small space.

## 2.2 Memory-Hardness

We say that a function $f_n$ on space parameter $n$ is *memory-hard* in the (sequential) random-oracle model [13] if, for all adversaries computing $f_n$ with high probability using space $S$ and $T$ random oracle queries, we have that $S \cdot T \in \Omega(n^2)$. This definition deserves a bit of elaboration. Following Dziembowski et al. [34] we say that an algorithm "uses space $S$" if the entire configuration of the Turing Machine (or RAM machine) computing the algorithm requires at least $S$ bits to describe. When, we say that an algorithm computes a function "with high probability," we mean that the probability that the algorithm computes the function is non-negligible as the output size of the random oracle and the space parameter $n$ tend to infinity. In practice, we care about the adversary's concrete success probability, so we avoid asymptotic notions of security wherever possible. In addition, as we discuss in the evaluation section (Sect. 6), the exact value of the constant hidden inside the $\Omega(\cdot)$ is important for practical purposes, so our analysis makes explicit and optimizes these constants.

A function that is memory-hard under this definition requires the adversary to use either a lot of working space or a lot of execution time to compute the function. Functions that are memory-hard in this way are not amenable to implementation in special-purpose hardware (ASIC), since the cost to power a unit of memory for a unit of time on an ASIC is the same as the cost on a commodity server. An important limitation of this definition is that it does not take into account parallel or multiple-instance attacks, which we discuss in Sect. 5.1.

## 2.3 Password-Independent Access Pattern

A first-class design goal of the Balloon algorithm is to have a memory access pattern that is *independent* of the password being hashed. (We allow the data-access pattern to depend on the salt, since the salts can be public.) As mentioned above, employing a password-independent access pattern reduces the risk that information about the password will leak to other users on the same machine via cache or other side-channels [23,54,77]. This may be especially important in cloud-computing environments, in which many mutually distrustful users share a single physical host [69].

Creating a memory-hard function with a password-independent access pattern presents a technical challenge: since the data-access pattern depends only upon the salt—which an adversary who steals the password file knows—the adversary can compute the entire access pattern in advance of a password-guessing attack. With the access pattern in hand, the adversary can expend a huge amount of effort to find an efficient strategy for computing the hash function in small space. Although this pre-computation might be expensive, the adversary can amortize its cost over billions of subsequent hash evaluations. A function that is memory-hard *and* that uses a password-independent data access pattern must be impervious to *all* small-space strategies for computing the function so that it maintains its strength in the face of these pre-computation attacks. (Indeed, as we discuss in Sect. 5.1, Alwen and Blocki show that in some models of computation, memory-hard functions with password-independent access patterns do not exist [3].)

### 2.4  Collision Resistance, etc.

If necessary, we can modify the Balloon function so that it provides the standard properties of second-preimage resistance and collision resistance [51]. It is possible to achieve these properties in a straightforward way by composing the Balloon function $B$ with a standard cryptographic hash function $H$ as

$$H_B(\text{passwd}, \text{salt}) := H(\text{passwd}, \text{salt}, B(\text{passwd}, \text{salt})).$$

Now, for example, if $H$ is collision-resistant, then $H_B$ must also be.[2] That is because any inputs $(x_p, x_s) \neq (y_p, y_s)$ to $H_B$ that cause $H_B(x_p, x_s) = H_B(y_p, y_s)$ immediately yield a collision for $H$ as:

$$(x_p, x_s, B(x_p, x_s)) \qquad \text{and} \qquad (y_p, y_s, B(y_p, y_s)),$$

no matter how the Balloon function $B$ behaves.

## 3  Balloon Hashing Algorithm

In this section, we present the Balloon hashing algorithm.

### 3.1  Algorithm

The algorithm uses a standard (non-memory-hard) cryptographic hash function $H : \mathbb{Z}_N \times \{0,1\}^{2k} \to \{0,1\}^k$ as a subroutine, where $N$ is a large integer. For the purposes of our analysis, we model the function $H$ as a random oracle [13].

---

[2] We are eliding important definitional questions about what it even means, in a formal sense, for a function to be collision resistant [16,70].

```
func Balloon(block_t passwd, block_t salt,
     int s_cost,           // Space cost (main buffer size)
     int t_cost):          // Time cost (number of rounds)
  int delta = 3            // Number of dependencies per block
  int cnt = 0              // A counter (used in security proof)
  block_t buf[s_cost]):    // The main buffer

  // Step 1. Expand input into buffer.
  buf[0] = hash(cnt++, passwd, salt)
  for m from 1 to s_cost-1:
    buf[m] = hash(cnt++, buf[m-1])

  // Step 2. Mix buffer contents.
  for t from 0 to t_cost-1:
    for m from 0 to s_cost-1:
      // Step 2a. Hash last and current blocks.
      block_t prev = buf[(m-1) mod s_cost]
      buf[m] = hash(cnt++, prev, buf[m])

      // Step 2b. Hash in pseudorandomly chosen blocks.
      for i from 0 to delta-1:
        block_t idx_block = ints_to_block(t, m, i)
        int other = to_int(hash(cnt++, salt, idx_block)) mod s_cost
        buf[m] = hash(cnt++, buf[m], buf[other])

  // Step 3. Extract output from buffer.
  return buf[s_cost-1]
```

**Fig. 1.** Pseudo-code of the Balloon hashing algorithm.

The Balloon algorithm uses a large memory buffer as working space and we divide this buffer into contiguous *blocks*. The size of each block is equal to the output size of the hash function $H$. Our analysis is agnostic to the choice of hash function, except that, to prevent pitfalls described in the full version of this paper, the internal state size of $H$ must be at least as large as its output size. Since $H$ maps blocks of $2k$ bits down to blocks of $k$ bits, we sometimes refer to $H$ as a *cryptographic compression function*.

The Balloon function operates in three steps (Fig. 1):

1. **Expand.** In the first step, the Balloon algorithm fills up a large buffer with pseudo-random bytes derived from the password and salt by repeatedly invoking the compression function $H$ on a function of the password and salt.
2. **Mix.** In the second step, the Balloon algorithm performs a "mixing" operation $r$ times on the pseudo-random bytes in the memory buffer. The user-specified round parameter $r$ determines how many rounds of mixing take place. At each mixing step, for each block $i$ in the buffer, the routine updates the contents of block $i$ to be equal to the hash of block $(i-1) \bmod n$, block $i$, and $\delta$ other blocks chosen "at random" from the buffer. (See Theorem 1 for an illustration of how the choice of $\delta$ affects the security of the scheme.)

Since the Balloon functions are deterministic functions of their arguments, the dependencies are not chosen truly at random but are sampled using a pseudorandom stream of bits generated from the user-specific salt.

3. **Extract.** In the last step, the Balloon algorithm outputs the last block of the buffer.

**Multi-core Machines.** A limitation of the Balloon algorithm as described is that it does not allow even limited parallelism, since the value of the $i$th block computed always depends on the value of the $(i-1)$th block. To increase the rate at which the Balloon algorithm can fill memory on a multi-core machine with $M$ cores, we can define a function that invokes the Balloon function $M$ times in parallel and XORs all the outputs. If $\mathsf{Balloon}(p, s)$ denotes the Balloon function on password $p$ and salt $s$, then we can define an $M$-core variant $\mathsf{Balloon}_M(p, s)$ as:

$$\mathsf{Balloon}_M(p, s) := \mathsf{Balloon}(p,\ s\|\text{``1''}) \oplus \cdots \oplus \mathsf{Balloon}(p,\ s\|\text{``}M\text{''}).$$

A straightforward argument shows that computing this function requires computing $M$ instances of the single-core Balloon function. Existing password hashing functions deploy similar techniques on multi-core platforms [18,37,60,61].

## 3.2  Main Security Theorem

The following theorem demonstrates that attackers who attempt to compute the Balloon function in small space must pay a large penalty in computation time. The complete theorem statement is given in the full version of this paper.

**Theorem 1 (informal).**  *Let $\mathcal{A}$ be an algorithm that computes the n-block r-round Balloon function with security parameter $\delta \geq 3$, where $H$ is modeled as a random oracle. If $\mathcal{A}$ uses at most $S$ blocks of buffer space then, with overwhelming probability, $\mathcal{A}$ must run for time (approximately) $T$, such that*

$$S \cdot T \geq \frac{r \cdot n^2}{8}.$$

*Moreover, under the stated conditions, one obtains the stronger bound:*

$$S \cdot T \geq \frac{(2^r - 1)n^2}{8} \qquad if \qquad \begin{cases} \delta = 3 \ and \ S < n/64 \ or, \\ \delta = 4 \ and \ S < n/32 \ or, \\ \delta = 5 \ and \ S < n/16 \ or, \\ \delta = 7 \ and \ S < n/8. \end{cases}$$

The theorem shows that, when the adversary's space usage falls below a certain threshold (parameterized by $\delta$), the computation time increases exponentially in the number of rounds $r$. For example, when $\delta = 7$ and the space $S$ is less than $n/8$, the time to evaluate Balloon is at least $2^r$ times the time to evaluate it with space $n$. Thus, attackers who attempt to compute the Balloon function in very small space must pay a large penalty in computation time.

```
v_1 = hash(input, "0")
v_2 = hash(input, "1")
v_3 = hash(v_1, v_2)
v_4 = hash(v_2, v_3)
v_5 = hash(v_3, v_4)
return v_5
```

**Fig. 2.** An example computation (left) and its corresponding data-dependency graph (right).

The proof of the theorem is given in the full version of this paper. Here we sketch the main ideas in the proof of Theorem 1.

**Proof idea.** The proof makes use of *pebbling arguments*, a classic technique for analyzing computational time-space trade-offs [42,48,59,63,72,78] and memory-hard functions [8,32,33,37]. We apply pebbling arguments to the *data-dependency graph* corresponding to the computation of the Balloon function (See Fig. 2 for an example graph). The graph contains a vertex for every random oracle query made during the computation of Balloon: vertex $v_i$ in the graph represents the response to the $i$th random-oracle query. An edge $(v_i, v_j)$ indicates that the input to the $j$th random-oracle query depends on the response of the $i$th random-oracle query.

The data-dependency graph for a Balloon computation naturally separates into $r + 1$ layers—one for each round of mixing (Fig. 3). That is, a vertex on level $\ell \in \{1, \dots, r\}$ of the graph represents the output of a random-oracle query made during the $\ell$th mixing round.

The first step in the proof shows that the data-dependency graph of a Balloon computation satisfies certain connectivity properties, defined below, with high probability. The probability is taken over the choice of random oracle $H$, which determines the data-dependency graph. Consider placing a pebble on each of a subset of the vertices of the data-dependency graph of a Balloon computation. Then, as long as there are "not too many" pebbles on the graph, we show that the following two properties hold with high probability:

– *Well-Spreadedness.* For every set of $k$ consecutive vertices on some level of the graph, at least a quarter of the vertices on the prior level of the graph are on unpebbled paths to these $k$ vertices.
– *Expansion.* All sets of $k$ vertices on any level of the graph have unpebbled paths back to at least $2k$ vertices on the prior level. The value of $k$ depends on the choice of the parameter $\delta$.

The next step is to show that every *graph-respecting* algorithm computing the Balloon function requires large space or time. We say that an adversary $\mathcal{A}$ is graph respecting if for every $i$, adversary $\mathcal{A}$ makes query number $i$ to the

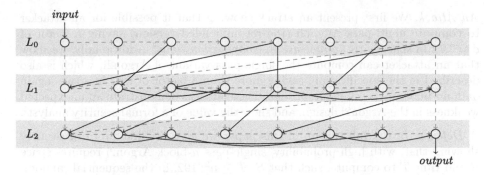

**Fig. 3.** The Balloon data-dependency graph on $n = 8$ blocks and $r = 2$ rounds, drawn with $\delta = 1$ for simplicity. (The real construction uses $\delta \geq 3$.) The dashed edges are fixed and the solid edges are chosen pseudorandomly by applying the random oracle to the salt.

random-oracle only after it has in storage all of the values that this query takes as input.[3]

We show, using the well-spreadedness and expansion properties of the Balloon data-dependency graph, that every graph-respecting adversary $\mathcal{A}$ must use space $S$ and time $T$ satisfying $S \cdot T \geq n^2/8$, with high probability over the choice of $H$. We use the graph structure in the proof as follows: fix a set of $k$ values that the adversary has not yet computed. Then the graph properties imply that these $k$ values have many dependencies that a space-$S$ adversary cannot have in storage. Thus, making progress towards computing the Balloon function in small space requires the adversary to undertake a huge amount of recomputation.

The final step uses a technique of Dwork, Naor, and Wee [32]. They use the notion of a graph *labeling* to convert a directed acyclic graph $G$ into a function $f_G$. They prove that if $G$ is a graph that is infeasible for time-$T$ space-$S$ graph-respecting pebbling adversaries to compute, then it is infeasible for time-$T'$ space-$S'$ arbitrary adversaries to compute the labeling function $f_G$, with high probability in the random-oracle model, where $T' \approx T$ and $S' \approx S$.

We observe that Balloon computes the function $f_G$ where $G$ is the Balloon data-dependency graph. We then directly apply the technique of Dwork, Naor, and Wee to obtain a upper bound on the probability that an arbitrary adversary can compute the Balloon function in small time and space.     □

## 4   Attacking and Defending Argon2

In this section, we analyze the Argon2i password hashing function [18], which won the recent Password Hashing Competition [56].

---

[3] This description is intentionally informal—see the full version of the paperor the precise statement.

*An Attack.* We first present an attack showing that it possible for an attacker to compute multi-pass Argon2i (the recommended version) saving a factor of $e \approx 2.72$ in space with *no increase in computation time.*[4] Additionally, we show that an attacker can compute the single-pass variant of Argon2i, which is also described in the specification, saving more than a factor of four in space, again *with no increase in computation time.* These attacks demonstrate an unexpected weakness in the Argon2i design, and show the value of a formal security analysis.

*A Defense.* In the full version of this paper we give the first proof of security showing that, with high probability, single-pass $n$-block Argon2i requires space $S$ and time $T$ to compute, such that $S \cdot T \geq n^2/192$, in the sequential random-oracle model. Our proof is relatively simple and uses the same techniques we have developed to reason about the Balloon algorithm. The time-space lower bound we can prove about Argon2i is weaker than the one we can prove about Balloon, since the Argon2i result leaves open the possibility of an attack that saves a factor of 192 factor in space with no increase in computation time. If Argon2i becomes a standard algorithm for password hashing, it would be a worthwhile exercise to try to improve the constants on both the attacks and lower bounds to get a clearer picture of its exact memory-hardness properties.

### 4.1   Attack Overview

Our Argon2i attacks require a linear-time pre-computation operation that is independent of the password and salt. The attacker need only run the pre-computation phase once for a given choice of the Argon2i public parameters (buffer size, round count, etc.). After running the pre-computation step once, it is possible to compute many Argon2i password hashes, on different salts and different passwords using our small-space computation strategy. Thus, the cost of the pre-computation is amortized over many subsequent hash computations.

The attacks we demonstrate undermine the security claims of the Argon2i (version 1.2.1) design documents [18]. The design documents claim that computing $n$-block single-pass Argon2i with $n/4$ space incurs a 7.3× computational penalty [18, Table 2]. Our attacks show that there is no computational penalty. The design documents claim that computing $n$-block three-pass Argon2i with $n/3$ space incurs a 16,384× computational penalty [18, Sect. 5.4]. We compute the function in $n/2.7 \approx n/3$ space with no computational penalty.

We analyze a idealized version the Argon2i algorithm, which is slightly simpler than that proposed in the Argon2i v1.2.1 specification [18]. Our idealized analysis *underestimates* the efficacy of our small-space computation strategy, so the strategy we propose is actually *more effective* at computing Argon2i than the analysis suggests. The idealized analysis yields an expected $n/4$ storage cost, but as Fig. 4 demonstrates, empirically our strategy allows computing

---

[4] We have notified the Argon2i designers of this attack and the latest version of the specification incorporates a design change that attempts to prevent the attack [19]. We describe the attack on the original Argon2i design, the winner of the password hashing competition [56].

single-pass Argon2i with only $n/5$ blocks of storage. This analysis focuses on the single-threaded instantiation of Argon2i—we have not tried to extend it to the many-threaded variant.

## 4.2   Background on Argon

At a high level, the Argon2i hashing scheme operates by filling up an $n$-block buffer with pseudo-random bytes, one 1024-byte block at a time. The first two blocks are derived from the password and salt. For $i \in \{3, \ldots, n\}$, the block at index $i$ is derived from two blocks: the block at index $(i - 1)$ and a block selected pseudo-randomly from the set of blocks generated so far. If we denote the contents of block $i$ as $x_i$, then Argon2i operates as follows:

$$x_1 = H(\text{passwd, salt} \parallel 1)$$
$$x_2 = H(\text{passwd, salt} \parallel 2)$$
$$x_i = H(x_{i-1}, x_{r_i}) \quad \text{where } r_i \in \{1, \ldots, i - 1\}$$

Here, $H$ is a non-memory-hard cryptographic hash function mapping two blocks into one block. The random index $r_i$ is sampled from a non-uniform distribution over $S_i = \{1, \ldots, i-1\}$ that has a heavy bias towards blocks with larger indices. We model the index value $r_i$ as if it were sampled from the uniform distribution over $S_i$. Our small-space computation strategy performs *better* under a distribution biased towards larger indices, so our analysis is actually somewhat conservative.

The single-pass variant of Argon2i computes $(x_1, \ldots, x_n)$ in sequence and outputs bytes derived from the last block $x_n$. Computing the function in the straightforward way requires storing every generated block for the duration of the computation— $n$ blocks total.

The multiple-pass variant of Argon2i works as above except that it computes $pn$ blocks instead of just $n$ blocks, where $p$ is a user-specified integer indicating the number of "passes" over the memory the algorithm takes. (The number of passes in Argon2i is analogous to number of rounds $r$ in Balloon hashing.) The default number of passes is three. In multiple-pass Argon2i, the contents of block $i$ are derived from the prior block and one of the most recent $n$ blocks. The output of the function is derived from the value $x_{pn}$. When computing the multiple-pass variant of Argon2i, one need only store the latest $n$ blocks computed (since earlier blocks will never be referenced again), so the storage cost of the straightforward algorithm is still roughly $n$ blocks.

Our analysis splits the Argon2i computation into discrete time steps, where time step $t$ begins at the moment at which the algorithm invokes the compression function $H$ for the $t$th time.

## 4.3   Attack Algorithm

Our strategy for computing $p$-pass Argon2i with fewer than $n$ blocks of memory is as follows:

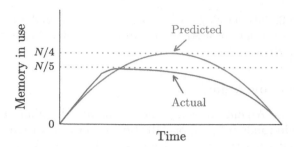

**Fig. 4.** Space used by our algorithm for computing single-pass Argon2i during a single hash computation.

- **Pre-computation Phase.** We run the entire hash computation once—on an arbitrary password and salt—and write the memory access pattern to disk. For each memory block $i$, we pre-compute the time $t_i$ after which block $i$ is never again accessed and we store $\{t_1, \ldots, t_{pn}\}$ in a read-only array. The total size of this table on a 64-bit machine is at most $8pn$ bytes.[5]

    Since the Argon2i memory-access pattern does not depend on the password or salt, it is possible to use this same pre-computed table for many subsequent Argon2i hash computations (on different salts and passwords).
- **Computation Phase.** We compute the hash function as usual, except that we delete blocks that will never be accessed again. After reading block $i$ during the hash computation at time step $t$, we check whether the current time $t \geq t_i$. If so, we delete block $i$ from memory and reuse the space for a new block.

    The expected space required to compute $n$-block single-pass Argon2i is $n/4$. The expected space required to compute $n$-block many-pass Argon2i tends to $n/e \approx 2.7$ as the number of passes tends to infinity. We analyze the space usage of the attack algorithm in detail in Appendix A.

## 5   Discussion

In this section, we discuss parallel attacks against memory-hard functions and compare Balloon to other candidate password-hashing functions.

### 5.1   Memory Hardness Under Parallel Attacks

The Balloon Hashing algorithm achieves the notion of memory-hardness introduced in Sect. 2.2: an algorithm for computing Balloon must, with high probability in the random-oracle model, use (roughly) time $T$ and space $S$ that satisfy

---

[5] On an FPGA or ASIC, this table can be stored in relatively cheap shared read-only memory and the storage cost can be amortized over a number of compute cores. Even on a general-purpose CPU, the table and memory buffer for the single-pass construction together will only require $8n + 1024(n/4) = 8n + 256n$ bytes when using our small-space computation strategy. Argon2i normally requires $1024n$ bytes of buffer space, so our strategy still yields a significant space savings.

$S \cdot T \in \Omega(n^2)$. Using the time-space product in this way as a proxy metric for computation cost is natural, since it approximates the area-time product required to compute the function in hardware [60].

As Alwen and Serbinenko [8] point out, there are two key limitations to the standard definition of memory-hardness in which we prove security. First, the definition yields a *single-instance* notion of security. That is, our definition of memory-hardness puts a lower-bound on the $ST$ cost of computing the Balloon function once, whereas in a password-guessing attack, the adversary potentially wants to compute the Balloon function *billions of times*.[6] Second, the definition treats a *sequential* model of computation—in which the adversary can make a single random-oracle query at each time step. In contrast, a password-guessing adversary may have access to thousands of computational cores operating *in parallel*.

To address the limitations of the conventional single-instance sequential adversary model, which we use for our analysis of the Balloon function, Alwen and Serbinenko introduce a new adversary model and security definition. Essentially, they allow the adversary to make many parallel random-oracle queries at each time step. In this "parallel random-oracle model" (pROM), they attempt to put a lower bound on the sum of the adversary's space usage over time: $\sum_t S_t \in \Omega(n^2)$, where $S_t$ is the number of blocks of space used in the $t$-th computation step. We call a function that satisfies this notion of memory-hardness in the pROM an *amortized memory-hard* function.[7]

To phrase the definition in different terms: Alwen and Serbinenko look for functions $f$ such that computing $f$ requires a large amount of working space *at many points* during the computation of $f$. In contrast, the traditional definition (which we use) proves the weaker statement that the adversary computing $f$ must use a lot of space *at some point* during the computation of $f$.

An impressive recent line of work has uncovered many new results in this model:

- Alwen and Blocki [3,4] show that, in the pROM, there *does not exist* a perfectly memory-hard function (in the amortized sense) that uses a password-independent memory-access pattern. In the sequential setting, Balloon and other memory-hard functions require space $S$ and time $T$ to compute such that $S \cdot T \in \Omega(n^2)$. In the parallel setting, Alwen and Blocki show that the best one can hope for, in terms of amortized space usage is $\Omega(n^2/\log n)$. Additionally, they give special-case attack algorithms for computing many candidate password-hashing algorithms in the pROM. Their algorithm

---

[6] Bellare, Ristenpart, and Tessaro consider a different type of multi-instance security [12]: they are interested in key-derivation functions $f$ with the property that finding $(x_1, \ldots, x_m)$ given $(f(x_1), \ldots, f(x_m))$ is roughly $m$ times as costly as inverting $f$ once. Stebila et al. [73] and Groza and Warinschi [40] investigate a similar multiple-instance notion of security for client puzzles [31] and Garay et al. [38] investigate related notions in the context of multi-party computation.

[7] In the original scrypt paper, Percival [60] also discusses parallel attacks and makes an argument for the security of scrypt in the pROM.

computes Balloon, for example, using an amortized time-space product of roughly $O(n^{7/4})$.[8]

- Alwen et al. [6] show that the amortized space-time complexity of the single-round Balloon function is at least $\tilde{\Omega}(n^{5/3})$, where the $\tilde{\Omega}(\cdot)$ ignores logarithmic factors of $n$. This result puts a limit on the effectiveness of parallel attacks against Balloon.
- Alwen et al. [5] construct a memory-hard function with a password-independent access pattern and that has an asymptotically optimal amortized time-space product of $S \cdot T \in \Omega(n^2/\log n)$. Whether this construction is useful for practical purposes will depend heavily on value of the constant hidden in the $\Omega(\cdot)$. In practice, a large constant may overwhelm the asymptotic improvement.
- Alwen et al. [6] prove, under combinatorial conjectures[9] that scrypt *is* near-optimally memory-hard in the pROM. Unlike Balloon, scrypt uses a data-dependent access pattern—which we would like to avoid—and the data-dependence of scrypt's access pattern seems fundamental to their security analysis.

As far as practical constructions go, these results leave the practitioner with two options, each of which has a downside:

*Option 1.* Use scrypt, which seems to protect against parallel attacks, but which uses a password-dependent access pattern and is weak in the face of an adversary that can learn memory access information. (We describe the attack in the full version of the paper)

*Option 2.* Use Balloon Hashing, which uses a password-independent access pattern and is secure against sequential attacks, but which is asymptotically weak in the face of a massively parallel attack.

A good practical solution is to hash passwords using a careful composition of Balloon and scrypt: one function defends against memory access pattern leakage and the other defends against massively parallel attacks. For the moment, let us stipulate that the pROM attacks on vanilla Balloon (and all other practical password hashing algorithms using data-independent access patterns) make these algorithms less-than-ideal to use on their own. Can we somehow combine the two constructions to get a "best-of-both-worlds" practical password-hashing algorithm? The answer is yes: compose a data-independent password-hashing algorithm, such as Balloon, with a data-dependent scheme, such as scrypt. To use the composed scheme, one would first run the password through the

---

[8] There is no consensus on whether it would be feasible to implement this parallel attack in hardware for realistic parameter sizes. That said, the fact that such pROM attacks exist at all are absolutely a practical concern.

[9] A recent addendum to the paper suggests that the combinatorial conjectures that underlie their proof of security may be false [7, Sect. 0].

data-independent algorithm and next run the resulting hash through the data-dependent algorithm.[10]

It is not difficult to show that the composed scheme is memory-hard against either: (a) an attacker who is able to learn the function's data-access pattern on the target password, or (b) an attacker who mounts an attack in the pROM using the parallel algorithm of Alwen and Blocki [3]. The composed scheme defends against the two attacks separately but does not defend against both of them simultaneously: the composed function does not maintain memory-hardness in the face of an attacker who is powerful enough to get access-pattern information *and* mount a massively parallel attack. It would be even better to have a practical construction that could protect against both attacks simultaneously, but the best known algorithms that do this [5,8] are likely too inefficient to use in practice.

The composed function is almost as fast as Balloon on its own—adding the data-dependent hashing function call is effectively as costly as increasing the round count of the Balloon algorithm by one.

## 5.2    How to Compare Memory-Hard Functions

If we restrict ourselves to considering memory-hard functions in the sequential setting, there are a number of candidate constructions that all can be proven secure in the random-oracle model: Argon2i [19],[11] Catena BRG, Catena DBG [37], and Balloon. There is no widely accepted metric with which one measures the quality of a memory-hard function, so it is difficult to compare these functions quantitatively.

In this section, we propose one such metric and compare the four candidate functions under it. The metric we propose captures the notion that a good memory-hard function is one that makes the attacker's job as difficult as possible given that the defender (e.g., the legitimate authentication server) still needs to hash passwords in a reasonable amount of time. Let $T_f(\mathcal{A})$ denote the expected running time of an algorithm $\mathcal{A}$ computing a function $f$ and let $ST_f(\mathcal{A})$ denote its expected space-time product. Then we define the quality $Q$ of a memory-hard function against a sequential attacker $\mathcal{A}_S$ using space $S$ to be the ratio:

$$Q[\mathcal{A}_S, f] = \frac{ST_f(\mathcal{A}_S)}{T_f(\text{Honest})}.$$

We can define a similar notion of quality in the amortized/parallel setting: just replace the quantity in the numerator (the adversary's space-time product) with the sum of a pROM adversary's space usage over time: $\sum_t S_t$ of $\mathcal{A}_S$.

---

[10] Our argument here gives some theoretical justification for the Argon2id mode of operation proposed in some versions of the Argon2 specification [19, Appendix B]. That variant follows a hashing with a password-independent access pattern by hashing with a password-dependent access pattern.

[11] We provide a proof of security for single-pass Argon2i in the full version of this paper.

We can now use the existing memory-hardness proofs to put lower bounds on the quality (in the sequential model) of the candidate memory-hard functions. We show in the full version of the paper that Argon2i has a sequential time-space lower bound of the form $S \cdot T \geq n^2/192$, for $S < n/24$. The $n$-block $r$-round Balloon function has a time-space lower-bound of the form $S \cdot T \geq (2^r - 1)n^2/8$ for $S < n/64$ when the parameter $\delta = 3$. The $n$-block Catena BRG function has a time-space lower bound of the form $S \cdot T \geq n^2/16$ (Catena BRG has no round parameter). The $r$-round $n$-block Catena DBG function has a claimed time-space lower bound of the form $S \cdot T \geq n(\frac{rn}{64S})^r$, when $S \leq n/20$. These lower-bounds yield the following quality figures against an adversary using roughly $n/64$ space:

$$Q[\mathcal{A}_S, \text{Balloon}_{(r=1)}] \geq \frac{n}{16}; \qquad Q[\mathcal{A}_S, \text{Balloon}_{(r>1)}] \geq \frac{(2^r - 1)n}{8(r + 1)}$$

$$Q[\mathcal{A}_S, \text{Catena-BRG}] \geq \frac{n}{32}; \qquad Q[\mathcal{A}_S, \text{Catena-DBG}] \geq \frac{r^r}{2r \log_2 n}$$

$$Q[\mathcal{A}_S, \text{Argon2i}] \geq \frac{n}{192}$$

From these quality ratios, we can draw a few conclusions about the protection these functions provide against one class of small-space attackers (using $S \approx n/64$):

– In terms of provable memory-hardness properties in the sequential model, one-round Balloon always outperforms Catena-BRG and Argon2i.
– When the buffer size $n$ grows and the number of rounds $r$ is held fixed, Balloon outperforms Catena-DBG as well.
– When the buffer size $n$ is fixed and the number of rounds $r$ grows large, Catena-DBG provides the strongest provable memory-hardness properties in the sequential model.
– For many realistic choices of $r$ and $n$ (e.g., $r = 5$, $n = 2^{18}$), $r$-round Balloon outperforms the other constructions in terms of memory-hardness properties.

# 6    Experimental Evaluation

In this section, we demonstrate experimentally that the Balloon hashing algorithm is competitive performance-wise with two existing practical algorithms (Argon2i and Catena), when all are instantiated with standard cryptographic primitives.

## 6.1    Experimental Set-Up

Our experiments use the OpenSSL implementation (version 1.0.1f) of SHA-512 and the reference implementations of three other cryptographic hash functions (Blake2b, ECHO, and SHA-3/Keccak). We use optimized versions of the underlying cryptographic primitives where available, but the core Balloon code is written entirely in C. Our source code is available at https://crypto.stanford.edu/balloon/ under the ISC open-source license. We used a workstation running an Intel Core i7-6700 CPU (Skylake) at 3.40 GHz with 8 GiB of RAM for our performance benchmarks. We compiled the code for our timing results with gcc version 4.8.5 using the -O3 option. We average all of our measurements over 32 trials. We compare the Balloon functions against Argon2i (v.1.2.1) [18] and Catena [37]. For comparison purposes, we implemented the Argon2i, Catena BRG, and Catena DBG memory-hard algorithms in C.

*On the Choice of Cryptographic Primitives.* The four memory-hard functions we evaluate (Argon2i, Balloon, Catena-BRG, Catena-DBG) are all essentially modes of operation for an underlying cryptographic hash function. The choice of the underlying hash function has implications for the performance and the security of the overall construction. To be conservative, we instantiate all of the algorithms we evaluate with the Blake2b as the underlying hash function [10].

Memory-hard functions going back at least as far as scrypt [60] have used reduced-round hash functions as their underlying cryptographic building block. Following this tradition, the Argon2i specification proposes using a new and very fast reduced-round hash function as its core cryptographic primitive. Since the Argon2i hash function does not satisfy basic properties of a traditional cryptographic hash function (e.g., it is not collision resistant), modeling it as a random oracle feels particularly problematic. Since our goal in this work is to analyze memory-hard functions with *provable* security guarantees, we instantiate the memory-hard functions we evaluate with traditional cryptographic hashes for the purposes of this evaluation.

That said, we stress that the Balloon construction is agnostic to the choice of underlying hash function—it is a mode of operation for a cryptographic hash function—and users of the Balloon construction may instantiate it with a faster reduced-round hash function (e.g., scrypt's BlockMix or Argon2i's compression function) if they so desire.

## 6.2  Authentication Throughput

The goal of a memory-hard password hash function is to *use as much working space as possible as quickly as possible* over the course of its computation. To evaluate the effectiveness of the Balloon algorithm on this metric, we measured the rate at which a server can check passwords (in hashes per second) for various buffer sizes on a single core.

Figure 5 shows the minimum buffer size required to compute each memory-hard function with high probability with no computational slowdown, for a variety of password hashing functions. We set the block size of the construction to be equal to the block size of the underlying compression function, to avoid the issues discussed in the full version of this paper. The charted results for Argon2i incorporate the fact that an adversary can compute many-pass Argon2i (v.1.2.1) in a factor of $e \approx 2.72$ less working space than the defender must allocate for the computation and can compute single-pass Argon2i with a factor of four less space (see Sect. 4). For comparison, we also plot the space usage of two

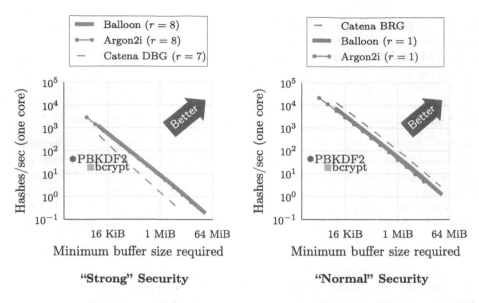

**Fig. 5.** The Balloon algorithm outperforms Argon2i and Catena DBG for many settings of the security parameters, and Balloon is competitive with Catena BRG. We instantiate Argon2i, Balloon, and Catena with Blake2b as the underlying cryptographic hash function.

non-memory-hard password hashing functions, bcrypt [66] (with cost $= 12$) and PBKDF2-SHA512 [43] (with $10^5$ iterations).

If we assume that an authentication server must perform 100 hashes per second per four-core machine, Fig. 5 shows that it would be possible to use one-round Balloon hashing with a 2 MiB buffer or eight-round Balloon hashing with a 256 KiB buffer. At the same authentication rate, Argon2i (instantiated with Blake2b as the underlying cryptographic hash function) requires the attacker to use a smaller buffer—roughly 1.5 MiB for the one-pass variant. Thus, with Balloon hashing we simultaneously get better performance than Argon2i and stronger memory-hardness guarantees.

## 6.3   Compression Function

Finally, Fig. 6 shows the result of instantiating the Balloon algorithm construction with four different standard cryptographic hash functions: SHA-3 [17], Blake2b [10], SHA-512, and ECHO (a SHA-3 candidate that exploits the AES-NI instructions) [14]. The SHA-3 function (with rate $= 1344$) operates on 1344-bit blocks, and we configure the other hash functions to use 512-bit blocks.

On the $x$-axis, we plot the buffer size used in the Balloon function and on the $y$-axis, we plot the rate at which the Balloon function fills memory, in bytes of written per second. As Fig. 6 demonstrates, Blake2b and ECHO outperform the SHA functions by a bit less than a factor of two.

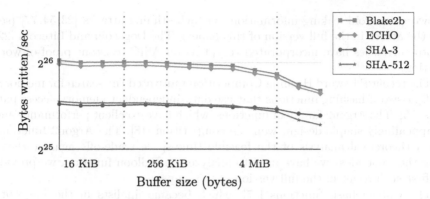

**Fig. 6.** Throughput for the Balloon algorithm when instantiated with different compression functions. The dotted lines indicate the sizes of the L1, L2, and L3 caches on our test machine.

## 7    Related Work

*Password Hashing.* The problem of how to securely store passwords on shared computer systems is nearly as old as the systems themselves. In a 1974 article, Evans et al. described the principle of storing passwords under a hard-to-invert function [35]. A few years later, Robert Morris and Ken Thompson presented the now-standard notion of password *salts* and explained how to store passwords under a moderately hard-to-compute one-way function to increase the cost of dictionary attacks [52]. Their DES-based "crypt" design became the standard for password storage for over a decade [49] and even has a formal analysis by Wagner and Goldberg [80].

In 1989, Feldmeier and Karn found that hardware improvements had driven the cost of brute-force password guessing attacks against DES crypt down by five orders of magnitude since 1979 [36,46]. Poul-Henning Kamp introduced the costlier md5crypt to replace crypt, but hardware improvements also rendered that design outmoded [27].

Provos and Mazières saw that, in the face of ever-increasing processor speeds, any fixed password hashing algorithm would eventually become easy to compute and thus ineffective protection against dictionary attacks. Their solution, bcrypt, is a password hashing scheme with a variable "hardness" parameter [66]. By periodically ratcheting up the hardness, a system administrator can keep the time needed to compute a single hash roughly constant, even as hardware improves. A remaining weakness of bcrypt is that it exercises only a small fraction of the CPU's resources—it barely touches the L2 and L3 caches during its execution [50]. To increase the cost of custom password-cracking hardware, Reinhold's HEKS hash [67] and Percival's popular scrypt routine consume an adjustable amount of storage space [60], in addition to time, as they compute a hash. Balloon, like scrypt, aims to be hard to compute in little space. Unlike scrypt, however, we require that our functions' data access pattern be independent of the password to avoid leaking information via cache-timing attacks [23,54,77] (see also the attack in the full version of this paper). The Dogecoin and Litecoin [22] crypto-currencies have incorporated scrypt as an ASIC-resistant proof-of-work function.

The recent Password Hashing Competition motivated the search for memory-hard password-hashing functions that use data-independent memory access patterns [56]. The Argon2 family of functions, which have excellent performance and an appealingly simple design, won the competition [18]. The Argon2 functions lack a theoretical analysis of the feasible time-space trade-offs against them; using the same ideas we have used to analyze the Balloon function, we provide the first such result in the full version of this paper.

The Catena hash functions [37], which became finalists in the Password Hashing Competition, are memory-hard functions whose analysis applies pebbling arguments to classic graph-theoretic results of Lengauer and Tarjan [48]. The Balloon analysis we provide gives a tighter time-space lower bounds than Catena's analysis can provide in many cases, and the Balloon algorithm outperforms the more robust of the two Catena algorithms (see Sect. 6). Biryokov and Khovratovich demonstrated a serious flaw in the security analysis of one of the Catena variants, and they provide a corresponding attack against that Catena variant [20].

The other competition finalists included a number of interesting designs that differ from ours in important ways. Makwa [64] supports offloading the work of password hashing to an untrusted server but is not memory-hard. Lyra [2] is a memory-hard function but lacks proven space-time lower bounds. Yescrypt [61] is an extension of scrypt and uses a password-dependent data access pattern.

Ren and Devadas [68] give an analysis of the Balloon algorithm using bipartite expanders, following the pebbling techniques of Paul and Tarjan [58]. Their results imply that an adversary that computes the $n$-block $r$-round Balloon function in $n/8$ space, must use at least $2^r n/c$ time to compute the function (for some constant $c$), with high probability in the random-oracle model. We prove the stronger statement that an adversary's space-time product must satisfy: $S \cdot T \in \Omega(n^2)$ for almost all values of $S$. Ren and Devadas also prove statements showing that algorithms computing the Balloon functions efficiently must use a certain amount of space at *many points* during their computation. Our time-space lower bounds only show that the adversary must use a certain amount of space a *some point* during the Balloon computation.

*Other Studies of Password Protection.* Concurrently with the design of hashing schemes, there has been theoretical work from Bellare et al. on new security definitions for password-based cryptography [12] and from Di Crescenzo et al. on an analysis of passwords storage systems secure against adversaries that can steal only a bounded number of bits of the password file [28]. Other ideas for modifying password hashes include the *key stretching* schemes of Kelsey et al. [44] (variants on iterated hashes), a proposal by Boyen to keep the hash iteration count (e.g., time parameter in bcrypt) secret [24], a technique of Canetti et al.

for using CAPTCHAs in concert with hashes [25], and a proposal by Dürmuth to use password hashing to do meaningful computation [29].

*Parallel Memory-Hardness.* In a recent line of work [3–6,8] has analyzed memory-hard functions from a number of angles in the parallel random-oracle model, introduced by Alwen and Serbinenko [8]. We discuss these very relevant results at length in Sect. 5.1.

*Memory-Bound Functions.* Abadi et al. [1] introduced memory-bound functions as more effective alternatives to traditional proofs-of-work in heterogeneous computing environments [11,31]. These functions require many cache misses to compute and, under the assumption that memory latencies are consistent across computing platforms, they are roughly as hard to compute on a computationally powerful device as on a computationally weak one. The theoretical analysis of memory-bound functions represented one of the first applications of pebbling arguments to cryptography [30,32].

*Proofs of Space.* Dziembowski et al. [33] and Ateniese et al. [9] study proofs-of-space. In these protocols, the prover and verifier agree on a large bitstring that the prover is supposed to store. Later on, the prover can convince the verifier that the prover has stored some large string on disk, even if the verifier does not store the string herself. Spacemint proposes building a cryptocurrency based upon a proof-of-space rather than a proof-of-work [55]. Ren and Devadas propose using the problem of pebbling a Balloon graph as the basis for a proof of space [68].

*Time-Space Trade-Offs.* The techniques we use to analyze Balloon draws on extensive prior work on computational time-space trade-offs. We use pebbling arguments, which have seen application to register allocation problems [72], to the analysis of the relationships between complexity classes [15,26,42,75], and to prior cryptographic constructions [32–34,37]. Pebbling has also been a topic of study in its own right [48,59]. Savage's text gives a clear introduction to graph pebbling [71] and Nordström surveys the vast body of pebbling results in depth [53].

# 8 Conclusion

We have introduced the Balloon password hashing algorithm. The Balloon algorithm is provably memory-hard (in the random-oracle model against sequential adversaries), exhibits a password-independent memory access pattern, and meets or exceeds the performance of the fastest heuristically secure schemes. Using a novel combinatorial pebbling argument, we have demonstrated that password-hashing algorithms can have memory-hardness proofs without sacrificing practicality.

This work raises a number of open questions:

- Are there efficient methods to defend against cache attacks on scrypt? Could a special-purpose ORAM scheme help [39]?

- Are there *practical* memory-hard functions with password-independent access patterns that retain their memory-hardness properties under parallel attacks [8]? The recent work of Alwen et al. [4] is promising, though it is still unclear whether the pROM-secure constructions will be competitive with Balloon for concrete settings of the parameters.
- Is it possible to build hardware that effectively implements the pROM attacks [3–5] against Argon2i and Balloon at realistic parameter sizes? What efficiency gain would this pROM hardware have over a sequential ASIC at attacking these constructions? Are these parallel attacks still practical in hardware when the function's memory-access pattern depends on the salt (as Balloon's access pattern does)?

**Acknowledgements.** We would like to our anonymous reviewers for their helpful comments. We also thank Josh Benaloh, Joe Bonneau, Greg Hill, Ali Mashtizadeh, David Mazières, Yan Michalevsky, Bryan Parno, Greg Valiant, Riad Wahby, Keith Winstein, David Wu, Sergey Yekhanin, and Greg Zaverucha for comments on early versions of this work. This work was funded in part by an NDSEG Fellowship, NSF, DARPA, a grant from ONR, and the Simons Foundation. Opinions, findings and conclusions or recommendations expressed in this material are those of the authors and do not necessarily reflect the views of DARPA.

# A    Details of the Attack on Argon2

In this section, we provide a detailed analysis of the attack on Argon2i that we introduced in Sect. 4.

The goal of the attack algorithm is to compute Argon2i in the same number of time steps as the naïve algorithm uses to compute the function, while using a constant factor less space than the naïve algorithm does. In this way, an attacker mounting a dictionary attack against a list of passwords hashed with Argon2i can do so at less cost (in terms of the space-time product) than the Argon2i specification claimed possible.

Argon2i has one-pass and many-pass variants and our attack applies to both; the many-pass variant is recommended in the specification. We first analyze the attack on the one-pass variant and then analyze the attack on the many-pass variant.

We are interested in the attack algorithm's expected space usage at time step $t$—call this function $S(t)$.[12]

**Analysis of One-Pass Argon2i.** At each step of the one-pass Argon2i algorithm, the expected space usage $S(t)$ is equal to the number of memory blocks generated so far minus the expected number of blocks in memory that will never

---

[12] As described in Sect. 4.2, the contents of block $i$ in Argon2i are derived from the contents of block $i-1$ and a block chosen at random from the set $r_i \xleftarrow{\text{R}} \{1, \ldots, i-1\}$. Throughout our analysis, all probabilities are taken over the random choices of the $r_i$ values.

be used after time $t$. Let $A_{i,t}$ be the event that block $i$ is never needed after time step $t$ in the computation. Then $S(t) = t - \sum_{i=1}^{t} \Pr[A_{i,t}]$.

To find $S(t)$ explicitly, we need to compute the probability that block $i$ is never used after time $t$. We know that the probability that block $i$ is never used after time $t$ is equal to the probability that block $i$ is not used at time $t+1$ *and* is not used at time $t+2$ *and* [...] *and* is not used at time $n$. Let $U_{i,t}$ denote the event that block $i$ is *unused* at time $t$. Then:

$$\Pr[A_{i,t}] = \Pr\left[\bigcap_{t'=t+1}^{n} U_{i,t'}\right] = \prod_{t'=t+1}^{n} \Pr[U_{i,t'}] \tag{1}$$

The equality on the right-hand side comes from the fact that $U_{i,t'}$ and $U_{i,t''}$ are independent events for $t' \neq t''$.

To compute the probability that block $i$ is not used at time $t'$, consider that there are $t' - 1$ blocks to choose from and $t' - 2$ of them are *not* block $i$: $\Pr[U_{i,t'}] = \frac{t'-2}{t'-1}$. Plugging this back into Eq. 1, we get:

$$\Pr[A_{i,t}] = \prod_{t'=t+1}^{n} \left(\frac{t'-2}{t'-1}\right) = \frac{t-1}{n-1}$$

Now we substitute this back into our original expression for $S(t)$:

$$S(t) = t - \sum_{i=1}^{t} \left(\frac{t-1}{n-1}\right) = t - \frac{t(t-1)}{n-1}$$

Taking the derivative $S'(t)$ and setting it to zero allows us to compute the value $t$ for which the expected storage is maximized. The maximum is at $t = n/2$ and the expected number of blocks required is $S(n/2) \approx n/4$.

**Larger in-degree.** A straightforward extension of this analysis handles the case in which $\delta$ random blocks—instead of one—are hashed together with the prior block at each step of the algorithm. Our analysis demonstrates that, even with this strategy, single-pass Argon2i is vulnerable to pre-computation attacks. The maximum space usage comes at $t^* = n/(\delta+1)^{1/\delta}$, and the expected space usage over time $S(t)$ is:

$$S(t) \approx t - \frac{t^{\delta+1}}{n^\delta} \quad \text{so} \quad S(t^*) \approx \frac{\delta}{(\delta+1)^{1+1/\delta}}n.$$

**Analysis of Many-Pass Argon2i.** One idea for increasing the minimum memory consumption of Argon2i is to increase the number of passes that the algorithm takes over the memory. For example, the Argon2 specification proposes taking three passes over the memory to protect against certain time-space trade-offs. Unfortunately, even after *many* passes over the memory, the Argon2i algorithm sketched above still uses many fewer than $n$ blocks of memory, in expectation, at each time step.

To investigate the space usage of the many-pass Argon2i algorithm, first consider that the space usage will be maximized at some point in the middle of its computation—not in the first or last passes. At some time step $t$ in the middle of its computation the algorithm will have at most $n$ memory blocks in storage, but the algorithm can delete any of these $n$ blocks that it will never need after time $t$.

At each time step, the algorithm adds a new block to the end of the buffer and deletes the first block. At any one point in the algorithm's execution, there will be at most $n$ blocks of memory in storage. If we freeze the execution of the Argon2i algorithm in the middle of its execution, we can inspect the $n$ blocks it has stored in memory. Call the first block "stored block 1" and the last block "stored block $n$."

Let $B_{i,t}$ denote the event that stored block $i$ is never needed after time $t$. Then we claim $\Pr[B_{i,t}] = (\frac{n-1}{n})^i$. To see the logic behind this calculation: notice that, at time $t$, the first stored block in the buffer can be accessed at time $t+1$ but by time $t+2$, the first stored block will have been deleted from the buffer. Similarly, the second stored block in the buffer at time $t$ can be accessed at time $t+1$ or $t+2$, but not $t+3$ (since by then stored block 2 will have been deleted from the buffer). Similarly, stored block $i$ can be accessed at time steps $(t+1)$, $(t+2)$, ..., $(t+i)$ but not at time step $(t+i+1)$.

The total storage required is then:

$$S(t) = n - \sum_{i=1}^{n} \mathrm{E}[B_{i,t}] = n - \sum_{i=1}^{n} \left(\frac{n-1}{n}\right)^i \approx n - n\left(1 - \frac{1}{e}\right).$$

Thus, even after many passes over the memory, Argon2i can still be computed in roughly $n/e$ space with no time penalty.

# References

1. Abadi, M., Burrows, M., Manasse, M., Wobber, T.: Moderately hard, memory-bound functions. ACM Trans. Internet Technol. **5**(2), 299–327 (2005)
2. Almeida, L.C., Andrade, E.R., Barreto, P.S.L.M., Simplicio Jr., M.A.: Lyra: password-based key derivation with tunable memory and processing costs. J. Cryptographic Eng. **4**(2), 75–89 (2014)
3. Alwen, J., Blocki, J.: Efficiently computing data-independent memory-hard functions. In: Robshaw, M., Katz, J. (eds.) CRYPTO 2016. LNCS, vol. 9815, pp. 241–271. Springer, Heidelberg (2016). doi:10.1007/978-3-662-53008-5_9
4. Alwen, J., Blocki, J.: Towards practical attacks on Argon2i and Balloon Hashing. Cryptology ePrint Archive, Report 2016/759 (2016). http://eprint.iacr.org/2016/759
5. Alwen, J., Blocki, J., Pietrzak, K.: The pebbling complexity of depth-robust graphs. Manuscript (Personal Communication) (2016)
6. Alwen, J., Chen, B., Kamath, C., Kolmogorov, V., Pietrzak, K., Tessaro, S.: On the complexity of scrypt and proofs of space in the parallel random oracle model. In: Fischlin, M., Coron, J.-S. (eds.) EUROCRYPT 2016. LNCS, vol. 9666, pp. 358–387. Springer, Heidelberg (2016). doi:10.1007/978-3-662-49896-5_13

7. Alwen, J., Chen, B., Kamath, C., Kolmogorov, V., Pietrzak, K., Tessaro, S.: On the complexity of scrypt and proofs of space in the parallel random oracle model. Cryptology ePrint Archive, Report 2016/100 (2016). http://eprint.iacr.org/
8. Alwen, J., Serbinenko, V.: High parallel complexity graphs and memory-hard functions. In: STOC, pp. 595–603 (2015)
9. Ateniese, G., Bonacina, I., Faonio, A., Galesi, N.: Proofs of space: when space is of the essence. In: Abdalla, M., Prisco, R. (eds.) SCN 2014. LNCS, vol. 8642, pp. 538–557. Springer, Heidelberg (2014). doi:10.1007/978-3-319-10879-7_31
10. Aumasson, J.-P., Neves, S., Wilcox-O'Hearn, Z., Winnerlein, C.: BLAKE2: simpler, smaller, fast as MD5. In: Jacobson, M., Locasto, M., Mohassel, P., Safavi-Naini, R. (eds.) ACNS 2013. LNCS, vol. 7954, pp. 119–135. Springer, Heidelberg (2013). doi:10.1007/978-3-642-38980-1_8
11. Back, A.: Hashcash-a denial of service counter-measure, May 1997. http://www.cypherspace.org/hashcash/. Accessed 9 Nov 2015
12. Bellare, M., Ristenpart, T., Tessaro, S.: Multi-instance security and its application to password-based cryptography. In: Safavi-Naini, R., Canetti, R. (eds.) CRYPTO 2012. LNCS, vol. 7417, pp. 312–329. Springer, Heidelberg (2012). doi:10.1007/978-3-642-32009-5_19
13. Bellare, M., Rogaway, P.: Random oracles are practical: a paradigm for designing efficient protocols. In: CCS, pp. 62–73. ACM (1993)
14. Benadjila, R., Billet, O., Gilbert, H., Macario-Rat, G., Peyrin, T., Robshaw, M., Seurin, Y.: SHA-3 proposal: ECHO. Submission to NIST (updated) (2009)
15. Bennett, C.H.: Time/space trade-offs for reversible computation. SIAM J. Comput. 18(4), 766–776 (1989)
16. Bernstein, D.J., Lange, T.: Non-uniform cracks in the concrete: the power of free precomputation. In: Sako, K., Sarkar, P. (eds.) ASIACRYPT 2013. LNCS, vol. 8270, pp. 321–340. Springer, Heidelberg (2013). doi:10.1007/978-3-642-42045-0_17
17. Bertoni, G., Daemen, J., Peeters, M., Van Assche, G.: Keccak sponge function family. Submission to NIST (Round 2) (2009)
18. Biryukov, A., Dinu, D., Khovratovich, D.: Argon2 design document (version 1.2.1), October 2015
19. Biryukov, A., Dinu, D., Khovratovich, D.: Argon2 design document (version 1.3), February 2016
20. Biryukov, A., Khovratovich, D.: Tradeoff cryptanalysis of memory-hard functions. In: Iwata, T., Cheon, J.H. (eds.) ASIACRYPT 2015. LNCS, vol. 9453, pp. 633–657. Springer, Heidelberg (2015). doi:10.1007/978-3-662-48800-3_26
21. Bitcoin wiki - mining comparison. https://en.bitcoin.it/wiki/Mining_hardware_comparison
22. Bonneau, J., Miller, A., Clark, J., Narayanan, A., Kroll, J.A., Felten, E.W.: SoK: research perspectives and challenges for Bitcoin and cryptocurrencies. In: Symposium on Security and Privacy. IEEE, May 2015
23. Bonneau, J., Mironov, I.: Cache-collision timing attacks against AES. In: Goubin, L., Matsui, M. (eds.) CHES 2006. LNCS, vol. 4249, pp. 201–215. Springer, Heidelberg (2006). doi:10.1007/11894063_16
24. Boyen, X.: Halting password puzzles. In: USENIX Security (2007)
25. Canetti, R., Halevi, S., Steiner, M.: Mitigating dictionary attacks on password-protected local storage. In: Dwork, C. (ed.) CRYPTO 2006. LNCS, vol. 4117, pp. 160–179. Springer, Heidelberg (2006). doi:10.1007/11818175_10
26. Chan, S.M.: Just a pebble game. In: IEEE Conference on Computational Complexity, pp. 133–143. IEEE (2013)

27. CVE-2012-3287: md5crypt has insufficient algorithmic complexity (2012). http://cve.mitre.org/cgi-bin/cvename.cgi?name=CVE-2012-3287. Accessed 9 Nov 2015
28. Di Crescenzo, G., Lipton, R., Walfish, S.: Perfectly secure password protocols in the bounded retrieval model. In: Halevi, S., Rabin, T. (eds.) TCC 2006. LNCS, vol. 3876, pp. 225–244. Springer, Heidelberg (2006). doi:10.1007/11681878_12
29. Dürmuth, M.: Useful password hashing: how to waste computing cycles with style. In: New Security Paradigms Workshop, pp. 31–40. ACM (2013)
30. Dwork, C., Goldberg, A., Naor, M.: On memory-bound functions for fighting spam. In: Boneh, D. (ed.) CRYPTO 2003. LNCS, vol. 2729, pp. 426–444. Springer, Heidelberg (2003). doi:10.1007/978-3-540-45146-4_25
31. Dwork, C., Naor, M.: Pricing via processing or combatting junk mail. In: Brickell, E.F. (ed.) CRYPTO 1992. LNCS, vol. 740, pp. 139–147. Springer, Heidelberg (1993). doi:10.1007/3-540-48071-4_10
32. Dwork, C., Naor, M., Wee, H.: Pebbling and proofs of work. In: Shoup, V. (ed.) CRYPTO 2005. LNCS, vol. 3621, pp. 37–54. Springer, Heidelberg (2005). doi:10.1007/11535218_3
33. Dziembowski, S., Faust, S., Kolmogorov, V., Pietrzak, K.: Proofs of space. In: Gennaro, R., Robshaw, M. (eds.) CRYPTO 2015. LNCS, vol. 9216, pp. 585–605. Springer, Heidelberg (2015). doi:10.1007/978-3-662-48000-7_29
34. Dziembowski, S., Kazana, T., Wichs, D.: One-time computable self-erasing functions. In: Ishai, Y. (ed.) TCC 2011. LNCS, vol. 6597, pp. 125–143. Springer, Heidelberg (2011). doi:10.1007/978-3-642-19571-6_9
35. Evans Jr., A., Kantrowitz, W., Weiss, E.: A user authentication scheme not requiring secrecy in the computer. Commun. ACM **17**(8), 437–442 (1974)
36. Feldmeier, D.C., Karn, P.R.: UNIX password security - ten years later. In: Brassard, G. (ed.) CRYPTO 1989. LNCS, vol. 435, pp. 44–63. Springer, Heidelberg (1990). doi:10.1007/0-387-34805-0_6
37. Forler, C., Lucks, S., Wenzel, J.: Memory-demanding password scrambling. In: Sarkar, P., Iwata, T. (eds.) ASIACRYPT 2014. LNCS, vol. 8874, pp. 289–305. Springer, Heidelberg (2014). doi:10.1007/978-3-662-45608-8_16
38. Garay, J., Johnson, D., Kiayias, A., Yung, M.: Resource-based corruptions and the combinatorics of hidden diversity. In: ITCS, pp. 415–428. ACM (2013)
39. Goldreich, O., Ostrovsky, R.: Software protection and simulation on oblivious RAMs. J. ACM **43**(3), 431–473 (1996)
40. Groza, B., Warinschi, B.: Revisiting difficulty notions for client puzzles and DoS resilience. In: Gollmann, D., Freiling, F.C. (eds.) ISC 2012. LNCS, vol. 7483, pp. 39–54. Springer, Heidelberg (2012). doi:10.1007/978-3-642-33383-5_3
41. Ho, S.: Costco, Sam's Club, others halt photo sites over possible breach, July 2015. http://www.reuters.com/article/2015/07/21/us-cyberattack-retail-idUSKCN0PV00520150721. Accessed 9 Nov 2015
42. Hopcroft, J., Paul, W., Valiant, L.: On time versus space. J. ACM (JACM) **24**(2), 332–337 (1977)
43. Kaliski, B.: PKCS #5: Password-based cryptography specification, version 2.0. IETF Network Working Group, RFC 2898, September 2000
44. Kelsey, J., Schneier, B., Hall, C., Wagner, D.: Secure applications of low-entropy keys. In: Okamoto, E., Davida, G., Mambo, M. (eds.) ISW 1997. LNCS, vol. 1396, pp. 121–134. Springer, Heidelberg (1998). doi:10.1007/BFb0030415
45. Kirk, J.: Internet address overseer ICANN resets passwords after website breach, August 2015. http://www.pcworld.com/article/2960592/security/icann-resets-passwords-after-website-breach.html. Accessed 9 Nov 2015

46. Klein, D.V.: Foiling the cracker: a survey of, and improvements to, password security. In: Proceedings of the 2nd USENIX Security Workshop, pp. 5–14 (1990)
47. Krantz, L.: Harvard says data breach occurred in June, July 2015. http://www.bostonglobe.com/metro/2015/07/01/harvard-announces-data-breach/pqzk9IPWLMiCKBl3IijMUJ/story.html. Accessed 9 Nov 2015
48. Lengauer, T., Tarjan, R.E.: Asymptotically tight bounds on time-space trade-offs in a pebble game. J. ACM **29**(4), 1087–1130 (1982)
49. Leong, P., Tham, C.: UNIX password encryption considered insecure. In: USENIX Winter, pp. 269–280 (1991)
50. Malvoni, K., Designer, S., Knezovic, J.: Are your passwords safe: energy-efficient bcrypt cracking with low-cost parallel hardware. In: USENIX Workshop on Offensive Technologies (2014)
51. Menezes, A.J., Van Oorschot, P.C., Vanstone, S.A.: Handbook of Applied Cryptography. CRC Press, Boca Raton (1996)
52. Morris, R., Thompson, K.: Password security: a case history. Commun. ACM **22**(11), 594–597 (1979)
53. Nordström, J.: New wine into old wineskins: a survey of some pebbling classics with supplemental results, March 2015. http://www.csc.kth.se/~jakobn/research/PebblingSurveyTMP.pdf. Accessed 9 Nov 2015
54. Osvik, D.A., Shamir, A., Tromer, E.: Cache attacks and countermeasures: the case of AES. In: Pointcheval, D. (ed.) CT-RSA 2006. LNCS, vol. 3860, pp. 1–20. Springer, Heidelberg (2006). doi:10.1007/11605805_1
55. Park, S., Pietrzak, K., Alwen, J., Fuchsbauer, G., Gazi, P.: Spacemint: a cryptocurrency based on proofs of space. Technical report, Cryptology ePrint Archive, Report 2015/528 (2015)
56. Password hashing competition. https://password-hashing.net/
57. Paterson, M.S., Hewitt, C.E.: Comparative schematology. In: Record of the Project MAC Conference on Concurrent Systems and Parallel Computation, pp. 119–127. ACM (1970)
58. Paul, W.J., Tarjan, R.E.: Time-space trade-offs in a pebble game. Acta Informatica **10**(2), 111–115 (1978)
59. Paul, W.J., Tarjan, R.E., Celoni, J.R.: Space bounds for a game on graphs. Math. Syst. Theor. **10**(1), 239–251 (1976)
60. Percival, C.: Stronger key derivation via sequential memory-hard functions. In: BSDCan, May 2009
61. Peslyak, A.: yescrypt, October 2015. https://password-hashing.net/submissions/specs/yescrypt-v2.pdf. Accessed 13 Nov 2015
62. Peterson, A.: E-Trade notifies 31,000 customers that their contact info may have been breached in 2013 hack, October 2015. https://www.washingtonpost.com/news/the-switch/wp/2015/10/09/e-trade-notifies-31000-customers-that-their-contact-info-may-have-been-breached-in-2013-hack/. Accessed 9 Nov 2015
63. Pippenger, N.: A time-space trade-off. J. ACM (JACM) **25**(3), 509–515 (1978)
64. Pornin, T.: The Makwa password hashing function, April 2015. http://www.bolet.org/makwa/. Accessed 13 Nov 2015
65. Privacy Rights Clearinghouse: Chronology of data breaches. http://www.privacyrights.org/data-breach. Accessed 9 Nov 2015
66. Provos, N., Mazières, D.: A future-adaptable password scheme. In: USENIX Annual Technical Conference, pp. 81–91 (1999)
67. Reinhold, A.: HEKS: a family of key stretching algorithms (Draft G), July 2001. http://world.std.com/~reinhold/HEKSproposal.html. Accessed 13 Nov 2015

68. Ren, L., Devadas, S.: Proof of space from stacked expanders. Cryptology ePrint Archive, Report 2016/333 (2016). http://eprint.iacr.org/
69. Ristenpart, T., Tromer, E., Shacham, H., Savage, S.: Hey, you, get off of my cloud: exploring information leakage in third-party compute clouds. In: CCS, pp. 199–212. ACM (2009)
70. Rogaway, P.: Formalizing human ignorance. In: Nguyen, P.Q. (ed.) VIETCRYPT 2006. LNCS, vol. 4341, pp. 211–228. Springer, Heidelberg (2006). doi:10.1007/11958239_14
71. Savage, J.E.: Models of Computation: Exploring the Power of Computing. Addison-Wesley, New York (1998)
72. Sethi, R.: Complete register allocation problems. SIAM J. Comput. 4(3), 226–248 (1975)
73. Stebila, D., Kuppusamy, L., Rangasamy, J., Boyd, C., Gonzalez Nieto, J.: Stronger difficulty notions for client puzzles and denial-of-service-resistant protocols. In: Kiayias, A. (ed.) CT-RSA 2011. LNCS, vol. 6558, pp. 284–301. Springer, Heidelberg (2011). doi:10.1007/978-3-642-19074-2_19
74. Takala, R.: UVA site back online after chinese hack, August 2015. http://www.washingtonexaminer.com/uva-site-back-online-after-chinese-hack/article/2570383. Accessed 9 Nov 2015
75. Tompa, M.: Time-space tradeoffs for computing functions, using connectivity properties of their circuits. In: STOC, pp. 196–204. ACM (1978)
76. Tracy, A.: In wake of T-Mobile and Experian data breach, John Legere did what all CEOs should do after a hack, October 2015. http://www.forbes.com/sites/abigailtracy/2015/10/02/in-wake-of-t-mobile-and-experian-data-breach-john-legere-did-what-all-ceos-should-do-after-a-hack/. Accessed 9 Nov 2015
77. Tromer, E., Osvik, D.A., Shamir, A.: Efficient cache attacks on AES, and countermeasures. J. Cryptology 23(1), 37–71 (2010)
78. Valiant, L.G.: Graph-theoretic arguments in low-level complexity. In: Gruska, J. (ed.) MFCS 1977. LNCS, vol. 53, pp. 162–176. Springer, Heidelberg (1977). doi:10.1007/3-540-08353-7_135
79. Vaughan-Nichols, S.J.: Password site LastPass warns of data breach, June 2015. http://www.zdnet.com/article/lastpass-password-security-site-hacked/. Accessed 9 Nov 2015
80. Wagner, D., Goldberg, I.: Proofs of security for the unix password hashing algorithm. In: Okamoto, T. (ed.) ASIACRYPT 2000. LNCS, vol. 1976, pp. 560–572. Springer, Heidelberg (2000). doi:10.1007/3-540-44448-3_43

# Linear Structures: Applications to Cryptanalysis of Round-Reduced Keccak

Jian Guo[1,2], Meicheng Liu[1,2,3](✉), and Ling Song[1,2,3]

[1] Cryptanalysis Taskforce, Temasek Laboratories@NTU, Singapore, Singapore
ntu.guo@gmail.com, meicheng.liu@gmail.com, songling@iie.ac.cn
[2] School of Physical and Mathematical Sciences,
Nanyang Technological University, Singapore, Singapore
[3] State Key Laboratory of Information Security, Institute of Information Engineering,
Chinese Academy of Sciences, Beijing 100093, People's Republic of China

**Abstract.** In this paper, we analyze the security of round-reduced versions of the Keccak hash function family. Based on the work pioneered by Aumasson and Meier, and Dinur *et al.*, we formalize and develop a technique named *linear structure*, which allows linearization of the underlying permutation of Keccak for up to 3 rounds with large number of variable spaces. As a direct application, it extends the best zero-sum distinguishers by 2 rounds without increasing the complexities. We also apply linear structures to preimage attacks against Keccak. By carefully studying the properties of the underlying Sbox, we show bilinear structures and find ways to convert the information on the output bits to linear functions on input bits. These findings, combined with linear structures, lead us to preimage attacks against up to 4-round Keccak with reduced complexities. An interesting feature of such preimage attacks is low complexities for small variants. As extreme examples, we can now find preimages of 3-round SHAKE128 with complexity 1, as well as the first practical solutions to two 3-round instances of Keccak challenge. Both zero-sum distinguishers and preimage attacks are verified by implementations. It is noted that the attacks here are still far from threatening the security of the full 24-round Keccak.

**Keywords:** Cryptanalysis · SHA-3 · Keccak · Preimage attacks · Zero-sum distinguishers

## 1 Introduction

The Keccak sponge function family [6] was designed by Bertoni *et al.* as one of the 64 proposals submitted to the SHA-3 competition [24] in October 2008. It won in October 2012 after intense competition, and was subsequently standardized by the U.S. National Institute of Standards and Technology (NIST) as *Secure Hash Algorithm-3* [25] (SHA-3) in August 2015. As such, Keccak has received intensive security analysis, since the design was made public in 2008, against the traditional security notions such as collision, preimage, and second-preimage

© International Association for Cryptologic Research 2016
J.H. Cheon and T. Takagi (Eds.): ASIACRYPT 2016, Part I, LNCS 10031, pp. 249–274, 2016.
DOI: 10.1007/978-3-662-53887-6_9

resistance, as well as distinguishers of the underlying permutations and securities under some message authentication code, stream cipher, and authenticated cipher modes.

Up to date, the best collision attacks are reduced up to 4 out of 24 rounds of KECCAK-224/256 with practical complexities [12,14], and up to 5 rounds of KECCAK-256 with theoretical complexities [13], by differential attacks. Practical preimage attacks are up to 2 rounds, by the approaches of meet-in-the-middle [23] and SAT solvers [21]. Theoretical preimage attacks work up to 7/8/9 rounds for KECCAK-224/256/512 respectively with small time complexity gains over bruteforce [3,11,20]. There were mainly two types of distinguishers against the underlying permutation of KECCAK (named KECCAK-$f$), *i.e.*, zero-sum distinguishers [2,9] and those involving high probability differentials [17,19]. These distinguishers work for 9 rounds in [2], 8 rounds in [19] with practical complexities, and up to 15 rounds with theoretical complexities bounded by $2^{800}$ (birthday bound) for the 1600-bit KECCAK-$f$ permutation. Besides these, there are also attacks in other security settings, we are not listing them all as they are less relevant with our work here.

To promote security analysis with practical complexities, the KECCAK team has been organizing the "KECCAK Crunchy Crypto Collision and Preimage Contest" [4] (we will call it KECCAK **Challenge** for short) and offering cash prizes for the winners. To make it feasible, the instances are set to be round-reduced variants of KECCAK with capacity $c = 160$ and the output truncated to 160 bits for collisions and 80 bits for preimages, so the theoretical complexities for both are $2^{80}$, which is relatively small but yet beyond PC's capability. Instances have been solved for up to 4 and 2 rounds for collisions and preimages, respectively.

**Our Contributions.** In this paper, we focus on security analysis of KECCAK with respective to two security notions, *i.e.*, distinguisher of round-reduced versions of the underlying permutation KECCAK-$f$, and preimage of round-reduced variants of the KECCAK hash function family. Firstly, we review the zero-sum distinguisher by Aumasson and Meier [2]. Zero-sum distinguishers finds a set of input to the permutation, whose sum is zero and the set of corresponding output sums to zero at the same time. This distinguisher makes use of the property of low algebraic degrees 2 and 3 of the Sbox and its inverse used in KECCAK-$f$, which is the only non-linear step of the round function. The attack starts from the middle of the permutation, and extends freely towards both forward and backward directions of the permutations. By setting up initial values of the middle starting point, one can bypass one round without increasing the algebraic degrees. Similar idea was extended to bypass one round by Dinur *et al.* [15] for key recovery attacks in keyed settings. In this paper, we formalize this idea as *linear structures* and extend the free starting rounds to 3 by combining properties of the linear layers and Sbox of the KECCAK-$f$ round function, and generally increase the attacked rounds of the zero-sum distinguishers by Aumasson and Meier by 2 without increasing the complexities. Notably, we extend the practically attacked rounds from 9 to 11. Furthermore, the 12-round KECCAK-$f$ permutations can be distinguished with complexity $2^{65}$ or $2^{82}$. This is of special interests since the 12-round KECCAK-$f$ permutation variants are used in the CAESAR candidates KEYAK [8] and KETJE [7]. Nevertheless, we stress here that this distinguisher does not affect the security of KEYAK or KETJE. A summary of the comparisons of our results with the previous ones is shown in Table 1. Our results are verified by an implementation of the 11-round distinguisher with time and data complexity $2^{33}$, with all the 1600 bits of the output summing to zero with certainty. Note that Table 1 does not include the distinguishers with complexities $\geq 2^{800}$, such as [9,16]. In the KECCAK reference [6, Page 61], the designers mentioned that "Only structural distinguishers on $f$ that have non-zero advantage below $2^{800}$ queries can possibly qualify as a threat for the security of a sponge function that uses it." This is a birthday bound with regard to the size of the permutation.

**Table 1.** Summary of distinguishers on the 1600-bit KECCAK-$f$ permutation, with complexities bounded by $2^{800}$

| #Rounds | inv+forw | Best Known | inv + forw | Improved | inv + forw | Further |
|---------|----------|------------|------------|----------|------------|---------|
| 7  | 3 + 4 | $2^{13}$ [19]    | 3 + 4 | $2^{10}$  | 2 + 5 | $2^9$    |
| 8  | 3 + 5 | $2^{18}$ [2,19]  | 3 + 5 | $2^{17}$  | 3 + 5 | $2^{10}$ |
| 9  | 4 + 5 | $2^{30a}$ [2]    | 4 + 5 | $2^{28}$  | 3 + 6 | $2^{17}$ |
| 10 | 4 + 6 | $2^{60b}$ [2]    | 4 + 6 | $2^{33}$  | 4 + 6 | $2^{28}$ |
| 11 | 5 + 6 | $2^{60c}$ [2]    | 4 + 7 | $2^{65}$  | 4 + 7 | $2^{33}$ |
| 12 | 5 + 7 | $2^{129}$ [2]    | 5 + 7 | $2^{82}$  | 4 + 8 | $2^{65}$ |
| 13 | 6 + 7 | $2^{244}$ [2]    | 5 + 8 | $2^{129}$ | 5 + 8 | $2^{82}$ |
| 14 | 6 + 8 | $2^{257}$ [2]    | 6 + 8 | $2^{244}$ | 5 + 9 | $2^{129}$ |
| 15 | 6 + 9 | $2^{513}$ [2]    | 6 + 9 | $2^{257}$ | -     | -        |

$^a$ Corrected: $2^{33}$. Note that the complexity $2^{30}$ estimated in [2] is based on the experiments made over a 25-dimensional space by the designers in [6], which shows the maximum degree over 25 variables of 4 rounds to be 15. We expect the maximum degree over 30 variables of 4 rounds to be 16, and thus we estimate the time complexity for 5 rounds to be $2^{33}$.
$^b$ Corrected: $2^{65}$.
$^c$ Corrected: $2^{82}$.

The second contributions of this paper are improved preimage attacks. In contrast to the meet-in-the-middle and SAT solver techniques used previously, we adopt the techniques of linear structures and find preimages by linearizing the KECCAK round functions and converting the preimage finding problems to that of solving systems of linear equations. This technique leads to attacks on up to 4-round KECCAK with reduced complexities. The complexities of 3-round preimages are so significantly reduced that enables us to find preimages of SHAKE128 (a variant of KECCAK[$r = 1344, c = 256$] adapted by SHA-3) practically, and to solve two of 3-round preimage instances and a near-preimage with only two bits difference of 4-round preimage instance of the KECCAK Challenge. The summary of our preimage attacks together with the previous best ones is shown in Table 2. Note that Table 2 does not include small optimizations of exhaustive search, such as [3,11]. In this table, by variant 128 we mean SHAKE128($M$, 128). Different with the attacks of [20] which outperform exhaustive search by a larger factor as the hash size becomes larger, our attacks outperform exhaustive search by a larger factor as the hash size (or the capacity) becomes smaller.

**Table 2.** Summary of preimage attacks on KECCAK reduced up to 4 rounds.

| #Rounds | Variant | Time | Reference |
|---------|---------|------|-----------|
| 2 | 128/224/256 | $2^{33}$ | [23] |
| 2 | 128/224/256 | 1 | Sect. 6.1 |
| 2 | 384 | $2^{129}$ | Sect. 6.1 |
| 2 | 512 | $2^{384}$ | Sect. 6.1 |
| 3 | 128 | $2^{26.6}$ | Sect. 6.2 |
| 3 | 128 | 1 | Sect. 6.4 |
| 3 | 224 | $2^{97}$ | Sect. 6.2 |
| 3 | 256 | $2^{192}$ | Sect. 6.2 |
| 3 | 384 | $2^{322}$ | Sect. 6.3 |
| 3 | 512 | $2^{482}$ | Sect. 6.3 |
| 3 | 512 | $2^{506}$ | [20] |
| 4 | 128 | $2^{106}$ | Sect. 6.4 |
| 4 | 224 | $2^{213}$ | Sect. 6.3 |
| 4 | 256 | $2^{251}$ | Sect. 6.3 |
| 4 | 224/256 | $2^{221}/2^{252}$ | [20] |
| 4 | 384/512 | $2^{378}/2^{506}$ | [20] |

Both improved zero-sum distinguishers and preimage attacks are possible thanks to the technique *linear structures*. By exploiting the properties of the Sbox used in KECCAK, we find ways to linearize both the Sbox itself and its inverse. Combining with properties of the linear layer of the KECCAK round function, we are able to find linear subspaces with large dimension by setting proper initial values. A nice property of these linear structures is that the algebraic degrees can be kept the same for up to 3 rounds, *i.e.*, output bits of 3-round KECCAK-$f$ can be expressed as linear combinations of input bits. As a special feature of the linear structure, complexities of our attacks reduce significantly when the targets are KECCAK instances with small capacities. In such cases, the number of required constraints derived from pre-set constants is small and the degree of freedom left is relatively large, and hence leads to faster attacks. As extreme examples, we can find preimages of 3-round SHAKE128($M$, 128) and solve the 3-round KECCAK[$r = 1440, c = 160, \ell = 80$] instance of KECCAK **Challenge** with complexity 1.

**Organization.** The rest of the paper is organized as follows. Section 2 gives the details of the KECCAK hash function family, followed by the properties of the Sbox in Sect. 3. The linear structure is introduced in Sect. 4. Its applications of zero-sum distinguishers and preimages attacks are presented in Sects. 5 and 6, respectively. Section 7 concludes the paper.

## 2 Definition of KECCAK

### 2.1 The Sponge Function

The KECCAK hash function follows the sponge construction, as depicted in Fig. 1. The message $M$ is padded and split into blocks of $r$ bits each. Beginning with an initial value (IV), the first $r$ bits of $b$-bit state is XORed with the message block, followed by the application of the permutation $f$. This step is repeated until all message blocks are processed. Then the first $r$ bits are outputted, $r$ more bits can be obtained after an additional application of $f$, and this process is repeated until all required digest bits are obtained. The number of iterations is determined by the requested number of digest bits $\ell$. Finally the output is truncated to its first $\ell$ bits.

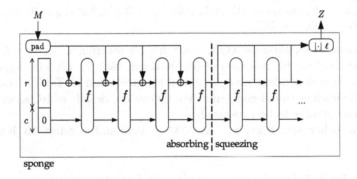

**Fig. 1.** The sponge function [5]

### 2.2 The KECCAK Hash Functions

To define the KECCAK hash function family, the designers give the details of the underlying permutation $f$, as well as parameters set of $(r, c, \ell)$. The IV is set to be all "0"s. The underlying permutation of the KECCAK hash function is chosen in a set of seven KECCAK-$f$ permutations, denoted KECCAK-$f[b]$, where $b \in \{25, 50, 100, 200, 400, 800, 1600\}$ is the width of the permutation. The default version of KECCAK-$f$ is of size $b = 1600$ bits, which can be represented as $5 \times 5$ 64-bit lanes as depicted in Fig. 2, denoted as $A[x, y]$ with $x$ for the index of column and $y$ for the index of row. In what follows, indexes of $x$ and $y$ are from the set $\{0, 1, 2, 3, 4\}$ and they are working in modulo 5 without other specification.

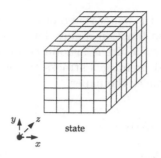

**Fig. 2.** The KECCAK state [6]

The underlying permutation KECCAK-$f$[1600] consists of 24 identical round functions up to a difference of constant addition. The round function R consists of five operations ($\theta$ goes first):

$$R = \iota \circ \chi \circ \pi \circ \rho \circ \theta.$$

$\theta : A[x, y] = A[x, y] \oplus \bigoplus_{j=0}^{4}(A[x-1, j] \oplus (A[x+1, j] \lll 1))$, for $x, y = 0, \ldots, 4$.
$\rho : A[x, y] = A[x, y] \lll r[x, y]$, for $x, y = 0, \ldots, 4$.
$\pi : A[y, 2x + 3y] = A[x, y]$, for $x, y = 0, \ldots, 4$.
$\chi : A[x, y] = A[x, y] \oplus ((\sim A[x+1, y]) \& A[x+2, y])$, for $x, y = 0, \ldots, 4$.
$\iota : A[0, 0] = A[0, 0] \oplus RC$.

Here "$\oplus$" denotes for bit-wise XOR, "$\lll$" for bit rotation towards MSB of the 64-bit word, "$\sim$" for bit negation of 64-bit word, "$\&$" for bit-wise logic AND, "$r[x, y]$" for lane dependent rotation constants presented in Table 3, and "RC" for round-dependent round constants. We ignore the details of RC since it does not affect our attacks to be presented.

Without other specifications, KECCAK-$f$ hereinafter refers to KECCAK-$f$ [1600].

**Table 3.** Rotation constants $r[x, y]$ in KECCAK $\rho$ operation.

|         | $x=0$ | $x=1$ | $x=2$ | $x=3$ | $x=4$ |
|---------|-------|-------|-------|-------|-------|
| $y=0$   | 0     | 1     | 62    | 28    | 27    |
| $y=1$   | 36    | 44    | 6     | 55    | 20    |
| $y=2$   | 3     | 10    | 43    | 25    | 39    |
| $y=3$   | 41    | 45    | 15    | 21    | 8     |
| $y=4$   | 18    | 2     | 61    | 56    | 14    |

## 2.3    Instances of KECCAK and SHA-3

The hash function KECCAK$[r, c, \ell]$ refers to the instance of the KECCAK sponge function family with parameters capacity $c$, bitrate $r$ and output length $\ell$.

The official versions of KECCAK have $r = 1600 - c$ and $c = 2\ell$, where $\ell \in \{224, 256, 384, 512\}$, called KECCAK-224, KECCAK-256, KECCAK-384, and KECCAK-512. The padding rule of KECCAK is of the form $M10^*1$, i.e., it pads a single bit "1" followed by a smallest non-negative number of "0", then a bit "1", such that the bit length of the padded message becomes multiple of $r$.

The SHA-3 standard takes mainly the four default instances of KECCAK with digest sizes $224, 256, 384, 512$. The only difference is the padding rule. These four SHA-3 instances pad the message with two bits "01" before applying the KECCAK padding rule, so the padded message becomes $M0110^*1$, i.e., it pads $M$ by three bits "011", followed by a smallest non-negative number of "0"s, then a bit "1", such that the padded message is of multiple of $r$ bits. Generally, all our analysis results in this paper on KECCAK applies to SHA-3, under the same parameters $(r, c, \ell)$, possibly with an increment of the complexities by at most $2^2$ due to the two extra padding bits.

The SHA-3 family also includes two extendable-output functions (XOFs), called SHAKE128 and SHAKE256. More exactly, these instances SHAKE128$(M, \ell)$ and SHAKE256$(M, \ell)$ are defined from KECCAK$[r = 1344, c = 256]$ and KECCAK $[r = 1088, c = 512]$ by appending a four-bit suffix "1111" to the message, for any output length $\ell$. Our preimage attacks on KECCAK-256 also applies to SHAKE256 $(M, 256)$. We will only consider preimage attacks on SHAKE128$(M, 128)$.

# 3  Properties of the Sbox $\chi$

In this section, we discuss the properties of the Sbox $\chi$, which will be used to construct distinguishers on KECCAK-$f$ permutation in Sect. 5, and to mount preimage attacks on KECCAK in Sect. 6.

## 3.1  Setting up Linear Equations from the Output of $\chi$

**Bilinear Structure.** We show in this section that given $t$ consecutive bits out of the 5 output bits of $\chi$, one can set up at least $t - 1$ linear equations on the 5 input bits due to the bilinear structure of the $\chi$. Hereinafter, we may also refer to the $\chi$ operation by Sbox.

The algebraic normal form of $\chi$ mapping 5-bit $a = a_0 a_1 a_2 a_3 a_4$ into 5-bit $b = b_0 b_1 b_2 b_3 b_4$ can be written as $b_i = a_i \oplus (a_{i+1} \oplus 1) \cdot a_{i+2}$, i.e.,

$$b_0 = a_0 \oplus (a_1 \oplus 1) \cdot a_2, \tag{1}$$

$$b_1 = a_1 \oplus (a_2 \oplus 1) \cdot a_3, \tag{2}$$

$$b_2 = a_2 \oplus (a_3 \oplus 1) \cdot a_4, \tag{3}$$

$$b_3 = a_3 \oplus (a_4 \oplus 1) \cdot a_0, \tag{4}$$

$$b_4 = a_4 \oplus (a_0 \oplus 1) \cdot a_1. \tag{5}$$

Then, we show that given two consecutive bits of the output of $\chi$, one linear equation can be set up on the input bits. Without loss of generality, assume that $b_0$ and $b_1$ are known. By (2), we have

$$b_1 \cdot a_2 = (a_1 \oplus (a_2 \oplus 1) \cdot a_3) \cdot a_2 = a_1 \cdot a_2$$

and thus according to (1) we obtain

$$b_0 = a_0 \oplus (b_1 \oplus 1) \cdot a_2. \tag{6}$$

Given three consecutive bits of the output of $\chi$, to say $b_0$, $b_1$ and $b_2$, an additional linear equation can be similarly set up:

$$b_1 = a_1 \oplus (b_2 \oplus 1) \cdot a_3. \tag{7}$$

Generally, the input $a$ and output $b$ of $\chi$ satisfy $F(a,b) = 0$ with $F(a,b) = aSb + Ta + Qb$, for some $5 \times 5$ binary matrices $S, T, Q$.

Given four output bits of $\chi$, any bit of the input can be represented as a linear function on the unknown bit of the output, and one can naturally set up four linear equations on the input bits by eliminating the unknown output bit. It is clear that given all the five output bits of $\chi$, the input bits are all determined. We summarize in Table 4 the number of linear equations on the input bits that can set up for given $t$ consecutive bits of the output of $\chi$.

**Table 4.** Number of linear equations obtained from the output of $\chi$

| #Known consecutive output bits | 2 | 3 | 4 | 5 |
|---|---|---|---|---|
| #Linear equations | | 1 | 2 | 4 | 5 |

## 3.2   Setting up More Linear Equations

As explained above, given $t$ bits of the output of $\chi$, for $t = 4$ or $5$, one can set up $t$ linear equations on the input of $\chi$, and for $t < 4$, one can set up $t - 1$ linear equations. Here we present two more methods for setting up one or more extra linear equations on input bits when less than 4 bits of the output are known.

The first method is to guess the value of an input bit. We obtain two extra linear equations at cost of doubling the operations needed. For example, if a single bit $b_0$ of the output is known, no linear equation could be set with previous methods. However, here we can guess the input bit $a_1$ so that the equation $b_0 = a_0 \oplus (a_1 \oplus 1) \cdot a_2$ becomes linear. Together with the guess of $a_1$ itself, we obtain in total two more linear equations. The cost is that there are 2 choices of the guess, so we obtained the two extra linear equations with the cost of an increase of time complexity by a factor of 2. This is generally true when the number of known output bits is less than 4, as summarized in Setting 1.

**Setting 1.** *When the number of known output bits is in the range $[1,3]$, a guess of an input bit leads to two extra linear equations on the input bits, by the cost of doubling the time complexity.*

The second method is to make use of the probabilistic equation $b_i = a_i$ which holds with probability 0.75, due to the fact that $b_i = a_i \oplus (a_{i+1} \oplus 1) \cdot a_{i+2}$ and $(a_{i+1} \oplus 1) \cdot a_{i+2}$ is 0 with probability 0.75 assuming uniformly distributed $a_{i+1}$ and $a_{i+2}$, as summarized in Setting 2. This method will result in time complexity increase by a factor $0.75^{-1} = 2^{0.415}$.

**Setting 2.** *$b_i = a_i$ of the $\chi$ holds with probability 0.75 when input bit $a_j$'s are uniformly distributed, for all $i \in \{0, \ldots, 4\}$.*

### 3.3   Linearizing the Inverse of $\chi$

The inverse $\chi^{-1} : b \mapsto a$ has algebraic degree 3, and its algebraic normal form can be written as

$$a_i = b_i \oplus b_{i+2} \oplus b_{i+4} \oplus b_{i+1} \cdot b_{i+2} \oplus b_{i+1} \cdot b_{i+4} \oplus b_{i+3} \cdot b_{i+4} \oplus b_{i+1} \cdot b_{i+3} \cdot b_{i+4}$$
$$= b_i \oplus (b_{i+1} \oplus 1) \cdot (b_{i+2} \oplus (b_{i+3} \oplus 1) \cdot b_{i+4}) \tag{8}$$

where $0 \le i \le 4$ and the indexes are operated on modulo 5, that is,

$$a_0 = b_0 \oplus (b_1 \oplus 1) \cdot (b_2 \oplus (b_3 \oplus 1) \cdot b_4), \tag{9}$$
$$a_1 = b_1 \oplus (b_2 \oplus 1) \cdot (b_3 \oplus (b_4 \oplus 1) \cdot b_0), \tag{10}$$
$$a_2 = b_2 \oplus (b_3 \oplus 1) \cdot (b_4 \oplus (b_0 \oplus 1) \cdot b_1), \tag{11}$$
$$a_3 = b_3 \oplus (b_4 \oplus 1) \cdot (b_0 \oplus (b_1 \oplus 1) \cdot b_2), \tag{12}$$
$$a_4 = b_4 \oplus (b_0 \oplus 1) \cdot (b_1 \oplus (b_2 \oplus 1) \cdot b_3). \tag{13}$$

It is obvious to note

**Setting 3.** *If there is a single unknown output bit $b_j$ of $\chi$ and all other output bits are constants, then all input bits $a_i$ can be expressed as linear combination of $b_j$.*

If we impose $b_3 = 0$ and $b_4 = 1$, then we have

$$a_0 = b_0 \oplus (b_1 \oplus 1) \cdot (b_2 \oplus 1),$$
$$a_1 = b_1,$$
$$a_2 = 1 \oplus b_2 \oplus (b_0 \oplus 1) \cdot b_1,$$
$$a_3 = 0,$$
$$a_4 = 1 \oplus (b_0 \oplus 1) \cdot b_1,$$

and thus all $a_i$'s are linear on $b_0$ and $b_2$. That's, for $b_3 = 0$, $b_4 = 1$ and any fixed $b_1$, the algebraic degree of $\chi^{-1}$ becomes 1.

If we further impose $b_1 = 1$, then we have

$$a_0 = b_0, \quad a_1 = 1, \quad a_2 = b_0 \oplus b_2, \quad a_3 = 0, \quad a_4 = b_0,$$

so all inputs bits $a_i$'s become linear combinations of $b_i$'s. Similar property holds when $b_1 = 0$. This is summarized as:

**Setting 4.** *When $b_{j+3} = 0$, $b_{j+4} = 1$, and $b_{j+1}$ is known (either 0 or 1), then all inputs bits $a_i$'s can be written as linear combinations of $b_i$'s, for all $j \in \{0, \ldots, 4\}$.*

# 4   The Linear Structures

In this section, we review the previous work, and formalize the idea of linear structure. We show linearization of KECCAK-$f$ permutation for up to 3 rounds. Our distinguisher and preimage attacks using linear structures depend directly on the space size of the variables of these linear structures, $i.e.$, more variable bits result in lower attack complexities. We show in details how the largest space size possible could be obtained in each scenario.

## 4.1   Techniques for Keeping 2 Rounds Being Linear

In [15], Dinur $et\ al.$ exploited a method for keeping the first round of KECCAK-$f$ being linear and used it to analyze the security of keyed variants of KECCAK. Here we restate and formalize their technique. Let $A[1,i]$, $i = 0, 1, 2, 3$, be variables and $A[1,4] = \bigoplus_{i=0}^{3} A[1,i] \oplus \alpha$ with any constant $\alpha$ so that variables in each column sum to a constant. Then, as shown in Fig. 3, we can see how the variables affect the internal state under the transformation of KECCAK-$f$ round function $\mathsf{R} = \iota \circ \chi \circ \pi \circ \rho \circ \theta$. In Fig. 3 and hereinafter, the 2-tuple number "$x, y$" denotes the position of a lane at the initial state, and we track its position under the $\pi$ function, where $0 \leq x, y \leq 4$. All bits of the lanes with orange slashes have algebraic degree 1, those of the lanes in orange have algebraic degree at most 1, and the other lanes are all constants. Note the algebraic degrees will $not$ be affected by the linear operations $\theta$, $\rho$, $\pi$, and $\iota$. The only non-linear operation is the $\chi$, and its degree is 2 or 3 in forward or backward directions, respectively. As shown in the third state in Fig. 3, each row contains a single bit of degree 1 and the other 4 bits are constants. Since the only possibility for $\chi$ to increase the algebraic degree is through two neighbouring bits due to the term $(a_{i+1} \oplus 1) \cdot a_{i+2}$, the algebraic degree of the state bits remains at most 1 after $\chi$, $i.e.$, after one round function $\mathsf{R}$. The size of free variables can be at most 4 lanes, $i.e.$, $64 \times 4 = 256$ bits.

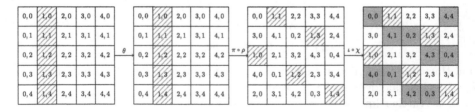

**Fig. 3.** Keeping the 1st forward round being linear with the degrees of freedom up to 256, with orange bits of degree at most 1, and white bits being constants.

Noting that the only nonlinear part of $\mathsf{R}$ is $\chi$ which operates on each 5-bit row. Since there is at most 1 variable in each row as in the first state in Fig. 3, the inverse function $\chi^{-1}$ is linear on these variables due to Setting 3. Thus, the first inverse round $R^{-1}$ is linear on these variables. This property was first used to construct zero-sum distinguishers on KECCAK-$f$ in [2].

**Increasing the Degrees of Freedom up to 512 for 2 Rounds.** Let $A[i,j]$ for $i = 0,2$ and $j = 0,1,2,3$ be variables and $A[i,4] = \bigoplus_{j=0}^{3} A[i,j] \oplus \alpha_i$ with constants $\alpha_i$ for $i = 0,2$. Figure 4 shows how the variables propagate in one round R for $\alpha_0 = 0$ and $\alpha_2 = \mathtt{0xff\cdots f}$. The bits of the lanes in gray (resp. lightgray) are set to all 1's (resp. 0's), and the bits of white lanes are set to arbitrary constants. The lanes with orange slashes or orange have algebraic degree at most 1 as above. Since there are at most two variables in each row input to $\chi$ and the variables are not adjacent, the outputs of $\chi$ are all linear on these variables. Therefore, the algebraic degree of the state bits in these variables remains 1 after the first round of KECCAK-$f$ permutation, and the size of free variables can achieve at most $64 \times 4 \times 2 = 512$. This is also true for other constants $\alpha_i$.

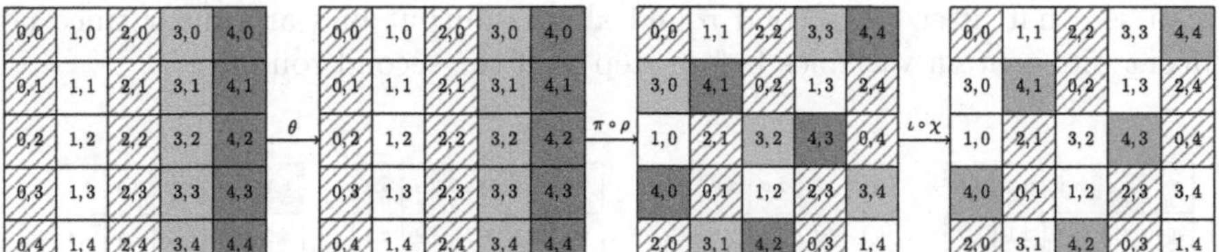

**Fig. 4.** Keeping the 1st forward round being linear with the degrees of freedom up to 512, with orange bits of degree at most 1, and gray, lightgray and white bits being values 1, 0, and arbitrary constants, respectively. (Color figure online)

To keep the algebraic degrees to be at most one when $\chi^{-1}$ is applied (inverting one round) to the 512 variables in the first state of Fig. 4, according to Setting 4, we restrict the bits of gray lanes to be all ones and the bits of lightgray lanes to be all zeros, where the bits in gray and lightgray lanes respectively correspond to $b_{i+4}$'s and $b_{i+3}$'s in Setting 4. Note that the step $\iota$ only adds a constant to the first lane and thus it does not affect the gray and lightgray lanes. In this case, the first inverse round $R^{-1}$ is linear on these 512 variables.

## 4.2 How to Keep 3 Rounds Being Linear

Based on the technique above, in this section we describe a technique for keeping an additional forward round of KECCAK-$f$ being linear.

Let $A[i,j]$ with $i = 0,2$ and $j = 0,1,2$ be variables. In what follows, we show how to impose some conditions on the input bits such that all the output bits after two rounds forward are linear. To make sure that the variables do not affect the values of the other bits after step $\theta$ of the first round, *i.e.*, keeping the sum of all columns to be zero constants, we impose the following $2 \times 64$ equations:

$$A[i,0] \oplus A[i,1] \oplus A[i,2] = 0, \quad i = 0,2.$$

The values of white lanes are set in such a way that the value of the gray and lightgray lanes remained unchanged after step $\theta$ of the first round, as shown in Fig. 5. The steps $\rho$ and $\pi$ are respectively shifts of the bits in the same lanes and permutations of the positions of the lanes. After the steps $\chi$ and $\iota$, the lane at column 0 and row 0 equals $A[0,0] \oplus A[2,2]_{\lll 43}$, the other lanes in orange remain unchanged up to constants, and the white lanes are all constants. To make sure that the variables do not propagate after step $\theta$ of the second round, we impose $3 \times 64$ more equations:

$$A[2,0]_{\lll 62} = A[0,0] \oplus A[2,2]_{\lll 43},$$
$$A[2,1]_{\lll 6} = A[0,1]_{\lll 36},$$
$$A[2,2]_{\lll 43} = A[0,2]_{\lll 3}.$$

Note that this result is still valid when constants are XORed to the above three equations. Since this linear system has in total $5 \times 64 = 320$ equations and $6 \times 64 = 384$ variables, there remains 64 degrees of freedom. As shown in Fig. 5, we can see that after the second round all the output bits are linear since no adjacent bits contain variables before step $\chi$ of the second round.

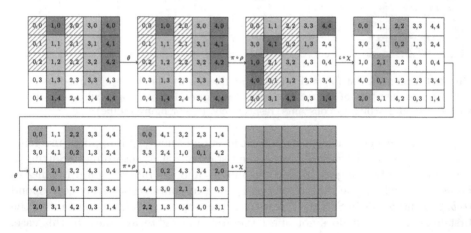

**Fig. 5.** Keeping the 2nd forward round being linear with degree of freedom up to 64

To ensure that the inverse function $\chi^{-1}$ is linear, we restrict the bits of lanes $A[4,j]$ with $j = 0, 1, 2$ to be all ones and the bits of lanes $A[3,j]$ with $j = 0, 1, 2$ to be all zeros as in Setting 4.

**Increasing the Degrees of Freedom to up to 128.** Similarly, we can increase the degrees of freedom from 64 to 128 by setting $A[i,j]$ with $i = 0, 2$ and $j = 0, 1, 2, 3$ be variables and imposing some conditions on the input bits as shown in Fig. 6. We build a linear system of $6 \times 64 = 384$ equations on $8 \times 64 = 512$ variables which has 128 degrees of freedom and satisfies that the output bits after the second round are all linear. To ensure that the inverse function $\chi^{-1}$ is

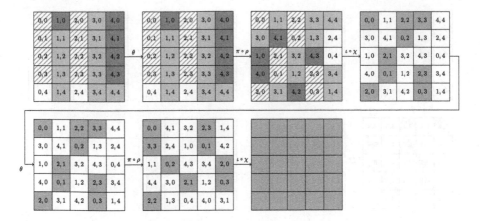

**Fig. 6.** Keeping the 2nd forward round being linear with the degree of freedom up to 128

linear, we restrict the bits of lanes $A[4, j]$ with $j = 0, 1, 2, 3$ to be all ones and the bits of lanes $A[3, j]$ with $j = 0, 1, 2, 3$ to be all zeros.

**Increasing the Degrees of Freedom to up to 194.** We further extend the degrees of freedom to 194 by setting $A[i, j]$ with $i = 0, 2$ and $j = 0, 1, \cdots, 4$ be variables and imposing some conditions on the input bits as shown in Fig. 7. We build a linear system of $7 \times 64 = 448$ equations on $10 \times 64 = 640$ variables which has 194 degrees of freedom and satisfies that the output bits after the second round are all linear. Note that there are two linear equations linearly dependent on the other equations, so the degree of freedom is 194 instead of 192. To ensure that the inverse function $\chi^{-1}$ is linear, we restrict the bits of lanes $A[4, j]$ with $j = 0, 1, \cdots, 4$ to be all ones and the bits of lanes $A[3, j]$ with $j = 0, 1, \cdots, 4$ to be all zeros.

In summary, we found linear structures of KECCAK-$f$ permutation reduced to 2 rounds with degree of freedom up to 512, and 3 rounds with degree of freedom up to 194.

## 5    Zero-Sum Distinguishers

A zero-sum distinguisher for a function is a method to find a set of values summing to zero such that their respective images also sum to zero. That is, it is a method to find a set $S$ such that $\sum_{x \in S} x = 0$ and $\sum_{x \in S} f(x) = 0$ for the function $f$. It is well known that the $d$-th order derivative of a polynomial with degree at most $d$ is a constant. For a Boolean function of algebraic degree at most $d$, its $d$-th order derivative is also a constant. Thus the outputs of a Boolean function of degree at most $d$ sum to zero when the inputs take over a linear space of dimension at least $d + 1$.

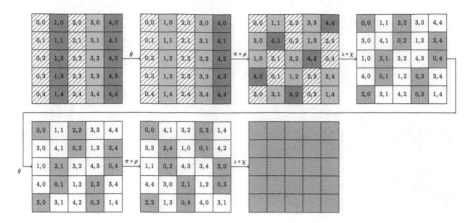

**Fig. 7.** Keeping the 2nd forward round being linear with the degree of freedom up to 194

The KECCAK-$f$ permutation is the core function of KECCAK and SHA-3. The known method for constructing zero-sum distinguishers on KECCAK-$f$ permutation, exploits the fact that adding a round in KECCAK-$f$ only doubles the degree of the algebraic expression of the output bits in terms of the input bits, and only triples the degree of the algebraic expression of the input bits in terms of the output bits. This is due to that the algebraic degree of one KECCAK-$f$ round is 2 and the algebraic degree of one inverse round is 3. The real zero-sum distinguisher starts from some middle round of the KECCAK-$f$ permutation, and extends $n$ rounds forward and $m$ rounds backward. So the algebraic degree of $n$ forward rounds $\mathsf{R}^n$ is bounded by $2^n$, and $m$ backward rounds $\mathsf{R}^{-m}$ by $3^m$. With a linear space $S_M$ from the middle round of size at least $2^{1+\max(2^n,3^m)}$, one can be ensured that both input and output sum to zero, i.e., $\sum_{x \in S_M} \mathsf{R}^n(x) = 0$ and $\sum_{x \in S_M} \mathsf{R}^{-m}(x) = 0$. The desired input space $S$ of the $(m+n)$-round distinguisher can be obtained by $S = \{\mathsf{R}^{-m}(x) \mid x \in S_M\}$.

The attack has been extended in two different directions, finding better bounds of the algebraic degrees of $\mathsf{R}^n$ and $\mathsf{R}^{-m}$ [9,16], and inserting rounds in the starting point in the middle [2]. Our improved zero-sum distinguisher is in line with the second approach. Aumasson and Meier showed in [2] that one round could be inserted for free. This is achieved by carefully choosing the set $S_M$ so that the algebraic degree keeps to be 1 after one round. It becomes obvious to note the linear structures presented in Sect. 4 could be used here to extend the number of free rounds to *three*, i.e., with linear structures as $S_M$ (similar to the way how *initial structures* are used in MITM preimage attacks [1,18]), the algebraic degrees of one backward round $\mathsf{R}^{-1}$ and two forward rounds $\mathsf{R}^2$ are kept to be 1.

$$\underset{\text{backward}}{\underbrace{\hspace{2cm}}^{m+1}} \Big| \underset{\text{forward}}{\underbrace{\hspace{2cm}}^{2+n}}.$$

As such, with the same complexity $2^{1+\max(2^n,3^m)}$, our improved distinguisher works for $(m+n+3)$ rounds, *i.e.*, $(m+1)$ rounds backward and $(n+2)$ rounds forward. In Table 1, we summarize our results with the best combinations of $m$ and $n$. Note the number of attacked rounds is limited by the size of $S_M$, a.k.a., the size of the linear structures. For instance, the largest space we found for 3-round linear structure is $2^{194}$, so the distinguisher works for all combinations of $m$ and $n$ such that $2^{1+\max(2^n,3^m)} \leq 2^{194}$. When $m = 4$ (5 rounds backward) and $n = 7$ (9 rounds forward), as stated in the last entry of the third column of Table 1, the attack applies to $m + n + 3 = 14$ rounds with time/data complexity $2^{1+\max(2^n,3^m)} = 2^{1+\max(2^7,3^4)} = 2^{129} \leq 2^{194}$.

As a trade-off, the size of linear structure could be larger for less rounds, *e.g.*, up to $2^{512}$ for 2 rounds. So the distinguisher works as below

$$\underbrace{\phantom{xxxxxx}}_{\text{backward}}^{m+1} \Big| \underbrace{\phantom{xxxxxx}}_{\text{forward}}^{1+n}.$$

While there is one free round less, we can afford larger complexities, *e.g.*, with $m = 5$ and $n = 8$, we can distinguish $m + n + 2 = 15$ rounds with complexity $2^{1+\max(2^n,3^m)} = 2^{257}$. Results of other choices of $(m,n)$ are listed in the second column of Table 1.

As a direct application to the 12-round KECCAK-$f$ permutation used in the CAESAR candidate KEYAK [8], the 3-round linear structure is large enough and the choice of $(m = 3, n = 6)$ results in attack complexity $2^{65}$. KETJE [7] uses a 12-round KECCAK-$f$ permutation reduced to 400 bits (denoted as KECCAK-$p[400, n_r = 12]$), by reducing the length of lanes from 64 to 16 bits. When we project the zero-sum distinguisher to this small variant, the maximum sizes of linear structures are reduced to $512/4 = 128$ and $192/4 = 48$ bits respectively for 2 and 3 rounds. While the size for the 3-round linear structure is insufficient for distinguishing 12 rounds, the 128-bit 2-round linear structure makes it eligible with complexity $2^{82}$. We note that though our distinguishers work for 12-round KECCAK-$f$, they do not result in attacks in settings of authenticated cipher against KEYAK or KETJE.

In summary our improved zero-sum distinguishers work for up to 15 rounds, and for up to 11 rounds with practical complexities.

**Experiments.** We have made an experiment for verifying our distinguishers on KECCAK-$f$ permutation reduced to 7 rounds in the forward direction. We use the structure with degrees of freedom up to 64 as shown in Sect. 4.2. Note that all the bits of the 7-round output have algebraic degree at most $2^{7-2} = 32$ for this structure. It is sufficient to use a 33 dimensional space. In our experiment, 31 out of those 64 variables are first randomly valued and fixed, then the outputs are summed over all the possible values of the rest input variables. It turns out that all the 1600 bits of this sum are zeros.

# 6    Preimage Attacks

In this section, we exploit algebraic techniques to mount preimage attacks on several variants of KECCAK based on the properties of the Sbox $\chi$ and the linear structures of KECCAK-$f$ permutation. The preimage attacks on SHA-3 are the same except that the time complexity may be at most $2^2$ larger in some cases due to the two extra padding bits. In general, here we find preimages of message with length $\leq r - 2$ bits by setting the $(r - 1)$-th bit of the input state to be 1 so that the padded message is one block, unless the degree of freedom is insufficient. We choose the message in such a way that the internal states of the first few rounds follow linear structures as presented in Sect. 4 and the $\chi$ of the last round is inverted by the methods presented in Sect. 3. To achieve smallest possible time complexities, we will use different linear structures, and different methods inverting the $\chi$ for each instance of KECCAK. Note, the first $r - 1$ bits of the input to KECCAK-$f$ can be chosen freely by choosing the proper message bit values. However, the last $c = b - r$ bits could not be chosen since there is no addition of message bits, so we can only choose "variables" of linear structures from the first $r-1$ bits, and this is why we must use different linear structures for different instances. In what follows, we present the preimage attacks by showing the choice of linear structures, ways to invert the Sbox, followed by a complexity analysis of each instance attacked. The basic idea of our attacks is to set up and solve linear equations. The complexity in this section is measured by the number of times for solving the linear system of equations.

## 6.1    Preimage Attacks on 2-Round KECCAK

First we discuss the preimage attacks on KECCAK reduced to 2 rounds. They follow 1-round linear structures, plus 1-round inversion of the Sbox. These attacks adopt some similar ideas of meet-in-the-middle [23], while they exploit the linear structures of KECCAK. For 2-round KECCAK-512, we execute the attack as follows (depicted in Fig. 8):

1. Invert the first 320 bits of a given hash value $h$ through $\chi^{-1} \circ \iota^{-1}$. Note these bits form the full output of the 64 Sboxes in the first row, so the corresponding input bits can be fully determined.
2. Randomly guess the values of the lanes in white of the state input to the first round, as shown in Fig. 8, where the 1024 bits of the lanes in lightgray are set to all zeros and the last bit of $A[3, 1]$ is set to 1 such that the state input to the first round satisfies the padding rule;
3. For each guess, we set $A[0, 1] = A[0, 0] + \alpha_0$ and $A[2, 1] = A[2, 0] + \alpha_2$ with random constants $\alpha_0, \alpha_2$, build a linear system between $A[0, 0]$, $A[2, 0]$ and the recovered 320 input bits of the $\chi$ in the second round, then solve this system and check whether the resulted hash value is correct.

Since $A[0, 0]$ and $A[2, 0]$ have 128 bits, so we have a complexity gain over brute-force of $2^{128}$, i.e., $2^{512-128} = 2^{384}$ for 2-round KECCAK-512 preimage attack.

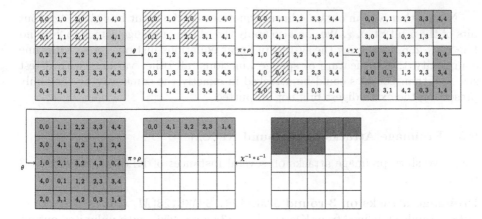

**Fig. 8.** Preimage attack on 2-round KECCAK-512

Note the degree of freedom in our setting is sufficient to find a preimage eventually. There are 128 bits from $A[0,0]$ and $A[2,0]$, 319 bits from white lanes, and 128 bits from $\alpha_0$ and $\alpha_2$, which sums to 575 bits, larger than the required 512 bits.

For the 2-round KECCAK-384, the attack is similar to that for KECCAK-512, except that we can construct linear structure from $r = 1600 - 2 \times 384 - 1 = 831$ bits instead of 575 bits for KECCAK-512. We can obtain a linear structure of 256-bit variables from $(A[0,0], A[0,1], A[2,0], A[2,1])$ with $A[0,2] = A[0,0] \oplus A[0,1] \oplus \alpha_0$ and $A[2,2] = A[2,0] \oplus A[2,1] \oplus \alpha_2$, hence a linear system of 256-bit equations, as shown in Fig. 9. For generating a message satisfying the padding rule, we just need a solution with the last bit of $A[2,2]$ being 1. Therefore, the time complexity of this attack is $2^{384-256+1} = 2^{129}$.

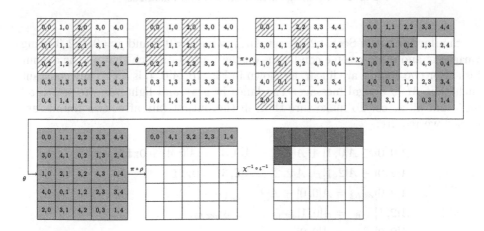

**Fig. 9.** Preimage attack on 2-round KECCAK-384

Noting that we can obtain 4 linear equations on the input bits given 4 output bits of the 5-bit Sbox $\chi$. We can also apply the above preimage attack to 2-round KECCAK-256, by solving the system of linear equations just once, *i.e.*, with time complexity 1. As a feature of sponge functions, all other variants with digest size less than 256 bits could be attacked in exactly the same way by randomly presetting the extra digest bits not outputted.

## 6.2  Preimage Attacks on 3-Round KECCAK

Next, we show preimage attacks on several instances of KECCAK reduced to 3 rounds.

**Preimage attacks on 3-round SHAKE128.** SHAKE128$(M, \ell)$ is an instance of SHA-3 standard defined from KECCAK$[r = 1344, c = 256]$, with unlimited output length $\ell$. We focus on the preimage attack on SHAKE128$(M, 128)$, denoted by SHAKE128 hereinafter for simplicity.

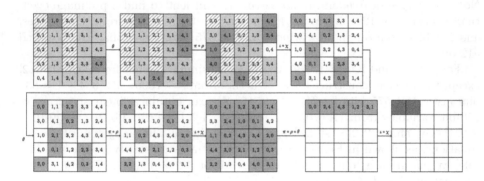

**Fig. 10.** Preimage attack on 3-round SHAKE128

Similar to that in Sect. 4.2, we set $A[i, j]$ with $i = 0, 2$ and $j = 0, 1, 2, 3$ being variables, and impose some conditions on the input bits such that all the output bits after two rounds are linear, as shown in Fig. 10. $A[0, 4]$ is set to any constant such that $M$ is a legal message. The lanes in gray and lightgray are set to all ones and all zeros. To make sure that all the output bits after two rounds are linear, we require:

$$A[0, 0] \oplus A[0, 1] \oplus A[0, 2] \oplus A[0, 3] = A[0, 4] \oplus \texttt{0xff} \cdots \texttt{f},$$
$$A[2, 0] \oplus A[2, 1] \oplus A[2, 2] \oplus A[2, 3] = \texttt{0xff} \cdots \texttt{f},$$
$$A[2, 0]_{\lll 62} = A[0, 0] \oplus A[2, 2]_{\lll 43},$$
$$A[2, 1]_{\lll 6} = A[0, 1]_{\lll 36} \oplus A[2, 3]_{\lll 15},$$
$$A[2, 2]_{\lll 43} = A[0, 2]_{\lll 3},$$
$$A[2, 3]_{\lll 15} = A[0, 3]_{\lll 41} \oplus A[2, 0]_{\lll 62}.$$

All these $6 \times 64$ linear equations are linearly independent and thus have $2^{128}$ solutions. We expect that there is one solution matching the given 128-bit hash value. Since $\pi \circ \rho \circ \theta$ is linear, the bits input to $\chi$ of the last round are all linear on the variables. For $\texttt{SHAKE128}$, the first two output bits of each 5 bits of the 64 Sboxes $\chi$ in the first row of the last round are known. According to the properties of $\chi$ as shown in Table 4, we can set up 1 linear equation for each Sbox, hence 64 linear equations in total between the input bits to the Sboxes of the last round and hash value. There are two methods to obtain extra 64 linear equations, as shown in Sect. 3.1, including guess-and-determine technique in Setting 1 and probabilistic linearization in Setting 2. For the former, we guess 32 bits input to $\chi$ of the last round and obtain 64 more linear equations, which will find the correct solution in $2^{32}$. For the latter, we exploit the probabilistic equations $b_i = a_i$'s each of which holds with probability 0.75. Since we have 64 probabilistic equations, the total probability of this system is $0.75^{64} = 2^{-26.6}$. We can expect a correct solution from $2^{26.6}$ such systems which can be obtained by changing the values of $A[0,4]$. Thus the complexity of this attack is $2^{26.6}$.

**Preimage attacks on 3-round KECCAK[$\mathbf{r = 1440, c = 160, \ell = 80}$].** Similar techniques as presented previously allow us to find solutions for the 3-round preimage challenge with width 1600 in the KECCAK $\texttt{Challenge}$ [4]. As shown in Fig. 11, we set the lanes with orange slashes of the first state to be variables. The 31st bit of $A[2,4]$ is set to 1 for ensuring that the state input to the first round complies with the padding. Finally, we get 161 degrees of freedom such that the bits input to $\chi$ of the last round are all linear. The sketch of the processing is shown in Fig. 11. According to the properties of $\chi$ as presented in Sect. 3.1, we can set up 16 linear equations between the bits input to the last $\chi$ and hash value. We can obtain extra $2 \times 64$ linear equations by guessing 64 bits input to the last $\chi$. Now, we build a linear system of $16 + 2 \times 64 = 144$ equations on 161 variables. Therefore, we immediately get correct solutions for any given hash value by solving this system. A solution for the 3-round preimage challenge with width 1600 is listed as below, where the message has length 1438 and each 64-bit word is expressed in hexadecimal.

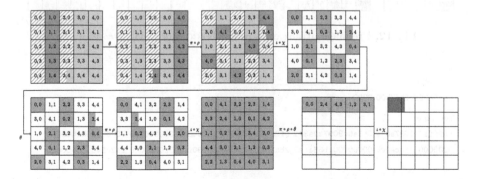

**Fig. 11.** Preimage attack on 3-round KECCAK[$r = 1440, c = 160, \ell = 80$]

Challenge:

```
e7cfc02846a32506 756c
```

Preimage:

```
01e0bc766796d36f ffffffffffffffff bd25fc21a299814e 0000000000000000 0000000000000000
cc85265f6f0e696a ffffffffffffffff 3a6f339c0eb075b9 0000000000000000 0000000000000000
d22ac7903b459dc2 ffffffffffffffff 903a19e9986a2ac7 0000000000000000 0000000000000000
539674b5f5e23187 ffffffffffffffff 1770d654e35ec89e 0000000000000000 0000000000000000
b326d6f339c0e9bf ffffffffffffffff d71d16ae
```

**Preimage attacks on 3-round** KECCAK$[r = 640, c = 160, \ell = 80]$. Similar techniques also allow us to find solutions for the 3-round preimage challenge with width 800. The sketch of the attack is shown in Fig. 12. To keep two rounds being linear, the six lanes with orange slashes of input state are expressed by 64 variables for any fixed values of auxiliary variables, and the two lanes with red grid, $A[3,0]$ and $A[4,3]$, are represented by 32 auxiliary variables. We set up 64 linear equations on 64 variables for a given 80-bit hash value by guessing 8 bits of the variables, and expect a correct preimage for $2^{16}$ tries. The time complexity of this attack is $2^{24}$. As a matter of fact, the time complexity can be further cut down to $2^7$ by applying a similar attack as described in Sect. 6.4. A solution for the 3-round preimage challenge with width 800 is listed as below, where the message has length 638.

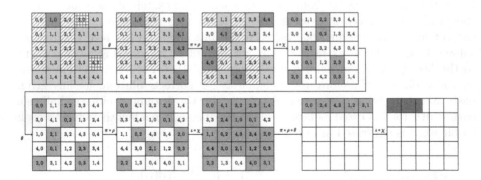

**Fig. 12.** Preimage attack on 3-round KECCAK$[r = 640, c = 160]$

Challenge:

```
0e668099c5b57b00 9302
```

Preimage:

```
ffffffff1097e68a 069e5c9097c2a342 9128124400000000 3bc3a3a300000000 0000000000000000
0000000056ace9cb 00000000cb56ace9 2ba3ccb200000000 990fc4d300000000 ff2c346d00000000
```

**Preimage Attacks on 3-Round** KECCAK-224 **and** KECCAK-256. Since the rates $r$ of KECCAK-224 and KECCAK-256 are much smaller than that of

SHAKE128, there are less choices of constant part when keeping two rounds being linear with as many degrees of freedom as possible. For KECCAK-256, we use 384 variables $A[i, j]$, $i = 0, 2, j = 0, 1, 2$, out of which there will be 64 independent variables after forcing the sum of variables in column $0, 2$ of the input to $\theta$ in the first round, and in column $0, 1, 2$ of the input to the $\theta$ in the second round to be constants as depicted in Fig. 13, *i.e.*, the size of this linear structure is $2^{64}$. However, it is insufficient to match a 256-bit hash value by 64-bit variables. To get enough choices for the state input to the first round, we set the constant part by using 128 auxiliary variables $A[3, 0]$ and $A[4, 2]$ such that the linear structure remains linear after two rounds for any fixed values of auxiliary variables. As depicted in Fig. 13, we required that the gray and lightgray lanes of the state after step $\theta$ of the first round are respectively ones and zeros. To achieve this, we first fix the values of $A[0, 3]$ and $A[3, 0]$, and then set up 192 linear equations, $\bigoplus_{j=0}^{4}(A[i - 1, j] \oplus (A[i + 1, j] \lll 1)) = \texttt{0xff} \cdots \texttt{f}, i = 1, 4$ and $\bigoplus_{j=0}^{4}(A[i - 1, j] \oplus (A[i + 1, j] \lll 1)) = 0, i = 3$, which implies that $A[4, 2]$ is determined by $A[3, 0]$. To make sure that the variables do not affect the other bits after step $\theta$ of the second round, we impose 192 more equations according to the value of $A[1, 2]$. Finally, the six lanes with orange slashes of input state can be expressed by 64 variables for any fixed values of auxiliary variables, and the two lanes with red grid can be represented by 64 auxiliary variables. As usual, we can set up 64 linear equations on these 64 variables for a given 256-bit hash value. Since there are $2^{64}$ choices for variable lanes, $2^{64}$ choices for auxiliary variable lanes, and $2^{128}$ choices for constant lanes, we have $2^{256}$ choices for the state input to the first round, and we expect a correct solution. The time complexity of this attack is $2^{192}$.

The preimage attack on KECCAK-224 is similar, as shown in Fig. 14. To keep two rounds being linear, the eight lanes with orange slashes of input state are expressed by 128 variables for any fixed values of auxiliary variables, and the four lanes with red grid are represented by 64 auxiliary variables. We set up 128 linear equations on 128 variables for a given 224-bit hash value (half solutions correspond to legal messages), and expect a correct preimage for $2^{97}$ tries. The time complexity of this attack is $2^{97}$.

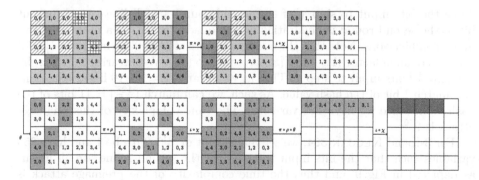

**Fig. 13.** Preimage attack on 3-round KECCAK-256

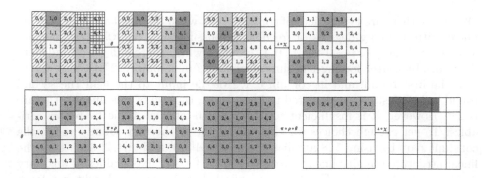

Fig. 14. Preimage attack on 3-round KECCAK-224

## 6.3 Preimage Attacks on 3-Round KECCAK-384/512 and 4-round KECCAK-224/256

For 3-round KECCAK-512, on one hand, we have 128 variables such that the bits input to step $\chi$ of the second round are all linear, as depicted in Fig. 8; on the other hand, we can directly inverse 320 bits through $\chi^{-1} \circ \iota^{-1}$ from a given hash value, each bit of which is a sum of 11 bits of the output of the second round. Since $\pi \circ \rho$ just permutate the positions of the bits and $\iota$ just add a constant to the first lane, they do not increase the nonlinear terms, and thus we neglect these steps in the last one and a half rounds.

$$M \xrightarrow[\text{1.5 rounds}]{\pi \circ \rho \circ \theta \circ R} A \xrightarrow{\iota \circ \chi} B \xrightarrow{\theta} C \xrightarrow{\pi \circ \rho} \mid \xleftarrow{\chi^{-1} \circ \iota^{-1}} h.$$

The expressions of $\theta$ and $\chi$ are given as follows,

$$\chi : \ B[x][y][z] = A[x][y][z] \oplus (A[x+1][y][z] \oplus 1) \cdot A[x+2][y][z],$$

$$\theta : \ C[x][y][z] = B[x][y][z] \oplus \bigoplus_{y'=0}^{4} B[x-1][y'][z] \oplus \bigoplus_{y'=0}^{4} B[x+1][y'][z-1]. \quad (14)$$

Since the bits input to step $\chi$ of the second round are all linear, each output bit of the second round is quadratic and the quadratic part is a product of two linear combinations. Note that the quadratic parts of $B[x][y][z]$ and $B[x-1][y][z]$ share a common factor $A[x+1][y][z]$ according to (14). We linearize $C[x][y][z]$ by guessing 10 bits input to step $\chi$. That is, we obtain $11 = 1 + 10$ linear equations and match 1 bit of the hash value. As such, we can match $\lfloor \frac{128}{11} \rfloor = 11$ bits of the hash value since we have 128 variables. The time complexity of this preimage attack is $2^{501}$.

For 3-round KECCAK-384, we set the last bit of $A[2,2]$ to be 1 and have 255 variables such that the bits input to step $\chi$ of the second round are all linear, as depicted in Fig. 9, and thus the time complexity of the preimage attack is $2^{384-\lfloor \frac{255}{11} \rfloor} = 2^{361}$.

For KECCAK-224/256, we cannot inverse the hash value through $\chi^{-1}$ as KECCAK-384/512, but we can set up the equations such as $a_0 = b_0$ for $b_1 = 1$ according to (6). Since we have 127 and 64 variables such that the bits input to step $\chi$ of the third round are all linear, as depicted in Figs. 14 and 13, the time complexities for 4-round KECCAK-224/256 are respectively $2^{213}$ and $2^{251}$.

**Improved Preimage Attacks on 3-Round KECCAK-384 and KECCAK-512.** In the above attacks, we assume that all the guessed linear combinations are linearly independent. It is possible to cut down the time complexity if elaborately choose linearly dependent ones, since there will be more degrees of freedom for guessing more linear combinations to match more bits of the hash value. For 3-round KECCAK-384/512, we can further improve the attacks by this method. Since we can inverse 320 bits of through $\chi^{-1} \circ \iota^{-1}$ from a given hash value, we can choose the bits which share a sum of one column (according to the property of $\theta$) or common linear parts in quadratic terms (according to the property of $\chi$).

By (14), $B[x - 1][y][z]$ and $B[x][y][z]$ are linear after guessing the value of $A[x + 1][y][z]$ for $0 \leq y \leq 4$. It is also true that $B[x + 1][y][z - 1]$ and $B[x + 2][y][z - 1]$ are linear after guessing the value of $A[x + 3][y][z - 1]$ for $0 \leq y \leq 4$. This means that after guessing the above 10 bits input to step $\chi$, we not only linearize $C[x][y][z]$, but also obtain an extra equation:

$$C[x + 1][y + 1][z] = B[x + 1][y + 1][z] \oplus \bigoplus_{y'=0}^{4} B[x][y'][z] \oplus \bigoplus_{y'=0}^{4} B[x + 2][y'][z - 1],$$

the quadratic part of which only appears in $B[x+1][y+1][z]$. Thus we can set up 2 extra linear equations and match one more bit of the hash value by guessing one more bit. Totally we set up 13 linear equations and match two bits of the given hash value.

Then we consider another two equations:

$$C[x + 2][y + 2][z - 1] = B[x + 2][y + 2][z - 1] \oplus \bigoplus_{y'=0}^{4} B[x + 1][y'][z - 1] \oplus \bigoplus_{y'=0}^{4} B[x + 3][y'][z - 2],$$

$$C[x + 3][y + 3][z - 1] = B[x + 3][y + 3][z - 1] \oplus \bigoplus_{y'=0}^{4} B[x + 2][y'][z - 1] \oplus \bigoplus_{y'=0}^{4} B[x + 4][y'][z - 2].$$

Again, we can set up another 8 linear equations and match two more bits of the hash value by guessing 6 more bits.

Generally, we can match $2\lfloor \frac{t-5}{8} \rfloor$ bits of a given hash value if we have $t$ variables. For 3-round KECCAK-384/512, we have 255 and 128 variables, and thus match 62 and 30 bits respectively. Therefore, the time complexities of this improved preimage attack are respectively $2^{322}$ and $2^{482}$ for 3-round KECCAK-384/512.

## 6.4 Improved Preimage Attacks on SHAKE128

The idea presented in Sect. 6.3 also applies to SHAKE128. In this section we extend it to improve the preimage attacks on SHAKE128.

Instead of linearizing 2 rounds forward, we linearize 2 rounds by combining one round forward and one round backward as discussed in Sect. 4.1, and we have 512 variables such that these two rounds are linear. To make sure that the state input to the first round corresponds to a legal message, we set up 262 linear equations such that the last 256 bits are all zeros and the following last 6 bits are all ones. Then there remains 250 degrees of freedom such that the bits input to step $\chi$ of the third round are all linear.

For 3-round SHAKE128, we set up 64 linear equations between these 250 variables and a given hash value as the same way done in Sect. 6.2, and then obtain extra $2 \times 64$ linear equations by guessing 64 bits input to step $\chi$ of the third round. Each solution of this linear system corresponds to a preimage of the given hash value. Therefore, the time complexity of this attack is 1.

For 4-round SHAKE128, given a 128-bit hash value, we expect 32 zeros and 32 ones among its last 64 bits ($b_1$'s), and thus we can set up a linear system, which matches 22 bits ($b_0$'s) of the hash value, by guessing 220 bits input to step $\chi$ of the third round. This attack gives a correct preimage in $2^{106}$.

## 6.5  Preimage Attacks on 4-Round KECCAK $[r = 1440, C = 160, \ell = 80]$

A similar attack as proposed in Sect. 6.4 also applies to KECCAK$[r = 1440, c = 160, \ell = 80]$. In stead, we use two rounds forward and one round backward for linearization. As shown in Sect. 4.2, we have 194 degrees of freedom for such 3-round linear structure. To make sure that the state input to the first round corresponds to a legal message, we set up 161 linear equations such that the last 161 bits are fixed. Then there remains 33 degrees of freedom such that the bits input to step $\chi$ of the fourth round are all linear. Given an 80-bit hash value of 4-round KECCAK$[r = 1440, c = 160, \ell = 80]$, we can set up 16 linear equations by (6), and set up 17 probabilistic equations using $b_i = a_i$. This attack gives a correct two-block preimage in $2^{47+17 \times 0.42} \approx 2^{54}$. We estimate that the computations of the whole attack need approximately $2^{20}$ CPU core hours. We run this attack in less than $2^{10}$ CPU core hours, and find a 78-bit matched preimage of length 2874 for the 4-round KECCAK preimage challenge with width 1600.

Message:
bc739847dd59b8f6 21e6f9016ae9292d 44c2f9f008f175fc fb1a9d7d2f5af0d9 c709f78dfa830460
0000000000000000 0000000000000000 0000000000000000 0000000000000000 0000000000000000
0000000000000000 0000000000000000 0000000000000000 0000000000000000 0000000000000000
0000000000000000 0000000000000000 0000000000000000 0000000000000000 0000000000000000
0000000000000000 0000000000000000 00000000
34d781770fae25d9 4bcdf7304704b1a0 aeb1cc6a3d9a4b9f 879b5b095e744910 09096232b744ac44
63faab93d1b6a3f5 7aca93b5c0c2afa0 f1b2772194934266 41e5a573d5efc16f 34e0e077bfb4ce43
48bb5cb11aa15738 3ecb466e4aa6fec3 4e3e5449626d5e2d ccec6be24c92d63b fb652d66cc6a4621
356d6bfdd56b1afb d9da9b8c0e366cd3 034ad6fdd9caa885 236ade6960c8edaf 03d6d60e45aeb00e
b8132036d4e20f33 8e4a29bbbd2c1cb8 8549b303

Output:

7d aa d8 07 b0 50 6c 9c 02 76

Challenge:

7d aa d8 07 f8 50 6c 9c 02 76

Difference:

-- -- -- -- 48 -- -- -- -- --

# 7 Conclusions

In conclusion, we have described the linear structures of KECCAK-$f$ and exploited them to analyze the security of KECCAK, including zero-sum distinguishers on KECCAK-$f$ permutation and preimage attacks on KECCAK. Our distinguishers work on KECCAK-$f$ reduced to up to 15 rounds, and are practical for up to 11 rounds. These results improve the previously best known distinguishers by two more rounds with the same complexities. Our preimage attacks work on all variants of KECCAK reduced to up to 4 rounds except for 4-round KECCAK-384/512, much faster than the exhaustive search. Specially, in terms of practical preimage attacks, we could find the preimage by solving a small linear system just once for 2-round KECCAK-224/256 and 3-round SHAKE128. With these techniques, we have found preimages for 3-round KECCAK Challenge with widths 1600 and 800, and a 78-bit matched preimage for 4-round KECCAK Challenge with width 1600. It will be interesting to see applications of linear structures to other KECCAK-like ciphers or functions.

**Acknowledgement.** We are grateful to Florian Mendel, Lei Wang, and anonymous reviewers of ASIACRYPT 2016 for their fruitful discussions and helpful comments. The second author was supported by the National Natural Science Foundation of China (Grant Nos. 61672516, 61303258, 61379139 and 11526215) and the Strategic Priority Research Program of the Chinese Academy of Sciences under Grant XDA06010701.

# References

1. Aoki, K., Guo, J., Matusiewicz, K., Sasaki, Y., Wang, L.: Preimages for step-reduced SHA-2. In: Matsui, M. (ed.) ASIACRYPT 2009. LNCS, vol. 5912, pp. 578–597. Springer, Heidelberg (2009). doi:10.1007/978-3-642-10366-7_34
2. Aumasson, J.P., Meier, W.: Zero-sum distinguishers for reduced Keccak-f and for the core functions of Luffa and Hamsi (2009). https://131002.net/data/papers/AM09.pdf
3. Bernstein, D.J.: Second Preimages for 6 (7?(8??)) Rounds of Keccak. NIST mailing list (2010)
4. Bertoni, G., Daemen, J., Peeters, M., Van Assche, G.: Keccak crunchy crypto collision and pre-image contest. http://keccak.noekeon.org/crunchy_contest.html
5. Bertoni, G., Daemen, J., Peeters, M., Van Assche, G.: Cryptographic sponge functions, January 2011. http://sponge.noekeon.org/CSF-0.1.pdf
6. Bertoni, G., Daemen, J., Peeters, M., Van Assche, G.: The Keccak reference, Version 3.0, January 2011. http://keccak.noekeon.org

7. Bertoni, G., Daemen, J., Peeters, M., Van Assche, G., Van Keer, R.: CAESAR submission: KETJE v1, March 2014. http://ketje.noekeon.org
8. Bertoni, G., Daemen, J., Peeters, M., Van Assche, G., Van Keer, R.: CAESAR submission: Keyak v2, December 2015. http://keyak.noekeon.org/
9. Boura, C., Canteaut, A., Cannière, C.: Higher-order differential properties of KECCAK and Luffa. In: Joux, A. (ed.) FSE 2011. LNCS, vol. 6733, pp. 252–269. Springer, Heidelberg (2011). doi:10.1007/978-3-642-21702-9_15
10. Canteaut, A. (ed.): FSE 2012. LNCS, vol. 7549. Springer, Heidelberg (2012)
11. Chang, D., Kumar, A., Morawiecki, P., Sanadhya, S.K.: 1st and 2nd preimage attacks on 7, 8 and 9 rounds of Keccak-224,256,384,512. In: SHA-3 Workshop, August 2014
12. Dinur, I., Dunkelman, O., Shamir, A.: New attacks on Keccak-224 and Keccak-256. In: Canteaut, A. (ed.) FSE 2012. LNCS, vol. 7549, pp. 442–461. Springer, Heidelberg (2012). doi:10.1007/978-3-642-34047-5_25
13. Dinur, I., Dunkelman, O., Shamir, A.: Collision attacks on up to 5 rounds of SHA-3 using generalized internal differentials. In: Moriai, S. (ed.) FSE 2013. LNCS, vol. 8424, pp. 219–240. Springer, Heidelberg (2014). doi:10.1007/978-3-662-43933-3_12
14. Dinur, I., Dunkelman, O., Shamir, A.: Improved practical attacks on round-reduced Keccak. J. Cryptol. **27**(2), 183–209 (2014)
15. Dinur, I., Morawiecki, P., Pieprzyk, J., Srebrny, M., Straus, M.: Cube attacks and cube-attack-like cryptanalysis on the round-reduced Keccak sponge function. In: Oswald, E., Fischlin, M. (eds.) EUROCRYPT 2015. LNCS, vol. 9056, pp. 733–761. Springer, Heidelberg (2015). doi:10.1007/978-3-662-46800-5_28
16. Duan, M., Lai, X.: Improved zero-sum distinguisher for full round Keccak-f permutation. Cryptology ePrint Archive, Report 2011/023 (2011). http://eprint.iacr.org/
17. Duc, Alexandre, Guo, Jian, Peyrin, Thomas, Wei, Lei: Unaligned rebound attack: application to Keccak. In: [10] 402–421
18. Guo, J., Ling, S., Rechberger, C., Wang, H.: Advanced meet-in-the-middle preimage attacks: first results on full tiger, and improved results on MD4 and SHA-2. In: Abe, M. (ed.) ASIACRYPT 2010. LNCS, vol. 6477, pp. 56–75. Springer, Heidelberg (2010). doi:10.1007/978-3-642-17373-8_4
19. Jean, J., Nikolić, I.: Internal differential boomerangs: practical analysis of the round-reduced Keccak-$f$ permutation. In: Leander, G. (ed.) FSE 2015. LNCS, vol. 9054, pp. 537–556. Springer, Heidelberg (2015). doi:10.1007/978-3-662-48116-5_26
20. Morawiecki, P., Pieprzyk, J., Srebrny, M.: Rotational Cryptanalysis of Round-Reduced Keccak. In: [22] 241–262
21. Morawiecki, P., Srebrny, M.: A SAT-based preimage analysis of reduced Keccak hash functions. Inf. Process. Lett. **113**(10–11), 392–397 (2013)
22. Morawiecki, P., Pieprzyk, J., Srebrny, M.: Rotational cryptanalysis of round-reduced KECCAK. In: Moriai, S. (ed.) FSE 2013. LNCS, vol. 8424, pp. 241–262. Springer, Heidelberg (2014). doi:10.1007/978-3-662-43933-3_13
23. Naya-Plasencia, M., Röck, A., Meier, W.: Practical analysis of reduced-round KECCAK. In: Bernstein, D.J., Chatterjee, S. (eds.) INDOCRYPT 2011. LNCS, vol. 7107, pp. 236–254. Springer, Heidelberg (2011). doi:10.1007/978-3-642-25578-6_18
24. NIST: SHA-3 COMPETITION (2007–2012). http://csrc.nist.gov/groups/ST/hash/sha-3/index.html
25. The U.S. National Institute of Standards and Technology: SHA-3 Standard: Permutation-Based Hash and Extendable-Output Functions . Federal Information Processing Standard, FIPS 202, 5th August 2015

# Randomness

# When Are Fuzzy Extractors Possible?

Benjamin Fuller[1]($\boxtimes$), Leonid Reyzin[2], and Adam Smith[3]

[1] University of Connecticut, Storrs, CT, USA
benjamin.fuller@uconn.edu
[2] Boston University, Boston, MA, USA
reyzin@cs.bu.edu
[3] Pennsylvania State University, University Park, PA, USA
asmith@cse.psu.edu

**Abstract.** Fuzzy extractors (Dodis et al., Eurocrypt 2004) convert repeated noisy readings of a high-entropy secret into the same uniformly distributed key. A minimum condition for the security of the key is the hardness of guessing a value that is similar to the secret, because the fuzzy extractor converts such a guess to the key.

We define *fuzzy min-entropy* to quantify this property of a noisy source of secrets. Fuzzy min-entropy measures the success of the adversary when provided with *only* the functionality of the fuzzy extractor, that is, the *ideal* security possible from a noisy distribution. High fuzzy min-entropy is necessary for the existence of a fuzzy extractor.

We ask: *is high fuzzy min-entropy a sufficient condition for key extraction from noisy sources?* If only computational security is required, recent progress on program obfuscation gives evidence that fuzzy min-entropy is indeed sufficient. In contrast, information-theoretic fuzzy extractors are not known for many practically relevant sources of high fuzzy min-entropy.

In this paper, we show that fuzzy min-entropy is *sufficient* for information theoretically secure fuzzy extraction. For every source distribution $W$ for which security is possible we give a secure fuzzy extractor.

Our construction relies on the fuzzy extractor knowing the precise distribution of the source $W$. A more ambitious goal is to design a single extractor that works for all possible sources. Our second main result is that this more ambitious goal is impossible: we give a family of sources with high fuzzy min-entropy for which no single fuzzy extractor is secure. We show three flavors of this impossibility result: for standard fuzzy extractors, for fuzzy extractors that are allowed to sometimes be wrong, and for secure sketches, which are the main ingredient of most fuzzy extractor constructions.

**Keywords:** Fuzzy extractors · Secure sketches · Information theory · Biometric authentication · Error-tolerance · Key derivation · Error-correcting codes

© International Association for Cryptologic Research 2016
J.H. Cheon and T. Takagi (Eds.): ASIACRYPT 2016, Part I, LNCS 10031, pp. 277–306, 2016.
DOI: 10.1007/978-3-662-53887-6_10

# 1    Introduction

Sources of reproducible secret random bits are necessary for many cryptographic applications. In many situations these bits are not explicitly stored for future use, but are obtained by repeating the same process (such as reading a biometric or a physically unclonable function) that generated them the first time. However, bits obtained this way present a problem: noise [4,8,12,14,19,30,31,33,37,39,43]. That is, when a secret is read multiple times, readings are close (according to some metric) but not identical. To utilize such sources, it is often necessary to remove noise, in order to derive the same value in subsequent readings.

The same problem occurs in the interactive setting, in which the secret channel used for transmitting the bits between two users is noisy and/or leaky [42]. Bennett, Brassard, and Robert [4] identify two fundamental tasks. The first, called information reconciliation, removes the noise without leaking significant information. The second, known as privacy amplification, converts the high entropy secret to a uniform random value. In this work, we consider the noninteractive version of these problems, in which these tasks are performed together with a single message.

The noninteractive setting is modeled by a primitive called a fuzzy extractor [13], which consists of two algorithms. The generate algorithm (Gen) takes an initial reading $w$ and produces an output key along with a nonsecret helper value $p$. The reproduce (Rep) algorithm takes the subsequent reading $w'$ along with the helper value $p$ to reproduce key. The correctness guarantee is that the key is reproduced precisely when the distance between $w$ and $w'$ is at most $t$.

The security requirement for fuzzy extractors is that key is uniform even to a (computationally unbounded) adversary who has observed $p$. This requirement is harder to satisfy as the allowed error tolerance $t$ increases, because it becomes easier for the adversary to guess key by guessing a $w'$ within distance $t$ of $w$ and running $\mathsf{Rep}(w', p)$.

**Fuzzy Min-Entropy.** We introduce a new entropy notion that precisely measures how hard it is for the adversary to guess a value within distance $t$ of the original reading $w$. Suppose $w$ is sampled from a distribution $W$. To have the maximum chance that $w'$ is within distance $t$ of $w$, the adversary would want to maximize the total probability mass of $W$ within the ball $B_t(w')$ of radius $t$ around $w'$. We therefore define *fuzzy min-entropy*

$$\mathrm{H}^{\mathsf{fuzz}}_{t,\infty}(W) \overset{\text{def}}{=} -\log \max_{w'} \Pr[W \in B_t(w')].$$

The security of the resulting key cannot exceed the fuzzy min-entropy (Proposition 1).

However, existing constructions do not measure their security in terms of fuzzy min-entropy; instead, their security is shown to be the min-entropy of $W$, denoted $\mathrm{H}_\infty(W)$, minus some loss, for error-tolerance, that is at least $\log |B_t|$.[1]

---

[1] We omit $w$ in the notation $|B_t|$ since, as with almost all previous work, we study metrics where the volume of the ball $B_t(w)$ does not depend on the center $w$.

Since (trivially) $H_\infty(W) - \log|B_t| \leq H_{t,\infty}^{\text{fuzz}}(W)$, it is natural to ask whether this loss is necessary. This question is particularly relevant when the gap between the two sides of the inequality is high.[2] As an example, iris scans appear to have significant $H_{t,\infty}^{\text{fuzz}}(W)$ (because iris scans for different people appear to be well-spread in the metric space [11]) but negative $H_\infty(W) - \log|B_t|$ [6, Sect. 5]. We therefore ask: *is fuzzy min-entropy sufficient for fuzzy extraction?* There is evidence that it may be sufficient when the security requirement is computational rather than information-theoretic—see Sect. 1.2. We provide an answer for the case of information-theoretic security in two settings.

**Contribution 1: Sufficiency of $H_{t,\infty}^{\text{fuzz}}(W)$ for a Precisely Known $W$.** It should be easier to construct a fuzzy extractor when the designer has *precise knowledge* of the probability distribution function of $W$. In this setting, we show that it is possible to construct a fuzzy extractor that extracts a key almost as long as $H_{t,\infty}^{\text{fuzz}}(W)$ (Theorem 1). Our construction crucially utilizes the probability distribution function of $W$ and, in particular, cannot necessarily be realized in polynomial time (this is similar, for example, to the interactive information-reconciliation feasibility result of [34]). This result shows that $H_{t,\infty}^{\text{fuzz}}(W)$ is a necessary and sufficient condition for building a fuzzy extractor for a given distribution $W$.

A number of previous works in the precise knowledge setting have provided efficient algorithms and tight bounds for specific distributions—generally the uniform distribution or i.i.d. sequences (for example, [20,26–28,38,41]). Our characterization unifies previous work, and justifies using $H_{t,\infty}^{\text{fuzz}}(W)$ as the measure of the quality of a noisy distribution, rather than cruder measures such as $H_\infty(W) - \log|B_t|$. Our construction can be viewed as a reference to evaluate the quality of efficient constructions in the precise knowledge setting by seeing how close they get to extracting all of $H_{t,\infty}^{\text{fuzz}}(W)$.

**Contribution 2: The Cost of Distributional Uncertainty.** Assuming precise knowledge of a distribution $W$ is often unrealistic for high-entropy distributions; they can never be fully observed directly and must therefore be modeled. It is imprudent to assume that the designer's model of a distribution is completely accurate—the adversary, with greater resources, would likely be able to build a better model. (In particular, the adversary has more time to build the model after a particular construction is deployed.) Because of this, existing designs work for a family of sources (for example, all sources of min-entropy at least $m$ with at most $t$ errors). The fuzzy extractor is designed given only knowledge of the family. The attacker may know more about the distribution than the designer. We call this the *distributional uncertainty* setting.

Our second contribution is a set of negative results for this more realistic setting. We provide two impossibility results for fuzzy extractors. Both demonstrate families $\mathcal{W}$ of distributions over $\{0,1\}^n$ such that each distribution in

---

[2] For nearly uniform distributions, $H_{t,\infty}^{\text{fuzz}}(W) \approx H_\infty(W) - \log|B_t|$. In this setting, standard coding based constructions of fuzzy extractors (using appropriate codes) yield keys of size approximately $H_{t,\infty}^{\text{fuzz}}(W)$.

the family has $\mathrm{H}_{t,\infty}^{\mathrm{fuzz}}$ linear in $n$, but no fuzzy extractor can be secure for most distributions in $\mathcal{W}$. Thus, a fuzzy extractor designer who knows only that the distribution comes from $\mathcal{W}$ is faced with an impossible task, even though our positive result, Theorem 1, shows that fuzzy extractors can be designed for each distribution in the family individually.

The first impossibility result (Theorem 2) assumes that Rep is perfectly correct and rules our fuzzy extractors for entropy rates as high as $\mathrm{H}_{t,\infty}^{\mathrm{fuzz}}(W) \approx 0.18n$. The second impossibility result (Theorem 3), relying on the work of Holenstein and Renner [25], also rules out fuzzy extractors in which Rep is allowed to make a mistake, but applies only to distributions with entropy rates up to $\mathrm{H}_{t,\infty}^{\mathrm{fuzz}}(W) \approx 0.07n$.

We also provide a third impossibility result (Theorem 4), this time for an important building block called "secure sketch," which is used in most fuzzy extractor constructions (in order to allow Rep to recover the original $w$ from the input $w'$). The result rules out secure sketches for a family of distributions with entropy rate up to $0.5n$, even if the secure sketches are allowed to make mistakes. Because secure sketches are used in most fuzzy extractors constructions, the result suggests that building a fuzzy extractor for this family will be very difficult. We define secure sketches formally in Sect. 7.

These impossibility results motivate further research into computationally, rather information-theoretically, secure fuzzy extractors (Sect. 1.2).

## 1.1   Our Techniques

**Techniques for Positive Results for a Precisely Known Distribution.** We now explain how to construct a fuzzy extractor for a precisely known distribution $W$ with fuzzy min-entropy. We begin with distributions in which all points in the support have the same probability (so-called "flat" distributions). Gen simply extracts a key from the input $w$ using a randomness extractor. Consider some subsequent reading $w'$. To achieve correctness, the string $p$ must permit Rep to disambiguate which point $w \in W$ within distance $t$ of $w'$ was given to Gen. Disambiguating multiple points can be accomplished by universal hashing, as long as the size of hash output space is slightly greater than the number of possible points. Thus, Rep includes into the public value $p$ a "sketch" of $w$ computed via a universal hash of $w$. To determine the length of that sketch, consider the heaviest (according to $W$) ball $B^*$ of radius $t$. Because the distribution is flat, $B^*$ is also the ball with the most points of nonzero probability. Thus, the length of the sketch needs to be slightly greater than the logarithm of the number of non-zero probability points in $B^*$. Since $\mathrm{H}_{t,\infty}^{\mathrm{fuzz}}(W)$ is determined by the weight of $B^*$, the number of points cannot be too high and there will be entropy left after the sketch is published. This remaining entropy suffices to extract a key.

For an arbitrary distribution, we cannot afford to disambiguate points in the ball with the greatest number of points, because there could be too many low-probability points in a single ball despite a high $\mathrm{H}_{t,\infty}^{\mathrm{fuzz}}(W)$. We solve this problem by splitting the arbitrary distribution into a number of nearly flat distributions we call "levels." We then write down, as part of the sketch, the level of the original reading $w$ and apply the above construction considering only points in that level. We call this construction *leveled hashing* (Construction 1).

**Techniques for Negative Results for Distributional Uncertainty.** We construct a family of distributions $\mathcal{W}$ and prove impossibility for a uniformly random $W \leftarrow \mathcal{W}$. We start by observing the following asymmetry: Gen sees only the sample $w$ (obtained via $W \leftarrow \mathcal{W}$ and $w \leftarrow W$), while the adversary knows $W$.

To exploit the asymmetry, in our first impossibility result (Theorem 2), we construct $\mathcal{W}$ so that conditioning on the knowledge of $W$ reduces the distribution to a small subspace (namely, all points on which a given hash function produces a given output), but conditioning on *only* $w$ leaves the rest of the distribution uniform on a large fraction of the entire space. An adversary can exploit the knowledge of the hash value to reduce the uncertainty about key, as follows.

The nonsecret value $p$ partitions the metric space into regions that produce a consistent value under Rep (preimages of each key under $\mathsf{Rep}(\cdot, p)$). For each of these regions, the adversary knows that possible $w$ lie at distance at least $t$ from the boundary of the region (else, the fuzzy extractor would have a nonzero probability of error). However, in the Hamming space, the vast majority of points lie near the boundary (this result follows by combining the isoperimetric inequality [21], which shows that the ball has the smallest boundary, with bounds on the volume of the interior of a ball, which show that this boundary is large). This allows the adversary to rule out so many possible $w$ that, combined with the adversarial knowledge of the hash value, many regions become empty, leaving key far from uniform.

For the second impossibility result (Theorem 3, which rules out even fuzzy extractors that are allowed a possibility of error), we let the adversary know some fraction of the bits of $w$. Holenstein and Renner [25] showed that if the adversary knows each bit of $w$ with sufficient probability, and bits of $w'$ differ from bits of $w$ with sufficient probability, then so-called "information-theoretic key agreement" is impossible. Converting the impossibility of information-theoretic key agreement to impossibility of fuzzy extractors takes a bit of technical work.

## 1.2   Related Settings

**Other Settings with Close Readings: $\mathrm{H}_{t,\infty}^{\mathrm{fuzz}}$ is Sufficient.** The security definition of fuzzy extractors can be weakened to protect only against computationally bounded adversaries [17]. In this computational setting, for most distance metrics a single fuzzy extractor can simultaneously secure all possible distributions by using virtual grey-box obfuscation for all circuits in $\mathrm{NC}^1$ [5]. This construction is secure when the adversary can rarely learn key with oracle access to the program functionality. The set of distributions with fuzzy min-entropy are exactly those where an adversary learns key with oracle access to the functionality with negligible probability. Thus, extending our negative result to the computational setting would have negative implications on the existence of obfuscation.

Furthermore, the functional definition of fuzzy extractors can be weakened to permit interaction between the party having $w$ and the party having $w'$. Such a weakening is useful for secure remote authentication [7]. When both interaction and computational assumptions are allowed, secure two-party computation can produce a key that will be secure whenever the distribution $W$ has fuzzy min-entropy. The two-party computation protocol needs to be secure without assuming authenticated channels; it can be built under the assumptions that collision-resistant hash functions and enhanced trapdoor permutations exist [3].

**Correlated Rather than Close Readings.** A different model for the problem of key derivation from noisy sources does not explicitly consider the distance between $w$ and $w'$, but rather views $w$ and $w'$ as samples of drawn from a correlated pair of random variables. This model is considered in multiple works, including [1,10,29,42]; recent characterizations of when key derivation is possible in this model include [35,40]. In particular, Hayashi et al. [22] independently developed an interactive technique similar to our non-interactive leveled hashing, which they called "spectrum slicing." To the best of our knowledge, prior results on correlated random variables are in the precise knowledge setting; we are unaware of works that consider the cost of distributional uncertainty.

# 2 Preliminaries

**Random Variables.** We generally use uppercase letters for random variables and corresponding lowercase letters for their samples. A repeated occurrence of the same random variable in a given expression signifies the same value of the random variable: for example $(W, \mathsf{SS}(W))$ is a pair of random variables obtained by sampling $w$ according to $W$ and applying the algorithm $\mathsf{SS}$ to $w$.

The *statistical distance* between random variables $A$ and $B$ with the same domain is $\mathbf{SD}(A,B) = \frac{1}{2}\sum_a |\Pr[A=a] - \Pr[B=b]| = \max_S \Pr[A \in S] - \Pr[B \in S]$.

**Entropy.** Unless otherwise noted logarithms are base 2. Let $(X,Y)$ be a pair of random variables. Define *min-entropy* of $X$ as $\mathrm{H}_\infty(X) = -\log(\max_x \Pr[X = x])$, and the *average (conditional)* min-entropy of $X$ given $Y$ as $\tilde{\mathrm{H}}_\infty(X|Y) = -\log(\mathbb{E}_{y \in Y} \max_x \Pr[X = x|Y = y])$ [13, Sect. 2.4]. Define Hartley entropy $H_0(X)$ to be the logarithm of the size of the support of $X$, that is $H_0(X) = \log|\{x|\Pr[X = x] > 0\}|$. Define average-case Hartley entropy by averaging the support size: $\tilde{H}_0(X|Y) = \log(\mathbb{E}_{y \in Y}|\{y|\Pr[X = x|Y = y] > 0\}|)$. For $0 < a < 1$, define the binary entropy $h_2(p) = -p \log p - (1-p) \log(1-p)$ as the Shannon entropy of any random variable that is 0 with probability $p$ and 1 with probability $1 - p$.

**Randomness Extractors.** We use randomness extractors [32], as defined for the average case in [13, Sect. 2.5].

**Definition 1.** *Let* $\mathcal{M}$, $\chi$ *be finite sets. A function* $\mathsf{ext} : \mathcal{M} \times \{0,1\}^d \to \{0,1\}^\kappa$ *a* $(\tilde{m}, \epsilon)$ *-average case extractor if for all pairs of random variables* $X, Y$ *over* $\mathcal{M}, \chi$ *such that* $\tilde{H}_\infty(X|Y) \geq \tilde{m}$, *we have*

$$\mathbf{SD}((\mathsf{ext}(X, U_d), U_d, Y), U_\kappa \times U_d \times Y) \leq \epsilon.$$

**Metric Spaces and Balls.** For a metric space $(\mathcal{M}, \mathsf{dis})$, the *(closed) ball of radius* $t$ *around* $w$ is the set of all points within radius $t$, that is, $B_t(w) = \{w' | \mathsf{dis}(w, w') \leq t\}$. If the size of a ball in a metric space does not depend on $w$, we denote by $|B_t|$ the size of a ball of radius $t$. We consider the Hamming metric over vectors in $\mathcal{Z}^n$ for some finite alphabet $\mathcal{Z}$, defined via $\mathsf{dis}(w, w') = |\{i | w_i \neq w'_i\}|$. $U_\kappa$ denotes the uniformly distributed random variable on $\{0,1\}^\kappa$.

We will use the following bounds on $|B_t|$ in $\{0,1\}^n$, see [2, Lemma 4.7.2, Eq. 4.7.5, p. 115] for proofs.

**Lemma 1.** *Let* $\tau = t/n$. *The volume* $|B_t|$ *of the ball of radius in* $t$ *in the Hamming space* $\{0,1\}^n$ *satisfies*

$$\frac{1}{\sqrt{8n\tau(1-\tau)}} \cdot 2^{nh_2(\tau)} \leq |B_t| \leq 2^{nh_2(\tau)}.$$

## 2.1 Fuzzy Extractors

In this section, we define fuzzy extractors, slightly modified from the work of Dodis et al. [13, Sect. 3.2]. First, we allow for error as discussed in [13, Sect. 8]. Second, in the *distributional uncertainty* setting we consider a general family $\mathcal{W}$ of distributions instead of families containing all distributions of a given min-entropy. Let $\mathcal{M}$ be a metric space with distance function $\mathsf{dis}$.

**Definition 2.** *An* $(\mathcal{M}, \mathcal{W}, \kappa, t, \epsilon)$-*fuzzy extractor with error* $\delta$ *is a pair of randomized procedures, "generate"* (Gen) *and "reproduce"* (Rep). *Gen on input* $w \in \mathcal{M}$ *outputs an extracted string* key $\in \{0,1\}^\kappa$ *and a helper string* $p \in \{0,1\}^*$. *Rep takes* $w' \in \mathcal{M}$ *and* $p \in \{0,1\}^*$ *as inputs.* (Gen, Rep) *have the following properties:*

1. *Correctness: if* $\mathsf{dis}(w, w') \leq t$ *and* (key, $p$) $\leftarrow$ Gen($w$), *then* $\Pr[\mathsf{Rep}(w', p) = \mathsf{key}] \geq 1 - \delta$.
2. *Security: for any distribution* $W \in \mathcal{W}$, *if* (Key, $P$) $\leftarrow$ Gen($W$), *then* $\mathbf{SD}((\mathsf{Key}, P), (U_\kappa, P)) \leq \epsilon$.

In the above definition, the errors must be chosen before $p$ is known in order for the correctness guarantee to hold.

**The Case of a Precisely Known Distribution.** If in the above definition we take $\mathcal{W}$ to be a one-element set containing a single distribution $W$, then the fuzzy extractor is said to be for a *precisely known distribution*. In this case, we need to require correctness only for $w$ that have nonzero probability. Note that we have no requirement that the algorithms are compact or efficient, and so the distribution can be fully known to them.

# 3    New Notion: Fuzzy Min-Entropy

The fuzzy extractor helper string $p$ allows everyone, including the adversary, to find the output of $\mathsf{Rep}(\cdot, p)$ on any input $w'$. Ideally, $p$ should not provide any useful information beyond this ability, and the outputs of $\mathsf{Rep}$ on inputs that are too distant from $w$ should provide no useful information, either. In this ideal scenario, the adversary is limited to trying to guess a $w'$ that is $t$-close to $w$. Letting $w'$ be the center of the maximum-weight ball in $W$ is optimal, we measure the quality of a source by (the negative logarithm of) this weight.

**Definition 3.** *The $t$-fuzzy min-entropy of a distribution $W$ in a metric space $(\mathcal{M}, \mathsf{dis})$ is:*

$$\mathrm{H}_{t,\infty}^{\mathbf{fuzz}}(W) = -\log \left( \max_{w'} \sum_{w \in \mathcal{M} \mid \mathsf{dis}(w,w') \leq t} \Pr[W = w] \right)$$

Fuzzy min-entropy measures the functionality provided to the adversary by $\mathsf{Rep}$ (since $p$ is public), and thus is a necessary condition for security. We formalize this statement in the following proposition.

**Proposition 1.** *Let $W$ be a distribution over $(\mathcal{M}, \mathsf{dis})$ with $\mathrm{H}_{t,\infty}^{\mathbf{fuzz}}(W) = m$. Let $(\mathsf{Gen}, \mathsf{Rep})$ be a $(\mathcal{M}, \{W\}, \kappa, t, \epsilon)$-fuzzy extractor with error $\delta$. Then*

$$2^{-\kappa} \geq 2^{-m} - \delta - \epsilon.$$

*If $\delta = \epsilon = 2^{-\kappa}$, then $\kappa$ cannot exceed $m + 2$. Additionally, if fuzzy min-entropy of the source is only logarithmic in a security parameter while the $\delta$ and $\epsilon$ parameters are negligible, then extracted key must be of at most logarithmic length.*

*Proof.* Let $W$ be a distribution where $\mathrm{H}_{t,\infty}^{\mathbf{fuzz}}(W) = m$. This means that there exists a point $w' \in \mathcal{M}$ such that $\Pr_{w \in W}[\mathsf{dis}(w, w') \leq t] = 2^{-m}$. Consider the following distinguisher $D$: on input $(\mathsf{key}, p)$, if $\mathsf{Rep}(w', p) = \mathsf{key}$, then output 1, else output 0.
$\Pr[D(\mathsf{Key}, P) = 1] \geq 2^{-m} - \delta$, while $\Pr[D(U_\kappa, P) = 1] = 1/2^{-\kappa}$. Thus,

$$\mathbf{SD}((\mathsf{Key}, P), (U_\kappa, P)) \geq \delta^D((\mathsf{Key}, P), (U_\kappa, P)) \geq 2^{-m} - \delta - 2^{-\kappa}. \qquad \square$$

Proposition 1 extends to the settings of computational security and interactive protocols. Fuzzy min-entropy represents an upper bound on the security from a noisy source. However, there are many distributions with fuzzy min-entropy with no known information-theoretically secure fuzzy extractor (or corresponding impossibility result).

We explore other properties of fuzzy min-entropy, not necessary for the proofs presented here, in the full version [18, Appendix E].

# 4  $\mathrm{H}_{t,\infty}^{\mathrm{fuzz}}(W)$ is Sufficient in the Precise Knowledge Setting

In this section, we build fuzzy extractors that extract almost all of $\mathrm{H}_{t,\infty}^{\mathrm{fuzz}}(W)$ for any distribution $W$. We reiterate that these constructions assume precise knowledge of $W$ and are not necessarily polynomial-time. They should thus be viewed as feasibility results. We begin with flat distributions and then turn to arbitrary distributions.

## 4.1  Warm-Up for Intuition: Fuzzy Extractor for Flat Distributions

Let $\mathrm{supp}(W) = \{w | \Pr[W = w] > 0\}$ denote the support of a distribution $W$. A distribution $W$ is *flat* if all elements of $\mathrm{supp}(W)$ have the same probability. Our construction for this case is quite simple: to produce $p$, Gen outputs a hash of its input point $w$ and an extractor seed; to produce key, Gen applies the extractor to $w$. Given $w'$, Rep looks for $w \in \mathrm{supp}(W)$ that is near $w'$ and has the correct hash value, and applies the extractor to this $w$ to get key.

The specific hash function we use is *universal*. (We note that universal hashing has a long history of use for information reconciliation, for example [4,34,36]. This construction is not novel; rather, we present it as a stepping stone for the case of general distributions).

**Definition 4** ([9]). *Let $F : \mathcal{K} \times \mathcal{M} \to R$ be a function. We say that $F$ is universal if for all distinct $x_1, x_2 \in \mathcal{M}$:*

$$\Pr_{K \leftarrow \mathcal{K}}[F(K, x_1) = F(K, x_2)] = \frac{1}{|R|} .$$

In our case, the hash output length needs to be sufficient to disambiguate elements of $\mathrm{supp}(W) \cap B_t(w')$ with high probability. Observe that there are at most $2^{\mathrm{H}_\infty(W) - \mathrm{H}_{t,\infty}^{\mathrm{fuzz}}(W)}$ such elements when $W$ is flat, so output length slightly greater (by $\log 1/\delta$) than $\mathrm{H}_\infty(W) - \mathrm{H}_{t,\infty}^{\mathrm{fuzz}}(W)$ will suffice. Thus, the output key length will be $\mathrm{H}_{t,\infty}^{\mathrm{fuzz}}(W) - \log 1/\delta - 2\log 1/\epsilon + 2$ (by using average-case leftover hash lemma, per [13, Lemmas 2.2b and 2.4]). As this construction is only a warm-up, so we do not state it formally and proceed to general distributions.

## 4.2  Fuzzy Extractor for Arbitrary Distributions

The hashing approach used in the previous subsection does not work for arbitrary sources. Consider a distribution $W$ consisting of the following balls: $B_t^1$ is a ball with $2^{\mathrm{H}_\infty(W)}$ points with total probability $\Pr[W \in B_t^1] = 2^{-\mathrm{H}_\infty(W)}$, $B_t^2, ..., B_t^{2^{-\mathrm{H}_\infty(W)}}$ are balls with one point each with probability $\Pr[W \in B_t^i] = 2^{-\mathrm{H}_\infty(W)}$. The above hashing algorithm writes down $\mathrm{H}_\infty(W)$ bits to achieve correctness on $B_t^1$. However, with probability $1 - 2^{-\mathrm{H}_\infty(W)}$ the initial reading is outside of $B_t^1$, and the hash completely reveals the point.

Instead, we use a layered approach: we separate the input distribution $W$ into nearly-flat layers, write down the layer from which the input $w$ came (i.e., the approximate probability of $w$) as part of $p$, and rely on the construction from the previous part for each layer. In other words, the hash function output is now variable-length, longer if probability of $w$ is lower. Thus, $p$ now reveals a bit more about $w$. To limit this information and the resulting security loss, we limit number of layers. As a result, we lose only $1 + \log H_0(W)$ more bits of security compared to the previous section. We emphasize that this additional loss is quite small: if $W$ is over $\{0,1\}^n$, it is only $1 + \log n$ bits (so, for example, only 11 bits if $W$ is 1000 bits long, and no more than 50 bits for any remotely realistic $W$). We thus obtain the following theorem.

**Theorem 1.** *For any metric space* $\mathcal{M}$, *distribution* $W$ *over* $\mathcal{M}$, *distance* $t$, *error* $\delta > 0$, *and security* $\epsilon > 0$, *there exists a* $(\mathcal{M}, \{W\}, \kappa, t, \epsilon)$-*known distribution fuzzy extractor with error* $\delta$ *for* $\kappa = \mathrm{H}_{t,\infty}^{\mathrm{fuzz}}(W) - \log H_0(W) - \log 1/\delta - 2\log 1/\epsilon + 1$. *(Note that the value* $\log H_0(W)$ *is doubly logarithmic in the size of the support of* $W$ *and is smaller than* $\log 1/\delta$ *and* $\log 1/\epsilon$ *for typical setting of parameters.)*

We provide the construction and the proof in Appendix A. The main idea is that providing the level information makes the distribution look nearly flat (the probability of points differs by at most a factor of two, which increases the entropy loss as compared to the flat case by only one bit). And the level information itself increases the entropy loss by $\log H_0(W)$ bits, because there are only $H_0(W)$ levels that contain enough weight to matter.

## 5  Impossibility of Fuzzy Extractors for Family with $\mathrm{H}_{t,\infty}^{\mathrm{fuzz}}$

In the previous section, we showed the sufficiency of $\mathrm{H}_{t,\infty}^{\mathrm{fuzz}}(W)$ for building fuzzy extractors when the distribution $W$ is precisely known. However, it may be infeasible to completely characterize a high-entropy distribution $W$. Traditionally, algorithms deal with this *distributional uncertainty* by providing security for a family of distributions $\mathcal{W}$. In this section, we show that distributional uncertainty comes at a real cost.

We demonstrate an example over the binary Hamming metric in which every $W \in \mathcal{W}$ has linear $\mathrm{H}_{t,\infty}^{\mathrm{fuzz}}(W)$ (which is in fact equal to $\mathrm{H}_\infty(W)$), and yet there is some $W \in \mathcal{W}$ where even for 3-bit output keys and high constant $\epsilon = \frac{1}{4}$. In fact, we show that the adversary need not work hard: even a uniformly random choice of distribution $W$ from $\mathcal{W}$ will thwart the security of any $(\mathsf{Gen}, \mathsf{Rep})$. The one caveat is that, for this result, we require $\mathsf{Rep}$ to be always correct (i.e., $\delta = 0$). As mentioned in the introduction, this perfect correctness requirement is removed in Sects. 6 and 7 at a cost of lower entropy rate and stronger primitive, respectively.

As basic intuition, the result is based on the following reasoning: $\mathsf{Gen}$ sees only a random sample $w$ from a random $W \in \mathcal{W}$, but not $W$. The adversary sees $W$ but not $w$. Because $\mathsf{Gen}$ does not know which $W$ the input $w$ came from, $\mathsf{Gen}$ must produce $p$ that works for many distributions $W$ that contain $w$ in their support. Such $p$ must necessarily reveal a lot of information. The adversary can combine information gleaned from $p$ with information about $W$ to narrow down the possible choices for $w$ and thus distinguish key from uniform.

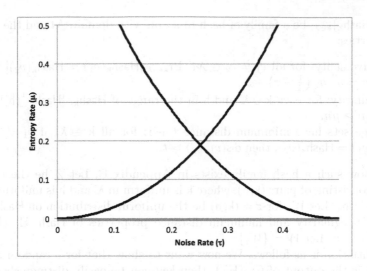

**Fig. 1.** The region of $\tau$ ($x$-axis) and $\mu$ ($y$-axis) pairs for which Theorem 2 applies is the region below both curves.

**Theorem 2.** *Let $\mathcal{M}$ denote the Hamming space $\{0,1\}^n$. There exists a family of distributions $\mathcal{W}$ over $\mathcal{M}$ such that for each element $W \in \mathcal{W}$, $\mathrm{H}_{t,\infty}^{\mathrm{fuzz}}(W) = \mathrm{H}_\infty(W) \geq m$, and yet any $(\mathcal{M}, \mathcal{W}, \kappa, t, \epsilon)$-fuzzy extractor with error $\delta = 0$ has $\epsilon > 1/4$.*

*This holds as long as $\kappa \geq 3$ and under the following conditions on the entropy rate $\mu = m/n$, noise rate $\tau = t/n$, and $n$:*

- *any $0 \leq \tau < \frac{1}{2}$ and $\mu > 0$ such that $\mu < 1 - h_2(\tau)$ and $\mu < 1 - h_2\left(\frac{1}{2} - \tau\right)$*
- *any $n \geq \max\left(\frac{2}{1-h_2(\tau)-\mu}, \frac{5}{1-h_2\left(\frac{1}{2}-\tau\right)-\mu}\right)$.*

Note that the conditions on $\mu$ and $\tau$ imply the result applies to any entropy rate $\mu \leq .18$ as long as $\tau$ is set appropriately and $n$ is sufficiently large (for example, the result applies to $n \geq 1275$ and $\tau = .6\sqrt{\mu}$ when $0.08 \leq \mu \leq .18$; similarly, it applies to $n \geq 263$ and $\tau = \sqrt{\mu}$ when $0.01 \leq \mu \leq 0.08$). The $\tau$ vs. $\mu$ tradeoff is depicted in Fig. 1.

*Proof (Sketch).* Here we describe the family $\mathcal{W}$ and provide a brief overview of the main proof ideas. We provide a full proof in Appendix B. We will show the theorem holds for an average member of $\mathcal{W}$. Let $Z$ denote a uniform choice of $W$ from $\mathcal{W}$ and denote by $W_z$ the choice specified by a particular value of $z$.

Let $\{\mathsf{Hash}_k\}_{k \in \mathcal{K}}$ be a family of hash function with domain $\mathcal{M}$ and the following properties:

- $2^{-a}$-universality: for all $v_1 \neq v_2 \in \mathcal{M}$, $\mathrm{Pr}_{k \leftarrow \mathcal{K}}[\mathsf{Hash}_k(v_1) = \mathsf{Hash}_k(v_2)] \leq 2^{-a}$, where $a = n \cdot h_2\left(\frac{1}{2} - \tau\right) + 3$.
- $2^m$-regularity: for each $k \in \mathcal{K}$ and h in the range of $\mathsf{Hash}_k$, $|\mathsf{Hash}_k^{-1}(h)| = 2^m$, where $m \geq \mu n$.
- preimage sets have minimum distance $t + 1$: for all $k \in \mathcal{K}$, if $v_1 \neq v_2$ but $\mathsf{Hash}_k(v_1) = \mathsf{Hash}_k(v_2)$, then $\mathrm{dis}(v_1, v_2) > t$.

We show such a hash family exists in Appendix B. Let $Z$ be the random variable consisting of pairs $(k, h)$, where $k$ is uniform in $\mathcal{K}$ and $h$ is uniform in the range of $\mathsf{Hash}_k$. Let $W_z$ for $z = (k, h)$ be the uniform distribution on $\mathsf{Hash}_k^{-1}(h)$. By the $2^m$-regularity and minimum distance properties of $\mathsf{Hash}$, $\mathrm{H}_\infty(W_z) = \mathrm{H}_{t,\infty}^{\mathtt{fuzz}}(W_z) = m$. Let $\mathcal{W} = \{W_z\}$.

The intuition is as follows. We now want to show that for a random $z \leftarrow Z$, if $(\mathsf{key}, p)$ is the output of $\mathsf{Gen}(W_z)$, then key can be easily distinguished from uniform in the presence of $p$ and $z$.

In the absence of information about $z$, the value $w$ is uniform on $\mathcal{M}$ (by regularity of $\mathsf{Hash}$). Knowledge of $p$ reduces the set of possible $w$ from $2^n$ to $2^{n \cdot h_2\left(\frac{1}{2} - \tau\right)}$, because, by correctness of $\mathsf{Rep}$, every candidate input $w$ to $\mathsf{Gen}$ must be such that all of its neighbors $w'$ of distance at most $t$ produce the same output of $\mathsf{Rep}(w', p)$. And knowledge of $z$ reduces the set of possible $w$ by another factor of $2^a$, because a hash value with a random hash function key likely gives fresh information about $w$.

# 6   Impossibility in the Case of Imperfect Correctness

The impossibility result in the previous section applies only to fuzzy extractors with perfect correctness. In this section, we build on the work of Holenstein and Renner [25] to show the impossibility of fuzzy extractors even when they are allowed to make mistakes a constant fraction $\delta$ (as much as 4%) of the time. However, the drawback of this result, as compared to the previous section, is that we can show impossibility only for a relatively low entropy rate of at most 7%. In Sect. 7, we rule out stronger primitives called secure sketches with nonzero error (which are used in most fuzzy extractor constructions), even for entropy rate as high as 50%.

**Theorem 3.** *Let $\mathcal{M}$ denote the Hamming space $\{0, 1\}^n$. There exists a family of distributions $\mathcal{W}$ over $\mathcal{M}$ such that for each element $W \in \mathcal{W}$, $\mathrm{H}_{t,\infty}^{\mathtt{fuzz}}(W) = \mathrm{H}_\infty(W) \geq m$, and yet any $(\mathcal{M}, \mathcal{W}, \kappa, t, \epsilon)$-fuzzy extractor with error $\delta \leq \frac{1}{25}$ has $\epsilon > \frac{1}{25}$.*

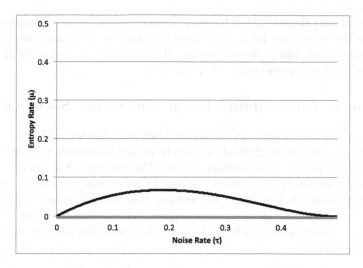

**Fig. 2.** The region of $\tau$ ($x$-axis) and $\mu$ ($y$-axis) pairs for which Theorem 3 applies is the region below this curve.

*This holds for any $\kappa > 0$ under the following conditions on the entropy rate $\mu = m/n$, noise rate $\tau = t/n$, and $n$:*

- *any $0 \leq \tau \leq \frac{1}{2}$ and $\mu$ such that $\mu < 4\tau(1-\tau)\left(1 - h_2\left(\frac{1}{4-4\tau}\right)\right)$*
- *any sufficiently large $n$ (as a function of $\tau$ and $\mu$)*

Note that the conditions on $\mu$ and $\tau$ imply that the result applies to any entropy rate $\mu \leq \frac{1}{15}$ as long as $\tau$ is set appropriately and $n$ is sufficiently large. The $\tau$ vs. $\mu$ tradeoff is depicted in Fig. 2.

*Proof (Proof Sketch).* We now describe the family $\mathcal{W}$ and provide an overview of the main ideas. The full proof is in Appendix C.

Similarly to the proof of Theorem 2, we will prove that any fuzzy extractor fails for an element $W_z$ of $\mathcal{W}$ chosen according to the distribution $Z$. In this case, $Z$ will not be uniform but rather binomial (with tails cut off). Essentially, $Z$ will contain each bit of $w$ with (appropriately chosen) probability $\beta$; given $Z = z$, the remaining bits of $w$ will be uniform and independent.

For a string $z \in \{0, 1, \perp\}^n$, denote by $info(z)$ the number of entries in $z$ that are not $\perp$: $info(z) = |\{i \text{ s.t } z_i \neq \perp\}|$. Let $W_z$ be the uniform distribution over all strings in $\{0, 1\}^n$ that agree with $z$ in positions that are not $\perp$ in $z$ (i.e., all strings $w \in \{0, 1\}^n$ such that for $1 \leq i \leq n$, either $z_i = \perp$ or $w_i = z_i$).

We will use $\mathcal{W}$ to prove the theorem statement. First, we show that every distribution $W_z \in \mathcal{W}$ has sufficient $\mathrm{H}_{t,\infty}^{\mathrm{fuzz}}$. Indeed, $z$ constrains $info(z)$ coordinates out of $n$ and leaves the rest uniform. Thus, $\mathrm{H}_{t,\infty}^{\mathrm{fuzz}}(W_z)$ is the same as $\mathrm{H}_{t,\infty}^{\mathrm{fuzz}}$ of the uniform distribution on the space $\{0, 1\}^{n-info(z)}$. Second, we now want to show that $\mathbf{SD}((\mathrm{Key}, P, Z), (U_\kappa, P, Z)) > \frac{1}{25}$. To show this, we use a result of Holenstein and Renner [25, Theorem 4]. Their result shows impossibility of interactive key agreement for a noisy channel where the adversary observes each bit with some probability. Several technical results are necessary to apply the result in our setting (presented in Appendix C).

# 7  Stronger Impossibility Result for Secure Sketches

Most fuzzy extractor constructions share the following feature with our construction in Sect. 4: $p$ includes information that is needed to recover $w$ from $w'$; both Gen and Rep simply apply an extractor to $w$. The recovery of $w$ from $w'$, known as information-reconciliation, forms the core of many fuzzy extractor constructions. The primitive that performs this information reconciliation is called *secure sketch*. In this section we show stronger impossibility results for secure sketches. First, we recall their definition from [13, Sect. 3.1] (modified slightly, in the same way as Definition 2).

**Definition 5.** *An* $(\mathcal{M}, \mathcal{W}, \tilde{m}, t)$-*secure sketch with error $\delta$ is a pair of randomized procedures, "sketch" (SS) and "recover" (Rec). SS on input $w \in \mathcal{M}$ returns a bit string $ss \in \{0,1\}^*$. Rec takes an element $w' \in \mathcal{M}$ and $ss \in \{0,1\}^*$. (SS, Rec) have the following properties:*

1. *Correctness:* $\forall w, w' \in \mathcal{M}$ *if* $\mathsf{dis}(w, w') \leq t$ *then* $\Pr[\mathsf{Rec}(w', \mathsf{SS}(w)) = w] \geq 1 - \delta$.
2. *Security: for any distribution* $W \in \mathcal{W}$, $\tilde{\mathrm{H}}_\infty(W|\mathsf{SS}(W)) \geq \tilde{m}$.

Secure sketches are more demanding than fuzzy extractors (secure sketches can be converted to fuzzy extractors by using a randomness extractors like in our Construction 1 [13, Lemma 4.1]). We prove a stronger impossibility result for them. Specifically, in the case of secure sketches, we can extend the results of Theorems 2 and 3 to cover imperfect correctness (that is, $\delta > 0$) and entropy rate $\mu$ up to $\frac{1}{2}$. Since most fuzzy extractor constructions rely on secure sketches, this result gives evidence that fuzzy extractors even with imperfect correctness and for high entropy rates are difficult to construct in the case of distributional uncertainty.

**Theorem 4.** *Let $\mathcal{M}$ denote the Hamming space $\{0,1\}^n$. There exists a family of distributions $\mathcal{W}$ over $\mathcal{M}$ such that for each element $W \in \mathcal{W}$, $\mathrm{H}_{t,\infty}^{\mathsf{fuzz}}(W) = \mathrm{H}_\infty(W) \geq m$, and yet any $(\mathcal{M}, \mathcal{W}, \tilde{m}, t)$-secure sketch with error $\delta$ has $\tilde{m} \leq 2$.*

*This holds under the following conditions on $\delta$, the entropy rate $\mu = m/n$, noise rate $\tau = t/n$, and $n$:*

- *any $0 \leq \tau < \frac{1}{2}$ and $\mu > 0$ such that $\mu < h_2(\tau)$ and $\mu < 1 - h_2(\tau)$*
- *any $n \geq \max\left(\frac{.5 \log n + 4\delta n + 4}{h_2(\tau) - \mu}, \frac{2}{1 - h_2(\tau) - \mu}\right)$*

Note that the result holds for any $\mu < 0.5$ as long as $\delta < (h_2(\tau) - \mu)/4$ and $n$ is sufficiently large. The $\tau$ vs. $\mu$ tradeoff is depicted in Fig. 3.

We provide the proof, which uses similar ideas to the proof of Theorem 2, in Appendix D.

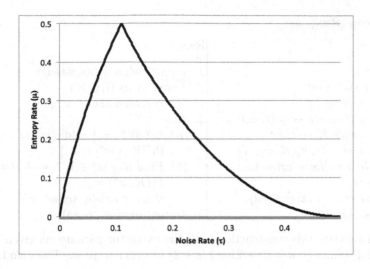

**Fig. 3.** The region of $\tau$ ($x$-axis) and $\mu$ ($y$-axis) pairs for which Theorem 4 applies is the region below both curves.

**Acknowledgements.** The authors are grateful to Gene Itkis and Yevgeniy Dodis for helpful discussions and to Thomas Holenstein for clarifying the results of [24,25]. The work of Benjamin Fuller was done while at MIT Lincoln Laboratory and Boston University and is sponsored in part by US NSF grants 1012910 and 1012798 and the United States Air Force under Air Force Contract FA8721-05-C-0002. Opinions, interpretations, conclusions and recommendations are those of the authors and are not necessarily endorsed by the United States Government. Leonid Reyzin is supported in part by US NSF grants 0831281, 1012910, 1012798, and 1422965, and The Institute of Science and Technology, Austria, where part of this work was performed. Adam Smith's work was supported in part by NSF awards 0747294, 0941553 and 1447700 and was performed partly while at Boston University's Hariri Institute for Computing and RISCS Center, and the Harvard Center for Research on Computation & Society.

# A    Proof of Theorem 1

We first provide a full description of the layered hashing construction.

**Construction 1.** *Let $W$ be a distribution over a metric space $\mathcal{M}$ with $\mathrm{H}_\infty(W) = m$.*

- *Let $\delta \leq \frac{1}{2}$ be the error parameter.*
- *Let $\ell = m + H_0(W) - 1$; round $\ell$ down so that $\ell - m$ is an integer (i.e., set $\ell = m + \lfloor(\ell - m)\rfloor$).*
- *For each $i = m, m+1, \ldots, \ell-1$, let $L_i = (2^{-(i+1)}, 2^{-i}]$ and let $F_i : \mathcal{K}_i \times \mathcal{M} \to R_i$ be a family of universal hash functions with $\log|R_i| = i + 1 - \mathrm{H}_{t,\infty}^{\mathrm{fuzz}}(W) + \log 1/\delta$. Let $L_\ell = (0, 2^{-\ell}]$.*
- *Let ext be an $(\tilde{m}, \epsilon)$-average-case extractor for $\tilde{m} = \mathrm{H}_{t,\infty}^{\mathrm{fuzz}}(W) - \log H_0(W) - \log 1/\delta - 1$ with output length $\kappa$.*

*Define* $\mathsf{Gen}_W, \mathsf{Rep}_W$ *as:*

| $\mathsf{Gen}_W$ | $\mathsf{Rep}_W$ |
|---|---|
| 1. *Input:* $w$. | 1. *Input:* $(w', p = (ss, \mathsf{seed}))$ |
| 2. *Find $i$ such that* $\Pr[W = w] \in L_i$. | 2. *Parse $ss$ as* $(i, y, K)$ |
| 3. *If $i = \ell$ then set $ss = (i, w, 0)$.* | 3. *If $i = \ell$ then set $w^* = y$.* |
| 4. *Else sample $K \leftarrow \mathcal{K}_i$ and set $ss = (i, F_i(K, w), K)$* | 4. *Else* |
| 5. *Sample a uniform extractor seed* $\mathsf{seed}$ | (a) *Let* $W^* = \{w^* | \mathsf{dis}(w^*, w') \le t \; \wedge \; \Pr[W = w^*] \in L_i\}$. |
| 6. *Output* $\mathsf{key} = \mathsf{ext}(w, \mathsf{seed})$, $p = (ss, \mathsf{seed})$. | (b) *Find any $w^* \in W^*$ such that* $F_i(K, w^*) = y$; *if none exists, set $w^* = \perp$.* |
|  | 5. *Output* $\mathsf{ext}(w^*, \mathsf{seed})$. |

We instantiate this construction with the extractor parameters given by [13, Lemma 2.4] (namely, $\kappa = \tilde{m} - 2 \log 1/\epsilon + 2$) in order to prove Theorem 1.

*Proof (Proof of Theorem 1).* We first argue **correctness**. Fix some $w, w'$ within distance $t$. When $\Pr[W = w] \in L_\ell$, then Rep is always correct, so let's consider only the case when $\Pr[W = w] \notin L_\ell$. The algorithm Rep will never output $\perp$ since at least the correct $w$ will match the hash. Thus, an error happens when another element $w^* \in W^*$ has the same hash value $F(K_i, w^*)$ as $F(K_i, w)$. Observe that the total probability mass of $W^*$ is less than $|W^*| \cdot 2^{-(i+1)}$ but greater than or equal to the maximum probability mass in a ball of radius $t$, $2^{-\mathrm{H}^{\mathsf{fuzz}}_{t,\infty}(W)}$. Therefore, $|W^*| \le 2^{i+1-\mathrm{H}^{\mathsf{fuzz}}_{t,\infty}(W)}$. Each element of $W^*$ has the same hash as $F(K, w)$ with probability at most $1/|R_i|$, and thus correctness with error $|W^*|/|R| \le \delta$ follows by the union bound.

**Security:** We now argue security of the construction. Let $W_i = \{w | \Pr[W = w] \in L_i\}$. For ease of notation, let us make the special case of $i = \ell$ as part of the general case, as follows: define $\mathcal{K}_\ell = \{0\}$, $F_\ell(0, w) = w$, and $R_\ell = W_\ell$. Also, denote by SS the randomized function that maps $w$ to $ss$. First, we set up the analysis by levels:

$$2^{-\tilde{\mathrm{H}}_\infty(W | \mathsf{SS}(W))} = \mathbb{E}_{ss} \max_w \Pr[W = w \mid \mathsf{SS}(W) = ss]$$

$$= \sum_{ss} \max_w \Pr[W = w \wedge \mathsf{SS}(W) = ss]$$

$$= \sum_{i=m}^{\ell} \sum_{K \in \mathcal{K}_i} \sum_{y \in R_i} \max_w \Pr[W = w \wedge \mathsf{SS}(W) = (i, y, K)]$$

$$\le \sum_{i=m}^{\ell} \sum_{K \in \mathcal{K}_i} \sum_{y \in R_i} \max_{w \in W_i} \Pr[W = w \wedge F_i(K, w) = y \wedge K \text{ output by Gen}].$$

We now pay the penalty of $|R_i|$ for the presence of $y$ (observe that removing the condition that $F_i(K, w) = y$ from the conjunction cannot reduce the probability):

$$2^{-\tilde{H}_\infty(W|SS(W))} \leq \sum_{i=m}^{\ell} \sum_{K \in \mathcal{K}_i} \sum_{y \in R_i} \max_{w \in W_i} \Pr[W = w \wedge K \text{ is chosen by } SS]$$

$$= \sum_{i=m}^{\ell} \sum_{K \in \mathcal{K}_i} |R_i| \cdot \max_{w \in W_i} \Pr[W = w \wedge K \text{ is chosen by } SS].$$

We now get rid of the key, because it is independent:

$$2^{-\tilde{H}_\infty(W|SS(W))} \leq \sum_{i=m}^{\ell} \sum_{K \in \mathcal{K}_i} |R_i| \cdot \max_{w \in W_i} \Pr[W = w] \cdot \frac{1}{|\mathcal{K}_i|}$$

$$= \sum_{i=m}^{\ell} |R_i| \cdot \max_{w \in W_i} \Pr[W = w]$$

$$\leq |R_\ell| \cdot 2^{-\ell} + \sum_{i=m}^{\ell-1} |R_i| \cdot 2^{-i}.$$

Finally, we add everything up, recalling that $|R_i|$ for $i < \ell$ is $2^{i+1-H_{t,\infty}^{fuzz}(W)+\log 1/\delta}$.

$$2^{-\tilde{H}_\infty(W|SS(W))} \leq 2^{H_0(W)} \cdot 2^{-\ell} + (\ell - m) \cdot 2^{1-H_{t,\infty}^{fuzz}(W)+\log 1/\delta}$$

(next line uses $\ell > m + H_0(W) - 2$)

$$< 2^{2-m} + (\ell - m) \cdot 2^{1-H_{t,\infty}^{fuzz}(W)+\log 1/\delta}$$

(next line uses $m \geq H_{t,\infty}^{fuzz}(W)$ and $\log 1/\delta \geq 1$)

$$\leq (\ell - m + 1) \cdot 2^{1-H_{t,\infty}^{fuzz}(W)+\log 1/\delta}$$

(next line uses $\ell \leq m + H_0(W) - 1$)

$$\leq H_0(W) \cdot 2^{1-H_{t,\infty}^{fuzz}(W)+\log 1/\delta}.$$

Taking the negative logarithm of both sides, we obtain $\tilde{m} \overset{\text{def}}{=} \tilde{H}_\infty(W|SS(W)) = H_{t,\infty}^{fuzz}(W) - \log H_0(W) - \log 1/\delta - 1$. Applying the $(\tilde{m}, \epsilon)$ randomness extractor gives us the desired result. □

# B  Proof of Theorem 2

*Proof.* As a reminder, we show the impossibility for an average member of $\mathcal{W}$. For completeness, we reiterate the family $\mathcal{W}$ introduced in the proof sketch.

Let $\{\mathsf{Hash}_k\}_{k \in \mathcal{K}}$ be a family of hash function with domain $\mathcal{M}$ and the following properties:

- $2^{-a}$-universality: for all $v_1 \neq v_2 \in \mathcal{M}$, $\Pr_{k \leftarrow \mathcal{K}}[\mathsf{Hash}_k(v_1) = \mathsf{Hash}_k(v_2)] \leq 2^{-a}$, where $a = n \cdot h_2\left(\frac{1}{2} - \tau\right) + 3$.
- $2^m$-regularity: for each $k \in \mathcal{K}$ and h in the range of $\mathsf{Hash}_k$, $|\mathsf{Hash}_k^{-1}(h)| = 2^m$, where $m \geq \mu n$.

294 B. Fuller et al.

- preimage sets have minimum distance $t + 1$: for all $k \in \mathcal{K}$, if $v_1 \neq v_2$ but $\mathsf{Hash}_k(v_1) = \mathsf{Hash}_k(v_2)$, then $\mathsf{dis}(v_1, v_2) > t$.

We demonstrate the existence of such a hash family in Lemma 4. Let $Z$ be the random variable consisting of pairs $(k, h)$, where $k$ is uniform in $\mathcal{K}$ and $h$ is uniform in the range of $\mathsf{Hash}_k$. Let $W_z$ for $z = (k, h)$ be the uniform distribution on $\mathsf{Hash}_k^{-1}(h)$. By the $2^m$-regularity and minimum distance properties of $\mathsf{Hash}$, $H_\infty(W_z) = H_{t,\infty}^{\mathsf{fuzz}}(W_z) = m$. Let $\mathcal{W} = \{W_z\}$.

We now want to show that for a random $z \leftarrow Z$, if $(\mathsf{key}, p)$ is the output of $\mathsf{Gen}(W_z)$, then $\mathsf{key}$ can be easily distinguished from uniform in the presence of $p$ and $z$. The intuition is as follows: in the absence of information about $z$, the value $w$ is uniform on $\mathcal{M}$ (by regularity of $\mathsf{Hash}$). Knowledge of $p$ reduces the set of possible $w$ from $2^n$ to $2^{n \cdot h_2\left(\frac{1}{2} - \tau\right)}$, because, by correctness of $\mathsf{Rep}$, every candidate input $w$ to $\mathsf{Gen}$ must be such that all of its neighbors $w'$ of distance at most $t$ produce the same output of $\mathsf{Rep}(w', p)$ (see Lemma 2). And knowledge of $z$ reduces the set of possible $w$ by another factor of $2^a$, because a hash value with a random hash function key likely gives fresh information about $w$ (see Lemma 3).

To formalize the intuition of the previous two sentences, view the sequence of events that we are trying to analyze as a game. The adversary chooses a uniform $k \in \mathcal{K}$ and uniform $h$ in the range of $\mathsf{Hash}_k$. A uniform $w$ from $\mathcal{M}$ s.t. $\mathsf{Hash}_k(w) = h$ then gets chosen, $(\mathsf{key}, p) = \mathsf{Gen}(w)$ gets computed, and the adversary receives $p$. The output of this game is $(k, h, w, p, \mathsf{key})$. Note that, by regularity of $\mathsf{Hash}_k$, $w$ is uniform in $\mathcal{M}$.

Consider now an alternative game. A uniform $w$ gets chosen from $\mathcal{M}$ and uniform key $k$ gets chosen from $\mathcal{K}$. $(\mathsf{key}, p) = \mathsf{Gen}(w)$ gets computed. The adversary receives $(k, h = \mathsf{Hash}_k(w), p)$. The output of the game is $(k, h, w, p, \mathsf{key})$.

The distributions of the adversary's views and the outputs in the two games are identical: indeed, in both games, three random variable are uniform and independent (i.e., $w$ is uniform in $\mathcal{M}$, $k$ is uniform in $\mathcal{K}$, and the random coins of $\mathsf{Gen}$ are uniform in their domain), and the rest are determined fully by these three. However, the second game is easier to analyze, which is what we now do.

The following lemma shows that the knowledge of $p$ and $\mathsf{key}$ reduces the entropy of $w$.

**Lemma 2.** *Suppose $\mathcal{M}$ is $\{0,1\}^n$ with the Hamming metric, $\kappa \geq 2$, $0 \leq t \leq n/2$, and $\epsilon \geq 0$. Suppose $(\mathsf{Gen}, \mathsf{Rep})$ is a $(\mathcal{M}, \mathcal{W}, \kappa, t, \epsilon)$-fuzzy extractor with error $\delta = 0$, for some distribution family $\mathcal{W}$ over $\mathcal{M}$. Let $\tau = t/n$. For any fixed $p$, there is a set $\mathsf{GoodKey}_p \subseteq \{0,1\}^\kappa$ of size at least $2^{\kappa-1}$ such that for every $\mathsf{key} \in \mathsf{GoodKey}_p$,*

$$\log |\{v \in \mathcal{M} | (\mathsf{key}, p) \in \mathsf{supp}(\mathsf{Gen}(v))\}| \leq n \cdot h_2\left(\frac{1}{2} - \tau\right) \leq n \cdot \left(1 - \frac{2}{\ln 2} \cdot \tau^2\right),$$

*and, therefore, for any distribution $D_\mathcal{M}$ on $\mathcal{M}$,*

$$H_0(D_\mathcal{M} | \mathsf{Gen}(D_\mathcal{M}) = (\mathsf{key}, p)) \leq n \cdot h_2\left(\frac{1}{2} - \tau\right) \leq n \cdot \left(1 - \frac{2}{\ln 2} \cdot \tau^2\right).$$

*Proof.* The set $\mathsf{GoodKey}_p$ consists of all keys for which $H_0(\mathcal{M}|\mathsf{Rep}(\mathcal{M},p) = \mathsf{key}) \leq 2^{n-\kappa+1}$.

The intuition is as follows. By perfect correctness of $\mathsf{Rep}$, the input $w$ to $\mathsf{Gen}$ has the following property: for all $w'$ within distance $t$ of $w$, $\mathsf{Rep}(w',p) = \mathsf{Rep}(w,p)$. Thus, if we partition $\mathcal{M}$ according to the output of $\mathsf{Rep}$, the true $w$ is $t$ away from the interior of a part. Interior sets are small, which means the set of possible of $w$ values is small. (We note that by perfect correctness, $\mathsf{Rep}$ has a deterministic output even if the algorithm is randomized, so this partition is well-defined.)

To formalize this intuition, fix $p$ and partition $\mathcal{M}$ according to the output of $\mathsf{Rep}(\cdot,p)$ as follows: let $Q_{p,\mathsf{key}} = \{w' \in \mathcal{M}|\mathsf{Rep}(w',p) = \mathsf{key}\}$. Note that there are $2^\kappa$ keys and thus $2^\kappa$ parts $Q_{p,\mathsf{key}}$. Let $\mathsf{GoodKey}_p$ by the set of keys for which these parts are not too large: $\mathsf{key} \in \mathsf{GoodKey}_p \Leftrightarrow |Q_{p,\mathsf{key}}| \leq 2 \cdot \mathcal{M}/2^\kappa = 2^{n-\kappa+1}$. Observe that $\mathsf{GoodKey}_p$ contains at least half the keys: $|\mathsf{GoodKey}_p| \geq 2^{\kappa-1}$ (if not, then $\cup_{\mathsf{key}}|Q_{p,\mathsf{key}}| > |\mathcal{M}|$). For the remainder of the proof we focus on elements in $\mathsf{GoodKey}_p$.

As explained above, if $w$ is the input to $\mathsf{Gen}$, then every point $w'$ within distance $t$ of $w$ must be in the same part $Q_{p,\mathsf{key}}$ as $w$, by correctness of $\mathsf{Rep}$. Thus, $w$ must come from the interior of some $Q_{p,\mathsf{key}}$, where interior is defined as

$$\mathsf{Inter}(Q_{p,\mathsf{key}}) = \{w \in Q_{p,\mathsf{key}}|\forall w' \text{ s.t. } \mathsf{dis}(w,w') \leq t, w' \in Q_{p,\mathsf{key}}\}.$$

We now use the isoperimetric inequality to bound the size of $\mathsf{Inter}(Q_{p,\mathsf{key}})$. Define a *near-ball*[3] centered at $x$ to be any set $S$ that is contained in a ball of some radius $\eta$ and contains the ball of radius $\eta - 1$ around $x$. The inequality of [16, Theorem 1] (the original result is due to Harper [21]) says that for any sets $A, B \subset \{0,1\}^n$, there are near-balls $X$ and $Y$ centered at $0^n$ and $1^n$, respectively, such that $|A| = |X|$, $|B| = |Y|$, and $\min_{a \in A, b \in B} \mathsf{dis}(a,b) \leq \min_{x \in X, y \in Y} \mathsf{dis}(x,y)$.

Letting $A$ be the $\mathsf{Inter}(Q_{p,\mathsf{key}})$ and $B$ be the complement of $Q_{p,\mathsf{key}}$ and applying this inequality, we get a near-ball $S_{p,\mathsf{key}}$ centered at $0^n$ and a near-ball $D$ centered at $1^n$, such that $|S_{p,\mathsf{key}}| = |\mathsf{Inter}(Q_{p,\mathsf{key}})|$, $|D| = 2^n - |Q_{p,\mathsf{key}}|$, and $\forall s \in S_{p,\mathsf{key}}, d \in D, \mathsf{dis}(s,d) > t$. Note that since $\mathsf{key} \in \mathsf{GoodKey}_p$ and $\kappa \geq 2$, we have $|Q_{p,\mathsf{key}}| \leq 2^{n-\kappa+1}$, and therefore $|D| \geq 2^{n-1}$.

Thus, $D$ includes all the strings of Hamming weight $\lceil n/2 \rceil$ (because it is centered at $1^n$ and takes up at least half the space), which means that the maximum Hamming weight of an element of $S_{p,\mathsf{key}}$ is $\lceil n/2 \rceil - t - 1 \leq n/2 - t$ (because each element of $S_{p,\mathsf{key}}$ is at distance more than $t$ from $D$). We can now use binary entropy to bound the size of $S_{p,\mathsf{key}}$ by Lemma 1:

$$|\mathsf{Inter}(Q_{p,\mathsf{key}})| = |S_{p,\mathsf{key}}| \leq |\{x|\mathsf{dis}(x,0) \leq n/2 - t\}| \leq 2^{n \cdot h_2\left(\frac{1}{2} - \frac{t}{n}\right)}.$$

The theorem statement follows by taking the logarithm of both sides and by observing (using Taylor series expansion at $\tau = 0$ and noting that the third derivative is negative) that $h_2\left(\frac{1}{2} - \tau\right) \leq 1 - \frac{2}{\ln 2} \cdot \tau^2$. $\qquad\square$

---

[3] In most statements of the isoperimetric inequality, this type of set is simply called a ball. We use the term *near*-ball for emphasis.

We now analyze how the entropy drops further when the adversary learns $\mathsf{Hash}_k(w)$. Let $\mathsf{K}$ denote the uniform distribution on $\mathcal{K}$. We defer the proof to the full version of this work [18, Lemma B.2].

**Lemma 3.** *Let $L$ be a distribution. Let $\{\mathsf{Hash}_k\}_{k\in\mathcal{K}}$ be a family of $2^{-a}$-universal hash functions on the support of $L$. Assume $\mathsf{k}$ is uniform in $\mathcal{K}$ and independent of $L$. Then*

$$\tilde{H}_0(L|\mathsf{K},\mathsf{Hash}_\mathsf{K}(L)) < \log(1 + |\operatorname{supp}(L)| \cdot 2^{-a}) \le \max(1, 1 + H_0(L) - a).$$

Let $\mathsf{M}$ denote the uniform distribution on $\mathcal{M}$. By Lemma 2, for any $p$, $H_0(\mathsf{M}|\mathsf{Gen}(\mathsf{M}) = (\mathsf{key}, p)$ s.t. $\mathsf{key} \in \mathsf{GoodKey}_p) \le n \cdot h_2\left(\frac{1}{2} - \frac{t}{n}\right) + \kappa$ (because there are most $2^\kappa$ keys in $\mathsf{GoodKey}_p$). Applying Lemma 3 (and recalling that $\kappa \ge 3$), we get that for any $p$,

$$\tilde{H}_0(\mathsf{M}|\mathsf{Gen}(\mathsf{M}) = (\mathsf{key}, p) \text{ s.t. } \mathsf{key} \in \mathsf{GoodKey}_p, \mathsf{K}, \mathsf{Hash}_\mathsf{K}(\mathsf{M}))$$
$$< \max\left(1, 1 + n \cdot h_2\left(\frac{1}{2} - \frac{t}{n}\right) + \kappa - a\right) \le \kappa - 2.$$

(Note carefully the somewhat confusing conditioning notation above, because we are conditioning on both events and variables. The event is $\mathsf{key} \in \mathsf{GoodKey}_p$ and the variables are $\mathsf{k}$ and $\mathsf{Hash}_k(\mathsf{M})$.)

By correctness, for a fixed $p$, $\mathsf{Rep}(w, p)$ can produce only one key—the same one that was produces during $\mathsf{Gen}(w)$. Since applying a deterministic function (in this case, $\mathsf{Rep}$) cannot increase $H_0$, we get that for each $p$,

$$\tilde{H}_0(\mathsf{key}|\mathsf{Gen}(\mathsf{M}) = (\mathsf{key}, p) \text{ s.t. } \mathsf{key} \in \mathsf{GoodKey}_p, \mathsf{K}, \mathsf{Hash}_\mathsf{K}(\mathsf{M})) < \kappa - 2.$$

Thus, on average over $z = (\mathsf{k}, \mathsf{h})$, over half the keys in $\mathsf{GoodKey}_p$ (i.e., over a quarter of all possible $2^\kappa$ keys) cannot be produced. Let $\mathsf{Implausible}$ be the set of triples $(\mathsf{key}, p, z = (\mathsf{k}, \mathsf{h}))$ such that $\Pr[\mathsf{Gen}(W_z) = (\mathsf{key}, p)] = 0$. Triples drawn by sampling $w$ from $W_z$ and computing $(p, \mathsf{key}) = \mathsf{Gen}(w)$ never come from this set. On other hand, random triples come $\mathsf{Implausible}$ at over quarter of the time. Thus, by definition of statistical distance, $\epsilon > \frac{1}{4}$.

It remains to show that the hash family with the desired properties exists.

**Lemma 4.** *For any $0 \le \tau < \frac{1}{2}$, $\mu > 0$, $\alpha$, and $n$ such that $\mu \le 1 - h_2(\tau) - \frac{2}{n}$ and $\mu \le 1 - \alpha - \frac{2}{n}$, there exists a family of hash functions $\{\mathsf{Hash}_k\}_{k\in\mathcal{K}}$ on $\{0,1\}^n$ that is $2^{-a}$-universal for $a = \alpha n$, $2^m$ regular for $m \ge \mu n$, and whose preimage sets have minimum distance $t + 1$ for $t = \tau n$.*

*Proof.* Let $\mathcal{C}$ be the set of all binary linear codes of rate $\mu$ (to be precise, dimension $m = \lceil \mu n \rceil$), length $n$, and minimum distance $t + 1$:

$$\mathcal{C} = \{C | C \text{ is a linear subspace of } \{0,1\}^n, \dim(C) = m, \min_{c \in C - \{0^n\}} \operatorname{dis}(c, 0^n) > t\}.$$

For each $C \in \mathcal{C}$, fix $H_C$, an $(n - m) \times n$ parity check matrix for $C$, such that $C = \ker H_C$. For $v \in \{0,1\}^n$, let the syndrome $\mathsf{syn}_C(v) = H_C \cdot v$. Let $\{\mathsf{Hash}_k\}_{k\in\mathcal{K}} = \{\mathsf{syn}_C\}_{C\in\mathcal{C}}$.

$2^m$ regularity follows from the fact that for each $h \in \{0,1\}^{n-\mu n}$, $\mathsf{Hash}_k^{-1}(h)$ is a coset of $C$, which has size $2^m$. The minimum distance property is also easy: if $v_1 \neq v_2$ but $\mathsf{syn}_C(v_1) = \mathsf{syn}_C(v_2)$, then $H_C(v_1 - v_2) = 0^n$, hence $v_1 - v_2 \in C - \{0^n\}$ and hence $\mathsf{dis}(v_1, v_2) = \mathsf{dis}(v_1 - v_2, 0) > t$.

We show $2^{-a}$-universality by first considering a slightly larger hash family. Let $\mathcal{K}'$ be the set of *all* $m$-dimensional subspaces of $\{0,1\}^n$; for each $C' \in \mathcal{K}'$, choose a parity check matrix $H_{C'}$ such that $C' = \ker H_{C'}$, and let $\mathsf{syn}_{C'}(v) = H_{C'} \cdot v$. Let $\{\mathsf{Hash}'_{k'}\}_{k' \in \mathcal{K}'} = \{\mathsf{syn}_{C'}\}_{C' \in \mathcal{K}'}$. This family is $2^{m-n}$-universal: for $v_1 \neq v_2$, $\Pr_{C' \in \mathcal{K}'}[H_{C'} \cdot v_1 = H_{C'} \cdot v_2] = \Pr_{C' \in \mathcal{K}'}[v_1 - v_2 \in \ker H_{C'} = C'] = \frac{2^m}{2^n}$, because $C'$ is a random $m$-dimensional subspace. Note that this family is not much bigger than our family $\{\mathsf{Hash}_k\}_{k \in \mathcal{K}}$, because, as long as $\mu < 1 - h_2(\tau)$, almost every subspace of $\{0,1\}^n$ of dimension $m$ has minimum distance $t+1$ for a sufficiently large $n$. Formally,

$$\Pr_{C' \in \mathcal{K}'}[C' \not\subseteq \mathcal{C}] = \Pr_{C' \in \mathcal{K}'}[\exists v_1 \neq v_2 \in C' \text{ s. t. } \mathsf{dis}(v_1, v_2) \leq t]$$

$$= \Pr_{C' \in \mathcal{K}'}[\exists v_1 \neq v_2 \in C' \text{ s. t. } \mathsf{dis}(v_1 - v_2, 0^n) \leq t]$$

$$= \Pr_{C' \in \mathcal{K}'}[\exists v \in C' - \{0^n\} \text{ s. t. } \mathsf{dis}(v, 0^n) \leq t]$$

$$\leq \sum_{v \in B_t(0^n) - \{0^n\}} \Pr_{C' \in \mathcal{K}'}[v \in C'] \leq 2^{nh_2(\tau)} \cdot \frac{2^m}{2^n} \leq \frac{1}{2}$$

(the penultimate inequality follows by Lemma 1 and the last one from $m \leq \mu n + 1$ and $\mu \leq 1 - h_2(\tau) - \frac{2}{n}$).

Since this larger family is universal and at most factor of two bigger than our family, our family is also universal:

$$\Pr_{C \in \mathcal{C}}[\mathsf{syn}_C(v_1) = \mathsf{syn}_C(v_2)] = \frac{|\{C \in \mathcal{C} | \mathsf{syn}_C(v_1) = \mathsf{syn}_C(v_2)\}|}{|\mathcal{C}|}$$

$$\leq \frac{|\{C \in \mathcal{K}' | \mathsf{syn}_C(v_1) = \mathsf{syn}_C(v_2)\}|}{|\mathcal{K}'|} \cdot \frac{|\mathcal{K}'|}{|\mathcal{C}|} \leq 2^{m-n+1}$$

Thus, we obtain the desired result as long as $m - n + 1 \leq -a$, which is implied by the condition $\mu \leq 1 - \alpha - \frac{2}{n}$ and the fact that $m \leq \mu n + 1$. $\square$

Applying Lemma 4 with $\alpha = h_2\left(\frac{1}{2} - \tau\right) + \frac{3}{n}$, we see that the largest possible $\mu$ is $\max_\tau \min\left(1 - h_2(\tau), 1 - h_2\left(\frac{1}{2} - \tau\right)\right) \approx 0.1887$. Using the quadratic approximation to $h_2\left(\frac{1}{2} - \tau\right)$ (see Lemma 2), we can let $\mu$ be a free variable and set $\tau = .6\sqrt{\mu}$, in which case both constraints will be satisfied for all $0 < \mu \leq .18$ and sufficiently large $n$, as in the theorem statement. This concludes the proof of Theorem 2. $\square$

## C  Proof of Theorem 3

*Proof.* Similarly to the proof of Theorem 2, we will prove that any fuzzy extractor fails for an average element of $\mathcal{W}$: letting $Z$ denote a choice of $W$ from $\mathcal{W}$, we will show that $\mathbf{SD}((\mathsf{Key}, P, Z), (U_\kappa, P, Z)) > \frac{1}{25}$.

For completeness, we reiterate the family of distributions introduced in the proof sketch. In this case, $Z$ will not be uniform but rather binomial (with tails cut off). Essentially, $Z$ will contain each bit of $w$ with (appropriately chosen) probability $\beta$; given $Z = z$, the remaining bits of $w$ will be uniform and independent.

For a string $z \in \{0, 1, \perp\}^n$, denote by $info(z)$ the number of entries in $z$ that are not $\perp$: $info(z) = |\{i \text{ s.t } z_i \neq \perp\}|$. Let $W_z$ be the uniform distribution over all strings in $\{0, 1\}^n$ that agree with $z$ in positions that are not $\perp$ in $z$ (i.e., all strings $w \in \{0, 1\}^n$ such that for $1 \leq i \leq n$, either $z_i = \perp$ or $w_i = z_i$).

Let $0 \leq \beta' \leq 1$ be a parameter (we will set it at the end of the proof). Let $Z'$ denote the distribution on strings in $\{0, 1, \perp\}^n$ in which each symbol is, independently of other symbols, $\perp$ with probability $1 - \beta'$, 0 with probability $\beta'/2$, and 1 with probability $\beta'/2$. Let $\beta = \beta' + \frac{1.4}{\sqrt{n}}$. Consider two distribution families: $\mathcal{W}' = \{W_z\}_{z \leftarrow Z'}$ and a smaller family $\mathcal{W} = \{W_z\}_{z \leftarrow Z}$, where $Z = Z' | info(Z') \leq \beta n$ (the second family is smaller because, although on average $info(Z') = \beta' n$, there is a small chance that $info(Z')$ is higher than even $\beta n$).

We will use $\mathcal{W}$ to prove the theorem statement. First, we will show that every distribution $W_z \in \mathcal{W}$ has sufficient $\mathrm{H}_{t,\infty}^{\mathrm{fuzz}}$. Indeed, $z$ constrains $info(z)$ coordinates out of $n$ and leaves the rest uniform. Thus, $\mathrm{H}_{t,\infty}^{\mathrm{fuzz}}(W_z)$ is the same as $\mathrm{H}_{t,\infty}^{\mathrm{fuzz}}$ of the uniform distribution on the space $\{0, 1\}^{n-info(z)}$. Let $a = n - info(z)$. By Lemma 1

$$\mathrm{H}_{t,\infty}^{\mathrm{fuzz}}(W_z) \geq a\left(1 - h_2\left(\frac{t}{a}\right)\right) \geq n(1 - \beta)\left(1 - h_2\left(\frac{t}{n(1-\beta)}\right)\right)$$
$$= n(1 - \beta)\left(1 - h_2\left(\frac{\tau}{1-\beta}\right)\right).$$

and therefore

$$\mu = (1 - \beta)\left(1 - h_2\left(\frac{\tau}{1-\beta}\right)\right). \tag{1}$$

Note that smaller $\beta$ gives a higher fuzzy entropy rate.

Second, we now want to show, similarly to the proof of Theorem 2, that $\mathbf{SD}((\mathsf{Key}, P, Z), (U_\kappa, P, Z)) > \frac{1}{25}$. We will do so by considering the family $\mathcal{W}$. Observe that by triangle inequality

$$\mathbf{SD}((\mathsf{Key}, P, Z), (U_\kappa, P, Z)) \geq \mathbf{SD}((\mathsf{Key}, P, Z'), (U_\kappa, P, Z'))$$
$$- \mathbf{SD}((\mathsf{Key}, P, Z'), (\mathsf{Key}, P, Z))$$
$$- \mathbf{SD}((U_\kappa, P, Z), (U_\kappa, P, Z'))$$
$$\geq \mathbf{SD}((\mathsf{Key}, P, Z'), (U_\kappa, P, Z')) - 2 \cdot \mathbf{SD}(Z', Z)$$
$$\geq \mathbf{SD}((\mathsf{Key}, P, Z'), (U_\kappa, P, Z')) - \frac{1}{25}.$$

The last line follows by Hoeffding's inequality [23],

$$\mathbf{SD}(Z', Z) = \Pr[info(Z') > \beta n] \leq \exp\left(-2n\left(\frac{1.4}{\sqrt{n}}\right)^2\right) < \frac{1}{50}.$$

Denote $\mathbf{SD}((\mathsf{Key}, P, Z'), (U_\kappa, P, Z'))$ by $\epsilon'$. To bound $\epsilon'$, we recall a result of Holenstein and Renner [25, Theorem 4] (we will use the version presented in [24, Lemma 4.4]). For a random variable $W$ with a values in $\{0,1\}^n$, let $W^{noisy}$ denote a noisy copy of $W$: namely, the random variable obtained by passing $W$ through a binary symmetric channel with error rate $\frac{1-\alpha}{2}$ (that is, $W_i^{noisy} = W_i$ with probability $\frac{1+\alpha}{2}$ and $W_i^{noisy} = 1 - W_i$ with probability $\frac{1-\alpha}{2}$, independently for each position $i$). Holenstein and Renner show that if $\alpha^2 \leq \beta$, then Shannon entropy of $\mathsf{Key}$ conditioned on $P$ and $W^{noisy}$ is greater than Shannon entropy of $\mathsf{Key}$ conditioned on $Z$ and $W^{noisy}$. Intuitively, this means that the Rep, when given $P$ and $W^{noisy}$, knows less about $\mathsf{Key}$ than the adversary (who knows $P$ and $Z$).

Recall the definitions of Shannon entropy $H_1(X) \stackrel{\text{def}}{=} \mathbb{E}_{x \leftarrow X} - \log \Pr[X = x]$ and conditional Shannon entropy $H_1(X|Y) \stackrel{\text{def}}{=} \mathbb{E}_{y \leftarrow Y} H_1(X|Y = y)$.

**Theorem 5** ([25, Theorem 4]; [24, Lemma 4.4]). *Suppose that $(P, \mathsf{Key})$ is a pair of random variables derived from $W$. If $\alpha^2 \leq \beta'$, then*

$$H_1(\mathsf{Key}|P, Z') \leq H_1(\mathsf{Key}|P, W^{noisy})$$

*where $H_1$ denotes Shannon entropy, $W^{noisy}$ is $W$ passed through a binary symmetric channel with error rate $\frac{1-\alpha}{2}$, and $Z'$ is $W$ passed through a binary erasure channel with erasure rate $1 - \beta'$.*

(For a reader interested in how our statement of Lemma 5 follows from [24, Lemma 4.4], note that what we call $\mathsf{Key}, P, W^{noisy}$, and $Z'$ are called $U, V, Y$, and $Z$, respectively, in [24]. Note also that we use only the part of the lemma that says that secret key rate $S_\rightarrow = 0$ when $\alpha^2 \leq \beta$, and the definition [24, Definition 3.1] of the notion $S_\rightarrow$ in terms of Shannon entropy.)

We now need to translate this bound on Shannon entropy to the language of statistical distance $\epsilon$ of the key from uniform, reliability $\delta$ of the procedure Rep, and key length $\kappa$, as used in the definition of fuzzy extractors. First, we will do this translation for the case of noisy rather than worst-case input to Rep.

**Corollary 1.** *Let $(W, W^{noisy}, Z')$ be a triple of correlated random variables such that*

- *$W$ and $W^{noisy}$ are uniform over $\{0,1\}^n$,*
- *$W^{noisy}$ is $W$ passed through a binary symmetric channel with error rate $\frac{1-\alpha}{2}$ (that is, each bit position of $W$ agrees with corresponding bit position of $W^{noisy}$ with probability $\frac{1+\alpha}{2}$), and*
- *$Z'$ is $W$ passed through a binary erasure channel with erasure rate $1 - \beta'$ (that is, each bit position of $Z'$ agrees with the corresponding bit position of $W$ with probability $\beta'$ and is equal to $\perp$ otherwise).*

*Suppose $\mathsf{Gen}(W)$ produces $(\mathsf{Key}, P)$ with $\mathsf{Key}$ of length $\kappa$. Suppose $\Pr[\mathsf{Rep}(W^{noisy}, P) = \mathsf{Key}] = 1 - \delta'$. Suppose further that $\mathbf{SD}((\mathsf{Key}, P, Z'), (U_\kappa, P, Z')) = \epsilon'$. If $\alpha^2 \leq \beta'$, then*

$$\kappa \leq \frac{h_2(\epsilon') + h_2(\delta')}{1 - \epsilon' - \delta'}.$$

*In other words, if $\alpha^2 \leq \beta'$, $\epsilon' \leq \frac{1}{12}$, and $\delta' \leq \frac{1}{12}$, then even a 1-bit Key is impossible to obtain.*

(We note that a similar result follows from [24, Theorem 3.17] if we set the variables $S_\rightarrow$, $\gamma$, and $m$ in that theorem to $0, \delta$, and $\kappa$, respectively. However, we could not verify the correctness of that theorem due to its informal treatment of what "$\epsilon$-close to uniform" means; it seems that the small correction term $-h_2(\epsilon)$, just like in our result, is needed on the right-hand side to make that theorem correct.)

*Proof.* Reliability allows us to bound the entropy of the key. By Fano's inequality [15, Sect. 6.2, p. 187], $H_1(\mathsf{Key}|P, W^{noisy}) \leq \kappa\delta' + h_2(\delta')$. Hence, by Theorem 5 (and the assumption that $\alpha^2 > \beta'$), we have

$$H_1(\mathsf{Key}|P, Z') \leq \kappa\delta' + h_2(\delta'). \tag{2}$$

We now need the following lemma, which shows that near-uniformity implies high entropy.

**Lemma 5.** *For a pair of random variables $(A, B)$ such that the statistical distance between $(A, B)$ and $U_\kappa \times B$ is $\epsilon$, then $H_1(A|B) \geq (1 - \epsilon)\kappa - h_2(\epsilon)$.*

*Proof.* Let $E$ denote a binary random variable correlated with $(A, B)$ as follows: when $A = a$ and $B = b$, then $E = 0$ with probability

$$\max(\Pr[(A, B) = (a, b)] - \Pr[U_\kappa \times B = (a, b)], 0).$$

Similarly, let $F$ denote a binary random variable correlated with $U_\kappa \times B$ as follows: when $U_\kappa = a$ and $B = b$, then $F = 0$ with probability

$$\max(\Pr[U_\kappa \times B = (a, b)] - \Pr[(A, B) = (a, b)], 0).$$

Note that $\Pr[E = 0] = \Pr[F = 0] = \epsilon$, by definition of statistical distance. Note also that $(A, B|E = 1)$ is the same distribution as $(U_\kappa \times B|F = 1)$. Since conditioning cannot increase Shannon entropy (by a simple argument — see, e.g., [2, Theorem 1.4.4]), we get

$$\begin{aligned} H_1(A|B) &\geq H_1(A|B, E) \\ &= \Pr[E = 1]H_1(A|B, E = 1) + \Pr[E = 0]H_1(A|B, E = 0) \\ &\geq (1 - \epsilon)H_1(A|B, E = 1) = (1 - \epsilon)H_1(U_\kappa|B, F = 1). \end{aligned}$$

To bound this latter quantity, note that (the first line follows from the chain rule $H_1(X) \leq H_1(X, Y) = H_1(X|Y) + H_1(Y)$ [2, Theorem 1.4.4])

$$\begin{aligned} \kappa = H_1(U_\kappa|B) &\leq H_1(U_\kappa|B, F) + H_1(F) \\ &= (1 - \epsilon)H_1(U_\kappa|B, F = 1) + \epsilon \cdot H_1(U_\kappa|B, F = 0) + h_2(\epsilon) \\ &\leq (1 - \epsilon)H_1(U_\kappa|B, F = 1) + \epsilon \cdot \kappa + h_2(\epsilon) \end{aligned}$$

Rearranging terms, we get $H_1(U_\kappa|B, F = 1) \geq \kappa - h_2(\epsilon)/(1 - \epsilon)$, and thus

$$H_1(A|B) \geq (1 - \epsilon)\kappa - h_2(\epsilon).$$

This concludes the proof of Lemma 5. □

Combining (2) and Lemma 5 (applied to $A = $ Key, $B = (P, Z')$, and $\epsilon = \epsilon'$), we get the claimed bound. This concludes the proof of Corollary 1. □

Next, we translate this result from the noisy-input-case to the worst-case input case. Set $\alpha = \sqrt{\beta'}$. Suppose $t \geq n \left( \frac{1-\sqrt{\beta'}}{2} + \frac{1.4}{\sqrt{n}} \right)$. By Hoeffding's inequality [23],

$$\Pr[\mathsf{dis}(W, W^{noisy}) > t] \leq \exp\left( -2n \left( \frac{1.4}{\sqrt{n}} \right)^2 \right) < \frac{1}{50}.$$

Thus, a fuzzy extractor that corrects $t$ errors with reliability $\delta$ implies that $\Pr[\mathsf{Rep}(W^{noisy}, P) = \mathsf{Key}] \geq 1 - \delta'$ for $\delta' = \delta + \frac{1}{50}$. Since $\delta \leq 1/25$, we have $\delta' < 1/12$ and Corollary 1 applies to gives us $\epsilon' > 1/12$ and $\epsilon > 1/12 - 1/25 > 1/25$ as long as $\kappa > 0$.

Finally, we work out the relationship between $\mu$ and $\tau$ and eliminate $\beta$, as follows. Recall that $\beta = \beta' + \frac{1.4}{\sqrt{n}}$; therefore $\sqrt{\beta} \leq \sqrt{\beta'} + \frac{1.2}{n^{1/4}}$, and it suffices to take $\tau \geq \frac{1-\sqrt{\beta}}{2} + \frac{2}{\sqrt[4]{n}}$. Thus, we can set any $\tau > \frac{1-\sqrt{\beta}}{2}$ as long as $n$ is sufficiently large. Solving for $\beta$ (that is, taking any $\beta > (1 - 2\tau)^2$) and substituting into Eq. 1, we can get any $\mu < 4\tau(1 - \tau) \left( 1 - h_2\left( \frac{1}{4-4\tau} \right) \right)$ for a sufficiently large $n$. □

# D  Proof of Theorem 4

*Proof.* Similarly to the proof of Theorem 2, we will prove that any secure sketch algorithm fails for an average element of $\mathcal{W}$: letting $Z$ denote a uniform choice of $W$ from $\mathcal{W}$, we will show that $\tilde{H}_\infty(W_Z|\mathsf{SS}(W_Z), Z) \leq 2$. The overall proof strategy is the same as for Theorem 2. We highlight only the changes here. Recall that $|B_t|$ denotes the volume of the ball of radius $t$ in the space $\{0, 1\}^n$. The parameters of the hash family are the same, except for universality: we require $2^{-a}$-universality for $a = (n - \log|B_t| + h_2(2\delta))/(1 - 2\delta)$.

We postpone the question of the existence of such a hash family until the end of the proof.

We can now state and the analogue of Lemma 2. This result is an extension of lower bounds from [13, Appendix C], which handles only the case of perfect correctness. It shows that the value of the sketch reduces the entropy of a uniform point by approximately $\log|B_t|$.

**Lemma 6.** *Let $\mathcal{M}$ denote the Hamming space $\{0, 1\}^n$ and $|B_t|$ denote the volume of a Hamming ball of radius $t$ in $\{0, 1\}^n$. Suppose $(\mathsf{SS}, \mathsf{Rec})$ is a $(\mathcal{M}, \mathcal{W}, \tilde{m}, t)$ secure sketch with error $\delta$, for some distribution family $\mathcal{W}$ over $\mathcal{M}$. Then for every $v \in \mathcal{M}$ there exists a set $\mathsf{GoodSketch}_v$ such that $\Pr[\mathsf{SS}(v) \in \mathsf{GoodSketch}_v] \geq 1/2$ and for any fixed $ss$,*

$$\log|\{v \in \mathcal{M}|ss \in \mathsf{GoodSketch}_v\}| \leq \frac{n - \log|B_t| + h_2(2\delta)}{1 - 2\delta},$$

and, therefore, for any distribution $D_{\mathcal{M}}$ over $\mathcal{M}$,

$$H_0(D_{\mathcal{M}}|ss \in \text{GoodSketch}_{D_{\mathcal{M}}}) \leq \frac{n - \log|B_t| + h_2(2\delta)}{1 - 2\delta}.$$

*Proof.* For any $v \in M$, define $\text{Neigh}_t(v)$ be the uniform distribution on the ball of radius $t$ around $v$ and let

$$\text{GoodSketch}_v = \{ss| \Pr_{v' \leftarrow \text{Neigh}_t(v)}[\text{Rec}(v', ss) \neq v] \leq 2\delta]\}.$$

We prove the lemma by showing two propositions.

**Proposition 2.** *For all $v \in \mathcal{M}$, $\Pr[\text{SS}(v) \in \text{GoodSketch}_v] \geq 1/2$.*

*Proof.* Let the indicator variable $1_{v',ss}$ be 1 if $\text{Rec}(v', ss) = v$ and 0 otherwise. Let $q_{ss}$ be the quality of the sketch on the ball $B_t(v)$:

$$q_{ss} = \Pr_{v' \leftarrow \text{Neigh}_t(v)}[\text{Rec}(v', ss) = v] = \mathop{\mathbb{E}}_{v' \in \text{Neigh}_t(v)} 1_{v',ss}.$$

By the definition of correctness for $(\text{SS}, \text{Rec})$, for all $v' \in B_t(v)$,

$$\Pr_{ss \leftarrow \text{SS}(v)}[\text{Rec}(v', ss) = v] \geq 1 - \delta.$$

Hence, $\mathbb{E}_{ss \leftarrow \text{Gen}(v)} 1_{v',ss} \geq 1 - \delta$. Therefore,

$$\mathop{\mathbb{E}}_{ss \leftarrow \text{Gen}(v)} q_{ss} = \mathop{\mathbb{E}}_{ss} \mathop{\mathbb{E}}_{v'} 1_{v',ss} = \mathop{\mathbb{E}}_{v'} \mathop{\mathbb{E}}_{ss} 1_{v',ss} \geq \mathop{\mathbb{E}}_{v'}(1 - \delta) = 1 - \delta.$$

Therefore, applying Markov's inequality to $1 - q_{ss}$, we get $\Pr[q_{ss} \geq 1 - 2\delta] = \Pr[1 - q_{ss} \leq 2\delta] \leq 1/2$.

$\square$

To finish the proof of Lemma 6, we will show that the set $\{v \in \mathcal{M}|ss \in \text{GoodSketch}_v\}$ forms a kind of error-correcting code, and then bound the size of the code.

**Definition 6.** *We say that a set $C$ is an $(t, \delta)$-Shannon code if there exists a (possibly randomized) function $\text{Decode}$ such that for all $c \in C$,*

$$\Pr_{c' \leftarrow \text{Neigh}_t(c)}[\text{Decode}(c') \neq c] \leq \delta.$$

The set $\{v \in \mathcal{M}|ss \in \text{GoodSketch}_v\}$ forms $(t, 2\delta)$ Shannon code if we set $\text{Decode}(y) = \text{Rec}(y, ss)$. We now bound the size of such a code.

**Proposition 3.** *If $C \subseteq \{0,1\}^n$ is a $(t, \delta)$-Shannon code, then*

$$\log|C| \leq \frac{n - \log|B_t| + h_2(\delta)}{1 - \delta}.$$

*Proof.* Let the pair of random variables $(X, Y)$ be obtained as follows: let $X$ be a uniformly chosen element of $C$ and $Y$ be a uniformly chosen element of the ball of radius $t$ around $Y$. By the existence of Decode and Fano's inequality [15, Sect. 6.2, p. 187], $H_1(X|Y) \leq h_2(\delta) + \delta \log |C|$. At the same time, $H_1(X|Y) = H_1(X) - H_1(Y) + H_1(Y|X)$ (because $H_1(X, Y) = H_1(X) + H_1(Y|X) = H_1(Y) + H_1(X|Y)$), and therefore $H_1(X|Y) \geq \log |C| - n + \log |B_t|$ (because $H_1(Y) \leq n$). Therefore, $\log |C| - n + \log |B_t| \leq h_2(\delta) + \delta \log |C|$, and the lemma follows by rearranging terms.

$\square$

Lemma 6 follows from Proposition 3. $\square$

We now show that entropy drops further when the adversary learns $\mathsf{Hash}_k(w)$. Let M denote the uniform distribution on $\mathcal{M}$ and K denote the uniform distribution on $\mathcal{K}$. Applying Lemma 3 to Lemma 6, we get that for any $ss$,

$$\tilde{H}_0(\mathsf{M}|ss \in \mathsf{GoodSketch}_\mathsf{M}, \mathsf{K}, \mathsf{Hash}_\mathsf{K}(\mathsf{M}))$$
$$< \max\left(1, 1 + \frac{n - \log |B_t| + h_2(2\delta)}{1 - 2\delta} - a\right). \tag{3}$$

To complete the proof, we will use this bound on $\tilde{H}_0$ as a bound on $\tilde{H}_\infty$, as justified by the following lemma (proof in the full version of this work [18, Lemma D.7]).

**Lemma 7.** *For any random variables $X$ and $Y$, $\tilde{H}_\infty(X|Y) \leq \tilde{H}_0(X|Y)$.*

We need just one more lemma before we can complete the result, an analogue of [13, Lemma 2.2b] for conditioning on a single value $Z = z$ rather than with $Z$ on average (we view conditioning on a single value as equivalent to conditioning on an event). The proof of this lemma is natural and is shown in the full version of this work [18, Lemma D.8].

**Lemma 8.** *For any pair of random variables $(X, Y)$ and event $\eta$ that is a (possibly randomized) function of $(X, Y)$, $\tilde{H}_\infty(X|\eta, Y) \geq \tilde{H}_\infty(X|Y) - \log 1/\Pr[\eta]$.*

Combining Lemmas 8 and 7 with Eq. 3, we get

$$\tilde{H}_\infty(W_Z|Z, \mathsf{SS}(W_Z)) = \tilde{H}_\infty(\mathsf{M}|\mathsf{SS}(\mathsf{M}), \mathsf{K}, \mathsf{Hash}_\mathsf{K}(\mathsf{M}))$$
$$\leq \log \frac{1}{\Pr[\mathsf{SS}(\mathsf{M}) \in \mathsf{GoodSketch}_\mathsf{M}]} +$$
$$\tilde{H}_\infty(\mathsf{M}|ss \text{ s.t. } ss = \mathsf{SS}(\mathsf{M}) \text{ and } ss \in \mathsf{GoodSketch}_\mathsf{M}, \mathsf{K}, \mathsf{Hash}_\mathsf{K}(\mathsf{M}))$$
$$\leq \log \frac{1}{\Pr[\mathsf{SS}(\mathsf{M}) \in \mathsf{GoodSketch}_\mathsf{M}]} +$$
$$\tilde{H}_0(\mathsf{M}|ss \text{ s.t. } ss = \mathsf{SS}(\mathsf{M}) \text{ and } ss \in \mathsf{GoodSketch}_\mathsf{M}, \mathsf{K}, \mathsf{Hash}_\mathsf{K}(\mathsf{M}))$$
$$< \log \frac{1}{\Pr[\mathsf{SS}(\mathsf{M}) \in \mathsf{GoodSketch}_\mathsf{M}]} + \max\left(1, 1 + \frac{n - \log |B_t| + h_2(2\delta)}{1 - 2\delta} - a\right).$$

We can have shown that $\tilde{H}_\infty(W_Z|Z, \mathsf{SS}(W_Z)) \leq 2$, because the first term of the above sum is at most 1 by Proposition 2 and the second term is 1 by our choice of $a$ as $a = \frac{n - \log|B_t| + h_2(2\delta)}{1 - 2\delta}$.

It remains to show that the desired hash family exists. Note in that (because $\delta < .25$) setting any $\alpha \geq 1 - h_2(\tau) + \frac{.5 \log n + 4\delta n + 2}{n}$ and choosing an $\alpha n$-universal hash function will be sufficient, because, by Lemma 1, $\log|B_t| \geq nh_2(\tau) - \frac{1}{2}\log n - 1$, and so

$$
\begin{aligned}
a = \frac{n - \log|B_t| + h_2(2\delta)}{1 - 2\delta} &\leq n \cdot \frac{1 - h_2(\tau) + (.5\log n + 1 + h_2(2\delta))/n}{1 - 2\delta} \\
&< n \cdot \left(1 - h_2(\tau) + \frac{.5\log n + 1 + h_2(2\delta)}{n} + 4\delta\right) \\
&\leq n \cdot \left(1 - h_2(\tau) + \frac{.5\log n + 4\delta n + 2}{n}\right) \\
&\leq n \cdot \alpha
\end{aligned}
$$

(the second inequality is true because for any $x < 1$ and $0 < y < .5$, $x/(1-y) < x + 2y$, because $x < (x+2y)(1-y)$, because $0 < y(2-x-2y)$; the third inequality follows from $h_2(2\delta) < 1$).

Such a hash family exists by Lemma 4 as long as $\mu \leq 1 - \alpha - 2/n \leq h_2(\tau) - (.5\log n + 4\delta n + 4)/n$ and $\mu \leq 1 - h_2(\tau) - 2/n$.                         $\square$

# References

1. Ahlswede, R., Csiszár, I.: Common randomness in information theory and cryptography - I: secret sharing. IEEE Trans. Inf. Theory **39**(4), 1121–1132 (1993)
2. Ash, R.: Information Theory. Intersciene Publishers, New York (1965)
3. Barak, B., Canetti, R., Lindell, Y., Pass, R., Rabin, T.: Secure computation without authentication. J. Cryptology **24**(4), 720–760 (2011)
4. Bennett, C.H., Brassard, G., Robert, J.M.: Privacy amplification by public discussion. SIAM J. Comput. **17**(2), 210–229 (1988)
5. Bitansky, N., Canetti, R., Kalai, Y.T., Paneth, O.: On virtual grey box obfuscation for general circuits. In: Garay, J.A., Gennaro, R. (eds.) CRYPTO 2014. LNCS, vol. 8617, pp. 108–125. Springer, Heidelberg (2014). doi:10.1007/978-3-662-44381-1_7
6. Blanton, M., Hudelson, W.M.P.: Biometric-based non-transferable anonymous credentials. In: Qing, S., Mitchell, C.J., Wang, G. (eds.) ICICS 2009. LNCS, vol. 5927, pp. 165–180. Springer, Heidelberg (2009). doi:10.1007/978-3-642-11145-7_14
7. Boyen, X., Dodis, Y., Katz, J., Ostrovsky, R., Smith, A.: Secure remote authentication using biometric data. In: Cramer, R. (ed.) EUROCRYPT 2005. LNCS, vol. 3494, pp. 147–163. Springer, Heidelberg (2005). doi:10.1007/11426639_9
8. Brostoff, S., Sasse, M.: Are passfaces more usable than passwords?: a field trial investigation. In: McDonald, S., Waern, Y., Cockton, G. (eds.) People and Computers, pp. 405–424. Springer, London (2000)
9. Carter, L., Wegman, M.N.: Universal classes of hash functions. J. Comput. Syst. Sci. **18**(2), 143–154 (1979)
10. Csiszár, I., Körner, J.: Broadcast channels with confidential messages. IEEE Trans. Inf. Theory **24**(3), 339–348 (1978)

11. Daugman, J.: Probing the uniqueness and randomness of iriscodes: results from 200 billion iris pair comparisons. Proc. IEEE **94**(11), 1927–1935 (2006)
12. Daugman, J.: How iris recognition works. IEEE Trans. Circ. Syst. Video Technol. **14**(1), 21–30 (2004)
13. Dodis, Y., Ostrovsky, R., Reyzin, L., Smith, A.: Fuzzy extractors: how to generate strong keys from biometrics and other noisy data. SIAM J. Comput. **38**(1), 97–139 (2008)
14. Ellison, C., Hall, C., Milbert, R., Schneier, B.: Protecting secret keys with personal entropy. Future Gener. Comput. Syst. **16**(4), 311–318 (2000)
15. Fano, R.: Transmission of Information: A Statistical Theory of Communications. MIT Press Classics, M.I.T. Press, New York (1961)
16. Frankl, P., Füredi, Z.: A short proof for a theorem of Harper about Hamming-spheres. Discrete Math. **34**(3), 311–313 (1981)
17. Fuller, B., Meng, X., Reyzin, L.: Computational fuzzy extractors. In: Sako, K., Sarkar, P. (eds.) ASIACRYPT 2013. LNCS, vol. 8269, pp. 174–193. Springer, Heidelberg (2013). doi:10.1007/978-3-642-42033-7_10
18. Fuller, B., Smith, A., Reyzin, L.: When are fuzzy extractors possible? IACR Cryptology ePrint Archive 2014, 961 (2014)
19. Gassend, B., Clarke, D., Van Dijk, M., Devadas, S.: Silicon physical random functions. In: Proceedings of the 9th ACM Conference on Computer and Communications Security, pp. 148–160. ACM (2002)
20. Hao, F., Anderson, R., Daugman, J.: Combining crypto with biometrics effectively. IEEE Trans. Comput. **55**(9), 1081–1088 (2006)
21. Harper, L.H.: Optimal numberings and isoperimetric problems on graphs. J. Comb. Theory **1**(3), 385–393 (1966)
22. Hayashi, M., Tyagi, H., Watanabe, S.: Secret key agreement: general capacity and second-order asymptotics. In: 2014 IEEE International Symposium on Information Theory, pp. 1136–1140. IEEE (2014)
23. Hoeffding, W.: Probability inequalities for sums of bounded random variables. J. Am. Stat. Assoc. **58**(301), 13–30 (1963)
24. Holenstein, T.: Strengthening key agreement using hard-core sets. Ph.D. thesis, ETH Zurich (May 2006), reprint as vol. 7 of ETH Series in Information Security and Cryptography, ISBN 3-86626-088-2, Hartung-Gorre Verlag, Konstanz (2006)
25. Holenstein, T., Renner, R.: One-way secret-key agreement and applications to circuit polarization and immunization of public-key encryption. In: Shoup, V. (ed.) CRYPTO 2005. LNCS, vol. 3621, pp. 478–493. Springer, Heidelberg (2005). doi:10.1007/11535218_29
26. Ignatenko, T., Willems, F.M.: Biometric security from an information-theoretical perspective. Found. Trends Commun. Inf. Theory **7**(2–3), 135–316 (2012)
27. Juels, A., Wattenberg, M.: A fuzzy commitment scheme. In: Sixth ACM Conference on Computer and Communication Security, pp. 28–36. ACM, November 1999
28. Linnartz, J.-P., Tuyls, P.: New shielding functions to enhance privacy and prevent misuse of biometric templates. In: Kittler, J., Nixon, M.S. (eds.) AVBPA 2003. LNCS, vol. 2688, pp. 393–402. Springer, Heidelberg (2003). doi:10.1007/3-540-44887-X_47
29. Maurer, U.M.: Secret key agreement by public discussion from common information. IEEE Trans. Inf. Theory **39**(3), 733–742 (1993)
30. Mayrhofer, R., Gellersen, H.: Shake well before use: intuitive and secure pairing of mobile devices. IEEE Trans. Mob. Comput. **8**(6), 792–806 (2009)
31. Monrose, F., Reiter, M.K., Wetzel, S.: Password hardening based on keystroke dynamics. Int. J. Inf. Secur. **1**(2), 69–83 (2002)

32. Nisan, N., Zuckerman, D.: Randomness is linear in space. J. Comput. Syst. Sci. **52**(1), 43–52 (1996)
33. Pappu, R., Recht, B., Taylor, J., Gershenfeld, N.: Physical one-way functions. Science **297**(5589), 2026–2030 (2002)
34. Renner, R., Wolf, S.: The exact price for unconditionally secure asymmetric cryptography. In: Cachin, C., Camenisch, J.L. (eds.) EUROCRYPT 2004. LNCS, vol. 3027, pp. 109–125. Springer, Heidelberg (2004). doi:10.1007/978-3-540-24676-3_7
35. Renner, R., Wolf, S.: Simple and tight bounds for information reconciliation and privacy amplification. In: Roy, B. (ed.) ASIACRYPT 2005. LNCS, vol. 3788, pp. 199–216. Springer, Heidelberg (2005). doi:10.1007/11593447_11
36. Skoric, B., Tuyls, P.: An efficient fuzzy extractor for limited noise. Cryptology ePrint Archive, Report 2009/030 (2009)
37. Suh, G.E., Devadas, S.: Physical unclonable functions for device authentication and secret key generation. In: Proceedings of the 44th Annual Design Automation Conference, pp. 9–14. ACM (2007)
38. Tuyls, P., Goseling, J.: Capacity and examples of template-protecting biometric authentication systems. In: Maltoni, D., Jain, A.K. (eds.) BioAW 2004. LNCS, vol. 3087, pp. 158–170. Springer, Heidelberg (2004). doi:10.1007/978-3-540-25976-3_15
39. Tuyls, P., Schrijen, G.-J., van Škorić, B., Geloven, J., Verhaegh, N., Wolters, R.: Read-proof hardware from protective coatings. In: Goubin, L., Matsui, M. (eds.) CHES 2006. LNCS, vol. 4249, pp. 369–383. Springer, Heidelberg (2006). doi:10.1007/11894063_29
40. Tyagi, H., Watanabe, S.: Converses for secret key agreement and secure computing. IEEE Trans. Inf. Theo. **61**(9) (2015)
41. Wang, Y., Rane, S., Draper, S.C., Ishwar, P.: A theoretical analysis of authentication, privacy and reusability across secure biometric systems. IEEE Trans. Inf. Forensics Secur. **6**(6), 1825–1840 (2012)
42. Wyner, A.D.: The wire-tap channel. Bell Syst. Tech. J. **54**(8), 1355–1387 (1975)
43. Zviran, M., Haga, W.J.: A comparison of password techniques for multilevel authentication mechanisms. Comput. J. **36**(3), 227–237 (1993)

# More Powerful and Reliable Second-Level Statistical Randomness Tests for NIST SP 800-22

Shuangyi Zhu[1,2,3], Yuan Ma[1,2(✉)], Jingqiang Lin[1,2], Jia Zhuang[1,2], and Jiwu Jing[1,2]

[1] Data Assurance and Communication Security Research Center,
Chinese Academy of Sciences, Beijing, China
{zhushuangyi,yma,linjq,jzhuang13,jing}@is.ac.cn
[2] State Key Laboratory of Information Security,
Institute of Information Engineering, Chinese Academy of Sciences, Beijing, China
[3] University of Chinese Academy of Sciences, Beijing, China

**Abstract.** Random number generators (RNGs) are essential for cryptographic systems, and statistical tests are usually employed to assess the randomness of their outputs. As the most commonly used statistical test suite, the NIST SP 800-22 suite includes 15 test items, each of which contains two-level tests. For the test items based on the binomial distribution, we find that their second-level tests are flawed due to the inconsistency between the assessed distribution and the assumed one. That is, the sequence that passes the test could still have statistical flaws in the assessed aspect. For this reason, we propose *Q-value* as the metric for these second-level tests to replace the original P-value without any extra modification, and the first-level tests are kept unchanged. We provide the correctness proof of the proposed Q-value based second-level tests. We perform the theoretical analysis to demonstrate that the modification improves not only the detectability, but also the reliability. That is, the tested sequence that dissatisfies the randomness hypothesis has a higher probability to be rejected by the improved test, and the sequence that satisfies the hypothesis has a higher probability to pass it. The experimental results on several deterministic RNGs indicate that, the Q-value based method is able to detect some statistical flaws that the original SP 800-22 suite cannot realize under the same test parameters.

**Keywords:** Statistical randomness test · NIST SP 800-22 · Random number generator · P-value

## 1 Introduction

As essential primitives, random number generators (RNGs) are important for cryptographic systems. The security of many cryptographic schemes and protocols is built on the perfect randomness of RNG outputs. RNGs are classified

© International Association for Cryptologic Research 2016
J.H. Cheon and T. Takagi (Eds.): ASIACRYPT 2016, Part I, LNCS 10031, pp. 307–329, 2016.
DOI: 10.1007/978-3-662-53887-6_11

into two types: pseudo/deterministic and true/non-deterministic random number generators (PRNGs and TRNGs, respectively). In general, TRNGs based on some random physical phenomenons, may be used directly as random bit sources or generate seeds for PRNGs, and PRNGs extend the seeds to produce deterministic long sequences.

For any type of RNG, statistical hypothesis tests have been widely employed to assess the quality of the RNG, which evaluate whether the output sequences fit with the given hypothesis (i.e., the sequence has perfect randomness) or not. In addition, statistical randomness tests are also used to evaluate the outputs of other cryptographic primitives such as hash functions and block ciphers, to preliminarily validate the indistinguishability of their outputs from random mapping. The commonly used statistical test suites, each of which is composed of a serial of test items, include Diehard [7] proposed by Marsaglia and SP 800-22 [11] standardized by US National Institute of Standard and Technology (NIST).

The most commonly used NIST SP 800-22 test suite is composed of 15 test items, and provides comprehensive evaluation for different randomness aspects of assessed sequences. For example, the Frequency Test assesses the uniformity of the sequence, and the Runs Test assesses the transform frequency of 0's and 1's. In the beginning of the testing process, the whole bit sequence is divided into $N$ blocks. In every test item, a test statistic value is computed for each data block. According to the assumed distribution of the test statistic value, 15 test items are divided into two types: *binomial distribution* based (binomial-based for short in this paper) and *chi-square distribution* based (chi-square based for short). Each test item uses its assumed distribution to compute the P-value, which roughly represents the probability that the block is random. A test item is considered to be passed when the computed P-value is larger than the *significance level*. Then, based on the computed $N$ P-values for $N$ blocks, each test item performs two-level tests: the first-level test and the second-level test, where passing the former is the premise to execute the latter. The second-level testing approach was found to increase the testing capability [9]. The first-level test focuses on the passing ratio of the $N$ P-values, and the second-level test further focuses on the uniformity of the $N$ P-values to assess whether the test statistic values follow the expected distribution, i.e., the standard normal distribution[1] or the chi-square distribution. In the remainder of this paper, *the test statistic* refers to the test statistic value that is assumed to follow the standard normal distribution in the binomial-based tests.

**Related Work.** Several papers on the NIST SP 800-22 test suite have been presented in literature. Among the test items, Kim *et al.* [6] analyzed the correctness of the Spectral Test and the Lemple-Ziv Test, and Hamano [4,5] adjusted the distribution parameters for the Spectral Test and corrected the Overlapping Test. Sulak *et al.* [14] found that the P-values for short sequences (less than 512 bits) follow a specific discrete distribution, rather than the assumed uniform distribution for long sequences. Pareschi *et al.* [9] investigated the reliability of the

---

[1] For a sufficiently large number of trials, the distribution of the binomial sum after normalizing, is closely approximated by a standard normal distribution [11].

second-level tests, and analyzed the sensitivity to the approximation errors introduced by the computation of P-values. Furthermore, as the sequence length is finite in practice and thus the set of possible statistic values is discrete, Pareschi *et al.* [10] provided the actual distributions of P-values for the Frequency Test, the Runs Test, and the Spectral Test, and evaluated the test errors for different testing methods based on P-values. In our preliminary work [15], we analyzed the correctness and the reliability of the second-level tests in the NIST SP 800-22 test suite.

**Our Contribution.** In this paper, we find that the P-values derived from the binomial-based tests are unqualified for the second-level tests of the NIST SP 800-22 suite, though they are proper to be used for the first-level tests. The P-values in the binomial-based tests are computed, using the absolute values of the test statistics. Therefore, the second-level tests on P-values do not exactly tell whether the test statistics follow the standard normal distribution or not, because we cannot learn from the P-values that the test statistics are positive or negative. In particular, even if P-values follow the uniform distribution on [0, 1], there still exists a non-ignorable probability that the test statistics are not aligned with the expected standard normal distribution; then, it fails to detect some imperfect random sequences.

We propose a new metric called *Q-value* in this paper,[2] for the second-level tests of the binomial-based tests (but not the chi-square based ones), to replace the original P-value without any extra modification. The Q-value is computed directly using the test statistics, rather than their absolute values. We prove that the uniformity of Q-values is equal to that the test statistics follow the standard normal distribution as expected. In the case that there exists some mean drifts of the assumed normal distribution, which is commonly caused by flawed generators, the Q-value based tests produce greater gaps than the P-value based ones under both the total variation distance (TVD) and the Kullback-Leibler divergence (KLD), i.e., Q-value is more sensitive to detect such drifts. Therefore, for the binomial-based tests, our Q-value based second-level tests have greater testing capability than the P-value based ones.

Furthermore, inspired by [10], we investigate the actual distributions of P-values and Q-values with a finite block length, for the binomial-based tests. The comparison in the Frequency Test shows that the distribution of Q-value is more smooth, i.e., it is closer to the uniform distribution on [0, 1]. Hence, the Q-value based second-level tests are more reliable, i.e., our improvement also decreases the probability of erroneously identifying an ideal generator as not random.

Finally, we perform the improved statistical tests on the outputs of several PRNGs. The experimental results demonstrate that the Q-value based second-level tests are able to detect some statistical flaws that the original SP 800-22 suite cannot detect under the same test parameters.

---

[2] The term of *q-value* is defined as a measure of significance in terms of the false discovery rate [12, 13], while in this paper we use Q-value as another definition.

**Organization.** The rest of this paper is organized as follows. In Sect. 2, we introduce the two-level statistical tests included in the NIST SP 800-22 test suite. In Sect. 3, we state the problem in the second-level tests of binomial-based tests. In Sect. 4, we propose Q-value based second-level statistical tests, and investigate the detectability and the reliability. In Sect. 5, we apply the statistical tests on several popular PRNGs to validate the effectiveness. Section 6 concludes the paper.

# 2   Two-Level Statistical Tests in SP 800-22

## 2.1   Statistical Hypothesis Testing for Randomness

Hypothesis testing is a commonly used method to assess whether the tested data fit with the null hypothesis that is denoted as $\mathcal{H}_0$. In the statistical hypothesis testing, a statistic value is chosen and used to determine whether $\mathcal{H}_0$ should be accepted or rejected. Under the null hypothesis, the theoretical reference distribution of this statistic value is figured out by mathematical methods. From this reference distribution, a confidence interval is determined based on a preset *confidence level* $\gamma$ (e.g., $\gamma = 0.99$), i.e., the probability that the statistic values are inside the confidence interval is $\gamma$.

The null hypothesis in statistical tests for randomness is that, the tested bit sequence is random. In the testing, the test statistic value is computed on the tested bit sequence, and then is compared to the bounds of the confidence interval. If the test statistic value lies outside the confidence interval, the null hypothesis that the sequence is random is rejected. Otherwise, $\mathcal{H}_0$ is accepted.

A randomness test suite may contain a serial of test items, which evaluate different aspects of randomness. These test items produce different confidence intervals based on the same confidence level. Then, *P-value* is employed as a unified metric for different test items, which is calculated using the test statistic. For a randomness test item, a P-value is the probability that a perfect random number generator would have produced a sequence less random than the tested sequence [11]. More specifically, the P-value is computed as the probability of obtaining a statistic value $S$ equal to or "more extreme" than the observed value $S_{obs}$ of the tested sequence. According to the definition of "more extreme" cases, the tests are generally divided into two categories: one-sided tests and two-sided tests.

In the NIST SP 800-22 test suite, a test is considered to be two-sided when $S$ is assumed to follow a normal distribution, and the P-value is computed as $2\min\{\Pr(S > S_{obs}), \Pr(S < S_{obs})\}$. A test is considered to be one-sided when $S$ is assumed to follow a chi-squared distribution, and the P-value is computed as $\Pr(S > S_{obs})$. Figure 1 shows the computations of P-values based on the observed values for the one-sided and two-sided tests included in the NIST SP 800-22 test suite, where the shaped areas are the P-values.

Then the test is performed by comparing P-value with a *significance level* denoted as $\alpha$, and $\alpha = 1 - \gamma$ where $\gamma$ is the confidence level. If P-value $p < \alpha$, then $\mathcal{H}_0$ is rejected and the tested sequence is considered to be non-random. If $p \geq \alpha$,

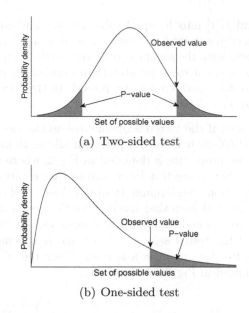

(a) Two-sided test

(b) One-sided test

Fig. 1. P-value in one-sided and two-sided tests

$\mathcal{H}_0$ is accepted and the sequence is considered to be random. When $\mathcal{H}_0$ is true and $p < \alpha$, $\mathcal{H}_0$ is erroneously rejected, which is called Type I Error. The probability of Type I Error is $\alpha$. On the contrary, the fact that, $p \geq \alpha$ when $\mathcal{H}_0$ is false, is called Type II Error. The significance level recommended by NIST is $\alpha = 0.01$.

## 2.2 Two-Level Tests

The current version of the NIST SP 800-22 test suite [11] is composed of 15 test items. According to the assumed distribution of the test statistic values, these test items are divided into two categories: the binomial-based (i.e., the two-sided tests) and the chi-square based (i.e., the one-sided tests). The Frequency (Monobit) Test, the Runs Test, the Spectral Test, Maurer's "Universal Statistical" Test, and the Random Excursions Variant Test belong to the binomial-based tests, and the others are chi-square based.

In the testing process, according to the test parameters, the whole tested bit sequence is partitioned into $N$ blocks, and each block contains $n$ bits. For each test item, the hypothesis testing, where the null hypothesis is that the tested sequence is random, is executed for each data block, and then $N$ P-values are obtained. Based on these P-values, the following two-level test is performed in each test item.

1. Count the number of the blocks whose P-values are equal or greater than $\alpha$, and compute the passing ratio. If the ratio lies in the confidence interval defined as $1 - \alpha \pm 3\sqrt{\frac{(1-\alpha)\alpha}{N}}$, the first-level test is passed;

2. Divide the interval $[0,1]$ into $K$ equal sub-intervals, and count each number of the P-values in each sub-interval. Perform a chi-square goodness-of-fit test on these $K$ numbers with the assumed uniform distribution, yielding another P-value $p_T$. If $p_T$ is equal to or greater than another significance level $\alpha_T$, the second-level test is considered to be passed. In the NIST SP 800-22 test suite, $K = 10$ and $\alpha_T = 0.0001$.

A test item is passed if the tested sequence passes the two-level test of this test item, and the SP 800-22 test suite is passed if all the 15 included test items are passed. The testing procedure is depicted in Fig. 2, where we use $N = 1000$ as an example. Note that, some test items are further composed of a serial of sub-items (such as the Non-Overlapping Template Test), and each sub-item can be treated as a separate test item that has its own P-values and $p_T$. In addition, for the Random Excursions and Random Excursions Variant Tests, the P-values are computed only if the tested sequence block meets specific criteria, so the number of available P-values may be less than $N$ for the $N$ sequence blocks. These details are omitted in Fig. 2 for simplicity.

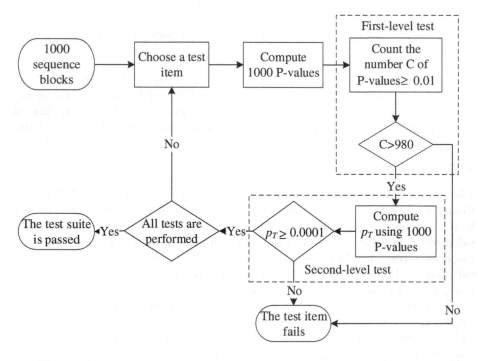

**Fig. 2.** The testing procedure of the NIST SP 800-22 test suite ($N = 1000$)

## 2.3   Frequency Test

We take the Frequency Test as an example to explain the P-value computation in the binomial-based tests. The bit block with length $n$ is denoted as

$\varepsilon = \{\varepsilon_1, \varepsilon_2, \ldots, \varepsilon_n\} \in \{0,1\}^n$. Then $S = \sum_{i=1}^{n}(2\varepsilon_i - 1)$ is computed. Under the null hypothesis, $S$ is assumed to follow a binomial distribution. As $n$ is always very large, the limiting binomial distribution is approximated as a normal distribution. Hence, $S$ is assumed to follow the normal distribution $\mathcal{N}(u, \sigma^2)$, where $u = 0$ and $\sigma^2 = n$. The test statistic $d = (S - u)/\sigma$ follows $\mathcal{N}(0,1)$. Then the P-value is computed using the cumulative distribution function (CDF) of the standard normal distribution $\Phi(\cdot)$ or the complementary error function $\mathrm{erfc}(\cdot)$:

$$p = 2(1 - \Phi(|d|)) = \mathrm{erfc}(\frac{|d|}{\sqrt{2}}),$$

where

$$\Phi(x) = \frac{1}{\sqrt{2\pi}} \int_{-\infty}^{x} e^{-\frac{\eta^2}{2}} d\eta,$$

$$\mathrm{erfc}(x) = \frac{2}{\sqrt{\pi}} \int_{x}^{\infty} e^{-\eta^2} d\eta.$$

In all binomial-based tests, the same formula is used to compute P-values based on the test statistics, while each test item has a unique formula to compute the test statistic.

## 2.4 Spectral Test

The Spectral Test is also known as the Discrete Fourier Transform (DFT) Test or the Fast Fourier Transform (FFT) Test. The purpose of this test is to detect periodic features (i.e., repetitive patterns that are near each other) in the tested sequence [11]. For tested sequence $\varepsilon = \{\varepsilon_1, \varepsilon_2, \ldots, \varepsilon_n\} \in \{0,1\}^n$, the observed value $N_1$ is assumed to follow $\mathcal{N}(u, \sigma^2)$, where $u$ is the expected number of frequency components that are beyond the 95 % threshold $T = \sqrt{(\ln \frac{1}{0.05})n}$. Then the test statistic $d$ is computed as:

$$d = \frac{N_1 - u}{\sigma},$$

where $u = 0.95n/2$, $\sigma^2 = 0.95 \cdot 0.05 \cdot n/c$, and $c = 4$ in the NIST SP 800-22 test suite.

## 3   Incompleteness of P-Value Based Second-Level Tests

In the binomial-based tests, the standard normal distribution should be used as the reference for the observed test statistics. However, we find that, when the computed P-values follow a uniform distribution, the test statistic values are aligned with the *half-normal distribution*[3], rather than the expected normal

---

[3] The half-normal distribution refers to the fold at the mean of the standard normal distribution in this paper.

distribution. We prove this observation using the following Lemma 1 [3] and Theorem 1. To ensure the continuity of the statistic values' CDF, we assume that the sequence block length $n$ is large enough in this section.

**Lemma 1.** *Let $F$ be a continuous CDF on $\mathbb{R}$ with inverse $F^{-1}$ defined by*

$$F^{-1}(z) = \inf\{x : F(x) = z, 0 < z < 1\},$$

*where inf means the infimum. If $Z$ is a uniform random variable on $[0,1]$, then $F^{-1}(Z)$ has distribution function $F$. Also, if a random variable $X$ has distribution function $F$, then $F(X)$ is uniformly distributed on $[0,1]$.*

*Proof.* The first statement follows after noting that for all $x \in \mathbb{R}$,

$$\Pr(F^{-1}(Z) \leq x) = \Pr(\inf\{y : F(y) = Z\} \leq x)$$
$$= \Pr(Z \leq F(x)) = F(x).$$

The second statement follows from the fact that for all $0 < z < 1$,

$$\Pr(F(X) \leq z) = \Pr(X \leq F^{-1}(z))$$
$$= F(F^{-1}(z)) = z. \qquad \square$$

**Theorem 1.** *Let $d$ be the test statistic in a binomial-based test, and let $p$ be the P-value computed in the test. The following two statements are equivalent: (1) $|d|$ follows the half-normal distribution, and (2) $p$ is uniformly distributed on $[0,1]$.*

*Proof.* Let $Y$ be a random variable following the half-normal distribution, and let $F_Y(\cdot)$ be the CDF of $Y$. On one hand, if $|d|$ follows the half-normal distribution, $F_Y(|d|)$ is a uniformly distributed variable on $[0,1]$ according to the second statement of Lemma 1. Since $p$ is computed as $1 - F_Y(|d|)$, $p$ is also uniformly distributed on $[0,1]$. On the other hand, if $p = 1 - F_Y(|d|)$ is uniformly distributed on $[0,1]$, $F_Y(|d|)$ also follows the uniform distribution on $[0,1]$. According to the first statement of Lemma 1 (by replacing $Z$ with $F_Y(|d|)$), $F_Y^{-1}(F_Y(|d|)) = |d|$, has the same CDF with $Y$, thus $|d|$ follows the half-normal distribution. $\square$

Obviously, the condition that $|d|$ follows the half-normal distribution is insufficient to deduce that $d$ follows the normal distribution. Therefore, for the second-level tests of the binomial-based tests, checking the uniformity of P-values is unqualified to assess whether $d$ satisfies the null hypothesis. Hence, the second-level tests in the binomial-based tests could fail to detect some imperfect random sequences or elaboratively constructed sequences.

*Remark.* As to the chi-square based tests in the NIST SP 800-22 test suite, we clarify that these tests do not have the mentioned problem. The chi-square based tests are one-sided, and their P-values are not computed from the absolute values of the test statistic values.

**Biased "Random" Sequence Construction.** Below we will construct a biased sequence, yet it passes the NIST SP 800-22 test suite with given test parameters.

1. Generate a random bit sequence with an appropriate length that passes the test suite. For example, use the Blum-Blum-Shub generator (BBS) [2] which is acknowledged as a good PRNG.
2. Perform the Frequency Test according to the test parameters $n$ and $N$: calculate the test statistic value $d_i$ of the $i$th block ($i = 1, ..., N$). For each $i$, if $d_i$ is less than zero (i.e., 0's are more than 1's), perform a bitwise NOT (negation) on the sequence block; otherwise, keep the block unchanged.

The processed sequence is significantly biased, as the number of 1's is larger than that of 0's for each block after processing. However, the processed sequence still has a very high probability to pass the test suite due to the following reasons.

– For the Frequency Test, the P-value for each block is unchanged since $|d|$ remains unchanged.
– For most test items, "0" and "1" have equal roles in the evaluation of randomness. For example, in the Block Frequency, Cumulative Sums, Runs, Spectral, Universal, Approximate Entropy, and Serial Tests, their P-values remain unchanged after processing.

The effectiveness of the construction is confirmed by the statistical testing for the original and processed BBS output sequences. The two test reports about the original and processed BBS outputs are presented in Appendix A. We emphasize that, the constructed sequence is elaborative, and changing the testing method (e.g., enlarging the block length adopted by the test) certainly can detect the bias. The goal of our construction is to demonstrate the incompleteness of the P-value based second-level test, rather than to construct a flawed sequence which can pass all the existing test methods. In practice, an undetectable flaw may occur in other manners more than the unbalance, or occur in the focused aspects of other binomial-based tests more than the Frequency Test.

## 4    Second-Level Tests Based on Q-Value

### 4.1    Q-Value

The bias in the constructed sequence above should be detected by the Frequency Test that assesses the balance of the tested sequence. In our construction experiment, as the P-value based second-level tests cannot assess the symmetry of the test statistics, the constructed sequence "bypasses" the Frequency Test, even the whole test suite. For this reason, we introduce *Q-value* to replace P-value in the second-level tests of the binomial-based tests, and Q-value is defined as

$$q = 1 - \Phi(d) = \frac{1}{2}\mathrm{erfc}(\frac{d}{\sqrt{2}}).$$

The relationship between $p$ and $q$ is

$$p = \begin{cases} 2q, & q \leq 0.5; \\ 2(1-q), & q > 0.5. \end{cases}$$

Referring to the proof of Theorem 1, we have Theorem 2 for Q-value.

**Theorem 2.** *Let d be the test statistic in a binomial-based test, and let q be the Q-value computed in the test. The following two statements are equivalent: (1) d follows the standard normal distribution, and (2) q is uniformly distributed on [0, 1].*

Checking the uniformity of Q-value is equal to assessing the distribution of $d$ rather than $|d|$. Therefore, we propose the Q-value based second-level tests to replace the original second-level tests for the binomial-based tests. In the testing process, the modification is only using $N$ Q-values rather than P-values to perform the chi-square goodness-of-fit test.

Different from P-value, Q-value is computed directly using the test statistics, rather than their absolute values. Hence, Q-value based tests are able to assess the symmetry (to zero) of the test statistics, and have greater testing capability. The constructed sequence in Sect. 3 cannot pass the Q-value based second-level test of the Frequency Test, because all the derived Q-values are not greater than 0.5.

## 4.2   Testing Capability on the Drift of Test Statistics

The second-level tests in the binomial-based tests are designed to assess the difference between the theoretical reference distribution (i.e., the standard normal distribution) and the observed distribution. In the practical testing on the output sequences of RNGs (rather than the elaboratively constructed sequences), we emphasize that both the P-value based and Q-value based tests can detect the statistical flaws when the observed distribution is quite different from the standard normal distribution. Hence, we focus on the case that the observed distribution is (or is similar to) a normal distribution, but the distribution parameters (such as the mean or the variance) drift from the ideal ones. Next, we compare the sensitivity to the drifts between the Q-value based test and the P-value based test.

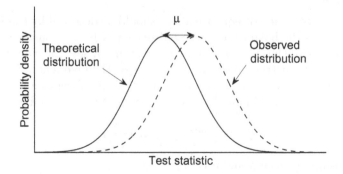

**Fig. 3.** The mean drift between the theoretical distribution and the observed one

**Mean Drift.** The mean drift is defined as the distance between the mean values of the theoretical distribution and the observed one. We assume that the test statistic $d$, which is computed by a formula on the tested data, follows the standard normal distribution. The mean drift with $\mu$ for the test statistic is depicted in Fig. 3.

Either an error in the computation formula of $d$ or the flawed data can cause a drift. For example, Kim *el al.* [6] improved the formula in the Spectral Test, which makes the the distribution of the calculated test statistics from good RNGs show better consistency with the theoretical reference distribution. The other case that, the tested data are flawed, is more common in the testing. Below we show the consequence if one uses a biased generator of a noticeable mean drift.

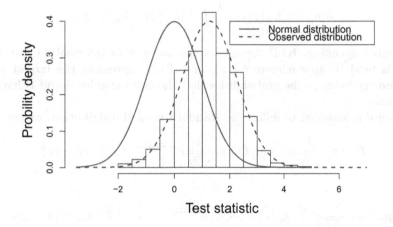

**Fig. 4.** The mean drift caused by the simulated biased sequence

The statistical flaw on the frequency (i.e., the tested sequence is biased) is common for imperfect RNGs, especially for those TRNGs where the physical phenomenons are not ideal. We assume that the flawed generator outputs a biased sequence with 50.2 % 1's. The generator is simulated by the R software [1], and the observed test statistics are computed with the parameters $n = 10^5$ and $N = 1000$. Due to the existence of the bias, the distribution of the observed test statistics has a mean drift from the expected standard normal distribution, as shown in Fig. 4. The histogram in Fig. 4 is plotted using the probability density values computed on the 1000 test statistics. The mean of the observed distribution drifts to 1.265. Knowing the inherent bias of the generator output sequence, one can optimize brute-force attacks to reduce the breaking complexity for cryptographic systems. Hence, it is important to detect the mean drift for the testing of RNGs.

**KLD and TVD.** We denote the probability distribution function (PDF) of the ideal test statistic $d$ as $f(x)$, and the PDF of $|d|$ as $g(x)$. When the mean of the observed test statistics $d_\mu$ has a drift $\mu$ ($\mu \neq 0$), the PDFs of $d_\mu$ and $|d_\mu|$ are represented as $f_\mu(x)$ and $g_\mu(x)$, respectively. If the deviation caused by $\mu$ between $f(x)$ and $f_\mu(x)$ is larger than that between $g(x)$ and $g_\mu(x)$, we say that $f(x)$ is more sensitive to the drift, and the statistical test based on $f(x)$ has greater testing capability on detecting the drift.

We choose Kullback-Leibler divergence (KLD) and total variation distance (TVD) as the measurements of the sensitivity. For PDFs $h_A(x)$ and $h_B(x)$ of two continuous random variables $A$ and $B$, the KLD between them is defined as

$$D_{\mathrm{KL}}(h_A(x) \| h_B(x)) = \int_{-\infty}^{\infty} h_A(x) \log \frac{h_A(x)}{h_B(x)} \mathrm{d}x, \tag{1}$$

and the TVD between them is defined as

$$\delta(h_A(x), h_B(x)) = \frac{1}{2} \int_{-\infty}^{\infty} |h_A(x) - h_B(x)| \mathrm{d}x. \tag{2}$$

Roughly speaking, KLD represents the amount of information lost when $h_B(x)$ is used to approximate $h_A(x)$, and TVD represents the largest possible difference between the probabilities that the two variables $A$ and $B$ have the same value.

When $d$ is assumed to follow the standard normal distribution, we get

$$f(x) = \frac{1}{\sqrt{2\pi}} e^{-\frac{x^2}{2}}, f_\mu(x) = \frac{1}{\sqrt{2\pi}} e^{-\frac{(x-\mu)^2}{2}}, x \in (-\infty, +\infty),$$

and

$$g(x) = \frac{2}{\sqrt{2\pi}} e^{-\frac{x^2}{2}}, g_\mu(x) = \frac{1}{\sqrt{2\pi}} (e^{-\frac{(x-\mu)^2}{2}} + e^{-\frac{(x+\mu)^2}{2}}), x \in [0, +\infty).$$

Then, substituting $f(x)$ and $f_\mu(x)$ into Eq. (1), we get the KLD between $f(x)$ and $f_\mu(x)$, as shown in Eq. (3).

$$D_{\mathrm{KL}}(f(x) \| f_\mu(x)) = \int_{-\infty}^{\infty} f(x) \log \frac{f(x)}{f_\mu(x)} \mathrm{d}x \tag{3}$$

$$= \int_{-\infty}^{\infty} \frac{1}{\sqrt{2\pi}} e^{-\frac{x^2}{2}} \log \frac{e^{-\frac{x^2}{2}}}{e^{-\frac{(x-\mu)^2}{2}}} \mathrm{d}x = \frac{\mu^2}{2}$$

By noting that $e^{-\frac{(x-\mu)^2}{2}} + e^{-\frac{(x+\mu)^2}{2}} \geq 2e^{-\frac{x^2+\mu^2}{2}}$, we get the KLD between $g(x)$ and $g_\mu(x)$, which is strictly smaller than $\mu^2/2$, as shown in Eq. (4).

$$D_{\mathrm{KL}}(g(x) \| g_\mu(x)) = \int_{-\infty}^{\infty} g(x) \log \frac{g(x)}{g_\mu(x)} \mathrm{d}x \tag{4}$$

$$= \int_{0}^{\infty} \frac{2}{\sqrt{2\pi}} e^{-\frac{x^2}{2}} \log \frac{2e^{-\frac{x^2}{2}}}{(e^{-\frac{(x-\mu)^2}{2}} + e^{-\frac{(x+\mu)^2}{2}})} \mathrm{d}x < \frac{\mu^2}{2}$$

By observing that $e^{-\frac{x^2}{2}}$ and $e^{-\frac{(x-\mu)^2}{2}}$ are symmetrical to $x = 0$ and $x = \mu$, respectively, we compare the result between $\delta(f(x), f_\mu(x))$ and $\delta(g(x), g_\mu(x))$, as shown in Eq. (5).

$$\delta(f(x), f_\mu(x)) = \frac{1}{2} \cdot \frac{1}{\sqrt{2\pi}} \int_{-\infty}^{\infty} |e^{-\frac{x^2}{2}} - e^{-\frac{(x-\mu)^2}{2}}| dx \tag{5}$$

$$= \frac{1}{2\sqrt{2\pi}} \int_{-\infty}^{\frac{-\mu}{2}} e^{-\frac{x^2}{2}} - e^{-\frac{(x-\mu)^2}{2}} dx + \frac{1}{2\sqrt{2\pi}} \int_{-\frac{\mu}{2}}^{\frac{\mu}{2}} e^{-\frac{x^2}{2}} - e^{-\frac{(x-\mu)^2}{2}} dx$$

$$+ \frac{1}{2\sqrt{2\pi}} \int_{\frac{\mu}{2}}^{\infty} e^{-\frac{(x-\mu)^2}{2}} - e^{-\frac{x^2}{2}} dx$$

$$= \frac{1}{2\sqrt{2\pi}} \int_{\frac{\mu}{2}}^{\infty} |(e^{-\frac{(x-\mu)^2}{2}} - e^{-\frac{x^2}{2}})| + |(e^{-\frac{x^2}{2}} - e^{-\frac{(x+\mu)^2}{2}})| dx$$

$$+ \frac{1}{2\sqrt{2\pi}} \int_{0}^{\frac{\mu}{2}} |2e^{-\frac{x^2}{2}} - e^{-\frac{(x-\mu)^2}{2}} - e^{-\frac{(x+\mu)^2}{2}}| dx$$

$$> \frac{1}{2\sqrt{2\pi}} \int_{0}^{\infty} |2e^{-\frac{x^2}{2}} - e^{-\frac{(x-\mu)^2}{2}} - e^{-\frac{(x+\mu)^2}{2}}| dx = \delta(g(x), g_\mu(x))$$

From Eqs. (3)–(5), we deduce $D_{KL}(f(x)\|f_\mu(x)) > D_{KL}(g(x)\|g_\mu(x))$ and $\delta(f(x), f_\mu(x)) > \delta(g(x), g_\mu(x))$.

The KLD and TVD results with $\mu = 0.5$ on the normal distribution $f(x)$ and the half-normal distribution $g(x)$ are also depicted in Fig. 5, where the distances represent the integral parts in Eqs. (1) and (2). We can see that the change caused by $\mu$ in the normal distribution is larger than that in the half-normal distribution, which means that the test based on the normal distribution is more sensitive to the drift. Thus, we conclude that Q-value based second-level tests are more powerful than the P-value based ones to detect the mean drift of the test statistics.

Regarding to the drift of the variance, we note that the testing capability of the two testing methods are identical, as their KLDs (or TVDs) are equal.

## 4.3 Testing Reliability Analysis Based on Actual Distribution

An asymptotic distribution refers to the limiting distribution when $n$ approaches infinity. The asymptotic distribution of P-value is the uniform distribution on $[0, 1]$. However, in the practical cases that $n$ is finite, the number of possible P-values is limited, i.e., the set of P-values is discrete. This fact makes the actual distribution of P-value is not a perfect uniform distribution on $[0, 1]$. When the number of blocks $N$ is very large, the inconsistency is revealed and the observed P-values do not follow the assumed uniform distribution, which makes these P-values fail the chi-square test in the second-level test. This decreases the reliability of the statistical tests, i.e., increases the probability of erroneously identifying an ideal generator as not random.

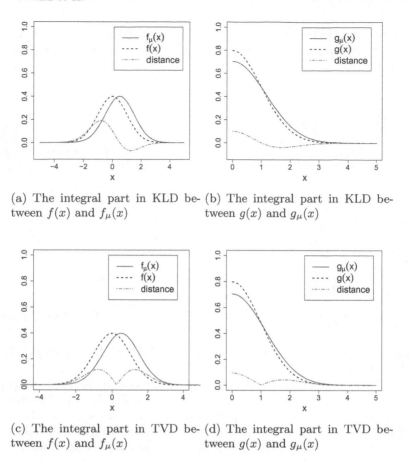

(a) The integral part in KLD be-  (b) The integral part in KLD between $f(x)$ and $f_\mu(x)$        tween $g(x)$ and $g_\mu(x)$

(c) The integral part in TVD be-  (d) The integral part in TVD between $f(x)$ and $f_\mu(x)$        tween $g(x)$ and $g_\mu(x)$

**Fig. 5.** The comparison of TVDs and KLDs with the drift $\mu = 0.5$

In order to investigate the reliability of the Q-value based second level tests, we deduce the actual distributions of Q-values for the binomial-based tests, and compare them with those of P-values. The actual distributions of P-values for the binomial-based tests have been analyzed in [10]. The actual distribution of Q-value is closer to the assumed uniform distribution, meaning that the Q-value based test has a lower probability that a sequence with perfect randomness fails the test, i.e., higher reliability.

**Actual Distribution.** As we mentioned in Sect. 2.3, each binomial-based test computes its normally distributed value $S \sim \mathcal{N}(u, \sigma^2)$. For an $n$-bit sequence block, the number of possible values of $S$ is denoted as $m$. The possible values are increasingly ordered as $\mathcal{S} = \{s_1, s_2, ..., s_m\}$, i.e., $s_{i-1} < s_i$ for $i = 2, ..., m$. Note that the variables $u, \sigma, m$ depend on the specific test item, such as the Frequency, Runs, Spectral, Universal, and Random Excursions Variant Tests.

As our goal is to provide a general conclusion for the binomial-based tests, we do not consider the specific values of these variables.

For simplicity, we consider a common situation that $m$ is odd and $S$ is symmetrical with respect to $u$. For each $s_i$, P-value $p_i = \mathrm{erfc}(\frac{|s_i - u|}{\sqrt{2}\sigma})$, and Q-value $q_i = \frac{1}{2}\mathrm{erfc}(\frac{s_i - u}{\sqrt{2}\sigma})$. The sets of possible P-values and possible Q-values are denoted as $\mathcal{P}$ and $\mathcal{Q}$, respectively. According to the symmetry of $S$, it is observed that $p_i = p_{m+1-i}$ and $q_i + q_{m+1-i} = 1$, thus the cardinality $|\mathcal{P}| = m/2 + 1$ and $|\mathcal{Q}| = m$.

The actual CDFs of P-value and Q-value are represented as:

$$F'_p(x) = \sum_{i=1}^{m} \Pr\{S = s_i\} U(x - p_i), \tag{6}$$

$$F'_q(x) = \sum_{j=1}^{m} \Pr\{S = s_j\} U(x - q_j), \tag{7}$$

where

$$U(x) = \begin{cases} 1, x \geq 0; \\ 0, x < 0. \end{cases}$$

Using the property $p_i = p_{m+1-i}$, $F'_p(x)$ is rewritten as:

$$F'_p(x) = 2 \sum_{i=1}^{(m-1)/2} \Pr\{S = s_i\} U(x - p_i) + \Pr\{S = s_{\frac{m+1}{2}}\} U(x - p_{\frac{m+1}{2}}). \tag{8}$$

In fact, these two CDFs are both stepladder-like functions. We compare the number, height, and width of the steps between $F'_p(x)$ and $F'_q(x)$. Note that $|\mathcal{Q}|$ is almost as twice as $|\mathcal{P}|$, so the number of steps in $F'_q(x)$ is approximately as twice as that in $F'_p(x)$. The coefficient of the step function in $F'_p(x)$ is as twice as that in $F'_q(x)$, so the maximum width and height of the step in $F'_p(x)$ are also as twice as those in $F'_q(x)$. Therefore, the actual distribution of Q-values is more smooth, and is closer to the uniform distribution than that of P-values.

It should be noted that we assume $S$ is symmetrical with respect to $u$, the mean of the asymptotic distribution. The assumption is appropriate for the Frequency Test; however, in other binomial-based tests, there may be a little deviation between $u$ and the mean of $S$. We leave the study on this case as our future work.

**Actual Distribution in the Frequency Test.** We take the Frequency Test as an example to demonstrate the difference between the distributions of P-value and Q-value. Without loss of generality, the length $n$ of the sequence block is assumed to be even. It is easy to figure out that $|\mathcal{S}| = n + 1$, $|\mathcal{P}| = \frac{n}{2} + 1$, $|\mathcal{Q}| = n+1$, and $u = 0$, $\sigma^2 = n$. Then, from Equation (8) we get the actual CDF of P-value:

$$F'_p(x) = 2 \sum_i \Pr\{S = s_i\} U(x - p_i) + \frac{2}{\sqrt{2\pi n}} U(x - 1),$$

where $\Pr\{S = s_i\} = 2^{-n}\binom{n}{i-1} \approx \frac{2}{\sqrt{2\pi n}}e^{\frac{-(2i-n-2)^2}{2n}}$, $p_i = \mathrm{erfc}(\frac{|2i-n-2|}{\sqrt{2n}})$, and $i \in \{1, 2, ..., \frac{n}{2}\}$.

From Eq. (7), we get the actual CDF of Q-value:

$$F'_q(x) = \sum_j \Pr\{S = s_j\}U(x - q_j),$$

where $\Pr\{S = s_j\} = 2^{-n}\binom{n}{j-1} \approx \frac{2}{\sqrt{2\pi n}}e^{\frac{-(2j-n-2)^2}{2n}}$, $q_j = \frac{1}{2}\mathrm{erfc}(\frac{2j-n-2}{\sqrt{2n}})$ and $j \in \{1, 2, ..., n+1\}$.

For the parameter $n = 200$, we plot the actual CDFs of P-value and Q-value, as shown in Fig. 6. It is observed that Q-value's actual CDF is closer to the uniform distribution than P-value's.

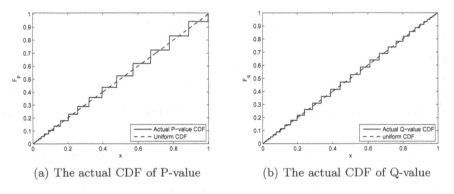

(a) The actual CDF of P-value          (b) The actual CDF of Q-value

**Fig. 6.** Actual CDF comparison between P-value and Q-value for the Frequency Test ($n = 200$)

Then, we compare the uniformity between actual P-values and Q-values through the chi-square goodness-of-fit test. Here we choose $n = 2^{20}$ and $K = 16$ to better express the difference between Q-values and P-values in the chi-square test. As shown in Eq. (9), the statistic value $\chi^2$ is computed using $O_i$ which is the number of P-values or Q-values in the $i$th sub-interval.

$$\chi^2 = \sum_{i=1}^{K} \frac{(O_i - N/K)^2}{N/K} \tag{9}$$

Using the Q-value and P-value CDFs with $n = 2^{20}$, we calculate two sets of $O_i$ based on P-values and Q-values, respectively. As expected, the set of $O_i$ based on Q-values shows better consistency with the uniform distribution than that based on P-values, as shown in Fig. 7. Therefore, we conclude that, under the same test parameters, the Q-value based second-level test has higher reliability than the P-value based one.

To verify the correctness of the derived actual CDF of Q-value, we test the BBS output sequence with test parameters $n = 2^{10}$ and $N = 100000$, and count

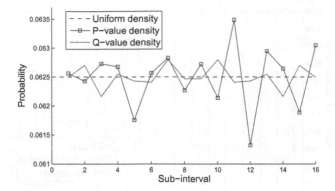

**Fig. 7.** The probability comparison between P-values and Q-values in each sub-interval ($K = 16$, $n = 2^{20}$)

the number of Q-values in each sub-interval. The experimental and theoretical counting results in each sub-interval are shown in Fig. 8, which shows good consistency between the theory and the experiment.

## 5    Statistical Tests on PRNGs

In this section, our experiments confirm that the Q-value based second-level tests have lower probabilities to erroneously identify good RNGs as not random, and also demonstrate that they have greater testing capability.

### 5.1    Experiment Setup

We choose several popular PRNGs including BBS, Linear Congruential Generator (LCG), Modular Exponentiation Generator (MODEXPG), and Micall-Schnorr Generator (MSG), and test their original output sequences using the NIST Statistical Test Suite (sts v2.1) [8] and our version using Q-values.

The test parameters adopted by each test item are the default values specified in the sts v2.1 toolkit. Also, we run the PRNG functions included in the toolkit to generate the output sequences, and the input parameters for these PRNGs are the default values fixed in the source code of sts v2.1, where the default seed of LCG is 23482349.

### 5.2    Statistical Testing

Using the recommended test parameters $n = 10^6$ and $N = 1000$, we perform statistical tests on the output sequences of these PRNGs. We only list the second-level test results (i.e., $p_t$'s) for the Frequency Test, the Runs Test, the Spectral Test, and the Universal Test, as shown in Table 1. We omit the results of the

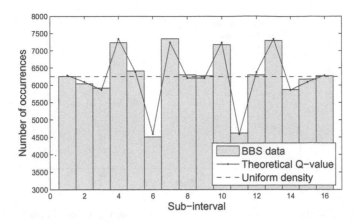

**Fig. 8.** Q-value comparison between the experimental and theoretical results in each sub-interval ($n = 2^{10}$, $N = 100000$, $K = 16$)

**Table 1.** Second-level test results for PRNGs ($n = 10^6$, $N = 1000$)

| PRNG | Second-level Test | Frequency | Runs | Spectral $(c = 4)$ | Spectral $(c = 3.8)$ | Universal |
|------|------|------|------|------|------|------|
| BBS | P-value | 0.6641 | 0.6350 | 0.5281 | 0.2480 | 0.4299 |
|     | Q-value | 0.4817 | 0.9379 | 0.0218 | 0.0113 | 0.4263 |
| MSG | P-value | 0.3899 | 0.1746 | 0.6642 | 0.9619 | 0.7734 |
|     | Q-value | 0.8055 | 0.1786 | 0.1825 | 0.6350 | 0.9996 |
| LCG | P-value | 0.8596 | 0.7075 | 0.4788 | 0.6392 | 0.8111 |
|     | Q-value | 0.4769 | 0.8905 | **0.0007** | **0.0026** | 0.4447 |
| MODEXPG | P-value | 0.0 | * | 0.4541 | 0.2636 | 0.2676 |
|         | Q-value | 0.0 | * | 0.1538 | 0.1107 | 0.7578 |

\* The first-level test fails.

Random Excursions Variant Test, for 18 different subitems are included in this item and all these subitems are passed.

All the tested sequences of these PRNGs pass the whole original SP 800-22 test suite, except for MODEXPG. For BBS, MSG, and LCG, the $p_t$'s of the three binomial-based tests are all greater than the preset threshold $\alpha_T = 0.0001$, thus these PRNGs pass both P-value based and Q-value based second-level tests. For MODEXPG, the Frequency Test fails either for P-value or Q-value. However, the Q-value's $p_t$ of LCG in the Spectral Test becomes very small (0.0007), which indicates that the test statistics are not well consistent with the standard normal distribution, though $p_t$ is still greater than $\alpha_T$.

In order to confirm the discovery in the Spectral Test, we plot the histograms using the probability density values computed on the 1000 test statistics from LCG or MSG, and compare them with the PDF of the standard normal distribution, as depicted in Fig. 9. The distribution of the test statistics of MSG has

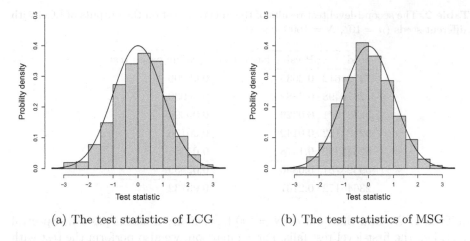

(a) The test statistics of LCG

(b) The test statistics of MSG

**Fig. 9.** Comparison between the standard normal distribution and the distribution of test statistics for LCG and MSG

better consistency with the standard normal distribution, while the distribution of the test statistics of LCG drifts to the right. As we analyzed in Sect. 4.2, the Q-value based test is more sensitive to the mean drift, thus detects the drift better than the P-value based test.

For the Spectral Test, Pareschi *et al.* [10] pointed out that the variance $\sigma^2 = 0.95 \cdot 0.05 \cdot n/c$ with $c = 3.8$, is closer to the ideal distribution than the original value ($c = 4$) in the NIST SP 800-22 test suite [4,6,11]. Here we emphasize that the modification only adjusts the variance of the test statistic value, rather than the mean. The reason why the tested sequence almost fails the Spectral Test is the asymmetry of the statistic values. Therefore, the mean drift (or the asymmetry) still exists after modifying the variance, thus the Q-value based test can still detect the drift. This is confirmed by the experiment, and the experimental results for $c = 3.8$ are also shown in Table 1.

## 5.3 Further Analysis on LCG

We repeat the Spectral Test on the output sequences of LCG with different seeds, and the $p_t$ results of the P-value and Q-valued based second-level tests are presented in Table 2. From Table 2, we confirm that the conflict in Table 1 is not a coincidence or individual example, as similar results are also obtained for other seeds. It is noted that the choice of the LCG parameters has an impact on the quality of the output, thus the output sequences derived from some seeds are possible to show better statistical properties, as shown in the latter rows of Table 2.

Although we get small $p_t$'s in the Q-value based second-level tests for LCG outputs, the sequence is still considered to pass the test ($p_t \geq \alpha_T = 0.0001$). Therefore, we further test the LCG outputs using a longer block length $n = 10^7$ to improve the testing capability, and the tested sequence is the same with that

**Table 2.** The second-level test results of the Spectral Test on the outputs of LCG with different seeds ($n = 10^6$, $N = 1000$, $c = 4$)

| Seed | P-value based test | Q-value based test |
|---|---|---|
| 73724612 | 0.3635 | 0.00006 |
| 12876498 | 0.2882 | 0.00030 |
| 52731971 | 0.0329 | 0.00096 |
| 92134122 | 0.0142 | 0.00106 |
| 82345342 | 0.1478 | 0.01581 |
| 59823781 | 0.6890 | 0.02959 |
| 23646172 | 0.2167 | 0.03732 |

in Table 1. We find that, out of $N = 100$ blocks only 2 blocks pass the Spectral Test, i.e., the first-level test fails. For comparison, we also perform the test with $n = 10^7$ and $N = 100$ on the same BBS output sequence in Table 1, and the test is still passed. The detailed test reports are presented in Appendix B.

It is reasonable to conclude that the Q-value based second-level tests improve the detectability under the same test parameters. In the process of increasing the block length to improve the testing capability, the Q-value based second-level tests discover statistical flaws sooner.

## 6   Conclusion

We investigate the testing capability of the second-level tests of the binomial-based tests in the NIST SP 800-22 test suite, and find that, the sequence that passes the tests could still have statistical flaws in the assessed aspect. Hence, we propose *Q-value* as the metric for the second-level tests to replace the original P-value without any extra modification. The Q-value based second-level test is applicable for all the five binomial-based tests, including the Frequency, Runs, Spectral, Universal, and Random Excursions Variant Tests. We provide the correctness proof of the proposed Q-value based second-level tests, and the distance analyses show that the modification improves the testing capability. Surprisingly, the comparison between the P-value's and Q-value's actual distributions indicates that the testing reliability is also improved. The experiments on several popular PRNGs demonstrate that the Q-value based second-level tests improve the detectability under the same test parameters. In the future, we will study the effectiveness of our method on TRNGs, and further analyze the properties of the Q-value based second-level tests.

**Acknowledgments.** We thank the anonymous reviewers of CHES 2016 and ASI-ACRYPT 2016, for their invaluable suggestions and comments to improve the quality and fairness of this paper. This work was partially supported by National Basic Research Program of China (973 Program No. 2013CB338001), National Natural Science Foundation of China (No. 61602476) and Strategy Pilot Project of Chinese Academy of Sciences (No. XDA06010702).

## A  Statistical Test Results on the Original and Processed BBS Output Sequences

See Tables 3 and 4.

**Table 3.** Statistical test report of the original BBS outputs ($n = 10^6$, $N = 10^3$)

| C1 | C2 | C3 | C4 | C5 | C6 | C7 | C8 | C9 | C10 | P-VALUE | PROPO | STATISTICAL TEST |
|----|----|----|----|----|----|----|----|----|-----|---------|-------|-------------------|
| 102 | 88 | 101 | 108 | 101 | 89 | 93 | 95 | 111 | 112 | 0.664168 | 995/1000 | Frequency |
| 86 | 109 | 99 | 91 | 89 | 106 | 106 | 94 | 116 | 104 | 0.474986 | 987/1000 | BlockFrequency |
| 88 | 102 | 85 | 106 | 112 | 98 | 88 | 109 | 103 | 109 | 0.463512 | 994/1000 | CumulativeSums |
| 97 | 96 | 83 | 104 | 103 | 94 | 107 | 109 | 103 | 104 | 0.807412 | 996/1000 | CumulativeSums |
| 89 | 115 | 89 | 103 | 102 | 95 | 100 | 102 | 93 | 112 | 0.635037 | 990/1000 | Runs |
| 98 | 90 | 97 | 101 | 116 | 102 | 99 | 93 | 104 | 100 | 0.883171 | 994/1000 | LongestRun |
| 103 | 98 | 80 | 92 | 102 | 96 | 116 | 94 | 108 | 111 | 0.371941 | 989/1000 | Rank |
| 107 | 108 | 83 | 99 | 109 | 101 | 90 | 94 | 96 | 113 | 0.528111 | 983/1000 | FFT |
| 97 | 104 | 101 | 118 | 84 | 86 | 112 | 94 | 97 | 107 | 0.319084 | 993/1000 | NonOverlappingTemplate |
| 103 | 101 | 104 | 106 | 112 | 94 | 90 | 95 | 90 | 105 | 0.841226 | 992/1000 | OverlappingTemplate |
| 114 | 118 | 104 | 101 | 98 | 93 | 93 | 97 | 84 | 98 | 0.429923 | 987/1000 | Universal |
| 107 | 98 | 97 | 89 | 95 | 99 | 101 | 101 | 106 | 107 | 0.965860 | 995/1000 | ApproximateEntropy |
| 62 | 58 | 67 | 60 | 68 | 61 | 53 | 63 | 60 | 52 | 0.906970 | 598/604 | RandomExcursions |
| 59 | 53 | 55 | 56 | 66 | 59 | 73 | 51 | 58 | 74 | 0.380976 | 600/604 | RandomExcursionsVariant |
| 90 | 96 | 81 | 105 | 109 | 96 | 104 | 117 | 109 | 93 | 0.323668 | 994/1000 | Serial |
| 81 | 97 | 91 | 104 | 112 | 100 | 105 | 103 | 113 | 94 | 0.484646 | 988/1000 | Serial |
| 111 | 95 | 100 | 107 | 113 | 88 | 97 | 97 | 97 | 95 | 0.779188 | 993/1000 | LinearComplexity |

**Table 4.** Statistical test report of the processed BBS outputs ($n = 10^6$, $N = 10^3$)

| C1 | C2 | C3 | C4 | C5 | C6 | C7 | C8 | C9 | C10 | P-VALUE | PROPO | STATISTICAL TEST |
|----|----|----|----|----|----|----|----|----|-----|---------|-------|-------------------|
| 102 | 88 | 101 | 108 | 101 | 89 | 93 | 95 | 111 | 112 | 0.664168 | 995/1000 | Frequency |
| 86 | 109 | 99 | 91 | 89 | 106 | 106 | 94 | 116 | 104 | 0.474986 | 987/1000 | BlockFrequency |
| 88 | 102 | 85 | 106 | 112 | 98 | 88 | 109 | 103 | 109 | 0.463512 | 994/1000 | CumulativeSums |
| 97 | 96 | 83 | 104 | 103 | 94 | 107 | 109 | 103 | 104 | 0.807412 | 996/1000 | CumulativeSums |
| 89 | 115 | 89 | 103 | 102 | 95 | 100 | 102 | 93 | 112 | 0.635037 | 990/1000 | Runs |
| 101 | 88 | 88 | 111 | 118 | 100 | 104 | 99 | 92 | 99 | 0.518106 | 993/1000 | LongestRun |
| 99 | 100 | 81 | 85 | 107 | 98 | 111 | 98 | 110 | 111 | 0.361938 | 990/1000 | Rank |
| 107 | 108 | 83 | 99 | 109 | 101 | 90 | 94 | 96 | 113 | 0.528111 | 983/1000 | FFT |
| 98 | 102 | 88 | 105 | 91 | 105 | 104 | 97 | 102 | 108 | 0.926487 | 994/1000 | NonOverlappingTemplate |
| 122 | 89 | 90 | 98 | 112 | 96 | 109 | 108 | 84 | 92 | 0.147815 | 991/1000 | OverlappingTemplate |
| 114 | 118 | 104 | 101 | 98 | 93 | 93 | 97 | 84 | 98 | 0.429923 | 987/1000 | Universal |
| 107 | 98 | 97 | 89 | 95 | 99 | 101 | 101 | 106 | 107 | 0.965860 | 995/1000 | ApproximateEntropy |
| 61 | 56 | 57 | 53 | 74 | 57 | 64 | 65 | 65 | 52 | 0.654467 | 597/604 | RandomExcursions |
| 56 | 60 | 58 | 55 | 76 | 54 | 69 | 51 | 54 | 71 | 0.280306 | 601/604 | RandomExcursionsVariant |
| 90 | 96 | 81 | 105 | 109 | 96 | 104 | 117 | 109 | 93 | 0.323668 | 994/1000 | Serial |
| 81 | 97 | 91 | 104 | 112 | 100 | 105 | 103 | 113 | 94 | 0.484646 | 988/1000 | Serial |
| 97 | 92 | 110 | 99 | 101 | 105 | 98 | 97 | 108 | 93 | 0.953089 | 992/1000 | LinearComplexity |

# B Statistical Test Results with the Longer Block Length on the LCG and BBS Output Sequences

See Tables 5 and 6.

**Table 5.** Statistical test report of the LCG outputs ($n = 10^7$, $N = 10^2$)

| C1 | C2 | C3 | C4 | C5 | C6 | C7 | C8 | C9 | C10 | P-VALUE | PROPO | STATISTICAL TEST |
|----|----|----|----|----|----|----|----|----|-----|---------|-------|------------------|
| 12 | 16 | 4 | 12 | 12 | 8 | 10 | 8 | 11 | 7 | 0.334538 | 100/100 | Frequency |
| 12 | 10 | 8 | 11 | 11 | 8 | 11 | 10 | 6 | 13 | 0.911413 | 100/100 | BlockFrequency |
| 12 | 13 | 14 | 6 | 6 | 11 | 8 | 7 | 13 | 10 | 0.494392 | 99/100 | CumulativeSums |
| 9 | 11 | 10 | 11 | 11 | 17 | 10 | 6 | 9 | 6 | 0.474986 | 100/100 | CumulativeSums |
| 9 | 10 | 7 | 12 | 13 | 8 | 13 | 12 | 9 | 7 | 0.834308 | 99/100 | Runs |
| 9 | 10 | 9 | 10 | 13 | 16 | 7 | 6 | 12 | 8 | 0.534146 | 100/100 | LongestRun |
| 14 | 10 | 12 | 7 | 11 | 7 | 9 | 12 | 9 | 9 | 0.867692 | 98/100 | Rank |
| 100 | 0 | 0 | 0 | 0 | 0 | 0 | 0 | 0 | 0 | **0.000000** | **2/100** | FFT |
| 6 | 11 | 10 | 10 | 6 | 15 | 12 | 11 | 14 | 5 | 0.319084 | 100/100 | NonOverlappingTemplate |
| 17 | 13 | 7 | 10 | 13 | 7 | 8 | 6 | 9 | 10 | 0.304126 | 97/100 | OverlappingTemplate |
| 6 | 8 | 7 | 8 | 12 | 10 | 7 | 17 | 12 | 13 | 0.289667 | 98/100 | Universal |
| 10 | 3 | 8 | 10 | 8 | 12 | 13 | 6 | 21 | 9 | 0.013569 | 99/100 | ApproximateEntropy |
| 8 | 14 | 7 | 13 | 3 | 9 | 6 | 13 | 9 | 8 | 0.213309 | 88/90 | RandomExcursions |
| 9 | 8 | 13 | 11 | 7 | 9 | 11 | 11 | 6 | 5 | 0.694743 | 89/90 | RandomExcursionsVariant |
| 5 | 9 | 10 | 9 | 4 | 12 | 8 | 18 | 10 | 15 | 0.066882 | 100/100 | Serial |
| 8 | 10 | 11 | 7 | 13 | 6 | 12 | 13 | 10 | 10 | 0.816537 | 100/100 | Serial |
| 7 | 11 | 6 | 8 | 13 | 11 | 14 | 7 | 9 | 14 | 0.514124 | 100/100 | LinearComplexity |

**Table 6.** Statistical test report of the BBS outputs ($n = 10^7$, $N = 10^2$)

| C1 | C2 | C3 | C4 | C5 | C6 | C7 | C8 | C9 | C10 | P-VALUE | PROPO | STATISTICAL TEST |
|----|----|----|----|----|----|----|----|----|-----|---------|-------|------------------|
| 12 | 10 | 7 | 6 | 11 | 6 | 15 | 7 | 8 | 18 | 0.096578 | 99/100 | Frequency |
| 11 | 8 | 7 | 11 | 8 | 7 | 6 | 12 | 14 | 16 | 0.350485 | 100/100 | BlockFrequency |
| 12 | 10 | 4 | 15 | 8 | 10 | 8 | 8 | 13 | 12 | 0.437274 | 99/100 | CumulativeSums |
| 12 | 5 | 10 | 12 | 9 | 4 | 11 | 13 | 11 | 13 | 0.437274 | 99/100 | CumulativeSums |
| 11 | 12 | 10 | 12 | 8 | 6 | 8 | 8 | 13 | 12 | 0.834308 | 100/100 | Runs |
| 7 | 16 | 14 | 11 | 5 | 7 | 13 | 9 | 5 | 13 | 0.122325 | 100/100 | LongestRun |
| 11 | 11 | 12 | 6 | 8 | 7 | 10 | 11 | 10 | 14 | 0.816537 | 99/100 | Rank |
| 16 | 13 | 13 | 6 | 14 | 10 | 7 | 7 | 5 | 9 | 0.162606 | 97/100 | FFT |
| 13 | 11 | 12 | 7 | 6 | 7 | 6 | 10 | 11 | 17 | 0.249284 | 97/100 | NonOverlappingTemplate |
| 18 | 12 | 10 | 11 | 8 | 9 | 11 | 7 | 7 | 7 | 0.334538 | 96/100 | OverlappingTemplate |
| 15 | 10 | 9 | 8 | 8 | 8 | 12 | 11 | 7 | 12 | 0.779188 | 100/100 | Universal |
| 9 | 11 | 11 | 6 | 7 | 8 | 10 | 7 | 8 | 23 | 0.010988 | 99/100 | ApproximateEntropy |
| 11 | 8 | 9 | 7 | 7 | 9 | 9 | 7 | 7 | 14 | 0.689019 | 88/88 | RandomExcursions |
| 9 | 8 | 8 | 6 | 10 | 7 | 4 | 11 | 17 | 8 | 0.105618 | 87/88 | RandomExcursionsVariant |
| 9 | 12 | 10 | 9 | 13 | 10 | 7 | 6 | 7 | 17 | 0.366918 | 100/100 | Serial |
| 7 | 20 | 8 | 8 | 9 | 9 | 8 | 8 | 9 | 14 | 0.108791 | 99/100 | Serial |
| 11 | 7 | 11 | 10 | 14 | 12 | 6 | 8 | 12 | 9 | 0.779188 | 100/100 | LinearComplexity |

# References

1. The R project for statistical computing. http://www.r-project.org
2. Blum, L., Blum, M., Shub, M.: A simple unpredictable pseudo-random number generator. SIAM J. Comput. **15**(2), 364–383 (1986)
3. Devroye, L.: Introduction. In: Devroye, L. (ed.) Non-Uniform Random Variate Generation, pp. 1–26. Springer, New York (1986)
4. Hamano, K.: The distribution of the spectrum for the discrete fourier transform test included in SP800-22. IEICE Trans. **88–A**(1), 67–73 (2005)
5. Hamano, K., Kaneko, T.: Correction of overlapping template matching test included in NIST randomness test suite. IEICE Trans. **90–A**(9), 1788–1792 (2007)
6. Kim, S., Umeno, K., Hasegawa, A.: Corrections of the NIST statistical test suite for randomness. IACR Cryptology ePrint Archive 2004, 18 (2004). http://eprint.iacr.org/2004/018
7. Marsaglia, G.: Diehard Battery of Tests of Randomness. http://www.stat.fsu.edu/pub/diehard/
8. NIST: Statistical test suite (sts 2.1). http://csrc.nist.gov/groups/ST/toolkit/rng/documents/sts-2.1.2.zip
9. Pareschi, F., Rovatti, R., Setti, G.: Second-level NIST randomness tests for improving test reliability. In: International Symposium on Circuits and Systems (ISCAS 2007), pp. 1437–1440 (2007)
10. Pareschi, F., Rovatti, R., Setti, G.: On statistical tests for randomness included in the NIST SP800-22 test suite and based on the binomial distribution. IEEE Trans. Inf. Forensics Secur. **7**(2), 491–505 (2012)
11. Rukhin, A., et al.: A statistical test suite for random and pseudorandom number generators for cryptographic applications. NIST Special Publication 800–22. http://csrc.nist.gov/publications/nistpubs/800-22-rev1a/SP800-22rev1a.pdf
12. Storey, J.D.: The positive false discovery rate: A bayesian interpretation and the q-value. Ann. Stat. **31**(6), 2013–2035 (2003)
13. Storey, J.D., Tibshirani, R.: Statistical significance for genomewide studies. Proc. Nat. Acad. Sci. **100**(16), 9440–9445 (2003)
14. Sulak, F., Doğanaksoy, A., Ege, B., Koçak, O.: Evaluation of randomness test results for short sequences. In: Carlet, C., Pott, A. (eds.) SETA 2010. LNCS, vol. 6338, pp. 309–319. Springer, Heidelberg (2010). doi:10.1007/978-3-642-15874-2_27
15. Zhuang, J., Ma, Y., Zhu, S., Lin, J., Jing, J.: Q-value test: a new method on randomness statistical test. J. Cryptologic Res. **3**(2), 192–201 (2016). (in Chinese)

# Authenticated Encryption

# Trick or Tweak: On the (In)security of OTR's Tweaks

Raphael Bost[1,2]([✉]) and Olivier Sanders[3]

[1] Direction Générale de l'Armement - Maîtrise de l'Information, Bruz, France
raphael_bost@alumni.brown.edu
[2] Université de Rennes 1, Rennes, France
[3] Orange Labs, Cesson-Sévigné, France

**Abstract.** Tweakable blockcipher (TBC) is a powerful tool to design authenticated encryption schemes as illustrated by Minematsu's Offset Two Rounds (OTR) construction. It considers an additional input, called tweak, to a standard blockcipher which adds some variability to this primitive. More specifically, each tweak is expected to define a different, independent pseudo-random permutation.

In this work we focus on OTR's way to instantiate a TBC and show that it does not achieve independence for a large amount of parameters. We indeed describe collisions between the input masks derived from the tweaks and explain how they result in practical attacks against this scheme, breaking privacy, authenticity, or both, using a single encryption query, with advantage at least 1/4.

We stress however that our results do not invalidate the OTR construction as a whole but simply prove that the TBC's input masks should be designed differently.

## 1 Introduction

Communications over an insecure channel usually rise the issue of confidentiality and authenticity of data exchanged through this channel. Although efficient solutions are known for each of these properties individually, their combination to ensure both is not obvious [BN00, Kra01] and has, in practice, resulted in security breaches (*e.g.* [Kra01, AP13]). Also, the combination of different constructions, potentially relying on different primitives, may reveal quite costly.

Designing an *authenticated encryption* (AE) scheme, which efficiently achieves both authenticity and confidentiality, has thus become a major topic in cryptography, with many past contributions [Dwo04, Dwo07, MV04, BRW04, Rog04, KR11]. Since the beginning of the CAESAR competition [CAE14], a large number of new constructions have been proposed, from blockcipher modes of operation [IMGM15, Min14, AFF+15, DN14, HKR15] to ad-hoc designs [Nik14], or sponge-based constructions [BDP+14, ABB+14]. Among the former, OTR [Min14] follows an approach based on tweakable blockciphers (TBC), a powerful primitive introduced by Liskov, Rivest and Wagner [LRW02].

© International Association for Cryptologic Research 2016
J.H. Cheon and T. Takagi (Eds.): ASIACRYPT 2016, Part I, LNCS 10031, pp. 333–353, 2016.
DOI: 10.1007/978-3-662-53887-6_12

## 1.1 Tweakable Blockcipher

Compared to a regular blockcipher, a TBC $\widetilde{E} : \mathcal{K} \times \mathcal{T} \times \{0,1\}^n \to \{0,1\}^n$ takes an additional input $T \in \mathcal{T}$, called a tweak, which adds some variability. As illustrated in [LRW02], a TBC enables simpler designs and security proofs for AE schemes, and can be instantiated from a blockcipher. To achieve efficiency, the design of the input masks must take into account the fact that the TBC is generally not used alone but rather in a mode of operation. In particular, the cost of changing the tweak must be much smaller than the cost of changing the key.

The now common constructions to build a TBC out of a block cipher are the Xor-Encrypt (XE) and Xor-Encrypt-Xor (XEX) constructions of [Rog04]. The principle of XE is to derive an input mask $\Delta$ from the tweak and xor it with the message before calling $E_K$ (XEX also xors this mask to the output). The efficiency comes from designing the input mask $\Delta$ in such a way that $\Delta_{i+1}$ (used to encrypt the $i$-th message block) can be easily derived from $\Delta_i$. For example, in OCB2 [Rog04], $\Delta_{i+1}$ is obtained from $\Delta_i$ by multiplying the latter by some elements of $\mathbb{F}_{2^n}$ (namely $X$ or $(X+1)$, where $X$ generates $\mathbb{F}_{2^n}^*$).

OTR's masks slightly differs from OCB2's one by using, among others, $\Delta_{i,0} = X^{i+1}\delta$ for the $2i-1$-th block and $\Delta_{i,1} = (X^{i+1}+1)\delta$ for the $2i$-th block (where $\delta$ is the encryption of the nonce). This approach is very well suited to the Feistel-based construction of OTR.

## 1.2 Our Contribution

However, we show in this paper that this solution is, at best, unsafe and even totally insecure in many cases. Indeed, the security of XE relies on the hardness of constructing collisions among the input masks $\Delta_i$.

This can easily be proven for OCB2 due to the form of $\Delta = X^i(X + 1)^j E_K(N)$. A collision in the offsets means that $X^i(1+X)^j = X^{i'}(1+X)^{j'}$ for some integers $i, i', j$ and $j'$, and so that $(1+X)^{j-j'} = X^{i'-i}$. This equation, along with the discrete logarithm of $X+1$ in base $X$, allows to define bounds on $i$ and $j$ excluding any collision. Unfortunately, this is no longer true for OTR due to the special form of its offsets. For example, if we just consider the input masks $\Delta_{i,0} = X^{i+1}\delta$ and $\Delta_{i,1} = (X^{i+1}+1)\delta$, it is impossible to formally exclude collisions: there are no algebraic reason why $X^i$ should differ from $X^j + 1$ for any $i, j \leq B$, for some bound $B$.

The simple fact that no formal proof can be provided should itself call for another design of the masks, nevertheless one might still wonder if these collisions are likely.

In this work, we investigate this issue and show that, for a large family of blocksize $n \leq 10000$ (OTR is defined for any blockcipher size $n \in \mathbb{N}^*$), standard choices of parameters lead to trivial collisions. Moreover, we show that the block sizes outside this family are not necessarily secure and need a specific, costly study to exclude collision for reasonable $B$. We focus on the most popular choices, namely $n = 64$ and $n = 128$, and present a collision for the former case when

$\mathbb{F}_{2^{64}}$ is generated, as usual, using the primitive pentanomial $P = X^{64} + X^4 + X^3 + X + 1$. We get similar results for $n = 128$ when $\mathbb{F}_{2^{128}}$ is generated by some specific primitive pentanomials. However, the latter do not include the usually used one, namely $P = X^{128} + X^7 + X^2 + X + 1$. We therefore study more thoroughly this case and propose a bound $B = 2^{45}$ excluding collisions. We do not claim that this bound is optimal but we provide evidence that collisions are likely to occur between $2^{45}$ and $2^{64}$.

In a second part, we describe concrete attacks against privacy and authenticity resulting from these collisions. They show that the latter do not simply invalidate the security proof but also completely break the security of the construction.

Finally, we describe some ways of constructing the input masks which prevent collisions. We therefore emphasize that our work does not question the intrinsic security of OTR seen as a TBC mode of operation, but simply shows that the instantiation of the TBC in [Min14] should be fixed. In particular, due to our attack, Minematsu modified the masks generation in the last version of the CAESAR submission, AES-OTRv3 [Min16].

## 2 Preliminaries

### 2.1 Basic Notations

For sake of clarity, we will use the same notations as the ones of [Min14]. The set of all finite-length binary strings, including the empty string $\epsilon$, is denoted by $\{0,1\}^*$. $\forall S \in \{0,1\}^*$, $|S|$ denotes the length of $S$ and $|S|_a = \max\{\lceil(|S|/a)\rceil, 1\}$. The concatenation of two binary strings $S$ and $T$ is written $ST$. $\forall S \in \{0,1\}^*, (S[1], \ldots, S[m]) \xleftarrow{n} S$ denotes the $n$-bit block partitioning of $S$, i.e. $S = S[1] \ldots S[m]$, where $|S[i]| = n$ for $i < m$ and $|S[m]| \leq n$ (we thus have $m = |S|_n$). The sequence of $a$ zeros is denoted by $0^a$. For all $n \in \mathbb{N}$ and $S$ such that $|S| \leq n$, $S_n$ denotes the padding $S10^{n-|S|-1}$ if $|S| < n$ and $S$ otherwise. In the following, we will omit the subscript $n$ if it is made obvious by the context. For a finite set $\mathcal{S}$, we write $S \xleftarrow{\$} \mathcal{S}$ if $S$ is uniformly chosen from $\mathcal{S}$.

### 2.2 Blockciphers and Tweakable Blockciphers

We review the standard definitions of blockciphers and tweakable blockciphers from [LRW02, Rog04]. A blockcipher is a function $E : \mathcal{K} \times \{0,1\}^n \to \{0,1\}^n$ where $n \in \mathbb{N}$, $\mathcal{K} \neq \emptyset$ is a finite set and $E(K,.) = E_K(.)$ is a permutation for each $K \in \mathcal{K}$. The PRF and PRP advantages of $E$ against adversary $\mathcal{A}$ are defined as:

$$\mathrm{Adv}_E^{\mathrm{prf}}(\mathcal{A}) = \mathbb{P}[K \xleftarrow{\$} \mathcal{K} : \mathcal{A}^{E_K(.)} \Rightarrow 1] - \mathbb{P}[\rho \xleftarrow{\$} \mathrm{Func}(n) : \mathcal{A}^{\rho(.)} \Rightarrow 1]$$

$$\mathrm{Adv}_E^{\mathrm{prp}}(\mathcal{A}) = \mathbb{P}[K \xleftarrow{\$} \mathcal{K} : \mathcal{A}^{E_K(.)} \Rightarrow 1] - \mathbb{P}[\pi \xleftarrow{\$} \mathrm{Perm}(n) : \mathcal{A}^{\pi(.)} \Rightarrow 1]$$

where $\mathrm{Func}(n)$ (resp. $\mathrm{Perm}(n)$) is the set of all the functions (resp. permutations) $\{0,1\}^n \to \{0,1\}^n$.

A tweakable blockcipher is a blockcipher with an additional public input. It is formalized as a function $\widetilde{E} : \mathcal{K} \times \mathcal{T} \times \{0,1\}^n \to \{0,1\}^n$ where $n \in \mathbb{N}$, $\mathcal{K}, \mathcal{T} \neq \emptyset$ are finite sets and $\widetilde{E}(K, T, .) = \widetilde{E}_K(T, .) = \widetilde{E}_K^T(.)$ is a permutation for each $K \in \mathcal{K}$ and $T \in \mathcal{T}$. The tweakable PRF and tweakable PRP advantages of $\widetilde{E}$ against adversary $\mathcal{A}$ is defined as:

$$\mathrm{Adv}_{\widetilde{E}}^{\widetilde{\mathrm{prf}}}(\mathcal{A}) = \mathbb{P}[K \xleftarrow{\$} \mathcal{K} : \mathcal{A}^{\widetilde{E}_K(\cdot,\cdot)} \Rightarrow 1] - \mathbb{P}[\widetilde{\rho} \xleftarrow{\$} \mathrm{Func}(\mathcal{T}, n) : \mathcal{A}^{\widetilde{\rho}(\cdot,\cdot)} \Rightarrow 1]$$

$$\mathrm{Adv}_{\widetilde{E}}^{\widetilde{\mathrm{prp}}}(\mathcal{A}) = \mathbb{P}[K \xleftarrow{\$} \mathcal{K} : \mathcal{A}^{\widetilde{E}_K(\cdot,\cdot)} \Rightarrow 1] - \mathbb{P}[\widetilde{\pi} \xleftarrow{\$} \mathrm{Perm}(\mathcal{T}, n) : \mathcal{A}^{\widetilde{\pi}(\cdot,\cdot)} \Rightarrow 1]$$

where $\mathrm{Func}(\mathcal{T}, n)$ (resp. $\mathrm{Perm}(\mathcal{T}, n)$) is the set of all mappings from $\mathcal{T}$ to functions (resp permutations) $\{0,1\}^n \to \{0,1\}^n$.

## 2.3 Authenticated Encryption

**Definition.** An authenticated encryption $\mathrm{AE}[\tau]$ having a $\tau$-bit tag consists of an encryption algorithm $\mathrm{AE}\text{-}\mathcal{E}_\tau$ and a decryption algorithm $\mathrm{AE}\text{-}\mathcal{D}_\tau$. The former takes as input a key $K \in \mathcal{K}_{ae}$, a nonce $N \in \mathcal{N}_{ae}$ and an associated data $A \in \mathcal{A}_{ae}$ along with a message $M \in \mathcal{M}_{ae}$ and outputs a ciphertext $C \in \mathcal{M}_{ae}$ as well as a tag $T_E \in \{0,1\}^\tau$. On input $(K, N, A, C, T_E)$, the latter outputs a plaintext $M$ such that $|M| = |C|$ or an error symbol $\perp$. The sets $\mathcal{K}_{ae}, \mathcal{N}_{ae}, \mathcal{A}_{ae}$ and $\mathcal{M}_{ae}$ are assumed to be non-empty and finite.

**Security Model.** The security properties expected from an authenticated encryption scheme are privacy and authenticity. The former informally requires that no adversary, even given access to encryption queries, is able to distinguish $\mathrm{AE}[\tau]$ from an oracle $\$$ returning a random pair $(C, T_E) \xleftarrow{\$} \{0,1\}^{|M|} \times \{0,1\}^\tau$ on input $(N, A, M)$. This is formally defined by the following advantage:

$$\mathrm{Adv}_{\mathrm{AE}[\tau]}^{\mathrm{priv}}(\mathcal{A}) = \Pr[K \xleftarrow{\$} \mathcal{K}_{ae} : \mathcal{A}^{\mathrm{AE}\text{-}\mathcal{E}_\tau} \to 1] - \Pr[\mathcal{A}^\$ \to 1].$$

We say an adversary $\mathcal{A}$ is *nonce-respecting* if it cannot submit two queries $(N_i, A_i, M_i)$ and $(N_j, A_j, M_j)$ with $N_i = N_j$ for $i \neq j$. In this paper, we will always consider nonce-respecting adversaries. It is claimed in [Min14] that $\mathrm{Adv}_{\mathrm{OTR}[\tau]}^{\mathrm{priv}}(\mathcal{A}) \leq \frac{6(q + \sigma_A + \sigma_M)^2}{2^n}$ where $q$ is the number of encryption queries and $(\sigma_A, \sigma_M) = (\sum_i^q |A_i|, \sum_i^q |M_i|)$.

Authenticity informally requires that no adversary, even with access to encryption and decryption queries, is able to produce a valid tuple $(N, A, C, T_E)$, *i.e.* one such that $\mathrm{AE}\text{-}\mathcal{D}_\tau(N, A, C, T_E) \neq \perp$. Obviously, $(N, A, C, T_E)$ must not have been previously returned by the encryption oracle. The authenticity notion is defined by the advantage:

$$\mathrm{Adv}_{\mathrm{AE}[\tau]}^{\mathrm{auth}}(\mathcal{A}) = \Pr[K \xleftarrow{\$} \mathcal{K}_{ae} : \mathcal{A}^{\mathrm{AE}\text{-}\mathcal{E}_\tau, \mathrm{AE}\text{-}\mathcal{D}_\tau} \text{ forges}]$$

where $\mathcal{A}$ forges if one of the decryption query $(N_i', A_i', C_i', T_{E,i}')$ does not return $\perp$. Notice that $N_i'$ may be equal to $N_j$ or $N_{i'}'$ for all $i, i'$ and $j$. It is claimed in [Min14] that $\mathrm{Adv}^{\mathrm{auth}}_{\mathrm{OTR}[\tau]}(\mathcal{A}) \leq \frac{6(q+q'+\sigma_A+\sigma_M+\sigma_{A'}+\sigma_{C'})^2}{2^n}$ where $q$ (resp. $q'$) is the number of encryption (resp. decryption) queries, $(\sigma_A, \sigma_M) = (\sum_i^q |A_i|, \sum_i^q |M_i|)$ and $(\sigma_{A'}, \sigma_{C'}) = (\sum_i^q |A'_i|, \sum_i^q |C'_i|)$.

## 2.4   Galois Field

For all non negative integers $n$, we denote by $\mathbb{F}_{2^n}$ the field with $2^n$ elements and by $\mathbb{F}^*_{2^n}$ its multiplicative group. To represent this field one [IK03, Rog04, Min14] usually selects the lexicographically first polynomial $P$ among the primitive polynomials of degree $n$ with coefficients in $\mathbb{F}_2$ having a minimum number of non-zero coefficients, and use $\mathbb{F}_2[X]/P(X)$ as a representation of $\mathbb{F}_{2^n}$. [Ser98] provides such polynomials for $n \leq 10000$. An element $a \in \mathbb{F}_{2^n}$ can then be written as a formal polynomial $b_1 X^{n-1} + \ldots + b_{n-1}X + b_n$ of degree $n-1$ or equivalently as a $n$-bit string $b_1 \ldots b_n$. In the following, we will use both notations interchangeably.

For any $a = b_1 X^{n-1} + \ldots + b_n$ and $c = b'_1 X^{n-1} + \ldots + b'_n$ in $\mathbb{F}_{2^n}$, the product $a \cdot c$ is $(\sum_{i=1}^n b_i X^{n-i})(\sum_{j=1}^n b'_j X^{n-j}) \bmod P(X)$. In particular, it is worthy to note that $a \cdot X$ can be computed very efficiently with a shift and a conditional xor, hence the interest of a low-weight polynomial $P$. For example, for $n = 119$, one would select $P(X) = X^{119} + X^8 + 1$ [Ser98], so $a \cdot X = (a << 1) \oplus 0^{110} b_1 0^7 b_1$.

The table in [Ser98] shows that, up to $n = 10000$, primitive trinomials exist for slightly over one half of the values of $n$. In this case, the field $\mathbb{F}_{2^n}$ is usually generated by $X^n + X^j + 1$ for some $j \in [1, n-1]$. Otherwise, the table shows that, for $n \leq 10000$, one can at least find an irreducible pentanomial. For example, for $n = 128$, one can use $P(X) = X^{128} + X^7 + X^2 + X + 1$.

## 3   Description of OTR

Before describing our attack, we recall the AE scheme of [Min14], $\mathrm{OTR}[E, \tau]$, parametrized by a keyed permutation $E_K : \{0,1\}^n \to \{0,1\}^n$, and a tag length $\tau \leq n$. Its encryption algorithm $\mathrm{OTR}\text{-}\mathcal{E}_{E,\tau}$ consists of an encryption core $EF_E$ and an authentication core $AF_E$ which processes the additional authenticated data. Since our attack applies on $EF_E$, we omit the description of $AF_E$ in Fig. 1 and assume that the string $A$ (authenticated data) is empty.

$EF_E$ can be seen as a variation of the tweakable blockcipher based authenticated encryption mode OCB [Rog04]. In OTR, tweakable blockciphers are instantiated using a two-rounds Feistel permutation where internal round functions are PRFs with tweak-dependent input masks. Algorithm 1 gives a formal description of the authenticated encryption algorithm $\mathbb{EF}[\widetilde{\rho}, \tau]$ that uses a tweakable random function $\widetilde{\rho}$. As defined in [Min14], the tweak space of $\widetilde{\rho}$ is $\mathcal{T} = (\{0,1\}^n \times \mathbb{N} \times \{0,1\}) \cup (\{*\} \times \{0,1\}^n \times \mathbb{N} \times \{0,1\} \times \{0,1\})$.[1]

An important theorem in the security proof of OTR is that, if $\widetilde{\rho}$ is a tweakable random function, then $\mathbb{EF}[\widetilde{\rho}, \tau]$ is a secure authenticated encryption scheme.

---

[1] We slightly changed the notations from [Min14] to give a more formal construction of the tweakable PRF.

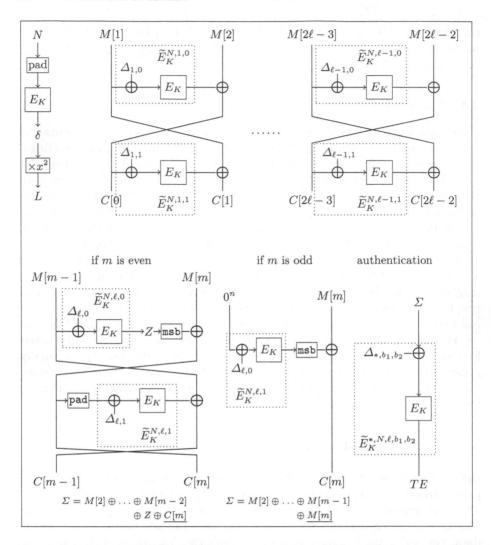

**Fig. 1.** Encryption core $EF_E$ of OTR for a message $M = M[1] \ldots M[m]$ and a blocksize $n$. The integer $\ell$ is defined as $\lceil \frac{m}{2} \rceil$. $\Delta_{i,b} = (X^{i+1} + b)\delta$, for $i = 1, \ldots, \ell$ and $b \in \{0,1\}$. $\Delta_{*,b_1,b_2} = [(X+1)X^{\ell+1} + X \cdot b_1 + b_1 + b_2]\delta$ with $b_1 = 0$ if $m$ is odd and 1 otherwise while $b_2 = 0$ if $|M[m]| < n$ and 1 otherwise. The dotted boxes represent the tweakable random functions of the $\mathbb{OTR}$ construction.

**Theorem 1 (Theorem 3 of [Min14]).** *Fix $\tau \in \{1, \ldots, n\}$. For any adversary $\mathcal{A}$, and tweakable random function $\tilde{\rho}$*

$$\mathrm{Adv}^{\mathrm{priv}}_{\mathrm{EF}[\tilde{\rho},\tau]}(\mathcal{A}) = 0.$$

*Moreover, for any adversary $\mathcal{A}$ making $q$ encryption queries and $q_v$ decryption queries,*

$$\mathrm{Adv}_{\mathrm{EF}[\widetilde{\rho},\tau]}^{\mathrm{auth}}(\mathcal{A}) \leq \frac{2q_v}{2^n} + \frac{q_v}{2^\tau}.$$

We refer to the original paper for the full proof of this theorem. Minematsu also instantiates $\widetilde{\rho}$ using the XE approach [Rog04]:

$$\widetilde{E}_K^{N,i,a}(P) = E_K(P + \Delta_{i,a}) \text{ with } \Delta_{i,a} = X^{i-1}L + a \cdot \delta$$
$$\widetilde{E}_K^{*,N,i,b_1,b_2}(P) = E_K(P + \Delta_{*,i,b_1,b_2}) \text{ with } \Delta_{*,i,b_1,b_2} = (X+1)(X^{i-1}L + b_1 \cdot \delta) + b_2 \cdot \delta$$

where $\delta = E_K(N)$ and $L = X^2\delta$. Once developed, the final expression of the $\Delta$ values is

$$\Delta_{i,a} = (X^{i+1} + a)\delta$$
$$\Delta_{*,i,b_1,b_2} = (X^{i+2} + X^{i+1} + b_1X + b_1 + b_2)\delta.$$

---

**Algorithm 1.** Description of $\mathbb{EF}[\widetilde{\rho}, \tau]$.

---

1: $\Sigma \leftarrow 0^n$
2: $(M[1], \ldots, M[m]) \xleftarrow{n} M$
3: $\ell \leftarrow \lceil m/2 \rceil$
4: **for** $i = 1$ **to** $\ell - 1$ **do**
5:      $C[2i-1] \leftarrow \widetilde{\rho}^{N,i,0}(M[2i-1]) \oplus M[2i]$
6:      $C[2i] \leftarrow \widetilde{\rho}^{N,i,1}(C[2i-1]) \oplus M[2i-1]$
7:      $\Sigma \leftarrow \Sigma \oplus M[2i]$
8: **end for**
9: **if** $m$ is even **then**
10:      $Z \leftarrow \widetilde{\rho}^{N,\ell,0}(M[m-1])$
11:      $C[m] \leftarrow \mathtt{msb}_{|M[m]|}(Z) \oplus M[m]$
12:      $C[m-1] \leftarrow \widetilde{\rho}^{N,\ell,1}(C[m]) \oplus M[m-1]$
13:      $\Sigma \leftarrow \Sigma \oplus Z \oplus C[m]$
14:      **if** $|M[m]| \neq n$ **then** $TE \leftarrow \widetilde{\rho}^{*,N,\ell,1,0}(\Sigma)$
15:      **else** $TE \leftarrow \widetilde{\rho}^{*,N,\ell,1,1}(\Sigma)$
16: **else**              $\triangleright$ *m is odd*
17:      $C[m] \leftarrow \mathtt{msb}_{|M[m]|}(\widetilde{\rho}^{N,\ell,0}(0^n)) \oplus M[m]$
18:      $\Sigma \leftarrow \Sigma \oplus M[m]$
19:      **if** $|M[m]| \neq n$ **then** $TE \leftarrow \widetilde{\rho}^{*,N,\ell,0,0}(\Sigma)$
20:      **else** $TE \leftarrow \widetilde{\rho}^{*,N,\ell,0,1}(\Sigma)$
21: **end if**
22: $C \leftarrow (C[1], \ldots, C[m])$
23: **return** $(C, TE)$

---

To finish the proof of security, [Min14] uses the Lemma 1, claiming the CPA security of the tweakable PRF $\widetilde{E}$, provided that $E$ is a perfect blockcipher (a random permutation):

**Lemma 2 (Lemma 1 of [Min14]).** *For any adversary $\mathcal{A}$ making $q$ queries,*

$$\mathrm{Adv}_{\widetilde{E}}^{\mathrm{prf}}(\mathcal{A}) \leq \frac{5q^2}{2^n}.$$

The proof of Lemma 1 relies on the fact that the masks $\Delta$ are assumed to be "differentially uniform" for any two distinct inputs. However, we show below that this is not the case for a large choice of parameters $n$, and that it actually completely breaks the security of OTR.

# 4   Collision in Masks Polynomials

## 4.1   Flaw in OTR's Proof

In [Min14], all possible masks $\Delta$ are regrouped in a set

$$\mathcal{S}_1(\delta) = \big\{ X^{i+1}\delta, (X^{i+1}+1)\delta, (X^{i+2}+X^{i+1})\delta, (X^{i+2}+X^{i+1}+X)\delta,$$
$$(X^{i+2}+X^{i+1}+1)\delta, (X^{i+2}+X^{i+1}+X+1)\delta \big\}_{i=1}$$

(no upper bound on $i$ is given but we can suppose that it is bounded by the maximum number of blocks one can query for an encryption, and that is it at most $2^{n/2}$) and it is claimed that for any $\Delta, \Delta' \in \mathcal{S}_1(\delta_1) \cup \mathcal{S}_1(\delta_2)$ such that $\Delta$ and $\Delta'$ are generated from two different expressions, and $d \in \{0,1\}^n$,

$$\Pr_{\delta_1, \delta_2 \xleftarrow{\$} \{0,1\}^n} [\Delta + \Delta' = d] \leq \frac{1}{2^n}$$

where the probability is taken over the random choices of $\delta_1$ and $\delta_2$. This is true if $\Delta \in \mathcal{S}_1(\delta_1)$ and $\Delta' \in \mathcal{S}_1(\delta_2)$, but not if both $\Delta$ and $\Delta'$ are generated from the same $\delta$.

Namely, suppose that there are two integers $i$ and $j \geq 2$ such that

$$X^i = X^j + 1 \tag{1}$$
$$\text{or } X^i = X^{j+1} + X^j + r(X) \tag{2}$$
$$\text{or } X^{i+1} + X^i = X^{j+1} + X^j + r(X) \tag{3}$$

with $r(X) \in \{0, 1, X, X+1\}$. Then we directly have a collision inside $\mathcal{S}_1(\delta)$ for any $\delta$. This problem is not highlighted in the proof and we will show that we can actually find (and use) such pairs of integers.

In the following, we will use the terms 'type-1', 'type-2', and 'type-3' for collisions satisfying, respectively, Eqs. (1), (2) and (3).

## 4.2   Finding Collisions

The problem with the polynomials considered above is that it seems impossible, given $n \in \mathbb{N}$ and a polynomial $P$ generating $\mathbb{F}_{2^n}$, to provide a formal argument excluding collisions for any $i, j \in [2, t]$ for some integer $2 < t \leq 2^{n/2}$. One can note that we do not consider collisions in the set $\{X^i\}_{i=2}^t$, as $X$ is a generator of $\mathbb{F}_{2^n}^*$ (since $P$ is primitive) and we chose $t \leq 2^{n/2}$.

Actually, we show that trivial collisions can be found when the definition polynomial $P$ has a special form, in particular when $P$ is a trinomial or a pentanomial.

**Case 1:** $\mathbb{F}_{2^n}$ is generated by a trinomial $P(X) = X^n + X^j + 1$.

As explained in [Ser98], this is the standard choice for a majority of values $n \leq 10000$. In such a case, a collision in $\mathcal{S}_1$ is trivially given by $P$ since $X^n = X^j + 1$ (this is thus a type-1 collision). Any encryption of a message $M$ of $m$ blocks such that $\lceil \frac{m}{2} \rceil \geq n - 1$ will then lead to the re-use of a mask and so to one of the attacks described in the next session.

One might argue that this can be avoided by generating $\mathbb{F}_{2^n}$ with a pentanomial instead of a trinomial. However, this unconventional choice will negatively impact the performances of the scheme and will not necessarily prevent collisions.

**Case 2:** $\mathbb{F}_{2^n}$ is generated by a pentanomial $P(X) = X^n + X^{j_1} + X^{j_2} + X^{j_3} + 1$. This case includes, for example, $n = 64$ and $n = 128$. Although there is no trivial collision as in the previous case, it is still necessary to check, for the chosen $n$ and $P$, that $\mathcal{S}_1$ only contains distinct elements, which requires a significant amount of computations and storage space. We here describe the most popular cases:

- $n = 64$. The lexicographically first primitive pentanomial of degree 64 is $X^{64} + X^4 + X^3 + X + 1$ [Ser98]. It leads to a type-2 collision since $X^{64} = X^4 + X^3 + X + 1$.
- $n = 128$. Here again, the pentanomial generating $\mathbb{F}_{2^{128}}$ may give an obvious collision. For example, setting $P = X^{128} + X^{68} + X^{67} + X + 1$ leads to a type-2 collision $X^{128} = X^{68} + X^{67} + X + 1$, and setting $P = X^{128} + X^{127} + X^{61} + X^{60} + 1$ leads to a type-3 collision $X^{128} + X^{127} = X^{61} + X^{60} + 1$. However, this is not the case with the lexicographically first primitive pentanomial of degree 128, $P = X^{128} + X^7 + X^2 + X + 1$, that one generally uses to define $\mathbb{F}_{2^{128}}$. The latter therefore needs a more thorough study that we defer to Sect. 6.

## 5   Practical Attacks

One may wonder if the collisions found in the input masks simply invalidate the security proofs of OTR. Unfortunately, this is not the case and we show below that any kind of collision leads to attacks breaking privacy and/or authenticity. We recall that, for sake of simplicity, authenticated data are assumed to be empty in the following attacks. Attacks for non-empty authenticated data can easily be derived from them.

### 5.1   Type-1 Collisions

A type-1 collision occurs when there are $i$ and $j$ such that $X^i = X^j + 1$. We can assume, without loss of generality, that $j < i$ (since $X^i = X^j + 1 \Leftrightarrow X^j = X^i + 1$).

**Breaking Authenticity.** To break authenticity, one can make a query on an arbitrary message $M = M[1] \ldots M[2i-3]$ for a nonce $N$, defining $\delta = E_K(N)$ and $L = X^2\delta$, and receive the ciphertext $C = C[1] \ldots C[2i-3]$ along with the tag $T = TE$.

The message $M$ has an odd number of blocks so $C[2i-3] = E_K(X^i\delta) \oplus M[2i-3]$.

Let $C' = C'[1] \ldots C'[2i-3]$ such that $C'[k] = C[k]$ for $k \notin \{2j-3, 2j-2, 2i-3\}$, $C'[2j-3] = 0^n$, $C'[2j-2] = M[2j-3] \oplus C[2i-3] \oplus M[2i-3]$ and $C'[2i-3] = C[2i-3] \oplus C[2j-3]$.

Then, the pair $(C', TE)$ is valid: OTR-$\mathcal{D}_{E,\tau}(N, \epsilon, C', T) = M'[1] \ldots M'[2i-3] \neq \perp$. Indeed, by construction, we have $M'[k] = M[k] \, \forall k \notin \{2j-3, 2j-2, 2i-3\}$. Moreover, we have

$$
\begin{aligned}
M'[2j-3] &= E_K(C'[2j-3] \oplus (X^j+1)\delta) \oplus C'[2j-2] \\
&= E_K(0^n \oplus (X^j+1)\delta) \oplus M[2j-3] \oplus C[2i-3] \oplus M[2i-3] \\
&= E_K((X^j+1)\delta) \oplus M[2j-3] \oplus E_K(X^i\delta) \\
&= M[2j-3]
\end{aligned}
$$

and

$$
\begin{aligned}
M'[2j-2] &= E_K(M'[2j-3] \oplus X^j\delta) \oplus C'[2j-3] \\
&= E_K(M[2j-3] \oplus X^j\delta) \oplus 0^n \\
&= C[2j-3] \oplus M[2j-2].
\end{aligned}
$$

Finally, we have $M'[2i-3] = M[2i-3] \oplus C[2j-3]$. Therefore:

$$
\Sigma' = \Sigma \oplus C[2j-3] \oplus C[2j-3] = \Sigma
$$

and the tag $TE$ remains valid for $C'$.

For an adversary $\mathcal{A}$ following this procedure,

$$
\mathrm{Adv}^{\mathrm{auth}}_{\mathrm{AE}[\tau]}(\mathcal{A}) = 1.
$$

**Breaking Privacy.** We describe here a way that an adversary $\mathcal{A}$ can use to break privacy with advantage almost $1/4$ with a single query. To break privacy, $\mathcal{A}$ queries the encryption oracle with a random nonce $N$ and a message $M = M[1] \ldots M[2i-2]$ such that $|M[2i-2]| = 1$ and $M[2j-3] = 010^{n-2}$. $\mathcal{A}$ will receive $C = C[1] \ldots C[2i-2]$ with $|C[2i-2]| = 1$. If $C[2i-2] = 1$ (which happens with probability $\frac{1}{2}$), $\mathcal{A}$ just picks its output bit at random (she does not try further up). Otherwise, we have $C[2i-2] = 010^{n-2} = M[2j-3]$.

As a consequence, we get the following:

$$
\begin{aligned}
M[2i-3] &= E_K(\underline{C[2i-2]} \oplus (X^i+1)\delta) \oplus C[2i-3] \\
&= E_K(M[2j-3] \oplus X^j\delta) \oplus C[2i-3] \\
&= C[2j-3] \oplus M[2j-2] \oplus C[2i-3]
\end{aligned}
$$

and $M[2j - 2] \oplus M[2i - 3] = C[2j - 3] \oplus C[2i - 3]$, which defines an efficient distinguisher between the random encryption oracle and the real encryption oracle. More formally,

$$\mathrm{Adv}^{\mathrm{priv}}_{\mathrm{AE}[\tau]}(\mathcal{A}) = \frac{1}{2}\left(1 - \frac{1}{2^n}\right) - \frac{1}{2}\cdot\frac{1}{2} = \frac{1}{4} - \frac{1}{2^{n+1}}.$$

## 5.2   Type-2 Collisions

A type-2 collision occurs when there are $i$ and $j$ such that $X^i = X^{j+1} + X^j + r(X)$ with $r(X) \in \{0, 1, X, X + 1\}$. We show below how one can break authenticity if $i \geq j$ and privacy if $i < j$.

**Breaking Privacy for $i < j$.** To break privacy, one submits a message $M = M[1]\ldots M[m] = 0^n \ldots 0^n M[2i - 3]M[2i - 2]0^n \ldots M[m - 1]0^{|M[m]|}$ where $m$, $|M[m]|$, $M[2i - 3]$,$M[2i - 2]$ and $M[m - 1]$ are defined as follows:

- If $r(X) = X + 1$, then one sets $m = 2(j - 1)$, $|M[m]| = n - 1$, $M[2i - 3] = M[2i - 2] \in \{0, 1\}^n$ and $M[m - 1] \in \{0, 1\}^n$.
  Since the last block of $M$ is $0^{n-1}$, the $n - 1$ most significant bits of $Z \oplus C[m]$ are $0^{n-1}$. Therefore, if the last bit of $Z$ is 1 (which occurs with probability $\frac{1}{2}$), $Z \oplus C[m] = 0^n$. Also, in this case, $\Sigma = M[2i - 2] = M[2i - 3]$. If the last bit of $Z$ is not 1, one simply submits new messages with different $M[m - 1]$ until this condition is fulfilled.
  The authentication tag $TE$ then verifies the following relation:

$$
\begin{aligned}
TE &= E_K(\Sigma \oplus \Delta_{*,m,1,0}) \\
   &= E_K(M[2i - 3] \oplus (X^{j+1} + X^j + X + 1)\delta) \\
   &= E_K(M[2i - 3] \oplus X^i\delta) \\
   &= C[2i - 3] \oplus M[2i - 2]
\end{aligned}
$$

  Therefore, $TE \oplus C[2i - 3] = M[2i - 2]$, which breaks privacy.
- If $r(X) = X$, then one sets $m = 2(j - 1)$, $|M[m]| = n$, $M[2i - 3] = M[2i - 2] \in \{0, 1\}^n$ and $M[m - 1] \in \{0, 1\}^n$. In such a case, $\Sigma = M[2i - 2] = M[2i - 3]$ and the previous attack still applies.
- If $r(X) = 1$, then one sets $m = 2(j - 1) - 1$, $|M[m]| = n$, $M[2i - 3] = M[2i - 2] \in \{0, 1\}^n$ and $M[m - 1] = 0^n$. Here again, $\Sigma = M[2i - 2] = M[2i - 3]$ so the equality $TE \oplus C[2i - 3] = M[2i - 2]$ still holds.
- Else, $r(X) = 0$. One then sets $m = 2(j - 1) - 1$, $|M[m]| = n - 1$, $M[2i - 3] \in \{0, 1\}^n$, $M[m - 1] = 0^n$ and $M[2i - 2]$ is equal to $M[2i - 3]$ except on the last bit. We then have:

$$
\begin{aligned}
\Sigma &= M[2i - 2] \oplus M[m] \\
       &= M[2i - 2] \oplus 0^{|M[m]|}1 \\
       &= M[2i - 3]
\end{aligned}
$$

and $TE \oplus C[2i - 3] = M[2i - 2]$, as before.

In all these cases, we have a distinguishing criteria between the truly random oracle and the real encryption oracle that can be trivially checked. An adversary $\mathcal{A}$ using this algorithm will break the privacy with advantage $\frac{1}{4} - \frac{1}{2^{n+1}}$ with a single encryption query.

**Breaking Authenticity for $i \geq j$.** The previous attacks against privacy shows that, for any $r(X)$, if there is a type-2 collision among the tweaks polynomials, with $i < j$, one can submit a message $M$ such that its encryption $(C, TE)$ satisfies the equation $TE = C[2i - 3] \oplus M[2i - 2]$. Informally, by taking this assertion backward, this means that one can compute a valid tag for some specific message from $C[2i - 3]$ and $M[2i - 2]$. The idea of the authenticity attacks is to query encryption for a message $M$ such that $|M| > 2in$ to get these two bitstrings and then to truncate it to make $TE$ a valid tag for a shorter message of size $\approx 2jn$.

More specifically, we distinguish the following cases:

- If $r(X) = X$, then $\Delta_{i-1,0} = \Delta_{*,j-1,1,1}$. $\mathcal{A}$ selects an integer $m > 2(i-1)$ and submits a message $M = M[1] \ldots M[m]$ such that $M[k] = 0^n$ for $k \in [1, 2(j-2)]$, $M[2j-3], M[2j-2] \in \{0,1\}^n$, $M[2i-2] = M[2i-3] = M[2j-2]$ and $M[k] \in \{0,1\}^n$ otherwise. Let $(C, TE)$ be the response to this encryption query. Then, the pair $(C', TE') \leftarrow (C[1] \ldots C[2j-4]C[2j-2]C[2j-3], C[2i-3] \oplus M[2i-2])$ is valid (recall that the last two blocks of $C$ are switched during the encryption process), and decrypts to $M' = M[1] \ldots M[2j-3]$. Indeed, if $M'$ is the decryption of $C'$, $M'[k] = M[k]$ for $k \leq 2j-2$, $\Sigma' = M'[2j-2]$, the valid tag for $C'$ should be

$$\widetilde{TE} = E_K(\Sigma' \oplus \Delta_{*,j-1,1,1})$$
$$= E_K(M'[2j-2] \oplus \Delta_{*,j-1,1,1})$$
$$= E_K(M[2i-3] \oplus \Delta_{i-1,0})$$
$$= C[2i-3] \oplus M[2i-2]$$
$$= TE'$$

This clearly breaks the authenticity of the scheme.

- If $r(X) = X + 1$ (and $\Delta_{i-1,0} = \Delta_{*,j-1,1,0}$), then one selects an integer $n > 2(i-1)$ and queries the message $M = M[1] \ldots M[m]$ such that $M[k] = 0^n$ for $k \in [1, 2(j-2)]$, $M[2j-3], M[2j-2] \in \{0,1\}^n$, $M[2i-2] = M[2i-3] = M[2j-2]$ and $M[k] \in \{0,1\}^n$ are arbitrary strings otherwise.
  With probability $\frac{1}{2}$, the last bit of $C[2j-3]$ is 1. In this case, $\mathtt{msb}_{n-1}(C[2j-3]) = C[2j-3]$. Let $(C', TE') = (C[1] \ldots C[2j-4]C[2j-2]\mathtt{msb}_{n-1}(C[2j-3]), C[2i-3] \oplus M[2i-2])$ and $M'$ the decryption of $C'$. Again, for $k < 2j-3$, $M'[k] = M[k]$, but we also have $M'[2j-3] = M[2j-3]$ and $Z' = C[2j-3] \oplus M[2j-2]$:

$$M'[2j-3] = E_K(C'[2j-2] \oplus \Delta_{j-1,1}) \oplus C'[2j-3]$$
$$= E_K(\mathtt{msb}_{n-1}(C[2j-3]) \oplus \Delta_{j-1,1}) \oplus C[2j-2]$$
$$= E_K(C[2j-3] \oplus \Delta_{j-1,1}) \oplus C[2j-2]$$
$$= M[2j-3]$$

$$Z' = E_K(M'[2j-3] \oplus \Delta_{j-1,0})$$
$$= E_K(M[2j-3] \oplus \Delta_{j-1,0})$$
$$= C[2j-3] \oplus M[2j-2]$$

As a direct consequence, we also have

$$\Sigma' = Z' \oplus \underline{C'[2j-2]} = C[2j-3] \oplus M[2j-2] \oplus \mathsf{msb}_{n-1}(C[2j-3])$$
$$= M[2j-2].$$

As a consequence, using similar equalities to the $r(X) = X$ case, we can show that the authentication tag for $C'$ should be $\widetilde{TE} = C[2i-3] \oplus M[2i-2] = TE'$. This attack produces a forgery with probability $\frac{1}{2}$.

- If $r(X) = 1$, $\Delta_{i-1,0} = \Delta_{*,j-1,0,1}$. $\mathcal{A}$ again selects $m \geq 2(i-2)$ and queries encryption of $M = M[1]\dots M[m]$ such that $M[k] = 0^n$ for $k \in [1, 2(j-1)]$, $M[2i-3] = 0^n$ and $M[k] \in \{0,1\}^n$ for $k > 2i-2$. Let $(C', TE') = (C[1]\dots C[2j-4]C[2j-3], C[2i-3] \oplus M[2i-2])$ and $M'$ its decryption. Once again, we have $M[k] = M'[k]$ for $k < 2j-3$. Moreover, as the number of blocks in $C'$ is odd,

$$M'[2j-3] = C'[2j-3] \oplus E_K(\Delta_{j-1,0})$$
$$= C[2j-3] \oplus E_K(M[2j-3] \oplus \Delta_{j-1,0})$$
$$= M[2j-2] = 0^n$$

and hence $\Sigma' = 0^n (= M[2i-3])$. Finally

$$TE' = C[2i-3] \oplus M[2i-2] = E_K(M[2i-3] \oplus \Delta_{i-1,0})$$
$$= E_K(\Sigma' \oplus \Delta_{*,j-1,0,1}) = \widetilde{TE}$$

where $\widetilde{TE}$ is the expected tag for $C'$. Again, we are able to produce a forgery.
- If $r(X) = 0$, then one proceeds as in the previous case except that $M[2i-3] = 0^{n-1}1$. We will still have $\Sigma' = M[2i-3]$ and the pair $(C', TE') = (C[1]\dots C[2j-4]\mathsf{msb}_{n-1}(C[2j-3]), C[2i-3] \oplus M[2i-2])$ is a valid forgery.

## 5.3   Type-3 Collisions

A type-3 collision occurs when there are $\ell$ and $\ell'$ such that $X^{\ell+2} + X^{\ell+1} = X^{\ell'+2} + X^{\ell'+1} + r(X)$, with $r(X) \in \{0, 1, X, X+1\}$. We assume, without loss of generality, that $\ell < \ell'$.

The input masks of the form $X^{k+2} + X^{k+1} + r(X)$ are the ones involved in the computation of the tag $TE$. So a type-3 collision informally means that the input mask used to compute $TE$ for a message of length $m'$ such that $\ell' = \lceil \frac{m'}{2} \rceil$ is the same than the one used to compute $TE$ for a truncated message of length $m$ verifying $\ell = \lceil \frac{m}{2} \rceil$. Again, this leads to a practical attack against authenticity.

**Breaking Authenticity.** As previously, the attack will slightly differ according to $r(X)$.

- If $r(X) = X$, $\Delta_{*,\ell,0,0} = \Delta_{*,\ell',1,1}$ $\mathcal{A}$ submits an encryption query for the message $M[1] \ldots M[2\ell]M[2\ell+1] \ldots M[2\ell'-1]M[2\ell']$ with $M[2\ell-1] = 0^n$, $M[2\ell]$ has its last bit set to 1 (in particular $\mathtt{msb}_{n-1}(M[2\ell]) = M[2\ell])$, and $M[i] = 0^n$ for $i \in [2\ell+1, 2\ell']$. Upon receiving $(C[1] \ldots C[2\ell'], TE)$, $\mathcal{A}$ forges $(C', TE') = (C[1] \ldots C[2\ell-2]\mathtt{msb}_{n-1}(C[2\ell-1]), TE)$, which is a valid ciphertext.

  Indeed, if $\Sigma$ is the checksum corresponding to $(C[1] \ldots C[2\ell'], TE)$ and $\Sigma'$ is the one corresponding to the forged ciphertext, we have:

$$\Sigma' = M[2] \oplus \ldots \oplus M[2\ell-2] \oplus \underline{\mathtt{msb}_{n-1}(E_K(\Delta_{\ell,0}))} \oplus C'[2\ell-1]$$
$$= M[2] \oplus \ldots \oplus M[2\ell-2] \oplus \underline{\mathtt{msb}_{n-1}(E_K(\Delta_{\ell,0}) \oplus C[2\ell-1])}$$
$$= M[2] \oplus \ldots \oplus M[2\ell-2] \oplus \underline{\mathtt{msb}_{n-1}(M[2\ell])}$$
$$= M[2] \oplus \ldots \oplus M[2\ell-2] \oplus M[2\ell]$$
$$= \Sigma$$

  Therefore, $\widetilde{TE} = E_K(\Sigma' \oplus \Delta_{*,\ell,0,0}) = E_K(\Sigma \oplus \Delta_{*,\ell',1,1}) = TE$, so the tag $TE$ is also valid for this truncated ciphertext $C'$.

- if $r(X) = X + 1$, one proceeds as in the previous case except that we take any value for $M[2\ell]$ and $(C', TE') = (C[1] \ldots C[2\ell-2]C[2\ell-1], TE)$: we don't have to play with the padding. Therefore, $\widetilde{TE} = E_K(\Sigma' \oplus \Delta_{*,\ell,0,1}) = E_K(\Sigma \oplus \Delta_{*,\ell',1,1}) = TE$, and $TE$ remains valid for this truncated ciphertext.

- If $r(X) = 1$, $\Delta_{*,\ell,0,0} = \Delta_{*,\ell',0,1}$, and $\mathcal{A}$ will proceed as in the first case $r(X) = X$, except that its first query will be with $M$ with an odd number of blocks. $\mathcal{A}$ will query $M = M[1] \ldots M[2\ell'-1]$ such that $M[2\ell-1] = 0^n$, $M[2\ell]$ has its last bit set to 1, and $M[i] = 0^n$ for $i \in [2\ell+1, 2\ell'-1]$. The forgery will be $(C', TE') = (C[1] \ldots C[2\ell-2]\mathtt{msb}_{n-1}(C[2\ell-1]), TE)$.

  The proof that $(C', TE')$ is a valid forgery proceeds exactly as for the $r(X) = X$ case.

- if $r(X) = 0$, $\Delta_{*,\ell,0,1} = \Delta_{*,\ell',0,1}$, and $\mathcal{A}$ submits an encryption query on $M = M[1] \ldots M[2\ell'-1]$ such that $M[2\ell-1] = 0^n$, and $M[i] = 0^n$ for $i \in [2\ell+1, 2\ell'-1]$. The forgery will be $(C', TE') = (C[1] \ldots C[2\ell-2]C[2\ell-1], TE)$. The validity of the forgery can be easily proven from the same arguments as before.

In every case, we are able to easily produce a valid forgery from a single encryption request. For an adversary $\mathcal{A}$ following this procedure,

$$\mathrm{Adv}_{\mathrm{AE}[\tau]}^{\mathrm{auth}}(\mathcal{A}) = 1.$$

# 6 Practical Security of OTR with 128 Bits Blocks

In the previous sections we exhibited tweak collisions on OTR breaking the security claim, in particular for non generic block sizes (sizes that are not divisible by 8) and for 64 bits block ciphers. These collisions allow the adversary to break privacy and/or authenticity of the scheme in two encryption/decryption requests with a small number of blocks. Here, we focus on the case $n = 128$.

Also, note that for the sake of breaking OTR, we are only interested in collisions before the birthday bound, *i.e.* collisions for which the maximum index $i$ of the polynomials defined by $\Delta_{i,a}$ or $\Delta_{*,i,b_1,b_2}$ is smaller than $2^{n/2}$. Higher order collisions are less interesting as OTR's proofs only guarantees security below the birthday bound.

## 6.1    Analytical Collisions

One strategy for quickly finding collisions could rely on the fact that $\mathbb{F}_{2^d} \subset \mathbb{F}_{2^{128}}$ for any $d$ dividing 128. Indeed, any relation $Y^i = Y^j + 1$ for some $Y \in \mathbb{F}_{2^d}$ gives us a type-1 collision $X^{a \cdot i} = X^{a \cdot j} + 1$ with $a$ such that $Y = X^a$ in $\mathbb{F}_{2^{128}}$. Such relations can easily be found in $\mathbb{F}_{2^d}$ for $d \in \{16, 32, 64\}$, for example by computing the discrete logarithm of $Y^j + 1$ in base $Y$. However, they do not lead to truly practical attacks because $Y^{2^d - 1} = 1$ (as any element of $\mathbb{F}_{2^d}$) which implies that $2^{128} - 1 | a \cdot (2^d - 1)$ (recall that $X$ generates $\mathbb{F}_{2^{128}}^*$) and so that $(2^{128} - 1)/(2^d - 1)$ divides $a$. Therefore, such relations will only give collisions for quite large indices $a \cdot i$ (since $a$ is at least greater than $2^{64} + 1$) and so beyond the birthday bound.

## 6.2    Searching for Collisions Exhaustively

We also tried to algorithmically and exhaustively find collisions among tweaks polynomials. This can be done easily on a desktop computer for $n = 64$, but not for $n = 128$.

Indeed, to check collisions for tweak polynomials of index less than $d$, we need at least $2d \cdot 128$ bits of memory: the index $i$ polynomials we are interested in are of the form $X^i(+1)$ and $X^i + X^{i-1}(+X)(+1)$, so to save memory, we can only store $X^i$ and $X^i + X^{i-1} \bmod P(X)$, and do the collision search on the 126 high degree bits. To exhibit a genuine collision, we then just have to recompute the different possibilities for the polynomials and find the matching ones. Also, for each polynomial, we have to store its 'index' $i$, adding $O(\log d)$ storage. So if we were to exhaustively search for all collisions for $d < 2^{64}$, we would need $2 \cdot 2^{64} \cdot 192$ bits, *i.e.* 24 exabytes.

On the computational point of view, the complexity of the algorithm is well-known, $O(d \log d)$, as we can generate all the $2d$ polynomials, sort them using the lexicographic order on their bits, and finally search a collision in $O(d)$.

It is also important to notice that the collision search is embarrassingly parallelizable: once generated, we can put the polynomials in some bins, depending on the value of the high degree bits, and limit the search to collisions inside each bin. This algorithm is described by Algorithm 2.

Algorithm 2 also offers a nice time/memory tradeoff: instead of keeping all bins in memory, we can instead limit ourself to the bins fitting in memory, and run the algorithms several times so that all the bins are spanned.

---

**Algorithm 2.** Our collision search algorithm

---

**for** $k = 0$ **to** $2^p - 1$ **do**                                               ▷ *In parallel*
  $S_k \leftarrow \emptyset$                                                          ▷ *Initialize bins*
**end for**
**for** $i = 0$ **to** $d$ **do**                                                      ▷ *In parallel*
  $\alpha_i \leftarrow X^i \bmod P$
  $k_\alpha \leftarrow \mathtt{msb}_p(\alpha_i)$
  $S_{k_\alpha} \leftarrow S_{k_\alpha} \cup (\alpha_i, i)$
  $\beta_i \leftarrow X^{i+1} + X^i \bmod P$
  $k_\beta \leftarrow \mathtt{msb}_p(\beta_i)$
  $S_{k_\beta} \leftarrow S_{k_\beta} \cup (\beta_i, i)$
**end for**
**for** $k = 0$ **to** $2^p - 1$ **do**                                               ▷ *In parallel*
  Lexicographically sort $S_k$
  Sequentially scan $S_k$ for a collision
**end for**

---

We coded this algorithm in C, using OpenMP and SSE instructions, and we were able to show that there is no collisions among the tweak polynomials of index less than $2^{45}$ for $\mathbb{F}_{2^{128}}$ defined by $X^{128} + X^7 + X^2 + X + 1$, proving Proposition 3, which fixes Lemma 1 of [Min14].

**Proposition 3.** *For any adversary $\mathcal{A}$ making $q$ queries on $\widetilde{E}$ as defined in Sect. 3, with tweak space $\mathcal{T} = \{0,1\}^{128} \times \{0, \ldots, 2^{45}\} \times \{0,1\} \cup \{*\} \times \{0,1\}^{128} \times \{0, \ldots, 2^{45}\} \times \{0,1\} \times \{0,1\}$,*

$$\mathrm{Adv}_{\widetilde{E}}^{\widetilde{\mathrm{prp}}}(\mathcal{A}) \leq 5q^2/2^{128}.$$

This exhaustive search took us around 15 CPU-years, using 3TB of RAM.

### 6.3 Probable Collision Before the Birthday Bound

The collisions exhibited earlier in the paper, for example for $n = 64$ or $n = 119$, use the special form of the polynomial. For the latter, we use the fact that it is a trinomial, directly giving a type-1 collision. For the former, as there are non zero coefficients of two consecutive degrees higher than 2, the polynomial gives a type-2 collision. One could wonder if, excepting these 'trivial' collisions, it is easy to find other before-birthday-bound collisions? Said otherwise, what is the repartition of the indices of colliding polynomials? We can also remember that if the tweak polynomials behaved randomly, we would expect a collision to be happening just before the birthday bound.

We ran experiments for $n = 16, 32$ and $64$, using (respectively) irreducible polynomials $X^{16} + X^5 + X^3 + X + 1$, $X^{32} + X^7 + X^3 + X^2 + 1$ and $X^{64} + X^4 + X^3 + X + 1$. They are summarized in Table 1.

If we were to extrapolate, we would expect a collision for $n = 128$ using irreducible polynomial $X^{128} + X^7 + X^4 + X + 1$ to also happen slightly before

**Table 1.** Lower indices of colliding tweak polynomials (excepted trivial ones).

| $n$ | 16 | 32 | 64 |
|---|---|---|---|
| Polynomial | $(X+1)X^{105} =$ $(X+1)X^{134}+X$ | $(X+1)X^{30115} =$ $X^{19743} + X$ | $X^{2242000936} =$ $X^{2302312163} + 1$ |
| log(degree) | 7.07 | 14.88 | 31.10 |

the birthday bound. We support this claim with a few experiments we ran on smaller fields. Figures 2, 3 and 4 show the repartition of the smallest collisions of tweak polynomials (*i.e.* the collision with the lowest index) depending on the choice of the irreducible polynomial chosen to define $\mathbb{F}_{2^n}$.

The graphs not only show that the first collision is extremely likely to happen before the birthday bound, but also that it should not happen too early before: we cannot really hope for gaining more than a few bits.

In this case the security proof of [Min14] is only invalidated by a small amount. However, we do not have any formal argument to fill the gap between $2^{45}$ and $2^{64}$.

# 7 Other Instantiations of Input Masks

The previous collisions do not exclude GF doublings to derive the offsets but simply show that this should be done differently. One of the most obvious solution consists in defining the input mask for the block $M[i]$ as $X^{i+2}\delta$ and $\Delta_*$ as

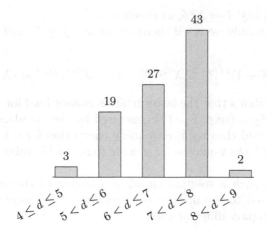

**Fig. 2.** Log of the lowest indices of colliding tweak polynomials for every $\mathbb{F}_{2^{16}}$ representations using the 94 degree 16 irreducible pentanomials over $\mathbb{F}_2$. In other words, among the 94 possible representations of $\mathbb{F}_{2^{16}}$, 3 leads to a collision between the $2^5$ first tweak polynomials, 19 to a collision between polynomials of indices $i$ and $j$ such that $\mathtt{max}(i,j) \in ]2^5, 2^6]$, and so on and so forth.

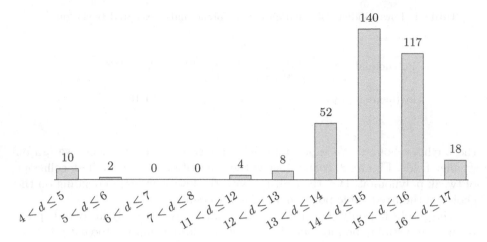

**Fig. 3.** Log of the lowest indices of colliding tweak polynomials for every $\mathbb{F}_{2^{32}}$ representations using the 351 degree 32 irreducible pentanomials over $\mathbb{F}_2$.

$X^m(X+1)^j\delta$ where $m$ is the number of blocks of $M$ and where $j$ would depend on some properties of $M$, namely the parity and the number of bits of $M[m]$.

More specifically, the tweakable random function $\widetilde{\rho}$ (see Sect. 3) can be instantiated as follows:

$$\widetilde{E}_K^{N,i,a}(P) = E_K(\Delta_{i,a} + P) \text{ with } \Delta_{i,a} = X^{2(i-1)+a}L$$

$$\widetilde{E}_K^{*,N,i,b_1,b_2}(P) = E_K(\Delta_{*,i,b_1,b_2} + P) \text{ with } \Delta_{*,i,b_1,b_2} = (X+1)^{1+b_2+2b_1}X^{2(i-1)}L$$

where $\delta = E_K(N)$ and $L = X^2\delta$, as previously.

A collision then only occurs if there are some $i, j \in \mathbb{N}^*$ and $a, b_1, b_2 \in \{0,1\}$ such that:

$$X^{2(i-1)+a} = (X+1)^{1+b_2+2b_1}X^{2(j-1)} \quad \Leftrightarrow \quad X^{2(i-j)+a} = (X+1)^{1+b_2+2b_1}$$

However, [Rog04] shows that the latter relation cannot hold for $i, j \leq 2^{115}$ (resp. $i, j \leq 2^{51}$) when $\mathbb{F}_{2^{128}}$ (resp. $\mathbb{F}_{2^{64}}$) is generated by the standard polynomial. A collision attack would thus require to query encryption for a huge message $M$, whose number of blocks would be far greater than the birthday bound, which is impossible.

Unfortunately, such a solution entails a doubling of the number of multiplications, compared to the original construction. It is therefore preferable to construct $\widetilde{\rho}$ in a slightly different way:

$$\widetilde{E}_K^{N,i,a}(P) = E_K(\Delta_{i,a} + P) \text{ with } \Delta_{i,a} = (X+1)^a X^{i-1}L$$

$$\widetilde{E}_K^{*,N,i,b_1,b_2}(P) = E_K(\Delta_{*,i,b_1,b_2} + P) \text{ with } \Delta_{*,i,b_1,b_2} = (X+1)^{2+b_2+2b_1}X^{i-1}L.$$

Here again, the argument of [Rog04] formally excludes any practical collision attack. The point is that, since $\Delta_{i,1} = \Delta_{i,0} \oplus \Delta_{i+1,0}$, almost one half of the

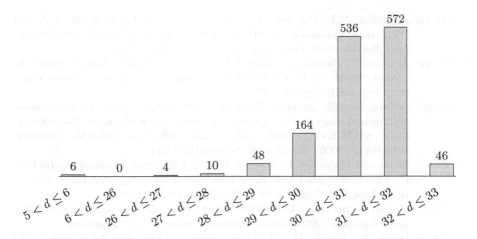

**Fig. 4.** Log of the lowest indices of colliding tweak polynomials for every $\mathbb{F}_{2^{64}}$ representations using the 1386 degree 64 irreducible pentanomials over $\mathbb{F}_2$.

offsets only require one xor to be computed. The cost is thus similar to the one of the original instantiation [Min14]. The last version of OTR [Min16] uses a similar method to generate tweaks and thus avoid our attack.

## 8  Conclusion

In this work, we have presented practical attacks against OTR resulting from collisions between the input masks. Although the occurrence of such collisions depend on both the blocksize $n$ and on the polynomial generating $\mathbb{F}_{2^n}$, we argue that the large number of parameters concerned calls for another design of the input masks. We have therefore proposed some ways to immunize OTR to these attacks which do not affect efficiency while being provably secure.

Our results thus do not question the intrinsic security of OTR but simply point out a flaw in the current instantiation.

**Acknowledgements.** We thank Jean-Gabriel Kammerer for helpful discussions on the implementation of the collision search algorithm, and Julien Devigne for his help. We also thank anonymous reviewers for their useful remarks.

## References

[ABB+14] Andreeva, E., Bilgin, B., Bogdanov, A., Luykx, A., Mendel, F., Mennink, B., Mouha, N., Wang, Q., Yasuda, K.: Primates. CAESAR 1st Round (2014). https://competitions.cr.yp.to/caesar-submissions.html

[AFF+15] Abed, F., Fluhrer, S., Forler, C., List, E., Lucks, S., McGrew, D., Wenzel, J.: Pipelineable on-line encryption. In: Cid, C., Rechberger, C. (eds.) FSE 2014. LNCS, vol. 8540, pp. 205–223. Springer, Heidelberg (2015). doi:10.1007/978-3-662-46706-0_11

[AP13] AlFardan, N.J., Paterson, K.G.: Lucky thirteen: breaking the TLS and DTLS record protocols. In: 2013 IEEE Symposium on Security and Privacy, pp. 526–540. IEEE Computer Society Press, May 2013

[BDP+14] Bertoni, G., Daemen, J., Michaël Peeters, G.V., Assche, R.K.: Caesar submission: keyak v1. In: CAESAR 1st Round (2014). https://competitions.cr.yp.to/round1/keyakv1.pdf

[BN00] Bellare, M., Namprempre, C.: Authenticated encryption: relations among notions and analysis of the generic composition paradigm. In: Okamoto, T. (ed.) ASIACRYPT 2000. LNCS, vol. 1976, pp. 531–545. Springer, Heidelberg (2000). doi:10.1007/3-540-44448-3_41

[BRW04] Bellare, M., Rogaway, P., Wagner, D.: The EAX mode of operation. In: Roy, B., Meier, W. (eds.) FSE 2004. LNCS, vol. 3017, pp. 389–407. Springer, Heidelberg (2004). doi:10.1007/978-3-540-25937-4_25

[CAE14] Caesar: competition for authenticated encryption: security, applicability and robustness (2014). http://competitions.cr.yp.to/caesar.html

[DN14] Datta, N., Nandi, M.: ELmE: a misuse resistant parallel authenticated encryption. In: Susilo, W., Mu, Y. (eds.) ACISP 2014. LNCS, vol. 8544, pp. 306–321. Springer, Heidelberg (2014). doi:10.1007/978-3-319-08344-5_20

[Dwo04] Dworkin, M.J.: Recommendation for block cipher modes of operation: the CCM mode for authentication and confidentiality, sp. 800–38c. Technical report, National Institute of Standards and Technology (2004)

[Dwo07] Dworkin, M.J.: Recommendation for block cipher modes of operation: galois/counter mode (GCM) and GMAC, spp. 800–38d. Technical report, National Institute of Standards and Technology (2007)

[HKR15] Hoang, V.T., Krovetz, T., Rogaway, P.: Robust authenticated-encryption AEZ and the problem that it solves. In: Oswald, E., Fischlin, M. (eds.) EUROCRYPT 2015. LNCS, vol. 9056, pp. 15–44. Springer, Heidelberg (2015). doi:10.1007/978-3-662-46800-5_2

[IK03] Iwata, T., Kurosawa, K.: OMAC: one-key CBC MAC. In: Johansson, T. (ed.) FSE 2003. LNCS, vol. 2887, pp. 129–153. Springer, Heidelberg (2003). doi:10.1007/978-3-540-39887-5_11

[IMGM15] Iwata, T., Minematsu, K., Guo, J., Morioka, S.: CLOC: authenticated encryption for short input. In: Cid, C., Rechberger, C. (eds.) FSE 2014. LNCS, vol. 8540, pp. 149–167. Springer, Heidelberg (2015). doi:10.1007/978-3-662-46706-0_8

[KR11] Krovetz, T., Rogaway, P.: The software performance of authenticated-encryption modes. In: Joux, A. (ed.) FSE 2011. LNCS, vol. 6733, pp. 306–327. Springer, Heidelberg (2011). doi:10.1007/978-3-642-21702-9_18

[Kra01] Krawczyk, H.: The order of encryption and authentication for protecting communications (or: How Secure Is SSL?). In: Kilian, J. (ed.) CRYPTO 2001. LNCS, vol. 2139, pp. 310–331. Springer, Heidelberg (2001). doi:10.1007/3-540-44647-8_19

[LRW02] Liskov, M., Rivest, R.L., Wagner, D.: Tweakable block ciphers. In: Yung, M. (ed.) CRYPTO 2002. LNCS, vol. 2442, pp. 31–46. Springer, Heidelberg (2002). doi:10.1007/3-540-45708-9_3

[Min14] Minematsu, K.: Parallelizable rate-1 authenticated encryption from pseudorandom functions. In: Nguyen, P.Q., Oswald, E. (eds.) EUROCRYPT 2014. LNCS, vol. 8441, pp. 275–292. Springer, Heidelberg (2014). doi:10.1007/978-3-642-55220-5_16

[Min16] Kazuhiko Minematsu. AES-OTR v3. Technical report, NEC Corporation (2016). https://groups.google.com/group/crypto-competitions/attach/1290d45334f8a3/AESOTR_v3.pdf?part=0.1

[MV04] McGrew, D.A., Viega, J.: The security and performance of the galois/-counter mode (GCM) of operation. In: Canteaut, A., Viswanathan, K. (eds.) INDOCRYPT 2004. LNCS, vol. 3348, pp. 343–355. Springer, Heidelberg (2004). doi:10.1007/978-3-540-30556-9_27

[Nik14] Nikolic, I.: Tiaoxin-346. CAESAR Submission (2014)

[Rog04] Rogaway, P.: Efficient instantiations of tweakable blockciphers and refinements to modes OCB and PMAC. In: Lee, P.J. (ed.) ASIACRYPT 2004. LNCS, vol. 3329, pp. 16–31. Springer, Heidelberg (2004). doi:10.1007/978-3-540-30539-2_2

[Ser98] Seroussi, G.: Table of low-weight binary irreducible polynomials. Technical report, HP (1998). http://www.hpl.hp.com/techreports/98/HPL-98-135.pdf?jumpid=reg_R1002_USEN

# Universal Forgery and Key Recovery Attacks on ELmD Authenticated Encryption Algorithm

Aslı Bay[1]($\boxtimes$), Oğuzhan Ersoy[2], and Ferhat Karakoç[1]

[1] TÜBİTAK BİLGEM, Gebze, Turkey
{asli.bay,ferhat.karakoc}@tubitak.gov.tr
[2] Electrical and Electronics Engineering Department,
Boğaziçi University, Istanbul, Turkey
oguzhan.ersoy@boun.edu.tr

**Abstract.** In this paper, we provide a security analysis of ELmD: a block cipher based Encrypt-Linear-mix-Decrypt authentication mode. As being one of the second-round CAESAR candidate, it is claimed to provide misuse resistant against forgeries and security against block-wise adaptive adversaries as well as 128-bit security against key recovery attacks. We scrutinize ElmD in such a way that we provide universal forgery attacks as well as key recovery attacks. First, based on the collision attacks on similar structures such as Marble, AEZ, and COPA, we present universal forgery attacks. Second, by exploiting the structure of ELmD, we acquire ability to query to the block cipher used in ELmD. Finally, for one of the proposed versions of ELmD, we mount key recovery attacks reducing the effective key strength by more than 60 bits.

**Keywords:** Authenticated encryption · CAESAR · ELmD · Forgery attack · Key recovery

## 1 Introduction

CAESAR competition [1] (Competition for Authenticated Encryption: Security, Applicability, and Robustness) has been announced in January 2013 aiming at fulfilling the needs of secure, efficient and robust authenticated encryption schemes. In total, 57 candidates are submitted to the competition. These schemes are released to crypto community for their security analysis and around 20 of them were eliminated in the first round of the competition in July 2015. Since then, around 30 candidates compete in the second round, and are being analyzed in terms of their security and efficiency.

ELmD is amongst the second-round CAESAR candidates designed by Datta and Nandi [5]. It is an Encrypt-Linear-mix-Decrypt block cipher authentication mode accepting associated data, and its structure is similar to some other

---

A. Bay—This author is financially supported by TÜBİTAK (BİDEB 2232, Project No. 115C119).

O. Ersoy—The work was done while this author was working at TÜBİTAK BİLGEM.

J.H. Cheon and T. Takagi (Eds.): ASIACRYPT 2016, Part I, LNCS 10031, pp. 354–368, 2016.
DOI: 10.1007/978-3-662-53887-6_13

authenticated encryption schemes such as AES-COPA [2], Marble [10], and SHELL [12]. ELmD is fully parallelizable and online, that is, $i^{th}$ block of ciphertext only depends on the first $i$ blocks of plaintext. As an optional property, it provides intermediate tag verification in order to fasten verification process and to be secure against block-wise adaptive adversaries. Designers of ELmD claim that the scheme provides nonce misuse resistance against forgery attacks. According to authors' assertion, ELmD provides 62.8-bit security for integrity (forgery attacks) and for privacy (distinguishing attacks). Indeed, they claim that ELmD provides 128-bit security against key recovery attacks that we disprove by applying partial-sum [7] and Demirci-Selçuk meet-in-the-middle attacks [6] on ELmD(6,6) where 6-round AES is used as the block cipher.

**Previous Results.** As far as we know, ELmD has been analyzed only by Zhang and Wu [13] in terms of both integrity and privacy. Very similar to our internal state recovery, they first find internal state parameter of ELmD by birthday attack and then they provide an almost universal forgery attack with a few queries. For breaking privacy, they propose a truncated differential analysis of reduced version of ELmD (ELmD(4, 4)) with $2^{123}$ time and memory complexities. In [13], the authors consider the internal parameter $L$ generated by only the encryption of zero with 4-round AES, i.e., $L = \text{AES}^4(0)$. However, both the usage of 4 rounds of encryption/decryption and the generation of the internal parameter $L$ with four AES rounds in ELmD are not acceptable in the proposal. Actually, after obtaining an input and output pair of 4-round AES (i.e., $L = \text{AES}^4(0)$), it is feasible to make a meet-in-the-middle analysis to recover the secret key. Previously, similar efforts are made to other CAESAR candidates COPA [11], Marble and AEZ in [8] to find state collisions beyond the birthday bound. Indeed, for AEZ and Marble [8], this attack is used for realizing a key recovery attack.

**Our Contribution:** In this paper, after obtaining the internal state parameter of ELmD, we make universal forgeries with a few queries to the oracle. Furthermore, by exploiting the structure of ELmD, we are able to query decryption oracle of the block cipher in ELmD. Finally, we mount key recovery attacks on ELmD(6,6) reducing effective key strength more than 60 bits.

Outline of the rest of the paper: In Sect. 2, a brief description of ELmD is given. Then in Sect. 3, we show how to recover internal state parameter $L$, and present universal forgery attacks on ELmD with a few queries to the oracle. In Sect. 4, we introduce novel methods to generate special plaintext pairs having relation between their ciphertexts and to query to the decryption oracle of the block cipher. By using chosen ciphertexts, in Sect. 5, key recovery attacks on ELmD(6,6) are presented. Section 6 concludes the paper.

# 2   Brief Description of ELmD

**Notation:** '$\oplus$': bitwise addition in modulo 2 (exclusive OR), '$\cdot$': field multiplication modulo the polynomial $p(x) = x^{128} + x^7 + x^2 + x + 1$ in $GF(2^{128})$. Also, $0^a$ denotes $a$-bit string of 0.

**Algorithm 1.** Processing associated data: IV generation

1: **Input:** $D, d, L$
2: **Output:** $IV$
3: **for** $i = 0$ to $d - 1$ **do**
4:     $DD_i = D_i \oplus 3 \cdot 2^i \cdot L$
5:     $Z_i = E_K(DD_i)$
6:     $(Y_i, W'_{i+1}) = \rho(Z_i, W'_i)$
7: **end for**
8: **if** $|D^*_d| = 128$ **then** $DD_d = D_d \oplus 3 \cdot 2^d \cdot L$
9: **else**  $DD_d = D_d \oplus 7 \cdot 3 \cdot 2^{d-1} \cdot L$
10: **end if**
11: $Z_d = E_K(DD_d)$
12: $(Y_d, W'_{d+1}) = \rho(Z_d, W'_d)$
13: $IV = W'_{d+1}$

**Algorithm 2.** Encryption and tag generation without producing intermediate tag ($t = 0$)

1: **Input:** $\ell, IV, M_1, \dots, M_\ell, L, |M^*_\ell|$
2: **Output:** $C_1, \dots, C_\ell, C_{\ell+1}$
3: $W_0 = IV$
4: $M_{\ell+1} = M_\ell$
5: **for** $i = 1$ to $\ell - 1$ **do**
6:     $MM_i = M_i \oplus 2^{i-1} \cdot L$
7:     $X_i = E_K(MM_i)$
8:     $(Y_i, W_i) = \rho(X_i, W_{i-1})$
9:     $CC_i = E_K^{-1}(Y_i)$
10:     $C_i = CC_i \oplus 3^2 \cdot 2^{i-1} \cdot L$
11: **end for**
12: **if** $|M^*_\ell| = 128$ **then** $MM_\ell = M_\ell \oplus 2^{\ell-1} \cdot L$ and $MM_{\ell+1} = M_{\ell+1} \oplus 2^\ell \cdot L$
13: **else**  $MM_\ell = M_\ell \oplus 7 \cdot 2^{\ell-2} \cdot L$ and $MM_{\ell+1} = M_{\ell+1} \oplus 7 \cdot 2^{\ell-1} \cdot L$
14: **end if**
15: **for** $i = \ell$ to $\ell + 1$ **do**
16:     $X_i = E_K(MM_i)$
17:     $(Y_i, W_i) = \rho(X_i, W_i)$
18: **end for**
19: $CC_\ell = E_K^{-1}(Y_\ell)$
20: $C_\ell = CC_\ell \oplus 3^2 \cdot 2^{\ell-1} \cdot L$
21: $CC^*_{\ell+1} = E_K^{-1}(Y_{\ell+1} \oplus 1)$
22: $C^*_{\ell+1} = CC^*_{\ell+1} \oplus 3^2 \cdot 2^\ell \cdot L$
23: **if** $|M^*_\ell| \neq 128$ **then** $C_{\ell+1} = trunc(C^*_{\ell+1})_{|M^*_\ell|}$
24: **else**  $C_{\ell+1} = C^*_{\ell+1}$
25: **end if**

ELmD is a block cipher based Encrypt-Linear-mix-Decrypt authentication mode proposed by Datta and Nandi [5] for CAESAR competition. In the proposal of ELmD, AES-128 [4] is used as the block cipher where the number of rounds can be either 10 or 6. Note that 6-round AES used in ELmD includes whitening-key layer and MixColumns operation at the last round. Hence from now on, $\text{AES}^{\text{rd}}$ denotes AES with **rd** rounds. For simplicity, $E_K$ is also used for AES-128 in the rest of the paper. In addition, $L$ is a key-depending mask which is generated in two ways; $L = \text{AES}^6(\text{AES}^6(0))$ when rd $= 6$ and $L = \text{AES}^{10}(0)$ when rd $= 10$.

The linear mixing function $\rho$ takes two inputs $t, x \in \{0,1\}^{128}$ and produces two outputs $t', y \in \{0,1\}^{128}$ as follows

$$\rho(x,t) = (y,t') : \quad y = x \oplus 3 \cdot t \text{ and } t' = x \oplus 2 \cdot t.$$

Associated data is used to generate IV (see Algorithm 1) which is an input to both encryption/decryption function of ELmD. Let pub and param be a public message number and the parameter set, respectively, which are both 64 bits, and $D = (D_1, \ldots, D_d^*)$ be an associated data. By construction, the designers of ELmD assign $D_0 = \text{pub}\|\text{param}$ and $W_0' = 0$. The last block of associated data is padded as $D_d = D_d^*\|10^*$ if $|D_d^*| \neq 128$, otherwise $D_d = D_d^*$.

ELmD has two versions, namely v1.0 and v2.0. ELmD v1.0 was modified by the generation of last message block in such a way that the XOR of previous messages added to this block. Also, **rd** is modified to ELmD(6,6) and ELmD(10,10).

Tagged ciphertext is generated as follows. Let $M = M_1\|M_2\|\cdots\|M_\ell^*$ be the message to be encrypted. Padding is performed as $M_\ell = (\oplus_{i=1}^{\ell-1}M_i) \oplus (M_\ell^*\|10^*)$

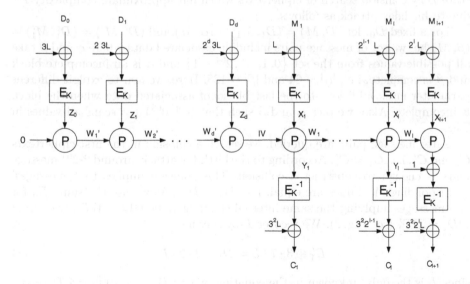

**Fig. 1.** Processing associated data and the generation of tagged ciphertext in ELmD when $|D_d| = |M_\ell| = n$

if $|M_\ell^*| < 128$, otherwise $M_\ell = (\oplus_{i=1}^{\ell-1} M_i) \oplus M_\ell^*$. ELmD has an intermediate tag option if it is needed, however for the simplicity we mention only tagged ciphertext generation without producing intermediate tags ($t = 0$) in Algorithm 2. ELmD encryption including processing associated data is depicted in Fig. 1.

ELmD decrypts and verifies a given tagged ciphertext pair in three steps. First of all, IV is produced by using pub, param, and $D$ as in Algorithm 1. Afterwards, the tagged ciphertext is decrypted as an inversion of Algorithm 2, and then tag is verified when $M_{\ell+1} = M_\ell$. Once the tag is verified, plaintext is released otherwise $\perp$ is returned.

# 3   Universal Forgery Attack on ELmD

In this section, we present universal forgery attacks on ELmD. First, we recover ELmD state $L$ by collision search of ciphertexts. Using $L$, we can make universal forgery attack on ELmD. Before going into details, we briefly describe the two main forgery models:

– **Existential Forgery** is the generation of a valid ciphertext and tag pair for an unspecified message which is not previously queried to an oracle.
– **Universal Forgery** is the generation of ciphertext and tag pair for a given message which is not previously queried to an oracle.

## 3.1   Recovering Internal State Parameter L

Similar to state recovery attacks of COPA and Marble [8,11], we recover ELmD state $L$ by collision search of ciphertexts which has approximate complexity $2^{65}$ due to birthday attack as follows.

For a fixed $D_0$, let $(D, M) = (D_1, M_1) = (\alpha, M)$ and $(D', M') = (D_1', M_1') = (\beta, M)$ be two set of message pairs including associated data where $\alpha$ and $\beta$ take all possible values from the set $\{0, 1, \ldots, 2^{64} - 1\}$ and $\alpha$ is an incomplete block and $\beta$ is complete, i.e., $|\alpha| = 64$ and $|\beta| = 128$. Here, we aim to exploit different parameter mask additions to the last blocks of associated data when the block is incomplete. Also, we pick $\alpha$ and $\beta$ such that $(\alpha \| 10^{63}) \oplus \beta$ scans all values in $\mathbb{F}_{2^{128}}$.

After message pairs are queried, we search a collision in the first ciphertexts $C_1$ and $C_1'$, i.e., $C_1 = C_1'$. According to the birthday attack, around $2 \cdot 2^{64}$ message pairs is enough to construct a collision. This collision implies that messages' corresponding IV values are equal, i.e., $IV = IV'$. As we use the same $D_0$ for two messages implying the same internal chaining value ($W_1' = W_1''$), we obtain $DD_1 = DD_1'$ (see Fig. 2). We recover $L$ by solving

$$D_1' \oplus 3 \cdot 7 \cdot L = D_1 \oplus 3 \cdot 2 \cdot L, \tag{1}$$

since $L$ is the only unknown in the equation, where $D_1 = \alpha \| 10^{63}$ and $D_1' = \beta$.

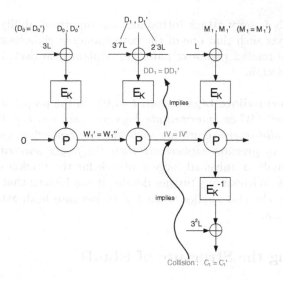

**Fig. 2.** Recovering $L$ by finding a collision in $(t = 0)$

## 3.2 Forgery

Once we recover $L$, we can make universal forgery attacks on ELmD by making a few queries to the oracle.

**A Universal Forgery Attack.** Let $(D, M) = (D_1, \ldots, D_{d-1}, D_d, M_1, \ldots, M_{\ell-1}, M_\ell)$ be targeted associated data and message pair with assigned $D_0 = \text{pub}\|\text{param}$, where $|D_d| = 128$. Compute $D'_d$ such that $D'_d\|10^* = D_d \oplus 2^d \cdot 3L \oplus 7 \cdot 2^{d-1} \cdot 3L$ and $|D'_d| < 128$. Note that because of the padding rule, we can always obtain $D'_d$ with $|D'_d| < 128$.

Query $(D', M) = (D_1, \ldots, D_{d-1}, D'_d, M_1, \ldots, M_{\ell-1}, M_\ell)$ with the same $D_0$ and obtain the corresponding ciphertext and tag pair as $(\widetilde{C}, \widetilde{T})$. Due to the choice of associated data, $D$ and $D'$ produce the same IV. Hence, the corresponding ciphertext and tag pair $(C, T)$ of $(D, M)$ is equal to that of $(D', M)$, i.e., $(C, T) = (\widetilde{C}, \widetilde{T})$. Note that the same attack also works for $|D_d| < 128$ case. In a similar manner, a $|D'_d| = 128$ block can be chosen where $D'_d = D_d\|10^* \oplus 2^d \cdot 3L \oplus 7 \cdot 2^{d-1} \cdot 3L$, and the rest of the attack is the same. Therefore, this forgery attack works for any associated data and message pair.

**Another Universal Forgery Attack.** Here we present another forgery for the same $(D_0, D, M)$ triple using only completed blocks. First, query $M_1 = D_0 \oplus 3L \oplus L$ without D, and obtain $C_1$. Then, query $(D', M)$ such that $D'_0 = D_0$, $D'_1 = C_1 \oplus 3^2L \oplus 2 \cdot 3L$, $D'_{i+2} = D_i \oplus 2^i \cdot 3L \oplus 2^{i+2} \cdot 3L$ for $i = 0, 1, \ldots, d$ and obtain ciphertext $C$ and tag $T$. It can be seen that this $(C, T)$ pair is also valid for $(D, M)$.

Note that this forgery attack introduces an important ability of generating a pair of plaintexts such that one of the corresponding ciphertext is half of the other one. These related plaintext pairs are explained in Sect. 4, and used for key recovery in Sect. 5.

**Forgery of Intermediate tags (when $t \neq 0$).** In the proposal of ELmD, the authors state that "*When intermediate tags are used i.e. $t \neq 0$, if the forger can compute a valid intermediate tag such that the ciphertext up to that is not identical to any of previous ciphertexts then the forger succeeds*". Once $L$ is known, we can make a universal forgery attack for the version of ELmD with intermediate tags. Without any further details, it can be seen that the previously given forgery attacks also applies when $t \neq 0$. Because both attacks only uses the associated data.

# 4    Exploiting the Structure of ELmD

In this section, we explore the block cipher used in ELmD by exploiting the general structure of the authenticated encryption algorithm where the bottom function is the decryption mode of the upper one. First, using the recovered $L$ value, we can obtain two types of plaintext pairs:

1. For any $P_1$ and $\mu$, $(P_1, P_2)$ pair such that $\mu \cdot E(P_1) = E(P_2)$.
2. For any $\Delta$, $(Q_1, Q_2)$ pair such that $E(Q_1) = E(Q_2) \oplus \Delta$.

Using these special plaintext pairs, we can obtain plaintext and corresponding ciphertext pairs of the encryption block cipher $E_K(\cdot)$ or $\text{AES}^{\text{rd}}$. Especially, we can query any ciphertext to the decryption mode of the cipher.

Following attacks are mostly explained for the maskless version of ELmD. Since we know the $L$ value, we can easily switch from $(D, M, C)$ triple to $(DD, MM, CC)$ triple and vice versa, where $D_i = DD_i \oplus 2^{i-1} \cdot 3L$, $M_i = MM_i \oplus 2^{i-1}L$ and $C_i = CC_i \oplus 2^{i-1} \cdot 3^2L$. In other words, we can query $(DD, MM)$ and obtain $CC$ values. For the simplicity, we usually use $(DD, MM, CC)$ triples. It is important to note that the last message block cannot be controlled since $MM_{\ell+1} = MM_\ell \oplus 2^{\ell-1}L \oplus 2^\ell L$.

## 4.1    2-Multiplicative Pairs: $(R_1, R_2)$ with $2 \cdot E(R_1) = E(R_2)$

Initially, for any given/fixed $D_0 = \mathsf{pub}\|\mathsf{param}$, we make a query for one block message $MM_1^1 = DD_0$ without an additional associated data and obtain the corresponding ciphertext and tag pair $(C^1, T^1)$. As seen in Fig. 3, $IV^1 = E_K(DD_0)$. Because of our message choice, $X_1^1$ is also equal to $IV^1$ and therefore $Y_1^1 = 2 \cdot IV^1$. Even without knowing $IV^1$ value, we obtain $CC_1^1$ such that $E_K(CC_1^1) = 2 \cdot IV^1 = 2 \cdot E_K(DD_0)$. Here, it is important to note that $D_0$ has a special structure and cannot take any 128-bit value. For any $R_1$, using the same $D_0$, query $DD_1^2 = CC_1^1, MM_1^2 = MM_2^2 = R_1$ and obtain the corresponding ciphertext and tag pair $(C^2, T^2)$. It can be seen that $IV^2 = \rho(IV^1, 2 \cdot IV^1) = 0$ and therefore $X_1^2 = W_1^2 = E_K(MM_1^2)$. $W_1^2 = E_K(MM_1^2) = X_2^2$ implies $Y_2^2 = 2 \cdot X_2^2$ and $E_K(CC_2^2) = 2 \cdot E_K(MM_1^2)$. As can be seen in Fig. 3, by setting $R_2 = CC_2^2$, we obtain $(R_1, R_2)$ pair such that $2 \cdot E(R_1) = E(R_2)$. The complexity to obtain $N$ such 2-multiplicative pairs is only $N + 1$ queries if the same $D_0 = \mathsf{pub} \| \mathsf{param}$ is used. Therefore, the complexity of getting a 2-multiplicative pair is approximately one block query.

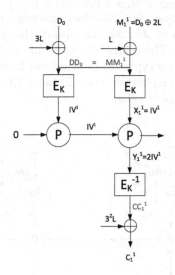

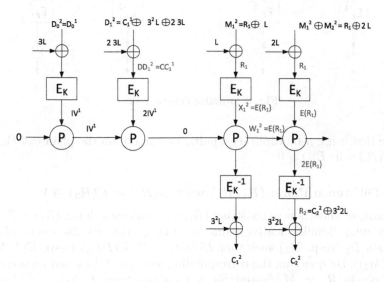

**Fig. 3.** 2-multiplicative pairs

## 4.2  $\mu$-multiplicative Pairs: $(P_1, P_2)$ with $\mu \cdot E(P_1) = E(P_2)$

Here, we present a method to generate $(P_1, P_2)$ pair satisfying $\mu \cdot E(P_1) = E(P_2)$ for any $P_1$ and $\mu$ values with the help of observations in the previous part. First, for a given $P_1$, we obtain the plaintext $R_2$ such that $2 \cdot E(P_1) = E(R_2)$. Also, we arrange associated data to make $IV = 0$.

Let $\mu' = 3^{-1}(\mu \oplus 1)$ where $3^{-1}$ represents the multiplicative inverse of $3$ in the given field. It can be seen that any $\mu' \in \mathbb{F}_{2^{128}}$ can be represented as $2^{127} \cdot m_1 \oplus 2^{126} \cdot m_2 \oplus \cdots \oplus 2 \cdot m_{127} \oplus m_{128}$ where $m_i \in \{1, 2\}$.

As shown in Fig. 4, by querying 129-block message with $MM_i = R_{m_i}$ for $i = 1, \dots, 128$ and $MM_{129} = P_1$, we can obtain the plaintext $P_2 = CC_{129}$ satisfying $E(P_2) = \mu \cdot E(P_1)$. The complexity to obtain any multiplicative pair of a given $P_i$ is about $2^7$ block encryptions. In other words, obtaining the plaintext of a given multiple of a given ciphertext costs $2^7$ block ELmD encryptions which is approximately $2^8$ block cipher calls.

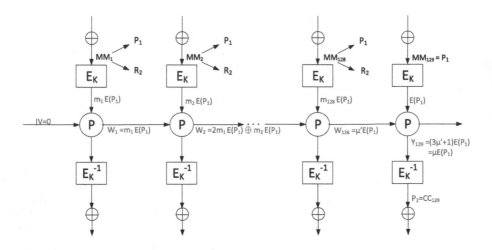

**Fig. 4.** $\mu$-multiplicative pairs

Note that using $\mu$-multiplicative pairs, we can obtain the plaintext $P_0$ satisfying $E(P_0) = 0 \cdot E(\cdot) = 0$.

## 4.3  1-Difference Pairs: $(R_1, R_2)$ with $E(R_1) = E(R_2) \oplus 1$

In this part, we show how to construct $(R_1, R_2)$ pairs such that $E(R_1) = E(R_2) \oplus 0^{127}1$ by using 2-multiplicative pairs (see Fig. 5). For any $D_0$ (resp. $M_1$), we can obtain $D_1$ (resp. $M_2$) such that $E(DD_1) = 2 \cdot E(DD_0)$ (resp. $E(MM_2) = 2 \cdot E(MM_1)$). By querying the corresponding associated data and message pair, we can obtain $R_1 = MM_3$ and $R_2 = CC_3$ satisfying $E(R_1) = E(R_2) \oplus 1$. The complexity to obtain a 1-difference pair is simply a query of 1 associated data block and 2 message blocks where associated data and message blocks are 2-multiplicative pairs.

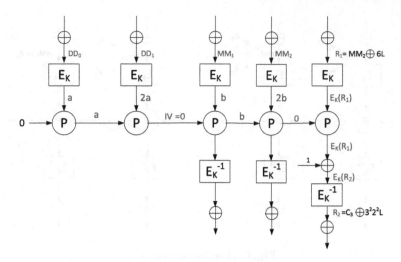

**Fig. 5.** 1-difference pairs

## 4.4 $\Delta$-difference Pairs: $(Q_1, Q_2)$ with $E(Q_1) = E(Q_2) \oplus \Delta$

First, we generate a 1-difference pair: $\{R_1, R_2\}$ where $E(R_1) = E(R_2) \oplus 0^{127}1$. Then, for any $\Delta$, compute $\delta = \delta_1 \| \delta_2 \| \cdots \| \delta_{128}$ such that $3 \cdot \delta = \Delta$ over the defined field.

We construct two messages $M, M'$ each containing 129 blocks with the same associated data $D$ such that

$$MM_i = R_1 \quad \text{and} \quad MM_i' = R_{\delta_i+1} \quad \text{for} \quad i = 1, 2, \ldots, 129$$

where $\delta_{129} = 0$.

As illustrated in Fig. 6, $129^{th}$ ciphertext blocks of $(D, M)$ and $(D, M')$ differ by $\Delta$. Here, we briefly, explain the differential path of two messages $(D, M)$ and $(D, M')$. As their associated data are equal, they will provide the same $IV$, that is $IV \oplus IV' = 0$. After processing of the first blocks of two messages $R_1$ and $R_{\delta_1+1}$ in the upper layer of encryption, we will get difference in $X_1$'s as $\Delta X_1 = X_1 \oplus X_1' = \delta_1$. Since $\Delta IV = 0$, $\Delta W_1 = W_1 \oplus W_1' = \delta_1$. For the second message blocks $R_1$ and $R_{\delta_2+1}$, we get $\Delta X_2 = X_2 \oplus X_2' = \delta_2$. Then, we have $\Delta W_2 = W_2 \oplus W_2' = 2\delta_1 + \delta_2$. Similarly, after the encryption of 128th blocks, we have $\Delta W_{128} = W_{128} \oplus W_{128}' = 2^{127}\delta_1 + 2^{126}\delta_2 + \cdots + \delta_{128} = \delta$. Finally, as we choose the last message blocks equal, we have $\Delta X_{129} = X_{129} \oplus X_{129}' = 0$. Since no difference is coming from upper encryption layer $\Delta Y_{129} = Y_{129} \oplus Y_{129}' = 3 \cdot \Delta W_{128} = 3 \cdot \delta = \Delta$. Hence, we obtain plaintexts $Q_1 = CC_{129}$ and $Q_2 = CC_{129}'$ having required ciphertext difference: $E(Q_1) = E(Q_2) \oplus \Delta$. Note that by changing the last message block, we can get several message pairs having desired ciphertext difference.

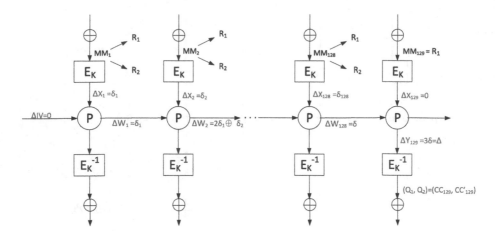

**Fig. 6.** $\Delta$-difference pairs

### 4.5  Querying Decryption Oracle of the Block Cipher

Here, we describe how to query inner block cipher of ELmD, $AES^{rd}$. Since, we can obtain any multiple of a given ciphertext in $\mu$-multiplicative pairs, it is obvious that any ciphertext can be queried, i.e., plaintext of a given ciphertext can be obtained, if the decryption of $0^{127}1$ is known.

First, using 1-difference pairs, we obtain a pair $(R_1, R_2)$ with $E(R_1) = E(R_2) \oplus 1$. Then, using $\mu$-multiplicative pairs, we acquire $R_3$ such that $3^{-1}E(R_1) = E(R_3)$. By querying associated data satisfying $IV = 0$ and message with $MM_1 = R_3$, $MM_2 = R_2$, we obtain $CC_2$ which is equal to decryption of 1, i.e., $E(CC_2) = 0^{127}1$. After obtaining decryption of 1, we can query any ciphertext with the help of $\mu$-multiplicative pairs.

This property enables us to mount a chosen ciphertext attack.

## 5  Key Recovery

The encryption function $E_K$ used in ElmD is either 6-round AES ($AES^6$) or 10-round AES ($AES^{10}$) depending on the application. For both versions of ELmD, the designers claim that ELmD provides 128 bits of security against plaintext and key recovery attacks. In this section, we show that this claim is not valid if the function $E_K$ is $AES^6$.

In Sect. 4, after recovering $L$ parameter, it is shown how to obtain corresponding plaintext for any given ciphertext in a time complexity of about $2^8$ encryption operations. As a result, we can mount attacks on 6-round AES with chosen ciphertexts. In [7], by using partial sums an attack on 6-round AES was given with a time and data complexities of $2^{44}$ and $2^{34.6}$, respectively in chosen plaintext scenario. This attack can be easily adapted to chosen ciphertext case because of the AES structure. MixColumns and AddRoundKey operations can be swapped with applying the inverse of MixColumns to the round key. As known, the inverse of AES without the MixColumns operation in the last round has the same structure with AES, the similar attack can be applied. Note that the MixColumn operations at the end of the cipher is not important because ciphertexts can be easily manipulated. The total time complexity of key recovery is $2^{65} + 2^8 \times 2^{34.6} + 2^{44} \approx 2^{65}$ which is dominated by the cost of recovery of $L$.

In addition, we propose a Demirci-Selçuk meet-in-the-middle attack [6] using the distinguisher on 3-round AES [9]. This attack also uses chosen ciphertexts. The time and data complexities of this attack is $2^{66}$ and $2^{33}$, respectively. With this attack the time complexity of key recovery attack on ELmD is $2^{65}+2^8 \times 2^{33} + 2^{66} \approx 2^{66.6}$ encryptions. Even though the time complexity is relatively higher than the previous attack, this attack uses relatively less data and illustrates Demirci-Selçuk MITM in a splice-and-cut [3] perspective.

While presenting the attack we use the following notation. $AES^6$ consists of 6 full rounds of AES with initial key whitening and supports a key size of 128 bits. One full round of AES is composed of SubBytes (SB), ShiftRows (SR), MixColumns (MC) and AddRoundKey (AK) operations [4]. The whitening key and $i$-th round key ($i \in \{1, 2, 3, 4, 5, 6\}$) are denoted by $k_0$ and $k_i$ respectively. We use $x_i$, $y_i$, $z_i$ and $w_i$ to represent the blocks in $i$-th round before the SubBytes, ShiftRows, MixColumns and AddRoundKey operations respectively where the input of the first round is $x_1 = P \oplus k_0$ and $P$ is the plaintext. In the case of swapping MixColumns and AddRoundKey operations we denote the round key as $u_i = MC^{-1}(k_i)$ and the state after round key addition as $\bar{w}_i$. Also $x_i^j$, $y_i^j$, $z_i^j$, $w_i^j$ and $\bar{w}_i^j$ denotes the blocks for $j$-th plaintext and $a(m, n, ..., l)$ are used for $m, n, ..., l$-th bytes of a block $a$. The orders of 128-bit blocks' bytes in $4 \times 4$ matrix of bytes is as conventional, that is the first row is composed of 0, 4, 8 and 12-th bytes of 128-bit block where 0-th byte is the left-most byte.

The attack is given in Algorithm 3 and depicted in Fig. 7. The number of bits guessed in the attack is 144 and the probability that a wrong guess passes the condition in Step 10 is $2^{-144}$. Thus, with the correct guess, a wrong one can be returned by the algorithm. In Step 3 in Algorithm 3, $2^{80} \times 19 \times \frac{10}{16 \times 6} \approx 2^{81}$ encryptions are performed by guessing 10 bytes. Note that this step can be done offline. For a ciphertext the time complexity of getting the corresponding plaintext is approximately $2^8$ encryptions as mentioned in Sect. 4. Thus the number of operations performed in Step 6 is $19 \times 2^{32} \times 2^8 = 2^{44.25}$ encryptions. In Step 9, 144-bit differences are computed performing $2^{64} \times 19 \times \frac{10}{16 \times 6} \approx 2^{65}$ encryption operations. As a result the time complexity of Algorithm 3 is $2^{81}$ offline and $2^{65}$ online encryptions. To store the 144-bit difference for possible $2^{80}$ values $2^{80} \times 144$-bit memory is required. Note that in the attack 12 bytes of $\bar{w}_6^j$ are fixed to a constant 0. Thus the attack needs $2^{32}$ chosen ciphertexts and corresponding plaintexts.

Notice that with this attack we obtain 4 bytes of $k_0$ so far. With slight modifications in the attack it can be seen easily that other 4 bytes of $k_0$ can be found. Remaining 64 bits of the key can be recovered by brute force. The

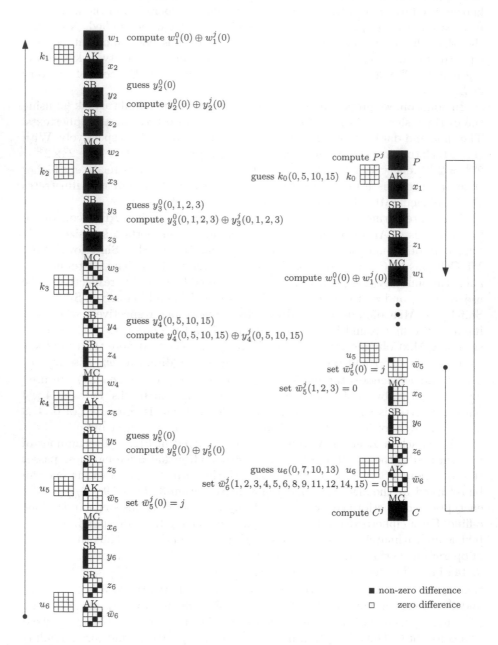

**Fig. 7.** Demirci-Selçuk MITM attack on 6-round AES. The offline and online steps are on the left-hand and right-hand sides, respectively.

**Algorithm 3.** Demirci-Selçuk MITM Attack on 6-round AES.

1: Take 19 different values for $\bar{w}_5^j(0,1,2,3)$ and $\bar{w}_6^j(1,2,3,4,5,6,8,9,11,12,14,15)$ such that $\bar{w}_5^j(0) = j$, and the other bytes are 0 for $0 \leq j \leq 18$.
2: **for** each possible values of $y_5^0(0)$, $y_4^0(0,5,10,15)$, $y_3^0(0,1,2,3)$ and $y_2^0(0)$ **do**
3:     Compute the difference $(w_1^0(0) \oplus w_1^1(0), w_1^0(0) \oplus w_1^2(0), ..., w_1^0(0) \oplus w_1^{18}(0))$ and store it in Table $T$.
4: **end for**
5: **for** each possible values of $u_6(0,7,10,13)$ **do**
6:     Compute $C^j$'s
7:     Find $P^j$'s by using the method in Sect. 4.
8:     **for** each possible values of $k_0(0,5,10,15)$ **do**
9:         Compute the difference $(w_1^0(0) \oplus w_1^1(0), w_1^0(0) \oplus w_1^2(0), ..., w_1^0(0) \oplus w_1^{18}(0))$ and find the difference in Table $T$.
10:         **if** a match found **then**
11:             Return $k_0(0,5,10,15)$ as the correct key
12:         **end if**
13:     **end for**
14: **end for**

total complexity of recovering 128-bit key will be $2 \times 2^{81} = 2^{82}$ offline and $2 \times 2^{65} + 2^{64} \approx 2^{66}$ online encryptions.

The memory and data complexities will be $2 \times 2^{80} \times 144$-bit memory and $2 \times 2^{32} = 2^{33}$ data respectively. Note that the offline time complexity can be reduced to $2^{74}$ by removing the guess of $y_5(0)$ from the offline step and adding 8-bit guess for $u_5(0)$ to online step. In that case the time complexity of online step will be $2^{74}$ encryption operations.

# 6    Conclusion

ELmD is an a block cipher based Encrypt-Linear-mix-Decrypt authentication mode submitted to CAESAR Competition. It is claimed to be strong against misuse forgery attacks, block-wise adaptive adversaries and key recovery attacks with 128-bit security. This work provides universal forgery attacks against ELmD. Furthermore, we disprove the 128-bit security claim of ELmD by applying two key recovery attacks, namely partial-sum and Demirci Selçuk meet-in-the-middle attacks with $2^{65}$ and $2^{66.6}$ time complexities, respectively.

# References

1. CAESAR - Competition for authenticated encryption: security, applicability, and robustness. http://competitions.cr.yp.to/caesar.html
2. Andreeva, E., Bogdanov, A., Luykx, A., Mennink, B., Tischhauser, E., Yasuda, K.: AES-COPA v. 1, submission to the CAESAR competition, March 2014. http://competitions.cr.yp.to/round1/aescopav1.pdf

3. Bogdanov, A., Khovratovich, D., Rechberger, C.: Biclique cryptanalysis of the full AES. In: Lee, D.H., Wang, X. (eds.) ASIACRYPT 2011. LNCS, vol. 7073, pp. 344–371. Springer, Heidelberg (2011). doi:10.1007/978-3-642-25385-0_19
4. Daemen, J., Rijmen, V.: The Design of Rijndael: AES - The Advanced Encryption Standard. Information Security and Cryptography. Springer, Heidelberg (2002). http://dx.doi.org/10.1007/978-3-662-04722-4
5. Datta, N., Nandi, M.: ELmD v2.0, submission to the CAESAR competition, August 2015. https://competitions.cr.yp.to/round2/elmdv20.pdf
6. Demirci, H., Selçuk, A.A.: A meet-in-the-middle attack on 8-round AES. In: Nyberg, K. (ed.) FSE 2008. LNCS, vol. 5086, pp. 116–126. Springer, Heidelberg (2008). doi:10.1007/978-3-540-71039-4_7
7. Ferguson, N., Kelsey, J., Lucks, S., Schneier, B., Stay, M., Wagner, D., Whiting, D.: Improved cryptanalysis of rijndael. In: Goos, G., Hartmanis, J., Leeuwen, J., Schneier, B. (eds.) FSE 2000. LNCS, vol. 1978, pp. 213–230. Springer, Heidelberg (2001). doi:10.1007/3-540-44706-7_15
8. Fuhr, T., Leurent, G., Suder, V.: Collision attacks against CAESAR candidates. In: Iwata, T., Cheon, J.H. (eds.) ASIACRYPT 2015. LNCS, vol. 9453, pp. 510–532. Springer, Heidelberg (2015). doi:10.1007/978-3-662-48800-3_21
9. Gilbert, H., Minier, M.: A collision attack on 7 rounds of rijndael. In: AES Candidate Conference, pp. 230–241 (2000)
10. Guo, J.: Marble specification version 1.0, submission to the CAESAR competition, March 2014. http://competitions.cr.yp.to/round1/marblev10.pdf
11. Lu, J.: On the security of the COPA and marble authenticated encryption algorithms against (almost) universal forgery attack. IACR Crypt. ePrint Arch. **2015**, 79 (2015). http://eprint.iacr.org/2015/079
12. Wang, L.: SHELL v2.0, submission to the CAESAR competition, August 2015. https://competitions.cr.yp.to/round2/shellv20.pdf
13. Zhang, J., Wu, W.: Security analysis of CAESAR second-round candidate: ELmD (2016). www.escience.cn/system/download/77967

# Statistical Fault Attacks on Nonce-Based Authenticated Encryption Schemes

Christoph Dobraunig[1], Maria Eichlseder[1], Thomas Korak[1], Victor Lomné[2],
and Florian Mendel[1]([⊠])

[1] Graz University of Technology, Graz, Austria
florian.mendel@iaik.tugraz.at
[2] ANSSI, Paris, France
victor.lomne@ssi.gouv.fr

**Abstract.** Since the first demonstration of fault attacks by Boneh et al. on RSA, a multitude of fault attack techniques on various cryptosystems have been proposed. Most of these techniques, like Differential Fault Analysis, Safe Error Attacks, and Collision Fault Analysis, have the requirement to process two inputs that are either identical or related, in order to generate pairs of correct/faulty ciphertexts. However, when targeting authenticated encryption schemes, this is in practice usually precluded by the unique nonce required by most of these schemes.

In this work, we present the first practical fault attacks on several nonce-based authenticated encryption modes for AES. This includes attacks on the ISO/IEC standards GCM, CCM, EAX, and OCB, as well as several second-round candidates of the ongoing CAESAR competition. All attacks are based on the Statistical Fault Attacks by Fuhr et al., which use a biased fault model and just operate on collections of faulty ciphertexts. Hereby, we put effort in reducing the assumptions made regarding the capabilities of an attacker as much as possible. In the attacks, we only assume that we are able to influence some byte (or a larger structure) of the internal AES state before the last application of MixColumns, so that the value of this byte is afterwards non-uniformly distributed.

In order to show the practical relevance of Statistical Fault Attacks and for evaluating our assumptions on the capabilities of an attacker, we perform several fault-injection experiments targeting real hardware. For instance, laser fault injections targeting an AES co-processor of a smartcard microcontroller, which is used to implement modes like GCM or CCM, show that 4 bytes (resp. all 16 bytes) of the last round key can be revealed with a small number of faulty ciphertexts.

**Keywords:** Fault attacks · Authenticated encryption · CAESAR · Differential Fault Attacks (DFA) · Statistical Fault Attacks (SFA)

## 1 Introduction

Fault attacks pose a serious threat for cryptographic implementations. For this kind of attacks, the analyzed device is operated outside its defined operating

© International Association for Cryptologic Research 2016
J.H. Cheon and T. Takagi (Eds.): ASIACRYPT 2016, Part I, LNCS 10031, pp. 369–395, 2016.
DOI: 10.1007/978-3-662-53887-6_14

370 C. Dobraunig et al.

conditions, which can lead to erroneous outputs. By analyzing the erroneous output data, secret information can be revealed. In the worst case, a single fault can reveal the entire secret key of a block cipher like AES, which has been shown to be feasible by many researchers in the last decade [7,33]. Popular techniques to inject faults include modifications of the power supply [50] or the clock source [6] by injecting glitches. Other methods, such as laser fault injection [45], have been proven even more powerful, because they additionally allow a precise localization of the fault injection.

While fault attacks on block ciphers and stream ciphers have received a great deal of attention from the scientific community, authenticated ciphers have been arguably less popular targets among researchers. At the same time, they describe an important class of cryptographic algorithms with many applications in information security. Authenticated encryption provides both confidentiality and authentication of data to two parties communicating via an insecure channel. This is essential for many applications such as SSL/TLS, IPSEC, SSH, or hard-disk encryption. In most applications, there is not much value in keeping the data secret without ensuring that it has not been intentionally or unintentionally modified. For this reason, in practical applications, block ciphers like AES are typically used mainly as a building block for an authenticated encryption scheme.

An authenticated encryption scheme is usually modeled as a function with four inputs: a unique nonce $N$, associated data $A$, plaintext $P$, and secret key $K$. It generates two outputs: the ciphertext $C$, and the authentication tag $T$:

$$\mathcal{E}(K, N, A, P) = (C, T).$$

The corresponding decryption algorithm takes the secret key $K$, nonce $N$, authenticated data $A$, ciphertext $C$, and tag $T$, and either outputs the plaintext $P$ if the verification tag is correct, or $\perp$ if the verification of the tag failed:

$$\mathcal{D}(K, N, A, C, T) \in \{P, \perp\}.$$

It is usually assumed (and typically essential for the security of the authenticated encryption scheme) that nonces never repeat for encryptions $\mathcal{E}$ under the same key $K$. We refer to such schemes as nonce-based authenticated encryption. While some schemes claim a certain level of robustness even in misuse settings (such as repeated nonces, or release of unverified plaintext), this does not mean that they are intended to be intentionally misused in practical implementations: repeating nonces always incurs a certain loss of security.

An interesting consequence of the unique nonce in the encryption procedure is the implicitly provided protection against several classes of fault attacks [11, 12,49]. In particular, Differential Fault Analysis (DFA) [11] is rendered almost impossible, since an attacker is unable to observe both the correct and the faulty output for the same input, if the attacker cannot fix the value of the nonce. Moreover, in contrast to nonce-based (but unauthenticated) encryption schemes (such as CBC, CTR, etc.), where the decryption procedure (with a fixed nonce) is still susceptible to DFA, this is not the case for nonce-based authenticated encryption schemes that only return the plaintext if the tag is correct. For this reason, all published fault attacks on authenticated encryption schemes so far are in settings where either the nonce is repeated, or unverified plaintext is released [42, 43].

These observations might lead to the impression that nonce-based authenticated encryption schemes are not susceptible to fault attacks and thus, no dedicated fault attack countermeasures might be necessary to protect the implemented scheme against these attacks. However, in this work, we show that this assumption is not true, and present the first fault attacks on authenticated encryption schemes that are not performed in some kind of misuse scenario. We show that countermeasures against fault attacks are essential for implementations of authenticated encryption schemes operating in hostile environments.

**Our Contribution.** We present fault attacks for a wide range of authenticated encryption schemes. Our attacks do not require any misuse scenario, such as nonce reuse or release of unverified plaintext. We focus our discussion on various AES-based schemes, including the ISO/IEC standards CCM [48], GCM [32], EAX [9], and OCB [40], as well as several second-round CAESAR [46] candidates. However, our analysis is applicable to a broader range of constructions and is not limited to AES-based schemes.

All our attacks are based on an enhancement of the Statistical Fault Attack (SFA) presented by Fuhr et al. [18], which requires only very limited assumptions about the attacker's capabilities: the ability to induce a fault that leads to a biased (non-uniform) distribution in certain bytes. In case of AES, we assume that the attacker is able to influence some byte (or a larger structure) of the internal state of AES before the last application of MixColumns, so that the value of this byte is non-uniformly distributed. Particularly, we do not have to rely on the exact position of a fault, the number of faults injected during a single encryption, or even the knowledge that a certain fault has happened at all in an individual encryption. All we need to do is to collect ciphertexts and estimate the distribution of a single byte for various key guesses.

In order to evaluate the assumptions on the capabilities of an attacker, we also perform fault-injection experiments targeting three different hardware platforms. In the first setting, clock glitch attacks on a GCM software implementation executed on an 8-bit microcontroller are performed. In addition, we evaluate implementations using AES co-processors on a smartcard chip and a general-purpose microcontroller by means of laser fault injection and clock tampering, respectively. In all three settings, 4 bytes of the last round key of AES could be successfully recovered with 30, 16, and 1 200 faulty ciphertexts, respectively. In all practical scenarios, the attack has to be repeated three more times to recover the full last round key (in case of AES-128).

**Outline.** The remainder of the paper is organized as follows. In Sect. 2, we give some background on fault attacks in general, recapitulate the work of Fuhr et al. [18] on SFA, and introduce our attack model. In Sect. 3, we show how SFA can be applied to various AES-based authenticated encryption schemes.

Finally, we present practical experiments and verify the practicality of SFA on three different hardware platforms in Sect. 4.

## 2   Background

In this section, we revisit the Statistical Fault Attacks on AES underlying our attacks. We start with a general overview of different types of fault attacks, and briefly describe the biased fault model in the attack of Fuhr et al. [18]. Finally, we discuss the modified, much more general biased fault model we use in this paper, and how to identify the best key candidates.

### 2.1   Fault Attacks

Since the seminal work of Boneh et al. [13], it has been shown that many cryptographic algorithms are susceptible to Fault Attacks (FA). Indeed, numerous papers have proposed FA on most cryptographic primitives, including symmetric ciphers (DES [11], AES [37], etc.) as well as asymmetric schemes (RSA [13], Elliptic Curve Cryptography [10], etc.).

Fault attacks induce a logical error by physical means in one of the intermediate variables of a cryptographic primitive, and exploit the erroneous result to get information on the secret key. The means to inject a logical error can consist in over- or under-powering the device during a short time period, tampering its clock, or injecting a light beam or an electro-magnetic field inside the device [7,31,45].

Several cryptanalytic methods have been developed to exploit erroneous results in order to retrieve the key. In Differential Fault Analysis (DFA) [11], the attacker runs a cryptographic function twice on the same input and introduces a fault near the end of one of the computations. Then, information on the key can be retrieved from the differences between the correct and the faulty output. The Safe Error Attack (SEA) [49] fixes part of the cryptographic secret to a known value. Then, the observation of a collision on the result of a correct and faulted computation for identical inputs leaks information on the secret. In Collision Fault Analysis [12], one runs a cryptographic operation on two related inputs, and introduces a fault near the beginning of one of the computations. The adversary then exploits cases where a collision on the outputs occurs.

A common requirement of all these fault attacks is the necessity of processing two inputs that are either identical or related, in order to generate pairs of correct/faulty ciphertexts. Therefore, the attacker needs to be able to control the input of a cryptographic operation, which classifies them as chosen-plaintext attacks. Some of these FA require only one pair of correct/faulty outputs obtained from the same input, whereas others require several pairs to retrieve the secret key.

## 2.2   Statistical Fault Attacks

In 2013, Fuhr et al. proposed a new type of fault attack, called Statistical Fault Attack (SFA) [18]. In contrast to most previous attacks, the adversary only requires a collection of faulty ciphertexts encrypted with the same key. Hence, SFA works with random and unknown plaintexts.

**Fault Model.** Unlike most traditional fault attacks, SFA requires a slightly different fault model. Assuming that intermediate variables get uniformly distributed towards the last rounds for secure cryptographic primitives like AES, an attacker has to be able to induce faults which change the distribution of some intermediate values to be non-uniform. In particular, Fuhr et al. considered the following three fault models:

(a) the stuck-at-0 fault model with probability 1,
(b) the stuck-at-0 fault model with probability 1/2,
(c) the stuck-at model to an unknown and random value $e$ with probability 1.

Using these non-uniform fault models, Fuhr et al. were able to show several attacks on AES based on simulations. Their attacks target the last 4 rounds with a small number of faulty ciphertexts and practical complexity.

**Description of the AES.** AES is a byte-oriented block cipher following the wide-trail design strategy. It operates on a state of $4 \times 4$ bytes and updates it in 10, 12, or 14 rounds, depending on the key size of 128, 192, or 256 bits. In each round (except the last one with no MixColumns), the following four transformations are applied.

SubBytes (SB): This step is the only non-linear transformation of the cipher. It is a permutation consisting of an S-box $S$ applied to each byte of the state.

ShiftRows (SR): This step is a byte transposition that cyclically shifts each row of the state by different offsets. Row $j$ is shifted right by $j$ byte positions.

MixColumns (MC): This step is a permutation operating on the state column by column. To be more precise, it is a left-multiplication by a $4 \times 4$ circular MDS matrix $M$ over $\mathbb{F}_{2^8}$.

AddRoundKey (AK): In this transformation, the state is modified by combining it with a round key with a bitwise xor operation.

**Attack Procedure and Complexity.** While Fuhr et al. proposed several attack variants, we will focus only on the attack that targets the $9^{\text{th}}$ round of AES. When changing the distribution of one byte of AES before the last MixColumns, they showed that with these fault models, 4 bytes of the last round key could be recovered with high probability using the Squared Euclidean Imbalance (SEI) distinguisher with only 6, 14, and 80 faulty ciphertexts, respectively. We briefly recount the attack below, but refer to [18] for a more detailed description.

If we denote our target state before the last MixColumns in the encryption to the $i^{\text{th}}$ ciphertext by $\tilde{S}_9^i$, we can express one byte of this state as a function of the ciphertext $\tilde{C}^i$, 4 bytes of the last round key $K_{10}$, and one byte of $\mathsf{MC}^{-1}(K_9)$, as follows. Our target state is

$$\tilde{S}_9^i = \mathsf{MC}^{-1}(\mathsf{SB}^{-1} \circ \mathsf{SR}^{-1}(\tilde{C}^i \oplus K_{10}) \oplus K_9)$$
$$= \mathsf{MC}^{-1}(\mathsf{SB}^{-1} \circ \mathsf{SR}^{-1}(\tilde{C}^i \oplus K_{10})) \oplus \mathsf{MC}^{-1}(K_9).$$

Each byte of $\tilde{S}_9^i$ can therefore be deduced using one hypothesis on 4 bytes of $K_{10}$ and on one particular byte of $\mathsf{MC}^{-1}(K_9)$. As shown by Fuhr et al., the xor with $\mathsf{MC}^{-1}(K_9)$ does not modify the distance of the biased distribution from uniform. Hence, it can be omitted in the attack. In other words, this allows to mount the attack on a modified $\tilde{S}_9^{i\prime}$:

$$\tilde{S}_9^{i\prime} = \mathsf{MC}^{-1} \circ \mathsf{SB}^{-1} \circ \mathsf{SR}^{-1}(\tilde{C}^i \oplus K_{10}).$$

This allows us to recover 4 bytes of the last round key $K_{10}$ by making $2^{32}$ hypotheses on their value and predicting one byte of $\tilde{S}_9^{i\prime}$. By repeating the attack 4 times, one can recover the complete last round key $K_{10}$.

## 2.3 A Generalized Fault Model

In this work, we want to go beyond specific fault models like in Sect. 2.2. The only assumption we make is that the attacker is able to influence some byte (or a larger structure) of the internal state of AES before the last MixColumns such that this value becomes clearly non-uniformly distributed. We make no assumptions about the details of this non-uniformity, nor do we require that the attacker knows the new distribution. To exploit this type of fault, the attacker will collect faulty (biased) ciphertexts, compute backwards to the target byte for different key guesses, and try to reject wrong key guesses that would result in an approximately uniform measured distribution of the biased target byte. In the remainder of this section, we discuss how to identify the non-uniform distribution for the wrong key guesses.

We do not consider the distribution on bit-level, but for example on byte-level. Exploiting such non-uniform distributions of multi-bit values (more specifically, distributions of several sums of single bits) has already been investigated in the context of multidimensional linear cryptanalysis [21]. However, the distributions in this context are typically very close to uniform, unlike the distributions we expect in the case of SFA. Unfortunately, as noted by Samajder and Sarkar [44], the state-of-the-art framework for multidimensional cryptanalysis is not suitable for handling distributions which are significantly different from uniform. On the positive side, testing the closeness of discrete distributions [41] is a well-established field of research. Here, the central challenge is to determine whether two discrete distributions are the same (or close to each other) with the help of as few samples as possible. In our case, we want to determine whether our given samples are distributed uniformly or not.

The algorithms needing the fewest samples to perform this task are based on an idea of Goldreich and Ron [19]. Their algorithm makes use of collisions between sampled values to test for uniformity, since the expected number of collisions is lowest for uniformly distributed samples. Hence, the further a distribution deviates from the uniform distribution, the more collisions and multi-collisions we expect.

Of course, it is possible to directly base the testing of the key hypothesis on uniformity testing. For instance, Batu et al. [8] present a test which requires $O\left(\epsilon^{-4} \cdot \sqrt{2^s} \cdot \log(1/\gamma)\right)$ samples for distributions over $2^s$-element sets. Their test accepts with probability $1 - \gamma$ if the samples come from a distribution with $\ell_1$-norm distance smaller than $\epsilon/\sqrt{3 \cdot 2^s}$ to the uniform distribution. It rejects with probability $1 - \gamma$ if the samples come from a distribution which is more than $\epsilon$ away from the uniform distribution.

However, for our use-case, an approach that ranks keys according to some metric, like the number of collisions, is more suitable than a binary decision whether the measured distribution is uniform or not. Significantly more samples are needed to clearly separate the distribution for the right key hypothesis from the wrong ones to enforce a binary decision, whereas for the ranking, it is usually sufficient if the right key is ranked somewhere among the top candidates. Since the uniformity tests of Batu et al. [8] and Paninski [36] are actually based on counting collisions, they also provide us with a starting point for a ranking algorithm. This algorithm ranks the key hypothesis according to the number of collisions, and gives multi-collisions a higher weight. In our experiments, this ranking algorithm performs as good as ranking based on the SEI.

Interestingly, the key ranking mechanism based on the SEI used in [18, 38] can also be linked to counting collisions. Let $s$ be the bitsize of our biased intermediate value $S_i = f^{-1}(\hat{K}, \tilde{C}_i)$, computed from the faulty ciphertext $\tilde{C}_i$ under the key hypothesis $\hat{K}$. Assuming that we have $N$ faulty ciphertexts, the SEI $d$ is calculated as

$$d(\hat{K}) = \sum_{\delta=0}^{2^s-1} \left( \frac{\#\{i \mid f^{-1}(\hat{K}, \tilde{C}_i) = \delta\}}{N} - \frac{1}{2^s} \right)^2 .$$

This distinguisher assigns high values to key hypotheses $\hat{K}$ that lead to distributions of intermediate values $S_i$ with many collisions. For instance, consider a sample size of $N = 2^s$ samples. Then, the SEI is essentially counting collisions, since only events that occur exactly once do not increase $d$. Moreover, since the deviation from uniform is squared, a greater deviation, or in our sense a multi-collision, contributes more to $d$.

To sum up, it turned out that the SEI cannot be outperformed in practice by a new ranking algorithm based on counting collisions, since the SEI is actually doing that. Hence, we decided to stick to the more common SEI to measure if the distribution of one byte value becomes clearly non-uniformly distributed. So for AES, the 4-byte key guesses of the last round key are ranked according to the resulting SEI of one byte before the last MixColums when decrypting faulty ciphertexts for one round. To be able to observe non-uniformness and to evaluate the SEI, we require the input to the block cipher to be different for each fault and the block cipher output to be known.

# 3   Statistical Fault Attacks on Authenticated Encryption

In this section, we evaluate the applicability of the Statistical Fault Attack to several authenticated encryption modes for AES. This includes the widely-used ISO/IEC-standardized modes like CCM [48], EAX [9], GCM [32] and OCB [40], as well as new authenticated encryption modes proposed in the CAESAR initiative [46]. For evaluating the applicability of the fault attacks to these authenticated encryption schemes, we only need very limited assumptions. As already stated in Sect. 2, we assume that the attacker is able to influence some byte (or a larger structure) of the internal state of AES before the last MixColumns operation in a way that this value becomes clearly non-uniformly distributed.

We classify the investigated authenticated encryption modes into three categories, as illustrated in Fig. 1:

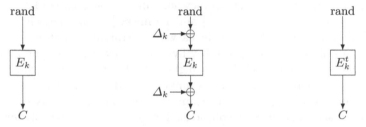

(a) Basic Construction    (b) XEX-like Construction    (c) Tweakable Block Cipher

**Fig. 1.** Classification of AES-based authenticated encryption schemes.

**Basic Construction.** The schemes in this category allow to directly observe the output of the block cipher. This includes schemes based on classical encryption schemes such as CTR [15], CBC [17], CFB [17], etc., but also schemes based on the XE construction [39], which masks the input of the block cipher using secret masks $\Delta_k$. More generally, we assume that the input to the block cipher is a secret random value, but the output is observable to the attacker.

**XEX-like Construction.** This construction is similar to XE, but unlike XE, both the input and the output of the block cipher are masked using secret, nonce-dependent masks $\Delta_k$. Constructions following the XEX construction [39] include for instance IAPM [28], OCB [40], and several of the CAESAR candidates.

**Tweakable Block Cipher.** The third category covers schemes that use a dedicated tweakable block cipher, which depends on a (typically nonce-dependent) tweak in addition to the secret key. Since the focus of this work is on

AES-based modes, we will restrict ourselves to constructions using the AES round function and following the TWEAKEY framework [27], such as for instance the CAESAR candidates KIASU [26] and Deoxys [24].

In the remainder of this section, we will discuss the applicability of Statistical Fault Attacks to schemes of these three categories in turn.

### 3.1    Application to the Basic Construction

In this construction, the output of the block cipher is directly known to the attacker, or can trivially be recovered by, say, xoring observable values with public values or constants. It is easy to see that in this case, the Statistical Fault Attack described in Sect. 2 can be applied in a straight-forward way to recover the secret key $k$. As an example, we discuss the application of Statistical Fault Attacks on AES in counter (CTR) modes as used in GCM, CCM and EAX (all standardized by ISO/IEC).

**Statistical Fault Attack on CCM, EAX and GCM.** As a representative example for the three modes, we will discuss the attack on CCM, which is shown in Fig. 2. As its name implies, the CTR-with-CBC-MAC mode (CCM) can be split into an encryption part using AES in counter mode to encrypt the plaintext $P$ and an authentication part using CBC-MAC to authenticate the nonce $N$, associated data $A$, and plaintext $P$, which generates the tag $T$. For clarity, we have substituted the first part of the CBC-MAC, where the associated data is processed, with its outcome $V$ in Fig. 2. Since the fault attack is solely performed on the encryption part, the following observations also hold for EAX and GCM that both use AES in CTR mode for encryption.

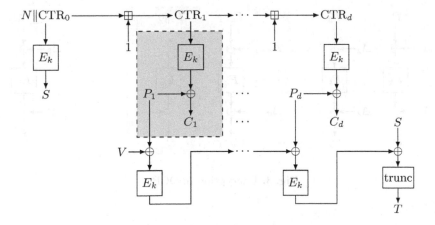

**Fig. 2.** The counter with CBC-MAC mode.

For the sake of simplicity, we restrict our fault attack to the encryption $E_k$ of the first plaintext block (marked by the dashed rectangle in Fig. 2). Let us recall the conditions of Sect. 2 that are necessary for the Statistical Fault Attack to work:

1. The inputs of the block cipher need to be different for each fault.
2. The block cipher output needs to be known.

Condition 1 is always fulfilled, since it is required that the nonce $N$ changes for each encryption and thus, the input to $E_k$ changes as well. Condition 2 is fulfilled assuming a known plaintext attack, where the plaintext block $P_1$ is known to the attacker. Then, one can compute the keystream part for encrypting this plaintext block by xoring it with $C_1$. The resulting keystream is the output of the block cipher $E_k$. To sum up, we are able to observe outputs of the block cipher $E_k$ for various inputs. Thus, we have the same preconditions as for the fault attack on plain AES described in Sect. 2. Hence, the attack can be applied to CCM (and any other scheme based on CTR mode) in a straight-forward way. We want to stress that the attacker does not require to know the input of the block cipher, it is just necessary that it changes. Therefore, the attack also applies to modes where the value of the counter is unknown, such as EAX.

**Statistical Fault Attack on OCB.** Although ISO/IEC-standard OCB is based on the XEX construction, we show that it is also vulnerable to the attack on the basic construction. The reason for this is that if the last plaintext block is incomplete, it is instead processed using the XE construction, as shown in Fig. 3. Therefore, the knowledge of this incomplete last plaintext and ciphertext block allows an attacker to compute the output of the block cipher $E_k$ and thus, the Statistical Fault Attack is again applicable.

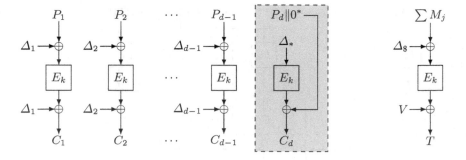

**Fig. 3.** Encryption in OCB.

**Application to Other Modes.** Besides CCM, EAX, GCM, and OCB, the fault attack discussed in this section also applies to several other authenticated encryption modes. For instance, to the CAESAR candidates Cloc [22] and

Silc [23], which are based on cipher-feed-back mode (CFB), where the ciphertext is the xor of the output of a block cipher $E_k$ and the plaintext blocks. Another example is AES-OTR [34], which uses a balanced two-round Feistel network for encryption. The round function of this network is AES in an XE mode. Since the balanced Feistel network has only two rounds, knowledge of the plaintext and ciphertext implies knowledge of the block cipher output. Thus, again, the Statistical Fault Attack is directly applicable.

## 3.2 Application to XEX-Like Constructions

In this construction, the output of the block cipher is masked with a secret value $\Delta_k$, which prevents a straightforward application of the basic attack. However, depending on how $\Delta_k$ is computed, the Statistical Fault Attack may nevertheless be applicable. In the simplest case, $\Delta_k$ is not nonce-dependent. This allows to repeatedly observe ciphertexts masked with a secret, but constant value $\Delta_k$. We demonstrate how to exploit this in an attack on the CAESAR candidate AES-COPA [4].

**Statistical Fault Attack on AES-COPA.** AES-COPA uses an XEX-like construction for encrypting the plaintext, which is shown in Fig. 4. The input $V$ of the plaintext processing is the result of a PMAC-like processing of the associated data $A$ and the nonce $N$. Thus, $V$ will change for different nonce values. Each processed ciphertext block requires two invocations of the block cipher $E_k$. AES-COPA masks both the input of the block cipher processing the plaintext blocks $P_j$, and the output of the block cipher that generate ciphertext blocks $C_j$. The masks are based on a secret value $L = E_k(0)$. We focus our attack on the block cipher call that generates $C_1$, as marked in Fig. 4.

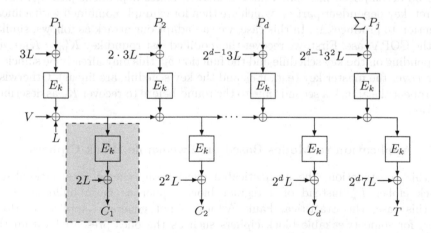

**Fig. 4.** Plaintext processing of AES-COPA, $L = E_k(0)$.

So far, only one of our two prerequisites for the SFA from Sect. 2 is fulfilled. We can vary the input of the block cipher calls by changing, for example, the nonce, associated data, or plaintext. However, the output of the block cipher is unknown, since it is masked with the secret value $\Delta_k = 2 \cdot E_k(0)$ to get $C_1$. To overcome this obstacle and since $\Delta_k$ solely depends on the secret key $k$, we consider $\Delta_k$ as a part of the key schedule to compute the last round key. Thus, instead of the last round key $K_{10}$ of AES, we get $K'_{10} := K_{10} \oplus (2 \cdot E_k(0))$ as the last round key.

Hence, instead of recovering the last round key $K_{10}$ of AES as in the attacks before, we now can recover $K'_{10}$ by using SFA as described in Sect. 2. For recovering $K'_{10}$, the complexity and the needed numbers of faults are the same as for the attack on AES itself. However, the knowledge of $K'_{10}$ does not directly lead to a key recovery attack of the master key $k$. Therefore, we need to perform the Statistical Fault Attack a second time. One option is to target again the first plaintext block and use our knowledge of $K'_{10}$ to now target the AES round key $K_9$. Alternatively, we repeat the attack for the second plaintext block to recover $K_{10} \oplus (4 \cdot E_k(0))$ and thus get $K_{10}$ by solving the resulting linear system. In both cases, the master key can then easily be recovered from $K_9$ and $K_{10}$, respectively.

**Application to Other Modes.** Besides COPA, other schemes that use a nonce-independent $\Delta_k$ and allow the Statistical Fault Attack include ELmD [14] and Shell [47]. In contrast, some schemes, such as IAPM, OCB, or some CAESAR candidates, also include the nonce in the computation of $\Delta_k$. All these schemes have in common that $\Delta_k$ changes unpredictably for each block cipher call, which prevents a straight-forward application of Statistical Fault Attacks.

Instead of relying on misuse settings like repeated nonces, we will have a closer look at how these schemes typically compute $\Delta_k$. In many cases, $\Delta_k$ can be decomposed into two values: a known, nonce-dependent part $\delta_N$, and a secret, key-dependent part $\delta_k$, which are then for example combined with a linear function to produce $\Delta_k$. In this case, we can adapt our attack as follows, similar to the COPA case. First, we recover the modified last round key $K'_{10} = K_{10} \oplus \delta_k$. Depending on the key schedule and the function $\delta_k$, this may already be sufficient to recover the master key (e.g., if $\delta_k$ and the key schedule are linear). Otherwise, we repeat the attack a second time to the round before to recover $K_9$ as described before.

### 3.3 Application to Modes Based on Tweakable Block Ciphers

In this construction, the authenticated encryption scheme uses a tweakable block cipher $E^t_k$ instead of a regular block cipher as basic building block. In this case, the Statistical Fault Attack is not generally applicable. However, for some tweakable block ciphers such as the ones presented within the TWEAKEY framework [27], we can adapt our attack. In particular, this is possible if the last subkeys of the tweakable block cipher can be described by the composition of two values, $\delta_t \oplus \delta_k$. We illustrate the working principle of the attack for the CAESAR candidate Deoxys [24], but the same attack is also applicable to KIASU [26], where the tweak $t$ is only xored to each round-key.

**Statistical Fault Attack on Deoxys.** Deoxys offers two modes of operation, both using two variants of the underlying tweakable block cipher Deoxys-BC. We focus on Deoxys$^{\neq}$-128-128, which uses Deoxys-BC-256 as underlying tweakable block cipher. As shown in Fig. 5, Deoxys$^{\neq}$ encrypts the individual plaintext blocks $P_j$ in an $\Theta$CB3-like [30] way. This ensures both the variation of the tweakable block cipher inputs, and knowledge of the outputs. However, since the tweak is partly defined by the nonce, we have to determine the influence of this nonce on the last round key that we want to recover using SFA. Thus, we have to have a closer look at the definition of the tweakable block cipher Deoxys-BC-256.

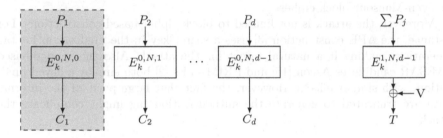

**Fig. 5.** Plaintext processing for Deoxys$^{\neq}$.

Figure 6 shows how Deoxys-BC-256 uses the round function $f$ of the AES, but computes different round keys $K_i$ based on the master key $k$ and tweak $t$. Here, $K_i$ is the xor sum of three values: a key-dependent round key $K_i^k$, a tweak-dependent round tweak $K_i^t$, and a round constant $c_i$. The values are updated using a simple byte permutation $h$. For instance, $K_0^k = k$, $K_0^t = t$, $K_1^k = 2\,h(k)$, $K_1^t = h(t)$, $K_r^k = 2\,h(2\,h(\ldots 2\,h(k)\ldots))$, and $K_r^t = h(h(\ldots h(t)\ldots))$.

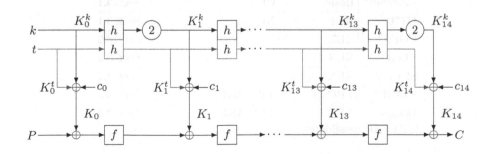

**Fig. 6.** Block cipher Deoxys-BC-256.

Since the value of the tweak used for encryption is publicly known, the varying part $K_i^t$ of the round keys $K_i$ can be easily calculated. The unknown parts $K_i^k$ of the round key are constant for multiple calls of the block cipher under the same key $k$. Hence, the last round key $K_{14}^k$ can be recovered with the SFA on AES described in Sect. 2.

## 3.4   Summary and Discussion of Results

We demonstrated in the previous sections that several authenticated encryption modes for AES are susceptible to Statistical Fault Attacks. A summary of the results is given in Table 1. However, Statistical Fault Attacks are applicable to a broader range of authenticated encryption schemes, and are not limited to AES-based modes. Natural targets for the attack include, for instance, the CAESAR candidates Joltik [25] and Scream [20], which also follow the TWEAKEY framework [27], or Prøst [29], which applies the modes of COPA [5] and OTR [35] to an Even-Mansour block cipher.

Moreover, the attack is not limited to block cipher based constructions. For instance, the APE construction [3] uses a secret key in the finalization for tag generation, making it a natural target for the attack. Also the sponge-based CAESAR candidates Ascon [16] and PRIMATEs [2] both employ a keyed finalization, with similar effects. However, the fact that large parts of the internal state are truncated to generate the authentication tag might complicate the attack.

**Table 1.** Statistical fault attacks on AES-based authenticated encryption modes in the nonce-respecting setting.

| Primitive | Classification | Comments | Reference |
|---|---|---|---|
| CCM | Basic | CTR | Sect. 3.1 |
| GCM | Basic | CTR | Sect. 3.1 |
| EAX | Basic | CTR | Sect. 3.1 |
| OCB | Basic | XE (incomplete blocks) | Sect. 3.1 |
| Cloc/Silc[a] | Basic | CFB | Sect. 3.1 |
| OTR[a] | Basic | XE | Sect. 3.1 |
| COPA[a] | XEX | | Sect. 3.2 |
| ELmD[a] | XEX | | Sect. 3.2 |
| SHELL[a] | XEX | | Sect. 3.2 |
| KIASU[a] | TBC | TWEAKEY | Sect. 3.3 |
| Deoxys[a] | TBC | TWEAKEY | Sect. 3.3 |

[a]CAESAR candidates.

# 4    Practical Verification/Implementation of the Attacks

In order to demonstrate the practical relevance of Statistical Fault Attacks and to validate the assumptions from previous sections, we performed three fault-injection experiments targeting real hardware.

An AES-GCM implementation executed on an off-the-shelf microcontroller served as target for the first experiment. In this context we used the ASM AES version from [1] to realize the block cipher. Due to the lack of embedded platforms implementing GCM or CCM completely in hardware, we put the focus of the following analysis on hardware AES co-processors available on a smart-card microcontroller and on a general-purpose microcontroller, respectively. The remaining parts for realizing the authenticated encryption modes are then implemented in software.

In all settings, the fault injections aim to induce a bias on at least one byte of the AES state before the last MixColumns transformation, and allow to reveal 32 bits of the last AES round key. For full key recovery, the attack has to be repeated three more times. The following list provides an overview of the fault-injection methods and the attack results for the three settings:

1. Clock tampering has been used to disturb the execution of the AES software implementation running on an ATxmega 256A3 general-purpose microcontroller. This setting allowed to reveal 4 bytes of the last round key with less than 30 faulted ciphertexts.
2. Laser fault injections on an AES co-processor on a smartcard microcontroller. Our experiments show that less than 16 faulty ciphertexts are sufficient to reveal 4 bytes of the last round key.
3. Clock tampering on a hardware AES co-processor implemented on a general-purpose microcontroller. In this setting, we need approximately 1 200 faulted ciphertexts for recovering 4 bytes of the last round key.

For all attacks, 4 bytes of the last round key can be recovered out of the faulted ciphertexts in less than one hour using an Intel Core i7 3770K. In the following, we give a detailed description and summary of the practical fault-injection attacks.

## 4.1    AES Software Implementation on an 8-Bit Microcontroller

In the following setting, we used clock glitches to provoke faults during an AES computation implemented in software on an 8-bit microcontroller. In particular, we used the ASM AES version from [1] for realizing the GCM AE mode.

For the clock-glitch experiments, we used a nominal clock frequency of 24 MHz ($T_{clk} = 41.7$ ns). According to [1], one 128-bit encryption requires 2 555 clock cycles. For simplicity, we used one general-purpose I/O pin of the microcontroller for indicating the start of the AES encryption. This trigger pin together with the knowledge of the length of the AES encryption procedure allows to find the correct time interval for inserting the clock glitch. Next to that, our results show that faults in consecutive clock cycles also lead to successful key recovery. As a consequence, this behavior allows to relax the precision prerequisite of the trigger information.

With the found parameters, we collected two sets, each containing 80 faulty ciphertexts. For the first set, a single clock glitch was inserted. For the second set, clock glitches in 50 consecutive clock cycles were inserted. Next, we performed SFA attacks using an increasing number of faulty ciphertexts on both sets individually. The results containing the set size $N$, the SEI value for the correct subkey ($\text{SEI}_c$), and the maximum SEI value of the wrong subkey guesses ($\text{SEI}_w$) were stored in two separate lists (one list for each set) in the format $[N, \text{SEI}_c, \max(\text{SEI}_w)]$. For this attack scenario, we started with $N = 4$ and increased $N$ in every iteration by 4.

Figure 7 displays the evolution of the SEI values for increasing number of ciphertexts in the single clock glitch setting. Values corresponding to the correct subkey are plotted in red, the maximum SEI values of the wrong subkey guesses are plotted in blue. With 30 faulty ciphertexts, $\text{SEI}_c$ exceeds $\max(\text{SEI}_w)$, which allows to reveal the correct subkey value.

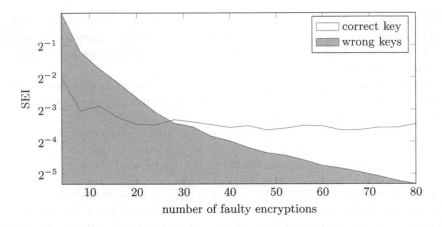

**Fig. 7.** SEI values for correct key ($\text{SEI}_c$) plotted against best SEI for a wrong key ($\max(\text{SEI}_w)$) for increasing number of faulty encryptions. Setup: AES software implementation, single clock glitch (Table 2). (Color figure online)

Figure 8 displays the evolution of the SEI values for an increasing number of ciphertexts for the setting with 50 consecutive clock glitches. In this setting, 24 ciphertexts are sufficient for $\text{SEI}_c$ to exceed $\max(\text{SEI}_w)$, which allows to reveal the correct subkey value.

Results of the fault attacks targeting the AES software implementations using clock glitches show that with 30 faulty ciphertexts, it is possible to reveal the 32-bit subkey if a single clock glitch is inserted. Furthermore, if the clock glitch is inserted in 50 consecutive clock cycles, approximately 25 faulty ciphertexts are sufficient for subkey recovery. We did not further investigate the approach

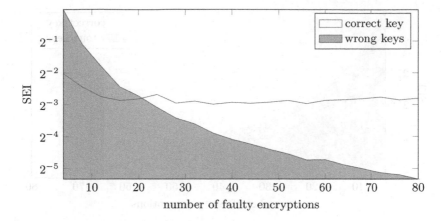

**Fig. 8.** Evolution of the SEI values with increasing number of faulty encryptions. Setup: AES software implementation, multiple clock glitches (Table 2).

of inserting the clock glitch in consecutive clock cycles because this is out of scope of the current work. Nevertheless, by carefully trimming the fault injection parameters, the number of faulty ciphertexts for successful subkey recovery could probably be further decreased.

## 4.2 AES Hardware Co-Processor of a Smartcard Microcontroller

In this experiment, we used a laser fault injection system to induce faults during encryptions of an AES Hardware co-processor of a smartcard microcontroller. This co-processor can easily be used as building block for realizing authenticated encryption modes like GCM or CCM on the smartcard.

The laser fault injection system consists of an infrared laser diode module and a microscope allowing to focus the laser spot depending on the microscope objective used. Here an objective with a $10\times$ magnification is used. The whole system is mounted on a motorized X-Y-Z stage.

As the smartcard microcontroller runs its own operating system, the only signal available for triggering the laser injection system is the sending of the encryption command through APDU command. Therefore, a temporal delay is added to postpone the laser injection during the AES encryption thanks to a remotely controllable pulse generator. Furthermore, as the smartcard microcontroller runs on its own internal clock network, an inherent temporal jitter is present due to the asynchronism between the laser injection system and the smartcard microcontroller clock network. These experimental conditions are very close to the ones present in real world scenarios.

By applying a spatial fault injection cartography, we have been able to find a spatial position where only one byte of the AES state is faulted. Furthermore, by trying different delays, we found a spatio-temporal setting where only 4 bytes of the ciphertext were faulted with a high reliability. By studying the indices of

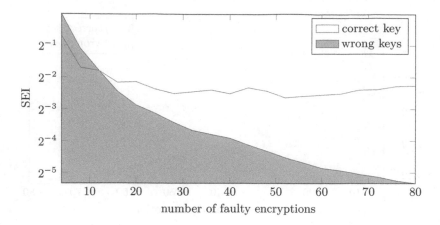

**Fig. 9.** Evolution of the SEI values with increasing number of faulty encryptions. Setup: AES hardware co-processor of a smartcard microcontroller, laser (Table 3). (Color figure online)

the faulted ciphertext bytes, we concluded that we successfully induced a fault on one byte of the AES state just before the last MixColumns. The fact that the hardware AES module can also be used outside of the context of authenticated encryption, i.e., for encrypting single plaintext blocks, simplified this profiling. However, if the stand-alone usage of the AES co-processor is not possible on the attacked platform, the search for the right fault injection parameters becomes more complicated, but is still feasible.

With the found parameters, we collected again 80 faulty ciphertexts. With the collected faulty ciphertexts, the same evaluation as in the previous section was conducted. We started again with an initial attack set size $N = 4$ and increased the size of the attack set by 4 in every iteration. The evolution of the SEI values with increasing set size is depicted in Fig. 9. Values corresponding to the correct subkey are plotted in red, the maximum SEI values of the wrong subkey guesses are plotted in blue.

As depicted on Fig. 9, $SEI_c$ already exceeds $\max(SEI_w)$ with only $N = 16$ ciphertexts. Therefore, this number of ciphertexts allows to retrieve 4 bytes of the correct last round key. This result validates the practicability of the fault model and even shows that laser-based fault injection systems are well suitable for this kind of attacks.

### 4.3   AES Co-Processor on a General-Purpose Microcontroller

In this setting, we use clock glitches to inject faults during the encryption procedure of an AES co-processor integrated on a general-purpose microcontroller. This co-processor can on the one hand be used as stand-alone block cipher to encrypt plaintext blocks, on the other hand it can be used in the context of AE for realizing a mode of operation like GCM or CCM. The co-processor in

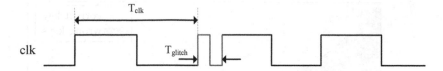

**Fig. 10.** Clock signal with intentionally inserted additional positive clock edge.

stand-alone mode allows profiling the hardware in order to find suitable fault-injection parameters. The target of the fault injection is the output of the byte substitution (SubBytes) in the $9^{th}$ AES round. The AES co-processor implements the SubBytes function with pure combinational logic. Since one column of the state is processed in a single clock cycle, this allows to create faults in 4 bytes of the state with a single clock glitch.

We define with $T_{glitch}$ the time interval between two subsequent positive clock edges in case of a clock glitch. This value is smaller compared to the nominal clock period $T_{clk}$, as illustrated in Fig. 10. If $T_{glitch}$ is smaller than the path delay of the combinational SubBytes block, the output value of this block has not settled to its correct, stable value. As a result, a wrong value is sampled by the registers at the output of the block, which leads to faults in the ciphertext.

For the clock glitch experiments, we used a nominal clock frequency of $10\,MHz$ ($T_{clk} = 100\,ns$). Preliminary fault experiments allowed to find the correct clock cycle (i.e., the delay between the start of the encryption and the targeted instruction) to disturb the SubBytes operation in the $9^{th}$ round before the MixColumns step. With $T_{glitch} = 10.2\,ns$, we achieved a fault probability of $99.5\,\%$.

With these parameters, we executed the AES encryption to receive $2\,000$ faulty ciphertexts. The increased number of ciphertexts was required because preliminary experiments revealed that the bias introduced with the clock glitch was significantly smaller compared to the bias introduced by the laser attack. With the collected faulty ciphertexts, the same evaluation as in the previous section was conducted. Due to a smaller bias, we started with an initial attack set size $N = 32$ and increased the size of the attack set by 32 in every iteration. The evolution of the SEI values with increasing set size is depicted in Fig. 11. Values corresponding to the correct subkey are again plotted in red, the maximum SEI values of the wrong subkey guesses are plotted in blue.

As depicted on Fig. 11, starting at $1\,200$ ciphertexts, $SEI_c$ exceeds $\max(SEI_w)$. This allows to reveal the correct subkey in an attack setting. Compared to the results presented in the previous section, the number of required ciphertexts is nearly 100 times higher, but the number is still practical and this amount of ciphertexts can be collected within minutes. However, the effort for performing clock-glitch attacks compared to laser fault attacks (e.g., preparing the fault-injection environment, finding good fault-injection parameters) is significantly smaller, which has to be taken into account.

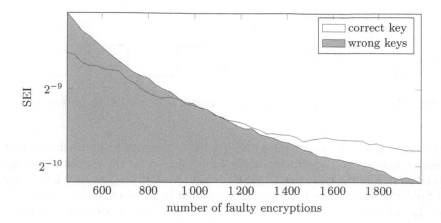

**Fig. 11.** Evolution of the SEI values with increasing number of faulty encryptions. Setup: AES co-processor on a general-purpose microcontroller, clock glitch (Table 4). (Color figure online)

## 4.4 Discussion and Remarks

The goal of the attacks presented in this section is a feasibility study proving that the assumed biased fault model is indeed valid on different platforms using different fault-injection mechanisms.

For the software implementation, a general-purpose I/O pin indicating the start of the AES encryption has been used, which allowed a precise fault injection using clock glitches. Real-world scenarios, like the second experiment targeting the smartcard microcontroller, typically do not allow the usage of a trigger pin. In such scenarios, other sources for synchronizing the fault-injection procedure can be applied, like spying the communication or the power profile. This can decrease the precision of the fault injections.

But it is important to note that the outcome of the SFA attack does not strictly rely on a precise fault injection. If only a subset of the received ciphertexts are affected by the expected fault pattern, the remaining ciphertexts (fault-free or fault hitting another location during the cipher rounds) are treated as noise. A more reliable fault injection process however minimizes the number of required ciphertexts for successful key recovery.

Furthermore, when the attacked platform allows the usage of the AES co-processor for stand-alone encryption (e.g., as in the previous experiments), one can easily perform a profiling step which simplifies the search for appropriate fault injection parameters. Nevertheless, if the AES co-processor can only be used in the context of the authenticated encryption mode, it is still possible to find the appropriate fault injection parameters. Of course, the number of attempts and the search space for the parameters increase, resulting in a more time-consuming setup phase for the fault injection.

With the practical results presented in this section, we showed that implementations of AES-based authenticated encryption modes on different hardware platforms are vulnerable to the proposed fault attacks introduced in this work.

## 5    Conclusion

In this work, we demonstrate for the first time that a wide range of nonce-based authenticated encryption schemes, including the widely used ISO/IEC standards CCM, GCM, EAX, and OCB, are susceptible to fault attacks. All our attacks need only very limited assumptions about the attacker's capabilities. To confirm these assumptions and to show the practical relevance of the attacks, we perform several fault-injection experiments targeting real hardware. This highlights the need for dedicated fault attack countermeasures for authenticated encryption schemes. Although our analysis focus only on AES-based constructions, we want to note that it is applicable to a broader range of authenticated encryption schemes. This is part of future work.

**Acknowledgments.** The authors would like to thank the organizers and participants of ASK 2015 that initiated this work and the anonymous reviewers for useful comments. The research leading to these results has received funding from the European Union's Horizon 2020 research and innovation programme under grant agreement No 644052 (HECTOR). Furthermore, this work has been supported in part by the Austrian Research Promotion Agency (FFG) under grant number 845589, by the Austrian Science Fund (project P26494-N15) and by the French ANR-14-CE28-0015 project.

# A   Data of Practical Verification/Implementation

**Table 2.** Evolution of the SEI values for correct key ($SEI_c$) and the best wrong key ($\max(SEI_w)$) for increasing number of faulty encryptions $N$. Setup: AES software implementation, single clock glitch (left) and multiple clock glitches (right).

| $N$ | $SEI_c$ | $\max(SEI_w)$ | $N$ | $SEI_c$ | $\max(SEI_w)$ |
|---|---|---|---|---|---|
| 4 | 0.25 | 1.00 | 4 | 0.25 | 1.00 |
| 8 | 0.12 | 0.43 | 8 | 0.18 | 0.46 |
| 12 | 0.13 | 0.30 | 12 | 0.15 | 0.29 |
| 16 | 0.11 | 0.22 | 16 | 0.14 | 0.18 |
| 20 | 0.09 | 0.16 | 20 | 0.14 | 0.15 |
| 24 | 0.09 | 0.12 | 24 | 0.16 | 0.12 |
| 28 | 0.10 | 0.09 | 28 | 0.13 | 0.09 |
| 32 | 0.10 | 0.09 | 32 | 0.13 | 0.08 |
| 36 | 0.09 | 0.07 | 36 | 0.13 | 0.07 |
| 40 | 0.08 | 0.06 | 40 | 0.13 | 0.06 |
| 44 | 0.09 | 0.05 | 44 | 0.13 | 0.05 |
| 48 | 0.08 | 0.05 | 48 | 0.13 | 0.05 |
| 52 | 0.08 | 0.05 | 52 | 0.14 | 0.04 |
| 56 | 0.09 | 0.04 | 56 | 0.13 | 0.04 |
| 60 | 0.09 | 0.04 | 60 | 0.14 | 0.04 |
| 64 | 0.08 | 0.04 | 64 | 0.14 | 0.03 |
| 68 | 0.08 | 0.03 | 68 | 0.14 | 0.03 |
| 72 | 0.08 | 0.03 | 72 | 0.15 | 0.03 |
| 76 | 0.08 | 0.03 | 76 | 0.14 | 0.03 |
| 80 | 0.09 | 0.03 | 80 | 0.14 | 0.02 |

**Table 3.** Evolution of the SEI values for correct key ($\text{SEI}_c$) and the best wrong key ($\max(\text{SEI}_w)$) for increasing number of faulty encryptions $N$. Setup: AES hardware co-processor of a smartcard microcontroller, laser.

| $N$ | $\text{SEI}_c$ | $\max(\text{SEI}_w)$ |
|---|---|---|
| 4 | 0.62 | 1.00 |
| 8 | 0.31 | 0.46 |
| 12 | 0.29 | 0.29 |
| 16 | 0.22 | 0.18 |
| 20 | 0.23 | 0.14 |
| 24 | 0.19 | 0.11 |
| 28 | 0.17 | 0.09 |
| 32 | 0.18 | 0.08 |
| 36 | 0.19 | 0.07 |
| 40 | 0.17 | 0.07 |
| 44 | 0.20 | 0.06 |
| 48 | 0.19 | 0.05 |
| 52 | 0.16 | 0.04 |
| 56 | 0.17 | 0.04 |
| 60 | 0.17 | 0.03 |
| 64 | 0.17 | 0.03 |
| 68 | 0.19 | 0.03 |
| 72 | 0.19 | 0.03 |
| 76 | 0.21 | 0.03 |
| 80 | 0.21 | 0.02 |

**Table 4.** Evolution of the SEI values for correct key ($SEI_c$) and the best wrong key ($\max(SEI_w)$) for increasing number of faulty encryptions $N$. Setup: AES co-processor on a general-purpose microcontroller, clock glitch.

| $N$ | $SEI_c$ | $\max(SEI_w)$ | $N$ | $SEI_c$ | $\max(SEI_w)$ |
|---|---|---|---|---|---|
| 32 | 0.02930 | 0.08203 | 1 024 | 0.00165 | 0.00164 |
| 64 | 0.01514 | 0.03369 | 1 056 | 0.00162 | 0.00162 |
| 96 | 0.01020 | 0.02040 | 1 088 | 0.00155 | 0.00154 |
| 128 | 0.00769 | 0.01489 | 1 120 | 0.00150 | 0.00151 |
| 160 | 0.00625 | 0.01125 | 1 152 | 0.00147 | 0.00145 |
| 192 | 0.00521 | 0.00971 | 1 184 | 0.00145 | 0.00140 |
| 224 | 0.00474 | 0.00817 | 1 216 | 0.00143 | 0.00136 |
| 256 | 0.00430 | 0.00693 | 1 248 | 0.00138 | 0.00137 |
| 288 | 0.00398 | 0.00620 | 1 280 | 0.00135 | 0.00130 |
| 320 | 0.00355 | 0.00535 | 1 312 | 0.00131 | 0.00128 |
| 352 | 0.00341 | 0.00492 | 1 344 | 0.00131 | 0.00125 |
| 384 | 0.00304 | 0.00448 | 1 376 | 0.00130 | 0.00122 |
| 416 | 0.00284 | 0.00416 | 1 408 | 0.00129 | 0.00120 |
| 448 | 0.00271 | 0.00388 | 1 440 | 0.00127 | 0.00117 |
| 480 | 0.00266 | 0.00359 | 1 472 | 0.00122 | 0.00113 |
| 512 | 0.00247 | 0.00330 | 1 504 | 0.00124 | 0.00111 |
| 544 | 0.00241 | 0.00315 | 1 536 | 0.00125 | 0.00107 |
| 576 | 0.00240 | 0.00297 | 1 568 | 0.00126 | 0.00106 |
| 608 | 0.00233 | 0.00280 | 1 600 | 0.00124 | 0.00104 |
| 640 | 0.00231 | 0.00264 | 1 632 | 0.00123 | 0.00103 |
| 672 | 0.00229 | 0.00250 | 1 664 | 0.00123 | 0.00101 |
| 704 | 0.00213 | 0.00238 | 1 696 | 0.00123 | 0.00100 |
| 736 | 0.00206 | 0.00227 | 1 728 | 0.00123 | 0.00098 |
| 768 | 0.00195 | 0.00219 | 1 760 | 0.00119 | 0.00097 |
| 800 | 0.00188 | 0.00215 | 1 792 | 0.00120 | 0.00095 |
| 832 | 0.00182 | 0.00202 | 1 824 | 0.00117 | 0.00093 |
| 864 | 0.00180 | 0.00195 | 1 856 | 0.00116 | 0.00089 |
| 896 | 0.00181 | 0.00190 | 1 888 | 0.00114 | 0.00087 |
| 928 | 0.00178 | 0.00180 | 1 920 | 0.00113 | 0.00088 |
| 960 | 0.00171 | 0.00173 | 1 952 | 0.00113 | 0.00087 |
| 992 | 0.00168 | 0.00172 | 1 984 | 0.00113 | 0.00084 |

# References

1. AVR crypto lib. http://avrcryptolib.das-labor.org. Accessed 13 Jan 2016
2. Andreeva, E., Bilgin, B., Bogdanov, A., Luykx, A., Mendel, F., Mennink, B., Mouha, N., Wang, Q., Yasuda, K.: PRIMATEs. Submission to the CAESAR Competition (Round 2). http://competitions.cr.yp.to/round2/primatesv102.pdf
3. Andreeva, E., Bilgin, B., Bogdanov, A., Luykx, A., Mennink, B., Mouha, N., Yasuda, K.: APE: authenticated permutation-based encryption for lightweight cryptography. In: Cid, C., Rechberger, C. (eds.) FSE 2014. LNCS, vol. 8540, pp. 168–186. Springer, Heidelberg (2015). doi:10.1007/978-3-662-46706-0_9
4. Andreeva, E., Bogdanov, A., Luykx, A., Mennink, B., Tischhauser, E., Yasuda, K.: AES-COPA. Submission to the CAESAR Competition (Round 2). http://competitions.cr.yp.to/round2/aescopav2.pdf
5. Andreeva, E., Bogdanov, A., Luykx, A., Mennink, B., Tischhauser, E., Yasuda, K.: Parallelizable and authenticated online ciphers. In: Sako, K., Sarkar, P. (eds.) ASIACRYPT 2013. LNCS, vol. 8269, pp. 424–443. Springer, Heidelberg (2013). doi:10.1007/978-3-642-42033-7_22
6. Balasch, J., Gierlichs, B., Verbauwhede, I.: An in-depth and black-box characterization of the effects of clock glitches on 8-bit MCUs. In: Fault Diagnosis and Tolerance in Cryptography - FDTC 2011, pp. 105–114. IEEE (2011)
7. Bar-El, H., Choukri, H., Naccache, D., Tunstall, M., Whelan, C.: The sorcerer's apprentice guide to fault attacks. In: Fault Diagnosis and Tolerance in Cryptography - FDTC 2004, pp. 330–342 (2004)
8. Batu, T., Fortnow, L., Rubinfeld, R., Smith, W.D., White, P.: Testing closeness of discrete distributions. J. ACM 60(1), 4 (2013)
9. Bellare, M., Rogaway, P., Wagner, D.: The EAX mode of operation. In: Roy, B., Meier, W. (eds.) FSE 2004. LNCS, vol. 3017, pp. 389–407. Springer, Heidelberg (2004). doi:10.1007/978-3-540-25937-4_25
10. Biehl, I., Meyer, B., Müller, V.: Differential fault attacks on elliptic curve cryptosystems. In: Bellare, M. (ed.) CRYPTO 2000. LNCS, vol. 1880, pp. 131–146. Springer, Heidelberg (2000). doi:10.1007/3-540-44598-6_8
11. Biham, E., Shamir, A.: Differential fault analysis of secret key cryptosystems. In: Kaliski Jr., B.S. (ed.) CRYPTO 1997. LNCS, vol. 1294, pp. 513–525. Springer, Heidelberg (1997). doi:10.1007/BFb0052259
12. Blömer, J., Krummel, V.: Fault based collision attacks on AES. In: Breveglieri, L., Koren, I., Naccache, D., Seifert, J.-P. (eds.) FDTC 2006. LNCS, vol. 4236, pp. 106–120. Springer, Heidelberg (2006). doi:10.1007/11889700_11
13. Boneh, D., DeMillo, R.A., Lipton, R.J.: On the importance of checking cryptographic protocols for faults. In: Fumy, W. (ed.) EUROCRYPT 1997. LNCS, vol. 1233, pp. 37–51. Springer, Heidelberg (1997). doi:10.1007/3-540-69053-0_4
14. Datta, N., Nandi, M.: ELmD. Submission to the CAESAR Competition (Round 2). http://competitions.cr.yp.to/round2/elmdv20.pdf
15. Diffie, W., Hellman, M.E.: Privacy and authentication: an introduction to cryptography. Proc. IEEE 67(3), 397–427 (1979)
16. Dobraunig, C., Eichlseder, M., Mendel, F., Schläffer, M.: Ascon. Submission to the CAESAR Competition (Round 2). http://competitions.cr.yp.to/round2/asconv11.pdf
17. Dworkin, M.: Recommendation for block cipher modes of operation. NIST Spec. Publ. 800(38A), 1–59 (2001)

18. Fuhr, T., Jaulmes, É., Lomné, V., Thillard, A.: Fault attacks on AES with faulty ciphertexts only. In: Fischer, W., Schmidt, J. (eds.) Fault Diagnosis and Tolerance in Cryptography - FDTC 2013, pp. 108–118. IEEE Computer Society, Washington, DC (2013)
19. Goldreich, O., Ron, D.: On testing expansion in bounded-degree graphs. Electron. Colloquium Comput. Complex. (ECCC) **7**(20), 1–6 (2000)
20. Grosso, V., Leurent, G.L., Standaert, F., Varici, K., Journault, A., Durvaux, F., Gaspar, L., Kerckhof, S.: SCREAM. Submission to the CAESAR Competition (Round 2). http://competitions.cr.yp.to/round2/screamv3.pdf
21. Hermelin, M., Cho, J.Y., Nyberg, K.: Multidimensional extension of Matsui's Algorithm 2. In: Dunkelman, O. (ed.) FSE 2009. LNCS, vol. 5665, pp. 209–227. Springer, Heidelberg (2009). doi:10.1007/978-3-642-03317-9_13
22. Iwata, T., Minematsu, K., Guo, J., Morioka, S., Kobayashi, E.: CLOC. Submission to the CAESAR Competition (Round 2). http://competitions.cr.yp.to/round2/clocv2.pdf
23. Iwata, T., Minematsu, K., Guo, J., Morioka, S., Kobayashi, E.: SILC. Submission to the CAESAR Competition (Round 2). http://competitions.cr.yp.to/round2/silcv2.pdf
24. Jean, J., Nikolic, I., Peyrin, T.: Deoxys. Submission to the CAESAR Competition (Round 2). http://competitions.cr.yp.to/round2/deoxysv13.pdf
25. Jean, J., Nikolic, I., Peyrin, T.: Joltik. Submission to the CAESAR Competition (Round 2). http://competitions.cr.yp.to/round2/joltikv13.pdf
26. Jean, J., Nikolic, I., Peyrin, T.: KIASU. Submission to the CAESAR Competition (Round 1). http://competitions.cr.yp.to/round1/kiasuv1.pdf
27. Jean, J., Nikolić, I., Peyrin, T.: Tweaks and keys for block ciphers: the TWEAKEY framework. In: Sarkar, P., Iwata, T. (eds.) ASIACRYPT 2014. LNCS, vol. 8874, pp. 274–288. Springer, Heidelberg (2014). doi:10.1007/978-3-662-45608-8_15
28. Jutla, C.S.: Encryption modes with almost free message integrity. In: Pfitzmann, B. (ed.) EUROCRYPT 2001. LNCS, vol. 2045, pp. 529–544. Springer, Heidelberg (2001). doi:10.1007/3-540-44987-6_32
29. Kavun, E.B., Lauridsen, M.M., Leander, G., Rechberger, C., Schwabe, P., Yalçin, T.: Prøst. Submission to the CAESAR Competition (Round 1). http://competitions.cr.yp.to/round1/proestv11.pdf
30. Krovetz, T., Rogaway, P.: The software performance of authenticated-encryption modes. In: Joux, A. (ed.) FSE 2011. LNCS, vol. 6733, pp. 306–327. Springer, Heidelberg (2011). doi:10.1007/978-3-642-21702-9_18
31. Maurine, P.: Techniques for EM fault injection: equipments and experimental results. In: Bertoni, G., Gierlichs, B. (eds.) Fault Diagnosis and Tolerance in Cryptography - FDTC 2012, pp. 3–4. IEEE Computer Society, Washington, DC (2012)
32. McGrew, D.A., Viega, J.: The security and performance of the galois/counter mode (GCM) of operation. In: Canteaut, A., Viswanathan, K. (eds.) INDOCRYPT 2004. LNCS, vol. 3348, pp. 343–355. Springer, Heidelberg (2004). doi:10.1007/978-3-540-30556-9_27
33. Tunstall, M., Mukhopadhyay, D., Ali, S.: Differential fault analysis of the advanced encryption standard using a single fault. In: Ardagna, C.A., Zhou, J. (eds.) WISTP 2011. LNCS, vol. 6633, pp. 224–233. Springer, Heidelberg (2011). doi:10.1007/978-3-642-21040-2_15
34. Minematsu, K.: AES-OTR. Submission to the CAESAR Competition (Round 2). http://competitions.cr.yp.to/round2/aesotrv2.pdf

35. Minematsu, K.: Parallelizable rate-1 authenticated encryption from pseudorandom functions. In: Nguyen, P.Q., Oswald, E. (eds.) EUROCRYPT 2014. LNCS, vol. 8441, pp. 275–292. Springer, Heidelberg (2014). doi:10.1007/978-3-642-55220-5_16

36. Paninski, L.: A coincidence-based test for uniformity given very sparsely sampled discrete data. IEEE Trans. Inf. Theory **54**(10), 4750–4755 (2008)

37. Piret, G., Quisquater, J.-J.: A differential fault attack technique against SPN structures, with application to the AES and KHAZAD. In: Walter, C.D., Koç, Ç.K., Paar, C. (eds.) CHES 2003. LNCS, vol. 2779, pp. 77–88. Springer, Heidelberg (2003). doi:10.1007/978-3-540-45238-6_7

38. Rivain, M.: Differential fault analysis on DES middle rounds. In: Clavier, C., Gaj, K. (eds.) CHES 2009. LNCS, vol. 5747, pp. 457–469. Springer, Heidelberg (2009). doi:10.1007/978-3-642-04138-9_32

39. Rogaway, P.: Efficient instantiations of tweakable blockciphers and refinements to modes OCB and PMAC. In: Lee, P.J. (ed.) ASIACRYPT 2004. LNCS, vol. 3329, pp. 16–31. Springer, Heidelberg (2004). doi:10.1007/978-3-540-30539-2_2

40. Rogaway, P., Bellare, M., Black, J.: OCB: a block-cipher mode of operation for efficient authenticated encryption. ACM Trans. Inf. Syst. Secur. **6**(3), 365–403 (2003)

41. Rubinfeld, R.: Taming big probability distributions. ACM Crossroads **19**(1), 24–28 (2012)

42. Saha, D., Chowdhury, D.R.: SCOPE: on the side channel vulnerability of releasing unverified plaintexts. In: Dunkelman, O., Keliher, L. (eds.) SAC 2015. LNCS, vol. 9566, pp. 417–438. Springer, Heidelberg (2016). doi:10.1007/978-3-319-31301-6_24

43. Saha, D., Kuila, S., Roy Chowdhury, D.: EscApe: diagonal fault analysis of APE. In: Meier, W., Mukhopadhyay, D. (eds.) INDOCRYPT 2014. LNCS, vol. 8885, pp. 197–216. Springer, Heidelberg (2014). doi:10.1007/978-3-319-13039-2_12

44. Samajder, S., Sarkar, P.: Another look at normal approximations in cryptanalysis. Cryptology ePrint Archive, Report 2015/679 (2015). http://ia.cr/2015/679

45. Skorobogatov, S.P., Anderson, R.J.: Optical fault induction attacks. In: Kaliski Jr., B.S., Koç, K., Paar, C. (eds.) CHES 2002. LNCS, vol. 2523, pp. 2–12. Springer, Heidelberg (2003). doi:10.1007/3-540-36400-5_2

46. The CAESAR committee: CAESAR: Competition for authenticated encryption: Security, applicability, and robustness (2014). http://competitions.cr.yp.to/caesar.html

47. Wang, L.: Shell. Submission to the CAESAR Competition (Round 2). http://competitions.cr.yp.to/round2/shellv20.pdf

48. Whiting, D., Ferguson, N., Housley, R.: Counter with CBC-MAC (CCM). RFC 3610 (2003)

49. Yen, S., Joye, M.: Checking before output may not be enough against fault-based cryptanalysis. IEEE Trans. Comput. **49**(9), 967–970 (2000). http://dx.doi.org/10.1109/12.869328

50. Zussa, L., Dutertre, J.M., Clediere, J., Tria, A.: Power supply glitch induced faults on FPGA: an in-depth analysis of the injection mechanism. In: On-Line Testing Symposium - IOLTS 2013, pp. 110–115. IEEE (2013)

# Authenticated Encryption
# with Variable Stretch

Reza Reyhanitabar[1($\boxtimes$)], Serge Vaudenay[2], and Damian Vizár[2]

[1] NEC Laboratories Europe, Heidelberg, Germany
reza.reyhanitabar@neclab.eu
[2] EPFL, Lausanne, Switzerland

**Abstract.** In conventional authenticated-encryption (AE) schemes, the *ciphertext expansion*, a.k.a. *stretch* or *tag length*, is a constant or a parameter of the scheme that must be *fixed* per key. However, using variable-length tags per key can be desirable in practice or may occur as a result of a misuse. The RAE definition by Hoang, Krovetz, and Rogaway (Eurocrypt 2015), aiming at the *best-possible* AE security, supports variable stretch among other strong features, but achieving the RAE goal incurs a particular inefficiency: *neither encryption nor decryption can be online.* The problem of enhancing the well-established nonce-based AE (nAE) model and the standard schemes thereof to support variable tag lengths per key, without sacrificing any desirable functional and efficiency properties such as *online* encryption, has recently regained interest as evidenced by extensive discussion threads on the CFRG forum and the CAESAR competition. Yet there is a lack of formal definition for this goal. First, we show that several recently proposed heuristic measures trying to augment the known schemes by inserting the tag length into the nonce and/or associated data *fail* to deliver any meaningful security in this setting. Second, we provide a formal definition for the notion of nonce-based variable-stretch AE (nvAE) as a natural extension to the traditional nAE model. Then, we proceed by showing a second modular approach to formalizing the goal by combining the nAE notion and a new property we call *key-equivalent separation by stretch* (*kess*). It is proved that (after a mild adjustment to the syntax) any nAE scheme which additionally fulfills the *kess* property will achieve the nvAE goal. Finally, we show that the nvAE goal is efficiently and provably achievable; for instance, by simple tweaks to off-the-shelf schemes such as OCB.

**Keywords:** Authenticated encryption · Variable-length tags · Robustness · Security definitions · CAESAR competition

## 1 Introduction

Authenticated encryption (AE) algorithms have recently faced an immense increase in popularity as appropriate cryptographic tools for providing *data* confidentiality (privacy) and integrity (together with authenticity) services simultaneously. The notion of AE, as a cryptographic scheme in its own right, was

© International Association for Cryptologic Research 2016
J.H. Cheon and T. Takagi (Eds.): ASIACRYPT 2016, Part I, LNCS 10031, pp. 396–425, 2016.
DOI: 10.1007/978-3-662-53887-6_15

originally put forward in several (partially) independent papers [3,4,20] and further evolved to notions of nonce-based AE (nAE) by Rogaway et al. [35], nonce-based AE with associated data (AEAD) by Rogaway [32,34], deterministic AE (DAE) and misuse-resistant AE (MRAE) by Rogaway and Shrimpton [36], online nonce-misuse resistant AE by Fleischmann et al. [14], AE under the release of unverified plaintext (AE-RUP) by Andreeva et al. [1], robust AE (RAE) by Hoang et al. [16], and online AE (OAE2) by Hoang et al. [17].

Providing *authenticity* requires any AE scheme to incur a non-zero ciphertext expansion or stretch, $\tau = |C| - |M|$, where $|M|$ and $|C|$ are the lengths of the plaintext and ciphertext in bits, respectively. Most standard AE schemes adopt a syntax in which the ciphertex is explicitly partitioned as $C = C_{core}||\text{Tag}$ with $C_{core}$ as the ciphertext core (decryptable to a putative plaintext) and Tag as the authentication tag (used for verifying the decrypted message). In this paper, we will use the terms *ciphertext expansion*, *stretch* and *tag length* interchangeably unless the syntax of an AE scheme (e.g. an RAE scheme) does not allow partitioning of the ciphertext to a core and a tag part, in which case we use the general term *stretch*.

THE PROBLEM. This paper investigates the problem of using an AE scheme with variable-length tags (variable stretch) under the same key. All the known security notions for AE schemes [1,14,17,32,34,36] and constructions thereof, with the exception of RAE [16], assume that the stretch $\tau$ is a constant or a scheme parameter which must be *fixed* per key, and security is proved under this assumption. A correct usage of such a scheme shall ensure that two instances of the same scheme with different stretches $\tau_1$ and $\tau_2$ always use two independently chosen keys $K_1$ and $K_2$. However, this rigid correct-use mandate may be violated in practice for different reasons.

First, AE schemes may be used with variable-length tags per key due to misuse and poorly engineered security systems. With the increasing scale of deployment of cryptography, various types of misuse of cryptographic tools (i.e. their improper use that leads to compromised security) occur routinely in practice [9,12,18,22,23,41]. Identifying potential ways of misuse and mitigating their impact by sound design is therefore of great importance, while waving such a potential misuse off because there have been no cases of occurrence is a dangerous practice. Prior "Disasters" [6] have shown that it's a question of when, not if, a misuse will eventually happen in applications of (symmetric-key) cryptographic schemes in practice.

The ongoing CAESAR competition [5] has explicitly listed a set of conventional confidentiality and integrity goals for AE, but has left "any additional security goals and robustness goals that the submitters wish to point out" as an option. Among the potential additional goals, *robustness* features, in particular, different flavours of misuse-resistance to nonce reuse [14,36] have attracted a lot of attention. While the recent focus has been mainly on nonce misuse, proper characterization and formalization of other potential misuse dimensions seems yet a challenge to be further investigated. The current literature lacks a systematic approach to formalizing an appropriate notion of AE with misuse-resistance to tag-length variation under the same key, *without sacrificing* interesting functional and efficiency features such as online encryption.

Second, there are use cases such as resource-constrained communication devices, where the support for variable-length tags is desired, but changing the key per tag length and renegotiating the system parameters is a costly process due to bandwidth and energy constraints. In those cases, supporting variable stretch per key while still being able to provide a "sliding scale" authenticity is deemed to be a useful functional and efficiency feature as pointed out by Struik [39]. For instance, de Meulenaer et al. demonstrate that in case of wireless sensor networks, communication-related energy consumption is substantially higher than the consumption caused by computation [10]. Sliding scale authenticity could significantly extend the lifetime of such sensors, especially if processed plaintexts are very short, while only a handful of them requires a very high level of authenticity.

The problem has appeared to be highly interesting from both theoretical and practical perspectives as evidenced by the relatively long CFRG forum thread on issues arising from variable-length tags in OCB [24], followed by ongoing discussions in the CAESAR competition mailing list [19], which in turn has motivated several second-round CAESAR candidates to be tweaked [19,25,28] with the aim of providing some *heuristic* measures for addressing the problem.

ISSUES ARISING FROM VARIABLE STRETCH PER KEY. Lack of support for variable-length tags per key in conventional AE models, in particular in the widely-used nAE security model, is not just a theoretical and definitional complaint, rather all known standard AE schemes such as the widely-deployed CCM, GCM, and OCB schemes do *misbehave* in one way or another if misused in this way [24,31,38]. Depending on the application scenario, the consequences of such a misbehavior may range from a degraded security level to a complete loss of security.

A CFRG forum discussion thread initiated by Manger [24], has raised the following concerns with an "Attacker changing tag length in OCB":

- OCB with different tag lengths are defined. Under the same key, shorter tags are simply truncation of longer tags. The tag length is not mixed into the ciphertext as it never affects any input to the underlying blockcipher. Consequently, given a valid output from e.g. the OCB algorithm with 128-bit tag it is trivial to produce a valid output for the OCB algorithm with 64-bit tag under the same key, by just dropping the last 8 bytes.
- An attacker wanting to change the associated data while keeping the same plaintext and the same tag length as applied by the originator (e.g. 128 bits) only has to defeat the shortest accepted tag length (e.g. 64 bits) and the differences between accepted tag lengths up to the targeted stretch. This is not fulfilled by OCB.
- Would OCB be better if the algorithms with different tag lengths could not affect each other? Perhaps restricting the nonce to <126 bits (instead of <128 bits) and encoding the tag length in 2 bits.

The CFRG discussions concluded by adopting Manger's suggested heuristic measure by designers of OCB: "just drop the tag length into the nonce" [31]. One may call this method *nonce stealing* for tag length akin to "nonce stealing" for associated data (AD), proposed by Rogaway [32] to convert an AE scheme to an AEAD scheme. The problem of variable-length tags per key has regained interest in recent CAESAR competition discussions. Nandi [27] has raised the question whether including the tag length in the associated data can resolve the problem. A natural extension would be combining both measures, i.e., including the tag length as part of both the nonce and the associated data.

But in the absence of a definitional and provable-security treatment of the problem of robustness to tag-length variation per key, the proposed heuristic measures and claims for added security in the tweaked schemes are informal, and only limited to showing lack of some specific type of misbehavior by the schemes.

RAE SOLVES THE PROBLEM, DO WE NEED ANOTHER DEFINITION? RAE aims to capture the "*best-possible*" AE security [16]. Similar to the MRAE and Pseudo-random Injection (PRI) notions [36] it targets robustness to nonce-misuse, but it also improves upon the prior notions by supporting variable stretch and hence sliding scale authenticity for any arbitrary stretch. However, the cost to pay for achieving such a strong goal is that any RAE scheme incurs a particular inefficiency: *neither encryption nor decryption can be online*. We also note that designing an *efficient* RAE scheme, e.g. AEZ [16], essentially entails designing an *efficient* tweakable block cipher with variable-length messages and tweaks at the first place followed by employing it in the encode-then-encipher paradigm, a task that has turned out to be non-trivial as evidenced by several non-ideal properties determined by recent attacks against the core cipher of prior AEZ versions by Fuhr et al. [15].

While RAE aims to facilitate the use of any stretch, even a small one, and promises to provide the best-possible security for any stretch even under nonce-reuse, our main aim in this paper is to provide an enhancement to the conventional AE models, in particular the popular nAE model, that just adds robustness to tag-length variation under the same key without sacrificing the highly desired *online-ness* feature. Unlike the RAE notion our aim is neither to facilitate/encourage using arbitrarily short tags nor to add nonce-misuse resistance to a scheme which does not already possess such a property. The core goal is to minimize/cut the interferences between instances of an AE scheme (e.g. OCB) using different tag lengths under the same key and to meaningfully achieve the best-possible authenticity in this setting without affecting/damaging the privacy property.

Intuitively, one aims to have an AE scheme that can guarantee $\tau_c$-bit authenticity to the recipient whenever a received ciphertext has a $\tau_c$-bit tag ($\tau_c$-bit stretch) irrespective of adversarial access to other instances of the same algorithm under the same key but different (shorter or longer) $\tau$-bit tags.

HEURISTIC MEASURES FAIL. We show in Sect. 3 that *in general*, several recently proposed heuristic measures, such as inserting the tag length into the nonce [31], into the associated data [27] or both methods combined, fail to capture the aforementioned intuition of a meaningful security in the variable-length tag setting. This is done by showing generic forgery attacks against these measures in a large class of nAE schemes (including e.g. GCM and OCB) that follow the "ciphertext translation" design paradigm of Rogaway [32]. The attacks have a much lower verification query complexity for $\tau$ bits of stretch than $2^\tau$. For example, an adversary having access to the instances of the same algorithm with 32-bit, 64-bit, 96-bit and 128-bit tags under the same key will only need a query complexity $O(2^{32})$ to forge a message with a 128-bit tag. The attacks are rather straightforward generalization of the tag-length misusing attack presented by the Ascon team on OMD version 1 [13].

OUR RESULTS. We formalize a security notion for nonce-based variable-stretch AE (nvAE). First we provide an all-in-one security definition to formulate the notion. Then we take an alternative modular approach for defining the notion by introducing a property, named *key-equivalent separation by stretch* (**kess**), that together with the conventional nAE security implies the nvAE security notion. While the former approach provides an easy-to-understand, stand-alone definition by directly capturing the whole aim of nvAE, the latter modular approach is easier to work with, at least for proving schemes nvAE-secure, in particular, when one tweaks an existing nAE-secure scheme and wants to establish the nvAE-security of the modified scheme by just proving its **kess** property rather than having to prove everything from scratch. We show that the nvAE goal is efficiently and provably achievable by application of simple tweaks to off-the-shelf popular schemes such as OBC, Minematsu's OTR [25] or OMD without sacrificing their desirable functional and efficiency features such as online encryption. Furthermore, we establish the relations (implications and separations) between different security notions in the conventional fixed-stretch AE setting and variable-stretch AE setting. A summary of the relations is depicted in Fig. 1.

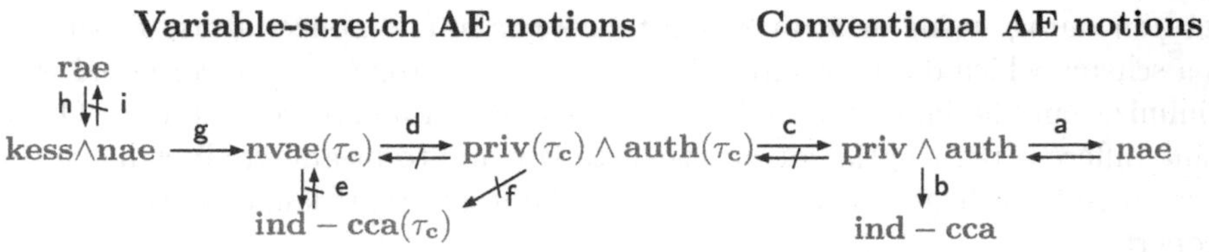

**Fig. 1.** Relations among notions for nonce-based AE with and without variable stretch. Previous works: a [36], b [3]. This paper: c (Remark 3, attacks in Sect. 3), d (Remark 3, Corollary 1), e (Theorem 1, Remark 2), f (Proposition 1), g (Theorem 2), h, i (Remark 4 together with [16]).

ORGANIZATION OF THE PAPER. In Sect. 2 we overview some of the prior AE definitions. Section 3 describes generic forgery attacks showing ineffectiveness of the heuristic measures of including the tag length in the nonce and/or associated data of a given nAE scheme to support variable-length tags per key. In Sect. 4 we provide formal definitions for the goal of AE with variable stretch per key, and Sect. 7 provides some discussions and remarks on the interpretation of the results of this work. In Sect. 6 we show how to efficiently achieve nvAE.

## 2  Preliminaries and Prior AE Definitions

NOTATIONS. For a set $S$ (either finite, or endowed with a natural definition of uniform distribution) we denote by $a \leftarrow\!\!\$\ S$ sampling an element of $S$ uniformly at random and storing it in the variable $a$. All strings are binary strings. We let $|X|$ denote the length of a string $X$, and $X\|Y$ the concatenation of two strings $X$ and $Y$. We let $\varepsilon$ denote the empty string of length 0. We let $\{0,1\}^*$ denote the set of all strings of arbitrary finite lengths (s.t. $\varepsilon \in \{0,1\}^*$) and we let $\{0,1\}^n$ denote the set of all strings of length $n$ for a positive integer $n$. We let $\mathbb{N}$ denote the set of all (positive) natural numbers and $\mathbb{N}_0 = \mathbb{N} \cup \{0\}$.

RESOURCE-PARAMETERIZED ADVERSARIAL ADVANTAGE. The insecurity of a scheme $\Pi$ in regard to a security property **xxx** is measured using the resource parameterized function $\mathbf{Adv}_{\Pi}^{\mathbf{xxx}}(\mathbf{r}) = \max_{\mathscr{A}}\{\mathbf{Adv}_{\Pi}^{\mathbf{xxx}}(\mathscr{A})\}$, where the maximum is taken over all adversaries $\mathscr{A}$ which use resources bounded by $\mathbf{r}$.

BLOCKCIPHERS AND TWEAKABLE BLOCKCIPHERS. Let $\mathrm{Perm}(n)$ be the set of all permutations over $n$-bit strings. Let $\mathrm{Perm}^{\mathcal{T}}(n) \subseteq \{\tilde{\pi} : \mathcal{T} \times \{0,1\}^n \to \{0,1\}^n\}$ be the set of all functions, s.t. for every $\tilde{\pi} \in \mathrm{Perm}^{\mathcal{T}}(n)$, $\tilde{\pi}(t,\cdot)$ is a permutation for every $t \in \mathcal{T}$ where $\mathcal{T}$ is a set of tweaks. We use $\tilde{\pi}^t(\cdot)$ and $\tilde{\pi}(t,\cdot)$ interchangeably. Let $E : \mathcal{K} \times \{0,1\}^n \to \{0,1\}^n$ be a blockcipher and let $\tilde{E} : \mathcal{K} \times \mathcal{T} \times \{0,1\}^n \to \{0,1\}^n$ be a tweakable blockcipher with a non-empty, finite $\mathcal{K} \subseteq \{0,1\}^*$. Let $D$ and $\tilde{D}$ denote the inverses of $E$ and $\tilde{E}$ respectively. Let $E_K(\cdot) = E(K,\cdot)$ and $\tilde{E}_K^t(\cdot) = \tilde{E}(K,t,\cdot)$. Let $\mathscr{A}$ be an adversary. Then:

$$\mathbf{Adv}_E^{\pm\mathrm{prp}}(\mathscr{A}) = \Pr\left[K \leftarrow\!\!\$\ \mathcal{K} : \mathscr{A}^{E_K,D_K} \Rightarrow 1\right] - \Pr\left[\pi \leftarrow\!\!\$\ \mathrm{Perm}(n) : \mathscr{A}^{\pi,\pi^{-1}} \Rightarrow 1\right]$$

$$\mathbf{Adv}_{\tilde{E}}^{\pm\widetilde{\mathrm{prp}}}(\mathscr{A}) = \Pr\left[K \leftarrow\!\!\$\ \mathcal{K} : \mathscr{A}^{\tilde{E}_K,\tilde{D}_K} \Rightarrow 1\right] - \Pr\left[\tilde{\pi} \leftarrow\!\!\$\ \mathrm{Perm}^{\mathcal{T}}(n) : \mathscr{A}^{\tilde{\pi},\tilde{\pi}^{-1}} \Rightarrow 1\right]$$

The resource parameterized advantage functions are defined accordingly, considering that the adversarial resources of interest here are the time complexity $(t)$ of the adversary and the total number of queries $(q)$ asked by the adversary.

In the following we recall the security notions for nonce-based AE (nAE) schemes with associated data (a.k.a. "AEAD" schemes) [32] and RAE schemes. We will simply use nAE to refer to any (nonce-based) AEAD scheme as all nAE schemes must now support associated data processing.

SYNTAX. We augment the syntax of original nAE schemes [32] to include a stretch variable. A scheme for authenticated encryption is a triplet $\Pi = (\mathcal{K}, \mathcal{E}, \mathcal{D})$ where $\mathcal{K} \subseteq \{0,1\}^*$ is the set of keys endowed with a (uniform) distribution and $\mathcal{E} : \mathcal{K} \times \mathcal{N} \times \mathcal{A} \times \mathcal{I}_T \times \mathcal{M} \to \mathcal{C}$ and $\mathcal{D} : \mathcal{K} \times \mathcal{N} \times \mathcal{A} \times \mathbb{N} \times \mathcal{C} \to \mathcal{M} \cup \{\bot\}$ are the encryption and decryption algorithm respectively, both deterministic and stateless. We call $\mathcal{N}$ nonce space, $\mathcal{A}$ AD space, $\mathcal{M}$ plaintext space, $\mathcal{C}$ ciphertext space, and $\mathcal{I}_T$ stretch space (i.e. the set of ciphertext expansion values that can be applied upon encryption) of $\Pi$, and we have that $\mathcal{N} \subseteq \{0,1\}^*$, $\mathcal{M} \subseteq \{0,1\}^*$, $\mathcal{A} \subseteq \{0,1\}^*$, $\mathcal{C} \subseteq \{0,1\}^*$ and $\mathcal{I}_T \subseteq \mathbb{N}$.

We insist that if $M \in \mathcal{M}$ then $\{0,1\}^{|M|} \subseteq \mathcal{M}$ (any reasonable AE scheme would certainly have this property). We additionally limit ourselves to *correct* and *tidy* (defined by Namprempre et al. [26]) schemes with *variable stretch*. Namely, the correctness means that for every $(K, N, A, \tau, M) \in \mathcal{K} \times \mathcal{N} \times \mathcal{A} \times \mathcal{I}_T \times \mathcal{M}$, if $\mathcal{E}(K, N, A, \tau, M) = C$ then $\mathcal{D}(K, N, A, \tau, C) = M$, and tidiness means that for every $(K, N, A, \tau, C) \in \mathcal{K} \times \mathcal{N} \times \mathcal{A} \times \mathcal{I}_T \times \mathcal{C}$, if $\mathcal{D}(K, N, A, \tau, C) = M \neq \bot$ then $\mathcal{E}(K, N, A, \tau, M) = C$. In both cases $|C| = |M| + \tau$ where $\tau$ denotes the *stretch*.

VARIATIONS IN SYNTAX. In the case of conventional nAE schemes, the expansion of ciphertexts is fixed to some constant value $\tau$; this is equivalent to setting $\mathcal{I}_T = \{\tau\}$. For such schemes, we omit stretch from the list of input arguments of both the encryption and the decryption algorithm. We sometimes create an ordinary nonce-based AE scheme $\Pi'$ from a nonce-based AE scheme with variable stretch $\Pi$ by fixing the expansion value for all queries to some value $\tau \in \mathcal{I}_T$. We will denote this as $\Pi' = \Pi[\tau]$.

TWO-REQUIREMENT SECURITY DEFINITION. The nAE notion was originally formalized by a two-requirement (privacy and authenticity) definition [4,32]. The privacy of a scheme $\Pi$ is captured by its indistinguishability from a random strings-oracle in a chosen plaintext attack with non-repeating nonces, while its authenticity is defined as adversary's inability to *forge* a new ciphertext, i.e. issue a decryption query returning $M \neq \bot$. The **priv** advantage of an adversary $\mathcal{A}$ against $\Pi$ is defined as $\mathbf{Adv}_{\Pi}^{\mathrm{priv}}(\mathcal{A}) = \Pr[\mathcal{A}^{\mathrm{priv}\text{-}R_\Pi} \Rightarrow 1] - \Pr[\mathcal{A}^{\mathrm{priv}\text{-}I_\Pi} \Rightarrow 1]$ and the **auth** advantage of $\mathcal{A}$ as $\mathbf{Adv}_{\Pi}^{\mathrm{auth}}(\mathcal{A}) = \Pr[\mathcal{A}^{\mathrm{auth}_\Pi} \text{ forges}]$ where the corresponding security games are defined in Fig. 2. In the following $x \leftarrow\!\!\$\ \mathcal{S}$ will denote sampling an element $x$ from a set $\mathcal{S}$ with uniform distribution.

ALL-IN-ONE SECURITY DEFINITION. Rogaway and Shrimpton introduced an alternative, all-in-one approach for defining the nAE security, and proved it to be equivalent to the two-requirement definition [36]. The all-in-one **nae** notion captures AE security as indistinguishability of the real encryption and decryption algorithms from a random strings oracle and an always-reject oracle in a nonce-respecting, chosen ciphertext attack. The **nae** advantage of an adversary $\mathcal{A}$ against a scheme $\Pi$ is defined as $\mathbf{Adv}_{\Pi}^{\mathrm{nae}}(\mathcal{A}) = \Pr[\mathcal{A}^{\mathrm{nae}\text{-}R_\Pi} \Rightarrow 1] - \Pr[\mathcal{A}^{\mathrm{nae}\text{-}I_\Pi} \Rightarrow 1]$ where the corresponding security games are defined in Fig. 3.

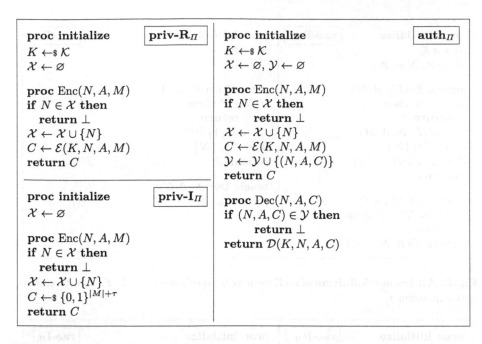

**Fig. 2. Two-requirement definition** of nAE security for a scheme $\Pi = (\mathcal{K}, \mathcal{E}, \mathcal{D})$ with ciphertext expansion $\tau$.

ROBUST AE. As mentioned in Sect. 1, the notion of robust AE (RAE) [16], aims to capture a very strong security goal. The RAE security is captured as indistinguishability of a scheme from a particular idealized primitive in an unrestricted chosen ciphertext attack. The **rae** advantage of an adversary $\mathscr{A}$ against a scheme $\Pi$ is defined as $\mathbf{Adv}_{\Pi}^{\mathrm{rae}}(\mathscr{A}) = \Pr[\mathscr{A}^{\mathrm{rae\text{-}R}_{\Pi}} \Rightarrow 1] - \Pr[\mathscr{A}^{\mathrm{rae\text{-}I}_{\Pi}} \Rightarrow 1]$ where the corresponding security games are defined in Fig. 4.

It is known that the strong RAE security of a scheme implies its nAE security. This can be easily verified by showing that $\mathbf{Adv}_{\Pi}^{\mathrm{priv}}(\mathscr{B}) \leq \mathbf{Adv}_{\Pi}^{\mathrm{rae}}(\mathscr{A})$ and $\mathbf{Adv}_{\Pi}^{\mathrm{auth}}(\mathscr{C}) \leq \mathbf{Adv}_{\Pi}^{\mathrm{rae}}(\mathscr{A}) + \frac{q_d}{2^{\tau}}$ for some adversaries $\mathscr{B}$ and $\mathscr{C}$ with the same resources as $\mathscr{A}$, $q_d$ the number of decryption queries and $\tau$ the amount of stretch in all queries. However, the robustness of RAE comes at the expense of efficiency; an RAE-secure AE scheme must be inherently "offline", i.e. it cannot encrypt a plaintext with constant memory while outputting ciphertext bits with constant latency, as every bit of the ciphertext must depend on every bit of plaintext.

STRETCH (IN)DEPENDENT ADVANTAGE. For some of the security notions we discuss, the adversarial advantage is trivially dependent on the value of stretch. The advantage for notions that capture integrity of ciphertexts will necessarily be high whenever stretch $\tau$ is low, as there is always a trivial attack that queries a random ciphertext with probability $2^{-\tau}$ of being successfully decrypted. This concerns the notions **auth** and **nae**. The notions that do not *directly* capture

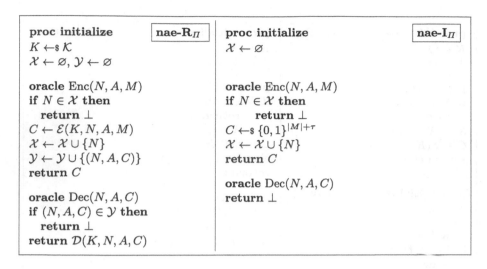

**Fig. 3. All-in-one definition** of nAE security for a scheme $\Pi = (\mathcal{K}, \mathcal{E}, \mathcal{D})$ with ciphertext expansion $\tau$.

| proc initialize $\quad$ rae-R$_\Pi$ | proc initialize $\quad$ rae-I$_\Pi$ |
|---|---|
| $K \leftarrow_\$ \mathcal{K}$ | for $N, A, \tau \in \mathcal{N} \times \{0,1\}^* \times \mathbb{N}$ do |
| | $\quad \pi_{N,A,\tau} \leftarrow_\$ \mathrm{Inj}(\tau)$ |
| proc Enc$(N, A, \tau, M)$ | proc Enc$(N, A, \tau, M)$ |
| return $\mathcal{E}(K, N, A, \tau, M)$ | return $\pi_{N,A,\tau}(M)$ |
| proc Dec$(N, A, \tau, C)$ | proc Dec$(N, A, \tau, C)$ |
| return $\mathcal{D}(K, N, A, \tau, C)$ | if $\exists M \in \{0,1\}^*$ s.t. $\pi_{N,A,\tau}(M) = C$ then |
| | $\quad$ return $M$ |
| | return $\perp$ |

**Fig. 4. RAE security.** Defining security for a robust AE scheme $\Pi = (\mathcal{K}, \mathcal{E}, \mathcal{D})$ with nonce space $\mathcal{N}$. $\mathrm{Inj}(\tau)$ denotes the set of all injective, $\tau$-expanding functions from $\{0,1\}^*$ to $\{0,1\}^{\geq \tau}$.

integrity of ciphertexts are not inherently impacted by the value of $\tau$. In particular, no trivial attack with advantage $2^{-\tau}$ exists for the notions **priv** or **rae**. Note that **rae** captures the integrity property indirectly; the idealized reference of RAE security itself will still yield to the trivial attack mentioned above.

## 3    Failure of Inserting Stretch into Nonce And/or AD

Using a generic forgery attack, we show that the recently proposed heuristic measures, namely, inclusion of the tag length in the nonce [31], in the AD [27] or in both nonce and AD fail when applied to a large class of nAE schemes (including

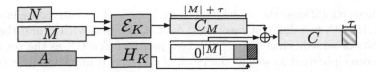

**Fig. 5. Ciphertext translation.** The message-only nAE encryption $\mathcal{E}$ produces an intermediate ciphertext $C_M$ with $\tau$ bits of stretch. The leftmost $\tau$ bits of the output of a keyed hash $H_K(A)$ are xored to the rightmost $\tau$ bits of $C_M$, forming the final ciphertext $C$.

| | | | |
|---|---|---|---|
| 1: $\Delta_A \leftarrow \varepsilon$; $A^* \leftarrow_\$ \mathcal{A}\backslash\{A\}$ | | 7: | $C_i \leftarrow C_i^* \oplus 0^{\lvert C_i \rvert - \tau_i} \lVert \Delta_A \rVert \delta$ |
| 2: **for** $i \leftarrow 1$ **to** $g$ **do** | | 8: | $M_i \leftarrow \mathrm{Dec}(N_i, A, \tau_i, C_i)$ |
| 3: $\quad$ pick fresh nonce $N_i$ | | 9: | **while** $M_i = \perp$ |
| 4: $\quad$ $C_i^* \leftarrow \mathrm{Enc}(N_i, A^*, \tau_i, M)$ | | 10: | $\Delta_A \leftarrow \mathsf{right}_{\tau_i}(C_i \oplus C_i^*)$ |
| 5: $\quad$ **do** | | 11: | **return** $N_g, A, C_g$ |
| 6: $\quad\quad$ pick fresh $\delta \in \{0,1\}^{\tau_i - \tau_{i-1}}$ | | | |

**Fig. 6. Ciphertext forgery** for a ciphertext translation-based AEAD scheme with associated data $A$ and message $M$ in presence of variable stretch. Here $\tau_0 = 0$.

e.g. GCM and OCB) that follow the "ciphertext translation" design paradigm of Rogaway [32] which is depicted in Fig. 5. The attack is not completely new, it is a rather straightforward generalization of the tag-length misusing attack originally proposed by the Ascon team on a specific algorithm, namely OMD version 1 [13] which also follows the ciphertext translation method.

THE ATTACK. We target a ciphertext translation-based AEAD scheme $\Pi$ that supports any amount of stretch from a set $\mathcal{I}_T = \{\tau_1, \ldots, \tau_r\}$ with $\tau_1 < \tau_2 < \ldots < \tau_r$. We assume oracle access to encryption and decryption algorithms, such that the amount of stretch can be chosen for every query independently. The goal is to forge a ciphertext for $A, M$ expanded by $\tau_g \in \mathcal{I}_T$ bits, with $g > 1$. The attack proceeds as in Fig. 6. We let $\mathsf{left}_i(X)$ and $\mathsf{right}_j(X)$ denote $i$ leftmost bits and $j$ rightmost bits of a string $X$ respectively.

The hash function $\Pi_K(\cdot)$ used to process AD must fulfil some mild conditions for the attack to work against the described heuristic countermeasures [27,31], namely:

- In case that the tag length is only injected into the nonce, the attack works with *arbitrary* $H_K(\cdot)$.
- For inclusion of the tag length in the AD or a combination of this method and nonce stealing, the attack works if $H_K(A) = H_{1_K}(A_1) \oplus H_{2_K}(A_2) \oplus \cdots \oplus H_{m_K}(A_m)$, for arbitrary functions $H_{i_K}$, $1 \leq i \leq m$, where $A = A_1 \lVert A_2 \rVert \cdots \lVert A_m$ for $A_j \in \{0,1\}^n$ for some positive integer $n$ (this is the case for both GCM and OCB). In this case, we must ensure that the block of AD that contains the amount of stretch $\tau$ is unchanged between $A$ and $A^*$.

Under these conditions, the attack will always succeed: whenever we encrypt a message $M$ with two different associated data $A, A^*$, first with $\tau_i$ and then with $\tau_j > \tau_i$ bits of stretch, then $C_i \oplus C_i^*$ will be a prefix of $C_j \oplus C_j^*$, as the xor cancels out the core ciphertext as well as the block of AD that is impacted by $\tau$ (if any).

The complexity of the attack in terms of verification queries will be $O(2^\mu)$ with $\mu = \max\{\tau_1, \tau_2 - \tau_1, \ldots, \tau_g - \tau_{g-1}\}$. For example, an adversary having access to the instances of the algorithm with 32-bit, 64-bit, 96-bit and 128-bit tags under the same key will only need a query complexity $O(2^{32})$ to forge a message with a 128-bit tag, which is in stark contrast with the expected $O(2^{128})$ query complexity.

# 4    Formalizing Nonce-Based AE with Variable Stretch

Defining a meaningful security notion for AE schemes with variable stretch under the same key has turned out to be a non-trivial task [24,31,38]. Allowing the adversary to choose the amount of stretch freely from a set $\mathcal{I}_T = \{\tau_{\min}, \ldots, \tau_{\max}\}$ will inevitably enable it to produce forgeries with a high probability $2^{-\tau_{\min}}$ by targeting the shortest allowed stretch; a forgery is sure to be found with at most $2^{\tau_{\min}}$ verification queries. This is inherent to *any* AE scheme.

Despite this limit to its *global* security guarantees, there is a meaningful security property which *can* be expected from an nvAE scheme by a user: the scheme must guarantee $\tau$ bits of security for ciphertexts with $\tau$ bits of stretch, regardless of adversarial access to other instances with the same key but other (shorter and/or longer) amount of stretch than $\tau$. For example, forging a ciphertext with $\tau$-bit stretch should require $\approx 2^\tau$ verification queries *with $\tau$-bit stretch*, regardless of the number of queries made under other different amounts of stretch.

This non-interference between different instances that use the same key but different stretch (tag length) is the intuition behind a formal definition for the notion of nonce-based, variable-stretch AE.

SECURITY DEFINITION. We define a security notion parameterized by the challenge stretch value $\tau_c \in \mathcal{I}_T$ as a natural extension to the notion of nAE. This is done in the compact all-in-one definition style of [36].

Let $\Pi = (\mathcal{K}, \mathcal{E}, \mathcal{D})$ be a nvAE scheme whose syntax is defined in Sect. 2. An **nvae**$(\tau_c)$ adversary $\mathscr{A}$ gets to interact with games **nvae**$(\tau_c)$-$\mathbf{R}_\Pi$ (left) and **nvae**$(\tau_c)$-$\mathbf{I}_\Pi$ (right) in Fig. 7, defining respectively the real and ideal behavior of such a scheme. The adversary has access to two oracles Enc and Dec determined by these games and its goal is to distinguish the two games.

The adversary must respect a *relaxed nonce-requirement*; it must use a unique pair of nonce and stretch for encryption queries. Compared to the standard nonce-respecting requirement in nAE schemes, here nonce may be reused provided that the stretch does not repeat simultaneously.

In the ideal game **nvae**$(\tau_c)\mathbf{I}_\Pi$, the encryption and decryption queries with $\tau_c$-bit stretch are answered in the same idealized way as in the "ideal" game of **nae** notion (Fig. 3 right). However, the queries with stretch other than $\tau_c$ are treated with the real encryption/decryption algorithm. This lets the adversary to issue arbitrary queries (e.g. repeated forgeries) for any stretch $\tau \neq \tau_c$ and leverage the information thus gathered to attack the challenge expansion. At the same time, only queries with $\tau_c$ bits of stretch can help the adversary to actually distinguish the two games, capturing the exact level of security for queries with $\tau_c$ bits of stretch in presence of variable stretch.

We measure the advantage of $\mathscr{A}$ in breaking the $\mathbf{nvae}(\tau_c)$ security of $\Pi$ as
$$\mathbf{Adv}_{\Pi}^{\mathbf{nvae}(\tau_c)}(\mathscr{A}) = \Pr[\mathscr{A}^{\mathbf{nvae}(\tau_c)\text{-}\mathbf{R}_\Pi} \Rightarrow 1] - \Pr[\mathscr{A}^{\mathbf{nvae}(\tau_c)\text{-}\mathbf{I}_\Pi} \Rightarrow 1].$$

ADVERSARIAL RESOURCES. The adversarial resources of interest for the $\mathbf{nvae}(\tau_c)$ notion are $(t, \mathbf{q_e}, \mathbf{q_d}, \boldsymbol{\sigma})$, where $t$ denotes the running time of the adversary, $\mathbf{q_e} = (q_e^\tau | \tau \in \mathcal{I}_T)$ denotes the vector that holds the number of encryption queries $q_e^\tau$ made with stretch $\tau$ for every stretch $\tau \in \mathcal{I}_T$, and $\mathbf{q_d} = (q_d^\tau | \tau \in \mathcal{I}_T)$ denotes the same for the decryption queries and $\boldsymbol{\sigma} = (\sigma^\tau | \tau \in \mathcal{I}_T)$ denotes the vector that holds the total amount of data $\sigma^\tau$ processed in all queries with stretch $\tau$ for every $\tau \in \mathcal{I}_T$.

Despite being focused on queries stretched by $\tau_c$ bits, we watch adversarial resources for every stretch $\tau \in \mathcal{I}_T$ in a detailed, vector-based fashion. This approach appears to be most flexible w.r.t. the security analysis. However, in a typical case we will be interested in the resources related to $\tau_c$ (i.e. $q_e^{\tau_c}, q_d^{\tau_c}, \sigma^{\tau_c}$) and cumulative resources of the adversary $q_e$, $q_d$, $\sigma$ with $q_e = \sum_{\tau \in \mathcal{I}_T} q_e^\tau$, $q_d = \sum_{\tau \in \mathcal{I}_T} q_d^\tau$ and $\sigma = \sum_{\tau \in \mathcal{I}_T} \sigma^\tau$.

*Remark 1 (Relation to nAE).* The notion of $\mathbf{nvae}(\tau_c)$ is indeed an extension of the classical all-in-one security notion for nonce-based AE schemes. If the scheme $\Pi$ is secure with some stretch-space $\mathcal{I}_T$, then it will be secure for any stretch-space $\mathcal{I}_T' \subseteq \mathcal{I}_T$, in particular for $\mathcal{I}_T' = \{\tau_c\}$. If a scheme has a stretch-space $\mathcal{I}_T = \{\tau_c\}$, then $\mathbf{nvae}(\tau_c)$ becomes the classical $\mathbf{nae}$ notion. It easily follows, that $\mathbf{nvae}(\tau_c)$ security of a scheme $\Pi$ tightly implies $\mathbf{nae}$ security of $\Pi[\tau_c]$.

Similar to the $\mathbf{nae}$ notion, the $\mathbf{nvae}(\tau_c)$ adversarial advantage will be trivially high if $\tau_c$ is low (due to successful forgeries). Yet, if the $\mathbf{nvae}(\tau_c)$ advantage of a scheme behaves "reasonably", we will call the scheme secure. We discuss the interpretation of the $\mathbf{nvae}(\tau_c)$ bounds in Appendix 7.

PARAMETERIZED CCA SECURITY. An $\mathbf{nae}$-secure AE scheme is also $\mathbf{ind} - \mathbf{cca}$-secure. This follows from the equivalence of the all-in-one and dual nAE notions and a well-known implication $\mathbf{priv} \wedge \mathbf{auth} \Rightarrow \mathbf{ind} - \mathbf{cca}$ established by Bellare and Namprempre [3]. It is natural to ask: *Does the $\mathbf{nvae}(\tau_c)$-security also provide a privacy guarantee against chosen ciphertext attacks?* We define a $\tau_c$-parameterized extension of the $\mathbf{ind} - \mathbf{cca}$ security notion and answer this question positively.

The parameterized $\mathbf{ind} - \mathbf{cca}(\tau_c)$ notion captures the exact privacy level guaranteed by an nvAE scheme for encryption queries stretched by $\tau_c$ bits, in presence of arbitrary queries with expansions $\tau \neq \tau_c$ and reasonable decryption queries stretched by $\tau_c$ bits. The notion is building on the intuition that privacy level of $\tau_c$-expanded queries should not be affected by the adversarial queries with other amounts of stretch.

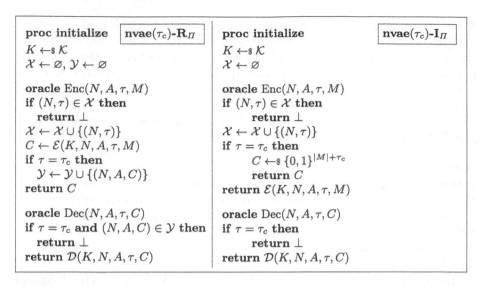

**Fig. 7. AE security with variable stretch.** Security games for defining AE security of a nonce-based AE scheme $\Pi = (\mathcal{K}, \mathcal{E}, \mathcal{D})$ with variable-stretch.

SECURITY DEFINITION. Let $\Pi = (\mathcal{K}, \mathcal{E}, \mathcal{D})$ be an nvAE with syntax defined in Sect. 2. We let an adversary $\mathscr{A}$ interact with the games $\mathbf{ind} - \mathbf{cca}(\tau_c)\text{-}\mathbf{R}_\Pi$ and $\mathbf{ind} - \mathbf{cca}(\tau_c)\text{-}\mathbf{I}_\Pi$ defined in Fig. 8 and its goal is to distinguish them. In the "ideal" game $\mathbf{ind} - \mathbf{cca}(\tau_c)\text{-}\mathbf{I}_\Pi$, the $\tau_c$-stretched encryption queries are answered with random strings while the decryption queries are processed with the real decryption algorithm. $\mathscr{A}$ must respect the relaxed nonce-requirement and is prevented to win the game trivially (i.e. by re-encrypting output of decryption query with $\tau_c$ bits of stretch and vice-versa). We measure $\mathscr{A}$'s advantage in breaking $\mathbf{ind} - \mathbf{cca}(\tau_c)$ security of $\Pi$ as $\mathbf{Adv}_\Pi^{\mathrm{ind-cca}(\tau_c)}(\mathscr{A}) = \Pr\left[\mathscr{A}^{\mathrm{ind-cca}(\tau_c)\text{-}\mathbf{R}} \Rightarrow 1\right] - \Pr\left[\mathscr{A}^{\mathrm{ind-cca}(\tau_c)\text{-}\mathbf{I}} \Rightarrow 1\right]$.
The adversarial resources of interest for the $\mathbf{ind} - \mathbf{cca}(\tau_c)$ notion are the same as for the $\mathbf{nvae}(\tau_c)$ notion, i.e. $(t, \mathbf{q_e}, \mathbf{q_d}, \boldsymbol{\sigma})$.

*Remark 2 (Relations to ind-cca and nvAE).* Similarly as in the case of $\mathbf{nvae}(\tau_c)$ and $\mathbf{nae}$, $\mathbf{ind} - \mathbf{cca}(\tau_c)$ security with some stretch space $\mathcal{I}_T$ implies $\mathbf{ind} - \mathbf{cca}(\tau_c)$ security with any stretch space $\mathcal{I}'_T \subseteq \mathcal{I}_T$, e.g. $\mathcal{I}'_T = \{\tau_c\}$. It follows that $\mathbf{ind} - \mathbf{cca}(\tau_c)$ security of a scheme $\Pi$ implies the classical $\mathbf{ind} - \mathbf{cca}$ security of $\Pi[\tau_c]$.

The notions of $\mathbf{ind} - \mathbf{cca}(\tau_c)$ and $\mathbf{nvae}(\tau_c)$ differ mainly in the way the "ideal" games treat the decryption queries expanded by $\tau_c$ bits. The impact of this difference is substantial; the $\mathbf{ind} - \mathbf{cca}(\tau_c)$ notion does not capture integrity of ciphertexts. E.g. a scheme that concatenates output of a length-preserving, nonce-based, ind-cca-secure encryption scheme (using encoding of the nonce and stretch as a "nonce") and an image of the nonce and stretch under a PRF would be secure in the sense of $\mathbf{ind} - \mathbf{cca}(\tau_c)$, but insecure in the sense of $\mathbf{nvae}(\tau_c)$.

| **proc initialize** $\boxed{\text{ind-cca}(\tau_c)\text{-}\mathbf{R}_\Pi}$ | **proc initialize** $\boxed{\text{ind-cca}(\tau_c)\text{-}\mathbf{I}_\Pi}$ |
|---|---|
| $K \leftarrow_\$ \mathcal{K}$ | $K \leftarrow_\$ \mathcal{K}$ |
| $\mathcal{V} \leftarrow \varnothing, \mathcal{X} \leftarrow \varnothing, \mathcal{Y} \leftarrow \varnothing$ | $\mathcal{V} \leftarrow \varnothing, \mathcal{X} \leftarrow \varnothing, \mathcal{Y} \leftarrow \varnothing$ |
| | |
| **oracle** $\text{Enc}(N, A, \tau, M)$ | **oracle** $\text{Enc}(N, A, \tau, M)$ |
| **if** $(N, \tau) \in \mathcal{X}$ **then return** $\bot$ | **if** $(N, \tau) \in \mathcal{X}$ **then return** $\bot$ |
| **if** $\tau = \tau_c$ **and** $(N, A, M) \in \mathcal{V}$ **then** | **if** $\tau = \tau_c$ **and** $(N, A, M) \in \mathcal{V}$ **then** |
|   **return** $\bot$ |   **return** $\bot$ |
| $\mathcal{X} \leftarrow \mathcal{X} \cup \{(N, \tau)\}$ | $\mathcal{X} \leftarrow \mathcal{X} \cup \{(N, \tau)\}$ |
| $C \leftarrow \mathcal{E}(K, N, A, \tau, M)$ | **if** $\tau = \tau_c$ **then** |
| **if** $\tau = \tau_c$ **then** |   $C \leftarrow_\$ \{0,1\}^{|M|+\tau_c}$ |
|   $\mathcal{Y} \leftarrow \mathcal{Y} \cup \{(N, A, C)\}$ |   $\mathcal{Y} \leftarrow \mathcal{Y} \cup \{(N, A, C)\}$ |
| **return** $C$ |   **return** $C$ |
| | **return** $\mathcal{E}(K, N, A, \tau, M)$ |
| | |
| **oracle** $\text{Dec}(N, A, \tau, C)$ | **oracle** $\text{Dec}(N, A, \tau, C)$ |
| **if** $\tau = \tau_c$ **and** $(N, A, C) \in \mathcal{Y}$ **then** | **if** $\tau = \tau_c$ **and** $(N, A, C) \in \mathcal{Y}$ **then** |
|   **return** $\bot$ |   **return** $\bot$ |
| $M \leftarrow \mathcal{D}(K, N, A, \tau, C)$ | $M \leftarrow \mathcal{D}(K, N, A, \tau, C)$ |
| **if** $\tau = \tau_c$ **and** $M \neq \bot$ | **if** $\tau = \tau_c$ **and** $M \neq \bot$ |
|   $\mathcal{V} \leftarrow \mathcal{V} \cup \{(N, A, M)\}$ |   $\mathcal{V} \leftarrow \mathcal{V} \cup \{(N, A, M)\}$ |
| **return** $M$ | **return** $M$ |

**Fig. 8. Parameterized ind-cca security.** Games for defining $\mathbf{ind} - \mathbf{cca}(\tau_c)$ security of a nonce-based AE scheme with variable-stretch $\Pi = (\mathcal{K}, \mathcal{E}, \mathcal{D})$.

We examine the relation between the two notions in the other direction in Theorem 1. We would like to stress that the result in Theorem 1 holds for *any* nvAE scheme, and in particular for any stretch space $\mathcal{I}_T$.

**Theorem 1 ($\mathbf{nvae}(\tau_c) \Rightarrow \mathbf{ind\text{-}cca}(\tau_c)$).** *Let* $\Pi = (\mathcal{K}, \mathcal{E}, \mathcal{D})$ *be an arbitrary nonce-based AE scheme with variable stretch. We have that*

$$\mathbf{Adv}_\Pi^{\mathbf{ind}-\mathbf{cca}(\tau_c)}(t, \mathbf{q_e}, \mathbf{q_d}, \boldsymbol{\sigma}) \leq 2 \cdot \mathbf{Adv}_\Pi^{\mathbf{nvae}(\tau_c)}(t', \mathbf{q_e}, \mathbf{q_d}, \boldsymbol{\sigma}),$$

*with* $t' = t + O(q)$ *and* $q = \sum_{\tau \in \mathcal{I}_T}(q_e^\tau + q_d^\tau)$.

*Proof.* Let $\mathscr{A}$ be an $\mathbf{ind} - \mathbf{cca}$ adversary with indicated resources. We define the game $\mathbf{ind} - \mathbf{cca}(\tau_c)\text{-}\mathbf{I}_\Pi^\perp$ as an intermediate step in the proof; it is exactly the same as $\mathbf{ind} - \mathbf{cca}(\tau_c)\text{-}\mathbf{I}_\Pi$, except that the decryption queries with $\tau_c$ bits of stretch are always answered with $\bot$. We have that

$$\mathbf{Adv}_\Pi^{\mathbf{ind}-\mathbf{cca}(\tau_c)}(\mathscr{A}) = \Pr[\mathscr{A}^{\mathbf{ind}-\mathbf{cca}(\tau_c)\text{-}\mathbf{R}_\Pi} \Rightarrow 1] - \Pr[\mathscr{A}^{\mathbf{ind}-\mathbf{cca}(\tau_c)\text{-}\mathbf{I}_\Pi^\perp} \Rightarrow 1]$$
$$+ \Pr[\mathscr{A}^{\mathbf{ind}-\mathbf{cca}(\tau_c)\text{-}\mathbf{I}_\Pi^\perp} \Rightarrow 1] - \Pr[\mathscr{A}^{\mathbf{ind}-\mathbf{cca}(\tau_c)\text{-}\mathbf{I}_\Pi} \Rightarrow 1].$$

We start by showing that $\Pr[\mathscr{A}^{\mathbf{ind}-\mathbf{cca}(\tau_c)\text{-}\mathbf{R}_\Pi} \Rightarrow 1] - \Pr[\mathscr{A}^{\mathbf{ind}-\mathbf{cca}(\tau_c)\text{-}\mathbf{I}_\Pi^\perp} \Rightarrow 1] \leq \mathbf{Adv}_\Pi^{\mathbf{nvae}(\tau_c)}(\mathscr{B})$ for an $\mathbf{nvae}(\tau_c)$ adversary $\mathscr{B}$ with the resources $(t', \mathbf{q_e}, \mathbf{q_d}, \boldsymbol{\sigma})$. The reduction of $\mathscr{A}$ to $\mathscr{B}$ is straightforward: $\mathscr{B}$ simply answers $\mathscr{A}$'s queries with its own oracles, making sure that the trivial win-preventing restrictions of

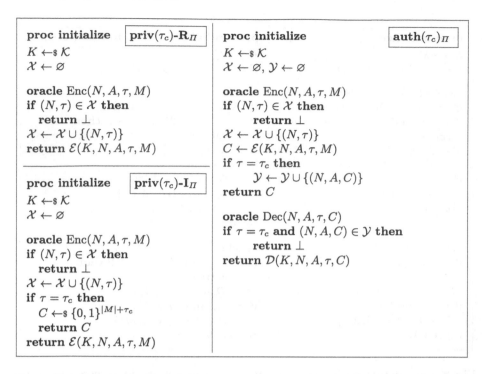

**Fig. 9. Dual nvAE security.** Security games for defining AE security of a nonce-based AE scheme $\Pi = (\mathcal{K}, \mathcal{E}, \mathcal{D})$ with variable-stretch.

$\text{ind} - \text{cca}(\tau_c)$ games are met. At the end of experiment, $\mathscr{B}$ outputs whatever $\mathscr{A}$ outputs. This ensures perfect simulation of both games for $\mathscr{A}$.

It remains to show that $\Pr[\mathscr{A}^{\text{ind}-\text{cca}(\tau_c)\text{-}\mathbf{I}_{\Pi}^{\perp}} \Rightarrow 1] - \Pr[\mathscr{A}^{\text{ind}-\text{cca}(\tau_c)\text{-}\mathbf{I}_{\Pi}} \Rightarrow 1] \leq \mathbf{Adv}_{\Pi}^{\text{nvae}(\tau_c)}(\mathscr{C})$ for an $\text{nvae}(\tau_c)$ adversary $\mathscr{C}$ with the resources $(t', \mathbf{q_e}, \mathbf{q_d}, \boldsymbol{\sigma})$. We reduce $\mathscr{A}$ to $\mathscr{C}$ as follows. $\mathscr{C}$ answers all $\mathscr{A}$'s queries directly with its own oracles (again making sure to enforce all the restrictions of $\text{ind} - \text{cca}(\tau_c)$ games), except for encryption queries expanded by $\tau_c$ bits. For those, $\mathscr{C}$ ignores its encryption oracle and answers with $|M| + \tau_c$ random bits if $\mathscr{A}$'s query has a fresh nonce-stretch pair an is not a re-encryption. At the end of experiment, $\mathscr{C}$ outputs the inverse of $\mathscr{A}$'s output. If $\mathscr{C}$ interacts with $\text{nvae}(\tau_c)\text{-}\mathbf{R}_{\Pi}$, then it perfectly simulates $\text{ind} - \text{cca}(\tau_c)\text{-}\mathbf{I}_{\Pi}$ for $\mathscr{A}$ while if $\mathscr{C}$ interacts with $\text{nvae}(\tau_c)\text{-}\mathbf{I}_{\Pi}$, then it perfectly simulates $\text{ind} - \text{cca}(\tau_c)\text{-}\mathbf{I}_{\Pi}^{\perp}$. $\square$

No Two-Requirement Notion. The equivalence of the two-requirement (privacy and authenticity) approach and all-in-one approach for defining AE security is among the best known results in AE [36]. One may wonder whether such an equivalence also holds in the setting of variable-stretch AE schemes for natural $\tau_c$-parameterized extensions of these notions. Surprisingly, we answer this question negatively. We consider the conventional privacy (ind-cpa\$) and authenticity (integrity of ciphertexts) notions for AE schemes [3, 32] and define the notions of $\tau_c$-privacy and $\tau_c$-authenticity as natural parameterized extensions of their conventional counterparts.

Let $\Pi = (\mathcal{K}, \mathcal{E}, \mathcal{D})$ be an nvAE scheme with syntax defined in Sect. 2. An adversary $\mathscr{A}$ against $\tau_c$-privacy of $\Pi$ interacts with games $\mathbf{priv}(\tau_c)\text{-}\mathbf{R}_\Pi$ (real scheme) and $\mathbf{priv}(\tau_c)\text{-}\mathbf{I}_\Pi$ (ideal behaviour) defined in Fig. 9, and tries to distinguish them. We measure $\mathscr{A}$'s advantage in breaking the $\tau_c$-privacy of $\Pi$ in a chosen plaintext attack as $\mathbf{Adv}_\Pi^{\mathbf{priv}(\tau_c)}(\mathscr{A}) = \Pr[\mathscr{A}^{\mathbf{priv}(\tau_c)\text{-}\mathbf{R}_\Pi} \Rightarrow 1] - \Pr[\mathscr{A}^{\mathbf{priv}(\tau_c)\text{-}\mathbf{I}_\Pi} \Rightarrow 1]$.

An adversary $\mathscr{A}$ that attacks the $\tau_c$-authenticity of $\Pi$ is left to interact with the game $\mathbf{auth}(\tau_c)_\Pi$ defined in Fig. 9 and its goal is to find a valid forgery (i.e. produce a decryption query returning $M \neq \bot$) with the target stretch of $\tau_c$ bits. We measure the advantage of $\mathscr{A}$ in breaking $\tau_c$-authenticity of $\Pi$ in a chosen ciphertext attack by $\mathbf{Adv}_\Pi^{\mathbf{auth}(\tau_c)}(\mathscr{A}) = \Pr\left[\mathscr{A}^{\mathbf{auth}(\tau_c)_\Pi} \text{ forges with } \tau_c\right]$. The adversarial resources of interest for the $\mathbf{priv}(\tau_c)$ and $\mathbf{auth}(\tau_c)$ notions are $(t, \mathbf{q_e}, \boldsymbol{\sigma})$ and $(t, \mathbf{q_e}, \mathbf{q_d}, \boldsymbol{\sigma})$ respectively, defined as for the notion of $\mathbf{nvae}(\tau_c)$ in the current Section.

*Remark 3 (Relations with the all-in-one nvAE, priv and auth notions).* As before, if a scheme $\Pi$ is $\mathbf{priv}(\tau_c)$ ($\mathbf{auth}(\tau_c)$) secure with stretch-space $\mathcal{I}_T$, then it will be secure for any stretch-space $\mathcal{I}_T' \subseteq \mathcal{I}_T$ including $\mathcal{I}_T' = \{\tau_c\}$, implying the $\mathbf{priv}$ ($\mathbf{auth}$) security of the scheme $\Pi[\tau_c]$.

We can easily verify that the $\mathbf{nvae}(\tau_c)$ security of a scheme $\Pi$ implies both the $\mathbf{priv}(\tau_c)$ security and the $\mathbf{auth}(\tau_c)$ of $\Pi$, by adapting the reductions for corresponding conventional notions [36] slightly. In Proposition 1, we show that the converse of this implication does not hold.

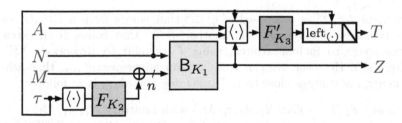

**Fig. 10.** The encryption algorithm of the scheme $\Pi_{\neg\text{cca}}$. $\langle \cdot \rangle$ is an efficiently computable, injective encoding scheme.

**Proposition 1.** *There exists a nonce-based AE scheme with variable stretch, that is secure in the sense of both the $\mathbf{priv}(\tau_c)$ notion and the $\mathbf{auth}(\tau_c)$ notion but insecure in the sense of $\mathbf{ind} - \mathbf{cca}(\tau_c)$ notion, i.e.*

$$\mathbf{priv}(\tau_c) \wedge \mathbf{auth}(\tau_c) \not\Rightarrow \mathbf{ind} - \mathbf{cca}(\tau_c),$$

*assuming the existence of secure tweakable blockciphers and PRFs.*

| **proc** $\mathcal{E}_{\neg\mathrm{cca}}(K, N, A, \tau, M)$ | **proc** $\mathcal{D}_{\neg\mathrm{cca}}(K, N, A, \tau, C)$ |
|---|---|
| Parse $K$ as $K_1, K_2, K_3$ | Parse $K$ as $K_1, K_2, K_3$ |
| $W \leftarrow M \oplus F(K_2, \langle\tau\rangle)$ | Parse $C$ as $Z\|T$ with $\|T\| = \tau$ |
| $Z \leftarrow \mathsf{B}(K_1, N, W)$ | **if** $\mathsf{left}_\tau(F'(K_3, \langle N, A, \tau, Z\rangle)) \neq T$ **then** |
| $T \leftarrow \mathsf{left}_\tau(F'(K_3, \langle N, A, \tau, Z\rangle))$ | $\qquad$ **return** $\perp$ |
| **return** $Z\|T$ | $W \leftarrow \mathsf{B}^{-1}(K_1, N, Z)$ |
| | **return** $W \oplus F(K_2, \langle\tau\rangle)$ |

**Fig. 11.** Encryption and decryption algorithms of the nonce-based, variable-stretch AE scheme $\Pi_{\neg\mathrm{cca}} = (\mathcal{K}_{\neg\mathrm{cca}}, \mathcal{E}_{\neg\mathrm{cca}}, \mathcal{E}_{\neg\mathrm{cca}})$. $\langle\cdot\rangle$ is an efficiently computable, injective encoding scheme.

To support the claim in Proposition 1, we define the nvAE scheme $\Pi_{\neg\mathrm{cca}} = (\mathcal{K}_{\neg\mathrm{cca}}, \mathcal{E}_{\neg\mathrm{cca}}, \mathcal{D}_{\neg\mathrm{cca}})$ constructed from an ind-cpa secure tweakable blockcipher $\mathsf{B} : \mathcal{K}_1 \times \mathcal{N} \times \{0,1\}^n \rightarrow \{0,1\}^n$ and two PRFs $F : \mathcal{K}_2 \times \{0,1\}^* \rightarrow \{0,1\}^n$ and $F' : \mathcal{K}_3 \times \{0,1\}^* \rightarrow \{0,1\}^m$. We define $\mathcal{K}_{\neg\mathrm{cca}} = \mathcal{K}_1 \times \mathcal{K}_2 \times \mathcal{K}_3$, $\mathcal{M}_{\neg\mathrm{cca}} = \{0,1\}^n$, $\mathcal{A}_{\neg\mathrm{cca}} = \{0,1\}^*$, $\mathcal{N}_{\neg\mathrm{cca}} = \mathcal{N}$ and the encryption and decryption algorithms as in Fig. 11. We require that $|\mathcal{I}_{T\,\neg\mathrm{cca}}| \geq 2$ and that $m \geq \max(\mathcal{I}_{T\,\neg\mathrm{cca}})$. The encryption algorithm $\mathcal{E}_{\neg\mathrm{cca}}$ is depicted in Fig. 10.

The scheme $\Pi_{\neg\mathrm{cca}}$ is by far no real-life AE construction (mainly due to its limited message space), its purpose is merely to act as a counter example. It can be verified, that $\mathbf{Adv}_{\Pi_{\neg\mathrm{cca}}}^{\mathbf{auth}(\tau_c)}(t, \mathbf{q_e}, \mathbf{q_d}, \boldsymbol{\sigma}) \leq \mathbf{Adv}_{F'}^{PRF}(t, q_e + q_d, \sigma) + q_d^{\tau_c}/2^{\tau_c}$; every forgery attempt equals to guessing $\tau_c$ bits of an output of $F'$, evaluated on a fresh input.[1] For privacy, we have that $\mathbf{Adv}_{\Pi_{\neg\mathrm{cca}}}^{\mathbf{priv}(\tau_c)}(t, \mathbf{q_e}, \mathbf{q_d}, \boldsymbol{\sigma}) \leq \mathbf{Adv}_F^{PRF}(t, q_e, \sigma) + \mathbf{Adv}_{F'}^{PRF}(t, q_e, \sigma) + \mathbf{Adv}_{\mathsf{B}}^{\widetilde{prp}}(t, q_e) + 2q_e^2/2^n$. Here $q_e = \sum_{\tau \in \mathcal{I}_T} q_e^\tau$, $q_d = \sum_{\tau \in \mathcal{I}_T} q_d^\tau$ and $\sigma = \sum_{\tau \in \mathcal{I}_T} \sigma^\tau$.

The term $2q_e^2/2^n$ is composed of $q_e^2/2^n$ that comes from a $RP$-$RF$ switch for the tweakable blockcipher and another $q_e^2/2^n$ that comes from extending the tweakspace to include stretch, using $F$ (similar to Rogaway's XE construction [33]). However, we can construct an adversary $\mathscr{A}_{\neg\mathrm{cca}}$, that achieves $\mathbf{ind} - \mathbf{cca}(\tau_c)$ advantage close to 1. The strategy of $\mathscr{A}_{\neg\mathrm{cca}}$ is as follows:

1. ask query $Z_1\|T_1 \leftarrow \mathrm{Enc}(N_1, A_1, \tau_c, M_1)$ with arbitrary $N_1, A_1, M_1$,
2. iterate through $T_1^* \in \{0,1\}^{\tau_{\min}}$ until $M_1^* \leftarrow \mathrm{Dec}(N_1, A_1, \tau_{\min}, Z_1\|T_1^*)$ returns $M_1^* \neq \perp$,
3. ask query $Z_2\|T_2 \leftarrow \mathrm{Enc}(N_2, A_2, \tau_c, M_2)$ with arbitrary $N_2, A_2, M_2$,
4. iterate through $T_2^* \in \{0,1\}^{\tau_{\min}}$ until $M_2^* \leftarrow \mathrm{Dec}(N_2, A_2, \tau_{\min}, Z_2\|T_2^*)$ returns $M_2^* \neq \perp$,
5. return 1 iff $M_1 \oplus M_1^* = M_2 \oplus M_2^*$ (otherwise return 0),

---

[1] Note that $\tau_c$ is an index rather than a power in $q_d^{\tau_c}$.

| proc initialize $\quad$ **kess-R$_\Pi$** | proc initialize $\quad$ **kess-I$_\Pi$** |
|---|---|
| $K \leftarrow_\$ \mathcal{K}$ | for $\tau \in \mathcal{I}_T$ do |
| $\mathcal{X} \leftarrow \varnothing$ | $\quad K_\tau \leftarrow_\$ \mathcal{K}$ |
| | |
| **oracle** Enc$(N, A, \tau, M)$ | **oracle** Enc$(N, A, \tau, M)$ |
| **if** $(N, \tau) \in \mathcal{X}$ **then** | **if** $(N, \tau) \in \mathcal{X}$ **then** |
| $\quad$ **return** $\bot$ | $\quad$ **return** $\bot$ |
| $\mathcal{X} \leftarrow \mathcal{X} \cup \{(N, \tau)\}$ | $\mathcal{X} \leftarrow \mathcal{X} \cup \{(N, \tau)\}$ |
| **return** $\mathcal{E}(K, N, A, \tau, M)$ | **return** $\mathcal{E}(K_\tau, N, A, \tau, M)$ |
| | |
| **oracle** Dec$(N, A, \tau, C)$ | **oracle** Dec$(N, A, \tau, C)$ |
| **return** $\mathcal{D}(K, N, A, \tau, C)$ | **return** $\mathcal{D}(K_\tau, N, A, \tau, C)$ |

**Fig. 12. Key-equivalent separation by stretch.** Games defining **kess** property of a nonce-based AE scheme $\Pi = (\mathcal{K}, \mathcal{E}, \mathcal{D})$ with variable stretch. Note that the independent keying for each $\tau \in \mathcal{I}_T$ in game **kess-I$_\Pi$** can be done by lazy sampling if needed.

where $\tau_{\min} = \min(\mathcal{I}_T \backslash \{\tau_c\})$. We have that $\mathbf{Adv}^{\mathrm{ind-cca}(\tau_c)}_{\Pi_{\neg\mathrm{cca}}}(\mathscr{A}_{\neg\mathrm{cca}}) = 1 - 2^{-n}$. As amount of stretch $\tau$ has no effect on the encryption by B, we can verify that

$$M_1 \oplus F(K_2, \langle\tau_c\rangle) = M_1^* \oplus F(K_2, \langle\tau_{\min}\rangle)$$
$$M_2 \oplus F(K_2, \langle\tau_c\rangle) = M_2^* \oplus F(K_2, \langle\tau_{\min}\rangle)$$

The final conditional statement verified by the adversary is always true for the real scheme. The probability of the same event in the "ideal" game is $2^{-n}$. As a consequence of Theorem 1 and Proposition 1, we can state Corollary 1.[2]

**Corollary 1.** *There exists a nonce-based AE scheme with variable stretch, that is secure in the sense of both the* $\mathbf{priv}(\tau_c)$ *notion and the* $\mathbf{auth}(\tau_c)$ *notion but insecure in the sense of* $\mathbf{nvae}(\tau_c)$ *notion, i.e.*

$$\mathbf{priv}(\tau_c) \wedge \mathbf{auth}(\tau_c) \not\Rightarrow \mathbf{nvae}(\tau_c)$$

KEY-EQUIVALENT SEPARATION BY STRETCH. The notion of $\mathbf{nvae}(\tau_c)$ captures the immediate intuition about the security goal one expects to achieve using a nonce-based AE scheme with variable stretch. We now introduce a modular approach to *achieving* the notion. Assume that an AE scheme is already known to be secure in the sense of the nAE model. What additional security property should such a scheme possess (i.e. on top of nAE-security) so that it can achieve the full aim of being a $\mathbf{nvae}(\tau_c)$-secure scheme? We formalize such a desirable property, naming it *key-equivalent separation by stretch* (**kess**), which captures the intuition that for each value of stretch the scheme should behave as if keyed with a fresh, independent secret key.

---

[2] The same attack strategy yields also $\mathbf{Adv}^{\mathrm{nvae}(\tau_c)}_{\Pi_{\neg\mathrm{cca}}}(\mathscr{A}_{\neg\mathrm{cca}}) = 1 - 2^{-n}$.

```
proc initialize                          oracle Enc(N, A, τ, M)
for τ ∈ I_T do                           if (N, τ) ∈ X then
    K_τ ←$ K                                  return ⊥
X ← ∅, Y ← ∅                             X ← X ∪ {(N, τ)}
                                         C ← E(K_τ, N, A, τ, M)
oracle Dec(N, A, τ, C)                   if τ = τ_c then
if τ = τ_c and (N, A, C) ∈ Y then            Y ← Y ∪ {(N, A, C)}
    return ⊥                             return C
return D(K_τ, N, A, τ, C)
```

**Fig. 13.** Security game $\mathbf{nvae}(\tau_c)\text{-}G_\Pi$.

Let $\Pi = (\mathcal{K}, \mathcal{E}, \mathcal{D})$ be an nvAE scheme with the syntax defined in Sect. 2. We let an adversary $\mathscr{A}$ that tries to break **kess** of $\Pi$ interact with games defined in Fig. 12. The goal of the adversary is to distinguish these two games. The advantage of $\mathscr{A}$ in breaking the **kess** property of the scheme $\Pi$ is measured by $\mathbf{Adv}_\Pi^{\mathbf{kess}}(\mathscr{A}) = \Pr\left[\mathscr{A}^{\mathbf{kess}\text{-}\mathbf{R}_\Pi} \Rightarrow 1\right] - \Pr\left[\mathscr{A}^{\mathbf{kess}\text{-}\mathbf{I}_\Pi)} \Rightarrow 1\right].$

The adversarial resources of interest for the **kess** notion are $(t, \mathbf{q_e}, \mathbf{q_d}, \boldsymbol{\sigma})$, as defined for the $\mathbf{nvae}(\tau_c)$ notion in the current Section.

We note that **kess** on its own says nothing about AE security of a scheme (e.g. identity "encryption" concatenated with $\tau$ zeroes achieves **kess**, but is far from **nae**-secure). However, we show in Theorem 2 that when combined with **nae** security, **kess** implies $\mathbf{nvae}(\tau_c)$ security. Informally, the **kess** notion takes care of interaction between queries with different values of stretch. Once this is done, we are free to argue that the queries with $\tau_c$ bits of stretch are "independent" of those with other values of stretch and will "inherit" the security level of $\Pi[\tau_c]$.

**Theorem 2. (kess $\wedge$ nae $\Rightarrow$ nvae($\tau_c$)).** *Let* $\Pi = (\mathcal{K}, \mathcal{E}, \mathcal{D})$ *be a nonce-based AE scheme with variable stretch. We have that*

$$\mathbf{Adv}_\Pi^{\mathbf{nvae}(\tau_c)}(t, \mathbf{q_e}, \mathbf{q_d}, \boldsymbol{\sigma}) \leq \mathbf{Adv}_\Pi^{\mathbf{kess}}(t', \mathbf{q_e}, \mathbf{q_d}, \boldsymbol{\sigma}) + \mathbf{Adv}_{\Pi[\tau_c]}^{\mathbf{nae}}(t'', q_e^{\tau_c}, q_d^{\tau_c}, \sigma^{\tau_c}),$$

*with* $t' = t + O(q)$ *and* $t'' = t + O(\sigma)$ *where* $q = \sum_{\tau \in \mathcal{I}_T}(q_e^\tau + q_d^\tau)$ *and* $\sigma = \sum_{\tau \in \mathcal{I}_T}(\sigma_e^\tau + \sigma_d^\tau)$.

*Proof.* Let $\mathscr{A}$ be an $\mathbf{nvae}(\tau_c)$ adversary with the indicated resources. Consider the security game $\mathbf{nvae}(\tau_c)\text{-}G$ defined in Fig. 13. We have that
$$\mathbf{Adv}_\Pi^{\mathbf{nvae}(\tau_c)}(\mathscr{A}) = \Pr[\mathscr{A}^{\mathbf{nvae}(\tau_c)\text{-}\mathbf{R}_\Pi} \Rightarrow 1] - \Pr[\mathscr{A}^{\mathbf{nvae}(\tau_c)\text{-}G_\Pi} \Rightarrow 1]$$
$$+ \Pr[\mathscr{A}^{\mathbf{nvae}(\tau_c)\text{-}G_\Pi} \Rightarrow 1] - \Pr[\mathscr{A}^{\mathbf{nvae}\text{-}\mathbf{I}_\Pi(\tau_c)} \Rightarrow 1].$$

We first show that $\Pr[\mathscr{A}^{\mathbf{nvae}(\tau_c)\text{-}\mathbf{R}_\Pi} \Rightarrow 1] - \Pr[\mathscr{A}^{\mathbf{nvae}(\tau_c)\text{-}G_\Pi} \Rightarrow 1] \leq \mathbf{Adv}_\Pi^{\mathbf{kess}}(\mathscr{B})$ for a **kess** adversary $\mathscr{B}$ with the resources $(t', \mathbf{q_e}, \mathbf{q_d}, \boldsymbol{\sigma})$. The $\mathbf{nvae}(\tau_c)$ adversary $\mathscr{A}$ can be straightforwardly reduced to $\mathscr{B}$. Any query of $\mathscr{A}$ is directly answered with $\mathscr{B}$'s own oracles, except for decryption queries with expansion of $\tau_c$ bits whose output is trivially known from previous encryption queries; here

$\mathscr{B}$ returns $\bot$ to $\mathscr{A}$. At the end, $\mathscr{B}$ outputs whatever $\mathscr{A}$ outputs. If $\mathscr{B}$ interacts with **kess-R**$_\Pi$ then it perfectly simulates **nvae**$(\tau_c)$-**R**$_\Pi$ for $\mathscr{A}$. If $\mathscr{B}$ interacts with **kess-I**$_\Pi$ then it perfectly simulates **nvae**$(\tau_c)$-**G**$_\Pi$.

We next show that $\Pr[\mathscr{A}^{\mathbf{nvae}(\tau_c)\text{-}G_\Pi} \Rightarrow 1] - \Pr[\mathscr{A}^{\mathbf{nvae}\text{-}I_\Pi(\tau_c)} \Rightarrow 1] \leq \mathbf{Adv}^{\mathbf{nae}}_{\Pi[\tau_c]}(\mathscr{C})$ for an **nae** adversary $\mathscr{C}$ with resources $(t'', q_e^{\tau_c}, q_d^{\tau_c}, \sigma^{\tau_c})$. $\mathscr{A}$ can be reduced to $\mathscr{C}$ in the following way. When $\mathscr{A}$ issues a query with expansion $\tau_c$, $\mathscr{C}$ answers it with its own oracles. For other amounts of stretch $\tau \neq \tau_c$, $\mathscr{C}$ first checks if there were previous queries with $\tau$ bits of stretch. If not, it samples a fresh key $K_\tau$. $\mathscr{C}$ then processes the query with the real (encryption or decryption) algorithm of $\Pi$ and the key $K_\tau$, making sure that encryption queries comply with the nonce requirement and are not re-encryptions. If $\mathscr{C}$ interacts with **nae-R**$_{\Pi[\tau_c]}$ then it perfectly simulates **nvae**$(\tau_c)$-**G**$_\Pi$ for $\mathscr{A}$. If $\mathscr{C}$ interacts with **nae-I**$_{\Pi[\tau_c]}$ then it perfectly simulates **nvae**$(\tau_c)$-**I**$_\Pi$. This yields the desired result. $\qquad\square$

*Remark 4.* An RAE secure scheme $\Pi$ will always have the **kess** property. To see why, note that replacing $\Pi$ by a collection of random injections in both the **kess-R**$_\Pi$ and **kess-I**$_\Pi$ games will not increase the advantage significantly, as that would contradict $\Pi$'s RAE security. After the replacement, the two games will be indistinguishable. On the other hand, **kess** property does not guarantee RAE security; the scheme OCBv described in Sect. 6 can serve as a counter-example, because it does not tolerate nonce reuse.

## 5   A Short Guide to NvAE

INTERPRETATION OF THE NVAE SECURITY ADVANTAGE. The notion of **nvae**$(\tau_c)$ is parameterized by a constant, but arbitrary amount of stretch $\tau_c$ from the stretch space $\mathcal{I}_T$ of the AE scheme $\Pi$ in question. In the **nvae**$(\tau_c)$-**I**$_\Pi$ security game, only queries expanded by $\tau_c$ bits will be subjected to "idealization". For all other expansions, we give the adversary complete freedom to ask any queries it wants (except for the nonce-requirement), but their behaviour is the same in both security games. An **nvae**$(\tau_c)$ security bound that assumes no particular value or constraint for $\tau_c$ will therefore tell us, what security guarantees can we expect from queries stretched by $\tau_c$ bits specifically, for any $\tau_c \in \mathcal{I}_T$.

Looking at the security bound itself, we are able to tell if there are any undesirable interactions between queries with different amounts of stretch. This is best illustrated by revisiting the problems and forgery attack from Sects. 1 and 3 in the **nvae**$(\tau_c)$ security model.

ATTACKS IN NVAE MODEL. With the formal framework defined, we revisit the heuristic attacks from Sect. 3 and analyse the advantage they achieve, as well as the resources they require. Consider the original, unmodified scheme OCB [21], that produces the tag by truncating an $n$-bit (with $n > \tau$) to $\tau$ bits. In case of simultaneous use of two (or more) amounts of stretch $\tau_1 < \tau_2$ with the same key, we can forge a ciphertext stretched by $\tau_1$ bits by $\tau_2$-bit-stretched ciphertext truncation. This would correspond to an attack with an **nvae**$(\tau_1)$ advantage of 1 and constant resources.

If the same scheme is treated with the heuristic measures, i.e.nonce-stealing, and encoding $\tau$ in AD, from Sect. 3 (let's call it hOCB), we consider the forgery attack from the same Section. Assume that there are four instances of hOCB, with $32, 64, 96$ and $128$ bit tags. To make a forgery with 128-bit tag, we have to find a forgery with 32 bits and then exhaustively search for three 32-bit extensions of this forgery. This gives us an **nvae**(128) advantage equal to 1, requiring 4 encryption queries, $3 \cdot 2^{32}$ verification queries with stretch other than 128 bits and $2^{32}$ verification stretched by 128 bits. The effort necessary for such a forgery is clearly smaller than we could hope for, especially in the amount of verification queries stretched by the challenge amount of bits (i.e. 128).

"GOOD" BOUNDS. After seeing examples of attacks, one may wonder: what kind of **nvae**($\tau_c$) security bound should we expect from a secure nvAE scheme? For every scheme, it must be always possible to guess a ciphertext with probability $2^{-\tau_c}$. Thus the bound must always contain a term of the form $c \cdot (q_d^{\tau_c})^{\alpha}/2^{\tau_c}$ for some positive constants $c$ and $\alpha$, or something similar.

Even though the security level for $\tau_c$-stretched queries should be independent of any other queries, it is usually unavoidable to have a gradual increase of advantage with every query made by the adversary. This increase can generally depend on all of the adversarial resources, but should not depend on $\tau_c$ itself.

An example of a secure scheme's **nvae**($\tau_c$) bound can be found in Theorem 4. It consist of the fraction $(q_d^{\tau_c} \cdot 2^{n-\tau_c})/(2^n - 1) \approx q_d^{\tau_c}/2^{\tau_c}$, advantage bounds for the used blockcipher and a birthday-type term that grows with the total amount of data processed. We see, that queries stretched by $\tau \neq \tau_c$ bits will not unexpectedly increase adversary's chances to break OCBv, and that the best attack strategy is indeed issuing decryption queries with $\tau_c$ bits of stretch.

# 6    Achieving AE with Variable Stretch

We demonstrate that the security of AE schemes in the sense of **nvae**($\tau_c$) notion is easily achievable by introducing a practical and secure scheme. Rather than constructing a scheme from the scratch, we modify an existing, well-established scheme and follow a modular approach to analyse its security in presence of variable stretch. The modification we propose is general enough to be applicable to most of the AE schemes based on a tweakable primitive (e.g. tweakable blockcipher).

OCB MODE FOR TWEAKABLE BLOCKCIPHER. The Offset Codebook mode of operation for a tweakable blockcipher ($\Theta$CB) is a nonce-based AE scheme proposed by Krovetz and Rogaway [21] (there are subtle differences from the prior versions of OCB [33,35]). It is parameterized by a tweakable blockcipher $\widetilde{E} : \mathcal{K} \times \mathcal{T} \times \{0,1\}^n \to \{0,1\}^n$ and a tag length $0 \leq \tau \leq n$. The tweak space of $\widetilde{E}$ is of the form $\mathcal{T} = \mathcal{N} \times \mathbb{N}_0 \times \{0,1,2,3\} \cup \mathbb{N}_0 \times \{0,1,2,3\}$ for a finite set $\mathcal{N}$. The encryption and the decryption algorithms of $\Theta$CB$[\widetilde{E}, \tau]$ are described in Fig. 14.

The security of $\Theta$CB is captured in Lemma 1.

```
101: Algorithm ℰ_K(N, A, M)              308:        Sum ← Sum ⊕ Ẽ_K^{m,1}(A_*‖10*)
102:    if N ∉ 𝒩 then                    309:        return Sum
103:        return ⊥
104:    M_1‖M_2⋯M_m‖M_*←M where
105:        each |M_i| = n and |M_*| < n   201: Algorithm 𝒟_K(N, A, C)
106:    Sum ← 0^n, C_* ← ε                202:    if N ∉ 𝒩 or |C| < τ then
107:    for i ← 1 to m do                 203:        return ⊥
108:        C_i ← Ẽ_K^{N,i,0}(M_i)        204:    C_1‖C_2⋯C_m‖C_*‖T←C where
109:        Sum ← Sum ⊕ M_i               205:        each where each |C_i| = n,
110:    if M_* = ε then                   206:        |C_*| < n and |T| = τ
111:        Final ← Ẽ_K^{N,m,2}(Sum)      207:    Sum ← 0^n, M_* ← ε
112:    else                             208:    for i ← 1 to m do
113:        Pad ← Ẽ_K^{N,m,1}(0^n)        209:        M_i ← D_K^{N,τ,i,0}(C_i)
114:        C_* ← M_* ⊕ left_{|M_*|}(Pad) 210:        Sum ← Sum ⊕ M_i
115:        Sum ← Sum ⊕ M_*‖10*           211:    if C_* = ε then
116:        Final ← Ẽ_K^{N,m,3}(Sum)      212:        Final ← Ẽ_K^{N,m,2}(Sum)
117:    Auth ← Hash_K(A)                  213:    else
118:    T ← left_τ(Final ⊕ Auth)          214:        Pad ← Ẽ_K^{N,m,1}(0^n)
119:    return C_1‖C_2‖⋯‖C_m‖C_*‖T        215:        M_* ← C_* ⊕ left_{|C_*|}(Pad)
                                          216:        Sum ← Sum ⊕ M_*‖10*
301: Algorithm HASH_K(A)                  217:        Final ← Ẽ_K^{N,m,3}(Sum)
302:    Sum ← 0^n                         218:    Auth ← Hash_K(A)
303:    A_1‖A_2⋯A_m‖A_*←A where           219:    T' ← left_τ(Final ⊕ Auth)
304:        each |A_i| = n and |A_*| < n  220:    if T = T' then
305:    for i ← 1 to m do                 221:        return C_1‖⋯‖C_m‖C_*‖T
306:        Sum ← Sum ⊕ Ẽ_K^{i,0}(A_i)    222:    else
307:    if A_* ≠ ε then                   223:        return ⊥
```

**Fig. 14.** Definition of $\Theta\text{CB}[\widetilde{E}, \tau]$.

**Lemma 1** *(Lemma 2 [21]). Let $\widetilde{E} : \mathcal{K} \times \mathcal{T} \times \{0,1\}^n \to \{0,1\}^n$ be a tweakable blockcipher with $\mathcal{T} = \mathcal{N} \times \mathbb{N}_0 \times \{0,1,2,3\} \cup \mathbb{N}_0 \times \{0,1,2,3\}$. Let $\tau \in \{0, \ldots, n\}$. Then we have that*

$$\mathbf{Adv}^{\text{priv}}_{\Theta\text{CB}[\widetilde{E},\tau]}(t, q_e, \sigma) \leq \mathbf{Adv}^{\pm\widetilde{\text{prp}}}_{\widetilde{E}}(t', q^p),$$

$$\mathbf{Adv}^{\text{auth}}_{\Theta\text{CB}[\widetilde{E},\tau]}(t, q_e, q_d, \sigma) \leq \mathbf{Adv}^{\pm\widetilde{\text{prp}}}_{\widetilde{E}}(t', q^a) + q_d \cdot \frac{2^{n-\tau}}{2^n - 1},$$

*where $q^p \leq \lceil \sigma/n \rceil + 2 \cdot q_e$, and $q^a \leq \lceil \sigma/n \rceil + 2 \cdot (q_e + q_d)$, and $t' = t + O(\sigma)$.*

Thanks to the results of [36,37], we can state as a corollary of Lemma 1 that
$$\mathbf{Adv}^{\text{nae}}_{\Theta\text{CB}[\widetilde{E},\tau]}(t, q_e, q_d, \sigma) \leq \mathbf{Adv}^{\pm\widetilde{\text{prp}}}_{\widetilde{E}}(t', (\lceil \sigma/n \rceil + 2 \cdot (q_e + q_d))) + q_d \frac{2^{n-\tau}}{2^n-1}.$$

OCB MODE WITH VARIABLE-STRETCH SECURITY. We introduce $\Theta\text{CBv}$ (variable-stretch-$\Theta\text{CB}$), a nonce-based AE scheme with variable stretch, obtained by slightly modifying $\Theta\text{CB}$.

The tweakable blockcipher mode of operation $\Theta\text{CBv}$ is parameterized only by a tweakable blockcipher $\widetilde{E} : \mathcal{K} \times \mathcal{T} \times \{0,1\}^n \to \{0,1\}^n$. The tweak $\mathcal{T}$ is different than the one needed for $\Theta\text{CB}$; it is of the form $\mathcal{T} = \mathcal{N} \times \mathcal{I}_T \times \mathbb{N}_0 \times \{0,1,2,3\} \cup \mathcal{I}_T \times \mathbb{N}_0 \times \{0,1,2,3\}$ where $\mathcal{I}_T \subseteq \{0,1,\ldots,n\}$ is the desired stretch-space of $\Theta\text{CBv}$.

```
101: Algorithm ℰ_K(N, A, τ, M)              308:        Sum ← Sum ⊕ Ẽ_K^{τ,m,1}(A_*‖10*)
102:    if N ∉ 𝒩 then                        309:        return Sum
103:        return ⊥
104:    M_1‖M_2···M_m‖M_*←M where
105:        each |M_i| = n and |M_*| < n     201: Algorithm 𝒟_K(N, A, τ, C)
106:    Sum ← 0^n, C_* ← ε                   202:    if N ∉ 𝒩 or |C| < τ then
107:    for i ← 1 to m do                    203:        return ⊥
108:        C_i ← Ẽ_K^{N,τ,i,0}(M_i)         204:    C_1‖C_2···C_m‖C_*‖T←C where
109:        Sum ← Sum ⊕ M_i                  205:        each |C_i| = n,
110:    if M_* = ε then                      206:        |C_*| < n and |T| = τ
111:        Final ← Ẽ_K^{N,τ,m,2}(Sum)       207:    Sum ← 0^n, M_* ← ε
112:    else                                 208:    for i ← 1 to m do
113:        Pad ← Ẽ_K^{N,τ,m,1}(0^n)         209:        M_i ← D_K^{N,τ,i,0}(C_i)
114:        C_* ← M_* ⊕ left_{|M_*|}(Pad)    210:        Sum ← Sum ⊕ M_i
115:        Sum ← Sum ⊕ M_*‖10*              211:    if C_* = ε then
116:        Final ← Ẽ_K^{N,τ,m,3}(Sum)       212:        Final ← Ẽ_K^{N,τ,m,2}(Sum)
117:    Auth ← Hash_K(A)                     213:    else
118:    T ← left_τ(Final ⊕ Auth)             214:        Pad ← Ẽ_K^{N,τ,m,1}(0^n)
119:    return C_1‖C_2‖···‖C_m‖C_*‖T         215:        M_* ← C_* ⊕ left_{|C_*|}(Pad)
                                             216:        Sum ← Sum ⊕ M_*‖10*
301: Algorithm HASH_K(A, τ)                  217:        Final ← Ẽ_K^{N,τ,m,3}(Sum)
302:    Sum ← 0^n                            218:    Auth ← Hash_K(A)
303:    A_1‖A_2···A_m‖A_*←A where            219:    T' ← left_τ(Final ⊕ Auth)
304:        each s|A_i| = n and |A_*| < n    220:    if T = T' then
305:    for i ← 1 to m do                    221:        return C_1‖···‖C_m‖C_*‖T
306:        Sum ← Sum ⊕ Ẽ_K^{τ,i,0}(A_i)     222:    else
307:    if A_* ≠ ε then                      223:        return ⊥
```

**Fig. 15.** Definition of ΘCBv[Ẽ]. Changes from ΘCB highlighted in red.

The encryption and decryption algorithms of ΘCBv are exactly the same as those of ΘCB, that they now allow incorporate variable stretch and that every call to Ẽ is now tweaked by τ, in addition to the other tweak components. Both algorithms are described in Fig. 15. An illustration of the encryption algorithm is depicted in Fig. 16.

Thanks to Theorem 2, establishing the **nvae**($τ_c$) security of ΘCBv requires little effort. The corresponding result is stated in Theorem 3.

**Theorem 3.** Let $\tilde{E}: \mathcal{K} \times \mathcal{T} \times \{0,1\}^n \to \{0,1\}^n$ be a tweakable blockcipher with $\mathcal{T} = \mathcal{N} \times \mathcal{I}_T \times \mathbb{N}_0 \times \{0,1,2,3\} \cup \mathcal{I}_T \times \mathbb{N}_0 \times \{0,1,2,3\}$. Then we have that

$$\mathbf{Adv}_{\Theta CBv[\tilde{E}]}^{nvae(τ_c)}(t, \mathbf{q_e}, \mathbf{q_d}, \sigma) \leq \mathbf{Adv}_{\tilde{E}}^{\pm\widetilde{prp}}(t', q) + \sum_{τ \in \mathcal{I}_T} \mathbf{Adv}_{\tilde{E}}^{\pm\widetilde{prp}}(t', q^τ)$$

$$+ \mathbf{Adv}_{\tilde{E}}^{\pm\widetilde{prp}}(t', q^{τ_c}) + q_d^{τ_c} \cdot \frac{2^{n-τ_c}}{2^n - 1}.$$

where $q^τ = \lceil σ^τ/n \rceil + 2 \cdot (q_e^τ + q_d^τ)$ for $τ \in \mathcal{I}_T$, and $q = \sum_{τ \in \mathcal{I}_T} q^τ$, and $t' = t + O(σ)$ with $σ = \sum_{τ \in \mathcal{I}_T} σ^τ$.

*Proof.* We observe that if we fix the expansion value to $τ_c$ in all queries, the nonce-based AE scheme $(\Theta CBv[\tilde{E}])[τ_c]$ that we get will be identical with the

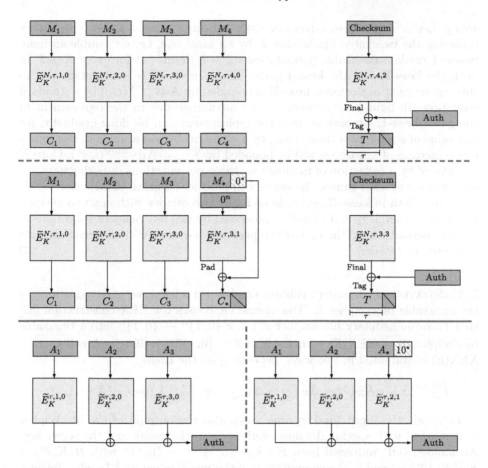

**Fig. 16.** Illustration of the encryption process of $\Theta$CBv (inspired by [21]) instantiated with a tweakable blokcipher $\widetilde{E} : \mathcal{K} \times \mathcal{T} \times \{0,1\}^n \to \{0,1\}^n$. The top half depicts the encryption of a message with four complete blocks (top) with Sum $= \bigoplus_{i=1}^4 M_i$ and the encryption of a message with three complete blocks and an incomplete block (bottom) with Sum $= \bigoplus_{i=1}^3 \oplus M_* \| 10^*$. The bottom half of the picture shows processing of associated data of three complete blocks (left) or two complete blocks and an incomplete block (right).

scheme $\Theta$CB$[\widetilde{E}, \tau_c]$. The result follows from this observation and the results of Lemmas 1 and 2 and Theorem 2. $\qquad \square$

**Lemma 2.** *Let* $\widetilde{E} : \mathcal{K} \times \mathcal{T} \times \{0,1\}^n \to \{0,1\}^n$ *be a tweakable blockcipher with* $\mathcal{T} = \mathcal{N} \times \mathcal{I}_T \times \mathbb{N}_0 \times \{0,1,2,3\} \cup \mathcal{I}_T \times \mathbb{N}_0 \times \{0,1,2,3\}$. *Then we have that*

$$\mathbf{Adv}^{\mathbf{kess}}_{\Theta\mathrm{CBv}[\widetilde{E}]}(t, \mathbf{q_e}, \mathbf{q_d}, \boldsymbol{\sigma}) \leq \mathbf{Adv}^{\pm\widetilde{\mathrm{prp}}}_{\widetilde{E}}(t', q) + \sum_{\tau \in \mathcal{I}_T} \mathbf{Adv}^{\pm\widetilde{\mathrm{prp}}}_{\widetilde{E}}(t', q^\tau)$$

*where* $q^\tau = \lceil \sigma^\tau/n \rceil + 2 \cdot (q_e^\tau + q_d^\tau)$ *for* $\tau \in \mathcal{I}_T$, *and* $q = \sum_{\tau \in \mathcal{I}_T} q^\tau$, *and* $t' = t + O(\sigma)$ *with* $\sigma = \sum_{\tau \in \mathcal{I}_T} \sigma^\tau$.

*Proof.* Let $\mathscr{A}$ be a **kess** adversary with indicated resources. We proceed by replacing the tweakable blockcipher $\widetilde{E}$ by an ideal one, i.e. we sample an independent random tweakable permutation $\widetilde{\pi}_K \leftarrow\$ \operatorname{Perm}^{\mathcal{T}}(n)$ for every $K \in \mathcal{K}$ in both the **kess-R** and the **kess-I** game. The increase of $\mathscr{A}$'s advantage due to this replacement in the game **kess-R** is bounded by $\mathbf{Adv}_{\widetilde{E}}^{\pm\widetilde{\mathrm{prp}}}(t,q)$ by a standard reduction. To bound the increase of $\mathscr{A}$'s advantage due to the replacement in the game **kess-I**, we observe that the replacement can be done gradually, for one value of stretch at a time. Thus, by a standard hybrid argument, the cumulative increase of advantage will be bounded by $\sum_{\tau\in\mathcal{I}_T}\mathbf{Adv}_{\widetilde{E}}^{\pm\widetilde{\mathrm{prp}}}(t,q^\tau)$. Once $\widetilde{E}$ is replaced by a collection of random tweakable permutations in both games, we observe that in both games, the games will produce identical distributions. This is because both in **kess-R** and in **kess-I**, any two queries with any two unequal amounts of stretch $\tau_1$ and $\tau_2$ will be processed by two independent collections of random permutations (thanks to the separation of queries with different amounts of stretch by tweaks). □

INSTANTIATING. $\widetilde{E}$. In order to obtain a real-world scheme, we need to instantiate the tweakable blockcipher $\widetilde{E}$. The scheme OCB uses the XEX construction [33] that turns an ordinary blockcipher $E : \mathcal{K} \times \{0,1\}^n \to \{0,1\}^n$ into a tweakable blockcipher $\widetilde{E} = \mathrm{XEX}[E]$ with $\widetilde{E} : \mathcal{K} \times \mathcal{T} \times \{0,1\}^n \to \{0,1\}^n$. A call to $\widetilde{E} = \mathrm{XEX}[E]$ is evaluated in two ways, depending on the tweak:

$$\widetilde{E}_K^{N,i,j}(X) = E_K(X \oplus \Delta_{N,i,j}) \oplus \Delta_{N,i,j}, \text{ or } \widetilde{E}_K^{i,j}(X) = E_K(X \oplus \Delta_{i,j}).$$

In each call, the input (and in some cases also the output) of the blockcipher $E$ is masked with special $\Delta$-values, derived from the tweak and the secret key. An almost XOR universal hash $H : \mathcal{K} \times \{0,1\}^{<n} \to \{0,1\}^n$ with $H(K,N) = E_K(N\|10^*)$ is used in the computation of the masking values.[3] In what follows, we silently represent binary strings and integers by element of $\mathrm{GF}(2^n)$ whenever needed and do the multiplications in this field with some fixed representation. E.g. $2^2 \cdot (0^{n-2}\|10)$ would return an $n$-bit string that represents the result of $x^2 \cdot x$ in $GF(2^n)$. The masking $\Delta$-values are computed as follows:

$$\Delta_{N,0,0} = H(K,N),$$
$$\Delta_{N,i+1,0} = \Delta_{N,i,0} \oplus L(\mathrm{ntz}(i+1)) \text{ for } i \geq 0,$$
$$\Delta_{N,i,j} = \Delta_{N,i,0} \oplus j \cdot L_* \text{ for } j \in \{0,1,2,3\},$$
$$\Delta_{0,0} = 0^n,$$
$$\Delta_{i+1,0} = \Delta_{i,0} \oplus L(\mathrm{ntz}(i+1)) \text{ for } i \geq 0,$$
$$\Delta_{i,j} = \Delta_{i,0} \oplus j \cdot L_* \text{ for } j \in \{0,1,2,3\},$$

where $L_* = E_K(0^n)$, $L(0) = 2^2 \cdot L_*$, $L(\ell) = 2 \cdot L(\ell - 1)$ for $\ell > 0$ and $\mathrm{ntz}(i)$ denotes the number of trailing zeros in the binary representation of the integer $i$, e.g. $\mathrm{ntz}(2) = 1$.

---

[3] A different AXU is used in the latest version of OCB [21], we opted for $E_K(\cdot)$ for the sake of simplicity.

**Lemma 3.** ([33]) *Let* $E : \mathcal{K} \times \{0,1\}^n \rightarrow \{0,1\}^n$ *be a blockcipher and* $\mathcal{T} = \mathcal{N} \times \mathbb{N}_0 \times \{0,1,2,3\} \cup \mathbb{N}_0 \times \{0,1,2,3\}$. *Let* $\mathscr{A}$ *be an adversary that runs in time at most* $t$, *asks at most* $q$ *queries, never asks queries with* $i$-*component exceeding* $2^{n-5}$ *and never asks decryption queries with tweaks from* $\mathbb{N}_0 \times \{0,1,2,3\}$. *Then*

$$\mathbf{Adv}^{\pm \widetilde{\mathrm{prp}}^{\mathcal{T}}}_{\mathrm{XEX}[E]}(\mathscr{A}) \leq \mathbf{Adv}^{\pm \mathrm{prp}}_{E}(\mathscr{B}) + \frac{9.5q^2}{2^n}$$

*for an adversary* $\mathscr{B}$ *that makes at most* $2q$ *queries and runs in time bounded by* $t + O(q)$.

EXTENDING THE TWEAKS WITH $\tau$. In order to instantiate $\Theta$CBv, we need to extend the tweaks of $\widetilde{E}$ with a fourth component: $\tau$. To this end, we propose XEX$'$, which is obtained by a slight modification of the XEX construction. Informally, we expand the domain of the "$j$-part" of tweaks and represent it as $\mathcal{I}_T \times \{0,1,2,3\}$, compensating for this by decreasing the maximal value of $i$.

The tweakable blockcipher $\widetilde{E}' = \mathrm{XEX}'[E]$ is defined as follows. We again use the AXU $H(K,N)$. We uniquely label each element of $\mathcal{I}_T$ by an integer with a bijection $\lambda : \mathcal{I}_T \rightarrow \{0,1,\dots,|\mathcal{I}_T|-1\}$. We define $m = \lceil \log_2 |\mathcal{I}_T| \rceil$, $L_* = E_K(0^n)$, $L_\tau = \lambda(\tau) \cdot 2^2 \cdot L_*$ for $\tau \in \mathcal{I}_T$, $L(0) = 2^{2+m} \cdot L_*$, and $L(\ell) = 2 \cdot L(\ell-1)$ for $\ell > 0$. The masking $\Delta$-values are computed as follows:

$$\Delta_{N,0,0,0} = H(K,N),$$
$$\Delta_{N,\tau,0,0} = \Delta_{N,0,0,0} \oplus L_\tau,$$
$$\Delta_{N,\tau,i+1,0} = \Delta_{N,\tau,i,0} \oplus L(\mathrm{ntz}(i+1)) \text{ for } i \geq 0,$$
$$\Delta_{N,\tau,i,j} = \Delta_{N,\tau,i,0} \oplus j \cdot L_* \text{ for } j \in \{0,1,2,3\},$$
$$\Delta_{\tau,0,0} = L_\tau,$$
$$\Delta_{\tau,i+1,0} = \Delta_{\tau,i,0} \oplus L(\mathrm{ntz}(i+1)) \text{ for } i \geq 0,$$
$$\Delta_{\tau,i,j} = \Delta_{\tau,i,0} \oplus j \cdot L_* \text{ for } j \in \{0,1,2,3\}.$$

A call to $\widetilde{E}'$ is evaluated as follows:

$$\widetilde{E}'^{N,\tau,i,j}_K(X) = E_K(X \oplus \Delta_{N,\tau,i,j}) \oplus \Delta_{N,\tau,i,j}, \text{ or } \widetilde{E}'^{\tau,i,j}_K(X) = E_K(X \oplus \Delta_{\tau,i,j}).$$

The security result for XEX$'$ construction is stated in Lemma 4.

**Lemma 4.** *Let* $E : \mathcal{K} \times \{0,1\}^n \rightarrow \{0,1\}^n$ *be a blockcipher and* $\mathcal{T} = \mathcal{N} \times \mathcal{I}_T \times \mathbb{N}_0 \times \{0,1,2,3\} \cup \mathcal{I}_T \times \mathbb{N}_0 \times \{0,1,2,3\}$ *for some finite, non-empty* $\mathcal{I}_T \subseteq \mathbb{N}_0$. *Let* $\mathscr{A}$ *be an adversary that runs in time at most* $t$, *asks at most* $q$ *queries, never asks queries with* $i$-*component exceeding* $2^{n-(5+\lceil \log_2 |\mathcal{I}_T| \rceil)}$ *and never asks decryption queries with tweaks from* $\mathcal{I}_T \times \mathbb{N}_0 \times \{0,1,2,3\}$. *Then*

$$\mathbf{Adv}^{\pm \widetilde{\mathrm{prp}}^{\mathcal{T}}}_{\mathrm{XEX}'[E]}(\mathscr{A}) \leq \mathbf{Adv}^{\pm \mathrm{prp}}_{E}(\mathscr{B}) + \frac{9.5q^2}{2^n}$$

*for an adversary* $\mathscr{B}$ *that makes at most* $2q$ *queries and runs in time bounded by* $t + O(q)$.

The treatment of $\tau$-tweak component in XEX′ construction is equivalent to a one where we would injectively encode $\tau, j$ into a single integer $j' = 2^2\tau + j$. Similar approach has been taken by Reyhanitabar et al. [29, 30], where it is shown that the essential properties of the masking values necessary for the security proof of [33] are preserved. The same arguments apply here, so we omit the proof of Lemma 4.

OCBv: PRACTICAL AE WITH VARIABLE STRETCH We define the blockcipher mode OCBv, a nonce based AE scheme with variable stretch. OCBv is only parameterized by a blockcipher $E$. It is obtained by instantiating the tweakable blockcipher in ΘCBv by the XEX′ costruction, i.e. OCBv[$E$] = ΘCBv[XEX′[$E$]] and its security is analysed in Theorem 4.

**Theorem 4.** *Let* $\widetilde{E} : \mathcal{K} \times \{0,1\}^n \to \{0,1\}^n$ *be a blockcipher. We have that*

$$\mathbf{Adv}_{\mathrm{OCBv}[E]}^{\mathrm{nvae}(\tau_c)}(t, \mathbf{q_e}, \mathbf{q_d}, \sigma) \leq \mathbf{Adv}_E^{\pm\mathrm{prp}}(t', 2q) + \sum_{\tau \in \mathcal{I}_T} \mathbf{Adv}_E^{\pm\mathrm{prp}}(t', 2q^\tau)$$

$$+ \mathbf{Adv}_E^{\pm\mathrm{prp}}(t', 2q^{\tau_c}) + \frac{28.5q^2}{2^n} + q_d^{\tau_c} \frac{2^{n-\tau_c}}{2^n - 1},$$

*where* $q^\tau = \lceil \sigma^\tau/n \rceil + 2 \cdot (q_e^\tau + q_d^\tau)$ *for* $\tau \in \mathcal{I}_T$, *and* $q = \sum_{\tau \in \mathcal{I}_T} q^\tau$ *and* $t' = t + O(\sigma)$ *with* $\sigma = \sum_{\tau \in \mathcal{I}_T} \sigma^\tau$.

If we further assume that the $\mathbf{Adv}_E^{\pm\mathrm{prp}}$ is non-decreasing w.r.t. both $q$ and $t$, then we can further simplify the bound to the form

$$\mathbf{Adv}_{\mathrm{OCBv}[E]}^{\mathrm{nvae}(\tau_c)}(t, \mathbf{q_e}, \mathbf{q_d}, \sigma) \leq (|\mathcal{I}_T| + 2) \cdot \mathbf{Adv}_E^{\pm\mathrm{prp}}(t', 2q) + \frac{28.5q^2}{2^n} + q_d^{\tau_c} \cdot \frac{2^{n-\tau_c}}{2^n - 1}.$$

*Proof.* The result in Theorem 4 follows from Theorem 3 and Lemma 4 by applying triangle inequality on the terms that arise from applying Lemma 4.    □

PERFORMANCE OF OCBv. The performance of OCBv can be expected to be very similar to that of OCB, as the two schemes only differ in the way the masking $\Delta$-values are computed. In addition to the operations necessary to compute $\Delta$-offsets in OCB, the computation of the $L_\tau$-values has to be done for OCBv. However, these can be precomputed at the initialization phase and stored, so the cost of their computation will be amortized over all queries. The only additional processing that remains after dealing with $L_\tau$-s is a single xor of a precomputed $L_\tau$ to a $\Delta$-value, necessary in every query. This is unlikely to impact the performance significantly.

# 7   Discussion

RELATION BETWEEN NVAE AND KESS+NAE. We define the **kess** property as useful, albeit strong property that facilitates modular security proofs of nvAE security for AE schemes whose nAE security has already been established. This is depicted as implication g in Fig. 1 and formally proven in Theorem 2. However, determining the exact nature of the relation in the reverse direction to implication g appears not to be straightforward, and we leave it as an open problem.

ACHIEVING NVAE SECURITY. In Sect. 6, we describe OCBv, a modified version of the OCB scheme for AEAD, that is provably secure in the sense of nvAE, and retains the desirable properties of OCB. Moreover, our transformation and analysis are generic enough to be applied to other schemes based on tweakable blockciphers, or other tweakable primitives (e.g. compression functions), which represents a large subset of current nAE schemes.

A natural problem to investigate would be to see if there exists a black-box transformation $\Gamma(\cdot)$, that would turn any nAE secure scheme $\Pi$ into an nvAE secure scheme $\Gamma(\Pi)$. A straightforward measure to take would be to derive a key $K'$ used internally with $\Pi$ from the key $K$ of $\Gamma(\Pi)$ as $K' = H(\tau, K)$ with a hash function $H$, as suggested by Struik [40]. This transformation can be easily proven secure, but only in random oracle model, and it makes the whole design unnecessarily complex. We leave the formal treatment of this question (in the standard model) as an open problem.

It is nevertheless possible to describe transformations that are applicable to large subsets of nAE secure schemes. One example is given in Sect. 6. Another such transformation is encoding $\tau$ in the nonce input of sponge-like modes. These either process all inputs in a single chain of permutation calls (e.g. Ketje [7], and Ascon [11]), or they use several such chains in parallel, but initialize all of them with nonce-dependent values (e.g. Keyak [8], and NORX [2]).

**Acknowledgments.** This work was partly supported by the EU H2020 TREDISEC project, funded by the European Commission under grant agreement no. 644412. Damian Vizár is supported in part by Microsoft Research under MRL Contract No. 2014-006 (DP1061305). We would like to thank the ASIACRYPT reviewers for their constructive comments. We would also like to thank Phillip Rogaway for an insightful discussion during CRYPTO 2015.

# References

1. Andreeva, E., Bogdanov, A., Luykx, A., Mennink, B., Mouha, N., Yasuda, K.: How to securely release unverified plaintext in authenticated encryption. In: Sarkar, P., Iwata, T. (eds.) ASIACRYPT 2014. LNCS, vol. 8873, pp. 105–125. Springer, Heidelberg (2014). doi:10.1007/978-3-662-45611-8_6
2. Aumasson, J.P., Jovanovic, P., Neves, S.: Norx. https://competitions.cr.yp.to/round2/norxv20.pdf
3. Bellare, M., Namprempre, C.: Authenticated encryption: relations among notions and analysis of the generic composition paradigm. In: Okamoto, T. (ed.) ASIACRYPT 2000. LNCS, vol. 1976, pp. 531–545. Springer, Heidelberg (2000). doi:10.1007/3-540-44448-3_41

4. Bellare, M., Rogaway, P.: Encode-then-encipher encryption: how to exploit nonces or redundancy in plaintexts for efficient cryptography. In: Okamoto, T. (ed.) ASI-ACRYPT 2000. LNCS, vol. 1976, pp. 317–330. Springer, Heidelberg (2000). doi:10. 1007/3-540-44448-3_24
5. Bernstein, D.J.: Cryptographic competitions: CAESAR. http://competitions.cr. yp.to
6. Bernstein, D.J.: Cryptographic competitions: Disasters. https://competitions.cr. yp.to/disasters.html
7. Bertoni, G., Daemen, J., Peeters, M., Assche, G.V., Keer, R.V.: Ketje. https:// competitions.cr.yp.to/round1/ketjev11.pdf
8. Bertoni, G., Daemen, J., Peeters, M., Assche, G.V., Keer, R.V.: Keyak. https:// competitions.cr.yp.to/round2/keyakv2.pdf
9. Borisov, N., Goldberg, I., Wagner, D.: Intercepting mobile communications: the insecurity of 802.11. In: MOBICOM, pp. 180–189 (2001)
10. De Meulenaer, G., Gosset, F., Standaert, F.X., Pereira, O.: On the energy cost of communication and cryptography in wireless sensor networks. In: 2008 IEEE International Conference on Wireless and Mobile Computing, Networking and Communications, pp. 580–585. IEEE (2008)
11. Dobraunig, C., Eichlseder, M., Mendel, F., Schlaffer, M.: Ascon. https:// competitions.cr.yp.to/round2/asconv11.pdf
12. Egele, M., Brumley, D., Fratantonio, Y., Kruegel, C.: An empirical study of cryptographic misuse in android applications. In: 2013 ACM SIGSAC Conference on Computer and Communications Security, CCS 2013, Berlin, Germany, 4–8 November 2013, pp. 73–84. ACM (2013)
13. Eichlseder, M.: Remark on variable tag lengths and OMD. crypto-competitions mailing list, 25 April 2014
14. Fleischmann, E., Forler, C., Lucks, S.: McOE: a family of almost foolproof on-line authenticated encryption schemes. In: Canteaut, A. (ed.) FSE 2012. LNCS, vol. 7549, pp. 196–215. Springer, Heidelberg (2012). doi:10.1007/978-3-642-34047-5_12
15. Fuhr, T., Leurent, G., Suder, V.: Collision attacks against CAESAR candidates. In: Iwata, T., Cheon, J.H. (eds.) ASIACRYPT 2015. LNCS, vol. 9453, pp. 510–532. Springer, Heidelberg (2015). doi:10.1007/978-3-662-48800-3_21
16. Hoang, V.T., Krovetz, T., Rogaway, P.: Robust authenticated-encryption AEZ and the problem that it solves. In: Oswald, E., Fischlin, M. (eds.) EUROCRYPT 2015. LNCS, vol. 9056, pp. 15–44. Springer, Heidelberg (2015). doi:10.1007/ 978-3-662-46800-5_2
17. Hoang, V.T., Reyhanitabar, R., Rogaway, P., Vizár, D.: Online authenticated-encryption and its nonce-reuse misuse-resistance. In: Gennaro, R., Robshaw, M. (eds.) CRYPTO 2015. LNCS, vol. 9215, pp. 493–517. Springer, Heidelberg (2015). doi:10.1007/978-3-662-47989-6_24
18. Hotz, G.: Console hacking 2010-ps3 epic fail. In: 27th Chaos Communications Congress (2010)
19. Iwata, T.: CLOC and SILC will be tweaked. crypto-competitions mailing list, 4 August 2015
20. Katz, J., Yung, M.: Unforgeable encryption and chosen ciphertext secure modes of operation. In: Goos, G., Hartmanis, J., Leeuwen, J., Schneier, B. (eds.) FSE 2000. LNCS, vol. 1978, pp. 284–299. Springer, Heidelberg (2001). doi:10.1007/ 3-540-44706-7_20
21. Krovetz, T., Rogaway, P.: The software performance of authenticated-encryption modes. In: Joux, A. (ed.) FSE 2011. LNCS, vol. 6733, pp. 306–327. Springer, Heidelberg (2011). doi:10.1007/978-3-642-21702-9_18

22. Langley, A.: Apple's SSL/TLS bug. Imperial Violet (2014)
23. Li, Y., Zhang, Y., Li, J., Gu, D.: iCryptoTracer: dynamic analysis on misuse of cryptography functions in iOS applications. In: Au, M.H., Carminati, B., Kuo, C.-C.J. (eds.) NSS 2014. LNCS, vol. 8792, pp. 349–362. Springer, Heidelberg (2014). doi:10.1007/978-3-319-11698-3_27
24. Manger, J.H.: [Cfrg] Attacker changing tag length in OCB. IRTFCFRGmailing list, 29 May 2013
25. Minematsu, K.: AES-OTR v2. crypto-competitions mailing list, 31 August 2015
26. Namprempre, C., Rogaway, P., Shrimpton, T.: Reconsidering generic composition. In: Nguyen, P.Q., Oswald, E. (eds.) EUROCRYPT 2014. LNCS, vol. 8441, pp. 257–274. Springer, Heidelberg (2014). doi:10.1007/978-3-642-55220-5_15
27. Nandi, M.: RE: CLOC and SILC will be tweaked. crypto-competitions mailing list, 5 August 2015
28. Reyhanitabar, R.: OMD version 2: a tweak for the 2nd round. crypto-competitions mailing list, 27 August 2015
29. Reyhanitabar, R., Vaudenay, S., Vizár, D.: Misuse-resistant variants of the OMD authenticated encryption mode. In: Chow, S.S.M., Liu, J.K., Hui, L.C.K., Yiu, S.M. (eds.) ProvSec 2014. LNCS, vol. 8782, pp. 55–70. Springer, Heidelberg (2014). doi:10.1007/978-3-319-12475-9_5
30. Reyhanitabar, R., Vaudenay, S., Vizár, D.: Boosting OMD for almost free authentication of associated data. In: Leander, G. (ed.) FSE 2015. LNCS, vol. 9054, pp. 411–427. Springer, Heidelberg (2015). doi:10.1007/978-3-662-48116-5_20
31. Rogaway, P.: Re: [Cfrg] Attacker changing tag length in OCB. IRTFCFRG mailing list, 3 June 2013
32. Rogaway, P.: Authenticated-encryption with associated-data. In: ACM CCS 2002, pp. 98 107 (2002)
33. Rogaway, P.: Efficient instantiations of tweakable blockciphers and refinements to modes OCB and PMAC. In: Lee, P.J. (ed.) ASIACRYPT 2004. LNCS, vol. 3329, pp. 16–31. Springer, Heidelberg (2004). doi:10.1007/978-3-540-30539-2_2
34. Rogaway, P.: Nonce-based symmetric encryption. In: Roy, B., Meier, W. (eds.) FSE 2004. LNCS, vol. 3017, pp. 348–358. Springer, Heidelberg (2004). doi:10.1007/978-3-540-25937-4_22
35. Rogaway, P., Bellare, M., Black, J., Krovetz, T.: OCB: a block-cipher mode of operation for efficient authenticated encryption. In: ACM CCS 2001, pp. 196–205 (2001)
36. Rogaway, P., Shrimpton, T.: A provable-security treatment of the key-wrap problem. In: Vaudenay, S. (ed.) EUROCRYPT 2006. LNCS, vol. 4004, pp. 373–390. Springer, Heidelberg (2006). doi:10.1007/11761679_23
37. Rogaway, P., Shrimpton, T.: Deterministic authenticated-encryption: a provable-security treatment of the key-wrap problem. In: IACR Cryptology ePrint Archive 2006, p. 221 (2006)
38. Rogaway, P., Wagner, D.: A critique of CCM. In: IACR Cryptology ePrint Archive 2003, p. 70 (2003)
39. Struik, R.: AEAD ciphers for highly constrained networks. In: DIAC 2013 presentation, 13 August 2013
40. Struik, R.: Re: [Cfrg] Attacker changing tag length in OCB. IRTFCFRG mailing list, 30 May 2013
41. Wu, H.: The misuse of rc4 in microsoft word and excel. Cryptology ePrint Archive, Report 2005/007 (2005). http://eprint.iacr.org/2005/007

# Block Cipher I

# Salvaging Weak Security Bounds for Blockcipher-Based Constructions

Thomas Shrimpton[1,2](✉) and R. Seth Terashima[1,2]

[1] Department of Computer and Information Science and Engineering,
University of Florida, Gainesville, USA
teshrim@ufl.edu, setht@qti.qualcomm.com
[2] Qualcomm Technologies, Inc., San Diego, USA

**Abstract.** The concrete security bounds for some blockcipher-based constructions sometimes become worrisome or even vacuous; for example, when a light-weight blockcipher is used, when large amounts of data are processed, or when a large number of connections need to be kept secure. Rotating keys helps, but introduces a "hybrid factor" $m$ equal to the number of keys used. In such instances, analysis in the ideal-cipher model (ICM) can give a sharper picture of security, but this heuristic is called into question when cryptanalysis of the real-world blockcipher reveals weak keys, related-key attacks, etc.

To address both concerns, we introduce a new analysis model, the ideal-cipher model under key-oblivious access (ICM-KOA). Like the ICM, the ICM-KOA can give sharp security bounds when standard-model bounds do not. Unlike the ICM, results in the ICM-KOA are less brittle to current and future cryptanalytic results on the blockcipher used to instantiate the ideal cipher. Also, results in the ICM-KOA immediately imply results in the ICM *and* the standard model, giving multiple viewpoints on a construction with a single effort. The ICM-KOA provides a conceptual bridge between ideal ciphers and tweakable blockciphers (TBC): blockcipher-based constructions secure in the ICM-KOA have TBC-based analogs that are secure under standard-model TBC security assumptions. Finally, the ICM-KOA provides a natural framework for analyzing blockcipher key-update strategies that use the blockcipher to derive the new key. This is done, for example, in the NIST CTR-DRBG and in the hardware RNG that ships on Intel chips.

## 1 Introduction

When a secret-key cryptographic primitive $\mathcal{E}$ is based upon a blockcipher $E$, a security proof for $\mathcal{E}$ will typically appeal to the pseudorandom-permutation (PRP) assumption—namely, that no efficient adversary can distinguish between the input-output behavior of the secretly (and randomly) keyed blockcipher $E_K$, and that of a truly random permutation $\pi$ with the same domain. When the proof states that the PRP-security of $E$ is a tight upperbound for the security of $\mathcal{E}$, one can derive from it useful messages for practice; e.g., how many calls to the blockcipher should be allowed before changing its key. When the upperbound

© International Association for Cryptologic Research 2016
J.H. Cheon and T. Takagi (Eds.): ASIACRYPT 2016, Part I, LNCS 10031, pp. 429–454, 2016.
DOI: 10.1007/978-3-662-53887-6_16

is not tight, the usefulness of any such messages can be unclear. In particular, when there is no known attack on the security of $\mathcal{E}$ whose success probability approaches the upperbound evidenced in the security proof. Such gaps are common when the security proof uses a "hybrid argument".

As an example, consider the following self-rekeying version of counter-mode encryption. (This is similar to the NIST CTR-DRBG [9] that underlies Intel's hardware RNG [11,19].) Let $\mathsf{CTR}[E]_K^N(\cdot)$ denote counter-mode encryption (over $n$-bit blockcipher $E$) under key $K$ and IV $N$. The scheme is initialized with a key $K_1$ that is random. To encrypt the $i$-th plaintext $X_i$, the scheme computes ciphertext $C_i \leftarrow \mathsf{CTR}[E]_{K_i}^0(X_i)$ using key $K_i$, and then computes a key $K_{i+1}$ for the next encryption call via $K_{i+1} \leftarrow \mathsf{CTR}[E]_{K_i}^{\lceil |X_i|/n \rceil + 1}(0^k)$. The standard proof would show that the security of this construction is (roughly) upperbounded by $m$ times the probability violating the PRP-security of $E$, where $m$ is the number of strings $X_i$ that are encrypted before the key is reinitialized to a fresh random, secret value. Such a bound can quickly become vacuous when the underlying blockcipher is lightweight and cannot be assumed to provide PRP-security comparable to blockciphers like AES, or in settings where frequent re-initialization (i.e., resetting to a fresh, random $K_1$) is difficult.

If this construction is analyzed instead in the ideal cipher model (ICM), the upperbound is considerably tighter, and nearly matched by an attack. This suggests that the multiplicative factor of $m$ in the standard-model result isn't "real", but rather an artifact of the proof technique. On the other hand ICM analysis provides only a security heuristic, and seems particularly inappropriate when the underlying blockcipher is known to have obvious non-ideal behavior for certain "weak" keys, or to suffer from related-key attacks.

Yet for constructions like this one, the presence of weak blockcipher keys is unlikely to be a real issue for the security of the construction: intuitively, if the initial key $K$ is *random*, then so should be the derived keys that follow it. Analysis in the ICM naturally captures this intuition, as the key $K_i$ is (essentially) independent of keys $K_1, K_2, \ldots, K_{i-1}$, and of the ciphertexts $C_1, C_2, \ldots, C_i$ that the construction outputs.

Moreover, observe that the construction doesn't actually need to know the *value* of any of the keys. It could carry out its duties if its access to $E$ was via an API that restricted it to refer to keys by handles, e.g., ask $(i, x, \text{"return"})$ and receive $E_{K_i}(x)$ in return, or $(i, x, \text{"key"})$ and cause the value $K_{i+1} = E_{K_i}(x)$ to be stored, receiving nothing in return. We refer to such an API as enforcing *key-oblivious access* (KOA) to $E$, and under this access model it is clear that the construction leaks nothing about the keys beyond what the blockcipher does. Said another way, the access model supports the intuition that if the initial key $K_1$ is secret, it and its successors remain so.

*The ICM under key-oblivious access.* We formalize all of this in a new model, the ICM under key-oblivious access (ICM-KOA). The construction has black-box access to the blockcipher via, roughly, the API just described. On the other hand, the adversary may query the ideal cipher freely, as in the traditional

ICM, capturing a real-world attacker's ability to compute (offline) blockcipher input-output pairs under any key it likes. Before we give more details about our formalism, let us explain what benefits it provides.

First, the ICM-KOA retains the power of ICM to give sharper bounds than those found under the standard-model PRP assumption. It can also expose important quantitative security distinctions among variants of a given blockcipher-based construction, where these would be hidden by a standard-model analysis. This may help to guide implementation decisions in practice. We also surface in our model the distinction between precomputation queries to the blockcipher, offline queries made to the blockcipher while attacking the construction, and online queries made to the construction under its secret keys.

Second, security results in the ICM-KOA imply comparable security results in the traditional ICM *and* results in the standard-model. The latter is possible precisely because the model guarantees that the blockcipher is called on random and secret keys. Thus a single effort yields multiple viewpoints on a given construction.

Third, while security proofs in this model are still heuristics, their value is more resilient to the discovery of weak keys and related-key attacks on the real blockcipher that is idealized. In fact, the formalism provides a clear path to analyzing the security of constructions when the blockcipher is modeled with explicit non-ideal behaviors. We leave this as interesting future work.

Finally, the ICM-KOA provides a conceptual bridge between ideal ciphers and tweakable blockciphers (TBC). This is pleasing because, intuitively, the strong-tweakable-PRP assumption suggests that a secure, secretly keyed TBC is computationally indistinguishable from an ideal cipher—both provide a *set* of random permutations (one permutation for each tweak or key, respectively). We show that blockcipher-based constructions that are secure in the ICM-KOA have TBC-based analogs that are secure in the standard model.

*Decomposing constructions into modes and schedulers.* We want our model to facilitate results for blockcipher-based constructions that may use many keys. So the ICM-KOA requires that constructions can be decomposed into two primitives, a *mode* $\mathcal{M}$ and a potentially stateful key-*scheduler* $\mathcal{S}$. Intuitively, the role of the mode is to affect the transformation of construction-inputs (e.g., plaintexts) into construction-outputs (e.g., ciphertexts), and the role of the scheduler is to determine what keys the mode must use during its execution. Many symmetric-key cryptographic primitives can be decomposed in this way, including encryption schemes and blockcipher-based PRFs, PRNGs, KDFs and MACs, whether or not rekeying strategies are applied to them.

Returning to our self-rekeying version of counter-mode encryption, we might decompose this into a mode $\mathcal{M}$ that, on input a key $K_i$ and a string $X$, computes $C \leftarrow \mathsf{CTR}[E]_{K_i}^0(X)$; and a scheduler $\mathcal{S}$ that (effectively) computes $K_{i+1} \leftarrow \mathsf{CTR}[E]_{K_i}^{\lceil |X|/n \rceil + 1}(0^k)$. Each will be forced to be oblivious of the actual key values by our model.

*Applying the ICM-KOA to constructions.* Given a blockcipher-based construction that admits decomposition, we define what it means for the construction to produce outputs that are indistinguishable from some reference-behavior-oracle in the ICM-KOA. To be clear, we do not claim that this is, on its own, an intuitive security goal. It is a new tool that provides a means to obtain strong bounds in the ICM that are backed by a guarantee that keys are kept random and secret. And because of this guarantee, we gain simultaneous results in the standard model. We illuminate the usefulness of the ICM-KOA via two case studies.

First we consider the NIST-CTR-DRBG. As the name suggests, it is a deterministic random-bit generator based on running a blockcipher in CTR mode. A result by Shrimpton and Terashima [19] shows that the standard-model security is around $q^2/2^k$, where $q$ is the number of calls the construction. For $k = 128$, this bound exceeds $2^{-40}$ when $q = 2^{44}$. This may seem safe; after all, this amounts to many terabytes of random bits. But the RNG has extremely high throughput—Intel reports 800 MB/s, which equates to 50 million queries per second—meaning the $q = 2^{44}$ limit in a little more than four days.

We analyze this in the ICM-KOA. For very little work, we recover the security bound from [19], and also get a much stronger bound in the ICM. The latter reveals the lack of a matching attack and shows that, barring cryptanalysis of AES *under random and secret keys*, we can permit on the order of $2^{70}$ queries before surpassing our $2^{-40}$ limit (assuming the adversary has resources for $2^{80}$ precomputation and $2^{80}$ offline queries). This translates to 750,000 years of runtime, and so is unlikely to be the limiting factor.

Next we consider three rekeying variants of CTR-mode, distinguished by how they choose IVs following a key change: (1) The IV is set to $0^n$; (2) the upper bits of the IV are unique for each key; (3) The IV is chosen randomly. In each case, we use the same key scheduler that sets $K_i \leftarrow E_{K_1}(i)$ (for $i > 1$). In the standard model, these three schemes all have the same security bound. Our analysis in the ICM-KOA uncovers significant quantitative differences their security bounds; in particular, we show how (1) succumbs to precomputation for shorter key lengths while (2) and (3) resist such attacks.

*Addressing hybrid-loss directly in the standard model.* Another, arguably more natural approach to avoiding a factor of $m$ hybrid-loss when analyzing a blockcipher-based construction that uses $m$ keys is to generalize the PRP notion to an $m$-PRP notion [18]. Here the adversary must distinguish between the collection of oracles $E_{K_1}(\cdot), E_{K_2}(\cdot), \ldots, E_{K_m}(\cdot)$ for random keys $K_1, \ldots, K_m$, and the collection $\pi_1(\cdot), \pi_2(\cdot), \ldots, \pi_m(\cdot)$ of random permutations. If a construction uses no more than $m$ blockcipher keys during the time that it is being attacked, reducing the construction's security to the blockcipher's $m$-PRP security can be done without a hybrid proof, and therefore does not incur a factor of $m$ loss.

But this may simply sweep problems under the rug: (1) it begs the question of how the $m$-PRP security of a given blockcipher relates to its PRP security (although we note that Hoang and Tessaro [12], building on the work of [18], have largely answered this question for key-alternating ciphers with independent round keys) (2) it doesn't directly model interesting scenarios where the keys are themselves derived from the $E$ using prior keys, particularly when, as with the NIST-RNG, the mode of operation is intertwined with key generation.

We explore this further in the full version of the paper. As one expects, the simplest result states that the $m$-PRP security of $E$ falls somewhere between its PRP-security and $m$ times that value. We go on to show that, under the assumption that a PRP-secure blockcipher $E$ exists: (1) there is a related blockcipher for which these upper- and lowerbounds on its $m$-PRP security are tight; and (2) there is a related blockcipher that is PRP-secure but not $m$-PRP-secure, for sufficiently large values of $m$. (Of course, these distinctions are not binary, but the quantitative results are reasonable for modest $m$). These results are mainly of theoretical importance, as no real blockcipher will resemble the ones used to prove them.

But we also give a result that sheds some light on how much of a gap exists between any particular blockcipher's PRP security and $m$-PRP security. Given a PRP-adversary $A$ for blockcipher $E$, the *best* $m$-PRP adversary $B[A]$ (that makes use of $A$ in a black-box fashion) will have an advantage between $\mathbf{Adv}_E^{prp}(A)$ and $m\mathbf{Adv}_E^{prp}(A)$; moreover, its location on this continuum can be computed from $\mathbf{Adv}_E^{prp}(A)$ and, interestingly, $A$'s false-positive rate when distinguishing a keyed instance of $E$ from a random permutation. When $A$'s false-positive and false-negative rates are similar, then $B[A]$'s advantage scales with $\sqrt{m}$, rather than $m$. Again, see the full version of this paper for details.

*Related Work.* Abdalla and Bellare [1] were the first to rigorously study the security of rekeyed symmetric-encryption schemes, under various rekeying strategies. Concretely, they show that CBC-mode over an $n$-bit blockcipher, consistently rekeyed after $2^{n/3}$ blocks, can have meaningful security bounds up to about $2^{2n/3}$ total message blocks. (Specifically, they show that $2^{2n/3}$ one-block messages can be encrypted.) Our KOA modeling captures their rekeyed encryption schemes. As one example, they consider a rekeying strategy that computes $(K_{i+1}, L_{i+1}) = (E(L_i, 0), E(L_i, 1))$; we would say the scheduler $S$ computes this $(K_{i+1}, L_{i+1})$, where $L_i$ (resp. $L_{i+1}$) is the current (resp. next) scheduler state.

There are a number of works that analyaze secretly keyed constructions in the ICM. Kilian and Rogaway [14] proved that the DESX construction is a secure SPRP in the ICM. Dai et al. [10] leverage the ICM to prove the security of multiple encryption. Lee [17] uses the ICM to consider key-length extension offered by cascade encryption (aka multiple encryption) and xor-cascade encryption (of which DESX is a simple example). Recently there have been a line of nice papers on the security of key-alternating ciphers (aka xor-cascade encryption), including [2,7,8,15,16], that perform their analysis in the public-random-permutation model, which is derivative of the ICM. The randomized message-authentication code RMAC was analyzed in the ICM [13].

The classic "Luby-Rackoff Backwards" paper by Bellare, Krovetz and Rogaway [4] addresses the construction of beyond birthday-bound secure PRFs from PRPs, but they are unable to do so in the standard model because of hybrid terms. Thus, their positive security results, which do show beyond-birthday-bound security of their constructions, are developed in the ICM, despite the presence of secret keys. It would be interesting to revisit their construction using the ICM-KOA.

Bellare, Boldyreva and Micali [3] consider multi-key security notions for public-key encryption, and show that, for left-or-right IND-CPA, the hybrid loss incurred by reducing from a multi-key instance to a single-key instance is inherent. Our discussion of the relationship between the PRP and $m$-PRP notions takes inspiration from that work, especially the construction of a cipher for which the bound is tight.

Bellare, Ristenpart and Tessaro [5] consider multi-instance (or multi-key) security notions, in which the attacker wins only if it breaks all of the instances. Their notions differ from ours, as it would suffice to break a single instance in our $m$-PRP notion.

Recent papers by Mouha and Luykx [18] and Hoang and Tessaro [12] consider the mutli-key security of key-alternating ciphers, demonstrating (in the random permutation model) that they do not suffer hybrid-like security losses. This work complements are own, which provides bounds for modes of operation that employ blockciphers with idealized behavior under random, secret keys.

*Roadmap.* Section 2 introduces the ICM with key-oblivious access. The central theorems are summarized up-front —that constructions (with certain properties) that are secure in the ICM-KOA are secure in both the ICM and standard models— and the bulk of the section is concerned with technical matters that support the formal theorem statements. The section ends by using the ICM-KOA framework to relate ideal ciphers and tweakable ciphers. Section 3 applies the results of Sect. 2 to various blockcipher-based constructions, including the NIST CTR-DRBG. Full proofs of all results are provided. Results on the relationship between the PRP and $m$-PRP standard-model notions will appear in the full version.

# 2    The ICM with Key-Oblivious Access

In this section, we formalize the notion of decomposing a construction into a mode (which carries out the cryptographic functionality) and a scheduler (which creates keys for the mode, as needed). We then define properties of modes and schedulers sufficient to imply results in both the standard model and the ICM. Roughly speaking:

- A mode and a scheduler constitute a *decomposition* of a construction if they preserve its black-box behavior.
- A mode is *compatible* with a scheduler if they query the underlying blockcipher on different points (and thus maintain an independence between keys and, e.g., ciphertexts).

- A decomposition has *dispersed inputs* if there are limits to how many blockcipher inputs an adversary can predict in advance.
- We quantify the computational resources consumed by the mode and scheduler using *mode efficiency*.

The first item and last items are straightforward, and the need for the second (in proofs) is intuitive after a moment's thought. Having dispersed inputs will help to make clear the impact of precomputation on security bounds. The coarser granularity of the standard model prevents it from benefiting from dispersed inputs, and we will demonstrate how this obscures the impact of precomputation.

The central theorems of this section, Theorems 1 and 2, have somewhat complicated statements. But, informally, they say the following:

**Theorems 1 and 2, informally.** If a decomposition (1) has these properties and (2) is difficult to distinguish from an appropriate reference oracle (e.g., an encryption oracle that returns random bits) when the underlying blockcipher is replaced by a random function that is inaccessible to the adversary, then the original construction is likewise hard to distinguish from the reference oracle *in both the standard model and in the ICM.*

We note that the "if" portion specifies indistinguishability when the blockcipher is treated as a random function that is inaccessible to the adversary. This isn't sweeping things under the rug: ICM-based proofs typically have to "decouple" the *actual* blockcipher used by the construction from the blockcipher available to the adversary using ad-hoc methods. Our informal theorem statement is merely surfacing this proof trick, and our model will allow us to enforce it cleanly.

The final significant contribution of this section is a result that uses the ICM-KOA framework to formalize a relationship between the ICM and TBCs.

## 2.1 Preliminaries

When $X, Y$ are strings, $X \parallel Y$ is the concatenation of those strings, and $X \oplus Y$ is their bitwise exclusive-or. When $\mathcal{X}$ is a set, $X \xleftarrow{\$} \mathcal{X}$ means to sample uniformly from $\mathcal{X}$ and assign the result to $X$. When $A$ is a randomized algorithm, then $X \xleftarrow{\$} A^{\mathcal{O}_1, \mathcal{O}_2, \ldots}(\sigma)$ means to provide $A$ with oracle (black-box) access to $\mathcal{O}_1, \mathcal{O}_2, \ldots$ and input $\sigma$, and to assign the result of its execution to $X$. An *adversary* is a randomized algorithm. The notation $A^{\mathcal{O}_1, \mathcal{O}_2, \ldots} \Rightarrow b$ refers to the event that an algorithm $A$, when provided the indicated oracles (if any), ends its execution with output $b$.

Fix integers $k, n > 0$. A function family $E \colon \{0,1\}^k \times \{0,1\}^n \to \{0,1\}^n$ is a blockcipher if, for all $K \in \{0,1\}^k$, the mapping $E_K(\cdot) = E(K, \cdot)$ is a permutation over $\{0,1\}^n$. We write $E_K^{-1}(\cdot)$ for the inverse of $E_K(\cdot)$. The set $\mathrm{Perm}\,(n)$ is the set of all permutations $\pi \colon \{0,1\}^n \to \{0,1\}^n$, and the set $\mathrm{BC}(k,n)$ is the set of all blockciphers $E \colon \{0,1\}^k \times \{0,1\}^n \to \{0,1\}^n$.

If $G$ is some game (in the sense of the game-playing framework of Bellare and Rogaway [6], where an adversary interacts with oracles) and $\mathcal{E}$ is some event, the notation $\Pr[\,G; \mathcal{C}\,]$ denotes the probability that the condition $\mathcal{C}$ will hold after $G$ terminates.

---

**Oracle** $\mathcal{M}[\mathcal{S}, E]_{(\mathsf{KM},\mathsf{KS})}(M)$:

$K_1 \xleftarrow{\$} \{0,1\}^k$
**return** $\mathcal{M}_{\mathsf{KM}}^{\mathsf{query},\mathsf{register}}(M)$

**Procedure** query$(i, X)$:

**if** $i > \mathsf{c}$ **then return** $\bot$
**return** $E(K_i, X)$

**Procedure** register():

$\mathsf{c} \leftarrow \mathsf{c} + 1$
$(i, X) \leftarrow \mathcal{S}_{\mathsf{KS}}^{\mathsf{query}}()$
$K_{\mathsf{c}} \leftarrow E(K_i, X)$

---

**Fig. 1.** A key-access manager exposes the query and register interfaces shown here. The oracle $\mathcal{M}[\mathcal{S}, E]_{(\mathsf{KM},\mathsf{KS})}$, to which attackers will have oracle access in security experiments, uses these interfaces and a to implement the mode $\mathcal{M}$ of a given decomposition $\hat{\mathcal{E}} = (\mathcal{M}, \mathcal{S}, \mathcal{K})$. Here, $\mathsf{c}$ is initially 1.

## 2.2 Decompositions and Their Associated Notions

Let $\mathcal{E} : \mathcal{K}_\mathcal{E} \times \mathcal{D} \to \mathcal{R}$ be some scheme (e.g., CTR mode) that makes black-box use of a blockcipher $E : \{0,1\}^k \times \{0,1\}^n \to \{0,1\}^n$. We write $\mathcal{E}_K^E$ for the construction being keyed by $K \in \mathcal{K}_\mathcal{E}$, with $E$ as a superscript to emphasize black-box access.

Our goal is to break $\mathcal{E}$ into a mode of operation and a key scheduler. A *decomposition* is a tuple $\hat{\mathcal{E}} = (\mathcal{M}, \mathcal{S}, \mathcal{K})$ of algorithms: a *mode* $\mathcal{M} : \mathcal{K}_\mathcal{M} \times \mathcal{D} \to \mathcal{R}$, a stateful but deterministic *scheduler* $\mathcal{S} : \mathcal{K}_\mathcal{S} \to \mathbb{N} \times \{0,1\}^n$, and a key-generation algorithm $\mathcal{K}$ that outputs values in $\mathcal{K}_\mathcal{M} \times \mathcal{K}_\mathcal{S}$. The mode $\mathcal{M}$ expects two oracles having the signatures of query and register, which are exposed as part of a *key-access manager* in Fig. 1. (Look ahead to World 1 of Fig. 3 for an illustration). The scheduler $\mathcal{S}$ expects oracle access to query, and is invoked by register.

A natural first attempt at defining key-oblivious access to an ideal cipher $E$ would be to choose set of keys $K_1, K_2, \ldots, K_m$ up front, and then give the mode $\mathcal{M}$ (e.g., CTR mode) being analyzed black-box access to some oracle $\mathcal{O}(i, X) := E(K_i, X)$ for $i \in [1..m]$. There would be no explicit scheduler, and the keys themselves would be independent of the blockcipher $E$. But we want to capture schemes that do use $E$ to derive the keys. For example, the Intel RNG [11] and the Abdalla and Bellare [1] constructions mentioned in the introduction. Hence we surface a key scheduler $\mathcal{S}$ as an explicit component of the decomposition, and must provide it with some kind of access to $E$. We cannot provide $\mathcal{S}$ *unfettered* access to $E$, however. If we did, then we would not be able to argue that $E$ is queried only under random (and secret) keys. Concretely, suppose $\mathcal{S}$ sets $K_i = E(C, E(C, K \oplus i))$, where $C$ is some constant and $K$ is some "master key"; this may be secure in the ICM, but if we instantiate $E$ with DES and $C$ is a one of the weak keys for DES, then we would have $K_i = K \oplus i$. The keys used by the mode of operation would be closely related, a scenario we wish to

**Table 1.** Symbols used in ICM-KOA security definitions.

| Symbol | Upperbound for number of... |
|--------|------------------------------|
| $q$ | Adversary queries |
| $m$ | Blockcipher keys used |
| $\sigma$ | $n$-bit blocks per adversary query |
| $\mu$ | Key aliases used to encipher any given block |
| $\nu$ | Blocks enciphered using any given key alias |

preclude. Thus we restrict the scheduler's access to $E$. Similar abuse from $\mathcal{M}$ must also be prevented.

The oracles in our key-access manager force both $\mathcal{S}$ and $\mathcal{M}$ to query the blockcipher via handles, values that are independent of the particular values of the keys. Moreover, when preparing to have a value assigned to the $m$th key $K_m$, the scheduler $\mathcal{S}$ can only request outputs of $E$ under keys $K_1$ through $K_{(m-1)}$. Note that $\mathcal{S}$ is not allowed to "know" the resulting value of $K_m$: instead, $\mathcal{S}$ outputs a pair $(i, X)$ and $K_m$ is assigned $E(K_i, X)$. We also force $\mathcal{M}$ to query $E$ using handles for keys.

We note that the syntax for both the mode $\mathcal{M}$ and the scheduler $\mathcal{S}$ provides them with what appear to be "master" keys KM and KS. This is to capture initial values (keys, IVs, etc.) provided to the blockcipher-based construction. We will not assume or demand that KM and KS are independent of each other, but allowing them to be distinct permits us to capture more general constructions.

**Definition 1 (Decompositions of schemes).** *Let $\mathcal{E} : \mathcal{K}_{\mathcal{E}} \times \mathcal{D} \to \mathcal{R}$ and $\hat{\mathcal{E}} = (\mathcal{M}, \mathcal{S}, \mathcal{K})$ be defined as above. For $K \in \mathcal{K}_{\mathcal{M}} \times \mathcal{K}_{\mathcal{S}}$, let $\mathcal{M}[\mathcal{S}, E]_K : \mathcal{D} \to \mathcal{R}$ be the procedure defined in Fig. 1; this procedure combines the mode of operation $\mathcal{M}$ with the key scheduler $\mathcal{S}$ and blockcipher $E$ in the natural way. We say $\hat{\mathcal{E}}$ is a faithful decomposition of $\mathcal{E}$ if, for any adversary $A$ and any $E \in \mathrm{BC}(k, n)$,*

$$k = n, \ \Pr\left[ A^{\mathcal{E}^E_{K'}, E, E^{-1}} \Rightarrow 1 \right] = \Pr\left[ A^{\mathcal{M}[\mathcal{S}, E]_K, E, E^{-1}} \Rightarrow 1 \right]. \ \textit{The probabilities are}$$

*over the choice of $K' \xleftarrow{\$} \mathcal{K}_{\mathcal{E}}$, $K \xleftarrow{\$} \mathcal{K}$ and the coins of $A$, $\mathcal{M}$, and $\mathcal{E}$.*

That is, the black-box behavior of $\mathcal{E}^E_{K'}$ must be identical to the black-box behavior of $\mathcal{M}[\mathcal{S}, E]_K$ (given the above distribution of keys) *for any blockcipher $E$* and *computationally unbounded* adversaries.

Note that by using blockcipher outputs as keys, this definition assumes for the sake of simplicity that the key size $k$ is equal to the blocksize $n$ (each key is the output of the blockcipher at some point). We note that our model could easily be extended to the case where $k \neq n$ by truncating or concatenating the keys produced, as required, at the expense of complicating notation. However, we will use both $k$ and $n$ in our definitions and security bounds in order to suggest how taking $k \neq n$ would impact our model and results.

*Compatible modes.* Our key-access manager formalism does not itself prevent a scheduler $\mathcal{S}$ from "cheating" by choosing non-random keys. For example, $\mathcal{S}$

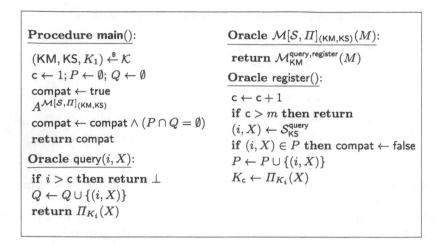

**Fig. 2.** Procedures and oracles for **Experiment** $\mathsf{COMPAT}^{\Pi}_{\hat{\mathcal{E}}}(A)$, where $\hat{\mathcal{E}} = (\mathsf{M}, \mathsf{S}, \mathcal{K})$. A mode $\mathcal{M}$ is $m$-compatible with a scheduler $\mathcal{S}$ if neither one queries the blockcipher on a point used to generate one of the first $m$ keys.

could use its query oracle to search for a point $(i, X)$ such that $E(K_i, X)$ ends in a zero, then output that point.

Informally, a scheduler $\mathcal{S}$ is *compatible* with a mode $\mathcal{M}$ if no adversary can cause either $\mathcal{S}$ or $\mathcal{M}$ to invoke query at a point $(i, X)$ used to generate a key $K_j = E(K_i, X)$. This ensures that both the $\mathcal{S}$ and $\mathcal{M}$ are oblivious to the actual values of each key.

We'll show that as long as each key alias $i$ is used significantly fewer than $2^{n/2}$ times, it follows that in both the ICM and the standard model there will be enough (computational) randomness in $E(K_i, X)$ for use as a cryptographic key. (This restriction results from the birthday paradox: since $E$ is being used to generate keys, we need it to behave like a random function, rather than random permutation.)

**Definition 2 (Compatible modes).** *Let $\hat{\mathcal{E}} = (\mathcal{M}, \mathcal{S}, \mathcal{K})$ be a decomposition over an $(k, n)$-bit blockcipher, $k = n$, and set $K \xleftarrow{\$} \mathcal{K}$. Let $m$ be a positive integer. Then $\mathcal{S}$ is $m$-compatible with $\mathcal{M}$ (with respect to $\mathcal{K}$) if for any keyed function $\Pi :$ $\{0, 1\}^k \times \{0, 1\}^n \to \{0, 1\}^n$, and any adversary $A$, $\Pr\left[\, \mathsf{COMPAT}^{\Pi}_{\hat{\mathcal{E}}}(A) \Rightarrow \mathsf{true} \right] = 1$, where Experiment $\mathsf{COMPAT}$ is defined in Fig. 2.*

Note that $\Pi$ need not be a blockcipher. This generality is required to make some of our later reductions work, and does not appear to exclude interesting modes.

Some other, arguably more natural definitions fail to capture our goal of preventing cheating schedulers. For example, suppose we instead query $\mathcal{S}_{\mathsf{KS}}$ to obtain keys $(K_1, K_2, \ldots, K_m)$ and require that no adversary with access to $E$ and $E^{-1}$ be able to distinguish these keys from truly random values. This definition proves *too* strict, as it excludes schedulers that deterministically derive $K_{i+1}$ from $K_i$.

It may then be tempting to instead allow schedulers to output keys directly (rather than $(i, X)$ pairs), and task an adversary $A$ to distinguish $\mathcal{M}[\mathcal{S}, E]_{(\mathsf{KM},\mathsf{KS})}$ from $\mathcal{M}[\$, E]_{\mathsf{KM},\mathsf{KS}}$, where $\$$ is a special oracle that samples and returns fresh random strings from $\{0,1\}^k$ on each invocation. This hides the keys from being directly observed by $A$, allowing $K_{i+1}$ to depend on $K_i$ deterministically. Such a definition, however, is too weak—it doesn't really depart from the familiar ICM. For example, if $\mathcal{S}_{\mathsf{KS}}$ sets $K_i = \mathsf{KS} \oplus i$ then the keys are not independent, yet $A$ is unlikely to be able to exploit this (in the ICM). One of our goals is that our security definition should imply security in the standard model, so this candidate also isn't acceptable.

*Dispersed inputs.* The next two definitions are used to measure some important combinatorial properties of decompositions. We will require several symbols to define the relevant parameters, and so provide Table 1 for reference.

**Definition 3 (Dispersed inputs).** *Let $k, n, \mu$ and $\sigma$ be non-negative integers, and let $\epsilon$ be positive. Let $F$ be a uniformly random function mapping $\{0,1\}^k \times \{0,1\}^n$ to $\{0,1\}^n$. A decomposition $\hat{\mathcal{E}}$ over an $(n, n)$-bit blockcipher has $(q, \sigma, \mu, \epsilon)$-dispersed inputs if for any adversary $A$ making $q$ queries, each no longer than $\sigma n$ bits,*

$$\Pr\left[\ \mathsf{COMPAT}_{\hat{\mathcal{E}}}^{F}(A)\ ;\ \max_{X}|\{i \mid (i, X) \in Q\}| > \mu\ \right] < \epsilon,$$

*where Experiment* $\mathsf{COMPAT}$ *is defined in Fig. 2, and $Q$ refers to the final value of the set so named constructed during this experiment (i.e., the set of points submitted to the* query *oracle).*

The condition states that no single input is evaluated under more than $\mu$ key aliases except with probability $\epsilon$. Small values of $\mu$ and $\epsilon$ limit the effectiveness of brute-force attacks by putting a cap on how many of the $m$ keys can be attacked in parallel with a single blockcipher invocation.

*Mode efficiency.* A final definition is used to bound the computational work done by $\mathcal{M}$ and $\mathcal{S}$ given restrictions on an adversary.

**Definition 4 (Mode efficiency).** *Let $\hat{\mathcal{E}}$ be a decomposition over an $(k, n)$-bit blockcipher $E$, with $k = n$. Let $\mathsf{COMPAT}$ be the experiment defined in Fig. 2, and let $A$ be any adversary making $q$ queries, each of length at most $\sigma n$ bits. We say $\hat{\mathcal{E}}$ is $(q, \sigma, m, \nu)$-efficient if after an execution of $\mathsf{COMPAT}_{\hat{\mathcal{E}}}^{E}(A)$, $\mathsf{c} < m$ and for each $i$, $|\{X \mid (i, X) \in P \cup Q\}| \leq \nu$. Here, $\mathsf{c}$, $P$, and $Q$ refer to the final values of the random variables constructed in the experiment's definition.*

That is, given such an adversary, the mode and scheduler will query the key manager using at most $m$ key aliases, and will use each alias to encipher at most $\nu$ blocks.

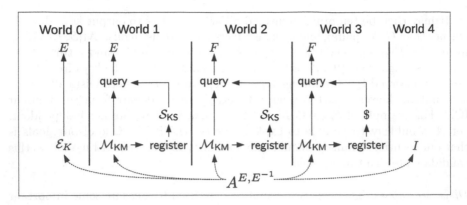

**Fig. 3.** Here, $F$ is an ideal cipher and $\mathcal{E}$ is some cryptographic scheme based on a (concrete) blockcipher $E$ that should be indistinguishable from some reference oracle $\mathcal{I}$. For example, $\mathcal{E}$ maybe an encryption scheme and $\mathcal{I}$ an oracle that returns a random string. From $A$'s perspective, World 0 = World 1 if $\hat{\mathcal{E}} = (\mathcal{M}, \mathcal{S}, \mathcal{K})$ is a *decomposition* of $\mathcal{E}$; World 1 ≈ World 2 if $\hat{\mathcal{E}}$ has *dispersed inputs* and $E$ is a PRP; World 2 ≈ World 3 if the scheduler $\mathcal{S}$ is *compatible* with the mode $\mathcal{M}$; World 3 ≈ World 4 if $\hat{\mathcal{E}}$ is indistinguishable from $\mathcal{I}$ in the ICM-KOA.

### 2.3   Generic Results About IND-KOA-ICM

We can now define what it means for a construction $\mathcal{E}$ to be indistinguishable from a reference oracle $\mathcal{I}$ in the ICM-KOA, the ICM, and the standard model. In general, we're interested in $\mathcal{I}$ that provide the desired idealized behavior of $\mathcal{E}$. For example, if $\mathcal{E}$ is an encryption algorithm, then we may want $\mathcal{I}$ to be the oracle that accepts a plaintext and outputs random bits.

We then show that ICM-KOA indistinguishability implies insecurity in both the ICM and the standard model, with a loss that is determined by the parameters of $\mathcal{E}$'s decomposition as surfaced by the efficiency and input-dispersion definitions. Figure 3 provides a graphical overview of how our key-access manager formalism will be used to argue indistinguishability of $\mathcal{E}$ and $\mathcal{I}$.

We emphasize that unlike most security definitions of this form, we do not claim that ICM-KOA indistinguishability offers an intuitive, compelling security goal on its own. Instead, it is a means to obtaining strong bounds in the ICM that are backed by a guarantee that keys are kept random and secret. And because of this guarantee, we gain simultaneous results in the standard model.

**Definition 5 (ICM-KOA indistinguishability).** *Let $\hat{\mathcal{E}} = (\mathcal{M}, \mathcal{S}, \mathcal{K})$ be a decomposition over an $(k,n)$-bit blockcipher, $k = n$, with $\mathcal{M}[\mathcal{S}, E]_K : \mathcal{D} \to \mathcal{R}$. Let $\mathcal{I} : \mathcal{D} \to \mathcal{R}$ be some reference scheme. Then the ICM-KOA-$\mathcal{I}$ advantage of an adversary $A$ is*

$$\mathbf{Adv}_{\hat{\mathcal{E}}}^{koa\text{-}ind\text{-}\mathcal{I}}(A) = \Pr\left[ A^{\mathcal{M}[F]_K, E, E^{-1}} \Rightarrow 1 \right] - \Pr\left[ A^{\mathcal{I}, E, E^{-1}} \Rightarrow 1 \right].$$

Here, $F \xleftarrow{\$} \text{Func}(k + n, n)$ and $\mathcal{M}[F]_K$ behaves identically to $\mathcal{M}[\mathcal{S}, F]_K$ (as defined in Fig. 1), except register assigns $K_c \xleftarrow{\$} \{0, 1\}^k$ instead of $K_c \leftarrow E_{K_i}(X)$.

Note that in this definition, the mode $\mathcal{M}$ does not interact with $E$, and so, without loss of generality, neither does $A$. ICM-KOA indistinguishability is only a useful notion for compatible decompositions with dispersed inputs, as these properties will allow us to "decouple" the ideal cipher used by the mode from the ideal cipher directly accessible by an adversary when proving results in the ICM.

**Definition 6 (ICM indistinguishability).** *Let $\hat{\mathcal{E}} = (\mathcal{M}, \mathcal{S}, \mathcal{K})$ be a decomposition over an $(k, n)$-bit blockcipher, $k = n$, where $\mathcal{M}[\mathcal{S}, E]_K : \mathcal{D} \to \mathcal{R}$. Let $\mathcal{I} : \mathcal{D} \to \mathcal{R}$ be some reference scheme (for example, an encryption algorithm with $\mathcal{D} = \mathcal{R} = \{0, 1\}^*$). Then the ICM-IND-$\mathcal{I}$ advantage of an adversary $A$ is*

$$\mathbf{Adv}_{\hat{\mathcal{E}}}^{icm\text{-}ind\text{-}\mathcal{I}}(A) = \Pr\left[ A^{\mathcal{M}[\mathcal{S}, E]_K, E, E^{-1}} \Rightarrow 1 \right] - \Pr\left[ A^{\mathcal{I}, E, E^{-1}} \Rightarrow 1 \right],$$

*where $K \xleftarrow{\$} \mathcal{K}$, and $E \xleftarrow{\$} \text{BC}(k, n)$ is an ideal cipher.*

*Precomputation, offline and online queries.* One benefit of the ICM-KOA model is that it can quantify the effectiveness of precomputation against specific modes. The following definition is general, but in it we have in mind $f_2 = E$, $f_3 = E^{-1}$ for some blockcipher $E$, while $f_1$ is an oracle for some blockcipher-based construction.

**Definition 7 (Precomputation, offline, and online queries).** *Let $A^{f_1, f_2, f_3}$ be an adversary. We say $A$ makes $q_P$ precomputation queries, $q_E$ offline queries, and $q$ online queries if*

- *$A$ makes $q_P$ combined queries to $f_2$ and $f_3$ before making its first query to $f_1$,*
- *and afterwards makes a combined $q_E$ queries to $f_2$ and $f_3$,*
- *while interleaving $q$ queries to $f_1$.*

*Relating the ICM-KOA and the ICM.* We now give the first of our two main model-implication results. Namely, that security in the ICM-KOA implies security in the ICM.

**Theorem 1 (ICM-KOA indistinguishability implies ICM indistinguishability).** *Let $\hat{\mathcal{E}} = (\mathcal{M}, \mathcal{S}, \mathcal{K})$ be a decomposition over an $(k, n)$-bit blockcipher with $k = n$, and let $\mathcal{I}$ be some reference scheme. Fix a positive integer $c$. Let $A$ be an adversary making $q_P$ precomputation queries, $q_E$ offline queries, and $q$ online queries, the latter of at most $\sigma n$ bits each. Suppose*

1. *$\mathcal{M}$ is compatible with $\mathcal{S}$,*
2. *$\hat{\mathcal{E}}$ is $(q, \sigma, m, \nu)$-efficient,*
3. *$\hat{\mathcal{E}}$ has $(q, \sigma, \mu, \epsilon)$-dispersed inputs, and*
4. *For any adversary $B$ making $q$ queries, $\mathbf{Adv}_{\hat{\mathcal{E}}}^{koa\text{-}ind\text{-}\mathcal{I}}(B) \leq \delta$.*

*Further suppose[1] that $q_E + q_P < 2^n$. Then*

$$\mathbf{Adv}_{\hat{\mathcal{E}}}^{icm\text{-}ind\text{-}\mathcal{I}}(A) \leq \delta + \frac{2q_E c\nu}{2^k(2^n - q_E - q_P)} + \frac{(q_E + q_P)m\nu}{2^{k+n}} + \frac{cm\nu^2}{2^n}$$

$$+ \frac{q_E(2\mu + c) + (q_P + m)\mu}{2^k} + \frac{m^{c+1}(1 + \nu^{c+1})}{2^{nc}(c+1)!} + 3\epsilon.$$

Although this general bound is complex, it simplifies substantially for various modes of operation. We will see this when we apply the general result to real constructions in Sect. 3. We note that the constant $c$ can be chosen more-or-less arbitrarily to minimize the bound. This permits the possibility of "beyond birthday-bound security" when $c > 1$. (The $cm\nu^2/2^n$ term gives a birthday bound with respect to the amount of data $\nu$ processed with a single key, but $m\nu$ blocks are enciphered in total.) Before proving this theorem, we give the following useful lemma.

**Lemma 1. ($c$-wise birthday bound).** *Let $c$, $q$, and $n$ be positive integers, with $c \leq q$. Let $X_1, \ldots, X_q$ be iid uniformly random $n$-bit strings. Then $\Pr[\exists S \subseteq \{1, \ldots, q\} \text{ s.t. } |S| = c, X_j = X_i \text{ for all } i, j \in S] \leq \frac{q^c}{2^{n(c-1)}c!}$.*

*Proof.* Fix some $x \in \{0,1\}^n$ and some $c$-sized index set $S \subseteq \{1, 2, \ldots, q\}$. Then $\Pr[\forall i \in S : x = X_i] = 2^{-cn}$. Since there are $2^n$ choices for $x$ and $\binom{q}{c} < q^c/c!$ choices for $S$, a union bound provides us with the desired upper bound.    □

*Proof (Theorem 1).* Let $F \xleftarrow{\$} \text{Func}(k + n, n)$. Then $\Pr[A^{\mathcal{M}[F]_K} \Rightarrow 1] - \Pr[A^{\mathcal{I}} \Rightarrow 1] \leq \delta$, where $K \xleftarrow{\$} \mathcal{K}$ and $\mathcal{M}[F]_K$ is defined as Definition 5.

Game $\text{G1}(A)$ (Fig. 4), which excludes the boxed statements, faithfully simulates $A^{\mathcal{M}[S,F]_K}$. In this figure, and for the remainder of the proof, $F$, $E$, and $E^{-1}$ (without subscripts) refer to oracles, while $F_K$ and $E_K$ (with subscripts) refer to the lazily-defined functions the game builds to help implement these oracles. We've moved the calls to register to the start of the game, without loss of generality.

In $\text{G1}(A)$, the behavior of $F$ is independent of the behavior of $E$ and $E^{-1}$. Consequently, the value of each key $K_i$ is information theoretically hidden from the adversary; the adversary can at best learn information about whether two key aliases correspond to the same key.

Recall that the difference between $\mathcal{M}[F]_K$ and $\mathcal{M}[S, F]_K$ is that the former's register procedure always assigns keys a uniformly random value that is independent of the other coins in the experiment. Hence, the oracle $\mathcal{M}[F]_K$ behaves identically to $\mathcal{M}[S, F]_K$ until there is some query input $(i, X)$ and some $S$ output $(j, X)$ with $K_i = K_j$.

Let us bound the probability of this happening during an execution of $A^{\mathcal{M}[F]_K}$. (The Fundamental Lemma of Game Playing implies that this probability is equal in both games; we are free to choose whichever best expedites

---

[1] The proof permits us to omit this final restriction by changing the first term in the bound to $2/2^k$.

the proof.) Fix one of the $m-1$ pairs $(j, X)$ output by $\mathcal{S}$. As $\mathcal{M}$ and $\mathcal{S}$ are compatible, query never receives an input $(j, X)$. Except with probability $\epsilon$, there are at most $\mu$ aliases $i$ such that query receives an input $(i, X)$. For each such alias $i$, $\Pr[K_i = K_j] = 1/2^k$; hence, some such alias exists with probability at most $\mu/2^k$. Taking a union bound over the $m-1$ pairs $(j, X)$ gives us $\Pr\left[A^{\mathcal{M}[F]_K} \Rightarrow 1\right] - \Pr\left[A^{\mathcal{M}[\mathcal{S}, F]_K} \Rightarrow 1\right] \leq \frac{m\mu}{2^k} + \epsilon$.

In Game G1, the $E$ and $E^{-1}$ oracles behave independently of the others. However, in Game G2, which includes the boxed statements, the $F$ and $E$ oracles have been coupled together (turning $F$ into a blockcipher). So $\Pr[\text{G2}(A) \Rightarrow 1] = \Pr\left[A^{\mathcal{M}[\mathcal{S}, E]_K, E, E^{-1}} \Rightarrow 1\right]$.

We therefore wish to bound $\Pr[\text{G1}(A) \Rightarrow 1] - \Pr[\text{G2}(A) \Rightarrow 1]$. The Fundamental Lemma of Game Playing allows us to do so by bounding the probability that one of the boolean "bad flags" of Fig. 4 is set during an execution of G1($A$).

Let $\mathcal{C}_c$ be the event that for some key $K$, $|\{i : K_i = K\}| > c$. By Lemma 1, $\Pr[\text{G1}(A) ; \mathcal{C}_c] \leq \frac{m^{c+1}}{2^{nc}(c+1)!}$.

Now, in Game G1($A$), $\mathsf{bad}_1$ is set on a particular query $(K, X)$ to $E$ only if the initial value for $Y$ is in $\text{Rng}(F_K)$:

$$\Pr[Y \in F_K \mid \neg\mathcal{C}_c] = \sum_{K_i} \Pr[K = K_i \mid \neg\mathcal{C}_c] \Pr[Y \in F_K \mid K = K', \neg\mathcal{C}_c]$$

$$\leq \sum_{K'} \frac{1}{2^k} \frac{|\text{Dom}(F_{K'})|}{2^n - q_E - q_P} \leq \frac{c\nu}{2^k(2^n - q_E - q_P)}.$$

Hence $\Pr[\text{G1}(A) ; \mathsf{bad}_1 \mid \neg\mathcal{C}_c] \leq \frac{q_E c\nu}{2^k(2^n - q_E - q_P)}$. A symmetric argument shows the same bound applies to $\Pr[\text{G1}(A) ; \mathsf{bad}_3 \mid \neg\mathcal{C}_c]$.

Similarly, $\mathsf{bad}_2$ is set on a particular query $(K, X)$ to $E$ only if $X \in \text{Dom}(F_K)$. Except with probability $\epsilon$, There are at most $\mu$ key aliases $i$ such that $X \in \text{Dom}(F_{K_i})$. Hence, $\Pr[\text{G1}(A) ; \mathsf{bad}_2] \leq \frac{q_E \mu}{2^k} + \epsilon$.

Note that $\mathsf{bad}_4$ is only set if the adversary makes a query $(K, Y)$ to $E^{-1}$ for some $Y \in \text{Rng}(F_K)$. Over the course of the game, the probability that there will exist some $Y' \in \{0, 1\}^n$ with $|\{(K, X) : F_K(X) = Y'\}| > c$ is at most $\frac{(m\nu)^c}{2^{n(c-1)}}$; i.e., except with this probability, $|\{K' : Y \in \text{Rng}(F_{K'})\}| \leq c$. (This follows from the fact that points in the range of each $F_K$ are uniform and mutually independent; see Lemma 1). Thus $\Pr[\text{G1}(A) ; \mathsf{bad}_4] \leq \frac{q_E c}{2^k} + \frac{(m\nu)^c}{2^{n(c-1)}}$.

To bound $\Pr[\text{G1}(A) ; \mathsf{bad}_5]$, consider a query $(i, X)$ to $F$. We sample a uniformly random $Y \xleftarrow{\$} \{0, 1\}^n$ and set $\mathsf{bad}_5$ if $Y \in \text{Rng}(E_{K_i})$ or $Y \in \text{Rng}(F_{K_i})$. Using an argument similar to that for our bound for $\mathsf{bad}_1$, $\Pr[Y \in \text{Rng}(E_{K_i})] \leq \frac{q_E + q_P}{2^{k+n}}$. Again fix a positive integer $c$. So as long as no key corresponds to more than $c$ aliases, $Y \in \text{Rng}(F_{K_i})$ with probability at most $c\nu/2^n$. Taking a union bound over each of $m\nu$ queries gives $\Pr[\text{G1}(A) ; \mathsf{bad}_5 \mid \neg\mathcal{C}_c] \leq \frac{(q_E + q_P)m\nu}{2^{k+n}} + \frac{cm\nu^2}{2^n}$.

Finally, we need to bound $\Pr[\text{G1}(A) ; \mathsf{bad}_6]$. This flag is set only if some $E$ or $E^{-1}$ query defines the point $E_K(X) = Y$ such that $K = K_i$ and $X = X'$, where

GAMES $G1$, $\boxed{G2}$

**Procedure main($A$):**

for $j = 1$ to $m - 1$ do
  register()
$b \leftarrow A^{f, E, E^{-1}}$
return $b$

**Oracle $F(i, X)$:**

if $X \in \text{Dom}\left(F_{K_i}\right)$ then
  return $F_{K_i}(X)$
$Y \xleftarrow{\$} \{0, 1\}^n$
if $Y \in \text{Rng}\left(E_{K_i}\right) \cup \text{Rng}\left(F_{K_i}\right)$ then
  $\text{bad}_5 \leftarrow$ true
  $\boxed{Y \xleftarrow{\$} \overline{\text{Rng}\left(E_{K_i}\right) \cup \text{Rng}\left(F_{K_i}\right)}}$
if $X \in \text{Dom}\left(E_{K_i}\right)$ then
  $\text{bad}_6 \leftarrow$ true
  $\boxed{Y \leftarrow E_{K_i}(X)}$
$F_{K_i}(X) \leftarrow Y$
return $Y$

**Oracle query($i, X$):**

if $i > \mathsf{c}$ then return $\perp$
return $F(i, X)$

**Oracle register():**

$(i, X) \leftarrow S_{\mathsf{SK}}^{\mathsf{query}}$
$K_{\mathsf{c}+1} \leftarrow F(i, X)$
$\mathsf{c} \leftarrow \mathsf{c} + 1$

**Oracle $E(K, X)$:**

$Y \xleftarrow{\$} \overline{\text{Rng}(E_K)}$
if $Y \in \text{Rng}(F_K)$ then
  $\text{bad}_1 \leftarrow$ true
  $\boxed{Y \xleftarrow{\$} \overline{\text{Rng}(E_K) \cup \text{Rng}(F_K)}}$
if $X \in \text{Dom}(F_K)$ then
  $\text{bad}_2 \leftarrow$ true
  $\boxed{Y \leftarrow F_K(X)}$
$E_K(X) \leftarrow Y$
return $E_K(X)$

**Oracle $E^{-1}(K, Y)$:**

$X \xleftarrow{\$} \overline{\text{Dom}(E_K)}$
if $X \in \text{Dom}(F_K)$ then
  $\text{bad}_3 \leftarrow$ true
  $\boxed{X \xleftarrow{\$} \overline{\text{Dom}(E_K) \cup \text{Dom}(F_K)}}$
if $Y \in \text{Rng}\left(F_{K_i}\right)$ then
  $\text{bad}_4 \leftarrow$ true
  $\boxed{X \leftarrow F_K^{-1}(Y)}$
$E_K^{-1}(Y) \leftarrow X$
return $E_K^{-1}(Y)$

**Oracle $f(M)$:**

return $\mathcal{M}_{\mathsf{KM}}^{\mathsf{query}, \mathsf{register}}(M)$

**Fig. 4.** In Game G2, $A$, $\mathcal{M}$, and $S$ access the same blockcipher (directly, through $\mathsf{query}_E$, and through $\mathsf{query}_F$, respectively). In Game G1, the behavior of $\mathsf{query}_F$ is decoupled from $E$ and $\mathsf{query}_E$, in effect giving the scheduler $S$ it's own blockcipher.

$(i, X')$ is some (future) $F$-query. Let us first consider a precomputation query that defines $E_K(X) = Y$. Then $\text{bad}_6$ will be triggered by this precomputation query only if $K$ is one of the at most $\mu$ keys under which $X$ is queried. Hence, the probability that some precomputation query will define a point on $E$ that triggers $\text{bad}_6$ is at most $q_P \mu / 2^k$.

Now let us consider an offline query that defines $E_K(X) = Y$. Except with probability $\epsilon$, there are at most $\mu$ key aliases $i$ that will be used to encipher $X$; the probability that one of these $\mu$ keys will be $K$ is at most $\frac{\mu}{2^k}$. Hence, the probability that some offline query will define a point on $E$ that triggers

$\mathsf{bad}_6$ is at most $q_E\mu/2^k$. Therefore $\Pr[\,\mathrm{G1}(A)\,;\,\mathsf{bad}_6\,] \leq \mu(q_E + q_P)/2^k + \epsilon$. The Fundamental Lemma of Game-Playing gives us:

$$\Pr[\,\mathrm{G1}(A) \Rightarrow 1\,] - \Pr[\,\mathrm{G2}(A) \Rightarrow 1\,]$$
$$\leq \Pr[\,\mathsf{bad}_1 \vee \mathsf{bad}_3 \vee \mathsf{bad}_5 \mid \neg\mathcal{C}_c\,] + \Pr[\,\mathcal{C}_c\,]$$
$$+ \Pr[\,\mathsf{bad}_2 \vee \mathsf{bad}_4 \vee \mathsf{bad}_6\,]$$
$$\leq \frac{2q_E c\nu}{2^k(2^n - q_E - q_P)} + \frac{(q_E + q_P)m\nu}{2^{k+n}} + \frac{cm\nu^2}{2^n} + \frac{m^{c+1}}{2^{nc}c + 1!}$$
$$+ \frac{2q_E\mu}{2^k} + \frac{q_E c}{2^k} + \frac{(m\nu)^c}{2^{n(c-1)}c!} + \frac{q_P\mu\epsilon}{2^k} + 3\epsilon$$
$$= \frac{2q_E c\nu}{2^k(2^n - q_E - q_P)} + \frac{(q_E + q_P)m\nu}{2^{k+n}} + \frac{cm\nu^2}{2^n}$$
$$+ \frac{q_E(2\mu + c) + q_P\mu}{2^k} + \frac{m^{c+1}(1 + \nu^{c+1})}{2^{nc}(c + 1)!} + 3\epsilon.$$

Collecting our results completes the proof.    □

*Relating the ICM-KOA to the standard model.* We now move on to a standard-model analogue. The indistinguishability advantage definition is the same, except now $A$ has an implicit description of $E$ rather than oracle access:

**Definition 8 (Standard model indistinguishability).** *Let $\mathcal{E} : \mathcal{K} \times \mathcal{D} \to \mathcal{R}$ be a scheme over an $(n, n)$-bit blockcipher and let $I : \mathcal{D} \to \mathcal{R}$ be some oracle. Let $E$ be an $(n, n)$-bit blockcipher. We define standard model indistinguishability advantage of an adversary $A$ (with respect to $\mathcal{E}$ and $I$) as:* $\mathbf{Adv}_{\mathcal{E};E}^{ind-I}(A) = \Pr[A^{\mathcal{M}[\mathcal{S},E]_K} \Rightarrow 1] - \Pr[A^I \Rightarrow 1]$, *where $K \xleftarrow{\$} \mathcal{K}$ is a random key and $E$ is an $(n, n)$-bit blockcipher.*

We now give the second of our two main model-implication results. Namely, that security in the ICM-KOA implies security in the standard model.

**Theorem 2 (ICM-KOA indistinguishability implies standard model indistinguishability).** *Let $\mathcal{E}$ be an $(k, n)$-bit blockcipher-based scheme, and let $\hat{\mathcal{E}} = (\mathcal{M}, \mathcal{S}, \mathcal{K})$ be a decomposition of $\mathcal{E}$. Suppose*

1. $\mathcal{M}$ *is compatible with* $\mathcal{S}$,
2. $\hat{\mathcal{E}}$ *is* $(q, \sigma, m, \nu)$-*efficient*,
3. *For any adversary $B'$ making $q$ queries,* $\mathbf{Adv}_{\hat{\mathcal{E}}}^{koa-ind-I}(B') \leq \delta$.

*Then for any adversary $A$ running in time $t$ and making $q$ queries, each at most $\sigma n$ bits in length, there exists some adversary $B$ running in time $t' \approx t$ and making $\nu$ queries such that $\mathbf{Adv}_{\mathcal{E};E}^{ind-I}(A) \leq m\mathbf{Adv}_E^{prf}(B) + \frac{m^2}{2^k} + \delta$.*

This theorem relates ICM-KOA security to the PRF security of the underlying blockcipher. This implies a relationship between ICM-KOA security and PRP security via the PRP-PRF switching lemma, at the expense of an additional $m\sigma^2/2^{n+1}$ term. This term beats the birthday bound by a factor of $m$.

**Oracle** $G_{(\mathsf{KM},\mathsf{KS})}(M)$:

$K_1 \xleftarrow{\$} \{0,1\}^k$
**return** $\mathcal{M}_{\mathsf{KM}}^{\mathsf{query},\mathsf{register}}(M)$

**Procedure** register():

$(i, X) \leftarrow \mathcal{S}_{\mathsf{KS}}^{\mathsf{query}}$
$K_{\mathsf{c}+1} \leftarrow R(i, X)$
$\mathsf{c} \leftarrow \mathsf{c} + 1$

**Procedure** query$(i, X)$:

**if** $i > \mathsf{c}$ **then return** $\bot$
**return** $R(i, X)$

**Fig. 5.** Replacing $E$ with a random function $R$

*Proof (Theorem 2).* We will use a game-playing proof. First $A$'s oracle will transition from $\mathcal{M}[\mathcal{S}, E]_K$ into $G$, where references to $E_{K_i}(X)$ are replaced with $R(i, X)$ for some random function $R$ (see Fig. 5).

This transition will itself involve a sequence of games. Define the oracle $G_\ell$ to be identical $\mathcal{M}[\mathcal{S}; E]_K$ for $K \xleftarrow{\$} \mathcal{K}$, except that query and register compute $R(i, X)$ in place of $E(K_i, X)$ when $i < \ell$. This gives us

$$\Pr\left[ A^{\mathcal{M}[\mathcal{S},E]_K} \Rightarrow 1 \right] - \Pr\left[ A^G \Rightarrow 1 \right]$$
$$\leq \sum_{j=0}^{m-1} \left( \Pr\left[ A^{G_{j+1}} \Rightarrow 1 \right] - \Pr\left[ A^{G_j} \Rightarrow 1 \right] \right).$$

Now in $G_{j+1}$, we have $K_{j+1} = R(i, X)$ for some $i \leq j$, where the compatibility condition ensures that this is the only time $R$ is evaluated at the point $(i, X)$. Consequently, $K_{j+1}$ is uniformly distributed and independent of the other coins of the experiment. It can therefore be freely discarded and replaced with some other value draw from this distribution without affecting the black-box behavior of $G_{j+1}$. Therefore from $A$ we can construct a PRF adversary $B_j$ with the property $\mathbf{Adv}_E^{\mathrm{prf}}(B_j) = \Pr\left[ A^{G_{j+1}} \Rightarrow 1 \right] - \Pr\left[ A^{G_j} \Rightarrow 1 \right]$. This is accomplished by having $B_j^f$ simulate $G_j$ for $A$, but using its own oracle to set query$(j+1, \cdot) = f(\cdot)$. So $B_j^f$ in behaves identically to either $G_j$ (when $f$ is $E_K$) or $G_{j+1}$ (when $f$ is a random function). We note that $B_j$ makes at most $\nu$ queries and has roughly the same running time as $A$.

Setting $B$ to be the $B_j$ with maximal advantage $(1 \leq j \leq m)$ gives us $\Pr\left[ A^{\mathcal{M}[\mathcal{S},E]_K} \Rightarrow 1 \right] - \Pr\left[ A^G \Rightarrow 1 \right] \leq m\mathbf{Adv}_E^{\mathrm{prf}}(B)$.

We observe that the $G$ and $\mathcal{M}[F]$ differ in behavior only when $K_i = K_j$ for some $i \neq j$, which happens with probability at most $m^2/2^k$. Hence, $\Pr\left[ A^G \Rightarrow 1 \right] - \Pr\left[ A^{\mathcal{M}[F]} \Rightarrow 1 \right] < m^2/2^k$.

Finally, by hypothesis $\Pr\left[A^{\mathcal{M}[F]} \Rightarrow 1\right] - \Pr\left[A^{\mathcal{I}} \Rightarrow 1\right] \leq \delta$. Combining these results provides the desired bound. □

## 2.4   Connection to TBC-based Constructions

A tweakable blockcipher $\widetilde{E}$ is a (strong) TPRP if a keyed instance of $\widetilde{E}$ is computationally indistinguishable from an ideal cipher. This suggests that there ought to be some formal relationship between TBCs and the ideal cipher model, but the fact that TBCs are a keyed construction means the two objects cannot be directly compared. However, the key managers we have introduced *are* keyed constructions that mediate access between modes of operation and an underlying cipher. They thus offer a means of bridging the conceptual gap between TBCs and ideal ciphers: specifically, the following theorem states that any mode of operation secure in the ICM-KOA can be transformed into a TBC-based construction secure in the standard model. In the following theorem statement, $\varepsilon$ denotes the empty string.

**Theorem 3 (Decompositions imply TBC-based constructions).** *Let $\mathcal{E}$ be a scheme over a $(k,n)$-bit blockcipher, and fix a decomposition $\hat{\mathcal{E}} = (\mathcal{M}, \mathcal{S}, \mathcal{K})$. Let be $\widetilde{E} : \{0,1\}^k \times \mathcal{T} \times \{0,1\}^n \to \{0,1\}^n$ be an n-bit TBC. Sample $K \xleftarrow{\$} \{0,1\}^k$ and $(\mathsf{KM}, \mathsf{KS}) \xleftarrow{\$} \mathcal{K}$.*

*Define an oracle $\mathcal{F}\langle\widetilde{E}_K\rangle_{\mathsf{KM}}$ as follows: On input $M$, the output of $\mathcal{F}\langle\widetilde{E}_K\rangle_{\mathsf{KM}}$ is the value returned by the oracle $\mathcal{M}[S,E]_{(\mathsf{KM},\varepsilon)}(M)$ in Fig. 1 when (1) the register procedure is replaced by a procedure register-nop that does nothing, and (2) the query procedure is modified so that, on input $(i,X)$, it returns $\widetilde{E}_K(i,X)$.[2] (This assumes that the maximum number of key aliases permitted by the mode is at most $|\mathcal{T}|$.) For any adversary $A$ running in time $t$ and making $q$ queries, each of length at most $\sigma n$ bits, there exists some adversary $B$ making $m\nu$ queries and running in time $t' \approx t$ such that*

$$\Pr\left[A^{\mathcal{F}\langle\widetilde{E}_K\rangle_{\mathsf{KM}}} \Rightarrow 1\right] - \Pr\left[A^{\mathcal{I}} \Rightarrow 1\right] \leq \mathbf{Adv}_{\widetilde{E}}^{\widetilde{\mathrm{prp}}}(B) + \frac{m\nu^2}{2^n} + \frac{m^2}{2^k} + \delta$$

*where $K \xleftarrow{\$} \mathcal{K}$.*

*Proof.* Let $\Pi \xleftarrow{\$} \mathrm{BC}(k,n)$ be an ideal cipher and $F \xleftarrow{\$} \mathrm{Func}(k+n,n)$ be a random function. By a standard reduction argument, there exists some adversary $B$ with the stated resources such that $\Pr\left[A^{\mathcal{F}\langle\widetilde{E}_K\rangle_{\mathsf{KM}}} \Rightarrow 1\right] - \Pr\left[A^{\mathcal{F}\langle\Pi\rangle_{\mathsf{KM}}} \Rightarrow 1\right] \leq \mathbf{Adv}_{\widetilde{E}}^{\widetilde{\mathrm{prp}}}(B)$. By the $m$ applications of the Switching Lemma, $\Pr\left[A^{\mathcal{F}\langle\Pi\rangle_{\mathsf{KM}}} \Rightarrow 1\right] - \Pr\left[A^{\mathcal{F}\langle F_K\rangle_{\mathsf{KM}}} \Rightarrow 1\right] \leq m\nu^2/2^n$. Finally, note that $\mathcal{F}\langle F_K\rangle_{\mathsf{KM}}$ and $\mathcal{F}[F]_{(\mathsf{KM},\varepsilon)}$ behave identically unless the $m$ random keys generated by the latter oracle's register procedure are not pairwise distinct, an event that happens with probability $m^2/2^k$. Collecting results completes the proof. □

---

[2] With these changes, the parameter $E$ is unused.

# 3    ICM-KOA Analysis of Constructions

We now put the ICM-KOA to work, using it to analyze example blockcipher-based constructions. We begin with the NIST-CTR-DRBG, as used in Intel's recent hardware random-number generator [11], whose standard-model security bounds [19] can become quite weak when an adversary is co-located on the same physical machine, due to the rate at which such an adversary can make queries. The weakness of these bounds is do to a hybrid-factor loss. Our ICM-KOA analysis yields considerably better bounds, and suggests that the multiplicative loss in the standard-model isn't "real".

Next, we give an example of when the standard-model fails to surface quantitative differences between the security of closely related schemes. In particular, we consider various rekeying and nonce-choice strategies for CTR mode. Although these schemes yield similar bounds in the standard model, we show that the best-possible black-box attacks tell quite a different story. These results are of particular importance when CTR is built over a lightweight blockciphers, where the standard-model security bounds for all of the strategies suggest that problems may arise quickly. Our ICM-KOA analysis (and the implied ICM results) offers a different viewpoint on these concerns, and identifies the best strategies from among the choices.

## 3.1    Analysis of NIST CTR-DRBG Generation Algorithm

As the name suggests, CTR-DRBG is a deterministic random-bit generator based on running a blockcipher in CTR mode. Here, we analyze its generation algorithm[3], specializing for the sake of simplicity to the case where AES-128 is used (so $n = k = 128$), and where 128 bits are requested on each invocation. This case is of special interest because these parameters are used inside of Intel's hardware random number generator.

Concretely, we consider the scheme ISK-RNG : $\{0,1\}^{2n} \times \{0,1\}^0 \to \{0,1\}^n$ over an $(n,n)$-bit blockcipher defined in Fig. 6. The system maintains an initially random internal state $(K, \mathsf{IV})$, and on each query computes $(R, K, \mathsf{IV}) \leftarrow (E_K(\mathsf{IV}), E_K(\mathsf{IV} + 1), E_K(\mathsf{IV} + 2))$, updating the state, and returns $R$. In order to decompose this into a model, we need the mode and scheduler to share the $\mathsf{IV}$ portion of the state. This is accomplished by using the initial $\mathsf{IV}$ as part of both the mode and scheduler key (these keys are not required to be independent).

We define Rand : $\{0,1\}^0 \to \{0,1\}^n$ to be the oracle that on each query samples $R \xleftarrow{\$} \{0,1\}^n$ and then returns $R$.

*Stronger than standard-model results desirable.* A result by Shrimpton and Terashima [19] shows, as one might expect, that the standard-model security bound for $q$ queries includes an $\mathcal{O}(q\mathbf{Adv}_E^{\mathrm{prp}}(B))$ term, where $B$ is an adversary making three queries. However, $B$ also has time $t$ to run, where $t$ is sufficient time to evaluate $E$ on $3q$ inputs. Hence even if $B$ conducts a naïve brute-force

---

[3] The specification also includes algorithms for, e.g., reseeding.

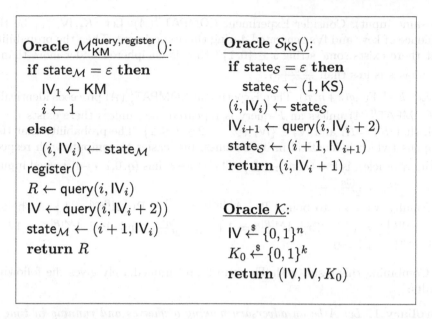

**Fig. 6.** The NIST CTR-DRBG decomposes into the mode and scheduler described above. The key-generation algorithm $\mathcal{K}$ ensures KM = KS.

attack, $\mathbf{Adv}_E^{\mathrm{prp}}(B) \approx 3q/2^k$. So the security bound becomes roughly $q^2/2^k$. For $k = 128$, this bound exceeds $2^{-40}$ when $q = 2^{44}$.

This may seem safe; after all, this amounts to many terabytes of random bits. But the RNG has extremely high throughput—Intel reports 800 MB/s, which equates to 50 million queries per second. This means an attacker who shares a physical machine with his target can reach the $q = 2^{44}$ limit in a little more than four days.

The following lemma provides a security bound for the ISK-RNG in the ICM-KOA. For very little work, we recover the security bound of Shrimpton and Terashima [19], and immediately also get a much stronger bound in the ICM. The ICM bound reveals the lack of a matching attack, and shows that barring cryptanalysis of AES *under random and secret keys*, we can permit on the order of $2^{70}$ queries before surpassing our $2^{-40}$ limit (assuming the adversary has resources for $2^{80}$ precomputation and $2^{80}$ offline queries). This translates to 750,000 years of ISK-RNG runtime, and so is unlikely to be the limiting factor.

**Lemma 2.** *For any positive integers $\mu$ and any adversary $A$ making at most $q$ online queries, ISK-RNG is $(q, 0, q, 3)$-efficient, has $(q, 0, c, \epsilon)$-dispersed inputs, and $\mathbf{Adv}_{ISK\text{-}RNG}^{koa\text{-}ind\text{-}Rand}(A) \leq \delta$, where $\delta = \frac{5q^2}{2^{2n}}$ and $\epsilon = \delta + \frac{(3q)^3}{2^{2n}3!}$.*

*Proof.* If $A$ makes $q$ queries (0 bits each), the RNG will make three queries using each of $q$ distinct key aliases. Hence $\hat{\mathcal{E}}$ is $(q, 0, q, 3)$-efficient.

Let $R : \{0,1\}^k \times \{0,1\}^n \rightarrow \{0,1\}^n$ be an oracle that samples and returns a fresh random string on each query (so $R$ may return different outputs on the same input). Consider Experiment $\mathsf{COMPAT}^R_{\hat{\mathcal{E}}}(A)$. Let $(K_i, \mathsf{IV}_i)_{i=1}^q$ be the sequence of keys and IVs generated during this experiment. Then the probability that there exists some string $x \in \{0,1\}^n$ that is enciphered under more than $c$ key aliases is less than $\frac{(3q)^{c+1}}{2^{nc}(c+1)!}$.

Let $F \xleftarrow{\$} \mathrm{Func}(k+n, n)$. Then Experiment $\mathsf{COMPAT}^F_{\hat{\mathcal{E}}}(A)$ proceeds identically to $\mathsf{COMPAT}^R_{\hat{\mathcal{E}}}(A)$ unless an $F$-query is repeated; i.e., unless there exists $i < j$ such that $K_j = K_i$ and $\mathsf{IV}_j \in \{\mathsf{IV}_i + \ell : -2 \leq \ell \leq 2\}$. The probability that this happens (which is identical in both games, but easier to compute with respect to the $R$ oracle), is less than $\frac{q^2}{2^k}\left(\frac{5}{2^n}\right)$. Therefore $\hat{\mathcal{E}}$ has $(q, 0, c, \epsilon)$-dispersed inputs for $\epsilon = \frac{5q^2}{2^{k+n}} + \frac{(3q)^{c+1}}{2^{nc}(c+1)!}$.

Finally, we need to bound $\Pr\left[A^{\mathcal{M}[F]_K} \Rightarrow 1\right] - \Pr\left[A^{\mathsf{Rand}} \Rightarrow 1\right]$. As before $\Pr\left[A^{\mathcal{M}[F]_K} \Rightarrow 1\right] - \Pr\left[A^{\mathcal{M}[R]_K} \Rightarrow 1\right] \leq \frac{5q^2}{2^{k+n}}$, and $\Pr\left[A^{\mathcal{M}[R]_K} \Rightarrow 1\right] - \Pr\left[A^{\mathsf{Rand}} \Rightarrow 1\right] = 0$.    $\square$

Combining this result with Theorem 2 and immediately gives the following results:

**Corollary 1.** *Let $A$ be an adversary making $q$ queries and running in time $t$. Then there exists an adversary $B$ making 3 queries and running in time $t' \approx t$ such that* $\mathbf{Adv}^{ind\text{-}Rand}_{ISK\text{-}RNG[E]}(A) \leq q\mathbf{Adv}^{prf}_E(B) + \frac{q^2}{2^n} + \frac{5q^2}{2^{2n}}$.

Note that up to a small constant factor, we've recovered, essentially the security bound from [19]. But we can do better:

**Corollary 2.** *Let $A$ be an adversary making $q_P$ precomputation queries, $q_E$ offline queries, and $q$ online queries, where $q_E + q_P < 2^{n-1}$. Then*

$$\mathbf{Adv}^{icm\text{-}ind\text{-}Rand}_{ISK\text{-}RNG}(A) \leq \frac{20q^2 + 24q_E + 3q(q_E + q_P) + 19q^3}{2^{2n}} + \frac{20q + 6q_E + 2q_P}{2^n}$$

Here we have set $c = 2$ for the sake of notational cleanliness.

Taking $q_E = q_P = 2^{80}$ allows the upper bound to stay below $2^{-40}$ even when $q = 2^{70}$, a substantial improvement over the previous $q = 2^{44}$ (which only applied to attackers with $q_P = 2^{44}$). This is a significantly stronger result than we could obtain in the standard model, and it retains the standard model's strength of only relying on random, secret keys. A brute-force attack on the key would obtain about the same success rate.

## 3.2    Analysis of CTR-mode Variants

We consider three variants on CTR mode, distinguished by how they choose IVs following a key change: (1) The IV is set to $0^n$; (2) the upper bits of the IV are unique for each key; (3) The IV is chosen randomly. In each case, we use the same key scheduler that sets $K_i \leftarrow E_{K_1}(i)$ (for $i > 1$). See Fig. 7. For simplicity, we consider the case where the key changes with each message. This models a situation where the counter state is retained between messages with the same key. The loss of adaptivity within the lifetime of a given key does not hamper a chosen-plaintext adversary in this context because the nature of CTR mode permits him to compute what a ciphertext would have been with a different plaintext. The variants are distinguished by the choice of iv-gen : $\mathbb{N} \to \{0,1\}^n$, which on input $i$ outputs some $\mathsf{IV}_i$. Define the reference scheme $\mathcal{R}[\text{iv-gen}]$ to be the stateful function that on its $i$th query $M$, computes $\mathsf{IV} \leftarrow \text{iv-gen}(i)$, samples $C \xleftarrow{\$} \{0,1\}^{|M|}$, and returns $(\mathsf{IV}, C)$.

**Theorem 4.** *Fix positive integers $\sigma$, $q$, and $b$ with $q < \sigma < 2^b$ and $b < n$. Let $\mathsf{const}(i) = 0^n$, let $\mathsf{unique}(i) = \langle i \rangle_b 0^{n-b}$ (where $\langle i \rangle_b$ is a $b$-bit encoding of $i$), and let $\mathsf{rand}(i)$ sample and return $R \xleftarrow{\$} \{0,1\}^n$ on each invocation. Let $A$ be an adversary making $q$ online queries, each at most $\sigma n$ bits long, $q_P$ precomputation queries, and $q_E$ offline queries. Then:*

$$\text{(1)} \quad \mathbf{Adv}_{\mathsf{CTR[const]}}^{ind\text{-}\mathcal{R}[const]}(A) \leq \frac{4q_E\sigma}{2^k(2^n - q_E - q_P)} + \frac{(q_E + q_P)q\sigma}{2^{k+n}} + \frac{2q\sigma^2}{2^n}$$
$$+ \frac{2q_E(q+1) + q_P q + 2q^2}{2^k} + \frac{q^3(1+\sigma^3)/6}{2^{2n}}$$

$$\text{(2)} \quad \mathbf{Adv}_{\mathsf{CTR[unique]}}^{ind\text{-}\mathcal{R}[unique]}(A) \leq \frac{4q_E\sigma}{2^k(2^n - q_E - q_P)} + \frac{(q_E + q_P)q\sigma}{2^{k+n}} + \frac{2q\sigma^2}{2^n}$$
$$+ \frac{6q_E + 2q_P + 2q}{2^k} + \frac{q^3(1+\sigma^3)/6}{2^{2n}}$$

$$\text{(3)} \quad \mathbf{Adv}_{\mathsf{CTR[rand]}}^{ind\text{-}\mathcal{R}[rand]}(A) \leq \frac{4q_E\sigma}{2^k(2^n - q_E - q_P)} + \frac{(q_E + q_P)q\sigma + (q\sigma)^2}{2^{k+n}} + \frac{2q\sigma^2}{2^n}$$
$$+ \frac{6q_E + 2q_P}{2^k} + \frac{q^3(1+4\sigma^3)/6}{2^{2n}}$$

*Proof.* Each decomposition is $(q, \sigma, q+1, \sigma)$-efficient. Sample $F \xleftarrow{\$} \mathsf{Func}(k+n, n)$. Let iv-gen $\in \{\text{const}, \text{unique}, \text{rand}\}$. Let bad be the event that during an execution $A^{\mathsf{CTR[iv\text{-}gen]}[F]}$, $\mathsf{CTR[iv\text{-}gen]}[F]$ repeats a query to $F$. Barring this event, the outputs of $\mathsf{CTR[iv\text{-}gen]}[F]$ are independent and uniformly random (with the possible exception of the $\mathsf{IV}$ component). Therefore $\Pr\left[ A^{\mathsf{CTR[iv\text{-}gen]}[F]} \Rightarrow 1 \right] - \Pr\left[ A^{\mathcal{R}[\text{iv-gen}]} \Rightarrow 1 \right] \leq \Pr[\text{bad}]$. We want to find an upper bound $\delta$ for $\Pr[\text{bad}]$, and do so for each method of generating the $\mathsf{IV}$. Specifically,

- When iv-gen = const, $\Pr[\text{bad}] \leq \Pr[\exists i \neq j : K_i = K_j] \leq q^2/2^k$
- When iv-gen = unique, $\Pr[\text{bad}] = 0$ because regardless of what value the keys have, the inputs never repeat.
- When iv-gen = rand, any two queries to $F$ collide with probability $1/2^{k+n}$ because both keys and IVs are uniform and independent. There are fewer than $(q\sigma)^2$ pairs of queries, so $\Pr[\text{bad}] < (q\sigma)^2/2^{k+n}$.

To apply Theorem 1 (with $c = 2$), we need to measure how much each variant disperses its inputs.

- CTR[const] has $(q, \sigma, q + 1, 0)$-dispersed inputs because $0^n$ is evaluated under each of the $q + 1$ keys.
- CTR[unique] has $(q, \sigma, 2, 0)$-dispersed inputs because each input is guaranteed to be used at most twice (including once by the scheduler).
- CTR[rand] has $(q, \sigma, c, (q\sigma)^{c+1}/2^{nc}(c + 1)!)$. The argument here follows that of Lemma 1, except each that we are interested in the probability that $x \in \{X_i, X_i + 1, \ldots, X_i + (\sigma - 1)\}$, instead of $x = X_i$, where $X_i$ plays the role of $\mathsf{IV}_i$.

Plugging these values into Theorem 1 gives us the previously stated bounds. □

---

**Oracle CTR[iv-gen]$_{\mathsf{KM}}^{\mathsf{query,register}}(M)$:**

if state$_\mathcal{M} = \varepsilon$ then
   $i \leftarrow 1$
else
   $i \leftarrow$ state$_\mathcal{M}$
register()
$\mathsf{IV}_i \leftarrow$ iv-gen$(i)$
$M_1 M_2 \cdots M_\ell \leftarrow_n M$
for $j = 1$ to $\ell$ do
   $C_j \leftarrow$ query$(i + 1, \mathsf{IV}_i + j) \oplus M_j$
state$_\mathcal{M} \leftarrow i + 1$
**return** $(\mathsf{IV}_i, C_1 C_2 \cdots C_\ell)$

**Oracle $\mathcal{S}_{\mathsf{KS}}()$:**

if state$_\mathcal{S} = \varepsilon$ then
   state$_\mathcal{S} \leftarrow$ KS
$i \leftarrow$ state$_\mathcal{S}$
state$_\mathcal{S} \leftarrow i + 1 \bmod 2^n$
**return** $(1, \text{state}_\mathcal{S})$

**Oracle $\mathcal{K}$:**

$K_1 \xleftarrow{\$} \{0, 1\}^k$
$V \xleftarrow{\$} \{0, 1\}^n$
**return** $(\varepsilon, V, K_1)$

**Fig. 7.** A general decomposition of CTR parameterized by the IV selection function, iv-gen.

---

*Interpretation.* Assume $q_P \gg q_E, q$. Using the const IV generation function permits $\sigma = 2^{n/3}, q = 2^{n/3}$ (up to constants) as long as $2^{k-n/3} \gg q_P$. This allows on the order of $2^{2n/3}$ $n$-bit blocks of data to be securely encrypted, beating the birthday bound. However, the constraint on $q_P$ may be worrisome for, e.g., $n = 64$, $k = 80$, which is only secure against adversaries for which $q_P \ll 2^{59}$. Using a predictable IV amplifies the effectiveness of precomputation because the adversary knows what precomputations will likely be helpful (in this case, finding preimages of $E_K(0^n)$). On the other hand, unique and rand also permit $\sigma = q = 2^{n/3}$, but the $\mathcal{O}(q_P q/2^k)$ term is now $\mathcal{O}(q_P/2^k)$. Precomputation is no longer nearly as much of a threat.

This $\mathcal{O}(q_P q/2^k)$ term for const corresponds to the following attack: Precompute $Y = E_K(0^n)$ for $q_P$ arbitrary keys $K$, and store each $K$ in a hash table using $Y$ as the hash table key. Encrypt the string $0^{2n}$ $q$ times, and perform a hash table lookup of the first $n$ bits of the ciphertext. This recovers the key if it happened to be one of the $q_P$ values used during precomputation. False positives can be all but eliminated by verifying the second $n$ bits of the ciphertext.

# References

1. Abdalla, M., Bellare, M.: Increasing the lifetime of a key: a comparative analysis of the security of re-keying techniques. In: Okamoto, T. (ed.) ASIACRYPT 2000. LNCS, vol. 1976, pp. 546–559. Springer, Heidelberg (2000). doi:10.1007/3-540-44448-3_42
2. Andreeva, E., Bogdanov, A., Dodis, Y., Mennink, B., Steinberger, J.P.: On the indifferentiability of key-alternating ciphers. In: Canetti, R., Garay, J.A. (eds.) CRYPTO 2013, Part I. LNCS, vol. 8042, pp. 531–550. Springer, Heidelberg (2013). doi:10.1007/978-3-642-40041-4_29
3. Bellare, M., Boldyreva, A., Micali, S.: Public-key encryption in a multi-user setting: security proofs and improvements. In: Preneel, B. (ed.) EUROCRYPT 2000. LNCS, vol. 1807, pp. 259–274. Springer, Heidelberg (2000). doi:10.1007/3-540-45539-6_18
4. Bellare, M., Krovetz, T., Rogaway, P.: Luby-Rackoff backwards: increasing security by making block ciphers non-invertible. In: Nyberg, K. (ed.) EUROCRYPT 1998. LNCS, vol. 1403, pp. 266–280. Springer, Heidelberg (1998). doi:10.1007/BFb0054132
5. Bellare, M., Ristenpart, T., Tessaro, S.: Multi-instance security and its application to password-based cryptography. In: Safavi-Naini, R., Canetti, R. (eds.) CRYPTO 2012. LNCS, vol. 7417, pp. 312–329. Springer, Heidelberg (2012). doi:10.1007/978-3-642-32009-5_19
6. Bellare, M., Rogaway, P.: The security of triple encryption and a framework for code-based game-playing proofs. In: Vaudenay, S. (ed.) EUROCRYPT 2006. LNCS, vol. 4004, pp. 409–426. Springer, Heidelberg (2006). doi:10.1007/11761679_25
7. Bogdanov, A., Knudsen, L.R., Leander, G., Standaert, F.-X., Steinberger, J.P., Tischhauser, E.: Key-alternating ciphers in a provable setting: encryption using a small number of public permutations (extended abstract). In: Pointcheval, D., Johansson, T. (eds.) EUROCRYPT 2012. LNCS, vol. 7237, pp. 45–62. Springer, Heidelberg (2012). doi:10.1007/978-3-642-29011-4_5
8. Chen, S., Steinberger, J.: Tight security bounds for key-alternating ciphers. In: Nguyen, P.Q., Oswald, E. (eds.) EUROCRYPT 2014. LNCS, vol. 8441, pp. 327–350. Springer, Heidelberg (2014). doi:10.1007/978-3-642-55220-5_19
9. Recommendation for random number generation using deterministic random bit generators. National Institute of Standards and Technology, NIST Special Publication 800–90A, U.S. Department of Commerce, January 2012
10. Dai, Y., Lee, J., Mennink, B., Steinberger, J.P.: The security of multiple encryption in the ideal cipher model. In: Garay, J.A., Gennaro, R. (eds.) CRYPTO 2014. LNCS, vol. 8616, pp. 20–38. Springer, Heidelberg (2014). doi:10.1007/978-3-662-44371-2_2
11. Hamburg, M., Kocher, P., Marson, M.E.: Analysis of Intel's Ivy Bridge digital random number generator (2012). http://www.cryptography.com/public/pdf/Intel_TRNG_Report_20120312.pdf

12. Hoang, V.T., Tessaro, S.: Key-alternating ciphers and key-length extension: exact bounds and multi-user security. In: Robshaw, M., Katz, J. (eds.) CRYPTO 2016. LNCS, vol. 9814, pp. 3–32. Springer, Heidelberg (2016). doi:10.1007/978-3-662-53018-4_1

13. Jaulmes, É., Joux, A., Valette, F.: On the security of randomized CBC-MAC beyond the birthday paradox limit a new construction. In: Daemen, J., Rijmen, V. (eds.) FSE 2002. LNCS, vol. 2365, pp. 237–251. Springer, Heidelberg (2002). doi:10.1007/3-540-45661-9_19

14. Kilian, J., Rogaway, P.: How to protect DES against exhaustive key search (an analysis of DESX). J. Cryptology 14(1), 17–35 (2001)

15. Lampe, R., Patarin, J., Seurin, Y.: An asymptotically tight security analysis of the iterated even-mansour cipher. In: Wang, X., Sako, K. (eds.) ASIACRYPT 2012. LNCS, vol. 7658, pp. 278–295. Springer, Heidelberg (2012). doi:10.1007/978-3-642-34961-4_18

16. Lampe, R., Seurin, Y.: Security analysis of key-alternating feistel ciphers. Cryptology ePrint Archive, Report 2014/151 (2014). http://eprint.iacr.org/2014/151

17. Lee, J.: Towards key-length extension with optimal security: cascade encryption and xor-cascade encryption. In: Johansson, T., Nguyen, P.Q. (eds.) EUROCRYPT 2013. LNCS, vol. 7881, pp. 405–425. Springer, Heidelberg (2013). doi:10.1007/978-3-642-38348-9_25

18. Mouha, N., Luykx, A.: Multi-key security: the even-mansour construction revisited. In: Gennaro, R., Robshaw, M. (eds.) CRYPTO 2015. LNCS, vol. 9215, pp. 209–223. Springer, Heidelberg (2015). doi:10.1007/978-3-662-47989-6_10

19. Shrimpton, T., Terashima, R.S.: A provable-security analysis of Intel's secure key RNG. In: Oswald, E., Fischlin, M. (eds.) EUROCRYPT 2015. LNCS, vol. 9056, pp. 77–100. Springer, Heidelberg (2015). doi:10.1007/978-3-662-46800-5_4

# How to Build Fully Secure Tweakable Blockciphers from Classical Blockciphers

Lei Wang[1,4]([✉]), Jian Guo[2], Guoyan Zhang[3], Jingyuan Zhao[4], and Dawu Gu[1]

[1] Department of Computer Science and Engineering,
Shanghai Jiao Tong University, Shanghai, China
wanglei_hb@sjtu.edu.cn, dwgu@sjtu.edu.cn
[2] Nanyang Technological University, Singapore, Singapore
guojian@ntu.edu.sg
[3] School of Computer Science and Technology, Shandong University, Jinan, China
guoyanzhang@sdu.edu.cn
[4] State Key Laboratory of Information Security,
Institute of Information Engineering, Chinese Academy of Sciences,
Beijing 100093, China
jingyuanzhao@live.com

**Abstract.** This paper focuses on building a tweakable blockcipher from a classical blockcipher whose input and output wires all have a size of $n$ bits. The main goal is to achieve full $2^n$ security. Such a tweakable blockcipher was proposed by Mennink at FSE'15, and it is also the only tweakable blockcipher so far that claimed full $2^n$ security to our best knowledge. However, we find a key-recovery attack on Mennink's proposal (in the proceeding version) with a complexity of about $2^{n/2}$ adversarial queries. The attack well demonstrates that Mennink's proposal has at most $2^{n/2}$ security, and therefore invalidates its security claim. In this paper, we study a construction of tweakable blockciphers denoted as $\widetilde{\mathbb{E}}[s]$ that is built on $s$ invocations of a blockcipher and additional simple XOR operations. As proven in previous work, at least two invocations of blockcipher with linear mixing are necessary to possibly bypass the birthday-bound barrier of $2^{n/2}$ security, we carry out an investigation on the instances of $\widetilde{\mathbb{E}}[s]$ with $s \geq 2$, and find 32 highly efficient tweakable blockciphers $\widetilde{E1}, \widetilde{E2}, \ldots, \widetilde{E32}$ that achieve $2^n$ provable security. Each of these tweakable blockciphers uses two invocations of a blockcipher, one of which uses a tweak-dependent key generated by XORing the tweak to the key (or to a secret subkey derived from the key). We point out the provable security of these tweakable blockciphers is obtained in the ideal blockcipher model due to the usage of the tweak-dependent key.

**Keywords:** Tweakable blockcipher · Full security · Ideal blockcipher · Tweak-dependent key

## 1 Introduction

Tweakable blockcipher, formalized by Liskov *et al.* [34,35], introduces an additional parameter called *tweak* to the classical blockcipher. More formally,

© International Association for Cryptologic Research 2016
J.H. Cheon and T. Takagi (Eds.): ASIACRYPT 2016, Part I, LNCS 10031, pp. 455–483, 2016.
DOI: 10.1007/978-3-662-53887-6_17

a classical blockcipher $E : \mathcal{K} \times \mathcal{M} \rightarrow \mathcal{M}$ is a family of permutations on $\mathcal{M}$ indexed by a secret key $k \in \mathcal{K}$. A tweakable blockcipher $\widetilde{E} : \mathcal{K} \times \mathcal{T} \times \mathcal{M} \rightarrow \mathcal{M}$ is a family of permutations on $\mathcal{M}$, indexed by two functionally distinct parameters: a key $k \in \mathcal{K}$ that is secret and used to provide the security, and a tweak $t \in \mathcal{T}$ that is public and used to provide the variability. The tweak is assumed to be known or even controlled by the adversary. $\widetilde{E}$ is considered secure if it with a secret key $k$ uniformly chosen from the key space $\mathcal{K}$ is indistinguishable from an ideal tweakable blockcipher $\widetilde{P} : \mathcal{T} \times \mathcal{M} \rightarrow \mathcal{M}$ that is a family of random permutations on $\mathcal{M}$ indexed by a public tweak $t \in \mathcal{T}$. As a more natural primitive for building modes of operation, tweakable blockcipher has found wide applications. Examples include encryption schemes [7,16,23,43,49,53], authenticated encryption [1,34,47,48], and disk encryption [24,25]. Moreover, many candidates of the ongoing cryptographic competition CAESAR [5] on authenticated encryption are based on tweakable blockciphers, e.g., Deoxys [29], Joltik [30], Scream [22], SHELL [51], etc.

There are mainly three approaches to design a tweakable blockcipher. The first one is from the scratch, including Hasty Pudding Cipher [50], Mercy [12] and Threefish (used in the hash function SKEIN [19]). Such designs usually have a drawback of lacking a security proof.

The second approach is to introduce the additional parameter tweak to generic constructions of blockcipher, including tweaking Luby-Rackoff cipher or Feistel cipher [20], tweaking Generalized Feistel cipher [44] and tweaking key-alternating cipher or (iterated) Even-Mansour [9–11,18,21,28,39]. These tweakable blockciphers except TWEAKEY framework in [28] are provably secure. In details, the designs in [11,18,20,39,44] have a provable security up to $2^{n/2}$ adversarial queries, often referred to as the *birthday-bound* security with respect to the $n$-bit block size of the underlying blockcipher (that is, the message space $\mathcal{M} = \{0,1\}^n$). To bypass the birthday-bound barrier and to achieve a higher security bound, Jean *et al.* proposed TWEAKEY framework [28] to construct ad-hoc tweakable blockciphers from key-alternating ciphers, and specified several TWEAKEY instances which are conjectured fully $2^n$ secure but lack formal security proofs. After that, Cogliati *et al.* designed several tweakable blockciphers[1] by tweaking Even-Mansour ciphers in [9,10], and these proposals are provably secure up to $2^{2n/3}$ adversarial queries.

The last and the most common approach is to start from a classical blockcipher and to use it as a black box to build a tweakable blockcipher, including LRW1 [34], LRW2 [34], variants and extensions of LRW2 such as XEX and CLRW2 [6,31,32,40,46,47], Minematsu's design [41] and Mennink's design [36]. Early proposals LRW1, LRW2, XEX and their variants [6,34,40,47] are limited to the birthday-bound security. After that, cryptographers considered the cascade of LRW2 in order to design tweakable blockciphers achieving beyond-birthday-bound security. One evaluation of LRW2 contains one invocation of a blockcipher, one invocation of a universal hash function, and each evaluation

---

[1] These tweakable blockciphers can be regarded as instances of TWEAKEY framework.

of LRW2 in the cascade construction requires an independent secret key. Landecker *et al.* proposed CLRW2 [32] that makes two evaluations of LRW2 (that is, two calls to a blockcipher, two invocations of a universal hash function, and two secret keys), and is proven secure up to $2^{2n/3}$ adversarial queries.[2] Lampe and Seurin analyzed the general case of the cascade of LRW2 [31]. For such a tweakable blockcipher making $s$ evaluations of LRW2 (that is, $s$ invocations of the underlying blockcipher and universal hash function, and $s$ secret keys), they proved that it has a security up to $2^{sn/(s+2)}$ queries (against adaptive chosen-ciphertext adversaries), and also conjectured that its security bound can be improved to $2^{sn/(s+1)}$ queries. Therefore by increasing the integer $s$, these tweakable blockciphers *asymptotically* approach full $2^n$ security, but meanwhile the efficiency gets worse as the necessary number of blockcipher invocations, universal hash function invocations, and the necessary key size linearly increase with $s$. Another direction to design a tweakable blockcipher achieving beyond-birthday-bound security is to use so-called *tweak-dependent key*. Roughly speaking, a tweak-dependent key is a key of an invocation of blockcipher in a tweakable blockcipher that is generated depending on the tweak. Liskov *et al.* suggested in [34] that changing the tweak should be less costly than changing the key from the efficiency concerns. Following it, early proposals of tweakable blockcipher avoided the usage of the tweak-dependent key. However, recently Jean *et al.* [28] pointed out that this suggestion is somewhat counter-intuitive from the security concern, because the adversary has full control on the tweak, but has very limited control on the key. They suggested that the tweak and the key should be treated comparably. In fact even before Jean *et al.*'s work, Minematsu [41] proposed a tweakable blockcipher built on two invocations of blockcipher, one of which uses a tweak-dependent key. His design is proven secure up to $\max\{2^{n/2}, 2^{n-|t|}\}$ adversarial queries, where $|t|$ is the bit size of the tweak (that is, the tweak space $\mathcal{T} = \{0,1\}^{|t|}$). Hence Minematsu's design is beyond-birthday-bound secure as long as the tweak is shorter than $n/2$ bits. A scheme XTX has been proposed to extend the tweak-length of any black-box tweakable blockcipher by using a universal hash function [42]. Recently Mennink [36] proposed two tweakable blockciphers $\widetilde{F}[1]$ and $\widetilde{F}[2]$ with the usage of the tweak-dependent key. $\widetilde{F}[1]$ consists of one invocation of blockcipher and one finite-field multiplication, and is proven secure up to $2^{2n/3}$ adversarial queries. $\widetilde{F}[2]$ makes two calls to blockcipher, and is *surprisingly* proven secure up to $2^n$ adversarial queries, that is achieving full security with very high efficiency. On the other hand, the security proof of Mennink's designs [36] are in the ideal blockcipher (information-theoretic) model, while other proposals [6,31,32,34,40,41,47] have security proofs in the standard (complexity-theoretic) model of assuming the underlying blockcipher as a pseudorandom permutation.

**Our Contributions.** In this paper, we focus on constructing tweakable blockciphers that achieve full $2^n$ security. This is mainly motivated by the scenarios where the blockciphers only have 32-, 48- or 64-bit block size, e.g., Simon and

---

[2] A flaw in the original proof was found and fixed by Procter [46].

Speck family of blockciphers [3] (refer to Sect. 4.2 for more discussions). As summarized above, so far there is only one tweakable blockcipher $\widetilde{F}[2]$ designed by Mennink [36] that claims full security. As a first contribution, we present a key-recovery attack on $\widetilde{F}[2]$ with a complexity of around $2^{n/2}$ adversarial queries, which invalidates the designer's security claim in [36]. Our attack has been verified by the designer [38]. Accordingly Mennink proposed a patch [37] to $\widetilde{F}[2]$ of the proceeding version, which can resist our key-recovery attack.

This paper designs tweakable blockciphers from classical blockciphers in the black-box way, that is following the above third design approach. We focus on a construction of tweakable blockcipher (see Fig. 2 as an example) denoted as $\widetilde{\mathbb{E}}[s]$ : $\mathcal{K} \times \mathcal{T} \times \mathcal{M} \to \mathcal{M}$, which consists of $s$ invocations of a blockcipher $E : \mathcal{K} \times \mathcal{M} \to \mathcal{M}$ and extra simple XOR operations. As a second and main contribution, we carry out a heuristic search to investigate the instances of $\widetilde{\mathbb{E}}[s]$, and successfully find 32 highly efficient tweakable blockciphers $\widetilde{E1}, \widetilde{E2}, \ldots$, and $\widetilde{E32}$ that achieve full $2^n$ security. Each of these tweakable blockcipher (see Figs. 6 and 7) makes two calls to the blockcipher $E$. In details, the first blockcipher call is to derive a secret subkey $y$ from the key $k$ such that $y = E(k, k)$, $y = E(k, 0)$ or $y = E(0, k)$. The second blockcipher call encrypts a plaintext $p$ (or decrypts a ciphertext $c$) with a tweak-dependent key, which is generated by XORing the tweak $t$ to the key $k$, the subkey $y$, or $k \oplus y$. In particular, we stress that by pre-computing and storing the subkey $y$, our tweakable blockciphers just need to make one blockcipher call for encrypting $(t, p)$ or decrypting $(t, c)$.

A comparison with previous tweakable blockciphers is detailed in Table 1. The main advantage of our designs is optimal $2^n$ provable security and high efficiency. From the security view, previous tweakable blockciphers except LRW2[s](with $s \to \infty$) and the patched $\widetilde{F}[2]$ (in ePrint version) have (at most) $2^{2n/3}$ provable security. From the efficiency view, LRW2[s] requires $s$ blockcipher calls, and $s$ universal hash function invocations, and hence the efficiency is significantly worse. Our designs also have an efficiency advantage compared with the patched $\widetilde{F}[2]$, as our designs require just one blockcipher call for encrypting a plaintext or decrypting a ciphertext when the subkey is pre-computed and stored.

**Organization.** The rest of the paper is organized as follows. Section 2 gives notations and definitions. Section 3 describes a key-recovery attack on Mennink's proposal. Section 4 presents the target construction, design goal and search strategy. We then write the search procedure and the found constructions in Sect. 5, and provide security proofs in Sect. 6. Finally we conclude the paper in Sect. 7.

# 2 Preliminaries

## 2.1 Notations

$\{0, 1\}^n$ denotes the set of all $n$-bit strings. For $a, b \in \{0, 1\}^n$, $a \oplus b$ denotes their bitwise exclusive-OR (XOR). For $a \in \{0, 1\}$ and $b \in \{0, 1\}^b$, $a \cdot b$ denotes the multiplication of $a$ and $b$, that is equal to $b$ if $a = 1$, and equal to 0 if

**Table 1.** Comparison of our designs with previous tweakable blockciphers: if we pre-compute and store the subkey, $\widetilde{E}1,\ldots,\widetilde{E}32$ require just one blockcipher call for encrypting a plaintext or decrypting a ciphertext.

| tweakable blockciphers | key size | security $(\log_2)$ | cost $E$ | $\otimes/h$ | tdk | reference |
|---|---|---|---|---|---|---|
| LRW1 | $n$ | $n/2$ | 1 | 0 | N | [34] |
| LRW2 | $2n$ | $n/2$ | 1 | 2 | N | [34] |
| XEX | $n$ | $n/2$ | 1 | 0 | N | [47] |
| LRW2[2] | $4n$ | $2n/3$ | 2 | 2 | N | [32] |
| LRW2[s] | $2sn$ | $sn/(s+2)$ | $s$ | $s$ | N | [31] |
| Min | $n$ | $\max\{n/2, n-\lvert t\rvert\}$ | 2 | 0 | Y | [41] |
| $\widetilde{F}[1]$ | $n$ | $2n/3$ | 1 | 1 | Y | [36] |
| $\widetilde{F}[2]$ | $n$ | $n/2$ | 2 | 0 | Y | [36] |
| patched $\widetilde{F}[2]$ | $n$ | $n$ | 2 | 0 | Y | [37] |
| $\widetilde{E}1,\ldots,\widetilde{E}32$ | $n$ | $n$ | 2 (1) | 0 | Y | Sect. 5 |

- $\otimes/h$ stands for multiplications or universal hashes;
- tdk stands for the tweak-dependent key. 'N' refers to not using tdk, and 'Y' refers to using tdk;
- $\lvert t\rvert$ stands for the bit length of the tweak;

$a = 0$. For a finite set $\mathcal{X}$, $x \xleftarrow{\$} \mathcal{X}$ denotes that an element $x$ is selected from $\mathcal{X}$ uniformly at random. $\lvert\mathcal{X}\rvert$ denotes the number of the elements in $\mathcal{X}$. Blockcipher is commonly denoted as $E : \mathcal{K} \times \mathcal{M} \to \mathcal{M}$, and tweakable blockcipher as $\widetilde{E} : \mathcal{K} \times \mathcal{T} \times \mathcal{M} \to \mathcal{M}$, where $\mathcal{K}$ is the key space, $\mathcal{T}$ is the tweak space, and $\mathcal{M}$ is the message space. Throughout this paper, we fix $\mathcal{K} = \mathcal{T} = \mathcal{M} = \{0,1\}^n$. Let $E(k, \cdot)$ and $E^{-1}(k, \cdot)$ be the encryption and the decryption of blockcipher $E$ with a key $k \in \mathcal{K}$ respectively. Let $E^{\pm}(k, \cdot)$ consist of both $E(k, \cdot)$ and $E^{-1}(k, \cdot)$. Sometimes we denote $E(k, \cdot)$, $E^{-1}(k, \cdot)$ and $E^{\pm}(k, \cdot)$ as $E_k(\cdot)$, $E_k^{-1}(\cdot)$ and $E_k^{\pm}(\cdot)$ respectively. Similarly we define notations $\widetilde{E}(k, \cdot, \cdot)$, $\widetilde{E}^{-1}(k, \cdot, \cdot)$, and $\widetilde{E}^{\pm}(k, \cdot, \cdot)$ for tweakable blockcipher $\widetilde{E}$, which can also be denoted as $\widetilde{E}_k(\cdot, \cdot)$, $\widetilde{E}_k^{-1}(\cdot, \cdot)$ and $\widetilde{E}_k^{\pm}(\cdot, \cdot)$, respectively. An input-output tuple of $E$ is commonly denoted as $(l, u, w)$ such that $w = E(l, u)$. An input-output tuple of $\widetilde{E}_k$ with $k \xleftarrow{\$} \mathcal{K}$ is denoted as $(t, p, c)$ such that $\widetilde{E}_k(t, p) = c$. Let Bloc be the set of all blockciphers with key space $\mathcal{K}$ and message space $\mathcal{M}$. A blockcipher $E$ is said to be an ideal blockcipher if it is selected from Bloc uniformly at random, that is $E \xleftarrow{\$}$ Bloc. Let $\widetilde{\text{Perm}}$ be the set of all functions $\widetilde{P} : \mathcal{T} \times \mathcal{M} \to \mathcal{M}$ such that for each $t \in \mathcal{T}$, $\widetilde{P}(t, \cdot)$ is a permutation on $\mathcal{M}$. A function $\widetilde{P}$ is said to be an ideal tweakable blockcipher if it is selected from $\widetilde{\text{Perm}}$ at random, that is $\widetilde{P} \xleftarrow{\$} \widetilde{\text{Perm}}$. Similarly we define notations $\widetilde{P}(\cdot, \cdot)$, $\widetilde{P}^{-1}(\cdot, \cdot)$ and $\widetilde{P}^{\pm}(\cdot, \cdot)$.

## 2.2   Tweakable Blockcipher and Security Definition

A *distinguisher* $\mathcal{D}$ is an algorithm that is given query access to one (or more) oracle of being either $\mathcal{O}$ or $\mathcal{Q}$, and outputs one bit. Its advantage in distinguishing these two primitives $\mathcal{O}$ and $\mathcal{Q}$ is defined as

$$\mathbf{Adv}(\mathcal{D}) = \left| \Pr\left[ \mathcal{D}^{\mathcal{O}} \Rightarrow 1 \right] - \Pr\left[ \mathcal{D}^{\mathcal{Q}} \Rightarrow 1 \right] \right|$$

A *tweakable blockcipher* with key space $\mathcal{K}$, tweak space $\mathcal{T}$ and message space $\mathcal{M}$ is a mapping $\widetilde{E} : \mathcal{K} \times \mathcal{T} \times \mathcal{M} \to \mathcal{M}$ such that for any key $k \in \mathcal{K}$ and any tweak $t \in \mathcal{T}$, $\widetilde{E}(k,t,\cdot)$ is a permutation over $\mathcal{M}$. The security of a tweakable blockcipher is defined via upper bounding the advantage of distinguisher $\mathcal{D}$ in the following game. $\mathcal{D}$ is given query access to oracles $(\mathcal{O}_1, E^{\pm})$: $\mathcal{O}_1$ is either $\widetilde{E}_k^{\pm}(\cdot,\cdot)$ with $k \xleftarrow{\$} \mathcal{K}$ or an ideal tweakable blockcipher $\widetilde{P}(\cdot,\cdot) \xleftarrow{\$} \widetilde{\mathrm{Perm}}$; $E^{\pm}$ is an ideal blockcipher (that is $E \xleftarrow{\$} \mathrm{Bloc}$) which is used as the underlying blockcipher of $\widetilde{E}$. The advantage of $\mathcal{D}$ in distinguishing $\widetilde{E}$ and $\widetilde{P}$ is defined as

$$\mathbf{Adv}_{\widetilde{E}}^{\widetilde{\mathrm{sprp}}}(\mathcal{D}) = \left| \Pr\left[ \mathcal{D}^{\widetilde{E}_k^{\pm}(\cdot,\cdot), E^{\pm}(\cdot,\cdot)} \Rightarrow 1 \right] - \Pr\left[ \mathcal{D}^{\widetilde{P}^{\pm}(\cdot,\cdot), E^{\pm}(\cdot,\cdot)} \Rightarrow 1 \right] \right|,$$

where the probabilities are taken over the choices of $k \xleftarrow{\$} \mathcal{K}$, $E \xleftarrow{\$} \mathrm{Bloc}$, $\widetilde{P} \xleftarrow{\$} \widetilde{\mathrm{Perm}}$, and $\mathcal{D}$'s coin (if any).

Throughout the paper, we consider information-theoretic distinguisher $\mathcal{D}$ such that $\mathcal{D}$ is computationally unbounded, but sorely limited by the number of queries to its oracles. We write

$$\mathbf{Adv}_{\widetilde{E}}^{\widetilde{\mathrm{sprp}}}(q) = \max_{\mathcal{D}}\{\mathbf{Adv}_{\widetilde{E}}^{\widetilde{\mathrm{sprp}}}(\mathcal{D})\},$$

where the maximum is taken over all distinguisher $\mathcal{D}$ that makes at most $q$ queries to its oracles.

A *tweak-dependent key* of a tweakable blockcipher is a key of an invocation of blockcipher which is generated depending on the tweak. In other words, changing the value of tweak leads to re-keying that blockcipher call. Liskov *et al.* suggested in [34] that changing the tweak should be less costly than changing the key. However, Jean *et al.* [28] pointed out that this suggestion is counter-intuitive, because the adversary has full control on the tweak, but has very limited control on the key. Indeed the tweak and the key should be treated comparably.

## 2.3   The H-Coefficient Technique

Our proof adopts the H-coefficient Technique [8,45], which is briefly introduced as follows. This paper considers information-theoretic distinguisher $\mathcal{D}$ that is computationally unbounded. Hence without loss of generality, we always assume $\mathcal{D}$ is deterministic. Suppose $\mathcal{D}$ interacts with $\mathcal{O}$ and $\mathcal{Q}$, and its advantage is defined in Sect. 2.2. A view $v$ is the query-response tuples that $\mathcal{D}$ receives when interacting with $\mathcal{O}$ or $\mathcal{Q}$. Let $X$ be the probability distribution of the view when

$\mathcal{D}$ interacts with $\mathcal{O}$, and $Y$ be the probability distribution of the view when $\mathcal{D}$ interacts with $\mathcal{Q}$. $\mathcal{V}$ is defined as the set of all attainable views $v$ while $\mathcal{D}$ interacting with $\mathcal{Q}$, that is $\mathcal{V} = \{v \mid \Pr[Y = v] > 0\}$.

The H-coefficient technique evaluates the upper bound of $\mathbf{Adv}(\mathcal{D})$ as follows. Firstly, partition $\mathcal{V}$ to two disjoint subsets $\mathcal{V}_{\text{good}}$ and $\mathcal{V}_{\text{bad}}$ such that $\mathcal{V} = \mathcal{V}_{\text{good}} \cup \mathcal{V}_{\text{bad}}$. Secondly, estimate a real value $\epsilon_{v_{\text{good}}}$ with $0 \leq \epsilon_{v_{\text{good}}} \leq 1$ such that for each view $v \in \mathcal{V}_{\text{good}}$, it has that

$$\frac{\Pr[X = v]}{\Pr[Y = v]} \geq 1 - \epsilon_{v_{\text{good}}}.$$

Moreover, compute the probability of $\mathcal{D}$ receiving a view from $\mathcal{V}_{\text{bad}}$ when interacting with $\mathcal{Q}$, that is $\Pr[Y \in \mathcal{V}_{\text{bad}}]$. Finally, conclude that the advantage of $\mathcal{D}$ is upper bounded as

$$\mathbf{Adv}(\mathcal{D}) \leq \epsilon_{v_{\text{good}}} + \Pr[Y \in \mathcal{V}_{\text{bad}}].$$

## 3    Key-Recovery Attack on Mennink's Design [36]

This section presents a key-recovery attack on Mennink's design in [36], which is depicted in Fig. 1. Let $\widetilde{E}_k : \mathcal{T} \times \mathcal{M} \to \mathcal{M}$ to denote Mennink's tweakable blockcipher with a secret key $k \in \mathcal{K}$ and $E : \mathcal{K} \times \mathcal{M} \to \mathcal{M}$ to denote its underlying blockcipher. The key-recovery attacker has query access to $\widetilde{E}_k^{\pm}(\cdot, \cdot)$ and $E^{\pm}(\cdot, \cdot)$. The attack procedure is detailed below.

At first step, the attacker recovers the value of $E(k, 0)$ by sending one query $(0, 0)$ to $\widetilde{E}_k^{-1}(\cdot, \cdot)$ to receive a plaintext $p$ such that $E(k, 0) = p$ holds. This is based on an observation for the case of tweak $t = 0$ and ciphertext $c = 0$.

- tweak $t = 0$ implies that the two blockcipher calls in $\widetilde{E}_k$ shares the same key value, and hence are identical permutation.
- ciphertext $c = 0$ implies that the outputs of two blockcipher calls in $\widetilde{E}_k$ are equal from $c = y_1 \oplus y_2 = 0$, that is $y_1 = y_2$.

When querying $(t = 0, c = 0)$ to $\widetilde{E}_k^{-1}(\cdot, \cdot)$, it has that $x_2 = t = 0$, and in turn the received plaintext $p = y_1 \oplus x_2 = y_1$, where $y_1$ is computed as $y_1 = E(k, 0)$. Hence the attacker gets the value of $E(k, 0)$ by sending one query $(0, 0)$ to $\widetilde{E}_k^{-1}(\cdot, \cdot)$.

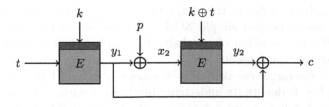

**Fig. 1.** Tweakable blockcipher in [36]

At second step, the attacker collects and stores a set of $E(k \oplus t, \text{const})$, where const is a fixed constant, for $2^{n/2}$ distinct tweak values $t$ by making $2^{n/2+1}$ queries to $\widetilde{E}_k(\cdot, \cdot)$. In details, for each tweak $t$, the attacker starts with recovering the value of $E(k,t)$ by sending one query $(0, E(k,0) \oplus t)$ to $\widetilde{E}_k(\cdot, \cdot)$ to receive ciphertext $c$ and computing $E(k,t) = c \oplus E(k,0)$. The reason is as follows. Note $y_1 = E(k,0)$. It has $x_2 = (E(k,0) \oplus t) \oplus y_1 = t$, which implies $y_2 = E(k,t)$. Also from $c = y_1 \oplus y_2$, it has that $y_2 = c \oplus y_1 = c \oplus E(k,0)$. Hence $E(k,t)$ is equal to $c \oplus E(k,0)$. Next and with a similar reason, the attacker recovers the value of $E(k \oplus t, \text{const})$ by sending one query $(t, E(k,t) \oplus \text{const})$ to $\widetilde{E}_k(\cdot, \cdot)$, and computing $E(k \oplus t, \text{const}) = c \oplus E(k,t)$. Overall, the attacker is able to recover the value of $E(k \oplus t, \text{const})$ for any tweak $t$, by sending two queries to $\widetilde{E}_k(\cdot, \cdot)$.

At third and the last step, the attacker selects $2^{n/2}$ distinct values $l$, queries $(l, \text{const})$ to $E(\cdot, \cdot)$ to receive $E(l, \text{const})$, and matches it to the set $\{E(k \oplus t, \text{const})\}$ stored at second step. If a match is found that is $E(k \oplus t, \text{const}) = E(l, \text{const})$, the attacker recovers the secret key $k$ as $k = l \oplus t$.

Now we evaluate the complexity and the success probability. The first step requires one query, the second step requires $2^{n/2+1}$ queries and the last step requires $2^{n/2}$ queries. Summing up, the total complexity is less than $2^{n/2+2}$ queries. Since there are $2^{n/2}$ distinct tweak values $t$ and $2^{n/2}$ distinct values $l$, the probability of existing a value of $t$ and a value of $l$ such that $t \oplus l = k$ is trivially computed as $1 - (1 - 2^{-n})^{2^n} \approx 1 - 1/e \approx 0.63$. Hence the success probability of recovering the key is about 0.63. Overall, the tweakable blockcipher designed by Mennink in [36] has at most around $2^{n/2}$ security, in other words, birthday-bound security, which is exponentially far lower than the designer's claim of full $2^n$ security.

**On proof flaw in [36].** In the proof, under the condition that the attacker cannot guess the key correctly (that is, (12a) defined in [36] is not set), it claimed that the distribution of output variable of the first blockcipher call, $y_1 = E(k,t)$, is independent from the second blockcipher call $y_2 = E(k \oplus t, x_2)$. This is a wrong claim. When tweak $t = 0$, both the two blockcipher calls share the same key, and therefore the distribution of their outputs are highly related.

## 4    Target Construction, Design Goal and Search Strategy

### 4.1    Tweakable Blockcipher $\widetilde{\mathbb{E}}[s]$

In this paper, we study a construction of tweakable blockcipher consisting of blockcipher calls and linear transformations. Furthermore, we restrict linear transformations to be just simple XOR operations for efficiency benefits. For a more generic construction of tweakable blockcipher from a classical blockcipher, we refer interested readers to [36].

We denote the target tweakable blockcipher as $\widetilde{\mathbb{E}}[s]$, which is built on $s$ blockcipher calls. Let $E$ denote its underlying blockcipher with $n$-bit block size and $n$-bit key size. Let $k$, $t$, $p$ and $c$ denote its key, tweak, plaintext and ciphertext, respectively, which are all $n$-bit long. Let $a_{i,j}$ and $b_{i,j}$ for $1 \leq i \leq s+1$ and

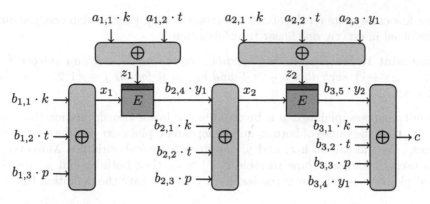

**Fig. 2.** Graphical view of $\widetilde{\mathbb{E}}[2]$ with key $k$, tweak $t$, plaintext $p$, ciphertext $c$, $a_{i,j} \in \{0,1\}$ and $b_{i,j} \in \{0,1\}$

---

**Algorithm 1.** Encryption of $\widetilde{E}[s](\cdot,\cdot,\cdot)$: '+' stands for addition operation in $GF(2^n)$, that is XOR operation.

---

**Input:** key $k$, plaintext $p$, tweak $t$, blockcipher $E(\cdot,\cdot)$, one-bit variables $a_{i,j}$'s and $b_{i,j}$'s

**Output:** ciphertext $c$

1. $x_1 = b_{1,1} \cdot k + b_{1,2} \cdot t + b_{1,3} \cdot p$
2. $z_1 = a_{1,1} \cdot k + a_{1,2} \cdot t$
3. **for** $i = 1$ **to** $s - 1$, **do**
4. $\quad y_i = E(z_i, x_i)$
5. $\quad x_{i+1} = b_{i+1,1} \cdot k + b_{i+1,2} \cdot t + b_{i+1,3} \cdot p + \sum_{j=4}^{i+3} b_{i+1,j} \cdot y_{j-3}$
6. $\quad z_{i+1} = a_{i+1,1} \cdot k + a_{i+1,2} \cdot t + \sum_{j=3}^{i+2} a_{i+1,j} \cdot y_{j-2}$
7.
8. **endfor**
9. $y_s = E(z_s, x_s)$
10. $c = b_{s+1,1} \cdot k + b_{s+1,2} \cdot t + b_{s+1,3} \cdot p + \sum_{j=4}^{s+3} b_{s+1,j} \cdot y_{j-3}$
11. **return** ciphertext $c$

---

$1 \leq j \leq i + 2$ be one-bit variables of being 0 or 1. The encryption procedure of $\widetilde{\mathbb{E}}[s]$ is provided in Algorithm 1. Each concrete instantiation of $\widetilde{\mathbb{E}}[s]$ is to determine the values of $a_{i,j}$'s and $b_{i,j}$'s. Moreover, a graphical view of $\widetilde{\mathbb{E}}[2]$ is depicted in Fig. 2 as an example, which is also useful for next sections. Throughout this paper, we always assume that all the $s$ blockcipher calls are indeed involved in the computation of the ciphertext $c$ from the key $k$, the tweak $t$ and the plaintext $p$ for $\widetilde{\mathbb{E}}[s]$.

A tweakable blockcipher must be invertible, namely plaintext $p$ should be efficiently decrypted from key $k$, tweak $t$ and ciphertext $c$. Such a requirement sets a few constraints on the above construction $\widetilde{\mathbb{E}}[s]$. Firstly, plaintext $p$ should be involved in exactly one linear transformation.

**Constraint 1.** *For $\widetilde{\mathbb{E}}[s]$ to be invertible, there should exist an integer $i \in \{1, 2, \ldots, s+1\}$ such that $b_{i,3} = 1$ and $b_{j,3} = 0$ for all $j \in \{1, 2, \ldots, s+1\}$ and $j \neq i$.*

Secondly, suppose plaintext $p$ is involved in the linear transformation that outputs $x_i$, then the values of both $x_i$ and $y_i$ depend on plaintext $p$ in the encryption process. We will call such $x_i$ and $y_i$ *plaintext-dependent* variables. Moreover, if $y_i$ is used to compute some variable $x_j$ $(j > i)$, then both $x_j$ and $y_j$ are also called plaintext-dependent variables. Iteratively, we have the definition below.

**Definition 1.** *For our target construction $\widetilde{\mathbb{E}}[s]$, internal variables $x_i$ and $y_i$ are said to be plaintext-dependent, if $x_i$ is computed depending on plaintext $p$ or a plaintext-dependent variable $y_j$ in the encryption process. Also we include plaintext $p$ as a plaintext-dependent variable.*

A plaintext-dependent variable cannot be used to produce any key value $z_j$.[3] Otherwise, the construction is not (efficiently) invertible, since one cannot compute $z_j$ without the knowledge of plaintext $p$.

**Constraint 2.** *For $\widetilde{\mathbb{E}}[s]$ to be invertible, if an internal state $y_i$ with $1 \leq i \leq s$ is a plaintext-dependent variable, the values of $a_{j,i+2}$'s for all $j \in \{i+1, i+2, \ldots, s\}$ must be 0.*

Moreover, the linear transformation to produce any internal state $x_i$ with $1 \leq i \leq s$ should have at most one input plaintext-dependent variable. Otherwise, one cannot efficiently inverse such a linear transformation in the decryption, because there are more than one unknown input variable.

**Constraint 3.** *For $\widetilde{\mathbb{E}}[s]$ to be invertible, the linear transformations to produce internal states $x_i$'s for all $i \in \{1, 2, \ldots, s+1\}$ must have at most one input variable that is plaintext-dependent.*

Summarizing up, an instantiation of $\widetilde{\mathbb{E}}[s]$ is efficiently invertible and therefore a valid tweakable blockcipher, as long as it satisfies the above three constraints. Nevertheless, additional conditions might be necessary from the concerns of security and efficiency. For example, it is important that all $s$ blockcipher invocations of $\widetilde{\mathbb{E}}[s]$ are indeed involved for computing ciphertext $c$ from the key $k$, the tweak $t$ and plaintext $p$. Here we omit such discussions for the general case, but leave them in next sections for specific case, *e.g.*, the instances of $\widetilde{\mathbb{E}}[2]$.

---

[3] Recall that all blockcipher calls are indeed involved in the computation of ciphertext $c$ from the key $k$, the tweak $t$ and plaintext $p$.

*Remarks.* It is interesting to note that many tweakable blockciphers proposed previously are instances of our target construction $\widetilde{\mathbb{E}}[2]$ in Fig. 2. For example, LRW1 construction designed by Liskov *et al.* in [34] is the instance with $b_{1,3} = a_{1,1} = b_{2,4} = b_{2,2} = a_{2,1} = b_{3,5} = 1$ and 0 for the other $a_{i,j}$'s and $b_{i,j}$'s. Minematsu's construction in [41] is the instance with $b_{1,2} = a_{1,1} = b_{2,3} = a_{2,3} = b_{3,5} = 1$ and 0 for the other $a_{i,j}$'s and $b_{i,j}$'s. Mennink's construction in [36] is the instance with $b_{1,2} = a_{1,1} = b_{2,4} = b_{2,3} = a_{2,1} = a_{2,2} = b_{3,5} = b_{3,4} = 1$ and 0 for the other $a_{i,j}$'s and $b_{i,j}$'s.

## 4.2   Design Goal

Our first and top-priority goal is *full $2^n$ provable security*, which has both theoretical and practical interests. A typical blockcipher nowadays such as AES [14] and SIMON [3] has a block size of 128 bits or 64 bits. In some constrained environment, the block size of lightweight blockciphers can be even shorter, *e.g.*, SIMON-48 [3]. Hence tweakable blockcipher constructions with merely a birthday-bound security may not be suited for various applications. Consequently other constructions providing higher security is definitely necessary. Particularly, designing tweakable blockciphers with optimal $2^n$ provable security is indeed a very interesting research topic.

Our second goal is *the minimum number of blockcipher calls*, which obviously comes from the efficiency concern. For our target construction, a blockcipher call is much more time-consuming than linear transformations which are merely XOR operations. Therefore the number of blockcipher calls dominates the overall efficiency of tweakable blockcipher. Besides, we also aim to optimize the efficiency of linear transformations under the condition of no security sacrifice, *i.e.*, erasing unnecessary input variables. In fact this is also the reason that we have limited the linear transformations to simple XORing variables when choosing the target construction $\widetilde{\mathbb{E}}[s]$.

Our third goal is *(comparably) high efficiency of changing a tweak*, which in particular should be more efficient than changing a key. It is motivated by the fact that tweak is changed more frequently than the key in applications. For instance, in most modes of operation such as OCB [48], tweak is changed for every plaintext block, while the secret key can be kept the same for up to birthday-bound number of plaintext blocks. Such a criteria of designing tweakable blockcipher has been suggested by Liskov *et al.* [35] and followed by several constructions in [6,31,32,40,46,47]. However, differently from those constructions, we allow to use tweak-dependent keys, in other words, changing a tweak leads to re-keying blockcipher. This is due to the above goals of security and efficiency. Indeed as shown in [31], without using tweak-dependent keys, an (almost) optimal secure tweakable blockcipher requires an unrestrained increase of blockcipher calls and the number of keys.

## 4.3  Search Strategy

In order to achieve the design goals listed in Sect. 4.2, we adopt a heuristic approach to search among the instances of $\widetilde{\mathbb{E}}[s]$.

- For the goal of full $2^n$ security, we should investigate the instances of $\widetilde{\mathbb{E}}[s]$ with $s \geq 2$. The reason is that Mennink in [36] proved any instance of $\widetilde{\mathbb{E}}[1]$ (that is with linear mixing) has at most $2^{n/2}$ security. It implies that at least 2 blockcipher calls are necessary to possibly bypass birthday-bound barrier and to reach full $2^n$ security.
- For the goal of minimum number of blockcipher calls, we start with analyzing the instances of $\widetilde{\mathbb{E}}[2]$. Moreover, we will not move to investigate the instances of $\widetilde{\mathbb{E}}[s+1]$, unless we have examined all the instances of $\widetilde{\mathbb{E}}[s]$ and none of them can achieve $2^n$ security. Once some instance of $\widetilde{\mathbb{E}}[s]$ is found with $2^n$ security, it is not needed to investigate the instances of $\widetilde{\mathbb{E}}[s']$ where $s' > s$.
- For the goal of high efficiency of changing a tweak, we should use the minimum number of tweak-dependent keys. Let $i$ denote the number of tweak-dependent keys. While searching among the instances of $\widetilde{\mathbb{E}}[s]$, we start with those with one tweak-dependent key. Moreover, we will not move to investigate the instances with $i + 1$ tweak-dependent keys, unless we have examined all the instances with $i$ tweak-dependent keys and none of them can achieve $2^n$ security. Once some instance of $\widetilde{\mathbb{E}}[s]$ with $i$ tweak-dependent keys is found with $2^n$ security, it is not needed to investigate the instances of $\widetilde{\mathbb{E}}[s]$ with $i'$ tweak-dependent keys, where $i' > i$.

Following the above search strategy, we start with investigating the instances of $\widetilde{\mathbb{E}}[2]$ with one tweak-dependent key, and find 32 such instances achieving full $2^n$ provable security. The search process is detailed in next section.

# 5  Search Among Instances of $\widetilde{\mathbb{E}}[2]$ with One Tweak-Dependent Key

To start with, we provide an observation that is used during the search: XORing tweak $t$ to plaintext $p$ and ciphertext $c$ does not have any impact to the security of tweakable blockcipher.

**Observation 1.** *For a tweakable blockcipher* $\widetilde{E} : \mathcal{K} \times \mathcal{T} \times \mathcal{M} \to \mathcal{M}$, *define a set of tweakable blockcipher* $\widetilde{E}[b_p, b_c] : \mathcal{K} \times \mathcal{T} \times \mathcal{M} \to \mathcal{M}$ *with* $b_p, b_c \in \{0, 1\}$ *as*

$$\widetilde{E}[b_p, b_c](k, t, p) := \widetilde{E}(k, t, p \oplus (b_p \cdot t)) \oplus (b_c \cdot t),$$

*for all* $k \in \mathcal{K}$, $t \in \mathcal{T}$ *and* $p \in \mathcal{M}$. *Each tweakable blockcipher* $\widetilde{E}[b_p, b_c]$ *provides the same security level as* $\widetilde{E}$, *that is* $\mathbf{Adv}_{\widetilde{E}[b_p, b_c]}^{\widetilde{\mathrm{sprp}}}(q) = \mathbf{Adv}_{\widetilde{E}}^{\widetilde{\mathrm{sprp}}}(q)$. *Thus, we do not use XORing tweak* $t$ *to plaintext* $p$ *and ciphertext* $c$ *for (slight) efficiency benefit.*

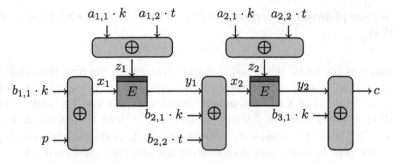

**Fig. 3.** Type I Constructions of $\widetilde{\mathbb{E}}[2]$

The proof of this observation is rather straightforward, and provided in full version of this paper [52].

Next, according to Constraint 1, we divide the instances of $\widetilde{\mathbb{E}}[2]$ into three types with respect to the place where the plaintext $p$ is injected.

**Type I:** $p$ is XORed to compute $x_1$, which sets $b_{1,3} = 1$, $b_{2,3} = 0$ and $b_{3,3} = 0$;
**Type II:** $p$ is XORed to compute $x_2$, which sets $b_{1,3} = 0$, $b_{2,3} = 1$ and $b_{3,3} = 0$;
**Type III:** $p$ is XORed to compute $x_3$, which sets $b_{1,3} = 0$, $b_{2,3} = 0$ and $b_{3,3} = 1$.

We search the instances of these types independently.

## 5.1  On the Instances of Type I

Constraint 2 sets $a_{2,3} = 0$, since $y_1$ is plaintext-dependent. Observation 1 sets $b_{1,2} = 0$ and $b_{3,2} = 0$. We set $b_{3,5} = 1$ such that the second blockcipher call is involved in $\widetilde{\mathbb{E}}[2]$.[4] Moreover, we set $b_{2,4} = 1$ in order to avoid overlap between the instances of Type I and of Type II, because if $b_{2,4} = 0$, the two blockcipher calls are parallel and indeed those instances are included in Type II. In turn, it implies that $x_2$ and $y_2$ are plaintext-dependent variables. Then Constraint 3 sets $b_{3,4} = 0$, because $y_2$ as a plaintext-dependent variable is already used to compute $c$. Putting all these together, the (simplified) construction of Type I is depicted in Fig. 3.

We investigate all the instances of Type I with one tweak-dependent key, which are divided into two cases depending on the position of the tweak-dependent key. More precisely, it depends on the values of $a_{1,2}$ and $a_{2,2}$.

**Case (1): $a_{1,2} = 1$ and $a_{2,2} = 0$.** $z_1$ is the tweak-dependent key. For these instances, the computation from internal variable $x_2$ to ciphertext $c$ is that

$$c = E(a_{2,1} \cdot k, x_2) \oplus b_{3,1} \cdot k.$$

---

[4] Otherwise, only one blockcipher call is actually involved, and such instances have at most $2^{n/2}$ security [36].

Hence, for any plaintext-ciphertext pair $(t, p, c)$ and $(t', p', c')$ with $(t, p) \neq (t', p')$, it has that

$$c = c' \implies x_2 = x_2'.$$

Exploiting this property, the attacker mainly focuses on the first blockcipher call and peels off the second blockcipher call, by using pairs of $(t, p, c)$ and $(t', p', c')$ with $c = c'$. Note that such a plaintext-ciphertext pair can be easily obtained by sending a query $(t, p)$ to $\widetilde{\mathbb{E}}[2]_k(\cdot, \cdot)$ to receive $c$, and then sending a query $(t', c)$ to $\widetilde{\mathbb{E}}[2]_k^{-1}(\cdot, \cdot)$ to receive $p'$.[5] Thanks to such plaintext-ciphertext pairs, the attacker gets to know and even control the internal difference $\Delta y_1 = b_{2,2} \cdot (t \oplus t')$. As a result, he can succeed to recover the key $k$ or to distinguish $\widetilde{\mathbb{E}}[2]$ from a random tweakable blockcipher $\widetilde{P}$ with a complexity of at most $O(2^{n/2})$ adversarial queries. The attack procedure is slightly different depending on the values of $a_{1,1}$ and $b_{1,1}$. Therefore, we further divide this case into four subcases, and describe the procedure for each subcase separately.

*Subcase (1.1): $a_{1,1} = 0$ and $b_{1,1} = 0$.* The key $k$ is not used in the first blockcipher $y_1 = E(t, p)$. Hence the attacker can get the value of $y_1$ by querying $(t, p)$ to $E(\cdot, \cdot)$. A distinguisher $\mathcal{D}$ is launched as follows. Firstly, $\mathcal{D}$ obtain a plaintext-ciphertext pair $(t, p, c)$ and $(t', p', c')$ with $c = c'$, and computes $\Delta y_1 = b_{2,2} \cdot (t \oplus t')$. Secondly, $\mathcal{D}$ queries $(t, p)$ and $(t', p')$ to $E(\cdot, \cdot)$ to receive $w$ and $w'$ respectively, and computes $\Delta w = w \oplus w'$. Finally, $\mathcal{D}$ outputs 1 if $\Delta y_1 = \Delta w$, and outputs 0 otherwise. The probability of $\mathcal{D}$ outputting 1 is 1 when interacting $\widetilde{\mathbb{E}}[2]$, and is $2^{-n}$ when interacting with $\widetilde{P}$. Thus, the advantage of $\mathcal{D}$ is $1 - 2^{-n}$. The complexity of $\mathcal{D}$ is 4 queries.

*Subcase (1.2): $a_{1,1} = 0$ and $b_{1,1} = 1$.* The first blockcipher call is $y_1 = E(t, p \oplus k)$. Its key $z_1$ is the tweak $t$, and can be controlled by the attacker. A key-recovery attack $\mathcal{A}$ is launched as follows. Firstly, $\mathcal{A}$ fixes a tweak value $t$ and a non-zero value $\Delta$. Secondly, $\mathcal{A}$ collects plaintext-ciphertext pairs $(t, p, c)$ and $(t', p', c')$ such that $t' = t \oplus \Delta$ and $c' = c$. Each pair has that

$$p \oplus p' = x_1 \oplus x_1' = E^{-1}(t, y_1) \oplus E^{-1}(t \oplus \Delta, y_1 \oplus b_{2,2} \cdot \Delta).$$

$\mathcal{A}$ stores $\{(p, p \oplus p')\}$ for $2^{n/2}$ distinct values of $p$, whose corresponding values of $y_1$ are also distinct. This needs $2^{n/2+1}$ queries. Thirdly, $\mathcal{A}$ selects $2^{n/2}$ distinct values $w$. For each $w$, he queries $(t, w)$ and $(t \oplus \Delta, w \oplus b_{2,2} \cdot \Delta)$ to $E^{-1}(\cdot, \cdot)$ to receive $u$ and $u'$ respectively, which has that

$$u \oplus u' = E^{-1}(t, w) \oplus E^{-1}(t \oplus \Delta, w \oplus b_{2,2} \cdot \Delta).$$

$\mathcal{A}$ matches $u \oplus u'$ to previously stored $p \oplus p'$. If a matched is found that implies $x_1 = p \oplus k = u$, the attacker computes the key $k$ as $k = u \oplus p$. The complexity

---

[5] Of course one may directly query $(t, c)$ and $(t, c' = c)$ to $\widetilde{\mathbb{E}}[2]_k^{-1}(\cdot, \cdot)$ to obtain such a pair. But the above approach allows the attacker to control the plaintext $p$, which is necessary in our attacks.

of $\mathcal{A}$ is around $2^{n/2+2}$ adversarial queries, and its success probability can be trivially computed as $1 - (1 - 2^{-n})^{2^n} \approx 1 - 1/e \approx 0.63$, since there are $2^{n/2}$ distinct values of $y_1$ and $2^{n/2}$ distinct values of $w$.

*Subcase (1.3):* $a_{1,1} = 1$ *and* $b_{1,1} = 0$. The first blockcipher call is $y_1 = E(t \oplus k, p)$. Its input $x_1$ is plaintext $p$, and can be controlled by the attacker. A key-recovery attack $\mathcal{A}$ is launched as follows. Firstly, $\mathcal{A}$ fixes a plaintext value $p$ and a non-zero value $\Delta$. Secondly, $\mathcal{A}$ collects plaintext-ciphertext pairs $(t, p, c)$ and $(t', p', c')$ such that $t' = t \oplus \Delta$ and $c' = c$. Each pair has that

$$p' = E^{-1}(t \oplus \Delta \oplus k, E(t \oplus k, p) \oplus b_{2,2} \cdot \Delta).$$

$\mathcal{A}$ stores $\{(t, p')\}$ for $2^{n/2}$ distinct values of $t$, which needs $2^{n/2+1}$ queries. Thirdly, $\mathcal{A}$ selects $2^{n/2}$ distinct values $l$. For each $l$, he queries $(l, p)$ to $E(\cdot, \cdot)$, receives $w$, and then queries $(l \oplus \Delta, w \oplus b_{2,2} \cdot \Delta)$ to $E^{-1}(\cdot, \cdot)$ to receive $u'$, which have that

$$u' = E^{-1}(l \oplus \Delta, E(l, p) \oplus b_{2,2} \cdot \Delta)$$

$\mathcal{A}$ matches $u'$ to previously stored $p'$. If a matched is found that implies $l = t \oplus k$, $\mathcal{A}$ computes the key $k$ as $k = l \oplus t$. The complexity of $\mathcal{A}$ is around $2^{n/2+2}$ adversarial queries. Similarly with the above subcases, its success probability can be computed as 0.63.

*Subcase (1.4):* $a_{1,1} = 1$ *and* $b_{1,1} = 1$. The first blockcipher call is $y_1 = E(t \oplus k, p \oplus k)$. XORing its inputs $x_1$ and $z_1$ is $x_1 \oplus z_1 = p \oplus t$, which can be controlled by the attacker. A key-recovery attack $\mathcal{A}$ is launched as follows. Firstly, $\mathcal{A}$ fixes a plaintext $p$ and a non-zero value $\Delta$. Secondly, $\mathcal{A}$ collects plaintext-ciphertext pairs $(t, p \oplus t, c)$ and $(t', p', c')$ with $t' = t \oplus \Delta$ and $c' = c$. Each pair has that

$$p' \oplus t = E^{-1}(t \oplus \Delta \oplus k, E(t \oplus k, p \oplus t \oplus k) \oplus b_{2,2} \cdot \Delta) \oplus k \oplus t$$

$\mathcal{A}$ stores $\{(t, p' \oplus t)\}$ for $2^{n/2}$ distinct values of $t$, which needs $2^{n/2+1}$ queries. Thirdly, $\mathcal{A}$ selects $2^{n/2}$ distinct values $l$. For each $l$, he queries $(l, p \oplus l)$ to $E(\cdot, \cdot)$, receives $w$, and then queries $(l \oplus \Delta, w \oplus b_{2,2} \cdot \Delta)$ to $E^{-1}(\cdot, \cdot)$ to receive $u'$, which have that

$$u' \oplus l = E^{-1}(l \oplus \Delta, E(l, p \oplus l) \oplus b_{2,2} \cdot \Delta) \oplus l$$

$\mathcal{A}$ matches $u' \oplus l$ to previously stored $p' \oplus t$. f a matched is found that implies $l = t \oplus k$, $\mathcal{A}$ computes the key $k$ as $k = l \oplus t$. The complexity of $\mathcal{A}$ is around $2^{n/2+2}$ adversarial queries, and its success probability can be trivially computed as 0.63 similarly with the above subcases.

Overall, we conclude that all the instances of Case (1) using one tweak-dependent key have at most around $2^{n/2}$ security.

**Case (2):** $a_{1,2} = 0$ **and** $a_{2,2} = 1$. $z_2$ is the tweak-dependent key. The analysis is highly similar with Case (1), which is written in full version of this paper [52]. In a high level, Case (2) can be regarded as the inverse of Case (1) by analyzing the decryption oracle $\widetilde{E}^{-1}$. Here we just provide the conclusion: all the instances of Case (2) using one tweak-dependent key have at most around $2^{n/2}$ security.

## 5.2  On the Instances of Type II

Observation 1 sets $b_{2,2} = 0$ and $b_{3,2} = 0$. We set $b_{3,5} = 1$ such that the second blockcipher call is involved in $\widetilde{\mathbb{E}}[2]$. The construction of Type II is depicted in Fig. 4. Similarly we also divide the instances of Type II into two cases depending on the position of the tweak-dependent key. More precisely, it depends on the values of $a_{1,2}$, $a_{2,2}$, and $a_{2,3}$ if $y_1$ is computed related to tweak $t$.

**Case (1): $a_{1,2} = 1$, $a_{2,2} = 0$, $a_{2,3} = 0$.** $z_1$ is the tweak-dependent key. The reason of setting $a_{2,3} = 0$ is that $y_1$ is computed depending on $t$ as

$$y_1 = E(a_{1,1} \cdot k \oplus t, b_{1,1} \cdot k \oplus b_{1,2} \cdot t).$$

We find the instances of this case have at most $2^{n/2}$ security based on the following observation. The computation from internal variable $y_1$ to ciphertext $c$ is that

$$c = E(a_{2,1} \cdot k, p \oplus b_{2,4} \cdot y_1 \oplus b_{2,1} \cdot k) \oplus b_{3,1} \cdot k \oplus b_{3,4} \cdot y_1,$$

which is not related to the tweak value. Therefore, for two distinct tweaks $t$ and $t'$ colliding on $y_1$ that is

$$E(a_{1,1} \cdot k \oplus t, b_{1,1} \cdot k \oplus b_{1,2} \cdot t) = E(a_{1,1} \cdot k \oplus t', b_{1,1} \cdot k' \oplus b_{1,2} \cdot t'),$$

it leads to the same ciphertext for any plaintext, more precisely,

$$\widetilde{\mathbb{E}}[2]_k(t, p) = \widetilde{\mathbb{E}}[2]_k(t', p), \quad \text{for} \quad \forall p \in \mathcal{M}.$$

Such a pair of tweaks can be found after trying $2^{n/2}$ distinct tweaks. Putting all together, a distinguisher $\mathcal{D}$ can be launched as follows. Firstly, $\mathcal{D}$ fixes a plaintext $p$. Secondly, he selects $2^{n/2}$ distinct tweak values $t$, queries $(t, p)$ to $\widetilde{\mathbb{E}}[2]_k(\cdot, \cdot)$ to search a collision among received ciphertexts. Let $t$ and $t'$ denote the corresponding tweaks for the colliding ciphertexts. Thirdly, $\mathcal{D}$ selects another plaintext $p'$ with $p' \neq p$, and queries $(t, p')$ and $(t', p')$ to $\widetilde{\mathbb{E}}[2]_k(\cdot, \cdot)$ and receives

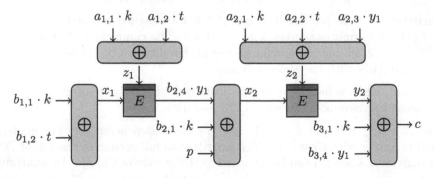

**Fig. 4.** Type II Construction of $\widetilde{\mathbb{E}}[2]$

ciphertexts $c'$ and $c''$ respectively. Finally, $\mathcal{D}$ outputs 1 if $c' \neq c''$, and outputs 0 otherwise. The complexity of $\mathcal{D}$ is around $2^{n/2}$ queries. When interacting with $\widetilde{\mathbb{E}}[2]$, $\mathcal{D}$ outputs 1, as long as he succeeds to find the colliding ciphertexts at second step, which has a probability of $1 - (1 - 2^{-n})^{2^{n-1}} \approx 0.4$. When interacting with a random tweakable blockcipher, the probability of $\mathcal{D}$ outputting 1 is obviously $2^{-n}$. Therefore, the advantage of $\mathcal{D}$ is computed as $0.4 - 2^{-n} \approx 0.4$.

**Case (2): $a_{1,2} = 0$.** We need to further set the values of $a_{2,2}$ and $a_{2,3}$ such that $z_2$ is a tweak-dependent key. There are two possible setting depending on the value of $b_{1,2}$. More precisely, if $b_{1,2} = 0$, then $y_1$ is computed unrelated to tweak $t$, and therefore $a_{2,2}$ must be 1. Otherwise, as long as one of $a_{2,2}$ and $a_{2,3}$ is not zero, $z_2$ is a tweak-dependent key. Accordingly we divide Case (2) to two subcases.

*Subcase (2.1): $b_{1,2} = 0$, $a_{2,2} = 1$.* A graphical view is provided in Fig. 5. Notably internal variable $y_1$ is computed as $y_1 = E(a_{1,1} \cdot k, b_{1,1} \cdot k)$, which is unrelated to tweak $t$. We refer to $y_1$ as a subkey derived from the key $k$ for those instances with $(a_{1,1}, b_{1,1}) \neq (0,0)$. Moreover, the computation from $p$ to $x_2$ is $x_2 = p \oplus b_{2,1} \cdot k \oplus b_{2,4} \cdot y_1$, and hence $\Delta x_2 = \Delta p$ always holds. Similarly, $\Delta y_2 = \Delta c$ always holds. In other words, for any plaintext-ciphertext pair $(t, p, c)$ and $(t', p', c')$, the internal variable differences $\Delta x_2$ and $\Delta y_2$ is known to the attacker. Due to these properties, we find several conditions on the instances of this subcase in order to possibly have a security beyond the birthday bound.

- $(a_{1,1}, b_{1,1}) \neq (0,0)$
  If $a_{1,1}, b_{1,1} = (0,0)$, it has that $y_1 = E(0,0)$. Then an attacker can query $(0,0)$ to $E(\cdot, \cdot)$, receive the value of $y_1$, and then peel off the first blockcipher call. As a result, the instances become essentially based on one blockcipher call in the view of the attacker. As proven in [36], the attacker can distinguish such instances from a random tweakable blockcipher with a complexity of at most $2^{n/2}$ adversarial queries.
- $(a_{2,1}, a_{2,3}) \neq (0,0)$
  If $(a_{2,1}, a_{2,3}) = (0,0)$, an attacker can fix the tweak $t$ to a constant and

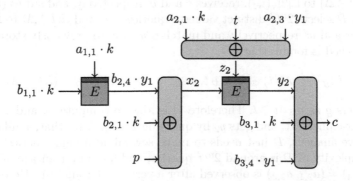

**Fig. 5.** Subcase (2.1) of Type II of $\widetilde{\mathbb{E}}[2]$

regard $b_{2,1} \cdot k \oplus b_{2,4} \cdot y_1$ and $b_{3,1} \cdot k \oplus b_{3,4} \cdot y_1$ as the pre- and post-whitening keys respectively. As a result, the instances become essentially one-step Even-Mansour blockcipher [17], and several attack procedures with a complexity of $2^{n/2}$ queries have been presented in [4,13,15].

- $(b_{2,1}, b_{2,4}) \neq (0,0)$ and $(b_{3,1}, b_{3,4}) \neq (0,0)$

  If $(b_{2,1}, b_{2,4}) = (0,0)$, it has $x_2 = p$. Then an attacker gets to know and control the value of $x_2$. A distinguisher $\mathcal{D}$ is launched as follows. Firstly, $\mathcal{D}$ fixes two distinct plaintexts $p$ and $p'$. Secondly, he selects $2^{n/2}$ distinct tweaks $t$. For each $t$, $\mathcal{D}$ queries $(t, p)$ and $(t, p')$ to $\widetilde{\mathbb{E}}[2]_k(\cdot, \cdot)$, receives ciphertexts $c$ and $c'$ respectively, and stores $(t, c \oplus c')$. Thirdly, $\mathcal{D}$ selects $2^{n/2}$ distinct values $l$. For each $l$, he queries $(l, p)$ and $(l, p')$ to $E(\cdot, \cdot)$, receives $w$ and $w'$ respectively, and matches $w \oplus w'$ to previously stored $c \oplus c'$ at second step. Once a matched is found, that is

$$E(a_{2,1} \cdot k \oplus t \oplus a_{2,3} \cdot y_1, p) \oplus E(a_{2,1} \cdot k \oplus t \oplus a_{2,3} \cdot y_1, p') = E(l, p) \oplus E(l, p'),$$

  $\mathcal{D}$ recovers $a_{2,1} \cdot k \oplus b_{2,3} \cdot y_1 = t \oplus l$. Finally, for any plaintext-ciphertext pair of $(t, p, c)$ and $(t', p', c')$, $\mathcal{D}$ can compute internal variables $z_2$ and $z_2'$, and query $(z_2, p)$ and $(z_2', p')$ to $E(\cdot, \cdot)$ to recover $y_2$ and $y_2'$, respectively. $\mathcal{D}$ outputs 1 if $c \oplus c' = y_2 \oplus y_2'$, and outputs 0 otherwise. The complexity of $\mathcal{D}$ is around $2^{n/2+2}$ queries. When interacting with $\widetilde{\mathbb{E}}[2]$, $\mathcal{D}$ outputs 1 as long as he recovers $a_{2,1} \cdot k \oplus b_{2,3} \cdot y_1$, which succeeds with a probability $1 - (1 - 2^{-n})^{2^n} \approx 1 - 1/e \approx$ 0.63. When interacting with a random tweakable blockcipher, $\mathcal{D}$ outputs 1 with a probability $2^{-n}$. Therefore the advantage of $\mathcal{D}$ is $0.63 - 2^{-n} \approx 0.63$.

  $(b_{3,1}, b_{3,4}) \neq (0,0)$ is observed after a very similar analysis. Just the attacker gets to know and control the value of $y_2$. Accordingly, he fixes two ciphertexts $c$ and $c'$, and queries $(t, c)$ and $(t, c')$ to $\widetilde{\mathbb{E}}[2]_k^{-1}(\cdot, \cdot)$ for distinct tweaks $t$. We omit the details.

- $(b_{2,1}, b_{2,4}) \neq (a_{2,1}, a_{2,3})$ and $(b_{3,1}, b_{3,4}) \neq (a_{2,1}, a_{2,3})$

  If $(b_{2,1}, b_{2,4}) = (a_{2,1}, a_{2,3})$, it has $b_{2,1} \cdot k \oplus b_{2,4} \cdot y_1 = a_{2,1} \cdot k \oplus a_{2,3} \cdot y_1$, which is denoted as $g$. Then $x_2 \oplus z_2 = g \oplus p \oplus g \oplus t = p \oplus t$. Hence an attacker gets to know and control $x_2 \oplus z_2$. A distinguisher $\mathcal{D}$ can be launched. Firstly, $\mathcal{D}$ fixes a non-zero $\Delta$. Secondly, he selects $2^{n/2}$ distinct tweaks $t$, queries $(t, p = t)$ and $(t, p' = t \oplus \Delta)$ to $\widetilde{\mathbb{E}}[2]_k(\cdot, \cdot)$, receives $c$ and $c'$ respectively, and stores $(t, c \oplus c')$. Thirdly, $\mathcal{D}$ selects $2^{n/2}$ distinct values $l$, queries $(l, l)$ and $(l, l \oplus \Delta)$ to $E(\cdot, \cdot)$ to receive $w$ and $w'$ respectively, and matches $w \oplus w'$ to previously stored $c \oplus c'$. If a matched is found, that is

$$E(g \oplus t, g \oplus t) \oplus E(g \oplus t, g \oplus t \oplus \Delta) = E(l, l) \oplus E(l, l \oplus \Delta),$$

  $\mathcal{D}$ recovers $g$ as $g = t \oplus l$. Therefore $\mathcal{D}$ is able to compute $x_2$ and $z_2$ for any plaintext-ciphertext, and gets $y_2$ by querying $E(\cdot, \cdot)$. After that, similarly with the above analysis, $\mathcal{D}$ just needs to make several additional queries. Overall, the complexity of $\mathcal{D}$ is around $2^{n/2}$ queries, and has an advantage of 0.63. $(b_{3,1}, b_{3,4}) \neq (a_{2,1}, a_{2,3})$ is observed after a very similar analysis. Here we omit the details.

Putting all these conditions together, there are 32 instances of this subcase left, which are denoted as $\widetilde{E}1$, $\widetilde{E}2$, ..., $\widetilde{E}32$ and have been depicted in Figs. 6 and 7. After further investigation, we find that these constructions achieve full $2^n$ provable security. The proof is presented in Sect. 6.

*Subcase (2.2):* $b_{1,2} = 1$, $(a_{2,2}, a_{2,3}) \neq (0,0)$. Interestingly, we notice that the instances of Subcase (2.1) has an efficiency advantage over the instances of Subcase (2.2). More precisely, if one pre-computes and stores internal variable $y_1$ as a subkey, an instance of Subcase (2.1) requires just one block-cipher call for encrypting $(t,p)$ or decrypting $(t,c)$, while the instances of Subcase (2.2) always need two blockcipher calls. Since we have found instances of Subcase (2.1) achieving full $2^n$ security, it is unnecessary to search among instances of Subcase (2.2). Nevertheless, we did investigate the instances of Subcase (2.2), and found 24 instances achieving full $2^n$ provable security. Here we omit the discussion on this subcase due to the limited space.

### 5.3  On the Instances of Type III

Clearly, plaintext and ciphertext are linearly related in this type of construction, and can be trivially distinguished by making two queries to $\widetilde{\mathbb{E}}[2]_k(\cdot, \cdot)$ with a fixed difference in plaintexts, *e.g.*, $(t,p)$ and $(t, p \oplus \Delta)$, and verifying $\Delta c = \Delta$.

## 6  Security Proof of $\widetilde{E}1, \ldots, \widetilde{E}32$

Let $\widetilde{E}$ be any tweakable blockcipher of $\widetilde{E}1$, $\widetilde{E}2$, ..., $\widetilde{E}32$, and $E$ denotes its underlying blockcipher. Let $\widetilde{P}$ be a random tweakable blockcipher that is $\widetilde{P} \xleftarrow{\$} \widetilde{\mathrm{Perm}}$. Let $(\mathcal{O}_1, \mathcal{O}_2)$ be either $(\widetilde{E}_k^{\pm}(\cdot, \cdot), E^{\pm}(\cdot, \cdot))$ with $k \xleftarrow{\$} \mathcal{K}$ or $(\widetilde{P}^{\pm}(\cdot, \cdot), E^{\pm}(\cdot, \cdot))$. Let $\mathcal{D}$ be a distinguisher interacting with $(\mathcal{O}_1, \mathcal{O}_2)$ that makes (at most) $q$ queries. We denote the number of $\mathcal{D}$'s queries to $\mathcal{O}_1$ and to $\mathcal{O}_2$ as $q_1$ and $q_2$ respectively: $q = q_1 + q_2$. Without loss of generality, we assume that $\mathcal{D}$ does not make duplicated queries to $\mathcal{O}_1$ or $\mathcal{O}_2$. We use views $v_1 = \{(t_1, p_1, c_1), \ldots, (t_{q_1}, p_{q_1}, c_{q_1})\}$ and $v_2 = \{(l_1, u_1, w_1), \ldots, (l_{q_2}, u_{q_2}, w_{q_2})\}$ to denote the transcripts, which are lists of query-responses, created by $\mathcal{D}$ interacting with $\mathcal{O}_1$ and $\mathcal{O}_2$, respectively. At the end of the interaction with $(\mathcal{O}_1, \mathcal{O}_2)$, the distinguisher $\mathcal{D}$ obtains a view $v = (v_1, v_2)$ before determining the output bit. Since $\mathcal{D}$ is computationally unbounded, without loss of generality we assume that $\mathcal{D}$ is deterministic. Therefore $\mathcal{D}$ computes its decision bit deterministically based on the view $v$. Accordingly, the probability distribution of the decision bit of $\mathcal{D}$ solely depends on the probability distribution of the view $v$.

Our proof adopts the H-coefficient technique [8,45], which has been introduced in Sect. 2.3. We use $X$ and $Y$ to denote the probability distribution on views when $\mathcal{D}$ interacts with $(\widetilde{E}_k^{\pm}(\cdot, \cdot), E^{\pm}(\cdot, \cdot))$ and interacts with $(\widetilde{P}^{\pm}(\cdot, \cdot), E^{\pm}(\cdot, \cdot))$, respectively. We use $\mathcal{V}$ to denote the set of attainable views $v$ when $\mathcal{D}$ interacts with $(\widetilde{P}^{\pm}(\cdot, \cdot), E^{\pm}(\cdot, \cdot))$, that is $\mathcal{V} = \{v \mid \Pr[Y = v] > 0\}$. Next,

474    L. Wang et al.

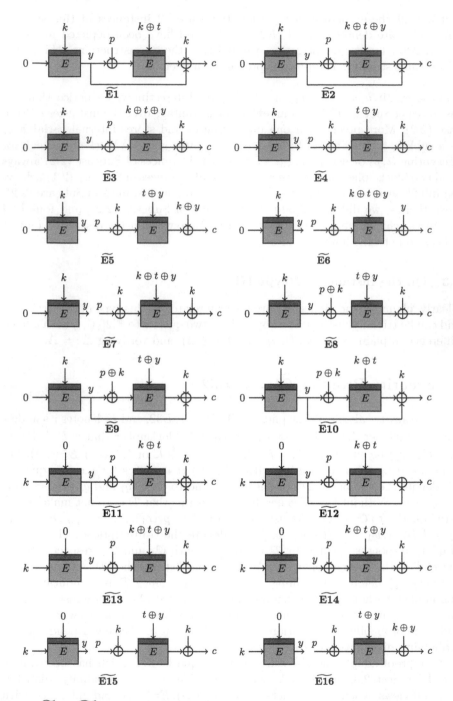

**Fig. 6.** $\widetilde{E}1$ to $\widetilde{E}16$ of the 32 efficient constructions: the internal variable $y$ is referred to as the subkey for these constructions.

How to Build Fully Secure Tweakable Blockciphers 475

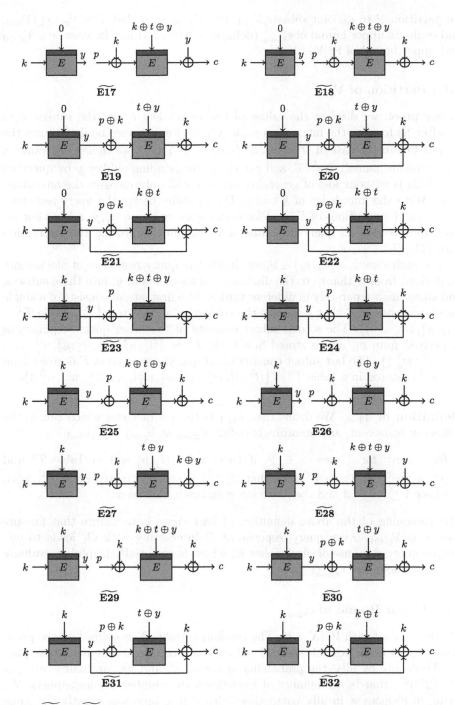

**Fig. 7.** $\widetilde{E17}$ to $\widetilde{E32}$ of the 32 efficient constructions: the internal variable $y$ is referred to as the subkey for these constructions.

we partition $\mathcal{V}$ to disjoint subsets $\mathcal{V}_{\text{bad}}$ and $\mathcal{V}_{\text{good}}$ such that $\mathcal{V} = \mathcal{V}_{\text{good}} \bigcup \mathcal{V}_{\text{bad}}$, and evaluate upper bound of $\epsilon_{v_{\text{good}}}$ (defined in Sect. 2.3) for the views $v \in \mathcal{V}_{\text{good}}$ and upper bound of $\Pr[Y \in \mathcal{V}_{\text{bad}}]$.

## 6.1    Partition of $\mathcal{V}$

In our proof, we disclose the values of the secret key $k$ and the subkey $y$ to $\mathcal{D}$, after he finishes the interaction with $(\mathcal{O}_1, \mathcal{O}_2)$ and before he determines the output bit. In the case of $(\widetilde{P}^{\pm}(\cdot, \cdot), E^{\pm}(\cdot, \cdot))$ as $(\mathcal{O}_1, \mathcal{O}_2)$, we choose the value of $k$ at random, namely $k \xleftarrow{\$} \mathcal{K}$, and get the corresponding subkey $y$ by querying $E^{\pm}$. This is without loss of generality since it will only increase the advantage of $\mathcal{D}$. With the knowledge of $k$ and $y$, $\mathcal{D}$ can easily derive the query-responses $(l, u, w)$'s of invocations of $E^{\pm}(\cdot, \cdot)$ for each query-response $(t_i, p_i, c_i)$ in view $v_1$. Therefore $\mathcal{D}$ gets all query-responses of blockcipher $E$ during the interaction with $(\mathcal{O}_1, \mathcal{O}_2)$.

For each view $v = (v_1, v_2) \in \mathcal{V}$, we divide the query-responses of blockcipher $E$, derived from it thanks to the disclosed values of $k$ and $y$, into three subsets, and store them separately in different tables. The first subset consists of a single query-response of $E$ that generates the subkey $y$, and is stored in a table $T^1 = \{(l_1^1, u_1^1, w_1^1 = y)\}$. The second subset consists of the other query-responses of $E$ derived from $v_1$, and is stored in a table $T^2 = \{(l_1^2, u_1^2, w_1^2), (l_2^2, u_2^2, w_2^2), \ldots, (l_{q_1}^2, u_{q_1}^2, w_{q_1}^2)\}$. The last subset consists of all query-responses of $E$ derived from $v_2$, and is stored in a table $T^3 = \{(l_1^3, u_1^3, w_1^3), (l_2^3, u_2^3, w_2^3), \ldots, (l_{q_2}^3, u_{q_2}^3, w_{q_2}^3)\}$.

**Definition of $\mathcal{V}_{\text{bad}}$.** We define that $\mathcal{V}_{\text{bad}}$ is the set of views which causes the following bad event, and accordingly define $\mathcal{V}_{\text{good}}$ as $\mathcal{V}_{\text{good}} = \mathcal{V} \backslash \mathcal{V}_{\text{bad}}$.

- *Bad event*: for a view $v \in \mathcal{V}$, if there exist $(l_j^i, u_j^i, w_j^i)$ in Table $T^i$ and $(l_{j'}^{i'}, u_{j'}^{i'}, w_{j'}^{i'})$ in Table $T^{i'}$ such that $(l_j^i, u_j^i) = (l_{j'}^{i'}, u_{j'}^{i'})$ or $(l_j^i, w_j^i) = (l_{j'}^{i'}, w_{j'}^{i'})$, where $1 \leq i, i' \leq 3$ and $i \neq i'$, we say $v$ causes a bad event.

The reasoning of the above definition of bad views is to ensure that for any view $v \in \mathcal{V}_{\text{good}}$, every query-response of $\mathcal{D}$ interacting with $\mathcal{O}_1$ leads to one unique query-response of blockcipher $E$, which is essentially helpful to evaluate the upper bound of $\epsilon_{v_{\text{good}}}$.

## 6.2    Upper Bound of $\epsilon_{v_{\text{good}}}$

Firstly, we deal with $\Pr[X = v]$. The random variable $X$ is defined on the probability space of all possible secret key $k$ and all possible underlying blockcipher $E$. We denote by $\text{all}_X$ the probability space of $X$, and its cardinality $|\text{all}_X|$ is $2^n \cdot (2^n!)^{2^n}$, that is the number of keys times the number of blockciphers. We write an element $\pi$ in $\text{all}_X$ compatible with $v$ if $\pi$ produces exactly the same responses for all queries in $v$. We denote by $\text{comp}_X(v)$ all the elements in $\text{all}_X$

compatible with view $v$. Since $k$ is chosen uniformly at random and $E$ is an ideal blockcipher, we have that

$$\Pr[X = v] = \frac{|\text{comp}_X(v)|}{|\text{all}_X|}.$$

Similarly, $Y$ is defined on the probability space of the key $k$, tweakable blockcipher $\tilde{\mathcal{P}}$ and blockcipher $E$. Define $\text{comp}_Y(v)$ and $\text{all}_Y$ accordingly, and then we have that

$$\Pr[Y = v] = \frac{|\text{comp}_Y(v)|}{|\text{all}_Y|}.$$

$\text{all}_Y$ is $2^n \cdot (2^n!)^{2^n} \cdot (2^n!)^{2^n}$, that is the number of keys times the number of tweakable blockciphers times the number of blockciphers.

Next is to compute $|\text{comp}_X(v)|$ and $|\text{comp}_Y(v)|$. Recall that the view $v$ contains the value of $k$, which is disclosed to $\mathcal{D}$ at the end of interaction, and then a set of input-outputs of underlying blockcipher $E$ are derived and separately stored in tables $T^1$, $T^2$ and $T^3$. Let $\alpha_i$ and $\beta_i$ denote the number of input-outputs $(l, u, w)$'s of $E$ with the value $i$ as the key value (that is $l = i$) in $T^2$ and $T^3$, respectively, for $0 \leq i \leq 2^n - 1$. Denote $i_t$ as the tweak value that produces $i$ as the key value (that is $z_2$ in Fig. 5) for the second blockcipher call in $\mathcal{O}_1$, and denote $\gamma_{i_t}$ the number of queries to $\mathcal{O}_1$ with tweak values as $i_t$. Since $v$ is a good view, there is no element collision between any two tables. Moreover, $\mathcal{D}$ does not make duplicate queries. Hence all input-outputs of $E$ in $T^1$, $T^2$ and $T^3$ are distinct. Therefore, it implies that $\gamma_{i_t} = \alpha_i$. The query-response $(l_1^1, u_1^1, w_1^1)$ of $E$ in $T^1$ has $l_1^1 = k$ or $l_1^1 = 0$.[6] Without loss of generality, we assume $l_1^1 = k$. Then, we get that

$$|\text{comp}_X(v)| = (2^n - \alpha_k - \beta_k - 1)! \cdot \prod_{i=0}^{k-1}(2^n - \alpha_i - \beta_i)! \cdot \prod_{i=k+1}^{2^n-1}(2^n - \alpha_i - \beta_i)!,$$

and

$$|\text{comp}_Y(v)| = \prod_{i=0}^{2^n-1}(2^n - \gamma_{i_t})! \cdot \left((2^n - \beta_k - 1)! \cdot \prod_{i=0}^{k-1}(2^n - \beta_i)! \cdot \prod_{i=k+1}^{2^n-1}(2^n - \beta_i)!\right)$$

$$= \prod_{i=0}^{2^n-1}(2^n - \alpha_i)! \cdot \left((2^n - \beta_k - 1)! \cdot \prod_{i=0}^{k-1}(2^n - \beta_i)! \cdot \prod_{i=k+1}^{2^n-1}(2^n - \beta_i)!\right)$$

$$= (2^n - \alpha_k)! \cdot (2^n - \beta_k - 1)! \cdot \prod_{i=0}^{k-1}((2^n - \alpha_i)! \cdot (2^n - \beta_i)!) \cdot \prod_{i=k+1}^{2^n-1}((2^n - \alpha_i)! \cdot (2^n - \beta_i)!).$$

From $(2^n - \alpha)! \cdot (2^n - \beta)! \leq (2^n - \alpha - \beta)! \cdot 2^n!$, we have that

$$|\text{comp}_Y(v)| \leq (2^n - \alpha_k - \beta_k - 1)! \cdot (2^n!)^{2^n} \cdot \prod_{i=0}^{k-1}(2^n - \alpha_i - \beta_i)! \cdot \prod_{i=k+1}^{2^n-1}(2^n - \alpha_i - \beta_i)!.$$

---

[6] More precisely, $\widetilde{E1}, \ldots, \widetilde{E10}, \widetilde{E23}, \ldots, \widetilde{E32}$ have $l_1^1 = k$, and the other tweakable blockciphers have $l_1^1 = 0$.

Then we compute

$$\frac{|\text{comp}_X(v)|}{|\text{comp}_Y(v)|} \geq \frac{(2^n - \alpha_k - \beta_k - 1)! \cdot \prod_{i=0}^{k-1}(2^n - \alpha_i - \beta_i)! \cdot \prod_{i=k+1}^{2^n-1}(2^n - \alpha_i - \beta_i)!}{(2^n - \alpha_k - \beta_k - 1)! \cdot (2^n!)^{2^n} \cdot \prod_{i=0}^{k-1}(2^n - \alpha_i - \beta_i)! \cdot \prod_{i=k+1}^{2^n-1}(2^n - \alpha_i - \beta_i)!}$$

$$= \frac{1}{(2^n!)^{2^n}}$$

Finally, we compute

$$\frac{\Pr[X = v]}{\Pr[Y = v]} = \frac{|\text{comp}_X(v)|}{|\text{comp}_Y(v)|} \times \frac{|\text{all}_Y|}{|\text{all}_X|}$$

$$\geq \frac{1}{(2^n!)^{2^n}} \times \frac{2^n \cdot (2^n!)^{2^n} \cdot (2^n!)^{2^n}}{2^n \cdot (2^n!)^{2^n}} = 1$$

which give that $\epsilon_{v_{\text{good}}} = 0$.

**Note.** We highlight that this upper bound of $\epsilon_{v_{\text{good}}} = 0$ is indeed shared by all these 32 constructions $\widetilde{E1}, \ldots, \widetilde{E32}$. Moreover, as long as every view in $\mathcal{V}_{\text{good}}$ does not cause the above bad event defined in Sect. 6.1, it always has that $\epsilon_{v_{\text{good}}} = 0$. Therefore, the advantage of all distinguishers making at most $q$ queries is upper bounded as

$$\mathbf{Adv}_{\widetilde{E}}^{\widetilde{\text{sprp}}}(q) \leq \Pr[Y \in \mathcal{V}_{\text{bad}}].$$

Thus, the remaining work is to evaluate $\Pr[Y \in \mathcal{V}_{\text{bad}}]$ for each construction of $\widetilde{E1}, \ldots, \widetilde{E32}$, separately.

### 6.3  Upper Bound of $\Pr[Y \in \mathcal{V}_{\text{bad}}]$

For each construction of $\widetilde{E1}$ to $\widetilde{E32}$, we give the exact definition of $\mathcal{V}_{\text{bad}}$ according to the specification, which also defines $\mathcal{V}_{\text{good}} = \mathcal{V} \backslash \mathcal{V}_{\text{bad}}$. We must ensure that every view $v \in \mathcal{V}_{\text{good}}$ does not cause the bad event defined in Sect. 6.1, such that the probability $\Pr[Y \in \mathcal{V}_{\text{bad}}]$ is upper bound of $\mathbf{Adv}_{\widetilde{E}}^{\widetilde{\text{sprp}}}(q)$. Due to the limited space, in this section we use $\widetilde{E1}$ as an example, and write the definitions of $\mathcal{V}_{\text{bad}}$ for the other constructions in full version of this paper [52].

$\mathcal{V}_{\text{bad}}$ *of* $\widetilde{E1}$ is defined as the set of views $v = (v_1, v_2)$ such that (at least) one of the following events occur:

**(1a).** $\exists (l_j, u_j, w_j) \in v_2$ such that $l_j = k$;
**(1b).** $\exists (t_i = 0, p_i, c_i) \in v_1$ such that $p_i = y$ or $c_i = k$;
**(1c).** $\exists (t_i, p_i, c_i) \in v_1$ and $(l_j, u_j, w_j) \in v_2$ such that $(l_j = k \oplus t_i, u_j = p_i \oplus y)$ or $(l_j = k \oplus t_i, w_j = c_i \oplus y \oplus k)$.

Since both $k$ and $y$ are selected uniformly at random from a set of size at least $2^n - q - 1$, we have that

$$\Pr\left[(1a)\right] \leq q/(2^n - q - 1);$$
$$\Pr\left[(1b)\right] \leq 2q/(2^n - q - 1);$$
$$\Pr\left[(1c)\right] \leq 2q^2/(2^n - q - 1)^2.$$

Therefore, we get that

$$\Pr\left[Y \in \mathcal{V}_{\text{bad}}\right] \leq \Pr\left[(1a)\right] + \Pr\left[(1b)\right] + \Pr\left[(1c)\right]$$
$$\leq \frac{3q}{2^n - q - 1} + \frac{2q^2}{(2^n - q - 1)^2}$$

Supposing $q < 2^{n-1}$, we have that

$$\Pr\left[Y \in \mathcal{V}_{\text{bad}}\right] \leq \frac{3q}{2^{n-1}} + \frac{2q^2}{(2^{n-1})^2} \leq \frac{5q}{2^{n-1}}.$$

Next, we look into the views in $\mathcal{V}_{\text{good}}$. A view in $\mathcal{V}_{\text{good}}$ implies that nonce of the three events $(1a)$, $(1b)$ and $(1c)$ occur. Then we have that

- $(1a)$ does not occur $\implies$ the tuple elements in $\mathcal{T}^1$ and in $\mathcal{T}^3$ do not collide;
- $(1b)$ does not occur $\implies$ the tuple elements in $\mathcal{T}^1$ and in $\mathcal{T}^2$ do not collide;
- $(1c)$ does not occur $\implies$ the tuple elements in $\mathcal{T}^2$ and in $\mathcal{T}^3$ do not collide;

where the notations $\mathcal{T}^1$, $\mathcal{T}^2$ and $\mathcal{T}^3$ are defined in Sect. 6.1. Combining them together, we can conclude that every view in $\mathcal{V}_{\text{good}}$ does not cause the bad event in Sect. 6.1. Hence $\epsilon_{v_{\text{good}}} = 0$ holds. Therefore it has that

$$\mathbf{Adv}_{\widetilde{E1}}^{\widetilde{\text{sprp}}}(q) \leq \frac{10q}{2^n}$$

## 6.4  Provable Security

Putting all together, we obtain the following theorem on the provable security of $\widetilde{E1}, \ldots, \widetilde{E32}$.

**Theorem 1.** *Let $\widetilde{E}$ be any tweakable blockcipher construction from the set of $\widetilde{E1}, \ldots, \widetilde{E32}$ depicted in Figs. 6 and 7. Let $q$ be an integer such that $q < 2^{n-1}$. Then the following bound holds.*

$$\mathbf{Adv}_{\widetilde{E}}^{\widetilde{\text{sprp}}}(q) \leq \frac{10q}{2^n}.$$

# 7  Conclusions and Discussions

This paper has proposed 32 tweakable blockcipher constructions that achieve full provable security via a minimum number of blockcipher calls, in the ideal blockcipher model. A direction of future work would be to investigate if such fully secure tweakable blockciphers can be constructed in the standard pseudo-random-permutation model with a constant number of blockcipher calls.

**On Key Check Value.** As highlighted in [27], ANSI X9.24-1 [2] suggests the use of the key check value KCV for the integrity verification of the blockcipher key, which may cause security loss for cryptographic primitives. In details, ANSI X9.24-1 suggests $KCV = E_k(0)$.[7] Moreover, KCV is a public value, and will be transmitted, sent or stored in clear. In other words, an attacker has chance to learn the value of KCV. It has a serious security impact to our constructions $\widetilde{E1}, \widetilde{E2}, \ldots, \widetilde{E10}$, whose subkey $y$ is computed as $y = E(k, 0)$. As we can see, $KCV = y$ holds, and hence an attacker can get the value of the subkey, and then is able to recover the key $k$ with a complexity of $2^{n/2}$ queries. We propose alternatives to these tweakable blockciphers when KCV is used: replace 0 by a non-zero constant const, and derive the subkey $y$ from the key $k$ as $y = E(k, const)$. On other hand, the usage of KCV has negligible impact to the security of the other tweakable blockcipher constructions $\widetilde{E11}, \ldots, \widetilde{E32}$.

**Acknowledgements.** Lei Wang and Dawu Gu are sponsored by the Natural Science Foundation of Shanghai (16ZR1416400), Major State Basic Research Development Program (973 Plan), the National Natural Science Foundation of China (61472250), and Innovation Plan of Science and Technology of Shanghai (14511100300). Guoyan Zhang is sponsored by National Natural Science Foundation of China (61602276). Jingyuan Zhao is sponsored by the National Science Foundation of China (no. 61379139) and the Strategic Priority Research Program of the Chinese Academy of Sciences (no. XDA06100701).

# References

1. Andreeva, E., Bogdanov, A., Luykx, A., Mennink, B., Tischhauser, E., Yasuda, K.: Parallelizable and authenticated online ciphers. In: Sako, K., Sarkar, P. (eds.) ASIACRYPT 2013. LNCS, vol. 8269, pp. 424–443. Springer, Heidelberg (2013). doi:10.1007/978-3-642-42033-7_22
2. ANSI: Retail Financial Services Symmetric Key Management Part 1: Using Symmetric Techniques. ANSI X9.24-1: 2009 (2009)
3. Beaulieu, R., Shors, D., Smith, J., Treatman-Clark, S., Weeks, B., Wingers, L.: SIMON and SPECK: Block Ciphers for the Internet of Things. Cryptology ePrint Archive, Report 2015/585 (2015). http://eprint.iacr.org/
4. Biryukov, A., Wagner, D.: Advanced slide attacks. In: Preneel, B. (ed.) EUROCRYPT 2000. LNCS, vol. 1807, pp. 589–606. Springer, Heidelberg (2000). doi:10.1007/3-540-45539-6_41
5. CAESAR Competition. http://competitions.cr.yp.to/caesar.html
6. Chakraborty, D., Sarkar, P.: A General construction of tweakable block ciphers and different modes of operations. In: Lipmaa, H., Yung, M., Lin, D. (eds.) Inscrypt 2006. LNCS, vol. 4318, pp. 88–102. Springer, Heidelberg (2006). doi:10.1007/11937807_8
7. Chakraborty, D., Sarkar, P.: HCH: A new tweakable enciphering scheme using the hash-counter-hash approach. IEEE Trans. Inf. Theory **54**(4), 1683–1699 (2008)

---

[7] More precisely, ANSI X9.24-1 suggests to use a few most significant bits of $E_k(0)$ as KCV.

8. Chen, S., Steinberger, J.: Tight security bounds for key-alternating ciphers. In: Nguyen, P.Q., Oswald, E. (eds.) EUROCRYPT 2014. LNCS, vol. 8441, pp. 327–350. Springer, Heidelberg (2014). doi:10.1007/978-3-642-55220-5_19

9. Cogliati, B., Lampe, R., Seurin, Y.: Tweaking even-mansour ciphers. In: Gennaro, R., Robshaw, M. (eds.) CRYPTO 2015. LNCS, vol. 9215, pp. 189–208. Springer, Heidelberg (2015). doi:10.1007/978-3-662-47989-6_9

10. Cogliati, B., Seurin, Y.: Beyond-birthday-bound security for tweakable even-mansour ciphers with linear tweak and key mixing. In: Iwata, T., Cheon, J.H. (eds.) ASIACRYPT 2015. LNCS, vol. 9453, pp. 134–158. Springer, Heidelberg (2015). doi:10.1007/978-3-662-48800-3_6

11. Cogliati, B., Seurin, Y.: On the provable security of the iterated even-mansour cipher against related-key and chosen-key attacks. In: Oswald, E., Fischlin, M. (eds.) EUROCRYPT 2015. LNCS, vol. 9056, pp. 584–613. Springer, Heidelberg (2015). doi:10.1007/978-3-662-46800-5_23

12. Crowley, P.: Mercy: A fast large block cipher for disk sector encryption. In: Goos, G., Hartmanis, J., Leeuwen, J., Schneier, B. (eds.) FSE 2000. LNCS, vol. 1978, pp. 49–63. Springer, Heidelberg (2001). doi:10.1007/3-540-44706-7_4

13. Daemen, J.: Limitations of the even-mansour construction. In: [26], pp. 495–498

14. Daemen, J., Rijmen, V.: The Design of Rijndael: AES - The Advanced Encryption Standard. Springer, Heidelberg (2002)

15. Dunkelman, O., Keller, N., Shamir, A.: Minimalism in cryptography: the even-mansour scheme revisited. In: Pointcheval, D., Johansson, T. (eds.) EUROCRYPT 2012. LNCS, vol. 7237, pp. 336–354. Springer, Heidelberg (2012). doi:10.1007/978-3-642-29011-4_21

16. Dworkin, M.: Recommendation for Block Cipher Modes of Operation: The XTS-AES Mode for Condentiality on Storage Devices. NIST Special Publication 800-38E (2010)

17. Even, S., Mansour, Y.: A construction of a cipher from a single pseudorandom permutation. In: [26], pp. 210–224

18. Farshim, P., Procter, G.: The related-key security of iterated even-mansour ciphers. In: [33], pp. 342–363

19. Ferguson, N., Lucks, S., Schneier, B., Whiting, D., Bellare, M., Kohno, T., Callas, J., Walker, J.: The SKEIN Hash Function Family. NIST SHA-3 Competition (2008)

20. Goldenberg, D., Hohenberger, S., Liskov, M., Schwartz, E.C., Seyalioglu, H.: On tweaking luby-rackoff blockciphers. In: Kurosawa, K. (ed.) ASIACRYPT 2007. LNCS, vol. 4833, pp. 342–356. Springer, Heidelberg (2007). doi:10.1007/978-3-540-76900-2_21

21. Granger, R., Jovanovic, P., Mennink, B., Neves, S.: Improved masking for tweakable blockciphers with applications to authenticated encryption. In: Fischlin, M., Coron, J.-S. (eds.) EUROCRYPT 2016. LNCS, vol. 9665, pp. 263–293. Springer, Heidelberg (2016). doi:10.1007/978-3-662-49890-3_11

22. Grosso, V., Leurent, G., Standaert, F., Varici, K., Journault, A., Durvaux, F., Gaspar, L., Kerckhof, S.: SCREAM Side-Channel Resistant Authenticated Encryption with Masking V3. CAESAR Competition Candidate (2015). http://competitions.cr.yp.to/round2/screamv3.pdf

23. Halevi, S.: EME*: Extending EME to handle arbitrary-length messages with associated data. In: Canteaut, A., Viswanathan, K. (eds.) INDOCRYPT 2004. LNCS, vol. 3348, pp. 315–327. Springer, Heidelberg (2004). doi:10.1007/978-3-540-30556-9_25

24. Halevi, S., Rogaway, P.: A tweakable enciphering mode. In: Boneh, D. (ed.) CRYPTO 2003. LNCS, vol. 2729, pp. 482–499. Springer, Heidelberg (2003). doi:10.1007/978-3-540-45146-4_28

25. Halevi, S., Rogaway, P.: A parallelizable enciphering mode. In: Okamoto, T. (ed.) CT-RSA 2004. LNCS, vol. 2964, pp. 292–304. Springer, Heidelberg (2004). doi:10.1007/978-3-540-24660-2_23

26. Imai, H., Rivest, R.L., Matsumoto, T. (eds.): ASIACRYPT 1991. LNCS, vol. 739. Springer, Heidelberg (1993). doi:10.1007/3-540-57332-1_17

27. Iwata, T., Wang, L.: Impact of ANSI X9.24-1:2009 key check value on ISO/IEC 9797-1:2011 MACs. In: Cid, C., Rechberger, C. (eds.) FSE 2014. LNCS, vol. 8540, pp. 303–322. Springer, Heidelberg (2015). doi:10.1007/978-3-662-46706-0_16

28. Jean, J., Nikolić, I., Peyrin, T.: Tweaks and keys for block ciphers: the TWEAKEY framework. In: Sarkar, P., Iwata, T. (eds.) ASIACRYPT 2014. LNCS, vol. 8874, pp. 274–288. Springer, Heidelberg (2014). doi:10.1007/978-3-662-45608-8_15

29. Jean, J., Nikolic, I., Peyrin, T.: Deoxys v1.3. CAESAR Competition Candidate (2015). http://competitions.cr.yp.to/round2/deoxysv13.pdf

30. Jean, J., Nikolic, I., Peyrin, T.: Joltik v1.3. CAESAR Competition Candidate (2015). http://competitions.cr.yp.to/round2/joltikv13.pdf

31. Lampe, R., Seurin, Y.: Tweakable blockciphers with asymptotically optimal security. In: Moriai, S. (ed.) FSE 2013. LNCS, vol. 8424, pp. 133–151. Springer, Heidelberg (2014). doi:10.1007/978-3-662-43933-3_8

32. Landecker, W., Shrimpton, T., Terashima, R.S.: Tweakable blockciphers with beyond birthday-bound security. In: Safavi-Naini, R., Canetti, R. (eds.) CRYPTO 2012. LNCS, vol. 7417, pp. 14–30. Springer, Heidelberg (2012). doi:10.1007/978-3-642-32009-5_2

33. Leander, G. (ed.): FSE 2015. LNCS, vol. 9054, pp. 428–448. Springer, Heidelberg (2015). doi:10.1007/978-3-662-48116-5_21

34. Liskov, M., Rivest, R.L., Wagner, D.: Tweakable block ciphers. In: Yung, M. (ed.) CRYPTO 2002. LNCS, vol. 2442, pp. 31–46. Springer, Heidelberg (2002). doi:10.1007/3-540-45708-9_3

35. Liskov, M., Rivest, R.L., Wagner, D.: Tweakable block ciphers. J. Cryptol. 24(3), 588–613 (2011)

36. Mennink, B.: Optimally secure tweakable blockciphers. In: [33], pp. 428–448

37. Mennink, B.: Optimally Secure Tweakable Blockciphers. IACR Cryptology ePrint Archive 2015 363 (2015). http://eprint.iacr.org/2015/363

38. Mennink, B.: Private communication (2015)

39. Mennink, B.: XPX: Generalized tweakable even-mansour with improved security guarantees. In: Robshaw, M., Katz, J. (eds.) CRYPTO 2016. LNCS, vol. 9814, pp. 64–94. Springer, Heidelberg (2016). doi:10.1007/978-3-662-53018-4_3

40. Minematsu, K.: Improved security analysis of XEX and LRW modes. In: Biham, E., Youssef, A.M. (eds.) SAC 2006. LNCS, vol. 4356, pp. 96–113. Springer, Heidelberg (2007). doi:10.1007/978-3-540-74462-7_8

41. Minematsu, K.: Beyond-birthday-bound security based on tweakable block cipher. In: Dunkelman, O. (ed.) FSE 2009. LNCS, vol. 5665, pp. 308–326. Springer, Heidelberg (2009). doi:10.1007/978-3-642-03317-9_19

42. Minematsu, K., Iwata, T.: Tweak-length extension for tweakable blockciphers. In: Groth, J. (ed.) IMACC 2015. LNCS, vol. 9496, pp. 77–93. Springer, Heidelberg (2015). doi:10.1007/978-3-319-27239-9_5

43. Minematsu, K., Matsushima, T.: Tweakable enciphering schemes from hash-sum-expansion. In: Srinathan, K., Rangan, C.P., Yung, M. (eds.) INDOCRYPT 2007. LNCS, vol. 4859, pp. 252–267. Springer, Heidelberg (2007). doi:10.1007/978-3-540-77026-8_19

44. Mitsuda, A., Iwata, T.: Tweakable pseudorandom permutation from generalized feistel structure. In: Baek, J., Bao, F., Chen, K., Lai, X. (eds.) ProvSec 2008. LNCS, vol. 5324, pp. 22–37. Springer, Heidelberg (2008). doi:10.1007/978-3-540-88733-1_2

45. Patarin, J.: A proof of security in $O(2^n)$ for the Xor of two random permutations. In: Safavi-Naini, R. (ed.) ICITS 2008. LNCS, vol. 5155, pp. 232–248. Springer, Heidelberg (2008). doi:10.1007/978-3-540-85093-9_22

46. Procter, G.: A Note on the CLRW2 Tweakable Block Cipher Construction. Cryptology ePrint Archive, Report 2014/111 (2014). http://eprint.iacr.org/2014/111

47. Rogaway, P.: Efficient instantiations of tweakable blockciphers and refinements to modes OCB and PMAC. In: Lee, P.J. (ed.) ASIACRYPT 2004. LNCS, vol. 3329, pp. 16–31. Springer, Heidelberg (2004). doi:10.1007/978-3-540-30539-2_2

48. Rogaway, P., Bellare, M., Black, J., Krovetz, T.: OCB: a block-cipher mode of operation for efficient authenticated encryption. In: Reiter, M.K., Samarati, P., (eds.) ACM CCS 2001, pp. 196–205. ACM (2001)

49. Sarkar, P.: Efficient tweakable enciphering schemes from (block-wise) universal hash functions. IEEE Trans. Inf. Theory 55(10), 4749–4760 (2009)

50. Schroeppel, R.: The Hasty Pudding Cipher. NIST AES Proposal (1998)

51. Wang, L.: SHELL v2.0. CAESAR Competition Candidate (2015). http://competitions.cr.yp.to/round2/shellv20.pdf

52. Wang, L., Guo, J., Zhang, G., Zhao, J., Gu, D.: How to Build Fully Secure Tweakable Blockciphers from Classical Blockciphers. Cryptology ePrint Archive, Report 2016/876 (2016). http://eprint.iacr.org/2016/876

53. Wang, P., Feng, D., Wu, W.: HCTR: A variable-input-length enciphering mode. In: Feng, D., Lin, D., Yung, M. (eds.) CISC 2005. LNCS, vol. 3822, pp. 175–188. Springer, Heidelberg (2005). doi:10.1007/11599548_15

# Design Strategies for ARX with Provable
# Bounds: SPARX and LAX

Daniel Dinu[(✉)], Léo Perrin, Aleksei Udovenko, Vesselin Velichkov,
Johann Großschädl, and Alex Biryukov

SnT, University of Luxembourg, Luxembourg City, Luxembourg
{daniel.dinu,leo.perrin,aleksei.udovenko,vesselin.velichkov,
johann.groszschaedl,alex.biryukov}@uni.lu

**Abstract.** We present, for the first time, a general strategy for designing ARX symmetric-key primitives with provable resistance against single-trail differential and linear cryptanalysis. The latter has been a long standing open problem in the area of ARX design. The *wide-trail design strategy* (WTS), that is at the basis of many S-box based ciphers, including the AES, is not suitable for ARX designs due to the lack of S-boxes in the latter. In this paper we address the mentioned limitation by proposing the *long trail design strategy* (LTS) – a dual of the WTS that is applicable (but not limited) to ARX constructions. In contrast to the WTS, that prescribes the use of small and efficient S-boxes at the expense of heavy linear layers with strong mixing properties, the LTS advocates the use of large (ARX-based) S-Boxes together with sparse linear layers. With the help of the so-called *long-trail argument*, a designer can bound the maximum differential and linear probabilities for any number of rounds of a cipher built according to the LTS.

To illustrate the effectiveness of the new strategy, we propose SPARX – a family of ARX-based block ciphers designed according to the LTS. SPARX has 32-bit ARX-based S-boxes and has provable bounds against differential and linear cryptanalysis. In addition, SPARX is very efficient on a number of embedded platforms. Its optimized software implementation ranks in the top 6 of the most software-efficient ciphers along with SIMON, SPECK, Chaskey, LEA and RECTANGLE.

As a second contribution we propose another strategy for designing ARX ciphers with provable properties, that is completely independent of the LTS. It is motivated by a challenge proposed earlier by Wallén and uses the differential properties of modular addition to minimize the maximum differential probability across multiple rounds of a cipher. A new primitive, called LAX, is designed following those principles. LAX partly solves the Wallén challenge.

**Keywords:** ARX · Block ciphers · Differential cryptanalysis · Linear cryptanalysis · Lightweight · Wide-trail strategy

© International Association for Cryptologic Research 2016
J.H. Cheon and T. Takagi (Eds.): ASIACRYPT 2016, Part I, LNCS 10031, pp. 484–513, 2016.
DOI: 10.1007/978-3-662-53887-6_18

# 1  Introduction

ARX, standing for Addition/Rotation/XOR, is a class of symmetric-key algorithms designed using only the following simple operations: modular addition, bitwise rotation and exclusive-OR. In contrast to S-box-based designs, where the only non-linear elements are the substitution tables (S-boxes), ARX designs rely on modular addition as the only source of non-linearity. Notable representatives of the ARX class include the stream ciphers Salsa20 [1] and ChaCha20 [2], the SHA-3 finalists Skein [3] and BLAKE [4] as well as several lightweight block ciphers such as TEA, XTEA [5], etc. Dinu et al. recently reported [6] that the most efficient software implementations on small processors belonged to ciphers from the ARX class: Chaskey-cipher [7] by Mouha et al., SPECK [8] by the American National Security Agency (NSA) and LEA [9] by the South Korean Electronic and Telecommunications Research Institute.[1]

For the mentioned algorithms, the choice of using the ARX paradigm was based on three observations[2]. First, getting rid of the table look-ups, associated with S-Box based designs, increases the resilience against side-channel attacks. Second, this design strategy minimizes the total number of operations performed during an encryption, allowing particularly fast software implementations. Finally, the computer code describing such algorithms is very small, making this approach especially appealing for lightweight block ciphers where the memory requirements are the harshest.

Despite the widespread use of ARX ciphers, the following problem has remained open up until now.

**Open Problem.** *Is it possible to design an ARX cipher that is provably secure against single-trail differential and linear cryptanalysis by design?*

To the best of our knowledge, there has only been one attempt at tackling this issue. In [10] Biryukov et al. have proposed several ARX constructions for which it is feasible to compute the exact maximum differential and linear probabilities over any number of rounds. However, these constructions are limited to 32-bit blocks. The general case of this problem, addressing any block size, has still remained without a solution.

More generally, the formal understanding of the cryptographic properties of ARX is far less satisfying than that of, for example, S-Box-based substitution-permutation networks (SPN). Indeed, the wide-trail strategy [11] (WTS) and the wide-trail argument [12] provide a way to design S-box based SPNs with provable resilience against differential and linear attacks. It relies on bounding the number of active S-Boxes in a differential (resp. linear) trail and deducing a lower bound on the best expected differential (resp. linear) probability.

---

[1] SPECK and the MAC Chaskey are being considered for standardization by ISO.
[2] For SPECK, we can only a guess it is the case as the designers have not published the rationale behind their algorithm.

*Our Contribution.* We propose two different strategies to build ARX-based block ciphers with provable bounds on the maximum expected differential and linear probabilities, thus providing a solution to the open problem stated above.

The first strategy is called the *Long Trail Strategy* (LTS). It borrows the idea of counting the number of active S-Boxes from the wide-trail argument but the overall principle is actually the opposite to the wide-trail strategy as described in [11]. While the WTS dictates the spending of most of the computational resources in the linear layer in order to provide good diffusion between small S-boxes, the LTS advocates the use of large and comparatively expensive S-Boxes in conjunction with cheaper and weaker linear layers. We formalize this method and describe the *long-trail argument* that can be used to bound the differential and linear trail probabilities of a block cipher built using this strategy.

Using this framework, we build a family of lightweight block ciphers called SPARX. All three instances in this family can be entirely specified using only three operations: addition modulo $2^{16}$, 16-bit rotations and 16-bit XOR. These ciphers are, to the best of our knowledge, the first ARX-based block ciphers for which the probability of both differential and linear trails are bounded. Furthermore, while one may think that these provable properties imply a performance degradation, we show that it is not the case. On the contrary, SPARX ciphers have very competitive performance on lightweight processors. In fact, the most lightweight version – SPARX-64 is in the top 3 for 16-bit micro-controllers according to the classification method presented in [6].

Finally, we propose the LAX construction, where bit rotations are replaced with a more general linear permutation. The bounds on the differential probability are expressed as a function of the branching number of the linear layer. We note that the key insight behind this construction has been published in [13], but its realization has been left as a challenge.

*Outline.* First, we introduce the notations and concepts used throughout the paper in Sect. 2. In Sect. 3, we describe how an ARX-based cipher with provable bounds can be built using an S-Box-based approach and how the method used is a particular case of the more general *Long Trail Strategy*. Section 4 contains the specification of the SPARX family of ciphers, the description of its design rationale and a discussion about the efficiency of its implementation on microcontrollers. The LAX structure is presented in Sect. 5. Finally, Sect. 6 concludes the paper.

## 2    Preliminaries

We use $\mathbb{F}_2$ to denote the set $\{0, 1\}$. Let $f : \mathbb{F}_2^n \to \mathbb{F}_2^n$, $(a, b) \in \mathbb{F}_2^n \times \mathbb{F}_2^n$ and $x \in \mathbb{F}_2^n$. We denote the probability of the differential trail $(a \xrightarrow{d} b)$ by $\Pr[f(x) \oplus f(x \oplus a) = b]$ and the correlation of the linear approximation $(a \xrightarrow{\ell} b)$ by $(2 \Pr[a \cdot x = b \cdot f(x)] - 1)$ where $y \cdot z$ is the scalar product of $y$ and $z$.

In an iterated block cipher, not all differential (respectively linear) trails are possible. Indeed, they must be coherent with the overall structure of the round function. For example, it is well known that a 2-round differential trail for the

AES with less than 4 active S-Boxes is impossible. To capture this notion, we use the following definition.

**Definition 1 (Valid Trail).** *Let $f$ be an $n$-bit permutation. A trail $a_0 \rightarrow \ldots \rightarrow a_r$ for $r$ rounds of $f$ is a valid trail if $Pr[a_i \rightarrow a_{i+1}] > 0$ for all $i$ in $[0, r-1]$. The set of all valid $r$-round differential (respectively linear) trails for $f$ is denoted $\mathcal{V}_\delta(f)^r$ (resp. $\mathcal{V}_\ell(f)^r$).*

We use the acronyms MEDCP and MELCC to denote resp. *maximum expected differential characteristic probability* and *maximum expected linear characteristic correlation* – a signature introduced earlier in [14]. The MEDCP of the keyed function $f_{k_i} : x \mapsto f(x \oplus k_i)$ iterated over $r$ rounds is defined as follows:

$$\text{MEDCP}(f^r) = \max_{(\Delta_0 \rightarrow \ldots \Delta_r) \in \mathcal{V}_\delta(f)^r} \prod_{i=0}^{r-1} \Pr[\Delta_i \xrightarrow{d} \Delta_{i+1}],$$

where $\Pr[\Delta_i \xrightarrow{d} \Delta_{i+1}]$ is the expected value of the differential probability of $\Delta_i \xrightarrow{d} \Delta_{i+1}$ for the function $f_k$ when $k$ is picked uniformly at random. $\text{MELCC}(f^r)$ is defined analogously. Note that $\text{MEDCP}(f^r)$ and $\left(\text{MEDCP}(f^1)\right)^r$ are *not* equal.

As designers, we thrive to provide upper bounds for both $\text{MEDCP}(f^r)$ and $\text{MELCC}(f^r)$. Doing so allows us to compute the number of rounds $f$ needed in a block cipher for the probability of all trails to be too low to be usable. In practice, we want $\text{MEDCP}(f^r) \ll 2^{-n}$ and $\text{MELCC}(f^r) \ll 2^{-n/2}$ where $n$ is the block size.

While this strategy is the best known, the following limitations must be taken into account by algorithm designers.

1. The quantities $\text{MEDCP}(f^r)$ and $\text{MELCC}(f^r)$ are relevant only if we make the *Markov assumption*, meaning that the differential and linear probabilities are independent in each round. This would be true if the subkeys were picked uniformly and independently at random but, as the master key has a limited size, it is not the case.
2. These quantities are averages taken over all possible keys: it is not impossible that there exists a weak key and a differential trail $T$ such that the probability of $T$ is higher than $\text{MEDCP}(f^r)$ for this particular key. The same holds for the linear probability.
3. These quantities deal with unique trails. However, it is possible that several differential trails share the same input and output differences, thus leading to a higher probability for said differential transition. This so-called *differential effect* can be leveraged to decrease the data complexity of differential attack. The same holds for linear attacks where several approximations may form a linear hull.

Still, this type of bound is the best that can be achieved in a generic fashion (to the best of our knowledge). In particular, this is the type of bound provided by the wide-trail argument used in the AES.

# 3    ARX-Based Substitution-Permutation Network

In this section, we present a general design strategy for building ARX-based block ciphers borrowing techniques from SPN design. The general idea is to build a SPN with ARX-based S-boxes instead of with S-boxes based on look-up tables (LUT). The proofs for the bound on the MEDCP and MELCC are inspired by the wide-trail argument introduced in the design of the AES [12]. However, because of the use of large S-Boxes, the method used relies on a different type of interaction between the linear and non-linear layers. We call the corresponding design strategy the *long trail strategy*. It is quite general and could be also applied in other contexts e.g. for non-ARX constructions.

First, we present possible candidates for the ARX-based S-Box and, along the way, identify the likely reason behind the choice of the rotation constants in SPECK-32. Then, we describe the long trail strategy in more details. Finally, we present two different algorithms for computing a bound for the MEDCP and MELCC of block ciphers built using a LT strategy. We also discuss how to ensure that the linear layer provides sufficient diffusion.

## 3.1    ARX-Boxes

**Definition 2 (ARX-box).** *An ARXbox is a permutation on m bits (where m is much smaller than the block size) which relies entirely on addition, rotation and XOR to provide both non-linearity and diffusion. An ARX-box is a particular type of S-Box.*

Possible constructions for ARX-boxes can be found in a recent paper by Biryukov et al. [10]. A first one is based on the MIX function of Skein [3] and is called MARX-2. The rotation amounts, namely $\{1, 2, 7, 3\}$, were chosen so as to minimize the differential and linear probabilities. The key addition is done over the full state. The second construction is called SPECKEY and consists of one round of SPECK-32 [8] with the key added to the full state instead of only to half the state as in the original algorithm. The two constructions MARX-2 and SPECKEY are shown in Fig. 1a and b. The differential and linear bounds for them are given in Table 1. While it is possible to choose the rotations used in SPECKEY in such a way as to slightly decrease the differential and linear bounds[3], such rotations are more expensive on small microcontrollers which only have instructions implementing rotations by 1 and by 8 (in both directions). We infer, although we cannot prove it, that the designers of SPECK-32 made similar observations.

## 3.2    Naive Approaches and Their Limitations

A very simple method to build ARX-based ciphers with provable bounds on MEDCP and MELCC is to use a SPN structure where the S-boxes are replaced

---

[3] Both can be lowered by a factor of 2 if we choose rotations $(9, 2), (9, 5), (11, 7)$ or $(7, 11)$ instead of $(7, 2)$.

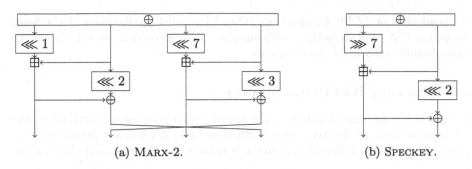

(a) MARX-2.                                                    (b) SPECKEY.

**Fig. 1.** Key addition followed by the candidate 32-bit ARX-boxes, MARX-2 and SPECKEY. The branch size is 8 bits for MARX-2, 16 bits for SPECKEY.

**Table 1.** Maximum expected differential characteristic probabilities (MEDCP) and maximum expected absolute linear characteristic correlations (MELCC) of MARX-2 and SPECKEY ($\log_2$ scale); $r$ is the number of rounds.

|         | $r$          | 1   | 2   | 3   | 4   | 5    | 6    | 7    | 8    | 9    | 10   |
|---------|--------------|-----|-----|-----|-----|------|------|------|------|------|------|
| MARX-2  | MEDCP($M^r$) | $-0$ | $-1$ | $-3$ | $-5$ | $-11$ | $-16$ | $-22$ | $-25$ | $-29$ | $-35$ |
|         | MELCC($M^r$) | $-0$ | $-0$ | $-1$ | $-3$ | $-5$  | $-8$  | $-10$ | $-13$ | $-15$ | $-17$ |
| SPECKEY | MEDCP($S^r$) | $-0$ | $-1$ | $-3$ | $-5$ | $-9$  | $-13$ | $-18$ | $-24$ | $-30$ | $-34$ |
|         | MELCC($S^r$) | $-0$ | $-0$ | $-1$ | $-3$ | $-5$  | $-7$  | $-9$  | $-12$ | $-14$ | $-17$ |

by ARX operations for which we can compute the MEDCP and MELCC. This is indeed the strategy we follow but care must be taken when actually choosing the ARX-based operations and the linear layer.

Let us for example build a 128-bit block cipher with an S-Box layer consisting in one iteration of SPECKEY on each 32-bit word and with an MDS linear layer, say a multiplication with the MixColumns matrix with elements in $GF(2^{32})$ instead of $GF(2^8)$. The MEDCP bound of such a cipher, computed using a classical wide-trail argument, would be equal to 1! Indeed, there exists probability 1 differentials for 1-round SPECKEY so that, regardless of the number of active S-Boxes, the bound would remain equal to 1. Such an approach is therefore not viable.

As the problem identified above stems from the use of 1-round SPECKEY, we now replace it with 3-round SPECKEY where the iterations are interleaved with the addition of independent round keys. The best linear and differential probabilities are no longer equal to 1, meaning that it is possible to build a secure cipher using the same layer as before provided that enough rounds are used. However, such a cipher would be very inefficient. Indeed, the MDS bound imposes that 5 ARX-boxes are active every 2 rounds, so that the MEDP bound is equal to $p_d^{5r/2}$ where $r$ is the number of rounds and $p_d$ is the best differential probability of the ARX-box (3-rounds SPECKEY). To push the bound below $2^{-128}$

we need at least 18 SPN rounds, meaning 54 parallel applications of the basic ARX-round! We will show that, with our alternative approach, we can obtain the same bounds with much fewer rounds.

### 3.3    The Long Trail Design Strategy

Informed by the shortcomings of the naive design strategies described in the previous section, we devised a new method to build ARX-based primitives with provable linear and differential bounds. It is based on the following observation.

**Observation 1 (Impact of Long Trails).** *Let $d(r)$ and $\ell(r)$ be the MEDCP and MELCC of some ARX-box iterated $r$ times and interleaved with the addition of independent subkeys. Then, in most cases:*

$$d(qr) \ll d(r)^q \text{ and } \ell(qr) \ll \ell(r)^q.$$

*In other words, in order to diminish the MEDCP and MELCC of a construction, it is better to allow long trails of ARX-boxes without mixing.*

For example, if we look at SPECKEY, the MEDCP for 3 rounds is $2^{-3}$ and that of 6 rounds is $2^{-15}$ which is far smaller than $(2^{-3})^2 = 2^{-6}$ (see Table 1). Similarly, the MELCC for 3 rounds is $2^{-1}$ and after 6 rounds it is $2^{-7} \ll (2^{-1})^2$.

In fact, a similar observation has been made by Nikolić when designing the CAESAR candidate family Tiaoxin [15]. It was later generalized to larger block sizes in [16], where Jean and Nikolić present, among others, the AES-based $\mathcal{A}_\oplus^2$ permutation family. It uses a partial S-Box layer where the S-Box consists of 2 AES rounds and a word-oriented linear layer in such a way that some of the S-Box calls can be chained within 2-round long trails. Thus, they may use the 4-round bound on the number of active 8-bit AES S-Boxes, which is 25, rather than twice the 2-round bound, which would be equal to 10 (see Table 2). Their work on this permutation can be interpreted as a particular case of the observation above.

**Definition 3 (Long Trail).** *We call Long Trail (LT) an uninterrupted sequence of calls to an ARX-box interleaved with key additions. No difference can be added into the trail from the outside. Such trails can happen for two reasons.*

1. *A Static Long Trail occurs with probability 1 because one output word of the linear layer is an unchanged copy of one of its input words.*

**Table 2.** Bound on the number of active 8-bit S-Boxes in a differential (or linear) trail for the AES.

| # R | 1 | 2 | 3 | 4 | 5 | 6 | 7 | 8 | 9 | 10 |
|---|---|---|---|---|---|---|---|---|---|---|
| # Active S-Boxes | 1 | 5 | 9 | 25 | 26 | 30 | 34 | 50 | 51 | 55 |

2. *A Dynamic Long Trail occurs within a specific differential trail because one output word of the linear layer consists of the XOR of one of its input words with a non-zero difference and a function of words with a zero difference. In this way the output word of the linear layer is again equal to the input word as in a Static LT, but here this effect has been obtained dynamically.*

**Definition 4 (Long Trail Strategy).** *The Long Trail Strategy is a design guideline: when designing a primitive with a rather weak but large S-Box (say, an ARX-based permutation), it is better to foster the existence of long trails rather than to have maximum diffusion in each linear layer.*

This design principle has an obvious caveat: although slow, diffusion is necessary! Unlike the WTS, in this context it is better to trade some of the power of the diffusion layer in favor of facilitating the emergence of long trails.

The long trail strategy is a method for building secure and efficient ciphers using a large but weak S-Box $S$ such that we can bound the MEDCP (and MELCC) of several iterations of $x \mapsto S(x \oplus k)$ with independent round keys. In this paper, we focus on the case where $S$ consists of ARX operations but this strategy could have broader applications such as, as briefly discussed above, the design of block ciphers operating on large blocks using the AES round function as a building block.

In a way, this design method is the direct opposite of the wide trail strategy as it is summarized by Daemen and Rijmen in [11] (emphasis ours):

> Instead of spending most of the resources on large S-boxes, the wide trail strategy aims at designing the round transformation(s) such that there are no trails with a low bundle weight. In ciphers designed by the wide trail strategy, *a relatively large amount of resources is spent in the linear step* to provide high multiple-round diffusion.

The long trail approach *minimizes* the amount of resources spent in the linear layer and does spend most of the resources on large S-Boxes. Still, as discussed in the next section, the method used to bound the MEDCP and MELCC in the long trail strategy is heavily inspired by the one used in the wide trail strategy.

**A Cipher Structure for the LT Strategy.** We can build block ciphers based on the long trail strategy using the following two-level structure. First, we must choose an S-Box layer operating on $w$ words in parallel. The composition of a key addition in the full state and the application of this S-Box layer is called a *round*. Several rounds are iterated and then a word-oriented linear mixing layer is applied to ensure diffusion between the words. The composition of $r$ rounds followed by the linear mixing layer is called a *step*[4], as described in Fig. 2. The encryption thus consists in iterating such steps. We used this design strategy to build a block cipher family, SPARX, which we describe in Sect. 4.

---

[4] This terminology is borrowed from the specification of LED [17] which also groups several calls of the round function into a step.

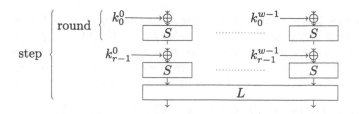

**Fig. 2.** A cipher structure for the LT strategy.

**Long Trail-Based Bounds.** In what follows we only discuss differential long trails for the sake of brevity. Linear long trails are treated identically.

**Definition 5 (Truncated LT Decomposition).** *Consider a cipher with a round function operating on $w$ words. A truncated differential trail is a sequence of values of $\{0,1\}^w$ describing whether an S-Box is active at a given round. The LT Decomposition of a truncated differential trail is obtained by grouping together the words of the differential trails into long trails and then counting how many active long trails of each length are present. It is denoted $\{t_i\}_{i\geq 1}$ where $t_i$ is equal to the number of truncated long trails with length $i$.*

*Example 1.* Consider a 64-bit block cipher using a 32-bit S-Box, one round of Feistel network as its linear layer and 4 steps without a final linear layer. Consider the differential trail $(\delta_0^L, \delta_0^R) \to (\delta_1^L, \delta_1^R) \to (0, \delta_2^R) \to (\delta_3^L, 0)$ (see Fig. 3 where the zero difference is dashed). Then this differential trail can be decomposed into 3 long trails represented in black, blue and red: the first one has length 1 and $\delta_0^R$ as its input; the second one has length 2 and $\delta_0^L$ as its input; and the third one has length 3 and $\delta_1^L$ as its input so that the LT decomposition of this trail is $\{t_1 = 1, t_2 = 1, t_3 = 1\}$. Using the terminology introduced earlier, the first two trails are Static LT, while the third one is Dynamic LT.

**Theorem 1 (Long Trail Argument).** *Consider a truncated differential trail $T$ covering $r$ rounds consisting of an S-Box layer with S-Box $S$ interleaved with key additions and some linear layer. Let $\{t_i\}_{i\geq 1}$ be the LT decomposition of $T$. Then the probability $p_D$ of any fully specified differential trail fitting in $T$ is upper-bounded by*

$$p_D \leq \prod_{i\geq 1} (\mathrm{MEDCP}(S^i))^{t_i}$$

*where $\mathrm{MEDCP}(S^i)$ is an upper-bound on the probability of a differential trail covering $i$ iterations of $S$.*

*Proof.* Let $\Delta_{i,s} \xrightarrow{d} \Delta_{j,s+1}$ denote any differential trail occurring at the S-Box level in one step, so that the S-Box with index $i$ at step $s$ sees the transition $\Delta_{i,s} \xrightarrow{d} \Delta_{j,s+1}$. By definition of a long trail, we have in each long trail a chain of differential trails $\Delta_{i_0,s_0} \xrightarrow{d} \Delta_{i_1,s_0+1} \xrightarrow{d} ... \xrightarrow{d} \Delta_{i_t,s_0+t}$ which, because of the lack

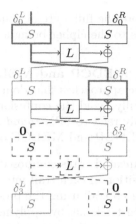

**Fig. 3.** An example of active LT decomposition.

of injection of differences from the outside, is a *valid trail* for $t$ iterations of the S-Box. This means that the probability of any differential trail following the same sequence of S-boxes as in this long trail is upper-bounded by $\mathrm{MEDCP}(S^t)$. We simply bound the product by the product of the bounds to derive the theorem. □

### 3.4   Choosing the Linear Layer: Bounding the MEDCP and MELCC while Providing Diffusion

In order to remain as general as possible, in this section we do not consider the details of a specific S-Box but instead we focus on fleshing out design criteria for the linear layer. All the information for the S-Box that is necessary to follow the explanation is the MEDCP and MELCC of its $r$-fold iterations including the key additions e.g. the data provided in Table 1 for our ARX-box candidates.

As the linear layers we consider may be weaker than usual designing SPN, it is also crucial that we ensure that ciphers built using such a linear layer are not vulnerable to integral attacks [18], in particular those based on the division property [19]. Incidentally, this gives us a criteria quantifying the diffusion provided by several steps of the cipher.

In this section, we propose two methods for bounding the MEDCP and MELCC of several steps of a block cipher. The first one is applicable to any linear layer but is relatively inefficient, while the second one works only for a specific subset of linear layers but is very efficient.

When considering truncated differential trails, it is hard to bound the probability of the event that differences in two or more words cancel each other in the linear layer i.e. the event that a Dynamic LT occurs. Therefore, for simplicity we assume that such cancellations happen *for free* i.e. with probability 1. Due to this simplification, we expect our bounds to be higher (i.e. looser) than the tight bounds. In other words, we *underestimate* the security of the cipher. Note that we also exclude the cases where the full state at some round has zero difference as the latter is impossible due to the cipher being a permutation.

**Algorithms for Bounding MEDCP and MELCC of a Cipher.** In this sub-section we propose generic approaches that do not depend on the number of rounds per step. In fact, to fully avoid the confusion between *rounds* and *steps* in what follows we shall simply refer to SPN *rounds*.

One way to bound the MEDCP and MELCC of a cipher is as follows:

1. Enumerate all possible truncated trails composed of active/inactive S-boxes.
2. Find an optimal decomposition of each trail into long trails (LT).
3. Bound the probability of each trail using the product of the MEDCP (resp. MELCC) of all active long trails i.e. by applying the Long Trail Argument (see Theorem 1) on the corresponding optimal trail decomposition.
4. The maximum bound over all trails is the final upper bound.

This approach is feasible only for a small number of rounds, because the number of trails grows exponentially. The algorithm is based on a recursive dynamic programming approach and has time complexity $O(wr^2)$, where $w$ is the number of S-Boxes applied in parallel in each S-Box layer and $r$ is the number of rounds.

As noted, the most complicated step in the above procedure is finding an optimal decomposition of a given truncated trail into long trails. The difficulty arises from the so-called *branching*: situation in which a long trail may be extended in more than one way. Recall that our definition of LT (cf. Definition 3) relies on the fact that there is no linear transformation on a path between two S-Boxes in a LT. The only transformations allowed are some XORs. Therefore, branching happens only when some output word of the linear layer receives two or more active input words without modifications. In order to cut off the branching effect (and thus to make finding the optimal decomposition of a LT feasible), we can put some additional linear functions that will modify the contribution of (some of) the input words. Equivalently, when choosing a linear layer we simply do not consider layers which cause branching of LTs. As we will show later, this restriction has many advantages.

To simplify our study of the linear layer, we introduce a matrix representation for it. In a block cipher operating on $w$ words, a linear layer may be expressed as a $w \times w$ block matrix. We will denote zero and identity sub-matrices by 0 and 1 respectively and an unspecified (arbitrary) sub-matrices by $L$. This information is sufficient for analyzing the high-level structure of a cipher. Using this notation, the linear layers to which we restrict our analysis have matrices where each column has at most one 1.

For the special subset of linear layers outlined above, we present an algorithm for obtaining MEDCP and MELCC bounds, that is based on a dynamic programming approach. Since there is no LT branching, any truncated trail consists of disjoint sequences of active S-Boxes. By Observation 1, we can treat each such sequence as a LT to obtain an optimal decomposition. Because of this simplification, we can avoid enumerating all trails by grouping them in a particular way.

We proceed round by round and maintain a set of best trails up to an equivalence relation, which is defined as follows. For all S-Boxes at the current last round $s$, we assign a number, which is equal to the length of the LT that covers this S-Box, or zero if the S-Box is not active. We say that two truncated trails for $s$ steps are equivalent if the tuples consisting of those numbers (current round $s$ and length of LT) are the same for both trails. This equivalence captures the possibility to replace some prefix of a trail by an equivalent one without breaking the validity of the trail or its LT decomposition. The total probability, however, can change. The key observation here is that from two equivalent trails we can keep only the one with the highest current probability. Indeed, if the optimal truncated trail for all $r$ rounds is an extension of the trail for $s$ rounds with lower probability, we can take the first $s$ rounds from the trail with higher probability without breaking anything and obtain a better trail, which contradicts the assumed optimality.

Due to page limit constraints, the pseudo-code for the algorithm is given in the full version of this paper [20].

This algorithm can be used to bound the probability of linear trails. Propagation of a linear mask through some linear layer can be described by multiplying the mask by the transposed inverse of the linear layer's matrix. In our matrix notation we can easily transpose the matrix but inversion is harder. However, we can build the linear trails bottom-up (i.e. starting from the last round): in this case we need only the transposed initial matrix. Our algorithm does not depend on the direction, so we obtain bounds on linear trails probabilities by running the algorithm on the transposed matrix using the linear bounds for the iterated S-box.

**Ensuring Resilience Against Integral Attacks.** As illustrated by the structural attack against SASAS and a recent generalization [21] to ciphers with more rounds, a SPN with few rounds may be vulnerable to integral attacks. This attack strategy has been further improved by Todo [19] who proposed the so-called *division property* as a means to track which bit should be fixed in the input to have a balanced output. He also described an algorithm allowing an attacker to easily find such distinguishers.

We implemented this algorithm to search for division-property-based integral trails covering as many rounds as possible. With it, for each matrix candidate we compute a maximum number of rounds covered by such a distinguisher. This quantity can then be used by the designer of the primitive to see if the level of protection provided against this type of attack is sufficient or not.

Tracking the evolution of the division property through the linear layer requires special care. In order to do this, we first make a copy of each word and apply the required XORs from the copy to the original words. Due to such state expansion, the algorithm requires both a lot of memory and time. In fact, it is even infeasible to apply on some matrices. To overcome this issue, we ran the algorithm with reduced word size. During our experiments, we observed that such an optimization may only result in longer integral characteristics and that this side effect occurs only for very small word sizes (4 or 5 bits). In light of this, we conjecture that the values obtained in these particular cases are upper bounds and are very close to the values which could be obtained without reducing the word size.

## 4   The SPARX Family of Ciphers

In this Section, we describe a family of block ciphers built using the framework laid out in the previous section. The instance with block size $n$ and key size $k$ is denoted SPARX-$n/k$.

### 4.1   High Level View

The plaintexts and ciphertexts consist of $w = n/32$ words of 32 bits each and the key is divided into $v = k/32$ such words. The encryption consists of $n_s$ steps, each composed of an ARX-box layer of $r_a$ rounds and a linear mixing layer. In the ARX-box layer, each word of the internal state undergoes $r_a$ rounds of SPECKEY, including key additions. The $v$ words in the key state are updated once $r_a$ ARX-boxes have been applied to one word of the internal state. The linear layers $\lambda_w$ for $w = 2, 4$ provide linear mixing for the $w$ words of the internal state.

This structure is summarized by the pseudo-code in Algorithm 1. The structure of one round is represented in Fig. 4, where $A$ is the 32-bit ARX-box consisting in one unkeyed SPECK-32 round. We also use $A^a$ to denote $a$ rounds of SPECKEY with the corresponding key additions (see Fig. 5a).

The different versions of SPARX all share the same definition of $A$. However, the permutations $\lambda_w$ and $K_v$ depend on the block and key sizes. The different members of the SPARX-family are specified below. The round keys can either be derived on the fly by applying $K_v$ on the key state during encryption or they can be precomputed and stored. The first option requires less RAM, while the second is faster. The only operations needed to implement any instance of SPARX are:

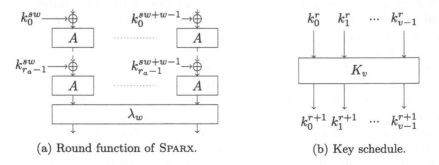

(a) Round function of SPARX.                    (b) Key schedule.

**Fig. 4.** A high level view of step $s$ of SPARX.

---

**Algorithm 1.** SPARX encryption
**Inputs** plaintext $(x_0, ..., x_{w-1})$; key $(k_0, ..., k_{v-1})$
**Output** ciphertext $(y_0, ..., y_{w-1})$

---

Let $y_i \leftarrow x_i$ for all $i \in [0, ..., w-1]$
**for all** $s \in [0, n_s - 1]$ **do**
    **for all** $i \in [0, w-1]$ **do**
        **for all** $r \in [0, r_a - 1]$ **do**
            $y_i \leftarrow y_i \oplus k_r$
            $y_i \leftarrow A(y_i)$
        **end for**
        $(k_0, ..., k_{v-1}) \leftarrow K_v\big((k_0, ..., k_{v-1})\big)$             ▷ Update key state
    **end for**
    $(y_0, ..., y_{w-1}) \leftarrow \lambda_w\big((y_0, ..., y_{w-1})\big)$           ▷ Linear mixing layer
**end for**
Let $y_i \leftarrow y_i \oplus k_i$ for all $i \in [0, ..., w-1]$            ▷ Final key addition
**return** $(y_0, ..., y_{w-1})$

---

- addition modulo $2^{16}$, denoted $\boxplus$,
- 16-bit exclusive-or (XOR), denoted $\oplus$, and
- 16-bit rotation to the left or right by $i$, denoted respectively $x \lll i$ and $x \ggg i$.

We claim that no attack using less than $2^k$ operations exists against SPARX-$n/k$ in neither the single-key nor in the related-key setting. We also faithfully declare that we have not hidden any weakness in these ciphers. SPARX is free for use and its source code is available in the public domain[5].

## 4.2   Specification

Table 3 summarizes the different SPARX instances and their parameters. The quantity $\min_{\text{secure}}(n_s)$ corresponds to the minimum number of steps for which we can prove that the MEDCP is below $2^{-n}$, that the MELCC is below $2^{-n/2}$ for the number of rounds per step chosen and for which we cannot find integral distinguishers covering this amount of steps.

**SPARX-64/128.** The lightest instance of SPARX is SPARX-64/128. It operates on two words of 32 bits and uses a 128-bit key. There are 8 steps and 3 rounds per step. As it takes 5 steps to achieve provable security against linear and differential attacks, our security margin is at least equal to 37 % of the rounds. Furthermore, while our long trail argument proves that 5 steps are sufficient to ensure that there are no single-trail differential and linear distinguishers, we do not expect this bound to be tight.

The linear layer $\lambda_2$ simply consists of a Feistel round using $\mathcal{L}$ as a Feistel function. The general structure of a step of SPARX-64/128 is provided in Fig. 5b. The

---

[5] See https://www.cryptolux.org/index.php/SPARX.

**Table 3.** The different SPARX instances.

|  | SPARX-64/128 | SPARX-128/128 | SPARX-128/256 |
|---|---|---|---|
| # State words $w$ | 2 | 4 | 4 |
| # Key words $v$ | 4 | 4 | 8 |
| # Rounds/Step $r_a$ | 3 | 4 | 4 |
| # Steps $n_s$ | 8 | 8 | 10 |
| Best Attack (# rounds) | 15/24 | 22/32 | 24/40 |
| $\min_{\text{secure}}(n_s)$ | 5 | 5 | 5 |

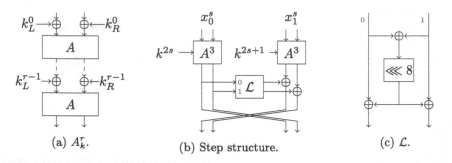

(a) $A_k^r$.    (b) Step structure.    (c) $\mathcal{L}$.

**Fig. 5.** A high level view of SPARX-64/128. Branches have a width of 16 bits (except for the keys in the step structure).

128-bit permutation used in the key schedule has a simple definition summarized in Fig. 6, where the counter $r$ is initialized to 0. It corresponds to the pseudo code given in Algorithm 2, where $(z)_L$ and $(z)_R$ are the 16-bit left and right halves of the 32-bit word $z$.

The $\mathcal{L}$ function is borrowed from NOEKEON [22] and can be defined using 16- or 32-bit rotations. It is defined as a Lai-Massey structure mapping a 32-bit value $x\|y$ to $x \oplus ((x \oplus y) \lll 8)\|y \oplus ((x \oplus y) \lll 8)$. Alternatively, it can be seen as a mapping of a 32-bit value $z$ to $z \oplus (z \lll^{32} 8) \oplus (z \ggg^{32} 8)$ where the rotations are over 32 bits.

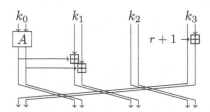

$$\begin{aligned}
&r \leftarrow r + 1 \\
&k_0 \leftarrow A(k_0) \\
&(k_1)_L \leftarrow (k_1)_L + (k_0)_L \mod 2^{16} \\
&(k_1)_R \leftarrow (k_1)_R + (k_0)_R \mod 2^{16} \\
&(k_3)_R \leftarrow (k_3)_R + r \mod 2^{16} \\
&k_0, k_1, k_2, k_3 \leftarrow k_3, k_0, k_1, k_2
\end{aligned}$$

**Fig. 6.** $K_4^{64}$ (used in SPARX-64/128).    **Algorithm 2.** Pseudo-code of $K_4^{64}$

**SPARX-128/128 and SPARX-128/256.** For use cases in which a larger block size can be afforded, we provide SPARX instances with a 128-bit block size and 128- or 256-bit keys. They share an identical step structure which is fairly similar to SPARX-64/128. Indeed, the linear layer relies again on a Feistel function except that $\mathcal{L}$ is replaced by $\mathcal{L}'$, a permutation of $\{0, 1\}^{64}$. Both SPARX-128/128 and SPARX-128/256 use 4 rounds per step but the first uses 8 steps while the last uses 10.

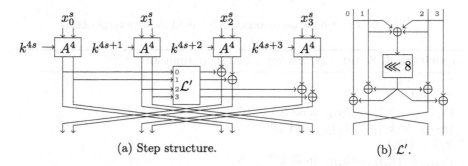

(a) Step structure.                              (b) $\mathcal{L}'$.

**Fig. 7.** The step structure of both SPARX-128/128 and SPARX-128/256.

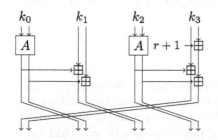

**Fig. 8.** The 128-bit permutation $K_4^{128}$ used in SPARX-128/128.

$r \leftarrow r + 1$
$k_0 \leftarrow A(k_0)$
$(k_1)_L \leftarrow (k_1)_L + (k_0)_L \mod 2^{16}$
$(k_1)_R \leftarrow (k_1)_R + (k_0)_R \mod 2^{16}$
$k_2 \leftarrow A(k_2)$
$(k_3)_L \leftarrow (k_3)_L + (k_2)_L \mod 2^{16}$
$(k_3)_R \leftarrow (k_3)_R + (k_2)_R + r \mod 2^{16}$
$k_0, k_1, k_2, k_3 \leftarrow k_3, k_0, k_1, k_2$

**Algorithm 3.** Pseudo-code of $K_4^{128}$

The Feistel function $\mathcal{L}'$ can be defined as follows. Let $a||b||c||d$ be a 64-bit word where each $a, ..., d$ is 16-bit long. Let $t = (a \oplus b \oplus c \oplus d) \lll 8$. Then $\mathcal{L}'(a||b||c||d) = c \oplus t \, || \, b \oplus t \, || \, a \oplus t \, || \, d \oplus t$. This function can also be expressed using 32-bit rotations. Let $x||y$ be the concatenation of two 32-bit words and $\mathcal{L}'_b$ denote $\mathcal{L}'$ without its final branch swap. Let $t = ((x \oplus y) \ggg^{32} 8) \oplus ((x \oplus y) \lll^{32} 8)$, then $\mathcal{L}'_b(x||y) = x \oplus t || y \oplus t$. Alternatively, we can use $\mathcal{L}$ to compute $\mathcal{L}'_b$ as follows: $\mathcal{L}'_b(x||y) = y \oplus \mathcal{L}(x \oplus y)||x \oplus \mathcal{L}(x \oplus y)$.

These two ciphers, SPARX-128/128 and SPARX-128/256, differ only by their number of steps and by their key schedule. The key schedule of SPARX-128/128 needs a 128-bit permutation $K_4^{128}$ described in Fig. 8 and Algorithm 3 while

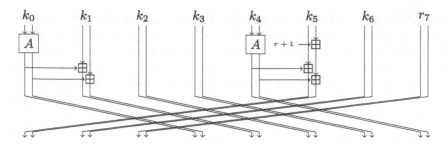

**Fig. 9.** The 256-bit permutation $K_8^{256}$ used in SPARX-128/256.

---

**Algorithm 4.** SPARX-128/256 key schedule permutation $K_8^{256}$.

---

$r \leftarrow r + 1$
$k_0 \leftarrow A(k_0)$
$(k_1)_L \leftarrow (k_1)_L + (k_0)_L \mod 2^{16}$
$(k_1)_R \leftarrow (k_1)_R + (k_0)_R \mod 2^{16}$
$k_4 \leftarrow A(k_4)$
$(k_5)_L \leftarrow (k_5)_L + (k_4)_L \mod 2^{16}$
$(k_5)_R \leftarrow (k_5)_R + (k_4)_R + r \mod 2^{16}$
$k_0, k_1, k_2, k_3, k_4, k_5, k_6, k_7 \leftarrow k_5, k_6, k_7, k_0, k_1, k_2, k_3, k_4$

---

SPARX-128/256 uses a 256-bit permutation $K_4^{256}$, which is presented in both Fig. 9 and Algorithm 4.

### 4.3  Design Rationale

**Choosing the ARX-box.** We chose the round function of SPECKEY/SPECK-32 over MARX-2 because of its superior implementation properties. Indeed, its smaller total number of operations means that a cipher using it needs to do fewer operations when implemented on a 16-bit platform. Ideally, we would have used an ARX-box with 32-bit operations but, at the time of writing, no such function has known differential and linear bounds (cf. Table 1) for sufficiently many rounds.

We chose to evaluate the iterations of the ARX-box over each branch rather than in parallel because such an order decreases the number of times each 32-bit branch must be loaded in CPU registers. This matters when the number of registers is too small to contain both the full key and the full internal state of the cipher and does not change anything if it is not the case.

**Mixing Layer, Number of Steps and Rounds per Step.** Our main approach for choosing the mixing layer was exhaustive enumeration of all matrices suitable for our long trail bounding algorithm from Sect. 3.4 and selecting the final matrix according to various criteria, which we will discuss later.

For SPARX-64/128, there is only one linear layer fulfilling our design criteria: one corresponding to a Feistel round. For such a structure, we found that the best integral covers 4 steps (without the last linear layer) and that, with 3 rounds per step, the MEDCP and MELCC are bounded by $2^{-75}$ and $2^{-38}$. These quantities imply that no single trail differential or linear distinguisher exists for 5 or more steps of SPARX-64/128.

For SPARX instances with 128-bit block we implemented an exhaustive search on a large subset of all possible linear layers. After some filtering, we arrived at roughly 3000 matrices. For each matrix we ran our algorithm from Sect. 3.4 to obtain bounds on MEDCP and MELCC for different values of the number of rounds per step $(r_a)$. We also ran the algorithm for searching integral characteristics described in Sect. 3.4.

Then, we analyzed the best matrices and found that there is a matrix which corresponds to a Feistel-like linear layer with the best differential/linear bound for $r_a = 4$. This choice also offered good compromise between other parameters, such as diffusion, strength of the ARX-box, simplicity and easiness/efficiency of implementation. It also generalizes elegantly the linear layer of SPARX-64/128. We thus settled for this Feistel-like function.

For more details on the selection procedure and other interesting candidates for the linear layer we refer the reader to the full version of this paper [20].

**The Linear Feistel Functions.** The linear layer obtained using the steps described above is only specified at a high level, it remains to define the linear Feistel functions $\mathcal{L}$ and $\mathcal{L}'$. The function $\mathcal{L}$ that we have chosen has been used in the Lai-Massey round constituting the linear layer of NOEKEON [22]. We reuse it here because it is cheap on lightweight processors as it only necessitates one rotation by 8 bits and 3 XORs. It also provides some diffusion as it has branching number 3. Its alternative representation using 32-bit rotations allows an optimized implementation on 32-bit processors.

Used for a larger block size, the Feistel function $\mathcal{L}'$ is a generalization of $\mathcal{L}$: it also relies on a Lai-Massey structure as well as a rotation by 8 bits. The reason behind these choices are the same as before: efficiency and diffusion. Furthermore, $\mathcal{L}'$ must also provide diffusion between the branches. While this is achieved by the XORs, we further added a branch swap in the bits of highest weight. This ensures that if only one 32-bit branch is active at the input of $\mathcal{L}'$ then two branches are active in its output. Indeed, there are two possibilities: either the output of the rotation is non-zero, in which case it gets added to the other branch and spreads to the whole state through the branch swap. Otherwise, the output is equal to 0, which means that the two 16-bit branches constituting the non-zero 32-bit branch hold the same non-zero value. These will then be spread over the two output 32-bit branches by the branch swap. The permutation $\mathcal{L}'$ also breaks the 32-bit word structure, which can help prevent the spread of integral patterns.

**Key Schedule.** The key schedules of the different versions of SPARX have been designed using the following general guidelines.

First, we look at criteria related to the implementation. To limit code size, components from the round function of SPARX are re-used in the key-schedule itself. To accommodate cases where the memory requirements are particularly stringent, we allow an efficient on-the-fly computation of the key.

We also consider cryptographic criteria. For example, we need to ensure that the keys used within each chain of 3 or 4 ARX-boxes are independent from one another. As we do not have enough entropy from the master key to generate truly independent round keys, we must also ensure that the round-keys are as different as possible from one another. This implies a fast mixing of the master key bits in the key schedule. Furthermore, in order to prevent slide attacks [23], we chose to have the round keys depend on the round index. Finally, since the subkeys are XOR-ed in the key state, we want to limit the presence of high probability differential pattern in the key update. Diffusion in the key state is thus provided by additions modulo $2^{16}$ rather than exclusive-or. While there may be high probability patterns for additive differences, these would be of little use because the key is added by an XOR to the state.

As with most engineering tasks, some of these requirements are at odds against each other. For example, it is impossible to provide extremely fast diffusion while also being extremely lightweight. Our designs are the most satisfying compromises we could find.

### 4.4   Security Analysis

**Single Trail Differential/Linear Attack.** By design and thanks to the long trail argument, we know that there is no differential or linear trail covering 5 steps (or more) with a useful probability for any instance of SPARX. Therefore, the 8 steps used by SPARX-64/128 and SPARX-128/128 and the 10 used by SPARX-128/256 are sufficient to ensure resilience against such attacks.

**Attacks Exploiting a Slow Diffusion.** We consider several attacks in this category, namely impossible and truncated differential attacks, meet-in-the middle attacks as well as integral attacks.

When we chose the linear layers, we ensured that they prevented division-property-based integral attacks, meaning that they provide good diffusion. Furthermore, the Feistel structure of the linear layer makes it easy to analyse and increases our confidence in our designs. In the case of 128-bit block sizes, the Feistel function $\mathcal{L}'$ has branching number 3 in the sense that if only one 32-bit branch is active then the two output branches are active. This prevents attacks trying to exploit patterns at the branch level. Finally, this Feistel function also breaks the 32-bit word structure through a 16-bit branch swap which frustrates the propagation of integral characteristics.

Meet-in-the-middle attacks are further hindered by the large number of key additions. This liberal use of the key material also makes it harder for an attacker to guess parts of it to add rounds at the top or at the bottom of, say, a differential characteristic.

**Best Attacks.** The best attacks we could find are integral attacks based on Todo's division property. The attack against SPARX-64/128 covers 15/24 rounds and recovers the key in time $2^{101}$ using $2^{37}$ chosen plaintexts and $2^{64}$ blocks of memory. For 22-round SPARX-128/128, we can recover the key in time $2^{105}$ using $2^{102}$ chosen plaintexts and $2^{72}$ blocks of memory. Finally, we attack 24-round SPARX-128/256 in time $2^{233}$, using $2^{104}$ chosen plaintexts and $2^{202}$ blocks of memory. A description of these attacks as well as the description of some time/data tradeoffs are provided in the full version of this paper [20].

### 4.5  Software Implementation

Next we describe how SPARX can be efficiently implemented on three resource constrained microcontrollers widely used in the Internet of Things (IoT), namely the 8-bit Atmel ATmega128, the 16-bit TI MSP430, and the 32-bit ARM Cortex-M3. We support the described optimization strategies with performance figures extracted from assembly implementations of SPARX-64/128 and SPARX-128/128 using the FELICS open-source benchmarking framework [24]. We use the same tool to get the most suitable implementations of SPARX for the two IoT-specific usage scenarios described in [6]. The first scenario uses a block cipher to encrypt 128 bytes of data using CBC mode, while the second encrypts 128 bits of data using a cipher in CTR mode. The most suitable implementation for a given usage scenario is selected using the *Figure of Merit (FOM)* defined in [6]:

$$\mathrm{FOM}(i_1, i_2, i_3) = \frac{p_{i_1, AVR} + p_{i_2, MSP} + p_{i_3, ARM}}{3},$$

where the performance parameter $p_{i,d}$ aggregates the code size, the RAM consumption, and the execution time for implementation $i$ according to the requirements of the usage scenario. The smaller the FOM value of an implementation in a certain use case, the better (more suitable) is the implementation for that particular use case. Finally, we compare the results of our implementations with the results available on the tool's website.[6]

**Implementation Aspects.** In order to efficiently implement SPARX on a resource constrained embedded processor, it is important to have a good understanding of its instruction set architecture (ISA). The number of general-purpose registers determines whether the entire cipher's state can be fitted into registers

---

[6] We submitted our implementations of SPARX to the FELICS framework. Up to date results are available at https://www.cryptolux.org/index.php/FELICS.

**Table 4.** Performance characteristics of the main components of SPARX

| Component | AVR | | MSP | | ARM | |
|---|---|---|---|---|---|---|
| | Cycles | Registers | Cycles | Registers | Cycles | Registers |
| $A$ | 16 | $4 + 1$ | 9 | 2 | 11 | $1 + 3$ |
| $A^{-1}$ | 19 | 4 | 9 | 2 | 12 | $1 + 3$ |
| $\lambda_2$ – 1-step | 24 | $8 + 1$ | 11 | $4 + 3$ | 5 | $2 + 1$ |
| $\lambda_2$ – 2-steps | 12 | 8 | 7 | $4 + 1$ | 3 | 2 |
| $\lambda_4$ – 1-step | 48 | $16 + 2$ | 36 | $8 + 1$ | 16 | $4 + 5$ |
| $\lambda_4$ – 2-steps | 24 | $16 + 2$ | 13 | $8 + 1$ | 12 | $4 + 4$ |

or whether a part of it has to be spilled to RAM. Memory operations are generally slower than register operations, consume more energy and increase the vulnerability of an implementation to side channel attacks [25]. Thus, the number of memory operations should be reduced as much as possible. Ideally the state should only be read from memory at the beginning of the cryptographic operation and written back at the end. Concerning the three targets we implemented SPARX for, they have 32 8-bit, 12 16-bit, and 13 32-bit general-purpose registers, which result in a total capacity of 256 bytes, 192 bytes, and 416 bytes for AVR, MSP, and ARM, respectively.

The SPARX family's simple structure consists only of three components: the ARX-box $A$ and its inverse $A^{-1}$, the linear layer $\lambda_2$ or $\lambda_4$ (depending on the version), and the key addition. The key addition (bitwise XOR) does not require additional registers and its execution time is proportional to the ratio between the operand width and the target device's register width. The execution time in cycles and the number of registers required to perform $A$, $A^{-1}$, $\lambda_2$, and $\lambda_4$ on each target device are given in Table 4.

The costly operation in terms of both execution time and number of required registers is the linear layer. The critical point is reached for the 128-bit linear layer $\lambda_4$ on MSP, which requires 13 registers. Since this requirement is above the number of available registers, a part of the state has to be saved onto the stack. Consequently, the execution time increases by 5 cycles for each **push** – **pop** instruction pair.

A 2-step implementation uses a simplified linear layer without the most resource demanding part – the branch swaps. It processes the result of the left branch after the first step as the right branch of the second step and similarly the result of the right branch after the first step as the left branch of the second step. This technique reduces the number of required registers and improves the execution time at the cost of an increase in code size. The performance gain is a factor of 2 on AVR, 2.7 on MSP, and 1.3 on ARM.

The linear transformations $\mathcal{L}$ and $\mathcal{L}'$ exhibit interesting implementation properties. For each platform there is a different optimal way to perform them. The

**Table 5.** Different trade-offs between the execution time and code size for encryption of a block using SPARX-64/128 and SPARX-128/128. Minimal values are given in bold.

| Implementation | Block size [bits] | AVR | | | MSP | | | ARM | | |
|---|---|---|---|---|---|---|---|---|---|---|
| | | Time [cyc.] | Code [B] | RAM [B] | Time [cyc.] | Code [B] | RAM [B] | Time [cyc.] | Code [B] | RAM [B] |
| 1-step rolled | 64 | 1789 | **248** | 2 | 1088 | **166** | 14 | 1370 | **176** | 28 |
| 1-step unrolled | 64 | 1641 | 424 | 1 | 907 | 250 | 12 | 1100 | 348 | **24** |
| 2-steps rolled | 64 | 1677 | 356 | 2 | 1034 | 232 | 10 | 1331 | 304 | 28 |
| 2-steps unrolled | 64 | **1529** | 712 | 1 | **853** | 404 | **8** | **932** | 644 | **24** |
| 1-step rolled | 128 | 4553 | **504** | 11 | 2809 | **300** | 26 | 3463 | **348** | 44 |
| 1-step unrolled | 128 | 4165 | 1052 | **10** | 2353 | 584 | 24 | 2784 | 884 | 40 |
| 2-steps rolled | 128 | 4345 | 720 | 11 | 2593 | 432 | 18 | 3399 | 620 | 40 |
| 2-steps unrolled | 128 | **3957** | 1820 | **10** | **2157** | 1004 | **16** | **2377** | 1692 | **36** |

optimal way to implement the linear layers on MSP is using the representations from Figs. 5c and 7b. On ARM the optimal implementation performs the rotations directly on 32-bit values. The function $\mathcal{L}$ can be executed on AVR using 12 XOR instructions and no additional registers. On the other hand, the optimal implementation of $\mathcal{L}'$ on AVR requires 2 additional registers and takes 24 cycles.[7]

The linear layer performed after the last step of SPARX can be dropped without affecting the security of the cipher, but it turns out that it results in poorer overall performances. The only case when this strategy helps is when top execution time is the main and only concern of an implementation. Thus we preferred to keep the symmetry of the step function and the overall balanced performance figures.

The salient implementation-related feature of SPARX family of ciphers is given by the simple and flexible structure of the step function depicted in Fig. 4, which can be implemented using different optimization strategies. Depending on specific constraints, such as code size, speed, or energy requirements to name a few, the rounds inside the step function can be rolled or unrolled; one or two step functions can be computed at once. The main possible trade-offs between the execution time and code size are explored in Table 5.

Except for the 1-step implementation of SPARX-128/128 on MSP, which needs RAM memory to save the cipher's state, all other RAM requirements are determined only by the process of saving the context onto the stack at the begging of the measured function. Thus, the RAM consumption of a pure assembly implementation would be zero, except for the 1-step rolled and unrolled implementations of SPARX-128/128 on MSP.

Due to the 16-bit nature of the cipher, performing $A$ and $A^{-1}$ on a 32-bit platform requires a little bit more execution time and more auxiliary registers than performing the same operations on a 16-bit platform. The process of packing

---

[7] For more details please see the implementations submitted to the FELICS framework (https://www.cryptolux.org/index.php/FELICS).

and unpacking a state register to extract and store back the two 16-bit branches of $A$ or $A^{-1}$ adds a performance penalty. The cost is amplified by the fact that the flexible second operand can not be used with a constant to extract the least or most significant 16 bits of a 32-bit register. Thus an additional masking register is required.

The simple key schedules of SPARX-64/128 and SPARX-128/128 can be implemented in different ways. The most efficient implementation turns out to be the one using the 1-iteration rolled strategy. Another interesting approach is the 4-iterations unrolled strategy, which has the benefit that the final permutation is achieved for free by changing the order in which the registers are stored in the round keys. This strategy increases the code size by up to a factor of 4, while the execution time is on average 25 % better.

Although we do not provide performance figures for SPARX-128/256, we emphasize that the only differences with respect to implementation aspects between SPARX-128/256 and SPARX-128/128 are the key schedules and the different number of steps.

**Evaluation and Comparison.** We evaluate the performance of our implementations of SPARX using FELICS in the two aforementioned usage scenarios. The key performance figures are given in the full version of this paper [20]. The balanced results are achieved using the 1-step implementations of SPARX-64/128 and SPARX-128/128.

**Table 6.** Top 10 best implementations in Scenario 1 (encryption key schedule + encryption and decryption of 128 bytes of data using CBC mode) ranked by the Figure of Merit (FOM) defined in FELICS. The results for all ciphers are the current ones from the Triathlon Competition at the moment of submission. The smaller the FOM, the better the implementation.

| Rank | Cipher | Block size | Key size | Scenario 1 FOM |
|---|---|---|---|---|
| 1 | SPECK | 64 | 128 | 5.0 |
| 2 | Chaskey-LTS | 128 | 128 | 5.0 |
| 3 | SIMON | 64 | 128 | 6.9 |
| 4 | RECTANGLE | 64 | 128 | 7.8 |
| 5 | LEA | 128 | 128 | 8.0 |
| 6 | SPARX | 64 | 128 | 8.6 |
| 7 | SPARX | 128 | 128 | 12.9 |
| 8 | HIGHT | 64 | 128 | 14.1 |
| 9 | AES | 128 | 128 | 15.3 |
| 10 | Fantomas | 128 | 128 | 17.2 |

Then we compare the performance of SPARX with the current results available on the Triathlon Competition at the time of submission.[8] As can be seen in Table 6 the two instances of SPARX perform very well across all platforms and rank very high in the FOM-based ranking. The forerunners are the NSA designs SIMON and SPECK, Chaskey, RECTANGLE and LEA, but, apart from RECTANGLE, none of them provides provable bounds against differential and linear cryptanalysis.

Besides the overall good performance figures in the two usage scenarios, the following results are worth mentioning:

- the execution time of SPARX-64/128 on MSP is in the top 3 of the fastest ciphers in both scenarios thanks to its 16-bit oriented operations;
- the code size of the 1-step rolled implementations of SPARX-64/128 and SPARX-128/128 on MSP is in the top 5 in both scenarios as well as in the small code size and RAM table for scenario 2;
- the 1-step rolled implementation of SPARX-64/128 breaks the previous minimum RAM consumption record on AVR in scenario 2;
- the execution time of the 2-steps implementation of SPARX-64/128 in scenario 2 is in the top 3 on MSP, in the top 5 on AVR, and in the top 7 on ARM; it also breaks the previous minimum RAM consumption records on AVR and MSP.

Given its simple and flexible structure as well as its very good overall ranking in the Triathlon Competition of lightweight block ciphers, the SPARX family of lightweight ciphers is suitable for applications on a wide range of resource constrained devices. The absence of look-up tables reduces the memory requirements and provides, according to [25], some intrinsic resistance against power analysis attacks.

# 5    Replacing Rotations with Linear Layers: The LAX Construction

In this section we outline an alternative strategy for designing an ARX cipher with provable bounds against differential and linear cryptanalysis. It is completely independent from the Long Trail Strategy outlined in the previous sections and uses the differential properties of modular addition to derive proofs of security.

## 5.1    Motivation

In his Master thesis [13] Wallén posed the challenge to design a cipher that uses only addition modulo-2 and GF(2)-affine functions, and that is provably resistant against differential and linear cryptanalysis [13, Sect. 5]. In this section we partially solve this challenge by proposing a construction with provable bounds against single-trail differential cryptanalysis (DC).

---

[8] Up to date results are available at https://www.cryptolux.org/index.php/FELICS.

## 5.2    Theoretical Background

**Definition 6** (xdp$^+$). *The XOR differential probability (DP) of addition modulo* $2^n$ *is defined as:*

$$\text{xdp}^+(\alpha, \beta \rightarrow \gamma) = 2^{-2n} \cdot \#\{(x,y) : ((x \oplus \alpha) + (y \oplus \beta)) \oplus (x + y) = \gamma\} \ ,$$

*where* $\alpha$, $\beta$ *and* $\gamma$ *are n-bit XOR differences and* $x$ *and* $y$ *are n-bit values.*

The XOR linear correlation of addition modulo $2^n$ (xlc$^+$) is defined in a similar way. Efficient algorithms for the computation of xdp$^+$ and xlc$^+$ have been proposed resp. in [26–29]. These results also reveal the following property. The magnitude of both xdp$^+$ and $|$xlc$^+|$ is inversely proportional to the number of bit positions at which the input/output differences (resp. masks) differ. For xdp$^+$, this fact is formally stated in the form of the following proposition.

**Proposition 1 (Bound on xdp$^+$).** *The differential probability* xdp$^+$ *is upper-bounded by* $2^{-k}$, *where* $k$ *is the number of bit positions, excluding the MSB, at which the bits of the differences are not equal:*

$$\text{xdp}^+(\alpha, \beta \rightarrow \gamma) \leq 2^{-k} : k = \#\{i : \neg(\alpha[i] = \beta[i] = \gamma[i]), 0 \leq i \leq w - 2\}$$

*Proof. Follows from [26, Alg. 2, Sect. 4].*

A similar proposition also holds for $|$xlc$^+|$ (see e.g. [10]). Proposition 1 provides the basis of the design strategy described in the following section.

## 5.3    The LAX Construction

LAX is a block cipher construction with $2n$-bit block and $n$-bit words. We investigate three instances of LAX designated by the block size: LAX-16, LAX-32 and LAX-64. A brief description of the round function of LAX-$2n$, shown in Fig. 10 (left), is given below.

Let $L$ be an $n \times n$ binary matrix that is (a) invertible and (b) has branch number $d > 2$. With $\ell(x)$ is denoted the multiplication of the $n$-bit vector $x$

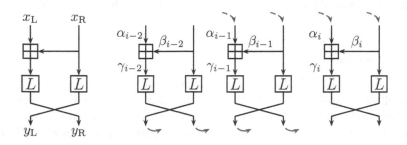

**Fig. 10. Left:** the round function of LAX; **Right:** three round differential of LAX.

by the matrix $L$: $\ell(x) = Lx$. Note that due to condition (b) it follows that $\forall x \neq 0 : h(x) + h(\ell(x)) \geq d$, where $h(x)$ is the Hamming weight of $x$.

The round function $\mathcal{A}(\cdot)$ of LAX-$2n$ maps a pair of $n$-bit words $(x_L, x_R)$ to a pair of $n$-bit words $(y_L, y_R)$ as follows (see Fig. 10 (left)):

$$(y_L, y_R) = \mathcal{A}(x_L, x_R) = (\ell(x_R),\ \ell(x_L \boxplus x_R))\ .$$

The matrix $L$ is chosen as the non-identity part of the generator matrix $G$ of a systematic $[2n, n, d]$ linear code over $\mathrm{GF}(2)$ such that $G = [I\ L]$. More specifically, the matrices $L$ for LAX-16, LAX-32 and LAX-64 are derived from the following codes respectively: $[16, 8, 5]$, $[32, 16, 8]$ and $[64, 32, 10]$. Note that the matrix of LAX-32 is the same as the one used in block cipher ARIA [30].

## 5.4 Bounds on the Differential Probability of LAX

**Lemma 1.** *For all differences $\alpha \neq 0$, the differential $(\alpha, \alpha \to \alpha)$ is impossible.*

*Proof.* Let $\mathrm{xdp}^+(\alpha, \beta \to \gamma) \neq 0$ for some differences $\alpha$, $\beta$ and $\gamma$. The statement of the lemma follows from the following two properties of $\mathrm{xdp}^+$ [26]. First, it must hold that $\alpha[0] \oplus \beta[0] \oplus \gamma[0] = 0$. Second, if $\alpha[i] = \beta[i] = \gamma[i]$ for some $0 \leq i \leq n-2$, then it must hold that $\alpha[i+1] \oplus \beta[i+1] \oplus \gamma[i+1] = \alpha[i]$. Since we want that $\alpha = \beta = \gamma$, from the first property it follows that $\alpha[0] = \beta[0] = \gamma[0] = 0$. Given that, due to the second property it follows that $\alpha[i] = \beta[i] = \gamma[i] = 0$, $\forall i \geq 1$. Therefore the only value of $\alpha$ for which $\mathrm{xdp}^+(\alpha, \beta \to \gamma) \neq 0$ and $\alpha = \beta = \gamma$ is $\alpha = 0$. □

**Theorem 2 (Differential bound on 3 rounds of LAX-$2n$).** *The maximum differential probability of any trail on 3 rounds of LAX-$2n$ is $2^{-(d-2)}$, where $d$ is the branch number of the matrix $L$.*

*Proof.* Let $(\alpha_{i-1}, \beta_{i-1}, \gamma_{i-1})$, $(\alpha_i, \beta_i, \gamma_i)$ and $(\alpha_{i+1}, \beta_{i+1}, \gamma_{i+1})$ be the input/output differences of the addition operations in three consecutive rounds of LAX-$2n$ and let $p_k = \mathrm{xdp}^+(\alpha_k, \beta_k \to \gamma_k)$ for $k \in \{i-1, i, i+1\}$ (see Fig. 10 (right)). We have to show that $p_{i-1} p_i p_{i+1} \leq 2^{-(d-2)}$ or, equivalently, that $\log_2 p_{i-1} + \log_2 p_i + \log_2 p_{i+1} \leq -(d-2)$. Denote with $h(x)$ the Hamming weight of the word $x$ and with $h^*(x)$ the Hamming weight of $x$, excluding the MSB. Note that $h^*(x) \leq h(x) - 1$. We consider two cases:

*Case 1:* $\beta_{i-1} \neq \gamma_{i-1}$. By Proposition 1 we have that $\log_2 p_{i-1} \leq -h^*(\beta_{i-1} \oplus \gamma_{i-1})$ and $\log_2 p_i \leq -h^*(\alpha_i \oplus \beta_i)$. Since $\beta_i = \ell(\gamma_{i-1})$ and $\alpha_i = \ell(\beta_{i-1})$ (see Fig. 10 (right)) and using the linearity of $\ell(\cdot)$ we have that $-h^*(\alpha_i \oplus \beta_i) = -h^*(\ell(\beta_{i-1} \oplus \gamma_{i-1}))$. As $\beta_{i-1} \neq \gamma_{i-1}$ it follows that $h^*(\beta_{i-1} \oplus \gamma_{i-1}) \neq 0$ and $h^*(\ell(\beta_{i-1} \oplus \gamma_{i-1})) \neq 0$. Thus we derive:

$$\log_2 p_{i-1} + \log_2 p_i \leq -h^*(\beta_{i-1} \oplus \gamma_{i-1}) - h^*(\ell(\beta_{i-1} \oplus \gamma_{i-1})).$$

From the properties of $L$ it follows that $-h(\beta_{i-1} \oplus \gamma_{i-1}) - h(\ell(\beta_{i-1} \oplus \gamma_{i-1})) \leq -d$ and so $-h^*(\beta_{i-1} \oplus \gamma_{i-1}) - h^*(\ell(\beta_{i-1} \oplus \gamma_{i-1})) \leq -(d-2)$. Therefore:

$$\log_2 p_{i-1} + \log_2 p_i \leq -(d-2).$$

510    D. Dinu et al.

*Case 2:* $\beta_{i-1} = \gamma_{i-1} \neq 0$. In this case $\alpha_i = \beta_i = \ell(\beta_{i-1}) = \ell(\gamma_{i-1})$. Due to Lemma 1 it follows that $\gamma_i \neq \beta_i$. Therefore we can apply the argument from *Case 1* on rounds $i$ and $i+1$ to derive the statement of the theorem in this case. □

### 5.5 Experimental Results

We have implemented the search algorithm proposed in [10] in order to find the probabilities of the best differential trails in LAX-16 and LAX-32. In Table 7, we compare the results to the theoretical bounds computed using Theorem 2.

**Table 7.** Best differential probabilities and best absolute linear correlations ($\log_2$ scale) for up to 12 rounds of LAX.

| | # Rounds | 1 | 2 | 3 | 4 | 5 | 6 | 7 | 8 | 9 | 10 | 11 | 12 |
|---|---|---|---|---|---|---|---|---|---|---|---|---|---|
| **LAX-16** | $p_{best}$ | +0 | −2 | −4 | −7 | −8 | −11 | −13 | −16 | −18 | −20 | −23 | −25 |
| | $c_{best}$ | +0 | +0 | −1 | −2 | −3 | −5 | −5 | −7 | −8 | −9 | −10 | −11 |
| | $p_{bound}$ | | | −3 | | | −6 | | | −9 | | | −12 |
| **LAX-32** | $p_{best}$ | +0 | −2 | −6 | −9 | −11 | −16 | −18 | −20 | −24 | −28 | −29 | −34 |
| | $c_{best}$ | +0 | +0 | +0 | −4 | −4 | −8 | −8 | −8 | −8 | −12 | −12 | −16 |
| | $p_{bound}$ | | | −6 | | | −12 | | | −18 | | | −24 |

Clearly the bound from Theorem 2 does not hold for the linear case. The problem is the "three-forked branch" in the LAX round function that acts as an XOR when the inputs are linear masks rather than differences. Thus, LAX only provides differential bounds and the full solution to the Wallén challenge still remains an open problem.

## 6  Conclusion

In this paper we presented, for the first time, a general strategy for designing ARX primitives with provable bounds against differential (DC) and linear cryptanalysis (LC) – a long standing open problem in the area of ARX design. The new strategy, called the Long Trail Strategy (LTS) advocates the use of large and computationally expensive S-boxes in combination with very light linear layers (the so-called Long Trail Argument). This makes the LTS to be the exact opposite of the Wide Trail Strategy (WTS) on which the AES (and many other SPN ciphers) are based. Moreover, the proposed strategy is not limited to ARX designs and can easily be applied also to S-box based ciphers.

To illustrate the effectiveness of the LTS we have proposed a new family of lightweight block ciphers, called SPARX, designed using the new approach. The family has three instances depending on the block and key sizes: SPARX-64/128, SPARX-128/128 and SPARX-128/256. With the help of the Long Trail Argument we prove resistance against single-trail DC and LC for each of the three instances of SPARX. In addition, we analyze the new constructions against a wide range of attacks such as impossible and truncated differentials, meet-in-the-middle and integral attacks. Our analysis did not find an attack covering 5 or more rounds of any of the three instances. The latter ensures a security margin of about 37 % of SPARX.

Beside (provable) security the members of the SPARX family are also very efficient. We have implemented them in software on three resource constrained microcontrollers widely used in the Internet of Things (IoT), namely the 8-bit Atmel ATmega128, the 16-bit TI MSP430, and the 32-bit ARM Cortex-M3. According to the FELICS open-source benchmarking framework our implementations of SPARX-64/128 and SPARX-128/128 rank respectively 6 and 7 in the list of top 10 most software efficient lightweight ciphers. In addition, the execution time of SPARX-64/128 on MSP is in the top 3 of this list. To the best of our knowledge, this paper is the first to propose a practical ARX design that has both arguments for provable security and competitive performance.

A secondary contribution of the paper is the proposal of an alternative strategy for ARX design with provable bounds against differential cryptanalysis. It is independent of the LTS and uses the differential properties of modular addition to derive proofs of security. As an illustration of this approach, the LAX family of constructions is described. The provable security of LAX against linear cryptanalysis is left as an open problem.

**Acknowledgements.** The work of Daniel Dinu and Léo Perrin is supported by the CORE project ACRYPT (ID C12-15-4009992) funded by the Fonds National de la Recherche, Luxembourg. The work of Aleksei Udovenko is supported by the Fonds National de la Recherche, Luxembourg (project reference 9037104). Vesselin Velichkov is supported by the Internal Research Project CAESAREA of the University of Luxembourg (reference I2R-DIR-PUL-15CAES). The authors thank Anne Canteaut for useful discussions regarding error correcting codes.

# References

1. Bernstein, D.J.: New Stream Cipher Designs: The eSTREAM Finalists. LNCS, vol. 4986. Springer, Heidelberg (2008)
2. Bernstein, D.J.: ChaCha, a variant of Salsa20. In: Workshop Record of SASC, vol. 8 (2008)
3. Niels, F., Lucks, S., Schneier, B., Whiting, D., Bellare, M., Kohno, T., Callas, J., Walker, J.: The Skein hash function family. Submission to NIST (round 3) (2010)
4. Aumasson, J.P., Henzen, L., Meier, W., Phan, R.C.W.: SHA-3 Proposal BLAKE (2010). https://131002.net/blake/blake.pdf
5. Needham, R.M., Wheeler, D.J.: Tea extensions. Technical report, Cambridge University, Cambridge, UK, October 1997
6. Dinu, D.D., Le Corre, Y., Khovratovich, D., Perrin, L., Großschädl, J., Biryukov, A.: Triathlon of lightweight block ciphers for the internet of things. In: NIST Workshop on Lightweight Cryptography 2015, National Institute of Standards and Technology (NIST) (2015)

7. Mouha, N., Mennink, B., Herrewege, A., Watanabe, D., Preneel, B., Verbauwhede, I.: Chaskey: an efficient MAC algorithm for 32-bit microcontrollers. In: Joux, A., Youssef, A. (eds.) SAC 2014. LNCS, vol. 8781, pp. 306–323. Springer, Heidelberg (2014). doi:10.1007/978-3-319-13051-4_19

8. Beaulieu, R., Shors, D., Smith, J., Treatman-Clark, S., Weeks, B., Wingers, L.: The SIMON and SPECK Families of Lightweight Block Ciphers. IACR Cryptology ePrint Archive **2013**, 404 (2013)

9. Hong, D., Lee, J.-K., Kim, D.-C., Kwon, D., Ryu, K.H., Lee, D.-G.: LEA: a 128-bit block cipher for fast encryption on common processors. In: Kim, Y., Lee, H., Perrig, A. (eds.) WISA 2013. LNCS, vol. 8267, pp. 3–27. Springer, Heidelberg (2014). doi:10.1007/978-3-319-05149-9_1

10. Biryukov, A., Velichkov, V., Le Corre, Y.: Automatic search for the best trails in ARX: application to block cipher SPECK. In: Peyrin, T. (ed.) FSE 2016. LNCS, vol. 9783, pp. 289–310. Springer, Heidelberg (2016). doi:10.1007/978-3-662-52993-5_15

11. Daemen, J., Rijmen, V.: The wide trail design strategy. In: Honary, B. (ed.) Cryptography and Coding 2001. LNCS, vol. 2260, pp. 222–238. Springer, Heidelberg (2001). doi:10.1007/3-540-45325-3_20

12. Daemen, J., Rijmen, V.: The Design of Rijndael: AES-the Advanced Encryption Standard. Springer, Heidelberg (2002)

13. Wallén, J.: On the Differential and Linear Properties of Addition. Master's thesis, Helsinki University of Technology (2003)

14. Keliher, L., Sui, J.: Exact maximum expected differential and linear probability for 2-round advanced encryption standard. IET Inf. Secur. **1**(2), 53–57 (2007)

15. Nikolić, I.: Tiaoxin-346. Submission to the CAESAR competition (2015)

16. Jean, J., Nikolić, I.: Efficient design strategies based on the AES round function. In: Peyrin, T. (ed.) FSE 2016. LNCS, vol. 9783, pp. 334–353. Springer, Heidelberg (2016). doi:10.1007/978-3-662-52993-5_17

17. Guo, J., Peyrin, T., Poschmann, A., Robshaw, M.: The LED block cipher. In: Preneel, B., Takagi, T. (eds.) CHES 2011. LNCS, vol. 6917, pp. 326–341. Springer, Heidelberg (2011). doi:10.1007/978-3-642-23951-9_22

18. Knudsen, L., Wagner, D.: Integral cryptanalysis. In: Daemen, J., Rijmen, V. (eds.) FSE 2002. LNCS, vol. 2365, pp. 112–127. Springer, Heidelberg (2002). doi:10.1007/3-540-45661-9_9

19. Todo, Y.: Structural evaluation by generalized integral property. In: Oswald, E., Fischlin, M. (eds.) EUROCRYPT 2015. LNCS, vol. 9056, pp. 287–314. Springer, Heidelberg (2015). doi:10.1007/978-3-662-46800-5_12

20. Dinu, D., Perrin, L., Udovenko, A., Velichkov, V., Großschädl, J., Biryukov, A.: Design Strategies for ARX with Provable Bounds: SPARX and LAX (Full Version).Cryptology ePrint Archive, to appear 2016. http://eprint.iacr.org/

21. Biryukov, A., Khovratovich, D.: Decomposition attack on SASASASAS. Cryptology ePrint Archive, Report 2015/646 (2015). http://eprint.iacr.org/

22. Daemen, J., Peeters, M., Van Assche, G., Rijmen, V.: Nessie proposal: NOEKEON. In: First Open NESSIE Workshop, pp. 213–230 (2000)

23. Biryukov, A., Wagner, D.: Slide attacks. In: Knudsen, L. (ed.) FSE 1999. LNCS, vol. 1636, pp. 245–259. Springer, Heidelberg (1999). doi:10.1007/3-540-48519-8_18

24. Dinu, D.D., Biryukov, A., Großschädl, J., Khovratovich, D., Le Corre, Y., Perrin, L.A.: FELICS-fair evaluation of lightweight cryptographic systems. In: NIST Workshop on Lightweight Cryptography 2015, National Institute of Standards and Technology (NIST) (2015)

25. Biryukov, A., Dinu, D., Großschädl, J.: Correlation power analysis of lightweight block ciphers: from theory to practice. In: Manulis, M., Sadeghi, A.-R., Schneider, S. (eds.) ACNS 2016. LNCS, vol. 9696, pp. 537–557. Springer, Heidelberg (2016). doi:10.1007/978-3-319-39555-5_29

26. Lipmaa, H., Moriai, S.: Efficient algorithms for computing differential properties of addition. In: Matsui, M. (ed.) FSE 2001. LNCS, vol. 2355, pp. 336–350. Springer, Heidelberg (2002). doi:10.1007/3-540-45473-X_28

27. Wallén, J.: Linear approximations of addition modulo $2^n$. In: Johansson, T. (ed.) FSE 2003. LNCS, vol. 2887, pp. 261–273. Springer, Heidelberg (2003). doi:10.1007/978-3-540-39887-5_20

28. Nyberg, K., Wallén, J.: Improved linear distinguishers for SNOW 2.0. In: Robshaw, M. (ed.) FSE 2006. LNCS, vol. 4047, pp. 144–162. Springer, Heidelberg (2006). doi:10.1007/11799313_10

29. Dehnavi, S.M., Rishakani, A.M., Shamsabad, M.R.M.: A more explicit formula for linear probabilities of modular addition modulo a power of two. Cryptology ePrint Archive, Report 2015/026 (2015). http://eprint.iacr.org/

30. Kwon, D., Kim, J., Park, S., Sung, S.H., Sohn, Y., Song, J.H., Yeom, Y., Yoon, E.-J., Lee, S., Lee, J., Chee, S., Han, D., Hong, J.: New block cipher: ARIA. In: Lim, J.-I., Lee, D.-H. (eds.) ICISC 2003. LNCS, vol. 2971, pp. 432–445. Springer, Heidelberg (2004). doi:10.1007/978-3-540-24691-6_32

# SCA and Leakage Resilience I

# Side-Channel Analysis Protection
# and Low-Latency in Action
## – Case Study of PRINCE and Midori –

Amir Moradi$^{(\boxtimes)}$ and Tobias Schneider

Horst Görtz Institute for IT-Security, Ruhr-Universität Bochum, Bochum, Germany
{amir.moradi,tobias.schneider-a7a}@rub.de

**Abstract.** During the last years, the industry sector showed particular interest in solutions which allow to encrypt and decrypt data within one clock cycle. Known as low-latency cryptography, such ciphers are desirable for pervasive applications with real-time security requirements. On the other hand, pervasive applications are very likely in control of the end user, and may operate in a hostile environment. Hence, in such scenarios it is necessary to provide security against side-channel analysis (SCA) attacks while still keeping the low-latency feature.

Since the single-clock-cycle concept requires an implementation in a fully-unrolled fashion, the application of masking schemes – as the most widely studied countermeasure – is not straightforward. The contribution of this work is to present and discuss about the difficulties and challenges that hardware engineers face when integrating SCA countermeasures into low-latency constructions. In addition to several design architectures, practical evaluations, and discussions about the problems and potential solutions with respect to the case study PRINCE (also compared with Midori), the final message of this paper is a couple of suggestions for future low-latency designs to – hopefully – ease the integration of SCA countermeasures.

## 1 Introduction

The need for integration of side-channel analysis (SCA) [29] countermeasures into pervasive security-enabled devices is known to both academia and industry. Such a demand has also been motivated by several practical key-recovery attacks on commercial applications, e.g., [2,22,31,34,41,54]. From another perspective, there are several important applications for which a low-latency encryption and instant response time is highly desirable, such as read/write access to encrypted memory modules, which should be preferably conducted in a single clock cycle (initially motivated by [27]). It is also expected that given the ongoing growth of pervasive computing, there will be many more future embedded systems that require low-latency encryption, especially applications with real-time requirements, e.g., in the automotive domain. Hence, such pervasive applications, where low-latency cryptography is required, should be protected against SCA threats.

© International Association for Cryptologic Research 2016
J.H. Cheon and T. Takagi (Eds.): ASIACRYPT 2016, Part I, LNCS 10031, pp. 517–547, 2016.
DOI: 10.1007/978-3-662-53887-6_19

Here in this work, we present the challenges one may face by integrating SCA countermeasures into implementations with low-latency target. Insertion of *hiding* techniques [32] in this scenario is either straightforward (such as noise generation or dual-rail logic) or ineffective (such as time randomization or shuffling) due to the fully unrolled architecture of low-latency implementations. Therefore, our focus is the integration of *masking* schemes into such designs. In particular, we concentrate on threshold implementation (TI) [40] as a provably-secure scheme against first-order SCA attacks. It should be noted that integration of ad hoc approaches, e.g., random pre-charging of [6], are out of our focus since we target solutions with provable security.

We should point out that it has previously been supposed that unrolled circuits – also the case of low-latency concept – are inherently secure against SCA attacks (see [6]). However, other practical results, e.g., in [36,51], showed that unrolling may make the attacks complicated since the common hypothetical power models (Hamming weight/distance) may not fit to the circuit's leakage anymore, but sophisticated yet first-order leakages can be exploited for key recovery.

As a known case study, PRINCE [13] (particularly designed as a low-latency cipher) is targeted in our investigations. We demonstrate design architectures and practical results with respect to the power consumption as well as SCA protection of different variants of implementations of PRINCE. In addition to several discussions about the SCA protection versus low-latency concept, we present a mixture of asynchronous circuit design methodology with threshold implementation which is expected to realize an SCA-protected self-timed design. Finally, having the PRINCE case study in mind, we give a couple of suggestions for the future low-latency cipher designs with the goal of mitigating the challenges, where SCA protection is desirable.

Furthermore, we consider the cipher Midori [3] which was designed with the goal of minimizing energy consumption. Since energy consumption and latency – to some extent – are proportional, we also provide a comparison between PRINCE and Midori with respect to latency when both are equipped with similar masking countermeasure.

## 2   Preliminaries

### 2.1   PRINCE

PRINCE [13] is a 64-bit block cipher that uses a 128-bit secret key $k$. The key expansion divides $k$ into two 64-bit parts as $k = (k_0 || k_1)$, and derives $k_0'$ from $k_0$ by a linear function as $(k_0 \ggg 1) \oplus (k_0 \gg 63)$. The subkeys $k_0$ and $k_0'$ are used as input and output whitening keys respectively, while $k_1$ is used as the round key for the core block cipher $\text{PRINCE}_{core}$ (see Fig. 1).

Each of the first five round functions $\mathcal{R}_i$ consists of **S-Layer** (by a 4-bit Sbox), **M'-Layer** (multiplication with a $64 \times 64$ matrix $M'$), **ShiftRows** (the same as the AES one but on 4-bit cells), **$RC_i$-add** (XORing the state with a 64-bit constant $RC_i$), and **$k_1$add** (XORing $k_1$ into the 64-bit state).

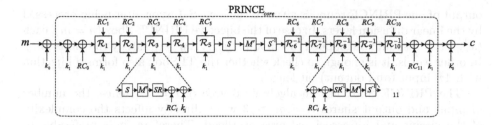

**Fig. 1.** A schematic view of PRINCE

The last five inverse round functions $\mathcal{R}_i^{-1}$ are formed by the inverse of the corresponding operations. It is noteworthy that $M'$ matrix is an involution, hence the inverse of $\mathbf{M'}$-**Layer** is itself. Further, due to its underlying FX-construction [26] as well as the $\alpha$-reflection, i.e., $RC_{i \in \{0,\dots,11\}} \oplus RC_{11-i} = \alpha$, the PRINCE encryption can turn to its decryption by swapping the whitening keys and XORing $\alpha$ to $k_1$,

$$\mathrm{PRINCE}^{\mathrm{Dec}}{}_{(k_0,k_0',k_1)} = \mathrm{PRINCE}^{\mathrm{Enc}}{}_{(k_0',k_0,k_1\oplus\alpha)}.$$

Note that $RC_0 = 0$, and $RC_{i \in \{1,\dots,5\}}$ as well as $\alpha$ are derived from the fraction part of $\pi = 3.141\dots$.

## 2.2   Threshold Implementation

Let us denote a 4-bit intermediate value of PRINCE, e.g., the Sbox input, as $x = \langle x_1, \dots, x_4 \rangle$. Under the $n - 1$ order Boolean masking concept, $x$ is represented by $(x^1, \dots, x^n)$, where $x = \bigoplus_{i=1}^{n} x^i$ and each $x^i$ similarly denotes a 4-bit vector $\langle x_1^i, \dots, x_4^i \rangle$.

The linear functions, such as $\mathbf{M'}$-**Layer**, can be simply applied to the shares of $x$ as $\mathsf{L}(x) = \bigoplus_{i=1}^{n} \mathsf{L}(x^i)$. Clearly, the non-linear functions, e.g., Sbox, cannot be trivially shared. Following the TI concept [8,40], the minimum number of shares to realize an Sbox to be secure against first-order attacks is $n = t + 1$, where $t$ denotes the algebraic degree of the Sbox. The shared Sbox should provide the output also in a shared form $(y^1, \dots, y^m)$, where $m \geq n$ when the Sbox is a bijection. Obviously, to ensure the *correctness* of the computation, we should have $\mathsf{S}(x) = y = \bigoplus_{i=1}^{m} y^i$.

Each output share $y^{j \in \{1,\dots,m\}}$ is given by a component function $\mathsf{f}^j(\cdot)$ over a subset of input shares. Defined as *non-completeness*, for first-order security each component function $\mathsf{f}^{j \in \{1,\dots,m\}}(\cdot)$ must be independent of at least one input share. The security of masking schemes (to some extent) depends on the uniform distribution of the masks. Therefore, the output of a TI Sbox must be also uniform, since it supplies other non-linear functions. For example, the Sbox output of one PRINCE round is given to the next **S-Layer** after being processed by the linear diffusion layers. In case of the bijective PRINCE Sbox ($n = m$), each $(x^1, \ldots, x^n)$ should be mapped to a unique $(y^1, \ldots, y^n)$ to satisfy the *uniformity*. In other words, it is enough to check whether the TI Sbox also forms a bijection with $4n$ input (and output) bit length.

The PRINCE Sbox has an algebraic degree of $t = 3$. Hence, the number of input and output shares $n = m > 3$ what directly affects the complexity of the circuit and its associated area overhead. Therefore, it is preferable to decompose the Sboxes into smaller non-linear functions each with maximum algebraic degree of 2, which enables staying with the minimum number of shares $n = m = 3$. Note that in this case, registers must be placed between the shared decomposed functions. Otherwise, the glitches propagate into cascaded shared non-linear circuits, and violate the *non-completeness* property. As an example, the authors of [42] presented a decomposition of the PRESENT [12] Sbox into two quadratic bijections g and f.

Above we briefly reviewed the TI concept. For detailed information, the interested reader is referred to the original articles [8,40].

## 3    Design Architectures

As stated before, PRINCE cipher has been designed with respect to low-latency feature. The goal was to achieve a short latency when the cipher is implemented in a fully-unrolled fashion. In other words, the implementation contains no sequential elements, e.g., register/flip-flop, and hence no clock.

In our investigations, in order to synthesize for an ASIC platform, we made use of Synopsys Design Compiler using the UMCL18G212T3 [49] ASIC standard cell library, i.e., UMC 0.18 μm. As a side note, such a standard library has not been covered by the original article [13], where Nangate 45 nm, UMC 90 nm, and UMC 130 nm technologies have been considered. Therefore, the performance figures which we report here are based on our syntheses. Since the area requirement, i.e., Gate Equivalence (GE), of an implementation varies depending on the desired latency, we give in Fig. 2 a curve of GE of the unrolled PRINCE implementation over the latency. We should stress that similar to the target of the seminal work [13], all our design architectures support both encryption and decryption. For the threshold implementations, the syntheses have been performed by keeping the hierarchy to avoid the combination of different shares (otherwise, first-order leakage is probable), and for the unrolled (unprotected) designs the hierarchy is avoided which allows the synthesizer to combine the cascaded circuits and reach the desired latency.

As stated in Sect. 2.2, in order to realize a masked hardware implementation, the masked non-linear functions (Sboxes) should be separated from each other by means of registers to avoid the propagation of glitches. Therefore, an unrolled architecture can never be properly masked. It is noteworthy that unrolled architectures already change the leakage characteristics of the device (see [6,51]). Hence, one may suppose that integration of masking into unrolled

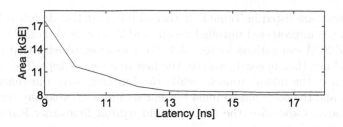

**Fig. 2.** Area versus latency of unrolled PRINCE

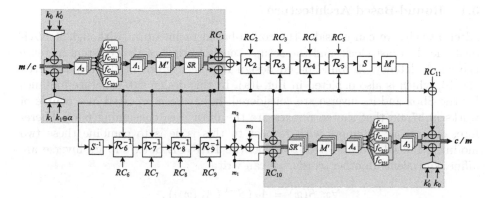

**Fig. 3.** Unrolled TI of PRINCE, only first and last round masked

architectures may complicate the device leakage in such a way that it becomes unexploitable. However, such a combination would definitely lead to first-order leakage detectable e.g., by $t$-test [18,24,44]. As a heuristic-based example, it can be supposed that masking the first and last rounds of PRINCE should suffice to protect against SCA attacks[1]. The PRINCE Sbox is a cubic 4-bit bijection, i.e., algebraic degree $t = 3$, and at least $n = m = 4$ shares are required. The PRINCE Sbox belongs to the class $\mathcal{C}_{231}$ (with respect to the category given in [10]), which needs three decomposition stages to be uniformly shared with 3 as well as 4 shares, while it can be uniformly shared in one stage with $n = m = 5$ shares. It has been given in Sect. 2.2 that the *uniformity* is required because the output of the shared Sbox feeds the next non-linear functions. Hence, the non-uniform output of the first cipher round does not play any role, if the second cipher round is not masked. Therefore, we can stay with $n = m = 4$ shares and make the (non-uniform) shared PRINCE Sbox in one stage by direct sharing [11]. We have implemented such a design, whose block diagram is shown in Fig. 3 and all the corresponding formulations are given in Appendix A. Further, its timing and

---

[1] In general, it is not a true statement since (*i*) the unmasking at the end of the first round (see Fig. 3) would anyway lead to (although hard-to-exploit) first-order leakage, and (*ii*) the adversary can set certain plaintext bits to a fixed value and target the second cipher round.

area overheads are listed in Table 1. It turned out that this design is 3–6 times larger than the unprotected unrolled design and 2–3 times slower. We deal with its practical SCA evaluations in Sect. 3.4. We should emphasize that except for the first and last (i.e., masked) rounds, the hierarchy is not kept. This allows the optimization of the middle rounds, while the functions over the shared signals (in the first and the last rounds) must be kept separate to avoid any combination over the shares. Otherwise, the design would exhibit first-order leakage at the first and/or last rounds.

### 3.1 Round-Based Architecture

Alternatively, we can consider the round-based architecture, although it obviously needs a fast clock, and the setup- and hold-time of the registers increase the whole latency. A round-based design has been given in the original article [13] which is also depicted in Fig. 4(a). In this design two separate modules for the Sbox and its inverse are considered. It has been reported in a couple of works [9,37,42] that shared Sboxes are the most area consuming part. Therefore, one of our attempts with respect to this issue is to combine these two modules. Indeed, we have realized that the PRINCE Sbox and its inverse are affine equivalent. In other words, we can write

$$\forall \boldsymbol{x}, S(\boldsymbol{x}) = A_2 \left( S^{-1} \left( A_1 \left( \boldsymbol{x} \right) \right) \right),$$

with $A_1$ and $A_2$ input- and output-affine transformations. In case of the PRINCE Sbox, there exists only one pair $(A_1, A_2)$, and $A_1$ and $A_2$ are the same. Hence, we can write $S = A \circ S^{-1} \circ A$, with $A$: B8A93021EDFC6574 as[2]

$$e = 1 + a + b + d, \qquad f = 1 + a, \qquad g = d, \qquad h = 1 + c,$$

with $\langle a, b, c, d \rangle$ the 4-bit input, $\langle e, f, g, h \rangle$ the 4-bit output, and $a$ and $e$ the least significant bits. Based on this findings, we developed another round-based architecture, shown in Fig. 4(b), where only one **S-Layer** module is instantiated.

More detailed information about the active data path of our developed round-based design at each cipher round is given in Appendix B (Fig. 17). Table 1 lists the differences between these two designs. Note that we constrained the syntheses of both designs with different latencies to obtain both fastest and smallest designs for fair comparisons. As stated, the objective is to make use of only one **S-Layer**, hence our design utilizes more multiplexers compared to the original round-based design. Further, we optimized the way that whitening keys $k_0$ and $k_0'$ are added to the state considering the fact that it should support both encryption and decryption. Another issue is how to deal with the round constants. As given in Sect. 2.1, the round constants have been randomly selected, hence a combinatorial circuit should realize the selection of $RC_i$ at each round. We have examined several cases, and the most optimized design (with respect to area) has been achieved by employing a multiplexer which selects one of the $RC_0$ to

---

[2] It also holds for $S^{-1} = A' \circ S \circ A'$, with $A'$: 5764FDCE1320B98A.

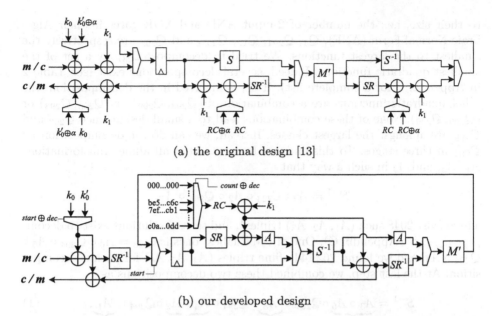

(a) the original design [13]

(b) our developed design

**Fig. 4.** Round-based designs

$RC_{11}$ by the round counter. The role of optimization was to assign 0111 as the round counter to the round number $i = 5$. Therefore, the order of the round counter 0010, 0011, ..., 0111, 1000, ..., 1100, 1101 can be reversed (required for decryption) by inverting the round counter bits (see Fig. 4(b)). We should also point out that in both original and our round-based designs, the state register is placed right after the multiplexer. That allows the synthesizer to combine them and make use of scan flip-flop, which is smaller than a sum of a multiplexer and a flip-flop [43].

As a side note, our round-based design is not necessarily the most optimized design. The tricks, that we used in our design, can be also applied in the original one (Fig. 4(a)). However, since our target is to instantiate one **S-Layer** to ease the threshold implementations, we consider our round-based architecture as the basis of the further designs.

## 3.2   Uniform Sharing of the Sbox

The uniform sharing of the cubic 4-bit bijection $\mathcal{C}_{231}$, to which the PRINCE Sbox belongs, with 3 shares can only be achieved by a three-stage quadratic decomposition [11]. As listed in [11], there exist 5 quadratic classes, $\mathcal{Q}_4$, $\mathcal{Q}_{12}$, $\mathcal{Q}_{293}$, $\mathcal{Q}_{294}$, and $\mathcal{Q}_{299}$, that can be uniformly shared in one stage[3]. With respect

---

[3] One more quadratic class $\mathcal{Q}_{300}$ exists, but needs two stages for a uniform sharing with 3 shares.

to their size, i.e., the number of 2-input AND and XOR gates in their Algebraic Normal Form (ANF), $\mathcal{Q}_4$, $\mathcal{Q}_{294}$, $\mathcal{Q}_{12}$, $\mathcal{Q}_{293}$, and $\mathcal{Q}_{299}$ are respectively the smallest to the largest functions. We tried to decompose $\mathcal{C}_{231}$ by a set of the smallest quadratic functions. Indeed, several decompositions exist (see Table 2 in Appendix C for a complete list). If $\mathcal{Q}_4$ is involved in the decomposition, the other quadratic functions are a combination of $(\mathcal{Q}_{293}, \mathcal{Q}_{299})$ or $(\mathcal{Q}_{293}, \mathcal{Q}_{293})$ or $(\mathcal{Q}_{299}, \mathcal{Q}_{299})$. None of these combinations lead to a small design since $\mathcal{Q}_{293}$ and $\mathcal{Q}_{299}$ are amongst the largest classes. Instead, we can do the decomposition by $\mathcal{Q}_{294}$ in three stages. To this end, we first extracted all affine transformations $A_1$, $A_2$, and $A_3$ in such a way that

$$S^{-1} = A_3 \circ \mathcal{C}_{223} \circ A_2 \circ \mathcal{Q}_{294} \circ A_1.$$

There exist 2048 such $(A_1, A_2, A_3)$ triples[4], and several solutions exist to decompose $\mathcal{C}_{223}$ (see Appendix C). One of the smallest ones is $\mathcal{C}_{223} = A_6 \circ \mathcal{Q}_{294} \circ A_5 \circ \mathcal{Q}_{294} \circ A_4$, and we found 262 144 affine triples $(A_4, A_5, A_6)$ for such a decomposition. At the last step, we combined these two decompositions as

$$S^{-1} = \underbrace{A_3 \circ A_6}_{A_{out}} \circ \mathcal{Q}_{294} \circ \underbrace{A_5}_{A_{m2}} \circ \mathcal{Q}_{294} \circ \underbrace{A_4 \circ A_2}_{A_{m1}} \circ \mathcal{Q}_{294} \circ \underbrace{A_1}_{A_{in}}, \tag{1}$$

and examined all $2048 \times 262\,144$ cases[5]. With respect to the size of the resulting affines, we considered the number of 2-input XOR gates as well as the Hamming weight of the constants. The smallest combination has been achieved as

$A_{in}$ : 8293C6D70A1B4E5F, $e = b$, $\quad f = a$, $\quad\quad g = c$, $\quad h = 1 + a + d$,

$A_{m1}$ : C480E6A2D591F7B3, $e = d$, $\quad f = c$, $\quad\quad g = 1 + b$, $\quad h = 1 + a$,

$A_{m2}$ : 08C43BF72AE619D5, $e = c$, $\quad f = c + d$, $\quad g = b$, $\quad h = a + b$,

$A_{out}$ : 21748BDE6530CF9A, $e = a + b$, $\quad f = 1 + a + c$, $\quad g = b + d$, $\quad h = c$. $\qquad(2)$

In order to share $\mathcal{Q}_{294}$ : 0123456789BAEFDC as

$$e = a + bd, \qquad f = b + cd, \qquad g = c, \qquad h = d,$$

we can follow the direct sharing [11], which has been applied in [38]. The component function $f_{\mathcal{Q}_{294}}^{i,j}(\langle a^i, b^i, c^i, d^i \rangle, \langle a^j, b^j, c^j, d^j \rangle) = \langle e, f, g, h \rangle$ has been defined in [38] as

$$e = a^i + b^i d^i + d^i b^j + b^i d^j \qquad\qquad g = c^i$$
$$f = b^i + c^i d^i + d^i c^j + c^i d^j \qquad\qquad h = d^i, \tag{3}$$

and it has been given that the three 4-bit output shares provided by $f_{\mathcal{Q}_{294}}^{2,3}(.,.)$, $f_{\mathcal{Q}_{294}}^{3,1}(.,.)$ and $f_{\mathcal{Q}_{294}}^{1,2}(.,.)$ make a uniform first-order sharing of $\mathcal{Q}_{294}$.

---

[4] It is the same for $S^{-1} = A_3 \circ \mathcal{Q}_{294} \circ A_2 \circ \mathcal{C}_{223} \circ A_1$.

[5] The result is a multiset, i.e., with repeated elements.

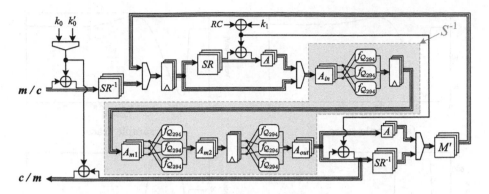

**Fig. 5.** Round-based first-order threshold implementation of PRINCE

Since the affine functions applied on all shares do not change the uniformity, our construction – given in Eq. (1) – in addition to the set of affines (Eq. (2)) and component function $f_{Q_{294}}$ (Eq. (3)) form a uniform first-order sharing of the PRINCE Sbox inverse. To the best of our knowledge, this is amongst the smallest construction which fulfills all the TI properties with $n = m = 3$ shares. Note that the shared quadratic functions should be separated by registers to avoid the propagation of glitches.

### 3.3    Implementation

Our construction of the first-order TI of PRINCE is depicted in Fig. 5. All operations except key and constant additions (and the **S-Layer**) are repeated three times. It suffices if the constant of the affine functions $A$, $A_{in}$, $A_{m1}$, $A_{m2}$, and $A_{out}$ are applied on only one share[6]. The key and the constants are not shared, which is the same scenario applied in several works, e.g., [7–9,37,42], and is adequate to resist against first-order attacks. Hence, the keys and constant are also applied on only one share.

Due to the registers integrated into the shared Sbox, the design realizes a pipeline with three stages. In other words, three consecutive (shared) inputs (plaintexts/ciphertexts) can be fed into the design, and after 40 clock cycles three outputs (ciphertexts/plaintexts) are consecutively given out. Thanks to the uniform sharing of the Sbox, excluding the masks required to share the input, the design does not require any fresh randomness during the computations. The performance figures of this design are also given in Table 1 for comparison purposes.

### 3.4    Practical Evaluations

For the practical investigations – rather than ASIC-based experiments or simulation – we ported the designs to an FPGA-based platform. We have used a

---

[6] It does not affect either the functionality or the uniformity.

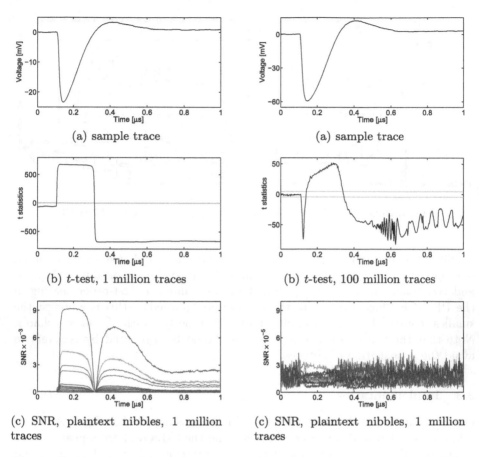

(a) sample trace

(a) sample trace

(b) *t*-test, 1 million traces

(b) *t*-test, 100 million traces

(c) SNR, plaintext nibbles, 1 million traces

(c) SNR, plaintext nibbles, 1 million traces

**Fig. 6.** Evaluation results, unrolled unprotected design

**Fig. 7.** Evaluation results, unrolled TI design (first and last rounds masked)

SAKURA-X board [1] with a Kintex-7 FPGA, particularly designed for SCA evaluations. In order to monitor the power consumption, we measured the voltage drop over a shunt resistor placed at the Vdd path of the Kintex FPGA. The power traces have been collected by means of a digital LeCroy oscilloscope at the sampling rate of 500 MS/s. Because of the low amplitude of the measured signal (due to the underlying low-power technology of Xilinx 7 series FPGAs), we employed an AC amplifier ZFL-1000LN+ from Mini-Circuits with 10 dB gain.

For SCA evaluation purposes, we applied the non-specific *t*-test (also known as *fixed versus random t*-test). This test procedure, originally called TVLA, has been proposed in [18], extended in [44], and [20,21] applied in e.g., [7,8,38]. The test – which compares the leakages associated to random inputs with that to a fixed input – can examine the existence of a detectable leakage, but cannot give any impression whether the leakage is exploitable. Hence, in case the *t*-test reports a first-order detectable leakage, we perform a signal-to-noise ratio (SNR)

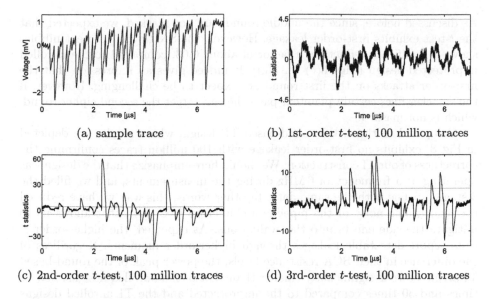

(a) sample trace

(b) 1st-order $t$-test, 100 million traces

(c) 2nd-order $t$-test, 100 million traces

(d) 3rd-order $t$-test, 100 million traces

**Fig. 8.** Evaluation results, round-based TI design

check. In such a check, the variance of the average leakage traces classified by e.g., the value of a plaintext nibble (divided by the variance of the noise) is examined [32]. It indeed can give an overview about the dependency of the average (first-order) leakages to the processed data. Here, we do not show any attack results, and only discuss about the existence of detectable leakages, and compare the amount of dependency of the leakages to the processed data.

For the unrolled unprotected design, Fig. 6 shows a sample power trace, the $t$-test result as well as the SNR over all 16 plaintext nibbles. Since the design is not masked, the $t$-test as expected shows a pretty strong first-order leakage. Along the same lines, the corresponding SNR exhibit a clear dependency between the traces and the plaintext nibbles. Hence, a successful key-recovery attack is expected (e.g., in [51]). We should here note that 23.3 mV power peak is relatively large[7] for this low-power FPGA. Since several gates are packed into one LUT, the equivalent design in ASIC can be more glitchy, and hence (probably) more energy consuming[8]. This **may** harden the development of fully unrolled (even low-latency) designs into low-energy, e.g., battery-powered, applications (see simulation-based results in [3,4]).

We have shown the corresponding results of the unrolled TI design in Fig. 7. As stated before, only the first and the last rounds are masked by means of four shares (see Fig. 3). During the measurements, the 4-share input as well as other 3 independent fresh random masks ($m_1$, $m_2$, $m_3$ for the last round) are given to the Kintex-7 FPGA. The output is also provided in a 4-share masked form.

---

[7] We showed real voltage values, i.e., output of the amplifier divided by 10.

[8] This is a guess by the authors and should be examined in practice.

As discussed before, since the middle rounds are not masked, we expected that the $t$-test exhibits first-order leakage. However, the SNR over plaintext nibbles shows a significant reduction, a factor of about 0.03, compared to the unrolled unprotected design (Figs. 6(c) vs. 7(c)). It indeed gives an impression that the first-order attacks on the first round are expect to be challenging. However, if the attacker fixes certain plaintext parts, he can target the second cipher round, which is not masked.

Compared to these, the round-based TI design, whose results are depicted in Fig. 8, exhibits no first-order leakage with 100 million traces confirming the correctness of our TI construction. We should here emphasize that the design was operating at a frequency of 6 MHz during the measurements, and we filled the 3-stage pipeline with the same data. In other words, this way we have reduced the algorithmic noise of the pipeline architecture (see [46]), that allows us to mitigate the side effects into the evaluations. As expected, the higher-order $t$-tests report detectable leakages through higher-order moments. Regardless of the difference in their SCA-resistance levels, the power peak of the round-based TI architecture is significantly smaller than that of the unrolled designs, i.e., 10 times and 30 times compared to the unprotected and the TI unrolled designs respectively.

## 4    Asynchronous Design

We have already discussed in Sect. 1 that low-latency concept is closely connected to the unrolled (single-cycle) architecture since the high clock rates (needed to rapidly run register-based designs) are not available or supported by many systems. For instance, in many FPGA designs clock rates above 200 MHz are often difficult to realize. In this settings, asynchronous circuits seem to be an alternative to this issue. With asynchronous circuit design, also known as self-timed and clock-less design, it is possible to realize circuits with high performance parameters in terms of their power, throughput, electromagnetic emissions, etc. [39,45]. Asynchronous design is not as well-established and widely-used as synchronous design methodology. Hence, the standard tools for asynchronous design are not available, or not widely known, or particularly customized for certain technologies.

Because the field of asynchronous circuit design covers a wide range, we focus only on certain concepts which are relevant to our case studies. In terms of PRINCE, consider the round-based synchronous architecture in Fig. 4(a). The maximum clock frequency is defined by the longest critical path (most likely when both Sbox and its inverse are active). However, such a path is not always active. In other words, in all clock cycles except the middle one the design can be clocked faster. If this design is realized by asynchronous design methodology, the end of the computation of one cipher round initiates the start of the next round. Hence, the design operates at its maximum speed, or let say with its lowest latency. In this case – similar to the unrolled architectures – the time when the computations are finished, i.e., the ciphertext is ready to be read, depends on the given inputs, but the maximum latency can be estimated.

**Table 1.** Performance figures of different PRINCE implementations. For each design, the first and the second row represents the smallest and the fastest variant respectively.

| Design | Area [GE] | Crit. Path [ns] | Clock # | Latency [ns] | Throughput [Gbps] | Power$^a$ Peak [mV] | DPA res. |
|---|---|---|---|---|---|---|---|
| Unrolled | 8 512 | 13 | 1 | 13 | 4.923 | 23.3 | |
| | 17 675 | 9 | | 9 | 7.111 | | |
| Unrolled TI | 48 012 | 38 | 1 | 38 | 1.684 | 59.5 | ~$^b$ |
| | 77 921 | 13.2 | | 13.2 | 4.848 | | |
| Round-based [13] | 2 809 | 5.6 | 13 | 72.8 | 0.879 | 1.7$^c$ | |
| | 4 698 | 1.5 | | 19.5 | 3.282 | | |
| Round-based ours | 2 286 | 4.9 | 14 | 68.6 | 0.933 | 1.5$^c$ | |
| | 4 663 | 2 | | 28 | 2.285 | | |
| Round-based TI | 9 292 | 4 | 40 | 160 | 1.143$^d$ | 2.3$^c$ 21$^e$ | ✓ |
| | 11 275 | 1.9 | | 76 | 2.406$^d$ | | |
| Round-based TI Asynchronous (simple Ack) | 25 701 | 11 | 40 × 2$^f$ | 800 | 0.208$^d$ | 53.7 | ✓ |
| | 31 936 | 5.4 | | 432 | 0.423$^d$ | | |
| Midori64 Round-based TI | 7 297 | 4 | 31 | 124 | 1.000$^g$ | 20.2$^e$ | ✓ |
| | 9 237 | 1.9 | | 58.9 | 2.105$^g$ | | |

$^a$measured from the FPGA implementations
$^b$only at the first and the last rounds
$^c$@ 6 MHz
$^d$considering the 3-stage pipeline
$^e$by controlled ring-oscillator clock
$^f$doubled due to pre-charge/evaluation phases
$^g$considering the 2-stage pipeline

As shown in Sect. 3, the round-based TI design can provide the first-order resistance, but it needs a clock with a frequency between 250 MHz and 500 MHz to achieve the highest throughput (see Table 1). Hence, our objective in this section is to realize the round-based TI design with an asynchronous design methodology.

**State of the Art.** We should emphasize that the asynchronous design has been previously applied as a sole SCA countermeasure. One of the earlier works [33] describes a smartcard chip which relies on self-timed circuits to provide protection against physical attacks. The authors proposed to solely use dual-rail encoding to reduce the threat of data-dependent power consumption but also noted the obvious difficulties of this approach, e.g., varying wire lengths. Furthermore, they highlighted the problem of timing leakage of asynchronous circuits and advise to minimize data dependent gate delays coupled with the insertion of dummy delays to reduce this leakage. Later in [23] the security of a similar self-timed circuit has thoroughly been tested in practice. The authors found that small imbalances in the dual-rail circuits cause data-dependent leakage which enables an attacker to perform a successful DPA on the asynchronous circuit. They showed that their asynchronous design alone is not sufficient to prevent SCA attacks, and that these imbalances need to be eliminated during the design process to increase the level of security. This is in line with [30, 53] where some of the difficulties, e.g., no global clock, with respect to performing DPA on asynchronous designs are described.

One of the first clock-less implementation of AES was presented in [52]. It also relies on power-balancing capability of dual-rail and the absence of a global clock to thwart DPA. The dual-rail circuits were found to be more secure than the single-rail one, however this is only based on simulation results and a thorough practical evaluation is missing.

Another approach to secure AES using clock-less circuits is presented in [14]. It again relies on an asynchronous style called quasi delay insensitive (QDI) which has a range of supply voltages. The authors noted the above-mentioned limitation of this implementation style with respect to SCA resistance [15]. Therefore, they proposed to lower the supply voltage to reduce the SNR and thwart DPA. However, [14] does not include practical experiments related to this approach. Further techniques [16,17] have been proposed to harden QDI against DPA based on the introduction of random timing and path swapping. However, their efficiency was only evaluated using electrical simulations.

More recently, an AES round function in Null Convention logic - another delay insensitive logic paradigm - has been proposed in [50] in which the SCA resistance has again been only evaluated with simulations.

It should be noted that in a majority of the aforementioned articles SCA resistance was not the sole motivation for asynchronous circuits. Other beneficial properties include a low-power consumption for embedded devices and some form of an integrated fault tolerant scheme.

What we want to examine here is not the application of asynchronous design to prevent SCA leakages. In short, we do not aim at e.g., realizing the round-based **unprotected** architecture with asynchronous methodology and examine its SCA resistance. Instead, our goal is to investigate the challenges and outcomes of implementing a correctly-masked design, e.g., round-based threshold implementation (Fig. 5), under the concept of asynchronous designs. Such an investigation is conducted with the goal of achieving a clock-less design while it is expected to still satisfy the desired first-order SCA protection due to its underlying uniform TI construction.

## 4.1  Fundamentals

Different parts of an asynchronous circuit need to communicate with each other. For example, the finish of one PRINCE round should initiate the next round. A couple of different handshaking protocols exist to establish such a communication.

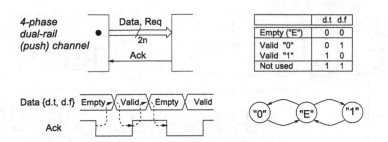

**Fig. 9.** A delay-insensitive 4-phase dual-rail protocol (taken from [45])

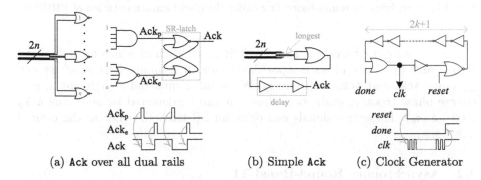

(a) Ack over all dual rails     (b) Simple Ack     (c) Clock Generator

**Fig. 10.** Exemplary circuits to generate the Ack and *clk* signals (a) and (b) for asynchronous designs, (c) for a synchronous design

*The 4-phase dual-rail protocol* encodes the data signals into two wires per bit (see Fig. 9). Each logical '1' or '0' is represented by {1,0} or {0,1} respectively, while {0,0} is known as "no data" (or "empty") and {1,1} as invalid. A transition from one valid coding to another is not allowed, unless an "empty" value is transmitted in-between, that forms a *return-to-zero* protocol. This protocol is very robust; two parties can communicate reliably regardless of delays in the wires, i.e., it is *delay-insensitive* [45].

This concept is very similar to the WDDL logic style [48], which has been designed to mitigate SCA leakages. The underlying dual-rail pre-charge logic is the same encoding; the valid encodings {0,1} and {1,0} are known as *evaluation phase* and the empty value {0,0} as *pre-charge phase*. A WDDL circuit is usually a synchronous design, where the evaluation/pre-charge phases are controlled by the clock signal (the same concept as in Fig. 9 by replacing the Ack signal with clock). This protocol is familiar to most digital designers, and avoids any glitches in the circuit hence achieving a low-power construction. However, it has a disadvantage due to the extra return-to-zero transitions that cost time and energy.

We can implement the combinatorial parts of a design based on the WDDL concept, and add extra logic to detect the end of the pre-charge as well as evaluation phase. This allows us to form the Ack signal (see Fig. 9). As shown in

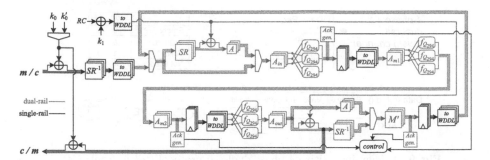

**Fig. 11.** Asynchronous round-based first-order threshold implementation of PRINCE

Fig. 10(a), we can integrate $n$ 2-input NOR gates, each of which for a dual-rail signal, and by means of an $n$-input AND and an $n$-input NOR gate[9] we can generate $\mathtt{Ack_p}$ and $\mathtt{Ack_e}$ respectively. When all $n$ dual-rail signals are in pre-charge phase (resp. in evaluation phase), it can be detected by observing $\mathtt{Ack_p}$ (resp. $\mathtt{Ack_e}$). These two signals can drive an SR-latch to generate the desired $\mathtt{Ack}$ signal.

## 4.2 Asynchronous Round-Based TI

WDDL combinatorial circuits (generally asynchronous circuits) are glitch free, i.e., each dual-rail signal changes only once at each pre-charge/evaluation phase. Threshold implementation has been developed mainly for glitchy circuits, and the registers should be placed between the non-linear shared functions to avoid the propagation of the glitches [40]. Hence, at the first glance it seems that it is not essential anymore to instantiate such registers if the circuit is glitch free.

Following this concept, we have implemented the round-based TI design presented in Fig. 11, and did not integrate registers between the shared $\mathcal{Q}_{294}$ functions. The state register is moved to the end of the round function, and the $\mathtt{Ack}$ signal is generated based on the state register input. By a couple of engineering tricks the design is mapped to our FPGA platform. We should here emphasize that Xilinx FPGAs are developed yet only for synchronous designs, and integration of asynchronous circuits is neither straightforward nor efficient. For example, each dual-rail WDDL gate should be implemented by a LUT [5].

Our design is a self-timed circuit, i.e., it does not require an external clock, and once the $\mathtt{reset}$ signal goes LO, the circuit starts the first evaluation phase, which is the first PRINCE round. Controlled by the internally-generated $\mathtt{Ack}$ signal, the end of the evaluation phase triggers the state register to save the cipher state and simultaneously the start of the pre-charge phase. As stated before, a disadvantage of such a concept is its required interleaved pre-charge/evaluation phases. Because we avoided the extra registers within the Sbox, the design does not form a pipeline anymore. Therefore, a full PRINCE is performed by 14

---

[9] Such large gates are made by cascading the smaller gates.

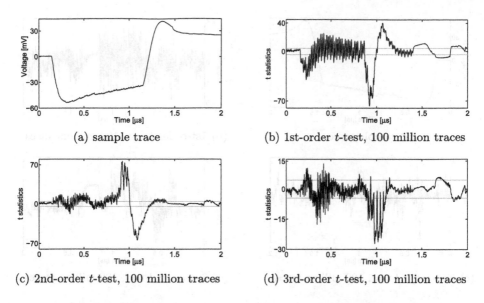

(a) sample trace

(b) 1st-order $t$-test, 100 million traces

(c) 2nd-order $t$-test, 100 million traces

(d) 3rd-order $t$-test, 100 million traces

**Fig. 12.** Evaluation results, asynchronous round-based TI design

(pre-charge, evaluation) cycles. Figure 12 shows a sample power trace of such a design, where the cipher rounds can be identified. However, the $t$-test indicates a pretty strong first-order leakage. Note that the design still realizes a uniform threshold implementation with 3 shares, and we have not used WDDL as an SCA countermeasure, rather as a 4-phase dual-rail protocol to enable detection of the end of the evaluation (and pre-charge) of the combinatorial circuit.

A more careful investigation about the detected first-order leakage clarified that although the circuit is glitch free, the non-linear circuits are cascaded. One of the component functions of the second non-linear circuit (the second shared $Q_{294}$ in Fig. 11) starts to evaluate when two output shares of the first non-linear circuit are both evaluated. Further, these two shares depend on all three shares of the Sbox input. Therefore, the start of the evaluation of the second non-linear circuit depends on all three input shares of the Sbox. This, which is a non-linear condition (i.e., when both two output shares of the first non-linear circuit are evaluated) is the reason for such a detectable first-order leakage (see [23] for a similar experience on an unmasked design). Although placing registers between the shared non-linear functions was initially introduced to avoid the propagation of glitches, it also synchronizes the start of their evaluation to be independent of the timing of the previous stage. As a result, the shared non-linear functions should also be isolated from each other even in asynchronous circuits.

If we isolate the shared non-linear circuits by means of registers, and trigger the registers to store when all 3 shares are evaluated, again the time of triggering the registers as well as the circuit which generates the Ack signals (Fig. 10(a)) depends on all 3 shares and leak through first-order moments. As a proof of concept, we have examined this issue by realizing the asynchronous round-based

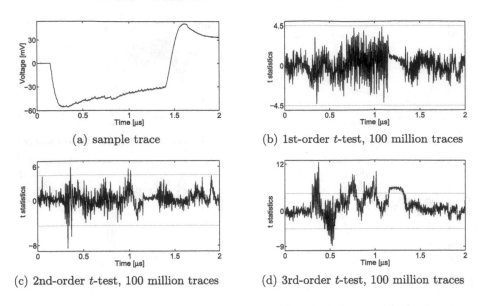

(a) sample trace

(b) 1st-order $t$-test, 100 million traces

(c) 2nd-order $t$-test, 100 million traces

(d) 3rd-order $t$-test, 100 million traces

**Fig. 13.** Evaluation results, asynchronous round-based TI design with simple Ack

TI design with registers in the Sbox module where the combinatorial parts are made by WDDL gates. In this case, the Ack signals are generated by observing the input of all three registers, i.e., when the pre-charge/evaluation of the entire circuit – pipeline with 3 stages – is completed. The evaluation of this construction has also showed detectable first-order leakage. So, we omit the corresponding results.

As a side note, the early propagation effect [47] of WDDL aggravates this issue. In the above explained experiments we have used the noEE version [5] of WDDL (available only for FPGAs), that avoids early propagation only in evaluation phase. We have also made use of its successor, AWDDL [35] (also only for FPGAs) which avoids early propagation in both phases. Regardless of its double area requirements, its utilization in our case slightly reduced the first-order leakage, but could not avoid it due to the known imbalances between the delay of dual rails. In other words, the time required for full pre-charge/evaluation phase of non-linear circuits still depends on three shares and hence on unshared input.

Therefore, the only solution which we could consider for a secure design is to simplify the Ack generator circuit. It means that if we generate the Ack signal based on only one share of one of the state registers, the start time of the next pre-charge/evaluation phase should be independent from the unshared values. However, such a circuit cannot guarantee that the pre-charge/evaluation of the other parts of the circuit are also finished. Therefore, we have found a path with the largest delay and connected the Ack generator circuit accordingly. To ensure the end of the pre-charge/evaluation of the other circuits, the generated Ack signal is delayed (see Fig. 10(b)).

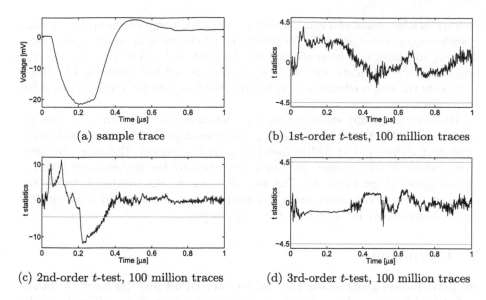

(a) sample trace

(b) 1st-order $t$-test, 100 million traces

(c) 2nd-order $t$-test, 100 million traces

(d) 3rd-order $t$-test, 100 million traces

**Fig. 14.** Evaluation results, round-based TI design clocked by a controlled ring oscillator

A sample trace as well as the $t$-test results are shown in Fig. 13, which confirms the prevention of first-order leakages. This construction is still a self-timed asynchronous circuit without external clock, but it is vastly customized. For instance, it does not operate at its maximum speed, and controlled mainly by a delayed periodic signal. Hence, we do not benefit from all the features of asynchronous methodology. If we ignore the low-power feature of this construction, it is not significantly different from the corresponding synchronous design with a high speed clock. As we listed in Table 1, the asynchronous design is much larger than its synchronous variant. Further, due to the interleaved (pre-charge, evaluation) phases, the latency of the asynchronous design is also not convincing.

The interleaved (pre-charge, evaluation) phases of 4-phase dual-rail protocols (e.g., WDDL which we used here) doubles the latency of the design. Alternatively, one can utilize a *2-phase dual-rail* protocol [45], where '1' and '0' values are encoded as signal transitions. Such protocols lead to faster but much more complex circuits. We have applied Level-Encoded Dual-Rail (LEDR) [19] concept and designed and evaluated the corresponding circuit, but due to the similarity of the results to that of the WDDL, their presentation is omitted. In short, the design was much bigger than its WDDL variant, but slightly faster. However, all issues with respect to isolation of non-linear functions as well as the Ack generator circuit hold true.

In this situation, where the operation of non-linear circuits **must** be isolated and independent of other non-linear parts, we believe that the synchronous design is favorable. For the remaining issue, i.e., absence of a fast clock in many applications where low-latency cryptography is required, we suggest to generate such a clock by means of a ring oscillator. Since the energy consumption of large clock-trees (operated at a high frequency) is not desirable in many applications, the ring oscillator can be controlled by the start and end of e.g., the encryption module. A schematic view of an exemplary circuit is depicted in Fig. 10(c). Obviously the ring oscillator should be adjusted based on the critical path delay of the circuit.

We have practically evaluated such a construction as well, whose results are shown by Fig. 14. As expected, higher power consumption peak compared to the same design operated at 6 MHz (see Fig. 8(a)) is observed. However, the first-order leakage is still avoided, and more interestingly the higher-order leakages are mitigated (Figs. 14 vs. 8). The reason is due to the overlap between the adjacent power peaks, which leads to higher amount of noise, and consecutively harder higher-order leakages to detect, e.g., in [38].

# 5    Discussion

We have discussed and shown that SCA-protected designs (by means of masking) should involve registers even in case of asynchronous designs. Therefore, the low-latency concept – with a perspective of unrolled architectures – is in contradiction with masking in hardware. As a result, round-based architectures are the only possible solution for applications, where provably-secure SCA protection is required. In this scenario, in order to achieve a low latency two parameters play the most important role: $(i)$ the latency of each cipher round, and $(ii)$ the number of rounds.

Obviously, the most challenging issue, which we faced, was uniform realization of the shared Sbox with 3 shares. In the seminal article [13], 8 different Sboxes (up to affine equivalent) are suggested for the PRINCE-family. However, all of them need at least a 3-stage decomposition to be able to uniformly shared with 3 shares. Such a decomposition, as shown in Sect. 3.3, leads to a pipeline round-based architecture with 3 stages. This – as stated above – increases the number of clock cycles required for each cipher round, and negatively affect the latency.

For the future designs, our first suggestion is to select Sboxes, whose uniform sharing needs a low number of stages. The extreme case is to apply quadratic Sboxes, which can be shared in one stage, but such a choice leads to higher number of rounds (see PrintCipher [28]), which affect the low-latency target as well. Hence, the trade-off here is to select either a quadratic Sbox, which needs more number of rounds, or a cubic TI-friendly Sbox which forms a pipeline, hence more number of clock cycles per round.

The second challenging issue was to deal with round constants. In case of PRINCE, the round constants have been selected from a semi-random source (fraction of $\pi = 3.141\ldots$). This design decision does not have any performance penalty in case of unrolled architecture, since a round constant just turns some XOR gates of the prior AddRoundKey to XNOR, i.e., for free[10]. However,

---

[10] 2-input XOR and XNOR gates need the same area [43].

for a round-based design, this leads to a relatively large combinatorial circuit since each round constant should be selected at each round based on the round counter[11]. Hence, it is advisable to systematically generate the round constants, e.g., by means of an LFSR. Note that if a large LFSR is chosen, the area required to save its state (by registers) has also a negative impact on the area overhead.

In case of PRINCE, due to its underlying $\alpha$-reflection structure, encryption and decryption circuits are very similar. **M′-Layer** of PRINCE is self-inverse, and the Sbox is affine equivalent to its inverse, but it consists of two different round functions. Such a construction makes the round-based architecture (required for SCA protection) more complicated as both round functions need to be implemented (see Fig. 4), which obviously increases the area requirements. Hence, it is preferred to have a design with a unique round function. In this case, achieving highly-similar encryption and decryption might be challenging.

## 5.1   Comparison to Midori

The Midori cipher has been introduced in [3] with the main goal of reducing the energy consumption. Based on the simulation results and the discussions given in [3,4], a round-based architecture is targeted to achieve the minimum energy consumption per bit. Further, it has been shown that the full latency of a round-based implementation of Midori outperforms that of other considered ciphers including PRINCE. Therefore, we considered Midori64 for comparison purposes[12].

Midori64 state is a 64-bit block, and its 4-bit Sbox (applied on all state nibbles) is an involution. Its linear layer includes an involutional MixColumn operation (made of a couple of XORs), and a ShiftCell which swaps the 4-bit cells of the state. It consists of 15 rounds, and respectively 15 round constants (each 16 bits) which are added to the LSB of the state nibbles. The 128-bit key is divided into two parts which are alternatively added to the sate at each round, and their XOR is used as a pre- and post-whitening key.

Figure 15 shows a round-based implementation of Midori64, which supports both encryption and decryption. Note that the authors of [3] proposed to apply the inverse of the linear operations, i.e., $\mathsf{ShiftCell}^{-1} \circ \mathsf{MixColumn}$, over the round keys and round constants for the decryption. However, we found our solution (see Fig. 15) which needs 64 extra 2-input XOR gates, cheaper than the original suggestion.

In order to realize its threshold implementation, the linear layers are simply repeated over the 3 shares, and a uniform representation of its Sbox is constructed. The Midori64 Sbox is affine equivalent to $\mathcal{C}_{266}$ class [11], which can be decomposed to two quadratic bijections with uniform TI. Amongst many possible solutions we selected $\mathcal{Q}_{12} \times \mathcal{Q}_{12}$ and found affine functions as

---

[11] In our round-based designs, the selection of the round constant followed by 64-bit XOR need an area of 265 GE.

[12] We are aware of the weakness reported in [25], but to be compatible with PRINCE, i.e., 64-bit block size, we excluded Midori128 in our investigations.

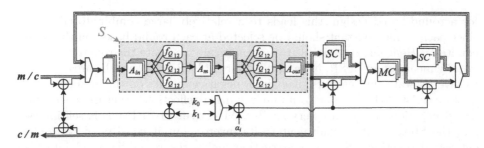

**Fig. 15.** Round-based first-order threshold implementation of Midori64

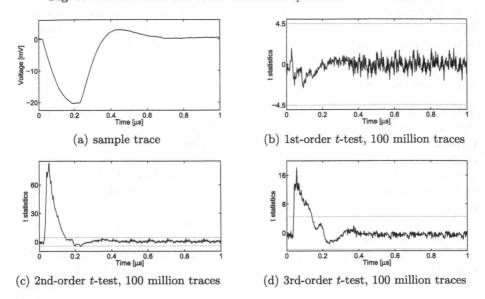

**Fig. 16.** Evaluation results, round-based TI design of **Midori64** clocked by a controlled ring oscillator

$$S = A_{out} \circ \mathcal{Q}_{12} \circ A_m \circ \mathcal{Q}_{12} \circ A_{in}.$$

There exist 147 456 such $(A_{in}, A_m, A_{out})$ triples, and we selected the following settings (with respect to the same criteria explained in Sect. 3.2):

$$
\begin{aligned}
A_{in} &: \texttt{0A1B82934E5FC6D7}, & e=b, & & f=a, & & g=d, & & h=a+c, \\
A_m &: \texttt{84B70C3F95A61D2E}, & e=b+d, & & f=b, & & g=a, & & h=1+a+c, \\
A_{out} &: \texttt{8A02DF57CE469B13}, & e=c, & & f=a, & & g=c+d, & & h=1+b. \quad (4)
\end{aligned}
$$

The sharing of $\mathcal{Q}_{12} : \texttt{0123456789CDEFAB}$ with

$$e=a, \qquad f=b+bd+cd, \qquad g=c+bd, \qquad h=d$$

can be derived by direct sharing [11]. If we define the component function $f_{\mathcal{Q}_{12}}^{i,j}(\langle a^i, b^i, c^i, d^i \rangle, \langle a^j, b^j, c^j, d^j \rangle) = \langle e, f, g, h \rangle$ as

$$e = a^i, \qquad\qquad\qquad\qquad g = b^i d^i + d^i b^j + b^i d^j,$$
$$f = b^i + b^i d^i + d^i b^j + b^i d^j + c^i d^i + d^i c^j + c^i d^j, \quad h = d^i, \tag{5}$$

we can form a uniform shared representation of $\mathcal{Q}_{12}$ by $f_{\mathcal{Q}_{12}}^{2,3}(.,.)$, $f_{\mathcal{Q}_{12}}^{3,1}(.,.)$ and $f_{\mathcal{Q}_{12}}^{1,2}(.,.)$, as shown in Fig. 15.

We have also practically examined its SCA resistance by the FPGA prototype. For comparison purposes we considered only a synchronous version, where the clock is provided by a controlled ring oscillator (with the same number of inverters as in the corresponding PRINCE design). The results (indicating first-order resistance and stronger leakage through higher-order moments compared to its corresponding PRINCE) are shown in Fig. 16, and the performance results are listed in Table 1.

## 5.2   Conclusions

We have presented the results of an extensive study on application of masking, particularly TI, on PRINCE considering its low-latency goal. As given in Table 1, the asynchronous design is around 2.8 times larger and around 2.6 times slower than its synchronous variant. Further, an overview about its power consumption (FPGA prototype) shows no advantage, even compared to the case when the synchronous design operates at a high frequency[13]. More importantly, we faced several issues regarding its detectable first-order leakage. Finally, the design, which could prevent the leakages, was not much structurally different to a synchronous design, whose clock is internally generated.

Based on Table 1, the fastest synchronous round-based TI needs 11 275 GE which is in the range of the unprotected unrolled design (8 512 - 17 675 GE). Although its critical path with 1.9 ns delay is around than 4 times shorter than that of the fastest unrolled design, its 40 clock cycle latency leads to 76 ns which is around 8 times more than 9 ns latency of the unrolled design. However, its underlying pipeline architecture compensates in terms of throughput to be between 2 and 3 times less than the unprotected unrolled designs.

Compared to the synchronous round-based TI of PRINCE, Midori64 is smaller and achieves lower latency (58.9 ns vs. 76 ns for the fastest designs), but their throughput are comparable considering the full capacity of the pipelines. We should emphasize that most of the suggestions (given above) can be seen in the design of Midori: (i) the Sbox is an involution and TI friendly, (ii) MixColumn is an involution, (iii) it consists of only one type of round function, and (iv) the round constants are short (16 bits per round) although they cannot be generated systematically. However, with respect to [25] our observation is that: there is still a gap to fill, i.e., a low-latency cipher, which in addition to the

---

[13] Note here the difference between power consumption of equivalent FPGA and ASIC circuits.

desired cryptographic strength, can easily deal with the challenges addressed in this article. In short, the candidate should still achieve a low latency when fully unrolled as well as in a round-based fashion, and at the same time its masked (TI) round-based variant is efficient in terms of area and latency for the applications, where provably-secure SCA protection is required.

**Acknowledgment.** The authors would like to acknowledge Ventzislav Nikov for his help with the decomposition process and Alexander Kühn for his help with implementation of different asynchronous variants of PRINCE on FPGA. The research in this work was supported in part by the DFG Research Training Group GRK 1817/1.

# A    Masked Unrolled Design (only First and Last Rounds)

To share the Sbox and its inverse with 4 shares, we represented the Sbox as $S = A_2 \circ C_{231} \circ A_1$ and its inverse as $S^{-1} = A_4 \circ C_{231} \circ A_3$ with $A_1$ : EF548932AB10CD76, $A_2$ : 08192A3B4C5D6E7F, $A_3$ : 92386DC7F45E0BA1, $A_4$ : 51736240FBD9C8EA, and $C_{231}$ : 0123468B59CEDA7F as

$$
\begin{aligned}
e &= a + d + ac + ad + bd + abc + bcd \\
f &= b + ac + bc + bd + abd \\
g &= c + d + bc + ad + cd + abd + bcd \\
h &= bc + ad + bd + cd + abd + acd + bcd.
\end{aligned}
$$

By applying direct sharing on $C_{231}$ we reach the component function $f_{C_{231}}^{i,j,k}(\langle a^i, b^i, c^i, d^i \rangle, \langle a^j, b^j, c^j, d^j \rangle, \langle a^k, b^k, c^k, d^k \rangle) = \langle e, f, g, h \rangle$ as

$$
\begin{aligned}
e =\, & a^i + d^i + a^ic^i + a^ic^j + a^ic^k + a^jc^i + a^id^i + a^id^j + a^id^k + a^jd^i + b^id^i + \\
& b^id^j + b^id^k + b^jd^i + a^ib^ic^i + a^ib^jc^k + a^ib^kc^j + a^jb^ic^k + a^jb^kc^i + \\
& a^kb^ic^j + a^kb^jc^i + a^ib^ic^j + a^ib^jc^j + a^ib^ic^k + a^ib^kc^k + a^jb^jc^i + \\
& a^jb^ic^i + a^ib^jc^i + a^ib^kc^i + a^jb^ic^j + b^ic^id^i + b^ic^jd^k + b^ic^kd^j + \\
& b^jc^id^k + b^jc^kd^i + b^kc^id^j + b^kc^jd^i + b^ic^id^j + b^ic^jd^j + b^ic^id^k + \\
& b^ic^kd^k + b^jc^jd^i + b^jc^id^i + b^ic^jd^i + b^ic^kd^i + b^jc^id^j
\end{aligned}
$$

$$
\begin{aligned}
f =\, & b^i + a^ic^i + a^ic^j + a^ic^k + a^jc^i + b^ic^i + b^ic^j + b^ic^k + b^jc^i + b^id^i + b^id^j + \\
& b^id^k + b^jd^i + a^ib^id^i + a^ib^jd^k + a^ib^kd^j + a^jb^id^k + a^jb^kd^i + a^kb^id^j + \\
& a^kb^jd^i + a^ib^id^j + a^ib^jd^j + a^ib^id^k + a^ib^kd^k + a^jb^jd^i + a^jb^id^i + \\
& a^ib^jd^i + a^ib^kd^i + a^jb^id^j
\end{aligned}
$$

$$g = c^i + d^i + b^i c^i + b^i c^j + b^i c^k + b^j c^i + a^i d^i + a^i d^j + a^i d^k + a^j d^i + c^i d^i + c^i d^j +$$
$$c^i d^k + c^j d^i + a^i b^i d^i + a^i b^j d^k + a^i b^k d^j + a^j b^i d^k + a^j b^k d^i + a^k b^i d^j +$$
$$a^k b^j d^i + a^i b^i d^j + a^i b^j d^j + a^i b^i d^k + a^i b^k d^k + a^j b^j d^i + a^j b^i d^i + a^i b^j d^i +$$
$$a^i b^k d^i + a^j b^i d^j + b^i c^i d^i + b^i c^j d^k + b^i c^k d^j + b^j c^i d^k + b^j c^k d^i + b^k c^i d^j +$$
$$b^k c^j d^i + b^i c^i d^j + b^i c^j d^j + b^i c^i d^k + b^i c^k d^k + b^j c^j d^i + b^j c^i d^i + b^i c^j d^i +$$
$$b^i c^k d^i + b^j c^i d^j$$

$$h = b^i c^i + b^i c^j + b^i c^k + b^j c^i + a^i d^i + a^i d^j + a^i d^k + a^j d^i + b^i d^i + b^i d^j + b^i d^k +$$
$$b^j d^i + c^i d^i + c^i d^j + c^i d^k + c^j d^i + a^i b^i d^i + a^i b^j d^k + a^i b^k d^j + a^j b^i d^k +$$
$$a^j b^k d^i + a^k b^i d^j + a^k b^j d^i + a^i b^i d^j + a^i b^j d^j + a^i b^i d^k + a^i b^k d^k + a^j b^j d^i +$$
$$a^j b^i d^i + a^i b^j d^i + a^i b^k d^i + a^j b^i d^j + a^i c^i d^i + a^i c^j d^k + a^i c^k d^j + a^j c^i d^k +$$
$$a^j c^k d^i + a^k c^i d^j + a^k c^j d^i + a^i c^i d^j + a^i c^j d^j + a^i c^i d^k + a^i c^k d^k + a^j c^j d^i +$$
$$a^j c^i d^i + a^i c^j d^i + a^i c^k d^i + a^j c^i d^j + b^i c^i d^i + b^i c^j d^k + b^i c^k d^j + b^j c^i d^k +$$
$$b^j c^k d^i + b^k c^i d^j + b^k c^j d^i + b^i c^i d^j + b^i c^j d^j + b^i c^i d^k + b^i c^k d^k + b^j c^j d^i +$$
$$b^j c^i d^i + b^i c^j d^i + b^i c^k d^i + b^j c^i d^j$$

By implementing four instances of this component function $f_{\mathcal{C}_{231}}^{2,3,4}(.,.,.)$, $f_{\mathcal{C}_{231}}^{3,4,1}(.,.,.)$, $f_{\mathcal{C}_{231}}^{4,1,2}(.,.,.)$, and $f_{\mathcal{C}_{231}}^{1,2,3}(.,.,.)$ we reach a correct, non-complete, but non-uniform sharing of $\mathcal{C}_{231}$. Note that the 64-bit masks $m_1$, $m_2$, and $m_3$ required to share the last input round are independent of the masks used to share the cipher input.

542    A. Moradi and T. Schneider

# B    Round-Based Designs

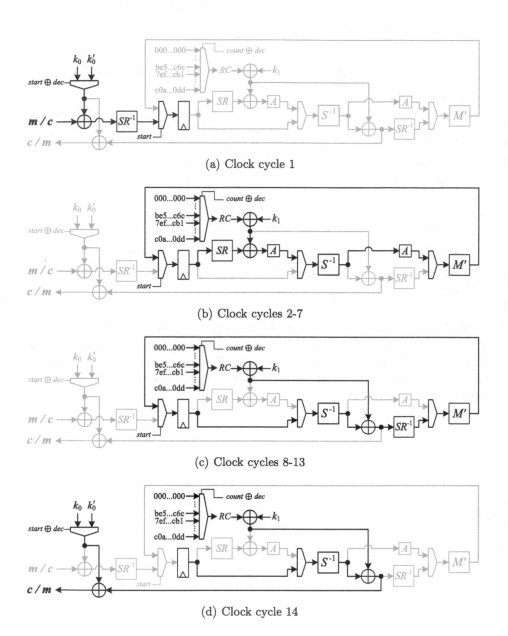

(a) Clock cycle 1

(b) Clock cycles 2-7

(c) Clock cycles 8-13

(d) Clock cycle 14

**Fig. 17.** Detailed active parts of our round-based architecture

# C    Decomposition of $\mathcal{C}_{231}$

**Table 2.** All possible ways to decompose $\mathcal{C}_{231}$ by selected quadratic bijections in three stages

| $\mathcal{C}_{231}$ | $\mathcal{C}_{150}$ | $\mathcal{C}_{151}$ | $\mathcal{C}_{158}$ | $\mathcal{C}_{159}$ | $\mathcal{C}_{168}$ | $\mathcal{C}_{171}$ | $\mathcal{C}_{172}$ | $\mathcal{C}_{214}$ | $\mathcal{C}_{215}$ | $\mathcal{C}_{223}$ | $\mathcal{C}_{262}$ | $\mathcal{C}_{266}$ | $\mathcal{C}_{296}$ | $\mathcal{C}_{297}$ |
|---|---|---|---|---|---|---|---|---|---|---|---|---|---|---|
| $\mathcal{Q}_4$ | | | × | × | | | | | | | | | | |
| $\mathcal{Q}_{12}$ | × | × | × | × | × | × | × | | | × | × | × | × | × |
| $\mathcal{Q}_{293}$ | × | × | × | × | × | × | × | × | | × | × | × | × | × |
| $\mathcal{Q}_{294}$ | × | × | × | × | × | × | × | | | × | × | × | | × |
| $\mathcal{Q}_{299}$ | × | × | × | × | × | × | × | | × | × | × | × | × | × |

| | |
|---|---|
| $\mathcal{C}_{150}$ : | $\mathcal{Q}_{12} \times \mathcal{Q}_{293}$ |
| $\mathcal{C}_{151}$ : | $\mathcal{Q}_{293} \times \mathcal{Q}_{12}$ |
| $\mathcal{C}_{158}$ : | $\mathcal{Q}_{299} \times \mathcal{Q}_{293}$ |
| $\mathcal{C}_{159}$ : | $\mathcal{Q}_{293} \times \mathcal{Q}_{299}$ |
| $\mathcal{C}_{168}$ : | $\mathcal{Q}_{293} \times \mathcal{Q}_{293}$ |

| | | |
|---|---|---|
| $\mathcal{C}_{171}$ : | $\mathcal{Q}_{293} \times \mathcal{Q}_{12}$ | $\mathcal{Q}_{294} \times \mathcal{Q}_{293}$ |
| $\mathcal{C}_{172}$ : | $\mathcal{Q}_{12} \times \mathcal{Q}_{293}$ | $\mathcal{Q}_{293} \times \mathcal{Q}_{294}$ |

| | | | | | | |
|---|---|---|---|---|---|---|
| $\mathcal{C}_{214}$ : | $\mathcal{Q}_4 \times \mathcal{Q}_{299}$ | $\mathcal{Q}_{12} \times \mathcal{Q}_{12}$ | $\mathcal{Q}_{12} \times \mathcal{Q}_{294}$ | $\mathcal{Q}_{12} \times \mathcal{Q}_{299}$ | $\mathcal{Q}_{293} \times \mathcal{Q}_4$ | $\mathcal{Q}_{293} \times \mathcal{Q}_{12}$ |
| | $\mathcal{Q}_{293} \times \mathcal{Q}_{294}$ | $\mathcal{Q}_{293} \times \mathcal{Q}_{299}$ | $\mathcal{Q}_{294} \times \mathcal{Q}_{12}$ | $\mathcal{Q}_{294} \times \mathcal{Q}_{294}$ | $\mathcal{Q}_{294} \times \mathcal{Q}_{299}$ | |

| | | | | | | |
|---|---|---|---|---|---|---|
| $\mathcal{C}_{215}$ : | $\mathcal{Q}_4 \times \mathcal{Q}_{293}$ | $\mathcal{Q}_{12} \times \mathcal{Q}_{12}$ | $\mathcal{Q}_{12} \times \mathcal{Q}_{293}$ | $\mathcal{Q}_{12} \times \mathcal{Q}_{294}$ | $\mathcal{Q}_{294} \times \mathcal{Q}_{12}$ | $\mathcal{Q}_{294} \times \mathcal{Q}_{293}$ |
| | $\mathcal{Q}_{294} \times \mathcal{Q}_{294}$ | $\mathcal{Q}_{299} \times \mathcal{Q}_4$ | $\mathcal{Q}_{299} \times \mathcal{Q}_{12}$ | $\mathcal{Q}_{299} \times \mathcal{Q}_{293}$ | $\mathcal{Q}_{299} \times \mathcal{Q}_{294}$ | |

| | | | | | |
|---|---|---|---|---|---|
| $\mathcal{C}_{223}$ : | $\mathcal{Q}_{12} \times \mathcal{Q}_{299}$ | $\mathcal{Q}_{293} \times \mathcal{Q}_{293}$ | $\mathcal{Q}_{293} \times \mathcal{Q}_{294}$ | $\mathcal{Q}_{294} \times \mathcal{Q}_{293}$ | $\mathcal{Q}_{294} \times \mathcal{Q}_{294}$ | $\mathcal{Q}_{299} \times \mathcal{Q}_{12}$ |
| | $\mathcal{Q}_{299} \times \mathcal{Q}_{299}$ | | | | | |

| | | | | |
|---|---|---|---|---|
| $\mathcal{C}_{262}$ : | $\mathcal{Q}_{12} \times \mathcal{Q}_{299}$ | $\mathcal{Q}_{294} \times \mathcal{Q}_{299}$ | $\mathcal{Q}_{299} \times \mathcal{Q}_{12}$ | $\mathcal{Q}_{299} \times \mathcal{Q}_{294}$ |
| $\mathcal{C}_{266}$ : | $\mathcal{Q}_{12} \times \mathcal{Q}_{12}$ | $\mathcal{Q}_{294} \times \mathcal{Q}_{299}$ | $\mathcal{Q}_{299} \times \mathcal{Q}_{294}$ | $\mathcal{Q}_{299} \times \mathcal{Q}_{299}$ |

| | | | | | |
|---|---|---|---|---|---|
| $\mathcal{C}_{296}$ : | $\mathcal{Q}_{12} \times \mathcal{Q}_{299}$ | $\mathcal{Q}_{293} \times \mathcal{Q}_{293}$ | $\mathcal{Q}_{294} \times \mathcal{Q}_{12}$ | $\mathcal{Q}_{299} \times \mathcal{Q}_{294}$ | $\mathcal{Q}_{299} \times \mathcal{Q}_{299}$ |
| $\mathcal{C}_{297}$ : | $\mathcal{Q}_{12} \times \mathcal{Q}_{294}$ | $\mathcal{Q}_{293} \times \mathcal{Q}_{293}$ | $\mathcal{Q}_{294} \times \mathcal{Q}_{299}$ | $\mathcal{Q}_{299} \times \mathcal{Q}_{12}$ | $\mathcal{Q}_{299} \times \mathcal{Q}_{299}$ |

# References

1. Side-channel attack user reference architecture. http://satoh.cs.uec.ac.jp/SAKURA/index.html
2. Balasch, J., Gierlichs, B., Verdult, R., Batina, L., Verbauwhede, I.: Power analysis of atmel cryptomemory – recovering keys from secure EEPROMs. In: Dunkelman, O. (ed.) CT-RSA 2012. LNCS, vol. 7178, pp. 19–34. Springer, Heidelberg (2012). doi:10.1007/978-3-642-27954-6_2
3. Banik, S., Bogdanov, A., Isobe, T., Shibutani, K., Hiwatari, H., Akishita, T., Regazzoni, F.: Midori: a block cipher for low energy. In: Iwata, T., Cheon, J.H. (eds.) ASIACRYPT 2015. LNCS, vol. 9453, pp. 411–436. Springer, Heidelberg (2015). doi:10.1007/978-3-662-48800-3_17

4. Banik, S., Bogdanov, A., Regazzoni, F.: Exploring energy efficiency of lightweight block ciphers. In: Dunkelman, O., Keliher, L. (eds.) SAC 2015. LNCS, vol. 9566, pp. 178–194. Springer, Heidelberg (2016). doi:10.1007/978-3-319-31301-6_10

5. Bhasin, S., Guilley, S., Flament, F., Selmane, N., Danger, J.: Countering early evaluation: an approach towards robust dual-rail precharge logic. In: Workshop on Embedded Systems Security - WESS 2010, p. 6. ACM (2010)

6. Bhasin, S., Guilley, S., Sauvage, L., Danger, J.-L.: Unrolling cryptographic circuits: a simple countermeasure against side-channel attacks. In: Pieprzyk, J. (ed.) CT-RSA 2010. LNCS, vol. 5985, pp. 195–207. Springer, Heidelberg (2010). doi:10.1007/978-3-642-11925-5_14

7. Bilgin, B., Gierlichs, B., Nikova, S., Nikov, V., Rijmen, V.: A more efficient AES threshold implementation. In: Pointcheval, D., Vergnaud, D. (eds.) AFRICACRYPT 2014. LNCS, vol. 8469, pp. 267–284. Springer, Heidelberg (2014). doi:10.1007/978-3-319-06734-6_17

8. Bilgin, B., Gierlichs, B., Nikova, S., Nikov, V., Rijmen, V.: Higher-order threshold implementations. In: Sarkar, P., Iwata, T. (eds.) ASIACRYPT 2014. LNCS, vol. 8874, pp. 326–343. Springer, Heidelberg (2014). doi:10.1007/978-3-662-45608-8_18

9. Bilgin, B., Gierlichs, B., Nikova, S., Nikov, V., Rijmen, V.: Trade-offs for threshold implementations illustrated on AES. IEEE Trans. CAD Integr. Circ. Syst. **34**(7), 1188–1200 (2015)

10. Bilgin, B., Nikova, S., Nikov, V., Rijmen, V., Stütz, G.: Threshold implementations of All 3×3 and 4×4 S-boxes. In: Prouff, E., Schaumont, P. (eds.) CHES 2012. LNCS, vol. 7428, pp. 76–91. Springer, Heidelberg (2012). doi:10.1007/978-3-642-33027-8_5

11. Bilgin, B., Nikova, S., Nikov, V., Rijmen, V., Tokareva, N., Vitkup, V.: Threshold implementations of small S-boxes. Crypt. Commun. **7**(1), 3–33 (2015)

12. Bogdanov, A., Knudsen, L.R., Leander, G., Paar, C., Poschmann, A., Robshaw, M.J.B., Seurin, Y., Vikkelsoe, C.: PRESENT: an ultra-lightweight block cipher. In: Paillier, P., Verbauwhede, I. (eds.) CHES 2007. LNCS, vol. 4727, pp. 450–466. Springer, Heidelberg (2007). doi:10.1007/978-3-540-74735-2_31

13. Borghoff, J., Canteaut, A., Güneysu, T., Kavun, E.B., Knezevic, M., Knudsen, L.R., Leander, G., Nikov, V., Paar, C., Rechberger, C., Rombouts, P., Thomsen, S.S., Yalçın, T.: PRINCE – a low-latency block cipher for pervasive computing applications. In: Wang, X., Sako, K. (eds.) ASIACRYPT 2012. LNCS, vol. 7658, pp. 208–225. Springer, Heidelberg (2012). doi:10.1007/978-3-642-34961-4_14

14. Bouesse, G., Renaudin, M., Witon, A., Germain, F.: A clock-less low-voltage AES crypto-processor. In: Proceedings of the 31st European Solid-State Circuits Conference, ESSCIRC 2005, pp. 403–406. IEEE (2005)

15. Bouesse, G.F., Renaudin, M., Dumont, S., Germain, F.: DPA on quasi delay insensitive asynchronous circuits: formalization and improvement. In: DATE, pp. 424–429. IEEE Computer Society (2005)

16. Bouesse, F., Renaudin, M., Sicard, G.: Improving DPA resistance of quasi delay insensitive circuits using randomly time-shifted acknowledgment signals. In: Reis, R., Osseiran, A., Pfleiderer, H.-J. (eds.) VLSI-SoC 2005. IIFIP, vol. 240, pp. 11–24. Springer, Heidelberg (2007). doi:10.1007/978-0-387-73661-7_2

17. Bouesse, F., Sicard, G., Renaudin, M.: Path swapping method to improve DPA resistance of quasi delay insensitive asynchronous circuits. In: Goubin, L., Matsui, M. (eds.) CHES 2006. LNCS, vol. 4249, pp. 384–398. Springer, Heidelberg (2006). doi:10.1007/11894063_30

18. Cooper, J., Demulder, E., Goodwill, G., Jaffe, J., Kenworthy, G., Rohatgi, P.: Test vector leakage assessment (TVLA) methodology in practice. In: International Cryptographic Module Conference (2013)

19. Dean, M.E., Williams, T.E., Dill, D.L.: Efficient self-timing with level-encoded 2-phase dual-rail (LEDR). In: Conference on Advanced Research in VLSI, pp. 55–70. MIT Press (1991)

20. Ding, A.A., Chen, C., Eisenbarth, T.: Simpler, faster, and more robust t-test based leakage detection. In: Standaert, F.-X., Oswald, E. (eds.) COSADE 2016. LNCS, vol. 9689, pp. 163–183. Springer, Heidelberg (2016). doi:10.1007/978-3-319-43283-0_10

21. Durvaux, F., Standaert, F.-X., Del Pozo, S.M.: Towards easy leakage certification. In: Gierlichs, B., Poschmann, A.Y. (eds.) CHES 2016. LNCS, vol. 9813, pp. 40–60. Springer, Heidelberg (2016). doi:10.1007/978-3-662-53140-2_3

22. Eisenbarth, T., Kasper, T., Moradi, A., Paar, C., Salmasizadeh, M., Shalmani, M.T.M.: On the power of power analysis in the real world: a complete break of the KEELOQ code hopping scheme. In: Wagner, D. (ed.) CRYPTO 2008. LNCS, vol. 5157, pp. 203–220. Springer, Heidelberg (2008). doi:10.1007/978-3-540-85174-5_12

23. Fournier, J.J.A., Moore, S., Li, H., Mullins, R., Taylor, G.: Security evaluation of asynchronous circuits. In: Walter, C.D., Koç, Ç.K., Paar, C. (eds.) CHES 2003. LNCS, vol. 2779, pp. 137–151. Springer, Heidelberg (2003). doi:10.1007/978-3-540-45238-6_12

24. Goodwill, G., Jun, B., Jaffe, J., Rohatgi, P.: A testing methodology for side channel resistance validation. In: NIST non-invasive attack testing workshop (2011). http://csrc.nist.gov/news_events/non-invasive-attack-testing-workshop/papers/08_Goodwill.pdf

25. Guo, J., Jean, J., Nikolić, I., Qiao, K., Sasaki, Y., Sim, S.M.: Invariant subspace attack against full Midori64. Cryptology ePrint Archive, Report 2015/1189 (2015). http://eprint.iacr.org/

26. Kilian, J., Rogaway, P.: How to protect DES against exhaustive key search. In: Koblitz, N. (ed.) CRYPTO 1996. LNCS, vol. 1109, pp. 252–267. Springer, Heidelberg (1996). doi:10.1007/3-540-68697-5_20

27. Knezevic, M., Nikov, V., Rombouts, P.: Low-latency encryption - Is "Lightweight = Light + Wait"? In: Prouff, E., Schaumont, P. (eds.) CHES 2012. LNCS, vol. 7428, pp. 426–446. Springer, Heidelberg (2012)

28. Knudsen, L., Leander, G., Poschmann, A., Robshaw, M.J.B.: PRINTCIPHER: a block cipher for ic-printing. In: Mangard, S., Standaert, F.-X. (eds.) CHES 2010. LNCS, vol. 6225, pp. 16–32. Springer, Heidelberg (2010). doi:10.1007/978-3-642-15031-9_2

29. Kocher, P., Jaffe, J., Jun, B.: Differential power analysis. In: Wiener, M. (ed.) CRYPTO 1999. LNCS, vol. 1666, pp. 388–397. Springer, Heidelberg (1999). doi:10.1007/3-540-48405-1_25

30. Kulikowski, K.J., Su, M., Smirnov, A.B., Taubin, A., Karpovsky, M.G., MacDonald, D.: Delay insensitive encoding and power analysis: a balancing act. In: ASYNC, pp. 116–125. IEEE Computer Society (2005)

31. Liu, J., Yu, Y., Standaert, F.-X., Guo, Z., Gu, D., Sun, W., Ge, Y., Xie, X.: Small tweaks do not help: differential power analysis of MILENAGE implementations in 3G/4G USIM cards. In: Pernul, G., Ryan, P.Y.A., Weippl, E. (eds.) ESORICS 2015. LNCS, vol. 9326, pp. 468–480. Springer, Heidelberg (2015). doi:10.1007/978-3-319-24174-6_24

32. Mangard, S., Oswald, E., Popp, T.: Power Analysis Attacks: Revealing the Secrets of Smart Cards. Springer, USA (2007)

33. Moore, S.W., Mullins, R.D., Cunningham, P.A., Anderson, R.J., Taylor, G.S.: Improving smart card security using self-timed circuits. In: ASYNC, pp. 211–218. IEEE Computer Society (2002)
34. Moradi, A., Barenghi, A., Kasper, T., Paar, C.: On the vulnerability of FPGA bitstream encryption against power analysis attacks: extracting keys from xilinx Virtex-II FPGAs. In: ACM Conference on Computer and Communications Security - CCS 2011, pp. 111–124. ACM (2011)
35. Moradi, A., Immler, V.: Early propagation and imbalanced routing, how to diminish in FPGAs. In: Batina, L., Robshaw, M. (eds.) CHES 2014. LNCS, vol. 8731, pp. 598–615. Springer, Heidelberg (2014). doi:10.1007/978-3-662-44709-3_33
36. Moradi, A., Mischke, O., Paar, C.: Practical evaluation of DPA countermeasures on reconfigurable hardware. In: HOST 2011, pp. 154–160. IEEE (2011)
37. Moradi, A., Poschmann, A., Ling, S., Paar, C., Wang, H.: Pushing the limits: a very compact and a threshold implementation of AES. In: Paterson, K.G. (ed.) EUROCRYPT 2011. LNCS, vol. 6632, pp. 69–88. Springer, Heidelberg (2011). doi:10.1007/978-3-642-20465-4_6
38. Moradi, A., Wild, A.: Assessment of hiding the higher-order leakages in hardware. In: Güneysu, T., Handschuh, H. (eds.) CHES 2015. LNCS, vol. 9293, pp. 453–474. Springer, Heidelberg (2015). doi:10.1007/978-3-662-48324-4_23
39. Myers, C.J.: Asynchronous Circuit Design. Wiley, New York (2001)
40. Nikova, S., Rijmen, V., Schläffer, M.: Secure hardware implementation of nonlinear functions in the presence of glitches. J. Cryptology **24**(2), 292–321 (2011)
41. Oswald, D., Paar, C.: Breaking mifare desfire MF3ICD40: power analysis and templates in the real world. In: Preneel, B., Takagi, T. (eds.) CHES 2011. LNCS, vol. 6917, pp. 207–222. Springer, Heidelberg (2011). doi:10.1007/978-3-642-23951-9_14
42. Poschmann, A., Moradi, A., Khoo, K., Lim, C., Wang, H., Ling, S.: Side-channel resistant crypto for less than 2, 300 GE. J. Cryptology **24**(2), 322–345 (2011)
43. Poschmann, A.Y.: Lightweight cryptography: cryptographic engineering for a pervasive world. Ph.D. thesis, Ruhr University Bochum (2009)
44. Schneider, T., Moradi, A.: Leakage assessment methodology — a clear roadmap for side-channel evaluations. In: Güneysu, T., Handschuh, H. (eds.) CHES 2015. LNCS, vol. 9293, pp. 495–513. Springer, Heidelberg (2015). doi:10.1007/978-3-662-48324-4_25
45. Spars, J., Furber, S.: Principles of Asynchronous Circuit Design: A Systems Perspective, 1st edn. Springer Publishing Company, Incorporated, USA (2010)
46. Standaert, F.-X., Örs, S.B., Preneel, B.: Power analysis of an FPGA: implementation of rijndael: is pipelining a DPA countermeasure? In: Joye, M., Quisquater, J.-J. (eds.) CHES 2004. LNCS, vol. 3156, pp. 30–44. Springer, Heidelberg (2004). doi:10.1007/978-3-540-28632-5_3
47. Suzuki, D., Saeki, M.: Security evaluation of DPA countermeasures using dual-rail pre-charge logic style. In: Goubin, L., Matsui, M. (eds.) CHES 2006. LNCS, vol. 4249, pp. 255–269. Springer, Heidelberg (2006). doi:10.1007/11894063_21
48. Tiri, K., Verbauwhede, I.: A logic level design methodology for a secure DPA resistant ASIC or FPGA implementation. In: Design, Automation and Test in Europe - DATE 2004, pp. 246–251. IEEE Computer Society (2004)
49. Virtual Silicon Inc.: 0.18 µm VIP standard cell library tape out ready, Part number: UMCL18G212T3, Process: UMC Logic 0.18 µm Generic II Technology: 0.18 µm, July 2004
50. Wu, J., Kim, Y., Choi, M.: Low-power side-channel attack-resistant asynchronous s-box design for AES cryptosystems. In: ACM Great Lakes Symposium on VLSI, pp. 459–464. ACM (2010)

51. Yli-Mäyry, V., Homma, N., Aoki, T.: Improved power analysis on unrolled architecture and its application to PRINCE block cipher. In: Güneysu, T., Leander, G., Moradi, A. (eds.) LightSec 2015. LNCS, vol. 9542, pp. 148–163. Springer, Heidelberg (2016). doi:10.1007/978-3-319-29078-2_9
52. Yu, A., Brée, D.S.: A clock-less implementation of the AES resists to power and timing attacks. In: ITCC (2), pp. 525–532. IEEE Computer Society (2004)
53. Yu, Z.C., Furber, S.B., Plana, L.A.: An investigation into the security of self-timed circuits. In: ASYNC, pp. 206–215. IEEE Computer Society (2003)
54. Zhou, Y., Yu, Y., Standaert, F.-X., Quisquater, J.-J.: On the need of physical security for small embedded devices: a case study with COMP128-1 implementations in SIM cards. In: Sadeghi, A.-R. (ed.) FC 2013. LNCS, vol. 7859, pp. 230–238. Springer, Heidelberg (2013). doi:10.1007/978-3-642-39884-1_20

# Characterisation and Estimation of the Key Rank Distribution in the Context of Side Channel Evaluations

Daniel P. Martin[1]([⊠]), Luke Mather[2], Elisabeth Oswald[1], and Martijn Stam[1]

[1] Department of Computer Science, University of Bristol,
Merchant Venturers Building, Woodland Road, Bristol BS8 1UB, UK
{dan.martin,elisabeth.oswald,martijn.stam}@bris.ac.uk
[2] HP Labs, Bristol, UK
luke.mather@bris.ac.uk

**Abstract.** Quantifying the side channel security of implementations has been a significant research question for several years in academia but also among real world side channel practitioners. As part of security evaluations, efficient key rank estimation algorithms were devised, which in contrast to analyses based on subkey recovery, give a holistic picture of the security level after a side channel attack. However, it has been observed that outcomes of rank estimations show a huge spread in precisely the range of key ranks where enumeration could lead to key recovery. These observations raise the question whether this is because of insufficient rank estimation procedures, or, if this is an inherent property of the key rank. Furthermore, if this was inherent, how could key rank outcomes be translated into practically meaningful figures, suitable to analysing the risk that real world side channel attacks pose? This paper is a direct response to these questions. We experimentally identify the key rank distribution and show that it is independent of different distinguishers and signal-to-noise ratios. Then we offer a theoretical explanation for the observed key rank distribution and determine how many samples thereof are required for a robust estimation of some key parameters. We discuss how this can be naturally integrated into real world side channel evaluation practices. We conclude our research by connecting non-parametric order statistics, in particular percentiles, in a practically meaningful way with business goals.

## 1 Introduction

To assess the outcome of an attack, researchers traditionally sought to determine the attack's success rate (SR). Standaert et al. [20] provided a formal definition for the SR and hypothesised that there is a link between attack outcomes (the success rate, assuming a single targeted intermediate value) and the leakage (measured in information theoretic terms in the same intermediate value). Further research aimed at characterising the SR, e.g. [18,21], or finding alternative

---

The research was carried out whilst L. Mather was employed at the University of Bristol.

J.H. Cheon and T. Takagi (Eds.): ASIACRYPT 2016, Part I, LNCS 10031, pp. 548–572, 2016.
DOI: 10.1007/978-3-662-53887-6_20

ways to predict differential power analysis (DPA) outcomes, e.g. [8]. These contributions brought much needed clarity about some aspects of the (interactions) between target functions and leakage models, but (necessarily) had to restrict themselves to considering attack outcomes for *a single subkey* only.

In practice however, the effort to reveal the *entire* secret key is the concern of most primacy: given a number of traces, and a computational budget for key enumeration, what is the likelihood to reveal the secret key? This question can be answered both by a generalised SR (which is closely connected to the key guessing entropy (GE, see [20]), this line of research has recently been developed further by Duc et al. [6].) or by computing the *rank* of the secret key. Consequently, fast methods to compute the rank of the secret key have become a hot topic [2,3,10, 14,24,25].

It is noteworthy that the first computationally efficient and accurate key estimation algorithm originated from an evaluation lab [10]. Their interest in the topic explains itself easily: assuming a sufficiently accurate method to estimate the true rank of the secret key, decisive leakage evaluations could be performed. However, the existing research brought to light an (unexpected) difficulty along the way: even though the aforementioned previous works sought to minimise the estimation error in key rank algorithms, the derived key ranks show a huge spread in exactly the range of ranks where enumeration is of practical importance. This opens up the question whether these ranks actually give meaningful information? And if so how would key rank computations be integrated in standardised security evaluations? The potential implication of these recent research results have prompted JHAS (JIL Hardware-related Attacks Subgroup, this industry led group essentially defines Common Criteria security evaluation practises for smart card products) to set up a specific working group that deals with the topic.

Our research offers answers to these questions: after introducing some background (Sect. 2) we improve the key rank algorithm of Martin et al. [14] to produce the (to date) most precise key ranking algorithm (Sect. 3). Using this high-precision ranking algorithm, we focus on the properties of the key rank distribution: we begin with an experimental exploration of the key rank, which we accompany and strengthen by a theoretical analysis. Then, drawing from carefully designed simulations, we justify some general observations about the key rank such as the independence of side channel distinguisher and trace characteristics. We evaluate statistical metrics for the purpose of quantifying the risk from side-channel attacks through an "evaluation through rank estimation" approach and relate it to (potential) business goals.

## 2   Side-Channel Evaluations and Key Rank

This section covers some basic notation related to differential power analysis (DPA) style attacks on modern blockciphers, as well as surveying the recent works on computing fast and accurate estimates for the key rank.

We use a bold type face to denote multi-dimensional variables. A key $\mathbf{k}$ can be partitioned into $m$ (independent) subkeys, which we denote as

$\mathbf{k} = (k^0, \dots, k^{m-1})$. We assume that all subkeys are of the same size (which holds in most scenarios in practice) and that each subkey can take one of $n$ possible values. As an example, for AES-128 typically the 128-bit key is subdivided in $m = 16$ subkeys of a byte ($n = 256$) each. The key to be recovered by the DPA attack is called the secret key and is denoted $\mathbf{sk} = (sk^0, \dots, sk^{m-1})$.

## 2.1  Standard DPA Model

In this paper we consider a standard DPA scenario as in Mangard et al. [13], which implies the attacks are single order and univariate. (Note that in higher order attacks the univariate targets still fit a standard DPA attack). An attacker has $N$ power measurements or traces $T_i$ corresponding to encryptions of $N$ known plaintexts $x_i \in \mathcal{X}$, $i = 1, \dots, N$ and wishes to recover the secret key $\mathbf{sk}$.

For each subkey ($j = 0, \dots, m-1$) we assume that each trace $T_i$ is condensed to a single point of interest $P_{i,j}$ and that this value $P_{i,j}$ decomposes additively as $P_{i,j} = P_{\exp} + P_{\text{noise}}$. Here $P_{\exp}$, called the *signal*, is a deterministic function of the value of the subkey $sk^j$ and the relevant input $x_i$, whereas $P_{\text{noise}}$, called the *noise*, is drawn at random according to some distribution that does not depend on any of the input values (including the secret key $\mathbf{sk}$). The signal-to-noise ratio (SNR) is then defined as the ratio of the variance in the signal (when ranging over secret keys and plaintexts) divided by the variance in the noise:[1]

$$\text{SNR} = \frac{Var(P_{\exp})}{Var(P_{\text{noise}})}.$$

The SNR is used to quantify the amount of leakage within a given measurement: the higher the SNR, the more information within the trace that can be exploited.

A distinguisher $D^j$ against the $j^{\text{th}}$ subkey takes as input the vector of condensed traces and corresponding plaintexts $(P_{i,j}, x_i)_{i=1,\dots,N}$ and outputs a distinguishing vector $\mathbf{D}^j \in \mathbb{R}^n$, which assigns a score for each possible hypothesis of the subkey under consideration. Without loss of generality, we will assume that the higher the score for a subkey hypothesis $k^j$, the more likely the distinguisher deems that secret key equals $k^j$. The distinguisher $D$ on the complete key, simply runs the subkey distinguishers $D^j$ for each subkey and outputs a list of distinguishing vectors $\mathbf{D} \in \mathbb{R}^{n \times m}$ (namely a distinguishing vector for each subkey).

## 2.2  Key Rank

The result of a side channel attack is a set of distinguishing vectors, which hold the information about subkeys (when studied individually) and the entire key

---

[1] Strictly speaking the SNR is defined relative to a subkey and should be indexed by $j$; however when we later refer to the SNR it will be the same for all subkeys. This is a simplifying assumption we make for our simulated data. It may not hold (nor do we require it to) on real devices.

(when studied jointly). To judge the potency of an attack, we need suitable metrics to express how well the distinguishing vectors enable key recovery.

Even though the ultimate goal is full key recovery, historically the emphasis has been on subkey recovery. The only relevant information in a subkey distinguishing vector $\mathbf{D}^j$ is the order it induces on possible subkey hypothesis, as a clever adversary would test the subkeys in order of likelihood (ignoring for a moment how one would test an individual subkey). The only information needed to identify the true subkey $sk^j$ in this ordering is its distinguishing score $d_{sk^j,j}$, leading to the following definition of subkey rank.

**Definition 1 (Subkey rank).** *Given the distinguishing vector $\mathbf{D}^j$, and the distinguishing score $d_{sk^j,j}$ for subkey $sk^j$, count the number of subkeys with score strictly larger than $d_{sk^j,j}$. We denote this $\mathrm{rank}^j_{sk^j}(\mathbf{D}^j)$.*

Extending subkey rank to a full key is based on the assumption that the distinguishing scores for individual subkeys can be added to give a meaningful score for the full score. For instance, given the distinguishing table $\mathbf{D} = (\mathbf{D}^0, \ldots, \mathbf{D}^{m-1})$ for the entire key, the score of secret key $\mathbf{sk}$ is computed as $W = \sum_{j=0}^{m-1} d_{sk^j,j}$ (where the notation $d_{sk^j,j}$ identifies the score corresponding to $sk^j$ in the distinguishing vector $\mathbf{D}^j$). In this case the actual values in the (subkey) distinguishing vectors becomes relevant.

**Definition 2 (Key rank).** *Given the distinguishing table $\mathbf{D}$, and the score $W$ of the secret key $\mathbf{sk}$, count the number of keys with score strictly larger than $W$. This is denoted $\mathrm{rank}_{\mathbf{sk}}(\mathbf{D})$.*

*Remark 1.* If multiple keys have the same score as the secret key, we assume that the latter is ranked first. This gives a conservative rank for a given distinguishing vector, as it will be the earliest an adversary would enumerate the key. For distinguishers that don't actually distinguish that well (e.g. because they do not exploit any leakage) this can lead to key ranks that significantly underestimate the remaining effort to recover the full key.

The key rank $\mathrm{rank}_{\mathbf{sk}}(\mathbf{D})$ of a single secret key given a specific distinguishing table is not particularly interesting on its own. To say something meaningful, we will consider the key rank as a random variable that is the outcome of the experiment in Fig. 1. Here a random key, random plaintexts, and (implicitly) random noise in the measurements $T_i$ are chosen, as a result of which the output of the experiment is a random variable. We will denote this random variable $\mathrm{keyrank}_D(N)$, which highlights the dependency on the number of traces $N$ and the distinguisher $D$ being used; obviously the experiment depends on the primitive under attack, and how it leaks, as well. The random variable $\mathrm{keyrank}^j_D(N)$ denotes the rank of the $j^{\mathrm{th}}$ subkey (that is, the experiment returns $\mathrm{rank}^j_{sk^j}(\mathbf{D}^j)$ instead).

**Definition 3 (Success rate).** *The success rate of a distinguisher $D$ as a function of the number of traces $N$ is defined as $SR_D(N) = \Pr[\mathrm{keyrank}_D(N) = 0]$, where the random variable $\mathrm{keyrank}_D(N)$ is defined by Fig. 1.*

**experiment Keyrank$_D(N)$:**

$\mathbf{sk} \xleftarrow{\$} \mathcal{K}$

$x_1, \ldots, x_N \xleftarrow{\$} \mathcal{X}$

For each $x_i$ capture trace $T_i$ leading to points
of interest $P_{i,j}$

For each subkey: $\mathbf{D}^j \leftarrow D^j((P_{i,j}, x_i)_{i=1,\ldots,N})$

$\mathbf{D} \leftarrow \{\mathbf{D}^j\}_{j=1}^m$

**return** $\text{rank}_{\mathbf{sk}}(\mathbf{D})$

**Fig. 1.** The key rank experiment leading to random variable keyrank$_D(N)$.

The success rate, or first-order success rate, captures how frequently the secret key $\mathbf{sk}$ is deemed (among) the most likely by the distinguisher. Given that the score for a full key is computed as the sum of its constituent subkey scores, a full key is deemed the most likely if, and only if, all its constituent subkeys are the most likely. Thus, when focusing on success rate, it suffices to look at the (first-order) *subkey* success rate.

Unfortunately, judging a distinguisher by its success rate only ignores key recovery attacks that include key enumeration as part of their strategy. One could look at higher-order success rates, where for the $M$-th order key recovery, the $M$ highest ranked key guesses are tested using a known plaintext–ciphertext pair, though this raises the question for which $M$ (and for realistic but large $M$, say $M = 2^{50}$ computing the $M$-th order success rate is a challenge on its own). Instead, we suggest to maintain the notion of keyrank$_D(N)$ as a random variable and we will investigate its distribution as a whole. This allows us to identify those properties of the distribution crucial to a holistic assessment of the potency of a side channel attack.

*Remark 2.* While the random variable keyrank$_D(N)$ is defined over the randomness of key, plaintexts, and the noise in the measurement, we emphasize that it is really the latter that matters. Indeed, one could equally consider key rank in a (non-adaptive) chosen plaintext setting and later on we will make the assumption that the randomness of the key is irrelevant (namely that conditioning the random variable keyrank$_D(N)$ on the key $\mathbf{sk}$ makes no difference). This does mean that if the leakage is noise-free, looking at key rank as a random variable is not that meaningful anymore. Instead, the leakage will allow an adversary to determine a set (containing $\mathbf{sk}$) of most likely keys it considers equiprobable; the relevant metric in this case is the size of this set, not the rank as we defined it (which will default to 0).

## 2.3  Theoretical Characterization of the Key Rank Distribution

When comparing DPA distinguishers, it is customary to assume a specific leakage model (e.g. the Hamming weight of some intermediate value with Gaussian noise added). When the subkey distinguishing vector is considered as a random variable (cf. Fig. 1), its distribution is known [18]: it takes the shape of a multi-variate normal distribution. Using order statistics, this leads to a characterization of the subkey rank distribution. This distribution is not particularly insightful in its algebraic form, but it can be numerically evaluated in time proportional to the $n$ (the size of the subkey space). However, extending this characterization into one for the full rank is not possible as the subkey rank distribution does not uniquely determine the full key rank distribution. One could attempt to use order statistics directly on the full key distinguishing table. However, even if this were possible, the resulting formulae are likely unwieldy in their algebraic form; moreover, numerical evaluation would this time be proportional in $n^m$ (i.e. the size of the full key space) which will be infeasible for any cryptosystem of relevance. This renders a full theoretical derivation of the key rank distribution moot. Instead, let us concentrate on typical statistics used to describe distributions, starting with the expected value, or the guessing entropy. Later we will hypothesise a candidate distribution.

**Guessing Entropy.** First defined by Massey [15], the guessing entropy captures the expected number of guesses (with an optimum strategy) to correctly guess the value of a random variable (in our scenario the secret key). This can be linked to the key rank by observing that the key rank is the number of guesses an optimal adversary would take to guess the secret key. Standaert et al. [20] first made this connection. We use the definition as given by Rivain [18].

**Definition 4 (Subkey guessing entropy).** *The subkey guessing entropy is defined as the expected value of the subkey rank, namely*

$$GE_D^j(N) = \mathbb{E}(\text{keyrank}_D^j(N)).$$

A key observation is that the guessing entropy is the *expected value* of the distribution of the subkey rank. Rivain found that the distribution of a *distinguishing vector* tends to a multivariate Gaussian [18], but the general distribution of the subkey rank itself has not been thoroughly explored.

Extending the guessing entropy metric into the context of a full key is simple—we now are required to find the expected value of the key rank.

**Definition 5 (Key guessing entropy).** *The key guessing entropy is defined as:*

$$GE_D(N) = \mathbb{E}(\text{keyrank}_D(N)).$$

**Ranking Entropy.** In this work we consider adversaries that would employ key enumeration as part of their attack strategy. This raises the question of how best to consider the relative strength of two adversaries that have different sized key enumeration budgets. Most differential attacks are chosen plaintext attacks, and thus the cost of checking the validity of a single key hypothesis is almost zero—a single call to an encryption or decryption. Thus, as in classical cryptanalysis, it is perhaps more useful to compare enumeration budgets in terms of *orders of magnitude*, i.e. consider the logarithm of (a function of) the key rank outcomes.

Recall that the guessing entropy $GE_D(N)$ is defined as $\mathbb{E}(\text{keyrank}_D(N))$ To consider the orders of magnitude in relation to the guessing entropy, the obvious approach would be to consider $\log(GE_D(N)) = \log(\mathbb{E}(\text{keyrank}_D(N)))$. We will later show that this approach is not satisfactory. For that reason, we introduce here an alternative, which we call the *ranking* entropy. The ranking entropy is defined as the expectation of the logarithm of the rank, that is it equals $\mathbb{E}(\mathfrak{R})$, where $\mathfrak{R} = \log(\text{keyrank}_D(N))$ (for brevity, we will henceforth refer to $\text{keyrank}_D(N)$ simply by $R$). Note that taking logarithms and expectation do not commute, so in general the ranking entropy will not equal the log of the guessing entropy.

Calculating either the guessing entropy or the ranking entropy directly appears to be a hard problem. Instead, for this and other statistics we will resort to sampling from the distribution by repeatedly running the experiment of Fig. 1 instead. This requires an algorithm to calculate the key rank.

## 2.4  Key Rank Estimation

We want to understand the distributional properties of the key rank for different distinguishers and leakage scenarios. A key tool for our empirical investigation is an efficient and highly accurate rank estimation algorithm. Finding the rank of a subkey is trivial after sorting the distinguishing vector. Unfortunately, for the full key this approach no longer works as sorting the complete distinguishing vector for the full key is at least as expensive as exhaustive search on the full key. For instance, in case of a typical attack on AES, the distinguishing table consists of 16 distinguishing vectors of dimension 256 each. A naive (but accurate) algorithm would be to compute the product distribution (i.e. list all combinations of all subkeys) in order to compute the rank of the secret key.

There have been a host of more advanced key rank *estimation* algorithms that return either an interval containing the actual rank or a point estimate of the rank. When comparing such algorithms, both the efficiency and the accuracy are relevant. Accuracy is measured in bits, where $b$ bits of accuracy means that if an algorithm says the key has rank $2^x$, the actual rank is in the range $2^{x\pm b}$. Below we give a brief overview of existing key rank estimation algorithms.[2]

Veyrat-Charvillon et al. [24] proposed the first non-trivial key rank algorithm. They represent the distinguishing scores in a multi-dimensional space, where each dimension represents an individual distinguishing vector (sorted in descending order). This space can naturally be divided into two parts; those keys with rank higher than the target key and those with a rank lower. Using the property that the 'frontier' between these two halves is convex, the rank of the key can be estimated to within 10 bits by repeatedly pruning the space.

---

[2]   A small technical caveat: we do not make a distinction between worst-case accuracy and the more fuzzy typical-case accuracy.

Glowacz et al. [10] construct an efficient rank algorithm based on the convolution of histograms. They utilise the property that if $H_1$ is a histogram of $S_1$ and $H_2$ is a histogram of $S_2$ then the convolution of $H_1$ and $H_2$ is a suitable approximation of $S_1 + S_2 = \{x_1 + x_2 | x_1 \in S_1, x_2 \in S_2\}$. By representing the distinguishing vectors as histograms and using this property they are able to estimate the rank of the key to within one bit of accuracy.

Duc et al. [6] propose a similar solution to that of Glowacz et al. [10]. They repeatedly 'merge' each set of data in (similar to the histogram convolution) and then down-sample the resulting data (this can be seen as the binning step in creating histograms). Additionally, they down-sample to a fixed number of samples after each 'merge', instead of just on the original data. While Duc et al. do not explicitly give a bound on the estimation error, the additional down-sampling implies it will be worse than that of Glowacz et al.'s algorithm.

Bernstein et al. [2] propose two key rank algorithms. The first adds a post-processing phase to the algorithm by Veyrat-Charvillon et al. [24], which tightens the accuracy to 5 bits. The second algorithm uses techniques similar to counting all $y$-smooth numbers less than $x$. By having an accuracy parameter they are able to get the bound arbitrarily tight, at the expense of runtime.

Martin et al. [14] propose a key rank algorithm based on the pseudo-polynomial time algorithm for the knapsack problem. After mapping the distinguishing scores to integer weights (such that larger distinguishing scores give smaller integers), they are able to efficiently count the number of keys with a weight less than the target key which directly corresponds to the rank of the key. Varying the size of the resulting integers allows them to make a trade-off between accuracy and runtime.

All-but-one of these algorithms are essentially interval estimates of the key rank; the only exception being the algorithm by Martin et al., which provides a point estimate. Clearly all works emphasised the need of an accurate rank estimation to ensure that the resulting key ranks are practically meaningful. In some of these papers, as well as in related work on key enumeration [25], some observations were made about the seemingly large variation of the key rank. Poussier et al. [16] compared a number of the interval-based algorithms to determine to what extent this variation was due to the algorithm being used (despite the researchers' best efforts to improve the accuracy of their algorithms, estimation introduces an error and with it variation). Our interest is not in the 'algorithmic' noise, but rather in the intrinsic distributional properties of the key rank itself.

For our empirical investigation into the key rank distribution, we opted for Martin et al.'s approach, as we found that it provides the best efficiency/accuracy tradeoff (it gives better accuracy than the algorithm by Glowacz et al. and is more efficient than the second algorithm by Bernstein et al.).

## 2.5  Summary Statistics

To be able to explore the characteristics of the key rank distribution further, we must sample from $R$ and estimate it—samples for $\mathfrak{R}$ are calculated by applying the logarithm to each sample from $R$. A first concern is to try to find the most appropriate estimators for the expected values of $R$ and $\mathfrak{R}$. However, these random variables have characteristics other than their mean (e.g. variance). To explore these, additional summary statistics—measures of location and spread—are necessary and we review the potential choices in the following.

*Estimates of the mean.* To compute the ranking entropy and the log of the guessing entropy (Sect. 2.3) we must estimate $\mathbb{E}(R)$ and $\mathbb{E}(\mathfrak{R})$. The arithmetic (or sample) mean of $N$ samples $x_1, x_2, \ldots, x_N$ is $\bar{x} = (x_1 + x_2 + \ldots + x_N)/N$. The law of large numbers states that the arithmetic mean over a large number of trials should be close to the expected value, and thus is the correct estimator for $\mathbb{E}(R)$.

When orders of magnitude are of concern, the arithmetic mean may not be suitable—consider a hypothetical scenario in which a DPA attack is evaluated 1024 times. In 1023 of the occasions, the rank of the key is 1, and in the one remaining occasion the rank of the key is $2^{32}$. The arithmetic mean in this case is (just over) $2^{22}$, which clearly misrepresents the strength of the attack. In this case, the geometric mean of $N$ samples $x_1, x_2, \ldots, x_N$ may be more appropriate. It is defined as:

$$\tilde{x} = \left(\prod_{i=1}^{N} x_i\right)^{\frac{1}{N}}$$

The logarithm of the geometric mean of $R$ is the arithmetic mean taken on $\mathfrak{R}$ ($\log \bar{R} = \tilde{\mathfrak{R}}$). Consequently, the geometric mean is a suitable estimator for the ranking entropy $\mathbb{E}(\mathfrak{R})$. With reference to our prior 'extreme example' the geometric mean would deliver a rank of (just over) 1—a better judgement on an adversary's "order of magnitude" ability.

*Standard deviation.* The estimated standard deviation

$$\hat{s}_X = \sqrt{\frac{1}{N}\sum_{i=1}^{N}(x_i - \bar{x})^2}$$

captures the degree of variation in a distribution. From the side-channel evaluation perspective, this will be of concern—if the standard deviation of $R$ is large, then the adversary has a higher probability of being "lucky" (or "unlucky"). A similar geometric standard deviation exists, such that the geometric standard deviation of $R$ is equivalent to the arithmetic standard deviation of $\mathfrak{R}$.

*Order statistics.* An alternative, non-parametric set of order statistics are the estimated percentiles of the distribution. The $P$-th percentile is the smallest value in an ordered sample such that $P$ percent of the data set is less than or equal to that value. More formally, the index $i$ in the ordered list of $N$ samples is

$$i = \left\lceil \frac{P}{100}N \right\rceil,$$

with the $P$-th percentile taken to be sample $x_i$.

The median (or 50th percentile) is a non-parametric measure of central tendency. In the case of our previous hypothetical scenario, the arithmetic mean was $2^{22}$, despite 99.9 % of the ranks being 1. In the same scenario, the median would report 1, a much more representative value for the strength of the adversary. The median (and percentiles in general) have already seen use as descriptive statistics in the context of key rankings in Veyrat-Charvillion et al. [23].

Finally, the minimum and maximum values observed within a sample may be important. In the side-channel context, these essentially correspond to estimates for the best and worst case scenario for the adversary (and vice-versa for the evaluator). The minimum value could also be associated with an indication of the min-entropy of the distribution (although we leave this as an avenue for future exploration).

The order of a set of samples from $R$ is invariant under logarithms, and thus the minimum, maximum and percentile values from $\mathfrak{R}$ can be computed by taking the logarithm of the values for the equivalent samples from $R$.

# 3   Accurate Estimation of the Rank Distribution

The ability to characterise the distribution of $R$ hinges on whether a sufficiently accurate estimation of an individual rank can be achieved. As previously established, the rank estimation algorithm of Martin et al. (hereafter, "KRE") is the optimal choice from the candidate set of algorithms for our experiments.

## 3.1   KRE Improvements

The KRE algorithm can be seen as having two components or steps: the first is a lossy conversation from floating point distinguishing scores to integer weights, and the second is an accurate counting method.

For the first step, the KRE algorithm takes a precision parameter, which is a number of bits $p$. Each of the distinguishing scores produced by a side-channel attack are then converted to positive integers of size at most $2^p$. A typical side-channel attack produces floating-point distinguishing scores which, assuming the use of a modern CPU, are highly likely to be computed using 64-bit floating point arithmetic. Thus for any $p < 64$, the conversion from raw distinguishing scores to integer values is lossy, and can theoretically 'collide' two different distinguishing scores together into the same integer value, losing some of the information produced by the side-channel attack.

The runtime of KRE is effectively exponential in $p$; for the same set of distinguishing vectors, ranking at precision $p + 1$ will take approximately twice as long as ranking at precision $p$. Using a variety of algorithmic and implementation improvements, we were able to accommodate a large increase in the precision retained by the algorithm. These improvements enabled us to perform rank calculations approximately 16 times faster than the previous work, allowing us to run experiments at a precision of up to $p = 23$ in the order of 1–2 min (depending on the 'true' rank being estimated).

These improvements included a modification to the first step (the "Map-ToWeight" function as described in [14]). We applied a linear shift to the integer weights such that the subkey with the smallest distinguishing score has an integer weight of 1, and thus typically lowers the integer weight of the correct key (which affects the run-time linearly). In addition to some optimisations at the level of the implementation, we also modified the recurrence relation to avoid all calls to the "left child" function. With these modifications, we were able to push our implementation to retain up to 23 bits of precision. Full details can be found in Appendix A.

## 3.2   KRE Precision

To provide a sanity check of how many bits of precision suffice for computing an 'exact' rank (similarly to the brief evaluation in [14]), we simulated a large number of DPA attacks and used the key rank estimation algorithm to estimate the rank of each attack using 8 to 23 bits of precision. Table 1 and Fig. 2 illustrate the average error between our best guess at the true key rank (which is obtained by taking the estimate at 23 bits) and the rank estimates at each level of precision. Each additional bit of precision used in the rank estimation algorithm can only increase accuracy (increasing the number of bits by one approximately doubles the weight of the target key; this will reduce the number of collisions when converting the distinguishing scores to integers and can not introduce new collisions).

As can be observed in the figure and table, the average error rapidly decreases between 8 and 14 bits of estimation precision. From 17 bits of precision onwards, the average error is within 3 decimal places, dropping as low as 4 decimal places at 20 bits of precision, and with each additional bit approximately halving the average error. Given our available computational budget for all our experiments, we selected 20 to be the precision used for the KRE algorithm. This allows us to both very accurately estimate ranks and to run a large amount of experiments.

**Table 1.** The average error, in bits, for increasing increments of precision used in the rank estimation algorithm. Average taken using 1091 DPA attacks with ranks spread across the range $2^0$ to $2^{128}$, using the geometric mean.

| Precision | Av. error (bits) | Precision | Av. error (bits) | Precision | Av. error (bits) |
|-----------|------------------|-----------|------------------|-----------|------------------|
| 8         | 0.302619         | 13        | 0.010231         | 18        | 0.000330         |
| 9         | 0.158402         | 14        | 0.005343         | 19        | 0.000154         |
| 10        | 0.082911         | 15        | 0.002756         | 20        | 0.000074         |
| 11        | 0.041216         | 16        | 0.001473         | 21        | 0.000033         |
| 12        | 0.020488         | 17        | 0.000641         | 22        | 0.000015         |

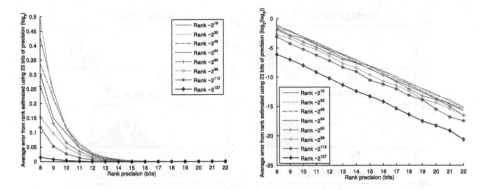

**Fig. 2.** (Left) Average error, in bits, from a 'true' rank taken to be the estimate as evaluated by the rank precision algorithm using 23 bits of precision. Rank estimates were evaluated using 8 to 22 bits of precision. Repeated DPA attacks were simulated using a random SNR, and placed into buckets if the estimated rank at 23 bits of precision was within 1 bit of $2^{16}, 2^{32}, \ldots, 2^{128}$. (Right) the same data, with the logarithm of the log-ranks applied.

# 4    Initial Exploratory Study

Now we have shown that we can estimate values from $R$ with a high degree of accuracy, we shift focus to exploring its distribution.

## 4.1    Visualising the Key Rank Distribution

As a first step we proceed to visualise the distribution of repeated key rank experiments at various depths. Histograms are an ideal tool for doing this; we hence run simulated experiments, using correlation power analysis (CPA, see [5]) as a distinguisher for attacking simulated Hamming-weight leakage with additive Gaussian noise with a low SNR of $2^{-7}$.

Figure 3 plots histograms of samples from $R$ across a range of different average rank values. In the middle range of rank values, the distribution appears to be appreciably normally distributed. However, we can observe non-normal behaviour at either end of the possible rank values, as can be seen in the top-left and bottom-right histograms. The bottom-right exhibits a much higher frequency of attacks of rank 0, producing a small additional peak at the left-tail of the distribution. Similarly, when the average rank is close to the maximum of $2^{128}$, the distribution is no longer symmetric, but is also without the additional peak. A review of statistical literature suggests that distributions that are 'clipped' in this way are defined as *truncated* distributions [9].

The x-axis of the histograms is log-scale: if the distribution of the logarithm of the ranks was indeed normal, then the distribution of the rank values themselves would be a log-normal distribution. The large skewness of a log-normal distribution would support our hypothesis that the arithmetic mean is not a

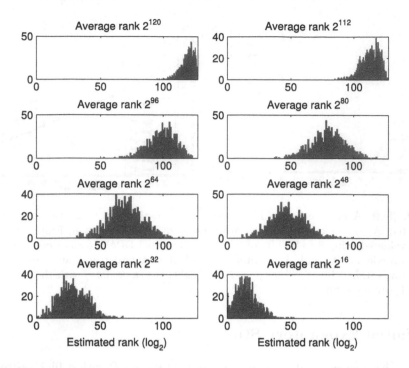

**Fig. 3.** Histograms for attacks with a (geometric) mean rank close to one of several values. Here the leakage is simulated Hamming-weight with Gaussian noise at an SNR of $2^{-7}$, with the attacker using CPA as a distinguisher.

suitable average, and rather the geometric mean is better suited. Our prediction of a log-normal distribution is supported by the central limit theorem, which implies that the product of positive random variables produces a log-normal distribution.

Given this information we conjecture that we have a delta-log-normal [1] distribution with truncation [9]. A delta-log-normal distribution is a distribution on a random variable $X$ such that $X$ is assigned value 0 with probability $\theta$ and follows the log-normal distribution with probability $1 - \theta$. In this particular context the value of $\theta$ would directly correspond to the success rate of an adversary for full key recovery (without enumeration). The log-normal distribution could then be parameterised separately using standard methods. Truncation corresponds to when a random variable can not be assigned a value passed a certain threshold. It is clear that the rank can only be assigned a value between 0 and $2^{128} - 1$, and thus must be truncated.

Whilst further research into this characterisation is a promising next-step, for the purposes of this work we instead pursue two questions of immediate importance: firstly, whether this shape and scale of distribution is consistent across the various contributory factors influencing the outcomes of side-channel attacks, and secondly whether the *non-parametric* order statistics outlined in

Sect. 2.5 can be used as a simple and efficient method for extracting meaningful conclusions without making any assumptions about the underlying distribution.

## 4.2  Is an Accurate Rank Distribution Estimation Viable?

Before further exploration of the candidacy of the summary statistics outlined in Sect. 2.5, we devised an experiment to determine how many repeat experiments are necessary to reliably estimate them. We kept the leakage model and SNR, as well as the distinguisher used by the adversary, constant but used randomly generated plaintexts, keys and Gaussian noise. We assumed a CPA attack using the Hamming-weight power model, and the leakage was simulated on the AES SubBytes target function, using the Hamming-weight leakage function and Gaussian noise. In the experiment, each statistic was estimated using increasing amounts of repeat experiments on simulated data.

The results in Fig. 4 exhibit the behaviour of the statistics. The maximum key rank values unsurprisingly exhibit the most variability—for key ranks above 80 we observe that the estimated values 'jump' at 50, 100, and 200 repeats where they stabilise. The other key ranks, hence those in the ranges were enumeration is within practical reach behave much more stable—from 25 repeats on they produce stable estimates, from 100 repeat experiments onwards the estimates have converged to the true value. The intuition behind the geometric mean being a sensible choice is sound, producing a line that is almost identical to that of the median, as expected under the assumption of a log-normal distribution. In fact, for all the experiments we pursued in this study, the geometric mean and median were nearly identical, and for simplicity we do not display it in future graphs. The unsuitability of the arithmetic mean (given orders of magnitude are a concern) is clear and consequently from here onwards we no longer calculate it.

*Resampling methods.* In the previous experiment, which was based on simulations, we were able to efficiently sample independent and mutually exclusive sets of key rank data. In practice this might not be possible as a single, large dataset might be available only. This situation is not uncommon and methods such as bootstrapping, jackknifing and k-fold cross validation are well understood [4,11,22] and therefore get employed in a variety of contexts. An important guideline though, irrespective of which resampling approach one chooses, is to pay attention to *randomly* selecting subsamples to avoid introducing a bias.

## 5  Characterising Rank Distributions

To understand whether the properties observed in the exploratory studies of Sect. 4 are common (or specific to the combination of distinguisher, leakage model and SNR), and to further explore characteristics of the distribution of $R$, we perform further simulated DPA attacks and vary the interesting parameters.

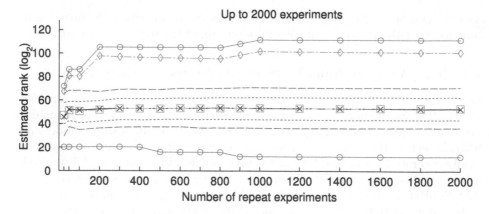

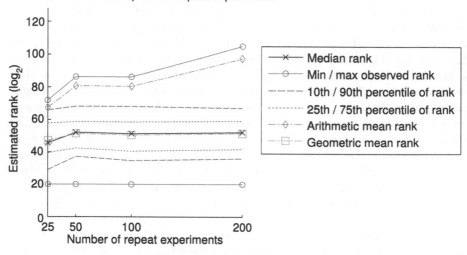

**Fig. 4.** Estimated summary statistics using increasing amounts of DPA attacks. Each DPA attack used a CPA-HW distinguisher on simulated AES SubBytes leakage using the Hamming-weight power model and Gaussian noise of an SNR $2^{-3}$.

## 5.1   SNR and Measurement Counts

The starting point of our simulated experiments was to consider whether both the measurement SNR and the quantity of trace measurements available affects the rank distribution. These two variables are clearly dependent; a very low SNR can be overcome by using more measurements, and at high SNR levels a successful attack can be created using fewer trace measurements. As a consequence we devised a set of experiments for which the rank distribution can be analysed as both these variables change. We assumed an 'optimal' adversary operating under commonly considered leakage conditions—namely, Hamming-weight leakage on the AES SubBytes operation with additive Gaussian noise, and where the adversary launches a CPA attack using the Hamming-weight as a power model.

We simulated data under a variety of SNRs, beginning with a low-noise scenario of SNR $2^{-1}$, up to a high-noise scenario with SNR $2^{-7}$. For each unique SNR, we simulated DPA attacks using increasing amounts of traces, beginning with a quantity for which the rank was approximately $2^{128}$, and increasing the number of traces until the vast majority of attacks produced a rank of 0. For each unique number of traces, we ran 1000 repeat attacks, and for each repeat generated the keys, plaintexts and additive noise at random.

Figure 5 visualises the summary statistics for attacks under the SNRs $2^{-7}$, $2^{-5}$ and $2^{-3}$. The general trends appear similar to those observed in our real world example. The variance observed is of most interest, both in terms of its magnitude and its consistency across multiple pairs of SNR and trace quantities.

Three main observations can be made:

1. The distribution appears to be at its widest in the middle range of ranks (e.g. when the rank is between $2^{40}$ and $2^{80}$), and variance minimises for very poor attacks (rank $\approx 2^{128}$) and very good attacks (rank $\approx 2^0$).
2. The maximum variance appears to be very large, with the difference between (for example) the 10th and 90th percentiles being in the order of up to 40 bits in some cases.
3. The exact level of SNR does not appear to affect the variance or shape of the distribution in any independent way—assuming the same distinguisher is used, at any given SNR, given sufficient traces to establish an average rank of $x$, the dispersion of the distribution will be very similar to that produced by attacks at any other SNR that have an average rank close to $x$.

To confirm these three intuitions, we plotted the estimated geometric standard deviation against the (geometric) mean rank (or equivalently the arithmetic mean and standard deviation of samples from $\mathfrak{R}$). The results can be seen in Fig. 6, where each line corresponds to results obtained for seven different SNRs. The shape and magnitude of each line very closely match, indicating that the behaviour is indeed consistent across all SNRs. The curves peak at an average rank of approximately $2^{64}$, suggesting that it is the 'true' *rank of the attack* that affects the variance, and *not* any characteristic of the leakage noise or quantity of data available (for a fixed key rank).

These three characteristics in tandem present an unfortunate problem for an evaluator and for the viability of the guessing entropy as a stand-alone metric. Not only is the variance very large, and thus an adversary may with non-negligible probability produce an attack far out-performing the average attack, but also the variance is largest in the range of key ranks that are of most interest to an evaluator. There is a threshold at which an adversary may be considered unrealistic (e.g. we might be confident that an adversary can enumerate $2^{54}$ keys, but not $2^{57}$), and unfortunately the distribution has the most variance here.

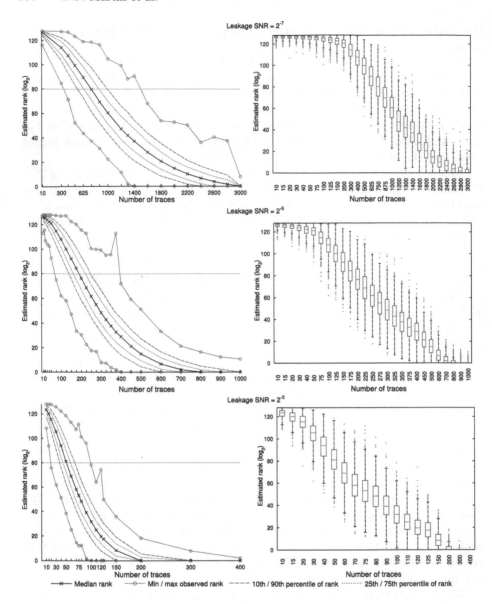

**Fig. 5.** (Left) Estimated ranks after 1,000 DPA attacks at SNRs $2^{-7}$, $2^{-5}$ and $2^{-3}$, using Hamming-weight CPA targeting simulated leakage on the AES SubBytes operation. (Right) Equivalent box-plots for using the same data as on the left. The central line in each box is the median, the box defines the inter-quartile range, the whiskers cover all samples not considered to be outlier values, and outliers are plotted individually.

## 5.2 Distinguishers and Higher-Order Attacks

A second consideration is whether the choice of distinguisher used by the adversary can change the characteristics of the rank distribution. Our previous

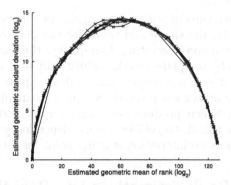

**Fig. 6.** A plot of the estimated geometric standard deviation against the geometric mean of the key rank, taken using 207,000 DPA attacks. Each line represents the standard deviation for attacks at the seven SNR values $2^{-7}, \ldots, 2^{-1}$. Each attack used simulated Hamming-weight leakage with CPA used as the distinguisher.

experiments used CPA as the distinguisher, and so to compare, we launched two additional types of attacks. Firstly, we tried reduced[3] template attacks on the simulated leakage. Secondly, we launched second-order attacks on a binary masked implementation of AES: the leakage sample corresponding to the mask value and the sample corresponding to the masked SubBytes operation were combined using the 'centre and multiply' method (see e.g [17]), and then a standard Hamming-weight CPA was launched. To enable a direct comparison with the standard CPA attacks, we ran the template attacks using data with an SNR of $2^{-7}$. For the second order attack, we reduced the SNR to $2^{-3}$ to alleviate the burden of having to use too many traces in the attack.

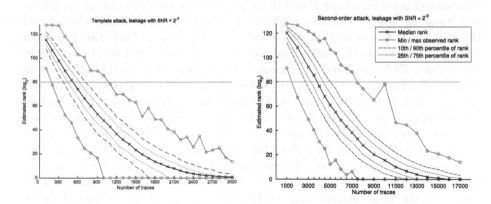

**Fig. 7.** Estimated ranks after template and second-order attacks, using simulated leakage on the AES SubBytes operation.

---

[3] Reduced in that we did not use multiple points or estimate a covariance matrix.

Figure 7 shows the results of our attacks. Again, we observe consistent behaviour as seen previously; the shape and trend for the percentiles is remarkably similar. We can observe one interesting discrepancy: the variance of the distribution produced by the template attacks, whilst still very large in the middle of the distribution and with a consistent shape, does appear to be *smaller* than that produced by any attack using correlation as a distinguisher (including the second-order attacks, which produce very similar rank variance to first-order ones). The reason for this distinguisher-specific dependency is unclear, and we leave this observation as an interesting starting-point for future research.

# 6   Embedding Rank Estimations into Real World Security Evaluations

In the previous sections we provided conclusive evidence that the key rank is random variable with inherently large variation. We showed that it is possible to meaningfully characterise average behaviour and spread using repeat experiments. A crucial questions remains though: how can this be integrated into practical side channel evaluations? In this section we discuss two radically different propositions for a solution. The first proposition is to employ some recent suggestions for short-cuts in evaluations; we find that these have limitations which restrict their practical use. The second proposition is a practical re-use of measurements for repeat experiments, which leads to practically meaningful results.

## 6.1   Bounding the Success Rate of an Adversary with Enumeration

In some recent work, Duc et al. [6] provided some bounds that relate the mutual information between a subkey and the leakage traces, as a function of the adversary's success rate, the number of shares (if used within a masking scheme) and the number of traces used within a side channel attack. They also present a construction relating the success rate, enumeration effort and number of traces (for a fixed SNR and number of masks), in the best case for the adversary in an extended version of their paper [7, Sect. 4.3c, Eq. 24, Algorithm 2]. We can interpret this as a lower bound on the key rank of the secret key at a given number of messages, by looking when the success rate first becomes non-zero.

Using code supplied by the authors of [6], we were able to evaluate this success rate bound in the context of idealised Hamming-weight information leakage. This data is shown in Fig. 8, re-using the simulated Hamming-weight CPA attack data under first-order and second-order attack conditions. As can be immediately seen from the large margin between the SR bound and the estimated ranks, this is a very loose bound—in the right hand graph, the SR bound is almost on top of the x- and y-axes. This supports their intuition (hinted at in [6]) that the theoretical bounds only tighten for a large number of masks, but cannot realistically approximate the performance of an adversary in the single or zero mask situation our work explores. Consequently this recent work, which hopes

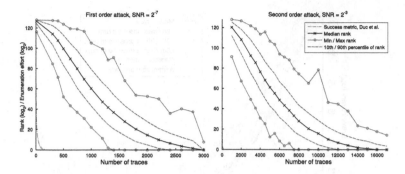

**Fig. 8.** Estimated key ranks compared to the success rate metric of Duc et al. [6] using simulated first and second order CPA attacks on Hamming-weight leakage.

to 'short-cut' the effort in evaluations, seems too inaccurate for the kind of implementations that are of immediate real world interest.

Using a different technique to Duc et al., but with the same intention to short-cut evaluation efforts somewhat, Ye et al. argue for an algorithm that allows to estimate the remaining effort of an adversary regarding enumeration and simultaneously provides the optimal guessing strategy. They suggest that their algorithm could be run once on a dataset. However, running their algorithm once can only deliver a single interval estimate of the key rank: repeat experiments would still produce a large variance which implies that any statement based on a single run is insufficient to determine the spread.

## 6.2 Real World Evaluation of a Challenging Target

We utilise an interesting real-world data set provided by Longo et al., which initially appeared at CHES 2015 [12], to illustrate how to integrate key rank into practical evaluations. We re-implemented and re-ran one of the attacks described by Longo et al. at CHES 2015 [12]. They illustrated several standard DPA attacks on a complex device, and we selected their most challenging one: an attack on a hardware AES implementation, utilising EM measurements. We refer the reader to the attack paper for full details, but note that we use an improved attack strategy communicated to us after correspondence with the authors [12].

The available dataset consisted of approximately one million EM traces. These were acquired in line of their 'standard' assessment approach for cryptographic devices, which as some of the authors are from a well known expert company, can be regarded as being in line with industry best practice: after initially identifying the source of the leakage, they gathered as many traces as they could afford (given some allotted time budget) for a given unknown secret key.

In the previous section of this paper we highlight the fact that estimations of key rank properties need repeat experiments. However, due to the EIS (Equal Images under different Subkeys) property that typical block ciphers have [19], it is not necessary to run these on different keys (since the results will be of

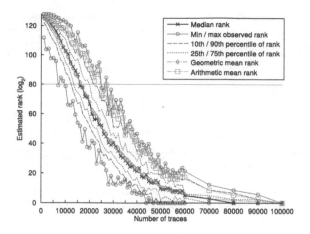

**Fig. 9.** Estimated key ranks after repeated DPA attacks on a set of 997,500 traces acquired from a BeagleBone Black device running AES-128 in hardware.

similar quality). Instead we can divide up any set of experiments into smaller subset to run repeat experiments, which is how we proceed. To analyse the distribution of $R$ produced by the attacker[4] we ran multiple DPA attacks using increasing amounts of traces from the data set. Figure 9 plots the trends of the minimum, maximum, various percentiles and geometric and arithmetic means for the estimated ranks as the number of traces available to the attacker increases.

The first attack reporting full key recovery uses approximately 45,000 traces, and we can immediately see from the graph that this should perhaps be considered a fortunate result for the attacker—at this number of traces, we observed some attacks of rank up to $\approx 2^{20}$. Also of interest is that the classical 80-bit security margin is first broken somewhere between 10,000 and 20,000 traces, and expected, considerable variance in the rank distribution, with a very large margin between the minimum and maximum values observed. The line for the arithmetic mean is again evidence that our intuition of computing statistics on $\Re$ is more meaningful—the line corresponding to the geometric mean (of $R$) is very close to the median.

**The Power of Percentiles.** Percentiles are particularly informative statistics in the evaluation context. Recall that percentiles give the value below which a specific percentage of observations (among the sampled observations) fall. We can relate this to business goals such as having no more than a certain percentage of devices be susceptible to a particular side channel adversary, as we show in the example below.

Consider our evaluation of the real world data-set before: we sampled from the rank distribution using repeated attacks for an increasing amount of traces.

---

[4] The attack is a Hamming-distance correlation power analysis on the input and output of the final round of encryption.

Risk can be assessed using these key rank samples. As an example, assume that the $10^{\text{th}}$ percentile of the estimated rank distribution is $2^{44}$ (in Fig. 9)—this indicates that of all the devices of that type sold into a market, 10 % would succumb to a full key recovery attack by an adversary using around 23,000 traces and with $2^{44}$ as an enumeration budget. An alternative but equivalent interpretation would be that 90 % of the devices are only vulnerable once the adversary's enumeration budget increases beyond $2^{44}$ (at 23,000 traces). Instead of phrasing attack scenarios around how many devices are vulnerable, one can focus on a single devices but many adversaries. For instance, if a series of fixed adversaries attacked the same device (using 23,000 traces and enumerating up to $2^{44}$ keys, then 10 % would succeed).

These examples demonstrate that percentiles are a very efficient and simple way to assess the spread of the rank distribution and report it in a meaningful way in business terms. The use of different percentiles then allows the evaluator to fine-tune these security margins.

# 7    Conclusion

One of our key findings is that the shape of the distribution of the key rank is *consistent*; these observations hold irrespective of the type of differential attack used (with the small but interesting observation that template attacks seem to produce key rank distributions with a slightly smaller variance than similar correlation based attacks) and SNR. We thereby confirm that it has a large variance in exactly the range in which the assumed enumeration capability of an adversary transitions from realistic to unrealistic.

In our efforts to explore suitable statistical measures to capture the practically important key rank characteristics we observe that the guessing entropy, defined as the *expected value* of the key rank, is not always meaningful. As an average, the guessing entropy cannot quantify *any* of the very large amounts of variance we observe. Consequently, additional metrics must be used, and a natural step is to instead consider non-parametric order statistics, which brings us to consider the usage of percentiles to connect side channel outcomes with business goals.

We additionally observe that the rank distribution $R$ follows some flavour of a truncated delta-log-normal distribution. However, in practice we typically are concerned with behaviour of adversaries in the log-domain (in $\mathfrak{R}$)—an evaluator tends to be more interested in the *magnitude* of an adversary's capabilities and not the exact value. Whilst the logarithm of the guessing entropy can be appropriately estimated using the geometric mean, it is perhaps easier to switch to considering guessing entropy defined using the logarithm of the ranks, and estimated using the arithmetic mean.

With regards to practical impacts, we observed that at least some repeat experiments are necessary for stable estimates of the geometric mean, median and percentiles. Whilst this appears to incur an overhead at first glance in terms of trace measurements, we explain that it is sound to simply 'split' any existing data set into smaller subsets with which the repeat runs can be conducted. Finally we show that caution is needed with regards to using short-cut formulas, and end by illustrating an approach to evaluating the security of a real world device using repeat rank experiments.

**Acknowledgements.** We would like to thank the authors of Longo et al. [12] for giving us access to their data set, and the authors of Duc et al. for giving us a preprint for their extended version of [6] as well as the corresponding Matlab code.

Dan Martin, Luke Mather, and Elisabeth Oswald were supported in part by EPSRC via the grants EP/I005226/1 and EP/N011635/1. This work was carried out using the computational facilities of the Advanced Computing Research Centre, University of Bristol http://www.bris.ac.uk/acrc/.

# A    KRE Optimisation

To allow the key rank algorithm of Martin et al. [14] to run with a precision of up to 23 bits we had to include several implementation and algorithmic tricks to bring down the runtime of the algorithm.

## A.1    Distinguishing Score to Integer Weight Conversion

When the distinguishing scores are converted to integer weights they are done in such a way that the largest distinguishing score results in value $2^p$. However it is possible that this leads to scenarios where the distinguishing scores are unnecessarily large—for example, if all the distinguishing scores have value 1 they will end up with value $2^p$. To counter this we subtract the minimum integer score from all scores to scale them back. This increases the efficiency of the algorithm since the runtime is linear in the weight of the key.

```
Algorithm KeyRank(m, n, W):
for j from m − 1 down to 0 do
  for w from 0 up to W − 1 do
    for i from n − 1 down to 0 do
      Rank[w] ← Rank[w] + OldRank[RightChild(j, w, i)]
    end for
  end for
  OldRank ← Rank
end for
return Rank[0]
```

**Fig. 10.** The resulting rank algorithm after adjusting the algorithm of Martin et al. [14]

## A.2    Recurrence Relation

One of the major changes to the algorithm was adjust the recurrence relation. The first step was to use the 'wide sort' given in the original paper as it had the smallest memory footprint. Using a three-dimensional coordinate system to index the graph, the single loop over the graph was replaced with three for loops, one for each integer in the representation. Using the combination of the wide sort and the triple index system, it can be noted given $(x, y, z)$, such that the left child is not reject, it will always return $(x, y + 1, z)$. This can be used to remove the majority of the memory copies and access by computing an entire partial weight within a subkey at once without having to work at an index at a time. The other advantage of the triple index system is that it greatly reduces the number of expensive operations required (such as mods) to calculate the child nodes. The resulting algorithm is given in Fig. 10.

# References

1. Aitchison, J.: On the distribution of a positive random variable having a discrete probability mass at the origin. J. Am. Stat. Assoc. **50**(271), 901–908 (1955)
2. Bernstein, D.J., Lange, T., van Vredendaal, C.: Tighter, faster, simpler side-channel security evaluations beyond computing power. IACR Cryptology ePrint Arch. **2015**, 221 (2015)
3. Bogdanov, A., Kizhvatov, I., Manzoor, K., Tischhauser, E., Witteman, M.: Fast and memory-efficient key recovery in side-channel attacks. IACR Cryptology ePrint Arch. **2015**, 795 (2015)
4. Bradley, E.: The Jackknife, the Bootstrap, and Other Resampling Plans. Society for Industrial and Applied Mathematics, Philadelphia (1982)
5. Brier, E., Clavier, C., Olivier, F.: Correlation power analysis with a leakage model. In: Joye, M., Quisquater, J.-J. (eds.) CHES 2004. LNCS, vol. 3156, pp. 16–29. Springer, Heidelberg (2004). doi:10.1007/978-3-540-28632-5_2
6. Duc, A., Faust, S., Standaert, F.-X.: Making masking security proofs concrete. In: Oswald, E., Fischlin, M. (eds.) EUROCRYPT 2015. LNCS, vol. 9056, pp. 401–429. Springer, Heidelberg (2015). doi:10.1007/978-3-662-46800-5_16
7. Duc, A., Faust, S., Standaert, F.: Making masking security proofs concrete - or how to evaluate the security of any leaking device (extended version). Cryptology ePrint Archive, Report 2014/119 (2015). http://eprint.iacr.org/
8. Fei, Y., Luo, Q., Ding, A.A.: A statistical model for DPA with novel algorithmic confusion analysis. In: Prouff, E., Schaumont, P. (eds.) CHES 2012. LNCS, vol. 7428, pp. 233–250. Springer, Heidelberg (2012). doi:10.1007/978-3-642-33027-8_14
9. Finney, D.J.: The truncated binomial distribution. Ann. Eugenics **14**(1), 319–328 (1947)
10. Glowacz, C., Grosso, V., Poussier, R., Schüth, J., Standaert, F.-X.: Simpler and more efficient rank estimation for side-channel security assessment. In: Leander, G. (ed.) FSE 2015. LNCS, vol. 9054, pp. 117–129. Springer, Heidelberg (2015). doi:10.1007/978-3-662-48116-5_6
11. Good, P.: Practitioner's Guide to Resampling Methods. CRC Press, Boca Raton (2012)

12. Longo, J., De Mulder, E., Page, D., Tunstall, M.: SoC it to EM: electromagnetic side-channel attacks on a complex system-on-chip. In: Güneysu, T., Handschuh, H. (eds.) CHES 2015. LNCS, vol. 9293, pp. 620–640. Springer, Heidelberg (2015). doi:10.1007/978-3-662-48324-4_31

13. Mangard, S., Oswald, E., Standaert, F.-X.: One for all - all for one: unifying standard DPA attacks. IET Inf. Secur. **5**(2), 100–110 (2011)

14. Martin, D.P., O'Connell, J.F., Oswald, E., Stam, M.: Counting keys in parallel after a side channel attack. In: Iwata, T., Cheon, J.H. (eds.) ASIACRYPT 2015. LNCS, vol. 9453, pp. 313–337. Springer, Heidelberg (2015). doi:10.1007/978-3-662-48800-3_13

15. Massey, J.L.: Guessing and entropy. In: IEEE International Symposium on Information Theory, p. 204 (1994)

16. Poussier, R., Grosso, V., Standaert, F.-X.: Comparing approaches to rank estimation for side-channel security evaluations. In: Homma, N., Medwed, M. (eds.) CARDIS 2015. LNCS, vol. 9514, pp. 125–142. Springer, Heidelberg (2016). doi:10.1007/978-3-319-31271-2_8

17. Prouff, E., Rivain, M., Bevan, R.: Statistical analysis of second order differential power analysis. IEEE Trans. Comput. **58**(6), 799–811 (2009)

18. Rivain, M.: On the exact success rate of side channel analysis in the Gaussian model. In: Avanzi, R.M., Keliher, L., Sica, F. (eds.) SAC 2008. LNCS, vol. 5381, pp. 165–183. Springer, Heidelberg (2009). doi:10.1007/978-3-642-04159-4_11

19. Schindler, W., Lemke, K., Paar, C.: A stochastic model for differential side channel cryptanalysis. In: Rao, J.R., Sunar, B. (eds.) CHES 2005. LNCS, vol. 3659, pp. 30–46. Springer, Heidelberg (2005). doi:10.1007/11545262_3

20. Standaert, F.-X., Malkin, T.G., Yung, M.: A unified framework for the analysis of side-channel key recovery attacks. In: Joux, A. (ed.) EUROCRYPT 2009. LNCS, vol. 5479, pp. 443–461. Springer, Heidelberg (2009). doi:10.1007/978-3-642-01001-9_26

21. Thillard, A., Prouff, E., Roche, T.: Success through confidence: evaluating the effectiveness of a side-channel attack. In: Bertoni, G., Coron, J.-S. (eds.) CHES 2013. LNCS, vol. 8086, pp. 21–36. Springer, Heidelberg (2013). doi:10.1007/978-3-642-40349-1_2

22. Tukey, J.: Bias and confidence in not quite large samples. Ann. Math. Stat. **29**, 614–623 (1958)

23. Veyrat-Charvillon, N., Gérard, B., Renauld, M., Standaert, F.-X.: An optimal key enumeration algorithm and its application to side-channel attacks. In: Knudsen, L.R., Wu, H. (eds.) SAC 2012. LNCS, vol. 7707, pp. 390–406. Springer, Heidelberg (2013). doi:10.1007/978-3-642-35999-6_25

24. Veyrat-Charvillon, N., Gérard, B., Standaert, F.-X.: Security evaluations beyond computing power. In: Johansson, T., Nguyen, P.Q. (eds.) EUROCRYPT 2013. LNCS, vol. 7881, pp. 126–141. Springer, Heidelberg (2013). doi:10.1007/978-3-642-38348-9_8

25. Ye, X., Eisenbarth, T., Martin, W.: Bounded, yet sufficient? how to determine whether limited side channel information enables key recovery. In: Joye, M., Moradi, A. (eds.) CARDIS 2014. LNCS, vol. 8968, pp. 215–232. Springer, Heidelberg (2015). doi:10.1007/978-3-319-16763-3_13

# Taylor Expansion of Maximum Likelihood Attacks for Masked and Shuffled Implementations

Nicolas Bruneau[1,2(✉)], Sylvain Guilley[1,3], Annelie Heuser[1], Olivier Rioul[1], François-Xavier Standaert[4], and Yannick Teglia[5]

[1] Institut Mines-Télécom, Télécom ParisTech,
CNRS LTCI Department Comelec, Paris, France
{nicolas.bruneau,sylvain.guilley,annelie.heuser,
olivier.rioul}@telecom-paristech.fr
[2] STMicroelectronics, AST Division, Rousset, France
[3] Secure-IC S.A.S., Rennes, France
[4] ICTEAM/ELEN/Crypto Group, Université catholique de Louvain,
Louvain-la-Neuve, Belgium
[5] Gemalto, Security Labs, La Ciotat, France

**Abstract.** The maximum likelihood side-channel distinguisher of a template attack scenario is expanded into lower degree attacks according to the increasing powers of the signal-to-noise ratio (SNR). By exploiting this decomposition we show that it is possible to build highly multivariate attacks which remain efficient when the likelihood cannot be computed in practice due to its computational complexity. The shuffled table recomputation is used as an illustration to derive a new attack which outperforms the ones presented by Bruneau et al. at CHES 2015, and so across the full range of SNRs. This attack combines two attack degrees and is able to exploit high dimensional leakage which explains its efficiency.

**Keywords:** Template attacks · Taylor expansion · Shuffled table recomputation

## 1 Introduction

In order to protect embedded systems against side-channel attacks, countermeasures need to be implemented. Masking and shuffling are the most investigated solutions for this purpose [18]. Intuitively, masking aims at increasing the order of the statistical moments (in the leakage distributions) that reveal sensitive information [8,15], while shuffling aims at increasing the noise in the adversary's

Annelie Heuser is a Google European Fellow in the field of Privacy and is partially founded by this fellowship.

Y. Teglia—Parts of this work have been done while the author was at STMicroelectronics.

J.H. Cheon and T. Takagi (Eds.): ASIACRYPT 2016, Part I, LNCS 10031, pp. 573–601, 2016.
DOI: 10.1007/978-3-662-53887-6_21

measurements [14]. As a result, an important challenge is to develop sound tools to understand the security of these countermeasures and their combination [31]. For this purpose, the usual strategy is to consider template attacks for which one can split the evaluation goals into two parts: offline profiling (building an accurate leakage model) and online attack (recovering the key using the leakage model). As far as profiling is concerned, standard methods range from non-parametric ones (e.g., based on histograms or kernels) of which the cost quite highly suffers from the curse of dimensionality (see e.g., [2] for an application of these methods in the context of non-profiled attacks) to parametric methods, typically exploiting the mixture nature of shuffled and masked leakage distributions [16,17,25,27,33], which is significantly easier if the masks (and permutations) are known during the profiling phase. Our premise in this paper is that an adversary is able to obtain such a mixture model via one of these means, and therefore we question its efficient exploitation during the online attack phase.

In this context, a starting observation is that the time complexity of template attacks exploiting mixture models increases exponentially with the number of masks (when masking) and permutation length (when shuffling [37]). So typically, the time complexity of an optimal template attack exploiting $Q$ traces against an implementation where each $n$-bit sensitive value is split into $\Omega$ shares and shuffled over $\Pi$ different positions is in $\mathcal{O}\left(Q \cdot (2^n)^{\Omega-1} \cdot \Pi!\right)$, which rapidly turns out to be intractable. In order to mitigate the impact of this high complexity, we propose a small, well-controlled and principled relaxation of the optimal distinguisher, based on its Taylor expansion (already mentioned in the field of side-channel analysis in [6,11]) of degree $L$. Such a simplification leads to various concrete advantages. First, when applied to masked implementations, it allows us to perform the (mixture) computations corresponding to the $(2^n)^\Omega$ factor in the complexity formula only once (thanks to precomputation) rather than $Q$ times. Second, when applied to shuffled implementations, it allows us to replace the $\Pi!$ factor in this formula by $\binom{\Pi}{\min(\lceil\frac{\Pi}{2}\rceil,L)} = \binom{\Pi}{L}$, thanks to the bounded degree $L$.

Additionally it can be noticed that an attacker will only build, during the offline profiling, the leakage models needed for the attack. By applying the Taylor expansion of the optimal distinguisher the complexity of the offline profiling is significantly reduced. In general the complexity of the offline profiling becomes equivalent to the complexity of the online attack.

The resulting "rounded template attacks" additionally carry simple intuitions regarding the minimum degree of the Taylor expansion needed for the attacks to succeed. Namely, this degree $L$ needs to be at least equal to the security order $O$ of the target implementation, defined as the smallest statistical moment in the leakage distributions that are key-dependent.

We then show that these attacks only marginally increase the data complexity (for a given success rate) when applied against a masked (only) implementation. More importantly, we finally exhibit that rounded template attacks are especially interesting in the context of high-dimensional higher-order side-channel attacks, and put forward the significant improvement of the attacks against the masked implementations with shuffled table recomputations from CHES 2015 [7].

**Introduction to Shuffled Table Recomputation.** Masking the linear parts of a block cipher is straightforward whereas protecting the non-linear parts is less obvious. To solve this issue different methods have been proposed. One can cite algebraic methods [3,30], using Global Look-Up Table (GLUT) [28] and table recomputation [1,8,10,19]. Table recomputation methods are often used in practice as they represent a good tradeoff between memory consumption and execution time since they precompute a masked substitution box (S-Box) that is stored in a table.

However, some attacks still manage to recover the mask during the table recomputation [6,36]. As a further protection the recomputation can be shuffled. This protection uses a random permutation which is drawn over $S_{2^n}$, the set of all the permutation of $\mathbb{F}_2^n$. Therefore, some random masks are uniformly drawn over $\mathbb{F}_2^n$ to ensure the security against first-order attacks.

**Contributions.** We show that the expansion of the likelihood allows attacks with a very high computational *efficiency*, while remaining very *effective* from a key recovery standpoint. This means that the expanded distinguisher requires only little more traces to reach a given success rate, while being much faster to compute.

We also show how to grasp in a multivariate setting several leakages of different orders. In particular, we present an attack on shuffled table recomputation which succeeds with less traces than [7]. Notice that the likelihood attack cannot be evaluated in this setting because it is computationally impossible to average over both the mask and the shuffle (the sole number of shuffles is $2^n! \approx 2^{1684}$ with $n = 8$).

Finally, we show that are our rounded version of the maximum likelihood allows better attacks than the state-of-the-art. Namely, our attack is better than the classical 2O-CPA and the recent attack of CHES'15 [7] in all noise variance settings.

**Outline.** The remainder of the paper is organized as follows. Section 2 provides the necessary notations and mathematical definitions. The theoretical foundation of our method is presented in Sect. 3. The case-study (shuffled table recomputation) is shown in Sect. 4. Section 5 evaluates the complexity of our method. The performance results are presented in Sect. 6. Conclusions and perspectives are presented in Sect. 7. Some technical results are deferred to the appendices.

# 2    Notations

## 2.1    Parameters

Randomization countermeasures consist in *masking* and *shuffling* protections. When evaluating randomized implementations, there are a number of important parameters to consider. First, the number of shares and the shuffle length in the scheme, next denoted as $\Omega$ and $\Pi$, are algorithmic properties of the counter-measure. These numbers generally influence the tradeoff between the implementation overheads and the security of the countermeasures. Second, the order of the implementation protected by a randomization countermeasure, next denoted as $O$, which is a statistical property of the implementation. It corresponds to the smallest key-dependent statistical moment in the leakage distributions. When only masking is applied and the masked implementation is "perfect" (meaning that the leakage of each share is independent of each other), the order $O$ equals to $\Omega$ at best. Finally, the number of dimensions (or dimensionality) used in the traces, next denoted as $D$, is a property of the adversary. In this respect, adversaries may sometimes be interested by using the lowest possible $D$ (since it makes the detection of POIs in the traces easier). But from the measurement complexity point of view, they have a natural incentive to use $D$ as large as possible. A larger dimension $D$ allows to increase the signal to noise ratio [5].

In summary, our notations are:

- $\Omega$: number of shares in the masking countermeasure,
- $\Pi$: length of the shuffling countermeasure,
- $O$: order of the implementation,
- $D$: dimensionality of the leakages.

**Examples.** Existing masking schemes combine these four values in a variety of manners. For example, in a perfect hardware masked implementation case with three shares, we may have $\Omega = 3$, $O = 3$ and $D = 1$ (since the three shares are manipulated in parallel). If this implementation is not perfect, we may observe lower order leakages (e.g. $\Omega = 3$, $O = 1$ and $D = 1$, that is a first-order leakage). And in order to prevent such imperfections, one may use a Threshold Implementation [24], in which case one share will be used to prevent glitches (so $\Omega = 3$, $O = 2$ and $D = 1$). If we move to the software case, we may then have more informative dimensions, e.g. $\Omega = 3$, $O = 3$, $D = 3$ if the adversary looks for a single triple of informative POIs. But we can also have a number of dimensions significantly higher than the order (which usually corresponds to stronger attacks). Let us also give an example of S-boxes masking with one mask, where the masking process of the S-box (often called recomputation) is shuffled. A permutation $\Phi$ of $\Pi = 2^n$ values is applied while computing the masked table. If the attacker ignores the recomputation step, he can carry out an attack on the already computed table. Hence parameters $\Omega = 2$, $O = 2$, $D = 2$ (also known as "second-order bivariate CPA"). But the attacker can also exploit the shuffled recomputation of the S-box in addition to a table look-up, as presented in [7]; the setting is thus highly multivariate: $\Omega = 2$, $\Pi = 2^n$, $O = 2$, $D = 2 \cdot 2^n + 1$. Interestingly, the paper [7] shows an attack at degree $L = 3$ which succeeds in less traces than attacks at minimal degree $L = O = 2$.

In general, a template attack based on mixture distributions (often used in parametric estimation) would require a summation over all random values of the countermeasure, that is $\mathcal{R}$, which consists in the set of masks and permutations. One can represent $\mathcal{R}$ as the Cartesian product of the set of mask and the set of permutations. Let us denote by $\mathcal{M}$ the set of mask and $\mathcal{S}$ the set of permutations. Then $\mathcal{R} = \mathcal{M} \times \mathcal{S}$. Therefore, the cardinality of $\mathcal{R}$ is $2^{n(\Omega-1)} \Pi!$.

Eventually, the security of a masked implementation depends on its order and noise level. More precisely, the security increases exponentially with the order (with the noise as basis) [12]. So for the designer, there is always an incentive to increase the noise and order. And for adversary, there is generally an incentive to use the largest possible $D$ (given the time constraints of his attack), so that he decreases the noise.

## 2.2 Model

We characterize the protection level in terms of the most powerful attacker, namely an attacker who knows everything about the design, except the masks and the noise. This means that we consider the case where the templates are known. How the attacker got the templates is related with *security by obscurity*, somehow he will know the model. Of course depending on the learning phase these estimations can be more or less accurate. For the sake of simplicity we assume in this paper the better scenario where all the estimations are exact[1].

Besides, we assume that the noise is independently distributed over each dimension. This is the least favorable situation for the attacker (as there is in this case the most noise entropy). For the sake of simplicity, we assume that the noise variance is equal to $\sigma^2$ at each point $d = 1, 2, \ldots, D$. This allows for a simple theoretical analysis. Let us give an index $q = 1, 2, \ldots, Q$ to each trace. For one trace $q$, the model is written as:

$$X = y(t, k^*, R) + N, \qquad (1)$$

where for notational convenience the dependency in $q$ and $d$ has been dropped. Here $X$ is a leakage measurement; $y = y(t, k^*, R)$ is the deterministic part of the model that depends on the correct key $k^*$, some known text (plaintext or ciphertext) $t$, and the unknown random values (masks and permutations) $R$. Each sample (of index $d$) of $N$ is a random noise, which follows a Gaussian distribution $p_N(z) = \frac{1}{\sqrt{2\pi\sigma^2}} \exp\left(-\frac{z^2}{2\sigma^2}\right)$.

Uppercase letters are generally used for random variables and the corresponding lowercase letters for their realizations. Bold symbols are used to denote vectors that have length $Q$, the number of measurements. Namely, $\mathbf{X}$ denotes a set of $Q$ random variables i.i.d. with the same law as $X$. So, $\mathbf{X}$ is a $Q \times D$ matrix; $\mathbf{R}$ denotes a set of random variables i.i.d. with the same law as $R$; $\mathbf{t}$ denotes the set of input texts of the measurements $\mathbf{X}$; $y(\mathbf{t}, k, \mathbf{R})$ denotes the set of leakage models, where $k$ is a key guess, $k^*$ being the correct key value.

Notations $\mathbf{X}_d$ and $\mathbf{X}^{(q)}$ are used to denote the $d$-th column and the $q$-th line of the matrix $\mathbf{X}$, respectively.

We are interested in attacks where each intermediate data is a $n$-bit vector. In particular, we target S-boxes, denoted by $S$. Regarding the transduction from

---

[1] We recall that, even if the templates are perfectly known, the online attack phase still requires $\mathcal{O}(Q \cdot 2^{n(\Omega-1)} \cdot \Pi!)$ computations.

the intermediate variable to the real-valued leakage, we take the example of the Hamming weight $w_H$ defined by $w_H(z) = \sum_{i=1}^{n} z_i$ where $z_i$ is the $i$th bit of $z$.

# 3    A Generic Log-Likelihood for Masked Implementations

In this section we derive a rounded version of Template Attack. Namely we expand a particular instantiation of the template attack the so-called optimal distinguisher using its Taylor Expansion. By rounding this expansion at the $L$th degree we are able to build a rounded version of the optimal distinguisher (later defined as $\text{ROPT}_L$). This attack features two advantages: it allows to combine different statistical moments and its complexity becomes manageable.

## 3.1    Maximum Likelihood (ML) Attack

The most powerful adversary knows exactly the leakage model (but the actual key, the masks, and the noise are unknown during the online step) and computes a likelihood. In the case of masking the optimal distinguisher which maximize the success rate is given by [6]:

**Theorem 1 (Maximum Likelihood).** *When the $y\,(t, k, R)$ are known and the Gaussian noise $N$ is i.i.d. across the queries (measurements) and independent across the dimension, then the optimal distinguisher is:*

$$\text{OPT}: \mathbb{R}^{DQ} \times \mathbb{R}^{DQ} \longrightarrow \mathbb{F}_2^n$$

$$(\mathbf{x}, y\,(\mathbf{t}, k, R)) \longmapsto \underset{k \in \mathbb{F}_2^n}{\operatorname{argmax}} \sum_{q=1}^{Q} \log \mathbb{E} \exp \frac{-\|x^{(q)} - y(t^{(q)}, k, R)\|^2}{2\sigma^2} \quad (2)$$

*where the expectation operator $\mathbb{E}$ is applied with respect to the random variable $R \in \mathcal{R}$, and the norm is the Euclidean norm $\|x^{(q)} - y(t^{(q)}, k, R)\|^2 = \sum_{d=1}^{D}(x_d^{(q)} - y_d(t^{(q)}, k, R))^2$.*

*Proof.* It is proven in [6] that the Maximum Likelihood distinguisher is:

$$\underset{k \in \mathbb{F}_2^n}{\operatorname{argmax}} \prod_{q=1}^{Q} \sum_{r \in \mathcal{R}} \mathbb{P}\,(r)\,p\left(x^{(q)}|y\left(t^{(q)}, k, r\right)\right).$$

Applying (1) for Gaussian noise and taking the logarithm yields (2).    □

In the sequel, we denote by $LL^{(q)} = \log \mathbb{E}_R \exp \frac{-\|x^{(q)} - y(t^{(q)}, k, R)\|^2}{2\sigma^2}$ the contribution of one trace $q$ of the Log-Likelihood full distinguisher $LL = \sum_{q=1}^{Q} LL^{(q)}$.

*Remark 1.* Notice that for each trace $q$, the Maximum Likelihood distinguisher involves a summation over $\#\mathcal{R}$ values, which correspond to $\#\mathcal{R}$ accesses to precharacterized templates.

If $D = 1$, then the signal-to-noise ratio (SNR) is defined in a natural way as the ratio between the variance of the model $Y$ and the variance of the noise $N$. But when the setup is multivariate, it is more difficult to quantify a notion of SNR. For this reason, we use the following quantity

$$\gamma = \frac{1}{2\sigma^2}, \tag{3}$$

which is actually proportional to an SNR, in lieu of SNR. In practice, we assume that $\gamma$ is small. It is indeed a condition for masking schemes to be efficient (see for instance [12]).

**Proposition 1 (Taylor Expansion of Optimal Attacks in Gaussian Noise).** *The attack consists in maximizing the sum over all traces $q = 1, \ldots, Q$ of*

$$\sum_{\ell=1}^{+\infty} \frac{\kappa_\ell}{\ell!} (-\gamma)^\ell, \tag{4}$$

*where $\kappa_\ell$ is the $\ell$th-order cumulant of the random variable $\|x - y(t, k, R)\|^2$, which can be found inductively from $\ell$th-order moments:*

$$\mu_\ell = \mathbb{E}_R\big(\|x - y(t, k, R)\|^{2\ell}\big), \tag{5}$$

*using the relation:*

$$\kappa_\ell = \mu_\ell - \sum_{\ell'=1}^{\ell-1} \binom{\ell-1}{\ell'-1} \kappa_{\ell'} \mu_{\ell-\ell'} \qquad (\ell \geq 1). \tag{6}$$

*Proof.* The log-likelihood can be expanded according to the increasing powers of the SNR as:

$$\log \mathbb{E} \exp\big(-\gamma \|x - y(t, k, R)\|^2\big) = \sum_{\ell=1}^{+\infty} \frac{\kappa_\ell}{\ell!} (-\gamma)^\ell, \tag{7}$$

where we have recognized the cumulant generating function [34]. The above relation (6) between cumulants and moments is well known [39].  □

**Definition 1.** *The Taylor expansion of the log-likelihood truncated to the $L$th degree $\mathrm{LL}_L$ in SNR is*

$$\mathrm{LL}_L = \sum_{\ell=1}^{L} (-1)^\ell \kappa_\ell \frac{\gamma^\ell}{\ell!}. \tag{8}$$

Put differently, we have $\mathrm{LL} = \mathrm{LL}_L + o(\gamma^L)$ (using the Landau notation). The optimal attack can now be "rounded" in the following way:

**Definition 2 (Rounded OPTimal Attack of Degree $L$ in $\gamma$).** *The rounded optimal $L$th-degree attack consists in maximizing over the key hypothesis the* *sum over all traces of the Lth order Taylor expansion* $\mathrm{LL}_L$ *in the SNR of the log-likelihood :*

$$\mathrm{ROPT}_L\colon \mathbb{R}^{DQ} \times \mathbb{R}^{DQ} \longrightarrow \mathbb{F}_2^n$$
$$(\mathbf{x}, y\,(\mathbf{t}, k, R)) \longmapsto \underset{k\in\mathbb{F}_2^n}{\mathrm{argmax}}\, \mathrm{LL}_L. \qquad (9)$$

**Proposition 2.** *If the degree $L$ is smaller than the order $O$ of the countermeasure then the attack fails to distinguish the correct key.*

*Proof.* One can notice that $\mu_\ell$ combines (by a product) a most $\ell$ terms following the formula:

$$\mu_\ell = \sum_{k_1+\ldots+k_D=\ell} \binom{\ell}{k_1,\ldots,k_D} \mathbb{E} \prod_{0<i<D+1} (x_i - y_i)^{2\cdot k_i},$$

with $k_1 + \ldots + k_d = \ell$. It implies that it exits at most $\ell$ different $k_i > 0$ and as a consequence there are at most $\ell$ different variables in the expectation. Therefore by definition of a perfect masking scheme $\mu_L$ does not depend on the key. As a consequence $\mathrm{LL}_L$ with $L < O$ neither depends on the key. $\qquad\square$

**Theorem 2.** *Let an implementation be secure at order $O$. The lowest-degree successful attack is the one at degree $L = O$ which maximizes $\mathrm{LL}_L$. This is equivalent to summing*

$$\mu_L = \mathbb{E}_R\big(\|x - y(t, k, R)\|^{2L}\big),$$

*over all traces and*

- *maximize the result over the key hypotheses, if $L$ is even;*
- *minimize the result over the key hypotheses, if $L$ is odd.*

*Proof.* Since $\kappa_\ell$ is independent of $k$ for all $\ell \leq L$, the first sensitive contribution to the log-likelihood is

$$(-1)^L \kappa_L \frac{\gamma^L}{L!}.$$

Now, $\kappa_L = \mu_L +$ lower order terms (which do not depend on the key as the implementation is secure at order $O$), and removing constants independent of $k$ the contribution to the log-likelihood reduces to $(-1)^L \mu_L$. $\qquad\square$

**Theorem 3 (Mixed Degree Attack).** *Assuming an implementation secure at order $O$, the next degree successful attack is the one at degree $L + 1 = O + 1$ which maximizes $\mathrm{LL}_{L+1}$. This is equivalent to summing*

$$\mu_L(1 + \gamma\mu_1) - \gamma\frac{\mu_{L+1}}{L+1},$$

*over all traces and*

- *maximize the result over the key hypotheses, if $L$ is even;*
- *minimize the result over the key hypotheses, if $L$ is odd.*

*Proof.* The $(L+1)$th-order term in the log-likelihood becomes

$$(-1)^L \kappa_L \frac{\gamma^L}{L!} + (-1)^{L+1} \frac{\kappa_{L+1}}{(L+1)!} \gamma^{L+1}.$$

Now from (6) we have, for $L > 0$

$$\kappa_{L+1} = \mu_{L+1} - (L+1)\mu_L\mu_1 + \text{ lower-order terms.}$$

Removing terms that do not depend on $k$, we obtain:

$$(-1)^L \gamma^L \left( \mu_L - \gamma(\frac{\mu_{L+1}}{L+1} - \mu_L\mu_1) \right).$$

Compared to a $L$th-degree attack, we see that $\mu_L$ is replaced by a corrected version:

$$\mu_L(1 + \gamma\mu_1) - \gamma\frac{\mu_{L+1}}{L+1},$$

where $\mu_1$ is independent of $k$. However, $\mu_1$ cannot be removed as it scales the relative contribution of $\mu_L$ and $\mu_{L+1}$ in the distinguisher. $\qquad\square$

*Remark 2.* In contrast to $\text{LL}_L$, implementing $\text{LL}_{L+1}$ requires knowledge of the SNR parameter $\gamma = 1/2\sigma^2$.

*Remark 3.* In general, when $L \geq O$ the rounded optimal attack $\text{ROPT}_L$ exploits all key dependent terms of degree $\ell$, where $O \leq \ell \leq L$, whereas an $LO$-CPA [8] or MCP-DPA [22] only exploit the term of degree $L$.

## 4    Case Study: Shuffled Table Recomputation

In this section we apply the $\text{ROPT}_L$ formula of Eq. (9) of Definition 2 to the particular case of a block cipher with a shuffled table recomputation stage. We show that in this scenario our new method allows to build a better attack than that from the state-of-the-art. By combining the second and the third cumulants we construct an attack which is better than:

- any second-order attack;
- the attack presented at CHES 2015. Following the notations of [7] we denote this attack by $\text{MVA}_{TR}$ (which stands for Multi-Variate Attack on Table Recomputation) in the rest of this article. This is a third-order attack that achieves better results than 2O-CPA when the noise level $\sigma$ is below a given threshold (namely $\sigma^2 \leq 2^{n-2} - n/2$).

### 4.1    Parameters of the Randomization Countermeasure

In order to validate our results we take as example a first order ($O = 2$), masking scheme where the sensitive variables are split into two shares ($\Omega = 2$). The nonlinear part of this scheme is computed using a table recomputation stage. This step is shuffled ($\Pi = 2^n$) for protection against some known attacks [26,36]. The beginning of this combined countermeasure is given in Algorithm 1. The table is recomputed in a random order from line 3 to line 7.

---

**Algorithm 1.** Beginning of computation of a block cipher masked by table recomputation in a random order

---

**input** : $t$, one byte of plaintext, and $k$, one byte of key
**output:** The application of AddRoundKey and SubBytes on $t$, i.e.,
$S[t \oplus k]$

// Table precomputation protected by shuffling ...............

1 $m \leftarrow_{\mathcal{R}} \mathbb{F}_2^n$, $m' \leftarrow_{\mathcal{R}} \mathbb{F}_2^n$     // Draw of random input and output masks
2 $\varphi \leftarrow_{\mathcal{R}} \mathbb{F}_2^n \rightarrow \mathbb{F}_2^n$     // Draw of random permutation of $\mathbb{F}_2^n$
3 **for** $\varphi(\omega) \in \{\varphi(0), \varphi(1), \ldots, \varphi(2^n - 1)\}$ **do**     // S-box masking
4 $\quad$ $z \leftarrow \varphi(\omega) \oplus m$     // Masked input
5 $\quad$ $z' \leftarrow S[\varphi(\omega)] \oplus m'$     // Masked output
6 $\quad$ $S'[z] = z'$     // Creating the masked S-box entry
7 **end**

// Masked computation ........................................

8 $t \leftarrow t \oplus m$     // Plaintext masking
9 $t \leftarrow t \oplus k$     // Masked AddRoundKey
10 $t \leftarrow S'[t]$     // Masked SubBytes
11 $t \leftarrow t \oplus m'$     // Demasking
12 **return** $t$

---

We used lower case letter (e.g., $m$, $\varphi$) for the realizations of random variables, written upper-case (e.g., $M$, $\Phi$). For the sake of simplicity in the rest of this case study, we assume that $m = m'$.

An overview of the leakages over time is given in Fig. 1.

We detail below the mathematical expression of these leakages. The randomization consists in one mask $M$ chosen randomly in $\{0, 1\}^n$, and one shuffle (random permutation of $\{0, 1\}^n$) denoted by $\Phi$. Thus, we denote $R = (M, \Phi)$, which is uniformly distributed over the Cartesian product $\{0, 1\}^n \times S_{2^n}$ (i.e. $\mathcal{M} = \{0, 1\}^n$ and $\mathcal{S} = S_{2^n}$), where $S_m$ is the symmetric group of $m$ elements. We have $D = 2^{n+1} + 2$ leakage models, namely:

- $X_0 = y_0(t, k, R) + N_0$ with $y_0(t, k, R) = w_H(M)$,
- $X_1 = y_1(t, k, R) + N_1$ with $y_1(t, k, R) = w_H(S[T \oplus k] \oplus M)$,
- $X_i = y_i(t, k, R) + N_i$, for $i = 2, \ldots, 2^n + 1$ with $y_i(t, k, R) = w_H(\Phi(i-2) \oplus M)$,
- $X_j = y_j(t, k, R) + N_j$, for $j = 2^n + 2, \ldots, 2^{n+1} + 1$ with $y_j(t, k, R) = w_H(\Phi(j - 2^n - 2))$.

We recall that we assume the noises $N$ are i.i.d. Clearly, there is a second-order leakage, as the pair $(X_0, X_1)$ does depend on the key. But there is also a large multiplicity of third-order leakages, such that $(X_1, X_i, X_{j=i+2^n})$, as will be analyzed in this case-study.

The following side-channel attacks are applied on a set of $Q$ realizations. Let us define $I$ and $J$ as $I = [\![2, 2^n + 1]\!]$ and $J = [\![2^n + 2, 2 \times 2^n + 1]\!]$. Then the maximal dimensionality is $D = 2 + 2 \times 2^n$, and we denote a sample $d$ as $d \in \{0, 1\} \cup I \cup J$. The $Q$ leaks (resp. models) at sample $d$ are denoted as $\mathbf{x}_d$ and $\mathbf{y}_d = y_d(\mathbf{t}, k, R)$.

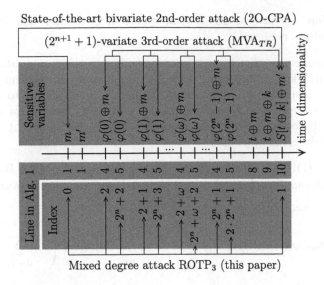

**Fig. 1.** Leakages of the shuffled table recomputation scheme

In order to simplify the notations we introduce

$$f_d^{(q)} = \left( x_d^{(q)} - y_d\left( t^{(q)}, k, R \right) \right)^2, \tag{10}$$

with $d \in \{0,1\} \cup I \cup J$. The $^{(q)}$ can be omitted where there is no ambiguity.

## 4.2   Second-Order Attacks

As any other high order masking scheme, our example can be defeated by High Order Attacks [8,20,29,38]. As our scheme is a first order masking scheme with two shares it can be defeated using a second order attack [8,20] which combines the leakages of the two shares using a *combination function* [8,20,25] such as the second order CPA (2O-CPA) with the centered product as combination function.

Using our notation it implies $D = 2$.

**Definition 3** (2O-CPA [29]). *We denote by* 2O-CPA *the* CPA *using the centered product as combination function. Namely:*

$$\text{2O-CPA: } \mathbb{R}^Q \times \mathbb{R}^Q \times \mathbb{R}^Q \longrightarrow \mathbb{F}_2^n$$
$$(\mathbf{x_0}, \mathbf{x_1}, \mathbf{y}) \longmapsto \underset{k \in \mathbb{F}_2^n}{\text{argmax}} \, \widehat{\rho}\left[\mathbf{x_0} \circ \mathbf{x_1}, \mathbf{y}\right], \tag{11}$$

*where* $\mathbf{y} = \mathbb{E}_M \left( y_0\left(\mathbf{t}, k, R\right) \circ y_1\left(\mathbf{t}, k, R\right) \right)$, $\circ$ *is the element wise product and* $\widehat{\rho}$ *is an estimator of the Pearson coefficient. It can be noticed that as the terms* $y_0\left(\mathbf{t}, k, R\right)$ *and* $y_1\left(\mathbf{t}, k, R\right)$ *only depend on* $M$ *the expectation is only computed over* $\mathcal{M}$.

*Remark 4.* Here we have assumed without loss of generality that the leakages and the model are centered.

An attacker can restrict himself in order to ignore the recomputation stage. Since such attacker ignores the table recomputation no random shuffle is involved. As a consequence the optimal distinguisher restricted to these leakages becomes computable. Nevertheless as we will see in Sect. 6 this approach is not the best. Indeed a lot of exploitable information is lost by not taking into account the table recomputation.

**Definition 4 (OPT$_{2O}$ Distinguisher — *Eq. (2) for $D = 2$*).** *We define by* OPT$_{2O}$ *the optimal attack which targets the mask and the masked sensitive value.*

$$\text{OPT}_{2O}: \quad \mathbb{R}^{2Q} \times \mathbb{R}^{2Q} \& \to \quad \mathbb{F}_2^n$$

$$(\mathbf{x}_d, y_d\,(t, k, R))_{d \in \{0,1\}} \mapsto \underset{k \in \mathbb{F}_2^n}{\text{argmax}} \sum_{q=1}^{Q} \log \mathbb{E} \exp \left( -\gamma \sum_{d \in \{0,1\}} f_d^{(q)} \right), \tag{12}$$

*with $f_d^{(q)}$ as defined in Eq. (10).*

### 4.3 Exploiting the Shuffled Table Recomputation Stage

It is known that the table recomputation step can be exploited to build better attacks than second order attacks [6,36]. Recently a new attack has been presented which remains better than the 2O-CPA even when the recomputation step is protected [7]. Let us recall the definition of this attack:

**Definition 5 (MVA$_{TR}$ [7]).** *The MultiVariate Attack (MVA) exploiting the leakage of the table recomputation (TR) is given by the function:*

$$\text{MVA}_{TR}: \mathbb{R}^{Q(2^{n+1}+1)} \times \mathbb{R}^Q \longrightarrow \mathbb{F}_2^n$$

$$(\mathbf{x}_d, \mathbf{y})_{d \in \{1\} \cup I \cup J} \mapsto \underset{k \in \mathbb{F}_2^n}{\text{argmax}} \, \widehat{\rho} \left[ \left( -\frac{1}{2} \sum_{i \in I, j=i+2^n} \mathbf{x_i} \circ \mathbf{x_j} \right) \circ \mathbf{x_1}, \mathbf{y} \right], \tag{13}$$

*where, like for Definition 3, $\mathbf{y} = \mathbb{E}_M (y_0\,(t, k, R) \circ y_1\,(t, k, R))$, $\circ$ is the element wise product and $\widehat{\rho}$ is an estimator of the Pearson coefficient.*

Let us now apply our new ROPT$_L$ on a block cipher protected with a shuffled table recomputation. In this case the lower moments are given by:

$$\mu_\ell = \mathbb{E}\left[ \left( \sum_d f_d \right)^\ell \right] = \mathbb{E}\left[ \left( \underbrace{f_0}_{S[t \oplus k] \oplus M} + \underbrace{f_1}_{M} + \sum_{i \in I} \underbrace{f_i}_{\Phi(\omega) \oplus M} + \sum_{j \in J} \underbrace{f_j}_{\Phi(\omega)} \right)^\ell \right].$$

**Proposition 3.** *The second degree rounded optimal attack on the table recomputation is:*

$$\text{ROPT}_2: \qquad \mathbb{R}^{2Q} \times \mathbb{R}^{2Q} \qquad \longrightarrow \mathbb{F}_2^n$$

$$(\mathbf{x}_d, y_d\,(\mathbf{t}, k, R))_{d \in \{0,1\}} \longmapsto \underset{k \in \mathbb{F}_2^n}{\text{argmax}} \sum_{q=1}^{Q} \mathbb{E}(f_0^{(q)} \times f_1^{(q)}). \qquad (14)$$

*Proof.* Combine Theorem 2 and Eq. (30) of Appendix A.2. □

*Remark 5.* The $\text{ROPT}_2$ which targets the second order moment happens not to take into account the terms of the recomputation stage. Naturally the only second order leakages are also the ones used by 2O-CPA and $\text{OPT}_{2O}$ distinguishers.

**Proposition 4.** *The third degree rounded optimal attack on the table recomputation is:*

$$\text{ROPT}_3: \ \mathbb{R}^{(2^{n+1}+2)Q} \times \mathbb{R}^{(2^{n+1}+2)Q} \ \longrightarrow \mathbb{F}_2^n$$

$$(\mathbf{x}_d, y_d\,(\mathbf{t}, k, R))_{d \in \{0,1\} \cup I \cup J} \longmapsto \underset{k \in \mathbb{F}_2^n}{\text{argmax}} \sum_{q=1}^{Q} \mu_2^{(q)}(1 + \gamma\mu_1^{(q)}) - \gamma\frac{\mu_3^{(q)}}{3},$$

$$(15)$$

*where the values of $\mu_1^{(q)}$, $\mu_2^{(q)}$ and, $\mu_3^{(q)}$ are respectively provided in Eq. (22) of Appendix A.1, Eq. (30) of Appendix A.2 and Eq. (33) of Appendix A.3.*

*Proof.* Combining Theorem 2 and Appendix A. □

**Proposition 5.** *To compute $\mu_1$, $\mu_2$ and $\mu_3$ an attacker does not need to compute the expectation over $S_{2^n}$.*

*Proof.* Proof given in Appendix A. □

## 5   Complexity

In this section we give the *time* complexity needed to *compute* OPT and $\text{ROPT}_L$. We also show that when $L \ll D$ the complexity of $\text{ROPT}_L$ remains manageable whereas the complexity of OPT is prohibitive. In this section all the complexities are computed for one key guess.

### 5.1   Complexity in the General Case

Let us first introduce an intermediate lemma.

**Lemma 1.** *The complexity of computing $\mu_\ell$ (for one trace) is lower than:*

$$\mathcal{O}\left(\binom{D+\ell-1}{\ell} \cdot 2^{(\Omega-1)n} \cdot \binom{\Pi}{\min\left(\lceil\frac{\Pi}{2}\rceil, \ell\right)}\right). \qquad (16)$$

*Proof.* See Appendix B.1.    □

**Proposition 6.** *The complexity of* OPT *is:*

$$\mathcal{O}\left(Q \cdot (2^n)^{\Omega-1} \cdot \Pi! \cdot D\right). \tag{17}$$

*The complexity of* $\mathrm{ROPT}_L$ *is lower than:*

$$\mathcal{O}\left(Q \cdot L \cdot \binom{D+L-1}{L} \cdot 2^{(\Omega-1)n} \cdot \binom{\Pi}{\min\left(\lceil \frac{\Pi}{2} \rceil, L\right)}\right). \tag{18}$$

*Proof.* The proof is given in Appendix B.2.    □

Proposition 6 allows to compare the complexity of the two attacks. One can notice that there are still terms with $\Pi!$ or $D!$ in $\mathrm{ROPT}_L$ such as $\binom{D+L-1}{L}$ or $\binom{\Pi}{\min(\lceil \frac{\Pi}{2} \rceil, L)}$. Nevertheless these two terms can be seen as constants where $L \ll D$. As a consequence we have the following remark.

***Important Remark.*** When the degree $L$ of the attack $\mathrm{ROPT}_L$ is such that $L \ll D$ the complexity of OPT is much higher than the complexity of $\mathrm{ROPT}_L$. Indeed the main term for OPT is $\Pi!$ whereas the one for $\mathrm{ROPT}_L$ is $2^{(\Omega-1)n}$.

**Proposition 7.** *The complexity of* $\mathrm{ROPT}_L$ *can be reduced to* $\mathcal{O}\left(Q \cdot L \cdot \binom{D+L-1}{L}\right)$ *with a precomputation in* $\mathcal{O}\left(L \cdot \binom{D+L-1}{L} \cdot 2^{(\Omega-1)n} \cdot \binom{\Pi}{\min(\lceil \frac{\Pi}{2} \rceil, L)}\right)$.

*Proof.* See Appendix B.3.    □

This means that for $Q$ large enough i.e. when $\gamma$ is low enough this computational "trick" allows a speed-up factor of $2^{(\Omega-1)n}\binom{\Pi}{\min(\lceil \frac{\Pi}{2} \rceil, L)}$. The idea is to output the values depending on the queries from the computation of the expectations. These expectations only depend on the model which can be computed only once.

## 5.2   Complexity of Our Case Study

Let us now compute the complexity of these two distinguishers applied to our case study. Of course an approach could be to use the formula of the previous Sect. 5.1. But one can notice that a lot of terms could be independent of the key and as consequence not needed in an attack. Another approach is to use the formula of the distinguisher.

**Proposition 8.** *The complexity of* OPT *is:*

$$\mathcal{O}\left(Q \cdot (2^n) \cdot 2^n! \cdot \left(2^{n+1} + 2\right)\right). \tag{19}$$

*The complexity of* $\mathrm{ROPT}_2$ *is:*

$$\mathcal{O}\left(Q \cdot 2^n\right). \tag{20}$$

*The complexity of* $\mathrm{ROPT}_3$ *is lower than:*

$$\mathcal{O}\left(Q \cdot 2^{4n}\right). \tag{21}$$

*Proof.* See Appendix B.4.                                                                □

*Remark 6.* As already mentioned an attacker can ignore the leakages of the table recomputation and only target the two shares. In such case the complexity of $OPT_{2O}$ (Definition 4) is $\mathcal{O}\left(Q \cdot (2^n)\right)$. With the result of Proposition 7 the complexity of $ROPT_2$ reduces to $\mathcal{O}(Q)$.

*Remark 7.* Using the result of Proposition 7 the complexity of $ROPT_3$ can be reduced to $\mathcal{O}\left(Q \cdot 2^{2n}\right)$ with a precomputation step of $\mathcal{O}\left(2^{2n}\right)$.

*Remark 8.* A summary of the complexity, and the computation time of the distinguishers are provided in Appendix B.5 in Table 1.

# 6  Simulation Results

In this section we validate in simulation the soundness of our approach for the case study described in Sect. 4.1. The results of these simulations are expressed in success rate (defined in [32] and denoted by SR). All simulations are computed using the Hamming weight model as a leakage model. As we assume an attacker with a perfect knowledge, the leakages are the model (denoted by $y$) plus some noise. The noise is Gaussian with a standard deviation of $\sigma$.

In Subsect. 6.1 we assume that the attacker does not take into account the table recomputation stage. He only targets the leakages of the mask and the masked share (the leakage of masked S-Box). Namely the leakages which occurs in lines 1 and 10 of Algorithm 1. This approach allows to compute the restricted version of the maximum likelihood. We compare the results of the maximum likelihood, our rounded version and the high order attacks.

In Subsect. 6.2 we present our main results. In this subsection the attacker can exploit the leakage of the mask, the masked share and all the leakages of the table recomputation. In this scenario we show that our rounded version of the optimal distinguisher outperforms all the attacks of the state-of-the-art.

## 6.1  Exploiting only Leakage of the Mask and the Masked Share

In this subsection all the attacks are computed using only the leakages of the line 1 and the line 10 of Algorithm 1.

In this case study we assume a perfect masking scheme with: $Y_0 = w_H(M)$ and $Y_1 = w_H(S[T \oplus k] \oplus M)$.

It can be seen in Fig. 2 that even for small noise ($\sigma = 1$, Fig. 2a) the 2O-CPA and $ROPT_2$ are equivalent. Indeed the two curves superimpose almost perfectly (in order to better highlight a difference, as many as 1000 attacks have been carried out for the estimation of the success rate). Moreover these two attacks are nearly equivalent to the optimal distinguisher (we recover here the results of [6]). We can notice that for both $\sigma = 1$ and $\sigma = 2$, $ROPT_4$ is not as good as $ROPT_2$. This means that the noise standard deviation is not large enough for approximations of higher degrees to be accurate. Indeed when the noise is not low enough the weight of each term of the decomposition can be such that some useful terms vanish due to the alternation of positive and negative terms in the Taylor expansion.

Let us recall that the decomposition of Eq. (8) is valid only for low $\gamma = 1/(2\sigma^2)$ i.e. high noise. The error term ( $o(\gamma^L)$ ) in the Taylor expansion gives the asymptotic evolution of this error when the noise increases but does not provide information about the error for a fixed value of noise variance. This means that the noise is too small for $ROPT_4$ to be a good approximation of OPT although $ROPT_2$ is nearly equivalent to OPT.

For $\sigma = 2$ the noise is high enough to have a good approximation of OPT by $ROPT_4$. For this noise all the attacks are close to OPT (Fig. 2b).

In the context where only the mask and the masked share are used it is equivalent to compute the 2O-CPA, $ROPT_2$ and OPT. As a consequence in the rest of this article only the 2O-CPA will be displayed.

To conclude our $ROPT_L$ is in this scenario at least as good as the HO-CPA of order $L$, which validates the optimality of state-of-the-art attacks against perfect masking schemes of order $O = L$.

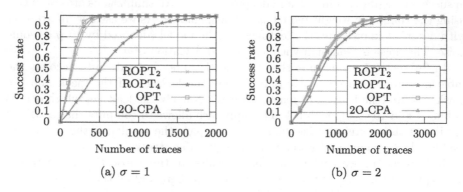

(a) $\sigma = 1$                    (b) $\sigma = 2$

**Fig. 2.** Bivariate attacks

## 6.2    Exploiting the Shuffled Table Recomputation

In this subsection the attacker can target the leakage of the mask, the masked share and all the leakages occurring during the table recomputation. As a consequence the attacks of Subsect. 6.1 remain possible. It has been shown in [6,33] that the 2O-CPA with the centered product becomes close to the $OPT_{2O}$ (the Maximum Likelihood) when the noise becomes high. It is moreover confirmed by our simulation results as it can be seen in Fig. 2. We choose as attack reference for the Fig. 3 the 2O-CPA and not the $OPT_{2O}$ because it performs similarly Fig. 2 and it is much faster to compute (see Table 1) which is mandatory for attacks with high noise (e.g. for $\sigma = 12$) which involve many traces.

Following the formulas provided previously empirical validations have been done. For $\sigma \leq 8$ the attacks have been redone 1000 times to compute the SR. For $\sigma > 8$ the attacks have been done 250 times. Results are plotted in Fig. 3.

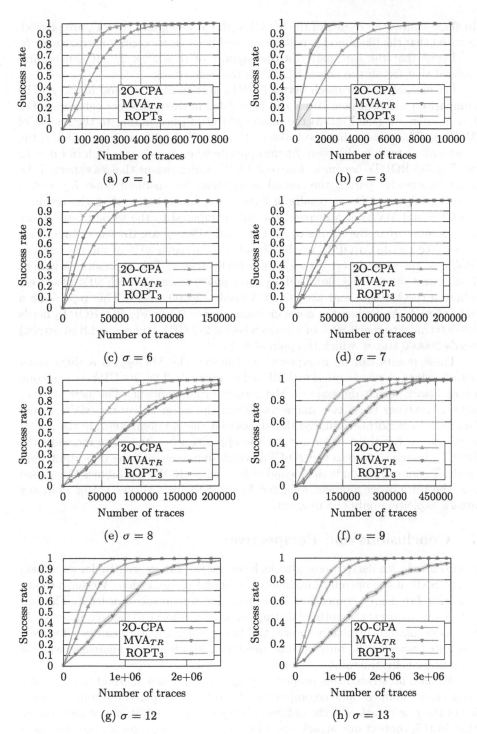

**Fig. 3.** Attack on shuffled table recomputation

In these figures the results of the 2O-CPA, the $MVA_{TR}$ and $ROPT_3$ are plotted. Noticed that the likelihood is not represented because we cannot average over $R$.

Recall that the cardinality of the support of $R$ is $2^n \times 2^n!$. It can be first noticed that for all the noises $ROPT_3$ is the best attack.

Let us analyze how much better $ROPT_3$ is than 2O-CPA and $MVA_{TR}$. The comparison with our new attack can be divided in three different categories. For low noise $\sigma = 3$ (see Fig. 3b) the results of $ROPT_3$ are similar to the results of $MVA_{TR}$. This means that the leakage of the shuffled table recomputation is the most leaking term in this case. At the opposite when the noise is high (for $\sigma = 12$ see Fig. 3g) $ROPT_3$ becomes close to 2O-CPA which means that as expected the most informative part is the second order term. For medium noise $7 \leq \sigma \leq 9$ (see Fig. 3d, e and f) the results of $ROPT_3$ are much better than the result of 2O-CPA and $MVA_{TR}$. Moreover, the gain compared to the second best attack is maximum when the results of 2O-CPA and $MVA_{TR}$ are the same. Indeed for $\sigma = 7$ (see Fig. 3d), $ROPT_3$ needs 35000 traces to reach 80 % of success whereas $MVA_{TR}$ (the second best attack) needs 60000 traces. This represents a gain of 71 %. For $\sigma = 8$ (see Fig. 3e), $ROPT_3$ needs 65000 traces to reach 80 % of success whereas the $MVA_{TR}$ and the 2O-CPA needs 120000 traces. This represents a gain of 85 %. And when the noise increases to $\sigma = 9$ (see Fig. 3f), $ROPT_3$ needs 120000 traces to reach 80 % of success whereas 2O-CPA (the second best attack) needs 200000 traces, which is a gain of 66 %.

These results can be interpreted as follows: The $MVA_{TR}$ is a third order attack which depends on the third order moment. The 2O-CPA is a second order attack which depends on the second order moment. The new $ROPT_3$ attack combines these two moments. When the noise is low the $MVA_{TR}$ and the $ROPT_3$ performs similarly; this shows that the dominant term in the Taylor expansion is the third order one. At the opposite when the noise increases the $ROPT_3$ becomes close to the 2O-CPA which indicates that the important term in the Taylor expansion is the second order one. As $ROPT_3$ combines the second and the third order moment weighted by the SNR it is always better than any attack exploiting only one moment.

# 7    Conclusions and Perspectives

In this article, we derived new attacks based on the $L$th degree Taylor expansion in the SNR of the optimal Maximum Likelihood distinguisher. We have shown that this $L$th degree truncation allows to target a moment of order $L$. The new attack outperforms the optimal distinguisher with respect to time complexity. In fact as we have theoretically shown, the Taylor approximation can be effectively computed whereas the fully optimal maximum likelihood distinguisher, was not computationally tractable.

We have illustrated this property by applying our new method in a complex scenario of "shuffled table recomputation" and have compared the time complexity of the new attack and the optimal distinguisher. In addition, we have shown that in this context our attack has a higher success rate than all the attacks of the state-of-art over all possible noise variances.

An open question is how to quantify the accuracy of the approximation $LL \longrightarrow LL_\ell$ as a function of the noise. In other words, what is the optimal degree of the Taylor expansion of the likelihood for a given SNR? Another interesting extension of this framework would be on hardware devices which are known to leak at various orders (see the real-world examples in [21–23]).

## A    Computation of the Moments

### A.1    Computation of $\mu_1$

There is no computational difficulty:

$$\mu_1 = \mathbb{E}(f_0) + \mathbb{E}(f_1) + \sum_{i \in I} \mathbb{E}(f_i) + \sum_{j \in J} \mathbb{E}(f_j). \tag{22}$$

Now, when there is no $\varphi$ in the R.V., then the expectation is only on $M$ (indeed, $\frac{1}{2^n!} \sum_{\varphi \in S_{2^n}} 1 = 1$). Thus,

$$E(f_0) = \frac{1}{2^n} \sum_{m \in \mathbb{F}_2^n} (x_0 - w_H(S[t \oplus k] \oplus m))^2 = \frac{1}{2^n} \sum_{m \in \mathbb{F}_2^n} (x_0 - w_H(m))^2, \tag{23}$$

which cannot further be simplified (in the simulations, it will be computed by the computer).

Similarly

$$E(f_1) = \frac{1}{2^n} \sum_{m \in \mathbb{F}_2^n} (x_1 - w_H(S[t \oplus k] \oplus m))^2 = \frac{1}{2^n} \sum_{m \in \mathbb{F}_2^n} (x_1 - w_H(m))^2. \tag{24}$$

When there is an expectation on $\Phi$, then at order one, it considers **only one value** $\Phi(\omega)$. It is uniformly distributed, hence one can replace the expectation on $\Phi$ by an expectation on one value of $\varphi$, we call $M'$. For instance:

$$\mathbb{E}(f_i) = \frac{1}{2^n!} \sum_{\varphi \in S_{2^n}} (x_i - w_H(\varphi(\omega)))^2$$

$$= \frac{1}{2^n} \sum_{m' \in \mathbb{F}_2^n} (x_i - w_H(m'))^2, \tag{25}$$

which can thus be computed with the same *average* method as $\mathbb{E}(f_0)$.

Lastly, when there is both $M$ and $\Phi(\omega)$, then whichever variable can absorb the other one, since both are uniformly distributed on $\mathbb{F}_2^n$. This means that:

$$\mathbb{E}(f_j) = \frac{1}{2^n} \sum_{m \in \mathbb{F}_2^n} \frac{1}{2^n!} \sum_{\varphi \in S_{2^n}} (x_j - w_H(\varphi(\omega) \oplus m))^2$$

$$= \frac{1}{2^{2n}} \sum_{m, m' \in \mathbb{F}_2^n} (x_j - w_H(m \oplus m'))^2$$

$$= \frac{1}{2^{2n}} \sum_{\widetilde{m}, m' \in \mathbb{F}_2^n} (x_j - w_H(\widetilde{m} \oplus m' \oplus m'))^2 \qquad \text{where } \widetilde{m} = m \oplus m' \quad (26)$$

$$= \frac{1}{2^n} \sum_{\widetilde{m} \in \mathbb{F}_2^n} (x_j - w_H(\widetilde{m}))^2, \qquad\qquad\qquad (27)$$

which is once again a similar computation as done for computing $\mathbb{E}(f_0)$.

## A.2  Computation of $\mu_2$

Recall that only the key dependent terms of $\mu_2$ are needed for ROPT$_2$ and ROPT$_3$.

   Notice that the square terms are computed as the non-square terms. For instance,

$$\mathbb{E}(f_0^2) = \frac{1}{2^n} \sum_{m \in \mathbb{F}_2^n} (x_0 - w_H(S[t \oplus k] \oplus m))^4 = \frac{1}{2^n} \sum_{m \in \mathbb{F}_2^n} (x_0 - w_H(m))^4, \quad (28)$$

which we drop since it does not depend on $k$. All in one, the only key-dependent term is:

$$\mathbb{E}(f_0 \times f_1) = \frac{1}{2^n} \sum_{m \in \mathbb{F}_2^n} (x_0 - w_H(S[t \oplus k] \oplus m))^2 (x_1 - w_H(m))^2, \quad (29)$$

which cannot be further simplified and will be computed by the computer. So, for the purpose of the attack, we have:

$$\mu_2 = \mathbb{E}(f_0 \times f_1) + \text{cst}. \qquad\qquad\qquad (30)$$

## A.3  Computation of $\mu_3$

We shall consider only terms which depend on the key, hence product of three terms, one of which (at least) is $f_0$. Obviously, $\mathbb{E}(f_0^3)$ does not depend on $k$, for the same reason as given in Eq. (28). But the two terms:

1. $\mathbb{E}(f_0^2 f_1)$ and
2. $\mathbb{E}(f_0 f_1^2)$

Notice that they are present $\binom{3}{2} = 3$ times each when developing the cube.

   **Interestingly**, those are **not** the only cases where $f_0$ and $f_1$ are selected.

$\mathbb{E}(f_0 f_1 f_j)$

$$= \frac{1}{2^n} \sum_{m \in \mathbb{F}_2^n} \frac{1}{2^n!} \sum_{\varphi \in S_{2^n}} (x_0 - w_H(S[t \oplus k] \oplus m))^2 (x_1 - w_H(m))^2 (x_j - w_H(\varphi(\omega) \oplus m))^2$$

$$= \frac{1}{2^n} \sum_{m \in \mathbb{F}_2^n} \frac{1}{2^n} \sum_{m' \in \mathbb{F}_2^n} (x_0 - w_H(S[t \oplus k] \oplus m))^2 (x_1 - w_H(m))^2 (x_j - w_H(m' \oplus m))^2$$

$$= \frac{1}{2^n} \sum_{m \in \mathbb{F}_2^n} (x_0 - w_H(S[t \oplus k] \oplus m))^2 (x_1 - w_H(m))^2 \frac{1}{2^n} \sum_{m' \in \mathbb{F}_2^n} (x_j - w_H(m' \oplus m))^2$$

$$= \frac{1}{2^n} \sum_{m \in \mathbb{F}_2^n} (x_0 - w_H(S[t \oplus k] \oplus m))^2 (x_1 - w_H(m))^2 \frac{1}{2^n} \sum_{\widetilde{m'} \in \mathbb{F}_2^n} (x_j - w_H(\widetilde{m'}))^2 \quad \text{(As in Eq. (26))}$$

$$= \mathbb{E}(f_0 f_1) \mathbb{E}(f_j).$$

Similarly, we have:

$$\mathbb{E}(f_0 f_1 f_i) = \mathbb{E}(f_0 f_1) \mathbb{E}(f_i).$$

Now, we consider products without $f_1$. Obviously, taking only $f_0$ and $f_i$ is not enough, since: $\mathbb{E}(f_0^2 f_i) = \mathbb{E}(f_0^2) \mathbb{E}(f_i)$ and $\mathbb{E}(f_0 f_i^2) = \mathbb{E}(f_0) \mathbb{E}(f_i^2)$ are key independent. The same goes for $\mathbb{E}(f_0^2 f_j)$ and $\mathbb{E}(f_0 f_j^2)$. We are left with $\mathbb{E}(f_0 f_i f_{i'})$, $\mathbb{E}(f_0 f_j f_{j'})$, and $\mathbb{E}(f_0 f_i f_j)$.

The term $\mathbb{E}(f_0 f_i f_{i'}) = \mathbb{E}(f_0) \mathbb{E}(f_i f_{i'}))$ does not depend on $k$, because there is no $M$ in $f_i$.

The term $\mathbb{E}(f_0 f_j f_{j'})$ can also factorize as $\mathbb{E}(f_0) \mathbb{E}(f_j f_{j'}))$, hence it does not depend on $k$. The reason is more subtle, so we detail it:

$$\mathbb{E}(f_0 f_j f_{j'}) = \frac{1}{2^n} \sum_{m \in \mathbb{F}_2^n} (x_0 - w_H(S[t \oplus k] \oplus m))^2$$

$$\times \frac{1}{2^n(2^n - 1)} \sum_{\substack{(m', m'') \in \mathbb{F}_2^n \times \mathbb{F}_2^n \\ \text{s.t. } m' \neq m''}} (x_j - w_H(m' \oplus m))^2 (x_{j'} - w_H(m'' \oplus m))^2.$$

Now, the second sum does not depend on $m$, as shown below:

$$\frac{1}{2^n(2^n - 1)} \sum_{\substack{(m', m'') \in \mathbb{F}_2^n \times \mathbb{F}_2^n \\ \text{s.t. } m' \neq m''}} (x_j - w_H(m' \oplus m))^2 (x_{j'} - w_H(m'' \oplus m))^2 =$$

$$\frac{1}{2^n} \sum_{m' \in \mathbb{F}_2^n} (x_j - w_H(m' \oplus m))^2 \frac{1}{2^n - 1} \sum_{m'' \in \mathbb{F}_2^n \setminus \{m'\}} (x_{j'} - w_H(m'' \oplus m))^2 =$$

$$\frac{1}{2^n} \sum_{\widetilde{m'} \in \mathbb{F}_2^n} (x_j - w_H(\widetilde{m'}))^2 \frac{1}{2^n - 1} \sum_{m'' \in \mathbb{F}_2^n \setminus \{\widetilde{m'} \oplus m\}} (x_{j'} - w_H(m'' \oplus m))^2 =$$

$$\frac{1}{2^n} \sum_{\widetilde{m'} \in \mathbb{F}_2^n} (x_j - w_H(\widetilde{m'}))^2 \frac{1}{2^n - 1} \sum_{\widetilde{m''} \in \mathbb{F}_2^n \setminus \{\widetilde{m''} \oplus m \oplus m\}} (x_{j'} - w_H(\widetilde{m''}))^2.$$

Consequently, the last case is $\mathbb{E}(f_0 f_i f_j)$. We can subdivide it into two cases: $j = i + 2^n$ and $j \neq i + 2^n$. When $j = i + 2^n$, the permutation $\Phi$ is evaluated at the same $\omega$ in $f_i$ and $f_j$. We denote by $M'$ the R.V. $\Phi(\omega)$, where $\omega = j - 2$. Hence:

$$\mathbb{E}(f_0 f_i f_{j=i+2^n}) =$$

$$\frac{1}{2^n} \sum_{m \in \mathbb{F}_2^n} (x_0 - w_H(S[t \oplus k] \oplus m))^2 \frac{1}{2^n} \sum_{m' \in \mathbb{F}_2^n} (x_i - w_H(m'))^2 (x_j - w_H(m' \oplus m))^2.$$

$$(31)$$

These terms (for all $j \in J$) correspond to the $\text{MVA}_{TR}$ attack published at CHES 2015 [7].

Eventually, there are the terms for $j \neq i - 2^n$. They are actually key dependent, hence must be kept. They are equal to:

$$\mathbb{E}(f_0 f_i f_{j \neq i+2^n}) = \frac{1}{2^n} \sum_{m \in \mathbb{F}_2^n} (x_0 - w_H(S[t \oplus k] \oplus m))^2$$

$$\times \frac{1}{2^n} \frac{1}{2^n - 1} \sum_{\substack{(m',m'') \in \mathbb{F}_2^n \times \mathbb{F}_2^n \\ \text{s.t. } m' \neq m''}} (x_i - w_H(m'))^2 (x_j - w_H(m'' \oplus m))^2.$$

Interestingly, without the constraint $m' \neq m''$, this quantity does not depend on the key. So, the leakage which is exploited here is due to the fact $\Phi$ is not a random function, but a bijection. As, in $\mu_3$, we are only interested in non constant terms, we can rewrite:

$$\mathbb{E}(f_0 f_i f_{j \neq i+2^n}) = \text{cst} - \frac{1}{2^n} \sum_{m \in \mathbb{F}_2^n} (x_0 - w_H(S[t \oplus k] \oplus m))^2$$

$$\times \frac{1}{2^n} \frac{1}{2^n - 1} \sum_{\substack{(m',m'') \in \mathbb{F}_2^n \times \mathbb{F}_2^n \\ \text{s.t. } m' = m''}} (x_i - w_H(m'))^2 (x_j - w_H(m'' \oplus m))^2$$

$$= \text{cst} - \frac{1}{2^n} \sum_{m \in \mathbb{F}_2^n} (x_0 - w_H(S[t \oplus k] \oplus m))^2$$

$$\times \frac{1}{2^n - 1} \sum_{m' \in \mathbb{F}_2} (x_i - w_H(m'))^2 (x_j - w_H(m' \oplus m))^2. \tag{32}$$

The non-constant term is similar to Eq. (31) provided a scaling by $-(2^n - 1)/2^n$ is done.

So, for the purpose of the attack, we have:

$$\mu_3 = \text{cst} + 3\mathbb{E}(f_0^2 f_1) + 3\mathbb{E}(f_0 f_1^2) + 3!\mathbb{E}(f_0 \times f_1)\left(\sum_{i \in I} \mathbb{E}(f_i) + \sum_{j \in J} \mathbb{E}(f_j)\right)$$

$$+ 3! \sum_{i=2}^{2^n+1} \mathbb{E}(f_0 f_i f_{j=i+2^n}) + 3! \sum_{i=2}^{2^n+1} \sum_{j \in \{2+2^n, \dots, 2^{n+1}+1\} \setminus \{i+2^n\}} \mathbb{E}(f_0 f_i f_j). \tag{33}$$

## B    Complexity Proofs

### B.1    Proof of Lemma 1

In order to prove Lemma 1 let us first introduce a preliminary result.

**Lemma 2.** *The quantity $\binom{\Pi}{\ell}$ is increasing if $\ell < \lceil \Pi/2 \rceil$ and its maximum is $\binom{\Pi}{\lceil \frac{\Pi}{2} \rceil}$.*

*Proof.*

$$\binom{\Pi}{\ell+1} = \frac{\Pi!}{(\Pi-\ell-1)!(\ell+1)!} = \frac{\Pi-\ell-1}{\ell+1}\binom{\Pi}{\ell},$$

and the factor $\frac{\Pi-\ell-1}{\ell+1}$ is strictly greater than 1. Indeed,

$$\frac{\Pi-\ell-1}{\ell+1} > 1 \iff \Pi > 2(\ell+1) \iff \ell < \lceil \Pi/2 \rceil.$$

□

Finally we can prove Lemma 1.

*Proof.* Let us first assume that one dimension leaks at most one element of the permutation. We can thus develop the expression of $\mu_\ell$, and we denote the complexity under the braces.

$$\mu_\ell = \mathbb{E}_R \left( \|x - y(t,k,R)\|^{2\ell} \right)$$

$$= \underbrace{\sum_{k_1+\ldots+k_D=\ell} \frac{\ell!}{\prod_{d=1}^{D} k_d!}}_{\binom{D+\ell-1}{\ell}} \underbrace{\mathbb{E}_R}_{2^{(\Omega-1)n}\left(\lceil\frac{\Pi}{2}\rceil\right)} \underbrace{\left( \prod_{d=1}^{D} f_d^{k_d} \right)}_{\min(D,\ell)}$$

As $k_1+\ldots+k_D = \ell$ there are at most $D$ indices $k_d, 1 \le d \le D$ such that $k_d \neq 0$. Hence there are at most $\min(D, \ell)$ elements in the product.

Each dimensions which leaks an element of the permutation can also leaks the masks. The worst case in terms of complexity is when all the permutation leakages depend also on the masks. Let us denote by $i$ such that $1 \le i \le \min(D, \ell)$ the number of those terms. Then the expectation is computed over $2^{(\Omega-1)n} \frac{\Pi!}{(\Pi-i)!}$. Nevertheless by taking into account the commutativity properties of the product one can only compute $2^{(\Omega-1)n}\binom{\Pi}{i}$.

By Lemma 2 we have that is value $\binom{\Pi}{i}$ is maximum with $\binom{\Pi}{\ell}$ when $\ell \le \lceil\frac{\Pi}{2}\rceil$. When $\ell > \frac{\Pi}{2} + 1$ the maximum is $\binom{\Pi}{\lceil\frac{\Pi}{2}\rceil}$.

Finally as there are $\binom{D+\ell-1}{\ell}$ elements in the sum.

The complexity of $\mu_\ell$ is lower than $\mathcal{O}\left( \binom{D+\ell-1}{\ell} 2^{(\Omega-1)n} \binom{\Pi}{\min(\lceil\frac{\Pi}{2}\rceil,\ell)} \right)$.    □

## B.2   Proof of Proposition 6

In order to prove Lemma 6 let us first introduce a preliminary result.

**Lemma 3.** *The quantity $\binom{D-1+\ell}{\ell}$ is increasing with $\ell$ if $D > 1$.*

*Proof.* We have that:

$$\binom{D-1+\ell+1}{\ell+1} = \frac{D+\ell}{\ell+1}\binom{D-1+\ell}{\ell},$$

where $\forall \ell, \frac{D+\ell}{\ell+1} > 1$ provided $D > 1$.    □

Finally let us prove Prop. 6.

*Proof. Complexity of OPT:*
Following Eq. (2) we have that the computation for a key guess of OPT is:

$$\underbrace{\sum_{q=1}^{Q} \log \underbrace{\mathbb{E}}_{\varPi! 2^{n(\varOmega-1)}} \exp \underbrace{\frac{-\|x - y(t,k,R)\|^2}{2\sigma^2}}_{D}}_{Q}. \tag{34}$$

We assume that the computation of the log and the exp is constant. As a consequence the complexity of the optimal distinguisher is $\mathcal{O}\left(Q \cdot (2^n)^{\varOmega-1} \cdot \varPi! \cdot D\right)$

*Complexity of* $\mathrm{ROPT}_L$: The computation of $\mathrm{ROPT}_L$ involves the computation of the $\mu_\ell$ with $\ell \leq L$ (Eqs. (2) and (1)). By Lemmas 1 and 3 all these terms have a complexity lower than $\mathcal{O}\left(\binom{D+L-1}{L} \cdot 2^{(\varOmega-1)n} \cdot \binom{\varPi}{\min(\lceil \frac{\varPi}{2} \rceil, L)}\right)$ (Eq. (16)).

As a consequence the complexity of $\mathrm{ROPT}_L$ is lower than

$$\mathcal{O}\left(Q \cdot L\binom{D+L-1}{L} \cdot 2^{(\varOmega-1)n} \cdot \binom{\varPi}{\min\left(\lceil \frac{\varPi}{2} \rceil, L\right)}\right). \tag{35}$$

□

## B.3   Proof of Proposition 7

*Proof.* Let us develop all the product in the term $\mu_\ell$ in order to compute the expectation in the minimum number of values.

$$\mu_\ell = \mathbb{E}_M\left(\left(\sum_{d=1}^{D}(x_{d,q} - y_d)^2\right)^\ell\right)$$

$$= \sum_{\substack{\ell_1,\ell_2,\dots,\ell_D \\ \sum_{d=1}^{D}=\ell}} \frac{\ell!}{\prod_{d=1}^{D}\ell_d!} \mathbb{E}_M\left((x_1 - y_1)^{2\ell_1} \cdots (x_D - y_D)^{2\ell_D}\right).$$

Moreover $(x_d - y_d(t,k,M))^{2\ell_d} = \sum_{i=0}^{2\ell_d}\binom{2\ell_d}{i}x_d^{2\ell_d-i}y_d(t,k,M)^i$

$$\mu_\ell = \sum_{\substack{\ell_1,\ell_2,\dots,\ell_D \\ \sum_{d=1}^{D}\ell_d=\ell}} \frac{\ell!}{\prod_{d=1}^{D}\ell_d!} \mathbb{E}_M\left(\prod_{d=1}^{D}\left(\sum_{i=0}^{2\ell_d}\binom{2\ell_d}{i}x_d^{2\ell_1-i}y_d(t,k,M)^i\right)\right)$$

$$= \sum_{\substack{\ell_1,\ell_2,\dots,\ell_D \\ \sum_{d=1}^{D}\ell_d=\ell}} \frac{\ell!}{\prod_{d=1}^{D}\ell_d!} \sum_{\substack{i_1 \leq 2\ell_1 \\ \vdots \\ i_D \leq 2\ell_D}} \prod_{d=1}^{D}\left(\binom{2\ell_d}{i_d}x_d^{2\ell_d-i_d}\right) \underbrace{\mathbb{E}_M\left(\prod_{d=1}^{D}y_d(t,k,M)^{i_d}\right)}_{\text{can be precomputed}}.$$

□

## B.4   Proof of Proposition 8

*Proof.* In our case study the size of the permutation is $\Pi = 2^n$.

Then the complexity of OPT is given by a straightforward application of Eq. (17).

From Eq. (14) we have that for $\text{ROPT}_2$ the computation for one key guess and one trace is given by $\mathbb{E}(f_0 \times f_1)$. In this equation the expectation is computed over $2^n$ values (Eq. (28)).

From Eq. (15) we have that for $\text{ROPT}_3$ the computation for one key guess and one trace is given by $\mu_2^{(q)}(1 + \gamma\mu_1^{(q)}) - \gamma\frac{\mu_3^{(q)}}{3}$. It can be seen in Eqs. (23), (24), (25) and (27) that the expectation of $\mu_1$ is computed over $2^n$ values. The dominant term in $\mu_3$ (Eq. (33)) is :

$$\sum_{i=2}^{2^n+1} \underbrace{\sum_{j \in \{2+2^n,\dots,2^{n+1}+1\} \setminus \{i+2^n\}} \underbrace{\mathbb{E}}_{2^{2n}}(f_0 f_i f_j)}_{2^{2n}}.$$

The expectation in this term is computed over $2^{2n}$ values (Eq. (32)). The sum is computed on less than $2^{2n}$.   □

## B.5   Time and Complexity

The times of the section are expressed in seconds. All the attacks have been run on Intel Xeon X5660 running at 2.67 GHz. All the implementations are mono-thread. The model of the simulations is the one describe in Sect. 6. For each distinguisher the attacks are computed 1000 times on 1000 traces.

**Table 1.** Time and complexity

| Attack | Dimension | Time (in seconds) | Computational complexity |
|---|---|---|---|
| 2O-CPA | 2 | 39 | $\mathcal{O}(Q)$ |
| $\text{ROPT}_2$ | 2 | 295 | $\mathcal{O}(Q)$ |
| $\text{OPT}_{2O}$ | 2 | 9473 | $\mathcal{O}(Q \cdot (2^n))$ |
| $\text{MVA}_{TR}$ | $2^{n+1}+1$ | 130 | $\mathcal{O}(Q \cdot 2^n)$ |
| $\text{ROPT}_3$ | $2^{n+1}+2$ | 2495 | $\mathcal{O}(Q \cdot 2^{2n})$ |
| OPT | $2^{n+1}+2$ | Not computable | $\mathcal{O}(Q \cdot (2^n) \cdot 2^n! \cdot (2^{n+1}+2))$ |

# C   Analysis of the DPAcontest

Recently an open implementation of a masking scheme with shuffling has been presented in the DPA contest v4.2 [35]. In this implementation the execution of the different states is performed in an random order.

An attacker can target the integrated leakages of the different states in order to counter the shuffling [9,31].

A better approach is to take into account the possible leakages of the permutation. In this case the optimal distinguisher will be not computable as it involves an expectation over 16! values. In this case the rounded optimal attack will reduced this complexity.

Let us defined the leakages of such implementations.

- $X_0 = y_0\,(t, k, R) + N_0$ with $y_0\,(t, k, R) = w_H(M)$,
- $X_1 = y_1\,(t, k, R) + N_1$ with $y_1\,(t, k, R) = w_H(S[\pi\,(T \oplus k)] \oplus M)$,
- $X_i = y_i\,(t, k, R) + N_i$, for $i = 2, \ldots, 18$ with $y_i\,(t, k, R) = w_H(\Phi(i - 2))$,

Then similarly to the Appendix A we have that:

$$\mu_1 = \mathbb{E}(f_0) + \mathbb{E}(f_1) + \sum_{i \in I} \mathbb{E}(f_i), \tag{36}$$

$$\mu_2 = \mathbb{E}(f_0 \times f_1) + \text{cst.} \tag{37}$$

Additionally as it is a low entropy masking scheme the secret key can leaked in an univariate high order attack. Depending on the number of masks involve in the masking scheme it could be at order 2, 3 or more. For simplicity let us assume it is at order 3. In such cases

$$\mu_3 = \mathbb{E}(f_1^3) + 3\mathbb{E}(f_0^2 f_1) + 3\mathbb{E}(f_0 f_1^2) + 3! \sum_{i=2}^{2^n+1} \mathbb{E}(f_0 f_1 f_i) + \text{cst.} \tag{38}$$

Of course an attacker can additionally exploit all the leakages of the different states in order to increase the success of the attacks.

In some particular low entropy masking schemes the same masks are reused several time or are linked by deterministic relations (e.g. the first version of the DPAcontest). In this context it could be interesting to combine the leakages of different states [4]. In this case our method could benefit of the multiple possible points combinations.

# References

1. Akkar, M.-L., Giraud, C.: An implementation of DES and AES, secure against some attacks. In: Koç, Ç.K., Naccache, D., Paar, C. (eds.) CHES 2001. LNCS, vol. 2162, pp. 309–318. Springer, Heidelberg (2001). doi:10.1007/3-540-44709-1_26
2. Batina, L., Gierlichs, B., Prouff, E., Rivain, M., Standaert, F.X., Veyrat-Charvillon, N.: Mutual information analysis: a comprehensive study. J. Cryptol. 24(2), 269–291 (2011)
3. Blömer, J., Guajardo, J., Krummel, V.: Provably secure masking of AES. In: Handschuh, H., Hasan, M.A. (eds.) SAC 2004. LNCS, vol. 3357, pp. 69–83. Springer, Heidelberg (2004). doi:10.1007/978-3-540-30564-4_5
4. Bruneau, N., Danger, J.-L., Guilley, S., Heuser, A., Teglia, Y.: Boosting higher-order correlation attacks by dimensionality reduction. In: Chakraborty, R.S., Matyas, V., Schaumont, P. (eds.) SPACE 2014. LNCS, vol. 8804, pp. 183–200. Springer, Heidelberg (2014). doi:10.1007/978-3-319-12060-7_13

5. Bruneau, N., Guilley, S., Heuser, A., Marion, D., Rioul, O.: Less is more dimensionality reduction from a theoretical perspective. In: Handschuh and Güneysu [13]

6. Bruneau, N., Guilley, S., Heuser, A., Rioul, O.: *Masks will fall off*. In: Sarkar, P., Iwata, T. (eds.) ASIACRYPT 2014. LNCS, vol. 8874, pp. 344–365. Springer, Heidelberg (2014). doi:10.1007/978-3-662-45608-8_19

7. Bruneau, N., Guilley, S., Najm, Z., Teglia, Y.: Multivariate high-order attacks of shuffled tables recomputation. In: Handschuh and Güneysu [13]

8. Chari, S., Jutla, C.S., Rao, J.R., Rohatgi, P.: Towards sound approaches to counteract power-analysis attacks. In: Wiener, M. (ed.) CRYPTO 1999. LNCS, vol. 1666, pp. 398–412. Springer, Heidelberg (1999). doi:10.1007/3-540-48405-1_26

9. Clavier, C., Coron, J.-S., Dabbous, N.: Differential power analysis in the presence of hardware countermeasures. In: Koç, Ç.K., Paar, C. (eds.) CHES 2000. LNCS, vol. 1965, pp. 252–263. Springer, Heidelberg (2000). doi:10.1007/3-540-44499-8_20

10. Coron, J.-S.: Higher order masking of look-up tables. In: Nguyen, P.Q., Oswald, E. (eds.) EUROCRYPT 2014. LNCS, vol. 8441, pp. 441–458. Springer, Heidelberg (2014). doi:10.1007/978-3-642-55220-5_25

11. Ding, A.A., Zhang, L., Fei, Y., Luo, P.: A statistical model for higher order DPA on masked devices. In: Batina, L., Robshaw, M. (eds.) CHES 2014. LNCS, vol. 8731, pp. 147–169. Springer, Heidelberg (2014). doi:10.1007/978-3-662-44709-3_9

12. Duc, A., Faust, S., Standaert, F.-X.: Making masking security proofs concrete. In: Oswald, E., Fischlin, M. (eds.) EUROCRYPT 2015. LNCS, vol. 9056, pp. 401–429. Springer, Heidelberg (2015). doi:10.1007/978-3-662-46800-5_16

13. Güneysu, T., Handschuh, H. (eds.): CHES 2015. LNCS, vol. 9293. Springer, Heidelberg (2015)

14. Herbst, C., Oswald, E., Mangard, S.: An AES smart card implementation resistant to power analysis attacks. In: Zhou, J., Yung, M., Bao, F. (eds.) ACNS 2006. LNCS, vol. 3989, pp. 239–252. Springer, Heidelberg (2006). doi:10.1007/11767480_16

15. Ishai, Y., Sahai, A., Wagner, D.: Private circuits: securing hardware against probing attacks. In: Boneh, D. (ed.) CRYPTO 2003. LNCS, vol. 2729, pp. 463–481. Springer, Heidelberg (2003). doi:10.1007/978-3-540-45146-4_27

16. Lemke-Rust, K., Paar, C.: Analyzing side channel leakage of masked implementations with stochastic methods. In: Biskup, J., López, J. (eds.) ESORICS 2007. LNCS, vol. 4734, pp. 454–468. Springer, Heidelberg (2007). doi:10.1007/978-3-540-74835-9_30

17. Lemke-Rust, K., Paar, C.: Gaussian mixture models for higher-order side channel analysis. In: Paillier, P., Verbauwhede, I. (eds.) CHES 2007. LNCS, vol. 4727, pp. 14–27. Springer, Heidelberg (2007). doi:10.1007/978-3-540-74735-2_2

18. Mangard, S., Oswald, E., Popp, T.: Power Analysis Attacks - Revealing the Secrets of Smart Cards. Springer, Heidelberg (2007)

19. Messerges, T.S.: Securing the AES finalists against power analysis attacks. In: Goos, G., Hartmanis, J., Leeuwen, J., Schneier, B. (eds.) FSE 2000. LNCS, vol. 1978, pp. 150–164. Springer, Heidelberg (2001). doi:10.1007/3-540-44706-7_11

20. Messerges, T.S.: Using second-order power analysis to attack DPA resistant software. In: Koç, Ç.K., Paar, C. (eds.) CHES 2000. LNCS, vol. 1965, pp. 238–251. Springer, Heidelberg (2000). doi:10.1007/3-540-44499-8_19

21. Moradi, A.: Statistical tools flavor side-channel collision attacks. In: Pointcheval, D., Johansson, T. (eds.) EUROCRYPT 2012. LNCS, vol. 7237, pp. 428–445. Springer, Heidelberg (2012). doi:10.1007/978-3-642-29011-4_26

22. Moradi, A., Standaert, F.X.: Moments-correlating DPA. IACR Cryptology ePrint Archive 2014, p. 409, 2 June 2014

23. Moradi, A., Wild, A.: Assessment of hiding the higher-order leakages in hardware. In: Güneysu, T., Handschuh, H. (eds.) CHES 2015. LNCS, vol. 9293, pp. 453–474. Springer, Heidelberg (2015). doi:10.1007/978-3-662-48324-4_23

24. Nikova, S., Rijmen, V., Schläffer, M.: Secure hardware implementation of nonlinear functions in the presence of glitches. J. Cryptol. **24**(2), 292–321 (2011)

25. Oswald, E., Mangard, S.: Template attacks on masking—resistance is futile. In: Abe, M. (ed.) CT-RSA 2007. LNCS, vol. 4377, pp. 243–256. Springer, Heidelberg (2006). doi:10.1007/11967668_16

26. Pan, J., Hartog, J.I., Lu, J.: You cannot hide behind the mask: power analysis on a provably secure $S$-Box implementation. In: Youm, H.Y., Yung, M. (eds.) WISA 2009. LNCS, vol. 5932, pp. 178–192. Springer, Heidelberg (2009). doi:10.1007/978-3-642-10838-9_14

27. Peeters, E., Standaert, F.-X., Donckers, N., Quisquater, J.-J.: Improved higher-order side-channel attacks with FPGA experiments. In: Rao, J.R., Sunar, B. (eds.) CHES 2005. LNCS, vol. 3659, pp. 309–323. Springer, Heidelberg (2005). doi:10.1007/11545262_23

28. Prouff, E., Rivain, M.: A generic method for secure SBox implementation. In: Kim, S., Yung, M., Lee, H.-W. (eds.) WISA 2007. LNCS, vol. 4867, pp. 227–244. Springer, Heidelberg (2007). doi:10.1007/978-3-540-77535-5_17

29. Prouff, E., Rivain, M., Bevan, R.: Statistical analysis of second order differential power analysis. IEEE Trans. Comput. **58**(6), 799–811 (2009)

30. Rivain, M., Prouff, E.: Provably secure higher-order masking of AES. In: Mangard, S., Standaert, F.-X. (eds.) CHES 2010. LNCS, vol. 6225, pp. 413–427. Springer, Heidelberg (2010). doi:10.1007/978-3-642-15031-9_28

31. Rivain, M., Prouff, E., Doget, J.: Higher-order masking and shuffling for software implementations of block ciphers. In: Clavier, C., Gaj, K. (eds.) CHES 2009. LNCS, vol. 5747, pp. 171–188. Springer, Heidelberg (2009). doi:10.1007/978-3-642-04138-9_13

32. Standaert, F.-X., Malkin, T.G., Yung, M.: A unified framework for the analysis of side-channel key recovery attacks. In: Joux, A. (ed.) EUROCRYPT 2009. LNCS, vol. 5479, pp. 443–461. Springer, Heidelberg (2009). doi:10.1007/978-3-642-01001-9_26

33. Standaert, F.-X., Veyrat-Charvillon, N., Oswald, E., Gierlichs, B., Medwed, M., Kasper, M., Mangard, S.: The world is not enough: another look on second-order DPA. In: Abe, M. (ed.) ASIACRYPT 2010. LNCS, vol. 6477, pp. 112–129. Springer, Heidelberg (2010). doi:10.1007/978-3-642-17373-8_7

34. Stuart, A., Ord, K.: Kendall's Advanced Theory of Statistics: Distribution Theory, 6th edn. Wiley-Blackwell, New York (1994). ISBN-10: 0470665300; ISBN-13: 978-0470665305

35. TELECOM ParisTech SEN research group. DPA Contest, 4th edn., 2013–2014. http://www.DPAcontest.org/v4/

36. Tunstall, M., Whitnall, C., Oswald, E.: Masking tables—an underestimated security risk. In: Moriai, S. (ed.) FSE 2013. LNCS, vol. 8424, pp. 425–444. Springer, Heidelberg (2014). doi:10.1007/978-3-662-43933-3_22

37. Veyrat-Charvillon, N., Medwed, M., Kerckhof, S., Standaert, F.-X.: Shuffling against side-channel attacks: a comprehensive study with cautionary note. In: Wang, X., Sako, K. (eds.) ASIACRYPT 2012. LNCS, vol. 7658, pp. 740–757. Springer, Heidelberg (2012). doi:10.1007/978-3-642-34961-4_44

38. Waddle, J., Wagner, D.: Towards efficient second-order power analysis. In: Joye, M., Quisquater, J.-J. (eds.) CHES 2004. LNCS, vol. 3156, pp. 1–15. Springer, Heidelberg (2004). doi:10.1007/978-3-540-28632-5_1
39. Weisstein, E.W.: Cumulant. From MathWorld A Wolfram Web Resource. http://mathworld.wolfram.com/Cumulant.html

# Unknown-Input Attacks in the Parallel Setting: Improving the Security of the CHES 2012 Leakage-Resilient PRF

Marcel Medwed[1(✉)], François-Xavier Standaert[2], Ventzislav Nikov[3], and Martin Feldhofer[1]

[1] NXP Semiconductors Austria, Gratkorn, Austria
marcel.medwed@gmail.com
[2] ICTEAM/ELEN/Crypto Group, Universite Catholique de Louvain, Louvain-la-Neuve, Belgium
[3] NXP Semiconductors Leuven, Leuven, Belgium

**Abstract.** In this work we present a leakage-resilient PRF which makes use of parallel block cipher implementations with unknown-inputs. To the best of our knowledge this is the first work to study and exploit unknown-inputs as a form of key-dependent algorithmic noise. It turns out that such noise renders the problem of side-channel key recovery intractable under very little and easily satisfiable assumptions. That is, the construction stays secure even in a noise-free setting and independent of the number of traces and the used power model. The contributions of this paper are as follows. First, we present a PRF construction which offers attractive security properties, even when instantiated with the AES. Second, we study the effect of unknown-input attacks in parallel implementations. We put forward their intractability and explain it by studying the inevitable model errors obtained when building templates in such a scenario. Third, we compare the security of our construction to the CHES 2012 one and show that it is superior in many ways. That is, a standard block cipher can be used, the security holds for all intermediate variables and it can even partially tolerate local EM attacks and some typical implementation mistakes or hardware insufficiencies. Finally, we discuss the performance of a standard-cell implementation.

## 1 Introduction

Countermeasures against side-channel attacks always imply implementation overheads and rely on physical assumptions. So designing such countermeasures comes with the equally important goals of maximizing security, while minimizing the overheads and relying on physical assumptions that are easy to fulfill by cryptographic engineers. Mainstream masking schemes (i.e. data randomization

F.-X. Standaert—Associate researcher of the Belgian Fund for Scientific Research (FNRS-F.R.S.). This work has been funded in part by the ERC project 280141 (acronym CRASH) and by the ARC project NANOSEC.

J.H. Cheon and T. Takagi (Eds.): ASIACRYPT 2016, Part I, LNCS 10031, pp. 602–623, 2016.
DOI: 10.1007/978-3-662-53887-6_22

based on secret sharing) are a typical example of this tradeoff, where security is exponential in the number of shares, performances are quadratic in the number of shares, and implementers need to guarantee that the leakages of the shares are independent and sufficiently noisy [7,10,15,26]. (Note that the condition of independent leakages is typically hard to guarantee, both in software and hardware implementations [2,8,17,18]). Threshold implementations are a specialization of masking that reduces the independence requirement (by ensuring that glitches do not harm the security of the masked implementations) [5,23], which can also lead to some performance gains with low number of shares [6,22].

At CHES 2012, a quite different tradeoff was introduced. Namely, and starting from the observation that leakage-resilience via re-keying alone is not sufficient to efficiently protect stateless symmetric cryptographic primitives such as block ciphers (later formalized in [3]), Medwed et al. proposed a tweaked construction of an AES-based leakage-resilient PRF, inspired from more formal works such as [1,9,11,28,32], which additionally requires that the AES is implemented in parallel and that its S-boxes have similar leakage models [20]. In this respect, and while the parallel implementation setting is easy to guarantee (and can even be emulated thanks to shuffling [14]), the "similar leakage assumption" turned out to be harder to evaluate. Later results showed that despite not easy to attack, such a solution may not be best suited the AES [4].

In this paper, we aim to improve the tradeoff between security, performance and physical assumptions for the CHES 2012 construction. For this purpose, our main ingredient is to replace the similar leakage assumption by an easier-to-guarantee requirement of unknown plaintexts. Interestingly, this requirement can be easily satisfied by exploiting a leakage-resilient stream cipher in order to generate these plaintexts (we use the efficient construction from [27] for this purpose). As a result, our contributions are as follows. We first describe our new construction of a leakage-resilient PRF based on unknown plaintexts. Second, we analyze its security in front of standard side-channel attacks where the adversary can observe noisy Hamming weight leakages (and compare it with the CHES 2012 proposal). Third, we evaluate the impact of implementation issues such as deviations from the Hamming weight leakages and leakages due to transitions between registers. Finally, we discuss alternative attack paths and put forward the good performances of our new construction. As part of our investigations, we also highlight the interesting security guarantees offered by the combination of unknown cipher inputs and parallel implementations for side-channel resistance, which is of independent interest.

Note that despite our design is inspired by previous constructions of leakage-resilient PRFs, the security guarantees we claim for it are significantly less formal/general. First, the only security we claim is key recovery security, as for the CHES 2012 PRF. This limitation is motivated by the discussions in [21,24], where it was shown that indistinguishability of the inputs is essentially impossible in a physically observable setting (excepted by artificially excluding some leakages of the analysis, which then reduces practical relevance). Second, it is worth emphasizing that our analyzes are only heuristic and based on concrete attacks. In this respect, the goal of our proposal is not to be proven secure under a formal model (in parts because it relies on non-standard implementation assumptions such as parallelism which make formal treatments much more challenging). By contrast, it is an attempt to implement a building block with bounded leakage. In other words, it is more an attempt to instantiate a way to fulfill the basic assumptions of leakage-resilient cryptography than an attempt to formally analyze a leakage-resilient primitive or functionality. For example, our unknown-input PRF could be a candidate for the leak-free block cipher required in [24].

## 2   Background: The CHES 2012 Leakage-Resilient PRF

We start with the description of the standard GGM PRF [13], depicted in the left part of Fig. 1, on which the CHES 2012 PRF is based. Let $F_k(x)$ denote the PRF indexed by $k$ and evaluated on $x$. Further, let the building blocks $E_{k^i}(p^i_j)$ denote the application of a block cipher $E$ to a plaintext $p^i_j$ under a key $k^i$ (the figure shows the example of $E = $ AES-128 with $1 \leq i \leq 128$ and $0 \leq j \leq 1$). Let also $x(i)$ denote the $i^{th}$ bit of $x$. The PRF first initializes $k^0 = k$ and then iterates as follows: $k^{i+1} = E_{k^i}(p^i_0)$ if $x(i) = 0$ and $k^{i+1} = E_{k^i}(p^i_1)$ if $x(i) = 1$. Eventually, the $(n+1)^{th}$ intermediate key $k^{128}$ is the PRF output as $F_k(x)$.

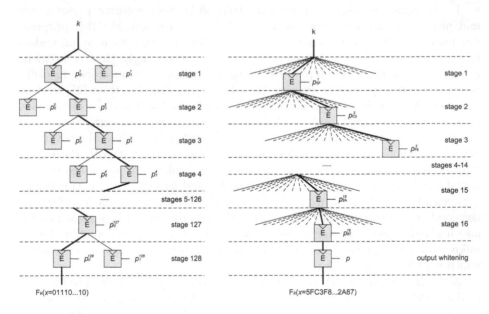

**Fig. 1.** Leakage-resilient PRFs: straight GGM (left) and efficient alternative (right).

In this basic version, the execution of the PRF guarantees that any side-channel adversary will at most observe the leakage corresponding to two plaintexts per intermediate key ($p^i_0$ and $p^i_1$). This implies 128 executions of the AES-128 to produce a single 128-bit output. A straightforward solution to trade improved performances for additional leakage is to increase the number of observable plaintexts per intermediate key. If one has $N_p$ such plaintexts per stage, the number of AES-128 executions to produce a 128-bit output is divided by $\log_2(N_p)$. However, as already discussed in [20], such a tradeoff scales badly and very rapidly decreases the side-channel security of an implementation (as it typically allows DPA with $N_p$ observable plaintexts).

To avoid this drawback, an efficient alternative (also proposed in [20]) is illustrated in the right part of Fig. 1. It can be viewed as a GGM construction with $N_p = 256$, but where the same set of 256 carefully chosen plaintexts is re-used in each PRF stage, excepted for the last stage where $N_p = 1$. In terms of efficiency, this proposal reduces the number of stages of a PRF based on the AES-128 to 17 (i.e. 16 plus one final whitening).

The security of this second construction is based on the combination of parallelism with carefully chosen plaintext values, in order to prohibit the application of standard divide-and-conquer strategies. For this purpose, plaintexts of the form $p_j = \{j - 1\}^{N_s}$, with $1 \leq j \leq N_p$ and $N_p$ being limited by the S-box input space were considered. Given that all S-boxes leak in parallel, the effect of this measure is that in a DPA attack, the predictions corresponding to the $N_s$ key bytes cannot be distinguished anymore, because these key bytes have to be targeted at the same time. As a result, and even when increasing $N_p$, not all the $N_s$ key bytes can be highly ranked by the attack. (We will re-detail this effect in Sect. 4.1, which is reflected by the higher guessing entropy of the targeted key bytes in Fig. 3). In [20], it was even shown that slight differences in the implementation – and therefore in the leakage of the $N_s$ S-boxes – are not easily exploitable. Eventually, if $N_s$ becomes sufficiently large, ordering the $N_s$ recovered subkeys has a cost of $N_s!$, meaning that even after seeing all leakages without noise, the adversary cannot fully recover the key.

Unfortunately, and despite conceptually appealing, this construction has several drawbacks which limit its applicability. First, the security parameter $N_s$ is defined by the number of S-boxes of the underlying block cipher. For some of the currently standardized block ciphers $N_s$ is not large enough (e.g. $N_s = 16$ for the AES-128, which corresponds to an insufficient $N_s! \approx 2^{44}$). Second, if intermediate values other than the first round's S-box outputs are targeted, the leakages might be sufficiently independent such that divide-and-conquer strategies work again. While this generally requires more computational power, recent results on multi-target attack DPA show that it is not out of reach [19]. (This is in fact the reason why attacks on the ciphertext need to be prevented by the whitening step in the CHES 2012 proposal). Finally, the size of the S-box defines the maximum value of $N_p$ and hence the maximum throughput.

## 3   New Leakage-Resilient PRF Construction

We now present a new construction which improves over the one in [20] in terms of performance and security, at the cost of higher memory requirements. For this purpose, we introduce a pre-computation step in which we generate $N_p$ secret,

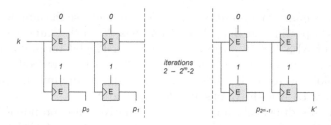

**Fig. 2.** Leakage-resilient PRG used to generate the $2^m$ secret plaintexts.

distinct plaintexts. This step can be seen in Fig. 2. It essentially uses the leakage-resilient PRG from [27] to generate $2^m$ secret plaintexts $p_0 \ldots p_{2^m-1}$ as well as an updated key $k'$. These secret plaintexts and updated key are then simply used in a tree-based PRF such as in the right side of Fig. 1. The output whitening step stays the same. By design, this new construction has the advantage (compared to the CHES 2012 one) that the plaintexts are secret and of no particular form. This implies that their number is not bounded by the S-box size, allowing for smaller trees of depth $128/m + 1$. From a security point of view, it also comes with interesting implications:

1. Since the plaintexts are unknown, a straight-forward unprofiled DPA is ruled out. Instead, an adversary has to build templates (for instance for the bivariate variable made of the plaintext and S-box output leakages).
2. For a similar reason, there is no straightforward way to verify a key candidate: for this purpose, one would not only need to recover the key but also at least one secret plaintext. In the worst case where the information leakages are not sufficient (i.e. if a successful attack requires additional key/plaintext enumeration [30]) this squares the attack time complexity.
3. As for the CHES 2012 construction using carefully chosen plaintexts, the adversary has no way of separating the leakages from the different subkeys. But contrary to this previous work, this feature now applies to any intermediate variable within the algorithm (not only to the first round leakages).

## 4    Security Analysis w.r.t. Basic Side-Channel Attacks

We now detail our security analysis against standard side-channel attacks and use the following notations. First, $k$ denotes a key, $k^*$ denotes a key candidate and $k_j$ the $j^{th}$ byte of a key. Next, $p_{i,j}$ is the $j^{th}$ byte of the $i^{th}$ plaintext out of $q$ ones that are available to an adversary. For $p$ and $k$, $j$ is assumed to be in the range $1, \ldots, N_s$ where $N_s = 16$ for the AES. Further, $t_i$ is a trace (aka leakage) vector, corresponding to the $i^{th}$ plaintext. A trace may contain several leakage points, denoted by $t_{i,j}$. $\mathsf{L}$ denotes the leakage function, e.g. the Hamming weight function in our examples below. Finally, $\mathsf{L}(\mathsf{S}(k_1 \oplus p_{2,1}))$ denotes the leakage of the S-box output corresponding to S-box 1 for the 2nd plaintext. The set of all plaintexts is denoted as $P$ and the set of all traces as $T$.

In a standard DPA attack, the adversary pursues a divide-and-conquer approach. That is, he first computes the correct subkeys as

$$\tilde{k}_j = \arg\max_{k_j^*} \Pr(k_j^* | p_{1,j} \dots p_{q,j}, t_{1,j} \dots t_{q,j}).$$

Here, $t_{i,j}$ denotes the sample within trace $i$ which only leaks about $k_j$. Afterwards
he combines these subkeys to $\tilde{k}$. The attack is successful if $\tilde{k} = k$.[1] In a parallel
hardware scenario, $t_i$ consists of a single leakage point that we approximate as:

$$t_{i,1} = \sum_{j=1}^{N_s} \mathsf{L}(\mathsf{S}(k_j \oplus p_{i,j})). \tag{1}$$

Nevertheless, even in this parallel scenario, an adversary can always target a
single key byte at a time by computing:

$$\tilde{k}_j = \arg\max_{k_j^*} \Pr(k_j^* | p_{1,j} \dots p_{q,j}, t_1 \dots t_q).$$

In this case, by just looking at a specific S-box or byte of the key, an adversary
neglects the other key bytes and their contribution to the leakage is interpreted
as (algorithmic) noise, which eventually averages out if plaintexts are uniformly
distributed. As already discussed in [20], for carefully chosen plaintexts, $p_{1,1} =
p_{1,j}$ for all $j = 1 \dots N$. Therefore, the equation becomes:

$$\tilde{k}_j = \arg\max_{k_j^*} \Pr(k_j^* | p_{1,1} \dots p_{q,1}, t_1 \dots t_q) \tag{2}$$

and all $\tilde{k}_j$ are the same. That is, since the probability condition is no longer
dependent on $j$, only one joint score vector can be obtained, which contains the
information about all the $N_s$ target key bytes at once. For unknown-plaintext
attacks, an adversary finally faces the problem of finding:

$$\tilde{k}_j = \arg\max_{k_j^*} \Pr(k_j^* | (t_{1,1}, t_{1,2}) \dots (t_{q,1}, t_{q,2})), \tag{3}$$

where he has no direct access to plaintext information, and therefore must extract
this information from the traces as well (reflected by the second sample of the
traces in the equation). We assume that this information is separately available
and that the traces take the form:

$$(t_{i,1}, t_{i,2}) = \left( \sum_{j=1}^{N_s} \mathsf{L}(p_{i,j}), \sum_{j=1}^{N_s} \mathsf{L}(\mathsf{S}(k_j \oplus p_{i,j})) \right). \tag{4}$$

This has the following important implications on the attack:

---

[1] If for some or all values of $j$ $\tilde{k}_j \neq k_j$, one may still find $k$ using key enumeration
techniques in the combination step, given that the bias is sufficiently high [30].

1. The adversary cannot apply a divide-and-conquer brute-force attack anymore. As in the case of carefully chosen plaintexts, also here the probability's condition becomes independent of $j$, which results in only a single score vector containing the information for all the $N_s$ subkeys.
2. Successful attacks have to be bi-variate ones, in which a second-order moment of the leakage distribution is exploited. This makes them more sensitive to noise. Furthermore, as for the CHES 2012 construction as well, $N_s - 1$ contributors for each leakage point represent key-dependent algorithmic noise, and cannot be averaged out like in the case of masking (as noted in [3]).[2]

In the following we present three experiments. In the first one we recap the security of the CHES 2012 scheme in order to allow for a later comparison. We do so by estimating the guessing entropy and the subkey rank distribution as a function of $N_s$ after seeing all possible traces. In the second experiment, we do the same for our improved proposal. This allows us to highlight the security improvement. In a third experiment we look at the model errors which are the reason for the security improvement. All experiments are carried out based on template attacks as this represents the most powerful side-channel adversary. We used discrete histograms (instead of continuous distributions) for our templates since the leakage function (aka power model) used in our experiments is also discrete and no noise is added. Hence, the number of bins is determined automatically and the histograms capture all the available information. Finally, we evaluate our metrics for increasing number of traces (with bounded number of plaintexts in the case of the CHES 2012 construction).

## 4.1    Security Based on Carefully Chosen Plaintexts

For the CHES 2012 scheme, the plaintexts are known, the target function is the AES S-box and the assumed power model is the Hamming weight model. Thus, the leakages are in the form of Eq. (1). Knowing this, we can generate a template $\mathcal{D}^i$ for each of the subkey candidates, assuming the plaintext to be zero. In our simulations we look at $N_s$ parallel AES S-boxes and the leaking variables are 8-bit valued. Therefore, each template is a histogram with $8 \cdot N_s + 1$ bins, starting at bin $\mathcal{D}^i(0)$ which indicates the probability that for a subkey $k = i$, the leakage sample has a value of 0. The templates are built according to Algorithm 1.

During the attack phase, the templates have been permuted according to the plaintext byte, that is, the probability for a certain leakage given a certain plaintext was calculated as $\Pr(t_1|p_{1,j}, k_j^*) = \mathcal{D}^{k_j^* \oplus p_{1,j}}(t_1)$.

The result of the known plaintext attack can be seen in Fig. 3. The left plot represents a scenario where the plaintexts were not carefully chosen and therefore, the S-boxes leak independently. This just serves as a reference for the right plot, where the actual CHES 2012 scheme with carefully chosen plaintexts was analyzed. The $y$-axes represent the average key rank of $k_1$ in $\log_2$-scale. A

---

[2] Note, that even if an implementation unintentionally compresses the distribution to a uni-variate one with an informative first-order moment, exploitations do not automatically become easier as discussed in Sect. 5.2.

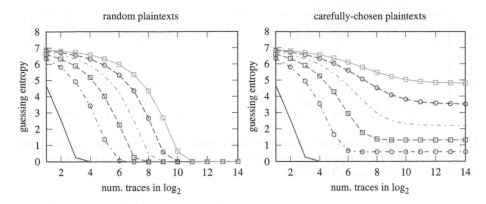

**Fig. 3.** Average guessing entropy after attacks with known plaintexts for $N_s = 1$ (blue, s/), 2 (green, dd/c), 4 (red, d/s), 8 (cyan, dd/), 16 (magenta, d/c), and 32 (yellow, s/s) with {s = solid, d = dashed, dd = dotted dashed}/{c = circle, s = square}. (Color figure online)

random guess would result in an average key rank of 128 and thus a 7 in $\log_2$-scale indicates that no information was retrieved via the side channel. Zero on the other hand indicates that the correct key was ranked first and thus it has been recovered with certainty. The $x$-axis shows the number of required traces to reach a certain average key rank, again in $\log_2$-scale. The different curves represent different numbers of parallel S-boxes ranging from 1 to 32 in powers of two. Each curve has been averaged over 10k attacks. On the right side we can observe a stagnation of the average rank at approximately $(N_s + 1)/2$ for $N_s \leq 8$ (in $\log_2$ this results in 0, 0.6, 1.3, and 2.2). As the adversary targets all subkeys at the same time, this is what one would expect intuitively. However, for 16 and 32 S-boxes, the average rank becomes higher, namely $\log_2(11.2) = 3.5$ (instead of 3.1) and $\log_2(27.1) = 4.8$ (instead of 4.0). This may look surprising, since due to the higher probability of collisions (i.e. repetitions within the $N_s$ subkey values for large values of $N_s$) the rank could be expected to be below $(N_s + 1)/2$. However, as the number of S-boxes increases, the key-dependent algorithmic noise also increases and starts to dominate, implying that incorrect keys start to be ranked amongst the most likely ones in this case.

Next to the average guessing entropy, it is also insightful to look at the rank distribution after seeing all possible leakages. This is done by analyzing the device's leakage distribution that we denote as $\mathcal{D}$. For instance, given two S-boxes and two subkeys $k_1$ and $k_2$, the exact leakage distribution of such device can be computed as $\mathcal{D} = conv(\mathcal{D}^{k_1}, \mathcal{D}^{k_2})$ (using convolutions reduces the complexity of computing $\mathcal{D}$ for an 8-bit S-box from $2^{8 \cdot N_s}$ for the naive approach to $(8 \cdot N_s + 1)^2$). The outcome of this experiment for 1000 random keys can be seen in Fig. 4. The plots show the PMF (in solid blue) and the CDF (in dotted dashed green) for the rank distribution after seeing all possible traces for $N_s = 16$ with carefully chosen plaintexts. The $x$-axis corresponds to the key ranks and the $y$-axis corresponds to

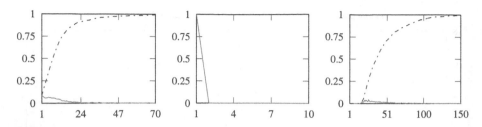

**Fig. 4.** Rank distributions for carefully-chosen plaintexts with $N_s = 16$. Average rank of subkey $k_1$ (left), average minimum rank amongst all $k_j$'s (middle) and average maximum rank amongst all $k_j$ (right).

the probabilities. This figure confirms the previous observations with additional intuitions. First for the left plot, since the median is at rank $\approx 8$ for each subkey in this case, an adversary would have a success rate of 0.5 to find the subkey within the $\approx 8$ most likely candidates. Next, in the middle plot, we show the distribution of the minimum rank within the 16 subkeys $k_j$. It can be clearly seen that the subkey ranked first is almost surely one of the correct ones. This is an important observation and will allow us to construct an advanced attack in Sect. 6.1. As for the distribution of the maximum rank within the 16 subkeys $k_j$ in the right plot, it can be seen that below rank 16, the success rate is almost zero since this can only happen (but is not given) if two subkeys are equal. Finally, in order to have a success rate of 0.5 to see the worst ranked $k_j$ (and therefore also seeing all other correct subkeys), the adversary would need to look at the first 37 most likely candidates.

## 4.2 Security Based on Unknown Plaintexts

For the unknown plaintext scenario we targeted leakages in the form of Eq. (4) and generated the templates as two-dimensional histograms. Each dimension has $8 \cdot N_s + 1$ bins, starting at bin $\mathcal{D}^i(0,0)$ which indicates the probability that for key $k = i$ both leakage samples have a value of 0. The templates are built according to Algorithm 2.

The left side of Fig. 5 again shows a reference result for independent noise. Since, in the unknown plaintext scenario, we cannot decouple the noise by simply randomizing the plaintexts, we had to use a trick. Namely, we only fixed $k_1$ and randomly drew $q$ different values for each $k_j$ with $j \in 2, \ldots, N_s$. It can be seen that a recovery for $N_s = 1$ S-boxes requires around $2^8$ traces, whereas for $N_s = 16$ around $2^{27}$ traces can be expected.

The right side of the figure represents the unknown-plaintext scenario (where the subkeys are constant over all traces within one instance of the experiment). It can be seen that key-dependent noise leads to a stagnation of the correct subkey's rank. This is similar to the carefully-chosen plaintext case and expected. However, the important difference compared with the previous experiment is that the stagnation does not take place at $y \approx \log_2((N_s + 1)/2)$ but much earlier. In

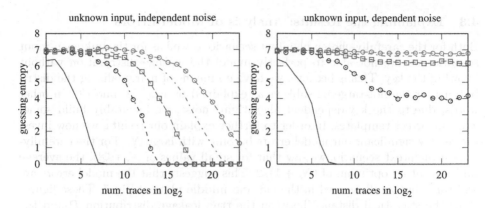

**Fig. 5.** Average guessing entropy after attacks with unknown plaintexts for $N_s =$ (blue, s/), 2 (green, dd/c), 4 (red, d/s), 8 (cyan, dd/), 16 (magenta, d/c). (Color figure online)

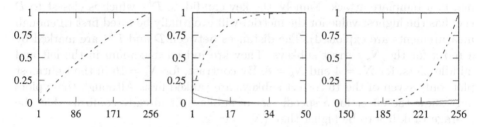

**Fig. 6.** Rank distributions for unknown plaintexts with $N_s = 16$. Average rank of subkey $k_1$ (left), average minimum rank amongst all $k_j$'s (middle) and average maximum rank amongst all $k_j$ (right)

order to get the full picture, we again look at the rank distributions in Fig. 6. First, we observe in the left plot that the subkey ranks (from 40 000 experiments) look close to uniformly distributed (which would be reflected by a straight line), with a median rank at $\approx 102$ (instead of 128 for the uniform distribution). For the minimum rank distribution (middle plot), the median rank is at $\approx 6$, which has to be compared to a value of 10 that would be obtained for a uniform distribution with $N_s = 16$. As for the median of the maximum rank, it moved to $\approx 240$ (whereas it would be at 245 for a uniform distribution with $N_s = 16$). In our experiments, the lowest maximum rank value found was 110. This essentially means that with a search complexity of $\binom{110}{16} \cdot 16! \approx 2^{107}$, the correct key is found with probability $\approx 1/40000 \approx 2^{-15}$. In fact, already for $N_s > 4$ and even when seeing all possible leakages in a noise-free Hamming weight scenario, the guessing entropy is close to 7 and the rank distribution close to uniform.

## 4.3    Explaining the Results: Analysis of Model Errors

Both for the carefully-chosen plaintext scenario as well as for the unknown-input scenario, we are not able to perfectly model the leakage distribution without knowing the key. This is because, we have no means of marginalizing the distributions for the not-targeted subkeys as explained by Eqs. (2) and (3). In other words, due to the key-dependent algorithmic-noise, we inevitably build somewhat incorrect templates. In order to further explain our results, we now investigate how significant our model errors become with large $N_s$. For the carefully-chosen plaintext scenario, we saw that for small values of $N_s$ ($\leq 8$), the average rank was at an optimum of $(N_s + 1)/2$. This suggests that the model errors are still tolerable, as illustrated in the left and middle plots of Fig. 7. These figures show the statistical distance between the true leakage distribution $\mathcal{D}$ and the models $\mathcal{D}^{k_j^*}$ for $N_s = 4$ and $N_s = 8$. For measuring the distance we computed one line of the mutual information matrix as defined in [10], corresponding to one used key. This metric was chosen because it directly reflects what will happen in a template attack. Namely, the key candidate $\mathcal{D}^{k_j^*}$ which is closest to $\mathcal{D}$ (i.e. has the highest value for the metric) will eventually be rated first (if enough measurements are exploited). The distances between $\mathcal{D}$ and $\mathcal{D}^{k_j}$ are marked by a red x for the $N_s$ correct subkeys. They are indeed maximum in the left and middle plots, for $N_s = 4$ and $N_s = 8$. By contrast, for $N_s = 16$ in the rightmost plot, only seven of the 16 correct subkeys are ranked first. Although these plots only show the effect for a specific set of subkeys, it already confirms that the average rank has to be higher than $(N_s + 1)/2$.

In the unknown plaintext scenario, we additionally need to estimate a second-order moment of a bi-variate distribution. From studies on masking, we know that such distributions are much more susceptible to noise [29]. Furthermore, in our case the relation between the leakage samples is not straightforward, as for affine or multiplicative masking [12]. Both circumstances suggest that the key dependent noise should cause more severe model errors and indeed, this

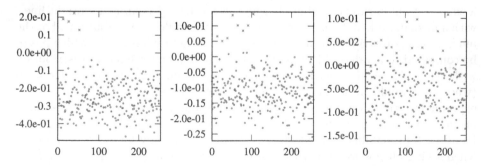

**Fig. 7.** Distance between $\mathcal{D}$ and $\mathcal{D}^{k_j^*}$ for carefully-chosen plaintexts. The device holds the subkeys $k_j$ marked by the red x. As the distance is measured by the entries of the mutual information matrix, a higher value on the y-axis indicates a smaller distance. From left to right the scenarios for $N_s = 4, 8,$ and 16 are depicted. (Color figure online)

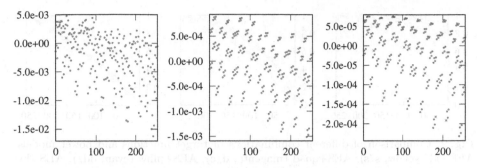

**Fig. 8.** Distance between $\mathcal{D}$ and $\mathcal{D}^{k_j^*}$ for unknown plaintexts. From left to right the scenarios for $N_s = 2, 8$, and 16 are depicted.

is what can be observed in Fig. 8. Be aware that this time, the leftmost plot depicts the case for $N_s = 2$ and even there already none of the correct subkeys is ranked first. As we move to higher values for $N_s$, it can also be seen that the distances themselves become much smaller. As a consequence, measurement noise (remember that until now all experiments were performed without noise) and the inability to calculate the templates will make attacks even harder, as will be discussed in Sect. 5.3.

## 5   Implementation and Attack Issues

The previous evaluations of our new construction assumed bi-variate noise-free Hamming weight leakages and perfectly calculated templates. In this section we want to address the violation of these assumptions in a real world implementation and attack scenario. We start by analyzing the deviation from Hamming weight leakages, then discuss the case of transition-based leakages (aka Hamming distance model), and finally look at the impact of a more realistic (bounded) template estimation phase.

### 5.1   Deviations from the Hamming Weight Leakage Function

In this section we show that the previous experiments based on Hamming weight leakages are appropriate and sufficient to argue about the security of our construction, even if such leakages are not accurately met in a real world application. We do so by exploring different power models. In particular, we choose power models with low and high resolution and with low and high non-linearity. As for the resolution, we choose the leakage functions to be the Hamming weight function (hw), the Hamming weight function plus quadratic terms (quad), and as the identity function (id). As for the non-linear leakage function we chose the Hamming weight function preceded by an AES S-box (nlhw). In addition, we target two kinds of S-boxes, the AES S-box (AES) and an identity function S-box (ID8). The latter one would correspond to directly attacking the key addition

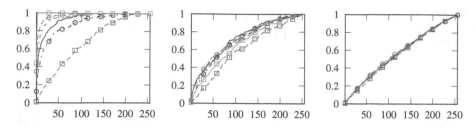

**Fig. 9.** Comparison of different combinations of target functions and power models: AES+id (yellow, s/s), AES+quad (magenta, d/c), AES+nlhw (cyan, dd/), AES+hw (blue, s/), ID8+nlhw (green, dd/c), ID8+hw (red, d/s). From left to right the scenarios for $N_s = 2, 4$, and 16 are depicted. (Color figure online)

layer of the AES. In Fig. 9 we compare the rank distributions for these various scenarios and $N_s = 2, 4$ and 16. For $N_s = 2$ it can be seen that the non-linearity of the target function is of higher importance than the one of the leakage function. The non-linearity of the leakage function only helps significantly for linear target functions (ID8), while for (AES) the impact is minor. Most importantly, we see that as soon as $N_s$ increases, the impact of all these combinations of leakage functions and targets vanishes, confirming our claims.

## 5.2    Distance-Based Leakages

In practice, a cryptographic implementations can be flawed because an adversary sees leakages which are not covered by the theoretical analysis. This can be due to glitches, early propagation, or most deadly for Boolean masking, unintentional distance-based leakage [2,8]. That is, a secret shared as $(s \oplus m, m)$ leaks via $HD(s \oplus m, m)$. Such leakage can occur if a register holding the first share is overwritten with the second share. Another scenario, where the adversary might get an advantage is if he can perform a normalized product combining before summing up the leakage points. This can be the case for a weakly shuffled software implementation which handles the key addition and the S-box operations together. Interestingly, we can show experimentally that none of these implementation issues represent a threat in the unknown-input case. From Fig. 10 (which contains the rank distributions of our construction in the context of uni-variate Hamming weight leakages corresponding to the XOR between the two intermediate values of our previous bi-variate distributions) we see that this Hamming distance case already performs badly for small values of $N_s$, whereas the normalized product combining still gives a slight advantage due to the reduced noise impact. Again, the higher the value of $N_s$ becomes, the more forgiving the scheme becomes w.r.t. implementation weaknesses.

## 5.3    Bounded Template Estimation

Besides the previously studied key-dependent algorithmic noise, another standard source of errors for templates is poor estimation. Usually, one exhaustively

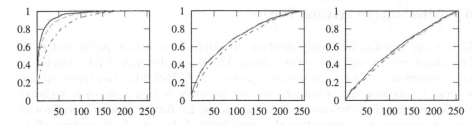

**Fig. 10.** Comparison of the rank distribution when attacking a standard bi-variate (red, d), a Hamming distance-based (green, dd) and a normalized product combining (blue, s) based leakage distribution. From left to right the scenarios for $N_s = 2, 4$, and 16 are depicted. (Color figure online)

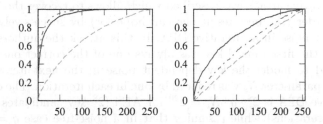

**Fig. 11.** Rank distribution for calculated templates with dependent noise (blue, s) and estimated templates with independent noise. For the dotted dashed green line the templates were estimated using $2^{26}$ traces, for the dashed red one using $2^{22}$ traces. $N_s = 2$ in the left plot and $N_s = 4$ in the right plot. (Color figure online)

acquires traces for all inputs. In practice, this is not possible as the number of inputs grows exponentially with $N_s$, but usually good enough if the number of traces is sufficiently large.[3] In Fig. 11 we can see that this is not the case for unknown-inputs. We compare the rank distribution for an attack with dependent noise to an attack with independent noise but with insufficiently sampled templates. For the left plot with $N_s = 2$, $2^{26}$ traces for template building yield a smaller error than key dependent noise, but still do not allow to recover the key with certainty as in the left plot of Fig. 5 where the templates where calculated. Using only $2^{22}$ traces already leads to a larger model error than dependent noise. Finally, for $N_s = 4$ the calculated templates for dependent noise already perform best.

---

[3] One could overcome this insufficiency by building the templates for the S-boxes independently and afterwards combine them like we did in our simulations. However, the errors for the $\mathcal{D}^i$'s will multiply when calculating the overall template and therefore the overall error will grow exponentially with $N_s$.

# 6    Alternative Attack Paths

In this last section, we finally mention two alternative attack paths that could be considered against our construction. These are iterative DPA attacks (as they represent the strongest attack against the CHES 2012 construction) and attacks to recover the plaintexts (as our security is based on their secrecy). While we leave their detailed analysis as a scope for further research, we provide concrete arguments showing that they have limited chances of success for realistic adversaries. Finally, we also discuss localized EM attacks.

## 6.1    Iterative DPA Attacks and Key Verification

In [20], an iterative attack was described which allows to recover the 16 subkeys up to their order (the best result one can hope for) by successively removing the dependent noise in an iterative DPA. In this attack the authors exploited the fact that the first ranked key was always one of the correct ones and thus could be used to model the key dependent noise in the next iteration. Thus virtually, the parameter $N_s$ was reduced by one in each iteration. The complexity of the iterative DPA is $2^8 \cdot q \cdot N_s = 2^{20}$ for AES ($2^8$ key candidates and $N_s = 16$, thus 16 iterations) while assuming that in a noise-free case $q = N_p = 2^8$ traces. Afterwards, the enumeration costs are $16! \approx 2^{44}$. Key verification during enumeration is straightforward since the plaintexts are known.

In the unknown-input case we could follow a similar strategy. In order to model the algorithmic noise, induced by already guessed subkeys, we would need to construct the templates freshly in every iteration. On top of that we cannot just take the first subkey candidate but need to exhaust the lists up to a certain threshold.[4] To estimate the effort of this, we multiply the medians of the ranks for the best ranked $k_j$ for $N_s = 1 \ldots 16$. The result is that with a probability of $2^{-16}$ we recover the correct key set after $\approx 2^{37}$ iterations. Each iteration comprises 16 template building and attack operations which in turn has a complexity of $\approx 2^{28}$ (at least $2^{20}$ traces and $2^8$ keys) each. Thus, investing around $2^{37+4+28} = 2^{69}$ one can recover the subkey bytes up to permutation. Ordering them costs again $16! \approx 2^{44}$. Note, that unlike for the carefully-chosen input case, the result of the iterative DPA is not conclusive and therefore has to be multiplied by the ordering effort. In fact, it would be even less complex to directly exhaust for the subkeys rather than the subkey set. Based on the medians for the actual ranks of $k_j$ for $j = N_s = 1 \ldots 16$ this would result in a complexity of $2^{90}$. Finally, after going through all this effort, one still has no means of verifying whether the correct key was found as one needs at least one secret plaintext to verify the key based on a known answer. As recovering a plaintext is as hard as recovering a key with the assumed unbounded data complexity and both need to be jointly verified, the effort squares.

---

[4] Be aware that key enumeration algorithms do not work here since the lists are dependent and thus no full key sorting according to probabilities is possible.

## 6.2   Attacks on the Plaintexts

Attacks on the key are restricted to $N_p$ traces in practice. As the plaintexts need to be precomputed, $N_p$ will take values between $2^4$ and $2^{16}$. As an adversary cannot launch a meaningful attack on the key with this restriction, he might instead target the plaintexts. This can be done by randomizing the PRF input for all iterations of the tree except for the last one. Hence, in the last iteration the key will be randomized, but the plaintext will be fixed. This in turn switches the role of the key and the plaintext in the attack and leads to a virtually unbounded data complexity. Recovering sufficient plaintexts following this strategy, a standard DPA on the key could be mounted.

Our previous analysis shows, that even with unlimited data complexity, one is far from recovering a key or a plaintext. Thus, for the standard DPA, the plaintext bytes for building the hypotheses are uncertain and have to be guessed. Let us first assume, that only one subkey byte is targeted. The adversary then needs to pick the plaintext byte for each trace from a set. Without side-channel information, this set would have a size of 256. From Fig. 6 we know that with a 50 % probability, the plaintext byte is contained in a set of 240 entries after an unbounded attack. Thus, overall in an attack where $r$ plaintexts are used, $2^8 \cdot 240^r$ hypotheses have to be built. Even then, the probability that the correct plaintext bytes are contained is only $2^{-r}$. Therefore, this seems to be a rather futile attack path.

## 6.3   Localized EM Attacks

We analyze the impact of localized EM attacks by reducing $N_s$. The simulation is performed for the generation of the secret plaintexts (two traces per key) and the PRF evaluation. For the latter, we look at attacks on the key (16 traces) and on the secret plaintexts (unlimited traces). Both scenarios are studied for the Hamming weight and for the ID leakage function.[5] Even though the same information can be extracted from Figs. 3, 5, and 9, we present a, for our purpose, more representative cross-section of these.

In Fig. 12 it can be seen that attacks on the key during the PRF evaluation are the least informative ones. Even for ID leakage, two parallel S-boxes are sufficient to raise the guessing entropy to a value close to seven. For recovering the secret plaintexts (unlimited traces), the situation is less clear, but remember from Sect. 6.2 that the complexity of such an attack grows exponentially with the number of plaintexts that need to be recovered. As a result two parallel S-boxes are sufficient in the HW case and for the ID case (notably uncommon in practice) three to four are required. Finally, attacking the plaintext generation seems the be the most promising strategy. Yet, already with $N_s = 2$ both scenarios lead to a considerable guessing entropy close to six. This is a quite positive result as we cannot do better than touching a key twice and therefore anyway need to

---

[5] As before, the ID leakage experiments were only carried out for up to 4 parallel S-boxes due to the prohibitive simulation complexity for larger values of $N_s$.

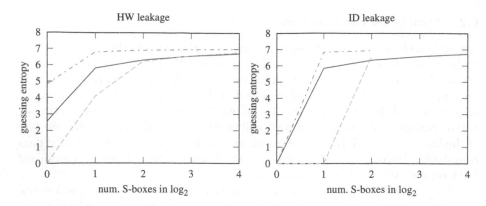

**Fig. 12.** Average guessing entropy after attacks with two known plaintexts (blue, s/), 16 unknown plaintexts (green, dd/c), unlimited unknown plaintexts (red, d/s). (Color figure online)

protect this part sufficiently. Note that, even if we require $N_s = 4$, this is a much stronger result than in [20] where $N_s \geq 24$ was suggested.

Now this leads us to the question how local EM attacks can be, that is, how few S-boxes can be targeted at once with an EM micro probe. Unfortunately, the only work studying such distinguishability so far aimed at a 90 nm FPGA implementation of a block cipher with 32 4-bit S-boxes [4]. From Fig. 5 in [4] one can see that the area in which leakage is observed covers approximately $1\,\text{mm}^2$. An AES implementation which suits our purposes in 40 nm technology on the other hand can be expected to cover only $10,7\,\text{kGE} * 0.71\,\mu\text{m}^2/GE \approx 7600\,\mu\text{m}^2$. Given that even in [4], the S-boxes could not be fully separated, we expect such an attack to be difficult. However, we leave this question to future research.

## 7    Implementation Figures

In this section we would like to discuss an implementation of the primitive and its performance. The AES coprocessor has been implemented using a 32-bit datapath with 4 S-boxes. All 32-bit operations (SubBytes, AddKey, MixCols) are performed in one cycle per column. ShiftRows and the key schedule are 128-bit operations and have been separated in order to minimize the power consumption. This results in a cost of 6 cycles per round and 66 in total. In addition the coprocessor features an IO register which allows to e.g. transfer data between the IO and the data or the key register within 4 cycles. The latter one has been implemented for a fast ciphertext to key transfer during the PRF evaluation. The total area of the coprocessor including the SFR interface accounts for 10.7 kGE.

Also, thanks to the IO register, loading data from the CPU to the coprocessor (32 cycles) can be done while the coprocessor is busy and the delay caused by the CPU between encryptions can be kept low. In total it takes therefore

2951 cycles to pre-compute the secret plaintexts and 2775 cycles for a PRF evaluation including a fault-protected final transformation. As a comparison, a fault protected AES takes four data operations (load key, load/unload data, compare) and two cipher operations, thus a total of 250 cycles. Hence, even though, one PRF evaluation takes 34 AES calls, in our architecture it only takes 11 times longer than a fault-protected AES. Since symmetric cryptographic operations are usually not the dominating part in an application, the overhead decreases with every abstraction layer. That is, when looking at the C function API level during a mutual authentication based on ISO-9798-4 (MAC based authentication), the overhead already decreases to a factor of 4. On an OS level or even transaction level (including communication overheads), the factor would decrease further.

Security wise, we addressed the need for parallelism against localized EM attacks in a hybrid way. On the one hand side, we implement a 32-bit datapath, that is, $N_s = 4$. On the other hand, we implement a four-fold shuffling, virtually resulting in $N_s = 16$. Thanks to this design, even if an adversary would be able to exploit localization, we still have time randomization as a backup. Furthermore, attacks on shuffling itself become more unlikely with increasing noise [31], which we take care of by widening the data path.

## 8    Conclusions

In this work we presented a leakage-resilient PRF which makes use of parallel block cipher implementations with unknown-inputs. To the best of our knowledge this is the first work to study and exploit this form of key-dependent algorithmic noise. It turns out that it renders the problem of side-channel key recovery intractable, even in a noise-free setting and independent of the number of traces and the used power model.

Thanks to this security improvement over the CHES 2012 construction, standardized algorithms like the AES can be used in our construction. Moreover, our analysis suggests that even localized EM attacks can be tolerated to some degree. That is, even if an EM probe would only catch the signal of 8 or 4 S-boxes, the attack would not suddenly become trivial. On top of that, we showed that opposed to the previous construction, the strong side-channel resistance holds throughout the entire algorithm and not only for the first round's S-box layer.

We also showed that these results hold even if actual implementations show leakage behaviors that significantly deviate from our experimental conditions. In fact, the security of our construction essentially relies on secret inputs and nothing else. Yet, and as usual, it will additionally benefit from any concrete limitation of the quality of the templates, e.g. due to a bounded number of traces for profiling and/or electrical noise.

Finally, and from a performance point of view, our new construction allows to use larger values for $N_p$ than the size of the S-box. In practice, it will be quite application specific whether large values for $N_p$ pay off (depending on the memory available for pre-computations). However, at least for block ciphers which use small S-boxes, like e.g. PRESENT, the construction should lead to a significant performance increase over the CHES 2012 one.

# A   Template building algorithms

In the algorithms below $conv(\cdot, \cdot)$ refers to the discrete convolution function.

---

**Algorithm 1.** Template construction for known inputs

---
**Require:** $M = 256$, $N_s$
**Ensure:** Templates
  //Build template for each key
  **for** $k = 0 \ldots M - 1$ **do**
    $\mathcal{D}^k = 0$
    $\mathcal{D}^k(\mathsf{L}(\mathsf{S}(k))) = 1$
  **end for**
  //Calculate the algorithmic noise contribution of an S-box as the marginal distribution
  $H = 0$
  **for** $k = 0 \ldots M - 1$ **do**
    $H + = \mathcal{D}^k / M$
  **end for**
  //Perform $N_s - 1$ convolutions with the marginal distribution
  **for** $k = 0 \ldots M - 1$ **do**
    **for** $i = 1 \ldots N_s - 1$ **do**
      $\mathcal{D}^k = conv(\mathcal{D}^k, H)$
    **end for**
  **end for**

---

---

**Algorithm 2.** Template construction for unknown-inputs

---
**Require:** $M = 256$, $N_s$
**Ensure:** Templates
  //Build template for each key
  **for** $k = 0 \ldots M - 1$ **do**
    **for** $p = 0 \ldots M - 1$ **do**
      $\mathcal{D}^k(\mathsf{L}(p), \mathsf{L}(\mathsf{S}(p \oplus k))) + = 1/M$
    **end for**
  **end for**
  //Calculate the algorithmic noise contribution of an S-box as the marginal distribution
  $H = 0$
  **for** $k = 0 \ldots M - 1$ **do**
    $H + = \mathcal{D}^k / M$
  **end for**
  //Perform $N_s - 1$ convolutions with the marginal distribution
  **for** $k = 0 \ldots M - 1$ **do**
    **for** $i = 1 \ldots N_s - 1$ **do**
      $\mathcal{D}^k = conv(\mathcal{D}^k, H)$
    **end for**
  **end for**

---

# References

1. Abdalla, M., Belaïd, S., Fouque, P.-A.: Leakage-resilient symmetric encryption via re-keying. In: Bertoni, G., Coron, J.-S. (eds.) CHES 2013. LNCS, vol. 8086, pp. 471–488. Springer, Heidelberg (2013). doi:10.1007/978-3-642-40349-1_27
2. Balasch, J., Gierlichs, B., Grosso, V., Reparaz, O., Standaert, F.-X.: On the cost of lazy engineering for masked software implementations. In: Joye and Moradi [16], pp. 64–81
3. Belaïd, S., Grosso, V., Standaert, F.-X.: Masking and leakage-resilient primitives: one, the other(s) or both? Crypt. Commun. **7**(1), 163–184 (2015)
4. Belaïd, S., De Santis, F., Heyszl, J., Mangard, S., Medwed, M., Schmidt, J.-M., Standaert, F.-X., Tillich, S.: Towards fresh re-keying with leakage-resilient PRFs: cipher design principles and analysis. J. Cryptographic Eng. **4**(3), 157–171 (2014)
5. Bilgin, B., Gierlichs, B., Nikova, S., Nikov, V., Rijmen, V.: Higher-order threshold implementations. In: Sarkar, P., Iwata, T. (eds.) ASIACRYPT 2014. LNCS, vol. 8874, pp. 326–343. Springer, Heidelberg (2014). doi:10.1007/978-3-662-45608-8_18
6. Bilgin, B., Gierlichs, B., Nikova, S., Nikov, V., Rijmen, V.: A more efficient AES threshold implementation. In: Pointcheval, D., Vergnaud, D. (eds.) AFRICACRYPT 2014. LNCS, vol. 8469, pp. 267–284. Springer, Heidelberg (2014). doi:10.1007/978-3-319-06734-6_17
7. Chari, S., Jutla, C.S., Rao, J.R., Rohatgi, P.: Towards sound approaches to counteract power-analysis attacks. In: Wiener, M. (ed.) CRYPTO 1999. LNCS, vol. 1666, pp. 398–412. Springer, Heidelberg (1999). doi:10.1007/3-540-48405-1_26
8. Coron, J.-S., Giraud, C., Prouff, E., Renner, S., Rivain, M., Vadnala, P.K.: Conversion of security proofs from one leakage model to another: a new issue. In: Schindler, W., Huss, S.A. (eds.) COSADE 2012. LNCS, vol. 7275, pp. 69–81. Springer, Heidelberg (2012). doi:10.1007/978-3-642-29912-4_6
9. Dodis, Y., Pietrzak, K.: Leakage-resilient pseudorandom functions and side-channel attacks on feistel networks. In: Rabin, T. (ed.) CRYPTO 2010. LNCS, vol. 6223, pp. 21–40. Springer, Heidelberg (2010). doi:10.1007/978-3-642-14623-7_2
10. Duc, A., Faust, S., Standaert, F.-X.: Making masking security proofs concrete or how to evaluate the security of any leaking device (extended version). Cryptology ePrint Archive, Report 2015/119 (2015). http://eprint.iacr.org/
11. Faust, S., Pietrzak, K., Schipper, J.: Practical leakage-resilient symmetric cryptography. In: Prouff and Schaumont [25], pp. 213–232
12. Fumaroli, G., Martinelli, A., Prouff, E., Rivain, M.: Affine masking against higher-order side channel analysis. In: Biryukov, A., Gong, G., Stinson, D.R. (eds.) SAC 2010. LNCS, vol. 6544, pp. 262–280. Springer, Heidelberg (2011). doi:10.1007/978-3-642-19574-7_18
13. Goldreich, O., Goldwasser, S., Micali, S.: How to construct random functions. J. ACM **33**(4), 792–807 (1986)
14. Grosso, V., Poussier, R., Standaert, F.-X., Gaspar, L.: Combining leakage-resilient prfs and shuffling - towards bounded security for small embedded devices. In: Joye and Moradi [16], pp. 122–136
15. Ishai, Y., Sahai, A., Wagner, D.: Private circuits: securing hardware against probing attacks. In: Boneh, D. (ed.) CRYPTO 2003. LNCS, vol. 2729, pp. 463–481. Springer, Heidelberg (2003). doi:10.1007/978-3-540-45146-4_27
16. Joye, M., Moradi, A. (eds.): CARDIS 2014. LNCS, vol. 8968. Springer, Heidelberg (2015)

17. Mangard, S., Popp, T., Gammel, B.M.: Side-channel leakage of masked CMOS gates. In: Menezes, A. (ed.) CT-RSA 2005. LNCS, vol. 3376, pp. 351–365. Springer, Heidelberg (2005). doi:10.1007/978-3-540-30574-3_24

18. Mangard, S., Pramstaller, N., Oswald, E.: Successfully attacking masked AES hardware implementations. In: Rao, J.R., Sunar, B. (eds.) CHES 2005. LNCS, vol. 3659, pp. 157–171. Springer, Heidelberg (2005). doi:10.1007/11545262_12

19. Mather, L., Oswald, E., Whitnall, C.: Multi-target DPA attacks: pushing DPA beyond the limits of a desktop computer. In: Sarkar, P., Iwata, T. (eds.) ASIACRYPT 2014. LNCS, vol. 8873, pp. 243–261. Springer, Heidelberg (2014). doi:10.1007/978-3-662-45611-8_13

20. Medwed, M., Standaert, F.-X., Joux, A.: Towards super-exponential side-channel security with efficient leakage-resilient PRFs. In: Prouff and Schaumont [25], pp. 193–212

21. Micali, S., Reyzin, L.: Physically observable cryptography (extended abstract). In: Naor, M. (ed.) TCC 2004. LNCS, vol. 2951, pp. 278–296. Springer, Heidelberg (2004). doi:10.1007/978-3-540-24638-1_16

22. Moradi, A., Poschmann, A., Ling, S., Paar, C., Wang, H.: Pushing the limits: a very compact and a threshold implementation of AES. In: Paterson, K.G. (ed.) EUROCRYPT 2011. LNCS, vol. 6632, pp. 69–88. Springer, Heidelberg (2011). doi:10.1007/978-3-642-20465-4_6

23. Nikova, S., Rijmen, V., Schläffer, M.: Secure hardware implementation of nonlinear functions in the presence of glitches. J. Cryptology 24(2), 292–321 (2011)

24. Pereira, O., Standaert, F.-X., Vivek, S.: Leakage-resilient authentication and encryption from symmetric cryptographic primitives. In: Ray, I., Li, N., Kruegel, C. (eds.) Proceedings of the 22nd ACM SIGSAC Conference on Computer and Communications Security, pp. 96–108. ACM, New York (2015)

25. Prouff, E., Schaumont, P. (eds.): CHES 2012. LNCS, vol. 7428. Springer, Heidelberg (2012)

26. Rivain, M., Prouff, E.: Provably secure higher-order masking of AES. In: Mangard, S., Standaert, F.-X. (eds.) CHES 2010. LNCS, vol. 6225, pp. 413–427. Springer, Heidelberg (2010). doi:10.1007/978-3-642-15031-9_28

27. Standaert, F.-X., Pereira, O., Yu, Y.: Leakage-resilient symmetric cryptography under empirically verifiable assumptions. In: Canetti, R., Garay, J.A. (eds.) CRYPTO 2013, Part I. LNCS, vol. 8042, pp. 335–352. Springer, Heidelberg (2013). doi:10.1007/978-3-642-40041-4_19

28. Standaert, F.-X., Pereira, O., Yu, Y., Quisquater, J.-J., Yung, M., Oswald, E.: Leakage resilient cryptography in practice. In: Sadeghi, A.-R., Naccache, D. (eds.) Towards Hardware-Intrinsic Security - Foundations and Practice. Information Security and Cryptography, pp. 99–134. Springer, Heidelberg (2010)

29. Standaert, F.-X., Veyrat-Charvillon, N., Oswald, E., Gierlichs, B., Medwed, M., Kasper, M., Mangard, S.: The world is not enough: another look on second-order DPA. In: Abe, M. (ed.) ASIACRYPT 2010. LNCS, vol. 6477, pp. 112–129. Springer, Heidelberg (2010). doi:10.1007/978-3-642-17373-8_7

30. Veyrat-Charvillon, N., Gérard, B., Renauld, M., Standaert, F.-X.: An optimal key enumeration algorithm and its application to side-channel attacks. In: Knudsen, L.R., Wu, H. (eds.) SAC 2012. LNCS, vol. 7707, pp. 390–406. Springer, Heidelberg (2013). doi:10.1007/978-3-642-35999-6_25

31. Veyrat-Charvillon, N., Medwed, M., Kerckhof, S., Standaert, F.-X.: Shuffling against side-channel attacks: a comprehensive study with cautionary note. In: Wang, X., Sako, K. (eds.) ASIACRYPT 2012. LNCS, vol. 7658, pp. 740–757. Springer, Heidelberg (2012). doi:10.1007/978-3-642-34961-4_44
32. Yu, Y., Standaert, F.-X.: Practical leakage-resilient pseudorandom objects with minimum public randomness. In: Dawson, E. (ed.) CT-RSA 2013. LNCS, vol. 7779, pp. 223–238. Springer, Heidelberg (2013). doi:10.1007/978-3-642-36095-4_15

# Block Cipher II

# A New Algorithm for the Unbalanced Meet-in-the-Middle Problem

Ivica Nikolić[1]($\boxtimes$) and Yu Sasaki[2]

[1] Nanyang Technological University, Singapore, Singapore
inikolic@ntu.edu.sg
[2] NTT Secure Platform Laboratories, Tokyo, Japan
sasaki.yu@lab.ntt.co.jp

**Abstract.** A collision search for a pair of $n$-bit unbalanced functions (one is $R$ times more expensive than the other) is an instance of the meet-in-the-middle problem, solved with the familiar standard algorithm that follows the tradeoff $TM = N$, where $T$ and $M$ are time and memory complexities and $N = 2^n$. By combining two ideas, unbalanced interleaving and van Oorschot-Wiener parallel collision search, we construct an alternative algorithm that follows $T^2 M = R^2 N$, where $M \leq R$. Among others, the algorithm solves the well-known open problem: how to reduce the memory of unbalanced collision search.

**Keywords:** Meet-in-the-middle · Tradeoff · Collision search

## 1 Introduction

Consider a collision search problem between two $n$-bit functions $f(x)$ and $g(x)$, in two similar scenarios. In the first case, assume $f(x)$ and $g(x)$ have the same cost (in terms of time complexity). In the second case, assume that $g(x)$ is only $2^{\frac{n}{10}}$ times more costly than $f(x)$. The state-of-the-art suggests we use two different time optimized algorithms for these two similar problems. For the first case we deploy Floyd's cycle finding algorithm [7] and produce a collision in $2^{\frac{n}{2}}$ time and *negligible memory*. For the second case, we store $2^{\frac{9n}{20}}$ images of $g(x)$, and with $2^{\frac{11n}{20}}$ evaluations of $f(x)$ find the collision – a process that requires a time equivalent[1] to $2^{\frac{11n}{20}}$ calls to $f(x)$ and a *memory of* $2^{\frac{9n}{20}}$. This sudden jump of memory from negligible to almost $2^{\frac{n}{2}}$, when the comparative cost of the functions has increased only by a small factor, indicates that the state-of-the-art algorithm is inefficient. We eliminate this inefficiency and show an alternative algorithm that relies on the more logical relation between the comparative cost of $g(x)$ to $f(x)$ and the memory: the smaller the comparative cost, the less memory is needed.

In the literature, the above second case is known as the meet-in-the-middle (MITM) problem, and it is solved with the described standard MITM algorithm.

---

[1] The $2^{\frac{9n}{20}}$ calls to $g(x)$ cost $2^{\frac{9n}{20}+\frac{n}{10}} = 2^{\frac{11n}{20}}$ calls to $f(x)$.

© International Association for Cryptologic Research 2016
J.H. Cheon and T. Takagi (Eds.): ASIACRYPT 2016, Part I, LNCS 10031, pp. 627–647, 2016.
DOI: 10.1007/978-3-662-53887-6_23

Many subproblems in cryptography can be modelled as MITM problems. In general, any collision search between two functions, which not necessary have the same domain and range, is a MITM problem. In such a form, this makes the MITM one of the most frequently occurring problems, and the MITM algorithms that solve the problems, one of the most widely used algorithms in cryptography.

The MITM problem has two instances. The first is the classical MITM as introduced by Diffie and Hellman [3] used for a key recovery in Double DES. It is a collision search problem between two functions with a range larger than a domain. The second instance aims at a collision search between two functions with a range not larger than a domain, but (usually[2]) with different weights. That is, one of the functions requires more time for execution. According to the previous naming convention, we call this instance an unbalanced MITM. In this paper we deal only with the unbalanced case. In the sequel, all references to the MITM problem implicitly assume the unbalanced MITM.

The algorithm that solves the unbalanced MITM allows a simple time-memory tradeoff. It is described with the curve $TM = N$, where $N = 2^n$, $T$ is the time complexity measured in accumulative cost of calls to the functions $f(x)$ and $g(x)$, while $M$ is the memory measured in blocks of certain size (comparable to $n$). By increasing time and reducing memory, solving certain MITM problems becomes feasible in practice, as usually, the memory is the bottleneck. Conversely, most theoretical applications require time optimized solutions, thus in these cases, the time is reduced and the memory is increased. Note, the time can be reduced only up to a certain bound, usually[3] defined as $\sqrt{N}$. If $T$ goes below the bound and $f(x), g(x)$ are random mappings, then a collision may not be found as the total number of pairs is below $N$.

**Our Contribution.** In our study of the unbalanced MITM problem, the MITM algorithms and the resulting tradeoffs, we include as a parameter the ratio $R$ of costs of the two function (e.g. in the above first scenario $R = 1$, while in the second $R = 2^{\frac{n}{10}}$). This is essential because $R$ defines how to balance the number of calls to $f(x)$ and to $g(x)$. In short, $f(x)$ can be evaluated $R$ times more frequently than $g(x)$, while maintaining the same time complexity.

Our new MITM algorithm relies on a combination of two ideas, both well known, but never combined together. The first idea is based on a selection function (we call this method *interleaving*) from the memoryless collision search of two functions. Floyd's algorithm can be used to find a collision between the two functions, by interleaving the calls to $f(x)$ and $g(x)$ during the detection of the cycle. That is, Floyd's algorithm is run for a function $F(x)$ that, based on some selection function, evaluates either $f(x)$ or $g(x)$ with equal probability. Thus a collision for $F(x)$ is an actual collision for $f(x)$ and $g(x)$ with a probability $\frac{1}{2}$ and consequently, the search has to be repeated twice. *Unbalanced interleaving*

---

[2] If the functions are balanced, then a collision can be found trivially with Floyd's cycle finding algorithm.

[3] As shown further, the bound is not universal, but depends on the comparative cost of the two functions.

happens when $F(x)$ evaluates one of the functions more frequently (e.g. $R$ times more) than the other. Then the collision search has to be repeated $R$ times. The second idea relies on van Oorschot-Wiener [22] multiple/parallel collision search based on Hellman's table. First, Hellman's table is built by storing the first and the last points of multiple chains produced from iterative evaluations of a function $F(x)$. Then, to find one collision, another chain for $F(x)$ is built. With the right choice of parameters, one chain in Hellman's table will collide with the newly constructed chain, which can be detected by the end point. Multiple collisions can be built by repeating the same process.

We combine these two ideas by constructing Hellman's table for the function $F(x)$, which is produced by unbalanced interleaving of $f(x)$ and $g(x)$, such that $f(x)$ is called $R$ times more often than $g(x)$. Then, the collision search for $F(x)$ is repeated $R$ times in order to obtain a single collision between $f(x)$ and $g(x)$. (A full description of the algorithm in a pseudo code is given in Appendix A.) Our analysis reveals that this new algorithm relies on the tradeoff

$$T^2 M = R^2 N$$

where $M \leq R$. It follows that, when $R$ tends to 1, then $M$ tends to 1, and $T$ tends to $2^{\frac{n}{2}}$. In other words, the closer the costs of the two functions, the less memory is required (and the time is closer to the case of balanced functions). In contrast, the standard MITM algorithm relies on the counterintuitive relation: the closer the costs of the two functions, the more memory is required.

We compare the new algorithm to the standard algorithm and show that the new is more memory effective and more time effective for certain values of $R$. In short, the new algorithm is more time effective when $MR^2 < N$, and more memory effective when $T > R^2$. A visual comparison of tradeoffs of the two algorithms is given in Appendix B.

We present a number of cases where the replacement of the standard algorithm with the new will lead either to a lower memory requirement or to a better time-memory tradeoff. In addition, we point out cases where such replacement will not work (e.g. known plaintext attacks on block ciphers). Finally, we show that some balanced collision search problems can be regarded as unbalanced, and thus with the use of the new algorithm, can be solved more efficiently (usually, will require less memory).

**Related Work.** The unbalanced MITM has been mentioned as a subproblem in a large number of papers. A few result provide an actual memoryless solution to the problem, for instance, Dunkelman et al. in [5]. The most extensive analysis in this direction has been done by Sasaki [21], who even considers unbalanced interleaving and comes to a conclusion that the time complexity of the memoryless unbalanced MITM is invariant of the interleaving factor. In short, all of the currently proposed memoryless algorithms for the unbalanced MITM provide the same time complexity, which is actually the precise point of our tradeoff curve with $M = 1$ and thus $T = R\sqrt{N}$.

Van Oorschot-Wiener multiple collision search [22] has been a fundamental tool in many research papers as well. Among the latest applications of this technique, we single out the memory efficient multicollision search by Joux and Lucks [12], the technique of dissection by Dinur et al. [4], the tradeoffs for the generalized birthday problem by Nikolić-Sasaki [20] and Khovratovich-Biryukov [1], the multi-user collisions by Fouque et al. [8], and others.

## 2    Preliminaries

### 2.1    Basics

Let $n$ be a positive integer, and $N = 2^n$. Let $f(x), g(x) : \{0,1\}^n \rightarrow \{0,1\}^n$ be two random functions (the range can be smaller than the domain, without affecting the presented analysis). Assume that the time $T_f$ required to compute $f(x)$ is not more than the time $T_g$ required to compute $g(x)$. Let $R = 2^\rho$ be the ratio of the costs of $g(x)$ to $f(x)$, that is, $R = 2^\rho = \frac{T_g}{T_f}$. Obviously, $R \geq 1$. We measure the time complexity of an algorithm in the number of equivalent calls/evaluations to $f(x)$. For instance, if an algorithm makes $u$ calls to $f(x)$ and $v$ calls to $g(x)$, then the time complexity is $u + R \cdot v$.

The MITM problem for $f(x), g(x)$, also known as the collision search problem between $f(x)$ and $g(x)$, consists in finding two $n$-bit values $a$ and $b$ such that $f(a) = g(b)$. This problem can be solved with the use of the MITM algorithm, referred further as the standard MITM algorithm or $\mathtt{MITM^{STD}}$. The algorithm works in two phases. First, in a hash table $L$ it stores $2^m$ pairs $(g(b_i), b_i)$ indexed by $g(b_i)$, where $b_i, i = 1, \ldots, 2^m$ are random values. Then, it keeps generating pairs $(a_j, f(a_j))$, where $a_j$ are random values, until for some $j$ the value of $f(a_j)$ collides with some $g(b_i)$ from the table $L$. As $f(x), g(x)$ are random, a collision will occur after around $2^{n-m}$ values of $f(a_j)$ have been generated.

The memory complexity of the standard algorithm is $M = 2^m$. It makes $2^m$ calls to $g(x)$ to create $L$, and $2^{n-m}$ calls to $f(x)$ to find the collision. According to the above notation, the time complexity $T$ of the algorithm is $T = R \cdot 2^m + 2^{n-m} = 2^{m+\rho} + 2^{n-m}$. For convenience, assume that $T = \max(2^{m+\rho}, 2^{n-m})$, as this reduces the actual time at most by a factor of two.

Let us focus on possible time-memory tradeoffs. When, $m + \rho \leq n - m$, then $T = 2^{n-m}$, and thus the standard algorithm allows the tradeoff $TM = 2^{n-m}2^m = 2^n = N$. On the other hand, when $m + \rho > n - m$, then $T = 2^{m+\rho}$, and thus $TM = 2^{m+\rho}2^m > 2^{n-m}2^m = N$. Obviously this option is worse and therefore further we assume that the memory satisfies $m + \rho \leq n - m$ or equivalently $RM^2 \leq N$ and focus on the tradeoff

$$TM = N. \tag{1}$$

From $RM^2 \leq N$, it follows that $N \geq RM^2 = R(\frac{N}{T})^2$, which leads to $T \geq \sqrt{RN}$.

## 2.2   Collisions Search with Interleaving

Let us consider the collision search problem between two $n$-bit functions $f(x)$ and $g(x)$. A memoryless approach to this problem is based on alteration of the well-known Floyd's cycle-finding algorithm that finds collisions for a single function, i.e. finds $(a, b)$ such that $f(a) = f(b)$.

In the case of a single function $f(x)$, Floyd's algorithm picks a random starting point $u$, assigns $v_0 = w_0 = u$, and iteratively produces values $v_i = f(v_{i-1}), w_i = f(f(w_{i-1}))$ until a collision between $v_i$ and $w_i$ is reached. This colliding value belongs to a cycle, and if the random point $u$ was chosen to be outside the cycle, then with an additional effort, the two colliding values $a$ and $b$ for $f(x)$ can be found: $a$ will be the value that turns the iteration into a cycle, while $b$ the value of the cycle. From the properties of random mappings, it follows that length of the cycle and the length of a chain that leads to a cycle is around $2^{\frac{n}{2}}$, thus the whole algorithm has a time complexity of around $2^{\frac{n}{2}}$ evaluations of $f(x)$ and it uses a negligible memory.

In the case of two functions $f(x), g(x)$, Floyd's algorithm still works and requires a small alteration. The trick is to *interleave* the evaluations of $f(x)$ and $g(x)$ with the use of a selection function $\sigma(x)$ which maps $n$-bit values to a single bit in a random fashion. That is, $\sigma(x)$ outputs 0 or 1, randomly and with equal probability. Define a function $F(x)$ as follows:

$$F(x) = \begin{cases} f(x) & \text{if } \sigma(x) = 0 \\ g(x) & \text{if } \sigma(x) = 1 \end{cases}$$

Then, with Floyd's algorithm find a colliding pair $(a, b)$ for $F(x)$. Obviously if $\sigma(a) \neq \sigma(b)$, then this translates to a collision between $f(x)$ and $g(x)$. Otherwise, repeat the collision search with another starting value. As a result, a colliding pair $(a, b)$ for $f(x), g(x)$ is found with around $2^{\frac{n}{2}}$ evaluations of both $f(x)$ and $g(x)$, and it requires a negligible memory.

We have assumed above that the cost of the two functions is the same, i.e. $R = 1$. However, if $g(x)$ is more costly than $f(x)$, then the time complexity of the above Floyd's algorithm is around $R \cdot 2^{\frac{n}{2}}$. An alternative way to find a collision between two unbalanced function is to use *unbalanced interleaving* as suggested by Sasaki [21]. That is, the selection function $\sigma(x)$ outputs 0 around $R$ times more often than 1. In such a case, a collision for $F(x)$ can be found in around $2^{\frac{n}{2}}$ calls to $f(x)$ and $\frac{2^{\frac{n}{2}}}{R}$ calls to $g(x)$, thus in time equivalent to around $2^{\frac{n}{2}}$ calls to $f(x)$ (recall that we measure the time complexity in calls to $f(x)$). However, a collision for $F(x)$ is an actual collision between $f(x)$ and $g(x)$ only with a probability of $\frac{1}{R}$, thus the collision search has to be repeated around $R$ times. This brings the total time complexity of producing a collision between $f(x)$ and $g(x)$ to $R \cdot 2^{\frac{n}{2}}$.

## 2.3   Multiple Collision Search

Consider the problem of finding multiple collisions for a function $f(x)$, i.e. pairs $(a_1, b_1), \ldots, (a_s, b_s)$ such that $f(a_i) = f(b_i)$ for $i = 1, \ldots, s$. By running Floyd's cycle finding algorithm $s$ times (each with a different starting point and a different reduction function), the required $s$ collisions are found in $s \cdot 2^{\frac{n}{2}}$ evaluations of $f(x)$ and with negligible memory. However, if $s$ is sufficiently large, then the parallel collisions search algorithm by Van Oorschot and Wiener [22] has favourable time complexity, but it requires non-negligible memory.

Let $M = 2^m$ be the available amount of memory. Van Oorschot-Wiener algorithm (given in a pseudo code in Algorithm 1 in Appendix A) starts by building a hash table $L_m$ that resembles Hellman's table from the well known time-memory tradeoffs [9]. Each entry in the table consists of two values: a random starting value $v_s$, and a value $v_e$ produced after $2^{\frac{n-m}{2}}$ iterative applications of $f(x)$ to $v_s$ (i.e. $v_0 = v_s, v_{i+1} = f(v_i), v_e = v_{2^{\frac{n-m}{2}}}$). The table $L_m$ has $2^m$ such entries[4] indexed by the values $v_e$, and thus it requires $2^m$ memory. It is built in $2^m 2^{\frac{n-m}{2}} = 2^{\frac{n+m}{2}}$ time. Note, collisions between iterations are prevented by the so-called matrix stopping rule[5]. It guarantees that if $M \cdot l^2 \leq 2^n$, where $l$ is the length of an iteration, then the number of collisions is negligible. In the above case $l = 2^{\frac{n-m}{2}}$, hence $M \cdot l^2 = 2^n$, thus the condition is fulfilled.

To find one collision for $f(x)$ with the use of $L_m$, choose a random value $w_0$ and build a chain composed of values $w_i$ where $w_{i+1} = f(w_i)$ (refer to Algorithm 2 in Appendix A). Each time a new value of $w_{i+1}$ is computed, check in $L_m$ if it coincides with one of $v_e$. If it does, then pick the corresponding starting value $v_s$ and the value $w_0$ and find the colliding pair. The table $L_m$ covers $2^{\frac{n+m}{2}}$ values. Hence, if the length of the chain is around $2^{n-\frac{n+m}{2}} = 2^{\frac{n-m}{2}}$, we can expect that some value of the chain will hit a value produced during the construction of the table. Obviously, such a hit can be detected once one of the consecutive points of the chain has coincided with some $v_e$ from $L_m$. As mentioned earlier, the average length of the chain at the moment of hit is $2^{\frac{n-m}{2}}$ and with an additional effort of not more than $2^{\frac{n-m}{2}}$ evaluations[6] of $f(x)$ such hit can be detected. Therefore, a collision can be found in around $2^{\frac{n-m}{2}}$ evaluations of $f(x)$. The procedure of generating $L_m$ and finding a collision is illustrated in Fig. 1.

To produce $s$ collisions, van Oorschot-Wiener algorithm requires $T = 2^{\frac{n+m}{2}} + s \cdot 2^{\frac{n-m}{2}}$ evaluations of $f(x)$. Floyd's algorithm needs $T = s \cdot 2^{\frac{n}{2}}$. Thus, roughly when the required number of collisions $s$ satisfies $s > \sqrt{M}$, then van Oorschot-Wiener algorithm has a lower time complexity than Floyd's algorithm. Conversely, if $s$ collisions are required, then it suffices to use $M < s^2$ memory in order to achieve better algorithm in terms of time complexity. For instance, when $s = 2^{\frac{n}{3}}$ and $M = 2^{\frac{n}{3}}$, then van Oorschot-Wiener algorithm requires only $2^{\frac{2n}{3}}$ time, while Floyd's algorithm requires $2^{\frac{n}{3}+\frac{n}{2}} = 2^{\frac{5n}{6}}$ time.

---

[4] Each built by starting from a different point $v_s$.

[5] The term *matrix stopping rule* has been introduced by Biryukov-Shamir [2]. In the original Hellman's paper [9] on TMTO, this rule was given without any particular name (see page 403, Remark 1).

[6] Because $v_e$ is produced from $v_s$ in $2^{\frac{n-m}{2}}$ iterations.

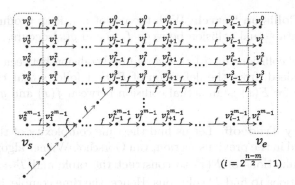

**Fig. 1.** $L_m$ generation and collision detection when $2^m$ memory is available. Only $v_s$ and $v_e$ are stored in $L_m$. Blue lines describe a collision detection performed after $L_m$ is generated. It will hit an intermediate value of one of the chains. By continuing the computation, it reaches $v_e$, thus the collided chain is identified. The exact colliding value can be detected after some re-computation of the two chains (refer to Algorithm 2 in Appendix A). (Color figure online)

## 3  A New Meet-in-the-Middle Algorithm

### 3.1  The Algorithm

To construct our new meet-in-the-middle algorithm, we combine the concepts of unbalanced interleaving and multiple collisions search.

**Specification of the Algorithm.** A description of the complete algorithm in a pseudo code is given in Algorithm 4 of the Appendix. In short, the algorithm can be defined as follows:

1. **Unbalanced Interleaving:** Define a function $F(x)$ as

$$F(x) = \begin{cases} f(x) & \text{if } \sigma(x) = 0 \\ g(x) & \text{if } \sigma(x) = 1 \end{cases}$$

   where the selection function $\sigma(x) : \{0,1\}^n \to \{0,1\}$ outputs 0 around $2^\rho$ times more frequently than 1. For instance, $\sigma(x)$ can be defined as[7]

$$\sigma(x) = \begin{cases} 1 & \text{if } \rho \text{ least significant bits of } x \text{ are zero} \\ 0 & \text{otherwise} \end{cases}$$

   Hence, $F(x)$ evaluates $f(x)$ around $2^\rho$ times more frequently than $g(x)$.

2. **Collision Table:** Based on van Oorschot-Wiener algorithm, create a table $L_m$ with $M = 2^m$ entries for the function $F(x)$.

---

[7] With such a definition, we assume that $f(x)$ and $g(x)$ are random.

3. **Multiple Collision Search:** With the use of $L_m$, keep producing collisions for $F(x)$, until actual collision between $f(x)$ and $g(x)$ occurs.

After around $2^\rho$ collisions for $F(x)$, the required collision between $f(x)$ and $g(x)$ will appear. Indeed, from the definition of $F(x)$ it follows that the probability that a collision for $F(x)$ is an actual collision between $f(x)$ and $g(x)$ is $2^{-\rho}$.

**Time-Memory Tradeoff.** Let us find the time complexity of the above algorithm. As stated in the previous section, van Oorschot-Wiener algorithm requires $T_1 = 2^{\frac{n+m}{2}}$ evaluations of $F(x)$ to construct the table and $T_2 = 2^\rho \cdot 2^{\frac{n-m}{2}} = 2^{\rho+\frac{n-m}{2}}$ evaluations to find $2^\rho$ collisions. Hence, the time complexity of our algorithm is $T = T_1 + T_2$. To simplify the analysis we assume that $T = \max(T_1, T_2)$ (we ignore the constant factor of 2). The required memory is $M = 2^m$.

Let us express the values of $T_1$ in terms of calls to $f(x)$ (recall that we measure the time cost in terms of the lighter function $f(x)$). In $T_1$, there are a total of $2^{\frac{n+m}{2}}$ evaluations of $F(x)$, out of which, around $2^{\frac{n+m}{2}}$ are to the function $f(x)$ and $2^{\frac{n+m}{2}}/2^\rho = 2^{\frac{n+m}{2}-\rho}$ to $g(x)$ which in turn are equivalent to $2^\rho \cdot 2^{\frac{n+m}{2}-\rho} = 2^{\frac{n+m}{2}}$ calls to $f(x)$. Thus $T_1 = 2^{\frac{n+m}{2}}$ calls to $f(x)$. In $T_2$, there are $2^{\rho+\frac{n-m}{2}}$ evaluations of $F(x)$, out of which around $2^{\rho+\frac{n-m}{2}}$ are to $f(x)$ and $2^{\rho+\frac{n-m}{2}}/2^\rho = 2^{\frac{n-m}{2}}$ to $g(x)$, which is equivalent to $2^{\rho+\frac{n-m}{2}}$ calls to $f(x)$. As a result, $T_2 = 2^{\rho+\frac{n-m}{2}}$ calls to $f(x)$. Therefore, the total time complexity $T$ expressed above as number of calls to $F(x)$ can be replaced with calls to $f(x)$.

Further we focus on $T = \max(T_1, T_2) = \max(2^{\frac{n+m}{2}}, 2^{\rho+\frac{n-m}{2}})$ and analyze the two cases:

1. Assume $2^{\frac{n+m}{2}} \le 2^{\rho+\frac{n-m}{2}}$ and thus $T = 2^{\rho+\frac{n-m}{2}}$. In this case, we obtain that

$$T^2 M = 2^{2\rho+n-m}2^m = 2^{2\rho}2^n 2^{-m}2^m = R^2 N \qquad (2)$$

2. Assume $2^{\frac{n+m}{2}} \ge 2^{\rho+\frac{n-m}{2}}$ and thus $T = 2^{\frac{n+m}{2}}$. Similarly, we end up with the tradeoff

$$T^2 = 2^{n+m} = N \cdot M \qquad (3)$$

At the point $2^{\frac{n+m}{2}} = 2^{\rho+\frac{n-m}{2}}$ the tradeoffs switch. This point is defined as

$$\frac{n+m}{2} = \rho + \frac{n-m}{2} \qquad (4)$$

$$m = \rho \qquad (5)$$

Hence, when the available memory $M$ is not more than $R$, the time $T$ and memory $M$ complexity of our meet-in-the-middle follows the tradeoff $T^2 M = R^2 N$. On the other hand, when $M \ge R$, then our tradeoff follows the curve $T^2 = N \cdot M$. In this case, we can see that when the memory increases, the time increases as well. Therefore this tradeoff is not beneficial and thus further in our discussion we focus only on the tradeoff $T^2 M = R^2 N$, where $M \le R$. In addition, the time is limited to $T = \sqrt{\frac{R^2 N}{M}} \ge \sqrt{\frac{R^2 N}{2^\rho}} = \sqrt{RN}$.

## 3.2  Comparison of Tradeoffs

Let us compare the new meet-in-the-middle algorithm $\mathrm{MITM}^{\mathrm{NEW}}$ to the standard meet-in-the-middle algorithm $\mathrm{MITM}^{\mathrm{STD}}$ in terms of time and memory complexities. A graphical comparison of the two tradeoffs is given in Appendix B.

**Time Comparison of the Tradeoffs.** Assume $\mathrm{MITM}^{\mathrm{NEW}}$ and $\mathrm{MITM}^{\mathrm{STD}}$ use the same amount of memory $M$ and we want to find the case when our algorithm has a lower time complexity than the standard. When $M \leq R$, then the time complexity of $\mathrm{MITM}^{\mathrm{NEW}}$ is $T_1 = R\sqrt{\frac{N}{M}}$, while of $\mathrm{MITM}^{\mathrm{STD}}$ is $T_2 = \frac{N}{M}$ and thus

$$R\frac{N^{\frac{1}{2}}}{M^{\frac{1}{2}}} = T_1 < T_2 = \frac{N}{M} \tag{6}$$

$$RM^{\frac{1}{2}} < N^{\frac{1}{2}} \tag{7}$$

$$R^2 M < N \tag{8}$$

From (8) and $M \leq R$ we can conclude that

**Fact 1.** *Let $R$ be the ratio of costs of $g(x)$ to $f(x)$, $M$ be the available memory, and let $M \leq R$. Then MITM$^{NEW}$ has a lower time complexity than MITM$^{STD}$ when*

$$M < \frac{N}{R^2}. \tag{9}$$

*Remark 1 (Necessary Condition).* From (9) it follows that the new algorithm may have a better time complexity only if $R < N^{\frac{1}{2}}$.

**Memory Comparison of the Tradeoffs.** Similarly, let us compare the memory complexities of the two algorithms when they use the same amount of time $T$. Assume $M \leq R$. Then the memory complexity of $\mathrm{MITM}^{\mathrm{NEW}}$ is $M_1 = \frac{R^2 N}{T^2}$, while of the $\mathrm{MITM}^{\mathrm{STD}}$ is $M_2 = \frac{N}{T}$, thus

$$\frac{R^2 N}{T^2} = M_1 < M_2 = \frac{N}{T} \tag{10}$$

$$T > R^2 \tag{11}$$

The condition $R \geq M_1 = \frac{R^2 N}{T^2}$ is equivalent to $T \geq \sqrt{RN}$. As a result we get

**Fact 2.** *Let $R$ be the ratio of costs of $g(x)$ to $f(x)$, $T$ be the available time, and let $T \geq \sqrt{NR}$. Then MITM$^{NEW}$ has a lower memory complexity than MITM$^{STD}$ when*

$$T > R^2. \tag{12}$$

*Remark 2 (Necessary condition).* From (9) and $T < N$ it follows that the new algorithm may have a better memory complexity only if $R < N^{\frac{1}{2}}$.

When used in analysis, often the parameters of the tradeoff are chosen in a way to minimize the time complexity. That is, the most used point of the curve in the tradeoff of the standard meet-in-the-middle algorithm is the one where the time complexity reaches the minimum. As mentioned in Sect. 2.1, this point is defined as $T = 2^{\frac{n+\rho}{2}} = \sqrt{NR}$ and $M = 2^{\frac{n-\rho}{2}} = \sqrt{\frac{N}{R}}$. As the condition $T \geq \sqrt{NR}$ of Fact 2 is satisfied, it follows that our $\texttt{MITM}^{\texttt{NEW}}$ will always use less memory than $\texttt{MITM}^{\texttt{STD}}$ as long as $T > R^2 = \frac{T^4}{N^2}$ or equivalently, $T < N^{\frac{2}{3}}$. This leads to

**Fact 3.** *Let $T < N^{\frac{2}{3}}$ be the minimal time complexity of $\texttt{MITM}^{\texttt{STD}}$, that uses $M_2 = \frac{N}{T}$ memory. Then, with the use of $\texttt{MITM}^{\texttt{NEW}}$, the memory complexity can be reduced to $M_1 = \frac{T^2}{N}$.*

*Proof.* From $T < N^{\frac{2}{3}}$ it follows that $R = \frac{T^2}{N} < N^{\frac{1}{3}}$ and $M_2 = \frac{N}{T} > N^{\frac{1}{3}}$. We choose $M_1 = R$, and use our $\texttt{MITM}^{\texttt{NEW}}$ to achieve $M_1 = \frac{R^2 N}{T^2} = \frac{T^4 N}{N^2 T^2} = \frac{T^2}{N} < \frac{N^{\frac{4}{3}}}{N} = N^{\frac{1}{3}}$.    $\square$

## 3.3    Practical Confirmation

We confirm the correctness of the new algorithm and the resulting tradeoff by implementing it and by running a series of computer experiments. In the experiments, the value of $N$ is in the range of $2^{32}$ to $2^{40}$, and the values of $R$ and $M$ vary (but comply to $M \leq R$). For each particular $N$, $R$, and $M$, we run 100 experiments, each with different $f(x)$ and $g(x)$, and measure the time complexity required to produce a collision between $f(x)$ and $g(x)$.

In Table 1 we report the measured time as the average of the 100 experiments. It is evident that the experimental time is very close to the expected time and differs roughly by a factor of four.

## 3.4    Additional Cases

Besides for the unbalanced MITM, $\texttt{MITM}^{\texttt{NEW}}$ can be used as well to solve a few other collision problems between balanced functions. Further we describe two potential applications. In Sect. 4 we provide concrete examples of these applications.

- **Reducing calls to one of the functions.** In certain applications, even though the costs of the two functions are the same ($R = 1$), it may be beneficial to reduce the number of calls to one of them. For instance, if $g(x)$ depends on a secret key $k$ thus is written as $g(k, x)$, then it has to be queried to get the result. Thus the number of calls to $g(x)$ corresponds to the data complexity $D$. If reducing $D$ is the priority, then the collision search becomes unbalanced.
- **Reduced domain of one of the functions.** So far, we have assumed that the ranges of the two functions are not larger than their domain. If one of the balanced functions has a domain smaller than the range, then $\texttt{MITM}^{\texttt{NEW}}$ can be

Table 1. Experimental verification of the new tradeoff.

| MITM space $N$ | Ratio $R$ | Memory $M$ | Expected time $T = \sqrt{R^2 N/M}$ | Experimental time $T$ |
|---|---|---|---|---|
| $2^{32}$ | $2^8$ | $2^6$ | $2^{21}$ | $2^{22.6}$ |
| $2^{32}$ | $2^4$ | $2^4$ | $2^{18}$ | $2^{20.0}$ |
| $2^{32}$ | $2^{12}$ | $2^{10}$ | $2^{23}$ | $2^{25.3}$ |
| $2^{32}$ | $2^{12}$ | $2^{12}$ | $2^{22}$ | $2^{24.0}$ |
| $2^{36}$ | $2^{10}$ | $2^8$ | $2^{24}$ | $2^{26.2}$ |
| $2^{36}$ | $2^{10}$ | $2^{10}$ | $2^{23}$ | $2^{24.7}$ |
| $2^{36}$ | $2^{12}$ | $2^{12}$ | $2^{24}$ | $2^{25.9}$ |
| $2^{40}$ | $2^6$ | $2^4$ | $2^{24}$ | $2^{26.1}$ |
| $2^{40}$ | $2^6$ | $2^6$ | $2^{23}$ | $2^{24.9}$ |
| $2^{40}$ | $2^8$ | $2^8$ | $2^{24}$ | $2^{25.7}$ |

used to find a collision. That is, a collision between $f(x) : \{0,1\}^n \to \{0,1\}^n$ and $g(x) : \{0,1\}^m \to \{0,1\}^n$, where $m < n$ and $f, g$ are balanced, can be found with the proposed algorithm.

## 3.5 Degenerate Cases

MITM$^{NEW}$ in an alternative to the MITM$^{STD}$, but in some cases it may not be applied or it may not follow the expected time-memory tradeoff curve. Let us take a closer look at such degenerate cases.

- **The ratio $R$ depends on the available memory.** An implicit assumption used in the above analysis is that the ratio $R$ of costs of the two function is fixed and invariant of the available memory. This may not always be the case, and one of the functions (most likely $g(x)$), may have execution time that depends on the available memory (the larger the memory, the shorter the time). In such a situation, the ratio $R$ becomes a function of the memory $M$, i.e. $R = R(M)$, and the curve becomes $T^2 M = R(M)^2 N$. This may limit the flexibility of choosing $M$, lead to another tradeoff, or even make the entire tradeoff invalid (recall that it is valid when $M \le R$, which becomes $M \le R(M)$ – this condition may not have a solution for $M > 0$).
- **Sets instead of functions.** MITM$^{NEW}$ makes calls to both $f(x)$ and $g(x)$, thus the functions must be computable. If one of the function is given as a set, then the algorithm will not function properly. Note, the naive idea of storing the set only leads to MITM$^{STD}$.
- **Known plaintext attack.** MITM$^{NEW}$ makes adaptive chosen queries to both $f(x)$ and $g(x)$. Thus attacks on block ciphers that are based on MITM$^{NEW}$ cannot be known plaintext attacks.

638    I. Nikolić and Y. Sasaki

# 4    Applications

Further we show applications of the MITM^NEW in three different cases: the first is the standard unbalanced MITM, while the remaining two are for the additional cases mentioned in Sect. 3.4.

## 4.1    The Case of Unbalanced Functions

Prior to presenting concrete applications of the MITM^NEW, we emphasize two points. First, MITM^NEW can be used to achieve better tradeoffs (for certain values of $M$ and $T$) in a lot of cases where MITM^STD has been applied. There are numerous such cases – listing and analyzing them is too tedious, and therefore we do not mention them. Second, when the amount of memory is not limited, then both MITM^NEW and MITM^STD have the same time complexity (both achieve the minimal possible theoretical time $T = \sqrt{RN}$). Hence, if the user is not concerned about the memory, then he/she can use either MITM^STD or MITM^NEW. We are ready now to proceed with concrete applications.

Iwamoto et al. [11] show that in narrow-pipe Merkle-Damgård hash functions, a collision attack for the compression function can be converted into a limited-birthday-distinguisher for the corresponding hash function. Recall that a collision[8] for a compression function $CF(h, m)$, where $h$ is the chaining value and $m$ is the message block, is a tuple $(h^*, m, m')$ such that $CF(h^*, m) = CF(h^*, m')$. On the other hand, a limited-birthday distinguisher for a hash function $H(M)$ is the following problem: given two sets $I, O$, find a message $M^*$ such that $H(M^* \oplus \delta_{in}) \oplus H(M^*) = \delta_{out}$, where $\delta_{in} \in I, \delta_{out} \in O$. It can be seen as a problem of finding a message that follows a certain truncated differential ($I, O$ are the truncated differences at the input and at the output, respectively).

Iwamoto et al. convert the collision into a limited-birthday distinguisher by placing the collision at the second block (refer to Fig. 2). That is, they first find multiple collisions $(h_1, m_1, m_1')$ for the compression function, store all $h_1$, and from the initial chaining value $h_0$, find a message $m_0$ that will produce a match with one of the stored $h_1$, i.e. find $m_0$ such that $CF(h_0, m_0) = h_1$. The complexity of the limited-birthday distinguisher in part depends on the complexity of producing collisions for the compression function. Thus, it is a classical example of an unbalanced MITM problem. Iwamoto et al. essentially use MITM^STD while we will switch to MITM^NEW.

**Application to LANE-256.** LANE-256 is a SHA-3 candidate hash function designed by Indesteege et al. [10] that has 256-bit state. A collision attack on the full compression function has been presented by Matusiewicz et al. [14]. Naya-Plasencia [19] has improved the attack – her collision search requires $2^{80}$ calls to the compression function and $2^{66}$ memory.

---

[8] Sometimes, it is called a semi-free-start collision for the compression function.

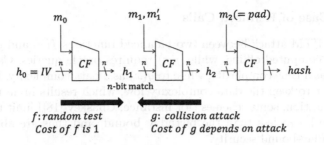

**Fig. 2.** Conversion from a collision attack on the compression function into a limited-birthday-distinguisher on hash function as shown in [11]. The third message block deals with padding. As the collision occurs on $h_2$, the third block preserves collision.

We use Iwamoto et al. conversion of the collision attack for the compression function into a limited-birthday distinguisher for the hash function. The resulting unbalanced MITM (on which the limited-birthday distinguishers relies on) consists of the two functions $f(x)$ and $g(x)$, such that $f(x)$ is equivalent to one compression function call (with a random message block), while $g(x)$ is equivalent to one collision for the compression function (according to Naya-Plasencia equivalent to $2^{80}$ calls and $2^{66}$ memory). Therefore, the ratio of costs is $R = 2^{80}$.

Iwamoto et al. [11] use $\texttt{MITM}^{\texttt{STD}}$ to find the complexity of the limited-birthday distinguisher for LANE-256. We use $\texttt{MITM}^{\texttt{NEW}}$ and show its advantage. From $N = 2^{256}$ and $R = 2^{80}$, it follows that $\texttt{MITM}^{\texttt{NEW}}$ can be described as $T^2 M = 2^{2*80+256} = 2^{416}$. For example, if we set the time complexity to be identical to [11], i.e. $T = 2^{169}$, then the memory complexity $M$ is reduced to $2^{416-2*169} = 2^{78}$, which improves the previous $2^{88}$ by a factor of $2^{10}$. If we set the memory complexity to the lowest possible $M = 2^{66}$ (Naya-Plasencia collision attack requires this much memory), then the previous $\texttt{MITM}^{\texttt{STD}}$ requires $T = 2^{190}$ (the tradeoff is $TM = 2^{256}$), while our $\texttt{MITM}^{\texttt{NEW}}$ requires $T = 2^{175}$.

**Application to AES-Miyaguchi-Preneel.** Iwamoto et al. show as well that in $2^{48}$ time they can find a collision for the compression function built upon 6-round AES in Miyaguchi-Preneel mode. Therefore, a limited-birthday distinguisher for the corresponding Merkle-Damgård hash function, is equivalent to an unbalanced MITM, where $R = 2^{48}$.

According to Fact 1, with parameters $N = 128$ and $R = 48$, $\texttt{MITM}^{\texttt{NEW}}$ has lower complexity than $\texttt{MITM}^{\texttt{STD}}$ if $M < \frac{N}{R^2} = 2^{32}, M \leq R = 2^{48}$, which reduces to $M < 2^{32}$. The time complexity of Iwamoto et al. result with $M = 2^{48}$ cannot be improved. However, $2^{48}$ memory may be too costly and it may be beneficial to reduce the time, when the available memory is much smaller. As suggested by the above condition, when the memory is limited up to $2^{32}$, $\texttt{MITM}^{\texttt{NEW}}$ gives better time than $\texttt{MITM}^{\texttt{STD}}$.

## 4.2    The Case of Reduced Calls

Consider a MITM attack between two balanced functions $f(x)$ and $g(x)$, where $f(x)$ can be computed offline, while $g(x)$ requires oracle queries. Calls to $f(x)$ are counted as a time complexity, while to $g(x)$ as a data complexity. In practice, we often want to keep the data complexity low, which results in an unbalanced MITM. In addition, some schemes (for instance, Chaskey [18]) limit the number of online queries to less than the birthday bound, and thus are able to prove beyond-birthday-bound security.

The best example that illustrates the importance of MITM$^{\text{NEW}}$ to these cases would be to use it to answer Dunkelman et al. [5] open problem about memoryless attack on Even-Mansour with $T$ time and $D = \frac{N}{T}$ data. However, this problem already has been solved partially by Fouque et al. [8]. They provide a solution that uses $M$ memory and $D$ data, such that $M < D$ and $MD^2 = N$. Interestingly, their approach also relies on van Oorschot-Wiener algorithm, but they do not use unbalanced interleaving. With MITM$^{\text{NEW}}$, we can obtain the same solution (thus we omit it from the paper). However, our approach is more generic than [8] – we show this by applying MITM$^{\text{NEW}}$ to key recovery attacks on tweakable block cipher constructions[9].

**Tweakable Block Cipher Mode-of-Operation.** The first example is a *Tweak-dependent Rekeying (TDR)* mode-of-operation proposed by Minematsu [17]. Let $E_K$ be a block cipher with $n$-bit state and $n$-bit key, and let $E_K^t$, where $t < \frac{n}{2}$, be a construction in which the first $n - t$ bits of the plaintext for $E_K$ are fixed to 0, namely the plaintext space is limited to $t$ bits. The TDR mode converts $E_K$ into a tweakable block cipher (uses $t$-bit tweak) with two $E_K$ calls: the first encrypts a tweak $Tw$ with $E_K^t$, used in the second call as a key:

$$K' \leftarrow E_K^t(Tw),$$
$$C \leftarrow E_{K'}(P).$$

Minematsu proves that the TDR mode achieves $O(\frac{2^n}{2^t})$ security. As $t < \frac{n}{2}$, the TDR mode achieves beyond-birthday-bound security. This bound is tight, as shown by the following attack that uses MITM$^{\text{STD}}$:

1. Fix $P$ to a randomly chosen value.
2. Choose $D$ random values of $Tw$, query $(P, Tw)$ to obtain the corresponding $C$, and store all $C$ in a table $L$.
3. Make $2^n/D$ guesses of $K'$, compute $C \leftarrow E_{K'}(P)$ and look for a match in $L$.

A match suggests a candidate for $K'$. With a negligibly small additional cost, the correct $K'$ can be verified. As the analysis relies on MITM$^{\text{STD}}$, it follows the tradeoff $TD = 2^n$. The required memory is identical to the data, i.e. $M = D$.

---

[9] In our understanding, these problems cannot be solved with the algorithm from [8].

To find a collision between steps (2) and (3), we can use $\texttt{MITM}^{\texttt{NEW}}-$ as in the above analysis[10], such collision will exist as long as $TD = 2^n$. The memory, however, can be reduced with $\texttt{MITM}^{\texttt{NEW}}$. The unbalanced MITM will make $T$ calls to $f(x)$ and $D$ calls to $g(x)$, if we set $R = \frac{T}{D}$. In such a case, the tradeoff becomes $T^2 M = \left(\frac{T}{D}\right)^2 2^n$, which is equivalent to $MD^2 = 2^n$. Thus, when the data $D$ satisfies $D > 2^{\frac{n}{3}}$, the new approach will require less memory. For instance, if $D = 2^{\frac{3n}{7}}$, then the standard (as given above in steps (2), (3)) will require $M = 2^{\frac{3n}{7}}$, while the new only $M = 2^{\frac{n}{7}}$ memory.

**Cryptanalysis on McOE-X.** At FSE 2009, Fleischmann et al. [6] propose a family of online authenticated encryption called McOE. Let $E_{K,Tw}$ be a tweakable block cipher under a key $K$ and a tweak $Tw$. Then, the ciphertext $C_i$ of the $i$-th message block $P_i$ of McOE is defined as follows:

$$t_i \leftarrow P_{i-1} \oplus C_{i-1},$$
$$C_i \leftarrow E_{K,t_i}(P_i).$$

McOE-X is an instance of the McOE family, such that $E_{K,Tw} = E_{K \oplus Tw}$.

Mendel et al. [15] show that the key of McOE-X can be recovered in $O(2^{\frac{n}{2}})$ time and data, or more general, in $T$ time and $D = \frac{2^n}{T}$ data, with $\texttt{MITM}^{\texttt{STD}}$.

1. Fix the message for the second block $P_1$ to a randomly chosen value.
2. Choose $D$ random values of the first message block $P_0$, query $P_0 \| P_1$ to obtain the corresponding $C_0 \| C_1$, and store them in a table $L$ along with $P_0 \oplus C_0$.
3. Make $2^n/D$ guesses of $K \oplus t_1$, denoted by $K'$, and compute $C_1 \leftarrow E_{K'}(P_1)$. Check for a match with $L$.

A match suggests that the $K$ can be computed as $P_0 \oplus C_0 \oplus K'$.

As in the case of TDR, with the use of $\texttt{MITM}^{\texttt{NEW}}$ we can reduce the memory requirement of Mendel et al. attack (which currently is $M = D$), while maintaining the same time $T$ and data $D$. Fleischmann et al. instantiate McOE-X with AES-128 as an underlying block cipher. Thus, according to Fact 3, Mendel et al. attack will have a lower memory complexity if $T < 2^{85.3}$ and if it relies on $\texttt{MITM}^{\texttt{NEW}}$ (rather than $\texttt{MITM}^{\texttt{STD}}$). (Considering that accessing $D$ data requires some computational cost of about $D$, limiting $T > D$ is reasonable. Then the range of $T$ becomes $2^{64} < T < 2^{85.3}$.) For instance, if $T = 2^{70}$, then $D = 2^{58}$, and thus Mendel et al. attack will require $2^{58}$ memory if it uses $\texttt{MITM}^{\texttt{STD}}$, and only $2^{12}$ memory if it relies on $\texttt{MITM}^{\texttt{NEW}}$. However, note that $\texttt{MITM}^{\texttt{NEW}}$ overweights $\texttt{MITM}^{\texttt{STD}}$ only if $D > 2^{42.7}$.

### 4.3   The Case of Reduced Domain

Let us apply $\texttt{MITM}^{\texttt{NEW}}$ to the case of a reduced domain. To do so, we focus on triple encryption $\overline{E_{k_1,k_2,k_3}}(P) = E_{k_3}(E_{k_2}(E_{k_1}(P))) = C$, where $E_k(P)$ is an

---

[10] We stress out that we are not showing a weakness of the TDR-mode, but a possible improvement in the memory requirement of the analysis that matches the proved security bound.

$n$-bit cipher with $n$-bit key $k$, and provide a key recovery given three pairs of known plaintext-ciphertext $(P_i, C_i), i = 1, 2, 3$.

First, let us reduce the key recovery to a collision search problem. For this purpose, we define two functions (below, $\|$ denotes concatenation)

$$F(k_1, k_2) = E_{k_2}(E_{k_1}(P_1)) \| E_{k_2}(E_{k_1}(P_2)),$$
$$G(k_3) = E_{k_3}^{-1}(C_1) \| E_{k_3}^{-1}(C_2)$$

Obviously, $F : \{0,1\}^{2n} \to \{0,1\}^{2n}$ and $G : \{0,1\}^n \to \{0,1\}^{2n}$, that is, $G$ *has a reduced domain*. A collision between $F$ and $G$ corresponds to a triplet of keys $(k_1, k_2, k_3)$ such that $\overline{E}_{k_1,k_2,k_3}(P_1) = C_1$ and $\overline{E}_{k_1,k_2,k_3}(P_2) = C_2$. We need to produce $2^n$ such collisions to get the final $\overline{E}_{k_1,k_2,k_3}(P_3) = C_3$, as on average there is only a single triplet of keys that encrypts the three plaintexts $P_1, P_2, P_3$, into the three ciphertexts $C_1, C_2, C_3$.

To find a single collision on $2n$ bits, we use $\mathtt{MITM}^{\mathtt{NEW}}$ with $R = 2^{\frac{n}{2}}$. This value is chosen to avoid collisions of chains in the Hellman's table. Recall that chains have length $\sqrt{\frac{2^{2n}}{M}}$. This ensures that the matrix stopping rule is fulfilled for points on which $F$ is evaluated: dimension of the domain of $F$ is $2n$, each chain has at most $\sqrt{\frac{2^{2n}}{M}}$ evaluations of $F$ and thus, $M \cdot \left(\sqrt{\frac{2^{2n}}{M}}\right)^2 \leq 2^{2n}$. When $R = 2^{\frac{n}{2}}$, then each chain has $\sqrt{\frac{2^{2n}}{M}}/2^{\frac{n}{2}}$ evaluations of $G$, thus the matrix stopping rule for $G$ (with domain of dimension $n$) is fulfilled as well because $M \cdot \left(\sqrt{\frac{2^{2n}}{M}}/2^{\frac{n}{2}}\right)^2 \leq 2^n$.

Therefore, the number of colliding chains is negligible. Note, as the ranges of the two functions are of dimensions $2n$ each, while the domain of $G$ has a dimension of only $n$, when building the chains, we need to use a reduction function for the inputs of $G$, which can be defined simply as a truncation of the $2n$-bit value to $n$ bits.

According to the tradeoff curve of $\mathtt{MITM}^{\mathtt{NEW}}$, we can produce a collision with complexities that follow $T^2 M = R^2 2^{2n} = 2^{3n}$. The condition of the tradeoff dictates that $M \leq R$, hence given a memory $M = 2^{\frac{n}{2}}$, we can get a collision in time $T_1 = 2^{\frac{5n}{4}}$. To get $2^n$ collisions we repeat $2^n$ times the whole collision search (rebuild a new Hellman's table with different reduction function). As a result, we can recover the whole $3n$-bit key in time $T = 2^{\frac{9n}{4}}$ and memory $M = 2^{\frac{n}{2}}$.

The standard MITM algorithm on triple encryption by Merkle and Hellman [16] follows $TM = 2^{3n}$, thus for $M = 2^{\frac{n}{2}}$ it requires time $T = 2^{\frac{5n}{2}}$, which is larger than our time. In addition, the dissection by Dinur et al. [4] used for attacks on multiple encryption, applies only when the number of encryption is at least four. Therefore, $\mathtt{MITM}^{\mathtt{NEW}}$ leads to the lowest time complexity attack on triple encryption with $M = 2^{\frac{n}{2}}$.

## 5  Conclusion

We have shown that one of the most common subproblems in cryptanalysis, the unbalanced meet-in-the-middle problem, can be solved with an alternative algorithm. The new algorithm relies on combination of two ideas: unbalanced interleaving and van Oorschot-Wiener multiple collision search. It follows the tradeoff $T^2 M = R^2 N$, where $R$ is the ratio of costs of the two functions. It outperforms the standard algorithm (with the tradeoff $TM = N$) in terms of time when $MR^2 < N$, and in terms of memory when $T > R^2$ (in both of the cases, assume that $M \leq R$).

The new algorithm follows a more intuitive relation between the ratio $R$ and the required memory $M$: the lower the ratio, the less memory is required. In fact, the complexity of the balanced collision search between two functions (solved with the Floyd's algorithm), can be described as a point of the tradeoff curve of the new algorithm ($R = 1, M = 1$ and thus $T^2 = N$). This is not the case with the standard algorithm ($M = 1$ will lead to $T = N$).

The new algorithm outperforms the standard algorithm in terms of time when $M \leq R$, $M \leq \sqrt{\frac{N}{R}}$ and $M < \frac{N}{R^2}$, and in terms of memory when $T \geq \sqrt{RN}$ and $T > R^2$.

In applications where minimizing the time complexity is the only concern, both the new and the standard algorithm behave the same ($T = \sqrt{RN}$). However, once the focus expands to memory as well as time, the new algorithm may provide significant advantage over the standard. As a general rule of the thumb, the new algorithm should be considered as the first choice in unbalanced meet-in-the-middle problems with $R < N^{\frac{1}{3}}$.

# A    Pseudo Code of Algorithms

---

**Algorithm 1.** Construction of table $L_m$

---

**procedure** CONSTRUCTL($f(x), n, m$)
    $L_m \leftarrow \emptyset$
    **for** $i=1$ to $2^m$ **do**
        $v_s \xleftarrow{\$} \{0,1\}^n$                                                 $\triangleright$ Generate random value
        $v_e \leftarrow v_s$
        **for** $j=1$ to $2^{\frac{n-m}{2}}$ **do**                               $\triangleright$ Iteratively apply $f(x)$
            $v_e \leftarrow f(v_e)$
        **end for**
        $L_m \leftarrow L_m \cup (v_s, v_e)$                                 $\triangleright$ Store $(v_s, v_e)$ in $L_m$
    **end for**
    **return** $L_m$
**end procedure**

---

---

**Algorithm 2.** Find collision with $L_m$

---

  **procedure** FINDCOLLISION($L_m, f(x), n, m$)

    $w_0 \xleftarrow{\$} \{0,1\}^n$                                                 ▷ Generate random value

    $w_i \leftarrow w_0$

    **length** $\leftarrow 0$

    **do**

        **length** $\leftarrow$ **length** $+ 1$

        $w_i \leftarrow f(w_i)$

        $v_s \leftarrow$ Find($L_m, w_i$)                            ▷ Check if $w_i$ is in $L_m$

    **while** $v_s = \emptyset$

    **for** $i=1$ to $2^{\frac{n-m}{2}} -$ **length do**           ▷ Align the two chains

        $v_s \leftarrow f(v_s)$

    **end for**

    **while** $f(v_s) \neq f(w_0)$ **do**                      ▷ Find the colliding pair

        $v_s \leftarrow f(v_s)$

        $w_0 \leftarrow f(w_0)$

    **end while**

    **return** $(v_s, w_0)$

  **end procedure**

---

**Algorithm 3.** Definition of $F(x)$

---

  **procedure** F($f, g, \rho$)

    **if** $x \% 2^\rho = 0$ **then**                          ▷ Least $\rho$ bits of $x$ are zeros

        **return** $g(x)$                               ▷ $F(x) = g(x)$

    **else**

        **return** $f(x)$                               ▷ $F(x) = f(x)$

    **end if**

  **end procedure**

---

**Algorithm 4.** New MITM Algorithm

---

  **procedure** MITM($n, m, \rho$)

    $L_m \leftarrow$ CONSTRUCTL(F($f, g, \rho$)$, n, m$)

    **do**

        $(a, b) \leftarrow$ FINDCOLLISION($L_m$, F($f, g, \rho$)$, n, m$))

    **while** $a \% 2^\rho > 0$ **and** $b \% 2^\rho > 0$

    **return** $(a, b)$

  **end procedure**

---

# B    Graphical Comparison of the Tradeoffs

A comparison of our tradeoff $T^2 M = R^2 N$ to the standard tradeoff $TM = N$ is given in Figs. 3, 4, and 5.

In Fig. 3, we can see that as long as $R < 2^{\frac{n}{2}}$, there is a range of values of $M$, where the time of MITM$^{\text{NEW}}$ is lower than the time of MITM$^{\text{STD}}$. For MITM$^{\text{NEW}}$, when $M > R$, the time remains the same as for the point $M = R$ (recall the tradeoff is valid as long as $M \leq R$). Note, a similar is true for MITM$^{\text{STD}}$ and is denoted

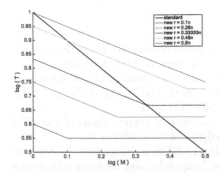

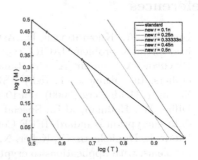

**Fig. 3.** Dependency of time on memory between $\mathtt{MITM}^{\mathtt{STD}}$ (in black) and $\mathtt{MITM}^{\mathtt{NEW}}$ (coloured), when $2^{\frac{n}{10}} \leq R \leq 2^{\frac{n}{2}}$.

**Fig. 4.** Dependency of memory on time between $\mathtt{MITM}^{\mathtt{STD}}$ (in black) and $\mathtt{MITM}^{\mathtt{NEW}}$ (coloured), when $2^{\frac{n}{10}} \leq R \leq 2^{\frac{n}{2}}$.

with coloured dots on the black line. For instance, when $R = 2^{\frac{n}{4}}$ (denoted in green), $\mathtt{MITM}^{\mathtt{STD}}$ is valid as long as $M \leq 2^{\frac{3n}{8}}$. For larger values of $M$, the standard tradeoff does not actually follow the black line (the time does not reduce), but the time remains the same as in the point $M = 2^{\frac{3n}{8}}$.

Similarly, in Fig. 4, we can see a range of values $T$ for which $\mathtt{MITM}^{\mathtt{NEW}}$ outperforms $\mathtt{MITM}^{\mathtt{STD}}$ in terms of memory. Note, both of the algorithms require minimal time of $T = \sqrt{RN}$. Therefore, the lines start from the point $T = \sqrt{RN}$.

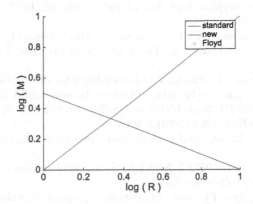

**Fig. 5.** Dependency of the memory on the ratio, when the time is minimal ($T = \sqrt{RN}$), between $\mathtt{MITM}^{\mathtt{NEW}}$ (in red) and $\mathtt{MITM}^{\mathtt{STD}}$ (in blue) (Color figure online).

Finally, in Fig. 5, we show the dependency on the memory of the comparative cost of the two functions, when the time is set to minimal, that is, when $T = \sqrt{RN}$. When $R = 1$, then the Floyd's algorithm requires no memory to find the collision (denoted with a yellow circle at the point $(0,0)$). However, once $R > 1$, the memory requirement of $\mathtt{MITM}^{\mathtt{STD}}$ immediately jumps to almost $2^{\frac{n}{2}}$ (in blue), whereas the memory of $\mathtt{MITM}^{\mathtt{NEW}}$ increases gradually (in red).

# References

1. Biryukov, A., Khovratovich, D.: Asymmetric proof-of-work based on the generalized birthday problem. IACR Cryptology ePrint Archive 2015, 946 (2015)
2. Biryukov, A., Shamir, A.: Cryptanalytic time/memory/data tradeoffs for stream ciphers. In: Okamoto, T. (ed.) ASIACRYPT 2000. LNCS, vol. 1976, pp. 1–13. Springer, Heidelberg (2000). doi:10.1007/3-540-44448-3_1
3. Diffie, W., Hellman, M.E.: Special feature exhaustive cryptanalysis of the NBS data encryption standard. IEEE Comput. **10**(6), 74–84 (1977)
4. Dinur, I., Dunkelman, O., Keller, N., Shamir, A.: Efficient dissection of composite problems, with applications to cryptanalysis, knapsacks, and combinatorial search problems. In: Safavi-Naini, R., Canetti, R. (eds.) CRYPTO 2012. LNCS, vol. 7417, pp. 719–740. Springer, Heidelberg (2012). doi:10.1007/978-3-642-32009-5_42
5. Dunkelman, O., Keller, N., Shamir, A.: Minimalism in cryptography: the even-mansour scheme revisited. In: Pointcheval, D., Johansson, T. (eds.) EUROCRYPT 2012. LNCS, vol. 7237, pp. 336–354. Springer, Heidelberg (2012). doi:10.1007/978-3-642-29011-4_21
6. Fleischmann, E., Forler, C., Lucks, S.: McOE: A family of almost foolproof on-line authenticated encryption schemes. In: Canteaut, A. (ed.) FSE 2012. LNCS, vol. 7549, pp. 196–215. Springer, Heidelberg (2012). doi:10.1007/978-3-642-34047-5_12
7. Floyd, R.W.: Nondeterministic algorithms. J. ACM **14**(4), 636–644 (1967)
8. Fouque, P.-A., Joux, A., Mavromati, C.: Multi-user collisions: applications to discrete logarithm, even-mansour and PRINCE. In: Sarkar, P., Iwata, T. (eds.) ASIACRYPT 2014. LNCS, vol. 8873, pp. 420–438. Springer, Heidelberg (2014). doi:10.1007/978-3-662-45611-8_22
9. Hellman, M.E.: A cryptanalytic time-memory trade-off. IEEE Trans. Inf. Theory **26**(4), 401–406 (1980)
10. Indesteege, S., Andreeva, E., De Canniere, C., Dunkelman, O., Käper, E., Nikova, S., Preneel, B., Tischhauser, E.: The LANE hash function, Submission to NIST (2008)
11. Iwamoto, M., Peyrin, T., Sasaki, Y.: Limited-birthday distinguishers for hash functions: collisions beyond the birthday bound can be meaningful. In: Sako, K., Sarkar, P. (eds.) ASIACRYPT 2013. LNCS, vol. 8270, pp. 504–523. Springer, Heidelberg (2013). doi:10.1007/978-3-642-42045-0_26
12. Joux, A., Lucks, S.: Improved generic algorithms for 3-collisions. In: Matsui [13], pp. 347–363
13. Matsui, M. (ed.): ASIACRYPT 2009. LNCS, vol. 5912. Springer, Heidelberg (2009). doi:10.1007/978-3-642-10366-7_21
14. Matusiewicz, K., Naya-Plasencia, M., Nikolić, I., Sasaki, Y., Schläffer, M.: Rebound attack on the full LANE compression function. In: Matsui [13], pp. 106–125
15. Mendel, F., Mennink, B., Rijmen, V., Tischhauser, E.: A simple key-recovery attack on McOE-X. In: Pieprzyk, J., Sadeghi, A.-R., Manulis, M. (eds.) CANS 2012. LNCS, vol. 7712, pp. 23–31. Springer, Heidelberg (2012). doi:10.1007/978-3-642-35404-5_3
16. Merkle, R.C., Hellman, M.E.: On the security of multiple encryption. Commun. ACM **24**(7), 465–467 (1981)
17. Minematsu, K.: Beyond-birthday-bound security based on tweakable block cipher. In: Dunkelman, O. (ed.) FSE 2009. LNCS, vol. 5665, pp. 308–326. Springer, Heidelberg (2009). doi:10.1007/978-3-642-03317-9_19

18. Mouha, N., Mennink, B., Herrewege, A., Watanabe, D., Preneel, B., Verbauwhede, I.: Chaskey: An efficient MAC algorithm for 32-bit microcontrollers. In: Joux, A., Youssef, A. (eds.) SAC 2014. LNCS, vol. 8781, pp. 306–323. Springer, Heidelberg (2014). doi:10.1007/978-3-319-13051-4_19

19. Naya-Plasencia, M.: How to improve rebound attacks. In: Rogaway, P. (ed.) CRYPTO 2011. LNCS, vol. 6841, pp. 188–205. Springer, Heidelberg (2011). doi:10.1007/978-3-642-22792-9_11

20. Nikolić, I., Sasaki, Y.: Refinements of the k-tree algorithm for the generalized birthday problem. In: Iwata, T., Cheon, J.H. (eds.) ASIACRYPT 2015. LNCS, vol. 9453, pp. 683–703. Springer, Heidelberg (2015). doi:10.1007/978-3-662-48800-3_28

21. Sasaki, Y.: Memoryless unbalanced meet-in-the-middle attacks: impossible results and applications. In: Boureanu, I., Owesarski, P., Vaudenay, S. (eds.) ACNS 2014. LNCS, vol. 8479, pp. 253–270. Springer, Heidelberg (2014). doi:10.1007/978-3-319-07536-5_16

22. van Oorschot, P.C., Wiener, M.J.: Parallel collision search with cryptanalytic applications. J. Cryptol. **12**(1), 1–28 (1999)

# Applying MILP Method to Searching Integral Distinguishers Based on Division Property for 6 Lightweight Block Ciphers

Zejun Xiang[1,2], Wentao Zhang[1,2(✉)], Zhenzhen Bao[1,2], and Dongdai Lin[1,2]

[1] State Key Laboratory of Information Security, Institute of Information
Engineering, Chinese Academy of Sciences, Beijing, China
{xiangzejun,zhangwentao,baozhenzhen,ddlin}@iie.ac.cn
[2] University of Chinese Academy of Sciences, Beijing, China

**Abstract.** Division property is a generalized integral property proposed by Todo at EUROCRYPT 2015, and very recently, Todo *et al.* proposed bit-based division property and applied to SIMON32 at FSE 2016. However, this technique can only be applied to block ciphers with block size no larger than 32 due to its high time and memory complexity. In this paper, we extend Mixed Integer Linear Programming (MILP) method, which is used to search differential characteristics and linear trails of block ciphers, to search integral distinguishers of block ciphers based on division property with block size larger than 32.

Firstly, we study how to model division property propagations of three basic operations (copy, bitwise AND, XOR) and an Sbox operation by linear inequalities, based on which we are able to construct a linear inequality system which can accurately describe the division property propagations of a block cipher given an initial division property. Secondly, by choosing an appropriate objective function, we convert a search algorithm under Todo's framework into an MILP problem, and we use this MILP problem appropriately to search integral distinguishers. As an application of our technique, we have searched integral distinguishers for SIMON, SIMECK, PRESENT, RECTANGLE, LBlock and TWINE. Our results show that we can find 14-, 16-, 18-, 22- and 26-round integral distinguishers for SIMON32, 48, 64, 96 and 128 respectively. Moreover, for two SP-network lightweight block ciphers PRESENT and RECTANGLE, we found 9-round integral distinguishers for both ciphers which are two more rounds than the best integral distinguishers in the literature [22,29]. For LBlock and TWINE, our results are consistent with the best known ones with respect to the longest distinguishers.

**Keywords:** MILP · Division property · Integral cryptanalysis · SIMON · SIMECK · PRESENT · RECTANGLE · LBlock · TWINE

## 1 Introduction

Programming problem is a mathematical optimization which aims to achieve the minimal or maximal value of an objective function under certain constraints, and

© International Association for Cryptologic Research 2016
J.H. Cheon and T. Takagi (Eds.): ASIACRYPT 2016, Part I, LNCS 10031, pp. 648–678, 2016.
DOI: 10.1007/978-3-662-53887-6_24

it has a wide range of applications from industry to academic community. Mixed Integer Linear Programming (MILP) is a kind of programming problem whose objective function and constraints are linear, and all or some of the variables involved in the problem are restricted to be integers. In recent years, MILP has found its applications in cryptographic community. Mouha *et al.* [11] and Wu *et al.* [21] applied MILP method to automatically count differential and linear active Sboxes for word-based block ciphers, which can be used to evaluate the resistance of block ciphers against differential and linear attacks. Later Sun *et al.* [13] extended this technique to count active Sboxes of SP-network block ciphers whose linear layer is a bit permutation.

Recently, this technique was improved [15] to search differential characteristics and linear trails with a minimal number of active Sboxes. They constructed the MILP model by a small number of linear inequalities chosen from the H-Representation of the convex hull of a set of points which are derived from the difference distribution (resp. linear approximation) table of Sbox. However, this method may result in invalid differential characteristics (resp. linear trails). Moreover, differential characteristic (resp. linear trail) with a minimal number of active Sboxes does not alway result in differential characteristic (resp. linear trail) with highest probability. To solve these problems, Sun *et al.* [14] encoded the probability of differentials (resp. linear approximations) of Sbox into the MILP model and they proved that it is always feasible to choose a set $\mathcal{L}$ of linear inequalities from the H-Representation of the convex hull of a set of points $A$, such that the feasible solutions of $\mathcal{L}$ are exactly the points in $A$. Thus, by adding $\mathcal{L}$ into the model and setting the probability as objective function, the MILP optimizer will always return (if the MILP problem can be solved in limited time) a valid differential characteristic (resp. linear trail) with highest probability.

Division property is a generalized integral property introduced by Todo [18] at EUROCRYPT 2015 to search integral distinguishers of block cipher structures which is the core part of integral cryptanalysis [4,7,8,10]. Todo studied propagation rules of division property through different block cipher operations and presented generalized algorithms to search integral distinguishers which only exploits the algebraic degree of nonlinear components of the block cipher. By using division property, Todo presented 10-, 12-, 12-, 14- and 14-round[1] integral distinguishers for SIMON32, 48, 64, 96 and 128 respectively. For PRESENT cipher a 6-round integral distinguisher was found. Later at CRYPTO 2015 Todo [17] proposed a full-round integral attack of MISTY1 based on a 6-round integral distinguisher. Sun *et al.* [12] revisited division property, and they studied the property of a set (multiset) satisfying certain division property. At CRYPTO 2016, Boura and Canteaut [6] proposed a new notion which they called parity set to study division property, based on which they found better integral distinguisher for PRESENT cipher.

---

[1] Since the round key is Xored into the state after the round function, we can easily extend one more round before the distinguisher by using the technique proposed in [20].

Very recently, Todo *et al.* [19] introduced bit-based division property at FSE 2016 which treats each bit independently in order to find better integral distinguishers. They applied this technique to SIMON32, and as a result a 14-round integral distinguisher for SIMON32 was found. However, as pointed out in [19], searching integral distinguisher by bit-based division property required much more time and memory. For a block cipher with block size $n$, the time and memory complexity is upper bounded by $2^n$. Thus, bit-based division property can only apply to block ciphers with block size at most 32. For block ciphers with a much larger block size, searching integral distinguisher by bit-based division property under Todo *et al.*'s framework would be computationally infeasible. Thus, Xiang *et al.* [24] proposed a state partition to get a tradeoff between the time-memory complexity and the accuracy of the integral distinguisher, and they improved distinguishers of SIMON48 and SIMON64 by one round for both variants.

## 1.1  Our Contributions

In this paper, we present a novel technique to search integral distinguishers based on bit-based division property by using MILP method. First we propose a new notion that we call *division trail* to illustrate division property propagation. We show that each division property propagation can be represented by division trails, furthermore, we have proved that it is sufficient to check the last vectors of all division trails in order to estimate whether a useful distinguisher exists. Based on this observation we construct a linear inequality system for a given block cipher such that all feasible solutions of this linear inequality system are exactly all the division trails. Thus, the constructed linear inequality system is sufficient to describe the division property propagations. Then, we study the stopping rule in division property propagation. The stopping rule determines whether the resulting division property can be propagated further to find a longer integral distinguisher. It is observed that for a division property propagation, if the resulting vectors for the first time contain all the vectors of Hamming weight one after propagating $r+1$ rounds, the propagation procedure should terminate and an $r$-round distinguisher can be derived. Hence, we set the sum of the coordinates of the last vector of $r$-round division trail as objective function. By combining this objective function and the linear inequality system derived from the division trails, we construct an MILP problem and present an algorithm to estimate whether $r$-round distinguisher exists given some initial division property. To illustrate our new technique, we run experiments (all the MILP problems in our experiments are solved by the openly available software Gurobi [1]) on SIMON, SIMECK, PRESENT, RECTANGLE, TWINE, LBlock:

1. For SIMON [3] family block ciphers, we first model division property propagations through *Copy*, *And* and *Xor* operations by linear inequalities, since those operations are the basic operations in SIMON family. By using these inequalities we construct an MILP problem and serve it in our search algorithm. As a result we found 14-, 16-, 18-, 22- and 26-round integral distinguishers for SIMON32, 48, 64, 96 and 128 respectively. For SIMON48, 64,

96 and 128, our results are 2, 1, 1 and 1 more rounds than the previous results in [27]. SIMECK [25] is a family of lightweight block ciphers whose round function is very similar to SIMON except the rotation constants. We applied our search technique to SIMECK and we found 15-, 18- and 21-round distinguishers for SIMECK32, 48 and 64 respectively.

2. PRESENT [5] and RECTANGLE [28] are two SP-network lightweight block ciphers whose linear layers are bit permutations. Unlike SIMON, these two ciphers are Sbox-based block ciphers. In [17,18], Sbox is treated as a whole, that is for an $n$-bit Sbox the input value to the Sbox is viewed as a value of $\mathbb{F}_2^n$. In this paper we study bit-based division property propagation of Sbox, and we present an algorithm to compute division trails of Sbox. We observed that, considering bit-based division property could preserve more integral property along with division property propagation through Sbox. By converting division trails of Sbox layer into a set of linear inequalities we construct MILP models for PRESENT and RECTANGLE, as a result, we found 9-round distinguishers for both ciphers which are two more rounds than the best integral distinguishers in the literature.

3. TWINE [16] and LBlock [23] are two generalized Feistel structure block ciphers. By modeling *Sbox*, *Copy* and *Xor* with linear inequalities, we apply our technique to these two ciphers and we found 16-round distinguishers which are in accordance with the results in [26].

Our results are listed in Table 1. All the ciphers explored above except SIMON32 have a block size larger than 32, and searching integral distinguishers by bit-based division property under Todo's framework is computationally infeasible for those ciphers. Note that all our experiments are conducted on a desktop and the consuming time varies from seconds to minutes which is very efficient, the details are listed in Table 1. Moreover, by converting the search algorithm into MILP problems, we can find better integral distinguishers for SIMON48/64/96/128, SIMECK48/64, PRESENT and RECTANGLE.

The rest of the paper is organized as follows: In Sect. 2 we introduce some basic background which will be used later. Section 3 studies how to model some basic operations and components used in block cipher, and to construct a linear inequality system to accurately describe the division property propagations. Section 4 studies the stopping rule and a search algorithm will be presented in this section. Section 5 shows some applications of the technique, and we conclude in Sect. 6.

## 2    Preliminaries

### 2.1    Notations

Let $\mathbb{F}_2$ denote the finite field with only two elements and $\mathbb{F}_2^n$ denote the $n$-bit string over $\mathbb{F}_2$. Let $\mathbb{Z}$ and $\mathbb{Z}^n$ denote the integer ring and the set of all vectors whose coordinates are integers respectively. For any $a \in \mathbb{F}_2^n$, let $a[i]$ denote the $i$-th bit of $a$, and the Hamming weight of $a$ is calculated as $\sum_{i=0}^{n-1} a[i]$. For any

**Table 1.** Results on some block ciphers.

| Cipher | Block size | Round (Previous) | Round (Sect. 5) | Data | Balanced bits | time |
|--------|-----------|------------------|-----------------|------|---------------|------|
| SIMON32 | 32 | 15 [19] | 14 | 31 | 16 | 4.1 s |
| SIMON48 | 48 | 14 [27] | 16 | 47 | 24 | 48.2 s |
| SIMON64 | 64 | 17 [27] | 18 | 63 | 22 | 6.7 m |
| SIMON96 | 96 | 21 [27] | 22 | 95 | 5 | 17.4 m |
| SIMON128 | 128 | 25 [27] | 26 | 127 | 3 | 58.4 m |
| SIMECK32 | 32 | 15 [19] | 15 | 31 | 7 | 6.5 s |
| SIMECK48 | 48 | 12 [18] | 18 | 47 | 5 | 56.6 s |
| SIMECK64 | 64 | 12 [18] | 21 | 63 | 5 | 3.0 m |
| PRESENT | 64 | 7 [22] | 9 | 60 | 1 | 3.4 m |
| RECTANGLE | 64 | 7 [28] | 9 | 60 | 16 | 4.1 m |
| LBlock | 64 | 16 [26] | 16 | 63 | 32 | 4.9 m |
| TWINE | 64 | 16 [26] | 16 | 63 | 32 | 2.6 m |

For SIMON and SIMECK family block ciphers, since the round key is Xored into the state after the round function, we can add one more round before the distinguishers using the technique in [20]. The results presented in the third and fourth columns have been added by one round.

$a = (a_0, \cdots, a_{m-1}) \in \mathbb{F}_2^{n_0} \times \cdots \times \mathbb{F}_2^{n_{m-1}}$, the vectorial Hamming weight of $a$ is defined as $W(a) = (w(a_0), \cdots, w(a_{m-1}))$ where $w(a_i)$ is the Hamming weight of $a_i$. Let $k = (k_0, k_1, \cdots, k_{m-1})$ and $k^* = (k_0^*, k_1^*, \cdots, k_{m-1}^*)$ be two vectors in $\mathbb{Z}^m$. Define $k \succeq k^*$ if $k_i \geq k_i^*$ holds for all $i = 0, 1, \cdots, m - 1$. Otherwise we write $k \not\succeq k^*$.

**Bit Product Function $\pi_u(x)$ and $\pi_u(x)$:** For any $u \in \mathbb{F}_2^n$, let $\pi_u(x)$ be a function from $\mathbb{F}_2^n$ to $\mathbb{F}_2$. For any $x \in \mathbb{F}_2^n$, define $\pi_u(x)$ as follows:

$$\pi_u(x) = \prod_{i=0}^{n-1} x[i]^{u[i]}$$

Let $\pi_u(x)$ be a function from $(\mathbb{F}_2^{n_0} \times \mathbb{F}_2^{n_1} \times \cdots \times \mathbb{F}_2^{n_{m-1}})$ to $\mathbb{F}_2$ for all $u \in (\mathbb{F}_2^{n_0} \times \mathbb{F}_2^{n_1} \times, \cdots, \times \mathbb{F}_2^{n_{m-1}})$. For any $u = (u_0, u_1, \cdots, u_{m-1}), x = (x_0, x_1, \cdots, x_{m-1}) \in (\mathbb{F}_2^{n_0} \times \mathbb{F}_2^{n_1} \times, \cdots, \times \mathbb{F}_2^{n_{m-1}})$, define $\pi_u(x)$ as follows:

$$\pi_u(x) = \prod_{i=0}^{m-1} \pi_{u_i}(x_i)$$

### 2.2 Division Property

Division property [18] is a generalized integral property which can exploit the properties hidden between traditional integral properties $\mathcal{A}$ and $\mathcal{B}$. Thus, by propagating division property we desire to get some better distinguishers. In the following we will introduce division property and present some propagation rules.

**Definition 1 (Division Property [17]).** *Let* $\mathbb{X}$ *be a multiset whose elements take a value of* $(\mathbb{F}_2^n)^m$, *and* $\boldsymbol{k}$ *be an m-dimensional vector whose coordinates take values between 0 and n. When the multiset* $\mathbb{X}$ *has the division property* $\mathcal{D}_{\boldsymbol{k}^{(0)},\boldsymbol{k}^{(1)},\cdots,\boldsymbol{k}^{(q-1)}}^{n,m}$, *it fulfills the following conditions: The parity of* $\pi_{\boldsymbol{u}}(\boldsymbol{x})$ *over all* $\boldsymbol{x} \in \mathbb{X}$ *is always even when*

$$\boldsymbol{u} \in \left\{ (u_0, u_1, \cdots, u_{m-1}) \in (\mathbb{F}_2^n)^m | W(\boldsymbol{u}) \not\succeq \boldsymbol{k}^{(0)}, \cdots, W(\boldsymbol{u}) \not\succeq \boldsymbol{k}^{(q-1)} \right\}$$

**Proposition 1 (Copy [17]).** *Denote* $\mathbb{X}$ *an input multiset whose elements belong to* $\mathbb{F}_2^n$, *and let* $x \in \mathbb{X}$. *The copy function creates* $(y_0, y_1)$ *from* $x$ *where* $y_0 = x, y_1 = x$. *Assuming the input multiset has division property* $\mathcal{D}_k^n$, *let* $\mathbb{Y}$ *be the corresponding output multiset, then* $\mathbb{Y}$ *has division property* $\mathcal{D}_{(0,k),(1,k-1),\cdots,(k,0)}^{n,2}$.

**Proposition 2 (Compression by And [24]).** *Denote* $\mathbb{X}$ *an input multiset whose elements belong to* $\mathbb{F}_2^n \times \mathbb{F}_2^n$, *let* $(x_0, x_1) \in \mathbb{X}$ *be an input to the compression function and denote the ouput value by* $y$ *where* $y = x_0 \& x_1$. *Let* $\mathbb{Y}$ *be the corresponding output multiset. If input multiset* $\mathbb{X}$ *has division property* $\mathcal{D}_{\boldsymbol{k}}^{n,2}$ *where* $\boldsymbol{k} = (k_0, k_1)$, *then the division property of* $\mathbb{Y}$ *is* $\mathcal{D}_k^n$ *where* $k = \max\{k_0, k_1\}$.

**Proposition 3 (Compression by Xor [17]).** *Denote* $\mathbb{X}$ *an input multiset whose elements belong to* $\mathbb{F}_2^n \times \mathbb{F}_2^n$, *let* $(x_0, x_1) \in \mathbb{X}$ *be an input to the compression function and denote the ouput value by* $y$ *where* $y = x_0 \oplus x_1$. *Let* $\mathbb{Y}$ *be the corresponding output multiset. If input multiset* $\mathbb{X}$ *has division property* $\mathcal{D}_{\boldsymbol{k}}^{n,2}$ *where* $\boldsymbol{k} = (k_0, k_1)$, *then the division property of* $\mathbb{Y}$ *is* $\mathcal{D}_{k_0+k_1}^n$.

**Proposition 4 (Substitution [17]).** *Denote* $\mathbb{X}$ *an input multiset whose elements belong to* $\mathbb{F}_2^{n_1}$, *let* $F$ *be a substitution function (Sbox) with algebraic degree* $d$ *and* $F$ *maps an element in* $\mathbb{F}_2^{n_1}$ *to an element in* $\mathbb{F}_2^{n_2}$, *denote* $\mathbb{Y}$ *the corresponding output multiset* $F(\mathbb{X})$. *Assuming the input multiset has division property* $\mathcal{D}_k^{n_1}$, *then the output multiset has division property* $\mathcal{D}_{\lceil \frac{k}{d} \rceil}^{n_2}$. *Moreover, if* $n_1 = n_2$ *and the substitution function is bijective, assuming the input multiset has division property* $\mathcal{D}_{n_1}^{n_1}$, *then the output multiset has division property* $\mathcal{D}_{n_1}^{n_1}$

For more details regarding division property we refer the readers to [17–19].

### 2.3   Modeling a Subset in $\{0, 1\}^n$ by Linear Inequalities

**Convex Hull** and **H-Representation**: The convex hull of a set $A$ of points is the smallest convex set that contains $A$, and the H-Representation of a convex set is a set of linear inequalities $\mathcal{L}$ corresponding to the intersection of some halfspaces such that the feasible solutions of $\mathcal{L}$ are exactly the convex set.

In [14,15] Sun *et al.* treat a differential $(x_{u-1}, \cdots, x_0) \rightarrow (y_{v-1}, \cdots, y_0)$ of an $u \times v$ Sbox as an $(u + v)$-dimensional vector $(x_{u-1}, \cdots, x_0, y_{v-1}, \cdots, y_0)$.

By computing the H-Representation of the convex hull of all possible input-output differential pairs of an Sbox, a set of linear inequalities will be returned to characterize the differential propagation. Moreover, they proved that for a given subset $A$ of $\{0,1\}^n$, it is always feasible to choose a set of linear inequalities $\mathcal{L}$ from the H-Representation of the convex hull of $A$, such that $A$ represents all feasible solutions of $\mathcal{L}$ restricted in $\{0,1\}^n$.

**Theorem 1** ([14]). *Let $A$ be a subset of $\{0,1\}^n$, and denote $\mathrm{Conv}(A)$ the convex hull of $A$. For any $x \in \{0,1\}^n$, $x \in \mathrm{Conv}(A)$ if and only if $x \in A$.*

Thus, they first computed a set of vectors $A$ which is composed of all differential pairs of a given Sbox, and then calculated the H-Representation of the convex hull of $A$ by using the inequality_generator() function in the Sage [2] software, and this will return a set of linear inequalities $\mathcal{L}$ which are the H-presentation of $\mathrm{Conv}(A)$. According to Theorem 1, $\mathcal{L}$ is an accurate description of the difference propagations of the given Sbox, that is, all feasible solutions of $\mathcal{L}$ restricted in $\{0,1\}^n$ are exactly $A$. Since $\mathcal{L}$ is the H-Representation of $\mathrm{Conv}(A)$, each possible differential characteristic corresponds to a point in $A$, thus, each possible differential characteristic satisfies the linear inequalities in $\mathcal{L}$. On the other hand, for any impossible differential characteristic $id$, there always exists at least one linear inequality in $\mathcal{L}$ such that $id$ does not satisfy this inequality. Otherwise, if $id$ satisfies all the inequalities in $\mathcal{L}$ which indicates $id$ belongs to $\mathrm{Conv}(A)$, and this is equivalent to $id \in A$.

Since $\mathcal{L}$ is an accurate description of $A$, adding all the linear inequalities in $\mathcal{L}$ into the MILP problem when searching differential characteristics of a block cipher, it will always return valid differential characteristics. However, the number of linear inequalities in the H-Representation of $\mathrm{Conv}(A)$ is often very large such that adding all the inequalities into the MILP model will make the problem computationally infeasible. Thus, Sun *et al.* [14] proposed a greedy algorithm (See Algorithm 1) to select a subset of $\mathcal{L}$ whose feasible solutions restricted in $\{0,1\}^n$ are exactly $A$. This algorithm can greatly reduce the number of inequalities required to accurately describe $A$.

In order to illustrate the procedure of this section, we present a toy example in Appendix A.

# 3   Modeling Division Property Propagations of Basic Operations and Sbox by Linear Inequalities

In [18] Todo introduced division property by using some vectors in $\mathbb{Z}^m$, and the propagation of division property through a round function of the block cipher is actually a transition of the vectors. Given an initial division property $\mathcal{D}_k^{n,m}$, let $f_r$ denote the round function of a block cipher, the division property of the state after one round $f_r$ can be computed from $\mathcal{D}_k^{n,m}$ by the rules introduced in [17,18], and denote the division property after one round $f_r$ by $\mathcal{D}_{\mathbb{K}}^{n,m}$ where $\mathbb{K}$ is a set of vectors in $\mathbb{Z}^m$. Thus, the division property propagation through $f_r$ is actually the transition from $k$ to the vectors in $\mathbb{K}$. Traditionally, if two

---

**Algorithm 1.** Select a subset of linear inequalities from $\mathcal{L}$

---

**Input** : $\mathcal{L}$: the set of all inequalities in the H-Representation of $\mathrm{Conv}(A)$ with
$A$ a subset of $\{0,1\}^n$
**Output:** A subset $\mathcal{L}^*$ of $\mathcal{L}$ whose feasible solutions restricted in $\{0,1\}^n$ are $A$

1 **begin**
2      $\mathcal{L}^* = \emptyset$
3      $B = \{0,1\}^n \setminus A$
4      $\bar{\mathcal{L}} = \mathcal{L}$
5      **while** $B \neq \emptyset$ **do**
6          $l \leftarrow$ The inequality in $\bar{\mathcal{L}}$ which maximizes the number of points in $B$
         that do not satisfy this inequality (choose the first one if there are
         multiple such inequalities).
7          $B^* \leftarrow$ The points in $B$ that do not satisfy $l$.
8          $\mathcal{L}^* = \mathcal{L}^* \cup \{l\}$
9          $\bar{\mathcal{L}} = \bar{\mathcal{L}} \setminus \{l\}$
10         $B = B \setminus B^*$
11      **end**
12      **return** $\mathcal{L}^*$
13 **end**

---

vectors $\mathbf{k}_1$ and $\mathbf{k}_2$ in $\mathbb{K}$ satisfying that $\mathbf{k}_1 \succeq \mathbf{k}_2$, then $\mathbf{k}_1$ is redundant and will be removed from $\mathbb{K}$. However, since the redundant vectors do not influence the division property, in this paper we do not remove redundant vectors in $\mathbb{K}$, that is for any vector derived from $\mathbf{k}$ by using the propagation rules we add this vector into $\mathbb{K}$. Moreover, for any vector $\bar{\mathbf{k}}$ in $\mathbb{K}$, we call that $\mathbf{k}$ can propagate to $\bar{\mathbf{k}}$ through $f_r$.

**Definition 2 (Division Trail).** *Let $f_r$ denote the round function of an iterated block cipher. Assume the input multiset to the block cipher has initial division property $\mathcal{D}_{\mathbf{k}}^{n,m}$, and denote the division property after $i$-round propagation through $f_r$ by $\mathcal{D}_{\mathbb{K}_i}^{n,m}$. Thus, we have the following chain of division property propagations:*

$$\{\mathbf{k}\} \stackrel{def}{=} \mathbb{K}_0 \xrightarrow{f_r} \mathbb{K}_1 \xrightarrow{f_r} \mathbb{K}_2 \xrightarrow{f_r} \cdots$$

*Moreover, for any vector $\mathbf{k}_i^*$ in $\mathbb{K}_i$ $(i \geq 1)$, there must exist an vector $\mathbf{k}_{i-1}^*$ in $\mathbb{K}_{i-1}$ such that $\mathbf{k}_{i-1}^*$ can propagate to $\mathbf{k}_i^*$ by division property propagation rules. Furthermore, for $(\mathbf{k}_0, \mathbf{k}_1, \cdots, \mathbf{k}_r) \in \mathbb{K}_0 \times \mathbb{K}_1 \times \cdots \times \mathbb{K}_r$, if $\mathbf{k}_{i-1}$ can propagate to $\mathbf{k}_i$ for all $i \in \{1, 2, \cdots, r\}$, we call $(\mathbf{k}_0, \mathbf{k}_1, \cdots, \mathbf{k}_r)$ an $r$-round **division trail**.*

**Proposition 5.** *Denote the division property of input multiset to an iterated block cipher by $\mathcal{D}_{\mathbf{k}}^{n,m}$, let $f_r$ be the round function. Denote*

$$\{\mathbf{k}\} \stackrel{def}{=} \mathbb{K}_0 \xrightarrow{f_r} \mathbb{K}_1 \xrightarrow{f_r} \mathbb{K}_2 \xrightarrow{f_r} \cdots \xrightarrow{f_r} \mathbb{K}_r$$

*the $r$-round division property propagation. Thus, the set of the last vectors of all $r$-round division trails which start with $\mathbf{k}$ is equal to $\mathbb{K}_r$.*

Generally, given an initial division property $\mathcal{D}_{\boldsymbol{k}}^{n,m}$, and if one would like to check whether there exists useful integral property after $r$-round encryption, we have to propagate the initial division property for $r$ rounds to get $\mathcal{D}_{\mathbb{K}_r}^{n,m}$ and check all the vectors in $\mathbb{K}_r$. According to Proposition 5, it is equivalent to find all $r$-round division trails which start with $\boldsymbol{k}$, and check the last vectors in the division trails to judge if any exploitable distinguisher can be extracted. Based on this observation, in the following we focus on how to accurately describe all division trails.

A linear inequality system will be adopted to describe division property propagations, that is we will construct a linear inequality system such that the feasible solutions represent all division trails. Since division property propagation is a deterministic procedure, the constructed linear inequality system must satisfy:

- For each division trail, it must satisfy all linear inequalities in the linear inequality system. That is each division trail corresponds to a feasible solution of the linear inequality system.
- Each feasible solution of the linear inequality system corresponds to a division trail. That is all feasible solutions of the linear inequality system do not contain any impossible division trail.

A linear inequality system satisfying the above two conditions is an accurate description of division property propagation. In the rest of the paper, we only consider bit-based division property. We start by modeling bit-based division property propagation of some basic operations and Sbox in block ciphers.

### 3.1    Modeling Copy, And and Xor

In this subsection, we show how to model bit-wise *Copy*, *And* and *Xor* operations by linear inequalities.

**Modeling Copy.** Copy operation is the basic operation used in Feistel block cipher. The left half of the input is copied into two equal parts, one of which is fed to the round function. Since we consider bit-based division property, the division property propagation of each bit is independent of each other. Thus, we consider only a single bit.

Let $\mathbb{X}$ be an input multiset whose elements take a value of $\mathbb{F}_2$. The copy function creates $y = (y_0, y_1)$ from $x \in \mathbb{X}$ where $y_0 = x$ and $y_1 = x$. Assuming the input multiset has division property $\mathcal{D}_k^1$, then the corresponding output multiset has division property $\mathcal{D}_{(0,k),\cdots,(k,0)}^1$ from Proposition 1. Since we consider bit-based division property, the input multiset division property $\mathcal{D}_k^1$ must satisfy $k \leq 1$. If $k = 0$, the output multiset has division property $\mathcal{D}_{(0,0)}^1$, otherwise if $k = 1$, the output multiset has division property $\mathcal{D}_{(0,1),(1,0)}^1$. Thus, $(0) \xrightarrow{copy} (0,0)$ is the only division trail given the initial division property $\mathcal{D}_0^1$, and $(1) \xrightarrow{copy} (0,1)$, $(1) \xrightarrow{copy} (1,0)$ are the two division trails given the initial division property $\mathcal{D}_1^1$.

Now we are ready to give a linear inequality description of these division trails. Denote $(a) \xrightarrow{copy} (b_0, b_1)$ a division trail of Copy function, the following inequality[2] is sufficient to describe the division propagation of Copy.

$$\begin{cases} a - b_0 - b_1 = 0 \\ a, b_0, b_1 \text{ are binaries} \end{cases} \tag{1}$$

Apparently, all feasible solutions of the inequalities in (1) corresponding to $(a, b_0, b_1)$ are $(0,0,0)$, $(1,0,1)$ and $(1,1,0)$, which are exactly the three division trails of Copy function described above.

**Modeling And.** Bit-wise And operation is a basic nonlinear function, it is the only nonlinear operation for SIMON family. Similar to the modeling procedure of Copy function, we can express its division property propagation as a set of linear inequalities.

Let $\mathbb{X}$ be an input multiset whose elements take a value of $\mathbb{F}_2 \times \mathbb{F}_2$. The And function creates $y = x_0 \& x_1$ from $x = (x_0, x_1) \in \mathbb{X}$. Assuming the input multiset has division property $\mathcal{D}_{\boldsymbol{k}}^{1,2}$ where $\boldsymbol{k} = (k_0, k_1)$, the division property of the corresponding output multiset is $\mathcal{D}_{k}^1$ where $k = \max\{k_0, k_1\}$ according to Proposition 2. Since we consider bit-based division property here, $\boldsymbol{k} = (k_0, k_1)$ must satisfy $0 \le k_0, k_1 \le 1$. Thus, there are four division trails for And function which are $(0,0) \xrightarrow{Xor} (0), (0,1) \xrightarrow{Xor} (1), (1,0) \xrightarrow{Xor} (1)$ and $(1,1) \xrightarrow{Xor} (1)$. Denote $(a_0, a_1) \xrightarrow{and} (b)$ a division trail of And function, the following linear inequalities are sufficient to describe this propagation features.

$$\begin{cases} b - a_0 \ge 0 \\ b - a_1 \ge 0 \\ b - a_0 - a_1 \le 0 \\ a_0, a_1, b \text{ are binaries} \end{cases} \tag{2}$$

It is easy to check that all feasible solutions of the inequalities in (2) corresponding to $(a_0, a_1, b)$ are $(0,0,0)$, $(0,1,1)$, $(1,0,1)$ and $(1,1,1)$, which are exactly the four division trails of And function described above.

**Modeling Xor.** Bit-wise Xor is another basic operation used in block ciphers. Similarly, a linear inequality system can be constructed to describe the division property propagation through Xor function.

Let $\mathbb{X}$ denote an input multiset whose elements take a value of $\mathbb{F}_2 \times \mathbb{F}_2$. The Xor function creates $y = x_0 \oplus x_1$ from $x = (x_0, x_1) \in \mathbb{X}$. Assuming the input multiset $\mathbb{X}$ has division property $\mathcal{D}_{\boldsymbol{k}}^{1,2}$ where $\boldsymbol{k} = (k_0, k_1)$, thus, the corresponding output multiset $\mathbb{Y}$ has division property $\mathcal{D}_{k_0+k_1}^1$. Since we consider bit-based

---

[2] In this paper we do not make a distinction between equality and inequality, since the MILP problem use both equalities and inequalities as constraints.

division property here, $\boldsymbol{k} = (k_0, k_1)$ must satisfy $0 \le k_0, k_1 \le 1$. Moreover, the element of $\mathbb{Y}$ takes a value in $\mathbb{F}_2$, the division property $\mathcal{D}^1_{k_0 + k_1}$ of $\mathbb{Y}$ must satisfy $k_0 + k_1 \le 1$. That is, if $(k_0, k_1) = (1, 1)$, the division property propagation will abort. Thus, there are three valid division trails: $(0, 0) \xrightarrow{Xor} (0)$, $(0, 1) \xrightarrow{Xor} (1)$ and $(1, 0) \xrightarrow{Xor} (1)$. Let $(a_0, a_1) \xrightarrow{Xor} (b)$ denote a division trail through Xor function, the following inequality can describe the division trail through Xor function.

$$\begin{cases} a_0 + a_1 - b = 0 \\ a_0, a_1, b \text{ are binaries} \end{cases} \tag{3}$$

We can check that all the feasible solutions of inequality (3) corresponding to $(a_0, a_1, b)$ are $(0, 0, 0)$, $(0, 1, 1)$ and $(1, 0, 1)$, which are exactly the division trails described above.

## 3.2   Modeling Sbox

Sbox is an important component of block ciphers, for a lot of block ciphers it is the only non-linear part. In [17,18], the Sbox is treated as a whole and the division property is considered while the element in the input multiset taking a value in $\mathbb{F}_2^n$ for an $n$-bit Sbox. In [19] Todo $et$ $al.$ introduced bit-based division property, but they only applied their technique to non-Sbox based ciphers SIMON and SIMECK. In this section, we study bit-based division property propagation through Sbox.

Assume we are dealing with an $n$-bit Sbox, the input and output of the Sbox are elements in $(\mathbb{F}_2)^n$. Suppose that the input multiset $\mathbb{X}$ has division property $\mathcal{D}^{1,n}_{\boldsymbol{k}}$ where $\boldsymbol{k} = (k_0, k_1, \cdots, k_{n-1})$, that is for any $\boldsymbol{u} \in (\mathbb{F}_2)^n$ the parity of $\pi_{\boldsymbol{u}}(\boldsymbol{x})$ over $\mathbb{X}$ is even only if $W(\boldsymbol{u}) \not\succeq \boldsymbol{k}$. Note that for bit-based division property it holds $W(\boldsymbol{u}) = \boldsymbol{u}$, thus, we do not make a distinction between $W(\boldsymbol{u})$ and $\boldsymbol{u}$ in the following. To compute the division property of the output multiset $\mathbb{Y}$, we first consider a naive approach.

**Previous Approach.** First by Concatenation function, each element in $\mathbb{X}$ can be converted into an element in $\mathbb{F}_2^n$. Denote output multiset of Concatenation function as $\mathbb{X}^*$, thus, the division property of $\mathbb{X}^*$ is $\mathcal{D}^n_{k_0 + k_1 + \cdots + k_{n-1}}$ according to Rule 5 in [17]. Secondly, we pass each element in $\mathbb{X}^*$ to the Substitution function Sbox, and denote the output multiset by $\mathbb{Y}^*$ whose elements take a value of $\mathbb{F}_2^n$. According to Proposition 4, the division property of $\mathbb{Y}^*$ is $\mathcal{D}^n_{\left\lceil \frac{k_0 + k_1 + \cdots + k_{n-1}}{d} \right\rceil}$ where $d$ is the algebraic degree of the Sbox. At last, for any value $y^* = y_0 || y_1 || \cdots || y_{n-1}$ in $\mathbb{Y}^*$, a Split function creates $y = (y_0, y_1, \cdots, y_{n-1})$ from $y^*$. Apparently, the output multiset of Split function equals to $\mathbb{Y}$. According to Rule 4 in [17], the division property of $\mathbb{Y}$ is $\mathcal{D}^{1,n}_{\boldsymbol{k}^0, \boldsymbol{k}^1, \cdots}$ where $\boldsymbol{k}^i = (k^i_0, k^i_1, \cdots, k^i_{n-1})$ $(i \ge 0)$ denote all solutions of $x_0 + x_1 + \cdots, x_{n-1} = \left\lceil \frac{k_0 + k_1 + \cdots + k_{n-1}}{d} \right\rceil$.

*Example:* Take the Sbox used in PRESENT as an example. The PRESENT Sbox is a $4 \times 4$ Sbox with algebraic degree three. Assume that the input multiset to the Sbox has division property $\mathcal{D}^{1,4}_{(0,1,1,1)}$. To compute the output multiset division property, we can proceed in three steps as described above: First by a concatenation function we convert the input multiset into another multiset $\mathbb{X}^*$ whose elements take a value in $\mathbb{F}^4_2$, thus the division property of $\mathbb{X}^*$ is $\mathcal{D}^4_3$. Secondly, make each value in $\mathbb{X}^*$ pass through the Sbox operation and this will result in a multiset $\mathbb{Y}^*$ with division property $\mathcal{D}^4_{\lceil \frac{3}{3} \rceil} = \mathcal{D}^4_1$. Finally, we split each value in $\mathbb{Y}^*$ into a value in $(\mathbb{F}_2)^4$, and we will get a multiset $\mathbb{Y}$ with division property $\mathcal{D}^{1,4}_{(0,0,0,1),(0,0,1,0),(0,1,0,0),(1,0,0,0)}$. Thus, we have obtained four division trails of Sbox: $(0,1,1,1) \xrightarrow{Sbox} (0,0,0,1)$, $(0,1,1,1) \xrightarrow{Sbox} (0,0,1,0)$, $(0,1,1,1) \xrightarrow{Sbox} (0,1,0,0)$ and $(0,1,1,1) \xrightarrow{Sbox} (1,0,0,0)$.

Note that only the algebraic degree is exploited to calculate the division trails of Sbox in this naive approach. From the example illustrated above, if the input multiset to the Sbox has division property $\mathcal{D}^{1,4}_{(0,1,1,1)}$, the corresponding output multiset does not balance on any of the four output bits. However, this is not actually true. Denote the input to PRESENT Sbox as $\boldsymbol{x} = (x_3, x_2, x_1, x_0)$, and the corresponding output as $\boldsymbol{y} = (y_3, y_2, y_1, y_0)$, the algebraic normal form (ANF) of PRESENT Sbox is listed as follows:

$$\begin{cases} y_3 = 1 \oplus x_0 \oplus x_1 \oplus x_3 \oplus x_1 x_2 \oplus x_0 x_1 x_2 \oplus x_0 x_1 x_3 \oplus x_0 x_2 x_3 \\ y_2 = 1 \oplus x_2 \oplus x_3 \oplus x_0 x_1 \oplus x_0 x_3 \oplus x_1 x_3 \oplus x_0 x_1 x_3 \oplus x_0 x_2 x_3 \\ y_1 = x_1 \oplus x_3 \oplus x_1 x_3 \oplus x_2 x_3 \oplus x_0 x_1 x_2 \oplus x_0 x_1 x_3 \oplus x_0 x_2 x_3 \\ y_0 = x_0 \oplus x_2 \oplus x_3 \oplus x_1 x_2 \end{cases} \tag{4}$$

Thus,

$$\boldsymbol{\pi}_{(0,0,0,1)}((y_3, y_2, y_1, y_0)) = y_0$$

and

$$\bigoplus_{x \in \mathbb{X}} \boldsymbol{\pi}_{(0,0,0,1)}((y_3, y_2, y_1, y_0))$$

$$= \bigoplus_{x \in \mathbb{X}} y_0$$

$$= \bigoplus_{x \in \mathbb{X}} (x_0 \oplus x_2 \oplus x_3 \oplus x_1 x_2)$$

$$= \bigoplus_{x \in \mathbb{X}} \boldsymbol{\pi}_{(0,0,0,1)}(\boldsymbol{x}) \oplus \bigoplus_{x \in \mathbb{X}} \boldsymbol{\pi}_{(0,1,0,0)}(\boldsymbol{x}) \oplus \bigoplus_{x \in \mathbb{X}} \boldsymbol{\pi}_{(1,0,0,0)}(\boldsymbol{x}) \oplus \bigoplus_{x \in \mathbb{X}} \boldsymbol{\pi}_{(0,1,1,0)}(\boldsymbol{x})$$

$$= 0 + 0 + 0 + 0$$

$$= 0$$

As illustrated above, the least significant bit $y_0$ of the output $\boldsymbol{y}$ is balanced. Similarly, we can check that $y_2$ and $y_0 y_2$ are all balanced. Furthermore, it can be observed that the expressions of $y_1$ and $y_3$ all contain monomial $x_0 x_1 x_2$

whose parity over $\mathbb{X}$ is undetermined according to the initial division property $\mathcal{D}_{(0,1,1,1)}^{1,4}$, thus $y_1$ and $y_3$ are not balanced. Based on these observations, the division property of $\mathbb{Y}$ should be $\mathcal{D}_{(0,0,1,0),(1,0,0,0)}^{1,4}$. In this case we obtain two division trails of PRESENT Sbox, and what is more important is that $y_0, y_2$ and $y_0 y_2$ are all balanced under this approach.

**Our Improved Approach.** Now we present a generalized algorithm to calculate division trails of an Sbox based on bit-based division property. In Algorithm 2, $\boldsymbol{x} = (x_{n-1}, \cdots, x_0)$ and $\boldsymbol{y} = (y_{n-1}, \cdots, y_0)$ denote the input and output to an $n$-bit Sbox respectively, and $y_i$ is expressed as a boolean function of $(x_{n-1}, \cdots, x_0)$.

---

**Algorithm 2.** Calculating division trails of an Sbox

**Input**  : The input division property of an $n$-bit Sbox $\mathcal{D}_k^{1,n}$ where
$\quad\quad\quad \boldsymbol{k} = (k_{n-1}, \cdots, k_0)$
**Output:** A set $\mathbb{K}$ of vectors such that the output multiset has division
$\quad\quad\quad$ property $\mathcal{D}_{\mathbb{K}}^{1,n}$

1 **begin**
2  $\quad \bar{\mathbb{S}} = \{\bar{\boldsymbol{k}} \mid \bar{\boldsymbol{k}} \succeq \boldsymbol{k}\}$
3  $\quad F(X) = \{\pi_{\bar{k}}(\boldsymbol{x}) \mid \bar{\boldsymbol{k}} \in \bar{\mathbb{S}}\}$
4  $\quad \bar{\mathbb{K}} = \emptyset$
5  $\quad$ **for** $\boldsymbol{u} \in (\mathbb{F}_2)^n$ **do**
6  $\quad\quad$ **if** $\pi_{\boldsymbol{u}}(\boldsymbol{y})$ contains any monomial in $F(X)$ **then**
7  $\quad\quad\quad \bar{\mathbb{K}} = \bar{\mathbb{K}} \cup \{\boldsymbol{u}\}$
8  $\quad\quad$ **end**
9  $\quad$ **end**
10 $\quad \mathbb{K} = \textbf{\textit{SizeReduce}}(\bar{\mathbb{K}})$
11 $\quad$ **return** $\mathbb{K}$
12 **end**

---

We explain Algorithm 2 line by line:

**Line 2–3** According to input division property $\mathcal{D}_k^{1,n}$, the parity of monomial $\pi_{\bar{k}}(\boldsymbol{x})$ with $\bar{\boldsymbol{k}} \succeq \boldsymbol{k}$ over $\mathbb{X}$ is undetermined, and we store these monomials in $F(X)$. Thus, the parity of any monomial that does not belong to $F(X)$ is zero.

**Line 4** Initialize $\mathbb{K}$ as an empty set.

**Line 5–9** For any possible $\boldsymbol{u}$, if boolean function $\pi_{\boldsymbol{u}}(\boldsymbol{y})$ contains any monomial in $F(X)$, the parity of $\pi_{\boldsymbol{u}}(\boldsymbol{y})$ over $\mathbb{X}$ is undetermined, and we store all these vectors in $\bar{\mathbb{K}}$.

**Line 10** *SizeReduce*() function removes all redundant vectors in $\bar{\mathbb{K}}$. Since we are interested in finding a set $\mathbb{K}$ such that for any $\boldsymbol{u} \in \{\boldsymbol{u} | \boldsymbol{u} \not\succeq \boldsymbol{k}$ for all $\boldsymbol{k} \in \mathbb{K}\}$, the parity of $\pi_{\boldsymbol{u}}(\boldsymbol{y})$ is zero. Note that for any vector $\boldsymbol{u} \in (\mathbb{F}_2)^n \backslash \bar{\mathbb{K}}$, the parity of $\pi_{\boldsymbol{u}}(\boldsymbol{y})$ is zero, thus, we must have $\{\boldsymbol{u} \mid \boldsymbol{u} \not\succeq \boldsymbol{k}$ for all

$k \in \mathbb{K}\} \subset (\mathbb{F}_2)^n \backslash \bar{\mathbb{K}}$, and if we let $\mathbb{K}= \boldsymbol{SizeReduce}(\bar{\mathbb{K}})$ it will meet this condition. Otherwise, if there exists a vector $\boldsymbol{u} \in \{\boldsymbol{u} \mid \boldsymbol{u} \not\succeq \boldsymbol{k}$ for all $\boldsymbol{k} \in \mathbb{K}\}$ such that $\boldsymbol{u} \notin (\mathbb{F}_2)^n \backslash \bar{\mathbb{K}}$, thus, we have $\boldsymbol{u} \in \bar{\mathbb{K}}$, which meants either $\boldsymbol{u} \in \mathbb{K}$ or there exists a vector $\boldsymbol{u}^* \in \mathbb{K}$ such that $\boldsymbol{u} \succeq \boldsymbol{u}^*$ since $\mathbb{K} = \boldsymbol{SizeReduce}(\bar{\mathbb{K}})$. In either case it won't happen $\boldsymbol{u} \in \{\boldsymbol{u} \mid \boldsymbol{u} \not\succeq \boldsymbol{k}$ for all $\boldsymbol{k} \in \mathbb{K}\}$, which leads to a contradiction. Therefore, $\mathbb{K}$ is sufficient to characterize the division property of output multiset.

**Line 11** Return $\mathbb{K}$ as output.

Given an Sbox and an initial division property $\mathcal{D}_k^{1,n}$, Algorithm 2 returns the output division property $\mathcal{D}_{\mathbb{K}}^{1,n}$. Thus for any vector $\boldsymbol{k}^* \in \mathbb{K}$, $(\boldsymbol{k}, \boldsymbol{k}^*)$ is a division trail of the Sbox. If we try all the $2^n$ possible input multiset division property, we will get a full list of division trails. Table 4 in Appendix B presents a complete list of all the 47 division trails of PRESENT Sbox.

Note that bit-based division property of an Sbox is closely related with Boura and Canteaut's work [6]. However, Boura and Canteaut's work is established on parity set, while our results are directly deduced from bit-based division property.

**Representing the Division Trails of Sbox as Linear Inequalities.** Each division trail of an $n$-bit Sbox can be viewed as a $2n$-dimensional vector in $\{0,1\}^{2n} \subset \mathbb{R}^{2n}$ where $\mathbb{R}$ is the real numbers field. Thus, all division trails form a subset $P$ of $\{0,1\}^{2n}$. Next, we compute the H-Representation of $\mathrm{Conv}(P)$ by using the inequality_generator() function in the Sage [2] software, and this will return a set of linear inequalities $\mathcal{L}$. However, $\mathcal{L}$ contains too many inequalities which will make the size of corresponding MILP problem too large to solve. Fortunately, we can select a subset $\mathcal{L}^*$ of $\mathcal{L}$ by Algorithm 1 such that the feasible solutions of $\mathcal{L}^*$ restricted in $\{0,1\}^{2n}$ are exactly $P$.

*Example:* PRESENT Sbox contains 47 division trails which forms a subset $P$ of $\{0,1\}^8$. By using the inequality_generator() function in the Sage software, a set of 122 linear inequalities will be returned. Furthermore, this set can be reduced by Algorithm 1 and we will get a set $\mathcal{L}^*$ of only 11 inequalities. The 11 inequalities for PRESENT Sbox are listed in Appendix C. In order to get the solutions of $\mathcal{L}^*$ restricted in $\{0,1\}^8$, we only need to specify that all variables can only take values in $\{0,1\}$.

So far, we have studied calculating and modeling division trails of basic operations and Sbox, thus, for block ciphers based on these operations and (or) Sbox, we can construct a set of linear inequalities which characterize one round division property propagation. By repeating this procedure $r$ times, we can get a linear inequality system $\mathcal{L}$ such that all feasible solutions of $\mathcal{L}$ are all $r$-round division trails.

### 3.3 Initial Division Property

Integral distinguisher search algorithm often has a given initial division property $\mathcal{D}_k^{1,n}$. Even though $\mathcal{L}$ is able to describe all division trails, we are interested in division trails starting from the given initial division property. Thus, we have to model the initial division property into the linear inequality system. Denote $(a_{n-1}^0, \cdots, a_0^0) \rightarrow \cdots \rightarrow (a_{n-1}^r, \cdots, a_0^r)$ an $r$-round division trail, $\mathcal{L}$ is thus a linear inequality system defined on variables $a_i^j$ ($i = 0, \cdots, n-1$. $j = 0, \cdots, r$) and some auxiliary variables. Let $\mathcal{D}_{\boldsymbol{k}}^{1,n}$ denote the initial input division property with $\boldsymbol{k} = (k_{n-1}, \cdots, k_0)$, we need to add $a_i^0 = k_i$ ($i = 0, \cdots n-1$) into $\mathcal{L}$, and thus all feasible solutions of $\mathcal{L}$ are division trails which start from vector $\boldsymbol{k}$.

# 4    Stopping Rule and Search Algorithm

In this section we first study the stopping rule in the search of integral distinguishers based on division property, and then we convert this stopping rule into an objective function of the MILP problem. At last, we propose an algorithm to determine whether an $r$-round integral distinguisher exists.

In the division property propagation, we note that only zero vector can propagate to zero vector. Thus if the given initial division property is $\mathcal{D}_{\boldsymbol{k}}^{1,n}$ with $\boldsymbol{k}$ a non-zero vector, and we denote the division property after $r$-round propagation by $\mathcal{D}_{\mathbb{K}_r}^{1,n}$, then it holds that $\mathbb{K}_r$ does not contain zero vector. In the following, we always assume $\boldsymbol{k} \neq \boldsymbol{0}$, since $\boldsymbol{k} = \boldsymbol{0}$ does not imply any integral property on the input multiset.

## 4.1    Stopping Rule

Let's first consider a set $\mathbb{X}$ with division property $\mathcal{D}_{\mathbb{K}}^{1,n}$. If $\mathbb{X}$ does not have any useful integral property, that is the Xor-sum of $\mathbb{X}$ does not balance on any bit, thus we have $\bigoplus_{\boldsymbol{x} \in \mathbb{X}} \pi_{\boldsymbol{u}}(\boldsymbol{x})$ is unknown for any unit vector $\boldsymbol{u} \in (\mathbb{F}_2)^n$. Since $\mathbb{X}$ has division property $\mathcal{D}_{\mathbb{K}}^{1,n}$, there must exist a vector $\boldsymbol{k} \in \mathbb{K}$ such that $\boldsymbol{u} \succeq \boldsymbol{k}$. Note that $\boldsymbol{u}$ is a unit vector, thus $\boldsymbol{u} = \boldsymbol{k}$, which means $\mathbb{K}$ contains all the $n$ unit vectors. On the other hand, if $\mathbb{K}$ contains all the $n$ unit vectors, then for any $\boldsymbol{0} \neq \boldsymbol{u} \in (\mathbb{F}_2)^n$ there must exist a unit vector $\boldsymbol{e} \in \mathbb{K}$ such that $\boldsymbol{u} \succeq \boldsymbol{e}$, that is $\bigoplus_{\boldsymbol{x} \in \mathbb{X}} \pi_{\boldsymbol{u}}(\boldsymbol{x})$ is unknown. Thus, $\mathbb{X}$ does not have any integral property.

**Proposition 6 (Set without Integral Property).** *Assume $\mathbb{X}$ is a multiset with division property $\mathcal{D}_{\mathbb{K}}^{1,n}$, then $\mathbb{X}$ does not have integral property if and only if $\mathbb{K}$ contains all the $n$ unit vectors.*

Denote the output division property after $i$-round encryption by $\mathcal{D}_{\mathbb{K}_i}^{1,n}$, and the initial input division property by $\mathcal{D}_{\boldsymbol{k}}^{1,n} \overset{def}{=} \mathcal{D}_{\mathbb{K}_0}^{1,n}$. If $\mathbb{K}_{r+1}$ for the first time contains all the $n$ unit vectors, the division property propagation should stop and an $r$-round distinguisher can be derived from $\mathcal{D}_{\mathbb{K}_r}^{1,n}$. In this case, $\mathbb{K}_r$ does not contain all $n$ unit vectors, thus we can always find a unit vector $\boldsymbol{e}$ such that $\boldsymbol{e} \notin \mathbb{K}_r$. Since $\boldsymbol{e}$ is a unit vector, it holds $\boldsymbol{e} \not\succeq \boldsymbol{k}$ for all $\boldsymbol{k} \in \mathbb{K}_r$. Therefore, the parity of $\pi_{\boldsymbol{e}}(\boldsymbol{x})$ over $r$-round outputs is even which is a zero-sum property, thus a balanced bit of the output is found. By repeating this process, all balanced bits can be found.

Based on this observation, we only need to detect whether $\mathbb{K}_r$ contains all unit vectors. According to Proposition 5, in order to check the vectors in $\mathbb{K}_r$, it is equivalent to check the last vectors of all $r$-round division trails. Denote $(a_{n-1}^0, \cdots, a_0^0) \rightarrow \cdots \rightarrow (a_{n-1}^r, \cdots, a_0^r)$ an $r$-round division trail, and let $\mathcal{L}$ denote a linear inequality system whose feasible solutions are all division trails which start with the given initial division property. It is clear that $\mathcal{L}$ is a linear inequality system defined on variables $a_i^j$ $(i = 0, \cdots, n-1.\ j = 0, \cdots, r)$ and some auxiliary variables. Thus, we can set the objective function as:

$$Obj : Min\{a_0^r + a_1^r + \cdots a_{n-1}^r\} \tag{5}$$

Now we get a complete MILP problem by setting $\mathcal{L}$ as constraints and $Obj$ as objective function. Note that $\mathbb{K}_i$ does not contain zero vector, in this case, the objective function will never take a value of zero, and the MILP problem will return an objective value greater than zero (if the MILP problem has feasible solutions). In the following we show how to determine whether $r$-round integral distinguisher exists based on this MILP problem.

### 4.2  Search Algorithm

Denote $\mathcal{L}$ a linear inequality description of all $r$-round division trails with the given initial input division property $\mathcal{D}_k^{1,n}$. Let the sum of the coordinates of the last vector in the division trail be the objective function $Obj$ as in Eq. (5). Denote $M(\mathcal{L}, Obj)$ the MILP problem composed of $\mathcal{L}$ and $Obj$. Algorithm 3 will return whether $r$-round integral distinguisher exists.

Our MILP problems are solved by the openly available MILP optimizer Gurobi [1], Algorithm 3 is presented with some Gurobi syntax. We denote the set of last vectors of all division trails by $\mathbb{K}_r$.

**Line 2** Initialize $\mathbb{S}$ as all possible output bit positions.

**Line 3–24** For an $n$-bit block cipher, check how many unit vectors there are in $\mathbb{K}_r$. Moreover, we remove the bit position marked by the unit vectors in $\mathbb{K}_r$ from $\mathbb{S}$, and return $\mathbb{S}$ as the output of the algorithm.

**Line 4** Check whether the MILP problem has a feasible solution. Note that the initial MILP problem always has feasible solutions. However, along with the execution of the procedure, it will add some constraints (Line 13) in the model which will possibly make the MILP problem unsolvable.

**Line 5** Optimize the MILP problem $M$ by Gurobi.

**Line 6–18** $M.ObjVal$ is Gurobi syntax which returns the current value of the objective function after $M$ has been optimized. $M.ObjVal = 1$ means we have found a division trail which ends up with a unit vector $e$, thus $e \in \mathbb{K}_r$. $M.getObjective()$ is a Gurobi function which returns the objective function of the model, which is $a_0^r + \cdots + a_{n-1}^r$ in our case. The functionality of Line 8–17 is to choose which variable of $(a_0^r, \cdots, a_{n-1}^r)$ is equal to one in $e$ and add a new constraint $var = 0$ into $M$, here $var$ denotes the variable taking a value of one. $obj.getVar(i)$ is used to return the $i$-th variable of $obj$ which is

---

**Algorithm 3.** Return whether $r$-round distinguisher exists

---

**Input** : $M = M(\mathcal{L}, Obj)$.
**Output:** A set $\mathbb{S}$ of balanced bit positions.

```
1  begin
2  |   S = {a_0^r, ⋯, a_{n-1}^r}
3  |   for i in range(0,n) do
4  |   |   if M has feasible solutions then
5  |   |   |   M.optimize()
6  |   |   |   if M.ObjVal = 1 then
7  |   |   |   |   obj = M.getObjective()
8  |   |   |   |   for i in range(0,n) do
9  |   |   |   |   |   var = obj.getVar(i)
10 |   |   |   |   |   val = var.getAttr('x')
11 |   |   |   |   |   if val = 1 then
12 |   |   |   |   |   |   S \ {var}
13 |   |   |   |   |   |   M.addConstr(var = 0)
14 |   |   |   |   |   |   M.update()
15 |   |   |   |   |   |   break
16 |   |   |   |   |   end
17 |   |   |   |   end
18 |   |   |   else
19 |   |   |   |   return S
20 |   |   |   end
21 |   |   else
22 |   |   |   return S
23 |   |   end
24 |   end
25 |   return S
26 end
```

---

$a_i^r$ in this case. $var.getAttr('x')$ retrieves the value of $var$ under the current solution. Line 12 removes $var$ from $\mathbb{S}$, since we have found $e \in \mathbb{K}_r$ whose nonzero position is $var$ which means $var$ can't be a balanced bit position. $M.addConstr(var = 0)$ adds a new constraints $var = 0$ into $M$, and this is used to rule out $e$ from $\mathbb{K}_r$. Line 14 updates the model since we have added a new constraint.

**Line 19** This step returns $\mathbb{S}$, the execution of this step means the objective value of $M$ is larger than one, that is we can no longer find a division trail with the last vector being a unit vector. In this case, we have found all unit vectors in $\mathbb{K}_r$ which represent undetermined bit positions, and thus we have ruled out all unbalanced bits and get an integral distinguisher.

**Line 22** $M$ do not have any feasible solutions means we have ruled out all units vectors of $\mathbb{K}_r$ and made $\mathbb{K}_r$ an empty set along with the execution. In this case, we can return $\mathbb{S}$ as output since we have checked all vectors.

**Line 25** If the for loop do not make the procedure exit, return $\mathbb{S}$ as output. Usually, in this case $\mathbb{S}$ is an empty set which means no distinguisher found.

Algorithm 3 always returns a set $\mathbb{S}$ indicating balanced bit positions. For a block cipher with a given initial division property $\mathcal{D}_k^{1,n}$, we can construct an $r$-round linear description of division property propagations and use Algorithm 3 to check whether a distinguisher exists. If for the first time the $(r+1)$-round model returns an empty set, then the longest distinguisher for the given initial division property is $r$-round.

# 5    Applications to SIMON, SIMECK, PRESENT, RECTANGLE, LBlock and TWINE

In this section, we show some applications of our technique. All the source codes are avaiable at https://github.com/xiangzejun/MILP_Division_Property. We applied our algorithm to SIMON, SIMECK, PRESENT, RECTANGLE, LBlock and TWINE block ciphers. The results are listed in Table 1. The *Round (Previous)* column and *Round (Sect. 5)* column list the number of rounds of the distinguishers of previous and our results. The *Data* column represents the number of active bits of the input pattern of the integral distinguisher, the data complexity of the distinguisher is determined by the initial input division property. *Balanced bits* column represents the number of balanced bits of the distinguisher we found. *Time* presents the time used by Algorithm 3 for searching the corresponding distinguishers, among which $s$ is short for second and $m$ is short for minute. All the experiments are conducted on the following platform: Intel Core i7-2600 CPU @3.40 GHz, 8.00G RAM, 64-bit Windows 7 system. Moreover, the distinguishes listed in Table 1 are presented in Appendix E. The table shows that we get improved distinguishers for SIMON48/64/96/128, SIMECK48/64, PRESENT and RECTANGLE. For SIMECK32, LBlock and TWINE our results are consistent with the previous best results. The result of SIMON32 is one round less than the result in [19]. However, we only use bit-based division property here, the 15-round distinguisher found in [19] for SIMON32 used bit-based division property using three subset. If bit-based division property is the only technique adopted, 14-round distinguisher is the longest distinguisher we can find.

## 5.1    Applications to SIMON and SIMECK

SIMON [3] is a family of lightweight block ciphers published by the U.S. National Security Agency (NSA) in 2013. SIMON adopts Fesitel structure and it has a very compact round function which only involves bit-wise And, Xor and circular shift operations. The structure of one round SIMON encryption is depicted in Fig. 1 where $S^i$ denotes left circular shift by $i$ bits.

**1-round Description of SIMON**: Denote one round division trail of SIMON$2n$ by $(a_0^i, \cdots, a_{n-1}^i, b_0^i, \cdots, b_{n-1}^i) \to (a_0^{i+1}, \cdots, a_{n-1}^{i+1}, b_0^{i+1}, \cdots, b_{n-1}^{i+1})$. In order to get a linear description of all possible division trails of one round SIMON, we introduce four vectors of auxiliary variables which are $(u_0^i, \cdots, u_{n-1}^i)$, $(v_0^i, \cdots, v_{n-1}^i)$, $(w_0^i, \cdots, w_{n-1}^i)$ and $(t_0^i, \cdots, t_{n-1}^i)$. We denote

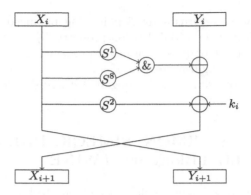

**Fig. 1.** Feistel structure of SIMON round function

$(u_0^i, \cdots, u_{n-1}^i)$ the input division property of $S^1$. Similarly, denote $(v_0^i, \cdots, v_{n-1}^i)$ and $(w_0^i, \cdots, w_{n-1}^i)$ the input division property of $S^8$ and $S^2$ respectively. Let $(t_0^i, \cdots, t_{n-1}^i)$ denote the output division property of bit-wise And operation. Subsection 3.1 has modeled Copy, And and Xor functions. According to Eq. (1), the following inequalities are sufficient to model the Copy operation used in SIMON$2n$:

$$\mathcal{L}_1 : a_j^i - u_j^i - v_j^i - w_j^i - b_j^{i+1} = 0 \text{ for } j \in \{0, 1, \cdots, n-1\}$$

Since we consider bit-based division property, division property propagation through circular shift is just a circular shift of the coordinates of the vector. Thus, the division property of the output of $S^1$ is $(u_1^i, \cdots, u_{n-1}^i, u_0^i)$. Similarly, the division property of the output of $S^8$ and $S^2$ are $(v_8^i, \cdots, v_6^i, v_7^i)$ and $(w_2^i, \cdots, w_0^i, w_1^i)$ respectively. We can model bit-wise And operation used in SIMON by the following inequalities according to Eq. (2):

$$\mathcal{L}_2 : \begin{cases} t_j^i - u_{j+1}^i \geq 0 & \text{for } j \in \{0, 1, \cdots, n-1\} \\ t_j^i - v_{j+8}^i \geq 0 & \text{for } j \in \{0, 1, \cdots, n-1\} \\ t_j^i - u_{j+1}^i - v_{j+8}^i \leq 0 & \text{for } j \in \{0, 1, \cdots, n-1\} \end{cases}$$

At last, the Xor operations in SIMON$2n$ can be modeled by the following inequalities according to Eq. (3):

$$\mathcal{L}_3 : a_j^{i+1} - b_j^i - t_j^i - w_{j+2}^i = 0 \text{ for } j \in \{0, 1, \cdots, n-1\}$$

So far, we have modeled all operations used in SIMON, and get an accurate description $\{\mathcal{L}_1, \mathcal{L}_2, \mathcal{L}_3\}$ of 1-round division trails. By repeating this procedure $r$ times, we can get a linear inequality system $\mathcal{L}$ for $r$-round division property propagation. Given some initial division property, we can add the corresponding constrains into $\mathcal{L}$ and estimate whether a useful distinguisher exists by Algorithm 3. The results for SIMON family are listed in Table 1.

For SIMON48/64/96/128, we found the best distinguishers so far. Note that by using bit-based division property under the framework of [19], it is computationally impractical to search distinguishers for these versions. Using Algorithm 3, distinguishers can be searched in practical time.

SIMECK [25] is a family of lightweight block cipher proposed at CHES 2015. The round function of SIMECK is very like SIMON except the rotation constants. We applied our technique to SIMECK, and 15-, 18- and 21-round distinguishers are found for SIMECK32, SIMECK48 and SIMECK64 respectively, which shows that SIMON has better security than SIMECK with respect to division property based integral cryptanalysis.

We found that the 14-round distinguisher of SIMON32 we found is the same as the 14-round distinguisher of SIMON32 in [19] based on bit-based division property. Surprisingly, the 15-round distinguisher for SIMECK32 in [19] is found by bit-based division property using three subsets, however, we also find the same distinguisher for SIMECK32 by only using bit-based division property.

In [9], the authors investigated the differential and linear behavior of SIMON family regarding rotation parameters, and they presented some interesting alternative parameters among which $(1, 0, 2)$ is optimal for the differential and linear characteristics with the restriction that the second rotation parameter is zero. In this paper, we investigated the integral property of this parameter by our technique. The results are listed in Table 2 ($h$ in the *time* column represents hour). The third column lists the rounds of the distinguishers we found. The results show that $(1, 0, 2)$ is a very bad choice with respect to division property based integral cryptanalysis.

**Table 2.** Results on SIMON(1,0,2).

| Cipher | Block size | Round | Data | Balanced bits | Time |
|---|---|---|---|---|---|
| SIMON32(1,0,2) | 32 | 20 | 31 | 1 | 34.1s |
| SIMON48(1,0,2) | 48 | 28 | 47 | 1 | 3.2m |
| SIMON64(1,0,2) | 64 | 36 | 63 | 1 | 10.3m |
| SIMON96(1,0,2) | 96 | 52 | 95 | 3 | 6.4h |
| SIMON128(1,0,2) | 128 | 68 | 127 | 3 | 24h |

## 5.2 Applications to PRESENT and RECTANGLE

PRESENT [5] and RECTANGLE [28] are two SP-network block ciphers, of which the linear layers are bit permutations. Figure 2 illustrates one round encryption of PRESENT.

**1-round Description of PRESENT**: Denote one round division trail of PRES-ENT by $(a_{63}^i, \cdots, a_0^i) \rightarrow (a_{63}^{i+1}, \cdots, a_0^{i+1})$. We first model the division

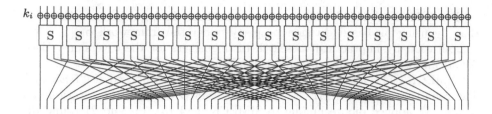

**Fig. 2.** One round SP structure of PRESENT

property propagation of Sbox layer. Denote the division property of the output of Sbox by $(b^i_{63}, \cdots, b^i_0)$. Subsection 3.2 has studied how to calculate the division trails of Sbox and model those trails by linear inequalities. Appendix C shows the 11 inequalities of PRESENT Sbox. For each of the 16 Sboxes of PRESENT, we introduce 11 inequalities and thus the Sbox layer of PRESENT can be modeled by $11 \times 16 = 176$ inequalities which is denoted by $\mathcal{L}_1$. The linear layer of PRESENT is a bit permutation, thus, the division property propagation through linear layer is just a permutation of the coordinates of the vector, that is

$$
\mathcal{L}_2 : \begin{cases} a^{i+1}_{16j \mod 63} = b^i_j & j \in \{0, 1, \cdots, 62\} \\ a^{i+1}_j = b^i_j & j = 63 \end{cases}
$$

Note that $\mathcal{L}_1$ is a linear inequality system defined on variables $(a^i_{63}, \cdots, a^i_0)$ and $(b^i_{63}, \cdots, b^i_0)$, we can use the equalities in $\mathcal{L}_2$ to replace the variables $(b^i_{63}, \cdots, b^i_0)$ in $\mathcal{L}_1$ in order to save auxiliary variables.

Now we have get a linear inequality system to describe one round division propagation of PRESENT. By repeating this procedure, an $r$-round linear inequality system can be constructed. For a given initial division property $\mathcal{D}^{1,64}_k$, we add this information into the linear inequality system and use Algorithm 3 to estimate whether there exists an integral distinguisher.

The result for PRESENT is listed in Table 1. We found a 9-round integral distinguisher for PRESENT which is two more rounds than the previous best results in [22].

The modeling procedure of RECTANGLE is very like to PRESENT, we only list the result here in Table 1. The previous longest integral distinguisher of RECTANGLE is found by the designers, and they gave a 7-round distinguisher. In this paper we find a 9-round distinguisher which is two more rounds.

## 5.3    Applications to LBlock and TWINE

This subsection applies our technique to two generalized Feistel block cipher LBlock and TWINE. The round function of these two ciphers are alike, and the round function composed of Copy, Sbox and Xor operations. We have showed how to model Copy and Xor operations in SIMON and Sbox in PRESENT, thus, we omit the details for these two ciphers due to the limit of space. The number of division trails and linear inequalities required to describe those division trails of LBlock and TWINE Sboxes are presented in Table 3. The $\sharp\{D.C\}$ column represents the number of division trails of the corresponding Sbox, and the $\sharp\{Ine\}$ column represents the number of linear inequalities we found to accurately describe the division trails. Note that we chose the first inequality in the sixth line of Algorithm 1, however, other choice rather than the first one may result in different set of inequlities.

**Table 3.** Sbox properties regrading division trails.

| Sbox | $\sharp\{D.C\}$ | $\sharp\{Ine\}$ |
| --- | --- | --- |
| PRESENT Sbox | 47 | 11 |
| RECTANGLE Sbox | 49 | 17 |
| LBlock S0 | 44 | 11 |
| LBlock S1 | 44 | 12 |
| Lblock S2 | 44 | 12 |
| LBlock S3 | 44 | 11 |
| LBlock S4 | 44 | 13 |
| LBlock S5 | 44 | 10 |
| LBlock S6 | 44 | 12 |
| LBlock S7 | 44 | 12 |
| TWINE Sbox | 47 | 11 |

Our experimental results regarding LBlock and TWINE are listed in Table 1. The distinguishers found in this paper are the same as the distinguishers found for these two ciphers in [26].

*Experiments.* To illustrate the validity of the technique proposed in this paper, we presented some integral distinguishers found by our technique with a small number of active bits, and we run experiments on these distinguishers. The experiments are presented at Appendix D. Our experiments showed that the distinguishers found by our technique are sound. Moreover, the results on PRESENT and RECTANGLE illustrate that our technique can find quite accurate distinguishers, that is the balanced bits found by Algorithm 3 are exactly in accordance with experimental results. For PRESENT cipher, we retrieved and improved the 5-round distinguisher found in [22], our technique found all the four balanced bits of the outputs of the fifth round given the same input pattern as in [22], while Wu *et al.* could only prove the balancedness of only one bit.

## 6  Summary and Discussion

In this paper we introduced a new technique to search integral distinguishers based on bit-based division property. We first proposed a new notion *division trail* and used this new notion to characterize the division property propagation, then we showed that it is sufficient to check the last vectors of all $r$-round division trails in order to estimate whether an $r$-round distinguisher exists.

Based on the observations on division trails, we proposed to construct a linear inequality system to characterize the division property propagations. We first studied how to model division property propagations of *Copy, And* and *Xor* operations by linear inequalities. For another basic component Sbox used in block ciphers, we studied the bit-based division property propagations for the first time, and we proposed an algorithm to compute the division trails of an Sbox. Moreover, we used those division trails to derive a set of linear inequalities whose feasible solutions are exactly all division trails. Thus, for a block cipher we can construct a linear inequality system whose solutions are all $r$-round division trails of the cipher, and we used this linear inequality system as constraints of the MILP problem. Then, the stopping rule in the search of integral distinguisher were studied and we converted it into an objective function of an MILP problem. To be specific, we set the sum of the coordinates of the last vector in an $r$-round division trail as objective function. Thus, we can get a complete MILP problem, based on which we presented an algorithm to estimate whether an $r$-round integral distinguisher exists by checking how many unit vectors are contained in the last vectors of all division trails.

We applied our technique to SIMON, SIMECK, PRESENT, RECTANGLE, LBlock and TWINE. For SIMON48/64/96/128, SIMECK48/64, PRESENT and RECTANGLE, we get much longer distinguishers than previous results based on division property in the open literature. Moreover, our results on PRESENT and RECTANGLE show that we can get better integral distinguishers by using the algebraic normal form of the Sboxes. Our results show that, by using our technique, we can search integral distinguishers based on bit-based division property in practical time for block ciphers with block size larger than 32, which is impractical under the traditional framework.

In [19], Todo *et al.* also introduced bit-based division property using three subsets, and they found 15-round distinguisher for SIMON32. However, we have not found a way to model this framework by an MILP problem at present. A surprising result is, by using our technique we also derived the 15-round distinguisher of SIMECK32 which are constructed by bit-based division property using three subsets [19]. We also used our technique on some Sbox-based block ciphers such as PRESENT and RECTANGLE, note that their linear layers are all bit permutations. However, this technique can be easily extended to arbitrary linear layers as pointed out by the reviewers, since any linear layer can be viewed as bit-level linear layer which can be treated as bit-wise copy and Xor.

**Acknowledgements.** We are very grateful to the anonymous reviewers. This work was supported by the National Natural Science Foundation of China (Grant No. 61379138), the "Strategic Priority Research Program" of the Chinese Academy of Sciences (Grant No. XDA06010701).

# A    An Example

Let's consider a simple example in this section. Suppose that $A = \{(0,1),(1,0),(1,1)\}$ is a subset of $\{0,1\}^2$ with three points, and we would like to get a linear inequality system $\mathcal{L}$ such that all feasible solutions of $\mathcal{L}$ restricted in $\{0,1\}^2$ are $A$.

We proceed by using inequality_generator() function in the Sage software to compute the H-Representation of $\mathrm{Conv}(A)$. The following is the source code.

```
Points = [[0,1],[1,0],[1,1]]
triangle = Polyhedron(vertices = Points)
for l in triangle.inequality_generator():
    print l
```

As a result, Sage returns three inequalities:

$$\mathcal{L} = \begin{cases} x + y - 1 \geq 0 \\ -y + 1 \geq 0 \\ -x + 1 \geq 0 \end{cases} \tag{6}$$

It is easy to check that the feasible solutions of $\mathcal{L}$ form a triangle with $A$ being its three vertices, and the set of all feasible solutions of $\mathcal{L}$ restricted in $\{0,1\}^2$ is exactly $A$. Thus, Eq. 6 is a description of $A$.

However, we can use Algorithm 1 to reduce the number of inequalities required. We apply Algorithm 1 to this example and we find that only one inequality is sufficient to accurately describe $A$:

$$\mathcal{L}^* = \{x + y - 1 \geq 0\} \tag{7}$$

It is easy to check that all solutions of $\mathcal{L}^*$ restricted in $\{0,1\}^2$ are $(0,1)$ ,$(1,0)$ and $(1,1)$ as expected.

# B    Division trails of PRESENT and RECTANGLE Sbox

Tables 4 and 5 present the division trails of PRESENT and RECTANGLE Sboxes respectively.

# C    Linear inequalities description of PRESENT and RECTANGLE Sbox

The following inequalities are the 11 inequalities used to describe PRESENT Sbox whose feasible solutions are exactly the 47 division trails of PRESENT

**Table 4.** Division trails of PRESENT Sbox

| Input $\mathcal{D}_k^{1,4}$ | Output $\mathcal{D}_{\mathbb{K}}^{1,4}$ |
|---|---|
| (0,0,0,0) | (0,0,0,0) |
| (0,0,0,1) | (0,0,0,1) (0,0,1,0) (0,1,0,0) (1,0,0,0) |
| (0,0,1,0) | (0,0,0,1) (0,0,1,0) (0,1,0,0) (1,0,0,0) |
| (0,0,1,1) | (0,0,1,0) (0,1,0,0) (1,0,0,0) |
| (0,1,0,0) | (0,0,0,1) (0,0,1,0) (0,1,0,0) (1,0,0,0) |
| (0,1,0,1) | (0,0,1,0) (0,1,0,0) (1,0,0,0) |
| (0,1,1,0) | (0,0,0,1) (0,0,1,0) (1,0,0,0) |
| (0,1,1,1) | (0,0,1,0) (1,0,0,0) |
| (1,0,0,0) | (0,0,0,1) (0,0,1,0) (0,1,0,0) (1,0,0,0) |
| (1,0,0,1) | (0,0,1,0) (0,1,0,0) (1,0,0,0) |
| (1,0,1,0) | (0,0,1,0) (0,1,0,0) (1,0,0,0) |
| (1,0,1,1) | (0,0,1,0) (0,1,0,0) (1,0,0,0) |
| (1,1,0,0) | (0,0,1,0) (0,1,0,0) (1,0,0,0) |
| (1,1,0,1) | (0,0,1,0) (0,1,0,0) (1,0,0,0) |
| (1,1,1,0) | (0,1,0,1) (1,0,1,1) (1,1,1,0) |
| (1,1,1,1) | (1,1,1,1) |

**Table 5.** Division trails of RECTANGLE Sbox

| Input $\mathcal{D}_k^{1,4}$ | Output $\mathcal{D}_{\mathbb{K}}^{1,4}$ |
|---|---|
| (0,0,0,0) | (0,0,0,0) |
| (0,0,0,1) | (0,0,0,1) (0,0,1,0) (0,1,0,0) (1,0,0,0) |
| (0,0,1,0) | (0,0,0,1) (0,0,1,0) (0,1,0,0) (1,0,0,0) |
| (0,0,1,1) | (0,0,0,1) (0,1,0,0) (1,0,1,0) |
| (0,1,0,0) | (0,0,0,1) (0,0,1,0) (0,1,0,0) (1,0,0,0) |
| (0,1,0,1) | (0,0,1,1) (0,1,0,0) (1,0,0,0) |
| (0,1,1,0) | (0,0,1,1) (0,1,0,0) (1,0,0,0) |
| (0,1,1,1) | (0,0,1,1) (0,1,0,0) (1,0,0,1) |
| (1,0,0,0) | (0,0,0,1) (0,0,1,0) (0,1,0,0) (1,0,0,0) |
| (1,0,0,1) | (0,0,1,1) (0,1,0,1) (0,1,1,0) (1,0,0,0) |
| (1,0,1,0) | (0,0,1,0) (0,1,0,1) (1,0,0,0) |
| (1,0,1,1) | (0,1,1,0) (1,0,1,1) (1,1,0,1) |
| (1,1,0,0) | (0,0,1,1) (0,1,0,0) (1,0,0,0) |
| (1,1,0,1) | (0,1,1,0) (1,0,1,0) (1,1,0,1) |
| (1,1,1,0) | (0,0,1,1) (0,1,0,1) (1,0,0,0) |
| (1,1,1,1) | (1,1,1,1) |

Sbox where $(a_3, a_2, a_1, a_0) \longrightarrow (b_3, b_2, b_1, b_0)$ denotes a division trail.

$$\mathcal{L}^* = \begin{cases} a_3 + a_2 + a_1 + a_0 - b_3 - b_2 - b_1 - b_0 \geq 0 \\ -a_2 - a_1 - 2a_0 + b_3 + b_1 - b_0 + 3 \geq 0 \\ -a_2 - a_1 - 2a_0 + 4b_3 + 3b_2 + 4b_1 + 2b_0 \geq 0 \\ -2a_3 - a_2 - a_1 + 2b_3 + 2b_2 + 2b_1 + b_0 + 1 \geq 0 \\ -2a_3 - a_2 - a_1 + 3b_3 + 3b_2 + 3b_1 + 2b_0 \geq 0 \\ -b_3 + b_2 - b_1 + b_0 + 1 \geq 0 \\ -2a_3 - 2a_2 - 2a_1 - 4a_0 + b_3 + 4b_2 + b_1 - 3b_0 + 7 \geq 0 \\ a_3 + a_2 + a_1 + a_0 - 2b_3 - 2b_2 + b_1 - 2b_0 + 1 \geq 0 \\ -4a_2 - 4a_1 - 2a_0 + b_3 - 3b_2 + b_1 + 2b_0 + 9 \geq 0 \\ -2a_0 - b_3 - b_2 - b_1 + 2b_0 + 3 \geq 0 \\ a_0 + b_3 - b_2 - 2b_1 - b_0 + 2 \geq 0 \\ a_3, a_2, a_1, a_0, b_3, b_2, b_1, b_0 \text{ are binaries} \end{cases} \qquad (8)$$

The following inequalities are the 17 inequalities used to describe RECTAN-GLE Sbox whose feasible solutions are exactly the 49 division trails of REC-TANGLE Sbox where $(a_3, a_2, a_1, a_0) \longrightarrow (b_3, b_2, b_1, b_0)$ denotes a division trail.

$$\mathcal{L}^* = \begin{cases} -a_3 - a_2 - 2a_1 - 3a_0 - 2b_3 + b_1 + 2b_0 + 6 \geq 0 \\ -b_3 - b_2 + b_0 + 1 \geq 0 \\ a_3 + a_2 + a_1 + a_0 - b_3 - b_2 - b_1 - b_0 \geq 0 \\ 3a_3 + a_2 - b_3 - 2b_2 - b_1 - 2b_0 + 2 \geq 0 \\ a_2 + a_0 - b_2 - 2b_1 - b_0 + 2 \geq 0 \\ -a_2 - a_1 - a_0 + b_3 + 2b_2 + 2b_0 + 1 \geq 0 \\ -2a_3 - a_1 - a_0 + b_3 + 2b_1 + b_0 + 2 \geq 0 \\ -3a_3 - a_2 - a_1 - 2a_0 + b_3 + 2b_2 + 2b_1 - b_0 + 4 \geq 0 \\ -a_2 - a_1 + b_3 + b_2 + b_1 + 1 \geq 0 \\ -3a_3 - a_2 - a_1 - 2a_0 + 3b_3 + 2b_2 + 2b_1 + b_0 + 2 \geq 0 \\ 2a_2 + 3a_1 - 3b_3 - b_2 - 2b_1 - b_0 + 3 \geq 0 \\ -a_3 - a_2 - a_0 + 2b_3 + 2b_2 + b_1 + b_0 \geq 0 \\ -2a_2 - a_1 - a_0 + 3b_3 + 4b_2 + 2b_1 + 2b_0 \geq 0 \\ a_3 + a_2 + a_1 + a_0 - 2b_3 - 2b_0 + 1 \geq 0 \\ 2a_0 - b_3 - b_2 - b_1 + 1 \geq 0 \\ 3a_3 - 4a_2 - a_1 - a_0 - 2b_3 - b_2 - 3b_1 + 2b_0 + 7 \geq 0 \\ a_3 + a_1 + a_0 + b_3 - 3b_2 - 2b_1 - 2b_0 + 3 \geq 0 \\ a_3, a_2, a_1, a_0, b_3, b_2, b_1, b_0 \text{ are binaries} \end{cases} \qquad (9)$$

# D    Experiments on PRESENT and RECTANGLE

For SIMON family block ciphers, we found a 14-round distinguisher of SIMON32 which is in accordance with the distinguisher presented in [19]. For Lblock and

TWINE the distinguisher found in this paper are in accordance with the distinguishers presented in [26]. Thus, we believe that the distinguishers found for SIMON, SIMECK, Lblock and TWINE are sound. In the following we only conduct some experiments on PRESENT and RECTANGLE.

*PRESENT*: We found the following 5-round distinguisher for PRESENT. If we fix the left most 60 bits as random constant and vary the right most 4 bits, then after five round encryption, the four right most bits of the state are balanced.

Input:(cccccccccccccccccccccccccccccccccccccccccccccccccccccccccccccaaaa)

Output:(????????????????????????????????????????????????????????????bbbb)

c: constant bit, a: active bit, ?: unknown bit, b: balanced bit

We run experiment on this distinguisher $2^{12}$ times. The experimental result returns the four right most bits as balanced bits which is in accordance with our theoretical result.

Note that in [22] Wu *el at.* found a 5-round distinguisher for PRESENT which has the same input pattern with the distinguisher presented here. However, they only proved that the right most bit is balanced. By using our technique, we can find all the four balanced bits.

*RECTANGLE*: We found the following 6-round distinguisher for RECTANGLE. The input of the distinguisher has 23 active bits, that is the right most six bits of the first, third and fourth rows, and the five right most bits of the second row are active. The output of six rounds encryption will be balanced on 40 bits, that is the first two rows, the two right most bits of the third row and the six left most bits of the last row.

$$
\text{Input}: \begin{pmatrix} \texttt{ccccccccccaaaaaa} \\ \texttt{cccccccccccaaaaa} \\ \texttt{ccccccccccaaaaaa} \\ \texttt{ccccccccccaaaaaa} \end{pmatrix} \longrightarrow \text{Output}: \begin{pmatrix} \texttt{bbbbbbbbbbbbbbbb} \\ \texttt{bbbbbbbbbbbbbbbb} \\ \texttt{??????????????bb} \\ \texttt{bbbbbb??????????} \end{pmatrix}
$$

We run experiment on this distinguisher $2^{10}$ times. The experimental result returns 40 balanced bits which is in accordance with our theoretical result.

# E   Integral Distinguishers listed in Table 1

For SIMON and SIMECK family block ciphers, all distinguisher can be extended one more round by the technique in [20].

## E.1   SIMON32's 13-Round Distinguisher

Input:(caaaaaaaaaaaaaaa,aaaaaaaaaaaaaaaa)

Output:(????????????????,bbbbbbbbbbbbbbbb)

## E.2  SIMON48's 15-Round Distinguisher

Input:(caaaaaaaaaaaaaaaaaaaaaaa,aaaaaaaaaaaaaaaaaaaaaaaa)
Output:(????????????????????????,bbbbbbbbbbbbbbbbbbbbbbbb)

## E.3  SIMON64's 17-Round Distinguisher

Input:(caaaaaaaaaaaaaaaaaaaaaaaaaaaaaaa,aaaaaaaaaaaaaaaaaaaaaaaaaaaaaaaa)
Output:(????????????????????????????????,bbbbbbbbbbb?????b?????bbbbbbbbbb)

## E.4  SIMON96's 21-Round Distinguisher

Input:(caaaaaaaaaaaaaaaaaaaaaaaaaaaaaaaaaaaaaaaaaaaaaaaa,
    aaaaaaaaaaaaaaaaaaaaaaaaaaaaaaaaaaaaaaaaaaaaaaaaa)
Output:(????????????????????????????????????????????????,
    b?b????b??????????????????????????????????b????b?)

## E.5  SIMON128's 25-Round Distinguisher

Input:(caaaaaaaaaaaaaaaaaaaaaaaaaaaaaaaaaaaaaaaaaaaaaaaaaaaaaaaaaaaaaaaa,
    aaaaaaaaaaaaaaaaaaaaaaaaaaaaaaaaaaaaaaaaaaaaaaaaaaaaaaaaaaaaaaaa)
Output:(????????????????????????????????????????????????????????????????,
    b?b?????????????????????????????????????????????????????????b?)

## E.6  SIMECK32's 14-Round Distinguisher

Input:(caaaaaaaaaaaaaaa,aaaaaaaaaaaaaaaa)
Output:(???????????????,bb???bb???bb???b)

## E.7  SIMECK48's 17-Round Distinguisher

Input:(caaaaaaaaaaaaaaaaaaaaaaa,aaaaaaaaaaaaaaaaaaaaaaaa)
Output:(????????????????????????,b???bb??????????????bb???)

## E.8  SIMECK64's 20-Round Distinguisher

Input:(caaaaaaaaaaaaaaaaaaaaaaaaaaaaaaa,aaaaaaaaaaaaaaaaaaaaaaaaaaaaaaaa)
Output:(????????????????????????????????,bb???b????????????????????b???b)

## E.9  PRESENT's 9-Round Distinguisher

Input:(aaaaaaaaaaaaaaaaaaaaaaaaaaaaaaaaaaaaaaaaaaaaaaaaaaaaaaaaaaaaaccccc)

Output:(???????????????????????????????????????????????????????????????b)

## E.10  RECTANGLE's 9-Round Distinguisher

$$
\text{Input}: \begin{pmatrix} \texttt{caaaaaaaaaaaaaaa} \\ \texttt{caaaaaaaaaaaaaaa} \\ \texttt{caaaaaaaaaaaaaaa} \\ \texttt{caaaaaaaaaaaaaaa} \end{pmatrix} \longrightarrow \text{Output}: \begin{pmatrix} \texttt{bbb?b?bbbbbbbbbb} \\ \texttt{?????????????b?b} \\ \texttt{????????????????} \\ \texttt{????????????????} \end{pmatrix}
$$

## E.11  LBlock's 16-Round Distinguisher

Input:(caaaaaaaaaaaaaaaaaaaaaaaaaaaaaaa,aaaaaaaaaaaaaaaaaaaaaaaaaaaaaaaa)

Output:(????????????????????????????????,bbbbbbbbbbbbbbbbbbbbbbbbbbbbbbbb)

## E.12  TWINE's 16-Round Distinguisher

Input:(caaaaaaaaaaaaaaaaaaaaaaaaaaaaaaa,aaaaaaaaaaaaaaaaaaaaaaaaaaaaaaaa)

Output:(????bbbb????bbbb????bbbb????bbbb,????bbbb????bbbb????bbbb????bbbb)

# References

1. http://www.gurobi.com/
2. http://www.sagemath.org/
3. Beaulieu, R., Shors, D., Smith, J., Treatman-Clark, S., Weeks, B., Wingers, L.: The SIMON and SPECK families of lightweight block ciphers. IACR Cryptology ePrint Archive **2013**, 404 (2013)
4. Biryukov, A., Shamir, A.: Structural cryptanalysis of SASAS. In: Pfitzmann, B. (ed.) EUROCRYPT 2001. LNCS, vol. 2045, pp. 395–405. Springer, Heidelberg (2001). doi:10.1007/3-540-44987-6_24
5. Bogdanov, A., Knudsen, L.R., Leander, G., Paar, C., Poschmann, A., Robshaw, M.J.B., Seurin, Y., Vikkelsoe, C.: PRESENT: an ultra-lightweight block cipher. In: Paillier, P., Verbauwhede, I. (eds.) CHES 2007. LNCS, vol. 4727, pp. 450–466. Springer, Heidelberg (2007). doi:10.1007/978-3-540-74735-2_31
6. Boura, C., Canteaut, A.: Another view of the division property. In: Robshaw, M., Katz, J. (eds.) CRYPTO 2016. LNCS, vol. 9814, pp. 654–682. Springer, Heidelberg (2016). doi:10.1007/978-3-662-53018-4_24
7. Daemen, J., Knudsen, L., Rijmen, V.: The block cipher Square. In: Biham, E. (ed.) FSE 1997. LNCS, vol. 1267, pp. 149–165. Springer, Heidelberg (1997). doi:10.1007/BFb0052343
8. Knudsen, L., Wagner, D.: Integral cryptanalysis. In: Daemen, J., Rijmen, V. (eds.) FSE 2002. LNCS, vol. 2365, pp. 112–127. Springer, Heidelberg (2002). doi:10.1007/3-540-45661-9_9

9. Kölbl, S., Leander, G., Tiessen, T.: Observations on the SIMON block cipher family. In: Gennaro, R., Robshaw, M. (eds.) CRYPTO 2015. LNCS, vol. 9215, pp. 161–185. Springer, Heidelberg (2015). doi:10.1007/978-3-662-47989-6_8

10. Lucks, S.: The saturation attack — a bait for twofish. In: Matsui, M. (ed.) FSE 2001. LNCS, vol. 2355, pp. 1–15. Springer, Heidelberg (2002). doi:10.1007/3-540-45473-X_1

11. Mouha, N., Wang, Q., Gu, D., Preneel, B.: Differential and linear cryptanalysis using mixed-integer linear programming. In: Wu, C.-K., Yung, M., Lin, D. (eds.) Inscrypt 2011. LNCS, vol. 7537, pp. 57–76. Springer, Heidelberg (2012). doi:10.1007/978-3-642-34704-7_5

12. Sun, B., Hai, X., Zhang, W., Cheng, L., Yang, Z.: New observation on division property. Science China Information Science (2016). http://eprint.iacr.org/2015/459

13. Sun, S., Hu, L., Song, L., Xie, Y., Wang, P.: Automatic security evaluation of block ciphers with S-bP structures against related-key differential attacks. In: Lin, D., Xu, S., Yung, M. (eds.) Inscrypt 2013. LNCS, vol. 8567, pp. 39–51. Springer, Heidelberg (2014). doi:10.1007/978-3-319-12087-4_3

14. Sun, S., Hu, L., Wang, M., Wang, P., Qiao, K., Ma, X., Shi, D., Song, L., Fu, K.: Towards finding the best characteristics of some bit-oriented block ciphers and automatic enumeration of (related-key) differential and linear characteristics with predefined properties. Technical report, Cryptology ePrint Archive, Report 2014/747 (2014)

15. Sun, S., Hu, L., Wang, P., Qiao, K., Ma, X., Song, L.: Automatic security evaluation and (Related-key) differential characteristic search: application to SIMON, PRESENT, LBlock, DES(L) and other bit-oriented block ciphers. In: Sarkar, P., Iwata, T. (eds.) ASIACRYPT 2014. LNCS, vol. 8873, pp. 158–178. Springer, Heidelberg (2014). doi:10.1007/978-3-662-45611-8_9

16. Suzaki, T., Minematsu, K., Morioka, S., Kobayashi, E.: *TWINE*: a lightweight block cipher for multiple platforms. In: Knudsen, L.R., Wu, H. (eds.) SAC 2012. LNCS, vol. 7707, pp. 339–354. Springer, Heidelberg (2013). doi:10.1007/978-3-642-35999-6_22

17. Todo, Y.: Integral cryptanalysis on full MISTY1. In: Gennaro, R., Robshaw, M. (eds.) CRYPTO 2015. LNCS, vol. 9215, pp. 413–432. Springer, Heidelberg (2015). doi:10.1007/978-3-662-47989-6_20

18. Todo, Y.: Structural evaluation by generalized integral property. In: Oswald, E., Fischlin, M. (eds.) EUROCRYPT 2015. LNCS, vol. 9056, pp. 287–314. Springer, Heidelberg (2015). doi:10.1007/978-3-662-46800-5_12

19. Todo, Y., Morii, M.: Bit-based division property and application to SIMON family. Cryptology ePrint Archive, Report 2016/285 (2016). http://eprint.iacr.org/

20. Wang, Q., Liu, Z., Varıcı, K., Sasaki, Y., Rijmen, V., Todo, Y.: Cryptanalysis of reduced-round SIMON32 and SIMON48. In: Meier, W., Mukhopadhyay, D. (eds.) INDOCRYPT 2014. LNCS, vol. 8885, pp. 143–160. Springer, Heidelberg (2014). doi:10.1007/978-3-319-13039-2_9

21. Wu, S., Wang, M.: Security evaluation against differential cryptanalysis for block cipher structures. IACR Cryptology ePrint Archive **2011**, 551 (2011)

22. Wu, S., Wang, M.: Integral attacks on reduced-round PRESENT. In: Qing, S., Zhou, J., Liu, D. (eds.) ICICS 2013. LNCS, vol. 8233, pp. 331–345. Springer, Heidelberg (2013). doi:10.1007/978-3-319-02726-5_24

23. Wu, W., Zhang, L.: LBlock: a lightweight block cipher. In: Lopez, J., Tsudik, G. (eds.) ACNS 2011. LNCS, vol. 6715, pp. 327–344. Springer, Heidelberg (2011). doi:10.1007/978-3-642-21554-4_19

24. Xiang, Z., Zhang, W., Lin, D.: On the division property of SIMON48 and SIMON64. International Workshop on Security (2016)
25. Yang, G., Zhu, B., Suder, V., Aagaard, M.D., Gong, G.: The Simeck family of lightweight block ciphers. In: Güneysu, T., Handschuh, H. (eds.) CHES 2015. LNCS, vol. 9293, pp. 307–329. Springer, Heidelberg (2015). doi:10.1007/978-3-662-48324-4_16
26. Zhang, H., Wu, W.: Structural evaluation for generalized feistel structures and applications to LBlock and TWINE. In: Biryukov, A., Goyal, V. (eds.) INDOCRYPT 2015. LNCS, vol. 9462, pp. 218–237. Springer, Heidelberg (2015). doi:10.1007/978-3-319-26617-6_12
27. Zhang, H., Wu, W., Wang, Y.: Integral attack against bit-oriented block ciphers. In: Kwon, S., Yun, A. (eds.) ICISC 2015. LNCS, vol. 9558, pp. 102–118. Springer, Heidelberg (2016). doi:10.1007/978-3-319-30840-1_7
28. Zhang, W., Bao, Z., Lin, D., Rijmen, V., Yang, B., Verbauwhede, I.: Rectangle: a bit-slice lightweight block cipher suitable for multiple platforms. Sci. China Inf. Sci. **58**(12), 1–15 (2015)
29. Zhang, W., Su, B., Wu, W., Feng, D., Wu, C.: Extending higher-order integral: an efficient unified algorithm of constructing integral distinguishers for block ciphers. In: Bao, F., Samarati, P., Zhou, J. (eds.) ACNS 2012. LNCS, vol. 7341, pp. 117–134. Springer, Heidelberg (2012). doi:10.1007/978-3-642-31284-7_8

# Reverse Cycle Walking and Its Applications

Sarah Miracle$^{(\boxtimes)}$ and Scott Yilek

University of St. Thomas, St. Paul, USA
{sarah.miracle,syilek}@stthomas.edu

**Abstract.** We study the problem of constructing a block-cipher on a "possibly-strange" set $S$ using a block-cipher on a larger set $T$. Such constructions are useful in format-preserving encryption, where for example the set $S$ might contain "valid 9-digit social security numbers" while $T$ might be the set of 30-bit strings. Previous work has solved this problem using a technique called cycle walking, first formally analyzed by Black and Rogaway. Assuming the size of $S$ is a constant fraction of the size of $T$, cycle walking allows one to encipher a point $x \in S$ by applying the block-cipher on $T$ a small *expected* number of times and $O(N)$ times in the worst case, where $N = |T|$, without any degradation in security. We introduce an alternative to cycle walking that we call *reverse cycle walking*, which lowers the worst-case number of times we must apply the block-cipher on $T$ from $O(N)$ to $O(\log N)$. Additionally, when the underlying block-cipher on $T$ is secure against $q = (1 - \epsilon)N$ adversarial queries, we show that applying reverse cycle walking gives us a cipher on $S$ secure even if the adversary is allowed to query all of the domain points. Such fully secure ciphers have been the the target of numerous recent papers.

**Keywords:** Format-preserving encryption · Small-domain block ciphers · Markov chains

## 1 Introduction

Suppose we have sets $S$ and $T$, with $S$ a subset of $T$. Typically, in this paper, the larger set $T$ will be $\{0, \ldots, 2^n - 1\}$ for some integer $n$, while the smaller set $S$ will be an arbitrary set for which we only assume we know how to efficiently test membership. The central problem we study in this paper is, given a cipher with domain $T$, how can we construct a cipher with domain $S$.

FORMAT-PRESERVING ENCRYPTION. The above problem arises when constructing *format preserving encryption* (FPE) [1,2,4] schemes for encrypting credit cards numbers, social security numbers, and other relatively short data objects. Suppose we have a customer database containing millions of US social security numbers (SSNs). SSNs are 9 decimal digit numbers with numerous additional restrictions (e.g., the first three digits may not be 666). Now suppose we later decide we need to encrypt the SSNs. One approach would be to use a standard block cipher like AES, representing the SSN as a 30-bit number and then padding

© International Association for Cryptologic Research 2016
J.H. Cheon and T. Takagi (Eds.): ASIACRYPT 2016, Part I, LNCS 10031, pp. 679–700, 2016.
DOI: 10.1007/978-3-662-53887-6_25

with 0 s before encrypting. The resulting ciphertext, however, would have a significantly different format from the original, unencrypted numbers. This could in turn require significant changes to the customer database, as well as to the hardware and software that process the SSNs. For this reason, it is desirable to have format-preserving encryption schemes, in which ciphertexts have the same format as plaintexts. A FPE scheme for SSNs would thus have ciphertexts that are 9 decimal digit numbers with the same restrictions as unencrypted SSNs.

CYCLE WALKING. A number of recent works [3,11,16–18] describe efficient, provably secure small-domain block ciphers for enciphering either bitstrings or, in most cases, points in the more general domain $\{0, \ldots, N-1\}$. This is already sufficient for many FPE applications. However, if the desired domain for a particular FPE application is not as simple as bitstrings of some length or integers up to $N$, then these ciphers alone are not sufficient. If we only assume that we can efficiently test membership in our target domain set $\mathcal{S}$,[1] then one approach to the problem is to find a cipher on a larger set $\mathcal{T}$ and transform it into a cipher on the smaller set $\mathcal{S}$. In the case of valid SSNs, for example, we might let the larger set $\mathcal{T}$ be 30-bit strings, since $10^9 < 2^{30}$ and we have many block ciphers that can encipher 30-bit strings. The canonical way to transform a cipher on the more general set $\mathcal{T}$ into a cipher on a subset $\mathcal{S}$ while maintaining the same level of security is to use *cycle walking*. Cycle walking is a folklore technique first formally analyzed by Black and Rogaway [4] that works as follows. Suppose $\pi$ is a permutation with domain $\mathcal{T}$ and we wish to use it to map a point $x \in \mathcal{S}$ to another point in $\mathcal{S}$. We first compute $\pi(x)$ and test if the result is in $\mathcal{S}$. If so, we map point $x$ to $\pi(x)$. If the result is not in $\mathcal{S}$, we apply $\pi$ again, computing $\pi(\pi(x))$ and again testing whether or not the result is in $\mathcal{S}$. We repeat this process until we get a point in $\mathcal{S}$. Let $\mathsf{CW}_\pi$ denote this cycle walking algorithm. Black and Rogaway showed that cycle walking maintains the security of $\pi$, and in particular showed that if cycle walking is applied to a CCA-secure cipher, then the resulting cipher is also CCA-secure.

If we are unlucky, we may have to apply $\pi$ numerous times before finally reaching a point in $\mathcal{S}$.[2] In fact, if we consider the *worst-case* running time of cycle walking, we might have to evaluate the permutation $\Theta(N)$ times. Yet, the *expected* running time is much better; if the size of $\mathcal{S}$ is at least half the size of $\mathcal{T}$ and if $\pi$ is a randomly-chosen permutation on $\mathcal{T}$, then the expected number of times $\mathsf{CW}_\pi$ will need to evaluate $\pi$ on a particular point is at most 2.

---

[1] If a set has an efficient way to rank and unrank elements, then instead one can apply the rank algorithm and then a cipher on $\{0, \ldots, |\mathcal{S}| - 1\}$. This is the case, for example, with regular languages described by a DFA [1]. For other languages, and even for regular languages described by a regular expression, the situation is more complicated. See [14,15] for more details. Nevertheless, in the current paper we are concerned with more general sets where only testing set membership is assumed to be efficient.

[2] We are guaranteed to eventually land back in the set $\mathcal{S}$, since permutations are made up of cycles and, if we don't hit another point in $\mathcal{S}$ first, we will cycle back around to the same point we started with.

It will be helpful to also examine the cycle structure of $\mathsf{CW}_\pi$ compared to $\pi$. Let $\mathcal{S}$ and $\mathcal{T}$ be as they were defined above, and let $\pi$ again be a permutation on the larger set $\mathcal{T}$. Recall that permutations are made up of disjoint cycles, so our chosen permutation $\pi$ is made up of disjoint cycles each with points from $\mathcal{T}$. Suppose that one of these cycles is $(t_1\ s_1\ s_2\ s_3\ t_2\ s_4\ s_5\ t_3\ t_4)$, where the $s$ points are all from $\mathcal{S}$ and the $t$ points are from $\mathcal{T} \setminus \mathcal{S}$. Now consider what happens when $\mathsf{CW}_\pi(s_3)$ is evaluated. Notice that $\pi(s_3) = t_2$, so we need to evaluate $\pi(\pi(s_3)) = s_4$. Thus, $\mathsf{CW}_\pi(s_3) = s_4$. In terms of the cycle structure, evaluating $\mathsf{CW}_\pi(s_3)$ corresponds to walking to the right in the cycle from $s_3$ until we hit another point in $\mathcal{S}$. Similarly, $\mathsf{CW}_\pi(s_5) = s_1$, which we can see since walking to the right from $s_5$ brings us to $t_3$, $t_4$, $t_1$ (after looping around to the front), and then finally $s_1$. We can thus determine the cycle structure of $\mathsf{CW}_\pi$ simply by erasing the $t$ points from all of the cycles in $\pi$, meaning the cycle above becomes $(s_1\ s_2\ s_3\ s_4\ s_5)$.

A CLOSER LOOK AT EXPECTED TIME. The small expected running time of cycle walking makes it an attractive option in practice for FPE. Yet, from a theoretical perspective, the fact that we do not know how to build permutations for arbitrary sets with worst-case running time better than $\Theta(N)$ is unsatisfying.

Finding alternative algorithms that do not run in expected time is not just an important theoretical question. From a practical perspective, in addition to the unpredictability of execution times potentially bothering practitioners, there is the danger that expected-time cryptographic algorithms can leak timing information that can be exploited by an adversary in an attack. Starting with the work of Kocher [12], there have been numerous examples of how such timing information can lead to subtle and damaging attacks on cryptographic protocols. Thus, generally, it would seem preferable to have cryptographic algorithms whose running time does not vary across different inputs.

Somewhat counter to this, Bellare, Ristenpart, Rogaway, and Stegers [1] analyzed the potential negative effects of the timing information leaked by cycle walking and concluded that the leakage is not damaging. Yet, their result is in a specific model where the adversary has access to the ciphertexts in addition to the number of cycle walking steps needed, which they call the cycle length.[3] This, however, does not preclude the possibility of other scenarios in which this timing information could be useful. As one simple example, suppose an adversary observes the time it takes to encipher and later learns the corresponding plaintext. If, at a later point, the adversary again observes the time it takes to encipher a point then this timing information can reveal whether or not the same point was enciphered without the adversary ever needing to observe any ciphertexts. Depending on the specific scenario and application, this information could be damaging.

REVERSE CYCLE WALKING. We now describe our main result: an alternative to cycle walking with substantially better worst-case running time that does not

---

[3] Specifically, they show that in a PRP security game the adversary gets no benefit from learning the cycle length in addition to the ciphertext on an encryption query.

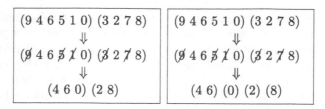

**Fig. 1.** Example of how one round of reverse 2-cycle walking differs from regular cycle walking. In this example, $\mathcal{T} = \{0, \ldots, 9\}$ and $\mathcal{S}$ are the even numbers in $\mathcal{T}$. **Left:** the effect regular cycle walking has on the cycle structure of the permutation. **Right:** the effect of one round of reverse 2-cycle walking on the cycle structure.

vary based on the input. Towards this, a first attempt might be to try to apply cycle walking, but somehow "cut-off" the algorithm if it is taking too long, since often with an expected-time algorithm one can simply stop the algorithm early and possibly introduce a small error. Unfortunately, it is not clear how to make this approach work. If we are evaluating $\mathsf{CW}_\pi(x)$ and walking through a long sequence of points in $\mathcal{T}\backslash\mathcal{S}$, we cannot just cut off the algorithm because we need to construct a permutation, and thus require a unique point in $\mathcal{S}$ to map $x$ to. Because of this difficulty, we introduce an alternative to cycle walking we call *reverse cycle walking*.

Let $\mathcal{S}$, $\mathcal{T}$, and $\pi$ be as they are defined above, and suppose again $\pi$ has a cycle $(t_1 \; s_1 \; s_2 \; s_3 \; t_2 \; s_4 \; s_5 \; t_3 \; t_4)$. As in traditional cycle walking, under reverse cycle walking $s_1$ is mapped to $s_2$ and $s_2$ is mapped to $s_3$. Where reverse cycle walking differs from traditional cycle walking is when a point in $\mathcal{S}$ is mapped outside of $\mathcal{S}$; this is the case for $s_3$, which is mapped under $\pi$ outside of $\mathcal{S}$ to $t_2$. To determine where $s_3$ should be mapped to, reverse cycle walking walks in the reverse direction, to the *left*, until a point outside of $\mathcal{S}$ is encountered; the last point encountered that is in $\mathcal{S}$ will be where $s_3$ is mapped. So in the case of the current cycle, if we wish to know where $s_3$ will be mapped, we walk to the left to $s_2$ and then finally to $s_1$. Since walking to the left any farther would result in a point outside of $\mathcal{S}$, reverse cycle walking stops here and maps $s_3$ to $s_1$. Similarly, $s_5$ would be mapped to $s_4$. The cycle structure that results from applying reverse cycle walking is thus $(s_1 \; s_2 \; s_3)(s_4 \; s_5)$.

Notice that reverse cycle walking, as just described, will still have poor worst-case running time and considerably better expected running time, much like traditional cycle walking. The main advantage now, though, is that we can consider variants of reverse cycle walking that "cut-off" the algorithm early and significantly reduce the worst-case running time. Specifically, when reverse $t$-cycle walking is applied to permutation $\pi$, any sequence of at most $t$ points from $\mathcal{S}$ that appear consecutively in a cycle of $\pi$ sandwiched between points from $\mathcal{T}\backslash\mathcal{S}$ will become a cycle. Any points in $\mathcal{S}$ that do not have this property are simply mapped to themselves.

For the rest of the paper, we focus on perhaps the simplest version of this idea, reverse 2-cycle walking. If $\pi$ has in some cycle $(\ldots t \; s \; s' \; t' \ldots)$, meaning two consecutive points from $\mathcal{S}$ sandwiched between points from $\mathcal{T} \setminus \mathcal{S}$, then under reverse 2-cycle walking (denoted $\mathsf{RCW}_\pi$) $(s\ s')$ becomes a cycle, meaning $\mathsf{RCW}_\pi(s) = s'$ and $\mathsf{RCW}_\pi(s') = s$. When reverse 2-cycle walking is applied to our example above where $\pi$ has cycle $(t_1\ s_1\ s_2\ s_3\ t_2\ s_4\ s_5\ t_3\ t_4)$, the resulting permutation will have cycles $(s_1)(s_2)(s_3)(s_4\ s_5)$. Notice that because $s_1$, $s_2$, and $s_3$ represented more than two consecutive points from $\mathcal{S}$, they were simply mapped to themselves. On the other hand, $s_4$ and $s_5$ were two consecutive points from $\mathcal{S}$ sandwiched between points outside of $\mathcal{S}$, so they are swapped. (For technical reasons, as we will see later in the paper, we will additionally flip a coin to see if these points are actually swapped or not.) The code of the reverse 2-cycle walking transformation can be found in Fig. 2 in Sect. 3. (Note that the transformation is an involution.) Another example illustrating how reverse 2-cycle walking compares to traditional cycle walking can be seen in Fig. 1.

WORST-CASE RUNNING TIME OF REVERSE 2-CYCLE WALKING. In our scenario above, even if $\pi$ is a random permutation on $\mathcal{T}$, $\mathsf{RCW}_\pi$ will clearly not be close to a random permutation on $\mathcal{S}$; many points are mapped to themselves (i.e., $\mathsf{RCW}_\pi(x) = x$). Thus, with reverse 2-cycle walking, we need to repeat the procedure for multiple rounds with independently chosen permutations $\pi$. The question then becomes how many rounds of RCW are needed before the resulting permutation on $\mathcal{S}$ is close to random.

To answer this question, we show that when $\pi$ is a randomly chosen permutation on $\mathcal{T}$ and when the size of $\mathcal{S}$ is a constant fraction of the size of $\mathcal{T}$, then reverse 2-cycle walking yields a *matching exchange process* (MEP), first defined and analyzed by Czumaj and Kutylowski [9]. A MEP proceeds in rounds to mix $N$ points, where in each round a random matching of some size is chosen and then a coin is flipped for each pair in the matching to decide whether its points should be swapped. Notice that this is exactly how multiple rounds of reverse 2-cycle walking proceed: in any given round, each point in $\mathcal{S}$ is either randomly paired with another point in $\mathcal{S}$, or it is mapped to itself and is not part of the matching for that round.

To analyze MEPs, Czumaj and Kutylowski used *non-Markovian delayed path coupling*, an extension of the well-known path coupling technique [5] in the area of Markov chains, to show that a matching exchange process will mix $N$ points in $O(\log N)$ rounds. Since we show reverse 2-cycle walking yields a MEP, directly applying their result gives us a way to construct an almost-random permutation on an arbitrary set with worst-case running time $\Theta(t(N) \cdot \log N)$, where $t(N)$ is the time it takes to apply permutation $\pi$ on $\mathcal{T}$. Recall that with traditional cycle walking, we get worst-case running time $\Theta(t(N) \cdot N)$, so our result is a significant improvement.

Since an asymptotic result is of limited practical value in the setting where cycle walking seems most useful, that of small-domain encryption for FPE, we also give concrete bounds relating the number or rounds of reverse 2-cycle walking to the CCA-advantage of an adversary attacking the encryption scheme. Unfortunately, because the Czumaj and Kutylowski paper targeted asymptotic results, their proof does not give explicit constants. To overcome this difficulty, we give new proofs of two key lemmas from CK's proof in order to minimize the constants for our setting where $N$ is perhaps $2^{30}$.

FULL SECURITY FROM REVERSE 2-CYCLE WALKING. Fully secure block ciphers, which are block ciphers that look like random permutations even to an adversary querying all $N$ domain points, have been the target of many recent papers [10,16,18] on small-domain encryption. While all of these recent results are based on a recursive shuffling technique from [8], we instead take a different approach and show that reverse 2-cycle walking can be used to achieve full security. In particular, we show that in certain situations we can take a cipher on a larger set $T$ that is *not* fully secure and apply reverse 2-cycle walking to get a fully secure cipher on the smaller set $S$.

To help explain this result in more detail, suppose we wish to construct a fully secure block cipher $E_{\text{full}}$ with domain $\{0, \ldots, N-1\}$ and further suppose we have another block cipher $E_{\text{part}}$ with a larger domain $\{0, \ldots, 2N-1\}$ and which is indistinguishable from a random permutation as long as the adversary only queries half the domain points. (Swap or Not [11] would be an example of such a cipher, which we call partially secure.) Notice that $E_{\text{part}}$, with domain size $2N$, will be secure against $N$ queries, which is the same *quantity* of queries we want $E_{\text{full}}$ to be secure against. But how should $E_{\text{full}}$ use $E_{\text{part}}$ to encipher points in $\{0, \ldots, N-1\}$? To encipher a point $x \in \{0, \ldots, N-1\}$, we could simply apply $E_{\text{part}}$ to $x$. But since $E_{\text{part}}$ has a larger domain, $E_{\text{part}}(x)$ might not be in $\{0, \ldots, N-1\}$. Czumaj [6] recently considered something similar and suggested using $E_{\text{part}}$ to shuffle all of the points $\{0, \ldots, 2N-1\}$ and then "remove" the points outside of $\{0, \ldots, N-1\}$. Unfortunately, it's not clear how to efficiently implement this "remove" step.

Another idea might be to use traditional cycle walking to always make sure we can map a point $x \in \{0, \ldots, N-1\}$ back into the same set. Unfortunately, proving this secure appears difficult, since in a reduction each of the $N$ adversarial queries made while attacking $E_{\text{full}}$ could result in many queries to $E_{\text{part}}$. Thus, in the reduction, our adversary attacking $E_{\text{part}}$ would likely need to query nearly all points in $\{0, \ldots, 2N-1\}$, many more queries than $E_{\text{part}}$ is assumed secure against.

Instead, we propose using reverse 2-cycle walking. Using our set names from earlier in the introduction, let $S = \{0, \ldots, N-1\}$ and let $T = \{0, \ldots, 2N-1\}$. Let $E_{\text{part}}$ be a block cipher with domain $T$. Then reverse 2-cycle walking has the following key property: if we evaluate $E_{\text{part}}$ on every point $x \in S$, then this gives us enough information to determine $\text{RCW}_{E_{\text{part}}}(x)$ for every $x \in S$. In other words, we never need to evaluate $E_{\text{part}}$ on any point outside of $\{0, \ldots, N-1\}$. This property allows the reduction to go through, giving us a fully secure cipher.

## 2    Preliminaries

NOTATION. For any set $\mathcal{X}$, let $\text{Perms}(\mathcal{X})$ be the set of all permutations $\pi : \mathcal{X} \to \mathcal{X}$. For sets $\mathcal{X}$ and $\mathcal{Y}$, let $\text{Funs}(\mathcal{X}, \mathcal{Y})$ be the set of all functions $f : \mathcal{X} \to \mathcal{Y}$. For set $\mathcal{X}$, let $x \leftarrow_{\$} \mathcal{X}$ denote choosing $x$ uniformly at random from $\mathcal{X}$.

MIXING TIME. The time a Markov chain $\mathcal{M}$ takes to converge to its stationary distribution $\mu$ is measured in terms of the distance between $\mu$ and $\mathcal{P}^t$, the distribution at time $t$. Let $\mathcal{P}^t(x, y)$ be the $t$-step transition probability and $\Omega$ be the state space. The *mixing time* of $\mathcal{M}$ is $\tau_{\mathcal{M}}(\epsilon) = \min\{t : ||\mathcal{P}^{t'} - \mu|| \leq \epsilon, \forall t' \geq t\}$, where $||\mathcal{P}^t - \mu|| = \max_{x \in \Omega} \frac{1}{2} \sum_{y \in \Omega} |\mathcal{P}^t(x, y) - \mu(y)|$ is the *total variation distance* at time $t$.

BLOCK CIPHERS AND THEIR SECURITY. A block cipher is a family of functions $E : \mathcal{K} \times \mathcal{M} \to \mathcal{M}$, with $\mathcal{K}$ a finite set called the key space and $\mathcal{M}$ a finite set called the domain or message space. For every $K \in \mathcal{K}$, the function $E_K(\cdot) = E(K, \cdot)$ is a permutation. Let $E^{-1} : \mathcal{K} \times \mathcal{M} \to \mathcal{M}$ be the inverse block cipher. We will typically let $N$ denote $|\mathcal{M}|$, the number of elements in the domain. Thus, when $\mathcal{M} = \{0, 1\}^n$, $N = 2^n$.

We will consider block cipher security against chosen-ciphertext attack (CCA), often referred to as strong-PRP security. Given block cipher $E :$ $\mathcal{K} \times \mathcal{M} \to \mathcal{M}$ and adversary $A$, the cca-advantage of $A$ against $E$ is defined to be

$$\mathbf{Adv}_E^{\mathrm{cca}}(A) = \mathbf{P}\left(A^{E(K, \cdot), E^{-1}(K, \cdot)} \Rightarrow 1\right) - \mathbf{P}\left(A^{\pi(\cdot), \pi^{-1}(\cdot)} \Rightarrow 1\right),$$

where the first probability is over the choice of $K$ and the coins of $A$, and the second probability is over the choice of $\pi$ from $\mathsf{Perms}(\mathcal{M})$ and the coins of $A$. In words, the adversary $A$ tries to determine which "world" he is in, where he is either in a world where he is given access to the block cipher and its inverse, or in a world where he is given access to a random permutation and its inverse. If an adversary $A$ is given oracle access to an algorithm $\mathcal{O}$ and its inverse $\mathcal{O}^{-1}$, we will sometimes write $A^{\pm \mathcal{O}(\cdot)}$ as shorthand for $A^{\mathcal{O}(\cdot), \mathcal{O}^{-1}(\cdot)}$.

PSEUDORANDOM FUNCTIONS. Let $F : \mathcal{K} \times \mathcal{X} \to \mathcal{Y}$ be a family of functions with key space $\mathcal{K}$. The prf-advantage of an adversary $A$ against $F$ is defined to be $\mathbf{Adv}_F^{\mathrm{prf}}(A) = \mathbf{P}\left(A^{F(K, \cdot)} \Rightarrow 1\right) - \mathbf{P}\left(A^{\rho(\cdot)} \Rightarrow 1\right)$, where the first probability is over the choice of key $K$ and the coins of $A$, and the second probability is over the choice of $\rho$ from $\mathsf{Funs}(\mathcal{X}, \mathcal{Y})$ and the coins of $A$. In words, the adversary tries to determine through oracle queries whether it is interacting with the keyed function $F$ or a random function chosen from all functions from $\mathcal{X}$ to $\mathcal{Y}$.

CYCLE WALKING. This paper focuses on the problem of using permutations on a set $\mathcal{T}$ to build a permutation on a smaller set $\mathcal{S} \subseteq \mathcal{T}$. Specifically, we will be interested in the scenario where $N_S = |\mathcal{S}|$ is a constant fraction of $N_T = |\mathcal{T}|$ (e.g., $2 \cdot N_S = N_T$). Black and Rogaway [3] analyzed a folklore technique for this called *cycle walking*, or *cycling*. Given a permutation $\pi$ on $\mathcal{T}$, let the cycle walking transformation of $\pi$ with target set $\mathcal{S}$ be function $\mathsf{CW}_\pi : \mathcal{S} \to \mathcal{S}$ defined as follows

Algorithm $\mathsf{CW}_\pi(x)$:
do
$\quad x \leftarrow \pi(x)$
while $(x \notin \mathcal{S})$
Return $x$

In words, cycle walking continues to apply permutation $\pi$ until it finally gets a point in set $\mathcal{S}$. Cycle walking can also be applied to block ciphers. Notationally, if $E : \mathcal{K} \times \mathcal{T} \rightarrow \mathcal{T}$ is a block-cipher on $\mathcal{T}$, then we will let $\bar{E} : \mathcal{K} \times \mathcal{S} \rightarrow \mathcal{S}$ be the block cipher that, on input $K$ and $x$, computes $\mathsf{CW}_{E_K}(x)$.

A key fact about cycle walking, argued by Black and Rogaway, is that if $\pi$ is a random permutation on $\mathcal{T}$, then $\mathsf{CW}_\pi$ will be a random permutation on $\mathcal{S}$. While this is an information-theoretic result, Black and Rogaway also briefly argued it can be used to show the cycle walking transformation preserves cca security as well, which we formalize as follows:

**Lemma 1 (Black-Rogaway).** *Let $\mathcal{S} \subseteq \mathcal{T}$ be such that $|\mathcal{S}| \geq (1/2)|\mathcal{T}|$ and let $E : \mathcal{K} \times \mathcal{T} \rightarrow \mathcal{T}$ be a block cipher on $\mathcal{T}$, and $\bar{E} : \mathcal{K} \times \mathcal{S} \rightarrow \mathcal{S}$ the block-cipher resulting from applying cycle walking to $E$ with target set $\mathcal{S}$. Let $A$ be an adversary making $q$ queries, then*

$$\mathbf{Adv}_{\bar{E}}^{\mathrm{cca}}(A) \leq \mathbf{Adv}_{E}^{\mathrm{cca}}(B)$$

*where adversary $B$ makes at most an expected $2q$ queries.*

As we explained in the introduction, cycle walking has small expected running time, but it has significantly worse worst-case running time. Additionally, the theorem above bounds the advantage of an adversary against $\bar{E}$ by the advantage of an adversary that makes and expected number of oracle queries.

## 3    Reverse 2-Cycle Walking

We now detail the reverse 2-cycle walking algorithm. Again, for sets $\mathcal{S} \subseteq \mathcal{T}$, let $N_S = |\mathcal{S}|$, $N_T = |\mathcal{T}|$, and assume $N_S$ is a constant fraction of $N_T$; $c \cdot N_S = N_T$ (e.g., $2 \cdot N_S = N_T$). Suppose we have permutation $\pi : \mathcal{T} \rightarrow \mathcal{T}$ with $\pi^{-1}$ its inverse. Also suppose we have a function $\mathsf{B} : \mathcal{S} \rightarrow \{0, 1\}$. The reverse 2-cycle walking transformation is a function $\mathsf{RCW}_{\pi,\mathsf{B}} : \mathcal{S} \rightarrow \mathcal{S}$ defined in Fig. 2.[4]

To understand the algorithm it is helpful to consider the cycle structure of $\mathsf{RCW}_{\pi,\mathsf{B}}$ (a permutation on $\mathcal{S}$) as compared to $\pi$. In $\mathsf{RCW}_{\pi,\mathsf{B}}$, points in $\mathcal{S}$ are mapped to themselves unless they are contained in a cycle (in $\pi$) where exactly two points in $\mathcal{S}$ are surrounded by points in $\mathcal{T} - \mathcal{S}$. For example, if $\pi$ contains the cycle $(s_1 s_2 t_1 t_2 s_3 s_4 t_4 s_5)$ where the $s_i$'s are in $\mathcal{S}$ and the $t_i$'s are in $\mathcal{T} - \mathcal{S}$, then the resulting permutation on $\mathcal{S}$ will contain the cycles $(s_1)(s_2)(s_3 s_4)(s_5)$. Note that for simplicity of analysis, if $\pi$ contains the cycle $(s_1 s_2)$ then the resulting permutation will contain the cycles $(s_1)(s_2)$, whereas $(s_1 s_2 t_1)$ will result in

---

[4] It should be noted that the pseudocode in Fig. 2 is written for ease of understanding and, if implemented exactly as written, could leak timing information about the input. An actual implementation would use standard techniques to ensure each path through the code results in the same number of operations. Additionally, if the underlying cipher $\pi$ has different timings in the forward and backward directions, then both $\pi(y)$ and $\pi^{-1}(z)$ would need to be computed regardless of the input point $x$.

Algorithm $\mathsf{RCW}_{\pi,\mathsf{B}}(x)$:

$u \leftarrow \bot \; ; \; v \leftarrow \bot$

$y \leftarrow \pi(x) \; ; \; z \leftarrow \pi^{-1}(x)$

if $y \in \mathcal{S}$ and $z \notin \mathcal{S}$ and $\pi(y) \notin \mathcal{S}$:

$\quad u \leftarrow x \; ; \; v \leftarrow y$

$\quad b \leftarrow \mathsf{B}(u)$

$\quad$ if $b = 1$ return $v$ else return $u$

else if $y \notin \mathcal{S}$ and $z \in \mathcal{S}$ and $\pi^{-1}(z) \notin \mathcal{S}$:

$\quad u \leftarrow z \; ; \; v \leftarrow x$

$\quad b \leftarrow \mathsf{B}(u)$

$\quad$ if $b = 1$ return $u$ else return $v$

else

$\quad$ return $x$

**Fig. 2.** The reverse 2-Cycle walking algorithm

$(s_1 s_2)$. Additionally, the function $\mathsf{B}$ has the effect of only including each 2-cycle in the final permutation on $\mathcal{S}$ with probability $1/2$. This is currently necessary for our analysis but we believe with further work this function can be removed.

When $\pi$ and $\mathsf{B}$ are clear from context, we will sometimes write just $\mathsf{RCW}(x)$. Note that each point $x$ is either mapped to itself (i.e., $\mathsf{RCW}(x) = x$), or is part of a 2-cycle (i.e., there is a $y \neq x$ s.t. $\mathsf{RCW}(x) = y$ and $\mathsf{RCW}(y) = x$). Notice also that the algorithm is its own inverse (an involution).

Given permutations $\pi_1, \pi_2, \ldots, \pi_k$ all on $\mathcal{T}$, and functions $\mathsf{B}_1, \ldots, \mathsf{B}_k$ from $\mathcal{S}$ to $\{0,1\}$, we denote by $\mathsf{RCW}^k_{(\pi_1,\ldots,\pi_k),(\mathsf{B}_1,\ldots,\mathsf{B}_k)}$ the composition $\mathsf{RCW}_{\pi_1,\mathsf{B}_1} \circ \ldots \circ \mathsf{RCW}_{\pi_k,\mathsf{B}_k}$. When the permutations $\pi_i$ and functions $\mathsf{B}_i$ are clear from the context, we will often write $\mathsf{RCW}^k$. The inverse of $\mathsf{RCW}^k$ will simply apply the rounds in reverse order, since the $\mathsf{RCW}$ algorithm above is its own inverse.

The rest of the paper focuses on the security of the reverse 2-cycle walking transformation. The next section gives an information theoretic result, bounding the mixing time of the Markov chain that results from applying a number of rounds of reverse 2-cycle walking where in each round we use a randomly chosen underlying permutation $\pi_i$ and function $\mathsf{B}_i$. In Sect. 5, we analyze the cca security of reverse 2-cycle walking when the underlying permutations on $\mathcal{T}$ are cca-secure and the round functions are implemented with a pseudorandom function. Finally, in Sect. 6, we show that reverse 2-cycle walking can be used to build fully secure block ciphers.

# 4    Bounding the Mixing Time

Here, we focus on how many rounds of reverse 2-cycle walking are needed before the resulting permutation on $\mathcal{S}$ is "close" to random. We consider the ideal case where at each round the underlying permutation $\pi_i$ and function $\mathsf{B}_i$ are chosen uniformly at random and bound the mixing time of the underlying Markov

S. Miracle and S. Yilek

chain. To do this, we use a technique called delayed path coupling, introduced by Czumaj and Kutylowski [9] to analyze what they call a matching exchange protocol. They are interested in studying a class of Markov chains for sampling permutations of $N$ points where at each step a number $\kappa$ is chosen according to some distribution, then a matching of size $\kappa$ is chosen uniformly from all matchings of size $\kappa$, and finally the points corresponding to each pair in the matching are each independently swapped (or not) with probability $1/2$. Assuming the expected size of the matching at each step is $\Theta(N)$ they show that after $\Theta(\log(N))$ steps the variation distance is $O(1/N)$. If you consider the effect of the RCW algorithm on all elements in $N_S$, a single step of the algorithm is equivalent to selecting a matching $M_i$ on $N_S$ (since we only consider 2-cycles) according to some distribution and then swapping each pair in the matching with probability $1/2$. Claim 1, which we prove below, implies that RCW is a matching exchange protocol with $\mathbf{E}\left[\,\kappa\,\right] \leq \frac{(c-1)^2 N_S}{c^3}$, where $c = N_T/N_S$. Given this, we can apply Czumaj and Kutylowski's results directly to bound the variation distance. Specifically, their result implies that there exist constants $k_1, k_2$ such that for $k = k_1 \log(N_S)$, $\|\nu_{\mathrm{rcw}^k} - \mu_s\| \leq \frac{k_2}{N_S}$ where $\nu_{\mathrm{rcw}^k}$ is the distribution after $k$ steps of the RCW algorithm and $\mu_s$ is the uniform distribution on permutations of the elements in $N_S$. However, their result does not explicitly compute the constants. Although we use many of the general ideas from their proof we not only give explicit constants but we provide new proofs of two key lemmas in order to provide a bound that is reasonable in our context and customized for the RCW algorithm. Despite these changes, we believe this is just a starting point and a further reduction of the constants is possible. We begin by providing an overview of the approach and then give a detailed proof focusing on our modifications. For additional information on the Markov chain analysis techniques used in this section please see [13,19].

We will first show that the Markov chain that results from repeatedly applying RCW is ergodic and it's stationary distribution is the uniform distribution. In a single step of the RCW algorithm there is a non-zero probability that we select any single transposition (i.e., $(s_i, s_j)(s_1)(s_2)\ldots$). It is well known that transpositions (swapping any two elements) connect the set of all permutations (see e.g., [13]) and thus RCW connects $\mathsf{Perms}(\mathcal{S})$ (the set of all permutations on $\mathcal{S}$). Additionally, RCW is aperiodic since there is a non-zero probability that no changes are made and thus ergodic. It is also relatively straightforward to see that RCW is symmetric (i.e., for all pairs of permutations $(x, y)$, $\mathcal{P}(x, y) = \mathcal{P}(y, x)$, where $\mathcal{P}(x, y)$ is the probability of moving from $x$ to $y$ in one step of RCW). Combining these implies that the stationary distribution of RCW is the uniform distribution as desired (see e.g., [13]).

In order to bound the mixing time of the matching exchange process, Czumaj and Kutylowski use a technique they call *delayed path coupling* which is an extension of coupling and path coupling, both well-known techniques in the Markov chain community. A *coupling* of a Markov chain $\mathcal{M}$ with state space $\Omega$ is a joint Markov process on $\Omega \times \Omega$ such that the marginals each agree with $\mathcal{M}$ and, once the two coordinates coalesce, they move in unison. The coupling time (or expected time until the two coordinates coalesce) can be used to upper bound the mixing time. *Path coupling*, introduced by Bubley and Dyer, simplifies this approach by considering only a subset $U$ of the joint state space $\Omega \times \Omega$ of a coupling [5]. By considering an appropriate metric $\Delta$ on $\Omega$, proving that the two marginal chains, if in a joint configuration in subset $U$, get no farther away in expectation after one iteration is sufficient to give a polynomial bound on the mixing time. For our argument we will define the distance between two configurations $\Delta(X, Y)$ as the minimum number of transpositions (swapping two points) needed to go from $X$ to $Y$ and $U$ as the set of all pairs of configurations that differ by a single transposition $\Delta(X, Y) = 1$. Using this definition of $U$ it is relatively straightforward to use path coupling to show that the mixing time is $O(N_S \log N_S)$. However for our application this bound is not sufficient and we require more complex techniques.

In delayed path coupling we consider the change in distance between two processes over more than just a single step. We bound the change in distance over $t = \Theta(\log(N_S))$ steps and use a non-Markovian coupling, allowing us to delay the coupling decisions based on future events. We will use the following delayed path coupling theorem due to Czumaj, Kanarek, Kutylowski and Lorys [7]. Let $\mathcal{M}$ be an ergodic Markov chain with statespace $\Omega$ (not necessarily $\mathsf{Perms}(\mathcal{S})$) and mixing time $\tau_{\mathcal{M}}(\epsilon)$ as defined in Sect. 2.

**Theorem 1 (Czumaj et al.).** *Let $\Delta$ be a metric defined on $\Omega \times \Omega$ which takes values in $\{0, \ldots, D\}$, let $U = \{(X, Y) \in \Omega \times \Omega : \Delta(X, Y) = 1\}$ and let $\delta$ be a positive integer. Let $(X_t, Y_t)_{t \in \mathbb{N}}$ be a coupling for $\mathcal{M}$, such that for every $(X_{t\delta}, Y_{t\delta}) \in U$ it holds that $\mathbf{E}\left[\Delta(X_{(t+1)\delta}, Y_{(t+1)\delta})\right] \leq \beta$ for some real $\beta < 1$. Then,*

$$\tau_{\mathcal{M}}(\epsilon) \leq \delta \cdot \left\lceil \frac{\ln(D * \epsilon^{-1})}{\ln \beta^{-1}} \right\rceil.$$

Czumaj and Kutylowski's use the distance metric $\Delta$ defined above (the minimum number of transpositions) and define a coupling $(X_t, Y_t)_{t=0}^T$ where $\Delta(X_0, Y_0) = 1$ (i.e., $X_0$ and $Y_0$ differ by a single transposition). They show that using their coupling, $\mathbf{E}\left[\Delta(X_T, Y_T)\right] \leq 1/N$, for $T = \Theta(\log N)$ which is sufficient to show the mixing time is $O(\log(N))$. We will use the same coupling to analyze the RCW algorithm and provide a brief overview here for completeness. Full details can be found in their paper [9]. Note that for ease of explanation, the matchings described here are the matchings actually applied at each step (i.e., the $B_i$'s are already incorporated into the description of the matchings). Let $M_1, M_2, \ldots M_T$ be the matchings defined by the coupling for the process $X$ and $N_1, N_2, \ldots N_T$ be the matchings for $Y$, so that applying these matchings at each step results in the coupling $(X_t, Y_t)_{t=0}^T$. We begin by choosing the permutations and corresponding matchings for $X$ at each step, $M_1, M_2, \ldots M_T$, according to the distribution given by the RCW algorithm thus ensuring that the marginals of $X$ agree with the RCW algorithm. Next using the matchings chosen for $X$ we will carefully select the matchings for $Y$, $N_1, N_2, \ldots N_T$ to ensure that by the end of $T$ steps the two processes will have coupled with probability $1 - 1/N_S$. Without loss of generality, assume that $X_0$ and $Y_0$ differ only by a transposition of points $x$ and $y$ (recall that $\Delta(X_0, Y_0) = 1$). If the matching $M_1$ contains the pair (or edge) $(x, y)$ then if we apply the same matching minus this pair to $Y_0$ then after one step, the process has coupled (e.g. $\Delta(X_1, Y_1) = 0$). However the probability that a matching contains this pair is only $\Theta(1/N_S)$ and thus not sufficient to obtain the bound we desire. In order to overcome this Czumaj and Kutylowski observe that if $(x, w)$ and $(y, z)$ are pairs in the matching $M_1$ then if we let $N_1 = M_1 - (x, w) - (y, z) + (x, z) + (y, w)$ then $X_1$ and $Y_1$ differ by a $(x, y)$ transposition. Conversely if we let $N_1 = M_1$ then $X_1$ and $Y_1$ differ by a $(w, z)$ transposition. Given this, if $M_2$ contains either $(x, y)$ or $(w, z)$ then we can choose $N_1, N_2$ so that $N_3 = M_3$ and the process has coupled. As Czumaj and Kutylowski do, we will call $(x, y)$ and $(w, z)$ good pairs and let $GP_t$ denote the set of good pairs at step $t$. The general idea behind the argument is to show that at every step the number of good pairs increases by a constant factor and thus after $\Theta(\log N_S)$ steps the number of good pairs is $\Omega(N_S)$. Given this, with high probability in another $\Theta(\log N_S)$ steps one of the matchings $M_t$ will contain a good pair and thus we can define corresponding matchings for $Y$ so that the process couples. We formally define a good pair as follows.

**Definition 1 (Czumaj, Kutylowski).** *Without loss of generality, assume $X_0$ and $Y_0$ differ by a $(x, y)$ transposition and let $GP_0 = \{(x, y)\}$. For each $(x, y) \in GP_{t-1}$:*

1. *If neither $x$ or $y$ is part of the matching $M_t$ then $(x, y) \in GP_t$.*
2. *If $(x, w) \in M_t$ and $y$ is not part of $M_t$ then $(w, y) \in GP_t$.*
3. *If $(y, w) \in M_t$ and $x$ is not part of $M_t$ then $(w, x) \in GP_t$.*
4. *If $(x, w), (y, z) \in M_t$ then if neither $w$ or $z$ are part of pairs in $GP_t$ then $(w, z) \in GP_t$ and $(x, y) \in GP_t$. Otherwise $(w, z) \in GP_t$.*

Using this strategy, Czumaj and Kutylowski formally give a coupling so that if a pair $(x, y)$ is a good pair at time $t$ and $M_t$ contains $(x, y)$ then $X_T = Y_T$. We use this coupling exactly and rely on their proof to show that it is indeed a valid coupling and the marginal distributions of $X$ and $Y$ agree with those given by RCW. Given this coupling, it remains to show that after a time $t_1$ the number of good pairs is large enough so that in the next $t_2$ steps one of the $t_2$ matchings will contain a good pair. We deviate from Czumaj and Kutylowski's approach in this analysis.

We begin by showing that after $t_1 = \Theta(\log N_S)$ steps the probability that there are less than $N_S/9$ good pairs is at most $.5N_S^{-2}$. Next, we show that after an additional $t_2 = \Theta(\log N_S)$ steps the probability that none of the matchings during those additional $t_2$ steps includes a good pair is at most $.5N_S^{-2}$. Combining these shows that using the given coupling, after $t_1 + t_2$ steps, with probability at most $N_S^{-2}$, the two processes remain at distance 1 and otherwise they are at distance 0. Thus, $\mathbf{E}\left[\Delta(X_{(t+1)\delta}, Y_{(t+1)\delta})\right] \leq N_S^{-2}$ for $\delta = t_1 + t_2$. Given this we can now apply the delayed path coupling theorem. Since $\Delta$ is the minimum number of transpositions to move from one configuration to another, $D$ (the maximum distance between two configurations) is at most $N_S$. This is due to the fact that by using a single transposition per point we can put each point in it's new location. Combining these and the delayed path coupling theorem gives the following bound on the mixing time.

**Theorem 2.** *For* $T \geq \max \left( 40\ln(2N_S^2), \frac{10\ln(N_S/9)}{\ln(1+.3(c-1)^4/c^6)} \right) + \frac{36c^3\ln(2N_S^2)}{(c-1)^2}$ *and* $N_S \geq 2^{10}$, *the mixing time* $\tau$ *of the RCW algorithm satisfies*

$$\tau(\epsilon) \leq T \cdot \left\lceil \frac{\ln(N_S/\epsilon)}{\ln N_S^2} \right\rceil.$$

*When* $\epsilon = 1/N_S$ *the bound simplifies to* $\tau(1/N_S) \leq T = \Theta(\ln(N_S))$

A straightforward manipulation of the bound on the mixing time gives us the following bound on the variation distance that will be useful in the remainder of the paper. Notice again that as long as the number of rounds of the RCW algorithm is at least $T = \Theta(\ln(N_S))$, the variation distance is less than $1/N_S$.

**Corollary 1.** *Let* $T = \max \left( 40\ln(2N_S^2), \frac{10\ln(N_S/9)}{\ln(1+.3(c-1)^4/c^6)} \right) + \frac{36c^3\ln(2N_S^2)}{(c-1)^2}$ *and* $N_S \geq 2^{10}$, *then*

$$\|\nu_{rcw^r} - \mu_s\| \leq N_S^{1-2r/T},$$

*where* $\nu_{rcw^r}$ *is the distribution after* $r$ *rounds of the RCW algorithm and* $\mu_s$ *is the uniform distribution on permutations of the elements in* $S$.

Our theorem does not explicitly condition on $E[\kappa] = \Theta(N_S)$ as in Czumaj and Kutylowski [9]. Instead our theorem applies only to the RCW algorithm and relies on more specific statements about the chain. For example, a key step in our analysis is to show that at each step of the RCW algorithm a particular point is part of a 2-cycle with constant probability (which implies that $E[\kappa] = \Theta(N_S)$). Let $c_x$ be the probability that point $x$ is part of a 2-cycle. We prove the following claim.

**Claim 1.**

$$c_x = \frac{(N_S - 1) \cdot (N_T - N_S)^2}{N_T \cdot (N_T - 1) \cdot (N_T - 2)} \geq \frac{(c-1)^2}{c^3}.$$

*where the probability is over the choice of* $\pi$ *and* $c = N_T/N_S$.

The point $x$ is part of a potential 2-cycle when the algorithm RCW is applied to $x$ and $u$ and $v$ are set; this happens in either the "if" of "else if" blocks of the RCW algorithm given in Sect. 3. The bit $b$ then determines whether or not $x$ and the point it gets paired with actually become part of a 2-cycle. To prove the claim, we need to consider how $u$ and $v$ can be set in the algorithm. There are two cases, corresponding to the "if" and "else if" blocks of the algorithm. First consider the "if" case. We need to determine the probability that $\pi(x) \in S \wedge \pi^{-1}(x) \notin S \wedge \pi(\pi(x)) \notin S$ and $B(x) = 1$ for a randomly chosen permutation $\pi$ on $T$ and B from $S$ to $\{0,1\}$. There are $N_S - 1$ choices for $\pi(x)$ (the minus one is since we don't want $x$ mapped to itself), $N_T - N_S$ choices for $\pi^{-1}(x)$, and $N_T - N_S$ choices for $\pi(\pi(x))$. This fixes three mappings, so there are then $(N_T - 3)!$ choices for how to map the remaining points. Thus, the probability we end up in the "if" case is

$$\frac{.5(N_S - 1) \cdot (N_T - N_S)^2 (N_T - 3)!}{N_T!} = \frac{(N_S - 1) \cdot (N_T - N_S)^2}{N_T \cdot (N_T - 1) \cdot (N_T - 2)}.$$

The argument for the "else if" case is almost identical, and gives the same probability. We lower bound the probability as follows.

$$\mathbf{P}(X_i = 1) = \frac{(N_S - 1) \cdot (N_T - N_S) \cdot (N_T - N_S)}{N_T \cdot (N_T - 1) \cdot (N_T - 2)} \geq \frac{(c - 1)^2}{c^3},$$

where $N_T = cN_S$. Note that using linearity of expectations over all points in $N_S$, this claim implies that the expected number of 2-cycles is $((c - 1)^2/c^3)N_S$.

Next, we prove the following lemma which shows that after $t_1$ steps there are linear number of good pairs.

**Lemma 2.** *Let $|GP_t|$ be the number of good pairs at step $t$, $N_S \geq 2^{10}$ and $t_1 = \max(40 \ln(2N_S^2), 10 \ln(N_S/9)/\ln(1 + .3(c - 1)^4/c^6)$ then*

$$\mathbf{P}(|GP_{t_1}| < N_S/9) \leq .5N_S^{-2}.$$

*Proof.* We start with one good pair at $t = 1$ and then at each step of the algorithm we say that a good pair $(x, y)$ *splits* if it creates a second good pair (this corresponds to the last case of Definition 1). We begin with bounding the probability that a good pair splits in the RCW algorithm. First, we assume that there are less than $N_S/9$ good pairs (if there are more than we're done). Since we have assumed that there are less than $N_S/9$ good pairs, there are at most $2N_S/9$ points in good pairs and at least $N_S - 2N_S/9 = (7/9)N_S$ points not in good pairs. Good pair $(x, y)$ splits when $x$ and $y$ are both matched to points that are not already in good pairs of which there are at least $(7/9)N_S$. Using this we can now extend the proof of Claim 1 to show the following where $c_p$ is the probability that a particular good pair splits:

$$c_p \geq \frac{(\frac{7}{9})^2(c - 1)^4(1 - 2^{-8})}{c^6}.$$

Let $(x, y)$ be a good pair. We want to lower bound that probability that point $x$ and $y$ are both part of potential 2-cycles $(x, w)$ and $(y, z)$ where $w$ and $z$ are not in good pairs. Since we are now interested in two points being part of potential 2-cycles, there are 4 different cases; the first case corresponds to $u$ and $v$ begin set for both $x$ and $y$ in the "if" block of the RCW algorithm. We need to determine the probability that $\pi(x) \in (\mathcal{S} - GP) \wedge \pi^{-1}(x) \notin \mathcal{S} \wedge \pi(\pi(x)) \notin \mathcal{S}$ and that $\pi(y) \in (\mathcal{S} - GP) \wedge \pi^{-1}(y) \notin \mathcal{S} \wedge \pi(\pi(y)) \notin \mathcal{S}$ and $\mathsf{B}(x) = \mathsf{B}(y) = 1$ for a randomly chosen permutation $\pi$ on $\mathcal{T}$ and $\mathsf{B}$ from $\mathcal{S}$ to $\{0, 1\}$. Since there are at least $(7/9)N_S$ points not in good pairs, there are $(7/9)N_S$ choices for $\pi(x)$, $N_T - N_S$ choices for $\pi^{-1}(x)$, and $N_T - N_S$ choices for $\pi(\pi(x))$. Given these mappings, there are $(7/9)N_S - 1$ choices for $\pi(y)$ (the minus one accounts for $\pi(x)$ which is already mapped to a point in $S - GP$), $N_T - N_S - 1$ choices for $\pi^{-1}(y)$, and $N_T - N_S - 1$ choices for $\pi(\pi(y))$. This fixes six mappings, so there are then $(N_T - 6)!$ choices for how to map the remaining points. Thus, the probability $x$ and $y$ are both mapped to points in $S$ that are not in good pairs in the "if" block of the algorithm is

$$.25\frac{(7/9)N_S \cdot (N_T - N_S)^2 \cdot ((7/9)N_S - 1) \cdot (N_T - N_S - 1)^2}{N_T \cdot (N_T - 1) \cdot (N_T - 2) \cdot (N_T - 3) \cdot (N_T - 4) \cdot (N_T - 5)}.$$

As in Claim 1, the argument for the other three cases is almost identical, and gives the same probability. We lower bound the probability as follows.

$$c_p \geq \frac{(\frac{7}{9}N_S)(\frac{7}{9}N_S - 1)(N_T - N_S)^2(N_T - N_S - 1)^2}{N_T(N_T - 1)(N_T - 2)(N_T - 3)(N_T - 4)(N_T - 5)} \geq \frac{(\frac{7}{9})^2(c - 1)^4(1 - 2^{-8})}{c^6},$$

where $N_T = cN_S$ and $N_S \geq 2^{10}$. Note that the restriction $N_S \geq 2^{10}$ could easily be loosened at the expense of a small constant factor in the bound.

By linearity of expectations, if we have $|GP_t|$ good pairs at step $t$, then the expected number of good pairs at step $t + 1$ is $\mathbf{E}\left[|GP_{t+1}|\right] = |GP_t| + c_p|GP_t|$. Let $G_t = (|GP_{t+1}| - |GP_t|)/|GP_t|$ be the fraction of good pairs that split between time $t$ and $t + 1$ (the growth rate). Thus, we have that $\mathbf{E}\left[G_t\right] = c_p$. Next, define an indicator random variable $Z_t$ that is 1 if $G_t \geq \mathbf{E}\left[G_t\right]/2 = c_p/2$ and 0 otherwise. Thus if $\sum_{t=0}^{t_1} Z_t \geq \frac{\ln n/9}{\ln(1+c_p/2)}$ then $|GP_{t_1}|$ is at least $(1 + c_p/2)^{(\ln N_S/9)/\ln(1+c_p/2)} = N_S/9$. This is due to the fact that each times $Z_t$ is one $|GP_t|$ increases at least by a factor of $1 + c_p/2$.

Next, we will show that for $t_1 = \max(40\ln(2N_S^2), 10\frac{\ln N_S/9}{\ln(1+c_p/2)})$,

$$\mathbf{P}\left(\sum_{t=0}^{t_1} Z_t < \frac{\ln N_S/9}{\ln(1 + c_p/2)}\right) < .5N_S^{-2}$$

which implies $\mathbf{P}\left(|GP_{t_1}| < N_S/9\right) \leq .5N_S^{-2}$. First, using Markov's inequality we will show that $\mathbf{P}\left(Z_t = 0\right) = \mathbf{P}\left(G_t \leq \mathbf{E}\left[G_t\right]/2\right) \leq 4/5$. Let $A = 3\mathbf{E}\left[G_t\right] - G_t$, then $\mathbf{P}\left(G_t \leq \mathbf{E}\left[G_t\right]/2\right) = \mathbf{P}\left(A \geq (3 - 1/2)\mathbf{E}\left[G_t\right] = 2.5\mathbf{E}\left[G_t\right]\right)$. By linearity of expectations, $\mathbf{E}\left[A\right] = \mathbf{E}\left[3\mathbf{E}\left[G_t\right] - G_t\right] = 2\mathbf{E}\left[G_t\right]$. Thus $\mathbf{P}\left(Z_t = 0\right) = \mathbf{P}\left(A \geq 2.5\mathbf{E}\left[G_t\right]\right) \leq \mathbf{E}\left[A\right]/2.5\mathbf{E}\left[G_t\right] = 4/5$. Next, we note that the $Z_i$'s are not independent since the probability $Z_i$ is 1 is determined by the number of good pairs. However, since we are assuming there are always at most $N_S/9$ good pairs, this process is stochastically lower bounded by a process with independent variables $X_1, \ldots X_{t_1}$ where each variable $X_i$ is 1 with probability $1/5$ and 0 with probability $4/5$. In the actual process, especially toward the beginning the $Z_i$'s are much more likely to be 1 because there are substantially fewer than $N_S/9$ good pairs. However throughout the process the probability is always at least $1/5$. Next, we will apply the Chernoff bound $\mathbf{P}\left(X < \mathbf{E}\left[X\right]/2\right) < \exp(-\mathbf{E}\left[X\right]/8)$

with $X = \sum_{t=0}^{t_1} X_t$ and $t_1 = \max(40\ln(2N_S^2), 10(\ln N_S/9)/\ln(1 + c_p/2))$. Our choice of $t_1$ implies that $\mathbf{E}[X] \geq (1/5)40\ln(2N_S^2) = 8\ln(2N_S^2)$. Therefore,

$$\mathbf{P}(X < \mathbf{E}[X]/2) < \exp(-\mathbf{E}[X]/8) <= \exp(-8\ln(2N_S^2)/8) = .5N_S^{-2}.$$

Again due to our choice of $t_1$, $\mathbf{E}[X] \geq (1/5)10\frac{\ln N_S/9}{\ln(1+c_p/2)} = 2\frac{\ln N_S/9}{\ln(1+c_p/2)}$. Combining these gives the desired result,

$$\mathbf{P}\left(X < \frac{\ln N_S/9}{\ln(1 + c_p/2)}\right) < \mathbf{P}(X < \mathbf{E}[X]/2) < .5N_S^{-2}.$$

$\qquad\square$

Finally, we consider the matchings during the next $t_2$ steps and show the probability that none of them includes a good pair is at most $.5N_S^{-2}$. We say that a pair $(x, y)$ is part of a *potential* matching if the RCW algorithm maps $x$ to $y$ regardless of the value of $B(x)$. Specifically, we prove the following lemma.

**Lemma 3.** *Let $t_2 = 36c^3\ln(2N_S^2)/(c-1)^2$ then conditioned on $|GP_{t_1}| \geq n/9$, the probability that the next $t_2$ potential matchings contain no edges from $GP_{t_1}$ is at most $.5N_S^{-2}$.*

*Proof.* First, consider a good pair $(x, y)$. We claim that the probability that $x$ is mapped to $y$ in one step of the RCW algorithm is $2 \cdot (c-1)^2/(c^3N_S)$. Again, there are two cases corresponding to the "if" and "else if" blocks of the algorithm. Consider the "if" case, we need to determine the probability that $\pi(x) = y \wedge \pi^{-1}(x) \notin S \wedge \pi(\pi(x)) \notin S$. There is one choice for $\pi(x)$, $N_T - N_S$ choices for $\pi^{-1}(x)$, and $N_T - N_S$ choices for $\pi(\pi(x))$. This fixes three mappings, resulting in $(N_T - 3)!$ choices for the remaining points. Thus, the probability $x$ is mapped to $y$ in the "if" case is

$$\frac{(N_T - N_S)^2 \cdot (N_T - 3)!}{N_T!} = \frac{(N_T - N_S)^2}{N_T \cdot (N_T - 1) \cdot (N_T - 2)} \leq \frac{(c-1)^2}{c^3 N_S}.$$

Again, the argument for the "else if" case is almost identical giving a factor of two in the probability that $x$ is mapped to $y$.

Let $H_t$ be the number of edges in the potential matching at time $t$ that correspond to good pairs. There are at least $N_S/9$ good pairs at time $t_1$ and each is in the potential matching with probability at least $2(c-1)^2/(c^3N_S)$. Thus, by linearity of expectations, for $t > t_1$ we have that

$$\mathbf{E}[H_t] \geq (N_S/9)(2(c-1)^2/(c^3N_S)) = 2(c-1)^2/(9c^3).$$

Since the potential matchings generated at each time step are independent we can now use a Chernoff bound to show that $\mathbf{P}\left(\sum_{t=t_1}^{t_1+t_2} H_t < 1\right) < .5N_S^{-2}$. Again, we will use the following form of the Chernoff bound; $\mathbf{P}(X < \mathbf{E}[X]/2) < \exp(-\mathbf{E}[X]/8)$. If we let $X = \sum_{t=t_1}^{t_1+t_2} H_t$ where $t_2 = 36c^3\ln(2N_S^2)/(c-1)^2$

then by linearity of expectations $\mathbf{E}[X] = \sum_{t=t_1}^{t_1+t_2} \mathbf{E}[H_t] \geq t_2 \cdot 2(c-1)^2/(9c^3) = 8\ln(2N_S^2)$. Applying the above Chernoff bound gives the following,

$$\mathbf{P}\left(\sum_{t=t_1}^{t_1+t_2} H_t < 4\ln(2N_S^2)\right) < \exp(-8\ln(2N_S^2)/8) = .5N_S^{-2}$$

Thus, since $\mathbf{P}\left(\sum_{t=t_1}^{t_1+t_2} H_t < 1\right) < \mathbf{P}\left(\sum_{t=t_1}^{t_1+t_2} H_t < 4\ln(2N_S^2)\right)$, we have that $\mathbf{P}\left(\sum_{t=t_1}^{t_1+t_2} H_t < 1\right) < .5N_S^{-2}$, as desired. $\qquad\square$

## 5    CCA Security

Let $\mathcal{S} \subseteq \mathcal{T}$ with $N_S$ and $N_T$ their sizes, respectively. Let $r$ be a positive integer called the repetition number. Let $E : \mathcal{K}_E \times \mathcal{T} \to \mathcal{T}$ be a block cipher with domain $\mathcal{T}$. Let $F : \mathcal{K}_F \times \{1,\ldots,r\} \times \mathcal{S} \to \{0,1\}$ be a pseudorandom function family.

We use reverse 2-cycle walking to define a new block cipher $\tilde{E} : \mathcal{K} \times \mathcal{S} \to \mathcal{S}$ as follows. The key space $\mathcal{K}$ is $\mathcal{K}_E^r \times \mathcal{K}_F$. Let $F_{i,K}(\cdot) = F(K,i,\cdot)$. Then $\tilde{E}$, on input key $K$ and point $x$, parses its key $K$ as $r$ block cipher keys $K_1,\ldots,K_r$ and a PRF key $K'$ and then computes

$$\mathsf{RCW}^r_{(E_{K_1},\ldots,E_{K_r}),(F_{1,K'},\ldots,F_{r,K'})}(x) .$$

The following theorem establishes the CCA security of block cipher $\tilde{E}$.

**Theorem 3.** *Let $E$, $F$, and $\tilde{E}$ be defined as above. Let $A$ be an adversary attacking $\tilde{E}$ and making $q$ queries. Then,*

$$\mathbf{Adv}^{\mathrm{cca}}_{\tilde{E}}(A) \leq r \cdot \mathbf{Adv}^{\mathrm{cca}}_E(B) + \mathbf{Adv}^{\mathrm{prf}}_F(C) + \Gamma ,$$

*with $B$ making $3 \cdot q$ queries, $C$ making $r \cdot q$ queries, and $\Gamma$ being the bound on variation distance from Corollary 1 that depends on $r$.*

*Proof.* We wish to bound the cca-advantage of an adversary $A$ attacking $\tilde{E}$ and making at most $q$ oracle queries. Thus we wish to bound

$$\mathbf{Adv}^{\mathrm{cca}}_{\tilde{E}}(A) = \mathbf{P}\left(A^{\pm\tilde{E}(K,\cdot)} \Rightarrow 1\right) - \mathbf{P}\left(A^{\pm\pi(\cdot)} \Rightarrow 1\right) .$$

We will start with the left term above, where $A$ is given access to oracles for $\tilde{E}$ and $\tilde{E}^{-1}$, and gradually change the oracles until they are simply random permutations on $\mathcal{S}$, bounding each oracle change accordingly.

Recall that $\pm\tilde{E}(K,\cdot)$ is really just a more compact way of writing

$$\pm\mathcal{O}_1(\cdot) = \pm\mathsf{RCW}^r_{(E_{K_1},\ldots,E_{K_r}),(F_{1,K'},\ldots,F_{r,K'})}(\cdot).$$

Our first oracle transition, from $\mathcal{O}_1$ to $\mathcal{O}_2$, replaces all of the block ciphers with random permutations on the same domain $\mathcal{T}$, turning the oracle into

$$\pm\mathcal{O}_2(\cdot) = \pm\mathsf{RCW}^r_{(\pi_1,\ldots,\pi_r),(F_{1,K'},\ldots,F_{r,K'})}(\cdot) ,$$

where each $\pi_i$ is a random permutation on $\mathcal{T}$. We can bound the difference using a hybrid argument and an adversary $B$ attacking the cca security of $E$.

The adversary $B$ is given an encryption algorithm and its inverse, which we denote by $\mathcal{O}_B$ and $\mathcal{O}_B^{-1}$. Adversary $B$ first chooses a random index $i \in \{1, \ldots, r\}$. Next, $B$ chooses $i-1$ keys $K_1, \ldots, K_{i-1}$ for block cipher $E$, and a PRF key $K'$. It then runs adversary $A$, simulating $A$'s oracle queries as follows.

On encryption query $x$ from $A$, $B$ first computes the mapping

$$x' = \mathsf{RCW}^{i-1}_{(E_{K_1}, \ldots, E_{K_{i-1}}), (F_{1,K'}, \ldots, F_{r,K'})}(x).$$

In words, $B$ applies $i-1$ rounds of the RCW algorithm with the keys $B$ chose earlier; let the result be $x'$. $B$ then uses its oracles to determine how $x'$ should be mapped in the $i$th step of RCW. Looking at the RCW algorithm in Sect. 3, we see that to determine this $B$ will need to query $\mathcal{O}(x')$, $\mathcal{O}^{-1}(x')$, and one of $\mathcal{O}(\mathcal{O}(x'))$ and $\mathcal{O}^{-1}(\mathcal{O}^{-1}(x'))$ depending on the results of the first two queries. Thus, for each encryption query $A$ makes, $B$ queries its own oracles three times, making either two forward and one inverse or one forward and two inverse queries. After $B$ determines how $x'$ should be mapped at the $i$th step (call the result $x''$), it computes how $x''$ should be mapped by steps $i+1$ through $r$ using the RCW algorithm with random permutations. To simulate these permutations, $B$ simply uses tables.

$B$ handles inverse queries from $A$ similarly, except it computes $\mathsf{RCW}^r$ in the reverse direction, using the same tables for random permutations in steps $i+1$ through $r$, using its own oracles at step $i$, and using the keys it chose for steps $1$ through $i-1$.

From the description of $B$, we can see that if $B$ is given as oracles a real block cipher and its inverse under some key, then $B$ simulates for $A$ the oracles $\pm\mathsf{RCW}^r_{(E_{K_1}, \ldots, E_{K_{i-1}}, E_{K_i}, \pi_{i+1}, \ldots, \pi_r)}$, while if $B$ is given as oracles a random permutation and its inverse, it simulates for $A$ oracles $\pm\mathsf{RCW}^r_{(E_{K_1}, \ldots, E_{K_{i-1}}, \pi_i, \ldots, \pi_r)}$.

Thus, it follows that

$$\mathbf{P}\left(A^{\pm\mathcal{O}_1(\cdot)} \Rightarrow 1\right) - \mathbf{P}\left(A^{\pm\mathcal{O}_2(\cdot)} \Rightarrow 1\right) \leq r \cdot \mathbf{Adv}_E^{\mathrm{cca}}(B) \tag{1}$$

where $B$ makes at most $3q$ oracle queries.

For our next oracle transition, from $\mathcal{O}_2$ to $\mathcal{O}_3$, we replace the PRF $F : \mathcal{K}_F \times \{1, \ldots, r\} \times \mathcal{S} \to \{0,1\}$ with a truly random function $\rho : \{1, \ldots, r\} \times \mathcal{S} \to \{0,1\}$. Similar to how we defined $F_{i,K}(\cdot) = F(K, i, \cdot)$, we let $\rho_i(\cdot) = \rho(i, \cdot)$. Thus, our oracle $\mathcal{O}_3$ becomes

$$\pm\mathcal{O}_3(\cdot) = \pm\mathsf{RCW}^r_{(\pi_1, \ldots, \pi_r), (\rho_1, \ldots, \rho_r)}(\cdot).$$

We can bound the change in advantage by the prf-advantage of an adversary $C$. The adversary $C$, given access to an oracle that is either the real pseudorandom function or a truly random function, runs $A$ and simulates its oracles by computing $\mathsf{RCW}^r$ using tables and random sampling to simulate the random permutations used by RCW, and using its own oracle to compute the bit $b$ used in each round. Since there are $r$ rounds of RCW and $A$ makes $q$ queries, $C$ will make $rq$ queries to its own oracle. Clearly, if $C$'s oracle is a real pseudorandom function, it simulates $\mathcal{O}_2$ for $A$, while if $C$'s oracle is a truly random function it simulates $\mathcal{O}_3$ for $A$. Thus,

$$\mathbf{P}\left(A^{\pm\mathcal{O}_2(\cdot)} \Rightarrow 1\right) - \mathbf{P}\left(A^{\pm\mathcal{O}_3(\cdot)} \Rightarrow 1\right) \leq \mathbf{Adv}_F^{\mathrm{prf}}(C) \qquad (2)$$

where $C$ makes at most $r \cdot q$ oracle queries.

At this point, we have $r$ rounds of RCW using only ideal components. For our last oracle transition, $\mathcal{O}_3$ to $\mathcal{O}_4$, we replace RCW entirely with a random permutation on $\mathcal{S}$. Thus,

$$\pm\mathcal{O}_4(\cdot) = \pm\pi(\cdot).$$

We are now in the information theoretic setting, and the maximum advantage of any adversary in distinguishing between $\mathsf{RCW}^r$ with ideal components and a random permutation on $\mathcal{S}$ is bounded in Corollary 1 in the previous section. Thus,

$$\mathbf{P}\left(A^{\pm\mathcal{O}_3(\cdot)} \Rightarrow 1\right) - \mathbf{P}\left(A^{\pm\mathcal{O}_4(\cdot)} \Rightarrow 1\right) \leq \Gamma \qquad (3)$$

where $\Gamma$ is the value the variation distance is bounded by in the Corollary.

We can thus bound the cca-advantage of $A$ as follows:

$$\mathbf{Adv}_{\tilde{E}}^{\mathrm{cca}}(A) \leq \left(\mathbf{P}\left(A^{\pm\mathcal{O}_1(\cdot)} \Rightarrow 1\right) - \mathbf{P}\left(A^{\pm\mathcal{O}_2(\cdot)} \Rightarrow 1\right)\right)$$
$$+ \left(\mathbf{P}\left(A^{\pm\mathcal{O}_2(\cdot)} \Rightarrow 1\right) - \mathbf{P}\left(A^{\pm\mathcal{O}_3(\cdot)} \Rightarrow 1\right)\right)$$
$$+ \left(\mathbf{P}\left(A^{\pm\mathcal{O}_3(\cdot)} \Rightarrow 1\right) - \mathbf{P}\left(A^{\pm\mathcal{O}_4(\cdot)} \Rightarrow 1\right)\right)$$

where recall that

$$\mathcal{O}_1 = \mathsf{RCW}^r_{(E_{K_1},\ldots,E_{K_r}),(F_{1,K'},\ldots,F_{r,K'})}$$
$$\mathcal{O}_2 = \mathsf{RCW}^r_{(\pi_1,\ldots,\pi_r),(F_{1,K'},\ldots,F_{r,K'})}$$
$$\mathcal{O}_3 = \mathsf{RCW}^r_{(\pi_1,\ldots,\pi_r),(\rho_1,\ldots,\rho_r)}$$
$$\mathcal{O}_4 = \pi$$

Substituting in Eqs. (1), (2), and (3) gives the bound from the theorem statement. □

## 6  Full Security via Reverse Cycle Walking

Let $\mathcal{S} = \{0,\ldots,N-1\}$ and $\mathcal{T} = \{0,\ldots,2N-1\}$. Thus, $N_T = |\mathcal{T}| = 2N$ and $N_S = |\mathcal{S}| = N$. Let $r$ be a positive integer called the repetition number.

Let $E : \mathcal{K} \times \mathcal{T} \to \mathcal{T}$ be a block cipher on $\mathcal{T}$ and let $E^{-1}$ be its inverse. Let $F : \mathcal{K}_F \times \{1, \ldots, r\} \times \mathcal{S} \to \mathcal{S}$ be a pseudorandom function family. Let $\tilde{E}$ be defined as it was in the previous section, using $r$ rounds of RCW with $E$ and $F$. The following theorem states that if $E$ is secure against cca adversaries making $N = (1/2)N_T$ queries, then $\tilde{E}$ is fully secure (i.e., secure against adversaries making $N = N_S$ queries). In other words, the RCW construction allows us to build a fully secure cipher out of a partially secure cipher on a larger set.

**Theorem 4.** *Let $\mathcal{S}$, $\mathcal{T}$, $E$, $F$, and $\tilde{E}$ be defined as above. Let $A$ be a cca adversary attacking $\tilde{E}$ and making $N = N_S$ queries. Then*

$$\mathbf{Adv}_{\tilde{E}}^{\mathrm{cca}}(A) \leq r \cdot \mathbf{Adv}_{E}^{\mathrm{cca}}(B) + \mathbf{Adv}_{F}^{\mathrm{prf}}(C) + \Gamma$$

*where $B$ makes $N = (1/2)N_T$ queries to its encryption oracle, $C$ makes $N \cdot r$ oracle queries, and $\Gamma$ is the bound on variation distance from Corollary 1 that depends on $r$.*

*Proof.* The proof is identical to the proof of Theorem 3 except for how adversary $B$ answers oracle queries from $A$ in the hybrid argument. As in the proof of Theorem 3, we let

$$\pm\mathcal{O}_1(\cdot) = \pm\mathsf{RCW}_{(E_{K_1}, \ldots, E_{K_r}), (F_{1,K'}, \ldots, F_{r,K'})}^r(\cdot),$$

and

$$\pm\mathcal{O}_2(\cdot) = \pm\mathsf{RCW}_{(\pi_1, \ldots, \pi_r), (F_{1,K'}, \ldots, F_{r,K'})}^r(\cdot).$$

We then use adversary $B$ attacking $E$ to argue that replacing $A$'s $\pm\mathcal{O}_1$ oracles with $\pm\mathcal{O}_2$ has little effect.

Let $B$'s oracles be denoted by $\mathcal{O}_B$ and $\mathcal{O}_B^{-1}$. As in the previous proof, $B$ begins by choosing a random index $i \in \{1, \ldots, r\}$, then $i - 1$ keys $K_1, \ldots, K_{i-1}$ for $E$ and a PRF key $K'$ for $F$.

This is where we encounter the major change from the adversary in the previous proof. At this point, adversary $B$ should query its own oracle $\mathcal{O}_B$ on all points $x \in \{0, \ldots, N-1\}$, recording the answers in a table. Specifically, $B$ sets $\mathrm{T}[x] = \mathcal{O}_B(x)$. $B$ then runs $A$, answer its oracle queries as follows.

On encryption query $x$ from $A$, $B$ computes

$$x' = \mathsf{RCW}_{(E_{K_1}, \ldots, E_{K_{i-1}}), (F_{1,K'}, \ldots, F_{r,K'})}^{i-1}(x).$$

using the block cipher keys and the PRF key it chose earlier. To determine how $x'$ should be mapped with the $i$th step of RCW, $B$ now has to use the answers it received from its oracle and stored in table T. Notice that $B$ can evaluate the boolean conditions in the "if" case of the RCW algorithm as follows: if $\mathrm{T}[x] \in \mathcal{S}$ and there is no $z \in \mathcal{S}$ s.t. $\mathrm{T}[z] = x$ and $\mathrm{T}[\mathrm{T}[x]] \notin \mathcal{S}$. Similarly, $B$ can evaluate the "else if" as follows: if $\mathrm{T}[x] \notin \mathcal{S}$ and there does exist $z \in \mathcal{S}$ s.t. $\mathrm{T}[z] = x$ and there does not exist $w \in \mathcal{S}$ s.t. $\mathrm{T}[w] = z$. In other words, the table T contains enough information to evaluate the RCW algorithm on any point in $\mathcal{S}$.

After $B$ computes how $x'$ is mapped in the $i$th RCW round, it uses tables to simulate random permutations for rounds $i + 1$ through $r$, just as it did in the proof of Theorem 3. Inverse queries from $A$ are handled similarly to in that proof, just with the $i$th round of RCW computed as above with table T.

The rest of the proof (i.e., bounding the change from using $F$ to using a truly random function) follows the exact steps as the proof in the previous section. □

# 7    Open Questions

There are a number of interesting open questions surrounding reverse cycle walking. We analyzed the security of reverse 2-cycle walking, but we explained in the introduction that the algorithm can be generalized to longer cycles. An interesting question is whether reverse $t$-cycle walking, for $t > 2$, leads to better bounds than we were able to prove here. Another interesting question is what is the optimal worst-case running time for strong pseudorandom permutations on general sets where only efficient membership testing is assumed. We were able to show a worst case running time of $\Theta(t(N)\log N)$, where $t(N)$ is the time to encipher a point in the larger set $\mathcal{T}$. We conjecture that this is in fact optimal.

**Acknowledgements.** We thank Tom Ristenpart for his very helpful comments on an earlier draft of this paper. We also thank the anonymous Asiacrypt reviewers for their detailed feedback.

# References

1. Bellare, M., Ristenpart, T., Rogaway, P., Stegers, T.: Format-preserving encryption. In: Jacobson, M.J., Rijmen, V., Safavi-Naini, R. (eds.) SAC 2009. LNCS, vol. 5867, pp. 295–312. Springer, Heidelberg (2009). doi:10.1007/978-3-642-05445-7_19

2. Bellare, M., Rogaway, P., Spies, T.: The FFX mode of operation for format-preserving encryption. Submission to NIST, February 2010

3. Black, J., Rogaway, P.: Ciphers with arbitrary finite domains. In: Preneel, B. (ed.) CT-RSA 2002. LNCS, vol. 2271, pp. 114–130. Springer, Heidelberg (2002). doi:10.1007/3-540-45760-7_9

4. Brightwell, M., Smith, H.: Using datatype-preserving encryption to enhance data warehouse security. In: National Information Systems Security Conference (NISSC) (1997)

5. Bubley, R., Dyer, M.E.: Faster random generation of linear extensions. In: Karloff, H.J. (ed.) 9th SODA, January 1998, pp. 350–354. ACM-SIAM (1998)

6. Czumaj, A.: Random permutations using switching networks. In: Servedio, R.A., Rubinfeld, R. (eds.) 47th ACM STOC, June 2015, pp. 703–712. ACM Press (2015)

7. Czumaj, A., Kanarek, P., Kutylowski, M., Lorys, K.: Delayed path coupling and generating random permutations via distributed stochastic processes. In: Tarjan, R.E., Warnow, T. (eds.) 10th SODA, January 1999, pp. 271–280. ACM-SIAM (1999)

8. Czumaj, A., Kanarek, P., Kutylowski, M., Lorys, K.: Fast generation of random permutations via networks simulation. In: European Symposium on Algorithms, pp. 246–260 (1996)

9. Czumaj, A., Kutylowski, M.: Delayed path coupling and generating random permutations. Random Struct. Algorithms **17**, 238–259 (2000)
10. Granboulan, L., Pornin, T.: Perfect block ciphers with small blocks. In: Biryukov, A. (ed.) FSE 2007. LNCS, vol. 4593, pp. 452–465. Springer, Heidelberg (2007). doi:10.1007/978-3-540-74619-5_28
11. Hoang, V.T., Morris, B., Rogaway, P.: An enciphering scheme based on a card shuffle. In: Safavi-Naini, R., Canetti, R. (eds.) CRYPTO 2012. LNCS, vol. 7417, pp. 1–13. Springer, Heidelberg (2012). doi:10.1007/978-3-642-32009-5_1
12. Kocher, P.C.: Timing attacks on implementations of diffie-hellman, RSA, DSS, and other systems. In: Koblitz, N. (ed.) CRYPTO 1996. LNCS, vol. 1109, pp. 104–113. Springer, Heidelberg (1996). doi:10.1007/3-540-68697-5_9
13. Levin, D.A., Peres, Y., Wilmer, E.L.: Markov Chains and Mixing Times. American Mathematical Society (2006)
14. Luchaup, D., Dyer, K.P., Jha, S., Ristenpart, T., Shrimpton, T.: LibFTE: a toolkit for constructing practical, format-abiding encryption schemes. In: Proceedings of the 23rd USENIX Security Symposium, pp. 877–891 (2014)
15. Luchaup, D., Shrimpton, T., Ristenpart, T., Jha, S.: Formatted encryption beyond regular languages. In: Ahn, G.J., Yung, M., Li, N. (eds.) ACM CCS 14, November 2014, pp. 1292–1303. ACM Press (2014)
16. Morris, B., Rogaway, P.: Sometimes-recurse shuffle. In: Nguyen, P.Q., Oswald, E. (eds.) EUROCRYPT 2014. LNCS, vol. 8441, pp. 311–326. Springer, Heidelberg (2014). doi:10.1007/978-3-642-55220-5_18
17. Morris, B., Rogaway, P., Stegers, T.: How to encipher messages on a small domain. In: Halevi, S. (ed.) CRYPTO 2009. LNCS, vol. 5677, pp. 286–302. Springer, Heidelberg (2009). doi:10.1007/978-3-642-03356-8_17
18. Ristenpart, T., Yilek, S.: The mix-and-cut shuffle: small-domain encryption secure against $N$ queries. In: Canetti, R., Garay, J.A. (eds.) CRYPTO 2013. LNCS, vol. 8042, pp. 392–409. Springer, Heidelberg (2013). doi:10.1007/978-3-642-40041-4_22
19. Sinclair, A.: Algorithms for Random Generation and Counting. Progress in Theoretical Computer Science. Birkhäuser, Boston (1993)

# Mathematical Analysis II

# Optimization of **LPN** Solving Algorithms

Sonia Bogos$^{(\boxtimes)}$ and Serge Vaudenay

EPFL, 1015 Lausanne, Switzerland
soniamihaela.bogos@epfl.ch
http://lasec.epfl.ch

**Abstract.** In this article we focus on constructing an algorithm that automatizes the generation of **LPN** solving algorithms from the considered parameters. When searching for an algorithm to solve an **LPN** instance, we make use of the existing techniques and optimize their use. We formalize an **LPN** algorithm as a path in a graph $G$ and our algorithm is searching for the optimal paths in this graph. Our results bring improvements over the existing work, i.e. we improve the results of the covering code from ASIACRYPT'14 and EUROCRYPT'16. Furthermore, we propose concrete practical codes and a method to find good codes.

## 1 Introduction

The Learning Parity with Noise (LPN) problem can be seen as a noisy system of linear equations in the binary domain. More specifically, we have a secret $s$ and an adversary that has access to an LPN oracle which provides him tuples of uniformly distributed binary vectors $v_i$ and the inner product between $s$ and $v_i$ to which some noise was added. The noise is represented by a Bernoulli variable with a probability $\tau$ to be 1. The goal of the adversary is to recover the secret $s$. The LPN problem is a particular case of the well-known Learning with Errors (LWE) [33] problem where instead of working in $\mathbb{Z}_2$ we extend the work to a ring $\mathbb{Z}_q$.

The LPN problem is attractive as it is believed to be resistant to quantum computers. Thus, it can be a good candidate for replacing the number-theoretic problems such as factorization and discrete logarithm (which can be easily broken by a quantum algorithm). Also, given its structure, it can be implemented in lightweight devices. The LPN problem is used in the design of the $HB$-family of authentication protocols [10,19,23,24,26,30] and several cryptosystems base their security on its hardness [1,14–16,20,25].

***Previous Work.*** LPN is believed to be hard. So far, there is no reduction from hard lattice problems to certify the hardness (like in the case of LWE). Thus, the best way to assess its hardness is by trying to design and improve algorithms that

**Electronic supplementary material** The online version of this chapter (doi:10. 1007/978-3-662-53887-6_26) contains supplementary material, which is available to authorized users.

J.H. Cheon and T. Takagi (Eds.): ASIACRYPT 2016, Part I, LNCS 10031, pp. 703–728, 2016.
DOI: 10.1007/978-3-662-53887-6_26

solve it. Over the years, the LPN problem was analyzed and there exist several solving algorithms. The first algorithm to target LPN is the BKW algorithm [6]. This algorithm can be described as a Gaussian elimination on blocks of bits (instead on single bits) where the secret is recovered bit by bit. Several improvements appeared afterwards [18, 28]. One idea that improves the algorithm is the use of the fast Walsh-Hadamard transform as we can recover several bits of the secret at once. In their work, Levieil and Fouque [28] provide an analysis with the level of security achieved by different LPN instances and propose secure parameters. Using BKW as a black-box, Lyubashevsky [29] presents an LPN solving algorithm useful for the case when the number of queries is restricted to an adversary. The best algorithm to solve LPN was presented at ASIACRYPT'14 [22] and it introduces the use of the covering codes to improve the performance. Some problems in the computation of complexities were reported [7, 36]. As discussed by Bogos et al. [7] and in the ASIACRYPT presentation [22], the authors used a too optimistic approximation for the bias introduced by their new reduction method, the covering codes. Some complexity terms are further missing (as discussed in Sect. 2.2) or are not in bit operations. Also, no method to construct covering codes were suggested. At EUROCRYPT'16, Zhang et al. [36] proposed a way to construct good codes by concatenating perfect codes and improved the algorithms. However, some other problem in complexities were reported [9]. The new LF(4) reduction technique introduced by Zhang et al. [36] was also shown to be incorrect [9].

For the case when the secret is sparse, i.e. its Hamming weight is small, the classical Gaussian elimination proves to give better results [7, 8, 11].

The LPN algorithms consist of two parts: one in which the size of the secret is reduced and one in which part of the secret is recovered. Once a part of the secret is recovered, the queries are updated and the algorithm restarts to recover the rest of the secret. When trying to recover a secret $s$ of $k$ bits, it is assumed that $k$ can be written as $a \cdot b$, for $a, b \in \mathbb{N}$ (i.e. secret $s$ can be seen as $a$ blocks of $b$ bits). Usually all the reduction steps reduce the size by $b$ bits and the solving algorithm recovers $b$ bits. While the use of the same parameter, i.e. $b$, for all the operations may be convenient for the implementation, we search for an algorithm that may use different values for each reduction step. We discover that small variations from the fixed $b$ can bring important improvements in the time complexity of the whole algorithm.

***Our Contribution.*** In this work we first *analyze the existing* LPN *algorithms and study the operations that are used in order to reduce the size of the secret. We adjust the expressions of the complexities of each step* (as in some works they were underestimated in the literature). For instance, the results from Guo et al. [22] and Zhang et al. [36] are displayed with corrections in Table 1.[1] (Details for this computation are provided as an additional material for this paper.)

Second, we *improve the theory behind the covering code reduction and show the link with perfect and quasi-perfect codes.* Using the average bias of covering

---

[1] As for [36], we only reported the results based on LF2 which are better than with LF1, as the LF(4) operation is incorrect [9].

**Table 1.** Time complexity to solve LPN (in bit operations). These complexities are based on the formulas from our paper with the most favorable covering codes we constructed from our pool, with adjusted data complexity to reach a failure probability bounded by 33 %. Originally claimed complexities by [22,36] are under parentheses.

| $(k,\tau)$ | ASIACRYPT'14 [22] | EUROCRYPT'16 [36] | Our results |
|---|---|---|---|
| $(512, 0.125)$ | $2^{86.96}(2^{79.9})$ (proceedings) $2^{81.90}(2^{79.7})$ (presentation)[a] | $2^{80.09}(2^{74.73})$ | $2^{78.84}$ |
| $(532, 0.125)$ | $2^{88.62}(2^{81.82})$ | $2^{82.17}(2^{76.90})$ | $2^{81.02}$ |
| $(592, 0.125)$ | $2^{97.71}(2^{88.07})$ | $2^{89.32}(2^{83.84})$ | $2^{87.57}$ |

[a] http://des.cse.nsysu.edu.tw/asiacrypt2014/doc/1-1_Solving LPN Using Covering Codes.pdf

codes allows us to use arbitrary codes and even random ones. Using the algorithm to construct optimal concatenated codes based on a pool of elementary ones allows us to improve complexities. (In Guo et al. [22], only a hypothetical code was assumed to be close to a perfect code; in Zhang et al. [36], only the concatenation of perfect codes are used; in Table 1, our computed complexities are based on the real codes that we built with our bigger pool to have a fair comparison.)

Third, we *optimize the order and the parameters used by the operations that reduce the size of the secret such that we minimize the time complexity required. We design a "meta-algorithm" that combines the reduction steps and finds the optimal strategy to solve* LPN. We *automatize the process of finding* LPN *solving algorithms, i.e. given a random* LPN *instance, our algorithm provides the description of the steps that optimize the time complexity.* In our formalization we call such algorithms "optimal chains". We perform a security analysis of LPN based on the results obtained by our algorithm and compare our results with the existing ones. We discover that we improve the complexity compared with the existing results [7,22,28,36], as shown in Table 1.

***Preliminaries and Notations.*** Given a domain $\mathcal{D}$, we denote by $x \xleftarrow{U} \mathcal{D}$ the fact that $x$ is drawn uniformly at random from $\mathcal{D}$. By $Ber_\tau$ we denote the Bernoulli distribution with parameter $\tau$. By $Ber_\tau^k$ we denote the binomial distribution with parameters $k$ and $\tau$. Let $\langle \cdot, \cdot \rangle$ denote the inner product, $\mathbb{Z}_2 = \{0,1\}$ and $\oplus$ denote the bitwise XOR. The Hamming weight of a vector $v$ is denoted by $\mathsf{HW}(v)$.

***Organization.*** In Sect. 2 we formally define the LPN problem and describe the main tools used to solve it. We carefully analyze the complexity of each step and show in footnote where it differs from the existing literature. Section 3 studies the failure probability of the entire algorithm and validates the use of the average bias in the analysis. Section 4 introduces the bias computation for perfect and quasi-perfect codes. We provide an algorithm to find good codes. The algorithm that searches the optimal strategy to solve LPN is presented in Sects. 5 and 6. We illustrate and compare our results in Sect. 7 and conclude in Sect. 8. We put in additional material details of our results: the complete list of the chains we obtain (for Tables 3 and 4), an example of complete solving algorithm, the random codes that we use for the covering code reduction, and an analysis of the results from [22, 36] to obtain Table 1.

## 2   LPN

### 2.1   LPN Definition

The LPN problem can be seen as a noisy system of equations in $\mathbb{Z}_2$ where one is asked to recover the unknown variables. Below, we present the formal definition.

**Definition 1 (LPN oracle).** *Let* $s \xleftarrow{U} \mathbb{Z}_2^k$, *let* $\tau \in ]0, \frac{1}{2}[$ *be a constant noise parameter and let* $\mathsf{Ber}_\tau$ *be the Bernoulli distribution with parameter* $\tau$. *Denote by* $D_{s,\tau}$ *the distribution defined as*

$$\{(v, c) \mid v \xleftarrow{U} \mathbb{Z}_2^k, c = \langle v, s \rangle \oplus d, d \leftarrow \mathsf{Ber}_\tau\} \in \mathbb{Z}_2^{k+1}.$$

*An* LPN *oracle* $\mathcal{O}_{s,\tau}^{\mathsf{LPN}}$ *is an oracle which outputs independent random samples according to* $D_{s,\tau}$.

**Definition 2 (Search LPN problem).** *Given access to an* LPN *oracle* $\mathcal{O}_{s,\tau}^{\mathsf{LPN}}$, *find the vector* $s$. *We denote by* $\mathsf{LPN}_{k,\tau}$ *the* LPN *instance where the secret has size* $k$ *and the noise parameter is* $\tau$. *Let* $k' \leq k$. *We say that an algorithm* $\mathcal{M}$ $(n, t, m, \theta, k')$-*solves the search* $\mathsf{LPN}_{k,\tau}$ *problem if*

$$\Pr[\mathcal{M}^{\mathcal{O}_{s,\tau}^{\mathsf{LPN}}}(1^k) = (s_1 \dots s_{k'}) \mid s \xleftarrow{U} \mathbb{Z}_2^k] \geq \theta,$$

*and* $\mathcal{M}$ *runs in time* $t$, *uses memory* $m$ *and asks at most* $n$ *queries from the* LPN *oracle.*

Remark that we consider here the problem of recovering only a part of the secret. Throughout the literature this is how the LPN problem is formulated. The reason for doing so is that the recovery of the first $k'$ bits dominates the overall complexity. Once we recover part of the secret, the new problem of recovering a shorter secret of $k - k'$ bits is easier.

The LPN problem has a decisional form where one has to distinguish between random vectors of size $k + 1$ and the samples from the LPN oracle. In this paper we are interested only in finding algorithms for the search version.

We define $\delta = 1 - 2\tau$. We call $\delta$ the bias of the error bit $d$. We have $\delta = E((-1)^d)$, with $E(\cdot)$ the expected value. We denote the bias of the secret bits by $\delta_s$. As $s$ is a uniformly distributed random vector, at the beginning we have $\delta_s = 0$.

## 2.2   Reduction and Solving Techniques

Depending on how many queries are given from the LPN oracle, the LPN solving algorithms are split in 3 categories. With a *linear number of queries*, the best algorithms are exponential, i.e. with $n = \Theta(k)$ the secret is recovered in $2^{\Theta(k)}$ time [31, 35]. Given a *polynomial number of queries* $n = k^{1+\eta}$, with $\eta > 0$, one can solve LPN with a sub-exponential time complexity of $2^{\mathcal{O}(\frac{k}{\log \log k})}$ [29]. When $\tau = \frac{1}{\sqrt{k}}$ we can improve this result and have a complexity of $e^{\frac{1}{2}\sqrt{k}(\ln k)^2 + \mathcal{O}(\sqrt{k}\ln k)}$ [8]. The complexity improves but remains in the sub-exponential range with a *sub-exponential number of queries*. For this category, we have the BKW [6], LF1, LF2 [28], FMICM [18] and the covering code algorithm [22, 36]. All these algorithms solve LPN with a time complexity of $2^{\mathcal{O}(\frac{k}{\log k})}$ and require $2^{\mathcal{O}(\frac{k}{\log k})}$ queries. In the special case when the noise is sparse, a simple Gaussian elimination can be used for the recovery of the secret [7, 11]. LF2, covering code or the Gaussian elimination prove to be the best one, depending on the noise level [7].

   All these algorithms have a common structure: given an $\mathsf{LPN}_{k,\tau}$ instance with a secret $s$, they reduce the original LPN problem to a new LPN problem where the secret $s'$ is of size $k' \leq k$ by applying several *reduction techniques*. Then, they recover $s'$ using a *solving method*. The queries are updated and the process is repeated until the whole secret $s$ is recovered. We present here the list of reduction and solving techniques used in the existing LPN solving algorithms. In the next section, we combine the reduction techniques such that we find the optimal reduction phases for solving different LPN instances.

   We assume for all the reduction steps that we start with $n$ queries, that the size of the secret is $k$, the bias of the secret bits is $\delta_s$ and the bias of the noise bits is $\delta$. After applying a reduction step, we will end up with $n'$ queries, size $k'$ and biases $\delta'$ and $\delta'_s$. Note that $\delta_s$ averages over all secrets although the algorithm runs with one target secret. As it will be clear below, the complexity of all reduction steps only depends on $k$, $n$, and the parameters of the steps but not on the biases. Actually, only the probability of success is concerned with biases. We see in Sect. 3 that the probability of success of the overall algorithm is not affected by this approach. Actually, we will give a formula to compute a value which approximates the average probability of success over the key based on the average bias.

   We have the following reduction steps:

- *sparse-secret* changes the secret distribution. In the formal definition of LPN, we take the secret $s$ to be a random row vector of size $k$. When other reduction steps or the solving phase depends on the distribution of $s$, one can transform an LPN instance with a random $s$ to a new one where $s$ has the same distribution as the initial noise, i.e. $s \leftarrow \mathsf{Ber}_\tau^k$. The reduction performs the following steps: from the $n$ queries select $k$ of them: $(v_{i_1}, c_{i_1}), \ldots, (v_{i_k}, c_{i_k})$ where the row vectors $v_{i_j}$, with $1 \leq j \leq k$, are linearly independent. Construct the matrix $M$ as $M = [v_{i_1}^T \cdots v_{i_k}^T]$ and rewrite the $k$ queries as $sM + d' = c'$, where

$d' = (d_{i_1}, \ldots, d_{i_k})$. With the rest of $n - k$ queries we do the following:

$$c'_j = \langle v_j (M^T)^{-1}, c' \rangle \oplus c_j = \langle v_j (M^T)^{-1}, d' \rangle \oplus d_j = \langle v'_j, d' \rangle \oplus d_j$$

We have $n - k$ new queries $(v'_j, c'_j)$ where the secret is now $d'$. In Guo et al. [22], the authors use an algorithm which is inappropriately called "the four Russians algorithm" [2]. This way, the complexity should be of $\mathcal{O}\left(\min_{\chi \in \mathbb{N}} \left(kn' \lceil \frac{k}{\chi} \rceil + k^3 + k\chi 2^\chi \right)\right)$.[2] Instead, the Bernstein algorithm [4] works in $\mathcal{O}\left(\frac{n' k^2}{\log_2 k - \log_2 \log_2 k} + k^2\right)$. We use the best of the two, depending on the parameters. Thus, we have:

---

*sparse-secret* : $k' = k$; $n' = n - k$; $\delta' = \delta$; $\delta'_s = \delta$

Complexity: $\mathcal{O}\left(\min_{\chi \in \mathbb{N}} \left(\frac{n' k^2}{\log_2 k - \log_2 \log_2 k} + k^2, kn' \lceil \frac{k}{\chi} \rceil + k^3 + k\chi 2^\chi \right)\right)$

---

- *xor-reduce*($b$) was first used by the LF2 algorithm. The queries are grouped in equivalence classes according to the values on $b$ random positions. In each equivalence class, we perform the xoring of every pair of queries. The size of the secret is reduced by $b$ bits and the new bias is $\delta^2$. The expected new number of queries is $E(\sum_{i<j} 1_{v_i}$ matches $v_j$ on the $b$-bit block$) = \frac{n(n-1)}{2^{b+1}}$ which improves previous results[3]. When $n \approx 1 + 2^{b+1}$, the number of queries are maintained. For $n > 1 + 2^{b+1}$, the number of queries will increase.

---

*xor-reduce*($b$) : $k' = k - b$; $n' = \frac{n(n-1)}{2^{b+1}}$; $\delta' = \delta^2$; $\delta'_s = \delta_s$

Complexity: $\mathcal{O}(k \cdot \max(n, n'))$

---

- *drop-reduce*($b$) is a reduction used only by the BKW algorithm. It consists in dropping all the queries that are not 0 on a window of $b$ bits. Again, these $b$ positions are chosen randomly. In average, we expect that half of the queries are 0 on a given position. For $b$ bits, we expect to have $\frac{n}{2^b}$ queries that are 0 on this window. The bias is unaffected and the secret is reduced by $b$ bits.

---

*drop-reduce*($b$) : $k' = k - b$; $n' = \frac{n}{2^b}$; $\delta' = \delta$; $\delta'_s = \delta_s$

Complexity: $\mathcal{O}(n(1 + \frac{1}{2} + \ldots + \frac{1}{2^{b-1}}))$

---

The complexity of $n(1 + \frac{1}{2} + \ldots + \frac{1}{2^{b-1}}) = \mathcal{O}(n)$ comes from the fact that we don't need to check all the $b$ bits: once we find a 1 we don't need to continue and just drop the corresponding query.

- *code-reduce*($k, k'$, params) is a method used by the covering code algorithm presented in ASIACRYPT'14. In order to reduce the size of the secret, one uses a linear code $[k, k']$ (which is defined by params) and approximates the $v_i$ vectors to the nearest codeword $g_i$. We assume that decoding is done in linear time

---

[2] But the $k^3 + k\chi 2^\chi$ terms is missing in [22].

[3] In Bogos et al. [7], the number of queries was approximated to $\frac{\frac{n}{2^b}\left(\frac{n}{2^b} - 1\right)}{2}$ which is less favorable.

for the code considered. (For the considered codes, decoding is indeed based on table look-ups.) The noisy inner product becomes:

$$\langle v_i, s \rangle \oplus d_i = \langle g_i' G, s \rangle \oplus \langle v_i - g_i, s \rangle \oplus d_i$$
$$= \langle g_i', sG^T \rangle \oplus \langle v_i - g_i, s \rangle \oplus d_i$$
$$= \langle g_i', s' \rangle \oplus d_i',$$

where $G$ is the generator matrix of the code, $g_i = g_i' G$, $s' = sG^T \in \{0,1\}^{k'}$ and $d_i' = \langle v_i - g_i, s \rangle \oplus d_i$. We denote $\mathsf{bc} = E((-1)^{\langle v_i - g_i, s \rangle})$ the bias of $\langle v_i - g_i, s \rangle$. We will see in Sect. 4 how to construct a $[k, k']$ linear code making $\mathsf{bc}$ as large as possible.

Here, $\mathsf{bc}$ averages the bias over the secret although $s$ is fixed by *sparse-secret*. It gives the correct average bias $\delta$ over the distribution of the key. We will see that it allows to approximate the expected probability of success of the algorithm.

By this transform, no query is lost.

> *code-reduce*$(k, k', \mathsf{params})$ : $k'$; $n' = n$; $\delta' = \delta \cdot \mathsf{bc}$
> $\delta_s'$ depends on $\delta_s$ and $G$
> Complexity: $\mathcal{O}(kn)$

The way $\delta_s'$ is computed is a bit more complicated than for the other types of reductions. However, $\delta_s$ only plays a role in the *code-reduce* reduction, and we will not consider algorithms that use more than one *code-reduce* reduction.

It is easy to notice that with each reduction operation the number of queries decreases or the bias is getting smaller. In general, for solving LPN, one tries to lose as few queries as possible while maintaining a large bias. We will study in the next section what is a good combination of using these reductions.

After applying the reduction steps, we assume we are left with an LPN$_{k', \delta'}$ instance where we have $n'$ queries. The original BKW algorithm was using a final solving technique based on majority decoding. Since the LF2 algorithm, we use a better solving technique based on the Walsh Hadamard Transform (WHT).

WHT recovers a block of the secret by computing the fast Walsh Hadamard transform on the function $f(x) = \sum_i 1_{v_i = x}(-1)^{\langle v_i, s \rangle \oplus d_i}$. The Walsh-Hadamard transform is

$$\hat{f}(\nu) = \sum_x (-1)^{\langle \nu, x \rangle} f(x) = \sum_i (-1)^{\langle v_i, s + \nu \rangle \oplus d_i}$$

For $\nu = s$, we have $\hat{f}(s) = \sum_i (-1)^{d_i}$. For a positive bias, we know that most of the noise bits are set to 0. It is the opposite when the bias is negative. So, $|\hat{f}(s)|$ is large and we suppose it is the largest value in the table of $\hat{f}$. Using again the Chernoff bounds, we need to have $n' = 8 \ln(\frac{2^{k'}}{\theta}) \delta'^{-2}$ [7] queries in order to bound the probability of guessing wrongly the $k'$-bit secret by $\theta$. We can improve further by applying directly the Central Limit Theorem and obtain a heuristic bound $\varphi(-\sqrt{\frac{n'}{2\delta'^{-2}-1}}) \leq 1 - (1-\theta)^{\frac{1}{2^{k'}-1}}$, where $\varphi(x) = \frac{1}{2} + \frac{1}{2}\text{erf}(\frac{x}{\sqrt{2}})$ and erf is the Gauss error function. We obtain that

$$\sqrt{n'} \geq -\sqrt{2\delta'^{-2}-1} \cdot \varphi^{-1}\left(1 - (1-\theta)^{\frac{1}{2^{k'}-1}}\right). \tag{1}$$

We can derive the approximation of Selçuk [34] that $n' \geq 4\ln(\frac{2^{k'}}{\theta})\delta'^{-2}$. We give the details of our results in Sect. 3. Complexity of the WHT$(k')$ is $\mathcal{O}(k'2^{k'}\frac{\log_2 n'+1}{2} + k'n')$ as we use the fast Walsh Hadamard Transform[4],[5].

---

WHT$(k')$;

Requires $\sqrt{n'} \geq -\sqrt{2\delta'^{-2}-1} \cdot \varphi^{-1}\left(1 - (1-\theta)^{\frac{1}{2^{k'}-1}}\right)$

Complexity: $\mathcal{O}(k'2^{k'}\frac{\log_2 n'+1}{2} + k'n')$

---

Given the reduction and the solving techniques, an LPN$_{k,\tau}$ solving algorithm runs like this: we start with a $k$-bit secret and with $n$ queries from the LPN oracle. We reduce the size of the secret by applying several reduction steps and we end up with $n'$ queries where the secret has size $k'$. We use one solving method, e.g. the WHT, and recover the $k'$-bit secret with a probability of failure bounded by $\theta$. We chose $\theta = \frac{1}{3}$. We have recovered a part of the secret. To fully recover the whole secret, we update the queries and start another chain to recover more bits, and so on until the remaining $k - k'$ bits are found. For the second part of the secret we will require for the failure probability to be $\theta^2$ and for the $i^{th}$ part it will be $\theta^i$. Thus, if we recover the whole secret in $i$ iterations, the total failure probability will be bounded by $\theta + \theta^2 + \cdots + \theta^i$. Given that we take $\theta = \frac{1}{3}$, we recover the whole secret with a success probability larger than 50 %. Experience shows that the time complexity for the first iteration dominates the total complexity.

As we can see in the formulas of each possible step, the computations of $k'$, $n'$, and of the complexity do not depend on the secret weight. Furthermore, the computation of biases is always linear. So, the correct average bias (over the distribution of the key made by the *sparse-secret* transform) is computed. Only the computation of the success probability is non-linear but we discuss about this in the next section. As it only matters in WHT, we will see in Sect. 3 that the approximation is justified.

---

[4] The second term $k'n'$ illustrates the cost of constructing the function $f$. In cases where $n' > 2^{k'}$ this is the dominant term and it should not be ignored. This was missing in several works [7,22]. For the instance LPN$_{592,0.125}$ from Guo et al. [22] this makes a big difference as $k' = 64$ and $n' = 2^{69}$; the complexity of WHT with the second term is $2^{75}$ vs $2^{70}$ [22]. Given that is must be repeated $2^{13}$ (as 35 bits of the secret are guessed), the cost of WHT is $2^{88}$.

[5] Normally, the values $\hat{f}(\nu)$ have an order of magnitude of $\sqrt{n'}$ so we have $\frac{1}{2}\log_2 n'$ bits.

## 3   On Approximating the Probability of Success

*Approximating n by using Central Limit Theorem.* In order to approximate
the number of queries needed to solve the LPN instance we consider when the
Walsh Hadamard Transform fails to give the correct secret. We first assume
that the bias is positive. We have a failure when for another $\bar{s} \neq s$, we have
that $\hat{f}(\bar{s}) > \hat{f}(s)$. Following the analysis from [7], we let $y = A'\bar{s}^T + c'^T$ and
$d' = A's^T + c'^T$. We have $\hat{f}(\bar{s}) = \sum_i (-1)^{y_i} = n' - 2.\mathsf{HW}(y)$ and similarly,
$\hat{f}(s) = n' - 2.\mathsf{HW}(d')$. So, $\hat{f}(\bar{s}) > \hat{f}(s)$ translates to $\mathsf{HW}(y) \leq \mathsf{HW}(d')$. Therefore

$$\Pr[\hat{f}(\bar{s}) > \hat{f}(s)] = \Pr\left[\sum_{i=1}^{n'} (y_i - d'_i) \leq 0\right].$$

For each $\bar{s}$, we take $y$ as a uniformly distributed random vector and we let $\delta'(s)$
be the bias introduce with a fixed $s$ for $d'_i$ (we recall that our analysis computes
$\delta' = E(\delta'(s))$ over the distribution of $s$). Let $X_1, \ldots, X_{n'}$ be random variable
corresponding to $X_i = y_i - d'_i$. Since $E(y_i) = \frac{1}{2}$, $E(d'_i) = \frac{1}{2} - \frac{\delta'(s)}{2}$ and $y_i$ and $d'_i$
are independent, we have that $E(X_i) = \frac{\delta'(s)}{2}$ and $\mathsf{Var}(X_i) = \frac{2 - \delta'(s)^2}{4}$. By using
the Central Limit Theorem we obtain that

$$\Pr[X_1 + \ldots + X_{n'} \leq 0] \approx \varphi(Z(s)) \text{ with } Z(s) = -\frac{\delta'(s)}{\sqrt{2 - \delta'(s)^2}}\sqrt{n'}$$

where $\varphi$ can be calculated by $\varphi(x) = \frac{1}{2} + \frac{1}{2}\mathsf{erf}(\frac{x}{\sqrt{2}})$ and erf is the Gauss error
function. For $\delta'(s) < 0$, the same analysis with $\hat{f}(\bar{s}) < \hat{f}(s)$ gives the same result.
Applying the reasoning for any $s' \neq s$ we obtain that the failure probability is

$$p(s) = 1 - (1 - \varphi(Z(s)))^{2^{k'}-1}, \text{if } \delta'(s) > 0$$
$$\text{and } p(s) = 1 - \frac{1}{2^{k'}}, \text{if } \delta'(s) \leq 0.$$

We deduce the following (for $\theta < \frac{1}{2}$)

$$p(s) \leq \theta \Leftrightarrow \sqrt{n'} \geq -\sqrt{2\delta'(s)^{-2} - 1}\varphi^{-1}\left(1 - (1-\theta)^{\frac{1}{2^{k'}-1}}\right) \text{ and } \delta'(s) > 0$$

As a condition for our WHT step, we adopt the inequality in which we replace
$\delta'(s)$ by $\delta'$. We give a heuristic argument below to show that it implies $E(p(s)) \leq$
$\theta$, which is what we want.

Note that if we use the approximation $\varphi(Z) \approx -\frac{1}{Z\sqrt{2\pi}}e^{-\frac{Z^2}{2}}$ for $Z \to -\infty$,
we obtain the condition $n' \geq 2(2\delta'^{-2} - 1)\ln(\frac{2^{k'}-1}{\theta})$. So, our analysis brings an
improvement of factor two over the Hoeffding bound method used by Bogos
et al. [7] that requires $n' \geq 8\delta'^{-2}\ln(\frac{2^{k'}}{\theta})$.

S. Bogos and S. Vaudenay

*On the validity of the using the bias average.* The above computation is correct when using $\delta'(s)$ but we use $\delta' = E(\delta'(s))$ instead. If no *code-reduce* step is used, $\delta'(s)$ does not depend on $s$ and we do have $\delta'(s) = \delta'$. However, when a *code-reduce* is used, the bias depends on the secret which is obtained after the *sparse-secret* step. For simplicity, we let $s$ denote this secret. The bias $\delta'(s)$ is actually of form $\delta'(s) = \delta^{2^x} \mathrm{bc}(s)$ where $x$ is the number of *xor-reduce* steps and $\mathrm{bc}(s)$ is the bias introduced by *code-reduce* depending on $s$. The values of $\delta'(s)$, $Z(s)$, and $p(s)$ are already defined above. We define $Z = -\frac{\delta'}{\sqrt{2-\delta'^2}}\sqrt{n'}$ and $p = 1 - (1 - \varphi(Z))^{2^{k'}-1}$. Clearly, $E(p(s))$ is the average failure probability over the distribution of the secret obtained after *sparse-secret*.

Our method ensures that $\delta' = E(\delta'(s))$ over the distribution of $s$. Since $\delta'$ is typically small (after a few *xor-reduce* steps, $\delta^{2^x}$ is indeed very small), we can consider $Z(s)$ as a linear function of $\delta'(s)$ and have $E(Z(s)) \approx Z$. This is confirmed by experiment. We make the **heuristic approximation** that

$$E\left(1 - (1 - \varphi(Z(s)))^{2^{k'}-1}\right) \approx 1 - (1 - \varphi(E(Z(s))))^{2^{k'}-1} \approx 1 - (1 - \varphi(Z))^{2^{k'}-1}$$

So, $E(p(s)) \approx p$.[6]

We did some experiments based on some examples in order to validate our heuristic assumption. Our results show indeed that $E(Z(s)) \approx Z$. There is a small gap between $E(p(s))$ and $p$ but this does not affect our results. Actually, we are in a phase transition region so any tiny change in the value of $n'$ makes $E(p(s))$ change a lot. We include our results in the additional material. Thus, ensuring that $p \leq \theta$ with the above analysis based on the average bias ensures that the expected failure probability to be bounded by $\theta$.

We also observed that the reduction *code-reduce* can introduce problems. More precisely, what can go wrong is that $s$ can have, with a given probability, a negative $\delta'(s)$ bias or a component in one of the concatenated codes giving a zero bias, making WHT to fail miserably.

## 4    Bias of the Code Reduction

In this section we present how to compute the bias introduced by a *code-reduce*. Recall that the reduction *code-reduce*$(k, k')$ introduces a new noise:

$$\langle v_i, s \rangle \oplus d_i = \langle g'_i, s' \rangle \oplus \langle v_i - g_i, s \rangle \oplus d_i,$$

where $g_i = g'_i G$ is the nearest codeword of $v_i$ and $s' = sG^T$. Note that $g_i$ is not necessarily unique, specially if the code is not perfect. We take $g_i = \mathrm{Decode}(v_i)$ obtained from an arbitrary decoding algorithm. Then the noise bc can be computed by the following formula:

---

[6]    Note that Zhang et al. [36] implicitly does the same assumption as they use the average bias as well.

$$\mathsf{bc} = E((-1)^{\langle v_i - g_i, s \rangle}) = \sum_{e \in \{0,1\}^k} \Pr[v_i - g_i = e]E((-1)^{\langle e,s \rangle})$$

$$= \sum_{w=0}^{k} \sum_{\substack{e \in \{0,1\}^k, \\ \mathsf{HW}(e)=w}} \Pr[v_i - g_i = e]\delta_s^w = E\left(\delta_s^{\mathsf{HW}(v_i - g_i)}\right)$$

for a $\delta_s$-sparse secret. (We recall that the *sparse-secret* reduction step randomizes the secret.) So, the probability space is over the distribution of $v_i$ *and* the distribution of $s$. Later, we consider $\mathsf{bc}(s) = E((-1)^{\langle v_i - g_i, s \rangle})$ over the distribution over $v_i$ only. (In the work of Guo et al. [22], only $\mathsf{bc}(s)$ is considered. In Zhang et al. [36], our bc was also considered.) In the last expression of bc, we see that the ambiguity in decoding does not affect bc as long as the Hamming distance $\mathsf{HW}(v_i - \mathsf{Decode}(v_i))$ is not ambiguous. This is a big advantage of averaging in bc as it allows to use non-perfect codes. From this formula, we can see that the decoding algorithm $v_i \to g_i$ making $\mathsf{HW}(v_i - g_i)$ minimal makes bc maximal. In this case, we obtain

$$\mathsf{bc} = E\left(\delta_s^{d(v_i, C)}\right), \tag{2}$$

where $C$ is the code and $d(v_i, C)$ denotes the Hamming distance of $v_i$ from $C$.

For a code $C$, the *covering radius* is $\rho = \max_v d(v, C)$. The *packing radius* is the largest radius $R$ such that the balls of this radius centered on all codewords are non-overlapping. So, the packing radius is $R = \lfloor \frac{D-1}{2} \rfloor$ where $D$ is the minimal distance. We further have $\rho \geq \lfloor \frac{D-1}{2} \rfloor$. A *perfect code* is characterized by $\rho = \lfloor \frac{D-1}{2} \rfloor$. A *quasi-perfect code* is characterized by $\rho = \lfloor \frac{D-1}{2} \rfloor + 1$.

**Theorem 1.** *We consider a $[k, k', D]$ linear code $C$, where $k$ is the length, $k'$ is the dimension, and $D$ is the minimal distance. For any integer $r$ and any positive bias $\delta_s$, we have*

$$\mathsf{bc} \leq 2^{k'-k} \sum_{w=0}^{r} \binom{k}{w} (\delta_s^w - \delta_s^{r+1}) + \delta_s^{r+1}$$

*where bc is a function of $\delta_s$ defined by (2). Equality for any $\delta_s$ such that $0 < \delta_s < 1$ implies that $C$ is perfect or quasi-perfect. In that case, the equality is reached when taking the packing radius $r = R = \lfloor \frac{D-1}{2} \rfloor$.*

By taking $r$ as the largest integer such that $\sum_{w=0}^{r} \binom{k}{w} \leq 2^{k-k'}$ (which is the packing radius $R = \lfloor \frac{D-1}{2} \rfloor$ for perfect and quasi-perfect codes), we can see that if a perfect $[k, k']$ code exists, it makes bc maximal. Otherwise, if a quasi-perfect $[k, k']$ code exists, it makes bc maximal.

*Proof.* Let decode be an optimal deterministic decoding algorithm. The formula gives us that

$$\mathsf{bc} = 2^{-k} \sum_{g \in C} \sum_{v \in \mathsf{decode}^{-1}(g)} \delta_s^{\mathsf{HW}(v-g)}$$

We define $\mathsf{decode}_w^{-1}(g) = \{v \in \mathsf{decode}^{-1}(g); \mathsf{HW}(v - g) = w\}$ and $\mathsf{decode}_{>r}^{-1}(g)$ the union of all $\mathsf{decode}_w^{-1}(g)$ for $w > r$. For all $r$, we have

$$\sum_{v \in \mathsf{decode}^{-1}(g)} \delta_s^{\mathsf{HW}(v-g)}$$

$$= \sum_{w=0}^{r} \binom{k}{w} \delta_s^w + \sum_{w=0}^{r} \left( \#\mathsf{decode}_w^{-1}(g) - \binom{k}{w} \right) \delta_s^w + \sum_{w>r} \delta_s^w \#\mathsf{decode}_w^{-1}(g)$$

$$\leq \sum_{w=0}^{r} \binom{k}{w} \delta_s^w + \sum_{w=0}^{r} \left( \#\mathsf{decode}_w^{-1}(g) - \binom{k}{w} \right) \delta_s^w + \delta_s^{r+1} \#\mathsf{decode}_{>r}^{-1}(g)$$

$$\leq \sum_{w=0}^{r} \binom{k}{w} \delta_s^w + \delta_s^{r+1} \left( \#\mathsf{decode}^{-1}(g) - \sum_{w=0}^{r} \binom{k}{w} \right)$$

where we used $\delta_s^w \leq \delta_s^{r+1}$ for $w > r$, $\#\mathsf{decode}_w^{-1}(g) \leq \binom{k}{w}$ and $\delta_s^w \geq \delta_s^{r+1}$ for $w \leq r$. We further have equality if and only if the ball centered on $g$ of radius $r$ is included in $\mathsf{decode}^{-1}(g)$ and the ball of radius $r + 1$ contains $\mathsf{decode}^{-1}(g)$. By summing over all $g \in C$, we obtain the result.

So, the equality case implies that the packing radius is at least $r$ and the covering radius is at most $r + 1$. Hence, the code is perfect or quasi-perfect. Conversely, if the code is perfect or quasi-perfect and $r$ is the packing radius, we do have equality.                                                                             □

So, for quasi-perfect codes, we can compute

$$\mathsf{bc} = 2^{k'-k} \sum_{w=0}^{R} \binom{k}{w} (\delta_s^w - \delta_s^{R+1}) + \delta_s^{R+1} \tag{3}$$

with $R = \lfloor \frac{D-1}{2} \rfloor$. For perfect codes, the formula simplifies to

$$\mathsf{bc} = 2^{k'-k} \sum_{w=0}^{R} \binom{k}{w} \delta_s^w \tag{4}$$

## 4.1   Bias of a Repetition Code

Given a $[k, 1]$ repetition code, the optimal decoding algorithm is the majority decoding. We have $D = k$, $k' = 1$, $R = \lfloor \frac{k-1}{2} \rfloor$. For $k$ odd, the code is perfect so $\rho = R$. For $k$ even, the code is quasi-perfect so $\rho = R + 1$. Using (3) we obtain

$$\mathsf{bc} = \begin{cases} \sum_{w=0}^{\frac{k-1}{2}} \frac{1}{2^{k-1}} \binom{k}{w} \delta_s^w & \text{if } k \text{ is odd} \\ \\ \sum_{w=0}^{\frac{k}{2}-1} \frac{1}{2^{k-1}} \binom{k}{w} \delta_s^w + \frac{1}{2^k} \binom{k}{k/2} \delta_s^{\frac{k}{2}} & \text{if } k \text{ is even} \end{cases}$$

We give below the biases obtained for some $[k, 1]$ repetition codes.

| $[k, 1]$ | Bias |
|---|---|
| $[1, 2]$ | $\frac{1}{2}\delta_s + \frac{1}{2}$ |
| $[3, 1]$ | $\frac{3}{4}\delta_s + \frac{1}{4}$ |
| $[4, 1]$ | $\frac{3}{8}\delta_s^2 + \frac{1}{2}\delta_s + \frac{1}{8}$ |
| $[5, 1]$ | $\frac{5}{8}\delta_s^2 + \frac{5}{16}\delta_s + \frac{1}{16}$ |
| $[6, 1]$ | $\frac{5}{16}\delta_s^3 + \frac{15}{32}\delta_s^2 + \frac{3}{16}\delta_s + \frac{1}{32}$ |
| $[7, 1]$ | $\frac{35}{64}\delta_s^3 + \frac{21}{64}\delta_s^2 + \frac{7}{64}\delta_s + \frac{1}{64}$ |
| $[8, 1]$ | $\frac{35}{128}\delta_s^4 + \frac{7}{16}\delta_s^3 + \frac{7}{32}\delta_s^2 + \frac{1}{16}\delta_s + \frac{1}{128}$ |
| $[9, 1]$ | $\frac{63}{128}\delta_s^4 + \frac{21}{64}\delta_s^3 + \frac{9}{64}\delta_s^2 + \frac{9}{256}\delta_s + \frac{1}{256}$ |
| $[10, 1]$ | $\frac{63}{256}\delta_s^5 + \frac{105}{256}\delta_s^4 + \frac{15}{64}\delta_s^3 + \frac{45}{512}\delta_s^2 + \frac{5}{256}\delta_s + \frac{1}{512}$ |

## 4.2  Bias of a Perfect Code

In previous work [22,36], the authors assume a perfect code. In this case, $\sum_{w=0}^{R} \binom{k}{w} = 2^{k-k'}$ and we can use (4) to compute bc. There are not so many binary linear codes which are perfect. Except the repetition codes with odd length, the only ones are the trivial codes $[k, k, 1]$ with $R = \rho = 0$ and bc $= 1$, the Hamming codes $[2^{\ell} - 1, 2^{\ell} - \ell - 1, 3]$ for $\ell \geq 2$ with $R = \rho = 1$, and the Golay code $[23, 12, 7]$ with $R = \rho = 3$.

For the Hamming codes, we have

$$\mathsf{bc} = 2^{-\ell} \sum_{w=0}^{1} \binom{2^{\ell} - 1}{w} \delta_s^w = \frac{1 + (2^{\ell} - 1)\delta_s}{2^{\ell}}$$

For the Golay code, we obtain

$$\mathsf{bc} = 2^{-11} \sum_{w=0}^{3} \binom{23}{w} \delta_s^w = \frac{1 + 23\delta_s + 253\delta_s^2 + 1771\delta_s^3}{2^{11}}$$

Formulae (2), (3) and (4) for bc are new. Previously [7,22], the value $\mathsf{bc}_w$ of $\mathsf{bc}(s)$ for any $s$ of Hamming weight $w$ was approximated to

$$\mathsf{bc}_w = 1 - 2\frac{1}{S(k, \rho)} \sum_{\substack{i \leq \rho, \\ i \text{ odd}}} \binom{w}{i} S(k - w, \rho - i),$$

where $w$ is the Hamming weight of the $k$-bit secret and $S(k', \rho)$ is the number of $k'$-bit strings with weight at most $\rho$. Intuitively the formula counts the number of $v_i - g_i$ that produce an odd number of xor with the 1's of the secret. (See [7,22].) So, Guo et al. [22] assumes a fixed value for the weight $w$ of the secret and considers the probability that $w$ is not correct. If $w$ is lower, the actual bias is larger but if $w$ is larger, the computed bias is overestimated and the algorithm fails.

For instance, with a $[3,1]$ repetition code, the correct bias is $\mathsf{bc} = \frac{3}{4}\delta_s + \frac{1}{4}$ following our formula. With a fixed $w$, it is of $\mathsf{bc}_w = 1 - \frac{w}{2}$ [7,22]. The probability of $w$ to be correct is $\binom{k}{w}\tau^w(1-\tau)^{k-w}$. We take the example of $\tau = \frac{1}{3}$ so that $\delta_s = \frac{1}{3}$.

| $w$ | $\mathsf{bc}_w$ | $\Pr[w]$ | $\Pr[w], \tau = \frac{1}{3}$ |
|---|---|---|---|
| 0 | 1 | $(1-\tau)^3$ | 0.2963 |
| 1 | $\frac{1}{2}$ | $3\tau(1-\tau)^2$ | 0.4444 |
| 2 | 0 | $3\tau^2(1-\tau)$ | 0.2222 |
| 3 | $-\frac{1}{2}$ | $\tau^3$ | 0.0370 |

So, by taking $w = 1$, we have $\delta = \mathsf{bc}_w = \frac{1}{2}$ but the probability of failure is about $\frac{1}{4}$. Our approach uses the average bias $\delta = \mathsf{bc} = \frac{1}{2}$.

## 4.3    Using Quasi-perfect Codes

If $C'$ is a $[k-1, k', D]$ perfect code with $k' > 1$ and if there exists some codewords of odd length, we can extend $C'$, i.e., add a parity bit and obtain a $[k, k']$ code $C$. Clearly, the packing radius of $C$ is at least $\lfloor \frac{D-1}{2} \rfloor$ and the covering radius is at most $\lfloor \frac{D-1}{2} \rfloor + 1$. For $k' > 1$, there is up to one possible length for making a perfect code of dimension $k'$. So, $C$ is a quasi-perfect, its packing radius is $\lfloor \frac{D-1}{2} \rfloor$ and its covering radius is $\lfloor \frac{D-1}{2} \rfloor + 1$.

If $C'$ is a $[k+1, k', D]$ perfect code with $k' > 1$, we can puncture it, i.e., remove one coordinate by removing one column from the generating matrix. If we chose to remove a column which does not modify the rank $k'$, we obtain a $[k, k']$ code $C$. Clearly, the packing radius of $C$ is at least $\lfloor \frac{D-1}{2} \rfloor - 1$ and the covering radius is at most $\lfloor \frac{D-1}{2} \rfloor$. For $k' > 1$, there is up to one possible length for making a perfect code of dimension $k'$. So, $C$ is a quasi-perfect, its packing radius is $\lfloor \frac{D-1}{2} \rfloor - 1$ and its covering radius is $\lfloor \frac{D-1}{2} \rfloor$.

Hence, we can use extended Hamming codes $[2^\ell, 2^\ell - \ell - 1]$ with packing radius 1 for $\ell \geq 3$, punctured Hamming codes $[2^\ell - 2, 2^\ell - \ell - 1]$ with packing radius 0 for $\ell \geq 3$, the extended Golay code $[24, 12]$ with packing radius 3, and the punctured Golay code $[22, 12]$ with packing radius 2.

There actually exist many constructions for quasi-perfect linear binary codes. We list a few in Table 2. We took codes listed in the existing literature [13, Table 1], [32, p. 122], [21, p. 47], [17, Table 1], [12, p. 313], and [3, Table 1]. In Table 2, $k$, $k'$, $D$, and $R$ denote the length, the dimension, the minimal distance, and the packing radius, respectively.

## 4.4    Finding the Optimal Concatenated Code

The linear code $[k, k']$ is typically instantiated by a concatenation of elementary codes for practical purposes. By "concatenation" of $m$ codes $C_1, \ldots, C_m$, we

**Table 2.** Perfect and quasi-perfect binary linear codes

| Name | Type | $[k, k', D]$ | $R$ | Comment | Ref. |
|------|------|--------------|-----|---------|------|
| | P | $[k, k, 1], \ k \geq 1$ | 0 | $[*, \ldots, *]$ | |
| r | P | $[k, 1, k], \ k$ odd | $\frac{k-1}{2}$ | Repetition code | |
| H | P | $[2^\ell - 1, 2^\ell - \ell - 1, 3], \ \ell \geq 3,$ | 1 | Hamming code | |
| G | P | $[23, 12, 7]$ | 3 | Golay code | |
| | QP | $[k, k-1, 1]$ | 0 | $[*, \ldots, *, 0]$ | |
| r | QP | $[k, 1, k], \ k$ even | $\frac{k}{2} - 1$ | Repetition code | |
| eG | QP | $[24, 12, 8]$ | 3 | Extended Golay code | |
| pG | QP | $[22, 12, 6]$ | 2 | Punctured Golay code | |
| eH | QP | $[2^\ell, 2^\ell - \ell - 1, 4], \ \ell \geq 2$ | 1 | Extended Hamming code | |
| | QP | $[2^\ell - 1, 2^\ell - \ell, 1], \ \ell \geq 2,$ | 0 | Hamming with an extra word | |
| pH | QP | $[2^\ell - 2, 2^\ell - \ell - 1, 2], \ \ell \geq 2$ | 0 | Punctured Hamming | |
| HxH | QP | $[2 * (2^\ell - 1), 2 * (2^\ell - \ell - 1)], \ \ell \geq 2$ | 1 | Hamming $\times$ Hamming | [13] |
| upack | QP | $[2^\ell - 2, 2^\ell - \ell - 2, 3], \ \ell \geq 3$ | 1 | Uniformly packed | [13] |
| 2BCH | QP | $[2^\ell - 1, (2^\ell - 1) - (2 * \ell)], \ \ell \geq 3$ | 2 | 2-e.c. BCH | [13] |
| Z | QP | $[2^\ell + 1, (2^\ell + 1) - (2 * \ell)], \ \ell > 3$ even | 2 | Zetterberg | [13] |
| rGop | QP | $[2^\ell - 2, (2^\ell - 2) - (2 * \ell)], \ \ell > 3$ even | 2 | Red. Goppa | [13] |
| iGop | QP | $[2^\ell, (2^\ell) - (2 * \ell)], \ \ell > 2$ odd | 2 | Irred. Goppa | [13] |
| Mclas | QP | $[2^\ell - 1, (2^\ell - 1) - 2 * \ell], \ \ell > 2$ odd | 2 | Mclas | [13] |
| S | QP | $[5, 2], [9, 5], [10, 5], [11, 6]$ | 1 | Slepian | [32] |
| S | QP | $[11, 4]$ | 2 | Slepian | [32] |
| FP | QP | $[15, 9], [21, 14], [22, 15], [23, 16]$ | 1 | Fontaine-Peterson | [32] |
| W | QP | $[19, 10], [20, 11], [20, 13], [23, 14]$ | 2 | Wagner | [32] |
| P | QP | $[21, 12]$ | 2 | Prange | [32] |
| FP | QP | $[25, 12]$ | 3 | Fontaine-Peterson | [32] |
| W | QP | $[25, 15], [26, 16], [27, 17], [28, 18],$ $[29, 19], [30, 20], [31, 20]$ | 1 | Wagner | [32] |
| GS | QP | $[13, 7], [19, 12]$ | 1 | GS85 | [21] |
| BBD | QP | $[7, 3, 3], [9, 4, 4], [10, 6, 3], [11, 7, 3],$ $[12, 7, 3], [12, 8, 3], [13, 8, 3],$ $[13, 9, 3], [14, 9, 3], [15, 10, 3],$ $[16, 10, 3], [17, 11, 4], [17, 12, 3],$ $[18, 12, 4], [18, 13, 3], [19, 13, 3],$ $[19, 14, 3], [20, 14, 4]$ | 1 | BBD08 | [3] |
| BBD | QP | $[22, 13, 5]$ | 2 | BBD08 | [3] |

mean the code formed by all $g_{i,1} \| \cdots \| g_{i,m}$ obtained by concatenating any set of $g_{i,j} \in C_j$. Decoding $v_1 \| \cdots \| v_m$ is based on decoding each $v_{i,j}$ in $C_j$ independently. If all $C_j$ are small, this is done by a table lookup. So, concatenated codes are easy to implement and to decode. For $[k, k']$ we have the concatenation of $[k_1, k'_1], \ldots, [k_m, k'_m]$ codes, where $k_1 + \cdots + k_m = k$ and $k'_1 + \cdots + k'_m = k'$. Let $v_{ij}, g_{ij}, s'_j$ denote the $j^{th}$ part of $v_i, g_i, s'$ respectively, corresponding to the concatenated $[k_j, k'_j]$ code. The bias of $\langle v_{ij} - g_{ij}, s_j \rangle$ in the code $[k_j, k'_j]$ is denoted by $\mathsf{bc}_j$. As $\langle v_i - g_i, s \rangle$ is the xor of all $\langle v_{ij} - g_{ij}, s_j \rangle$, the total bias introduced by this operation is computed as $\mathsf{bc} = \prod_{j=1}^{k'} \mathsf{bc}_j$ and the combination $\mathsf{params} = ([k_1, k'_1], \ldots, [k_m, k'_m])$ is chosen such that it gives the highest bias.

The way these $\mathsf{params}$ are computed is the following: we start by computing the biases for all elementary codes. I.e. we compute the biases for all codes from

Table 2. We may add random codes that we found interesting. (For these, we use (2) to compute bc.)[7] Next, for each $[i,j]$ code we check to see if there is a combination of $[i-n,j-m]$, $[n,m]$ codes that give a better bias, where $[n,m]$ is either a repetition code, a Golay code or a Hamming code. We illustrate below the algorithm to find the optimal concatenated code. This algorithm was independently proposed by Zhang et al. [36] (with perfect codes only).

---

**Algorithm 1.** Finding the optimal params and bias

---

1: **Input:** $k$
2: **Output:** table for the optimal bias for each $[i,j]$ code, $1 \leq j < i \leq k$

3: initialize all bias$(i,j) = 0$
4: initialize bias$(1,1) = 1$
5: initialize the bias for all elementary codes
6: **for all** $j : 2$ to $k$ **do**
7:     **for all** $i : j+1$ to $k$ **do**
8:         **for all** elementary code $[n,m]$ **do**
9:             **if** $|\text{bias}(i-n,j-m) \cdot \text{bias}(n,m)| > |\text{bias}(i,j)|$ **then**
10:                 bias$(i,j) = \text{bias}(i-n,j-m) \cdot \text{bias}(n,m)$
11:                 params$(i,j) = \text{params}(i-n,j-m) \cup \text{params}(n,m)$

---

Using $\mathcal{O}(k)$ elementary codes, this procedure takes $\mathcal{O}(k^3)$ time and we can store all params for any combination $[i,j]$, $1 \leq j < i \leq k$ with $\mathcal{O}(k^2)$ memory.

## 5    The Graph of Reduction Steps

Having in mind the reduction methods described in Sect. 2, we formalize an LPN solving algorithm in terms of finding the best chain in a graph. The intuition is the following: in an LPN solving algorithm we can see each reduction step as an edge from a $(k, \log_2 n)$ instance to a new instance $(k', \log_2 n')$ where the secret is smaller, $k' \leq k$, we have more or less number of queries and the noise has a different bias. For example, a $\chi or\text{-}reduce(b)$ reduction turns an $(k, \log_2 n)$ instance with bias $\delta$ into $(k', \log_2 n')$ with bias $\delta'$ where $k' = k - b$, $n' = \frac{n(n-1)}{2^{b+1}}$ and $\delta' = \delta^2$. By this representation, the reduction phase represents a chain in which each edge is a reduction type moving from LPN with parameters $(k,n)$ to LPN with parameters $(k',n')$ and that ends with an instance $(k_i, n_i)$ used to recover the $k_i$-bit length secret by a solving method. The chain terminates by the fast Walsh-Hadamard solving method.

We formalize the reduction phase as a chain of reduction steps in a graph $G = (V,E)$. The set of vertices $V$ is composed of $V = \{1, \ldots, k\} \times L$ where $L$ is a set of real numbers. For instance, we could take $L = \mathbb{R}$ or $L = \mathbb{N}$. For efficiency reasons, we could even take $L = \{0, \ldots, \eta\}$ for some bound $\eta$. Every

---

[7] The random codes that we used are provided as an additional material to this paper.

vertex saves the size of the secret and the logarithmic number of queries; i.e. a vertex $(k, \log_2 n)$ means that we are in an instance where the size of the secret is $k$ and the number of queries available is $n$. An edge from one vertex to another is given by a reduction step. An edge from $(k, \log_2 n)$ to a $(k', \log_2 n')$ has a label indicating the type of reduction and its parameters (e.g. *xor-reduce*$(k - k')$ or *code-reduce*$(k, k', \mathsf{params})$). This reduction defines some $\alpha$ and $\beta$ coefficients such that the bias $\delta'$ after reduction is obtained from the bias $\delta$ before the reduction by

$$\log_2 \delta'^2 = \alpha \log_2 \delta^2 + \beta$$

where $\alpha, \beta \in \mathbb{R}$.

We denote by $\lceil \lambda \rceil_L$ the smallest element of $L$ which is at least equal to $\lambda$ and by $\lfloor \lambda \rfloor_L$ the largest element of $L$ which is not larger than $\lambda$. In general, we could use a rounding function $\mathsf{Round}_L(\lambda)$ such that $\mathsf{Round}_L(\lambda)$ is in $L$ and approximates $\lambda$.

The reduction steps described in Subsect. 2.2 can be formalized as follows:

- *sparse-secret*: $(k, \log_2 n) \rightarrow (k, \mathsf{Round}_L(\log_2(n - k)))$ and $\alpha = 0, \beta = 0$
- *xor-reduce*$(b)$: $(k, \log_2 n) \rightarrow (k - b, \mathsf{Round}_L(\log_2\left(\frac{n(n-1)}{2^{b+1}}\right)))$ and $\alpha = 2, \beta = 0$
- *drop-reduce*$(b)$: $(k, \log_2 n) \rightarrow (k - b, \mathsf{Round}_L(\log_2\left(\frac{n}{2^b}\right)))$ and $\alpha = 1, \beta = 0$
- *code-reduce*$(k, k', \mathsf{params})$: $(k, \log_2 n) \rightarrow (k', \log_2 n)$ and $\alpha = 1, \beta = \log_2 \mathsf{bc}^2$, where bc is the bias introduced by the covering code reduction using a $[k, k']$ linear code defined by params.

Below, we give the formal definition of a reduction chain.

**Definition 3 (Reduction chain).** *Let*

$$\mathcal{R} = \{\textit{sparse-secret}, \textit{xor-reduce}(b), \textit{drop-reduce}(b), \textit{code-reduce}(k, k', \mathsf{params})\}$$

*for $k, k', b \in \mathbb{N}$. A **reduction chain** is a sequence*

$$(k_0, \log_2 n_0) \xrightarrow{e_1} (k_1, \log_2 n_1) \xrightarrow{e_2} \ldots \xrightarrow{e_i} (k_i, \log_2 n_i),$$

*where the change $(k_{j-1}, \log_2 n_{j-1}) \rightarrow (k_j, \log_2 n_j)$ is performed by one reduction from $\mathcal{R}$, for all $0 < j \leq i$.*

*A chain is **simple** if it is accepted by the automaton from Fig. 1.*

*Remark:* Restrictions for simple chains are modelled by the automaton in Fig. 1. We restrict to simple chains as they are easier to analyze. Indeed, *sparse-secret* is only used to raise $\delta_s$ to make *code-reduce* more effective. And, so far, it is hard to analyze sequences of *code-reduce* steps as the first one may destroy the uniform and high $\delta_s$ for the next ones. This is why we exclude multiple *code-reduce* reductions in a simple chain. So, we use up to one *sparse-secret* reduction, always one before *code-reduce*. And *sparse-secret* occurs before $\delta$ decreases. For convenience, we will add a state of the automaton to the vertex in $V$.

**Definition 4 (Exact chain).** *An **exact chain** is a simple reduction chain for $L = \mathbb{R}$. I.e. $\mathsf{Round}_L$ is the identity function.*

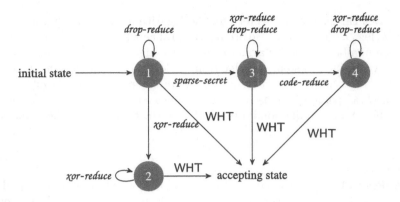

**Fig. 1.** Automaton accepting simple chains

A chain which is not exact is called **rounded**.

For solving LPN we are interested in those chains that end with a vertex $(k_i, \log_2 n_i)$ which allows to call a WHT solving algorithm to recover the $k_i$-bit secret. We call these chains valid chains and we define them below.

**Definition 5 (Valid reduction chain).** *Let*

$$(k_0, \log_2 n_0) \xrightarrow{e_1} (k_1, \log_2 n_1) \xrightarrow{e_2} \cdots \xrightarrow{e_i} (k_i, \log_2 n_i)$$

*be a reduction chain with $e_j = (\alpha_j, \beta_j, .)$. Let $\delta_j$ be the bias corresponding to the vertex $(k_j, \log_2 n_j)$ iteratively defined by $\delta_0 = \delta$ and $\log_2 \delta_j^2 = \alpha_j \log_2 \delta_{j-1}^2 + \beta_j$ for $j = 1, \ldots, i$. We say the chain is a $\theta$-valid reduction chain if $n_i$ satisfies (1) from p. 8 for $\delta' = \delta_i$ and $n' = n_i$.*

The *time complexity* of a chain $(e_1, \ldots, e_i)$ is simply the sum of the complexity of each reduction step $e_1, e_2, \ldots, e_i$ and WHT. We further define the **max-complexity** of a chain which is the maximum of the complexity of each reduction step and WHT. The max-complexity is a good approximation of the complexity. Our goal is to find a chain with optimal complexity. What we achieve is that, *given a set $L$, we find a rounded chain with optimal max-complexity up to some given precision.*

## 5.1    Towards Finding the Best LPN Reduction Chain

In this section we present the algorithm that helps finding the optimal valid chains for solving LPN. As aforementioned, we try to find the valid chain with optimal max-complexity for solving an $\mathsf{LPN}_{k,\tau}$ instance in our graph $G$.

The first step of the algorithm is to construct the directed graph $G = (V, E)$. We take the set of vertices $V = \{1, \ldots, k\} \times L \times \{1, 2, 3, 4\}$ which indicate the size of the secret, the logarithmic number of queries and the state in the automaton in Fig. 1. Each edge $e \in E$ represents a reduction step and is labelled with the following information: $(k_1, \log_2 n_1, st) \overset{\alpha,\beta,t}{\rightarrow} (k_2, \log_2 n_2, st')$ where $t$ is one of the reduction steps and $\alpha$ and $\beta$ save information about how the bias is affected by this reduction step.

The graph has $\mathcal{O}(k \cdot |L|)$ vertices and each vertex has $\mathcal{O}(k)$ edges. So, the size of the graph is $\mathcal{O}(k^2 \cdot |L|)$.

Thus, we construct the graph $G$ with all possible reduction steps and from it we try to see what is the optimal simple rounded chain in terms of max-complexity. We present in Algorithm 2 the procedure to construct the graph $G$ that contains all possible reduction steps with a time complexity bounded by $2^\eta$ (As explained below, Algorithm 2 is not really used).

The procedure of finding the optimal valid chain is illustrated in Algorithm 3. The procedure of finding a chain with upper bounded max-complexity is illustrated in Algorithm 4.

---

**Algorithm 2.** Construction of graph $G$

---

1: **Input:** $k, \tau, L, \eta$
2: **Output:** graph $G = (V, E)$ containing all the reduction steps that have a complexity smaller than $2^\eta$

3: $V = \{1, \ldots, k\} \times L \times \{1, \ldots, 4\}$
4: $E$ is the set of all $((i, \eta_1, st), (j, \eta_2, st'))$ labelled by $(\alpha, \beta, t)$ such that there is a $st \overset{t}{\rightarrow} st'$ transition in the automaton and for
5: $t = sparse\text{-}secret$:
6: **for all** $\eta : 1$ such that $\mathsf{lcomp} \leq \eta$ **do** set the edge
7:     where $i = k$, $(j, \eta_2) = (i, \mathsf{Round}_L(\log_2(2^{\eta_1} - i)))$, $\alpha = 1$, $\beta = 0$, $\mathsf{lcomp} = \min_x \log_2(\frac{(2^{\eta_1}-i)i^2}{log_2 i - log_2 log_2 i} + i^2, i(2^{\eta_1} - i)\lceil\frac{i}{2^x}\rceil + i^3 + i2^{x+2^x})$

8: $t = xor\text{-}reduce$:
9: **for all** $(i, \eta_1, b)$ such that $b \geq 1$ and $\mathsf{lcomp} \leq \eta$ **do** set the edge
10:     where $(j, \eta_2) = (i - b, \mathsf{Round}_L(\eta_1 - 1 + \log_2(\frac{2^{\eta_1}}{2^b} - 1)))$, $\alpha = 2$, $\beta = 0$, $\mathsf{lcomp} = \log_2 i + \max(\eta_1, \eta_2)$

11: $t = drop\text{-}reduce$:
12: **for all** $(i, \eta_1, b)$ such that $b \geq 1$ and $\mathsf{lcomp} \leq \eta$ **do** set the edge
13:     where $(j, \eta_2) = (i - b, \mathsf{Round}_L(\eta_1 - b))$, $\alpha = 1$, $\beta = 0$, $\mathsf{lcomp} = \log_2 b + \eta_1$

14: $t = code\text{-}reduce$:
15: **for all** $(i, \eta_1, j)$ such that $j < i$ and $\mathsf{lcomp} \leq \eta$ **do** set the edge
16:     where $\eta_2 = \eta_1$, $\alpha = 1$, $\beta = \log_2 bc^2$, $\mathsf{lcomp} = \log_2 i + \eta_1$, bc is the bias from the optimal $[i, j]$ code

---

Algorithm 4 receives as input the parameters $k$ and $\tau$ for the LPN instance, the parameter $\theta$ which represents the bound on the failure probability in recovering the secret. Parameter $\eta$ represents an upper bound for the logarithmic complexity of each reduction step. Given $\eta$, we build the graph $G$ which contains all possible reductions with time complexity smaller than $2^\eta$ (Step 4). Note that we don't really call Algorithm 2. Indeed, we don't need to store the edges of the graph. We rather keep a way to enumerate all edges going to a given vertex (in Step 11) by using the rules described in Algorithm 2.

---

**Algorithm 3.** Search for a rounded chain with optimal max-complexity

---

1: **Input**: $k, \tau, \theta$, precision
2: **Output**: a valid simple rounded chain in which rounding uses a given precision

3: set found = bruteforce                              ▷ found is the best found algorithm
4: set increment = $k$
5: set $\eta = k$                                        ▷ $2^\eta$ is a bound on the max-complexity
6: **repeat**
7:     set increment $\leftarrow \frac{1}{2}$increment
8:     define $L = \{0, \text{precision}, 2 \times \text{precision}, \ldots\} \cap [0, \eta - \text{increment}]$
9:     run (out, success) = Search$(k, \tau, \theta, L, \eta - \text{increment})$ with Algorithm 4
10:    **if** success **then**
11:        set found = out
12:        set $\eta = \eta - \text{increment}$
13: **until** increment $\leq$ precision
14: output found

---

For each vertex, we iteratively define $\Delta^{st}$ and $\mathsf{Best}^{st}$, the best reduction step to reach a vertex and the value of the corresponding error bias. The best reduction step is the one that maximizes the bias. We define these values iteratively until we reach a vertex from which the WHT solving algorithm succeeds with complexity bounded by $2^\eta$. Once we have reached this vertex, we construct the chain by going backwards, following the Best pointers.

We easily prove what follows by induction.

**Lemma 1.** *At the end of the iteration of Algorithm 4 for $(j, \eta_2, st')$, $\Delta^{st'}_{j,\eta_2}$ is the maximum of $\log_2 \delta^2$, where $\delta$ is the bias obtained by an $\mathsf{Round}_L$-rounded simple chain from a vertex of form $(k, \eta_1, 0)$ to $(j, \eta_2, st')$ with max-complexity bounded by $2^\eta$ ($\Delta^{st'}_{j,\eta_2} = -\infty$ if there is no such chain).*

**Lemma 2.** *If there exists a simple $\mathsf{Round}_L$-rounded chain $c$ ending on state $(k_j, \eta_j, st_j)$ and max-complexity bounded by $2^\eta$, there exists one $c'$ such that $\Delta^{st_i}_{i,\eta_i} = \log_2 \delta_i^2$ at each step.*

*Proof.* Let $c''$ be a simple chain ending on $(k_j, \eta_j, st_j)$ with $\Delta^{st_j}_{j\eta_j} = \log_2 \delta_j^2$. Let $(k_{j-1}, \eta_{j-1}, st_{j-1})$ be the preceding vertex in $c''$. We apply Lemma 2 on this vertex by induction to obtain a chain $c'''$. Since the complexity of the last edge does not depend on the bias and $\alpha \geq 0$ in the last edge, we construct the chain $c'$, by concatenating $c'''$ with the last edge of $c''$.                                    $\square$

**Theorem 2.** *Algorithm 4 finds a $\theta$-valid simple $\mathsf{Round}_L$-rounded chain for $\mathsf{LPN}_{k,\tau}$ with max-complexity bounded by $2^\eta$ if there exists one.*

*Proof.* We use Lemma 2 and the fact that increasing $\delta^2$ keeps constraint (1) valid.                                    $\square$

**Algorithm 4.** Search for a best LPN reduction chain with max-complexity bounded to $\eta$

---

1: **Input:** $k, \tau, \theta, L, \eta$
2: **Output:** a valid simple rounded chain with max-complexity bounded to $\eta$

3: $\delta = 1 - 2\tau$
4: Construct the graph $G$ using Algorithm 2 with parameters $k, \tau, L, \eta$
5: **for all** $\eta_1 \in L$ **do**
6:      set $\Delta^0_{k,\eta_1} = \log_2 \delta^2$, $\text{Best}^0_{k,\eta_1} = \perp$
7:      set $\Delta^{st}_{k,\eta_1} = -\infty$, $\text{Best}^{st}_{k,\eta_1} = \perp$    $\triangleright$ $\Delta^{st}$ stores the best bias for a vertex $(k, \eta_1, st)$ in a chain, and $\text{Best}^{st}$ is the edge ending to this vertex in this chain
8: **for** $j : k$ downto 1 **do**                                   $\triangleright$ Search for the optimal chain
9:      **for** $\eta_2 \in L$ in decreasing order **do**
10:          set $\Delta^{st}_{j,\eta_2} = 0$, $\text{Best}^{st} = \perp$ for all $st$
11:          **foreach** st' and each edge $e$ to $(j, \eta_2, st')$
12:              set $(i, \eta_1, st)$ to the origin of $e$ and $\alpha$ and $\beta$ as defined by $e$
13:              **if** $\alpha \Delta^{st}_{i,\eta_1} + \beta \geq \Delta^{st'}_{j,\eta_2}$ **then set** $\Delta^{st'}_{j,\eta_2} = \alpha \Delta^{st}_{i,\eta_1} + \beta$, $\text{Best}^{st} = e$
14:          **end foreach**
15:          **if** $\eta_2 > 1 - \Delta^{st'}_{j,\eta_2} + 2\log_2\left(-\varphi^{-1}(1 - (1-\theta)^{\frac{1}{2^{j}-1}})\right)$ and $j + \log_2 j \leq \eta$ **then**
16:              Construct the chain $c$ ending by $\text{Best}^{st'}_{j,\eta_2}$ and output $(c, \text{true})$
17: output $(\perp, \text{false})$

---

If we used $L = \mathbb{R}$, Algorithm 4 would always find a valid simple chain with bounded max-complexity when it exists. Instead, we use rounded chains and hope that rounding still makes us find the optimal chain.

So, we build Algorithm 3. In this algorithm, we look for the minimal $\eta$ for which Algorithm 4 returns something by a divide and conquer algorithm. First, we set $\eta$ as being in the interval $[0, k]$ where the solution for $\eta = k$ corresponds to a brute-force search. Then, we cut the interval in two pieces and see if the lower interval has a solution. If it does, we iterate in this interval. Otherwise, we iterate in the other interval. We stop once the amplitude of the interval is lower than the requested precision. The complexity of Algorithm 3 is of $\log_2 \frac{k}{\text{precision}}$ calls to Algorithm 4.

**Theorem 3.** *Algorithm 3 finds a $\theta$-valid simple $\text{Round}_L$-rounded chain for $\text{LPN}_{k,\tau}$ with parameter* precision, *with optimal rounded max-complexity, where the rounding function approximates* $\log_2$ *up to* precision *if there exists one.*

*Proof.* Algorithm 3 is a divide-and-conquer algorithm to find the smallest $\eta$ such that Algorithm 4 finds a valid simple $\text{Round}_L$-rounded chain of max-complexity bounded by $2^\eta$.             $\square$

We can see that the complexity of Algorithm 4 is of $\mathcal{O}\left(k^2 \cdot |L|\right)$ iterations as vertices have $k$ possible values for the secret length and $|L|$ possible values for the logarithmic number of equations. So, it is linear in the size of the graph. Furthermore, each type of edge to a fixed vertex has $\mathcal{O}(k)$ possible origins. The memory complexity is $\mathcal{O}\left(k \cdot |L|\right)$, mainly to store the $\Delta_{k,\eta}$ and $\mathsf{Best}_{k,\eta}$ tables. We also use Algorithm 1 which has a complexity $\mathcal{O}(k^3)$ but we run it only once during precomputation. Algorithm 3 sets $|L| \sim \frac{k}{\mathsf{precision}}$. So, the complexity of Algorithm 3 is $\mathcal{O}\left(k^3 + \frac{k^3}{\mathsf{precision}} \times \log \frac{k}{\mathsf{precision}}\right)$.

## 6    Chains with a Guessing Step

In order to further improve our valid chain we introduce a new reduction step to our algorithm. As it is done in previous works [5,22], we guess part of the bits of the secret. More precisely, we assume that $b$ bits of the secret have a Hamming weight smaller or equal to $w$. The influence on the whole algorithm is more complicated: it requires to iterate the WHT step $\sum_{i=0}^{w}\binom{w}{i}$ times. The overall complexity must further be divided by $\sum_{i=0}^{w}\binom{w}{i}\left(\frac{1-\delta_s}{2}\right)^i\left(\frac{1+\delta_s}{2}\right)^{w-i}$. Note that this generalized *guess-secret* step was used in Guo et al. [22].

We formalize this step as following:

- *guess-secret*$(b,w)$ guesses that $b$ bits of the secret have a Hamming weight smaller or equal to $w$. The $b$ positions are chosen randomly. The number of queries remains the same, the noise is the same and the size of the secret is reduced by $b$ bits. Thus, for this step we have

---

*guess-secret*$(b,w)$ : $k' = k - b$; $n' = n$; $\delta' = \delta$; $\delta'_s = \delta$

Complexity: $\mathcal{O}(nb)$ (included in *sparse-secret*) and

the Walsh transform has to be iterated $\sum_{i=0}^{w}\binom{w}{i}$ times and

the complexity of the whole algorithm is divided by

$\sum_{i=0}^{w}\binom{w}{i}\left(\frac{1-\delta_s}{2}\right)^i\left(\frac{1+\delta_s}{2}\right)^{w-i}$

---

This step may be useful for a sparse secret, i.e. $\tau$ is small, as then we reduce the size of the secret with a very small cost. In order to accommodate this new step we would have to add a transition from state 3 to state 3 in the automaton that accepts the simple chains (See Fig. 1).

To find the optimal chain using *guess-secret*$(b,w)$, we have to make a loop over all possible $b$ and all possible $w$. We run the full search $\mathcal{O}(k^2)$ times. The total complexity is thus $\mathcal{O}\left(\frac{k^5}{\mathsf{precision}} \times \log \frac{k}{\mathsf{precision}}\right)$.

## 7    Results

We illustrate in this section the results obtained by running Algorithm 4 for different LPN instances taken from Bogos et al. [7]. They vary from taking $k = 32$ to $k = 768$, with the noise levels: $0.05, 0.1, 0.125, 0.2$ and $0.25$. In Table 3 we display the logarithmic time complexity we found for solving LPN without using *guess-secret*.[8]

---

[8] Complete results are provided as an additional material to this paper.

Optimization of LPN Solving Algorithms    725

**Table 3.** Logarithmic time complexity on solving LPN without *guess-secret*

| $\tau$ | $k$ | | | | | | |
|---|---|---|---|---|---|---|---|
| | 32 | 48 | 64 | 100 | 256 | 512 | 768 |
| 0.05 | $13.89_{0.1}^{11.26}$ | $14.52_{0.1c}^{12.94}$ | $16.04_{0.1c}^{14.43}$ | $20.47_{0.1c}^{18.46}$ | $36.75_{0.1c}^{34.45}$ | $57.77_{0.1c}^{55.09}$ | $76.63_{0.1c}^{74.03}$ |
| 0.1 | $15.04_{0.1}^{12.70}$ | $18.58_{0.1}^{16.43}$ | $21.58_{0.1c}^{19.38}$ | $27.61_{0.1c}^{25.39}$ | $46.75_{0.1c}^{44.22}$ | $73.68_{0.1c}^{70.92}$ | $98.97_{0.1c}^{96.04}$ |
| 0.125 | $15.66_{0.1}^{13.52}$ | $19.29_{0.1}^{17.00}$ | $22.94_{0.1}^{20.50}$ | $28.91_{0.1}^{26.30}$ | $49.90_{0.1c}^{47.35}$ | $78.85_{0.1c}^{76.22}$ | $105.89_{0.1c}^{103.01}$ |
| 0.2 | $17.01_{0.1}^{14.80}$ | $21.25_{0.1}^{19.23}$ | $24.42_{0.1}^{22.00}$ | $32.06_{0.1}^{29.75}$ | $56.31_{0.1c}^{53.82}$ | $89.04_{0.1c}^{86.38}$ | $121.04_{0.1c}^{118.18}$ |
| 0.25 | $18.42_{0.1}^{16.30}$ | $22.34_{0.1}^{20.43}$ | $26.86_{0.1}^{24.58}$ | $32.94_{0.1}^{30.75}$ | $59.47_{0.1}^{56.88}$ | $94.66_{0.1c}^{91.97}$ | $127.35_{0.1c}^{124.63}$ |

Entry of form $a_{c...}^b$: $a = \log_2$ complexity, $b = \log_2$ max-complexity, $c =$ precision. Subscript $c$ means that a *code-reduce* is used.

**Table 4.** Logarithmic time complexity on solving LPN with *guess-secret*

| $\tau$ | $k$ | | | | | | |
|---|---|---|---|---|---|---|---|
| | 32 | 48 | 64 | 100 | 256 | 512 | 768 |
| 0.05 | $11.85_{0.1cg13o}^{10.90}$ | $13.01_{0.1cg23o}^{12.52}$ | $14.44_{0.1cg38o}^{13.74}$ | $17.20_{0.1cg75o}^{16.19}$ | $30.13_{0.1cg178o}^{28.02}$ | $49.56_{1cg417o}^{47.29}$ | $68.15_{1cg682o}^{65.98}$ |
| 0.1 | $12.41_{0.1cg23o}^{11.65}$ | $15.23_{0.1cg37o}^{14.25}$ | $17.71_{0.1cg52o}^{16.76}$ | $24.02_{0.1cg77o}^{22.14}$ | $45.99_{0.1cg100o}^{43.49}$ | $73.68_{1cg2}^{71.09}$ | $99.21_{1cg5}^{96.34}$ |
| 0.125 | $13.30_{0.1cg26o}^{12.40}$ | $16.49_{0.1cg39o}^{15.49}$ | $20.57_{0.1cg36o}^{18.61}$ | $27.14_{0.1cg47o}^{24.80}$ | $49.90_{0.1c}^{47.35}$ | $78.97_{1cg1}^{76.24}$ | $106.18_{1cg4}^{103.42}$ |
| 0.2 | $17.01_{0.1o}^{14.80}$ | $21.25_{0.1o}^{19.23}$ | $24.42_{0.1}^{22.00}$ | $32.06_{0.1}^{29.75}$ | $56.34_{0.1cg1}^{53.82}$ | $89.28_{1cg3}^{86.79}$ | $121.12_{1cg1}^{118.57}$ |
| 0.25 | $18.42_{0.1}^{16.30}$ | $22.34_{0.1}^{20.43}$ | $26.86_{0.1}^{24.58}$ | $32.94_{0.1}^{30.75}$ | $59.47_{0.1}^{56.88}$ | $94.85_{1cg2}^{92.36}$ | $127.63_{1cg3}^{125.01}$ |

Entry of form $a_{c...}^b$: $a = \log_2$ complexity, $b = \log_2$ max-complexity, $c =$ precision. Subscript $c$ means that a *code-reduce* is used. Subscript $o$ means that a only 1 bit of the secret is found by WHT. Subscript $gb$ means that a *guess-secret*$(b, \cdot)$ is used.

*Sequence of chains.* If we analyze in more details one of the chains that we obtained, e.g. the chain for LPN$_{512,0.125}$, we can see that it first uses a *sparse-secret*. Afterwards, the secret is reduced by applying 5 times the *xor-reduce* and one *code-reduce* at the end of the chain. With a total complexity of $2^{79.46}$ and $\theta < 33\%$ it recovers 64 bits of the secret.

$$(512, 63.3) \xrightarrow{sparse-secret} (512, 63.3) \xrightarrow{xor-reduce(59)} (453, 66.6) \xrightarrow{xor-reduce(65)}$$
$$(388, 67.2) \xrightarrow{xor-reduce(66)} (322, 67.4) \xrightarrow{xor-reduce(66)} (256, 67.8) \xrightarrow{xor-reduce(67)}$$
$$(189, 67.6) \xrightarrow{code-reduce} (64, 67.6) \xrightarrow{WHT}$$

The code used is a $[189, 64]$ concatenation made of ten random codes: one instance of a $[18, 6]$ code, five instances of a $[19, 6]$ code, and four instances of a $[19, 7]$ code. By manually tuning the number of equations without rounding, we can obtain with $n = 2^{63.299}$ a complexity of $2^{78.84}$. This is the value from Table 1.

*On the guess-secret reduction.* Our results show that the *guess-secret* step does not bring any significant improvement. If we compare Table 3 with Table 4 we can see that in few cases the guess step improves the total complexity. For $k \geq 512$, some results are not better than Table 3. This is most likely due to the lower precision used in Table 4.

We can see several cases where, at the end of a chain with *guess-secret*, only one bit of the secret is recovered by WHT. If only 1 bit of the secret is recovered by non-bruteforce methods, the next chain for $LPN_{k-1,\tau}$ will have to be run several times, given the *guess-secret* step used in the chain for $LPN_{k,\tau}$. Thus, it might happen that the first chain does not dominate the total complexity. So, our strategy to use sequences of chains has to be revised, but most likely, the final result will not be better than sequences of chains without *guess-secret*. So, we should rather avoid these chains ending with 1 bit recovery.

There is no case where a *guess-secret* without a chain ending with 1 bit brings any improvement.

*Comparing the results.* For practical values we compare our results with the previous work [7,22,28,36].

From the work of ASIACRYPT'14 [22] and EUROCRYPT'16 [36] we have that $LPN_{512,0.125}$ can be solved in time complexity of $2^{79.9}$ (with more precise complexity estimates). The comparison was shown in Table 1 in Introduction. We do better, provide concrete codes and we even remove the *guess-secret* step with an optimized use of a code. Thus, the results of Algorithm 4 improve all the existing results on solving LPN.

# 8    Conclusion

In this article we have proposed an algorithm for creating reduction chains with the optimal max-complexity. The results we obtain bring improvements to the existing work and to our knowledge we have the best algorithm for solving $LPN_{512,0.125}$. We believe that our algorithm could be further adapted and automatized if new reduction techniques would be introduced.

As future works, we could look at applications to the LWE problem. Kirchner and Fouque [27] improve the LWE solving algorithms by refining the modulus switching. We could also look at ways to keep track of biases of secret bits bitwise, in order to allow cascades of *code-reduce* steps.

# References

1. Alekhnovich, M.: More on average case vs approximation complexity. In: Proceedings of the 44th Symposium on Foundations of Computer Science (FOCS 2003), 11–14 October 2003, Cambridge, MA, USA, pp. 298–307. IEEE Computer Society (2003)
2. Arlazarov, V.L., Dinic, E.A., Kronrod, M.A., Faradzev, I.A.: On economical construction of the transitive closure of a directed graph. Sov. Math. Dokl. **11**, 1209–1210 (1970)
3. Baicheva, T.S., Bouyukliev, I., Dodunekov, S.M., Fack, V.: Binary and ternary linear quasi-perfect codes with small dimensions. IEEE Trans. Inf. Theory **54**(9), 4335–4339 (2008)
4. Bernstein, D.J.: Optimizing linear maps modulo 2. http://binary.cr.yp.to/linearmod2-20090830.pdf

5. Bernstein, D.J., Lange, T.: Never trust a bunny. In: Hoepman, J.-H., Verbauwhede, I. (eds.) RFIDSec 2012. LNCS, vol. 7739, pp. 137–148. Springer, Heidelberg (2013). doi:10.1007/978-3-642-36140-1_10

6. Blum, A., Kalai, A., Wasserman, H.: Noise-tolerant learning, the parity problem, and the statistical query model. In: Frances Yao, F., Luks, E.M. (eds.) Proceedings of the Thirty-Second Annual ACM Symposium on Theory of Computing, 21–23 May 2000, Portland, OR, USA, pp. 435–440. ACM (2000)

7. Bogos, S., Tramèr, F., Vaudenay, S.: On solving LPN using BKW and variants - implementation and analysis. Crypt. Commun. 8(3), 331–369 (2016)

8. Bogos, S., Vaudenay, S.: How to sequentialize independent parallel attacks? In: Iwata, T., Cheon, J.H. (eds.) ASIACRYPT 2015. LNCS, vol. 9453, pp. 704–731. Springer, Heidelberg (2015). doi:10.1007/978-3-662-48800-3_29

9. Bogos, S., Vaudenay, S.: Observations on the LPN Solving Algorithm from Eurocrypt2016. Cryptology ePrint Archive, Report 2016/451 (2016). https://eprint. iacr.org/2016/451

10. Bringer, J., Chabanne, H., Dottax, E.: HB$^{++}$: a lightweight authentication protocol secure against some attacks. In: Second International Workshop on Security, Privacy and Trust in Pervasive and Ubiquitous Computing (SecPerU 2006), 29 June 2006, Lyon, France, pp. 28–33. IEEE Computer Society (2006)

11. Carrijo, J., Tonicelli, R., Imai, H., Nascimento, A.C.A.: A novel probabilistic passive attack on the protocols HB and HB$^+$. IEICE Trans. 92–A(2), 658–662 (2009)

12. Cohen, G., Honkala, I., Litsyn, S., Lobstein, A.: Covering Codes. North-Holland Mathematical Library, Elsevier Science, Amsterdam (1997)

13. Cohen, G.D., Karpovsky, M.G., Mattson Jr., H.F., Schatz, J.R.: Covering radius - survey and recent results. IEEE Trans. Inf. Theory 31(3), 328–343 (1985)

14. Damgård, I., Park, S.: Is public-key encryption based on LPN practical? IACR Cryptology ePrint Arch. 2012, 699 (2012)

15. Döttling, N., Müller-Quade, J., Nascimento, A.C.A.: IND-CCA secure cryptography based on a variant of the LPN problem. In: Wang, X., Sako, K. (eds.) ASIACRYPT 2012. LNCS, vol. 7658, pp. 485–503. Springer, Heidelberg (2012). doi:10. 1007/978-3-642-34961-4_30

16. Duc, A., Vaudenay, S.: HELEN: a public-key cryptosystem based on the LPN and the decisional minimal distance problems. In: Youssef, A., Nitaj, A., Hassanien, A.E. (eds.) AFRICACRYPT 2013. LNCS, vol. 7918, pp. 107–126. Springer, Heidelberg (2013). doi:10.1007/978-3-642-38553-7_6

17. Etzion, T., Mounits, B.: Mounits.: quasi-perfect codes with small distance. IEEE Trans. Inf. Theory 51(11), 3938–3946 (2005)

18. Fossorier, M.P.C., Mihaljević, M.J., Imai, H., Cui, Y., Matsuura, K.: An algorithm for solving the LPN problem and its application to security evaluation of the HB protocols for RFID authentication. In: Barua, R., Lange, T. (eds.) INDOCRYPT 2006. LNCS, vol. 4329, pp. 48–62. Springer, Heidelberg (2006). doi:10.1007/11941378_5

19. Gilbert, H., Robshaw, M.J.B., Seurin, Y.: HB$^\#$: increasing the security and efficiency of HB$^+$. In: Smart, N. (ed.) EUROCRYPT 2008. LNCS, vol. 4965, pp. 361–378. Springer, Heidelberg (2008). doi:10.1007/978-3-540-78967-3_21

20. Gilbert, H., Robshaw, M.J.B., Seurin, Y.: How to encrypt with the LPN problem. In: Aceto, L., Damgård, I., Goldberg, L.A., Halldórsson, M.M., Ingólfsdóttir, A., Walukiewicz, I. (eds.) ICALP 2008. LNCS, vol. 5126, pp. 679–690. Springer, Heidelberg (2008). doi:10.1007/978-3-540-70583-3_55

21. Graham, R.L., Sloane, N.J.A.: On the covering radius of codes. IEEE Trans. Inf. Theory 31(3), 385–401 (1985)

22. Guo, Q., Johansson, T., Löndahl, C.: Solving LPN using covering codes. In: Sarkar, P., Iwata, T. (eds.) ASIACRYPT 2014. LNCS, vol. 8873, pp. 1–20. Springer, Heidelberg (2014). doi:10.1007/978-3-662-45611-8_1

23. Hopper, N.J., Blum, M.: Secure human identification protocols. In: Boyd, C. (ed.) ASIACRYPT 2001. LNCS, vol. 2248, pp. 52–66. Springer, Heidelberg (2001). doi:10.1007/3-540-45682-1_4

24. Juels, A., Weis, S.A.: Authenticating pervasive devices with human protocols. In: Shoup, V. (ed.) CRYPTO 2005. LNCS, vol. 3621, pp. 293–308. Springer, Heidelberg (2005). doi:10.1007/11535218_18

25. Kiltz, E., Masny, D., Pietrzak, K.: Simple chosen-ciphertext security from low-noise LPN. In: Krawczyk, H. (ed.) PKC 2014. LNCS, vol. 8383, pp. 1–18. Springer, Heidelberg (2014). doi:10.1007/978-3-642-54631-0_1

26. Kiltz, E., Pietrzak, K., Cash, D., Jain, A., Venturi, D.: Efficient authentication from hard learning problems. In: Paterson, K.G. (ed.) EUROCRYPT 2011. LNCS, vol. 6632, pp. 7–26. Springer, Heidelberg (2011). doi:10.1007/978-3-642-20465-4_3

27. Kirchner, P., Fouque, P.-A.: An improved BKW algorithm for LWE with applications to cryptography and lattices. In: Gennaro, R., Robshaw, M. (eds.) CRYPTO 2015. LNCS, vol. 9215, pp. 43–62. Springer, Heidelberg (2015). doi:10.1007/978-3-662-47989-6_3

28. Levieil, É., Fouque, P.-A.: An improved LPN algorithm. In: Prisco, R., Yung, M. (eds.) SCN 2006. LNCS, vol. 4116, pp. 348–359. Springer, Heidelberg (2006). doi:10.1007/11832072_24

29. Lyubashevsky, V.: The parity problem in the presence of noise, decoding random linear codes, and the subset sum problem. In: Chekuri, C., Jansen, K., Rolim, J.D.P., Trevisan, L. (eds.) APPROX/RANDOM - 2005. LNCS, vol. 3624, pp. 378–389. Springer, Heidelberg (2005). doi:10.1007/11538462_32

30. Lyubashevsky, V., Masny, D.: Man-in-the-middle secure authentication schemes from LPN and weak PRFs. In: Canetti, R., Garay, J.A. (eds.) CRYPTO 2013. LNCS, vol. 8043, pp. 308–325. Springer, Heidelberg (2013). doi:10.1007/978-3-642-40084-1_18

31. May, A., Meurer, A., Thomae, E.: Decoding random linear codes in $\tilde{\mathcal{O}}(2^{0.054n})$. In: Lee, D.H., Wang, X. (eds.) ASIACRYPT 2011. LNCS, vol. 7073, pp. 107–124. Springer, Heidelberg (2011). doi:10.1007/978-3-642-25385-0_6

32. Peterson, W.W., Weldon, E.J.: Error-Correcting Codes. MIT Press, Cambridge (1972)

33. Regev, O.: On lattices, learning with errors, random linear codes, and cryptography. In: Gabow, H.N., Fagin, R. (eds.) Proceedings of the 37th Annual ACM Symposium on Theory of Computing, Baltimore, MD, USA, 22–24 May 2005, pp. 84–93. ACM (2005)

34. Selçuk, A.A.: On probability of success in linear and differential cryptanalysis. J. Cryptology 21(1), 131–147 (2008)

35. Stern, J.: A method for finding codewords of small weight. In: Cohen, G.D., Wolfmann, J. (eds.) Coding Theory 1988. LNCS, vol. 388, pp. 106–113. Springer, Heidelberg (1989). doi:10.1007/BFb0019850

36. Zhang, B., Jiao, L., Wang, M.: Faster algorithms for solving LPN. In: Fischlin, M., Coron, J.-S. (eds.) EUROCRYPT 2016. LNCS, vol. 9665, pp. 168–195. Springer, Heidelberg (2016). doi:10.1007/978-3-662-49890-3_7

# The Kernel Matrix Diffie-Hellman Assumption

Paz Morillo[1]([✉]), Carla Ràfols[2], and Jorge L. Villar[1]

[1] Universitat Politècnica de Catalunya, Barcelona, Spain
{paz.morillo,jorge.villar}@upc.edu
[2] Universitat Pompeu Fabra, Barcelona, Spain
carla.rafols@upf.edu

**Abstract.** We put forward a new family of computational assumptions, the Kernel Matrix Diffie-Hellman Assumption. Given some matrix $\mathbf{A}$ sampled from some distribution $\mathcal{D}$, the kernel assumption says that it is hard to find "in the exponent" a nonzero vector in the kernel of $\mathbf{A}^\top$. This family is a natural computational analogue of the Matrix Decisional Diffie-Hellman Assumption (MDDH), proposed by Escala *et al.* As such it allows to extend the advantages of their algebraic framework to computational assumptions.

The $k$-Decisional Linear Assumption is an example of a family of decisional assumptions of strictly increasing hardness when $k$ grows. We show that for any such family of MDDH assumptions, the corresponding Kernel assumptions are also strictly increasingly weaker. This requires ruling out the existence of some black-box reductions between flexible problems (*i.e.*, computational problems with a non unique solution).

**Keywords:** Matrix assumptions · Computational problems · Black-box reductions · Structure preserving cryptography

## 1 Introduction

It is commonly understood that cryptographic assumptions play a crucial role in the development of secure, efficient protocols with strong functionalities. For instance, upon referring to the rapid development of pairing-based cryptography, X. Boyen [8] says that "it has been supported, in no small part, by a dizzying array of tailor-made cryptographic assumptions". Although this may be a reasonable price to pay for constructing new primitives or improve their efficiency, one should not lose sight of the ideal of using standard and simple assumptions. This is an important aspect of provable security. Indeed, Goldreich [16], for instance, cites "having clear definitions of one's assumptions" as one of the three main ingredients of good cryptographic practice.

There are many aspects to this goal. Not only it is important to use clearly defined assumptions, but also to understand the relations between them: to see,

Work supported by the Spanish research project MTM2013-41426-R and by a Sofja Kovalevskaja Award of the Alexander von Humboldt Foundation and the German Federal Ministry for Education and Research.

J.H. Cheon and T. Takagi (Eds.): ASIACRYPT 2016, Part I, LNCS 10031, pp. 729–758, 2016.
DOI: 10.1007/978-3-662-53887-6_27

for example, if two assumptions are equivalent or one is weaker than the other. Additionally, the definitions should allow to make accurate security claims. For instance, although technically it is correct to say that unforgeability of the Waters' signature scheme [42] is implied by the DDH Assumption, defining the CDH Assumption allows to make a much more precise security claim.

A notable effort in reducing the "dizzying array" of cryptographic assumptions is the work of Escala *et al.* [11]. They put forward a new family of decisional assumptions in a prime order group $\mathbb{G}$, the *Matrix Diffie-Hellman* Assumption ($\mathcal{D}_{\ell,k}$-MDDH). It says that, given some matrix $\mathbf{A} \in \mathbb{Z}_q^{\ell \times k}$ sampled from some distribution $\mathcal{D}_{\ell,k}$, it is hard to decide membership in Im $\mathbf{A}$, the subspace spanned by the columns of $\mathbf{A}$, in the exponent. Rather than as new assumption, it should be seen as an algebraic framework for decisional assumptions which includes as a special case the widely used $k$-Lin family.

This framework has some obvious conceptual advantages. For instance, it allows to explain all the members of the $k$-Lin assumption family (and also others, like the uniform assumption, appeared previously in [13,14,41]) as a single assumption and unify different constructions of the same primitive in the literature (e.g., the Naor-Reingold PRF [36] and the Lewko-Waters PRF [29] are special cases of the same construction instantiated with the 1-Lin and the 2-Lin Assumption, respectively). Another of its advantages is that it avoids arbitrary choices and instead points out to a trade-off between efficiency and security (a scheme based on any $\mathcal{D}_{\ell,k}$-MDDH Assumption can be instantiated with many different assumptions, some leading to stronger security guarantees and others leading to more efficient schemes). But follow-up work has also illustrated other possibly less obvious advantages. For instance, Herold *et al.* [21] have used the Matrix Diffie-Hellman abstraction to extend the model of composite-order to prime-order transformation of Freeman [13] and to derive efficiency improvements which were proven to be impossible in the original model.[1] We believe this illustrates that the benefits of conceptual clarity can translate into concrete improvements as well.

The security notions for cryptographic protocols can be classified mainly in hiding and unforgeability ones. The former typically appear in encryption schemes and commitments and the latter in signature schemes and soundness in zero-knowledge proofs. Although it is theoretically possible to base the hiding property on computational problems, most of the practical schemes achieve this notion either information theoretically or based on decisional assumptions, at least in the standard model. Likewise, unforgeability naturally comes from computational assumptions (typically implied by stronger, decisional assumptions). Thus, a natural question is if one can find a computational analogue of their MDDH Assumption which can be used in "unforgeability type" of security notions.

---

[1] More specifically, we are referring to the lower bounds on the image size of a projecting bilinear map of [39] which were obtained in Freeman model [13]. The results of [21] by-passed this lower bounds allowing to save on pairing operations for projecting maps in prime order groups.

Most computational problems considered in the literature are search problems with a unique solution like the discrete logarithm or CDH. But unforgeability actually means the inability to produce one among many solutions to a given problem (*e.g.*, in many signature schemes or zero knowledge proofs). Thus, unforgeability is more naturally captured by a *flexible computational problem*, namely, a problem which admits several solutions[2]. This maybe explains why several new flexible assumptions have appeared recently when considering "unforgeability-type" security notions in structure-preserving cryptography [2]. Thus a useful computational analogue of the MDDH Assumption should not only consider problems with a unique solution but also flexible problems which can naturally capture this type of security notions.

## 1.1   Our Results

In the following $\mathcal{G} = (\mathbb{G}, q, \mathcal{P})$, being $\mathbb{G}$ some group in additive notation of prime order $q$ generated by $\mathcal{P}$, that is, the elements of $\mathbb{G}$ are $\mathcal{Q} = a\mathcal{P}$ where $a \in \mathbb{Z}_q$. They will be denoted as $[a] := a\mathcal{P}$. This notation naturally extends to vectors and matrices as $[\boldsymbol{v}] = (v_1\mathcal{P}, \ldots, v_n\mathcal{P})$ and $[\mathbf{A}] = (A_{ij}\mathcal{P})$.

**Computational Matrix Assumptions.** In our first attempt to design a computational analogue of the MDDH Assumption, we introduce the *Matrix Computational DH Assumption*, (MCDH) which says that, given a uniform vector $[\boldsymbol{v}] \in \mathbb{G}^k$ and some matrix $[\mathbf{A}]$, $\mathbf{A} \leftarrow \mathcal{D}_{\ell,k}$ for $\ell > k$, it is hard to extend $[\boldsymbol{v}]$ to a vector in $\mathbb{G}^\ell$ in the image of $[\mathbf{A}]$, Im$[\mathbf{A}]$. Although this assumption is natural and is weaker than the MDDH one, we argue that it is equivalent to CDH.

We then propose the *Kernel Matrix DH Assumption* ($\mathcal{D}_{\ell,k}$-KerMDH). This new flexible assumption states that, given some matrix $[\mathbf{A}]$, $\mathbf{A} \leftarrow \mathcal{D}_{\ell,k}$ for some $\ell > k$, it is hard to find a vector $[\boldsymbol{v}] \in \mathbb{G}^\ell$ in the kernel of $\mathbf{A}^\top$. We observe that for some special instances of $\mathcal{D}_{\ell,k}$, this assumption has appeared in the literature in [2,18,19,27,32] under different names, like *Simultaneous Pairing, Simultaneous Double Pairing (SDP in the following), Simultaneous Triple Pairing, 1-Flexible CDH, 1-Flexible Square CDH*. Thus, the new KerMDH Assumption allows us to organize and give a unified view on several useful assumptions. This suggests that the KerMDH Assumption (and not the MCDH one) is the right computational analogue of the MDDH framework. Indeed, for any matrix distribution the $\mathcal{D}_{\ell,k}$-MDDH Assumption implies the corresponding $\mathcal{D}_{\ell,k}$-KerMDH Assumption. As a unifying algebraic framework, it offers the advantages mentioned above: it highlights the algebraic structure of any construction based on it, and it allows writing many instantiations of a given scheme in a compact way.

**The Power of Kernel Assumptions.** At Eurocrypt 2015, our KerMDH Assumptions were applied to design simpler QA-NIZK proofs of membership in

---

[2] In the cryptographic literature we sometimes find the term "strong" as an alternative to "flexible", like the Strong RSA or the Strong DDH.

linear spaces [26]. They have also been used to give more efficient constructions of structure preserving signatures [25], to generalize and simplify the results on quasi-adaptive aggregation of Groth-Sahai proofs [17] (given originally in [24]) and to construct a tightly secure QA-NIZK argument for linear subspaces with unbounded simulation soundness in [15]. The power of a KerMDH Assumption is that it allows to guarantee uniqueness. This has been used by Kiltz and Wee [26], for instance, to compile some secret key primitives to the public key setting. Indeed, Kiltz and Wee [26] modify a hash proof system (which is only designated verifier) to allow public verification (a QA-NIZK proof of membership). In a hash proof system for membership in some linear subspace of $\mathbb{G}^n$ spanned by the columns of some matrix $[\mathbf{M}]$, the public information is $[\mathbf{M}^\top \mathbf{K}]$, for some secret matrix $\mathbf{K}$, and given the proof $[\boldsymbol{\pi}]$ that $[\boldsymbol{y}]$ is in the subspace, verification tests if $[\boldsymbol{\pi}] \stackrel{?}{=} [\boldsymbol{y}^\top \mathbf{K}]$.

The core argument to compile this to a public key primitive is that given $([\mathbf{A}], [\mathbf{KA}])$, $\mathbf{A} \leftarrow \mathcal{D}_{\ell,k}$ and any pair $[\boldsymbol{y}], [\boldsymbol{\pi}]$, the previous test is equivalent to $e([\boldsymbol{\pi}^\top], [\mathbf{A}]) = e([\boldsymbol{y}^\top], [\mathbf{KA}])$, under the $\mathcal{D}_{\ell,k}$-KerMDH Assumption. Indeed,

$$e([\boldsymbol{\pi}^\top], [\mathbf{A}]) = e([\boldsymbol{y}^\top], [\mathbf{KA}]) \Longleftrightarrow e([\boldsymbol{\pi}^\top - \boldsymbol{y}^\top \mathbf{K}], [\mathbf{A}]) = [\mathbf{0}] \stackrel{\mathcal{D}_{\ell,k}\text{-KerMDH}}{\Longrightarrow}$$
$$\Longrightarrow [\boldsymbol{\pi}] = [\boldsymbol{y}^\top \mathbf{K}]. \tag{1}$$

That is, although potentially there are many possible proofs which satisfy the public verification equation (left hand side of Eq. (1)), the $\mathcal{D}_{\ell,k}$-KerMDH Assumption guarantees that only one of them is efficiently computable, so verification gives the same guarantees as in the private key setting (right hand side of Eq. (1)). This property is also used in a very similar way in [15] and also in the context of structure preserving signatures in [25]. In Sect. 5 we use it to argue that, of all the possible openings of a commitment, only one is efficiently computable, i.e. to prove computational soundness of a commitment scheme. Moreover, some previous works, notably in the design of structure preserving cryptographic primitives [1–3,31], implicitly used this property for one specific KerMDH Assumption: the Simultaneous (Double) Pairing Assumption.

On the other hand, we have already discussed the importance of having a precise and clear language when talking about cryptographic assumptions. This justifies the introduction of a framework specific to computational assumptions, because one should properly refer to the assumption on which security is actually based, rather than just saying "security is based on an assumption weaker than $\mathcal{D}_{\ell,k}$-MDDH". A part from being imprecise, a problem with such a statement is that might lead to arbitrary, not optimal choices. For example, the signature scheme of [30] is based on the SDP Assumption but a slight modification of it can be based on the $\mathcal{L}_2$-KerMDH Assumption. If the security guarantee is "the assumption is weaker than 2-Lin" then the modified scheme achieves shorter public key and more efficient verification with no loss in security. Further, the claim that security is based on the MDDH decisional assumptions when only computational ones are necessary might give the impression that a certain tradeoff is in place when this is not known to be the case. For instance, Jutla and

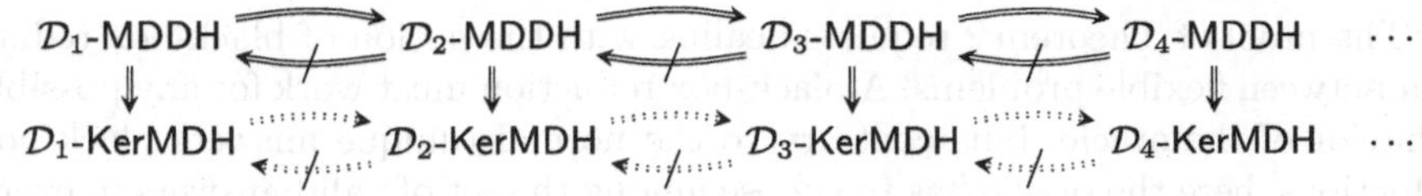

$$\mathcal{D}_1\text{-MDDH} \rightleftarrows_{/} \mathcal{D}_2\text{-MDDH} \rightleftarrows_{/} \mathcal{D}_3\text{-MDDH} \rightleftarrows_{/} \mathcal{D}_4\text{-MDDH} \quad \cdots$$

$$\downarrow \qquad\qquad \downarrow \qquad\qquad \downarrow \qquad\qquad \downarrow$$

$$\mathcal{D}_1\text{-KerMDH} \dashrightarrow_{/} \mathcal{D}_2\text{-KerMDH} \dashrightarrow_{/} \mathcal{D}_3\text{-KerMDH} \dashrightarrow_{/} \mathcal{D}_4\text{-KerMDH} \quad \cdots$$

**Fig. 1.** Implication and separation results between Matrix Assumptions (dotted arrows correspond to the new results).

Roy [24] construct constant-size QA-NIZK arguments of membership in linear spaces under what they call the "Switching Lemma", which is proven under a certain $\mathcal{D}_{k+1,k}$-MDDH Assumption. However, a close look at the proof reveals that in fact it is based on the corresponding $\mathcal{D}_{k+1,k}$-KerMDH Assumption[3]. For these assumptions, prior to our work, it was unclear whether the choice of larger $k$ gives any additional guarantees.

**Strictly Increasing Families of Kernel Assumptions.** An important problem is that it is not clear whether there are increasingly weaker families of KerMDH Assumptions. That is, some decisional assumptions families parameterized by $k$ like the $k$-Lin Assumption are known to be strictly increasingly weaker. The proof of increasing hardness is more or less immediate and the term *strictly* follows from the fact that every two $\mathcal{D}_{\ell,k}$-MDDH and $\mathcal{D}_{\widetilde{\ell},\widetilde{k}}$-MDDH problems with $\widetilde{k} < k$ are separated by an oracle computing a $k$-linear map. For the computational case, increasing hardness is also not too difficult, but nothing is known about *strictly* increasing hardness (see Fig. 1). This means that, as opposed to the decisional case, prior to our work, for protocols based on KerMDH Assumptions there was no-known tradeoff between larger $k$ (less efficiency) and security.

In this paper, we prove that the families of matrix distributions in [11], $\mathcal{U}_{\ell,k}$, $\mathcal{L}_k$, $\mathcal{SC}_k$, $\mathcal{C}_k$ and $\mathcal{RL}_k$, as well as a new distribution we propose in Sect. 6, the *circulant* family $\mathcal{CI}_{k,d}$, define families of kernel problems with increasing hardness. For this we show a tight reduction from the smaller to the larger problems in each family. Our main result (Theorem 2) is to prove that the hardness of these problems is *strictly* increasing. For this, we prove that there is no blackbox reduction from the larger to the smaller problems in the multilinear generic group model. These new results correspond to the dotted arrows in Fig. 1.

Having in mind that the computational problems we study in the paper are defined in a generic way, that is without specifying any particular group, the generic group approach arises naturally as the setting for the analysis of their hardness and reducibility relations. Otherwise, we would have to rely on specific properties of the representation of the elements of particular group families, not captured by the generic model.

---

[3] To see this, note that in the proof of their "Switching Lemma" on which soundness is based, they use the output of the adversary to decide if $f \stackrel{?}{\in} \mathrm{Im}\,\mathbf{A}$, $\mathbf{A} \leftarrow \mathcal{RL}_k$, by checking whether $[f]$ is orthogonal to the adversary's output (Eq. (1), proof of Lemma 1, [24], full version), and where $\mathcal{RL}_k$ is the matrix distribution of Sect. 2.3.

The proof of Theorem 2 requires dealing with the notion of black-box reduction between flexible problems. A black-box reduction must work for any possible behavior of the oracle, but, contrary to the normal (unique answer) black-box reductions, here the oracle has to choose among the set of valid answers in every call. Ruling out the existence of a reduction implies that for any reduction there is an oracle behavior for which the reduction fails. This is specially subtle when dealing with multiple oracle calls. We think that the proof technique we introduce to deal with these issues can be considered as a contribution in itself and can potentially be used in future work.

Combining the black-box techniques and the generic group model is not new in the literature. For instance Dodis et al. [10] combine the black-box reductions and a generic model for the group $\mathbb{Z}_n^*$ to show some uninstantiability results for FDH-RSA signatures.

Theorem 2 supports the intuition that there is a tradeoff between the size of the matrix—which typically results in less efficiency—and the hardness of the KerMDH Problems, and justifies the generalization of several protocols to different choices of $k$ given in [17, 24–26].

**Applications.** The discussion of our results given so far should already highlight some of the advantages of using the new Kernel family of assumptions and the power of these new assumptions, which have already been used in compelling applications in follow-up work in [17, 25, 26]. To further illustrate the usefulness of the new framework, we apply it to the study of trapdoor commitments. First, we revisit the Pedersen commitment [38] to vectors of scalars and its extension to vectors of group elements of Abe *et al.* [2] in bilinear groups. We unify these two constructions and we generalize to commit vectors of elements at each level $\mathbb{G}_r$, for any $0 \leq r \leq m$ under the extension of KerMDH Assumptions to the ideal $m$-graded encodings setting. In particular, when $m = 2$ we recover in a single construction as a special case both the original Pedersen and Abe *et al.* commitments.

The (generalized) Pedersen commitment maps vectors in $\mathbb{G}_r$ to vectors in $\mathbb{G}_{r+1}$, is perfectly hiding and computationally binding under any Kernel Assumption. In Sect. 5.2 we use it as a building block to construct a "group-to-group" commitment, which maps vectors in $\mathbb{G}_r$ to vectors in the same group $\mathbb{G}_r$. These commitments were defined in [3] because they are a good match to Groth-Sahai proofs. In [3], two constructions were given, one in asymmetric and the other in symmetric bilinear groups. Both are optimal in terms of commitment size and number of verification equations. Rather surprisingly, we show that both constructions in [3] are special instances of our group-to-group commitment for some specific matrix distributions.

**A New Family of MDDH Assumptions of Optimal Representation Size.** We also propose a new interesting family of Matrix distributions, the circulant matrix distribution, $\mathcal{CI}_{k,d}$, which defines new MDDH and KerMDH assumptions. This family generalizes the Symmetric Cascade Distribution $(\mathcal{SC}_k)$ defined in [11] to matrices of size $\ell \times k$, $\ell = k + d > k + 1$. We prove that it has optimal representation size $d$ independent of $k$ among all matrix distributions of the same size. The case $\ell > k + 1$ typically arises when one considers commitments/encryption in which the message is a vector of group elements instead of a single group element and the representation size typically affects the size of the public parameters.

We prove the hardness of the $\mathcal{CI}_{k,d}$-KerMDH Problem, by proving that the $\mathcal{CI}_{k,d}$-MDDH Problem is generically hard in $k$-linear groups. Analyzing the hardness of a family of decisional problems (depending on a parameter $k$) can be rather involved, specially when an efficient $k$-linear map is supposed to exist. This is why in [11], the authors gave a practical criterion for generic hardness when $\ell = k + 1$ in terms of irreducibility of some polynomials involved in the description of the problem. This criterion was used then to prove the generic hardness of several families of MDDH Problems. To analyze the generic hardness of the $\mathcal{CI}_{k,d}$-MDDH Problem for any $d$, the techniques in [11] are not practical enough, and we need some extensions of these techniques for the case $\ell > k + 1$, recently introduced in [20]. However, we could not avoid the explicit computation of a large (but well-structured) Gröbner basis of an ideal associated to the matrix distribution. The new assumption can be used to instantiate the commitment schemes of Sect. 5 with shorter public parameters and improved efficiency.

## 2 Preliminaries

For $\lambda \in \mathbb{N}$, we write $1^\lambda$ for the string of $\lambda$ ones. For a set $S$, $s \leftarrow S$ denotes the process of sampling an element $s$ from $S$ uniformly at random. For an algorithm $\mathcal{A}$, we write $z \leftarrow \mathcal{A}(x, y, \ldots)$ to indicate that $\mathcal{A}$ is a (probabilistic) algorithm that outputs $z$ on input $(x, y, \ldots)$. For any two computational problems $\mathbb{P}_1$ and $\mathbb{P}_2$ we recall that $\mathbb{P}_1 \Rightarrow \mathbb{P}_2$ denotes the fact that $\mathbb{P}_1$ reduces to $\mathbb{P}_2$, and then '$\mathbb{P}_1$ is hard' $\Rightarrow$ '$\mathbb{P}_2$ is hard'. Thus, we will use '$\Rightarrow$' both for computational problems and for the corresponding hardness assumptions.

Let Gen denote a cyclic group instance generator, that is a probabilistic polynomial time (PPT) algorithm that on input $1^\lambda$ returns a description $\mathcal{G} = (\mathbb{G}, q, \mathcal{P})$ of a cyclic group $\mathbb{G}$ of order $q$ for a $\lambda$-bit prime $q$ and a generator $\mathcal{P}$ of $\mathbb{G}$. We use additive notation for $\mathbb{G}$ and its elements are $a\mathcal{P}$, for $a \in \mathbb{Z}_q$ and will be denoted as $[a] := a\mathcal{P}$. The notation extends to vectors and matrices in the natural way as $[\boldsymbol{v}] = (v_1\mathcal{P}, \ldots, v_n\mathcal{P})$ and $[\mathbf{A}] = (A_{ij}\mathcal{P})$. For a matrix $\mathbf{A} \in \mathbb{Z}_q^{\ell \times k}$, Im $\mathbf{A}$ denotes the subspace of $\mathbb{Z}_q^\ell$ spanned by the columns of $\mathbf{A}$. Thus, Im$[\mathbf{A}]$ is the corresponding subspace of $\mathbb{G}^\ell$.

### 2.1 Multilinear Maps

In the case of groups with a bilinear map, or more generally with a $k$-linear map for $k \geq 2$, we consider a generator producing the tuple $(e_k, \mathbb{G}_1, \mathbb{G}_k, q, \mathcal{P}_1, \mathcal{P}_k)$, where $\mathbb{G}_1, \mathbb{G}_k$ are cyclic groups of prime-order $q$, $\mathcal{P}_i$ is a generator of $\mathbb{G}_i$ and $e_k$ is a non-degenerate efficiently computable $k$-linear map $e_k : \mathbb{G}_1^k \to \mathbb{G}_k$, such that $e_k(\mathcal{P}_1, \ldots, \mathcal{P}_1) = \mathcal{P}_k$. We actually consider graded encodings which offer a richer structure. For any fixed $k \geq 1$, let $\mathsf{MGen}_k$ be a PPT algorithm that on input $1^\lambda$ returns a description of a graded encoding $\mathcal{MG}_k = (e, \mathbb{G}_1, \ldots, \mathbb{G}_k, q, \mathcal{P}_1, \ldots, \mathcal{P}_k)$, where $\mathbb{G}_1, \ldots, \mathbb{G}_k$ are cyclic groups of prime-order $q$, $\mathcal{P}_i$ is a generator of $\mathbb{G}_i$ and $e$ is a collection of non-degenerate efficiently computable bilinear maps $e_{i,j} : \mathbb{G}_i \times \mathbb{G}_j \to \mathbb{G}_{i+j}$, for $i + j \leq k$, such that $e(\mathcal{P}_i, \mathcal{P}_j) = \mathcal{P}_{i+j}$. For simplicity we will omit the subindexes of $e$ when they become clear from the context. Sometimes $\mathbb{G}_0$ is used to refer to $\mathbb{Z}_q$. For group elements we use the following implicit notation: for all $i = 1, \ldots, k$, $[a]_i := a\mathcal{P}_i$. The notation extends in a natural way to vectors and matrices and to linear algebra operations. We sometimes drop the index when referring to elements in $\mathbb{G}_1$, i.e., $[a] := [a]_1 = a\mathcal{P}_1$. In particular, it holds that $e([a]_i, [b]_j) = [ab]_{i+j}$.

Additionally, for the asymmetric case, let $\mathsf{AGen}_2$ be a PPT algorithm that on input $1^\lambda$ returns a description of an asymmetric bilinear group $\mathcal{AG}_2 = (e, \mathbb{G}, \mathbb{H}, \mathbb{T}, q, \mathcal{P}, \mathcal{Q})$, where $\mathbb{G}, \mathbb{H}, \mathbb{T}$ are cyclic groups of prime-order $q$, $\mathcal{P}$ is a generator of $\mathbb{G}$, $\mathcal{Q}$ is a generator of $\mathbb{H}$ and $e : \mathbb{G} \times \mathbb{H} \to \mathbb{T}$ is a non-degenerate, efficiently computable bilinear map. In this case we refer to group elements as: $[a]_G := a\mathcal{P}$, $[a]_H := a\mathcal{Q}$ and $[a]_T := ae(\mathcal{P}, \mathcal{Q})$.

### 2.2   A Generic Model for Groups with Graded Encodings

In this section we describe a (purely algebraic) generic model for the graded encodings functionality, in order to obtain meaningful results about the hardness and separations of computational problems. The model is an adaptation of Maurer's generic group model [33,34] including the $k$-graded encodings, but in a completely algebraic formulation that follows the ideas in [5,12,20]. Since the $k$-graded encodings functionality implies the $k$-linear group functionality, the former gives more power to the adversaries or reductions working within the corresponding generic model. This in particular means that non-existential results proven in the richer $k$-graded encodings generic model also imply the same results in the $k$-linear group generic model. Therefore, in this paper we consider the former model. Due to the space limitations, we can only give a very succinct description of the model. See the full version of the paper [35] for a detailed and more formal description.

In a first approach we consider Maurer's model adapted to the graded encodings functionality, but still not phrased in a purely algebraic language. In this model, an algorithm $\mathcal{A}$ does not deal with proper group elements in $[y]_a \in \mathbb{G}_a$, but only with labels $(Y, a)$, and it has access to an additional oracle internally performing the group operations, so that $\mathcal{A}$ cannot benefit from the particular way the group elements are represented. Namely, on start all the group elements $[x_1]_{a_1}, \ldots, [x_\alpha]_{a_\alpha}$ in the input intended for $\mathcal{A}$ are replaced by the labels $(X_1, a_1), \ldots, (X_\alpha, a_\alpha)$. Then, $\mathcal{A}$ actually receives as input the set of labels, and possibly some other non-group elements (i.e., that do not belong to any of the groups $\mathbb{G}_1, \ldots, \mathbb{G}_k$), denoted as $\widetilde{x}$, and considered as a bit string. For each group $\mathbb{G}_a$ two additional labels $(0, a), (1, a)$, corresponding to the neutral element and the generator, are implicitly given to $\mathcal{A}$. Then $\mathcal{A}$ can adaptively make the following queries to an oracle implementing the $k$-graded encodings:

- GroupOp$((Y_1, a), (Y_2, a))$: group operation in $\mathbb{G}_a$ for two previously issued labels in $\mathbb{G}_a$ resulting in a new label $(Y_3, a)$ in $\mathbb{G}_a$.
- GroupInv$((Y, a))$: similarly for group inversion in $\mathbb{G}_a$.
- GroupPair$((Y_1, a), (Y_2, b))$: bilinear map for two previously issued labels in $\mathbb{G}_a$ and $\mathbb{G}_b$, $a + b \leq k$, resulting in a new label $(Y_3, a + b)$ in $\mathbb{G}_{a+b}$.
- GroupEqTest$((Y_1, a), (Y_2, a))$: test two previously issued labels in $\mathbb{G}_a$ for equality of the corresponding group elements, resulting in a bit ($1 = $ equality).

In addition, the oracle performs the actual computations with the group elements, and it uses them to answer the GroupEqTest queries. Every badly formed query (for instance, containing a label not previously issued by the oracle or as an input to $\mathcal{A}$) is answered with a special rejection symbol $\perp$. Following the usual step in generic group model proofs (see for instance [5,11,20]), we use polynomials as labels to group elements. Namely, labels in $\mathbb{G}_a$ are polynomials in $\mathbb{Z}_q[\boldsymbol{X}]$, where the algebraic variables $\boldsymbol{X} = (X_1, \ldots, X_\alpha)$ are just formal representations of the group elements in the input of $\mathcal{A}$. Now the oracle computes the new labels using the natural polynomial operations: GroupOp$((Y_1, a), (Y_2, a)) = (Y_1 + Y_2, a)$, GroupInv$((Y, a)) = (-Y, a)$ and GroupPair$((Y_1, a), (Y_2, b)) = (Y_1 Y_2, a + b)$. It is easy to see that for any valid label $(Y, a)$, $\deg Y \leq a$.[4]

The output of $\mathcal{A}$ consists only of some labels $(Y_1, b_1), \ldots, (Y_\beta, b_\beta)$ (given at some time by the oracle) corresponding to group elements $[y_1]_{b_1}, \ldots, [y_\beta]_{b_\beta}$, along with some non-group elements, denoted as $\widetilde{y}$. Therefore, for any fixed random tape of $\mathcal{A}$ and any choice of the non-group elements $\widetilde{x}$, there exist polynomials $Y_1, \ldots, Y_\beta \in \mathbb{Z}_q[\boldsymbol{X}]$ of degrees upper bounded by $b_1, \ldots, b_\beta$ respectively, with coefficients known to $\mathcal{A}$. Notice that $\mathcal{A}$ itself can predict all answers given by the oracle except for some GroupEqTest queries. In particular, some GroupEqTest queries trivially result in 1, due to the group structure (*e.g.*, GroupOp$((Y, a), $ GroupInv$((Y, a)))$ is the same as $(0, a)$), or due to the (known) *a priori* constraints in the input group elements (*i.e.*, the definition of the problem instance given to $\mathcal{A}$). The answers to nontrivial GroupEqTest queries (*i.e.*, queries that cannot be trivially predicted by $\mathcal{A}$) are the only effective information $\mathcal{A}$ can receive from the generic group oracle.

We now introduce a "purely algebraic" version of the generic model. For that, we need to assume that the distribution of $\boldsymbol{x}$ can be sampled by evaluating a polynomial map $f$ of constant degree at a random point.[5] This is not an actual restriction in our context since all Matrix Diffie-Hellman problems fulfil this requirement. In the "purely algebraic" model we redefine the oracle GroupEqTest to answer 1 if and only if $\mathcal{A}$ can itself predict the positive answer. Namely GroupEqTest$((Y_1, a), (Y_2, a)) = 1$ if and only if $Y_1 \circ f = Y_2 \circ f$ as polynomials over $\mathbb{Z}_q$. With this change the behavior of $\mathcal{A}$ can only differ

---

[4] It clearly holds for the input group elements (since $\deg Y = 1$), and the inequality is preserved by GroupOp, GroupInv and GroupPair.

[5] A formal definition of this notion is given in the full version of the paper.

negligibly from the original,[6] meaning that generic algorithms perform almost equally in Maurer's model and its purely algebraic version. But now, any generic algorithm is just modelled by a set of polynomials. As we need to handle elements in different groups, we will use the shorter vector notation $[\boldsymbol{x}]_{\boldsymbol{a}} = ([x_1]_{a_1}, \ldots, [x_\alpha]_{a_\alpha}) = (x_1 \mathcal{P}_{a_1}, \ldots, x_\alpha \mathcal{P}_{a_\alpha}) \in \mathbb{G}_{a_1} \times \cdots \times \mathbb{G}_{a_\alpha}$. Note that the length of a vector of indices $\boldsymbol{a}$ is denoted by a corresponding Greek letter $\alpha$. We will also use a tilde to denote variables containing only non-group elements (*i.e.*, elements not in any of $\mathbb{G}_1, \ldots, \mathbb{G}_k$).

**Lemma 1.** *Let $\mathcal{A}$ be an algorithm in the (purely algebraic) generic multilinear group model. Let $([\boldsymbol{x}]_{\boldsymbol{a}}, \widetilde{x})$ and $([\boldsymbol{y}]_{\boldsymbol{b}}, \widetilde{y})$ respectively be the input and output of $\mathcal{A}$. Then, for every choice of $\widetilde{x}$ and any choice of the random tape of $\mathcal{A}$, there exist polynomials $Y_1, \ldots, Y_\beta \in \mathbb{Z}_q[\boldsymbol{X}]$ of degree upper bounded by $b_1, \ldots, b_\beta$ such that $\boldsymbol{y} = \boldsymbol{Y}(\boldsymbol{x})$, for all possible $\boldsymbol{x} \in \mathbb{Z}_q^n$, where $\boldsymbol{Y} = (Y_1, \ldots, Y_\beta)$. Moreover, $\widetilde{y}$ does not depend on $\boldsymbol{x}$.*

The proof of the lemma comes from the above discussion.

As usually, the proposed generic model reduces the analysis of the hardness of some problems to solving a merely algebraic problem related to polynomials. As an example, consider a computational problem $\mathcal{P}$ which instances are entirely described by some group elements in the base group $\mathbb{G}_1$, $[\boldsymbol{x}] \leftarrow \mathcal{P}.\mathsf{InstGen}(1^\lambda)$, and its solutions are also described by some group elements $[\boldsymbol{y}]_{\boldsymbol{b}} \in \mathcal{P}.\mathsf{Sol}([\boldsymbol{x}])$. We also assume that $\mathcal{P}.\mathsf{InstGen}$ just samples $\boldsymbol{x}$ by evaluating polynomial functions of constant degree at a random point. Then, $\mathcal{P}$ is hard in the purely algebraic generic multilinear group model if and only if for all (randomized) polynomials $Y_1, \ldots, Y_\beta \in \mathbb{Z}_q[\boldsymbol{X}]$ of degrees upper bounded by $b_1, \ldots, b_\beta$ respectively,

$$\Pr([\boldsymbol{y}]_{\boldsymbol{b}} \in \mathcal{P}.\mathsf{Sol}([\boldsymbol{x}]) : [\boldsymbol{x}] \leftarrow \mathcal{P}.\mathsf{InstGen}(1^\lambda), \boldsymbol{y} = \boldsymbol{Y}(\boldsymbol{x})) \in negl(\lambda)$$

where $\boldsymbol{Y} = (Y_1, \ldots, Y_m)$ and the probability is computed with respect the random coins of the instance generator and the randomized polynomials.[7] In a few words, this means that the set $\mathcal{P}.\mathsf{Sol}([\boldsymbol{x}])$ cannot be hit by polynomials of the given degree evaluated at $\boldsymbol{x}$.

This model extends naturally to algorithms with oracle access (*e.g.*, black-box reductions) but only when the oracles fit well into the generic model. Let us consider the algorithm $\mathcal{A}^{\mathcal{O}}$, with oracle access to $\mathcal{O}$. A completely arbitrary oracle (specified in the plain model) could have access to the internal representation of the group elements, and then it could leak some information about the group elements that is outside the generic group model. Thus, we will impose the very

---

[6] As a standard argument used in proofs in the generic group model, the difference between the original model and its purely algebraic reformulation amounts to a negligible probability, which is typically upper-bounded by using Schwartz-Zippel Lemma and the union bound, as shown for instance in [5,12,20].

[7] We can similarly deal with problems with non-group elements both in the instance description and the solution, but this would require a more sophisticated formalization, in which both the polynomials and the non-group elements in the solution could depend on the non-group elements in the instance, but in an efficient way.

limiting constraint that the oracles are also "algebraic", meaning that the oracle's input/output behavior respects the one-wayness of the graded encodings, and it only performs polynomial operations on the input labels.

**Definition 1.** *Let $([u]_d, \widetilde{u})$ and $([v]_e, \widetilde{v})$ respectively be a query to an oracle $\mathcal{O}$ and its corresponding answer, where $\widetilde{u}$ and $\widetilde{v}$ contain the respective non-group elements. The oracle $\mathcal{O}$ is called algebraic if for any choice of $\widetilde{u}$ there exist polynomials $V_1, \ldots, V_\epsilon \in \mathbb{Z}_q[U, R]$, $R = (R_1, \ldots, R_\rho)$, of constant degree (in the security parameter) such that*

- *for the specific choice of $\widetilde{u}$, $v_i = V_i(u, r)$, $i = 1, \ldots, \epsilon$, for all $u \in \mathbb{Z}_q^\epsilon$ and $r \in \mathbb{Z}_q^\rho$, where $r = (r_1, \ldots, r_\rho)$ are random parameters defined and uniformly sampled by the oracle,*
- *$\widetilde{v}$ does not depend on $u, r$ (thus, $r$ can only have influence in the group elements in the answer),*
- *$V_j$ does not depend on any $U_i$ such that $e_j < d_i$ (in order to preserve the one-wayness of the graded encodings).*

The parameters $r$ capture the behavior of an oracle solving a problem with many solutions (called here a "flexible" problem). They could be independent or not across different oracle calls, depending on whether the oracle is stateless or stateful. For technical reasons we consider only the stateless case with uniform sampling. Observe that the first two requirements in the definition mean that $v$ depends algebraically on $u, r$ and no extra information about $u, r$ can be leaked through $\widetilde{v}$. Removing any of these requirements from the definition results in that a generic algorithm using such an oracle will no longer be algebraically generic. Also notice that after a call to an algebraic oracle, there is no guarantee that labels $(Y, a)$ fulfil the bound $\deg Y \leq a$.

Although the notion of algebraic oracle looks very limiting (*e.g.*, it excludes a Discrete Logarithm oracle, as it destroys the one-wayness property of the graded encodings, but oracles solving CDH or the Bilinear Computational Diffie-Hellman problem fit well in the definition), it is general enough for our purposes. We will need the following generalization of Lemma 1:

**Lemma 2.** *Let $\mathcal{A}^{\mathcal{O}}$ be an oracle algorithm in the (purely algebraic) generic multilinear group model, making a constant number of calls $Q$ to an algebraic oracle $\mathcal{O}$. Let $([x]_a, \widetilde{x})$ and $([y]_b, \widetilde{y})$ respectively be the input and output of $\mathcal{A}$. Then, for every choice of $\widetilde{x}$ and the random tape, there exist polynomials of constant degree $Y_1, \ldots, Y_\beta \in \mathbb{Z}_q[X, R_1, \ldots, R_Q]$, such that $y = Y(x, r_1, \ldots, r_Q)$, for all possible inputs, where $Y = (Y_1, \ldots, Y_\beta)$, and $r_1, \ldots, r_Q$ are the parameters introduced in Definition 1 for the $Q$ queries. Moreover, $\widetilde{y}$ does not depend on $x$ or $r_1, \ldots, r_Q$.*

The proof of this lemma is given in Appendix A.

## 2.3   The Matrix Decisional Diffie-Hellman Assumption

We recall here the definition of the decisional assumptions introduced in [11], which are the starting point of our flexible computational matrix problems.

**Definition 2.** *[11], Let $\ell, k \in \mathbb{N}$ with $\ell > k$. We call $\mathcal{D}_{\ell,k}$ a matrix distribution if it outputs (in polynomial time, with overwhelming probability) matrices in $\mathbb{Z}_q^{\ell \times k}$ of full rank $k$. We denote $\mathcal{D}_k := \mathcal{D}_{k+1,k}$.*

**Definition 3 ($\mathcal{D}_{\ell,k}$-MDDH Assumption).** *[11] Let $\mathcal{D}_{\ell,k}$ be a matrix distribution. The $\mathcal{D}_{\ell,k}$-Matrix Diffie-Hellman ($\mathcal{D}_{\ell,k}$-MDDH) Problem is telling apart the two probability distributions $(\mathbb{G}, q, \mathcal{P}, [\mathbf{A}], [\mathbf{A}\mathbf{w}])$ and $(\mathbb{G}, q, \mathcal{P}, [\mathbf{A}], [\mathbf{z}])$, where $\mathbf{A} \leftarrow \mathcal{D}_{\ell,k}, \mathbf{w} \leftarrow \mathbb{Z}_q^k, \mathbf{z} \leftarrow \mathbb{Z}_q^\ell$.*

*We say that the $\mathcal{D}_{\ell,k}$-Matrix Diffie-Hellman ($\mathcal{D}_{\ell,k}$-MDDH) Assumption holds relative to* Gen *if the corresponding problem is hard, that is, if for all PPT adversaries $\mathcal{A}$, the advantage*

$$\mathbf{Adv}_{\mathcal{D}_{\ell,k}, \mathsf{Gen}}(\mathcal{A}) = \Pr[\mathcal{A}(\mathcal{G}, [\mathbf{A}], [\mathbf{A}\mathbf{w}]) = 1] - \Pr[\mathcal{A}(\mathcal{G}, [\mathbf{A}], [\mathbf{z}]) = 1] \in negl(\lambda),$$

*where the probability is taken over $\mathcal{G} = (\mathbb{G}, q, \mathcal{P}) \leftarrow \mathsf{Gen}(1^\lambda)$, $\mathbf{A} \leftarrow \mathcal{D}_{\ell,k}, \mathbf{w} \leftarrow \mathbb{Z}_q^k, \mathbf{z} \leftarrow \mathbb{Z}_q^\ell$ and the coin tosses of adversary $\mathcal{A}$.*

In the case of asymmetric bilinear groups or symmetric $k$-linear groups, we similarly say that the $\mathcal{D}_{\ell,k}$-MDDH Assumption holds relative to AGen$_2$ or MGen$_k$, respectively. In the former we specify if the assumption holds in the left ($\mathcal{A}$ receives $[\mathbf{A}]_G$, $[\mathbf{A}\mathbf{w}]_G$ or $[\mathbf{z}]_G$), or in the right ($\mathcal{A}$ receives $[\mathbf{A}]_H$, $[\mathbf{A}\mathbf{w}]_H$ or $[\mathbf{z}]_H$).

**Definition 4.** *A matrix distribution $\mathcal{D}_{\ell,k}$ is hard if the corresponding $\mathcal{D}_{\ell,k}$-MDDH problem is hard in the generic $k$-linear group model.*

Many different matrix distributions appear in the literature. Namely, the cascade $\mathcal{C}_k$ and symmetric cascade $\mathcal{SC}_k$ distributions were presented in [11], while the uniform $\mathcal{U}_{\ell,k}$, the linear $\mathcal{L}_k$, the randomized linear $\mathcal{RL}_k$ and the square polynomial $\mathcal{P}_{\ell,2}$ distributions were implicitly used in some previous works. We give their explicit definitions in Appendix B.

## 3    The Matrix Diffie-Hellman Computational Problems

In this section we introduce two families of search problems naturally related to the Matrix Decisional Diffie-Hellman problems. In the first family, given a matrix $[\mathbf{A}]$, where $\mathbf{A} \leftarrow \mathcal{D}_{\ell,k}$, and the first $k$ components of a vector $[\mathbf{z}]$, the problem is completing it so that $\mathbf{z} \in \operatorname{Im} \mathbf{A}$.

**Definition 5 ($\mathcal{D}_{\ell,k}$-MCDH).** *Given a matrix distribution $\mathcal{D}_{\ell,k}$, such that the upper $k \times k$ submatrix of $\mathbf{A} \leftarrow \mathcal{D}_{\ell,k}$ has full rank with overwhelming probability, the computational matrix Diffie-Hellman Problem is given $([\mathbf{A}], [\mathbf{z}_0])$, with $\mathbf{A} \leftarrow \mathcal{D}_{\ell,k}, \mathbf{z}_0 \leftarrow \mathbb{Z}_q^k$, compute $[\mathbf{z}_1] \in \mathbb{G}^{\ell-k}$ such that $(\mathbf{z}_0 \| \mathbf{z}_1) \in \operatorname{Im} \mathbf{A}$.*

The full-rank condition ensures the existence of solutions to the $\mathcal{D}_{\ell,k}$-MCDH problem instance. Thus, we tolerate the existence of a negligible fraction of unsolvable problem instances. Indeed, all known interesting matrix distributions fulfil this requirement. Notice that CDH and the computational $k$-Lin problems are particular examples of MCDH problems. Namely, CDH is exactly $\mathcal{L}_1$-MCDH and the computational $k$-Lin problem is $\mathcal{L}_k$-MCDH. Indeed, the $\mathcal{L}_1$-MCDH problem is given $[1], [a], [z_1]$, compute $[z_2]$ such that $(z_1, z_2)$ is collinear with $(1, a)$, or equivalently, $z_2 = z_1 a$, which is solving the CDH problem. All MCDH problems have a unique solution and they appear naturally in some scenarios using MDDH problems. For instance, the one-wayness of the encryption scheme in [11] is equivalent to the corresponding MCDH assumption.

There is an immediate relation between any MCDH problem and its decisional counterpart. Not surprisingly, for any matrix distribution $\mathcal{D}_{\ell,k}$, $\mathcal{D}_{\ell,k}$-MDDH $\Rightarrow$ $\mathcal{D}_{\ell,k}$-MCDH.

We are not going to study the possible reductions between MCDH problems, due to the fact that, essentially, any MCDH problem amounts to computing some polynomial on the elements of $\mathbf{A}$, and it is then equivalent to CDH ([4,23]), although the tightness of the reduction depends on the degree of the polynomial.

In the second family of computational problems, given a matrix $[\mathbf{A}]$, where $\mathbf{A} \leftarrow \mathcal{D}_{\ell,k}$, the problem is finding $[\boldsymbol{x}]$ such that $\boldsymbol{x} \in \ker \mathbf{A}^\top \backslash \{\mathbf{0}\}$. It is notable that some computational problems in the literature are particular cases of this second family.

**Definition 6** ($\mathcal{D}_{\ell,k}$-KerMDH). *Given a matrix distribution $\mathcal{D}_{\ell,k}$, the Kernel Diffie-Hellman Problem is given $[\mathbf{A}]$, with $\mathbf{A} \leftarrow \mathcal{D}_{\ell,k}$, find a nonzero vector $[\boldsymbol{x}] \in \mathbb{G}^\ell$ such that $\boldsymbol{x}$ is orthogonal to $\mathrm{Im}\,\mathbf{A}$, that is, $\boldsymbol{x} \in \ker \mathbf{A}^\top \backslash \{\mathbf{0}\}$.*

Definition 6 naturally extends to asymmetric bilinear groups. There, given $[\mathbf{A}]_H$, the problem is to find $[\boldsymbol{x}]_G$ such that $\boldsymbol{x} \in \ker \mathbf{A}^\top \backslash \{\mathbf{0}\}$. A solution can be obviously verified by checking if $e([\boldsymbol{x}^\top]_G, [\mathbf{A}]_H) = [\mathbf{0}]_T$. We can also consider an extension of this problem in which the goal is to solve the same problem but giving the solution in a different group $\mathbb{G}_r$, in some ideal graded encoding $\mathcal{MG}_m$, for some $0 \le r \le \min(m, k-1)$. The case $r = 1$ corresponds to the previous problem defined in a $m$-linear group.

**Definition 7** ($(r, m, \mathcal{D}_{\ell,k})$-KerMDH). *Given a matrix distribution $\mathcal{D}_{\ell,k}$ over a $m$-linear group $\mathcal{MG}_m$ and $r$ an integer $0 \le r \le \min(m, k-1)$, the $(r, m, \mathcal{D}_{\ell,k})$-KerMDH Problem is to find $[\boldsymbol{x}]_r \in \mathbb{G}_r^\ell$ such that $\boldsymbol{x} \in \ker \mathbf{A}^\top \backslash \{\mathbf{0}\}$.*

When the precise degree of multilinearity $m$ is not an issue, we will write $(r, \mathcal{D}_{\ell,k})$-KerMDH instead of $(r, m, \mathcal{D}_{\ell,k})$-KerMDH, for any $m \ge r$. We excluded the case $r \ge k$ because the problem is easy.

**Lemma 3.** *For all integers $k \le r \le m$ and for all matrix distributions $\mathcal{D}_{\ell,k}$, the $(r, m, \mathcal{D}_{\ell,k})$-KerMDH Problem is easy.*

The kernel problem is also harder than the corresponding decisional problem, in multilinear groups.

**Lemma 4.** *In a $m$-linear group, $\mathcal{D}_{\ell,k}$-MDDH $\Rightarrow (r, m, \mathcal{D}_{\ell,k})$-KerMDH for any matrix distribution $\mathcal{D}_{\ell,k}$ and for any $0 \le r \le m - 1$. In particular, for $m \ge 2$, $\mathcal{D}_{\ell,k}$-MDDH $\Rightarrow \mathcal{D}_{\ell,k}$-KerMDH.*

The proofs of Lemmas 3, and 4 can be found in the full version of this paper [35].

### 3.1 The Kernel DH Assumptions in the Multilinear Maps Candidates

We have shown that for any hard matrix distribution $\mathcal{D}_{\ell,k}$ the $\mathcal{D}_{\ell,k}$-KerMDH problem is generically hard in $m$-linear groups. We emphasize that all our results refer to generic, ideal multilinear maps (in fact, to graded encodings, which have more functionality). Our aim is only to give necessary condition for the assumptions to hold in candidate multilinear maps. The status of current candidate multilinear maps is rather uncertain, e.g. it is described in [28] as "break-and-repair mode". Thus, it is hard to argue if our assumptions hold in any concrete instantiation and we leave this as an open question for further investigation.

### 3.2 A Unifying View on Computational Matrix Problems

In this section we show how some computational problems in the cryptographic literature are unified as particular instances of KerMDH problems. Their explicit definitions are given in Appendix C. It is straightforward to see that Find-Rep [9] Assumption is just $(0, \mathcal{U}_{\ell,1})$-KerMDH, the Simultaneous Double Pairing Assumption (SDP) [2] is $\mathcal{RL}_2$-KerMDH, the Simultaneous Triple Pairing [18] Assumption is $\mathcal{U}_2$-KerMDH, the Simultaneous Pairing [19] Assumption is $\mathcal{P}_{\ell,2}$-KerMDH. The Double Pairing (DP) [18] Assumption corresponds to $\mathcal{U}_1$-KerMDH in an asymmetric bilinear setting. On the other hand, the 1-Flexible Diffie-Hellman (1-FlexDH) [32] Assumption is $\mathcal{C}_2$-KerMDH, the 1-Flexible Square Diffie-Hellman (1-FlexSDH) [27] Assumption is $\mathcal{SC}_2$-KerMDH, and the $\ell$-Flexible Diffie-Hellman ($\ell$-FlexDH) [32] Assumption for $\ell > 1$ is the only one which is not in the KerMDH family. However, $\ell$-FlexDH $\Rightarrow \mathcal{C}_{\ell+1}$-KerMDH. Getting the last three results requires a bit more work, and they are proven in the full version [35].

## 4 Reduction and Separation of Kernel Diffie-Hellman Problems

In this section we prove that the most important matrix distribution families $\mathcal{U}_{\ell,k}$, $\mathcal{L}_k$, $\mathcal{SC}_k$, $\mathcal{C}_k$ and $\mathcal{RL}_k$ (see Appendix B) define families of KerMDH problems with **strictly** increasing hardness, as we precisely state in Theorem 2, at the end of the section. By 'strictly increasing' we mean that (1) there are known reductions of the smaller problems to the larger problems (in terms of $k$) within each family, and (2) there are no black-box reductions in the other way in the multilinear generic group model. This result shows the necessity of using $\mathcal{D}_{\ell,k}$-KerMDH Assumptions for $k > 2$. A similar result is known for the corresponding $\mathcal{D}_{\ell,k}$-MDDH problems. Indeed, one can easily prove a separation between large and small decisional problems. Observe that any efficient $m$-linear map can efficiently solve any $\mathcal{D}_{\ell,k}$-MDDH problem with $k \leq m - 1$, and therefore every two $\mathcal{D}_{\ell,k}$-MDDH and $\mathcal{D}_{\widetilde{\ell},\widetilde{k}}$-MDDH problems with $\widetilde{k} < k$ are separated by an oracle computing a $k$-linear map. However, when dealing with the computational $\mathcal{D}_{\ell,k}$-KerMDH family, no such a trivial argument is known to exist.

Actually, a $m$-linear map does not seem to help to solve any $\mathcal{D}_{\ell,k}$-KerMDH problem with $k > 1$. Furthermore, the $m$-linear map seems to be useless for any (reasonable) reduction between KerMDH problems defined over the same group. Indeed, all group elements involved in the problem instances and their solutions belong to the base group $\mathbb{G}$, and the result of computing any $m$-linear map is an element in $\mathbb{G}_m$, where no efficient map from $\mathbb{G}_m$ back to $\mathbb{G}$ is supposed to exist.

## 4.1 Separation

In this section we firstly show the non-existential part of Theorem 2. Namely, we show that there is no black-box reduction in the generic group model (described in Sect. 2.2) from $\mathcal{D}_{\ell,k}$-KerMDH to $\mathcal{D}_{\widetilde{\ell},\widetilde{k}}$-KerMDH for $k > \widetilde{k}$, assuming that the two matrix distributions $\mathcal{D}_{\ell,k}$ and $\mathcal{D}_{\widetilde{\ell},\widetilde{k}}$ are hard (see Definition 4). Before proving the main result we need some technical lemmas and also a new geometrical notion defined on a family of subspaces of a vector space, named $t$-*Elusiveness*.

In the first lemma we show that the natural (black-box, algebraic) reductions between KerMDH problems have a very special form. Observe that a black-box reduction to a flexible problem must work for any adversary solving it. In particular, the reduction should work for **any** solution given by this adversary, or for **any** probability distribution of the solutions given by it. Informally, the lemma states that the output of a successful reduction can always be computed in essentially two ways: (1) By just applying a (randomized) linear map to the answer given by the adversary in the last call. Therefore, all possibly existing previous calls to the adversary are just used to prepare the last one. (2) By just ignoring the last call to the adversary and using only the information gathered in the previous ones.

Let $\mathcal{R}^{\mathcal{O}}$ be a black-box reduction of $\mathcal{D}_{\ell,k}$-KerMDH to $\mathcal{D}_{\widetilde{\ell},\widetilde{k}}$-KerMDH, in the purely algebraic generic multilinear group model, discussed in Sect. 2.2, for some matrix distributions $\mathcal{D}_{\ell,k}$ and $\mathcal{D}_{\widetilde{\ell},\widetilde{k}}$. Namely, $\mathcal{R}^{\mathcal{O}}$ solves $\mathcal{D}_{\ell,k}$-KerMDH with a non-negligible probability by making $Q \geq 1$ queries to an oracle $\mathcal{O}$ solving $\mathcal{D}_{\widetilde{\ell},\widetilde{k}}$-KerMDH with probability one. As we aim at ruling out the existence of some reductions, we just consider the best possible case any black-box reduction must be able to handle. Now we split the reduction as $\mathcal{R}^{\mathcal{O}} = (\mathcal{R}_0^{\mathcal{O}}, \mathcal{R}_1)$, where the splitting point is the last oracle call, as shown in Fig. 2. We actually use the same splitting in the proof of Lemma 2 in Appendix D. More formally, on the input of $[\mathbf{A}]$, for $\mathbf{A} \leftarrow \mathcal{D}_{\ell,k}$, and after making $Q - 1$ oracle calls, $\mathcal{R}_0^{\mathcal{O}}$ stops by outputting the last query to $\mathcal{O}$, that is a matrix $[\widetilde{\mathbf{A}}]$, where $\widetilde{\mathbf{A}} \in \mathcal{D}_{\widetilde{\ell},\widetilde{k}}$, together with some state information $s$ for $\mathcal{R}_1$. Next, $\mathcal{R}_1$ resumes the execution from $s$ and the answer $[\boldsymbol{w}] \in \mathbb{G}^{\widetilde{\ell}}$ given by the oracle, and finally outputs $[\boldsymbol{v}] \in \mathbb{G}^{\ell}$. Without loss of generality, we assume that both stages $\mathcal{R}_0^{\mathcal{O}}$ and $\mathcal{R}_1$ receive the same random tape, \$ ($\mathcal{R}_1$ can redo the computations performed by $\mathcal{R}_0^{\mathcal{O}}$).

**Lemma 5.** *There exists an algebraic oracle $\mathcal{O}$ (in the sense of Definition 1), that solves the $\mathcal{D}_{\ell,k}$-KerMDH Problem with probability one.*

All the proofs in Sect. 4 are given in Appendix D.

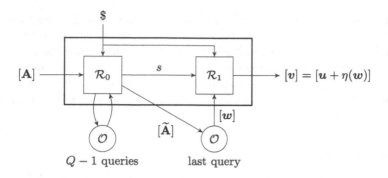

**Fig. 2.** Splitting of the black-box reduction.

Lemma 2 applied to $\mathcal{R}_0^{\mathcal{O}}$ (and using also Lemma 5) implies that only the group elements in $s$ can depend on $\mathbf{A}$. Indeed, the non-group elements in $s$ can only depend on the random tape \$. Now, from Lemma 1 applied to $\mathcal{R}_1$, we know that its output $[\boldsymbol{v}]$ is determined by a polynomial map of total degree at most one in the input group elements (*i.e.*, $\widetilde{\mathbf{A}}$ and the group elements in $s$), and the coefficients of this polynomial can only depend on \$, and the non-group elements in $s$, which in turn only depend on \$. Therefore, splitting the polynomial map into two parts, for every fixed \$ and every fixed oracle behavior in the first $Q-1$ oracle calls there exists a vector $u \in \mathbb{Z}_q^{\ell}$ and a linear map $\eta : \mathbb{Z}_q^{\tilde{\ell}} \to \mathbb{Z}_q^{\ell}$ such that we can write $\boldsymbol{v} = \boldsymbol{u} + \eta(\boldsymbol{w})$, where $\boldsymbol{u}$ actually depends on the group elements in $s$. The important fact here is that $\eta$ can only depend on \$, but not on $\mathbf{A}$.

**Lemma 6.** *Let* $\mathcal{R}^{\mathcal{O}} = (\mathcal{R}_0^{\mathcal{O}}, \mathcal{R}_1)$ *be a black-box reduction from* $\mathcal{D}_{\ell,k}$*-KerMDH to* $\mathcal{D}_{\tilde{\ell},\tilde{k}}$*-KerMDH, in the purely algebraic generic multilinear group model, making* $Q \geq 1$ *calls to an oracle* $\mathcal{O}$ *solving the latter with probability one. If* $\mathcal{R}^{\mathcal{O}}$ *succeeds with a non negligible probability* $\varepsilon$ *then, for every possible behavior of the oracle, either* $\Pr(\eta(\boldsymbol{w}) \in S') > negl$ *or* $\Pr(\boldsymbol{u} \in S') > negl$, *where* $S' = \ker \mathbf{A}^{\top} \backslash \{\mathbf{0}\}$, $[\mathbf{A}]$ *is the input of* $\mathcal{R}^{\mathcal{O}}$, *and its output is written as* $[\boldsymbol{u} + \eta(\boldsymbol{w})]$, *for some* $\boldsymbol{u}$ *only depending on the state output by* $\mathcal{R}_0^{\mathcal{O}}$, $[\boldsymbol{w}]$ *is the answer to the* $Q$*-th oracle query, and* $\eta : \mathbb{Z}_q^{\tilde{\ell}} \to \mathbb{Z}_q^{l}$ *is a (randomized) linear map that only depends on the random tape of* $\mathcal{R}^{\mathcal{O}}$.

The following property of the hard matrix distributions allows us to prove that indeed in the last lemma $\Pr(\eta(\boldsymbol{w}) \in S \backslash \{\mathbf{0}\}) \in negl$.

**Definition 8 (*t*-Elusiveness).** *A family of subspaces* $\mathcal{S}$ *of a vector space* $X$ *over the finite field* $\mathbb{Z}_q$ *is called* $t$*-elusive for some* $t < \dim X$ *if for all* $t$*-dimensional subspaces* $F \subset X$, $\Pr(F \cap S \neq \{\mathbf{0}\}) \in negl$, *where the probability is computed with respect to the choice of* $S \in \mathcal{S}$. *A matrix distribution* $\mathcal{D}_{\ell,k}$ *is called* $t$*-elusive if the family* $\{\ker \mathbf{A}^{\top}\}_{\mathbf{A} \in \mathcal{D}_{\ell,k}}$ *is* $t$*-elusive.*

**Lemma 7.** *If a matrix distribution* $\mathcal{D}_{\ell,k}$ *is hard (as given in Definition 4) then* $\mathcal{D}_{\ell,k}$ *is* $k$*-elusive.*

In the next theorem we use the $k$-elusiveness to prove that $\Pr(\boldsymbol{u} \in \ker \mathbf{A}^\top \backslash \{\mathbf{0}\}) > negl$ for all possible behaviors of the oracle in the first $Q - 1$ calls. This actually implies that the reduction can directly output $\boldsymbol{u}$, and only $Q - 1$ oracle calls are actually needed. Therefore, by the descent method we show that no successful reduction exists unless $\mathcal{D}_{\ell,k}$-KerMDH is easy.

**Theorem 1.** *Let* $\mathcal{D}_{\ell,k}$ *be* $k$-*elusive. If there exists a black-box reduction in the purely algebraic generic multilinear group model from* $\mathcal{D}_{\ell,k}$-KerMDH *to another problem* $\mathcal{D}_{\widetilde{\ell},\widetilde{k}}$-KerMDH *with* $\widetilde{k} < k$, *then* $\mathcal{D}_{\ell,k}$-KerMDH *is easy.*

Now we consider the contrapositive statement, that directly applies to the known families of hard matrix distributions.

**Corollary 1.** *If a matrix distribution family* $\{\mathcal{D}_{\ell,k}\}$ *is hard then for any* $\mathcal{D}_{\ell,k}$ *and* $\mathcal{D}_{\widetilde{\ell},\widetilde{k}}$ *in the family with* $k > \widetilde{k}$ *there is no black-box reduction in the generic group model from* $\mathcal{D}_{\ell,k}$-KerMDH *to* $\mathcal{D}_{\widetilde{\ell},\widetilde{k}}$-KerMDH.

*Proof.* Since all $\mathcal{D}_{\ell,k}$-MDDH problems in the family are generically hard on a $k$-linear group, we know that $\mathcal{D}_{\ell,k}$ is $k$-elusive by Lemma 7, and also $\mathcal{D}_{\ell,k}$-KerMDH is hard in that group (otherwise, any solution to $\mathcal{D}_{\ell,k}$-KerMDH can be used to solve $\mathcal{D}_{\ell,k}$-MDDH). By the above theorem, no black-box reduction in the generic group model from $\mathcal{D}_{\ell,k}$-KerMDH to $\mathcal{D}_{\widetilde{\ell},\widetilde{k}}$-KerMDH can exist for $k > \widetilde{k}$.

### 4.2  Increasing Families of KerMDH Problems

Most matrix distributions, like $\mathcal{U}_{\ell,k}$, $\mathcal{L}_k$, $\mathcal{SC}_k$, $\mathcal{C}_k$ and $\mathcal{RL}_k$, are indeed families parameterized by their size $k$. The negative results in Corollary 1 prevent us from finding reductions from larger to smaller KerMDH problems. Nevertheless, we provide here some examples of (tight) reductions going in the other way, within each of the previous families.

**Lemma 8.** $\mathcal{U}_{\widetilde{\ell},\widetilde{k}}$-KerMDH $\Rightarrow \mathcal{U}_{\ell,k}$-KerMDH *for any* $\widetilde{k} \le k$, $\widetilde{\ell} > \widetilde{k}$ *and* $\ell > k$.

*Proof.* We divide the proof into two steps: Firstly, assume that $\widetilde{\ell} = \widetilde{k} + 1$, $k \ge \widetilde{k}$, $\ell \ge k + 1$. Given an instance $[\widetilde{\mathbf{A}}]$, with $\widetilde{\mathbf{A}} \leftarrow \mathcal{U}_{\widetilde{k}+1,\widetilde{k}}$, we choose a full-rank matrix $\mathbf{L} \in \mathbb{Z}_q^{\ell \times (k+1)}$ and compute $[\mathbf{A}] = \mathbf{L}([\widetilde{\mathbf{A}}] \oplus [\mathbf{I}])$, where $\mathbf{I}$ is the identity matrix of size $(k - \widetilde{k}) \times (k - \widetilde{k})$ and $\oplus$ operation denotes diagonal block matrix concatenation. That is

$$U \oplus V = \begin{pmatrix} U & 0 \\ 0 & V \end{pmatrix}.$$

Clearly, the probability distribution of the new matrix is statistically close to the uniform distribution in $\mathbb{Z}_q^{\ell \times k}$. Any vector $[\boldsymbol{x}]$, obtained from a solver of $\mathcal{U}_{\ell,k}$-KerMDH, such that $\boldsymbol{x} \in \ker \mathbf{A}^\top \backslash \{\mathbf{0}\}$ can be transformed into $[\widetilde{\boldsymbol{x}}]$ such that $\widetilde{\boldsymbol{x}} \in \ker \widetilde{\mathbf{A}}^\top \backslash \{\mathbf{0}\}$ with overwhelming probability,[8] by just letting $[\widetilde{\boldsymbol{x}}]$ to

---

[8] Actually, $\widetilde{\boldsymbol{x}} = \mathbf{0}$ depends on the $(\widetilde{k} + 1)$-th column of $\mathbf{L}$ which is independent of $\mathbf{A}$.

be the first $\widetilde{k} + 1$ components of $\mathbf{L}^\top [\boldsymbol{x}]$. Thus, we have built a tight reduction $\mathcal{U}_{\widetilde{k}+1,\widetilde{k}}$-KerMDH $\Rightarrow \mathcal{U}_{\ell,k}$-KerMDH.

The second step, $k = \widetilde{k}$, $\widetilde{\ell} > \ell = \widetilde{k} + 1$, is simpler. Given an instance $[\widetilde{\mathbf{A}}]$, with $\widetilde{\mathbf{A}} \leftarrow \mathcal{U}_{\widetilde{\ell},\widetilde{k}}$, define the matrix $[\mathbf{A}]$ to be the upper $\widetilde{k} + 1$ rows of $[\widetilde{\mathbf{A}}]$. Clearly $\mathbf{A}$ follows the uniform distribution in $\mathbb{Z}_q^{(\widetilde{k}+1)\times\widetilde{k}}$. Now, any vector $[\boldsymbol{x}]$ such that $\boldsymbol{x} \in \ker \mathbf{A}^\top \backslash \{\mathbf{0}\}$ can be transformed into $[\widetilde{\boldsymbol{x}}]$ such that $\widetilde{\boldsymbol{x}} \in \ker \widetilde{\mathbf{A}}^\top \backslash \{\mathbf{0}\}$, by just padding $\boldsymbol{x}$ with $\widetilde{\ell} - \widetilde{k} - 1$ zeros. Thus, $\mathcal{U}_{\widetilde{\ell},\widetilde{k}}$-KerMDH $\Rightarrow \mathcal{U}_{\widetilde{k}+1,\widetilde{k}}$-KerMDH. By concatenating the two tight reductions we obtain the general case.

**Lemma 9.** *For* $\mathcal{D}_k = \mathcal{L}_k$, $\mathcal{SC}_k$, $\mathcal{C}_k$ *and* $\mathcal{RL}_k$, $\quad \mathcal{D}_k$-KerMDH $\Rightarrow \mathcal{D}_{k+1}$-KerMDH.

*Proof.* We start with the case $\mathcal{D}_k = \mathcal{L}_k$. Observe that given a matrix $\widetilde{\mathbf{A}} \leftarrow \mathcal{L}_k$, with parameters $a_1, \ldots, a_k$, we can build a matrix $\mathbf{A}$ following the distribution $\mathcal{L}_{k+1}$, by adding an extra row and column to $\widetilde{\mathbf{A}}$ corresponding to new random parameter $a_{k+1} \in \mathbb{Z}_q$. Moreover, given $\boldsymbol{x} = (x_1, \ldots, x_{k+2}) \in \ker \mathbf{A}^\top \backslash \{\mathbf{0}\}$, the vector $\widetilde{\boldsymbol{x}} = (x_1, \ldots, x_k, x_{k+2})$ is in $\ker \widetilde{\mathbf{A}}^\top \backslash \{\mathbf{0}\}$ (except for a negligible probability due to the possibility that $a_{k+1} = 0$ and $\widetilde{\boldsymbol{x}} = \mathbf{0}$, while $\boldsymbol{x} \neq \mathbf{0}$). The reduction consists of choosing a random $a_{k+1}$, then building $[\mathbf{A}]$ from $[\widetilde{\mathbf{A}}]$ as above, and finally obtaining $[\widetilde{\boldsymbol{x}}]$ from $[\boldsymbol{x}]$ by deleting the $(k+1)$-th coordinate.

Similarly, from a matrix $\widetilde{\mathbf{A}} \leftarrow \mathcal{SC}_k$, with parameter $a$, we can obtain a matrix $\mathbf{A}$ following $\mathcal{SC}_{k+1}$ by adding a new row and column to $\widetilde{\mathbf{A}}$. Now given $\boldsymbol{x} = (x_1, \ldots, x_{k+2}) \in \ker \mathbf{A}^\top \backslash \{\mathbf{0}\}$, it is easy to see that the vector $\widetilde{\boldsymbol{x}} = (x_1, \ldots, x_{k+1})$ is always in $\ker \widetilde{\mathbf{A}}^\top \backslash \{\mathbf{0}\}$.

$\mathcal{C}_k$-KerMDH $\Rightarrow \mathcal{C}_{k+1}$-KerMDH and $\mathcal{RL}_k$-KerMDH $\Rightarrow \mathcal{RL}_{k+1}$-KerMDH are proven using the same ideas.

By combining Corollary 1 with the explicit reductions given above, we can now state our main result in this section.

**Theorem 2.** *The matrix distribution families* $\{\mathcal{U}_{\ell,k}\}$, $\{\mathcal{L}_k\}$, $\{\mathcal{SC}_k\}$, $\{\mathcal{C}_k\}$ *and* $\{\mathcal{RL}_k\}$ *define families of* KerMDH *problems with **strictly** increasing hardness. Namely, for any* $\mathcal{D}_{\ell,k}$ *and* $\mathcal{D}_{\widetilde{\ell},\widetilde{k}}$ *belonging to one of the previous families, such that* $\widetilde{k} < k$,

1. *there exists a tight reduction,* $\mathcal{D}_{\widetilde{\ell},\widetilde{k}}$-KerMDH $\Rightarrow \mathcal{D}_{\ell,k}$-KerMDH,
2. *there is no black-box reduction in the generic group model in the opposite direction.*

## 5   Applications

We have already mentioned that the Kernel Matrix Diffie-Hellman Assumptions have already found applications in follow-up work, more concretely: (a) to generalize and improve previous constructions of QA-NIZK proofs for linear spaces [26], (b) to construct more efficient structure preserving signatures starting from affine algebraic MACS [25], (c) to improve and generalize aggregation of Groth-Sahai proofs [17] or (d) to construct a tightly secure QA-NIZK argument for linear subspaces with unbounded simulation soundness [15].

As a new application, we use our new framework to abstract two constructions of trapdoor commitments. See for instance [3] for the formal definition of a trapdoor commitment scheme $C = (\mathsf{K}, \mathsf{Comm}, \mathsf{Vrfy}, \mathsf{TrapdoorEquiv})$ and Sect. 6 for a discussion on the advantages of instantiating these commitments with the new circulant matrix distribution.

## 5.1 Generalized Pedersen Commments in Multilinear Groups

In a group $(\mathbb{G}, q, \mathcal{P})$ where the discrete logarithm is hard, the Pedersen commitment is a statistically hiding and computationally binding commitment to a scalar. It can be naturally generalized to several scalars. Abe *et al.* [2] show how to do similar Pedersen type commitments to vectors of group elements. With our new assumption family we can write both the Pedersen commitment and the commitment of [2] as a single construction and generalize it to (ideal) graded encodings.

- $\mathsf{K}(1^\lambda, d, m)$: Let $\mathcal{MG}_m = (e, \mathbb{G}_1, \mathbb{G}_2, \dots, \mathbb{G}_m, q, \mathcal{P}_1, \dots, \mathcal{P}_m) \leftarrow \mathsf{MGen}_m(1^\lambda)$. Sample $\mathbf{A} \leftarrow \mathcal{D}_{k+d,k}$. Let $\overline{\mathbf{A}}$ be the first $k$ rows of $\mathbf{A}$ and $\underline{\mathbf{A}}$ the remaining $d$ rows and $\mathbf{T} := \underline{\mathbf{A}}\,\overline{\mathbf{A}}^{-1}$ (w.l.o.g. we can assume $\overline{\mathbf{A}}$ is invertible). Output $ck := (\mathcal{MG}_m, [\mathbf{A}]_1), tk := (\mathbf{T})$.
- $\mathsf{Comm}(ck, [\boldsymbol{v}]_r)$: To commit to a vector $[\boldsymbol{v}]_r \in \mathbb{G}_r^d$, for any $r < m$, pick $\boldsymbol{s} \leftarrow \mathbb{Z}_q^k$, and output $[\boldsymbol{c}]_{r+1} := e([(\boldsymbol{s}^\top \| \boldsymbol{v}^\top)]_r, [\mathbf{A}]_1) = [(\boldsymbol{s}^\top \| \boldsymbol{v}^\top)\mathbf{A}]_{r+1} \in \mathbb{G}_{r+1}^k$, and the opening $Op = ([\boldsymbol{s}]_r)$.
- $\mathsf{Vrfy}(ck, [\boldsymbol{v}]_r, Op)$: Given a message $[\boldsymbol{v}]_r$ and opening $Op = ([\boldsymbol{s}]_r)$, this algorithm outputs 1 if $[\boldsymbol{c}]_{r+1} = e([(\boldsymbol{s}^\top \| \boldsymbol{v}^\top)]_r, [\mathbf{A}]_1)$.
- $\mathsf{TrapdoorEquiv}(ck, tk, [\boldsymbol{c}]_{r+1}, [\boldsymbol{v}]_r, Op, [\boldsymbol{v}']_r)$: On a commitment $[\boldsymbol{c}]_{r+1} \in \mathbb{G}_{r+1}^k$ to message $[\boldsymbol{v}]_r$ with opening $Op = ([\boldsymbol{s}]_r)$, compute: $[\boldsymbol{s}']_r := [\boldsymbol{s}]_r + \mathbf{T}^\top[(\boldsymbol{v} - \boldsymbol{v}')]_r \in \mathbb{G}_r^k$. Output $Op' = ([\boldsymbol{s}']_r)$ as the opening of $[\boldsymbol{c}]_{r+1}$ to $[\boldsymbol{v}']_r$.

The analysis is almost identical to [2]. The correctness of the trapdoor opening is straightforward. The hiding property of the commitment is unconditional, while the soundness (at level $r$) is based on the $(r, m, \mathcal{D}_{\ell,k})$-KerMDH Assumption. Indeed, given two messages $[\boldsymbol{v}]_r, [\boldsymbol{v}']_r$ with respective openings $[\boldsymbol{s}]_r, [\boldsymbol{s}']_r$, it obviously follows that $[\boldsymbol{w}] := [((\boldsymbol{s} - \boldsymbol{s}')^\top \| (\boldsymbol{v} - \boldsymbol{v}')^\top)]_r$ is a nonzero element in the kernel (in $\mathbb{G}_r$) of $\mathbf{A}^\top$, i.e. $e([\boldsymbol{w}^\top]_r, [\mathbf{A}]_1) = [\mathbf{0}]_{r+1}$.

Notice that the Pedersen commitment (to multiple elements) is for messages in $\mathbb{G}_0$ and $\mathbf{A} \leftarrow \mathcal{U}_{d+1,1}$ and soundness is based on the $(0, m, \mathcal{U}_{d+1,1})$-KerMDH. The construction proposed in [2] is for an asymmetric bilinear group $\mathcal{AG}_2$, and in this case messages are vectors in the group $\mathbb{H}$ and the commitment key consists of elements in $\mathbb{G}$, i.e. $ck = (\mathcal{AG}_2, [\mathbf{A}]_G), \mathbf{A} \leftarrow \mathcal{U}_{d+1,1}$. Further, a previous version of the commitment scheme of [2] in symmetric bilinear groups (in [18]) corresponds to our construction with $\mathbf{A} \leftarrow \mathcal{U}_{2+d,2}$.

## 5.2    Group-to-Group Commitments

The commitments of the previous section are "shrinking" because they map a vector of length $d$ in the group $\mathbb{G}_r$ to a vector of length $k$, for some $k$ independent of and typically smaller than $d$. Abe *et al.* [3] noted that in some applications it is useful to have "group-to-group" commitments, *i.e.* commitments which are defined in the same group as the vector message. The motivation for doing so in the bilinear case is that these commitments are better compatible with Groth-Sahai proofs.

There is a natural construction of group-to-group commitments which uses the generalized Pedersen commitment of Sect. 5.1, which is denoted as Ped.C = $(\widetilde{\mathsf{K}}, \widetilde{\mathsf{Comm}}, \widetilde{\mathsf{Vrfy}}, \mathsf{TrapdoorEquiv})$ in the following.

- $\mathsf{K}(1^\lambda, d, m)$: Run $(\widetilde{ck}, \widetilde{tk}) \leftarrow \widetilde{\mathsf{K}}(1^\lambda, m, d)$, output $ck = \widetilde{ck}$ and $tk = \widetilde{tk}$.
- $\mathsf{Comm}(ck, [\boldsymbol{v}]_r)$: To commit to a vector $[\boldsymbol{v}]_r \in \mathbb{G}_r^d$, for any $0 < r < m$, pick $[\boldsymbol{t}]_{r-1} \leftarrow [\mathbb{G}]_{r-1}^k$. Let $([\widetilde{\boldsymbol{c}}]_r, \widetilde{Op} = ([\boldsymbol{s}]_{r-1})) \leftarrow \widetilde{\mathsf{Comm}}(ck, [\boldsymbol{t}]_{r-1})$ and output $c := ([\boldsymbol{t} + \boldsymbol{v}]_r, [\widetilde{\boldsymbol{c}}]_r)$ and the opening $Op = ([\boldsymbol{s}]_r)$.
- $\mathsf{Vrfy}(ck, c, [\boldsymbol{v}]_r, Op)$: On input $c = ([\boldsymbol{y}]_r, [\widetilde{\boldsymbol{c}}]_r)$, this algorithm computes $[\widetilde{\boldsymbol{c}}]_{r+1}$ and outputs 1 if $[\boldsymbol{t}]_r := [\boldsymbol{y} - \boldsymbol{v}]_r$ satisfies that $1 \leftarrow \widetilde{\mathsf{Vrfy}}(ck, [\widetilde{\boldsymbol{c}}]_{r+1}, [\boldsymbol{t}]_r, [\boldsymbol{s}]_r)$, else it outputs 0.
- $\mathsf{TrapdoorEquiv}(ck, tk, c, [\boldsymbol{v}]_r, Op, [\boldsymbol{v}']_r)$: On a commitment $c = ([\boldsymbol{y}]_r, [\widetilde{\boldsymbol{c}}]_r)$ with opening $Op = ([\boldsymbol{s}]_r)$, if $[\boldsymbol{t}]_r := [\boldsymbol{y} - \boldsymbol{v}]_r$ and $[\boldsymbol{t}']_r := [\boldsymbol{y} - \boldsymbol{v}']_r$, this algorithm computes $[\widetilde{\boldsymbol{c}}]_{r+1}$ and runs the algorithm $\widetilde{Op} \leftarrow \widetilde{\mathsf{TrapdoorEquiv}}(ck, tk, [\widetilde{\boldsymbol{c}}]_{r+1}, [\boldsymbol{t}]_r, [\boldsymbol{s}]_r, [\boldsymbol{t}']_r)$, and outputs $\widetilde{Op}$.

A commitment is a vector of size $k + d$ and an opening is of size $k$. The required security properties follow easily from the properties of the generalized Pedersen commitment.

**Theorem 3.** *C is a perfectly hiding, computationally binding commitment.*

*Proof.* Since the generalized Pedersen commitment is perfectly hiding, then $([\boldsymbol{t} + \boldsymbol{v}]_r, \widetilde{\mathsf{Comm}}(\widetilde{ck}, [\boldsymbol{t}]_{r-1}))$ perfectly hides $[\boldsymbol{v}]_r$ because $[\boldsymbol{t}]_r$ acts as a one-time pad. Similarly, it is straightforward to see that the computationally binding property of $C$ follows from the computationally binding property of the generalized Pedersen commitment.

Interestingly, this construction explains the two instantiations of "group-to-group" commitments given in [3] (see the full version [35] for more details).

## 6    A New Matrix Distribution and Its Applications

Both of our commitment schemes of Sect. 5 base security on some $\mathcal{D}_{k+d,k}$-KerMDH assumptions, where $d$ is the length of the committed vector. When $d > 1$, the only example of $\mathcal{D}_{k+d,k}$-MDDH Assumption considered in [11] is the one corresponding to the uniform matrix distribution $\mathcal{U}_{k+d,k}$, which is the weakest MDDH Assumption of size $(k + d) \times k$. Another natural assumption for $d > 1$ is the one associated to the matrix distribution resulting from sampling from an arbitrary hard distribution $\mathcal{D}_{k+1,k}$ (e.g., $\mathcal{L}_k$) and adding $d - 1$ new random rows. Following the same ideas in the proof of Lemma 8, it is easy to see that the resulting $\mathcal{D}_{k+d,k}$-MDDH assumption is equivalent to the original $\mathcal{D}_{k+1,k}$-MDDH assumption. However, for efficiency reasons, we would like to have a matrix distributions with an even smaller representation size. This motivates us to introduce a new family of matrix distributions, the $\mathcal{CI}_{k,d}$ family.

**Definition 9 (Circulant Matrix Distribution).** *We define $\mathcal{CI}_{k,d}$ as*

$$
\mathbf{A} = \begin{pmatrix}
a_1 & & & 0 \\
\vdots & a_1 & & \\
a_d & \vdots & \ddots & \\
1 & a_d & & a_1 \\
& 1 & \ddots & \vdots \\
& & \ddots & a_d \\
0 & & & 1
\end{pmatrix} \in \mathbb{Z}_q^{(k+d)\times k}, \qquad \text{where } a_i \leftarrow \mathbb{Z}_q
$$

Matrix $\mathbf{A}$ is such that each column can be obtained by rotating one position the previous column, which explains the name. Notice that when $d = 1$, $\mathcal{CI}_{k,d}$ is exactly the symmetric cascade distribution $\mathcal{SC}_k$, introduced in [11]. It can be shown that the representation size of $\mathcal{CI}_{k,d}$, which is the number of parameters $d$, is the optimal among all hard matrix distributions $\mathcal{D}_{k+d,k}$ defined by linear polynomials in the parameters. A similar argument shows that the circulant assumption is also optimal in the sense that it has a minimal number of nonzero entries among all hard matrix distributions $\mathcal{D}_{k+d,k}$. It can also be proven that $\mathcal{CI}_{k,d}$-MDDH holds generically in $k$-linear groups, which implies the hardness of the corresponding KerMDH problem. To prove the generic hardness of the assumption, we turn to a result of Herold [20, Theorem 5.15 and corollaries]. It states that if all matrices produced by the matrix distribution are full-rank, $\mathcal{CI}_{k,d}$ is a hard matrix distribution. Indeed, an algorithm solving the $\mathcal{CI}_{k,d}$-MDDH problem in the generic $k$-linear group model must be able to compute a polynomial in the ideal $\mathfrak{H} \subset \mathbb{Z}_q[a_1, \ldots, a_d, z_1, \ldots, z_{k+d}]$ generated by all the $(k + 1)$-minors of $\mathbf{A}\|z$ as polynomials in $a_1, \ldots, a_d, z_1, \ldots, z_{k+d}$. Although this ideal can actually be generated using only a few of the minors, we need to build a Gröbner basis of $\mathfrak{H}$ to reason about the minimum degree a nonzero polynomial in $\mathfrak{H}$ can have. We show that, carefully selecting a monomial order, the set of all $(k + 1)$-minors of $\mathbf{A}\|z$ form a Gröbner basis, and all these minors have total degree exactly $k + 1$. Therefore, all nonzero polynomials in $\mathfrak{H}$ have degree at least $k + 1$, and then they cannot be evaluated by any algorithm in the generic $k$-linear group model. The full proof of both properties of $\mathcal{CI}_{k,d}$ can be found in the full version [35].

As for other matrix distribution families, we can combine Corollary 1 and the techniques used in Lemma 9 to show that for any fixed $d \geq 1$ the $\mathcal{CI}_{k,d}$-KerMDH problem family has strictly increasing hardness.

**Theorem 4.** *For any $d \geq 1$ and for any $k, \widetilde{k}$ such that $\widetilde{k} < k$*

1. *there exists a tight reduction, $\mathcal{CI}_{\widetilde{k},d}$-KerMDH $\Rightarrow \mathcal{CI}_{k,d}$-KerMDH,*
2. *there is no black-box reduction in the generic group model in the opposite direction.*

The new assumption gives new instantiations of the commitment schemes of Sect. 5 with public parameters of size $d$, independent of $k$. Further, because the matrix $\mathbf{A} \leftarrow \mathcal{CI}_{k,d}$ has a many zero entries, the number of exponentiations computed by the Commit algorithm, and the number of pairings of the verification algorithm is $kd$—as opposed to $k(k+d)$ for the uniform assumption. This seems to be optimal—but we do not prove this formally.

**Acknowledgements.** The authors thank E. Kiltz and G. Herold for improving this work through very fruitful discussions. Also G. Herold gave us the insight and guidelines to prove the hardness of the circulant matrix distribution.

## A    Deferred Proofs from Sect. 2.2

**Lemma 2.** *Let $\mathcal{A}^{\mathcal{O}}$ be an oracle algorithm in the (purely algebraic) generic multilinear group model, making a constant number of calls $Q$ to an algebraic oracle $\mathcal{O}$. Let $([\boldsymbol{x}]_a, \widetilde{x})$ and $([\boldsymbol{y}]_b, \widetilde{y})$ respectively be the input and output of $\mathcal{A}$. Then, for every choice of $\widetilde{x}$ and the random tape, there exist polynomials of constant degree $Y_1, \ldots, Y_\beta \in \mathbb{Z}_q[\boldsymbol{X}, \boldsymbol{R}_1, \ldots, \boldsymbol{R}_Q]$, such that $\boldsymbol{y} = \boldsymbol{Y}(\boldsymbol{x}, \boldsymbol{r}_1, \ldots, \boldsymbol{r}_Q)$, for all possible inputs, where $\boldsymbol{Y} = (Y_1, \ldots, Y_\beta)$, and $\boldsymbol{r}_1, \ldots, \boldsymbol{r}_Q$ are the parameters introduced in Definition 1 for the $Q$ queries. Moreover, $\widetilde{y}$ does not depend on $\boldsymbol{x}$ or $\boldsymbol{r}_1, \ldots, \boldsymbol{r}_Q$.*

*Proof.* We proceed by induction in $Q$. The first step, $Q = 0$, follows immediately from Lemma 1, because $\mathcal{A}^{\mathcal{O}}$ is just an algorithm (without oracle access). For $Q \geq 1$, we split $\mathcal{A}^{\mathcal{O}}$ into two sections $\mathcal{A}_0^{\mathcal{O}}$ and $\mathcal{A}_1$, separated exactly at the last query point (see Fig. 3). Let $([\boldsymbol{z}]_c, \widetilde{z})$ be the state information (group and non-group elements) that $\mathcal{A}_0^{\mathcal{O}}$ passes to $\mathcal{A}_1$, $([\boldsymbol{u}]_d, \widetilde{u})$ be the $Q$-th query to $\mathcal{O}$, and $([\boldsymbol{v}]_d, \widetilde{v})$ be its corresponding answer. We assume that $\mathcal{A}_0^{\mathcal{O}}$ and $\mathcal{A}_1$ receive the same random tape, \$, (perhaps introducing some redundant computations in $\mathcal{A}_1$). Observe that the output of $\mathcal{A}_0^{\mathcal{O}}$ consists of $([\boldsymbol{z}]_c, \widetilde{z})$ and $([\boldsymbol{u}]_c, \widetilde{u})$.

By the induction assumption, for any choice of $\widetilde{x}$ and \$, there exist some polynomials of constant degree $Z_1, \ldots, Z_\gamma \in \mathbb{Z}_q[\boldsymbol{X}, \boldsymbol{R}_1, \ldots, \boldsymbol{R}_{Q-1}]$ and $U_1, \ldots, U_\delta \in \mathbb{Z}_q[\boldsymbol{X}, \boldsymbol{R}_1, \ldots, \boldsymbol{R}_{Q-1}]$ such that $\boldsymbol{z} = \boldsymbol{Z}(\boldsymbol{x}, \boldsymbol{r}_1, \ldots, \boldsymbol{r}_{Q-1})$, where $\boldsymbol{Z} = (Z_1, \ldots, Z_\gamma)$, and $\boldsymbol{u} = \boldsymbol{U}(\boldsymbol{x}, \boldsymbol{r}_1, \ldots, \boldsymbol{r}_{Q-1})$, where $\boldsymbol{U} = (U_1, \ldots, U_\delta)$, for all possible $\boldsymbol{x} \in \mathbb{Z}_q^\alpha$ and $\boldsymbol{r}_1, \ldots, \boldsymbol{r}_{Q-1} \in \mathbb{Z}_q^\rho$. Moreover, $\widetilde{z}$ and $\widetilde{u}$ only depend on $\widetilde{x}$ and \$.

Now, the algorithm $\mathcal{A}_1$ receives as input $([\boldsymbol{z}]_c, \widetilde{z})$ and $([\boldsymbol{v}]_e, \widetilde{v})$. By Definition 1, $\boldsymbol{v}$ also depend polynomially on $\boldsymbol{u}$ and $\boldsymbol{r}_Q$. Namely, for every choice

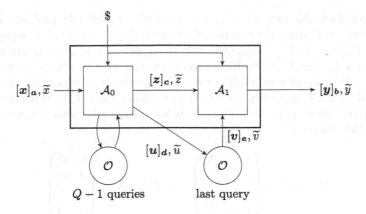

**Fig. 3.** Splitting of the oracle algorithm in Lemma 2.

of $\widetilde{u}$, there exist polynomials of constant degree $V_1, \ldots, V_\epsilon \in \mathbb{Z}_q[U, R_Q]$ such that $v = V(u, r_Q)$, where $V = (V_1, \ldots, V_\epsilon)$, while $\widetilde{v}$ only depends on $\widetilde{u}$.

Since $\mathcal{A}_1$ is just an algorithm without oracle access, by Lemma 1, for any choice of $\widetilde{v}$, $\widetilde{z}$ and \$, there exist polynomials of constant degree $Y_1, \ldots, Y_\beta \in \mathbb{Z}_q[V, Z]$ such that $y = Y(v, z)$, where $Y = (Y_1, \ldots, Y_\beta)$, for all $v \in \mathbb{Z}_q^\epsilon$ and $z \in \mathbb{Z}_q^\gamma$, while $\widetilde{y}$ only depends on $\widetilde{v}$, $\widetilde{z}$ and \$. By composition of all the previous polynomials, we show that $y$ depend polynomially on $x$ and $r_1, \ldots, r_Q$, where the polynomials depend only on \$ and $\widetilde{x}$. Indeed

$$y = Y(V(U(x, r_1, \ldots, r_{Q-1}), r_Q), Z(x, r_1, \ldots, r_{Q-1}))$$

and all the polynomials involved depend only on $\widetilde{x}$, $\widetilde{z}$, $\widetilde{u}$, $\widetilde{v}$ and \$, but all in turn only depend on $\widetilde{x}$ and \$. In addition, for the same reason, $\widetilde{y}$ only can depend on $\widetilde{x}$ and \$, which concludes the proof.

# B  Examples of Matrix Distributions

Some particular families of matrix distributions were presented in [11]. Namely,

$$\mathcal{SC}_k : \mathbf{A} = \begin{pmatrix} a & & & 0 \\ 1 & \ddots & & \\ & \ddots & \ddots & a \\ 0 & & 1 \end{pmatrix} \qquad \mathcal{C}_k : \mathbf{A} = \begin{pmatrix} a_1 & & & 0 \\ 1 & \ddots & & \\ & \ddots & \ddots & a_k \\ 0 & & 1 \end{pmatrix} \qquad \mathcal{L}_k : \mathbf{A} = \begin{pmatrix} a_1 & & & 0 \\ & \ddots & & \\ 0 & & & a_k \\ 1 & \cdots & & 1 \end{pmatrix},$$

where $a, a_i \leftarrow \mathbb{Z}_p$, and $\mathcal{U}_{\ell,k}$ which is simply the uniform distribution in $\mathbb{Z}_p^{\ell \times k}$. The $\mathcal{SC}_k$-MDDH Assumption is the Symmetric Cascade Assumption, the $\mathcal{C}_k$-MDDH Assumption is the Cascade Assumption, which were proposed for the first time. $\mathcal{U}_{\ell,k}$-MDDH is the Uniform Assumption, which appeared under other names in [7,37]. $\mathcal{L}_k$-MDDH is the Decisional Linear Assumption [6,22,40]. For instance, we can consider the case $k = 2$, in which the $\mathcal{L}_2$-MDDH problem is given $([1], [a_1], [a_2])$, tell apart the two distributions $([1], [a_1], [a_2], [w_1 a_1], [w_2 a_2], [w_1 + w_2])$ and $([1], [a_1], [a_2], [z_1], [z_2], [z_3])$, where $a_1, a_2, w_1, w_2, z_1, z_2, z_3$ are random. This is exactly the 2-Lin Problem, since we can always set $z_1 = w_1 a_1$ and $z_2 = w_2 a_2$. We also give examples of matrix distributions which did not appear in [11] but that are implicitly used in the problems 2 and 4 in Appendix C. The Randomized Linear and the Square Polynomial distributions are respectively given by the matrices

$$\mathcal{RL}_k : \mathbf{A} = \begin{pmatrix} a_1 & & 0 \\ & \ddots & \\ 0 & & a_k \\ b_1 & \cdots & b_k \end{pmatrix} \qquad \mathcal{P}_{\ell,2} : \mathbf{A} = \begin{pmatrix} a_1 & a_1^2 \\ a_2 & a_2^2 \\ \vdots & \vdots \\ a_\ell & a_\ell^2 \end{pmatrix}$$

where $a_i \leftarrow \mathbb{Z}_q$ and $b_i \leftarrow \mathbb{Z}_q^\times$. Jutla and Roy [24] referred to $\mathcal{RL}_k$-MDDH Assumption as the $k$-lifted Assumption.

## C  Flexible Problems That Fit into the New Framework

In this section we recall some computational problems in the cryptographic literature that we unify as particular instances of KerMDH problems. These problems are listed below, as they appear in the cited references. In the following, all parameters $a_i$ and $b_i$ are assumed to be randomly chosen in $\mathbb{Z}_q$.

1. Find-Rep [9]: Given $([a_1], \ldots, [a_\ell])$, find a nonzero tuple $(x_1, \ldots, x_\ell)$ such that $x_1 a_1 + \ldots + a_\ell x_\ell = 0$.
2. Simultaneous Double Pairing (SDP) [2]: Given the two tuples, $([a_1], [b_1])$ and $([a_2], [b_2])$, find a nonzero tuple $([x_1], [x_2], [x_3])$ such that $x_1 b_1 + x_2 a_1 = 0$, $x_1 b_2 + x_3 a_2 = 0$.
3. Simultaneous Triple Pairing [18]: Given the two tuples, $([a_1], [a_2], [a_3])$ and $([b_1], [b_2], [b_3])$, find a nonzero tuple $([x_1], [x_2], [x_3])$ such that $x_1 a_1 + x_2 a_2 + x_3 a_3 = 0$, $x_1 b_1 + x_2 b_2 + x_3 b_3 = 0$.
4. Simultaneous Pairing [19]: Given $([a_1], [a_2], \ldots, [a_\ell])$ and $([a_1^2], [a_2^2], \ldots, [a_\ell^2])$, find a nonzero tuple $([x_1], \ldots, [x_\ell])$ such that $\sum_{i=1}^\ell x_i a_i = 0$, $\sum_{i=1}^\ell x_i a_i^2 = 0$.
5. 1-Flexible Diffie-Hellman (1-FlexDH) [32]: Given $([1], [a], [b])$, find a triple $([r], [ra], [rab])$ with $r \neq 0$.
6. 1-Flexible Square Diffie-Hellman (1-FlexSDH) [27]: Given $([1], [a])$, find a triple $([r], [ra], [ra^2])$ with $r \neq 0$.
7. $\ell$-Flexible Diffie-Hellman ($\ell$-FlexDH) [32]: Given $([1], [a], [b])$, find a $(2\ell + 1)$-tuple $([r_1], \ldots, [r_\ell], [r_1 a], [r_1 r_2 a], \ldots, [(\prod_{i=1}^\ell r_i)a], [(\prod_{i=1}^\ell r_i)ab])$ such that $r_j \neq 0$ for all $j = 1, \ldots, \ell$.
8. Double Pairing (DP) [18]: In an asymmetric group $(\mathbb{G}, \mathbb{H}, \mathbb{T})$, given a pair of random elements $([a_1]_H, [a_2]_H) \in \mathbb{H}^2$, find a nonzero tuple $([x_1]_G, [x_2]_G)$ such that $[x_1 a_1 + x_2 a_2]_T = [0]_T$.

# D   Deferred Proofs from Sect. 4

**Lemma 5.** *There exists an algebraic oracle $\mathcal{O}$ (in the sense of Definition 1), that solves the $\mathcal{D}_{\ell,k}$-KerMDH Problem with probability one.*

*Proof.* Observe that $\mathcal{D}_{\ell,k}$ only uses group elements both in the instance description and in the solution to the problem. In addition, the problem (input/output relation) can be described by a polynomial map. Indeed, one can use the $k$-minors of $\mathbf{A}$, which are just polynomials of degree $k$, to obtain a basis of $\ker \mathbf{A}^\top$. Then the oracle can use parameters $r_1, \ldots, r_{\ell-k}$ as the coefficients of an arbitrary linear combination of the basis vectors. Sampling these parameters uniformly results in an oracle answer uniformly distributed in $\ker \mathbf{A}^\top$.

**Lemma 6.** *Let $\mathcal{R}^{\mathcal{O}} = (\mathcal{R}_0^{\mathcal{O}}, \mathcal{R}_1)$ be a black-box reduction from $\mathcal{D}_{\ell,k}$-KerMDH to $\mathcal{D}_{\widetilde{\ell},\widetilde{k}}$-KerMDH, in the purely algebraic generic multilinear group model, making $Q \geq 1$ calls to an oracle $\mathcal{O}$ solving the latter with probability one. If $\mathcal{R}^{\mathcal{O}}$ succeeds with a non negligible probability $\varepsilon$ then, for every possible behavior of the oracle, either $\Pr(\eta(\boldsymbol{w}) \in S') > \mathrm{negl}$ or $\Pr(\boldsymbol{u} \in S') > \mathrm{negl}$, where $S' = \ker \mathbf{A}^\top \backslash \{\mathbf{0}\}$, $[\mathbf{A}]$ is the input of $\mathcal{R}^{\mathcal{O}}$, and its output is written as $[\boldsymbol{u} + \eta(\boldsymbol{w})]$, for some $\boldsymbol{u}$ only depending on the state output by $\mathcal{R}_0^{\mathcal{O}}$, $[\boldsymbol{w}]$ is the answer to the $Q$-th oracle query, and $\eta : \mathbb{Z}_q^{\widetilde{l}} \to \mathbb{Z}_q^l$ is a (randomized) linear map that only depends on the random tape of $\mathcal{R}^{\mathcal{O}}$.*

*Proof.* Let us denote $S = \ker \mathbf{A}^\top$, where $[\mathbf{A}]$ is the input to $\mathcal{R}^{\mathcal{O}}$, and $S' = S \backslash \{\mathbf{0}\}$. Analogously, $\widetilde{S} = \ker \widetilde{\mathbf{A}}^\top$, where $[\widetilde{\mathbf{A}}]$ is the $Q$-th oracle query, and $\widetilde{S}' = \widetilde{S} \backslash \{\mathbf{0}\}$. From the discussion preceding the lemma, we know that $\boldsymbol{u}$ and $\eta$ are well-defined and fulfil the required properties. In particular, $\eta$ depends only on the random tape, \$, of $\mathcal{R}^{\mathcal{O}}$. As a black-box reduction, $\mathcal{R}^{\mathcal{O}}$ is successful means that it is successful for every possible behavior of the oracle in its $Q$ queries, with a success probability at least $\varepsilon$. We arbitrarily fix its behavior in the first $Q - 1$ queries. Concerning the last one, for all $\boldsymbol{w} \in \widetilde{S}'$, $\Pr(\boldsymbol{u} + \eta(\boldsymbol{w}) \in S') > \varepsilon$, where the probability is computed with respect to \$ and the randomness of $[\mathbf{A}]$. Now, defining

$$p_{\boldsymbol{w}} = \Pr(\boldsymbol{u} \in S \wedge \boldsymbol{u} + \eta(\boldsymbol{w}) \in S')$$

$$r_{\boldsymbol{w}} = \Pr(\boldsymbol{u} \notin S \wedge \boldsymbol{u} + \eta(\boldsymbol{w}) \in S')$$

we have $p_{\boldsymbol{w}} + r_{\boldsymbol{w}} > \varepsilon$. But not all $r_{\boldsymbol{w}}$ can be non-negligible since the corresponding events are disjoint. Indeed, for any vector $\boldsymbol{w} \neq \mathbf{0}$ and any different $\alpha_1, \alpha_2 \in \mathbb{Z}_q^\times$,

$$\boldsymbol{u} + \eta(\alpha_1 \boldsymbol{w}) \in S, \ \boldsymbol{u} + \eta(\alpha_2 \boldsymbol{w}) \in S \quad \Rightarrow \quad (\alpha_2 - \alpha_1)\boldsymbol{u} \in S \quad \Rightarrow \quad \boldsymbol{u} \in S$$

and then $\sum_{\alpha \in \mathbb{Z}_q^\times} r_{\alpha \boldsymbol{w}} \leq 1$. Thus, there exists $\alpha_m$ such that $r_{\alpha_m \boldsymbol{w}} \leq \frac{1}{q-1}$, which implies $p_{\alpha_m \boldsymbol{w}} > \varepsilon - \frac{1}{q-1}$. Now, we split $p_{\alpha_m \boldsymbol{w}}$, depending on whether $\boldsymbol{u} \in S'$ or $\boldsymbol{u} = \mathbf{0}$,

$$p_{\alpha_m \boldsymbol{w}} = \Pr(\boldsymbol{u} = \mathbf{0} \wedge \eta(\boldsymbol{w}) \in S') + \Pr(\boldsymbol{u} \in S' \wedge \boldsymbol{u} + \eta(\alpha_m \boldsymbol{w}) \in S')$$
$$\leq \Pr(\eta(\boldsymbol{w}) \in S') + \Pr(\boldsymbol{u} \in S')$$

and conclude that either $\Pr(\boldsymbol{u} \in S') > negl$ or for all nonzero $\boldsymbol{w} \in \widetilde{S}'$, $\Pr(\eta(\boldsymbol{w}) \in S') > negl$. However, which one is true could depend on the particular behavior of the oracle in the first $Q - 1$ calls.

The next lemma is needed in other subsequent proofs.

**Lemma 10.** *Consider integers* $l = k + d$, $\widetilde{l} = \widetilde{k} + \widetilde{d}$ *such that* $k, d, \widetilde{k}, \widetilde{d} > 0$ *and* $k > \widetilde{k}$. *Let* $\eta : \mathbb{Z}_q^{\widetilde{l}} \to \mathbb{Z}_q^l$ *be a linear map. Then, there exists a subspace* $F$ *of* $\operatorname{Im}\eta$ *of dimension at most* $k$ *such that for all* $\widetilde{d}$*-dimensional subspaces* $\widetilde{S}$ *of* $\mathbb{Z}_q^{\widetilde{l}}$, *either* $\widetilde{S} \subset \ker \eta$ *or* $\dim F \cap \eta(\widetilde{S}) \geq 1$.

*Proof.* If $\operatorname{rank}\eta \leq k$ it suffices to take $F = \operatorname{Im}\eta$. Indeed, if $\widetilde{S} \not\subset \ker \eta$, i.e., $\eta(\widetilde{S}) \neq \{\mathbf{0}\}$, then $\dim F \cap \eta(\widetilde{S}) = \dim \eta(\widetilde{S}) \geq 1$. Otherwise, $\operatorname{rank}\eta > k$, let $F$ a subspace of $\operatorname{Im}\eta$ of dimension $k$, using the Grassman's formula,

$$\dim F \cap \eta(\widetilde{S}) = \dim F + \dim \eta(\widetilde{S}) - \dim(F + \eta(\widetilde{S})) \geq k + \dim \eta(\widetilde{S}) - \operatorname{rank}\eta$$
$$\geq k + \dim \widetilde{S} - \dim \ker \eta - \operatorname{rank}\eta = k + \widetilde{d} - \widetilde{l} = k - \widetilde{k} \geq 1$$

**Lemma 7.** *If a matrix distribution* $\mathcal{D}_{\ell,k}$ *is hard (as given in Definition 4) then* $\mathcal{D}_{\ell,k}$ *is* $k$*-elusive.*

*Proof.* By definition, given a non-$k$-elusive matrix distribution $\mathcal{D}_{\ell,k}$, there exists a $k$-dimensional vector subspace $F \subset \mathbb{Z}_q^\ell$ such that $\Pr_{\mathbf{A} \leftarrow \mathcal{D}_{\ell,k}}(F \cap \ker \mathbf{A}^\top \neq \{\mathbf{0}\}) = \varepsilon > negl$. $F$ can be efficiently computed from the description of $\mathcal{D}_{\ell,k}$ with standard tools from linear algebra.

Let $\mathbf{M} \in \mathbb{Z}_q^{k \times \ell}$ be a maximal rank matrix such that $\operatorname{Im}\mathbf{M}^\top = F$. Then, $\dim(F \cap \ker \mathbf{A}^\top) = \dim(\operatorname{Im}\mathbf{M}^\top \cap \ker \mathbf{A}^\top) \leq \dim \ker(\mathbf{A}^\top \mathbf{M}^\top) = \dim \ker(\mathbf{MA})^\top = \dim \ker(\mathbf{MA})$, as $\mathbf{MA}$ is a $k \times k$ square matrix. Thus, we know that

$$\Pr_{\mathbf{A} \leftarrow \mathcal{D}_{\ell,k}} (\operatorname{rank}(\mathbf{MA}) < k) \geq \varepsilon$$

Now we show how to solve the $\mathcal{D}_{\ell,k}$-MDDH problem with advantage almost $\varepsilon$ on some $k$-linear group $\mathbb{G}$, by means of a $k$-linear map. Let $[(\mathbf{A} \| \boldsymbol{z})]$ be an instance of the $\mathcal{D}_{\ell,k}$-MDDH problem. In a 'real' instance $\boldsymbol{z} = \mathbf{A}\boldsymbol{x}$ for a uniformly distributed vector $\boldsymbol{x} \in \mathbb{Z}_q^k$, while in a 'random' instance, $\boldsymbol{z}$ is uniformly distributed $\mathbb{Z}_q^\ell$. A distinguisher can efficiently compute $[\mathbf{MA}]$ and $[\mathbf{M}\boldsymbol{z}]$. Observe that in a 'real' instance $\operatorname{rank}(\mathbf{MA} \| \mathbf{M}\boldsymbol{z}) = \operatorname{rank}(\mathbf{MA} \| \mathbf{MA}\boldsymbol{x}) = \operatorname{rank}(\mathbf{MA})$, while in a 'random' instance $\mathbf{M}\boldsymbol{z}$ is uniformly distributed in $\mathbb{Z}_q^k$. Therefore, for a 'random' instance there is a non-negligible probability, greater than $\varepsilon - \frac{1}{q}$, that $\operatorname{rank}(\mathbf{MA}) < k$ and $\operatorname{rank}(\mathbf{MA} \| \mathbf{M}\boldsymbol{z}) = \operatorname{rank}(\mathbf{MA}) + 1$, because $\mathbf{M}\boldsymbol{z} \in \operatorname{Im}(\mathbf{MA})$ occurs only with a negligible probability $< \frac{1}{q}$. Then, the distinguisher can efficiently tell apart the two cases because with a $k$-linear map at hand computing the rank of a $k \times k$ or a $k \times k + 1$ matrix can be done efficiently.

**Theorem 1.** *Let* $\mathcal{D}_{\ell,k}$ *be* $k$*-elusive. If there exists a black-box reduction in the purely algebraic generic multilinear group model from* $\mathcal{D}_{\ell,k}$*-KerMDH to another problem* $\mathcal{D}_{\widetilde{\ell},\widetilde{k}}$*-KerMDH with* $\widetilde{k} < k$, *then* $\mathcal{D}_{\ell,k}$*-KerMDH is easy.*

*Proof.* Let us assume the existence of the claimed reduction, $\mathcal{R}^{\mathcal{O}} = (\mathcal{R}_0^{\mathcal{O}}, \mathcal{R}_1)$, making $Q \geq 1$ oracle queries, where $Q$ is minimal, and with a success probability $\varepsilon$. Then, by Lemma 6, its output can be written as $[\boldsymbol{u} + \eta(\boldsymbol{w})]$, where $\eta : \mathbb{Z}_q^{\tilde{l}} \to \mathbb{Z}_q^l$ is a (randomized) linear map that does not depend on the particular choice of the matrix $\mathbf{A}$ in the $\mathcal{D}_{\ell,k}$-KerMDH input instance, but only on the random tape of the reduction. Let us denote as above $S = \ker \mathbf{A}^\top$, and $S' = S \backslash \{\mathbf{0}\}$. Analogously, $\tilde{S} = \ker \tilde{\mathbf{A}}^\top$, where $\tilde{\mathbf{A}} \leftarrow \mathcal{D}_{\tilde{\ell},\tilde{k}}$ and $\tilde{S}' = \tilde{S} \backslash \{\mathbf{0}\}$.

We now prove that in Lemma 6, for any possible behavior of the oracle in the first $Q - 1$ calls, there exists a particular behavior in the last call such that $\Pr(\eta(\boldsymbol{w}) \in S')$ is negligible. Namely, the $Q$-th query is answered by $\mathcal{O}$ by choosing a uniformly distributed $\boldsymbol{w} \in \tilde{S}'$ (as required to be algebraic, according to Definition 1). Indeed, $\Pr(\eta(\boldsymbol{w}) \in S') = \Pr(\eta(\boldsymbol{w}) \in S) - \Pr(\eta(\boldsymbol{w}) = 0)$. Now, developing the second term,

$$\begin{aligned}
\Pr(\eta(\boldsymbol{w}) = 0) &= \Pr(\eta(\boldsymbol{w}) = 0 \mid \tilde{S} \subset \ker \eta) \Pr(\tilde{S} \subset \ker \eta) \\
&\quad + \Pr(\eta(\boldsymbol{w}) = 0 \mid \tilde{S} \not\subset \ker \eta) \Pr(\tilde{S} \not\subset \ker \eta) \\
&= \Pr(\tilde{S} \subset \ker \eta) + \Pr(\boldsymbol{w} \in \tilde{S} \cap \ker \eta \mid \tilde{S} \not\subset \ker \eta) \Pr(\tilde{S} \not\subset \ker \eta) \\
&= \Pr(\tilde{S} \subset \ker \eta) + negl
\end{aligned}$$

where the last equality uses that the probability that a vector uniformly distributed in $\tilde{S}'$ belongs to a proper subspace of $\tilde{S}'$ is negligible. Analogously,

$$\begin{aligned}
\Pr(\eta(\boldsymbol{w}) \in S) &= \Pr(\eta(\boldsymbol{w}) \in S \mid \eta(\tilde{S}) \subset S) \Pr(\eta(\tilde{S}) \subset S) \\
&\quad + \Pr(\eta(\boldsymbol{w}) \in S \mid \eta(\tilde{S}) \not\subset S) \Pr(\eta(\tilde{S}) \not\subset S) \\
&= \Pr(\eta(\tilde{S}) \subset S) + \Pr(\boldsymbol{w} \in \tilde{S} \cap \eta^{-1}(S) \mid \eta(\tilde{S}) \not\subset S) \Pr(\eta(\tilde{S}) \not\subset S) \\
&= \Pr(\eta(\tilde{S}) \subset S) + negl
\end{aligned}$$

Thus, $\Pr(\eta(\boldsymbol{w}) \in S') = \Pr(\eta(\tilde{S}) \subset S) - \Pr(\tilde{S} \subset \ker \eta) + negl$. Now, using Lemma 10, we know that there exists a subspace $F$ of dimension at most $k$ such that if $\tilde{S} \not\subset \ker \eta$, then $\dim F \cap \eta(\tilde{S}) \geq 1$. Therefore $\Pr(\eta(\tilde{S}) \subset S) - \Pr(\tilde{S} \subset \ker \eta) \leq \Pr(\eta(\tilde{S}) \subset S \wedge \dim F \cap \eta(\tilde{S}) \geq 1) \leq \Pr(\dim F \cap S \geq 1)$. Due to the $k$-elusiveness of $\mathcal{D}_{\ell,k}$, from Lemma 7, the last probability is negligible. Namely, it is upper bounded by $\mathbf{Adv}_{\mathcal{D}_{\ell,k}\text{-MDDH}} + \frac{1}{q}$, where $\mathbf{Adv}_{\mathcal{D}_{\ell,k}\text{-MDDH}}$ denotes the advantage of a distinguisher for the $\mathcal{D}_{\ell,k}$-MDDH problem. By Lemma 6,

$$\Pr(\boldsymbol{u} \in S \backslash \{\mathbf{0}\}) > \varepsilon - \frac{1}{q-1} - \mathbf{Adv}_{\mathcal{D}_{\ell,k}\text{-MDDH}} - \frac{1}{q},$$

for any possible behavior of the oracle in the first $Q - 1$ calls. Therefore, we can modify the reduction $\mathcal{R}$ to output $\boldsymbol{u}$, without making the $Q$-th oracle call. The modified reduction is also successful, essentially with the same probability $\varepsilon$, with only $Q-1$ oracle calls, which contradicts the assumption that $Q$ is minimal. In summary, if the claimed reduction exists then there also exists an algorithm (a "reduction with $Q = 0$") directly solving $\mathcal{D}_{\ell,k}$-KerMDH without the help of any oracle and with the same success probability.

# References

1. Abe, M., Chase, M., David, B., Kohlweiss, M., Nishimaki, R., Ohkubo, M.: Constant-size structure-preserving signatures: generic constructions and simple assumptions. In: Wang, X., Sako, K. (eds.) ASIACRYPT 2012. LNCS, vol. 7658, pp. 4–24. Springer, Heidelberg (2012). doi:10.1007/978-3-642-34961-4_3
2. Abe, M., Fuchsbauer, G., Groth, J., Haralambiev, K., Ohkubo, M.: Structure-preserving signatures and commitments to group elements. In: Rabin, T. (ed.) CRYPTO 2010. LNCS, vol. 6223, pp. 209–236. Springer, Heidelberg (2010). doi:10.1007/978-3-642-14623-7_12
3. Abe, M., Haralambiev, K., Ohkubo, M.: Group to group commitments do not shrink. In: Pointcheval, D., Johansson, T. (eds.) EUROCRYPT 2012. LNCS, vol. 7237, pp. 301–317. Springer, Heidelberg (2012). doi:10.1007/978-3-642-29011-4_19
4. Bao, F., Deng, R.H., Zhu, H.F.: Variations of Diffie-Hellman problem. In: Qing, S., Gollmann, D., Zhou, J. (eds.) ICICS 2003. LNCS, vol. 2836, pp. 301–312. Springer, Heidelberg (2003). doi:10.1007/978-3-540-39927-8_28
5. Barthe, G., Fagerholm, E., Fiore, D., Mitchell, J., Scedrov, A., Schmidt, B.: Automated analysis of cryptographic assumptions in generic group models. In: Garay, J.A., Gennaro, R. (eds.) CRYPTO 2014. LNCS, vol. 8616, pp. 95–112. Springer, Heidelberg (2014). doi:10.1007/978-3-662-44371-2_6
6. Boneh, D., Boyen, X., Shacham, H.: Short group signatures. In: Franklin, M. (ed.) CRYPTO 2004. LNCS, vol. 3152, pp. 41–55. Springer, Heidelberg (2004). doi:10.1007/978-3-540-28628-8_3
7. Boneh, D., Halevi, S., Hamburg, M., Ostrovsky, R.: Circular-secure encryption from decision Diffie-Hellman. In: Wagner, D. (ed.) CRYPTO 2008. LNCS, vol. 5157, pp. 108–125. Springer, Heidelberg (2008). doi:10.1007/978-3-540-85174-5_7
8. Boyen, X.: The uber-assumption family. In: Galbraith, S.D., Paterson, K.G. (eds.) Pairing 2008. LNCS, vol. 5209, pp. 39–56. Springer, Heidelberg (2008). doi:10.1007/978-3-540-85538-5_3
9. Brands, S.: Untraceable off-line cash in wallet with observers. In: Stinson, D.R. (ed.) CRYPTO 1993. LNCS, vol. 773, pp. 302–318. Springer, Heidelberg (1994). doi:10.1007/3-540-48329-2_26
10. Dodis, Y., Haitner, I., Tentes, A.: On the instantiability of hash-and-sign RSA signatures. In: Cramer, R. (ed.) TCC 2012. LNCS, vol. 7194, pp. 112–132. Springer, Heidelberg (2012). doi:10.1007/978-3-642-28914-9_7
11. Escala, A., Herold, G., Kiltz, E., Ràfols, C., Villar, J.: An algebraic framework for Diffie-Hellman assumptions. In: Canetti, R., Garay, J.A. (eds.) CRYPTO 2013. LNCS, vol. 8043, pp. 129–147. Springer, Heidelberg (2013). doi:10.1007/978-3-642-40084-1_8
12. Escala, A., Herold, G., Kiltz, E. et al.: An algebraic framework for Diffie-Hellman assumptions. J. Cryptol. 1–47 (2015). doi:10.1007/s00145-015-9220-6
13. Freeman, D.M.: Converting pairing-based cryptosystems from composite-order groups to prime-order groups. In: Gilbert, H. (ed.) EUROCRYPT 2010. LNCS, vol. 6110, pp. 44–61. Springer, Heidelberg (2010). doi:10.1007/978-3-642-13190-5_3
14. Galindo, D., Herranz, J., Villar, J.: Identity-based encryption with master key-dependent message security and leakage-resilience. In: Foresti, S., Yung, M., Martinelli, F. (eds.) ESORICS 2012. LNCS, vol. 7459, pp. 627–642. Springer, Heidelberg (2012). doi:10.1007/978-3-642-33167-1_36
15. Gay, R., Hofheinz, D., Kiltz, E., Wee, H.: Tightly CCA-secure encryption without pairings. In: Fischlin, M., Coron, J.-S. (eds.) EUROCRYPT 2016. LNCS, vol. 9665, pp. 1–27. Springer, Heidelberg (2016). doi:10.1007/978-3-662-49890-3_1

16. Goldreich, O.: On post-modern cryptography. Cryptology ePrint Archive, Report 2006/461 (2006). http://eprint.iacr.org/2006/461
17. González, A., Hevia, A., Ràfols, C.: QA-NIZK arguments in asymmetric groups: new tools and new constructions. In: Iwata, T., Cheon, J.H. (eds.) ASIACRYPT 2015. LNCS, vol. 9452, pp. 605–629. Springer, Heidelberg (2015). doi:10.1007/978-3-662-48797-6_25
18. Groth, J.: Homomorphic trapdoor commitments to group elements. Cryptology ePrint Archive, Report 2009/007 (2009). http://eprint.iacr.org/2009/007
19. Groth, J., Lu, S.: A non-interactive shuffle with pairing based verifiability. In: Kurosawa, K. (ed.) ASIACRYPT 2007. LNCS, vol. 4833, pp. 51–67. Springer, Heidelberg (2007). doi:10.1007/978-3-540-76900-2_4
20. Herold, G.: Applications of classical algebraic geometry to cryptography. Ph.D. thesis, Ruhr-Universität Bochum (2014)
21. Herold, G., Hesse, J., Hofheinz, D., Ràfols, C., Rupp, A.: Polynomial spaces: a new framework for composite-to-prime-order transformations. In: Garay, J.A., Gennaro, R. (eds.) CRYPTO 2014. LNCS, vol. 8616, pp. 261–279. Springer, Heidelberg (2014). doi:10.1007/978-3-662-44371-2_15
22. Hofheinz, D., Kiltz, E.: Secure hybrid encryption from weakened key encapsulation. In: Menezes, A. (ed.) CRYPTO 2007. LNCS, vol. 4622, pp. 553–571. Springer, Heidelberg (2007). doi:10.1007/978-3-540-74143-5_31
23. Joux, A., Rojat, A.: Security ranking among assumptions within the uber assumption framework. Cryptology ePrint Archive, Report 2013/291 (2013). http://eprint.iacr.org/2013/291
24. Jutla, C.S., Roy, A.: Switching lemma for bilinear tests and constant-size NIZK proofs for linear subspaces. In: Garay, J.A., Gennaro, R. (eds.) CRYPTO 2014. LNCS, vol. 8617, pp. 295–312. Springer, Heidelberg (2014). doi:10.1007/978-3-662-44381-1_17
25. Kiltz, E., Pan, J., Wee, H.: Structure-preserving signatures from standard assumptions, revisited. In: Gennaro, R., Robshaw, M. (eds.) CRYPTO 2015. LNCS, vol. 9216, pp. 275–295. Springer, Heidelberg (2015). doi:10.1007/978-3-662-48000-7_14
26. Kiltz, E., Wee, H.: Quasi-adaptive NIZK for linear subspaces revisited. In: Oswald, E., Fischlin, M. (eds.) EUROCRYPT 2015. LNCS, vol. 9057, pp. 101–128. Springer, Heidelberg (2015). doi:10.1007/978-3-662-46803-6_4
27. Laguillaumie, F., Paillier, P., Vergnaud, D.: Universally convertible directed signatures. In: Roy, B. (ed.) ASIACRYPT 2005. LNCS, vol. 3788, pp. 682–701. Springer, Heidelberg (2005). doi:10.1007/11593447_37
28. Lepoint, T.: Zeroizing attacks on multilinear maps. In: ECRYPT-CSA Workshop on Tools for Asymmetric Cryptanalysis (2015). http://cryptool.hgi.rub.de/program.html
29. Lewko, A.B., Waters, B.: Efficient pseudorandom functions from the decisional linear assumption and weaker variants. In: Al-Shaer, E., Jha, S., Keromytis, A.D. (eds.) ACM CCS 2009, pp. 112–120. ACM Press, Chicago (2009)
30. Libert, B., Peters, T., Joye, M., Yung, M.: Linearly homomorphic structure-preserving signatures and their applications. In: Canetti, R., Garay, J.A. (eds.) CRYPTO 2013. LNCS, vol. 8043, pp. 289–307. Springer, Heidelberg (2013). doi:10.1007/978-3-642-40084-1_17
31. Libert, B., Peters, T., Joye, M., Yung, M.: Non-malleability from malleability: simulation-sound quasi-adaptive NIZK proofs and CCA2-secure encryption from homomorphic signatures. In: Nguyen, P.Q., Oswald, E. (eds.) EUROCRYPT 2014. LNCS, vol. 8441, pp. 514–532. Springer, Heidelberg (2014). doi:10.1007/978-3-642-55220-5_29

32. Libert, B., Vergnaud, D.: Multi-use unidirectional proxy re-signatures. In: Ning, P., Syverson, P.F., Jha, S. (eds.) ACM CCS 2008, pp. 511–520. ACM Press, Alexandria (2008)
33. Maurer, U.M.: Towards the equivalence of breaking the diffie-hellman protocol and computing discrete logarithms. In: Desmedt, Y.G. (ed.) CRYPTO 1994. LNCS, vol. 839, pp. 271–281. Springer, Heidelberg (1994). doi:10.1007/3-540-48658-5_26
34. Maurer, U.: Abstract models of computation in cryptography. In: Smart, N.P. (ed.) Cryptography and Coding 2005. LNCS, vol. 3796, pp. 1–12. Springer, Heidelberg (2005). doi:10.1007/11586821_1
35. Morillo, P., Ràfols, C., Villar, J.L.: Matrix computational assumptions in multilinear groups. Cryptology ePrint Archive, Report 2015/353 (2015). http://eprint.iacr.org/2015/353
36. Naor, M., Reingold, O.: Number-theoretic constructions of efficient pseudo-random functions. In: 38th FOCS, pp. 458–467. IEEE Computer Society Press, Miami Beach, Florida, 19–22 October 1997 (1997)
37. Naor, M., Segev, G.: Public-key cryptosystems resilient to key leakage. In: Halevi, S. (ed.) CRYPTO 2009. LNCS, vol. 5677, pp. 18–35. Springer, Heidelberg (2009). doi:10.1007/978-3-642-03356-8_2
38. Pedersen, T.P.: Non-interactive and information-theoretic secure verifiable secret sharing. In: Feigenbaum, J. (ed.) CRYPTO 1991. LNCS, vol. 576, pp. 129–140. Springer, Heidelberg (1992). doi:10.1007/3-540-46766-1_9
39. Seo, J.H.: On the (Im)possibility of projecting property in prime-order setting. In: Wang, X., Sako, K. (eds.) ASIACRYPT 2012. LNCS, vol. 7658, pp. 61–79. Springer, Heidelberg (2012). doi:10.1007/978-3-642-34961-4_6
40. Shacham, H.: A cramer-shoup encryption scheme from the linear assumption and from progressively weaker linear variants. Cryptology ePrint Archive, Report 2007/074 (2007). http://eprint.iacr.org/2007/074
41. Villar, J.L.: Optimal reductions of some decisional problems to the rank problem. In: Wang, X., Sako, K. (eds.) ASIACRYPT 2012. LNCS, vol. 7658, pp. 80–97. Springer, Heidelberg (2012). doi:10.1007/978-3-642-34961-4_7
42. Waters, B.: Efficient identity-based encryption without random oracles. In: Cramer, R. (ed.) EUROCRYPT 2005. LNCS, vol. 3494, pp. 114–127. Springer, Heidelberg (2005). doi:10.1007/11426639_7

# Cryptographic Applications of Capacity Theory: On the Optimality of Coppersmith's Method for Univariate Polynomials

Ted Chinburg[1]([⊠]), Brett Hemenway[1], Nadia Heninger[1], and Zachary Scherr[2]

[1] University of Pennsylvania, Philadelphia, USA
ted@math.upenn.edu
[2] Bucknell University, Lewisburg, USA

**Abstract.** We draw a new connection between Coppersmith's method for finding small solutions to polynomial congruences modulo integers and the capacity theory of adelic subsets of algebraic curves. Coppersmith's method uses lattice basis reduction to construct an auxiliary polynomial that vanishes at the desired solutions. Capacity theory provides a toolkit for proving when polynomials with certain boundedness properties do or do not exist. Using capacity theory, we prove that Coppersmith's bound for univariate polynomials is optimal in the sense that there are *no* auxiliary polynomials of the type he used that would allow finding roots of size $N^{1/d+\epsilon}$ for *any* monic degree-$d$ polynomial modulo $N$. Our results rule out the existence of polynomials of any degree and do not rely on lattice algorithms, thus eliminating the possibility of improvements for special cases or even superpolynomial-time improvements to Coppersmith's bound. We extend this result to constructions of auxiliary polynomials using binomial polynomials, and rule out the existence of any auxiliary polynomial of this form that would find solutions of size $N^{1/d+\epsilon}$ unless $N$ has a very small prime factor.

**Keywords:** Coppersmith's method · Lattices · Polynomial congruences · Capacity theory · RSA

## 1 Introduction

Coppersmith's method [Cop97, Cop01] is a celebrated technique in public-key cryptanalysis for finding small roots of polynomial equations modulo integers. In the simplest case, one is given a degree-$d$ monic polynomial $f(x)$ with integer coefficients, and one wishes to find the integers $r$ modulo a given integer $N$ for which $f(r) \equiv 0 \bmod N$. When $N$ is prime, this problem can be efficiently solved in polynomial time, but for composite $N$ of unknown factorization, no efficient method is known in general. In fact, such an algorithm would immediately break

© International Association for Cryptologic Research 2016
J.H. Cheon and T. Takagi (Eds.): ASIACRYPT 2016, Part I, LNCS 10031, pp. 759–788, 2016.
DOI: 10.1007/978-3-662-53887-6_28

the RSA cryptosystem, by allowing one to decrypt ciphertexts $c$ by finding roots of the polynomial $f(x) = x^e - c \bmod N$.

While it appears intractable to solve this problem in polynomial time, Coppersmith showed that one can efficiently find all *small* integers $r$ such that $f(r) \equiv 0 \bmod N$. More precisely, he proved the following result in [Cop97]:

**Theorem 1 (Coppersmith 1996).** *Suppose one is given a modulus $N$ and a monic polynomial $f(x) = x^d + f_{d-1}x^{d-1} + \cdots + f_1 x + f_0$ in $\mathbb{Z}[x]$. One can find all $r \in \mathbb{Z}$ such that*

$$|r| \leq N^{1/d} \quad \text{and} \quad f(r) \equiv 0 \bmod N \tag{1}$$

*in polynomial time in $\log(N) + \sum_i \log |f_i|$.*

The algorithm he developed to prove this result has applications across public-key cryptography, including cryptanalysis of low public exponent RSA with fixed-pattern or affine padding [Cop97], the security proof of RSA-OAEP [Sho01], and showing that the least significant bits of RSA are hardcore [SPW06]. We discuss these applications in more detail in Sect. 2.3. If the exponent $1/d$ in the bound in Eq. 1 could be increased, it would have immediate practical impact on the security of a variety of different cryptosystems.

In followup work, [Cop01, Sect. 4] Coppersmith speculates about possible improvements of this exponent $1/d$. The main conclusion of [Cop01, Sect. 4] is that "We have tried to abuse this method to obtain information that should otherwise be hard to get, and we always fail." We discuss these obstructions in more detail in Sect. 2.2.

Later, the hardness of finding roots of $f(x)$ of size $N^{1/d+\epsilon}$ for $\epsilon > 0$ was formalized as a concrete cryptographic hardness assumption [SPW06].

Coppersmith's proof of Theorem 1 relies on constructing a polynomial $h(x)$ such that any small integer $r$ satisfying $f(r) \equiv 0 \bmod N$ is a root of $h(x)$ over the integers. He finds such an auxiliary polynomial $h(x)$ by constructing a basis for a lattice of polynomials, and then by using the Lenstra-Lenstra-Lovasz lattice basis reduction algorithm [LLL82] to find a "small" polynomial in this lattice. The smallness condition ensures that any small integer $r$ satisfying $f(r) \equiv 0 \bmod N$ must be a root of $h(x)$. The algorithm then checks which rational roots $r$ of $h(x)$ have the desired properties.

*Our Results.* In this paper, our main result is that one cannot increase the exponent $1/d$ in Coppersmith's theorem by using auxiliary polynomials of the kind he considers. We give a formal proof that there do not exist polynomials of the kind required to extend Coppersmith's theorem by the same method, *regardless* of the polynomial $p(x)$, the modulus $N$, and the method used to find them. This is a much more general statement than previous partial results along these lines, and in particular it applies to the settings of most interest to cryptographers. This eliminates possible improvements to the method using improvements in lattice algorithms or shortest vector bounds. We obtain our results by drawing a new connection between this family of cryptographic techniques and results

from the *capacity theory* of adelic subsets of algebraic curves. We will use fundamental results of Cantor [Can80] and Rumely [Rum89, Rum13] about capacity
theory to prove several results about such polynomials.

In particular, we will prove in Theorem 6 a stronger form of the following
result. This result shows that there are *no* polynomials of the type used by
Coppersmith that could lead to an improvement of the bound in (1) from $N^{1/d}$
to $N^{1/d+\epsilon}$ for any $\epsilon > 0$.

**Theorem 2 (Optimality of Coppersmith's Theorem).** *Let $f$ be a monic
polynomial of degree $d$. Suppose $\epsilon > 0$. There does not exist a non-zero polynomial
$h(x) \in \mathbb{Q}[x]$ of the form*

$$h(x) = \sum_{i,j \geq 0} a_{i,j}\, x^i\, (f(x)/N)^j \tag{2}$$

*with $a_{i,j} \in \mathbb{Z}$ such that $|h(z)| < 1$ for all $z$ in the complex disk $\{z \in \mathbb{C} : |z| \leq
N^{(1/d)+\epsilon}\}$. Furthermore, if $\epsilon > \ln(2)/\ln(N)$ there is no such $h(x)$ such that
$|h(z)| < 1$ for all $z$ in the real interval $[-N^{1/d+\epsilon}, N^{1/d+\epsilon}]$.*

Note that in order for Coppersmith's method to run in polynomial time, $h(x)$
should have degree bounded by a polynomial in $\ln(N)$. Theorem 2 says that when
$\epsilon > 0$ there are no polynomials of *any* degree satisfying the stated bounds. We
can thus eliminate the possibility of an improvement to this method with even
superpolynomial running time.

In [Cop01], Coppersmith already noted that it did not appear possible to
improve the exponent $1/d$ in his result by searching for roots in the real interval
$[-N^{1/d+\epsilon}, N^{1/d+\epsilon}]$ instead of in the complex disk of radius $N^{1/d+\epsilon}$. The last
statement in Theorem 2 quantifies this observation, since $\ln(2)/\ln(N) \to 0$ as
$N \to \infty$.

Coppersmith also notes that since the binomial polynomials

$$b_i(x) = x \cdot (x-1) \cdots (x-i+1)/i!$$

take integral values on integers, one could replace $x^i$ in (2) by $b_i(x)$ and
$(f(x)/N)^j$ by $b_j(f(x)/N)$. Coppersmith observed (backed up by experiments)
that this leads to a small improvement on the size of the root that can be found,
and a speedup for practical computations. The improvement is proportional to
the degree of the auxiliary polynomial $h(x)$ that is constructed, and is thus
limited for a polynomial-time algorithm.

We show that the exponent $1/d$ in Coppersmith's theorem still cannot be
improved using binomial polynomials, but for a different reason. Our results
come in two parts. First, we show that the exact analogue of Theorem 2 is false in
the case of integral combinations of binomial polynomials. In fact, there are such
combinations that have all the properties required in the proof of Coppersmith's
theorem. The problem is that these polynomials have very large degree, and in
fact, they vanish at *every* small integer, not just the solutions of the congruence.
This is formalized in the following theorem, which is a simplified version of
Theorem 9.

**Theorem 3 (Existence of Binomial Auxiliary Polynomials).** *Suppose $\delta$ is any positive real number. For all sufficiently large integers $N$ there is a non-zero polynomial of the form $h(x) = \sum_i a_i \, b_i(x)$ with $a_i \in \mathbb{Z}$ such that $|h(z)| < 1$ for all $z$ in the complex disk $\{z \in \mathbb{C} : |z| \le N^\delta\}$.*

Second, we show that the existence of these polynomials still does not permit cryptographically useful improvements to Coppersmith's bound beyond $N^{1/d}$. This is because if one is able to use binomial polynomials of small degree to obtain such an improvement, then the modulus $N$ must have a small prime factor. In that case, it would have been more efficient to factor $N$ and use the factorization to find the roots. More precisely, we will show in Theorem 11 a stronger form of the following result:

**Theorem 4 (Negative Coppersmith Theorem for Binomial Polynomials).** *Let $f$ be a monic polynomial of degree $d$. Suppose $\epsilon > 0$ and that $M$ and $N$ are integers with $1.48774N^\epsilon \ge M \ge 319$. If there is a non-zero polynomial $h(x)$ of the form*

$$h(x) = \sum_{0 \le i,j \le M} a_{i,j} \, b_i(x) \, b_j(f(x)/N) \tag{3}$$

*with $a_{i,j} \in \mathbb{Z}$ such that $|h(z)| < 1$ for $z$ in the complex disk $\{z \in \mathbb{C} : |z| \le N^{1/d+\epsilon}\}$, then $N$ must have a prime factor less than or equal to $M$. In particular, this will be the case for all large $N$ if we let $M = \ln(N)^c$ for some fixed integer $c > 0$.*

Note that the integer $M$ quantifies "smallness" in Theorem 4 in two ways. First, it is a bound on the degree of the binomial polynomials that are allowed to be used to create auxiliary polynomials. But then if a useful auxiliary polynomial exists, then $N$ must have a factor of size less than or equal to $M$. As a special case of Theorem 4, if $N = pq$ is an RSA modulus with two large equal sized prime factors, then any auxiliary polynomial of the form in (3) that can find roots of size $N^{1/d+\epsilon}$ must involve binomial terms with $i$ or $j$ at least $1.48774N^\epsilon$.

Note that Coppersmith's theorem in its original form is not sensitive to whether or not $N$ has small prime factors. Theorem 4 shows that the existence of useful auxiliary polynomials *does* depend on whether $N$ has such small factors.

The paper is organized in the following way. In Sect. 2.1 we begin by recalling Coppersmith's algorithm for finding small solutions of polynomial congruences. In Sect. 2.3 we recall some mathematical hardness assumptions and we discuss their connection to the security of various cryptosystems and Coppersmith's algorithm. In Sect. 3 we review some basic notions from algebraic number theory, and we recall some results of Cantor [Can80] and Rumely [Rum89,Rum13] on which our work is based. At the end of Sect. 3 we prove Theorem 6, which implies Theorem 2. We state and prove Theorems 9 and 11 in Sect. 4; these imply Theorems 3 and 4. One of the goals of this paper is to provide a framework for using capacity theory to show when these auxiliary polynomials do or do not exist. We give an outline in Sect. 5 of how one proves these types of results. In the conclusion we summarize the implications of our results and discuss possible directions for future research.

## 2   Background and Related Work

Given a polynomial $f(x) = x^d + f_{d-1}x^{d-1} + \cdots + f_1 x + f_0 \in \mathbb{Z}[x]$ and a prime $p$ we can find solutions $x \in \mathbb{Z}$ to the equation

$$f(x) \equiv 0 \bmod p \tag{4}$$

in randomized polynomial time using e.g. Berlekamp's algorithm or the Cantor-Zassenhaus algorithm [Ber67,CZ81]. While it is "easy" to find roots of $f(x)$ in the finite field $\mathbb{Z}/p\mathbb{Z}$ and over $\mathbb{Z}$ as well, there is no known efficient method to find roots of $f(x)$ modulo $N$ for large composite integers $N$ unless one knows the factorization of $N$.

### 2.1   Coppersmith's Method

Although finding roots of a univariate polynomial, $f(x)$, modulo $N$ is difficult in general, if $f(x)$ has a "small" root, then this root can be found efficiently using Coppersmith's method [Cop97].

Coppersmith's method for proving Theorem 1 works as follows. We follow the exposition in [Cop01], which incorporates simplifications due to Howgrave-Graham [HG97]. Suppose $\epsilon > 0$ and that $f(x)$ has a root $r \in \mathbb{Z}$ with $|r| \leq N^{1/d-\epsilon}$ and $f(r) \equiv 0 \bmod N$. He considers the finite rank lattice $\mathcal{L}$ of rational polynomials in $\mathbb{Q}[x]$ of the form

$$h_{ij}(x) = \sum_{0 \leq i+dj < t} a_{i,j}\, x^i\, (f(x)/N)^j$$

where $t \geq 0$ is an integer parameter to be varied and all $a_{i,j} \in \mathbb{Z}$. Here $\mathcal{L}$ is a finite rank lattice because the denominators of the coefficients of $h_{ij}(x)$ are bounded and $h_{ij}(x)$ has degree bounded by $t$.

If we evaluate any polynomial $h_{ij} \in \mathcal{L}$ at a root $r$ satisfying $f(r) \equiv 0 \bmod N$, $h_{ij}(r)$ will be an integer.

Concretely, one picks a basis for a sublattice of $\mathcal{L} \in \mathbb{Q}^{t-1}$ by taking a suitable set of polynomials $\{h_{ij}(x)\}_{i,j}$ and representing each polynomial by its coefficient vector. Coppersmith's method applies the LLL algorithm to this sublattice basis to find a short vector representing a specific polynomial, $h_\epsilon(x)$ in $\mathcal{L}$. He shows that the fact that the vector of coefficients representing $h_\epsilon(x)$ is short implies that $|h_\epsilon(x)| < 1$ for all $x \in \mathbb{C}$ with $|x| \leq N^{1/d-\epsilon}$, and that for sufficiently large $t$, the LLL algorithm will find a short enough vector. Because $h_\epsilon(x)$ is an integral combination of terms of the form $x^i(f(x)/N)^j$, this forces $h(r) \in \mathbb{Z}$ because $f(r)/N \in \mathbb{Z}$. But $|r| \leq N^{1/d-\epsilon}$ forces $|h_\epsilon(r)| < 1$. Because 0 is the only integer less than 1 in absolute value, we see $h_\epsilon(r) = 0$. So $r$ is among the zeros of $h_\epsilon(x)$, and as discussed earlier, there is an efficient method to find the integer zeros of a polynomial in $\mathbb{Q}[x]$. One then lets $\epsilon \to 0$ and does a careful analysis of the computational complexity of this method.

The bound in Theorem 1 arises from cleverly choosing a subset of the possible $\{h_{ij}\}$ as a lattice basis so that one can bound the determinant of the lattice as tightly as possible, then using the LLL algorithm in a black-box way on the resulting lattice basis.

## 2.2    Optimality of Coppersmith's Theorem

Since Coppersmith's technique uses the LLL algorithm [LLL82] to find the specific polynomial $h(x)$ in the lattice $\mathcal{L}$, it is natural to think that improvements in lattice reduction techniques or improved bounds on the length of the shortest vector in certain lattices might improve the bound $N^{1/d}$ in Theorem 1.

Such an improvement would be impossible in polynomial time for arbitrary $N$, since the polynomial $f(x) = x^d$ has exponentially many roots modulo $N = p^d$ of absolute value $N^{1/d+\epsilon}$, but this does not rule out the possibility of improvements for cases of cryptographic interest, such as polynomial congruences modulo RSA moduli $N = pq$.

Coppersmith [Cop01] finds "cause for pessimism" in extending his technique. This pessimism comes from a specific example where the modulus $N$ is equal to $q^3$ the cube of a prime $q$. He observes that there are exponentially many small solutions to the congruence in question for such moduli, so his method cannot be expected to work in a black box manner for all moduli. He explains "we expect trouble whenever $q^2$ divides $N$ and $p(x)$ has repeated roots mod $q$." Since RSA moduli are square-free, Coppersmith's counterexample does not apply to RSA moduli. In general, Coppersmith's pessimism comes from examples where the discriminant of $f(x)$ and $N$ share a prime factor—in which case we can factor $N$ using a simple GCD calculation. Thus Coppersmith's counterexamples will never apply to any hard-to-factor modulus $N$. Coppersmith left open the possibility that his method could be improved for the applications of most interest to cryptographers. More explicitly, after discussing the above examples, he supposes he is not in the "unfavorable situation" in which the discriminant of $p(x)$ and $N$ have a common factor, and he discusses a "discriminant attack" which might work in this case. To say that the discriminant of $p(x)$ and $N$ have no common factor is the same as saying there are integer polynomials $D(x)$ and $E(x)$ together with an integer $F$ such that $D(x)p(x) + E(x)p'(x) + FN = 1$. Coppersmith wrote "Perhaps $D, E, F$ can be incorporated into the construction of the lattice $L$, in such a way that the bound $B$ can be improved to $N^{1/d+\epsilon}$. But I don't see how to do it." Our results show that such an improvement is impossible.

Aono, Agrawal, Satoh, and Watanabe [AASW12] showed that Coppersmith's lattice basis construction is optimal under the heuristic assumption that the lattice behaves as a random lattice; however they left open whether improved lattice bounds or a non-lattice-based approach to solving this problem could improve the $N^{1/d}$ bound.

## 2.3    Cryptanalytic Applications of Coppersmith's Theorem

Theorem 1 has many immediate applications to cryptanalysis, particularly the cryptanalysis of RSA. May [May07] gives a comprehensive survey of cryptanalytic applications of Coppersmith's method. In this paper, we focus on Coppersmith's method applied to univariate polynomials modulo integers. We highlight several applications of the univariate case below.

The RSA assumption posits that it is computationally infeasible to invert the map $x \mapsto x^d \bmod N$, i.e., it is infeasible to find roots of $f(x) = x^d - c \bmod N$. Because of their similar structure, almost all of the cryptographically hard problems (some of which are outlined below) based on factoring can be approached using Coppersmith's method (Theorem 1).

*Low public exponent RSA with stereotyped messages:* A classic example listed in Coppersmith's original paper [Cop97] is decrypting "stereotyped" messages encrypted under low public exponent RSA, where an approximation to the solution is known in advance. The general RSA map is $x \mapsto x^e \bmod N$. For efficiency purposes, $e$ can be chosen to be as small as 3, so that a "ciphertext" is $c_0 = x_0^3 \bmod N$. Suppose we know some approximation to the message $\tilde{x}_0$ to the message $x_0$. Then we can set

$$f(x) = (\tilde{x}_0 + x)^3 - c.$$

Thus $f(x)$ has a root (modulo $N$) at $x = x_0 - \tilde{x}_0$. If $|x_0 - \tilde{x}_0| < N^{1/3}$ then this root can be found using Coppersmith's method.

*Security of RSA-OAEP:* The RSA function $x \mapsto x^e \bmod N$ is assumed to be a one-way trapdoor permutation. Optimal Asymmetric Encryption Padding (OAEP) is a general method for taking a one-way trapdoor permutation and a random oracle [BR93], and creating a cryptosystem that achieves security against adaptive chosen ciphertext attacks (IND-CCA security).

Instantiating the OAEP protocol with the RSA one-way function yields RSA-OAEP, a standard cryptosystem. When the public exponent is $e = 3$, Shoup used Coppersmith's method to show that RSA-OAEP is secure against an adaptive chosen-ciphertext attack (in the random oracle model) [Sho01].

*Hard-core bits of the RSA Function:* Repeated iteration of the RSA function has been proposed as candidate for a pseudo random generator. In particular, we can create a stream of pseudo random bits by picking an initial "seed", $x_0$ and calculating the series

$$x_i \mapsto x_{i+1}$$
$$x_i \mapsto x_i^e \bmod N$$

At each iteration, the generator will output the $r$ least significant bits of $x_i$. For efficiency reasons, we would like $r$ to be as large as possible while still maintaining the provable security of the generator.

When we output only 1 bit per iteration, this was shown to be secure [ACGS88, FS00], and later this was increased to allow the generator to output any $\log \log(N)$ consecutive bits [HN04]. The maximum number of bits that can be safely outputted by such a generator is tightly tied to the approximation $\tilde{x}$ necessary for recovering $x$ from $x^e \bmod N$. Thus a bound on our ability to find small roots of $f(x) = (x - \tilde{x})^e - c \bmod N$ immediately translates into bounds on the maximum number of bits that can be safely outputted at each step of the RSA pseudo random generator.

In order to construct a provably secure pseudo random generator that outputs $\Omega(n)$ pseudo random bits for each multiplication modulo $N$, [SPW06] assume there is no probabilistic polynomial time algorithm for solving the $\left(\frac{1}{d} + \epsilon, d\right)$-SSRSA problem.

**Definition 1 (The $(\delta, d)$-SSRSA Problem [SPW06]).** *Given a random $n$ bit RSA modulus, $N$ and a polynomial $f(x) \in \mathbb{Z}[x]$ with $\deg(f) = d$, find a root $x_0$ such that $|x_0| < N^\delta$.*

Coppersmith's method solves the $\left(\frac{1}{d}, d\right)$-SSRSA Problem. Our results show that Coppersmith's method cannot be used to solve the $\left(\frac{1}{d} + \epsilon, d\right)$-SSRSA problem. Note that our results do not prove that the $\left(\frac{1}{d} + \epsilon, d\right)$-SSRSA problem is intractable—doing so would imply there is no polynomial-time algorithm for factoring—but instead we show that the best available class of techniques cannot be extended.

**Extensions to Coppersmith's Method.** Coppersmith's original work also considered the problem of finding small solutions to polynomial equations in two variables over the integers and applied his results to the problem of factoring RSA moduli $N = pq$ when half of the most or least significant bits of one of the factors $p$ is known [Cop97]. Howgrave-Graham gave an alternate formulation of this problem by finding approximate common divisors of integers using similar lattice-based techniques, and obtained the same bounds for factoring with partial information [HG01]. May [May10] gives a unified formulation of Coppersmith and Howgrave-Graham's results to find small solutions to polynomial equations modulo unknown divisors of integers. Later work by Jutla [Jut98] and Jochemsz and May [JM06] has generalized Coppersmith's method to multivariate equations, and Herrmann and May [HM08] obtained results for multivariate equations modulo divisors.

As we will show in the next section, existing results in capacity theory can be used to directly address the case of auxiliary polynomials for Coppersmith's method for univariate polynomials modulo integers. Adapting these results to the other settings of Coppersmith's method listed above is a direction for future research.

# 3   Capacity Theory for Cryptographers

In this section, we begin by recalling from [Can80, Rum89, Rum13] some background about arithmetic capacity theory, which is the tool we will use to prove our main results.

Classically, capacity theory arose from the following problem in electrostatics. How will a unit charge distribute itself so as to minimize potential energy if it is constrained to lie within a compact subset $E_\infty$ of $\mathbb{C}$ which is stable under complex conjugation? Define the *capacity* $\gamma(E_\infty)$ to be $e^{-V(E_\infty)}$, where $V(E_\infty)$ is the so-called Robbin's constant giving the minimal potential energy of a unit charge distribution on $E_\infty$.

It was discovered by Fekete and Szegő [Fek23, FS55] that the distribution of small charges on such an $E$ is related to the possible locations of zeros of monic integral polynomials. Heuristically, these zeros behave in the same way as charges that repel one another according to an inverse power law.

The *nth transfinite diameter* of a set $E_\infty$ is

$$d_n(E_\infty) = \sup_{z_1 \ldots z_n \in E} \prod_{i<j} |z_i - z_j|^{1/\binom{n}{2}}.$$

Then we can give a second definition of the capacity of $E_\infty$ as follows. It can be shown that this definition of capacity is equivalent to the definition via electrostatics.

**Definition 2 (Capacity of a Set via the Transfinite Diameter).**

$$\gamma(E_\infty) = \lim_{n \to \infty} d_n(E_\infty)$$

Let $z_1, \ldots, z_n$ be the conjugates of a degree-$n$ algebraic integer. Then they are the roots of the monic irreducible polynomial $f(x) = \prod_{i=1}^{n}(x - z_i) \in \mathbb{Z}[x]$. The discriminant of $f(x)$ is the non-zero rational integer $\Delta f(x) = \prod_{i<j}(z_i - z_j)^2$. Therefore the $n$th transfinite diameter of a set $E_\infty$ that contains the $z_i$ satisfies

$$d_n(E_\infty) \geq \prod_{i<j} |z_i - z_j|^{\frac{2}{n(n-1)}} = |\Delta f(x)|^{\frac{1}{n(n-1)}} \geq 1$$

Thus $d_n(E_\infty) \geq 1$ if $E$ contains all conjugates of a degree-$n$ algebraic integer. Since $E_\infty$ is bounded, only finitely many algebraic integers of degree $n$ have all their conjugates in $E$. Thus if there are infinitely many algebraic integers with all conjugates in $E_\infty$ then $\gamma(E_\infty) \geq 1$. The restriction that the discriminant of a monic integral polynomial without multiple zeros must be a non-zero integer prevents all the zeros from being too close to one another. Since the discriminant of the polynomial has absolute value at least 1, the potential energy is not positive.

The capacity can also be defined using the Chebyshev constant. Consider the set of degree-$n$ polynomials bounded on $E_\infty$:

$$b_n = \sup \left\{ |r| \mid \exists p(x) = rx^n + \cdots + p_0 \in \mathbb{R}[x] \text{ s.t. } \sup_{z \in E_\infty} |p(z)| \leq 1 \right\}.$$

**Definition 3 (Capacity of a set via the Chebyshev Constant).**

$$\gamma(E_\infty) = \lim_{n \to \infty} b_n^{-1/n}$$

A final equivalent definition of the capacity is the sectional capacity (see [Chi91, RLV00]). Consider the set of polynomials with real coefficients whose evaluations are bounded on $E_\infty$:

$$F_n = \{ p(x) \in \mathbb{R}[x] \mid \deg p(x) \leq n, \sup_{z \in E} |p(z)| < 1 \}$$

$F_n$ is a convex symmetric subset of $\mathbb{R}^{n+1}$.

**Definition 4 (Sectional Capacity).**

$$\log \gamma(E) = \lim_{n \to \infty} \frac{-2 \log \text{Vol}(F_n)}{n^2}$$

If $\gamma(E) < 1$ then for large $n$, we have $\log \text{Vol}(F_n) \approx (-n^2/2) \log \gamma(E) > (n+1) \log 2$. If $\text{Vol}(F_n) > 2^{n+1}$ then by Minkowski's theorem there must be a non-zero polynomial $p(x) \in F_n \cap \mathbb{Z}[x]$. Consider again $z_1, \ldots, z_n$ that are conjugates of some degree-$n$ algebraic integer in $E_\infty$. We have $|p(z_1)|, \ldots, |p(z_n)| < 1$, so $\text{Norm}(p(z_1)) = \prod_i |p(z_i)| < 1$, where Norm is the norm from $\mathbb{Q}(z_1)$ to $\mathbb{Q}$. But $\text{Norm}(p(z_1))$ is a rational integer, so $\text{Norm}(p(z_1)) = 0$ and $p(z_1) = 0$. Therefore the zeros of $p(x)$ include all algebraic integers with conjugates in this set, and thus $p(x)$ must vanish at all such elements in $E$.

These intuitions are behind the following striking result of Fekete and Szegő from [Fek23, FS55].

**Theorem 5 (Fekete and Szegő).** *Let $E_\infty$ be a compact subset of $\mathbb{C}$ closed under complex conjugation.*

- *If $\gamma(E_\infty) < 1$, then there are only finitely many irreducible monic polynomials with integer coefficients which have all of their roots in $E_\infty$.*
- *Conversely, if $\gamma(E_\infty) > 1$, then for every open neighborhood $U$ of $E_\infty$ in $\mathbb{C}$, there are infinitely many irreducible monic polynomials with integer coefficients having all their roots in $U$.*

The first case corresponds to the case in which the minimal potential energy $V(E_\infty)$ is positive, consistent with the physical intuition.

The work of Fekete and Szegő was vastly generalized by Cantor [Can80] to adelic subsets of the projective line, and by Rumely [Rum89, Rum13] to adelic subsets of arbitrary smooth projective curves over global fields. Their methods are based on potential theory, as in electrostatics. In [Chi91], Chinburg suggested sectional capacity theory, which applies to arbitrary regular projective varieties of any dimension and not just to curves. Sectional capacity theory was based on ideas from Arakelov theory, with the geometry of numbers and Minkowski's theorem being the primary tools. In [RLV00], Rumely, Lau and Varley showed that the limits hypothesized in [Chi91] do exist under reasonable hypotheses; this is a deep result.

This paper is the first application of capacity theory that we are aware of to cryptography. We will show that capacity theory is very suited to studying the kind of auxiliary polynomials used in the proof of Coppersmith's theorem. Before we begin, however, we review some number theory.

### 3.1    $p$-adic Numbers

For any prime $p$, and any $n \in \mathbb{Z}$, we define the $p$-*adic* valuation of $n$, to be the supremum of the integers $e$ such that $p^e | n$, *i.e.*,

$$v_p(n) = \begin{cases} \max\{e \in \mathbb{Z} : p^e \mid n\} & \text{if } n \neq 0 \\ \infty & \text{if } n = 0 \end{cases}$$

This is then extended to rational numbers in the natural way. If $a, b \in \mathbb{Z}$ and $a, b \neq 0$, then

$$v_p\left(\frac{a}{b}\right) = v_p(a) - v_p(b).$$

The $p$-adic valuation gives rise to a $p$-adic absolute value $| \ |_p : \mathbb{Q} \to \mathbb{R}$ given by

$$|x|_p = \begin{cases} p^{-v_p(x)} & \text{if } x \neq 0 , \\ 0 & \text{if } x = 0 . \end{cases} \tag{5}$$

It is straightforward to check that the $p$-adic absolute value is multiplicative and satisfies a stronger form of the triangle inequality:

$$|xy|_p = |x|_p \cdot |y|_p \quad \text{and} \quad |x + y|_p \leq \max\left(|x|_p, |y|_p\right) \quad \text{for} \quad x, y \in \mathbb{Q}. \tag{6}$$

The $p$-adic absolute value defines a metric on $\mathbb{Q}$. The $p$-adic numbers, $\mathbb{Q}_p$, are defined to be the completion of $\mathbb{Q}$ with respect to this metric. This is similar to the construction of $\mathbb{R}$ as the completion of $\mathbb{Q}$ with respect to the Euclidean absolute value $| \ | : \mathbb{Q} \to \mathbb{R}$.

Elements of $\mathbb{Q}_p$ are either 0 or expressed in a unique way as a formal infinite sum

$$\sum_{i=k}^{\infty} a_i p^i$$

in which $k \in \mathbb{Z}$, each $a_i$ lies in $\{0, 1, \ldots, p-1\}$ and $a_k \neq 0$. Such a sum converges to an element of $\mathbb{Q}_p$ because the sequence of integers $\{s_j\}_{j=k}^{\infty}$ defined by $s_j = \sum_{i=k}^{j} a_i p^i$ forms a Cauchy sequence with respect to the metric $| \ |_p$. One can add, subtract and multiply such sums by treating $p$ as a formal variable, performing operations in the resulting formal power series ring in one variable over $\mathbb{Z}$, and by then carrying appropriately. In fact, $\mathbb{Q}_p$ is a field, since multiplication is commutative and it is possible to divide elements by non-zero elements of $\mathbb{Q}_p$.

A field $L$ is algebraically closed if every non-constant polynomial $g(x) \in L[x]$ has a root in $L$. This implies that $g(x)$ factors into a product of linear polynomials in $L[x]$, since one can find in $L$ roots of quotients of $g(x)$ by products of previously found linear factors. For example, $\mathbb{C}$ is algebraically closed, but $\mathbb{Q}$ is certainly not.

In general, given a field $F$ there are many algebraically closed fields $L$ containing $F$. For example, given one such $L$, one could simply label the elements of $L$ by the elements of some other set, or one could put $L$ inside a larger algebraically closed field. Given one $L$, the set $\overline{F}$ of elements $\alpha \in L$ which are roots in $L$ of some polynomial in $F[x]$ is called the algebraic closure of $F$ in $L$. The set $\overline{F}$ is in fact an algebraically closed field. For a given $F$, the algebraic closure $\overline{F}$ will depend on the algebraically closed field $L$ which one chooses in this construction. But if one were to use a different field $\tilde{L}$, say, then the algebraic closure of $F$ in $\tilde{L}$ is isomorphic to $\overline{F}$ by a (non-unique) isomorphism which is the identity on $F$. So we often just fix one algebraic closure $\overline{F}$ of $F$.

For instance, if $F = \mathbb{Q}$, then $L = \mathbb{C}$ is algebraically closed, so we can take $\overline{\mathbb{Q}}$ to be the algebraic closure of $\mathbb{Q}$ in $\mathbb{C}$. The possible field embeddings $\tau : \overline{\mathbb{Q}} \to L = \mathbb{C}$ come from pre-composing with a field automorphism of $\overline{\mathbb{Q}}$.

However, for each prime $p$, there is another alternative. The field $\mathbb{Q}_p$ is not algebraically closed, but as noted above, we can find an algebraically closed field containing it and then construct the algebraic closure $\overline{\mathbb{Q}}_p$ of $\mathbb{Q}_p$ inside this field. Now we have $\mathbb{Q} \subset \mathbb{Q}_p \subset \overline{\mathbb{Q}}_p$, and $\overline{\mathbb{Q}}_p$ is algebraically closed. So we could take $L = \overline{\mathbb{Q}}_p$ and consider the algebraic closure $\overline{\mathbb{Q}}'$ of $\mathbb{Q}$ inside $\overline{\mathbb{Q}}_p$. We noted above that all algebraic closures of $\mathbb{Q}$ are isomorphic over $\mathbb{Q}$ in many ways. The possible isomorphisms of $\overline{\mathbb{Q}}$ (as a subfield of $\mathbb{C}$, for example) with $\overline{\mathbb{Q}}'$ (as a subfield of $\overline{\mathbb{Q}}_p$) correspond to the field embeddings $\sigma : \overline{\mathbb{Q}} \to \overline{\mathbb{Q}}_p$. Each such $\sigma$ gives an isomorphism of $\overline{\mathbb{Q}}$ with $\overline{\mathbb{Q}}'$ which is the identity map on $\mathbb{Q}$. Note here that $\overline{\mathbb{Q}}_p$ is much larger than $\overline{\mathbb{Q}}$, since $\overline{\mathbb{Q}}_p$ (and in fact $\mathbb{Q}_p$ as well) is uncountable while $\overline{\mathbb{Q}}$ is countable.

Each $\alpha \in \overline{\mathbb{Q}}$ is a root of a unique monic polynomial $m_\alpha(x) \in \mathbb{Q}[x]$ of minimal degree, and $m_\alpha(x)$ is irreducible. We will later need to discuss the image of such an $\alpha$ under all the field embeddings $\tau : \overline{\mathbb{Q}} \to \mathbb{C}$ and under all field embeddings $\sigma : \overline{\mathbb{Q}} \to \overline{\mathbb{Q}}_p$ as $p$ varies. The possible values for $\tau(\alpha)$ and $\sigma(\alpha)$ are simply the different roots of $m_\alpha(x)$ in $\mathbb{C}$ and $\overline{\mathbb{Q}}_p$, respectively.

*Example 1.* If $\alpha = \sqrt{7}$ then $m_\alpha(x) = x^2 - 7$. The possibilities for $\tau(\alpha)$ are the positive real square root 2.64575... and the negative real square root $-2.64575...$ of 7. When $p = 3$, it turns out that $x^2 - 7$ already has two roots $\alpha_1$ and $\alpha_2$ in the 3-adic numbers $\mathbb{Q}_3 \subset \overline{\mathbb{Q}}_3$. These roots are

$$\alpha_1 = 1 + 1 \cdot 3 + 1 \cdot 3^2 + 0 \cdot 3^3 + \cdots \quad \text{and} \quad \alpha_2 = 2 + 1 \cdot 3 + 1 \cdot 3^2 + 2 \cdot 3^3 + \cdots .$$

These expansions result from choosing 3-adic digits so that the square of the right hand side of each equality is congruent to 1 modulo an increasing power of 3. This is the 3-adic counterpart of finding the decimal digits of the two real square roots of 7. So the possibilities for $\sigma(\alpha)$ under all embeddings $\sigma : \overline{\mathbb{Q}} \to \overline{\mathbb{Q}}_3$ are $\alpha_1$ and $\alpha_2$.

Basic facts about integrality and divisibility are naturally encoded using $p$-adic absolute values:

**Fact 1.** As above, let $\overline{\mathbb{Q}}_p$ denote an algebraic closure of $\mathbb{Q}_p$. There is a unique extension of $|\;|_p : \mathbb{Q}_p \to \mathbb{R}$ to an absolute value $|\;|_p : \overline{\mathbb{Q}}_p \to \mathbb{R}$ for which (6) holds for all $x, y \in \overline{\mathbb{Q}}_p$.

**Fact 2.** The set $\overline{\mathbb{Z}}$ of algebraic integers is the set of all $\alpha \in \overline{\mathbb{Q}}$ for which $m_\alpha(x) \in \mathbb{Z}[x]$. In fact, $\overline{\mathbb{Z}}$ is a ring, so that adding, subtracting and multiplying algebraic integers produces algebraic integers. One can speak of congruences in $\overline{\mathbb{Z}}$ by saying $\alpha \equiv \beta \bmod \gamma \overline{\mathbb{Z}}$ if $\alpha - \beta = \gamma \cdot \delta$ for some $\delta \in \overline{\mathbb{Z}}$.

**Fact 3.** If $r \in \mathbb{Q}$ then $|r|_p \leq 1$ for all primes $p$ if and only if $r \in \mathbb{Z}$. More generally, an element $\alpha \in \overline{\mathbb{Q}}$ is in $\overline{\mathbb{Z}}$ if and only if for all primes $p$ and all field embeddings $\sigma : \overline{\mathbb{Q}} \to \overline{\mathbb{Q}}_p$ one has $|\sigma(\alpha)|_p \leq 1$.

**Fact 4.** Suppose $\alpha \in \overline{\mathbb{Z}}$ and $|\tau(\alpha)| < 1$ for all embeddings $\tau : \overline{\mathbb{Q}} \to \mathbb{C}$. Then in fact, $\alpha = 0$. To see why, note that $m_\alpha(0) \in \mathbb{Z}$ is $\pm 1$ times the product of the complex roots of $m_\alpha(x)$. These roots all have the form $\tau(\alpha)$, so $|m_\alpha(0)| < 1$. Then $m_\alpha(0) \in \mathbb{Z}$ forces $m_\alpha(0) = 0$. Because $m_\alpha(x)$ is monic and irreducible this means $m_\alpha(x) = x$, so $\alpha = 0$.

**Fact 5.** If $N = pq$ for distinct primes $p$ and $q$, then $|N|_p = \frac{1}{p}$, $|N|_q = \frac{1}{q}$, and $|N|_{p'} = 1$ for all other primes $p'$.

**Fact 6.** If $a, b \in \mathbb{Z}$, then

$$a|b \qquad \Leftrightarrow \qquad |b|_p \le |a|_p \quad \forall p$$

Thus $a|b$ is the statement that $b$ is in the $p$-adic disc of radius $|a|_p$ centered at $0$ for all $p$. More generally, if $\alpha, \beta \in \overline{\mathbb{Z}}$ then $\alpha$ divides $\beta$ in $\overline{\mathbb{Z}}$ if $\beta = \delta \cdot \alpha$ for some $\delta \in \overline{\mathbb{Z}}$. This is so if and only if $|\sigma(\beta)|_p \le |\sigma(\alpha)|_p$ for all primes $p$ and all field embeddings $\sigma : \overline{\mathbb{Q}} \to \overline{\mathbb{Q}}_p$.

## 3.2   Auxiliary Functions

The original question Coppersmith considered was this: Given an integer $N \ge 1$, a polynomial $f(x)$, and a bound $X$, can we find all integers $z \in \mathbb{Z}$ such that $|z| \le X$ and $f(z) \equiv 0 \bmod N$?

When $X$ is sufficiently small in comparison to $N$, Coppersmith constructed a non-zero auxiliary polynomial of the form

$$h(x) = \sum_{i,j} a_{i,j} x^i (f(x)/N)^j, \qquad a_{i,j} \in \mathbb{Z} \qquad (7)$$

satisfying $|h(z)| < 1$ for every $z \in \mathbb{C}$ with $|z| \le X$. As noted in Sect. 2.1, this boundedness property forces the set of $z \in \mathbb{Z}$ satisfying $|z| \le X$ and $f(z) \equiv 0 \bmod N$ to be among the roots of $h(x)$. In fact, the roots of the $h(x)$ include all algebraic integers $z \in \overline{\mathbb{Z}}$ satisfying

$$f(z) \equiv 0 \bmod N \cdot \overline{\mathbb{Z}} \quad \text{and} \quad |\sigma(z)| \le X \quad \text{for all embeddings} \quad \sigma : \overline{\mathbb{Q}} \to \mathbb{C}. \quad (8)$$

The reason is as follows. For $z \in \overline{\mathbb{Z}}$, the condition that $f(z) \equiv 0 \bmod N\overline{\mathbb{Z}}$ is equivalent to the condition that $f(z)/N \in \overline{\mathbb{Z}}$. Therefore, for any $h(x)$ in the form of Eq. 7, we have $h(z) \in \overline{\mathbb{Z}}$ whenever $f(z) \equiv 0 \bmod N\overline{\mathbb{Z}}$. If $h(x)$ further satisfies $|h(z)| < 1$ for all $z \in \mathbb{C}$ with $|z| \le X$, then the property that $|\sigma(z)| \le X$ for all embeddings $\sigma : \overline{\mathbb{Q}} \to \mathbb{C}$, means that $|h(\sigma(z))| < 1$ as well. Fact 4 therefore tells us that $h(z) = 0$.

Capacity theory can be used for solving the problem of deciding whether there exist non-zero auxiliary polynomials $h(x)$ which include among its roots the set of $z \in \overline{\mathbb{Z}}$ satisfying Eq. 8. The basic idea, which will be given in detail in Sect. 3.3, is that capacity theory gives one a way of deciding whether the set of algebraic integers satisfying Eq. 8 is finite or infinite.

When this set is infinite then there cannot exist *any* rational function $h(x)$ of any kind vanishing on the $z \in \overline{\mathbb{Z}}$ satisfying (8), and in particular no $h(x)$ of the form in (7) will exist satisfying the desired properties. If, on the other hand, this set is finite then there *will* exist an auxiliary polynomial $h(x)$ vanishing on the $z \in \overline{\mathbb{Z}}$ satisfying (8), and in fact Coppersmith explicitly constructed such a polynomial using the LLL algorithm. As we will see, the boundary for finite versus infinite occurs when $X = N^{1/d}$ where $d$ is the degree of $f(x)$.

### 3.3    When Do Useful Auxiliary Polynomials Exist?

In this section, we use capacity theory to give a characterization of when auxiliary polynomials $h(x)$ of the kind discussed in Sect. 3.2 exist. We will use the work of Cantor in [Can80] to show the following result.

**Theorem 6 (Existence of an Auxiliary Polynomial).** *Let $d$ be the degree of $f(x)$. Define $S(X)$ to be the set of all algebraic integers $z \in \overline{\mathbb{Z}}$ such that*

$$f(z) = 0 \bmod N\overline{\mathbb{Z}} \quad \text{and} \quad |\sigma(z)| \leq X \quad \text{for all embeddings} \quad \sigma : \overline{\mathbb{Q}} \to \mathbb{C}.$$

*There exists a polynomial $h(x) \in \mathbb{Q}[x]$ whose roots include every element of $S(X)$ if $X < N^{1/d}$. If $X > N^{1/d}$ there is no rational function $h(x) \in \mathbb{Q}(x)$ whose zero set contains $S(X)$ because $S(X)$ is infinite.*

We break the proof into a sequence of steps.

1. Since $f(x) \in \mathbb{Z}[x]$, and embeddings fix integers, then if $z \in \overline{\mathbb{Z}}$ we have $f(z) \in \overline{\mathbb{Z}}$, and $\sigma(f(x)) = f(\sigma(x))$ for all embeddings $\sigma : \overline{\mathbb{Q}} \to \overline{\mathbb{Q}}_p$.
2. Suppose $N = p_1^{e_1} \cdots p_k^{e_k}$ and $x \in \mathbb{Z}$, then by Fact 6

$$f(z) \equiv 0 \bmod N \Leftrightarrow |f(z)|_{p_i} \leq \left(\frac{1}{p_i}\right)^{e_i} \quad \forall i \in [k]$$

$$\Leftrightarrow |f(z)|_{p_i} \leq |N|_{p_i} \quad \forall i \in [k]$$

Similarly, if $z \in \overline{\mathbb{Z}}$ then

$$f(z) = 0 \bmod N\overline{\mathbb{Z}} \Leftrightarrow |\sigma(f(z))|_{p_i} = |f(\sigma(z))|_{p_i} \leq |N|_{p_i}$$

for all $i \in [k]$ and for all embeddings $\sigma : \overline{\mathbb{Q}} \to \overline{\mathbb{Q}}_p$.
3. For all primes, $p$, define the set of elements in $\overline{\mathbb{Q}}_p$ that solve the congruence in Eq. 8 $p$-adically:

$$E_p \stackrel{\text{def}}{=} \left\{ z \in \overline{\mathbb{Q}}_p \mid |f(z)|_p \leq |N|_p \right\} = f^{-1}\left(\left\{ z \in \overline{\mathbb{Q}}_p \mid |z|_p \leq |N|_p \right\}\right),$$

and similarly define the set of elements with bounded complex absolute value

$$E_\infty \stackrel{\text{def}}{=} \{ z \in \mathbb{C} \mid |z| \leq X \}$$

Let

$$\mathbb{E} \stackrel{\text{def}}{=} E_\infty \times \prod_{p \in \text{primes}} E_p$$

This specifies the set of $p$-adic and complex constraints on our solutions. Furthermore, $\mathbb{E}$ satisfies all of the conditions in [Rum89] for $\mathbb{E}$ to have a well-defined capacity $\gamma(\mathbb{E}) = \gamma(\mathbb{E}, \{\infty\})$ relative to the point $\infty$ on $\mathbb{P}^1$, and for the computations below to be valid. Note, one requirement in this case is that for all but finitely many primes $p$, $E_p$ is the integral closure $\overline{\mathbb{Z}}_p$ of $\mathbb{Z}_p$ in $\overline{\mathbb{Q}}_p$. We will compute the capacity of $\mathbb{E}$, a measurement of the size of $\mathbb{E}$.

4. We now define the *local capacities* $\gamma_p(E_p)$ and $\gamma_\infty(E_\infty)$ as well as the global capacity $\gamma(\mathbb{E})$. Suppose $0 \leq r \in \mathbb{R}$. We have $p$-adic and complex discs of radius $r$ defined by

$$D_p(a,r) = \left\{ z \in \overline{\mathbb{Q}}_p \,\middle|\, |z - a|_p \leq r \right\} \quad \text{for} \quad a \in \overline{\mathbb{Q}}_p$$

and

$$D_\infty(a,r) = \{ z \in \mathbb{C} \mid |z - a| \leq r \} \quad \text{for} \quad a \in \mathbb{C}.$$

**Fact 7 (Capacity of a Disc).** For $v = p$ and $v = \infty$, one has local capacity

$$\gamma_v(D_v(a,r)) = r$$

If $v = p$, $a = 0$ and $r = |N|_p$ is the $p$-adic absolute value of an integer $N \geq 1$, then $D_v(0, |N|_p) \cap \mathbb{Z}_p$ is just $N\mathbb{Z}_p$. We will need later the fact that the $p$-adic capacity of $N\mathbb{Z}_p$ is

$$\gamma_p(N\mathbb{Z}_p) = p^{-1/(p-1)} |N|_p$$

In a similar way, suppose $v = \infty$. The capacity of the real interval $[-r, r]$ is

$$\gamma_\infty([-r,r]) = r/2$$

**Fact 8 (Capacity of Polynomial Preimage).** If $f(x) \in \mathbb{Z}[x]$ is a monic degree $d$ polynomial, and $S$ is a subset of $\overline{\mathbb{Q}}_p$ if $v = p$ or of $\mathbb{C}$ if $v = \infty$ for which the capacity $\gamma_v(S)$ is well defined, then $\gamma_v(f^{-1}(S))$ is well defined and

$$\gamma_v\left(f^{-1}(S)\right) = \gamma_v(S)^{1/d}$$

Facts 7 and 8 show that

$$\gamma_p(E_p) = \gamma_p(D_p(0, |N|_p))^{1/d} = |N|_p^{1/d} \quad \text{and} \quad \gamma_\infty(E_\infty) = \gamma_p(D_\infty(0, X)) = X.$$

**Fact 9 (Capacity of a Product).**

$$\gamma(\mathbb{E}) = \gamma_\infty(E_\infty) \cdot \prod_{p \in \text{primes}} \gamma_p(E_p)$$

So

$$\gamma(\mathbb{E}) = X \cdot \prod_{p \in \text{primes}} |N|_p^{1/d} = X \cdot \prod_{i=1}^{k} p_i^{-e_i/d} = X \cdot N^{-1/d}.$$

5. Computing the capacity of our sets of interest tells us whether there exists a polynomial mapping the components of $\mathbb{E}$ into discs of radius 1. This allows us to apply the following theorem, due to Cantor [Can80], which tells us when an auxiliary polynomial exists.

**Theorem 7 (Existence of an Auxiliary Polynomial).** *If*

$$\mathbb{E} = E_\infty \times \prod_{p \in primes} E_p$$

*then there exists a non-zero auxiliary polynomial $h(x) \in \mathbb{Q}[x]$ satisfying*

$$h(E_p) \subset D_p(0,1) \quad \forall p$$

*and*

$$h(E_\infty) \subset \{z \in \mathbb{C} \mid |z| < 1\}$$

*if $\gamma(\mathbb{E}) < 1$, and no such polynomial exists if $\gamma(\mathbb{E}) > 1$.*

Once we have set up this framework, we are now ready to prove Theorem 6.

*Proof (Proof of Theorem 6).* Suppose first that $X < N^{1/d}$. Then by Fact 9, $\gamma(\mathbb{E}) < 1$. By Fact 7, there exists a polynomial $h(x) \in \mathbb{Q}[x]$ with $|h(z)|_p \le 1$ for all $p$ and $z \in E_p$, and $|h(z)| < 1$ for all $z \in E_\infty$. Suppose $z \in S(X)$. Then $f(z)/N \in \mathbb{Z}$, so Fact 3 says that for all primes $p$ and embeddings $\sigma : \overline{\mathbb{Q}} \to \overline{\mathbb{Q}_p}$ one has

$$|\sigma(f(z)/N)|_p \le 1$$

Since $f(x) \in \mathbb{Z}[x]$ and $N \in \mathbb{Z}$, we have $\sigma(f(z)) = f(\sigma(z))$ and $\sigma(N) = N$. So

$$|f(\sigma(z))|_p = |\sigma(f(z))|_p = \left|\frac{\sigma(f(z))}{\sigma(N)}\right|_p \cdot |\sigma(N)|_p = |\sigma(f(z)/N)|_p \cdot |N|_p \le |N|_p \,.$$

Therefore $\sigma(z) \in E_p$. Hence $|h(\sigma(z))|_p \le 1$, where $\sigma(h(z)) = h(\sigma(z))$ since $h(x) \in \mathbb{Q}[x]$. Because $p$ was an arbitrary prime, this means $h(z)$ is an algebraic integer, i.e. $h(z) \in \overline{\mathbb{Z}}$ by Fact 3. On the other hand, $z \in S(X)$ implies $|\sigma(z)| \le X$ so $|\sigma(h(z))| = |h(\sigma(z))| < 1$ for all $\sigma : \overline{\mathbb{Q}} \to \mathbb{C}$. Thus $h(z)$ is an algebraic integer such that $|\sigma(h(z))| < 1$ for all $\sigma : \overline{\mathbb{Q}} \to \mathbb{C}$, so by Fact 4, $h(z) = 0$ as claimed. When $X > N^{1/d}$, $S(X)$ is infinite by [Can80, Theorem 5.1.1]. $\quad\square$

To try to prove stronger results about small solutions of congruences, Copper-smith also considered auxiliary polynomials with absolute value less than 1 on a real interval which is symmetric about 0. We can quantify his observation that this does not lead to an improvement of the exponent $1/d$ in Theorem 1 by the following result.

**Theorem 8.** *Let $S'(X)$ be the subset of all $z \in S(X)$ such that $\sigma(z)$ lies in $\mathbb{R}$ for every embedding $\sigma : \overline{\mathbb{Q}} \to \mathbb{C}$. There exists a polynomial $h(x) \in \mathbb{Q}[x]$ whose roots include every element of $S'(X)$ if $X < 2N^{1/d}$. If $X > 2N^{1/d}$ there is no non-zero rational function $h(x) \in \mathbb{Q}(x)$ whose zero set contains $S'(X)$ because $S'(X)$ is infinite.*

*Proof (Proof of Theorem 8).* To prove the Theorem 8, one just replaces the complex disc $E_\infty = \{z \in \mathbb{C} : |z| \le X\}$ by the real interval $E'_\infty = \{z \in \mathbb{R} : |z| \le X\}$. Letting $\mathbb{E}' = \prod_p E_p \times E'_\infty$, we find $\gamma(\mathbb{E}') = 2 \cdot \gamma(\mathbb{E})$ because $\gamma(E'_\infty) = 2\gamma(E_\infty)$. So $\gamma(\mathbb{E}') < 1$ if $X < 2N^{1/d}$ and we find as above that there is a polynomial $h(x) \in \mathbb{Q}[x]$ whose roots contain every element of $S(X)'$. If $X > 2N^{1/d}$ then $\gamma(\mathbb{E}') > 1$ and $S(X)'$ is infinite by the main result of [Rum13], so $h(x)$ cannot exist.

# 4    Lattices of Binomial Polynomials

In this section, we will answer the question of whether Coppersmith's theorem can be improved using auxiliary polynomials that are combinations of binomial polynomials. The results we proved in Sect. 3 showed that it is impossible to improve the bounds for auxiliary polynomials of the form $h(x) = \sum_{i,j \ge 0} a_{i,j} x^i (f(x)/N)^j$.

Recall that if $i \ge 0$ is an integer, the binomial polynomial $b_i(x)$ is

$$b_i(x) = x \cdot (x-1) \cdots (x - i + 1)/i!.$$

Based on a suggestion by Howgrave-Graham and Lenstra, Coppersmith considered in [Cop01] auxiliary polynomials constructed from binomial polynomials; that is, of the form

$$h(x) = \sum_{i,j \ge 0} a_{i,j} b_i(x) b_j(f(x)/N). \tag{9}$$

He found that he was unable to improve the bound of $N^{1/d}$ using this alternate lattice. In this section we will prove some sharper forms of Theorems 3 and 4 that explain why this is the case.

Following the method laid out in Sect. 3, we find that capacity theory cannot rule out the existence of such polynomials. One of the key differences is that monomials send algebraic integers to algebraic integers, while binomial polynomials do not because of the denominators. Therefore, we are no longer able to use the same sets $E_p$ as in the previous section.

In fact, if one uses the lattice of binomial polynomials of the form (9), then for *any* disk in $\mathbb{C}$ there *do exist* auxiliary polynomials that have the required boundedness properties. This is in contrast to the situation for polynomials constructed from the monomial lattice. In Theorem 9, we exhibit, for any disk, an explicit construction of such a polynomial. However, since this polynomial is constructed with $j = 0$ in (9), it tells us nothing about the solution to the inputs to Coppersmith's theorem.

Theorem 11 shows that even if one manages to find an auxiliary polynomial in the lattice given by (9) that does give nontrivial information about the solutions to the inputs to Coppersmith's theorem, this polynomial will still not be useful. Either this polynomial must have degree so large that the root-finding step does not run in polynomial time, or $N$ must have a small prime factor. For this reason, for $N$ that has only large prime factors, using auxiliary polynomials constructed using binomial polynomials will not lead to an improvement in the $N^{1/d}$ bound in Coppersmith's method.

**Theorem 9 (Existence of Bounded Binomial Polynomials).** *Suppose $\delta$ is any positive real number. Suppose $c > 1$. For all sufficiently large integers $N$, there is a non-zero polynomial of the form*

$$h(x) = \sum_{0 \leq i \leq cN^\delta} a_i \, b_i(x) \tag{10}$$

*with $a_i \in \mathbb{Z}$ such that $|h(z)| < 1$ for all $z$ in the complex disk $\{z \in \mathbb{C} : |z| \leq N^\delta\}$.*

**Theorem 10 (Explicit Construction for Theorem 9).** *Let $q_0$ be the unique positive real number such that*

$$4\arctan(q_0/2) = q_0 \left( 2\ln(2) - \ln\left( \frac{4}{q_0^2} + 1 \right) \right) \tag{11}$$

*Suppose $c > q_0 = 3.80572...$, then one can exhibit an explicit $h(x)$ of the kind in (9) in the following way. Choose any constant $c'$ with $q_0 < c' < c$. Then for sufficiently large $N$ and all integers $t$ in the range $c'N^\delta/2 < t \leq cN^\delta/2 - 1/2$, the function*

$$h(x) = b_{2t+1}(x + t)$$

*will have the properties in (i).*

**Theorem 11 (Negative Coppersmith Theorem for Binomial Polynomials).** *Suppose $\epsilon > 0$ and that $M$ and $N$ are positive integers. Suppose further that*

$$N^\epsilon > \prod_{p \leq M} p^{1/(p-1)} \tag{12}$$

*where the product is over the primes $p$ less than or equal to $M$. This condition holds, for example, if $1.48774N^\epsilon \geq M \geq 319$. If there is a non-zero polynomial $h(x)$ of the form*

$$h(x) = \sum_{0 \leq i,j \leq M} a_{i,j} \, b_i(x) \, b_j(f(x)/N) \tag{13}$$

*with $a_{i,j} \in \mathbb{Z}$ such that $|h(z)| < 1$ for $z$ in the complex disk $\{z \in \mathbb{C} : |z| \leq N^{(1/d)+\epsilon}\}$, then $N$ must have a prime factor less than $M$.*

## 4.1   Proof of Theorems 9 and 10

The proof of Theorem 9 comes in several parts. We first use capacity theory to show that non-zero polynomials of the desired kind exist. This argument does not give any information about the degree of the polynomials, however. So we then use an explicit geometry of numbers argument to show the existence of a non-zero polynomial of a certain bounded degree which is of the desired type. Finally, we give an explicit construction of an $h(x)$. This $h(x)$ has a somewhat larger degree than the degree which the geometry of numbers argument shows can be achieved. It would be interesting to see if the LLL algorithm would lead to a polynomial time method for constructing a lower degree polynomial than the explicit construction.

In this section we assume the notations of Theorem 9. The criterion that $h(x)$ be a polynomial of the form

$$h(x) = \sum_i a_i b_i(x)$$

with $a_i \in \mathbb{Z}$ is an *extrinsic* property, which will be discussed in more detail in Step 1 of Sect. 5.1. In short, this extrinsic property arises because $h(x)$ must have a particular form. We need to convert this to an *intrinsic* criterion, in this case observing that these polynomials take $\mathbb{Z}_p$ to $\mathbb{Z}_p$. The key to doing so is the following result of Polya:

**Theorem 12 (Polya).** *The set of polynomials $h(x) \in \mathbb{Q}[x]$ which have integral values on every rational integer $r \in \mathbb{Z}$ is exactly the set of integral combinations $\sum_i a_i b_i(x)$ of binomial polynomials $b_i(x)$.*

**Corollary 1.** *The set of polynomials $h(x) \in \mathbb{Q}[x]$ which are integral combinations $\sum_i a_i b_i(x)$ of binomial polynomials $b_i(x)$ is exactly the set of $h(x)$ such that $|h(z)|_p \leq 1$ for all $z \in \mathbb{Z}_p$ and all primes $p$.*

The corollary follows because $\mathbb{Z}$ is dense in $\mathbb{Z}_p$.

Our main goal in the proof of Theorem 9 is to show there are $h(x) \neq 0$ as in Corollary 1 such that $|h(z)| < 1$ for $z$ in the complex disk $E_\infty = \{z \in \mathbb{C} : |z| \leq N^\delta\}$. We break reaching this goal into steps.

**Applying Capacity Theory Directly.** In view of Corollary 1, the natural adelic set to consider would be

$$\mathbb{E} = \prod_p E_p \times E_\infty \quad \text{with} \quad E_p = \mathbb{Z}_p \quad \text{for all} \quad p \tag{14}$$

However, this choice does not meet the criteria for $\gamma(\mathbb{E})$ to be well defined, because it is not true that $E_p = \overline{\mathbb{Z}}_p$ for all but finitely many $p$. However, for all $Y \geq 2$, the adelic set

$$\mathbb{E}' = \prod_{p \leq Y} \mathbb{Z}_p \times \prod_{p > Y} \overline{\mathbb{Z}}_p \times E_\infty \tag{15}$$

does satisfy the criteria for $\gamma(\mathbb{E})$ to be well defined. One has

$$\gamma_p(\mathbb{Z}_p) = p^{-1/(p-1)}, \quad \gamma_p(\overline{\mathbb{Z}}_p) = 1 \quad \text{and} \quad \gamma_\infty(E_\infty) = N^\delta.$$

So

$$\ln \gamma(\mathbb{E}') = \ln \left( \prod_{p \leq Y} \gamma_p(\mathbb{Z}_p) \times \gamma_\infty(E_\infty) \right) = -\sum_{p \leq Y} \frac{\ln(p)}{p-1} + \ln(N^\delta) \tag{16}$$

Here as $Y \to \infty$, the quantity $-\sum_{p \leq Y} \frac{\ln(p)}{p-1}$ diverges to $-\infty$. So for all sufficiently large $Y$ we have $\gamma(\mathbb{E}') < 1$. We then find as before that Cantor's work produces a non-zero polynomial $h(x) \in \mathbb{Q}[x]$ such that for all $v$ and all elements $z$ of the $v$-component of $\mathbb{E}'$ one has $|h(z)|_v \leq 1$, with $|h(z)| < 1$ if $v = \infty$. In particular, $|h(z)|_p \leq 1$ for all primes $p$ and all $z \in \mathbb{Z}_p \subset \overline{\mathbb{Z}}_p$. So Corollary 1 shows $h(x)$ is an integral combination of binomial polynomials such that $|h(z)| < 1$ if $z \in \mathbb{C}$ and $|z| \leq N^\delta$.

**Using the Geometry of Numbers to Control the Degree of Auxiliary Polynomials.** Minkowski's theorem says that if $L$ is a lattice in a Euclidean space $\mathbb{R}^n$ and $C$ is a convex symmetric subset of $\mathbb{R}^n$ of volume at least equal to $2^n$ times the generalized index $[L : \mathbb{Z}^n]$, there must be a non-zero element of $L \cap C$. To apply this to construct auxiliary polynomials, one takes $C$ to correspond to a suitably bounded set of polynomials with real coefficients, and $L$ to correspond to those polynomials with rational coefficients of the kind one is trying to construct.

In the case at hand, suppose $1 \leq r \in \mathbb{R}$. Let $\mathbb{Z}[x]_{\leq r}$ be the set of integral polynomials of degree $\leq r$, and let $L_{\leq r}$ be the $\mathbb{Z}$-span of $\{b_i(x) : 0 \leq i \leq r, i \in \mathbb{Z}\}$. To show the first statement of Theorem 9, it will suffice to show that if $c > 1$, then for sufficiently large $r = N^\delta > 0$, there is a non-zero $f(x) \in L_{\leq cr}$ such that $|f(z)| < 1$ for $z \in \mathbb{C}$ such that $|z| \leq r$.

Let $m = \lfloor cr \rfloor$ be the largest integer less than or equal to $cr$. By considering leading coefficients, we have

$$\ln[L_{\leq m} : \mathbb{Z}[x]_{\leq m}] = \ln \prod_{i=0}^{m} i! = m^2 \ln(m)/2 \cdot (1 + o(1))$$

where $o(1) \to 0$ as $m \to \infty$. Let $C$ be the set of polynomials with real coefficients of the form

$$\sum_{i=0}^{m} q_i(x/r)^i \quad \text{with} \quad |q_i| \leq 1/(m+2).$$

We consider $C$ as a convex symmetric subset of $\mathbb{R}^{m+1}$ by mapping a polynomial to its vector of coefficients. Then

$$\ln \operatorname{vol}(C) = (m+1) \cdot (\ln(2) - \ln(m+2)) - \sum_{i=0}^{m} i \ln(r) = -\ln(r)m^2/2 \cdot (1 + o(1)).$$

Since $\mathbb{Z}[x]_{\leq m}$ maps to a lattice in $\mathbb{R}^{m+1}$ with covolume 1, we find

$$\ln \operatorname{vol}(C) - \ln \operatorname{vol}(\mathbb{R}^{m+1}/L_{\leq m}) \geq (\ln(m) - \ln(r))m^2/2 \cdot (1 + o(1)) = \ln(c) \cdot m^2/2 \cdot (1 + o(1)).$$

Since $\ln(c) > 0$, for sufficiently large $m$, the right hand side is greater than $2 \ln(m+1)$. Hence Minkowski's Theorem produces a non-zero $f(x) \in L_{\leq m}$ in $C$. One has

$$|f(z)| \leq \sum_{i=0}^{m} |z/r|^i/(m+2) < 1$$

if $z \in \mathbb{C}$ and $|z| < r$, so we have proved Theorem 9.

**An Explicit Construction.** Theorem 10 concerns the polynomials $b_{2t+1}(x+t)$ when $t > 0$ is an integer. This polynomial takes integral values at integral $x$, so it is an integral combination of the polynomials $b_i(x)$ with $0 \le i \le 2t+1$ by Polya's Theorem 12. To finish the proof of Theorem 10, it will suffice to show the following. Let $q_0$ be the unique positive solution of the Eq. (11), and suppose $q > q_0$. Let $D(r)$ be the closed disk $D(r) = \{z \in \mathbb{C} : |z| \le r\}$. We will show that if $r$ is sufficiently large, then

$$|b_{2t+1}(z+t)| < 1 \quad \text{if} \quad 2t \ge qr \quad \text{and} \quad z \in D(r). \tag{17}$$

We have

$$b_{2t+1}(z+t) = \frac{\prod_{j=0}^{2t}(z+t-j)}{(2t+1)!} = \frac{\prod_{j=-t}^{t}(z-j)}{(2t+1)!} = \pm \frac{z \cdot \prod_{j=1}^{t}(z^2-j^2)}{(2t+1)!}$$

For $j \ge 0$ and $z \in D(r)$ we have

$$|-r^2-j^2| = r^2+j^2 \ge |z^2-j^2|.$$

So

$$\sup(\{b_{2t+t}(z+t) : z \in D(r)\}) = \frac{r \cdot \prod_{j=1}^{t}(r^2+j^2)}{(2t+1)!}.$$

Taking logarithms gives

$$\ln\sup(\{b_{2t+t}(z+t) : z \in D(r)\}) = \ln(r) + \sum_{j=1}^{t}\ln(r^2+j^2) - \ln((2t+1)!). \tag{18}$$

We now suppose $t \ge r$, so $\xi = r/t \le 1$. Then

$$\sum_{j=1}^{t}\ln(r^2+j^2) = t\ln(t^2) + t \cdot \frac{1}{t}\sum_{j=1}^{t}\ln(\xi^2+(j/t)^2)$$

$$= 2t\ln(t) + t \cdot \int_0^1 \ln(\xi^2+s^2)ds + o(t) \tag{19}$$

as $t \to \infty$. By integration by parts,

$$\int \ln(\xi^2+s^2)ds = s\ln(\xi^2+s^2) - 2s + 2\xi\arctan(s/\xi). \tag{20}$$

By Stirling's formula,

$$\ln((2t+1)!) = (2t+1)\ln(2t+1) - (2t+1) + o(t) = 2t\ln(t) + 2t\ln(2) - 2t + o(t). \tag{21}$$

Since $\ln(r) = o(t)$, we get from (18), (20) and (21) that

$$\ln(\sup\{b_{2t+t}(z+t) : z \in D(r)\}) = t \cdot (\ln(\xi^2+1) + 2\xi\arctan(\xi^{-1}) - 2\ln(2)) + o(t). \tag{22}$$

Writing $q = 2t/r = 2/\xi \geq 2$ and multiplying both sides of (22) by $q > 0$, we see that if

$$f(q) = q \ln\left(\frac{4}{q^2} + 1\right) + 4\arctan(q/2) - 2\ln(2)q < 0$$

then for sufficiently large $t$ the supremum on the left in (18) is negative and we have the desired bound. Here from $q \geq 2$ we have

$$f'(q) = \ln(1/q^2 + 1/4) \leq \ln(1/2) < 0 < f(2) \quad \text{and} \quad \lim_{q \to +\infty} f(q) = -\infty.$$

So there is a unique positive real number $q_0$ with $f(q_0) = 0$, and $f(q) < 0$ for $q > q_0$. This establishes (17) and finishes the proof of part (ii) of Theorem 9.

## 4.2   Proof of Theorem 11

The proof of Theorem 11 uses a feedback procedure. The feedback in this case is that if $N$ has no small prime factor $p$, then for all small primes $p$ we can increase the set $E_p$. This is described in more detail in Sect. 5.2.

Let $M$ be a positive integer and suppose $\epsilon > 0$. Suppose that there is a polynomial of the form

$$h(x) = \sum_{0 \leq i,j \leq M} a_{i,j} b_i(x) b_j(f(x)/N) \tag{23}$$

such that $a_{i,j} \in \mathbb{Z}$ and $|h(z)| < 1$ for all $z \in \mathbb{C}$ such that $|z| \leq N^{1/d+\epsilon}$. We show that if $M$ satisfies one of the inequalities involving $N$ in the statement of Theorem 11, then $N$ must have a prime divisor bounded above by $M$. We will argue by contradiction. Thus we need to show that the following hypothesis cannot hold:

**Hypothesis 1.** *No prime* $p \leq M$ *divides* $N$, *and either (12) holds or* $1.48774N^\epsilon \geq M \geq 319$.

The point of the proof is to show that Hypothesis 1 leads to $h(x)$ having small sup norms on all components of an adelic set $\mathbb{E}$ which has capacity larger than 1. The reason that the hypothesis that no prime $p \leq M$ divides $N$ enters into the argument is that this guarantees that $f(z)/N$ will lie in the $p$-adic integers $\mathbb{Z}_p$ for all $z \in \mathbb{Z}_p$ when $p \leq M$. This will lead to being able to take the component of $\mathbb{E}$ at such $p$ to be $\mathbb{Z}_p$. The $p$-adic capacity of $\mathbb{Z}_p$ is $p^{-1/(p-1)}$, as noted in Fact 7. This turns out to be relatively large when one applies various results from analytic number theory to get lower bounds on capacities.

To start a more detailed proof, let $p$ be a prime and suppose $0 \leq i, j \leq M$.

**Lemma 1.** *If* $p \leq M$ *set* $E_p = \mathbb{Z}_p$. *Then* $|h(z)|_p \leq 1$ *if* $z \in E_p$ *and the capacity* $\gamma_p(E_p)$ *equals* $p^{-1/(p-1)}|N|_p$.

*Proof.* If $p \leq M$ and $x \in \mathbb{Z}_p$, then $b_i(x) \in \mathbb{Z}_p$ since $\mathbb{Z}$ is dense in $\mathbb{Z}_p$ and $b_i(x) \in \mathbb{Z}$ for all $x \in \mathbb{Z}$. Furthermore, $f(x)/N \in \mathbb{Z}_p$ for $x \in \mathbb{Z}_p$ since we have assumed $N$ is prime to $p$ and $f(x) \in \mathbb{Z}[x]$. Therefore $b_j(f(x)/N) \in \mathbb{Z}_p$ for all $j$. Since the coefficients $a_{i,j}$ in (23) are integers, we conclude $|h(z)|_p \leq 1$. We remarked earlier in Fact 7 that $\gamma_p(\mathbb{Z}_p) = p^{-1/(p-1)}$. Since $p \leq M$, we have supposed that $p$ does not divide $N$. So $|N|_p = 1$, and we get $\gamma_p(E_p) = \gamma(\mathbb{Z}_p) = p^{-1/(p-1)}|N|_p$.

**Lemma 2.** *If $p > M$ set $E_p = f^{-1}(N\overline{\mathbb{Z}}_p)$. Then $|h(z)|_p \leq 1$ if $z \in E_p$ and $\gamma_p(E_p) = |N|_p^{-1/p}$.*

*Proof.* We first note that $0 \leq i, j \leq M < p$ implies that $|i!|_p = |j!|_p = 1$. Recall that $\overline{\mathbb{Z}}_p = \{x \in \overline{\mathbb{Q}}_p : |x|_p \leq 1\}$. If $x \in f^{-1}(N\overline{\mathbb{Z}}_p)$ then $x \in \overline{\mathbb{Z}}_p$ since $f(x)$ is monic with integral coefficients. So

$$|b_i(x)|_p = \frac{|x \cdot (x-1) \cdots (x-i+1)|_p}{|i!|_p} \leq 1$$

and

$$|b_j(f(x)/N)|_p = \frac{|f(x)/N \cdot (f(x)/N - 1) \cdots (f(x)/N - j + 1)|_p}{|j!|_p} \leq 1$$

since $x-k$ and $f(x)/N-k$ lie in $\overline{\mathbb{Z}}_p$ for all integers $k$ and $|i!|_p = |j!|_p = 1$. Because the $a_{i,j}$ in (2) are integral, we conclude $|h(z)|_p \leq 1$ if $z \in E_p = f^{-1}(N\overline{\mathbb{Z}}_p)$. The capacity $\gamma_p(E_p)$ is $|N|_p^{-1/p}$ by Fact 8.

**Lemma 3.** *Set $E_\infty = \{z \in \mathbb{C} : |z| \leq N^{1/d+\epsilon}\}$. Then $|h(z)|_\infty < 1$ if $z \in E_\infty$ and $\gamma_\infty(E_\infty) = N^{1/d+\epsilon}$.*

*Proof.* This first statement was one of our hypotheses on $h(x)$, while $\gamma_\infty(E_\infty) = N^{1/d+\epsilon}$ by Fact 7.

We conclude from these Lemmas and Fact 9 that when

$$\mathbb{E} = \prod_p E_p \times E_\infty$$

we have

$$\gamma(\mathbb{E}) = \left( \prod_{p \leq M} p^{-1/(p-1)} \right) \times \left( \prod_{\text{all } p} |N|_p^{1/d} \right) \times N^{1/d+\epsilon} = \left( \prod_{p \leq M} p^{-1/(p-1)} \right) N^\epsilon.$$

(24)

Here

$$\ln \left( \prod_{p \leq M} p^{-1/(p-1)} \right) = -\sum_{p \leq M} \frac{\ln(p)}{p-1}.$$

and it follows from [RS62, Theorem 6, p. 70] that if $M \geq 319$ then

$$-\sum_{p \leq M} \frac{\ln(p)}{p-1} = -\sum_{p \leq M} \frac{\ln(p)}{p} - \sum_{p \leq M} \frac{\ln(p)}{p(p-1)}$$

$$\geq -\sum_{p \leq M} \frac{\ln(p)}{p} - \sum_{p} \sum_{n=2}^{\infty} \frac{\ln(p)}{p^n}$$

$$\geq -\ln(M) + \gamma - \frac{1}{\ln(M)} \tag{25}$$

where $\gamma = 0.57721...$ is Euler's constant.

Hence (24) gives

$$\ln(\gamma(\mathbb{E})) = -\sum_{p \leq M} \frac{\ln(p)}{p-1} + \epsilon \ln(N) \geq -\ln(M) + \gamma - \frac{1}{\ln(M)} + \epsilon \ln(N). \tag{26}$$

The right hand side is positive if

$$N^\epsilon \cdot e^{\gamma - 1/\ln(M)} > M. \tag{27}$$

Since we assumed $M \geq 319$, we have $e^{\gamma - 1/\ln(M)} \geq 1.497445...$ and so (27) will hold if

$$1.48744 \cdot N^\epsilon > M \tag{28}$$

In any case, if the left hand side of (26) is positive then $\gamma(\mathbb{E}) > 1$. However, we have shown that $h(x)$ is a non-zero polynomial in $\mathbb{Q}[x]$ such that $|h(x)|_v \leq 1$ for all $v$ when $x \in E_v$ with strict inequality when $v = \infty$. By Cantor's Theorem 7, such an $h(x)$ cannot exist because $\gamma(\mathbb{E}) > 1$. The contradiction shows that Hypothesis 1 cannot hold, and this completes the proof of Theorem 11.

# 5   A Field Guide for Capacity-Theoretic Arguments

The proofs in Sects. 3 and 4 illustrate how capacity theory can be used to show the nonexistence and existence of polynomials with certain properties. This paper is a first step toward building a more general framework to apply capacity theory to cryptographic applications. In this section, we step back and summarize how capacity theory can be used in general to show either that auxiliary polynomials with various desirable properties do or do not exist.

The procedure for applying capacity theory to such problems allows for feedback between the type of polynomials one seeks and the computation of the relevant associated capacities. If it turns out that the capacity theoretic computations are not sufficient for a definite conclusion, they may suggest additional hypotheses either on the polynomials or on auxiliary parameters which would be useful to add in order to arrive at a definitive answer. They may also suggest some alternative proof methods which will succeed even when capacity theory used as a black box does not.

## 5.1   Showing Auxiliary Polynomials Exist

To use capacity theory to show that polynomials $h(x) \in \mathbb{Q}[x]$ with certain properties exist, one can follow these steps:

**Step 1.** State the conditions on $h(x)$ which one would like to achieve. These can be of an intrinsic or an extrinsic nature.

    (a) Intrinsic conditions have the following form:

        (i) For each prime $p$, one should give a subset $E_p$ of $\overline{\mathbb{Q}}_p$. For all but finitely many $p$, $E_p$ must be the set $\overline{\mathbb{Z}}_p$.

        (ii) One should give a subset $E_\infty$ of $\mathbb{C}$.

        (iii) The set of polynomials $h(x) \in \mathbb{Q}[x]$ one seeks are all polynomials such that $|h(z)|_p \le 1$ for all primes $p$ and all $z \in E_p$ and $|h(w)| < 1$ if $w \in E_\infty$.

    (b) To state conditions on $h(x)$ extrinsically, one writes down the type of polynomial expressions one allows. For example, one might require $h(x)$ to be an integral combination of integer multiples of specified polynomials, e.g. monomials in $x$ as in Theorem 6. Suppose one uses such an extrinsic description, and one is trying to show the existence of $h(x)$ of this form using capacity theory. It is then necessary to come up with an intrinsic description of the above kind with the property that any $h(x)$ meeting the intrinsic conditions must have the required extrinsic description. We saw another example of this in Sect. 4 on binomial polynomials; see also Step 5 below.

**Step 2.** Suppose we have stated an intrinsic condition on $h(x)$ as in parts (i), (ii) and (iii) of Step 1(a). One then needs to check that the adelic set $\mathbb{E} = \prod_p E_p \times E_\infty$ satisfies certain standard hypotheses specified in [Can80, Rum89, Rum13]. These ensure that the capacity

$$\gamma(\mathbb{E}) = \prod_p \gamma_p(E_p) \cdot \gamma_\infty(E_\infty) \tag{29}$$

is well defined. One then needs to employ [Can80, Rum89, Rum13] to find an upper bounds the $\gamma_p(E_p)$, on $\gamma_\infty(E_\infty)$ and then on $\gamma(\mathbb{E})$. This may also require results from analytic number theory concerning the distribution of primes. When using this method theoretically, there may be an issue concerning the computational complexity of finding such upper bounds. However, if $E_p$ and $E_\infty$ have a simple form (e.g. if they are disks), explicit formulas are available. Notice that the requirement in part (i) of Step 1 that $E_p = \overline{\mathbb{Z}}_p$ for all but finitely many $p$ forces $\gamma_p(E_p) = 1$ for all but finitely many $p$. So the product on the right side of (29) is well defined as long as $\gamma_\infty(E_\infty)$ and $\gamma_p(E_p)$ are for all $p$.

**Step 3.** If the computation in Step 2 shows $\gamma(\mathbb{E}) < 1$, capacity theory guarantees that there is some non-zero polynomial $h(x) \in \mathbb{Q}[x]$ which satisfies the bounds in part (iii) of Step 1. However, one has no information at this point about the degree of $h(x)$.

**Step 4.** Suppose that Step 2 shows $\gamma(\mathbb{E}) < 1$ and that we want to show there is an $h(x)$ as in Step 3 satisfying a certain bound on its degree. There are three levels of looking for such degree bounds.

    a. The most constructive method is to present an explicit construction of an $h(x)$ which one can show works. We did this in the previous section in the case of integral combinations of binomial polynomials.

    b. The second most constructive method is to convert the existence of $h(x)$ into the problem of finding a short vector in a suitable lattice of polynomials and to apply the LLL algorithm. One needs to show that the LLL criteria are met once one considers polynomials of a sufficiently large degree, and that a short vector will meet the intrinsic criteria on $h(x)$. We will return in later papers to the general question of when $\gamma(\mathbb{E}) < 1$ implies that there is a short vector problem whose solution via LLL will meet the intrinsic criteria. This need not always be the case. The reason is that in the geometry of numbers, one can find large complicated convex symmetric sets which are very far from being generalized ellipsoids. However, in practice, the statement that $\gamma(\mathbb{E}) < 1$ makes it highly likely that the above LLL approach will succeed.

    c. Because of the definition of sectional capacity in [Chi91,RLV00], the following approach is guaranteed to succeed by $\gamma(\mathbb{E}) < 1$. Minkowski's Theorem in the geometry of numbers will produce (in a non-explicit manner) a polynomial $h(x)$ of large degree $m$ which meets the intrinsic criteria. One can estimate how large $m$ must be by computing certain volumes and generalized indices. We illustrate such computations in Sect. 4 in the case of intrinsic conditions satisfied by integral combination of binomial polynomials.

**Step 5.** It can happen that the most natural choices for $E_p$ and $E_\infty$ in step 1 above do not satisfy all the criteria for the capacity of $\mathbb{E} = \prod_p E_p \times E_\infty$ to be well defined. One can then adjust these choices slightly. To obtain more control on the degrees of auxiliary functions, one can try an explicit Minkowski argument of the kind use in the proof of the positive result concerning integral combinations of binomial polynomials in Theorem 9 above.

## 5.2 Showing Auxiliary Polynomials Do Not Exist

To use capacity theory to show that polynomials $h(x) \in \mathbb{Q}[x]$ with certain properties do not exist, one can follow these steps:

**Step 1.** Specify the set of properties you want $h(x)$ to have. Then show that the following is true for every $h(x)$ with these properties:

    (i) For each prime $p$, exhibit a set $E_p$ of $\overline{\mathbb{Q}}_p$ such that $|h(z)|_p \le 1$ if $z \in E_p$. For all but finitely many $p$, $E_p$ must be the set $\overline{\mathbb{Z}}_p$.

    (ii) Exhibit a closed subset $E_\infty$ of $\mathbb{C}$ such that $|h(z)| < 1$ if $z \in E_\infty$. It is important that $h(x) \in \mathbb{Q}[x]$ with the desired properties meet the criteria in (i) and (ii).

**Step 2.** As before, one needs to check that the adelic set $\mathbb{E} = \prod_p E_p \times E_\infty$ satisfies certain standard hypotheses specified in [Can80, Rum89, Rum13]. These ensure that the capacity

$$\gamma(\mathbb{E}) = \prod_p \gamma_p(E_p) \cdot \gamma_\infty(E_\infty) \tag{30}$$

is well defined. One then needs to find a lower bound on $\gamma(\mathbb{E})$ using lower bounds on the $\gamma_p(E_p)$ and on $\gamma_\infty(E_\infty)$. One may also require information from analytic number theory, e.g. on the distributions of prime numbers less than a given bound.

**Step 3.** If the computation in Step 2 shows $\gamma(\mathbb{E}) > 1$, capacity theory guarantees that there is no non-zero polynomial $h(x) \in \mathbb{Q}[x]$ which satisfies the intrinsic conditions (i) and (ii) of Step 1. This means there do not exist of polynomials $h(x)$ having the original list of properties.

**Step 4.** Suppose that in Step 3, we cannot show $\gamma(\mathbb{E}) > 1$ due to the fact that the sets $E_p$ and $E_\infty$ in Step 1 are not sufficient large. One can now change the original criteria on $h(x)$, or take into account some additional information, to try to enlarge the sets $E_p$ and $E_\infty$ for which Step 1 applies. We saw in the previous section how this procedure works in the case of integral combinations of certain products of binomial polynomials. For example, if one assumes that certain other parameters (e.g. the modulus of a congruence) have no small prime factors, one can enlarge the sets $E_p$ in Step 1 which are associated to small primes.

# 6   Conclusion

In this work, we drew a new connection between two disparate research areas: lattice-based techniques for cryptanalysis and capacity theory. This connection has benefits for researchers in both areas.

- **Capacity Theory for cryptographers:** We have shown that techniques from capacity theory can be used to show that the bound obtained by Coppersmith's method in the case of univariate polynomials is optimal and the best available class of techniques for solving these types of problems cannot be extended. This has implications for cryptanalysis, and the tightness of cryptographic security reductions.
- **Cryptography for capacity theorists:** Capacity theory provides a method for calculating the conditions under which certain auxiliary polynomials exist. Coppersmith's method provides an efficient algorithm for *finding* these auxiliary polynomials. Until this time, capacity theory has not addressed the computational complexity actually *producing* auxiliary functions.

We used capacity theory to answer three questions of Coppersmith in [Cop01]

1. Can the exponent $1/d$ be improved (possibly through improved lattice reduction techniques)? No, the desired auxiliary polynomial simply does not exist.

2. Does restricting attention to the real line $[-N^{-1/d}, N^{1/d}]$ instead of the complex disk $|z| \leq N^{1/d}$ improve the situation? No.
3. Does considering lattices based on binomial polynomials improve the situation? No, these lattices have the desired auxiliary polynomials, but for RSA moduli, their degree is too large to be useful.

   Since Coppersmith's method is one of the primary tools in asymmetric cryptanalysis, these results give an indication of the security of many factoring-based cryptosystems.

   This paper lays a foundation for several directions of future work. Coppersmith's study of small integral solutions of equations in two variables and bivariate equations modulo $N$ [Cop97] is related to capacity theory on curves, as developed by Rumely in [Rum89, Rum13]. The extension of Coppersmith's method to multivariate equations [JM06, Jut98] is connected to capacity theory on higher dimensional varieties, as developed in [Chi91, RLV00, CMBPT15]. Multivariate problems raise deep problems in arithmetic geometry about the existence of finite morphisms to projective spaces which are bounded on specified archimedean and non-archimedean sets. Interestingly, Howgrave-Graham's extension of Coppersmith's method to find small roots of modular equations modulo *unknown* moduli [HG01, May10] appears to pertain to joint capacities of many adelic sets, a topic which has not been developed to our knowledge in the capacity theory literature. It is an intriguing question whether capacity theory can be extended to help us understand the limitations of these more general variants of Coppersmith's method.

**Acknowledgements.** This material is based upon work supported by the National Science Foundation under grants CNS-1513671, DMS-1265290, DMS-1360767, CNS-1408734, CNS-1505799, by the Simons Foundation under fellowship 338379, and a gift from Cisco.

# References

[AASW12]  Aono, Y., Agrawal, M., Satoh, T., Watanabe, O.: On the optimality of lattices for the Coppersmith technique. In: Susilo, W., Mu, Y., Seberry, J. (eds.) ACISP 2012. LNCS, vol. 7372, pp. 376–389. Springer, Heidelberg (2012). doi:10.1007/978-3-642-31448-3_28

[ACGS88]  Alexi, W., Chor, B., Goldreich, O., Schnorr, C.-P.: RSA and Rabin functions: certain parts are as hard as the whole. SIAM J. Comput. **17**(2), 194–209 (1988)

[Ber67]  Berlekamp, E.R.: Factoring polynomials over finite fields. Bell Syst. Tech. J. **46**(8), 1853–1859 (1967)

[BR93]  Bellare, M., Rogaway, P.: Random oracles are practical: a paradigm for designing efficient protocols. In: CCS 1993, pp. 62–73. ACM Press (1993)

[Can80]  Cantor, D.G.: On an extension of the definition of transfinite diameter and some applications. J. Reine Angew. Math. **316**, 160–207 (1980)

[Chi91]  Chinburg, T.: Capacity theory on varieties. Compositio Math. **80**(1), 75–84 (1991)

[CMBPT15] Chinburg, T., Moret-Bailly, L., Pappas, G., Taylor, M.J.: Finite morphisms to projective space and capacity theory. J. fur die Reine und.
Angew. Math. (2015)

[Cop97] Coppersmith, D.: Small solutions to polynomial equations, and low exponent RSA vulnerabilities. J. Cryptology 10(4), 233–260 (1997)

[Cop01] Coppersmith, D.: Finding small solutions to small degree polynomials.
Crypt. Lattices 2146, 20–31 (2001)

[CZ81] Cantor, D.G., Zassenhaus, H.: A new algorithm for factoring polynomials
over finite fields. Math. Comput. 36(154), 587–592 (1981)

[Fek23] Fekete, M.: Über die verteilung der wurzeln bei gewissen algebraischen
gleichungen mit ganzzahligen koeffizienten. Math. Z. 17(1), 228–249
(1923)

[FS55] Fekete, M., Szegö, G.: On algebraic equations with integral coefficients
whose roots belong to a given point set. Math. Z. 63(1), 158–172 (1955)

[FS00] Fischlin, R., Schnorr, C.-P.: Stronger security proofs for RSA and Rabin
bits. J. Cryptology 13(2), 221–244 (2000)

[HG97] Howgrave-Graham, N.: Finding small roots of univariate modular equa-
tions revisited. In: Darnell, M. (ed.) Cryptography and Coding 1997.
LNCS, vol. 1355, pp. 131–142. Springer, Heidelberg (1997). doi:10.1007/
BFb0024458

[HG01] Howgrave-Graham, N.: Approximate integer common divisors. In:
Silverman, J.H. (ed.) CaLC 2001. LNCS, vol. 2146, pp. 51–66. Springer,
Heidelberg (2001). doi:10.1007/3-540-44670-2_6

[HM08] Herrmann, M., May, A.: Solving linear equations modulo divisors: on
factoring given any bits. In: Pieprzyk, J. (ed.) ASIACRYPT 2008.
LNCS, vol. 5350, pp. 406–424. Springer, Heidelberg (2008). doi:10.1007/
978-3-540-89255-7_25

[HN04] Håstad, J., Nåslund, M.: The security of all RSA and discrete log bits. J.
ACM (JACM) 51(2), 187–230 (2004)

[JM06] Jochemsz, E., May, A.: A strategy for finding roots of multivariate polyno-
mials with new applications in attacking RSA variants. In: Lai, X., Chen,
K. (eds.) ASIACRYPT 2006. LNCS, vol. 4284, pp. 267–282. Springer,
Heidelberg (2006). doi:10.1007/11935230_18

[Jut98] Jutla, C.S.: On finding small solutions of modular multivariate polynomial
equations. In: Nyberg, K. (ed.) EUROCRYPT 1998. LNCS, vol. 1403, pp.
158–170. Springer, Heidelberg (1998). doi:10.1007/BFb0054124

[LLL82] Lenstra, H.W., Lenstra, A.K., Lovász, L.: Factoring polynomials with
rational coeficients. Math. Ann. 261(4), 515–534 (1982)

[May07] May, A.: Using LLL-reduction for solving RSA, factorization problems:
a survey. In: Conference Proceedings of the Conference in Honor of the
25th Birthday of the LLL Algorithm, pp. 1–34 (2007)

[May10] May, A.: Using LLL-reduction for solving RSA and factorization problems
the LLL algorithm. In: Nguyen, P.Q., Vallée, B. (eds.) The LLL Algorithm
Information Security and Cryptography, Chap. 10, pp. 315–348. Springer,
Heidelberg (2010)

[RLV00] Rumely, R., Lau, C.F., Varley, R.: Existence of the sectional capacity.
Mem. Am. Math. Soc. 145(690), viii+130 (2000)

[RS62] Rosser, J.B., Schoenfeld, L.: Approximate formulas for some functions of
prime numbers. Ill. J. Math. 6, 64–94 (1962)

[Rum89] Rumely, R.S.: Capacity Theory on Algebraic Curves. LNM, vol. 1378.
Springer, Heidelberg (1989). doi:10.1007/BFb0084525

[Rum13]  Rumely, R.: Capacity Theory with Local Rationality. Mathematical Surveys and Monographs, vol. 193. American Mathematical Society, Providence (2013). The strong Fekete-Szegö theorem on curves

[Sho01]  Shoup, V.: OAEP reconsidered. In: Kilian, J. (ed.) CRYPTO 2001. LNCS, vol. 2139, pp. 239–259. Springer, Heidelberg (2001). doi:10.1007/3-540-44647-8_15

[SPW06]  Steinfeld, R., Pieprzyk, J., Wang, H.: On the provable security of an efficient RSA-based pseudorandom generator. In: Lai, X., Chen, K. (eds.) ASIACRYPT 2006. LNCS, vol. 4284, pp. 194–209. Springer, Heidelberg (2006). doi:10.1007/11935230_13

# A Key Recovery Attack on MDPC with CCA Security Using Decoding Errors

Qian Guo[✉], Thomas Johansson, and Paul Stankovski

Department of Electrical and Information Technology,
Lund University, Lund, Sweden
{qian.guo,thomas.johansson,paul.stankovski}@eit.lth.se

**Abstract.** Algorithms for secure encryption in a post-quantum world are currently receiving a lot of attention in the research community, including several larger projects and a standardization effort from NIST. One of the most promising algorithms is the code-based scheme called QC-MDPC, which has excellent performance and a small public key size. In this work we present a very efficient key recovery attack on the QC-MDPC scheme using the fact that decryption uses an iterative decoding step and this can fail with some small probability. We identify a dependence between the secret key and the failure in decoding. This can be used to build what we refer to as a distance spectrum for the secret key, which is the set of all distances between any two ones in the secret key. In a reconstruction step we then determine the secret key from the distance spectrum. The attack has been implemented and tested on a proposed instance of QC-MDPC for 80 bit security. It successfully recovers the secret key in minutes.

A slightly modified version of the attack can be applied on proposed versions of the QC-MDPC scheme that provides IND-CCA security. The attack is a bit more complex in this case, but still very much below the security level. The reason why we can break schemes with proved CCA security is that the model for these proofs typically does not include the decoding error possibility.

**Keywords:** CCA-security · Key-recovery attack · Post-quantum cryptography · QC-MDPC · Reaction attack

## 1 Introduction

Given the existence of a large quantum computer, cryptosystems based on factoring or discrete logarithm will no longer be secure, as a quantum computer is able to solve both problems in polynomial time [33]. However, it is not yet known to what extent a future quantum computer can be used to successfully solve other types of problems. New algorithms for secure encryption in a post-quantum world (when large quantum computers exist) are currently receiving

Supported by the Swedish Research Council (Grants No. 2015-04528).

© International Association for Cryptologic Research 2016
J.H. Cheon and T. Takagi (Eds.): ASIACRYPT 2016, Part I, LNCS 10031, pp. 789–815, 2016.
DOI: 10.1007/978-3-662-53887-6_29

a lot of attention in the research community, including several larger projects and a standardization effort from NIST [9]. It is often mentioned that the new schemes could be from one of the areas: lattice-based, code-based, hash-based and multi-variate [5].

For code-based schemes, the basic construction is the McEliece public-key cryptosystem (PKC) [28], based on the hardness of decoding a random linear code. The general idea is to transform polynomially solvable instance of the problem into something that looks like a random instance. In this case we transform the generator matrix of a code with simple and efficient decoding to a generator matrix for a randomly looking code. Not knowing the inverse of this transformation, the attacker is facing a presumably hard problem, namely, decoding the random code.

The McEliece PKC has been extensively analyzed over a period of more than thirty years, and is still regarded secure in its original form using Goppa codes. Several other underlying codes have been proposed, but many of them have been broken [30]. A problem with the original McEliece construction is the size of the public key. McEliece proposed to use the generator matrix of a linear code as public key. The public key for the originally proposed parameters is roughly 500 Kbits. Although this can be managed today, it has motivated various attempts to decrease the key sizes but most of them have been unsuccessful.

Recently however, a very interesting version of the McEliece PKC was proposed, the QC-MDPC scheme [29]. This is a McEliece PKC that uses so-called moderate density parity check codes (MDPC codes) in quasi-cyclic (QC) form. The quasi-cyclic form allows us to represent a matrix by its first row, which leads to a small public key. As the MDPC codes have a random component, there is no need for scrambling and permutation matrices. Instead, the generator matrix is presented in systematic form. The QC-MDPC proposal with suitable parameters is yet unbroken and it is particularly interesting because of its simplicity and smaller key size.

An European initiative, PQCRYPTO, sponsored by the European Commission under its Horizon 2020 Program ICT-645622, is ≫developing cryptology that resists the unmatched power of quantum computers≪. In September 2015 this group of researchers published a report entitled "Initial Recommendation of long-term secure post-quantum systems" [1], where they recommended several algorithms as being ready for use and several others that warrant further study and may be recommended in coming years. This report recommends the QC-MDPC scheme for further study, confirming its competitiveness as a post-quantum candidate.

Many papers on its implementation have appeared since the introduction of the QC-MDPC scheme. In [15] and [24], the QC-MDPC McEliece is implemented in hardware using the same parameters that we attack in this paper. Implementation with side-channel protection is considered in [25].

## 1.1 Attack Models and Previous Work

In code-based public-key cryptography, one is typically concerned with two types of attacks: structural attacks and decoding attacks. Structural attacks aim to recover the secret code - *key recovery*, while the decoding attacks target an intercepted ciphertext and tries to recover the transmitted plaintext - *message recovery*. The plain versions of code-based schemes are designed to be secure in the chosen plaintext attack (CPA) model and it is known that chosen ciphertext attacks (CCA) can break them. To achieve security against adaptive chosen ciphertext attacks (CCA2), the schemes need to be converted. There are several standard conversions to achieve CCA2 security from CPA security, [4, 21], and basically the decoding problem is changed in such a way that the noise added in the encryption is no longer in control by Alice who is encrypting.

The standard attacks on the original McEliece scheme can be applied on the QC-MDPC scheme. These attacks are decoding attacks using *information set decoding algorithms*, typically improved versions of the Stern algorithm [3]. These attacks are message recovery attacks and can be applied in a few different scenarios, one of them being the "decoding one-out-of-many" [20,32]. The family of MDPC codes have parity checks of moderate weight (low but not very low). In a structural attack, one can thus consider the dual code, which is given from the generator matrix, and search for low weight codewords in the dual code. This is again done by the same type of algorithms as above. Being well known attacks, the instantiation of QC-MDPC schemes make sure that the computational complexity for these attacks are well beyond the selected security limit. More details can be found in for example [30,31].

For a plain QC-MDPC scheme without CCA2 conversion we can identify a few attacks that require more than the CPA assumption. Using a *partially known plaintext attack* [7], the attacker can reduce the code dimension in the decoding and thus achieve a lower complexity for the information set decoding. In a *resend attack*, Alice is resending the same message twice, or possible two related messages. Also in this case we can efficiently find the message [6]. A *reaction attack* [14] is a weaker version of a chosen ciphertext attack. The attacker sends an intercepted ciphertext with a modification (for example adding a single bit) and observes the reaction of the recipient (but not the result of decoding). Again, one can in certain cases efficiently find the message corresponding to the intercepted ciphertext. It is worth noting that all these attacks are message recovery attacks.

As mentioned before, to achieve a stronger security notion, the QC-MDPC scheme (as any McEliece PKC) can use a CCA2 conversion [21,26]. In this case, the above attacks are no longer possible. So to summarize the current state-of-the-art regarding attacks, for the plain schemes we have possibly some message recovery attacks using the model of reaction attacks. For CCA2 secure versions, we have no known successful attacks.

## 1.2   Contributions

Our basic scenario is the following. Bob has publicly announced his public key and Alice is continuously sending messages to him using the QC-MDPC scheme. Occasionally, Bob will suffer from a decoding error and will tell Alice, who may retransmit or simply discard sending that message. After sending a number of messages, Alice will be able to recover Bob's secret key using our proposed attack.

We present a very efficient **key recovery attack** on the QC-MDPC scheme using the fact that decryption uses an iterative decoding step and this can fail with some small probability. We identify a dependence between the secret key and the failure in decoding. This can be used to build what we call a distance spectrum for the secret key, which is the set of all distances between any two ones in the secret key. In a reconstruction step we then determine the secret key from the distance spectrum. The attack has been implemented and tested on a proposed instance of QC-MDPC for 80 bit security. It successfully recovers the secret key in minutes.

A slightly modified version of the attack can be applied on proposed versions of the QC-MDPC scheme that provides CCA2 security. The attack is a bit more complex in this case, but still very much below the security level. The reason why we can break schemes with proved CCA2 security is that the model for these proofs typically does not include the decoding error possibility. A similar situation has been identified and analyzed for the lattice-based scheme NTRU (NTRUEncrypt) [18,19].

The paper is organized as follows. We give some background in Sect. 2 and describe the QC-MDPC scheme in Sect. 3. We then present an overview of our new attack in Sect. 4 and give some related analysis in Sect. 5. In Sect. 6 we consider the case when we have a CCA2 converted version and demonstrate that a modified version of the attack is still valid. Section 7 presents some results from implementing the different steps of the attack. Finally, we conclude the paper in Sect. 8.

## 2   Background in Coding Theory and Public-Key Cryptography

Let us start by reviewing some basics from coding theory and how it can be applied to public-key cryptography through the McEliece PKC.

**Definition 1 (Linear codes).** *An $[n, k]$ linear code $\mathcal{C}$ over a finite field $\mathbb{F}_q$ is a linear subspace of $\mathbb{F}_q^n$ of dimension $k$.*

**Definition 2 (Generator matrix).** *A $k \times n$ matrix $\mathbf{G}$ with entries from $\mathbb{F}_q$ having rowspan $\mathcal{C}$ is a generator matrix for the $[n, k]$ linear code $\mathcal{C}$.*

Equivalently, $\mathcal{C}$ is the kernel of an $(n - k) \times n$ matrix $\mathbf{H}$ called a *parity-check matrix* of $\mathcal{C}$. We then have $\mathbf{cH}^\mathrm{T} = \mathbf{0}$, if and only if $\mathbf{c} \in \mathcal{C}$, where $\mathbf{H}^\mathrm{T}$ denotes the transpose of $\mathbf{H}$.

A code $\mathcal{C}$ can be represented by different generator matrices. An important representation is the systematic form, i.e., when each input symbol are in one-to-one correspondence with a position in the codeword. Then, one can find a $k \times k$ submatrix of $\mathbf{G}$ forming the identity matrix. After a row permutation we can consider $\mathbf{G}$ in the form $\mathbf{G} = (\mathbf{I}\,\mathbf{P})$. If $\mathbf{G}$ has the form $\mathbf{G} = (\mathbf{I}\,\mathbf{P})$, then $\mathbf{H} = (-\mathbf{P}^\mathsf{T}\,\mathbf{I})$.

The Hamming weight $w_\mathrm{H}(\mathbf{x})$ of a vector in $\mathbf{x} \in \mathbb{F}_q^n$ is the number of nonzero entries in the vector. The minimum (Hamming) distance of the code $\mathcal{C}$ is defined as $d \stackrel{\text{def}}{=} \min_{\mathbf{x},\mathbf{y}\in\mathcal{C}} w_\mathrm{H}(\mathbf{x}-\mathbf{y})$, where $\mathbf{x} \neq \mathbf{y}$. Continuing, we only consider the binary case $q = 2$.

**Definition 3 (Quasi-cyclic codes).** *An $[n,k]$-quasi-cylic (QC) code $\mathcal{C}$ is a linear block code such that for some integer $n_0$, every cyclic shift by $n_0$ is again a codeword.*

In particular, if $n = n_0 k$, then a generator matrix of the form

$$\mathbf{G} = (\mathbf{I}\,\mathbf{P}_0\,\mathbf{P}_1 \cdots \mathbf{P}_{n_0-1})$$

is a useful way to represent a QC code, where $\mathbf{P}_i$ is a $k \times k$ cyclic matrix, i.e. the rows (or columns) of $\mathbf{P}$ is obtained by cyclic rotation of the first row one step. Also, the algebra of $k \times k$ binary circulant matrices is isomorphic to the algebra of polynomials modulo $x^k + 1$ over $\mathbb{F}_2$, allowing an alternative description.

Another useful class of codes is the low-density parity-check code (LDPC) defined as a linear code that admits a sparse parity-check matrix $\mathbf{H}$, where sparsity means that each row of $\mathbf{H}$ has at most $w$ ones, for some small $w$. This sparse matrix can be represented in the form of a bipartite graph, that consists of $n - k$ upper nodes (named "check node") representing the $n - k$ parity equations and $n$ lower nodes (named "variable node") representing the $n$ codeword bits. A variable node is connected to a check node if the variable is present in that parity check. Each check node is then connected to $w$ variable nodes. We call this graph representation a "Tanner" graph, which is a frequently used term in work on iterative decoding algorithms.

## 2.1 McEliece Cryptosystem

In 1978 McEliece showed how a public key cryptosystem (PKC) could be constructed using tools from coding theory. We shortly describe the original McEliece PKC here. This scheme uses three matrices $\mathbf{G}, \mathbf{S}, \mathbf{P}$, where $\mathbf{G}$ is a $k \times n$ generator matrix of a binary $[n, k, 2t + 1]$ linear code. The original and still secure proposal in [28] is to use Goppa codes (see [13,23]). Then $\mathbf{S}$ a $k \times k$ random binary non-singular matrix (called the scrambling matrix), and $\mathbf{P}$ is an $n \times n$ random permutation matrix (called the permutation matrix). As designers we compute the new $k \times n$ matrix $\mathbf{G}' = \mathbf{SGP}$. The scheme works as follows:

- Private Key: $(\mathbf{G}, \mathbf{S}, \mathbf{P})$.
- Public Key: $(\mathbf{G}', t)$

- Encryption: A message $\mathbf{m}$ is mapped to a ciphertext $\mathbf{c}$ by $\mathbf{c} = \mathbf{m}\mathbf{G}' + \mathbf{e}$, where $\mathbf{c}$ is the $n$-bit ciphertext, $\mathbf{m}$ is the $k$-bit plaintext and $\mathbf{e}$ an $n$-bit error vector with (Hamming) weight $t$.
- Decryption: Use an efficient decoding algorithm for Goppa codes to decode $\mathbf{c}$ to find the error $\mathbf{e}\mathbf{P}^{-1}$, recover $\mathbf{m}\mathbf{S}$ and thus $\mathbf{m}$.

Knowing the description of the selected Goppa code allows efficient decoding, as there are many decoding algorithms for this problem running in polynomial time. But knowing only the public key, the attacker is facing a decoding problem for a code that looks like a random code, a presumably difficult problem.

## 3    The QC-MDPC Public Key Encryption Scheme

In [29] a new version of the McEliece PKC was proposed. It has a surprisingly simple description and does not use permutation and scrambling matrices as in the original McEliece construction, as well as in other generalizations [2,22] proposed. The idea is to use codes that allow iterative decoding. In coding theory, this usually involves low-density parity check codes, but for an encryption scheme this is not secure. The reason is that LDPC codes have parity-checks with very small Hamming weight (like 3-5) and these parity-checks in a given LDPC code correspond to codewords in the dual code. Since a basis of the dual code can be computed, it is computationally easy to find low-weight codewords in the dual code and hence the low-weight parity checks. The solution proposed in [29] is to increase the weight of the parity checks to a larger value, which is still small in comparison with the dimension of the code. This makes the task of finding low-weight codewords in the dual code much more costly. In this way, key-recovery attacks by algorithms searching for low weight codewords can be avoided.

The family of such codes with increased parity-check weight is called *Moderate-Density Parity-Check* codes (MDPC codes), and they can be decoded with the same decoding algorithms used to decode LDPC codes. The quasi-cyclic variant of MDPC codes are called QC-MDPC codes. These are of special interest, since the quasi-cyclic property allows us to represent the code to be used, by a single row of the generator matrix. Since the public key is the generator matrix, this gives us very compact keys. We will go through the different steps of the QC-MDPC public key cryptosystem as proposed in [29].

Let $r = n - k$.

### 3.1    Generation of Public-Key

1. Choose an $[n, n - r]$ code in the QC-MDPC family described by the parity-check matrix $\mathbf{H} \in \mathbb{F}_2^{r \times n}$, $n = n_0 r$, such that

$$\mathbf{H} = \left( \mathbf{H}_0 \ \mathbf{H}_1 \ \cdots \ \mathbf{H}_{n_0 - 1} \right),$$

where each $\mathbf{H}_i$ is a circulant $r \times r$ matrix with weight $w_i$ in each row and with $\hat{w} = \sum w_i$.

2. Generate the public key $\mathbf{G} \in \mathbb{F}_2^{(n-r) \times n}$ from $\mathbf{H}$ as,

$$\mathbf{G} = (\mathbf{I} \, \mathbf{P}),$$

where

$$\mathbf{P} = \begin{pmatrix} \mathbf{P}_0 \\ \mathbf{P}_1 \\ \vdots \\ \mathbf{P}_{n_0-2} \end{pmatrix} = \begin{pmatrix} \left(\mathbf{H}_{n_0-1}^{-1}\mathbf{H}_0\right)^{\mathrm{T}} \\ \left(\mathbf{H}_{n_0-1}^{-1}\mathbf{H}_1\right)^{\mathrm{T}} \\ \vdots \\ \left(\mathbf{H}_{n_0-1}^{-1}\mathbf{H}_{n_0-2}\right)^{\mathrm{T}} \end{pmatrix}.$$

Again, the QC-MDPC construction has no need for permutation or scrambling matrices.

**Encryption.** Let $\mathbf{m} \in \mathbb{F}_2^{(n-r)}$ be the plaintext. Multiply $\mathbf{m}$ with the public key $\mathbf{G}$ and add noise within the correction radius $t$ of the code, i.e., $\mathbf{c} = \mathbf{m}\mathbf{G} + \mathbf{e}$, where $w_{\mathrm{H}}(\mathbf{e}) \leq t$. The parameter $t$ is obtained from the error correcting capability of the decoding algorithm for the MDPC code [29]. The error vector is uniformly chosen among all binary $n$-tuples with $w_{\mathrm{H}}(\mathbf{e}) \leq t$.

### 3.2 Decryption

Let $\mathbf{c} \in \mathbb{F}_2^n$ be a received ciphertext. Given the secret low-weight parity check matrix $\mathbf{H}$, a low-complexity decoding procedure is used to obtain the plaintext $\mathbf{m}$.

The authors of [29] propose a variant of Gallager's bit-flipping algorithm [12] as the decoding procedure of MDPC codes. Here some details on this bit-flipping procedure are presented, which are vital to the proposed key recovery attack in the next section. The decoding algorithm works as follows:

1. Compute the syndrome, $\mathbf{s} = \mathbf{c}\mathbf{H}^T$. Since $\mathbf{m}\mathbf{H}^T = \mathbf{0}$, the syndrome is equivalently expressed as $\mathbf{s} = \mathbf{e}\mathbf{H}^T$. Now consider the Tanner graph for $\mathbf{H}$ and set the initial value in each variable node to 0. Create a counter with an initial value 0 for each variable node.
2. Run through all parity-check equations (rows of $\mathbf{H}$ and check nodes in the graph) and for every variable node connected to an unsatisfied check node, increase its corresponding counter by one.
3. Run through all variable nodes and flip its value if its counter satisfies a certain constraint—which usually is that the counter surpasses a threshold.
4. Check if all the equations are satisfied; if not, reset all the counters to 0 and go to Step 2. The procedure will stop if all the parity-checks are satisfied or if the limit on the maximum number of iterations is reached.

This iterative decoding algorithm commonly used with LDPC codes has an error-correction capability that increases linearly with the length of the code. The good performance of LDPC codes is due to the low-weight parities as the error-correction capability also decreases linearly with the weight of the parity-checks. MDPC codes have slightly higher parity-check weight than LDPC codes and one should anticipate that this influences the error-correction capability.

As expected, the actual performance of this procedure on MDPC codes is relatively poor compared with that on LDPC codes. Along the path of the work [29], researchers also proposed other variants [15,27] that reduce the decoding error probability further via changing the flipping threshold or introducing more rounds to handle the detected decoding errors. The reduced error probability, however, is still large compared with the corresponding security level[1].

### 3.3  Proposed Parameters

The authors of [29] proposed the parameters found in Table 1 for a QC-MDPC scheme with 80-bit, 128-bit and 256-bit security level.

**Table 1.** Some proposed QC-MDPC instances with key size and security level.

| Parameters | | | | | Key size | Security |
|---|---|---|---|---|---|---|
| $n$ | $r$ | $\hat{w}$ | $t$ | $n_0$ | | |
| 9602 | 4801 | 90 | 84 | 2 | 4801 | 80 |
| 19714 | 9857 | 142 | 134 | 2 | 9857 | 128 |
| 65542 | 32771 | 274 | 264 | 2 | 32771 | 256 |

Results from actual implementations of the QC-MDPC scheme [15,27] and also a QC-MDPC Niederreiter variant [26] were recently published. They all demonstrated excellent efficiency in terms of computational complexity and key sizes for encryption and decryption on constrained platforms such as embedded micro-controllers and FPGAs using the proposed parameters.

A European initiative, PQCRYPTO, sponsored by the European Commission under its Horizon 2020 Program ICT-645622, is ≫developing cryptology that resists the unmatched power of quantum computers≪. In September 2015 this group of researchers published a report entitled "Initial Recommendation of long-term secure post-quantum systems", where they recommended several algorithms as being ready in 2015 and several others that warrant further study and may be recommended in coming years. This report recommends the QC-MDPC scheme for further study, confirming its competitiveness as a post-quantum candidate.

## 4  A Key-Recovery Attack

In this section we describe our new attack against the plain QC-MDPC scheme as it has been proposed in [29] and described in the previous section.

---

[1] As in NTRUEncrypt [16,17], a secure approach is to require the decoding error probability to be less than $2^{-\kappa}$ for the $\kappa$-bit security.

## 4.1   Attack Model

The basic scenario for the attack is the following. Alice continuously sends messages to Bob using the QC-MDPC scheme and Bob's public key. Occasionally, a decoding error will occur and Bob will show a different reaction to report this decoding failure. The information will then be detected and collected. After repeating the procedure a number of times, Alice will be capable of recovering Bob's secret key using our proposed attack.

In terms of a security model definition, the attack is called a *reaction attack*. In previous work, resend and reaction attacks on McEliece PKC have appeared [14]. However, they have targeted message recovery only and there has been no key recovery attack in this model before.

The McEliece PKCs in their plain form have computational security against *chosen plaintext attacks* (CPAs), but are known to be insecure against *chosen ciphertext attacks* (CCAs). The reaction attack is an attack model in-between since it only requires the reaction of the decryption device (whether there was a decryption error) and not the result of decryption.

## 4.2   Attack Description

Continuing, we assume that the rate of the code is $R = k/n = 1/2$, corresponding to $n_0 = 2$. Also, let $w_0 = w_1 = w$. Attacks for other parameters follow in a similar fashion.

The key-recovery attack on QC-MDPC aims at finding the secret matrix $\mathbf{H}_0$, given only the public-key matrix $\mathbf{P}$. From $\mathbf{H}_0$, the remaining part of $\mathbf{H}$ can easily be recovered from $\mathbf{P}$ using basic linear algebra. Being a cyclic matrix, recovering $\mathbf{H}_0$ is equivalent to recovering its first row vector, denoted $\mathbf{h}_0$.

The key idea is to examine the decoding procedure for different error patterns. In particular, we will be interested in having Alice pick error patterns from special subsets. Let $\Psi_d$ be the set of all binary vectors of length $n = 2r$ having exactly $t$ ones, where all the $t$ ones are placed as random pairs[2] with distance $d$ in the first half of the vector. The second half of the vector is an all-0 vector. Formally, we select from the set $\Psi_d$, which guarantees repeated ones at distance $d$ at least $t/2$ times, where

$$\Psi_d = \{\mathbf{v} = (\mathbf{e}, \mathbf{f}) \mid w_{\mathrm{H}}(\mathbf{f}) = 0, \text{ and } \exists \text{ distinct } s_1, s_2, \ldots, s_t, \text{ s.t. } \mathbf{e}_{s_i} = 1, \text{ and}$$
$$s_{2i} = (s_{2i-1} + d) \bmod r \text{ for } i = 1, \ldots, \frac{t}{2}\}.$$

Alice will now send $M$ messages to Bob, using QC-MDPC with the error selected from the subset $\Psi_d$ of all possible error vectors of weight $t$. When there is a decoding error with Bob, she will record this and after $M$ messages she will be able to compute an empirical decoding error probability for the subset $\Psi_d$. Furthermore she will do this for $d = 1, 2, \ldots, U$ for some suitable upper bound $U$.

---

[2] We assume that $t$ is an even number for the ease of description; otherwise, we just pick $\frac{t-1}{2}$ random pairs and randomly choose another position to fulfill the constraint on the error weight.

---

**Algorithm 1** – Computing the distance spectrum

---

**Input:** parameters $n, r, w$ and $t$ of the underlying QC-MDPC code, number of decoding trials $M$ per distance.
**Output:** distance spectrum $D(\mathbf{h_0})$.

> **for** *all distances d* **do**
> > Try $M$ decoding trials using the designed error pattern
> > Perform statistical test to decide multiplicity $\mu(d)$
> > **if** $\mu(d) > 0$ **then**
> > > Add $d$ with multiplicity $\mu(d)$ to distance spectrum $D(\mathbf{h_0})$

---

The main observation of the paper is that there is a strong correlation between the decoding error probability for error vectors from $\Psi_d$ and the existence of a distance $d$ between two ones in the secret vector $\mathbf{h_0}$. Namely, if there exists two ones in $\mathbf{h_0}$ at distance $d$, the decoding error probability is much smaller than if distance $d$ does not exist between two ones. We will give an explanation to this in the next section.

So after sending $M \times U$ messages, we look at the decoding error probability for each $\Psi_d$ and classify each $d$, $d = 1, 2, \ldots, U$ as "does not exist in $\mathbf{h_0}$" (called CASE-0) or "existing in $\mathbf{h_0}$" (called CASE-1). This gives us what we call a distance spectrum for $\mathbf{h_0}$, denoted $D(\mathbf{h_0})$. It is given as

$$D(\mathbf{h_0}) = \{d : 1 \leq d \leq U, d \text{ classified as existing in } \mathbf{h_0}\}.$$

Also, since a distance $d$ can appear many times in the distance spectrum of a given bit pattern $\mathbf{c}$, we will let the multiplicity of $d$ in $\mathbf{c}$ be denoted $\mu_{\mathbf{c}}(d)$ in the sequel, or simply $\mu(d)$ when $\mathbf{c}$ is clearly defined from the context.

As an example, for the bit pattern $\mathbf{c} = 0011001$ we have $U = 3$ and

$$D(\mathbf{c}) = \{1, 3\},$$

with distance multiplicities $\mu(1) = 1, \mu(2) = 0$ and $\mu(3) = 2$.

The procedure for computing the distance spectrum is specified in Algorithm 1.

The final step is to do a reconstruction of $\mathbf{h_0}$ from knowing the distance spectrum $D(\mathbf{h_0})$. This is done through an iterative procedure. Start by assigning the first two ones in a length $i_0$ vector in position 0 and $i_0$, where $i_0$ is the smallest value in $D(\mathbf{h_0})$. Then put the third one in a position and test if the two distances between this third one and the previous two ones both appear in the distance spectrum. If they do not, we test the next position for the third bit. If they do, we move to test the fourth bit and its distances to the previous three ones, etc. After reconstruction, we have restored $\mathbf{h_0}$. The reconstruction procedure is illustrated in Fig. 1 and detailed in Algorithm 2.

For the above example with bit pattern $\mathbf{c} = 0011001$, the careful reader will note that this algorithm will reconstruct $\mathbf{c}$ as $1100100$ – with a rotation. However, this rotation is a non-issue in practice in our application.

---

**Algorithm 2** – Key recovery from distance spectrum

**Input:** distance spectrum $D(\mathbf{h}_0)$, partial secret key $\mathbf{h}_0$, current depth $l$.
**Output:** recovered secret key $\mathbf{h}_0$ or message "No such secret key exists".
**Initial recursion parameters:** distance spectrum $D(\mathbf{h}_0)$, empty set for secret key, current depth 0.

> **if** $l = w$ **then**
> $\quad\lfloor$ **return** $\mathbf{h}_0$ /* secret key found */
> **for** *all potential key bits $i$* **do**
> $\quad\lfloor$ **for** *all distances to key bit $i$ exist in $D(\mathbf{h}_0)$* **do**
> $\qquad$ Add key bit $i$ to secret key $\mathbf{h}_0$
> $\qquad$ Make recursive call with parameters $D(\mathbf{h}_0), \mathbf{h}_0$ and $l+1$
> $\qquad$ **if** *recursive call finds solution* $\mathbf{h}_0$ **then**
> $\qquad\quad\lfloor$ **if** $\mathbf{h}_0$ *is the secret key* **then**
> $\qquad\qquad\lfloor$ **return** $\mathbf{h}_0$ /* secret key found */
> $\qquad\lfloor$ Remove key bit $i$ from secret key $\mathbf{h}_0$
> **return** "No such secret key exists"

---

In addition, the reconstruction procedure may also find some key pattern $\mathbf{h}'$ with the same distance spectrum $D(\mathbf{h}_0)$ as $\mathbf{h}_0$. The algorithm will then discard it and recursively try other key patterns, which provides an exhaustive search process.

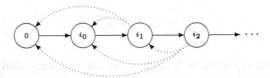

**Fig. 1.** The reconstruction process. Vertices represent nonzero bit positions in the bit pattern, solid arrows show the search order from left to right, dotted arrows show that for a newly determined bit position, its distances to all previous nonzero bit positions should all be in the distance spectrum.

# 5   Analysis

In this section we present an intuitive explanation why the proposed attack can recover the secret key from the decoding errors of a bit-flipping-type iterative decoder, and we also give some theoretical analysis on this new algorithm.

## 5.1   An Explanation for the Distinguishing Procedure

The authors in [27] pointed out that the employed iterative bit-flipping variants will stop in quite a small number of iterations (i.e., around 3 to 5 iterations on average), and further iterations have little effect on improving the success probability. Therefore, the behavior of the error variables in the first iteration plays a vital role in the decoding process: if almost all the variables flip from a wrong to a right value, the decoder will correct the errors quickly, otherwise it is more probable to fail.

We thus focus on the flipping behavior of the error bits in the first iteration for different input error patterns from the sets $\Psi_d$, containing random pairs of ones with distance $d$.

**Table 2.** The relation between the number of nonzero $h_{ij}e_i$'s and that of correctly changed counters in the first decoding iteration.

| # $(h_{ij}e_i = 1)$ | #(right change) | #(wrong change) |
|---|---|---|
| 0 | $w$ | 0 |
| 1 | 1 | $w - 1$ |
| 2 | $w - 2$ | 2 |
| 3 | 3 | $w - 3$ |
| $\vdots$ | $\vdots$ | $\vdots$ |

First, we present more observations on the first round of the bit-flipping process. Given the $j^{\text{th}}$ parity-check equation, i.e.,

$$\sum_{i=1}^{n} h_{ij}e_i = s_j,$$

for $1 \le j \le r$, this equation will affect $w$ counters corresponding to the error variables with a nonzero coefficient $h_{ij}$. The value of the syndrome bit $s_j$ determines if the equation is satisfied or not, since value for all the error variables $e_i$'s are initially set to 0. That is, if $s_j = 0$, then the parity-check equation holds and no counters are increased for this check node. On the other hand, if $s_j = 1$, all the $w$ counters for variable nodes included in this parity check are incremented.

Obviously, in the iterative decoding we do not want the counter for an error variable $e_i$ to increment if $e_i = 0$, and vice versa; we do want it to increment if $e_i = 1$. So we can consider whether the counter is correctly or erroneously changed.[3]

As a result, the number of nonzero terms of the form $h_{ij}e_i$'s in a parity-check equation determines the number of correctly changed counters in the first iteration, and the numbers are shown in Table 2. For example, in an equation, if there is no nonzero terms $h_{ij}e_i$, then $s_j = 0$ and the initial values of the $e_i$'s in this check are all correct; But since $s_j = 0$ none of their counters are incremented and hence all counters are correctly changed.

---

[3] Here "change" means increasing by 1 or preserving the value.

If there is only one nonzero term $h_{ij}e_i$ in the parity check, then $s_j = 1$ and the equation is unsatisfied. Every counter corresponding to an error variable in this equation will be increased, but only one variable is actually in error. Hence we are changing $w - 1$ counters erroneously and only one correctly. For two nonzero term $h_{ij}e_i$ in the parity check, $s_j = 0$ and it follows in the same way as before that $w - 2$ counters are correctly changed and two of them erroneously, etc.

According to the above observation, it is desirable to have a small even number (like 0, 2, ...) of nonzero terms $h_{ij}e_i$ when evaluating parity-check equations for having the best chances of success in decoding. We can observe that if we look at all the $r$ parity checks in $\mathbf{H}$, we will create a total of exactly $t \cdot w$ nonzero terms $h_{ij}e_i$ in the parity checks all together. For a randomly selected weight $t$ error, we can view this as putting $t \cdot w$ different objects in $r$ buckets and counting the number of objects in each bucket. An even number of objects in a bucket will be helpful in decoding, while an odd number of objects will act in opposite.

Now let us consider errors selected from our special error set $\Psi_d$. If the secret vector $\mathbf{h}_0$ contains two ones with distance $d$ inbetween (CASE-1), then, due to the many inserted pairs of distance $d$ in the error vector, we have "artificially" created a number of ($\geq \frac{t}{2}$) check equations where we know that we have at least two nonzero terms $h_{ij}e_i$ in the parity check. This "artificial" creation of pairs of nonzero terms $h_{ij}e_i$ in the same check equation changes the distribution of the number of nonzero terms $h_{ij}e_i$ in parity checks. If the secret vector $\mathbf{h}_0$ does not contains two ones with distance $d$ inbetween (CASE-0), then the same phenomenon does not appear.

In Table 3, we present a precise evaluation of the corresponding distributions of an instance using the suggested QC-MDPC parameters for 80-bit security where the weight of $\mathbf{h}_0$ is assumed to be exactly 45. These results are obtained by a heavy simulation using 1000 different random keys and 480100 valid error patterns for each key. In CASE-1, the probability of being 0 is higher and that of being 1 lower, which are both preferred for the decoding purpose. Also owing to that the probabilities of being other values larger than 1 are of a similar magnitude for the both cases, this table verifies the influences of the "artificially" created pairs.

**Table 3.** The distinct distributions of the number of nonzero terms $h_{ij}e_i$'s for the error patterns from $\Psi_d$ using the QC-MDPC parameters for 80-bit security and assuming that the weight of $\mathbf{h}_0$ is exactly 45.

| # ($h_{ij}e_i = 1$) | Probability | |
|---|---|---|
| | CASE-0 | CASE-1 |
| 0 | 0.4485 | 0.4534 |
| 1 | 0.3663 | 0.3602 |
| $\geq 2$ | 0.1852 | 0.1864 |

Since this algorithm iterates further and many quite short (e.g., length-4) cycles[4] appear in the corresponding Tanner graph, it is challenging to determine the variation of the decoding error probability caused by the different distributions in the first round, via presenting some precise theoretical estimations. On the other hand, several thousands of parity-check equations (e.g., 4801 equations in the 80-bit security case) exist, making the overall differences substantial. In addition, more correct values in the initial round will contribute positively in the following iterations. These facts explain why some significant differences can be detected in our experiments, and why they imply a successful key-recovery attack in real time.

### 5.2    Complexity Analysis

We now derive a complexity estimation for the key-recovery attacks. Making use of the obtained experimental results for certain key parameters, we can then approximate the concrete time complexity (shown in Sect. 7). This complexity consists of two parts: that of building the distance spectrum and that of reconstructing the secret polynomial. We analyze separately.

**The Complexity for Building the Distance Spectrum.** It is shown in experiments that the error rates for the different distances clearly separate into intervals according to multiplicity. When these intervals are disjoint, it is possible to determine the complete distance spectrum of the secret key fully and without error. In general, for a well-designed error pattern, the error probabilities increase with decreasing multiplicities, as sketched in Fig. 2a.

The distinguishing procedure involves $U$ groups of decoding tests, each of them consisting of $M$ decoding trails. Thus, overall $U \times M$ decoding data would be collected, implying that the complexity is of order $\mathcal{O}(MU)$. Here $M$ and $U$ are two algorithmic parameters that depend on the targeting security parameters $n, r, \hat{w}, t$. A reasonable upper bound for $U$ is $\lfloor \frac{r}{2} \rfloor$, since this is the number of possible (modular) distances given the block size $r$.

On the other hand, it is non-trivial to determine the minimal value of $M$ that is sufficient to execute a successful distinguishing. The experimental results suggest that the error rate for the error pattern using a distance $d$ with multiplicity $\mu(d)$ can be approximated by a Gaussian distribution with mean $m_{\mu(d)}$ and variance $\sigma^2_{\mu(d)}$; we can thus model this problem as a hypothesis testing problem determining whether $\mu(d)$ is zero or not.

Figure 2, which consists of two sub-figures, describes the approximated distributions of the error probability when performing the proposed reaction attack. With adequate decoding trials to make the widths of these Gaussian distributions "narrow" enough, we draw roughly the shape of the probability density function of the decoding error probability (in Fig. 2a). On the contrary, Fig. 2b records with the precision in magnitude the empirical distribution when performing the proposed reaction attack on the QC-MDPC parameters for 80-bit

---

[4] See Table 4 for more details.

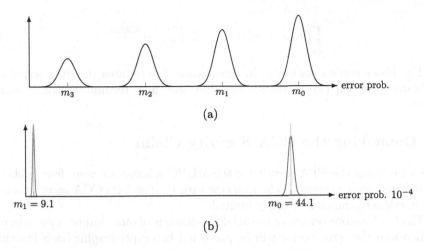

**Fig. 2.** Classification of distance multiplicities based on decoding error probability. (a): Distribution shape in general. (b): Empirical distribution using $M = 100,000$ decoding trials for each distance (proposed parameters for 80-bit security with $t = 84$).

security with error weight 84, where $100,000$ decoding trails are exploited for each distance $d$. In this figure, only the groups of multiplicity 0 and multiplicity 1 are depicted as the remaining groups are of a much smaller magnitude. More data can be found in Table 5.

**The Complexity for Reconstruction.** We show that the algorithm will return the correct key soon on average. Since this algorithm builds an enumeration tree to search for the possible solutions in a depth-first way, the time complexity can be represented by its paths to the leaves in the tree. Later we present a rough estimation of this number.

Suppose $n_s$ is the size of the distance spectrum $D(\mathbf{h}_0)$, $n_t$ the number of possible distances[5] required to be tested, and $\alpha$ the ratio between them, i.e. $n_s/n_t$. In the beginning, we chose the smallest distance in the spectrum and determine two positions 0 and $i_0$; this can be viewed as the root of the tree. Then we extend the tree to choose another position $i_1$. We know that the distances $i_1$ and $i_1 - i_0$ should be both in the distance spectrum; among the $n_s$ possible distances for $i_1$, thus, we can expect an $\alpha$ fraction of them are valid and there exist $n_s\alpha$ nodes in the first level. Similarly for one node in the first level, there are $n_s\alpha^2$ child nodes on average in the second level since the distances $i_2$, $i_2 - i_0$ and $i_2 - i_1$ should be all in the distance spectrum. Etc.

Since the average child number of a node after quite few steps[6] (denoting this number $\phi + 1$) will be less than 1, we can deduce a loose estimation on the average number of possible paths as

---

[5] A reasonable setting for $n_t$ is $T$, the number of distances bounded by $\frac{r}{2}$.

[6] The average child number of a node in the $l^{\text{th}}$ level drops exponentially in $l$.

$$\prod_{i=1}^{\phi} n_s \alpha^i = n_t^{\phi} \alpha^{\frac{\phi(\phi+3)}{2}} \leq \left(\frac{r}{2}\right)^{\phi} \alpha^{\frac{\phi(\phi+3)}{2}}. \tag{1}$$

The above results state that in expectation, the number of paths tested can be bounded by Eq. (1). In reality, the algorithm may terminate soon if we are lucky.

## 6    Debunking the CCA Security Claim

When targeting the CPA security of the MDPC scheme, we were free to choose the injected error patterns. When we now turn to attack its CCA-secure version, this freedom of choice is severely limited.

The CCA-secure version of the MDPC scheme is of more importance as in real applications the error vector will be protected by cryptographic hash functions after conversions (e.g., [21]) for making the MDPC scheme semantically secure.

The fundamental idea of the attack is as follows. We randomly generate $T$ plaintext-ciphertext pairs. We then form subsets of those with desired error patterns. In particular, we will be interested in error patterns that contain occurrences of distance $d$ between error bits, where $d$ is a length in the distance spectrum to be tested. Our simulations show that these error patterns can be used to efficiently distinguish whether a certain distance $d$ appears in the distance spectrum of the targeted secret polynomial.

We present the algorithm in two versions to match different levels of detail. The high level description is presented as Algorithm 3.

---

**Algorithm 3** – Breaking the CCA security of the converted MDPC scheme.

---

**Input:** number $T$ of ciphertexts to generate.
**Output:** distance spectrum $\mathbf{s}$ for the secret key $K$.

> Generate a collection $\Sigma$ of $T$ ciphertexts
> Record decryptability for each $c$ in $\Sigma$
> $\mathbf{s} \leftarrow$ storage for distance spectrum of secret key
> **for** *all distances $d$* **do**
> > $\Sigma_d \leftarrow \{c \in \Sigma \mid \mu_c(d) \geq 1\}$
> > $\mathbf{s}[d] \leftarrow$ multiplicity classification from decryptability rate in $\Sigma_d$
>
> **return s**

---

The description seems to suggest that we need lots of storage for handling ciphertexts, but this is not the case. An efficient implementation requires virtually no storage. To see this, consider the alternative description in Algorithm 4.

In Algorithm 4 we successively check the decryptability of ciphertexts and use these observations to obtain better and better estimates of decoding error probabilities related to all possible distances in the distance spectrum of the secret key.

---

**Algorithm 4** – Breaking the CCA security of the converted MDPC scheme. Detailed description.

---

**Input:** number $T$ of ciphertexts to generate.
**Output:** distance spectrum for the secret key $K$.

 $\mathbf{a} \leftarrow$ zero-initialized vector of length $\frac{r}{2}$ /* count decoding failures per distance */
 $\mathbf{b} \leftarrow$ zero-initialized vector of length $\frac{r}{2}$ /* count total samples per distance */
 $i \leftarrow 0$
 **while** $i < T$ **do**
  Generate ciphertext $c$
  $\mathbf{s}_{err} \leftarrow$ distance spectrum of ciphertext error
  $\ell \leftarrow$ decryptability of $c$ /* 0 for successful decryption, 1 for decryption failure */
  **for** *all distances $d$* **do**
   **if** $s_{err}[d] \geq 1$ **then**
    $\mathbf{a}[d] \leftarrow \mathbf{a}[d] + \ell$
    $\mathbf{b}[d] \leftarrow \mathbf{b}[d] + 1$
  $i \leftarrow i + 1$
 $\mathbf{s}_{key} \leftarrow$ vector of length $\frac{r}{2}$ /* distance spectrum of secret key */
 **for** *all distances $d$* **do**
  $\mathbf{s}_{key}[d] \leftarrow$ multiplicity classification from estimated error rate $\frac{\mathbf{a}[d]}{\mathbf{b}[d]}$
 **return** $\mathbf{s}_{key}$

---

The vector slots of $\mathbf{a}$ and $\mathbf{b}$ are used to represent the decoding error probabilities, so that $\frac{\mathbf{a}[d]}{\mathbf{b}[d]}$ is an approximation of the decoding error probability over all error patterns with distance spectrums containing distance $d$. This subset of error patterns is denoted $\Sigma_d$ in Algorithm 3.

Each ciphertext updates several entries in $\mathbf{a}$ and $\mathbf{b}$, and we need to observe the decryptability of sufficiently many ciphertexts in order to obtain probability estimates that are reliable enough for correct multiplicity classification.

For each ciphertext we utilize the nonzero (other thresholds are also possible) entries in the corresponding distance spectrum. Letting $\alpha$ denote the average fraction of nonzero entries in such a distance spectrum, one can see that the total number of iterations in the inner loop (per ciphertext) is about $\frac{\alpha r}{2}$.

The output of Algorithm 4 is the distance spectrum of the secret key, so the careful reader will note that the key recovery method described in Algorithm 2 needs to be applied as a final step for full key recovery. However, in terms of complexities, this additional step comes for free.

The time complexity of Algorithm 4 is precisely $T$ if we count the number of observed ciphertexts. If we count low-level operations, as defined by the inner loop of Algorithm 4, the time complexity is $T \times \frac{r}{2}$.

## 6.1   An Explanation of How Sample Collection Works

The precise nature of Algorithm 4 can easily and very conveniently be understood by modeling the sampling procedure as a generalized version of the coupon collector's problem. In the original coupon collector's problem, using the balls-and-bins paradigm, we randomly throw balls into $u$ bins until all bins are non-empty. We need to throw around $u \log u$ balls before we achieve this goal.

In the generalized problem, we keep throwing balls until all bins each contain at least $b$ balls. The time complexity for this (see [11]) is

$$J(u, b) = u \left( \log u + (b-1) \log \log u + \gamma - \log (b-1)! + o(1) \right). \tag{2}$$

It is even possible to arbitrarily bound the probability of failure by adding a linear number of samples (balls) according to

$$\lim_{t \to \infty} \mathbf{Pr} \left[ \mathcal{X}_{(u,b)} < u \log u + (b-1) u \log \log u + tu \right] = e^{-\frac{e^{-t}}{(b-1)!}},$$

where $\mathcal{X}_{(u,b)}$ is a statistical variable that represents the number of throws needed to fill up $u$ bins so that all of them contain at least $b$ balls.

In the CCA case we collect error patterns, but not all error patterns are useful. Instead, we form different subsets of useful error patterns denoted $\Sigma_d$ in Algorithm 3. We successively check the decryptability of ciphertexts and use these observations to obtain better and better estimates of decoding error probabilities related to all possible distances in the distance spectrum of the secret key.

For each ciphertext we then utilize the nonzero (other thresholds are also possible) entries in the corresponding distance spectrum, and each such nonzero entry corresponds to a ball. With $\alpha$ denoting the average fraction of nonzero entries in such a distance spectrum, one can see that the total number of balls we collect per ciphertext is about $\frac{\alpha r}{2}$.

In Algorithm 4, the bins are represented by the vector slots of $\mathbf{a}$ and $\mathbf{b}$, so there are $u = \frac{r}{2}$ bins. Each observed error pattern generates $\alpha$ balls, and each ball updates an entry in $\mathbf{a}$ and $\mathbf{b}$. We need at least $b$ balls in each bin in order to obtain probability estimates that are reliable enough for computing the distance spectrum of the secret key. The value $b$ determines the number $T$ of ciphertexts that we need to generate, since $b$ and $T$ are strongly related according to

$$\frac{\alpha r T}{2} \approx J \left( \frac{r}{2}, b \right). \tag{3}$$

It may also be noted that it is not immediately clear how to analytically derive $b$ or $T$ directly from the security parameters. For our results, we have determined $T$ explicitly by simulation, as described in Sect. 7.

# 7    Implementations and Numerical Results

We have conducted several simulation tests to verify the behaviors of the error rates related to different multiplicities and different error shapes. The following implementation results are all obtained by employing QC-MDPC with the proposed parameters for 80-bit security [29] and the original Gallager's bit-flipping algorithm [12], i.e., Decoder $\mathcal{B}$ in [27].

In the CPA case, we consider two different error weights. Error weight $t = 84$ is what is proposed for 80-bit security, but we also consider the case $t = 90$ here. This is motivated by security models that allow injection of more errors, where additional errors are not explicitly detected. For the CCA case, only results with $t = 84$ are stated.

Results for the CPA case are presented in Sect. 7.1, and the results for the CCA case are presented in Sect. 7.2. A discussion on the employment of other decoders follows in Sect. 7.3.

Before introducing the main implementation results, we show the probability distributions for distance multiplicities in the first polynomial when considering the QC-MDPC scheme with $n_0 = 2$ (see Table 4).

**Table 4.** Probability distributions for distance multiplicities in the first polynomial (of two), generated uniformly with total weight $t = 84$ and $t = 90$. The polynomial length is 4801, while the total vector length is 9602.

| multiplicity | $t = 84$ | | | $t = 90$ / key with $\hat{w} = 90$ | | |
|---|---|---|---|---|---|---|
| | probability | accumulated | accumulated | probability | accumulated | accumulated |
| 0 | 0.6955724 | 0.6955724 | 1.0000000 | 0.6589889 | 0.6589889 | 1.0000000 |
| 1 | 0.2524958 | 0.9480683 | 0.3044275 | 0.2748075 | 0.9337965 | 0.3410106 |
| 2 | 0.0458487 | 0.9939170 | 0.0519316 | 0.0573330 | 0.9911295 | 0.0662031 |
| 3 | 0.0055425 | 0.9994596 | 0.0060829 | 0.0079677 | 0.9990972 | 0.0088701 |
| 4 | 0.0005018 | 0.9999614 | 0.0005403 | 0.0008287 | 0.9999260 | 0.0009024 |
| 5 | 0.0000362 | 0.9999977 | 0.0000385 | 0.0000688 | 0.9999949 | 0.0000737 |
| 6 | 0.0000021 | 0.9999998 | 0.0000022 | 0.0000047 | 0.9999997 | 0.0000049 |
| 7 | 0.0000001 | 1.0000000 | 0.0000001 | 0.0000002 | 1.0000000 | 0.0000002 |

The vector is of length 9602, and is generated uniformly with weight 84 (or 90). These probability distributions are mainly of importance for the following two reasons.

- When the vector is viewed as a key vector, the data in the right part (corresponding to $t = 90$) show the distance multiplicity distributions of a random key, from which not only the size of its distance spectrum can be estimated, but some other vital information may also be revealed. For example, since about 6.6 percent of the distances are of multiplicity 2 or more when $t = 90$, quite a few length-4 cycles[7] will appear in the Tanner graph corresponding to the secret key.
- When the vector is viewed as an error vector, these data can be utilized to simulate the random error obtained from a CCA2-secure QC-MDPC scheme. We will explain this further in Sect. 7.2.

---

[7] A distance with multiplicity of 2 or more implies that there exists at least one length-4 cycle.

## 7.1   CPA Case

As described in Sect. 5.2, the time complexity of attacking the CPA-secure version consists of two parts: that of constructing the distance spectrum and of key reconstruction. From Table 5, we can see that for the MDPC parameters targeting 80-bit security, it is sufficient to choose $M$ to be 100, 000 to make the decoding error rates well-separated according to the multiplicity; this value can be even reduce to 10, 000 if an error with weight $t = 90$ is allowed to be used. Setting the number of different groups for decoding test as 2400, we derive that the time complexity for Alice to know the distance spectrum of the secret key is bounded by that of calling the decoder about 240, 000, 000 (or 24, 000, 000) times for solely the information whether the decoding succeeds, when the error weight $t$ is 84 (or 90). In the security model of a reaction attack, the decoding results (success or fail) are presumably provided to the adversary; therefore, the decoding cost is excluded from the time complexity, implying that the time complexity for constructing the distance spectrum can be estimated as $2^{28}$ (or $2^{25}$) operations for $t = 84$ (or 90).

**Table 5.** Decoding error rates when using the original Gallager's bit-flipping algorithm (Decoder $\mathcal{B}$ in [27]) and the designed error pattern $\Psi_d$ with $t = 84$ and $t = 90$. The number of decoding trials in a group is $M = 100, 000$ and $M = 10, 000$, respectively.

| multiplicity | $t = 84$ | | $t = 90$ | |
|---|---|---|---|---|
| | error rate | $\sigma$ | error rate | $\sigma$ |
| 0 | 0.0044099 | 0.00003868 | 0.415395 | 0.000830 |
| 1 | 0.0009116 | 0.00001304 | 0.248642 | 0.000729 |
| 2 | 0.0001418 | 0.00000475 | 0.121623 | 0.000529 |
| 3 | 0.0000134 | 0.00000112 | 0.048330 | 0.000299 |

For the MDPC parameters targeting 80-bit security, the weight of the secret key is set to be 90. By checking Table 4, therefore, the empirical ratio $\alpha$ can be approximated as 0.341 and thus $\phi$ is 6. We on average test no more than $2^{25.5}$ paths, costing less than $2^{35}$ operations since most of the invalid paths will be detected and removed soon (less than 20 steps). We implemented this algorithm, which performed quite well in practice — for most of the instances, the algorithm succeeded in minutes.

## 7.2   CCA Case

Next in turn is the CCA case and truly uniform error patterns with a certain weight. We have used Algorithm 4 for our simulation runs. One such simulation for the QC-MDPC scheme with the proposed parameters for 80-bit security (with $t = 84$) can be seen in Figure 3. Here we plot the number of utilized ciphertexts

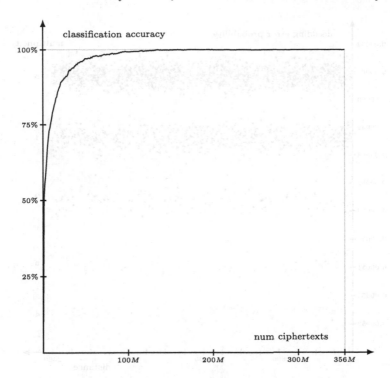

**Fig. 3.** CCA algorithm for a QC-MDPC McEliece instance for 80-bit security with $t = 84$. The distance spectrum of the key is fully recovered (no errors) after observing 356M ciphertexts. The graph shows the worst case out of ten full simulations.

vs. the fraction of correctly classified distance spectrum entries, resulting in a simple visualization of the algorithm efficiency.

The simulations suggest that $T = 356$ million observed ciphertexts are sufficient for fully determining the entire distance spectrum without any errors. That is, after we have observed 356 million ciphertexts, the distance spectrum remains stable and correct.

It should be noted that we ran ten independent simulation runs, and the result presented in Fig. 3 was the worst case simulation result. The 356 million ciphertexts estimate is therefore a conservative high probability estimate. That is, in all simulations, the multiplicity classifications were 100 % correct and stable after 356 million ciphertexts. For comparison, the best case yielded full distance spectrum recovery after 203 million ciphertexts.

The same simulation is shown in Fig. 4, providing a more detailed view of how the algorithm works. Each dot represents the estimated decoding error probability for one particular distance, and every possible distance has been plotted in the same graph.

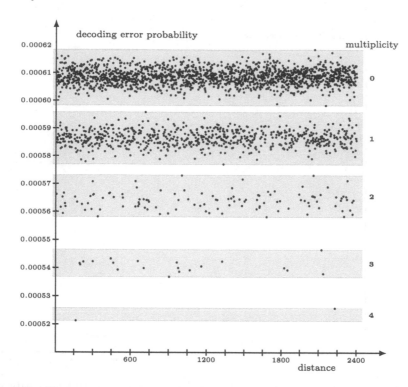

**Fig. 4.** Classification intervals for the $t = 84$ worst-case simulation after 356M cipher-texts. All 2400 data points plotted.

Now, have a closer look at the classification intervals that have been high-lighted with a grey background. These intervals span from the minimum to the maximum estimated decoding error probability *per multiplicity*.

As Fig. 4 shows the state of affairs at the end of the simulation, we can see that the distance dots are clearly separated into different and fully disjoint classification intervals depending on their multiplicity. The classification intervals are generally overlapping during the earlier parts of the simulation.

We can use the classification intervals and analyze their successive widths in the simulation. This is very useful, because we are only able to compute a perfect error-free distance spectrum when all of these intervals are mutually disjoint. In this way we can derive a reliable estimate for the value $T$ for different problem instances. The reader may note that there is some (little) room for improvement here. In the worst case simulation out of the ten we have performed, the classification intervals became disjoint after 268 million ciphertexts, which sets a lower bound for error-free distance spectrum recovery. However, this bound can be lowered even further if we allow errors in the recovered distance spectrum, or if we tweak the algorithm in other ways, and so on, but such improvements are out of scope here.

In a live scenario, the distance multiplicities are unknown, so it is not possible to compute the successive classification intervals and check when they are disjoint. Predetermined probability values cannot be used, since the decoding error probabilities differ significantly between instances (different keys). However, the general shape of the probability distribution in Fig. 2a is known and represents a "side view" of Fig. 4, so it can be seen that the multiplicity classification problem is not very difficult in practice.

The classification procedure we have used in our experiments is quite simplistic. In our simulations, we computed the current (estimated) decoding error probabilities $m_0, m_1, \ldots$ per multiplicity from the simulation values, and then computed boundary mid-points $\frac{m_0+m_1}{2}, \frac{m_1+m_2}{2}, \ldots$ and checked when the intervals were fully separated into these mid-point regions. In a live scenario one could efficiently achieve the same effect by using a simple clustering technique (counting dots in small intervals) to first accurately estimate $m_0, m_1, \ldots$, and then continue as we have done.

To conclude the simulation results, the total time complexity of Algorithm 4 is at most 356 million observed ciphertexts. If we count low-level operations instead, as specified in Sect. 6, then the total time complexity is about $T \times \frac{r}{2} = 2^{39.7}$ for the proposed security parameters for 80-bit security using the Gallager's original bit-flipping decoder. Compared with this complexity figure, the key reconstruction part is negligible.

## 7.3   Some Discussions

We discuss more about the decoding procedure employed in the implementation.

**Using Other Decoding Techniques.** The employed decoding algorithm in implementation is Gallager's bit-flipping algorithm, which is chosen not only because its relatively higher error-probability makes the implementation easier, but also because it is the original iterative decoding algorithm for LDPC settling the framework and the principle for the later improved decoders [15,27]. Hence, it is reasonable to assume that replacing the Gallager's bit-flipping decoder by another more advanced decoder may increase the attack complexity by a factor of around $2^e$, if the error probability is reduced with a factor of $2^{-e}$. However, the attack complexity is still far less than the claimed security level.

For example, the best implementations of bit-flipping-type decoders found in literature with respect to the decoding performance are the ones from [8,10] both claiming a decoding error probability less than $10^{-8}$ for the 80-bit secure QC-MDPC parameter set. These decoders improve upon the original Gallager's decoder by a factor of about $2^{15.6}$. Therefore, we might estimate the time complexity for attacking the 80-bit CPA (or CCA2) -secure version as $2^{43.6}$ (or $2^{55.3}$) operations, if these two decoders are instead implemented.

Moreover, some decoders (including the one in [29]) decrease the error probability by restarting the decoding process in the same decoding framework but only employing different thresholds, when an error is detected. On one side, since each calling of the bit-flipping algorithm might contribute to the variation between CASE-0 and CASE-1, the effects on the distinct decoding error probabilities may accumulate after more and more decoding rounds, implying that the estimation in the last paragraph is conservative. On the other hand, via some side-channel attacks, an adversary might get the information of the initial errors occurred, thereby reducing the problem to that of using a less powerful decoder like the one being implemented. It is definitely beneficial to design a countermeasure to withstand this type of attack.

We conjecture that this attack also works for the MDPC scheme employing a soft-decision decoding implementation.

**Moving to a Higher Security Level.** The QC-MDPC scheme using the suggested 80-bit secure parameter set is frequently discussed and implemented in literature due to its applications in power-constraint devices, for which we choose it as a study case. However, for the long-term security purpose, the bottom line nowadays is to achieve 128-bit security. There is no evidence that the scheme with the suggested 128-bit secure parameters will be invulnerable to the proposed reaction attack, and frankly speaking, the situation is even worse due to the larger gap between the current state-of-the-art implementation[8] of bit-flipping-type decoders and the required decoding performance[9] with respect to security.

**Higher Error Probability.** Another meaningful observation is that when utilizing the designed highly unbalanced error pattern, the error probability is higher than when harnessing a uniform distribution in the valid set of errors. This increased error probability jeopardizes the security of the MDPC scheme by boosting the proposed reaction attack further. Since we employ the same implementation of Gallager's bit-flipping decoder as in [15], the enlarged failure probability is mainly due to the specific error pattern used: all the $t$ error positions are gathered together in the first part of the error vector. We show the numerical results in Table 6.

**Table 6.** The comparison of failure rates among different error patterns using the QC-MDPC parameters for the 80-bit security.

| Error weight | All valid errors [15] | This work | |
|---|---|---|---|
| | | multiplicity 0 | multiplicity 1 |
| 84 | 0.00051 | 0.00441 | 0.00091 |
| 90 | 0.24080 | 0.41539 | 0.24864 |

---

[8] With respect to the decoding performance, the best known implementation using the suggested 80-bit secure parameter set outperforms the one using the suggested 128-bit secure parameters ($10^{-8}$ in [8,10] vs. $10^{-7}$ in [29]).

[9] One should decrease the decoding error probability for thwarting the proposed reaction attack within $2^{128}$ operations.

# 8    Conclusions and Future Work

In this paper, we have presented a reaction-type attack against the QC-MDPC public key encryption scheme. This novel attack exploits the strong correlation between certain structures in the secret key and the decoding error probability when errors with arranged patterns are employed. It then rebuilds the secret polynomial efficiently by executing a reconstruction procedure, therefore breaking the QC-MDPC scheme. With a slight modification, it can also be applied to the CCA2 converted version of the scheme to break its claimed security level against CCA2 attack. This (weaker) reaction attack can break the proved (stronger) CCA2 security because the decoding error probability is excluded in the proof models.

There are several research directions to be further investigated. A natural one is to design a countermeasure to protect the QC-MDPC scheme against this new attack. The most secure approach is to amend the employed iterative decoder to reduce the decoding error probability to be less than $2^{-\kappa}$ for $\kappa$-bit security. This is a challenging task due to the large between the state-of-the-art and the desired error levels, and also due to the lack of a precise theoretical error bound on these iterative algorithms. That is, changing to a more powerful decoder can enhance its security, but it is doubtful to claim that it can reach a quite high security level. Moreover, when moving to a decoder with improved performance via running itself more times with different algorithmic parameters if an error is detected, some types of side-channel attacks — like timing attacks — should be useful to know the original errors in the initial round, which can be used for a faster reaction attack.

Other directions include characterizing the strong and weak keys to resist this attack, deriving precise bounds on the decoding error probability for various error patterns given a security parameter, and designing more advanced reconstruction algorithms to handle more errors in the distance spectrum, etc. It would be fascinating if one can extend this attack to break the CCA-secure version of cryptosystems based on the LPN and LWE problems.

# References

1. Augot, D., Batina, L., Bernstein, D.J., Bos, J., Buchmann, J., Castryck, W., Dunkelman, O., Güneysu, T., Gueron, S., Hülsing, A., et al.: Initial recommendations of long-term secure post-quantum systems (2015). http://pqcrypto.eu.org/docs/initial-recommendations.pdf
2. Baldi, M., Chiaraluce, F., Garello, R., Mininni, F.: Quasi-cyclic low-density parity-check codes in the McEliece cryptosystem. In: Proceedings of IEEE International Conference on Communications, ICC 2007, Glasgow, Scotland, 24–28, pp. 951–956. IEEE (2007). http://dx.doi.org/10.1109/ICC.2007.161
3. Becker, A., Joux, A., May, A., Meurer, A.: Decoding random binary linear codes in $2^{\frac{n}{20}}$: how $1 + 1 = 0$ improves information set decoding. In: Pointcheval, D., Johansson, T. (eds.) EUROCRYPT 2012. LNCS, vol. 7237, pp. 520–536. Springer, Heidelberg (2012). doi:10.1007/978-3-642-29011-4_31

4. Bellare, M., Desai, A., Pointcheval, D., Rogaway, P.: Relations among notions of security for public-key encryption schemes. In: Krawczyk, H. (ed.) CRYPTO 1998. LNCS, vol. 1462, pp. 26–45. Springer, Heidelberg (1998). doi:10.1007/BFb0055718

5. Bernstein, D.J., Buchmann, J., Dahmen, E.: Post-Quantum Cryptography. Springer, Heidelberg (2009)

6. Berson, T.A.: Failure of the McEliece public-key cryptosystem under message-resend and related-message attack. In: Kaliski, B.S. (ed.) CRYPTO 1997. LNCS, vol. 1294, pp. 213–220. Springer, Heidelberg (1997). doi:10.1007/BFb0052237

7. Canteaut, A., Sendrier, N.: Cryptanalysis of the original Mceliece cryptosystem. In: Ohta, K., Pei, D. (eds.) ASIACRYPT 1998. LNCS, vol. 1514, pp. 187–199. Springer, Heidelberg (2000). doi:10.1007/3-540-49649-1_16

8. Chaulet, J., Sendrier, N.: Worst case QC-MDPC decoder for McEliece cryptosystem. In: IEEE International Symposium on Information Theory, ISIT 2016, Barcelona, Spain, 10–15 July 2016, pp. 1366–1370. IEEE (2016). http://dx.doi.org/10.1109/ISIT.2016.7541522

9. Chen, L., Jordan, S., Liu, Y.K., Moody, D., Peralta, R., Perlner, R., Smith-Tone, D.: Report on post-quantum cryptography. National Institute of Standards and Technology Internal Report 8105 (2016)

10. Chou, T.: QcBits: constant-time small-key code-based cryptography. In: Gierlichs, B., Poschmann, A.Y. (eds.) CHES 2016. LNCS, vol. 9813, pp. 280–300. Springer, Heidelberg (2016). doi:10.1007/978-3-662-53140-2_14

11. Flajolet, P., Sedgewick, R.: Analytic Combinatorics. Cambridge University Press, New York (2009)

12. Gallager, R.G.: Low-Density Parity-Check Codes. Ph.D. thesis, MIT Press, Cambridge (1963)

13. Goppa, V.D.: A new class of linear correcting codes. In: Problemy Peredachi Informatsii vol. 6, pp. 24–30 (1970)

14. Hall, C., Goldberg, I., Schneier, B.: Reaction attacks against several public-key cryptosystem. In: Varadharajan, V., Mu, Y. (eds.) ICICS 1999. LNCS, vol. 1726, pp. 2–12. Springer, Heidelberg (1999). doi:10.1007/978-3-540-47942-0_2

15. Heyse, S., Maurich, I., Güneysu, T.: Smaller keys for code-based cryptography: QC-MDPC McEliece implementations on embedded devices. In: Bertoni, G., Coron, J.-S. (eds.) CHES 2013. LNCS, vol. 8086, pp. 273–292. Springer, Heidelberg (2013). doi:10.1007/978-3-642-40349-1_16

16. Hoffstein, J., Pipher, J., Schanck, J.M., Silverman, J.H., Whyte, W., Zhang, Z.: Choosing Parameters for NTRUEncrypt. Cryptology ePrint Archive, Report 2015/708 (2015). http://eprint.iacr.org/

17. Hoffstein, J., Pipher, J., Silverman, J.H.: NTRU: A ring-based public key cryptosystem. In: Buhler, J.P. (ed.) ANTS 1998. LNCS, vol. 1423, pp. 267–288. Springer, Heidelberg (1998). doi:10.1007/BFb0054868

18. Howgrave-Graham, N., Nguyen, P.Q., Pointcheval, D., Proos, J., Silverman, J.H., Singer, A., Whyte, W.: The impact of decryption failures on the security of NTRU encryption. In: Boneh, D. (ed.) CRYPTO 2003. LNCS, vol. 2729, pp. 226–246. Springer, Heidelberg (2003). doi:10.1007/978-3-540-45146-4_14

19. Howgrave-Graham, N., Silverman, J.H., Singer, A., Whyte, W.: NTRU Cryptosystems: NAEP: Provable Security in the Presence of Decryption Failures. IACR Cryptology ePrint Archive 2003, 172 (2003)

20. Johansson, T., Jönsson, F.: On the complexity of some cryptographic problems based on the general decoding problem. IEEE Trans. Inf. Theory 48(10), 2669–2678 (2002)

21. Kobara, K., Imai, H.: Semantically secure McEliece public-key cryptosystems - conversions for McEliece PKC. In: Kim, K. (ed.) PKC 2001. LNCS, vol. 1992, pp. 19–35. Springer, Heidelberg (2001). doi:10.1007/3-540-44586-2_2

22. Löndahl, C., Johansson, T.: A new version of McEliece PKC based on convolutional codes. In: Chim, T.W., Yuen, T.H. (eds.) ICICS 2012. LNCS, vol. 7618, pp. 461–470. Springer, Heidelberg (2012). doi:10.1007/978-3-642-34129-8_45

23. MacWilliams, F.J., Sloane, N.J.A.: The Theory of Error Correcting Codes, vol. 16. Elsevier, Amsterdam (1977)

24. von Maurich, I., Güneysu, T.: Lightweight code-based cryptography: QC-MDPC McEliece encryption on reconfigurable devices. In: Proceedings of the conference on Design, Automation & Test in Europe, p. 38. European Design and Automation Association (2014)

25. von Maurich, I., Güneysu, T.: Towards side-channel resistant implementations of QC-MDPC McEliece encryption on constrained devices. In: Mosca, M. (ed.) PQCrypto 2014. LNCS, vol. 8772, pp. 266–282. Springer, Heidelberg (2014). doi:10.1007/978-3-319-11659-4_16

26. von Maurich, I., Heberle, L., Güneysu, T.: IND-CCA secure hybrid encryption from QC-MDPC niederreiter. In: Takagi, T. (ed.) PQCrypto 2016. LNCS, vol. 9606, pp. 1–17. Springer, Heidelberg (2016). doi:10.1007/978-3-319-29360-8_1

27. Maurich, I.V., Oder, T., Güneysu, T.: Implementing QC-MDPC McEliece encryption. ACM Trans. Embed. Comput. Syst. (TECS) 14(3), 44 (2015)

28. McEliece, R.J.: A public-key cryptosystem based on algebraic coding theory. DSN Prog. Rep. 42–44, 114–116 (1978)

29. Misoczki, R., Tillich, J.P., Sendrier, N., Barreto, P.S.: MDPC-McEliece: New McEliece variants from moderate density parity-check codes. In: 2013 IEEE International Symposium on Information Theory Proceedings (ISIT), pp. 2069–2073. IEEE (2013)

30. Overbeck, R., Sendrier, N.: Code-based cryptography. In: Bernstein, D.J., Buchmann, J., Dahmen, E. (eds.) Post-Quantum Cryptography, pp. 95–145. Springer, Heidelberg (2009)

31. Repka, M., Zajac, P.: Overview of the Mceliece cryptosystem and its Security. Tatra Mountains Math. Publ. 60(1), 57–83 (2014)

32. Sendrier, N.: Decoding one out of many. In: Yang, B.-Y. (ed.) PQCrypto 2011. LNCS, vol. 7071, pp. 51–67. Springer, Heidelberg (2011). doi:10.1007/978-3-642-25405-5_4

33. Shor, P.W.: Algorithms for quantum computation: discrete logarithms and factoring. In: 35th Annual Symposium on Foundations of Computer Science, 20–22 November 1994, Santa Fe, New Mexico, USA, pp. 124–134. IEEE Press (1994)

# SCA and Leakage Resilience II

# A Tale of Two Shares: Why Two-Share Threshold Implementation Seems Worthwhile—and Why It Is Not

Cong Chen[✉], Mohammad Farmani, and Thomas Eisenbarth

Worcester Polytechnic Institute, Worcester, MA, USA
{cchen3,mfarmani,teisenbarth}@wpi.edu

**Abstract.** This work explores the possibilities for practical Threshold Implementation (TI) with only two shares in order for a smaller design that needs less randomness but is still first-order leakage resistant. We present the first two-share Threshold Implementations of two lightweight block ciphers—Simon and Present. The implementation results show that two-share TI improves the compactness but usually further reduces the throughput when compared with first-order resistant three-share schemes. Our leakage analysis shows that two-share TI can retain perfect first-order resistance. However, the analysis also exposes a strong second-order leakage. All results are backed up by simulation as well as analysis of actual implementations.

**Keywords:** Threshold implementation · Paired t-test · Lightweight cryptography · FPGA

## 1 Motivation

Protecting cryptographic hardware against side channel analysis is a difficult task and usually incurs significant area overheads. Especially masking schemes aimed at hardware have been found to be flawed or prone to implementation errors that leave the countermeasure at least partially insecure [13,20,23].

Threshold Implementation (TI) has become a popular masking scheme for hardware implementations in the recent years, due to several advantages over competing schemes. Unlike secure logic styles [20,32], it does not require a change of the design flow. TI is fairly simple to apply to a wide range of ciphers, and its implementation is not very error-prone, if a known set of requirements and best practices is followed. Another advantage is that TI actually keeps the promise of reliable first-order side-channel resistance. It also provides good protection against higher-order attacks [6,24].

However, like most other masking schemes, TI incurs large area and time overheads, and often consumes huge amounts of randomness for remasking, which can make practical application cumbersome. So far the best results have an area overhead of approximately three while consuming at least two times the

© International Association for Cryptologic Research 2016
J.H. Cheon and T. Takagi (Eds.): ASIACRYPT 2016, Part I, LNCS 10031, pp. 819–843, 2016.
DOI: 10.1007/978-3-662-53887-6_30

combined plaintext and key size of randomness per encryption. Such overheads— the significant increase in area as well as the need for a high-performance random number generator—make TI an expensive choice, too expensive for a broad range of practical applications. Reparaz et al. [27] generalized TI to provide protection against higher-order attacks. The work mentioned the feasibility of reducing the number of shares to $d+1$, where $d$ is the desired protection order, suggesting that two shares are sufficient for first-order side channel protection. A first evaluation of using $d+1$ shares for AES was performed by De Cnudde et al. in [15].

*Our contribution.* In this work we explore the practical implications of reducing the number of shares of threshold implementations to only two shares (2-TI). Such a reduction of shares enables implementations that only incur an area over- head of two and at the same time can also reduce the need of minimally required randomness by a factor of two, making the incurred cost more bearable and thus allowing side channel protection for a much wider range of applications. Reduc- ing the number of shares is easily possible by applying the non-completeness requirement of TI at the bit-level rather than the state-level, as done by prevail- ing implementations.

While the feasibility of this approach has already been discussed in [27] and recently been practically verified in [15], this work explores the practical aspects, the benefits—and ramifications—of applying threshold implementation with only two shares to modern ciphers. Our case study focuses on applying 2-TI on two lightweight block ciphers, Present [7] and Simon [2]. Lightweight ciphers are usually a good target for TI, as the algebraic depth of their nonlinear func- tions is usually low. Low algebraic depth allows for cheap and effective masking while keeping the need for additional randomness low. In fact, our designs do not require remasking during the round functions, while a comparable masked implementation of AES requires more than 8,000 fresh random bits during one block encryption [15].

Our study shows that two-share TI is first order secure and also reduces the size of the sequential logic in hardware implementations. The 2-TI-conversion of nonlinear functions is more cumbersome and usually requires at least one additional pipeline stage, with negative impact on implementation size and/or performance. However, we also expose a strong second-order leakage in both of the designs and argue that this is inherent to two-share TI implementations. We show that these leakages exist both in the theoretical model and can also be quickly exposed by leakage detection tests. We validate the exploitability of the observed leakages by side channel key recovery attacks.

The remaining work is structured as follows: Relevant terminologies and methods are explained in Sect. 2. The theoretical discussion of two-share TI is given in Sect. 3 and two practical implementations of Simon and Present are introduced in Sects. 4 and 5. Sections 6 and 7 present implementation results and the outcome of the leakage analysis and we conclude at Sect. 8.

# 2   Preliminaries

## 2.1   Lightweight Cryptography

For many embedded applications, area and hence power or energy minimal implementations of cryptography are highly desirable. This has led to a rich literature on hardware-minimal crypto cores, which often rely on the numerous proposed "lightweight" block cipher designs, such as Present, Katan, or Simon and Speck. These lightweight ciphers as well as the area-minimal implementations share one common characteristic: *Serialization*.

Serialized implementations are very common for minimizing area of hardware implementations at the expense of increased run time. Area-critical functions are identified and broken into subfunctions that can be applied repeatedly, in an iterative manner, to achieve the same outcome. A typical example for block ciphers is the S-box layer, which due to its high nonlinearity usually is difficult to minimize in hardware. A classical area-optimized implementation of an S-box based cipher only features a single S-box, which is iteratively applied to different parts of the intermediate state. All modern block ciphers support this *vertical* type of serialization by using a single S-box (unlike DES which uses 8 different S-boxes). Similar techniques are also applied to decrease the size of large S-boxes (or in general functions of great algebraic complexity), by breaking them into subfunctions that are concatenated. Examples include implementations that compute the AES S-box by exploiting tower field representations by Canright [9] or the Present S-box into mappings of algebraic degree 2, which eases side-channel protection and decreases the size, at the cost of doubling the computation time [26]. We will refer to this serialization as *horizontal*. While vertical serialization is determined by the cipher at implementation time (usually determined by the number of S-boxes), the exploitable horizontal serialization is determined by the algebraic complexity of the nonlinear layer.

Typical vertical serialization parameters for hardware minimal implementations are ranging from data path sizes of 8 bit for AES, 4 bit for Present down to 1 bit for e.g. Simon or Katan. That is, as little as one bit of the cipher state are updated per cycle. Serial data paths increase the latency of the crypto core significantly. However, they also allow to reduce the combinational logic of the crypto core to low single-digit percentages of the entire design [14,29]. That means, in applications where the latency is not critical, the area of a cipher is almost entirely determined by the registers storing the key and state. As a result, significant area-improvements can only be achieved by breaking the memory barrier, for example by externalizing key storage (cf. Ktantan [14]), or, for FPGAs, hiding state and key in dedicated bulk memory such as block RAMs [19] or shift registers [1]. Since the remainder of the work uses Present and Simon for proof-of-concept implementations, we provide more details on these two ciphers here.

## 2.2    Present

Present is a hardware-oriented block cipher proposed in 2007, optimized for low area footprint [7]. It is a substitution-permutation network featuring a 4 × 4 bit S-box and a permutation layer consisting only of bit shifts, making it low cost in hardware. It features a block size of 64 bits and a key size of 80 or 128 bits, and has 31 rounds. Present has been optimized for many application scenarios, but the area-minimal implementations with a 4-bit data-path. It has also been standardized as a lightweight cryptographic block cipher as ISO/IEC 29192-2:2012. Each round of Present cipher consists of three steps including a key-addition layer, a substitution layer which is a non-linear function, and a permutation layer. In the first step, the round key which is consisted of left most significant 64 bits of the key is xored with the 64-bit current state. In the next step, the Present S-box is used which is a non-linear 4-bit to 4-bit function shown in the following table in hexadecimal notation.

| $x$ | 0 | 1 | 2 | 3 | 4 | 5 | 6 | 7 | 8 | 9 | $A$ | $B$ | $C$ | $D$ | $E$ | $F$ |
|---|---|---|---|---|---|---|---|---|---|---|---|---|---|---|---|---|
| $S(x)$ | $C$ | 5 | 6 | $B$ | 9 | 0 | $A$ | $D$ | 3 | $E$ | $F$ | 8 | 4 | 7 | 1 | 2 |

The substitution layer can be performed with 16 parallel S-box or using only one S-box 16 times which depends on the application requirement. In the last step, the permutation is applied to all the 64-bit data which is just a rewiring.

At the same time, the key is updated in the key schedule part. The key can be 80-bit or 120-bit; however we use 80-bit key in this paper. In each round the 64 left most bits of the current key, $k_{79}k_{78}k_{77}...k_{17}k_{16}$, is used in addroundkey. After using the round key, the 80-bit key register is updated by shifting, using S-box, and xoring with round-counter. More details about the specification of the Present is provided in [7].

## 2.3    Simon

Simon is a lightweight block cipher proposed by NSA in 2013 [2]. Simon implements a Feistel structure that accepts two $n$-bit words as input plaintext, with $n \in \{16, 24, 32, 48, 64\}$. For each input size $2n$, Simon has a set of allowable key sizes ranging from 64 bits to 256 bits. The number of rounds in Simon ranges from 32 to 72 rounds. Simon128/128, which can be seen as a drop-in replacement for AES-128, accepts 128 bits of plaintext at a word size of 64 bits and 128 bits of key. It generates a ciphertext after 68 rounds. The Simon128/128 parameter set will be used throughout this work, though the implementation strategies apply to other parameter sets in a natural way.

We denote the input words of round $i$ as $l_i$ and $r_i$. Then the output words are given as:

$$r_{i+1} = l_i$$
$$l_{i+1} = r_i + l_i^2 + (l_i^1 * l_i^8) + k_i \tag{1}$$

The upper index in $l_i^s$ indicates left circular shift by $s$ bits. The addition and the multiplication are in $GF(2)$ and equivalent to bitwise XOR and AND operations, respectively. Given the initial key words $k_0$ and $k_1$ (and possibly $k_2$ and $k_3$, depending on the key size), which are also used as first round keys, the subsequent round keys are computed as:

$$
\begin{aligned}
k_{i+2} &= k_i + k_{i+1}^{-3} + k_{i+1}^{-4} + c_i \quad \text{Two and Three Words} \\
k_{i+4} &= k_i + k_{i+1} + k_{i+1}^{-1} + k_{i+3}^{-3} + k_{i+3}^{-4} + c_i \quad \text{Four Words}
\end{aligned}
\tag{2}
$$

where $c_i$ is a round constant.

## 2.4 Masking

Masking is a common technique to prevent side channel leakage [10]. Sensitive states of a cryptographic implementation are split into shares by adding randomness. In an additive masking scheme, a variable $x$ is split into $s$ shares $x_i$ with $i \in \{0, 1, \ldots, s-1\}$ by choosing $x_{i>0}$ uniformly at random and $x_0 = x + \sum_{i=1}^{s-1} x_i$. These shares are then processed separately, ensuring that the sensitive state is never presented in the system, and—more importantly—that processed states are independent of the secret.

## 2.5 Threshold Implementation

Threshold Implementation (TI) was proposed by Nikova et al. [25] as a side-channel countermeasure to address the common problem of *glitches* that resulted in leakage for many other theoretically sound countermeasure techniques when applied to hardware. The original proposal only deals with protection against first-order side-channel leakages. Threshold Implementation has found widespread adoption in the academic community: several implementations of symmetric [5,6,24,26,31] and even asymmetric crypto algorithms [11,28] have been successfully protected with TI. Recently, TI has been expanded to protect against higher-order attacks as well [4], though potential pitfalls of the scheme in the multivariate setting have been pointed out and fixed in [27].

TI combines a set of three requirements with a constructive description of how to convert an algorithm into a side-channel resistant implementation in the presence of glitches. Sensitive states are converted into a shared representation by applying an additive Boolean masking, i.e., adding randomness. Functions $F(\cdot)$ are converted meeting the requirements of correctness, uniformity, and non-completeness.

– **Uniformity** requires all intermediate states (shares) to be uniformly distributed. Uniformity is intended to ensures the mean leakages to be state-independent, a key requirement to thwart first-order DPA. To ensure uniformity in a circuit it suffices to ensure uniformity for the output share of each function, as well as for the inputs of the circuit.

- **Non-Completeness** requires subfunctions $f_i$ of a shared function $F$ to be independent of at least one input share for first-order SCA resistance. That is, a function $F(x)$ shall be split into subfunctions $f_i(x_{j\neq i})$. This requirement was updated in [4] to require any $d$ subfunctions to be independent of at least one input share to achieve $d$-th order SCA resistance. Non-completeness ensures that the final circuit is not affected by glitches. Since glitches can only occur in subfunctions $f_i$, and each subfunction has insufficient knowledge to reconstruct a secret state (since it has no knowledge of at least one share $x_i$), no leakage can be caused by glitches.
- **Correctness** simply states that applying the subfunctions to a valid shared input must always yield a valid sharing of the correct output.

In the classic approach, a function of algebraic degree $t$ can be implemented using at least $t+1$ input shares for first order side-channel resistance, and $td+1$ for $d$-th order resistance [4,27]. In practice, virtually all implementations try to keep the number of shares low, i.e. for first order-protected designs at or close to 3. As a consequence, implementations of algebraically more complex functions need to be broken into algebraically simpler subfunctions. The described TI conversion always ensures correctness and non-completeness. Uniformity can be either achieved by using more input shares or by adding randomness during the computation. As a result, many of the published implementations, in order to reduce the size of the circuit, consume lots of randomness, up to thousands of bits per encrypted block.

## 2.6    Leakage Detection

A side channel leakage detection method based on Welch's t-test has been recently gaining popularity due to its simplicity, efficiency and reliability. The test procedures have been well studied in [12,30] and is often referred to as Test Vector Leakage Assessment (TVLA) test. Unlike other attacks or leakage models used for key recovery, TVLA only returns a confidence level to reject the leakage-free hypothesis and fail the device under test. Essentially, a t-statistic is calculated using two sets of leakage samples as:

$$t = \frac{\mu_A - \mu_B}{\sqrt{(\sigma_A^2/N_A) + (\sigma_B^2/N_B)}} \tag{3}$$

where $A$ and $B$ denote the two sets and $N_j$ denotes the number of traces in set $j \in \{A, B\}$. $\mu_j$ and $\sigma_j$ are the sample mean and sample variance respectively. The two sets of measurements are obtained with either fixed versus random plaintext (in a *non-specific t-test*) or random versus random plaintext (in a *specific t-test*). In our work we use the *non-specific t-test* since it does not depend on any intermediate value and power model. When the value of $t$ exceeds a certain threshold, the null hypothesis can be rejected with a small Type I error probability $p$. In this paper, we follow the threshold of $\pm 4.5$ used in [18,22].

An improved methodology based on paired t-test was suggested in [16]. The test uses matched pairs from the two sets of measurements. The advantage of this methodology is that common noise to both measurements can be rejected, making the test much more robust to slow changes of operating points in long measurement campaigns. When $n$ such pairs of measurements are obtained, we have $n$ difference measurements $D = L_A - L_B$ where $L_A$ is a random variable representing samples from set $A$ while $L_B$ from set $B$. The paired difference cancels the noise variation and makes it easier to detect nonzero population difference. Now, the null hypothesis becomes mean difference $\mu_D = 0$ instead of $\mu_A = \mu_B$. Let $\bar{D}$ and $s_D^2$ denote the sample mean and sample variances of the paired differences $D_1, ..., D_n$. The paired t-test statistic is calculated as:

$$t_p = \frac{\bar{D}}{\sqrt{\frac{s_D^2}{n}}}, \tag{4}$$

The null hypothesis of non-leakage is also rejected if $|t_p|$ exceeds the threshold of 4.5.

With respect to higher order leakage detection, the original traces should be preprocessed as explained in [30]. For example in a second order t-test, the traces - at each sample points independently - are mean free squared beforehand. Usually, the global mean of all samples at each time point is used. However, as suggested in [16], a moving average which is the average of neighboring traces around each trace is used instead to mitigate the environmental effects. In our experiments, we apply both tests, the classic TVLA test as well as the paired T-test, the latter one with moving averages for higher-order analysis.

# 3   Threshold Implementation with Two Shares

While the constructive approach by Nikova et al. allows to implement any $d$-th order algebraic functions in a straightforward way, actual implementations requiring to share functions of degree greater than 2 have put significant effort into keeping the number of shares as close as possible to three, which is perceived as the minimum possible to implement nonlinear functions, until [27][1]. In particular, [21] discussed the efficient implementation of 4-bit S-boxes with three shares. Similarly, the current TIs of AES utilize the algebraic structure of the AES S-box and four [24] or variable with up to five shares [6] to implement the S-box on a small area.

A natural question is: *Why to stop at three shares?* If small area is desirable, using similar techniques as the ones used by the above papers could enable TIs with just two shares, further reducing the area footprint as well as the need for randomness. This approach was already discussed in [27]. The approach is straightforward for the linear operations of an implementation, and has already

---

[1] It should be noted that [31] also proposed a two-share TI version of Simon, with the requirement of manually preventing glitches for two parts of the equation.

been widely used in several TIs for those parts [3,6,11]. The simplest nonlinear operation is a simple two-input and: $c = ab$ which can be processed with two shares as

$$c_0 = a_0 b_0 \qquad c_1 = a_1 b_1 \qquad c_2 = a_0 b_1 \qquad c_3 = a_1 b_0 \qquad (5)$$

This equation is in violation of the common interpretation of the non-completeness requirement, since $c_2$ and $c_3$ mix inputs from shares with different indices. However, non-completeness is not violated as long as $a$ and $b$ are statistically independent.

Equation (5) suggests a 4-share output, which is undesirable for a minimal implementation. To keep the number of shares low, the four shares $c_i$ can be recombined in the next cycle, e.g. $c_0' = c_0 + c_2$ and $c_1' = c_1 + c_3$. However, since the recombination would violate non-completeness, it must happen after a register-stage in the next clock cycle. In other words, a pipelining stage becomes necessary, increasing the register count and the delay of the output. The share proliferation gets worse for higher-degree algebraic functions, as stated in [27]. However, hardware-minimal implementations break higher-order algebraic functions into degree-minimal building blocks anyway, making share proliferation a theoretical concern only.

To also ensure uniformity and thus gain an implementable basic nonlinear building block, we implement $z = ab + c$ in two pipeline stages as

$$z_0' = a_0 b_0 + c_0 \qquad z_1' = a_1 b_1 + c_1 \qquad z_0 = z_0' + a_0 b_1 \qquad z_1 = z_1' + a_1 b_0 \qquad (6)$$

Note that $z_i'$ and $z_i$ are computed in separate cycles. Conveniently, the $z_i'$ and $z_i$ are uniform. Furthermore, this computation order only needs to store 2 intermediate states (unlike Eq. (5)). However, this assumes that the inputs are available in two subsequent clock cycles, which is a valid assumption in many serialized implementations. Either way, the resulting pipelining of the nonlinear function increases area overhead of that function, and also introduces a latency according to the number of pipeline stages needed. Most of this latency can be hidden if the data path of the implementation is small enough.

## 3.1    Potential Pitfalls

*Share rotation.* In [26] it was suggested to rotate the shares in every step to achieve increased side channel resistance. With two shares, this is highly dangerous: if $s_0$ overwrites $s_1$, the resulting leakage is likely to depend on both shares, hence has a direct dependence on the secret itself. In general, any register updates must be handled with great care.

*Increased Higher-order leakage.* The observed higher order leakage can be explained by the significant dependende of the variance on the value of the share $x$. For a simple example we compare a 2-sharing $S_2$ and a 3-sharing $S_3$ of a bit $x$ into $S_2(x) = \langle x_0, x_1 \rangle$ and $S_3(x) = \langle x_0, x_1, x_2 \rangle$ respectively. We further assume

**Table 1.** Comparison of leakage for a 2-sharing ($S_2$) and 3-sharing ($S_3$) of a bit $x$ in a Hamming weight model. The 2-sharing ($S_2$) shows a leakage in the variance $\sigma(S_2)$.

| $x$ | $S_2(x)$ | $S_3(x)$ | $wt(S_2)$ | $wt(S_3)$ | $\mu(S_2)$ | $\mu(S_3)$ | $\sigma(S_2)$ | $\sigma(S_3)$ |
|---|---|---|---|---|---|---|---|---|
| 0 | $\{00, 11\}$ | $\{000, 011, 101, 110\}$ | $\{0, 2\}$ | $\{0, 2, 2, 2\}$ | 1 | $3/2$ | **2** | 1 |
| 1 | $\{01, 10\}$ | $\{001, 010, 100, 111\}$ | $\{1, 1\}$ | $\{1, 1, 1, 3\}$ | 1 | $3/2$ | **0** | 1 |

a Hamming weight ($wt(\cdot)$) leakage on the shares. Table 1 lists the possible states and the resulting means and variances for both sharings.

As proper TI sharings of $x$, the mean leakage $\mu(S_i)$ is independent of the value of $x$. However, the variance of $S_2$ depends on $x$, in particular var($S_2(x = 0)$) $= 2 \neq 0 = $ var($S_2(x = 1)$). This is not true for the 3-sharing $S_3$, where the variances in both cases are identical as well. This is a strong indication why 2-sharings may have a strong second-order leakage. This was also observed for partial 2-share implementations in [3] and will be demonstrated for full 2-share implementations in the analysis of our reference implementations in Sect. 7.

# 4   Application to Simon

Threshold Implementations of Simon with three shares have been proposed in [31] to counteract first-order side channel attacks. Moreover, their bit-serialized implementation only consumes 87 slices on Spartan-3 xc3s50 FPGA which renders it the smallest threshold implementation of a block cipher. The authors also discussed how the requirement of *non-completeness* shuts the door on a two-share hardware implementation of Simon but not on software implementations.

In this section, we at first apply serialization technique in order to realize a two-share TI Simon on hardware. The leakage detection analysis and implementation results will be presented in Sects. 6 and 7.

## 4.1   Simon with Two Shares

We follow the notation used in [31] to describe the cipher. The input plaintext is initially split into two shares as:

$$
\begin{aligned}
r[a]_0 &= m[p][1] \\
l[a]_0 &= m[p][2] \\
r[b]_0 &= m[p][1] + r_0 \\
l[b]_0 &= m[p][2] + l_0
\end{aligned}
\tag{7}
$$

Where $r$ and $l$ represents the two input words, $a$ and $b$ denote two shares of the variables and subscript $i$ indicates the round of encryption. $m[p][1]$ and $m[p][2]$ are two fresh random values that mask the plaintext in the very beginning of the algorithm and no more random numbers are needed for the rest operations. Then, the round function is denoted as:

$$
\begin{aligned}
r[a]_{i+1} &= l[a]_i \\
l[a]_{i+1} &= r[a]_i + l[a]_i^2 + l[a]_i^1 * l[a]_i^8 + l[a]_i^1 * l[b]_i^8 + k[a]_i \\
r[b]_{i+1} &= l[b]_i \\
l[b]_{i+1} &= r[b]_i + l[b]_i^2 + l[b]_i^1 * l[b]_i^8 + l[b]_i^1 * l[a]_i^8 + k[b]_i
\end{aligned}
\tag{8}
$$

Where the superscripts $1, 2, 8$ on $l[*]_i$ represent left circular shift by corresponding numbers of bits. (Notice that both addition and multiplication are in $GF(2)$). Obviously, the computations of $l[a]_{i+1}$ and $l[b]_{i+1}$, if directly mapped into combinational circuits, are not *non-complete* since the two shares $l[a]_i^8$ and $l[b]_i^8$ are present in the same circuit and glitches may still cause leakage. We can serialize the above equations by enforcing them being executed in two steps rather than one. That is, we first compute the intermediate values $l[a]_{i+1,int}$ and $l[b]_{i+1,int}$ using only half of the terms in the equations as follows:

$$
\begin{aligned}
l[a]_{i+1,int} &= r[a]_i + l[a]_i^2 + l[a]_i^1 * l[a]_i^8 \\
l[b]_{i+1,int} &= r[b]_i + l[b]_i^2 + l[b]_i^1 * l[b]_i^8
\end{aligned}
\tag{9}
$$

Then, the round outputs can be further calculated as:

$$
\begin{aligned}
l[a]_{i+1} &= l[a]_{i+1,int} + l[a]_i^1 * l[b]_i^8 + k[a]_i \\
l[b]_{i+1} &= l[b]_{i+1,int} + l[b]_i^1 * l[a]_i^8 + k[b]_i
\end{aligned}
\tag{10}
$$

The serialization not only retains both *correctness* and *uniformity* but achieves *non-completeness* as well. In Eq. (9), the inputs $r[a]_i$, $l[a]_i^2$, $r[b]_i$ and $l[b]_i^2$ are all uniform and therefore the output intermediates are also uniform. Each function is independent of one share of every input and hence is *non-complete*. Similarly, Eq. (10) also satisfies the three requirements. Correctness can be easily proved by substituting $l[a]_{i+1,int}$ and $l[b]_{i+1,int}$ with Eq. (9). The uniformity of inputs $k[a]_i$ and $k[b]_i$ makes the outputs uniform too. Moreover, each function is independent of one share of every input and thus the functions are *non-complete* as well. One may argue that $l[a]_i^1$ and $l[b]_i^8$ (or $l[b]_i^1$ and $l[a]_i^8$) are two shares of $l_i$ with different rotations and may leak information of $l_i$. However, the multiplication between them is in $GF(2)$ and is equivalent with bitwise AND operation. Further, in order to ensure the *non-completeness*, "Keep Hierarchy" property of synthesize tool (ISE with XST) is enabled to separate the LUTs for $AND$.

## 4.2    Round-Based Implementation

Figure 1 depicts the structure of a FPGA implementation which contains two copies of the same data-path which consists of two registers $L_j$ and $R_j$ and the combinational circuits for round functions. Specifically, two clock cycles are taken to process each round operation. In the first clock cycle, the round inputs

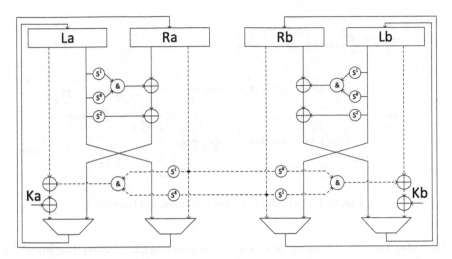

**Fig. 1.** Data-path of the simon with two shares. Solid line: First clock cycle; Dashed line: Second clock cycle

are evaluated with Eq. (9) and then the intermediates are overwritten back into the registers as illustrated by the solid lines in the figure. Note that $r[j]_{i+1} = l[j]_i$ is stored in $R_j$ while $l[a]_{i+1,int}$ is in $L_j$. Then, in the second clock cycle, Eq. (10) is evaluated as shown by the dashed line but remember that since $l[j]_i$ is now stored in $R_j$ and hence no extra buffer is needed for it.

The sharing of key schedule is not presented here since it consists of linear operations only and is trivial to implement.

### 4.3  Bit-Serialized Implementation

In order to fairly compare with the bit-serialized 3-TI Simon introduced in [31] and achieve a even smaller size of Simon implementation, a bit-serialized 2-TI Simon is constructed as depicted in the Fig. 2 (Only one share is shown).

Our design originates from the FIFO-based 3-TI bit-serialized in [31] but introduces new features in order for a 2-TI architecture.

First of all, the round function is adjusted according to Eqs. 9 and 10. (Note that both equations are evaluated in bits instead of the whole word in this case.) Therefore, as shown in the **LUT** part of Fig. 2, a one-bit register is inserted to hold the intermediate value $l[a]_{i+1,int}$ so that $l[a]_i^8$ and $l[b]_i^8$ will not be combined to cause leakages mistakenly.

Second, due to the insertion of this register, it will take two clock cycles for **LUT** to perform round operation for each bit. However, by using pipeline technique, the overall throughput will not be scarified too much. In fact, the 2-TI architecture processes all 64 bits within 65 clock cycles which is only one more than 3-TI in [31]. In order to achieve this, the FIFOs and shifted registers are designed to work as following.

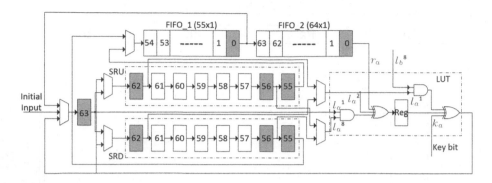

**Fig. 2.** Data-path of the bit-serialized 2-TI Simon

- Initially, the 128-bit block is stored in register #63, Shifted Registers Up (SRU) #62 to #55, FIFO_1 and FIFO_2.
- Once Encryption started, the values are right shifted and in the mean time bits in register #63, #62 and #56 as well as bit 0 in FIFO_2 are fed into **LUT** for logic operation.
- The output will be written back to Shifted Registers Down (SRD). Note that the valid outputs are generated since the second clock cycle. And then, after 64 clock cycles, the first 63 output bits are stored in Shifted Registers Down (SRD) #62 to #55, FIFO_1 and FIFO_2. In the last (65th) clock cycle, the final output bit will be written in register #63. Therefore, the whole round operation is done within 65 clock cycles.

## 5 Application to Present

In this section, we apply two-share Threshold Implementation to the Present cipher. In [21], the authors presented the 3-TI Present S-box. To achieve this, they decomposed the non-linear S-box of degree 3 into the combination of two quadratic functions—$G$ function—plus some linear functions, and then implement them with three shares. We follow their idea to use the same decomposition but then implement them with 2-TI while still retaining *uniformity*, *non-completeness*, and *correctness*. According to [21], the S-box of Present can be decomposed as:

$$S(X) = A(G(G(G(BX \oplus c)) \oplus d) \tag{11}$$

Where $G(.)$, $A$, $B$, and the constant vectors of $c, d$ are given as follows:

$$\begin{aligned}
G(x, y, z, w) &= (g_3, g_2, g_1, g_0) \\
g_3 &= x + yz + yw \\
g_2 &= w + xy \\
g_1 &= y \\
g_0 &= z + yw
\end{aligned} \tag{12}$$

$$A = \begin{bmatrix} 1&0&1&0 \\ 0&1&0&0 \\ 1&0&0&0 \\ 1&0&1&1 \end{bmatrix}, \quad B = \begin{bmatrix} 1&1&0&0 \\ 0&1&1&0 \\ 0&0&1&0 \\ 0&1&0&1 \end{bmatrix}, \quad c = \begin{bmatrix} 0&0&0&1 \end{bmatrix}, \quad d = \begin{bmatrix} 0&1&0&1 \end{bmatrix} \tag{13}$$

## 5.1   Present with Two Shares

A 2-sharing scheme of $G(.)$ can be expressed as follows546:

$$G_0(x_0, y_0, z_0, w_0, x_1, y_1, z_1, w_1) = (g_{03}, g_{02}, g_{01}, g_{00})$$
$$g_{03} = x_0 + y_0 z_0 + y_0 z_1 + y_0 w_0 + y_0 w_1$$
$$g_{02} = w_0 + x_0 y_0 + x_1 y_0 \tag{14}$$
$$g_{01} = y_0$$
$$g_{00} = z_0 + y_0 w_0 + y_0 w_1$$

$$G_1(x_0, y_0, z_0, w_0, x_1, y_1, z_1, w_1) = (g_{13}, g_{12}, g_{11}, g_{10})$$
$$g_{13} = x_1 + y_1 z_0 + y_1 z_1 + y_1 w_0 + y_1 w_1$$
$$g_{12} = w_1 + x_0 y_1 + x_1 y_1 \tag{15}$$
$$g_{11} = y_1$$
$$g_{10} = z_1 + y_1 w_0 + y_1 w_1$$

The above sharing satisfies both *correctness* and *uniformity* when the input shares are uniformly distributed. However, *non-completeness* is not fulfilled since two shares of the same inputs are fed into the same functions in some of the above equations.

As before, we serialize the computations into two steps in order to achieve *non-completeness* as illustrated in the following equations.

$$G_0^1(x_0, y_0, z_0, w_0) = (g_{03}^1, g_{02}^1, g_{01}^1, g_{00}^1)$$
$$g_{03}^1 = x_0 + y_0 z_0 + y_0 w_0$$
$$g_{02}^1 = w_0 + x_0 y_0 \tag{16}$$
$$g_{01}^1 = y_0$$
$$g_{00}^1 = z_0 + y_0 w_0$$

$$G_0^2(x_1, y_0, z_1, w_1, g_{03}^1, g_{02}^1, g_{01}^1, g_{00}^1) = (g_{03}^2, g_{02}^2, g_{01}^2, g_{00}^2)$$
$$g_{03}^2 = g_{03}^1 + y_0 z_1 + y_0 w_1$$
$$g_{02}^2 = g_{02}^1 + x_1 y_0 \tag{17}$$
$$g_{01}^2 = g_{01}^1$$
$$g_{00}^2 = g_{00}^1 + y_0 w_1$$

$$G_1^1(x_1, y_1, z_1, w_1) = (g_{13}^1, g_{12}^1, g_{11}^1, g_{10}^1)$$
$$g_{13}^1 = x_1 + y_1 z_1 + y_1 w_1$$
$$g_{12}^1 = w_1 + x_1 y_1 \tag{18}$$
$$g_{11}^1 = y_1$$
$$g_{10}^1 = z_1 + y_1 w_1$$

$$G_1^2(x_0, y_1, z_0, w_0, g_{13}^1, g_{12}^1, g_{11}^1, g_{10}^1) = (g_{13}^2, g_{12}^2, g_{11}^2, g_{10}^2)$$
$$g_{13}^2 = g_{13}^1 + y_1 z_0 + y_1 w_0$$
$$g_{12}^2 = g_{12}^1 + x_0 y_1 \tag{19}$$
$$g_{11}^2 = g_{11}^1$$
$$g_{10}^2 = g_{10}^1 + y_1 w_0$$

The superscript indicates the level of the circuit. Until now, we achieved a *correct*, *non-complete* and *uniform* two-share implementation of $G(.)$. The conversion of the remaining linear operations is discussed next.

### 5.2  Hardware Implementation

As depicted in Fig. 3, in order to provide the *non-completeness* to the design, we use registers to separate the two parts of the $G$. The second part of the shares ($G_0^2$ and $G_1^2$) use not only the outputs of the first part of the shares ($G_0^1$ and $G_1^1$) but also some of their inputs as well (depicted in Fig. 3). One 6-bit register and two 4-bit registers are used before the second part of the G module, to store the inputs $x_0$, $x_1$, $z_0$, $z_1$, $w_0$, and $w_1$; and the outputs of the first part of the G module, respectively.

In Fig. 4, the S-box architecture is depicted which includes two G modules, and functions $BX + c_0$ and $AX + d_0$ for the first share as well as functions $BX + c_1$ and $AX + d_1$ for second share in which $c_0 + c_1 = c$ and $d_0 + d_1 = d$. Furthermore, due to *non-completeness*, we use another row of registers in between two $G(.)$ functions in the S-box. One may argue that registers should also be inserted between non-linear functions (e.g. $G(.)$) and linear functions (e.g. $AX + d_0$), since when they are merged the two shares of certain variables may be combined again which fails the *non-completeness* requirement. While this is true in general cases, our design avoids this problem as $G_0^2$ and $G_1^2$ are both independent of one share of the inputs and hence any linear combination of $g_{13}^2, g_{12}^2, g_{11}^2, g_{10}^2$ or $g_{03}^2, g_{02}^2, g_{01}^2, g_{00}^2$ still satisfies *non-completeness*.

Figure 5 shows the whole Present cipher with two shares. The design includes two control inputs namely key_load and data_load. If key_load is high, at the rising edge of the clock signal, the 80-bit input key shares-Key A and Key B- are copied to the registers Key A and Key B respectively. When the data_load signal is high, at the rising edge of the clock signal, 64 right-most significant bits of the input shares (data_in A[63:0], data_in B[63:0]) are copied to state registers. It is worth mentioning that when the data_load is set, i.e. loading new two shares of plaintext into the state registers results in a

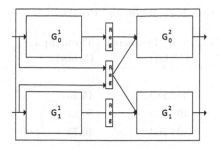

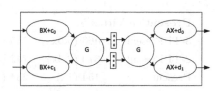

**Fig. 3.** Hardware architecture of the 2-share G module

**Fig. 4.** Hardware architecture of the 2-share S-box module

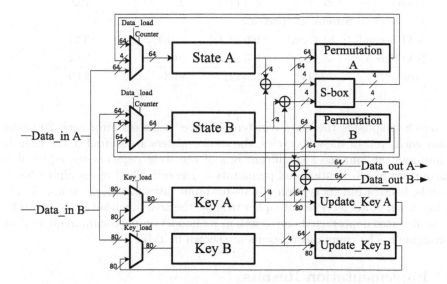

**Fig. 5.** Hardware architectures of the 2-shares Present Cipher.

reset of the state machine. That why this design does not have a reset signal. When the two-share keys and two-share plaintexts are loaded, both key_load and data_load must be set to zero. After that, it takes 31 rounds in order to Data_out A and Data_out B have a valid ciphertexts. In each round, the S-box and permutation operations respectively operate the inputs to update the state registers for the next round. Considering the hardware design, each $G(.)$ function needs one cycle and then every S-box needs four clock cycles to compute table lookup. According to the Fig. 5, each 64-bit input stored in the State register needs to use S-box 16 times. Hence, it needs 4 clock cycles for the first S-box due to its latency, plus 15 clock cycles for other 15 S-boxes in pipeline, also one more clock cycle for the permutation operation. Therefore, we need 20 cycles for each round of the Present cipher. Hence, we define another control signal, 'counter',

**Table 2.** Implementation results of two-share Simon and Present.

| Design | Slice (Regs) | Slice (LUTs) | Max. Frequency (MHz) | Throughput (Mbps) |
|---|---|---|---|---|
| **Present** on Virtex 5 | | | | |
| 3-TI Present | 466 (3.0x) | 715 (3.1x) | 397.289 | 45.567 |
| 2-TI Present | 370 (2.4x) | 742 (3.2x) | 490.252 | 50.61 |
| Present | 154 (1x) | 234 (1x) | 394.563 | 40.73 |
| Round-based **Simon** on Virtex 5 | | | | |
| 3-TI Simon | 777 (2.8x) | 1302 (2.8x) | 414 | 779 |
| 2-TI Simon | 520 (1.9x) | 1169 (2.5x) | 382 | 360 |
| Simon | 272 (1x) | 473 (1x) | 421 | 792 |
| Bit-serialized **Simon** on Spartan 3 | | | | |
| 3-TI Simon [31] | 61 (2.0x) | 160 (2.2x) | 109.4 | 3.21 |
| 2-TI Simon | 55 (1.8x) | 135 (1.9x) | 91.1 | 2.64 |
| Simon [1] | 30 (1x) | 72 (1x) | 91.4 | 2.69 |

in which it updates the state registers and Key registers after each 20 cycles. After each cycle of these 20 cycles, the state registers are shifted to the right by 4 bits and the four most significant bits of the state registers are replaced by the outputs of substitution and permutation network. The Present cipher has 31 rounds, hence a full encryption of a 64-bit input takes 620 clock cycles. We also design an unprotected Present cipher to show the area overhead of the protected Present versus unprotected one as well as its impact on maximum frequency and throughput. The comparison results are shown in Table 2.

# 6    Implementation Results

Table 2 summarizes the overhead and performance of two-share implementations of both ciphers. Note that we only implement Simon128/128 and Present64/80 as an example to show the advantage of two-share scheme. All the designs are implemented in Verilog and synthesized for Virtex-5 (xc5vlx50) or Spartan-3 (xc3s50) using XST.

For round-based Simon, we have three different implementations: unprotected, 2-TI and 3-TI. In terms of slice registers used, two-share TI implementation costs twice as much as the unprotected one and one third less than the 3-TI implementation. This is not surprising since increasing by one share will consume one more copy of registers to store the new share. Similarly, number of LUTs also increases. However, each round operation in 2-TI costs double clock cycles and therefore the throughput is greatly reduced compared with the other two designs.

We also implement bit-serialized 2-TI Simon to compare with the currently smallest block cipher designs for FPGAs, as given in [1], as well as its first-order protected 3-TI version from [31]. As shown in Table 2, our 2-TI design reduces the area overhead when compared to the 3-TI by about 13 %, i.e., cannot quite reach the optimal reduction of 33 % due to the pipelining overhead and the unaffected control logic. Nevertheless, this yields the smallest first-order protected block cipher design for FPGAs with the same parameters as AES-128.

With respect to Present, we have three implementations: Unprotected, Regular 3-TI, and the new 2-TI Present. In terms of slice registers used, regular 3-TI implementation used more than three times of the unprotected one. This is because we should use extra registers to guarantee the *non-completeness* of first-order resistant three-share Present cipher. Also, two-share implementation costs more than two times of unprotected Present because of the same reason mentioned before. Moreover, it is worth mentioning that the 2-TI first order resistant implementation uses less registers than 3-TI. For example, we use extra registers in $G(.)$ function as explained in Sect. 5. These registers help reducing the critical path, which explains the speed-up and resulting increase in throughput for 2-TI Present.

# 7   Leakage Analysis

In this section, we extend the discussion of a strong second-order leakage of two-share TI scheme, which was already described in Sect. 3.1, using simulation based leakage and the measurements from our reference implementations.

## 7.1   Theoretical Analysis

First we discuss the strong second-order leakage of two-share TI scheme using two-share Present S-box look-up as a target, namely the key-dependent intermediate value $y = S(x \oplus k)$ where $x, y, k$ are 4-bit input plaintext, S-box output and sub-key receptively.

*Synthetic samples and leakage model.* First, we generate noise free synthetic leakage samples of the 2-TI Present S-box based on Hamming weight model. As shown in Sect. 5, a 2-TI S-box processes two shares (4 bits for each share) in parallel and hence we use the Hamming weight of both output shares (8 bits in total) as the synthetic leakage samples. Further, in order for a second order analysis, the synthetic data should be center-and-then-squared. With respect to the leakage model, we use the Hamming weight of the regular S-box output which equals the bitwise XOR between the two output shares in the 2-TI S-box.

*First-order analysis.* We perform first-order *non-specific paired t-test* on the synthetic data and attempt to exploit any leakage using classic CPA as well. For this purpose, 1 million synthetic leakage samples for random input plaintext are generated as well as another 1 million for fixed inputs. The result of t-test using the 2 million samples is shown in Fig. 6(a) where the t value is less than 2 as the

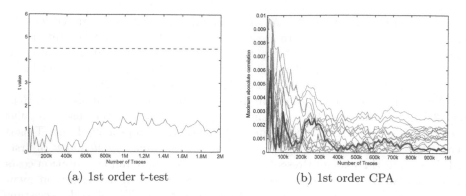

(a) 1st order t-test                    (b) 1st order CPA

**Fig. 6.** First-order leakage analysis of synthetic data. Left: first-order paired t-test. Right: first-order CPA; Red line corresponds to the correct key guess

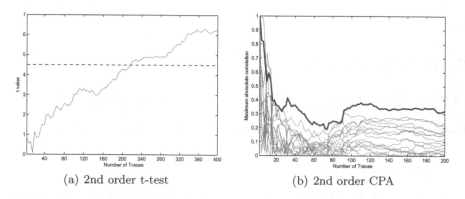

(a) 2nd order t-test                    (b) 2nd order CPA

**Fig. 7.** Second-order leakage analysis of synthetic data. Left: second-order paired t-test. Right: second-order CPA; Red line corresponds to the correct key guess

number of traces (synthetic samples) increases to 2 million. Then, a classic first-order CPA is performed on the 1 million samples associated with the random inputs using the above-mentioned leakage model. The results in Fig. 6(b) shows the correct key cannot be distinguished from the wrong key hypotheses with as much as 1 million samples and the attacks fail.

*Second-order analysis.* Then, we proceed with second-order *non-specific paired t-test* and CPA. For this purpose, 200 synthetic leakage samples for random input plaintext are generated as well as another 200 for fixed inputs. Figure 7(a) shows that t value exceed 4.5 with only a couple of hundreds of samples while classic CPA can recover the correct key with less than a hundred samples as shown in Fig. 7(b).

In summary, the theoretical analyses also show the first-order resistance of 2-TI scheme but reveals a strong second-order leakage. This strong second-order leakage is caused by the differing variances, as pointed out in Sect. 3.1. Note that we use perfect Hamming weight model for synthetic data without adding any noise. Hence, the CPA with a Hamming weight model can efficiently recover the key because it captures the leakage well. In fact, CPA on a perfect Hamming weight leakage is comparable to a profiled attack, in the absence of noise. But in the real world, actual leakages are more complex and CPA with Hamming weight model will not be as efficient as in this synthetic scenario. In the following we will conduct analysis on practical implementations to show this.

## 7.2   Practical Analysis

Next, we discuss the leakage analysis results for the two-share implementations of round-based Simon and Present. First, we apply the *non-specific* paired t-test method from [16] to detect any data-dependent leakage. Fixed (F) and random (R) measurements are interleaved using the FRRF pattern. Also, leakage detection tests are performed on round-based 3-TI Simon in order to compare with 2-TI and show the first-order leakage resistance of two-share scheme. Then, classic CPA is performed in order to exploit the second-order leakage detected by t-test and the results comply with the simulations in Sect. 7.1.

The analyzed implementations are ported into a Virtex-5 xc5vlx50 FPGA on the SASEBO-GII board clocked at 3 MHz. Measurements are taken using a Tektronix DPO-5104 oscilloscope which collects measurements with sample rate of 100 MS/s. The oscilloscope features a *FastFrame* functionality that can capture encryptions in bulk and thus 10 million measurements for each implementation can be taken in several hours.

*Round-based 2-TI Simon.* For two-share Simon implementation, 10 million measurements are collected, yielding 5 million fixed-random pairs. Each measurement contains 5000 time samples, covering the 68 rounds of Simon. The first-order paired t-test is performed using $n = 5000, 10000, 15000, \ldots$ pairs. Figure 8(a) shows the first order t-test result on the two-share Simon. The maximum absolute $t$ value across the 5000 time samples remains below the threshold of 4.5 with 10 million traces. We conclude that the two-share Simon implementation is resistant against first-order DPA and thus a validly implemented threshold implementation.

The results of the second order paired t-test are shown in Fig. 8(b). The step size is reduced to $n = 100, 200, \ldots$ to magnify the relevant area: The $t$ value of the second order analysis grows beyond 4.5 with about 500 traces. That is, a second order leakage is detectable with just hundreds of traces.

*Round-based 3-TI Simon.* In order to practically compare the performance of 2-TI and 3-TI in resisting first-order and second-order leakage, the paired t-test is also applied to 10 million FRRF measurements from a round-based 3-TI Simon. Figure 9(a) shows similar result as in Fig. 8(a) and the $t$ value is below the threshold of 4.5. The comparison shows again that the first-order resistance of 2-TI is solid as a 3-TI. However, 3-TI exhibits resistance against second-order analysis as shown in Fig. 9(b) and the $t$ value is still below 4.5 with

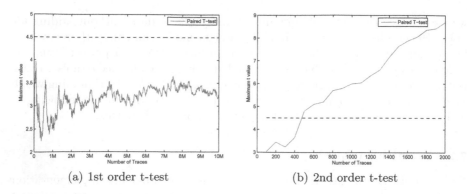

Fig. 8. Leakage detection results for the two-share implementation of Simon for first order (left) and second order (right) leakage over the number of traces. Note that the dimensions change for both axes.

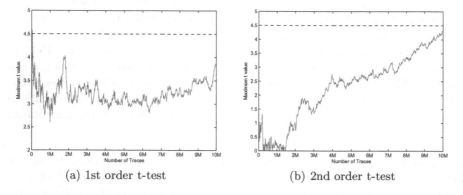

Fig. 9. Leakage detection results for the three-share implementation of Simon for first order (left) and second order (right) leakage over the number of traces.

10 million traces. That is, given more than 1000x as many measurements as for the 2-TI case, the leakage is just barely detectable. The results comply with the simulation analyses in Subsects. 3.1 and 7.1 and validate the weakness of 2-TI.

*2-TI Present.* As before, 10 million traces are captured for the two-share Present implementation, and then analyzed using paired t-test. The first order t-statistic is still below 4.5 with 10 million measurements, as shown in Fig. 10(a). The second order $t$-statistics exceeds the threshold with about 6000 traces as shown in Fig. 10(b). Again, the results suggest that two-share TI holds the promise of first order resistance, but fares terribly on the second order resistance.

*Exploiting the Uncovered Leakages.* In order to practically exploit this strong second-order leakage, a classic CPA [6,8,10] is performed on the measurements (center-and-then-squared) associated with the 5 million random plaintexts.

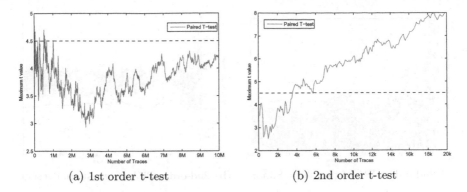

(a) 1st order t-test                            (b) 2nd order t-test

**Fig. 10.** Leakage detection results for the two-share implementation of Present for first order (left) and second order (right) leakage over the number of traces. Note that the dimensions change for both axes.

For *2-TI Simon*, the targeted operations occurred in the first clock cycle of the third round of encryption where shared values in registers $L_a$ and $L_b$ overwrite $R_a$ and $R_b$ respectively (see Fig. 1). The leakage model used is Hamming distance between registers $L$ and $R$ as in a plain or unprotected implementation. The reason why third round is chosen is because of the weak non-linearity of single Simon round operation (only one $AND$) and attacking third round would relieve the effect of "ghost peaks" [8]. Moreover, in order to reduce the computational complexity, we follow the divide-and-conquer approach and only attack the most significant four bits in $L$ and $R$ which are dependent on 10 bits in $k_0$ and 4 bits in $k_1$. Therefore, $2^{14}$ key hypotheses are required for the attack. To further reduce the complexity, we assume the knowledge of the relevant 4 bits in $k_1$ is known and only 10 bits in $k_0$ are aimed at to recover. Figure 11(a) shows the max correlation for each key hypothesis over the number of traces. The practical second-order attack successfully recovers the correct key with more than 3 million measurements even though ghost keys still exist. Note that these results can be significantly improved by using a profiled attack, predicting more bits, and by using a pruning technique as e.g. done in [17], which is always an option for ciphers with a low algebraic depth per round. Nevertheless, the results validate the second-order leakage of two-share TI detected by the t-test and it can be practically exploited.

We also performed the same second-order CPA on 5 million random traces (center-and-then-squared) on *2-TI Present*, targeting at the S-box output to exploit the leakage. Recall our 2-TI Present in which the 64-bit state registers are right rotated by 4 bits per clock cycle so that the least significant nibble is continuously fed into the S-box look-up and output is written back to the most significant nibble after 4 clock cycles. Therefore, a Hamming distance leakage occurs between consecutive output nibbles. In this attack, we use the Hamming distance power model between the first two consecutive S-box outputs which depends on the least significant key byte and thus $2^8$ key hypotheses are required.

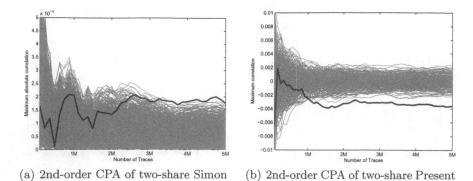

(a) 2nd-order CPA of two-share Simon    (b) 2nd-order CPA of two-share Present

**Fig. 11.** Second-order CPA. Max correlation for each key hypothesis over the number of traces.

The max correlations per key hypothesis over number of traces are shown in Fig. 11(b) and the results show that correct key can be successfully recovered with more than 1 million traces which demonstrates the practical exploitability of detected leakage.

The results from both validate our simulation analyses for the idealized case from Sects. 3.1 and 7.1, which suggests strong second-order leakage. The difference in sensitivity for the two implementations stems from their differing design strategies: 2-TI Simon is round based and does not use pipelining. Hence, it maximizes the leakage for the fixed-vs-random test: the entire state that is processed per cycle is constant in the fixed case and varies in the other case. For 2-TI Present, the implementation is serialized, with a 4-bit datapath, hence, a much smaller part of the implementation is updated per cycle, making the leakage less pronounced.

Moreover, unlike the theoretical analysis results in Sect. 7.1 where the number of traces needed for successful second-order t-test and CPA are of the same order magnitude, a lot more traces are needed for practical second-order CPA with Hamming distance model to exploit the leakage detected by t-test with only hundreds to thousands of traces. This is mainly because: (1) Practical implementation don't leak a perfect Hamming weight or Hamming distance leakage; (2) Noises also render the practical attacks inefficient.

While two-share TI shows potential in preventing first order leakage with less overhead, its poor performance on second order leakage resistance compared with three-sharing makes it less worthwhile.

## 8    Conclusion

This work presents the first practical threshold implementations using only two shares. We showed that lightweight ciphers have several features making them good targets for threshold implementations. Furthermore, we explain how using two shares can actually yield smaller cipher implementations that need less randomness and still show perfect first order resistance. While moving to two shares makes implementing the nonlinear functions of a cipher more cumbersome, resulting in either a loss in throughput, increase in circuit size, or even both, it allows to reduce the overhead of the sequential part of the implementation by only doubling the state and key size. Since the area of low-area crypto implementations usually depends mainly on the sequential part, significant improvements are possible. In fact, the presented bit-serialized two-share implementation of Simon is the smallest side-channel protected 128-bit block cipher implementation for FPGAs. To this end, we presented the first two-share threshold implementations of Simon and Present, which feature perfect first-order resistance.

However, these findings are of limited practical impact, as two-share TI features strong second-order leakage. Hence, on one hand, the results highlight that provable resistance against a "low" order of attack might be meaningless in practice. On the other hand, the previously observed feature that three-share TI not only keeps the promised first-order resistance, but also fails gracefully for higher order analysis, is undervalued and may deserve further analysis.

**Acknowledgments.** This work is supported by the National Science Foundation under grant CNS-1261399 and grant CNS-1314770.

# References

1. Aysu, A., Gulcan, E., Schaumont, P.: SIMON says: break area records of block ciphers on FPGAs. IEEE Embed. Syst. Lett. **6**(2), 37–40 (2014)
2. Beaulieu, R., Shors, D., Smith, J., Treatman-Clark, S., Weeks, B., Wingers, L.: The SIMON and SPECK families of lightweight block ciphers. IACR Cryptology ePrint Arch. **2013**, 404 (2013)
3. Bilgin, B., Gierlichs, B., Nikova, S., Nikov, V., Rijmen, V.: Trade-offs for threshold implementations illustrated on AES. IEEE Trans. Comput. Aided Des. Integr. Circuits Syst. **34**(7), 1188–1200 (2015)
4. Bilgin, B., Gierlichs, B., Nikova, S., Nikov, V., Rijmen, V.: Higher-order threshold implementations. In: Sarkar, P., Iwata, T. (eds.) ASIACRYPT 2014. LNCS, vol. 8874, pp. 326–343. Springer, Heidelberg (2014). doi:10.1007/978-3-662-45608-8_18
5. Bilgin, B., Daemen, J., Nikov, V., Nikova, S., Rijmen, V., Assche, G.: Efficient and first-order DPA resistant implementations of KECCAK. In: Francillon, A., Rohatgi, P. (eds.) CARDIS 2013. LNCS, vol. 8419, pp. 187–199. Springer, Heidelberg (2014). doi:10.1007/978-3-319-08302-5_13
6. Bilgin, B., Gierlichs, B., Nikova, S., Nikov, V., Rijmen, V.: A more efficient AES threshold implementation. In: Pointcheval, D., Vergnaud, D. (eds.) AFRICACRYPT 2014. LNCS, vol. 8469, pp. 267–284. Springer, Heidelberg (2014). doi:10.1007/978-3-319-06734-6_17
7. Bogdanov, A., Knudsen, L.R., Leander, G., Paar, C., Poschmann, A., Robshaw, M.J.B., Seurin, Y., Vikkelsoe, C.: PRESENT: an ultra-lightweight block cipher. In: Paillier, P., Verbauwhede, I. (eds.) CHES 2007. LNCS, vol. 4727, pp. 450–466. Springer, Heidelberg (2007). doi:10.1007/978-3-540-74735-2_31

8. Brier, E., Clavier, C., Olivier, F.: Correlation power analysis with a leakage model. In: Joye, M., Quisquater, J.-J. (eds.) CHES 2004. LNCS, vol. 3156, pp. 16–29. Springer, Heidelberg (2004). doi:10.1007/978-3-540-28632-5_2

9. Canright, D.: A very compact S-Box for AES. In: Rao, J.R., Sunar, B. (eds.) CHES 2005. LNCS, vol. 3659, pp. 441–455. Springer, Heidelberg (2005). doi:10.1007/11545262_32

10. Chari, S., Jutla, C.S., Rao, J.R., Rohatgi, P.: Towards sound approaches to counteract power-analysis attacks. In: Wiener, M. (ed.) CRYPTO 1999. LNCS, vol. 1666, pp. 398–412. Springer, Heidelberg (1999). doi:10.1007/3-540-48405-1_26

11. Chari, S., Jutla, C.S., Rao, J.R., Rohatgi, P.: Towards sound approaches to counteract power-analysis attacks. In: Wiener, M. (ed.) CRYPTO 1999. LNCS, vol. 1666, pp. 398–412. Springer, Heidelberg (1999). doi:10.1007/3-540-48405-1_26

12. Cooper, J., DeMulder, E., Goodwill, G., Jaffe, J., Kenworthy, G., Rohatgi, P.: Test vector leakage assessment (TVLA) methodology in practice. In: International Cryptographic Module Conference (2013). http://icmc-2013.org/wp/wp-content/uploads/2013/09/goodwillkenworthtestvector.pdf

13. Coron, J.-S., Prouff, E., Rivain, M.: Side channel cryptanalysis of a higher order masking scheme. In: Paillier, P., Verbauwhede, I. (eds.) CHES 2007. LNCS, vol. 4727, pp. 28–44. Springer, Heidelberg (2007). doi:10.1007/978-3-540-74735-2_3

14. Cannière, C., Dunkelman, O., Knežević, M.: KATAN and KTANTAN — a family of small and efficient hardware-oriented block ciphers. In: Clavier, C., Gaj, K. (eds.) CHES 2009. LNCS, vol. 5747, pp. 272–288. Springer, Heidelberg (2009). doi:10.1007/978-3-642-04138-9_20

15. De Cnudde, T., Reparaz, O., Bilgin, B., Nikova, S., Nikov, V., Rijmen, V.: Masking AES with $d + 1$ shares in hardware. In: Gierlichs, B., Poschmann, A.Y. (eds.) CHES 2016. LNCS, vol. 9813, pp. 194–212. Springer, Heidelberg (2016). doi:10.1007/978-3-662-53140-2_10

16. Ding, A.A., Chen, C., Eisenbarth, T.: Simpler, Faster, and More Robust T-test Based Leakage Detection. In: Constructive Side-Channel Analysis and Secure Design - 7th International Workshop, COSADE 2016, Graz, Austria, April 14–15, 2016, Revised Selected Papers, pp. 163–183. http://dx.doi.org/10.1007/978-3-319-43283-0_10

17. Eisenbarth, T., Kasper, T., Moradi, A., Paar, C., Salmasizadeh, M., Shalmani, M.T.M.: On the power of power analysis in the real world: a complete break of the KeeLoq code hopping scheme. In: Wagner, D. (ed.) CRYPTO 2008. LNCS, vol. 5157, pp. 203–220. Springer, Heidelberg (2008). doi:10.1007/978-3-540-85174-5_12

18. Goodwill, G., Jun, B., Jaffe, J., Rohatgi, P.: A Testing Methodology for Sidechannel Resistance Validation. Non-Invasive Attack Testing Workshop (2011). http://www.cryptography.com/public/pdf/a-testing-methodology-for-side-channel-resistance-validation.pdf

19. Kavun, E.B., Yalcin, T.: RAM-based ultra-lightweight FPGA implementation of PRESENT. In: 2011 International Conference on Reconfigurable Computing and FPGAs (ReConFig), pp. 280–285. IEEE (2011)

20. Kirschbaum, M., Popp, T.: Evaluation of a DPA-resistant prototype chip. In: Computer Security Applications Conference, ACSAC 2009, Annual, pp. 43–50, December 2009

21. Kutzner, S., Nguyen, P.H., Poschmann, A., Wang, H.: On 3-share threshold implementations for 4-Bit S-boxes. In: Prouff, E. (ed.) COSADE 2013. LNCS, vol. 7864, pp. 99–113. Springer, Heidelberg (2013). doi:10.1007/978-3-642-40026-1_7

22. Leiserson, A.J., Marson, M.E., Wachs, M.A.: Gate-level masking under a path-based leakage metric. In: Batina, L., Robshaw, M. (eds.) CHES 2014. LNCS, vol. 8731, pp. 580–597. Springer, Heidelberg (2014). doi:10.1007/978-3-662-44709-3_32

23. Moradi, A., Mischke, O.: How far should theory be from practice? In: Prouff, E., Schaumont, P. (eds.) CHES 2012. LNCS, vol. 7428, pp. 92–106. Springer, Heidelberg (2012). doi:10.1007/978-3-642-33027-8_6

24. Moradi, A., Poschmann, A., Ling, S., Paar, C., Wang, H.: Pushing the limits: a very compact and a threshold implementation of AES. In: Paterson, K.G. (ed.) EUROCRYPT 2011. LNCS, vol. 6632, pp. 69–88. Springer, Heidelberg (2011). doi:10.1007/978-3-642-20465-4_6

25. Nikova, S., Rechberger, C., Rijmen, V.: Threshold implementations against side-channel attacks and glitches. In: Ning, P., Qing, S., Li, N. (eds.) ICICS 2006. LNCS, vol. 4307, pp. 529–545. Springer, Heidelberg (2006). doi:10.1007/11935308_38

26. Poschmann, A., Moradi, A., Khoo, K., Lim, C.W., Wang, H., Ling, S.: Side-Channel resistant crypto for less than 2,300 GE. J. Cryptology 24(2), 322–345 (2011)

27. Reparaz, O., Bilgin, B., Nikova, S., Gierlichs, B., Verbauwhede, I.: Consolidating masking schemes. In: Gennaro, R., Robshaw, M. (eds.) CRYPTO 2015. LNCS, vol. 9215, pp. 764–783. Springer, Heidelberg (2015). doi:10.1007/978-3-662-47989-6_37

28. Reparaz, O., Sinha Roy, S., Vercauteren, F., Verbauwhede, I.: A masked ring-LWE implementation. In: Güneysu, T., Handschuh, H. (eds.) CHES 2015. LNCS, vol. 9293, pp. 683–702. Springer, Heidelberg (2015). doi:10.1007/978-3-662-48324-4_34

29. Rolfes, C., Poschmann, A., Leander, G., Paar, C.: Ultra-lightweight implementations for smart devices – security for 1000 gate equivalents. In: Grimaud, G., Standaert, F.-X. (eds.) CARDIS 2008. LNCS, vol. 5189, pp. 89–103. Springer, Heidelberg (2008). doi:10.1007/978-3-540-85893-5_7

30. Schneider, T., Moradi, A.: Leakage assessment methodology – a clear roadmap for side-channel evaluations. In: Güneysu, T., Handschuh, H. (eds.) CHES 2015. LNCS, vol. 9293, pp. 495–513. Springer, Heidelberg (2015). doi:10.1007/978-3-662-48324-4_25

31. Shahverdi, A., Taha, M., Eisenbarth, T.: Silent simon: a threshold implementation under 100 slices. In: 2015 IEEE International Symposium on Hardware Oriented Security and Trust (HOST), pp. 1–6, May 2015

32. Tiri, K., Verbauwhede, I.: A logic level design methodology for a secure DPA resistant ASIC or FPGA implementation. In: Proceedings of the Conference on Design, Automation and Test in Europe - vol. 1, DATE 2004, p. 10246 (2004). http://dl.acm.org/citation.cfm?id=968878.969036

# Cryptographic Reverse Firewall via Malleable Smooth Projective Hash Functions

Rongmao Chen[1,2(✉)], Yi Mu[1], Guomin Yang[1], Willy Susilo[1], Fuchun Guo[1], and Mingwu Zhang[3]

[1] School of Computing and Information Technology, Centre for Computer and Information Security Research, University of Wollongong, Wollongong, Australia
{rc517,ymu,gyang,wsusilo,fuchun}@uow.edu.au
[2] College of Computer, National University of Defense Technology, Changsha, China
[3] School of Computers, Hubei University of Technology, Wuhan, China
csmwzhang@gmail.com

**Abstract.** Motivated by the revelations of Edward Snowden, post-Snowden cryptography has become a prominent research direction in recent years. In Eurocrypt 2015, Mironov and Stephens-Davidowitz proposed a novel concept named *cryptographic reverse firewall* (CRF) which can resist exfiltration of secret information from an arbitrarily compromised machine. In this work, we continue this line of research and present generic CRF constructions for several widely used cryptographic protocols based on a new notion named *malleable smooth projective hash function*. Our contributions can be summarized as follows.

- We introduce the notion of malleable smooth projective hash function, which is an extension of the smooth projective hash function (SPHF) introduced by Cramer and Shoup (Eurocrypt'02) with the new properties of *key malleability* and *element rerandomizability*. We demonstrate the feasibility of our new notion using graded rings proposed by Benhamouda et al. (Crypto'13), and present an instantiation from the $k$-linear assumption.
- We show how to generically construct CRFs via malleable SPHFs in a modular way for some widely used cryptographic protocols. Specifically, we propose generic constructions of CRFs for the unkeyed message-transmission protocol and the oblivious signature-based envelope (OSBE) protocol of Blazy, Pointcheval and Vergnaud (TCC'12). We also present a new malleable SPHF from the linear encryption of valid signatures for instantiating the OSBE protocol with CRFs.
- We further study the two-pass oblivious transfer (OT) protocol and show that the malleable SPHF does not suffice for its CRF constructions. We then develop a new OT framework from graded rings and show how to construct OT-CRFs by modifying the malleable SPHF framework. This new framework encompasses the DDH-based OT-CRF constructions proposed by Mironov and Stephens-Davidowitz (Eurocrypt'15), and yields a new construction under the $k$-linear assumption.

**Keywords:** Cryptographic reverse firewall · Malleable smooth projective hash function · Oblivious signature-based envelope · Oblivious transfer

© International Association for Cryptologic Research 2016
J.H. Cheon and T. Takagi (Eds.): ASIACRYPT 2016, Part I, LNCS 10031, pp. 844–876, 2016.
DOI: 10.1007/978-3-662-53887-6_31

# 1   Introduction

In the last couple of years, the revelations of Edward Snowden [18,22] showed that the intelligence agencies successfully gained access to a massive collection of user sensitive data by undermining security mechanisms via a broad range of techniques, e.g., by subverting cryptographic protocols and actively deploying security weaknesses in the implementations of cryptosystems. The disclosures of Snowden have reawakened the cryptographic research community to the seriousness of the undermining of cryptographic solutions and standards [6–8,13,23,24], and led to a new research direction known as post-Snowden cryptography. The research problem could be generally summarized by the following question: "How to achieve meaningful security for cryptographic protocols in the presence of an adversary that may arbitrarily tamper with the victim's machine?"

**Cryptographic Reverse Firewall.** Motivated by the aforementioned question, Mironov and Stephens-Davidowitz [21] recently proposed a novel notion named cryptographic reverse firewall (CRF) aiming at providing strong security against inside vulnerabilities such as security backdoors. Informally, a CRF is a machine that sits at the boundary between the user's computer and the outside world. It plays as the role of an autonomous intermediary that intercepts and modifies the machine's incoming and outgoing messages to provide security protections even if the user's machine is compromised. A cryptographic protocol equipped with a correctly implemented CRF can guarantee that its security is preserved even if it is run on a compromised machine and the CRF could also resist exfiltration of secret information from the tampered machine. More specifically, Mironov and Stephens-Davidowitz defined three desirable properties for an honestly implemented CRF:

- *Functionality Maintaining.* A CRF should not break the functionality (i.e., correctness) of an honestly implemented protocol.
- *Security Preservation.* A protocol with a CRF should provide the same security guarantee as the properly implemented protocol regardless of how the underlying machine behaves.
- *Exfiltration Resistance.* A CRF should resist exfiltration so that a compromised implementation cannot leak any information to the outside world.

The above three properties deserve further interpretation. A good cryptographic protocol should be functional and secure regardless of the existence of the CRF when the protocol implementation is correct. That is, the user does not rely solely on the CRF for security but only requires it to preserve security. In particular, the CRF shares no secret with the protocol party, and thus even if the CRF is not functioning, an honestly implemented protocol would remain secure. This is one significant difference between the CRF and the prior work. On the other hand, when the protocol implementation is tampered but the CRF is implemented correctly, the CRF could provide the user with the desired security guarantee. In short, a protocol with CRF satisfies the security requirement as long as either the protocol implementation is not tampered or the CRF is implemented correctly.

The CRF could be viewed as a modern take on a line of work that received considerable attention in the 80s and 90s [10, 28]. It provides a general framework for building cryptographic schemes that remain secure when run on a compromised machine. The use of rerandomization to "sanitize" messages by the CRF is seemingly similar to the prior work, e.g., divertible protocols [10] and collusion-free protocols [3,19]. As summarized by Mironov and Stephens-Davidowitz in [21], the CRF is a generalization of these prior notions and models.

**Motivations of This Work.** In this work, we further explore the construction of CRFs. Unlike prior work that relies on concrete techniques and thus appears complicated, our goal is to develop generic paradigms for constructing CRFs in a conceptually simple and modular way. From a theoretical point of view, a generic paradigm can modularly explain concrete CRF constructions and their underlying design principles. From a practical point of view, a generic CRF construction based on abstract building blocks enables more concrete instantiations to be built for better security and/or efficiency. In fact, our work (partially) answers an open question raised by Mironov and Stephens-Davidowitz in [21]. Particularly, they stated that *"the "holy grail" would be a full characterization of functionalities and security properties for which reverse firewall exists"*.

## 1.1 Overview of Our Contributions

We introduce the notion of *malleable smooth projective hash function*, which is a new extension of the conventional SPHF. A malleable SPHF is a special SPHF which is of additional properties, namely projection key malleablility and element re-randomizability. Using this notion, we obtain generic CRF constructions for some widely used cryptographic protocols. Before we describe our results, we present an overview of the malleable smooth projective hash function.

**Malleable Smooth Projective Hash Function.** We first briefly recall the classical definition of the smooth projective hash function (SPHF) (also known as hash proof system) introduced by Cramer and Shoup [12].

CLASSICAL DEFINITION. An SPHF requires the existence of a domain $\mathcal{X}$ and an underlying $\mathcal{NP}$ language $\mathcal{L}$, where elements of $\mathcal{L}$ form a subset of $\mathcal{X}$, i.e., $\mathcal{L} \subset \mathcal{X}$. The key property of SPHF is that the hash value of any element $C \in \mathcal{L}$ can be computed by using either a secret hashing key hk, or a public projection key hp with the witness to the fact that $C \in \mathcal{L}$. However, the projection key gives almost no information about the hash value of any element in $\mathcal{X} \backslash \mathcal{L}$. Moreover, we say that the subset membership problem is hard if the distribution of $\mathcal{L}$ is computationally indistinguishable from $\mathcal{X} \backslash \mathcal{L}$.

NEW PROPERTIES. In addition to the above properties of a regular SPHF, we define two new properties for a malleable SPHF as follows.

- **Projection Key Malleability.** This property captures that,
  - *Key Indistinguishability*: any projection key hp can be re-randomized to an independent projection key $\widetilde{\text{hp}}$ using a uniformly chosen randomness $\widetilde{r}$; and
  - *Projection Consistency*: the hash value difference of any element due to the above key re-randomization is computable using $\widetilde{r}$.
- **Element Re-randomizability.** This property captures that,
  - *Element Indistinguishability*: any element $C$ can be re-randomized to another independent element $\widetilde{C}$ using a uniformly chosen witness $\widetilde{w}$; and
  - *Rerandomization Consistency*: the hash value difference between $C$ and $\widetilde{C}$ under the same hashing key is computable using the associated projection key with $\widetilde{w}$; and
  - *Membership Preservation*: the re-randomization of an element does not change its membership (i.e., $\widetilde{C} \in \mathcal{L} \Longleftrightarrow C \in \mathcal{L}$).

A SIMPLE EXAMPLE. We provide a very simple example of our new notion. We remark that such a simple example is just for a quick understanding of the properties captured by our malleable SPHF. The construction would be more complicated from other assumptions. The basic SPHF below is exactly the one of Cramer and Shoup for the DDH language in [12]. Let $g_1, g_2$ be two generators of a cyclic group $\mathbb{G}$ of prime order $p$. Let $\mathcal{X} = \mathbb{G}^{1 \times 2}$ and $\mathcal{L} = \{(g_1^r, g_2^r) \in \mathcal{X} \mid r \in \mathbb{Z}_p\}$. The hashing key is $\text{hk} = (\alpha_1, \alpha_2) \xleftarrow{\$} \mathbb{Z}_p^2$ and the associated projection key is $\text{hp} = g_1^{\alpha_1} g_2^{\alpha_2}$. For any element $C = (u_1, u_2) \in \mathcal{X}$, the hash value under hk is $\text{hv} = u_1^{\alpha_1} u_2^{\alpha_2}$.

- Choose $\widetilde{r} = (\beta_1, \beta_2) \xleftarrow{\$} \mathbb{Z}_p^2$, and compute $\widetilde{\text{hp}} = \text{hp} \cdot (g_1^{\beta_1} g_2^{\beta_2}) = g_1^{\alpha_1 + \beta_1} g_2^{\alpha_2 + \beta_2}$. $\widetilde{\text{hp}}$ is independent from hp and its associated hashing key is $\widetilde{\text{hk}} = (\alpha_1 + \beta_1, \alpha_2 + \beta_2)$. The hash value of element $C$ under $\widetilde{\text{hk}}$ is $\widetilde{\text{hv}} = u_1^{\alpha_1 + \beta_1} u_2^{\alpha_2 + \beta_2} = \text{hv} \cdot u_1^{\beta_1} u_2^{\beta_2}$, and hence the hash value difference is computable using $\widetilde{r}$.
- Choose $\widetilde{w} = \eta \xleftarrow{\$} \mathbb{Z}_p$ and compute $\widetilde{C} = (u_1 g_1^\eta, u_2 g_2^\eta)$. The hash value of $\widetilde{C}$ under hk is $\widetilde{\text{hv}} = (u_1 g_1^\eta)^{\alpha_1} (u_2 g_2^\eta)^{\alpha_2} = \text{hv} \cdot (\text{hp})^\eta$, and hence the hash value difference is computable using $\widetilde{w}$ (with hp). One can easily verify that $\widetilde{C} \in \mathcal{L} \Longleftrightarrow C \in \mathcal{L}$.

MORE CONSTRUCTIONS OF MALLEABLE SPHFs. To illustrate the feasibility of our new notion, we propose a generic construction of malleable SPHFs based on graded rings [9], which could be viewed as a common formalization for cyclic groups, bilinear groups, and multilinear groups. We rigorously prove that under some conditions, graded ring implies malleable SPHFs. Particularly, we rely on Katz and Vaikuntanathan [17] type SPHFs (KV-SPHF) where the projection key is independent from the element, as in many cases the linkability between the projection key and the element would make it difficult for a CRF to resist exfiltration and meanwhile maintain functionality. We will make this point clearer in our CRF constructions. We then provide a malleable SPHF instantiation of our generic framework from the $k$-linear assumption.

**Generic CRF Constructions via Malleable SPHFs.** We show how to generically construct CRFs via malleable SPHFs for some widely used protocols. Essentially, our CRF constructions rely on the *key indistinguishability* and the *element indistinguishability* properties of the underlying malleable SPHF for the security preservation and exfiltration resistance, and rely on the *projection consistency, rerandomization consistency* and *membership preservation* of the malleable SPHF for the functionality maintaining.

MESSAGE TRANSMISSION PROTOCOL. We first show as a warm up CRF constructions for the unkeyed message-transmission protocol. That is, both the sender and receiver have neither a shared secret key nor each other's public key. We remark that our framework can be seen as a generic construction of semantically secure public-key encryption scheme (with trusted setup) that is both key malleable and re-randomizable defined in [14], and hence provides a more intuitive way to build two-round message-transmission protocols with CRFs. The idea we illustrate via this simple protocol acts as a steppingstone toward other more complicated protocols.

OBLIVIOUS SIGNATURE-BASED ENVELOPE PROTOCOL. We also study the CRF constructions for another useful protocol, namely Oblivious Signature-Based Envelope (OSBE), which was proposed by Li, Du and Boneh [20] and later enhanced by Blazy, Pointcheval and Vergnaud [11]. An OSBE protocol allows a user Alice to send an envelope, which encapsulates her private message, to another user Bob in such a way that Bob will be able to recover the private message if and only if Bob has possessed a credential, e.g., a signature on an agreed-upon message from the certification authority. OSBE has been found useful in a growing number of protocols and applications such as Secret Handshakes [5] and Password-Based Authenticated Key-Exchange [15]. We show that the SPHF-based construction of OSBE in [11] is CRF-ready if the underlying SPHF is malleable. Surprisingly, we find that their proposed OSBE instantiation from linear encryption of Waters signature [25] could be extended to be malleable for the CRF instantiations. One should note that the extension does not strictly follow the aforementioned generic framework of constructing malleable SPHF from graded rings. This also shows more possibilities for constructing malleable SPHFs.

**CRF Constructions for Oblivious Transfer Protocol.** Another major contribution of our work is the CRF construction for the oblivious transfer (OT) protocol, which has been widely adopted as a basic tool by many cryptographic systems. Although our CRF constructions are inspired by our generic framework of malleable SPHF from graded rings, there is some substantive difference between them.

In this work, we start with the OT framework of Halevi and Kalai [16], which relies on a special SPHF. The basic idea is that: (1) the receiver picks and sends to the sender two elements $C_b \in \mathcal{L}, C_{1-b} \in \mathcal{X} \backslash \mathcal{L}$ ($b \in \{0,1\}$ is the choice bit);

(2) the sender generates two hashing key pairs and computes the hash values of $C_0$ and $C_1$ (using the secret hashing keys) to conceal its two message $M_0$ and $M_1$ respectively, and then sends the two concealed messages with projection keys to the receiver; (3) the receiver recovers $M_b$ by computing the hash value of $C_b$ (using the projection key with the witness to the fact $C_b \in \mathcal{L}$). Noting that a malicious receiver might choose both $C_b$ and $C_{1-b}$ from the language $\mathcal{L}$, the underlying SPHF is required to be verifiably smooth such that the sender can verify at least one of $(C_0, C_1)$ is not in the language.

DIFFICULTIES. It seems that we could extend the underlying SPHF of the HK-OT construction to be malleable so that the framework could admit CRFs. However, we found that it is actually not the case and the extension is not trivial at all.

– *The required* SPHF *here is not a classical one as it must be verifiably smooth.* Under the HK-OT framework, this is usually guaranteed by the verifiable linkability between $C_0$ and $C_1$ chosen by the receiver. However, a tampered implementation of the receiver may leak secret information to the outside world via the linkability. A desirable CRF for the receiver should be able to rerandomize $(C_0, C_1)$ to a uniform tuple $(\widetilde{C_0}, \widetilde{C_1})$ to resist exfiltration. However, the rerandomization would break the linkability of the tuple and lead to protocol failure.
– *The receiver freshly generates the element basis underlying the* SPHF *at the beginning of each protocol session, which means we have to deal with an untrusted setup.* Since the element basis (e.g., $g_1, g_2 \in \mathbb{G}$ for the DDH tuple generation) is chosen by the receiver per session, a tampered receiver may maliciously choose some "bad" basis in order to compromise the security or leak secret information to the outside. Therefore, the CRF should be able to rerandomize the element basis to preserve security and resist exfiltration, while still maintain the protocol functionality. This, unfortunately, could not be trivially realized by the malleable SPHF.

OUR SOLUTION. In order to resolve the problem, we first propose a special OT construction from graded rings. Particularly, the receiver sends to the sender only one element, based on which the sender could generate an element pair so that the verifiable smoothness can be guaranteed by the sender itself. We then propose CRF constructions for such an OT protocol. Our central idea mainly follows the generic framework of malleable SPHF from graded rings except that we require the receiver's CRF could also rerandomize the element basis chosen by the receiver. We show that the CRF could still achieve all the properties when the transformation matrix for rerandomizing the element basis meets some requirements. The modified semi-generic framework narrows the possible instantiations of the HK-OT framework. However, we show that the CRF construction following our framework not only captures the prior work [21], which is the only known OT-CRF to date, but also can yield new constructions under weaker assumptions. In particular, we present new CRF constructions based on the $k$-linear assumption, which is weaker than the DDH assumption underlying the OT-CRF construction in [21].

## 1.2   Related Work

**Comparisons with Other SPHF Variants.** SPHF was originally introduced by Cramer and Shoup [12]. Since its introduction, it has been widely used for constructions of many cryptographic primitives, including authenticated key exchange [15,17], oblivious transfer [16], zero-knowledge arguments [1,2,9] and so on. Here we mainly introduce the work that are closely related to our notion of malleable SPHF. Hoeteck Wee defined a notion of homomorphic SPHF for achieving key-dependent message security [26]. That is, the combination of hash values of two elements equal to the hash value of the combination of these two elements. One may note that their notion is somewhat similar to the sub-property of *rerandomization consistency* captured by the element re-randomizability of our malleable SPHF. However, their definition is solely based on the secret hashing key while ours uses the projection key to calculate the hash value difference. We should clarify that our defined property is not always the case especially for those SPHFs where the projection key depends on the element. Yang et al. [27] introduced the notion of updatable hash proof system (UHPS) for constructing public key encryption schemes that are secure against continuous memory attacks. The UHPS requires that the secret hashing key could be updated homomorphically. In fact, they mainly consider a special case in which a secret hashing key can be freshly updated while the associated projection key keeps the same.

**Other CRF Constructions.** Mironov and Stephens-Davidowitz [21] showed how to construct CRFs for a 1-out-of-2 oblivious protocol based on the DDH assumption and also proposed a protocol for private function evaluation. They also provided a generic way to prevent a tampered machine from leaking information to an eavesdropper via any protocol. Ateniese, Magri, and Venturi [4] continued the study on signatures and constructed the CRF to protect signatures schemes against algorithm substitution attacks. Recently, Dodis, Mironov and Stephens-Davidowitz [14] considered CRF constructions for message-transmission protocols. They proposed a rich collection of solutions that vary in efficiency, security, and setup assumptions in the classical setting. It is worth noting that the studied message-transmission protocol in our work belongs to the so-called unkeyed setting in their work. Our framework can be viewed as a generic construction of the semantically secure public-key encryption scheme (with a trusted setup) that is both key malleable and re-randomizable defined in [14].

## 2   Preliminaries

### 2.1   Cryptographic Reverse Firewalls

In general, a cryptographic protocol $\mathcal{P}$ must satisfy functionality (i.e., correctness) requirement $\mathcal{F}$, which places constraints on the output of the parties executing $\mathcal{P}$ for particular input, and security requirement $\mathcal{S}$, which places constraints on the message distribution conditioned on specific input. Below we briefly recall the definition of reverse firewalls from [21]. We refer the reader to [21] for more detailed discussions.

**Definition 1 (Cryptographic Reverse Firewall (CRF)).** *A cryptographic reverse firewall is a stateful algorithm $W$ that takes as input its state and a message and outputs an updated state and message. For simplicity, we do not write the state of $W$ explicitly. For a party $P$ and reverse firewall $W$, we define $W \circ P$ as the "composed" party where $W$ is applied to the incoming and outgoing messages of $P$. When the composed party engages in a protocol, the state of $W$ is initialized to the public parameters. If $W$ is meant to be composed with a party $P$, we call it a reverse firewall for $P$.*

One should note that $W$ has access to all public parameters, but not the private input or the output of $P$. In reality, $W$ can be regarded as an "active router" that sits at the boundary between $P$'s private network and the outside world and modifies the messages that $P$ sends and receives. The party $P$ of course does not want a reverse firewall to ruin its protocol's functionality when its internal implementation is correct. Following [21] we require that reverse firewalls should be "stackable", which means the composition of multiple reverse firewalls $W \circ W \circ \cdots \circ W \circ P$ should still maintain the functionality of the protocol. The following definition captures this property.

**Definition 2 (Functionality-maintaining CRFs).** *For any reverse firewall $W$ and any party $P$, let $W^1 \circ P = W \circ P$, and for $k \geq 2$, let $W^k \circ P = W \circ (W^{k-1} \circ P)$. For a protocol $P$ that satisfies some functionality requirements $\mathcal{F}$, we say that a reverse firewall $W$ maintains $\mathcal{F}$ for $P$ in $\mathcal{P}$ if $W^k \circ P$ maintains $\mathcal{F}$ for $P$ in $\mathcal{P}$ for any polynomial bounded $k \geq 1$. When $\mathcal{F}, P, \mathcal{P}$ are clear, we simply say that $W$ maintains functionality.*

Following the notations in [21], we use $\overline{P}$ to represent arbitrary adversarial implementations of party $P$ and $\widehat{P}$ to represent the functionality-maintaining adversarial implementations. For a protocol $\mathcal{P}$ with party $P$, we write $\mathcal{P}_{P \to \widehat{P}}$ to represent the protocol where the role of party $P$ is replaced by party $\widehat{P}$.

A reverse firewall should also preserve the security of the underlying protocol, even in the presence of compromise. The strongest notion requires that the protocol in which party $P$ is replaced with $W \circ \overline{P}$ for an arbitrarily corrupted party $\overline{P}$ still preserves the security while the weaker notion only considers tampered implementations that maintain functionality. The below definition captures this property.

**Definition 3 (Security-preserving CRFs).** *For a protocol $\mathcal{P}$ that satisfies some security requirements $\mathcal{S}$ and functionality $\mathcal{F}$ and a reverse firewall $W$,*

- *$W$ strongly preserves $\mathcal{S}$ for $P$ in $\mathcal{P}$ if the protocol $\mathcal{P}_{P \to W \circ \overline{P}}$ satisfies $\mathcal{S}$; and*
- *$W$ weakly preserves $\mathcal{S}$ for $P$ in $\mathcal{P}$ if the protocol $\mathcal{P}_{P \to W \circ \widehat{P}}$ satisfies $\mathcal{S}$.*

*When $\mathcal{P}, \mathcal{F}, \mathcal{S}, P$ are clear, we simple say that $W$ strongly preserves security or weakly preserves security.*

As introduced in [21], we also need the notion of exfiltration resistance. Intuitively, a reverse firewall is exfiltration resistant if "no corrupted implementation

of P can leak information through the firewall." We define this notion using the
game LEAK which is presented in Fig. 1. Intuitively, the game asks the adversary
to distinguish between a tampered implementation and an honest implementation. An exfiltration-resistant reverse firewall therefore prevents an adversary
from even learning whether a party has been compromised, let alone leaking
information.

$$
\begin{aligned}
&\textbf{Proc. } \mathsf{LEAK}(\mathcal{P}, \mathsf{P}_1, \mathsf{P}_2, \mathcal{W}, \ell)\\
&(\overline{\mathsf{P}}_1, \overline{\mathsf{P}}_2, I) \leftarrow \mathcal{A}(1^\ell)\\
&b \xleftarrow{\$} \{0,1\}\\
&\text{If } b = 1, \mathsf{P}^* \leftarrow \mathcal{W} \circ \overline{\mathsf{P}}_1\\
&\text{Else}, \mathsf{P}^* \leftarrow \mathcal{W} \circ \mathsf{P}_1\\
&\mathcal{T}^* \leftarrow \mathcal{P}_{\mathsf{P}_1 \to \mathsf{P}^*, \mathsf{P}_2 \to \overline{\mathsf{P}}_2}(I)\\
&b^* \leftarrow \mathcal{A}(\mathcal{T}^*, \mathsf{st}_{\overline{\mathsf{P}}_2})\\
&\text{Output } (b = b^*)
\end{aligned}
$$

**Fig. 1.** $\mathsf{LEAK}(\mathcal{P}, \mathsf{P}_1, \mathsf{P}_2, \mathcal{W}, \ell)$, the exfiltration resistance security game for a reverse
firewall $\mathcal{W}$ for party $\mathsf{P}_1$ in protocol $\mathcal{P}$ against party $\mathsf{P}_2$. $\mathcal{A}$ is the adversary, $\ell$ the
security parameter, $\mathsf{st}_{\overline{\mathsf{P}}_2}$ the state of $\overline{\mathsf{P}}_2$ after the run of the protocol, $I$ valid input for
$\mathcal{P}$, and $\mathcal{T}^*$ is the transcript of running protocol $\mathcal{P}_{\mathsf{P}_1 \to \mathsf{P}^*, \mathsf{P}_2 \to \overline{\mathsf{P}}_2}(I)$.

The advantage of any adversary $\mathcal{A}$ in the game LEAK is defined as

$$
\mathsf{Adv}^{\mathsf{LEAK}}_{\mathcal{A}, \mathcal{W}}(\ell) = \Pr[\mathsf{LEAK}(\mathcal{P}, \mathsf{P}_1, \mathsf{P}_2, \mathcal{W}, \ell) = 1] - 1/2.
$$

**Definition 4 (Exfiltration-resistant CRFs).** *For a protocol $\mathcal{P}$ that satisfies
functionality $\mathcal{F}$ and a reverse firewall $\mathcal{W}$,*

- *$\mathcal{W}$ is strongly exfiltration-resistant for party $\mathsf{P}_1$ against party $\mathsf{P}_2$ in protocol $\mathcal{P}$
  if for any PPT adversary $\mathcal{A}$, $\mathsf{Adv}^{\mathsf{LEAK}}_{\mathcal{A}, \mathcal{W}}(\ell)$ is negligible in the security parameter
  $\ell$; and*
- *$\mathcal{W}$ is weakly exfiltration-resistant for party $\mathsf{P}_1$ against party $\mathsf{P}_2$ in protocol $\mathcal{P}$,
  if for any PPT adversary $\mathcal{A}$, $\mathsf{Adv}^{\mathsf{LEAK}}_{\mathcal{A}, \mathcal{W}}(\ell)$ is negligible in the security parameter
  $\ell$ provided that $\overline{\mathsf{P}}_1$ maintains $\mathcal{F}$ for $\mathsf{P}_1$.*

*When $\mathcal{P}, \mathcal{F}, \mathsf{P}_1$ are clear, we simple say that $\mathcal{W}$ is strongly exfiltration-
resistant against $\mathsf{P}_2$ or weakly exfiltration-resistant against $\mathsf{P}_2$. In the special case
when $\mathsf{P}_2$ is empty, we say that $\mathcal{W}$ is exfiltration-resistant against eavesdroppers.*

### 2.2 Smooth Projective Hash Function

An SPHF is based on a domain $\mathcal{X}$ and an $\mathcal{NP}$ language $\mathcal{L}$, where $\mathcal{L}$ contains
a subset of the elements of the domain $\mathcal{X}$, i.e., $\mathcal{L} \subset \mathcal{X}$. An SPHF system over
a language $\mathcal{L} \subset \mathcal{X}$, onto a set $\mathcal{Y}$, is defined by the following five algorithms
(SPHFSetup, HashKG, ProjKG, Hash, ProjHash):

- SPHFSetup($1^\ell$) : The SPHFSetup algorithm takes as input a security parameter $\ell$, generates the *global parameters* param and the description of an $\mathcal{NP}$ language $\mathcal{L}$, outputs pp = $(\mathcal{L}, \text{param})$ as the public parameter.
- HashKG(pp) : The HashKG algorithm generates a *hashing key* hk;
- ProjKG(pp, hk, $C$) : The ProjKG algorithm derives the *projection key* hp from the hashing key hk and possibly an element $C$;
- Hash(pp, hk, $C$) : The Hash algorithm takes as input an element $C$ and the hashing key hk, outputs the hash value hv $\in \mathcal{Y}$;
- ProjHash(pp, hp, $C$, $w$) : The ProjHash algorithm takes as input the projection key hp and an element $C$ with the witness $w$ to the fact that $C \in \mathcal{L}$, outputs the hash value hv $\in \mathcal{Y}$.

SPHFs could be classified into two types according to whether ProjKG takes an element as input. The Gennaro and Lindell [15] type (GL-SPHF) allows hp to depend on $C$ while the Katz and Vaikuntanathan [17] type (KV-SPHF) does not. As shown later, our proposed new SPHF falls in the KV-SPHF category.

An SPHF should satisfy the following two properties.

**Correctness.** Formally, for any element $C \in \mathcal{L}$ with $w$ the witness, we have

$$\Pr\left[ \text{hv} \neq \text{hv}' : \begin{array}{l} \text{pp} \xleftarrow{\$} \text{SPHFSetup}(1^\ell); \\ \text{hk} \xleftarrow{\$} \text{HashKG}(\text{pp}); \text{hp} \leftarrow \text{ProjKG}(\text{pp}, \text{hk}); \\ \text{hv} \leftarrow \text{Hash}(\text{pp}, \text{hk}, C); \\ \text{hv}' \leftarrow \text{ProjHash}(\text{pp}, \text{hp}, C, w) \end{array} \right] \leq \text{negl}(\ell).$$

**Smoothness.** For any $C \in \mathcal{X} \backslash \mathcal{L}$, the following two distributions are statistically indistinguishable,

$$\mathcal{V}_1 = \{(\text{pp}, C, \text{hp}, \text{hv}) | \text{hv} = \text{Hash}(\text{hk}, C')\}, \mathcal{V}_2 = \{(\text{pp}, C, \text{hp}, \text{hv}) | \text{hv} \xleftarrow{\$} \mathcal{Y}\}.$$

That is, $\text{Adv}_{\text{SPHF}}^{\text{smooth}}(\ell) = \sum_{v \in \mathcal{Y}} |\Pr_{\mathcal{V}_1}[\text{hv} = v] - \Pr_{\mathcal{V}_2}[\text{hv} = v]| \leq \text{negl}(\ell)$.

It is required that one could efficiently sample elements from the set $\mathcal{X}$. That is, one could run a polynomial time algorithm SampYes(pp) to sample an element $(C, w)$ from $\mathcal{L}$ where $w$ is the witness to the membership $C \in \mathcal{L}$ and another polynomial time algorithm SampNo(pp) to sample an element $C$ from $\mathcal{X} \backslash \mathcal{L}$. The subset membership problem between $\mathcal{L}$ and $\mathcal{X}$ is usually required to be difficult, which is defined as follows.

**Definition 5 (Hard Subset Membership Problem).** *The subset membership problem* (SMP) *is hard on* $(\mathcal{X}, \mathcal{L})$ *for an* SPHF *that consists of* (SPHFSetup, HashKG, ProjKG, Hash, ProjHash), *if for any PPT adversary* $\mathcal{A}$,

$$\text{Adv}_{\mathcal{A}, \text{SPHF}}^{\text{SMP}}(\ell) = \Pr\left[ b' = b : \begin{array}{l} \text{pp} \xleftarrow{\$} \text{SPHFSetup}(1^\ell); \\ \text{hk} \xleftarrow{\$} \text{HashKG}(\text{pp}); \text{hp} \leftarrow \text{ProjKG}(\text{pp}, \text{hk}); \\ b \xleftarrow{\$} \{0, 1\}; (C_0, w) \xleftarrow{\$} \text{SampYes}(\text{pp}); \\ C_1 \xleftarrow{\$} \text{SampNo}(\text{pp}); \\ b' \leftarrow \mathcal{A}(\text{pp}, \text{hk}, \text{hp}, C_b) \end{array} \right] - \frac{1}{2} \leq \text{negl}(\ell).$$

# 3   Malleable Smooth Projective Hash Function

## 3.1   Definition

A malleable SPHF is defined by a tuple of algorithms (SPHFSetup, HashKG, ProjKG, Hash, ProjHash, MaulK, MaulH, ReranE, ReranH) which work as follows:

- SPHFSetup, HashKG, ProjKG, Hash, ProjHash are the same as in the classical SPHF;
- MaulK$(pp, hp, \widetilde{r})$. The MaulK algorithm takes as input a projection key hp and randomness $\widetilde{r}$, outputs a new projection key $\widetilde{hp}$;
- MaulH$(pp, hp, \widetilde{r}, C)$. The MaulH algorithm takes as input a projection key hp, the randomness $\widetilde{r}$ and an element $C$, outputs the hash value $\widetilde{hv}$;
- ReranE$(pp, C, \widetilde{w})$. The ReranE algorithm takes as input an element $C$ and the randomness $\widetilde{w}$, outputs a new element $\widetilde{C}$;
- ReranH$(pp, hp, C, \widetilde{w})$. The ReranH algorithm takes as input the projection key hp, an element $C$ and the randomness $\widetilde{w}$, outputs the hash value $\widetilde{hv}$;

We describe two randomness sampling algorithms named SampR and SampW. One could run SampR(pp) to sample $\widetilde{r}$ from the distribution of randomness using which we generate the hashing key. The algorithm SampW(pp) can be used to sample $\widetilde{w}$ from the witness distribution of the language.

Now we are ready to describe the properties of a malleable SPHF. In addition to the properties captured by a classical SPHF, a malleable SPHF also satisfies the following new properties which are essential in our constructions of CRFs.

**Definition 6 (Projection Key Malleability).** *A smooth projective hash function is projection key-malleable if the following properties hold.*

- **Key Indistinguishability.** *For any PPT adversary* $\mathcal{A} = (\mathcal{A}_1, \mathcal{A}_2)$,

$$
\mathsf{Adv}^{Key\text{-}Ind}_{\mathcal{A}, \mathsf{MSPHF}}(\ell) = \Pr \left[ b' = b : 
\begin{array}{l}
pp \xleftarrow{\$} \mathsf{SPHFSetup}(1^\ell); \\
(hp_1, hp_2, st) \leftarrow \mathcal{A}_1(pp); \\
b \xleftarrow{\$} \{0,1\}; \widetilde{r} \xleftarrow{\$} \mathsf{SampR}(pp); \\
\widetilde{hp} \leftarrow \mathsf{MaulK}(pp, hp_b, \widetilde{r}); \\
b' \leftarrow \mathcal{A}_2(pp, st, hp_1, hp_2, \widetilde{hp})
\end{array}
\right] - \frac{1}{2} \leq \mathsf{negl}(\ell).
$$

- **Projection Consistency.** *For any element* $C \in \mathcal{X}$,

$$
\Pr \left[ hv \neq hv' : 
\begin{array}{l}
pp \xleftarrow{\$} \mathsf{SPHFSetup}(1^\ell); \\
hk \xleftarrow{\$} \mathsf{HashKG}(pp); hp \leftarrow \mathsf{ProjKG}(pp, hk); \\
\widetilde{r} \xleftarrow{\$} \mathsf{SampR}(pp); \widetilde{hp} \leftarrow \mathsf{MaulK}(pp, hp, \widetilde{r}); \\
hv \leftarrow \mathsf{Hash}(pp, \widetilde{hk}, C); \\
\widetilde{hv} \leftarrow \mathsf{MaulH}(pp, hp, \widetilde{r}, C); \\
hv' \leftarrow \mathsf{Hash}(pp, hk, C) * \widetilde{hv}
\end{array}
\right] \leq \mathsf{negl}(\ell).
$$

*where* $\widetilde{hk}$ *is the associated hashing key of* $\widetilde{hp} \leftarrow \mathsf{MaulK}(pp, hp, \widetilde{r})$ *and* $*$ *denotes the operation between two hash values in* $\mathcal{Y}$.

**Definition 7 (Element Re-randomizability).** *A smooth projective hash function is element-rerandomizable if the followings hold.*

– **Element Indistinguishability.** *For any PPT adversary* $\mathcal{A} = (\mathcal{A}_1, \mathcal{A}_2)$,

$$\mathsf{Adv}_{\mathcal{A},\mathsf{MSPHF}}^{Element\text{-}Ind}(\ell) = \Pr\left[b' = b : \begin{array}{l} \mathsf{pp} \xleftarrow{\$} \mathsf{SPHFSetup}(1^\ell); \\ (C_1, C_2, \mathsf{st}) \leftarrow \mathcal{A}_1(\mathsf{pp}); \\ b \xleftarrow{\$} \{0,1\}; \widetilde{w} \xleftarrow{\$} \mathsf{SampW}(\mathsf{pp}); \\ \widetilde{C} \leftarrow \mathsf{ReranE}(\mathsf{pp}, C_b, \widetilde{w}); \\ b' \leftarrow \mathcal{A}_2(\mathsf{pp}, \mathsf{st}, C_1, C_2, \widetilde{C}) \end{array}\right] - \frac{1}{2} \leq \mathsf{negl}(\ell).$$

– **Rerandomization Consistency.** *For any element* $C \in \mathcal{X}$,

$$\Pr\left[\mathsf{hv} \neq \mathsf{hv}' : \begin{array}{l} \mathsf{pp} \xleftarrow{\$} \mathsf{SPHFSetup}(1^\ell); \\ \mathsf{hk} \xleftarrow{\$} \mathsf{HashKG}(\mathsf{pp}); \mathsf{hp} \leftarrow \mathsf{ProjKG}(\mathsf{pp}, \mathsf{hk}); \\ \widetilde{w} \xleftarrow{\$} \mathsf{SampW}(\mathsf{pp}); \widetilde{C} \leftarrow \mathsf{ReranE}(\mathsf{pp}, C, \widetilde{w}); \\ \mathsf{hv} \leftarrow \mathsf{Hash}(\mathsf{pp}, \mathsf{hk}, \widetilde{C}); \\ \widetilde{\mathsf{hv}} \leftarrow \mathsf{ReranH}(\mathsf{pp}, \mathsf{hp}, C, \widetilde{w}); \\ \mathsf{hv}' \leftarrow \mathsf{Hash}(\mathsf{pp}, \mathsf{hk}, C) * \widetilde{\mathsf{hv}} \end{array}\right] \leq \mathsf{negl}(\ell).$$

– **Membership Preservation.** *For any element* $C \in \mathcal{X}$, *let* $\widetilde{C} \leftarrow \mathsf{ReranE}(\mathsf{pp}, C, \widetilde{w})$ *where* $\widetilde{w} \xleftarrow{\$} \mathsf{SampW}(\mathsf{pp})$, *we have* $\widetilde{C} \in \mathcal{L}$ *if and only if* $C \in \mathcal{L}$.

**Definition 8 (Malleable SPHF).** *An SPHF is malleable if it is projection key-malleable and element-rerandomizable.*

### 3.2  Malleable SPHFs from Graded Rings

In this section, we show that under some conditions, the SPHF framework from graded rings proposed by Benhamouda et al. [9] could be extended into malleable SPHF. The main goal of this part is to demonstrate the feasibility of our definition. We remark that malleable SPHFs can be constructed using other approaches.

**Graded Rings.** Benhamouda et al. [9] proposed a generic framework for SPHFs using a new notion named graded rings, which is a common formalization for cyclic groups, bilinear groups, and even multilinear groups. The graded ring provides a practical way to manipulate elements of various groups involved in pairings and more generally, in multi-linear maps. Before describing their SPHF framework, we briefly recall the notion of graded rings. The notation $\oplus$ and $\odot$ correspond to the addition operation and the multiplication operation, respectively. For simplicity, here we focus on cyclic groups and symmetric bilinear groups. Let $\mathbb{G}, \mathbb{G}_T$ be two multiplicative groups with the same prime order $p$ with a symmetric bilinear map $e : \mathbb{G} \times \mathbb{G} \to \mathbb{G}_T$.

- For any $a, b \in \mathbb{Z}_p$, $a \oplus b = a + b$, $a \odot b = a \cdot b$;
- For any $u_1, v_1 \in \mathbb{G}$, $u_1 \oplus v_1 = u_1 \cdot v_1$, $u_1 \ominus v_1 = u_1 \cdot v_1^{-1}$, and for any $c \in \mathbb{Z}_p$, $c \odot u_1 = u_1^c$;
- For any $u_T, v_T \in \mathbb{G}_T$, $u_T \oplus v_T = u_T \cdot v_T$, $u_T \ominus v_T = u_T \cdot v_T^{-1}$, and for any $c \in \mathbb{Z}_p$, $c \odot u_T = u_T^c$;
- For any $u_1, v_1 \in \mathbb{G}$, $u_1 \odot v_1 = e(u_1, v_1) \in \mathbb{G}_T$.

That is, $\oplus$ and $\odot$ correspond to the addition and the multiplication of the exponents. The notations could be extended in a natural way when it comes to the case of vectors and matrices.

We are now ready to describe the framework of SPHF introduced in [9]. For a language $\mathcal{L}$ which is specified by the parameter aux, suppose there exist two positive integers $m$ and $n$, a function $\Gamma : \mathcal{X} \longmapsto \mathbb{G}^{m \times n}$ (for generating the element basis) and a function $\Theta_{\mathsf{aux}} : \mathcal{X} \longmapsto \mathbb{G}^{1 \times n}$, such that for any element $C \in \mathcal{X}$,

$$(C \in \mathcal{L}) \Longleftrightarrow (\exists \boldsymbol{\lambda} \in \mathbb{Z}_p^{1 \times m} \text{ s.t.}, \Theta_{\mathsf{aux}}(C) = \boldsymbol{\lambda} \odot \Gamma(C)).$$

In other words, $C \in \mathcal{L}$ if and only if $\Theta_{\mathsf{aux}}(C)$ is a linear combination of the rows in $\Gamma(C)$. Here it is required that the one who knows the witness $w$ of the membership $C \in \mathcal{L}$ can efficiently compute the above linear combination $\boldsymbol{\lambda}$. This requirement seems somewhat strong but is actually verified by very expressive languages [9].

With the above notations, the hashing key in an SPHF is a vector $\mathsf{hk} := \boldsymbol{\alpha} = (\alpha_1, ..., \alpha_n)^\mathsf{T} \overset{\$}{\leftarrow} \mathbb{Z}_p^n$ and the projection key for an element $C$ is $\mathsf{hp} := \boldsymbol{\gamma}(C) = \Gamma(C) \odot \boldsymbol{\alpha} \in \mathbb{G}^k$. Then the hash value computation for an element $C$ is:

$$\mathsf{Hash}(\mathsf{pp}, \mathsf{hk}, C) := \Theta_{\mathsf{aux}}(C) \odot \boldsymbol{\alpha}, \qquad \mathsf{ProjHash}(\mathsf{pp}, \mathsf{hp}, C, w) := \boldsymbol{\lambda} \odot \boldsymbol{\gamma}(C).$$

Intuitively, if $C \in \mathcal{L}$ with $\boldsymbol{\lambda}$, then we have,

$$\mathsf{Hash}(\mathsf{pp}, \mathsf{hk}, C) = \Theta_{\mathsf{aux}}(C) \odot \boldsymbol{\alpha} = \boldsymbol{\lambda} \odot \Gamma(C) \odot \boldsymbol{\alpha} = \boldsymbol{\lambda} \odot \boldsymbol{\gamma}(C) = \mathsf{ProjHash}(\mathsf{pp}, \mathsf{hp}, C, w).$$

This guarantees the correctness of the SPHF. As for the smoothness property, we can see that for any element $C \notin \mathcal{L}$ and a projection key $\mathsf{hp} = \boldsymbol{\gamma}(C) = \Gamma(C) \odot \boldsymbol{\alpha}$, the vector $\Theta_{\mathsf{aux}}(C)$ is not in the linear span of $\Gamma(C)$, and thus its hash value $\mathsf{hv} = \mathsf{Hash}(\mathsf{pp}, \mathsf{hk}, C) = \Theta_{\mathsf{aux}}(C) \odot \boldsymbol{\alpha}$ is independent from $\mathsf{hp} = \Gamma(C) \odot \boldsymbol{\alpha}$. We refer the readers to [9] for a more detailed analysis. One can note that if the function $\Gamma : \mathcal{X} \longmapsto \mathbb{G}^{m \times n}$ is a constant function, the corresponding SPHF is of KV-SPHF type, otherwise it is of GL-SPHF type.

**A Simple Example.** We illustrate this framework for the DDH language. Let $g_1, g_2$ be two generators of a cyclic group $\mathbb{G}$ of prime order $p$. Let $\mathcal{X} = \mathbb{G}^{1 \times 2}$ and $\mathcal{L} = \{(u_1, u_2) \mid r \in \mathbb{Z}_p, \text{ s.t.}, u_1 = g_1^r, u_2 = g_2^r\}$. For any $C = (u_1, u_2) \in \mathcal{L}$, $\Theta_{\mathsf{aux}}(C) = C$, $\Gamma(C) = (g_1, g_2)$ and the witness for $C \in \mathcal{L}$ is $w = r$ and here $\boldsymbol{\lambda} = w = r$. The hashing key is $\mathsf{hk} = \boldsymbol{\alpha} = (\alpha_1, \alpha_2)^\mathsf{T} \overset{\$}{\leftarrow} \mathbb{Z}_p^2$ and the projection key is $\mathsf{hp} = \boldsymbol{\gamma}(C) = \Gamma(C) \odot \boldsymbol{\alpha} = g_1^{\alpha_1} g_2^{\alpha_2} \in \mathbb{G}$. We then have

$$\mathsf{Hash}(\mathsf{pp}, \mathsf{hk}, C) = \Theta_{\mathsf{aux}}(C) \odot \boldsymbol{\alpha} = (u_1, u_2) \odot (\alpha_1, \alpha_2)^\mathsf{T} = u_1^{\alpha_1} u_2^{\alpha_2},$$

$$\mathsf{ProjHash}(\mathsf{pp}, \mathsf{hp}, C, w = r) = \boldsymbol{\lambda} \odot \boldsymbol{\gamma}(C) = r \odot (g_1^{\alpha_1} g_2^{\alpha_2}) = (g_1^{\alpha_1} g_2^{\alpha_2})^r.$$

This is exactly the original SPHF of Cramer and Shoup for the DDH language in [12].

**Generic Construction of Malleable SPHFs.** With the above definitions, we present a generic framework for constructing malleable SPHF based on graded rings.

- $\mathsf{SPHFSetup}(1^\ell)$. Output pp which defines the set $\mathcal{X}$ and the language $\mathcal{L}$ with the positive integers $m$ and $n$, and functions $\Gamma$ and $\Theta_{\mathsf{aux}}$.
- $\mathsf{HashKG}(\mathsf{pp})$. Sample $\boldsymbol{\alpha} \xleftarrow{\$} \mathbb{Z}_p^n$ and output $\mathsf{hk} = \boldsymbol{\alpha}$.
- $\mathsf{ProjKG}(\mathsf{pp}, \mathsf{hk}, C)$. Output $\mathsf{hp} = \boldsymbol{\gamma}(C) = \Gamma(C) \odot \boldsymbol{\alpha} \in \mathbb{G}^k$.
- $\mathsf{Hash}(\mathsf{pp}, \mathsf{hk}, C)$. Output $\mathsf{hv} = \Theta_{\mathsf{aux}}(C) \odot \boldsymbol{\alpha}$.
- $\mathsf{ProjHash}(\mathsf{pp}, \mathsf{hp}, C, w)$. Output $\mathsf{hv} = \boldsymbol{\lambda} \odot \boldsymbol{\gamma}(C)$ where $\boldsymbol{\lambda}$ is derived from $w$.
- $\mathsf{MaulK}(\mathsf{pp}, \mathsf{hp}, \widetilde{r})$. To re-randomize a projection key $\mathsf{hp} = \boldsymbol{\gamma}(C)$ using the randomness $\widetilde{r}$, compute and output $\widetilde{\mathsf{hp}}$ as:

$$\Delta\mathsf{hp} = \Gamma(C) \odot \widetilde{r}, \qquad \widetilde{\mathsf{hp}} = \boldsymbol{\gamma}(C) \oplus \Delta\mathsf{hp}.$$

- $\mathsf{MaulH}(\mathsf{pp}, \mathsf{hp}, \widetilde{r}, C)$. Output $\widetilde{\mathsf{hv}} = \Theta_{\mathsf{aux}}(C) \odot \widetilde{r}$.
- $\mathsf{ReranE}(\mathsf{pp}, C, \widetilde{w})$. To re-randomize an element $C$ using the random witness $\widetilde{w}$, derive $\widetilde{\boldsymbol{\lambda}}$ from $\widetilde{w}$, compute and output $\widetilde{C}$ as:

$$\Delta C = \widetilde{\boldsymbol{\lambda}} \odot \Gamma(C), \qquad \widetilde{C} = \Theta_{\mathsf{aux}}(C) \oplus \Delta C.$$

- $\mathsf{ReranH}(\mathsf{PP}, \mathsf{hp}, C, \widetilde{w})$. Derive $\widetilde{\boldsymbol{\lambda}}$ from $\widetilde{w}$ and output $\widetilde{\mathsf{hv}} = \widetilde{\boldsymbol{\lambda}} \odot \boldsymbol{\gamma}(C)$.

For the above construction, we have the following theorem.

**Theorem 1.** *The above generic construction is a malleable smooth projective hash function if the following conditions hold:*

a. $\Theta : \mathcal{X} \longmapsto \mathbb{G}^{1 \times n}$ *is an identity function; (Diverse Group [12])*
b. $\Gamma : \mathcal{X} \longmapsto \mathbb{G}^{k \times n}$ *is a constant function; (KV-SPHF type)*
c. *The subset membership problem between $\mathcal{L}$ and $\mathcal{X}$ is hard.*

*Proof.* It should be clear that the construction is an SPHF as it is exactly the graded ring-based SPHF framework proposed in [9]. Below we show that it is *projection key-malleable* and *element-rerandomizable*.

PROJECTION KEY MALLEABILITY. For any $\widetilde{r} = (r_1, ..., r_n)^\mathsf{T} \xleftarrow{\$} \mathsf{SampR}(\mathsf{pp})$, any element $C \in \mathcal{X}$, we have that

$$\begin{aligned}
\mathsf{MaulK}(\mathsf{pp}, \mathsf{hp}, \widetilde{r}) &= \boldsymbol{\gamma}(C) \oplus (\Gamma(C) \odot \widetilde{r}) \\
&= \Gamma(C) \odot \boldsymbol{\alpha} \oplus (\Gamma(C) \odot \widetilde{r}) \\
&= \Gamma(C) \odot (\boldsymbol{\alpha} \oplus \widetilde{r}) = \widetilde{\mathsf{hp}}.
\end{aligned}$$

One can easily notice that the new projection key $\widetilde{\mathsf{hp}}$ is independent of $\mathsf{hp}$, as the randomness $\widetilde{r}$ is uniformly chosen and $\Gamma$ is a constant function. Therefore, for any PPT adversary $\mathcal{A}$, we have that $\mathsf{Adv}_{\mathcal{A},\mathsf{MSPHF}}^{\mathsf{Key\text{-}Ind}}(\ell)$ is negligible. Moreover, the associated hashing key of $\widetilde{\mathsf{hp}}$ is $\widetilde{\mathsf{hk}} = \widetilde{\alpha} = \alpha \oplus \widetilde{r} = (\alpha_1 + r_1, ..., \alpha_n + r_n)^{\mathsf{T}} \in \mathbb{Z}_p^n$. Therefore, we have

$$\begin{aligned}
\mathsf{Hash}(\mathsf{pp}, \widetilde{\mathsf{hk}}, C) &= \Theta_{\mathsf{aux}}(C) \odot \widetilde{\alpha} = \Theta_{\mathsf{aux}}(C) \odot (\alpha \oplus \widetilde{r}) \\
&= \Theta_{\mathsf{aux}}(C) \odot \alpha \oplus \Theta_{\mathsf{aux}}(C) \odot \widetilde{r} \\
&= \mathsf{Hash}(\mathsf{pp}, \mathsf{hk}, C) \oplus \mathsf{MaulH}(\mathsf{pp}, \mathsf{hp}, \widetilde{r}, C).
\end{aligned}$$

This shows the projection consistency and thus the projection key is malleable.

ELEMENT RE-RANDOMIZABILITY. For any randomness $\widetilde{w}$, and any element $C \in \mathcal{X}$, we have that, $\mathsf{ReranE}(\mathsf{pp}, C, \widetilde{w}) = \Theta_{\mathsf{aux}}(C) \oplus (\widetilde{\lambda} \odot \Gamma(C)) = \widetilde{C}$. Due to the uniformly chosen randomness $\widetilde{w}$ (which derives $\widetilde{\lambda}$) and the hard subset membership problem, we have that $\widetilde{C}$ is computationally independent of $C$. Particularly, $\widetilde{\lambda} \odot \Gamma(C)$ could be viewed as a random chosen element from $\mathcal{L}$ as $\Gamma$ is a constant function (i.e., $\Gamma(C) = \Gamma(\widetilde{C})$). Therefore, for any PPT adversary $\mathcal{A}$, if $\mathsf{Adv}_{\mathcal{A},\mathsf{MSPHF}}^{\mathsf{Element\text{-}Ind}}(\ell)$ is non-negligible, we could use $\mathcal{A}$ to break the hard subset membership problem, which is a contradiction. Noting that here we require $\Theta$ to be an identity function, i.e., $\Theta_{\mathsf{aux}}(\widetilde{C}) = \widetilde{C}$, we have

$$\begin{aligned}
\mathsf{Hash}(\mathsf{pp}, \mathsf{hk}, \widetilde{C}) &= \Theta_{\mathsf{aux}}(\widetilde{C}) \odot \alpha = \widetilde{C} \odot \alpha \\
&= (\Theta_{\mathsf{aux}}(C) \oplus \widetilde{\lambda} \odot \Gamma(C))) \odot \alpha \\
&= \Theta_{\mathsf{aux}}(C) \odot \alpha \oplus \widetilde{\lambda} \odot \Gamma(C) \odot \alpha \\
&= \Theta_{\mathsf{aux}}(C) \odot \alpha \oplus \widetilde{\lambda} \odot \gamma(C) \\
&= \mathsf{Hash}(\mathsf{pp}, \mathsf{hk}, C) \oplus \mathsf{ReranH}(\mathsf{pp}, \mathsf{hp}, C, \widetilde{w}).
\end{aligned}$$

The above illustrates the *rerandomization consistency*. Below we show that the element rerandomization is also *membership-preserving*. Given any element $C \in \mathcal{L}$ with the witness $C = \lambda$, for any randomness $\widetilde{w}$ that derives $\widetilde{\lambda}$, we have that,

$$\begin{aligned}
\mathsf{ReranE}(\mathsf{pp}, C, \widetilde{w}) &= \Theta_{\mathsf{aux}}(C) \oplus (\widetilde{\lambda} \odot \Gamma(C)) \\
&= \lambda \odot \Gamma(C) \oplus (\widetilde{\lambda} \odot \Gamma(C)) \\
&= (\lambda \oplus \widetilde{\lambda}) \odot \Gamma(C) \\
&= \lambda' \odot \Gamma(\widetilde{C}) = \Theta_{\mathsf{aux}}(\widetilde{C}) = \widetilde{C}.
\end{aligned}$$

The above holds due to the fact that $\Theta$ is an identity function, i.e., $\Theta_{\mathsf{aux}}(\widetilde{C}) = \widetilde{C}$ and $\Gamma$ is a constant function, i.e., $\Gamma(C) = \Gamma(\widetilde{C})$. The witness to the fact $\widetilde{C} \in \mathcal{L}$ is $\lambda' = \lambda \oplus \widetilde{\lambda}$. For any element $C \in \mathcal{X} \backslash \mathcal{L}$, the vector $\Theta_{\mathsf{aux}}(C)$ is not in the linear span of $\Gamma(C)$. Therefore, for any $\widetilde{w}$, let $\widetilde{C} = \mathsf{ReranE}(\mathsf{pp}, C, \widetilde{w}) = \Theta_{\mathsf{aux}}(C) \oplus (\widetilde{\lambda} \odot \Gamma(C))$, we trivially have that $\Theta_{\mathsf{aux}}(\widetilde{C}) = \widetilde{C}$ is not in the linear span of $\Gamma(C)$ and thus $\widetilde{C} \in \mathcal{X} \backslash \mathcal{L}$.

**Instantiation from the $k$-Linear Assumption.** We instantiate the above framework based on the $k$-Linear ($k$-Lin) assumption. Let $\mathbb{G}$ be a group with prime order $p$ and $g$ a generator. The $k$-Lin assumption asserts that $g_{k+1}^{r_1+\cdots+r_k}$ is pseudo-random given $g_1,\cdots,g_{k+1},g_1^{r_1},\cdots,g_k^{r_k}$ where $g_1,\cdots,g_{k+1} \xleftarrow{R} \mathbb{G}, r_1,\cdots,r_k \xleftarrow{R} \mathbb{Z}_p$. Note that the DDH assumption is equivalent to the 1-Lin assumption.

We show how to construct a malleable SPHF from $k$-Lin assumption. The language is defined as,

$$\mathcal{L} = \{(c_1,\cdots,c_k)|\exists(r_1,\cdots,r_k) \in \mathbb{Z}_p^k, \text{s.t.}, c_1 = g_1^{r_1},\cdots,c_k = g_k^{r_k}, c_{k+1} = g_{k+1}^{\sum_{i=1}^k r_i})\}.$$

For any $C = (c_1,\cdots,c_{k+1})$, we have $\Theta_{\mathsf{aux}}(C) = C$ and

$$\Gamma(C) = \begin{pmatrix} g_1 & 1 & \cdots & 1 & g_{k+1} \\ 1 & g_2 & \cdots & 1 & g_{k+1} \\ \vdots & \vdots & \ddots & \vdots & \vdots \\ 1 & 1 & \cdots & g_k & g_{k+1} \end{pmatrix} \in \mathbb{G}^{k\times(k+1)}.$$

For any $C \in \mathcal{L}$ with witness $\boldsymbol{\lambda} = \boldsymbol{w} = (r_1,\cdots,r_k)$, we have, $\Theta_{\mathsf{aux}}(C) = (g_1^{r_1},\ \cdots,\ g_k^{r_k},\ g_{k+1}^{\sum_{i=1}^k r_i}) = \boldsymbol{\lambda} \odot \Gamma(C)$. Let $\mathsf{pp} = (\mathbb{G},p,g_1,\cdots,g_{k+1})$, $\widetilde{\boldsymbol{r}} = (\beta_1,\cdots,\beta_{k+1})^\mathsf{T}$ and $\widetilde{\boldsymbol{\lambda}} = \widetilde{\boldsymbol{w}} = (\eta_1,\cdots,\eta_k)$. The instantiation is as follows:

- $\mathsf{HashKG}(\mathsf{pp}) : \mathsf{hk} = \boldsymbol{\alpha} = (\alpha_1,\cdots,\alpha_{k+1})^\mathsf{T} \xleftarrow{\$} \mathbb{Z}_p^k$;
- $\mathsf{ProjKG}(\mathsf{pp},\mathsf{hk},C) : \mathsf{hp} = \boldsymbol{\gamma}(C) = \Gamma(C) \odot \boldsymbol{\alpha} = (g_1^{\alpha_1}g_{k+1}^{\alpha_{k+1}},\cdots,g_k^{\alpha_k}g_{k+1}^{\alpha_{k+1}})^\mathsf{T}$;
- $\mathsf{Hash}(\mathsf{pp},\mathsf{hk},C) : \mathsf{hv} = (c_1,\cdots,c_{k+1}) \odot (\alpha_1,\cdots,\alpha_{k+1})^\mathsf{T} = \prod_{i=1}^k c_i^{\alpha_i}$;
- $\mathsf{ProjHash}(\mathsf{pp},\mathsf{hp},C,\boldsymbol{w}) : \mathsf{hv} = \boldsymbol{\lambda} \odot \boldsymbol{\gamma}(C) = \prod_{i=1}^k (g_i^{\alpha_i}g_{k+1}^{\alpha_{k+1}})^{r_i}$;
- $\mathsf{MaulK}(\mathsf{pp},\mathsf{hp},\widetilde{\boldsymbol{r}}) : \widetilde{\mathsf{hp}} = \boldsymbol{\gamma}(C) \oplus (\Gamma(C) \odot \widetilde{\boldsymbol{r}}) = (g_1^{\alpha_1}g_{k+1}^{\alpha_{k+1}},\cdots,g_k^{\alpha_k}g_{k+1}^{\alpha_{k+1}})^\mathsf{T} \oplus (g_1^{\beta_1}g_{k+1}^{\beta_{k+1}},\cdots,g_k^{\beta_k}g_{k+1}^{\beta_{k+1}})^\mathsf{T} = (g_1^{\alpha_1+\beta_1}g_{k+1}^{\alpha_{k+1}+\beta_{k+1}},\cdots,g_k^{\alpha_k+\beta_k}g_{k+1}^{\alpha_{k+1}+\beta_{k+1}})^\mathsf{T}$;
- $\mathsf{MaulH}(\mathsf{pp},\mathsf{hp},\widetilde{\boldsymbol{r}},C) : \widetilde{\mathsf{hv}} = \Theta_{\mathsf{aux}}(C) \odot \widetilde{\boldsymbol{r}} = (c_1,\cdots,c_{k+1}) \odot (\beta_1,\cdots,\beta_{k+1})^\mathsf{T} = c_1^{\beta_1} \cdot c_2^{\beta_2}\cdots c_{k+1}^{\beta_{k+1}} = \prod_{i=1}^{k+1} c_i^{\beta_i}$;
- $\mathsf{ReranE}(\mathsf{pp},C,\widetilde{\boldsymbol{w}}) : \widetilde{C} = \Theta_{\mathsf{aux}}(C) \oplus (\widetilde{\boldsymbol{\lambda}} \odot \Gamma(C)) = (c_1 g_1^{\eta_1},\cdots,c_k g_k^{\eta_k},c_{k+1}g_{k+1}^{\sum_{i=1}^k \eta_i})$;
- $\mathsf{ReranH}(\mathsf{pp},\mathsf{hp},C,\widetilde{\boldsymbol{w}}) : \widetilde{\mathsf{hv}} = \widetilde{\boldsymbol{\lambda}} \odot \boldsymbol{\gamma}(C) = (\eta_1,\cdots,\eta_d) \odot (g_1^{\alpha_1}g_{k+1}^{\alpha_{k+1}},\cdots,g_k^{\alpha_k}g_{k+1}^{\alpha_{k+1}})^\mathsf{T} = \prod_{i=1}^k (g_i^{\alpha_i}g_{k+1}^{\alpha_{k+1}})^{\eta_i}$.

It is easy to verify that the above instantiation is a malleable SPHF as it satisfies all the conditions of Theorem 1.

*Remark.* Note that the function $\Theta_{\mathsf{aux}}$ is required to be an identity function in our framework. That is, the above generic construction is on diverse groups [12]. However, we remark that such a requirement is not necessary. We will show later (Sect. 4.2) a concrete malleable SPHF which demonstrates that instantiating malleable SPHF from graded rings can be done in different ways.

# 4  Generic Construction of CRFs via Malleable SPHFs

## 4.1  Warm-Up: Message-Transmission Protocol with CRFs

A message transmission protocol (MTP) enables one party, Alice, to securely communicate a message to another party, Bob. Here we focus on the unkeyed setting for message transmission. That is, both Alice and Bob have neither a shared secret key nor each other's public key. Specifically, the protocol does not assume a public-key infrastructure. It simply lets Bob send a randomly chosen public key as the first message and thereafter Alice sends an encryption of her message under Bob's public key as the second message. Since neither the sender nor the receiver can be authenticated in this setting, the strongest security guarantee is semantic security against passive adversaries. That is, the adversary should not be able to distinguish the protocol transcripts for transferring two different plaintexts which are chosen by the adversary. We remark that our framework can be seen as a generic construction of semantically secure public-key encryption that is both key malleable and re-randomizable defined in [14], and hence provides a more intuitive way to build two-round message-transmission protocols with CRFs. We show a two-round MTP constructed using SPHF in Fig. 2.

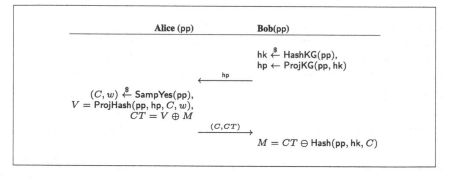

**Fig. 2.** Generic construction of two-round MTP from SPHF

**Theorem 2.** *The construction of MTP in Fig. 2 is correct and semantically secure.*

It should be clear that the protocol functionality is ensured by the correctness of the SPHF and the security is guaranteed by the pseudo-randomness of the SPHF, which is implied by the smoothness and the hardness of the subset membership problem.

**CRF for the Receiver.** In reality, a tampered implementation of Bob (the receiver) might choose an insecure public key so that an eavesdropper will be able to read Alice's plaintext. The key could also act as a channel to leak some secrets to Alice or an eavesdropper. Even assuming that the protocol is semantically secure, without the CRF, the compromised implementation of Bob can still leak some secret information to the outside. It is thus desirable for the CRF to resist exfiltration. Figure 3 shows the reverse firewall for Bob. The idea is that the CRF re-randomizes the public key chosen by Bob before it is sent to the outside world. To maintain the protocol functionality, it also intercepts Bob's incoming messages and converts Alice's ciphertext under the re-randomized key to that under Bob's original public key. The CRF should also preserve the semantic security of the protocol regardless of how Bob behaves. A computationally bounded adversary learns nothing about Alice's input plaintext from the transcript between Alice and Bob's CRF, even when the original public key chosen by Bob is insecure.

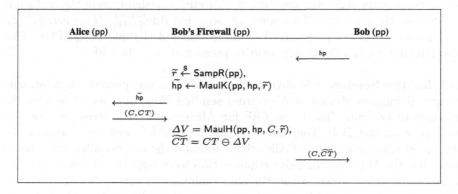

**Fig. 3.** Bob's CRF for the protocol shown in Fig. 2

**Theorem 3.** *The* CRF *for Bob shown in Fig. 3 maintains functionality and strongly preserves security for Bob, and strongly resists exfiltration against Alice, provided that the underlying* SPHF *is projection key-malleable.*

*Proof.* We verify that our construction satisfies the following properties.

*Functionality Maintaining.* For any ciphertext $(C, CT)$,

$$
\begin{aligned}
\widetilde{CT} = CT \ominus \Delta V &= CT \ominus \mathsf{MaulH}(\mathsf{pp}, \mathsf{hp}, C, \widetilde{r}) \\
&= M \oplus \mathsf{ProjHash}(\mathsf{pp}, \widetilde{\mathsf{hp}}, C, w) \ominus \mathsf{MaulH}(\mathsf{pp}, \mathsf{hp}, C, \widetilde{r}) \\
&= M \oplus \mathsf{Hash}(\mathsf{pp}, \widetilde{\mathsf{hk}}, C) \ominus \mathsf{MaulH}(\mathsf{pp}, \mathsf{hp}, C, \widetilde{r}) \\
&= M \oplus \mathsf{Hash}(\mathsf{pp}, \mathsf{hk}, C).
\end{aligned}
$$

The above holds due to the *projection consistency* of the *projection key malleability* in the underlying SPHF. Therefore, Bob is able to recover Alice's plaintext by computing $M = \widetilde{CT} \ominus \mathsf{Hash}(\mathsf{pp}, \mathsf{hk}, C)$.

*Strong Security Preservation and Strong Exfiltration Resistance.* It suffices to show that the CRF strongly resists exfiltration. Suppose there exists an adversary who has non-negligible advantage $\mathsf{Adv}^{\mathsf{LEAK}}_{\mathcal{A},\mathcal{W}}(\ell)$ in the game LEAK. We then show how to build an adversary $\mathcal{B}$ to break the *key indistinguishability* captured by the *projection key malleability* of the underlying SPHF by running $\mathcal{A}$. Recall that in the game LEAK, $\mathcal{A}$ would provide two parties $(\overline{\mathsf{P}}_1, \overline{\mathsf{P}}_2)$ which represent its chosen tampered implementations of Bob and Alice. $\mathcal{B}$ first runs the protocol between the honest party Bob and $\overline{\mathsf{P}}_2$, and obtains the output of Bob as $\mathsf{hp}_0$. $\mathcal{B}$ then runs again the protocol between $\overline{\mathsf{P}}_1$ and $\overline{\mathsf{P}}_2$, and obtains the output of $\overline{\mathsf{P}}_1$ as $\mathsf{hp}_1$. It then sends $(\mathsf{hp}_0, \mathsf{hp}_1)$ as the challenge projection keys for the key indistinguishability game, and receives the challenge re-randomized projection key $\widetilde{\mathsf{hp}}$. Finally, it forwards $\widetilde{\mathsf{hp}}$ to $\mathcal{A}$ as part of the challenge transcript $\mathcal{T}^*$ of the game LEAK and outputs the guess $b'$ of $\mathcal{A}$ as its guess. It is easy to see that the above behaviours of $\mathcal{B}$ are computationally indistinguishable from the real game LEAK from the view of $\mathcal{A}$. Therefore, we have that $\mathsf{Adv}^{\mathsf{Key\text{-}Ind}}_{\mathcal{B},\mathsf{MSPHF}}(\ell) \geq \mathsf{Adv}^{\mathsf{LEAK}}_{\mathcal{A},\mathcal{W}}(\ell)$, which contradicts the projection key malleability of the underling SPHF. This also trivially implies the strong security preservation of the CRF. $\qquad\square$

**CRF for the Sender.** It is obvious that a CRF cannot prevent an arbitrarily tampered implementation of Alice from sending Bob some secret besides the message to be sent. That is, no CRF for Alice can achieve strong exfiltration resistance against Bob. Therefore, the "best possible" security is against the corrupted implementations of Alice that maintain the functionality. One should note that the MTP functionality requires Bob to recover the plaintext message of Alice. In other words, a functionality-maintaining corruption of Alice can only send the given input but no other message. Formally, we have the following theorem for the CRF depicted in Fig. 4.

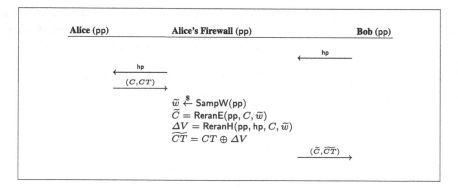

**Fig. 4.** Alice's CRF for the protocol shown in Fig. 2

**Theorem 4.** *The CRF for Alice shown in Fig. 4 maintains functionality and strongly preserves security for Alice, and weakly resists exfiltration against Bob, provided that the SPHF is element-rerandomizable.*

*Proof.* We verify that our construction satisfies the following properties.

*Functionality Maintaining.* One could easily have,

$$\begin{aligned}
\widetilde{CT} &= CT \oplus \Delta V = CT \oplus \mathsf{ReranH}(\mathsf{pp}, \mathsf{hp}, C, \widetilde{w}) \\
&= M \oplus \mathsf{ProjHash}(\mathsf{pp}, \mathsf{hp}, C, w) \oplus \mathsf{ReranH}(\mathsf{pp}, \mathsf{hp}, C, \widetilde{w}) \\
&= M \oplus \mathsf{Hash}(\mathsf{pp}, \mathsf{hk}, C) \oplus \mathsf{ReranH}(\mathsf{pp}, \mathsf{hp}, C, \widetilde{w}) \\
&= M \oplus \mathsf{Hash}(\mathsf{pp}, \mathsf{hk}, \widetilde{C}).
\end{aligned}$$

The above holds by the *rerandomization consistency* as the underlying SPHF is element re-randomizable. Bob is thus able to recover Alice's plaintext by computing $M = \widetilde{CT} \ominus \mathsf{Hash}(\mathsf{pp}, \mathsf{hk}, \widetilde{C})$.

*Strong Security Preservation and Weak Exfiltration Resistance.* For any tampered implementation of Alice that maintains functionality, suppose there exists an adversary who has non-negligible advantage $\mathsf{Adv}^{\mathsf{LEAK}}_{\mathcal{A},\mathcal{W}}(\ell)$ in the game LEAK. We then show how to build an adversary $\mathcal{B}$ to break the *element indistinguishability* captured by the *element re-randomizability* of the underlying SPHF by running $\mathcal{A}$. Recall that in the game LEAK, $\mathcal{A}$ would provide two parties $(\overline{\mathsf{P}}_1, \overline{\mathsf{P}}_2)$ which represent its chosen tampered implementations of Alice and Bob. Note that the tampered implementation of Alice is functionality-maintaining. $\mathcal{B}$ first runs the protocol between honest party Alice and $\overline{\mathsf{P}}_2$, and obtains the output of Alice as $(C_0, CT_0)$. $\mathcal{B}$ then runs again the protocol between $\overline{\mathsf{P}}_1$ and $\overline{\mathsf{P}}_2$, and obtains the output of $\overline{\mathsf{P}}_1$ as $(C_1, CT_1)$. It then sends $(C_0, C_1)$ as the challenge elements for the element indistinguishability game, and receives the challenge re-randomized element $\widetilde{C}$. It computes $\widetilde{CT} = M \oplus \mathsf{Hash}(\mathsf{pp}, \mathsf{hk}, \widetilde{C})$ and then forwards $(\widetilde{C}, \widetilde{CT})$ to $\mathcal{A}$ as part of the challenge transcript $\mathcal{T}^*$ of the game LEAK and outputs the guess $b'$ of $\mathcal{A}$ as its guess in the element indistinguishability game. It is easy to see that the above behaviours of $\mathcal{B}$ are computationally indistinguishable from the real game LEAK from the view of $\mathcal{A}$. Therefore, we have that $\mathsf{Adv}^{\mathsf{Element\text{-}Ind}}_{\mathcal{B},\mathsf{MSPHF}}(\ell) \geq \mathsf{Adv}^{\mathsf{LEAK}}_{\mathcal{A},\mathcal{W}}(\ell)$, which contradicts the *element re-randomizability* of the underling SPHF. Therefore, the CRF weakly resists exfiltration against Bob and of course against any eavesdropper. This also trivially implies the security preservation of the firewall. □

### 4.2   Oblivious Signature-Based Envelope with CRFs

In this section, we introduce the CRF constructions for the oblivious signature-based envelope protocol with an instantiation from the language of encryption of signature. Formally, an OSBE protocol involves: a sender, holding a string $P$, and a receiver holding a credential. The protocol *functionality* requires that at the end of protocol, the receiver could receive $P$ if and only if he/she possesses a certificate/signature on a predefined message $M$. The *security notion* asserts that the sender cannot determine whether the receiver owns the valid credential (*obliviousness*) and no other party learns anything about $P$ (*semantic security*).

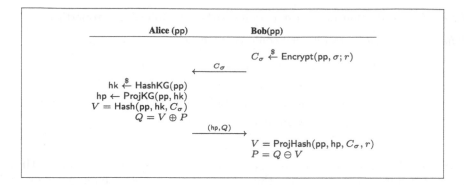

**Fig. 5.** Blazy-Pointcheval-Vergnaud OSBE framework [11]

**Blazy-Pointcheval-Vergnaud OSBE Framework** [11]. Noting that the original OSBE requires a secure channel during the execution to protect against eavesdroppers, Blazy, Pointcheval and Vergaud [11] clarified and enhanced the security models of OSBE by considering the security for both the sender and the receiver against the authority. Their new notion, namely *semantic security w.r.t. the authority*, requires that the authority who plays as the eavesdropper on the protocol, learns nothing about the private message of the sender. They showed how to generically build a 2-round OSBE scheme that can achieve the defined strong security in the standard model with a Common Reference String (CRS). We first recall a slightly modified version of their general framework, which is illustrated in Fig. 5. In particular, without loss of generality, we assume that the string $P$ is in the hash value space of the underlying SPHF. The main idea of the BPV-OSBE framework relies on the SPHF from the language defined by the encryption of valid signatures. Let $pp = (PP, ek, vk, M)$ where $PP$ is the collection of global parameters for the signature scheme, the encryption scheme and the SPHF system, $ek$ is the public key of the encryption scheme, $vk$ is the verification key of the signature scheme and $M$ is the predefined message. Suppose Encrypt is the encryption algorithm of the encryption scheme and Ver is the verification algorithm of the signature scheme. The language of the underlying SPHF is then defined as $\mathcal{L} = \{C_\sigma \mid \exists r, \sigma, \text{s.t.}, C_\sigma = \text{Encrypt}(pp, \sigma; r) \wedge \text{Ver}(pp, \sigma, M) = 1\}$. We then have that the subset membership problem is hard due to the security of the encryption scheme. Readers are referred to [11] for the detailed analysis of protocol correctness and security.

CRF **for the Receiver.** An tampered implementation of the receiver might produce a ciphertext $C_\sigma$ that either enables an eavesdropper to read Alice's message $P$, or acts as a channel to leak some secrets to the outsider (Alice or an eavesdropper). A CRF for Bob (denoted by $\mathcal{W}_B$) should be able to re-randomize the ciphertext $C_\sigma$ while still preserves the protocol functionality. It is also a requirement for $\mathcal{W}_B$ to preserve the protocol security, i.e., obliviousness, semantic security and semantic security w.r.t the authority. Regarding exfiltration, $\mathcal{W}_B$

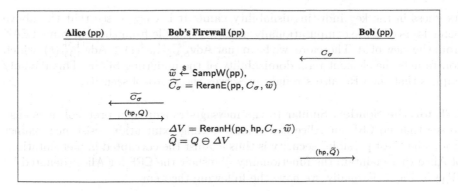

**Fig. 6.** Bob's CRF for the OSBE protocol shown in Fig. 5

should prevent the compromised Bob from using $C_\sigma$ as a channel to leak secrets. Figure 6 depicts the firewall $\mathcal{W}_B$ in the OSBE protocol.

**Theorem 5.** *The* CRF *for Bob shown in Fig. 6 maintains functionality and strongly preserves security for Bob, and strongly resists exfiltration against Alice, provided that the underlying* SPHF *is element-rerandomizable.*

*Proof.* We verify that our construction satisfies the following properties.
*Functionality Maintaining.* Due to the *rerandomization consistency* of the *element re-randomizability*, we have

$$\widetilde{Q} = Q \ominus \Delta V$$
$$= Q \ominus \mathsf{ReranH}(\mathsf{pp}, \mathsf{hp}, C_\sigma, \widetilde{w})$$
$$= P \oplus \mathsf{Hash}(\mathsf{pp}, \mathsf{hk}, \widetilde{C_\sigma}) \ominus \mathsf{ReranH}(\mathsf{pp}, \mathsf{hp}, C_\sigma, \widetilde{w})$$
$$= P \oplus \mathsf{Hash}(\mathsf{pp}, \mathsf{hk}, C_\sigma).$$

Bob is thus able to recover $P$ by computing $P = \widetilde{Q} \ominus \mathsf{ProjHash}(\mathsf{pp}, \mathsf{hk}, C_\sigma, r)$.

*Strong Security Preservation and Strong Exfiltration Resistance.* The strong exfiltration resistance follows from the fact that $\widetilde{C_\sigma}$ is independent of the original ciphertext $C_\sigma$ chosen by Bob who might be arbitrarily compromised. Precisely, suppose there exists an adversary who has non-negligible advantage $\mathsf{Adv}_{\mathcal{A},\mathcal{W}}^{\mathsf{LEAK}}(\ell)$ in the game LEAK. We then show how to build an adversary $\mathcal{B}$ to break the *element indistinguishability* captured by the *element re-randomizability* of the underlying SPHF by running $\mathcal{A}$. Recall that in the game LEAK, $\mathcal{A}$ would provide two parties $(\overline{\mathsf{P}}_1, \overline{\mathsf{P}}_2)$ which represent its chosen tampered implementations of Bob and Alice. $\mathcal{B}$ first runs the protocol between the honest party Bob and $\overline{\mathsf{P}}_2$, and obtains the output of Bob as $C_0$. $\mathcal{B}$ then runs again the protocol between $\overline{\mathsf{P}}_1$ and $\overline{\mathsf{P}}_2$, and obtains the output of $\overline{\mathsf{P}}_1$ as $C_1$. It then sends $(C_0, C_1)$ as the challenge elements for the element indistinguishability game, and receives the challenge re-randomized element $\widetilde{C_\sigma}$. Finally, it forwards $\widetilde{C_\sigma}$ to $\mathcal{A}$ as part of the challenge transcript $\mathcal{T}^*$ of the game LEAK and outputs the guess $b'$ of $\mathcal{A}$ as its guess in the key indistinguishability game. It is easy to see that the above behaviours of $\mathcal{B}$ are computationally indistinguishable from the real game LEAK from the view of $\mathcal{A}$. Therefore, we have that $\mathsf{Adv}_{\mathcal{B},\mathsf{MSPHF}}^{\mathsf{Element\text{-}Ind}}(\ell) \geq \mathsf{Adv}_{\mathcal{A},\mathcal{W}}^{\mathsf{LEAK}}(\ell)$, which contradicts the element-rerandomizability of the underling SPHF. This trivially implies that the CRF also strongly preserves the protocol security. $\qquad\square$

**CRF for the Sender.** Similar to the message-transmission protocol, it is easy to see that no CRF for Alice can achieve strong exfiltration resistance against Bob. The "best possible" security is thus against the corrupted implementations of Alice that maintain the functionality. We show the CRF for Alice (denoted by $\mathcal{W}_A$) in Fig. 7. Formally, we have the following theorem.

**Theorem 6.** *The* CRF *for Alice shown in Fig. 7 maintains functionality and strongly preserves security for Alice, and weakly resists exfiltration against Bob, provided that the underlying* SPHF *is projection key-malleable.*

*Proof.* We verify that our construction satisfies the following properties.

*Functionality Maintaining.* Due to the *projection consistency* of the *projection key-malleability* of the underlying SPHF, we have

$$\begin{aligned}
\widetilde{Q} &= Q \oplus \Delta V = Q \oplus \mathsf{MaulH}(\mathsf{pp},\mathsf{hp},C_\sigma,\widetilde{r}) \\
&= P \oplus \mathsf{Hash}(\mathsf{pp},\mathsf{hk},C_\sigma) \oplus \mathsf{MaulH}(\mathsf{pp},\mathsf{hp},C_\sigma,\widetilde{r}) \\
&= P \oplus \mathsf{Hash}(\mathsf{pp},\widetilde{\mathsf{hk}},C_\sigma).
\end{aligned}$$

In the above, $\widetilde{\mathsf{hk}}$ is the associated key of projection key $\widetilde{\mathsf{hp}} \leftarrow \mathsf{MaulK}(\mathsf{pp},\mathsf{hp},\widetilde{r})$. We can see that Bob can recover $P$ by computing $P = \widetilde{Q} \ominus \mathsf{ProjHash}(\mathsf{pp},\widetilde{\mathsf{hp}},C_\sigma,r)$.

*Strong Security Preservation and Weak Exfiltration Resistance.* For any tampered implementation of Alice that maintains functionality, suppose there exists an adversary who has non-negligible advantage $\mathsf{Adv}_{\mathcal{A},\mathcal{W}}^{\mathsf{LEAK}}(\ell)$ in the game LEAK. We then show how to build an adversary $\mathcal{B}$ to break the *key indistinguishability* captured by the *projection key-malleability* of the underlying MSPHF by running $\mathcal{A}$. Recall that in the game LEAK, $\mathcal{A}$ would provide two parties $(\overline{\mathsf{P}}_1,\overline{\mathsf{P}}_2)$ which represent its chosen tampered implementations of Alice and Bob. Note that the tampered implementation of Alice is functionality-maintaining. $\mathcal{B}$ first runs the protocol between honest party Alice and $\overline{\mathsf{P}}_2$, and obtains the output of Alice as $(\mathsf{hp}_0,Q_0)$. $\mathcal{B}$ then runs again the protocol between $\overline{\mathsf{P}}_1$ and $\overline{\mathsf{P}}_2$, and obtains the output of $\overline{\mathsf{P}}_1$ as $(\mathsf{hp}_1,Q_1)$. It then sends $(\mathsf{hp}_0,\mathsf{hp}_1)$ as the challenge projection key for the key indistinguishability game, and receives the challenge re-randomized projection key $\widetilde{\mathsf{hp}}$. It computes $\widetilde{Q} = P \oplus \mathsf{ProjHash}(\mathsf{pp},\widetilde{\mathsf{hp}},C_\sigma,r)$, and then forwards $(\widetilde{\mathsf{hp}},\widetilde{Q})$ to $\mathcal{A}$ as part of the challenge transcript $\mathcal{T}^*$ of the game LEAK and outputs the guess $b'$ of $\mathcal{A}$ as its guess in the key indistinguishability game. It is easy to see that the above behaviours of $\mathcal{B}$ are computationally indistinguishable from the real game LEAK from the view of $\mathcal{A}$. Therefore, we have that $\mathsf{Adv}_{\mathcal{B},\mathsf{MSPHF}}^{\mathsf{Key\text{-}Ind}}(\ell) \geq \mathsf{Adv}_{\mathcal{A},\mathcal{W}}^{\mathsf{LEAK}}(\ell)$, which contradicts the *projection key-malleability* of the underling MSPHF. Therefore, the firewall weakly resists exfiltration against

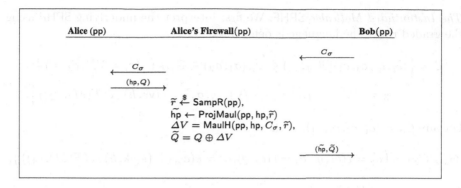

**Fig. 7.** Alice's CRF for the OSBE protocol shown in Fig. 5

Bob and of course against any eavesdropper. This also trivially implies the security preservation of the CRF. □

**Instantiation from the Linear Encryption of Valid Signatures.** In the work [11], an efficient OSBE protocol is proposed by combining the linear encryption scheme, the Waters signature [25] and an SPHF on the language of linear ciphertexts. Here we show how to extend the instantiated SPHF to be malleable for the CRF constructions. It is worth noting that the introduced malleable SPHF here could also be represented by graded ring but does not follow the generic framework proposed in Sect. 3.2 (i.e., $\Theta_{\text{aux}}$ is not an identity function). We first recall the SPHF proposed in the work [11]. Let $\mathbb{G}, \mathbb{G}_T$ be two multiplicative groups with the same prime order $p$. Let $g$ be the generator of $\mathbb{G}$ and $I$ be the identity element of $\mathbb{G}_T$. A symmetric bilinear map is a map $e : \mathbb{G} \times \mathbb{G} \to \mathbb{G}_T$ such that $e(u^a, v^b) = e(u, v)^{ab}$ for all $u, v \in \mathbb{G}$ and $a, b \in \mathbb{Z}_p$. It is worth noting that $e$ can be efficiently computed and $e(g, g) \neq 1_{\mathbb{G}_T}$.

*Linear Encryption of Waters Signatures.* Let $h \xleftarrow{\$} \mathbb{G}$ and $\mathbf{u} = (u_0, ..., u_k) \xleftarrow{\$} \mathbb{G}^{k+1}$ which defines the Waters hash of a message $M = (M_1, ..., M_k) \in \{0, 1\}^k$ as $\mathcal{F}(M) = u_0 \prod_{i=1}^{k} u_i^{M_i}$. The verification key is $\mathsf{vk} = g^z$ and the associated signing key is $\mathsf{sk} = h^z$ where $z \xleftarrow{\$} \mathbb{Z}_p$. The signature on a message $M$ is $\sigma = (\sigma_1 = \mathsf{sk} \cdot \mathcal{F}(M)^s, \sigma_2 = g^s)$ for some random $s \xleftarrow{\$} \mathbb{Z}_p$. It can be verified by checking $e(g, \sigma_1) = e(\mathsf{vk}, h) \cdot e(\mathcal{F}(M), \sigma_2)$. The linear encryption public key is $\mathsf{ek} = (Y_1 = g^{y_1}, Y_2 = g^{y_2})$ and the secret key is $\mathsf{dk} = (y_1, y_2) \xleftarrow{\$} \mathbb{Z}_p^2$. The ciphertext of a Waters signature $\sigma = (\sigma_1, \sigma_2)$ is $C_\sigma = (c_1 = Y_1^{r_1}, c_2 = Y_2^{r_2}, c_3 = g^{r_1+r_2} \cdot \sigma_1, c_4 = \sigma_2)$, where $(r_1, r_2) \xleftarrow{\$} \mathbb{Z}_p^2$.

*The Instantiated Malleable* SPHF. We first interpret the underlying SPHF using the graded ring. The language is defined as,

$$\mathcal{L} = \Big\{(c_1, c_2, c_3, c_4) | \exists (r_1, r_2) \in \mathbb{Z}_p^2, (\sigma_1, \sigma_2) \in \mathbb{G}_1^2, \text{s.t.}, \big(c_1 = Y_1^{r_1}, c_2 = Y_2^{r_2},$$

$$c_3 = g^{r_1 + r_2} \cdot \sigma_1, c_4 = \sigma_2\big) \bigwedge \big(e(g, \sigma_1) = e(\mathsf{vk}, h) \cdot e(\mathcal{F}(M), \sigma_2)\big)\Big\}.$$

For any $C_\sigma = (c_1, c_2, c_3, c_4)$, we have

$$\Theta_{\mathsf{aux}}(C_\sigma) = \Big(c_1' = e(c_1, g), c_2' = e(c_2, g), c_3' = e(c_3, g)/(e(\mathsf{vk}, h) \cdot e(\mathcal{F}(M), c_4))\Big),$$

and $\Gamma(C_\sigma) = \begin{pmatrix} Y_1 & 1 & g \\ 1 & Y_2 & g \end{pmatrix} \in \mathbb{G}^{2 \times 3}$. We can see that if $C_\sigma \in \mathcal{L}$ with witness $w = (r_1, r_2)$, let $\boldsymbol{\lambda} = (g^{r_1}, g^{r_2})$, we have,

$$\Theta_{\mathsf{aux}}(C_\sigma) = \Big(e(c_1, g), e(c_2, g), e(c_3, g)/(e(\mathsf{vk}, h) \cdot e(\mathcal{F}(M), c_4))\Big)$$
$$= \big(e(Y_1^{r_1}, g), e(Y_2^{r_2}, g), e(g^{r_1 + r_2}, g)\big)$$
$$= \boldsymbol{\lambda} \odot \Gamma(C_\sigma).$$

Let $\mathsf{pp} = (\mathbb{G}, p, g, Y_1, Y_2, \mathbf{u})$, $\widetilde{\boldsymbol{r}} = (\beta_1, \beta_2, \beta_3)^\mathsf{T}$ and $\widetilde{\boldsymbol{\lambda}} = \widetilde{\boldsymbol{w}} = (\eta_1, \eta_2, \eta_3)$. The instantiation is as follows:

- $\mathsf{HashKG}(\mathsf{pp})$ : $\mathsf{hk} = \boldsymbol{\alpha} = (\alpha_1, \alpha_2, \alpha_3)^\mathsf{T} \xleftarrow{\$} \mathbb{Z}_p^3$;
- $\mathsf{ProjKG}(\mathsf{pp}, \mathsf{hk}, C)$ : $\mathsf{hp} = \boldsymbol{\gamma}(C_\sigma) = \Gamma(C_\sigma) \odot \boldsymbol{\alpha} = (Y_1^{\alpha_1} g^{\alpha_3}, Y_2^{\alpha_2} g^{\alpha_3})^\mathsf{T}$;
- $\mathsf{Hash}(\mathsf{pp}, \mathsf{hk}, C)$ : $\mathsf{hv} = \Theta_{\mathsf{aux}}(C_\sigma) \odot \boldsymbol{\alpha} = (c_1', c_2', c_3') \odot (\alpha_1, \alpha_2, \alpha_3)^\mathsf{T} = e(c_1, g)^{\alpha_1} \cdot e(c_2, g)^{\alpha_2} \cdot \big(e(c_3, g)/(e(\mathsf{vk}, h) \cdot e(\mathcal{F}(M), c_4))\big)^{\alpha_3}$;
- $\mathsf{ProjHash}(\mathsf{pp}, \mathsf{hp}, C, \boldsymbol{w})$ : $\mathsf{hv} = \boldsymbol{\lambda} \odot \boldsymbol{\gamma}(C_\sigma) = (g^{r_1}, g^{r_2}) \odot (Y_1^{\alpha_1} g^{\alpha_3}, Y_2^{\alpha_2} g^{\alpha_3})^\mathsf{T} = e\big((Y_1^{\alpha_1} g^{\alpha_3})^{r_1} \cdot (Y_2^{\alpha_2} g^{\alpha_3})^{r_2}, g\big)$;
- $\mathsf{MaulK}(\mathsf{pp}, \mathsf{hp}, \widetilde{\boldsymbol{r}})$ : $\widetilde{\mathsf{hp}} = \boldsymbol{\gamma}(C) \oplus (\Gamma(C) \odot \widetilde{\boldsymbol{r}}) = \boldsymbol{\gamma}(C_\sigma) \oplus (\Gamma(C_\sigma) \odot \boldsymbol{\Delta r}) = (Y_1^{\alpha_1} g^{\alpha_3}, Y_2^{\alpha_2} g^{\alpha_3}) \oplus (Y_1^{\beta_1} g^{\beta_3}, Y_2^{\beta_2} g^{\beta_3}) = ((Y_1^{\alpha_1 + \beta_1} g^{\alpha_3 + \beta_3}, Y_2^{\alpha_2 + \beta_2} g^{\alpha_3 + \beta_3}))^\mathsf{T}$;
- $\mathsf{MaulH}(\mathsf{pp}, \mathsf{hp}, \widetilde{\boldsymbol{r}}, C)$ : $\mathsf{hv} = \Theta_{\mathsf{aux}}(C) \odot \widetilde{\boldsymbol{r}} = (c_1', c_2', c_3') \odot (\beta_1, \beta_2, \beta_3)^\mathsf{T} = e(c_1, g)^{\beta_1} \cdot e(c_2, g)^{\beta_2} \cdot \big(e(c_3, g)/(e(\mathsf{vk}, h) \cdot e(\mathcal{F}(M), c_4))\big)^{\beta_3}$;
- $\mathsf{ReranE}(\mathsf{pp}, C, \widetilde{\boldsymbol{w}})$ : $\widetilde{C} = C_\sigma \oplus (Y_1^{\eta_1}, Y_2^{\eta_2}, g^{\eta_1 + \eta_2} \mathcal{F}(M)^{\eta_3}, g^{\eta_3}) = (c_1 \cdot Y_1^{\eta_1}, c_2 \cdot Y_2^{\eta_2}, c_3 \cdot g^{\eta_1 + \eta_2} \mathcal{F}(M)^{\eta_3}, c_4 \cdot g^{\eta_3})$;
- $\mathsf{ReranH}(\mathsf{pp}, \mathsf{hp}, C, \widetilde{\boldsymbol{w}})$ : $\mathsf{hv} = (g^{\eta_1}, g^{\eta_2}) \odot \Gamma(C_\sigma) = (g^{\eta_1}, g^{\eta_2}) \odot (Y_1^{\alpha_1} g^{\alpha_3}, Y_2^{\alpha_2} g^{\alpha_3}) = e\big((Y_1^{\alpha_1} g^{\alpha_3})^{\eta_1} \cdot (Y_2^{\alpha_2} g^{\alpha_3})^{\eta_2}, g\big)$.

**Theorem 7.** *The above construction is a malleable smooth projective hash function.*

*Proof.* We verify that our construction satisfies the following properties. Note that the constructions of both MaulK and MaulH follow the framework proposed in Sect. 3.2. According to Theorem 1, we have that our constructed SPHF is projection key-malleable. Note that in our construction, $C_\sigma' = C_\sigma \oplus (Y_1^{\eta_1}, Y_2^{\eta_2},$

$g^{\eta_1+\eta_2}\mathcal{F}(M)^{\eta_3}, g^{\eta_3})$, one can easily observe the rerandomization is *element-indistinguishable* due to the 2-Lin assumption. Particularly, we have that $(Y_1^{\eta_1}, Y_2^{\eta_2}, g^{\eta_1+\eta_2})$ is a linear tuple w.r.t $(Y_1, Y_2, g)$. If any adversary can distinguish the rerandomized element, we can use it as a subroutine to break the 2-Lin assumption. We then prove that the element rerandomization is *membership-preserving*. Suppose $C_\sigma = \big(c_1 = Y_1^{r_1}, c_2 = Y_2^{r_2}, c_3 = g^{r_1+r_2}\cdot\sigma_1, c_4 = \sigma_2\big) \in \mathcal{L}$. We have that after it is rerandomized,

$$
\begin{aligned}
\widetilde{C_\sigma} &= C_\sigma \oplus (Y_1^{\eta_1}, Y_2^{\eta_2}, g^{\eta_1+\eta_2}\mathcal{F}(M)^{\eta_3}, g^{\eta_3}) \\
&= (c_1 \cdot Y_1^{\eta_1}, c_2 \cdot Y_2^{\eta_2}, c_3 \cdot g^{\eta_1+\eta_2}\mathcal{F}(M)^{\eta_3}, c_4 \cdot g^{\eta_3}) \\
&= \big(Y_1^{r_1+\eta_1}, \ Y_2^{r_2+\eta_2}, \ g^{r_1+r_2+\eta_1+\eta_2}\cdot\sigma_1\cdot\mathcal{F}(M)^{\eta_3}, \ \sigma_2\cdot g^{\eta_3}\big) \\
&\stackrel{\mathsf{def}}{=} \big(\widetilde{c}_1, \widetilde{c}_2, \widetilde{c}_3, \widetilde{c}_4\big)
\end{aligned}
$$

Since $\Gamma$ is a constant function, we know that, $\Gamma(\widetilde{C_\sigma}) = \Gamma(C_\sigma) = \begin{pmatrix} Y_1 & 1 & g \\ 1 & Y_2 & g \end{pmatrix}$. Let $\widetilde{\boldsymbol{\lambda}} = (g^{r_1+\eta_1}, g^{r_2+\eta_2})$, we then obtain:

$$
\begin{aligned}
\Theta_{\mathsf{aux}}(\widetilde{C_\sigma}) &= \Big(e(\widetilde{c}_1, g), e(\widetilde{c}_2, g), e(\widetilde{c}_3, g)/(e(\mathsf{vk}, h)\cdot e(\mathcal{F}(M), \widetilde{c}_4))\Big) \\
&= \Big(e(Y_1^{r_1+\eta_1}, g), e(Y_2^{r_2+\eta_2}, g), \frac{e(g^{r_1+r_2+\eta_1+\eta_2}\cdot\sigma_1\cdot\mathcal{F}(M)^{\eta_3}, g)}{e(\mathsf{vk}, h)\cdot e(\mathcal{F}(M), \sigma_2\cdot g^{\eta_3})}\Big) \\
&= \big(e(Y_1^{r_1+\eta_1}, g), e(Y_2^{r_2+\eta_2}, g), e(g^{r_1+r_2+\eta_1+\eta_2}, g)\big) \\
&= \widetilde{\boldsymbol{\lambda}} \odot \Gamma(\widetilde{C_\sigma}).
\end{aligned}
$$

This shows that $\widetilde{C_\sigma} \in \mathcal{L}$. If $C_\sigma \notin \mathcal{L}$, we trivially have that $\widetilde{C_\sigma} \notin \mathcal{L}$.

We then justify the rerandomization consistency. For any hashing key $\mathsf{hk} = \boldsymbol{\alpha} = (\alpha_1, \alpha_2, \alpha_3)^\top \xleftarrow{\$} \mathbb{Z}_p^2$, we have that,

$$
\begin{aligned}
\mathsf{Hash}(\mathsf{pp}, \mathsf{hk}, \widetilde{C_\sigma}) &= \Theta_{\mathsf{aux}}(\widetilde{C_\sigma}) \odot \boldsymbol{\alpha} \\
&= \Big(e(\widetilde{c}_1, g), e(\widetilde{c}_2, g), \frac{e(\widetilde{c}_3, g)}{e(\mathsf{vk}, h)\cdot e(\mathcal{F}(M), \widetilde{c}_4)}\Big) \odot (\alpha_1, \alpha_2, \alpha_3)^\top \\
&= (c_1', c_2', c_3') \odot (\alpha_1, \alpha_2, \alpha_3)^\top \oplus (g^{\eta_1}, g^{\eta_2}) \odot (Y_1^{\alpha_1}g^{\alpha_3}, Y_2^{\alpha_2}g^{\alpha_3}) \\
&= \Theta_{\mathsf{aux}}(C_\sigma) \odot \boldsymbol{\alpha} \oplus e\big((Y_1^{\alpha_1}g^{\alpha_3})^{\eta_1}\cdot(Y_2^{\alpha_2}g^{\alpha_3})^{\eta_2}, g\big) \\
&= \mathsf{Hash}(\mathsf{pp}, \mathsf{hk}, C_\sigma) \oplus \mathsf{ReranH}(\mathsf{pp}, \mathsf{hp}, C_\sigma, \widetilde{w}).
\end{aligned}
$$

# 5     Oblivious Transfer with Reverse Firewall

## 5.1     A New OT Framework from Graded Rings

Oblivious transfer forms a central primitive in modern cryptography. It is a protocol between the sender, holding two message $M_0$ and $M_1$, and a receiver holding a choice bit $b$. The OT *functionality* requires that at the end of the protocol, the receiver can learn the message $M_b$. The *security requirement* is

| $\mathsf{Sampl}(\boldsymbol{\Gamma}, b)$: | $\mathsf{PairG}(\boldsymbol{\Gamma}, C_0)$: |
|---|---|
| $\boldsymbol{w} \overset{\$}{\leftarrow} \mathsf{SampW}(\mathsf{pp})$ | Parse $\boldsymbol{\Gamma}$ as $(\Gamma_1, ..., \Gamma_n)$ |
| $C := \lambda(\boldsymbol{w}) \odot \boldsymbol{\Gamma}$ | set $\boldsymbol{\Gamma}' = (\mathbf{1}_{\mathbb{G}}, ..., \mathbf{1}_{\mathbb{G}}, \Gamma_n)_{1\times n}$ |
| Parse $\boldsymbol{\Gamma}$ as $(\Gamma_1, ..., \Gamma_n)$ | set $\boldsymbol{e} = (0_{\mathbb{Z}_p}, ..., 0_{\mathbb{Z}_p}, 1_{\mathbb{Z}_p})_{1\times m}$ |
| Set $\boldsymbol{e} = (0_{\mathbb{Z}_p}, ..., 0_{\mathbb{Z}_p}, b_{\mathbb{Z}_p})_{1\times m}$ | $\Delta C := \boldsymbol{e} \odot \boldsymbol{\Gamma}'$ |
| $\Delta C := \boldsymbol{e} \odot (\mathbf{1}_{\mathbb{G}}, ..., \mathbf{1}_{\mathbb{G}}, \Gamma_n)_{1\times n}$ | $C_1 := C_0 \ominus \Delta C$ |
| $C_0 := C \oplus \Delta C$ | return $C_1$ |
| Return $(C_0, \boldsymbol{w})$ | *Note:* $\mathbf{1}_{\mathbb{G}}$ is a $m \times 1$ matrix of $1_{\mathbb{G}}$ |

**Fig. 8.** Definitions of algortihms Sampl, PairG.

| Alice $(\mathsf{pp}, M_0, M_1)$ | | Bob$(\mathsf{pp}, b)$ |
|---|---|---|
| | | $\boldsymbol{\Gamma} \overset{\$}{\leftarrow} \mathsf{SampB}(\mathsf{pp})$ |
| | | $(C_0, \boldsymbol{w}) \overset{\$}{\leftarrow} \mathsf{Sampl}(\boldsymbol{\Gamma}, b)$ |
| | $\xleftarrow{(\boldsymbol{\Gamma}, C_0)}$ | |
| $C_1 = \mathsf{PairG}(\boldsymbol{\Gamma}, C_0)$, | | |
| $\mathsf{hk}_0 = \boldsymbol{\alpha_0} \overset{\$}{\leftarrow} \mathbb{Z}_p^n$, | | |
| $\mathsf{hp}_0 = \gamma_0 = \boldsymbol{\Gamma} \odot \boldsymbol{\alpha_0}$ | | |
| $\mathsf{hk}_1 = \boldsymbol{\alpha_1} \overset{\$}{\leftarrow} \mathbb{Z}_p^n$, | | |
| $\mathsf{hp}_1 = \gamma_1 = \boldsymbol{\Gamma} \odot \boldsymbol{\alpha_1}$ | | |
| $(V_i)_{i=0}^1 \leftarrow (C_i \odot \boldsymbol{\alpha_i})_{i=0}^1$ | | |
| $(CT_i)_{i=0}^1 \leftarrow (V_i \oplus M_i)_{i=0}^1$ | | |
| | $\xrightarrow{(\gamma_i, CT_i)_{i=0}^1}$ | |
| | | $V_b = \lambda(\boldsymbol{w}) \odot \gamma_b$ |
| | | $M_b = CT_b \ominus V_b$ |

**Fig. 9.** OT Protocols from graded rings.

that the receiver learns nothing about $M_{1-b}$ (*sender security*), and the sender learns nothing about the receiver's choice $b$ (*receiver security*). We introduce a variant of the HK-OT [16] framework in the context of graded rings. Essentially, we follow the generic framework of (malleable) SPHF from graded rings (shown in Sect. 3.2). The modified semi-generic framework narrows the possible instantiations of the HK-OT framework. However, as we will show later, the CRF construction following our framework not only captures the prior work [21], which is the only known OT-CRF to date, but also yields new constructions under weaker assumptions.

Before introducing our framework, we define two new algorithms Sampl, PairG depicted in Fig. 8. For the sake of clarity, we use $\boldsymbol{\lambda} = \lambda(w)$ to represent the derivation of $\boldsymbol{\lambda}$ from the witness $w$. We require $\Theta_{\mathsf{aux}}$ to be an identity function and $\Gamma$ to be a constant function. That is, we only consider the KV type SPHF on diverse groups. As before, the subset membership problem must also be hard. Note that these are exactly the same conditions (Theorem 1) for our malleable SPHF construction presented in Sect. 3.2. Our graded ring-based OT framework is shown in Fig. 9. Suppose the element basis (denoted by $\boldsymbol{\Gamma} = (\Gamma_1, ..., \Gamma_n) \in$

$\mathbb{G}^{m \times n})$ is chosen by the receiver using the algorithm named SampB. It is worth noting that for the sake of simplicity, we assume without loss of generality the receiver (even the tampered implementation) would not trivially choose $\Gamma_i = \mathbf{1}_\mathbb{G}$ for any $i \in [1, n]$, since such an attempt can be easily detected in reality. One can note that:

- $b = 0$: $C_0 \in \mathcal{L}$ as $C_0 = \lambda(\boldsymbol{w}) \odot \boldsymbol{\Gamma}$ and $C_1 \notin \mathcal{L}$ as $C_1$ is not a linear span of $\boldsymbol{\Gamma}$.
- $b = 1$: $C_0 \notin \mathcal{L}$ as $C_0$ is not a linear span of $\boldsymbol{\Gamma}$ and $C_1 \in \mathcal{L}$ as $C_1 = \lambda(\boldsymbol{w}) \odot \boldsymbol{\Gamma}$.

Formally, we have the following result for the above framework.

**Theorem 8.** *The generic construction of OT shown in Fig. 9 is correct and secure.*

The protocol functionality (correctness) follows from the fact that $C_b \in \mathcal{L}$ and the sender security is guaranteed as $C_{1-b} \notin \mathcal{L}$. The receiver security is due to the hardness of the subset membership problem.

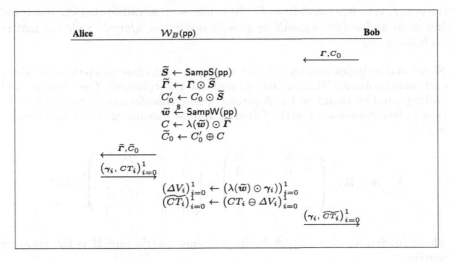

**Fig. 10.** Bob's CRF for the OT protocol in Fig. 9

## 5.2 Constructions of CRFs

**CRF for the Receiver.** The construction of the receiver CRF (denoted by $\mathcal{W}_B$) under our OT framework is shown in Fig. 10. The algorithm SampS outputs a *transformation matrix* (denoted by $\widetilde{\boldsymbol{S}} \in \mathbb{Z}_p^{n \times n}$) for the element basis $\boldsymbol{\Gamma}$. We denote the output of PairG($\boldsymbol{\Gamma}, \widetilde{C_0}$) as $\widetilde{C_1}$ and it should be clear that:

- $b = 0$: $\widetilde{C_0} = \lambda(\boldsymbol{w}) \odot \boldsymbol{\Gamma} \odot \widetilde{\boldsymbol{S}} \oplus \widetilde{w} \odot \widetilde{\boldsymbol{\Gamma}} = (\lambda(\boldsymbol{w}) \oplus \lambda(\widetilde{w})) \odot \widetilde{\boldsymbol{\Gamma}}$. $\widetilde{C_1} = (\lambda(\boldsymbol{w}) \odot \boldsymbol{\Gamma} \ominus \Delta C) \odot \widetilde{\boldsymbol{S}} \oplus \widetilde{w} \odot \widetilde{\boldsymbol{\Gamma}} = (\lambda(\boldsymbol{w}) \oplus \lambda(\widetilde{w})) \odot \widetilde{\boldsymbol{\Gamma}} \ominus \Delta C \odot \widetilde{\boldsymbol{S}}$, where $\Delta C = (0_{\mathbb{Z}_p}, ..., 0_{\mathbb{Z}_p}, 1_{\mathbb{Z}_p})_{1 \times m} \odot (\mathbf{1}_\mathbb{G}, ..., \mathbf{1}_\mathbb{G}, \Gamma_n)_{1 \times n}$.

– $b = 1$: $\widetilde{C_0} = (\lambda(\boldsymbol{w}) \odot \boldsymbol{\Gamma} \oplus \Delta C) \odot \widetilde{\boldsymbol{S}} \oplus \widetilde{w} \odot \widetilde{\boldsymbol{\Gamma}} = (\lambda(\boldsymbol{w}) \oplus \lambda(\widetilde{w})) \odot \widetilde{\boldsymbol{\Gamma}} \oplus \Delta C \odot \widetilde{\boldsymbol{S}}$, where
$\Delta C = (0_{\mathbb{Z}_p}, ..., 0_{\mathbb{Z}_p}, 1_{\mathbb{Z}_p})_{1 \times m} \odot (1_{\mathbb{G}}, ..., 1_{\mathbb{G}}, \Gamma_n)_{1 \times n}$. $\widetilde{C_1} = \lambda(\boldsymbol{w}) \odot \boldsymbol{\Gamma} \odot \widetilde{\boldsymbol{S}} \oplus \widetilde{w} \odot \widetilde{\boldsymbol{\Gamma}} = (\lambda(\boldsymbol{w}) \oplus \lambda(\widetilde{w})) \odot \widetilde{\boldsymbol{\Gamma}}$.

That is, $\widetilde{C_b} \in \mathcal{L}$ and thus $\mathcal{W}_B$ maintains the protocol functionality:

$$
\begin{aligned}
\widetilde{CT_b} &= CT_b \ominus \Delta V_b \\
&= M_b \oplus (\widetilde{C_b} \odot \boldsymbol{\alpha}_b) \ominus (\lambda(\widetilde{w}) \odot \gamma_b) \\
&= M_b \oplus (\lambda(\boldsymbol{w}) \oplus \lambda(\widetilde{w})) \odot \widetilde{\boldsymbol{\Gamma}} \odot \boldsymbol{\alpha}_b) \ominus (\lambda(\widetilde{w}) \odot \widetilde{\boldsymbol{\Gamma}} \odot \boldsymbol{\alpha}_b) \\
&= M_b \oplus (\lambda(\boldsymbol{w}) \odot \widetilde{\boldsymbol{\Gamma}} \odot \boldsymbol{\alpha}_b) \\
&= M_b \oplus (\lambda(\boldsymbol{w}) \odot \gamma_b).
\end{aligned}
$$

*Discussions on $\widetilde{S}$.* It is a trivial observation that $\mathcal{W}_B$ could strongly resist exfiltration if $\widetilde{\boldsymbol{\Gamma}}$ is independent from $\boldsymbol{\Gamma}$ as this also results in a random element $\widetilde{C}$ (by uniformly sampling $\widetilde{w}$). Precisely, let $\boldsymbol{\Gamma} = (\Gamma_1, ..., \Gamma_n)$. An ideal transformation matrix $\widetilde{S}$ should transfer each $\Gamma_i$ to another random $\widetilde{\Gamma}_i$ and for any $i, j \in [1, n]$ and $i \neq j$, $\widetilde{\Gamma}_i$ is independent from $\widetilde{\Gamma}_j$. To realize such a transformation, one could either *shear and uniformly scale* or *globally and non-uniformly scale* the matrix $\boldsymbol{\Gamma}$ as follows:

– *Shear and uniform scaling.* Choose a column and then independently shear each other column. Then uniformly scale all the columns. The shearing and scaling could be in any order. A corresponding transformation matrix for this type of transformation has the following format (assuming the chosen column is $\Gamma_1$):

$$
\widetilde{\boldsymbol{S}} = \mathbf{A} \odot \mathbf{B} = \begin{pmatrix} \alpha & 0 & \cdots & 0 \\ 0 & \alpha & \cdots & 0 \\ \vdots & \vdots & \ddots & \vdots \\ 0 & 0 & \cdots & \alpha \end{pmatrix} \odot \begin{pmatrix} 1 & \beta_2 & \cdots & \beta_n \\ 0 & 1 & \cdots & 0 \\ \vdots & \vdots & \ddots & \vdots \\ 0 & 0 & \cdots & 1 \end{pmatrix} \in \mathbb{Z}_p^{n \times n},
$$

where $(\alpha, \beta_2, ..., \beta_n) \xleftarrow{\$} \mathbb{Z}_p^n$, $\mathbf{A}$ is the a scaling matrix and $\mathbf{B}$ is the shearing matrix.

– *Globally non-uniform scaling.* Independently scale each column. A corresponding transformation matrix for this type of transformation has the following shape:

$$
\widetilde{\boldsymbol{S}} = \begin{pmatrix} \alpha_1 & 0 & \cdots & 0 \\ 0 & \alpha_2 & \cdots & 0 \\ \vdots & \vdots & \ddots & \vdots \\ 0 & 0 & \cdots & \alpha_n \end{pmatrix} \in \mathbb{Z}_p^{n \times n},
$$

where $(\alpha_1, ..., \alpha_n) \xleftarrow{\$} \mathbb{Z}_p^n$.

The first type has been used by Mironov and Stephens-Davidowitz in their OT-CRF construction [21]. One can note the second type of transformation is more efficient and thus can improve the efficiency. We will show the details in Sect. 5.3.

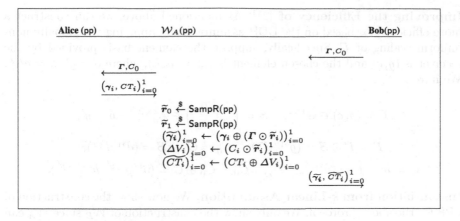

**Fig. 11.** Alice's CRF for the OT protocol in Fig. 9

**CRF for the Sender.** Figure 11 depicts the construction of CRF for the sender (denoted by $\mathcal{W}_A$). One may note that the construction is exactly part of the garded ring-based construction of malleable SPHF shown in Sect. 3.2. Therefore, according to Theorem 1, one could easily see that $\mathcal{W}_A$ maintains functionality, weakly resist exfiltration against Bob and strongly resist exfiltration against an eavesdropper. The composed firewall $\mathcal{W}_B \circ \mathcal{W}_A$ also weakly preservers security against Bob.

## 5.3    Instantiations

Due to the space limitation, the hardness assumptions and security analysis are given in the full version.

**Capturing the OT-CRF in [21].** Below we show that our framework indeed encompasses the construction in [21]. Precisely, in [21] the basis chosen by the receiver is $(g, c)$ and the chosen element is $C_0 = (d, h)$, where $d = g^y, h = c^y g^b$. We have that.

$$\boldsymbol{\varGamma} = (g, c), \quad \widetilde{\boldsymbol{S}} = \begin{pmatrix} \alpha & \alpha x' \\ 0 & \alpha \end{pmatrix}, \quad \widetilde{\boldsymbol{w}} = y',$$

$$\widetilde{\boldsymbol{\varGamma}} = \boldsymbol{\varGamma} \odot \widetilde{\boldsymbol{S}} = (g^\alpha, c^\alpha g^{\alpha x'}), \quad C_0' = C_0 \odot \widetilde{\boldsymbol{S}} = (d^\alpha, h^\alpha d^{\alpha x'}),$$

$$C = \widetilde{\boldsymbol{w}} \odot \widetilde{\boldsymbol{\varGamma}} = (g^{\alpha y'}, c^{\alpha y'} g^{\alpha x' y'}), \quad \widetilde{C_0} = C_0' \oplus C = (d^\alpha g^{\alpha y'}, h^\alpha d^{\alpha x'} c^{\alpha y'} g^{\alpha x' y'}).$$

One can note that the transformation of $\boldsymbol{\varGamma}$ adopted here is via shearing and uniform scaling as:

$$\widetilde{\boldsymbol{S}} = \begin{pmatrix} \alpha & \alpha x' \\ 0 & \alpha \end{pmatrix} = \begin{pmatrix} \alpha & 0 \\ 0 & \alpha \end{pmatrix} \odot \begin{pmatrix} 1 & x' \\ 0 & 1 \end{pmatrix}.$$

It is clear that other parts of protocol also follow the above framework.

**Improving the Efficiency of** [21]. As mentioned above, we can construct a more efficient $\mathcal{W}_B$ based on the DDH assumption by applying the globally non-uniform scaling of $\boldsymbol{\Gamma}$. Specifically, suppose the element basis provided by the receiver is $(g, c)$ and the chosen element is $C_0 = (d, h)$, where $d = g^y, h = c^y c^b$. We have

$$\boldsymbol{\Gamma} = (g, c) \in \mathbb{G}^{1 \times 2}, \quad \widetilde{\boldsymbol{S}} = \begin{pmatrix} s_1 & 0 \\ 0 & s_2 \end{pmatrix} \in \mathbb{Z}_p^{2 \times 2}, \quad \widetilde{\boldsymbol{w}} = y',$$

$$\widetilde{\boldsymbol{\Gamma}} = \boldsymbol{\Gamma} \odot \widetilde{\boldsymbol{S}} = (g^{s_1}, c^{s_2}), \quad C_0' = C_0 \odot \widetilde{\boldsymbol{S}} = (d^{s_1}, h^{s_2}),$$

$$C = \widetilde{\boldsymbol{w}} \odot \widetilde{\boldsymbol{\Gamma}} = (g^{s_1 y'}, c^{s_2 y'}), \quad \widetilde{C_0} = C_0' \oplus C = (d^{s_1} g^{s_1 y'}, h^{s_2} c^{s_2 y'}).$$

**Instantiation from $k$-Linear Assumption.** We now show the construction of CRF for the above protocol. We only show the construction of $\mathcal{W}_B$ since $\mathcal{W}_A$ can be easily obtained from the $k$-linear assumption based instantiation of malleable SPHF shown in Sect. 3.2. Specifically, we have

$$\boldsymbol{\Gamma} = \begin{pmatrix} g_1 & 1 & \cdots & 1 & g_{k+1} \\ 1 & g_2 & \cdots & 1 & g_{k+1} \\ \vdots & \vdots & \ddots & \vdots & \vdots \\ 1 & 1 & \cdots & g_k & g_{k+1} \end{pmatrix}, \quad \widetilde{\boldsymbol{S}} = \begin{pmatrix} s_1 & 0 & \cdots & 0 \\ 0 & s_2 & \cdots & 0 \\ \vdots & \vdots & \ddots & \vdots \\ 0 & 0 & \cdots & s_{k+1} \end{pmatrix},$$

$$\widetilde{\boldsymbol{\Gamma}} = \boldsymbol{\Gamma} \odot \widetilde{\boldsymbol{S}} = \begin{pmatrix} g_1^{s_1} & 1 & \cdots & 1 & g_{k+1}^{s_{k+1}} \\ 1 & g_2^{s_2} & \cdots & 1 & g_{k+1}^{s_{k+1}} \\ \vdots & \vdots & \ddots & \vdots & \vdots \\ 1 & 1 & \cdots & g_k^{s_k} & g_{k+1}^{s_{k+1}} \end{pmatrix}, \quad C_0' = C_0 \odot \widetilde{\boldsymbol{S}} = (c_1^{s_1}, c_2^{s_2}, ..., c_{k+1}^{s_{k+1}}),$$

$$\widetilde{\boldsymbol{w}} = (r_1', r_2', ..., r_{k+1}') \in \mathbb{Z}_p^k, \quad C = \widetilde{\boldsymbol{w}} \odot \widetilde{\boldsymbol{\Gamma}} = (g_1^{s_1 r_1'}, g_2^{s_2 r_2'}, ..., g_{k+1}^{s_{k+1} \sum_{i=1}^{k} r_i'}),$$

$$\widetilde{C_0} = C_0' \oplus C = (c_1^{s_1} g_1^{s_1 r_1'}, c_2^{s_2} g_2^{s_2 r_2'}, \cdots, c_k^{s_k} g_k^{s_k r_k'}, c_{k+1}^{s_{k+1}} g_{k+1}^{s_{k+1} \sum_{i=1}^{k} r_i'}).$$

## 6    Conclusion

In this work, we presented generic CRF constructions for several widely used cryptographic protocols based on a new notion named *malleable smooth projective hash function*, which is an extension of the SPHF with new properties. We showed how to generically construct CRFs via malleable SPHFs in a modular way. Specifically, we proposed generic constructions of CRFs for the unkeyed message-transmission protocol and the OSBE protocol. We further studied the OT protocol and developed a new OT framework from graded rings and showed how to construct OT-CRFs via a modified version of the malleable SPHF framework.

**Acknowledgements.** We would like to thank the anonymous reviewers for their invaluable comments on a previous version of this paper. Dr. Guomin Yang is supported by the Australian Research Council Discovery Early Career Researcher Award (Grant No. DE150101116). Dr. Mingwu Zhang is supported by the National Natural Science Foundation of China (Grant No. 61370224 and Grant No. 61672010).

# References

1. Abdalla, M., Benhamouda, F., Pointcheval, D.: Disjunctions for hash proof systems: new constructions and applications. In: Oswald, E., Fischlin, M. (eds.) EUROCRYPT 2015. LNCS, vol. 9057, pp. 69–100. Springer, Heidelberg (2015). doi:10.1007/978-3-662-46803-6_3
2. Abdalla, M., Chevalier, C., Pointcheval, D.: Smooth projective hashing for conditionally extractable commitments. In: Halevi, S. (ed.) CRYPTO 2009. LNCS, vol. 5677, pp. 671–689. Springer, Heidelberg (2009). doi:10.1007/978-3-642-03356-8_39
3. Alwen, J., Shelat, A., Visconti, I.: Collusion-free protocols in the mediated model. In: Wagner, D. (ed.) CRYPTO 2008. LNCS, vol. 5157, pp. 497–514. Springer, Heidelberg (2008). doi:10.1007/978-3-540-85174-5_28
4. Ateniese, G., Magri, B., Venturi, D.: Subversion-resilient signature schemes. In: ACM CCS, pp. 364–375 (2015)
5. Balfanz, D., Durfee, G., Shankar, N., Smetters, D.K., Staddon, J., Wong, H.: Secret handshakes from pairing-based key agreements. In: S&P, pp. 180–196 (2003)
6. Bellare, M., Hoang, V.T.: Resisting randomness subversion: fast deterministic and hedged public-key encryption in the standard model. In: Oswald, E., Fischlin, M. (eds.) EUROCRYPT 2015. LNCS, vol. 9057, pp. 627–656. Springer, Heidelberg (2015). doi:10.1007/978-3-662-46803-6_21
7. Bellare, M., Jaeger, J., Kane, D.: Mass-surveillance without the state: Strongly undetectable algorithm-substitution attacks. In: ACM CCS, pp. 1431–1440 (2015)
8. Bellare, M., Paterson, K.G., Rogaway, P.: Security of symmetric encryption against mass surveillance. In: Garay, J.A., Gennaro, R. (eds.) CRYPTO 2014. LNCS, vol. 8616, pp. 1–19. Springer, Heidelberg (2014). doi:10.1007/978-3-662-44371-2_1
9. Benhamouda, F., Blazy, O., Chevalier, C., Pointcheval, D., Vergnaud, D.: New techniques for SPHFs and efficient one-round PAKE protocols. In: Canetti, R., Garay, J.A. (eds.) CRYPTO 2013. LNCS, vol. 8042, pp. 449–475. Springer, Heidelberg (2013). doi:10.1007/978-3-642-40041-4_25
10. Blaze, M., Bleumer, G., Strauss, M.: Divertible protocols and atomic proxy cryptography. In: Nyberg, K. (ed.) EUROCRYPT 1998. LNCS, vol. 1403, pp. 127–144. Springer, Heidelberg (1998). doi:10.1007/BFb0054122
11. Blazy, O., Pointcheval, D., Vergnaud, D.: Round-optimal privacy-preserving protocols with smooth projective hash functions. In: Cramer, R. (ed.) TCC 2012. LNCS, vol. 7194, pp. 94–111. Springer, Heidelberg (2012). doi:10.1007/978-3-642-28914-9_6
12. Cramer, R., Shoup, V.: Universal hash proofs and a paradigm for adaptive chosen ciphertext secure public-key encryption. In: Knudsen, L.R. (ed.) EUROCRYPT 2002. LNCS, vol. 2332, pp. 45–64. Springer, Heidelberg (2002). doi:10.1007/3-540-46035-7_4
13. Dodis, Y., Ganesh, C., Golovnev, A., Juels, A., Ristenpart, T.: A formal treatment of backdoored pseudorandom generators. In: Oswald, E., Fischlin, M. (eds.) EUROCRYPT 2015. LNCS, vol. 9056, pp. 101–126. Springer, Heidelberg (2015). doi:10.1007/978-3-662-46800-5_5
14. Dodis, Y., Mironov, I., Stephens-Davidowitz, N.: Message transmission with reverse firewalls—secure communication on corrupted machines. In: Robshaw, M., Katz, J. (eds.) CRYPTO 2016. LNCS, vol. 9814, pp. 341–372. Springer, Heidelberg (2016). doi:10.1007/978-3-662-53018-4_13
15. Gennaro, R., Lindell, Y.: A framework for password-based authenticated key exchange. In: Biham, E. (ed.) EUROCRYPT 2003. LNCS, vol. 2656, pp. 524–543. Springer, Heidelberg (2003). doi:10.1007/3-540-39200-9_33

16. Halevi, S., Kalai, Y.T.: Smooth projective hashing and two-message oblivious transfer. J. Crypt. **25**(1), 158–193 (2012)

17. Katz, J., Vaikuntanathan, V.: Round-optimal password-based authenticated key exchange. In: Ishai, Y. (ed.) TCC 2011. LNCS, vol. 6597, pp. 293–310. Springer, Heidelberg (2011). doi:10.1007/978-3-642-19571-6_18

18. Larson, J., Perlroth, N., Shane, S.: Revealed: The NSAs Secret Campaign to Crack, Undermine Internet Security. Pro-Publica, New York (2013)

19. Lepinski, M., Micali, S., Shelat, A.: Collusion-free protocols. In: Proceedings of the 37th Annual ACM Symposium on Theory of Computing, Baltimore, MD, USA, 22–24 May 2005, pp. 543–552 (2005)

20. Li, N., Du, W., Boneh, D.: Oblivious signature-based envelope. In: PODC, pp. 182–189 (2003)

21. Mironov, I., Stephens-Davidowitz, N.: Cryptographic reverse firewalls. In: Oswald, E., Fischlin, M. (eds.) EUROCRYPT 2015. LNCS, vol. 9057, pp. 657–686. Springer, Heidelberg (2015). doi:10.1007/978-3-662-46803-6_22

22. Perlroth, N., Larson, J., Shane, S.: NSA Able to Foil Basic Safeguards of Privacy on Web. The New York Times (2013)

23. Rogaway, P.: The moral character of cryptographic work. IACR Crypt. ePrint Arch. **2015**, 1162 (2015)

24. Russell, A., Tang, Q., Yung, M., Zhou, H.: Cliptography: clipping the power of kleptographic attacks. IACR Crypt. ePrint Arch. **2015**, 695 (2015)

25. Waters, B.: Efficient identity-based encryption without random oracles. In: Cramer, R. (ed.) EUROCRYPT 2005. LNCS, vol. 3494, pp. 114–127. Springer, Heidelberg (2005). doi:10.1007/11426639_7

26. Wee, H.: KDM-security via homomorphic smooth projective hashing. In: Cheng, C.-M., Chung, K.-M., Persiano, G., Yang, B.-Y. (eds.) PKC 2016. LNCS, vol. 9615, pp. 159–179. Springer, Heidelberg (2016). doi:10.1007/978-3-662-49387-8_7

27. Yang, R., Xu, Q., Zhou, Y., Zhang, R., Hu, C., Yu, Z.: Updatable hash proof system and its applications. In: Pernul, G., Ryan, P.Y.A., Weippl, E. (eds.) ESORICS 2015. LNCS, vol. 9326, pp. 266–285. Springer, Heidelberg (2015). doi:10.1007/978-3-319-24174-6_14

28. Young, A., Yung, M.: The dark side of "Black-Box" cryptography or: should we trust capstone? In: Koblitz, N. (ed.) CRYPTO 1996. LNCS, vol. 1109, pp. 89–103. Springer, Heidelberg (1996). doi:10.1007/3-540-68697-5_8

# Efficient Public-Key Cryptography
# with Bounded Leakage and Tamper Resilience

Antonio Faonio[1](✉) and Daniele Venturi[2]

[1] Department of Computer Science, Aarhus University, Aarhus, Denmark
afaonio@gmail.com
[2] Department of Information Engineering and Computer Science,
University of Trento, Trento, Italy

**Abstract.** We revisit the question of constructing public-key encryption and signature schemes with security in the presence of bounded leakage and tampering memory attacks. For signatures we obtain the first construction in the standard model; for public-key encryption we obtain the first construction free of pairing (avoiding non-interactive zero-knowledge proofs). Our constructions are based on generic building blocks, and, as we show, also admit efficient instantiations under fairly standard number-theoretic assumptions.

The model of bounded tamper resistance was recently put forward by Damgård *et al.* (Asiacrypt 2013) as an attractive path to achieve security against arbitrary memory tampering attacks without making hardware assumptions (such as the existence of a protected self-destruct or key-update mechanism), the only restriction being on the number of allowed tampering attempts (which is a parameter of the scheme). This allows to circumvent known impossibility results for unrestricted tampering (Gennaro *et al.*, TCC 2010), while still being able to capture realistic tampering attacks.

**Keywords:** Public-key encryption · Signatures · Related-key attacks · Tampering · Leakage

## 1 Introduction

Motivated by the proliferation of memory tampering attacks and fault injection [11,13,46], a recent line of research—starting with the seminal work of Bellare and Kohno [8] on the related-key attack (RKA) security of blockciphers—aims at designing cryptographic primitives that provably resist such attacks. Briefly, memory tampering attacks allow an adversary to modify the secret key of a targeted cryptographic scheme, and later violate its security by observing the effect of such changes at the output. In practice such attacks can be implemented by several means, both in hardware and software.

This paper is focused on designing public-key primitives—i.e., public-key encryption (PKE) and signature schemes—with provable security guarantees against memory tampering attacks. In this setting, the modified secret key might

© International Association for Cryptologic Research 2016
J.H. Cheon and T. Takagi (Eds.): ASIACRYPT 2016, Part I, LNCS 10031, pp. 877–907, 2016.
DOI: 10.1007/978-3-662-53887-6_32

be the signing key of a certification authority or of an SSL server, or the decryption key of a user. Informally, security of a signature scheme under tampering attacks can be cast as follows. The adversary is given a target verification key $vk$ and can observe signatures of adaptively chosen messages both under the original secret key $sk$ and under related keys $sk' = T(sk)$, derived from $sk$ by applying efficient tampering functions $T$ chosen by the adversary; the goal of the adversary is to forge a signature on a "fresh message" (i.e., a message not asked to the signing oracle) under the original verification key. Tamper resistance of PKE schemes under chosen-ciphertext attacks (CCA) can be defined similarly, the difference being that the adversary is allowed to observe decryption of adaptively chosen ciphertexts under related secret keys $sk'$, and its goal is now to violate semantic security.

*Unrestricted tampering.* The best we could hope for would be, of course, to allow the adversary to make any polynomial number of arbitrary, efficiently computable, tampering queries. Unfortunately, this type of "unrestricted tampering" is easily seen to be impossible without making further assumptions, as observed for the first time by Gennaro et al. [29]. The attack of [29] is simple enough to recall it here. The first tampering attempt defines $sk_1'$ to be equal to $sk$ with the first bit set to zero, so that verifying a signature under $sk_1'$ essentially allows to learn the first bit $b_1$ of the secret key with overwhelming probability. The second tampering attempt defines $sk_2'$ to be equal to $sk$ with the second bit set to zero, and with the first bit equal to $b_1$, and so on. This way each tampering attempt can be exploited to reveal one bit of the secret key, yielding a total security breach after $s(\kappa)$ queries, where $s(\kappa)$ is the bit-length of the secret key as a function of the security parameter.[1]

A possible way out to circumvent such an attack is to rely on the so-called self-destruct feature: Find a way how to detect tampering with high probability, and completely erase the memory or "blow-up the device" whenever tampering is detected. While this is indeed a viable approach, it has some shortcomings (at it can, e.g., be exploited for carrying out denial-of-service attacks), and so finding alternatives is an important research question. One natural such alternative is to simply restrict the power of the tampering functions $T$, in such a way that carrying out the above attack simply becomes impossible. This approach led to the design of several public-key primitives resisting an arbitrary polynomial number of *restricted* tampering attempts. All these schemes share the feature that the secret key belongs to some finite field, and the set of allowed modifications consist of all linear or affine functions, or all polynomials of bounded degree, applied to the key [7,10,54].

*Bounded tampering.* Unfortunately, the approach of restricting the tampering class only offers a partial solution to the problem; the main reason for this is

---

[1] A similar attack works for PKE schemes, and more generally for a large class of cryptographic primitives that can be tested for malfunctioning [29]; one can also make the above attack completely stateless.

that it is not a priori clear how the above mentioned algebraic relations capture realistic tampering attacks (where, e.g., a chip is shot with a laser). Motivated by this shortcoming, in a recent work, Damgård et al. [18] suggested the model of *bounded* tampering, where one assumes an upper-bound $\tau \in \mathbb{N}$ on the total number of tampering attempts the adversary is allowed to ever make; apart from this, and from the fact that the tampering functions $T$ should be efficiently computable, there is no further restriction on the adversarial tampering. Arguably, such form of tamper-proof security is sufficient to capture realistic attacks in which tampering might anyway destroy the device under attack or it could be detected by auxiliary hardware countermeasures; moreover, this model allows to analyze the security of cryptographic primitives already "in the wild," *without* the need to modify the implementation to include, e.g., a self-destruct feature.

An important parameter in the model of bounded tampering is the so-called tampering rate $\rho(\kappa) := \tau(\kappa)/s(\kappa)$ defined to be the ratio between the number of allowed tampering attempts and the size $s(\kappa)$ of the secret key in bits. The attack of Gennaro et al. [29] shows that necessarily $\rho(\kappa) \leq 1 - 1/p(\kappa)$ for some polynomial $p(\cdot)$. The original work of [18] shows how to obtain signature schemes and PKE schemes tolerating linear tampering rate $\rho(\kappa) = O(1/\kappa)$. However, the signature construction relies on the so-called Fiat–Shamir heuristic [28], whose security can only be proven in the random oracle model; the PKE construction can be instantiated in the standard model, but requires an untamperable common reference string (CRS), being based on (true simulation-extractable) non-interactive zero-knowledge (NIZK) [20].

In a follow-up work [19], the same authors show that resilience against bounded tampering can be obtained via a generic transformation yielding tampering rate $\rho(\kappa) = O(1/\sqrt[3]{\kappa^2})$; however, the transformation only gives a weaker form of security against non-adaptive (or semi-adaptive [19]) tampering attacks.

## 1.1  Our Contribution

In this work we improve the current state of the art on signature schemes and PKE schemes provably resisting bounded memory tampering. In the case of signatures, we obtain the first constructions in the standard model based on generic building blocks; as we argue, this yields concrete signature schemes tolerating tampering rate $\rho(\kappa) = O(1/\kappa)$ under standard complexity assumptions such as the Symmetric External Diffie-Hellman (SXDH) [12,52] and the Decisional Linear (DLIN) [35,53] assumptions. In the case of PKE, we obtain a direct, pairing-free, construction based on certain hash-proof systems [17], yielding concrete PKE schemes tolerating tampering rate $\rho(\kappa) = O(1/\kappa)$ under a particular instantiation of the Refined Subgroup Indistinguishability (RSI) assumption [45].

More precisely, we show that *already existing* schemes can be proved secure against bounded tampering. We do not view this as a limitation of our result, as it confirms the perspective that the model of bounded tamper resilience allows to make statements about cryptographic primitives already used "in the wild" (that might have already been implemented and adopted in applications). Additionally, our security arguments are non-trivial, requiring significant modifications

**Table 1.** Comparing known constructions of public-key primitives with security against related-key attacks (without self-destruct and key updating mechanisms). The value "∞" under the column "tampering rate" means that the scheme supports an arbitrary polynomial number of tampering queries. [a]Only achieves security against non-adaptive tampering.

| Reference | Type | Attack class | Model | Tampering rate | Assumption |
|---|---|---|---|---|---|
| BCM11 [7] | Sig./PKE | Linear | Standard | ∞ | DDHI [1] |
| Wee12 [54] | PKE | Linear | Standard | ∞ | BDDH/LWE |
| BPT12 [10] | Sig./PKE | Affine | Random Oracle | ∞ | BDH |
| | | Polynomial | Standard | ∞ | EDBDH |
| DFMV13 [18] | Sig. | Any | Random Oracle | $O(1/\kappa)$ | DLOG/Factoring |
| | PKE | Any | Standard | $O(1/\kappa)$ | SXDH/DLIN |
| BMT14 [9] | Sig. | Affine | Standard | ∞ | DLOG |
| | | Exponentiation | Standard | ∞ | RSA |
| | | Addition | Standard | ∞ | LWE |
| DFMV15 [19] | Sig./PKE | Any | Standard[a] | $O(1/\sqrt[3]{\kappa^2})$ | OWF/TDP |
| JW15 [37] | Sig./PKE | Poly-size Circuits | Standard | ∞ | OWF/TDP |
| QLY$^+$15 [51] | Sig./PKE | Polynomial | Standard | ∞ | DDH/DCR |
| Ours Sect. 3 | Sig. | Any | Standard | $O(1/\kappa)$ | SXDH/DLIN |
| Ours Sect. 4 | PKE | Any | Standard | $O(1/\kappa)$ | RSI |

to the original proofs (more on this below). In what follows we explain our contributions and techniques more in details. We refer the reader to Table 1 for a summary of our results and a comparison with previous work.

*Signatures.* We prove that the leakage-resilient signature scheme by Dodis *et al.* [20] is secure against bounded tampering attacks. The scheme of [20] satisfies the property that it remains unforgeable even given bounded leakage on the signing key. The main idea for showing security against bounded tampering, is to reduce tampering to leakage. Notice that this is non-trivial, because in the tampering setting the adversary is allowed to see polynomially many signatures corresponding to each of the tampered secret keys (which are at most $\tau$), and this yields a total amount of key-dependent information which is much larger than the tolerated leakage.

We now explain how to overcome this obstacle. The scheme exploits a so-called leakage-resilient hard relation $R$; such a relation satisfies the property that, given a statement $y$ generated together with a witness $x$, it is unfeasible to compute a witness $x^*$ for $(x^*, y) \in R$; moreover the latter holds even given bounded leakage on $x$. The verification key of the signature scheme consists of a random $y$, while the secret key is equal to $x$, where $(x, y)$ is a randomly generated pair belonging to the relation $R$. In order to sign a message $m$, one simply outputs a non-interactive zero-knowledge proof of knowledge $\pi$ of $x$, where the message $m$ is used as a label in the proof. Verification of a signature can be done by verifying the accompanying proof.

In the security proof, by the zero-knowledge property, we can replace real proofs with simulated proofs. Moreover, by the proof of knowledge property, we can actually extract a valid witness $x^*$ for $(x^*, y)$ from the adversarial forgery $\pi^*$; note that, since the forger gets to see simulated proofs, the extractability requirement must hold even after seeing proofs generated via the zero-knowledge simulator. Finally, we can transform a successful forger for the signature scheme into an adversary breaking the underlying leakage-resilient relation; the trick is that the reduction can leak the statement $y'$ corresponding to any tampered witness $x' = T(x)$, which allows to simulate an arbitrary polynomial number of signature queries corresponding to $x'$ by running several independent copies of the zero-knowledge simulator upon input $y'$. Thus bounded tamper resilience follows by bounded leakage resilience.

A subtle technicality in the above argument is that the statement $y'$ must be efficiently computable as a function of $x'$. We call a relation $R$ satisfying this property a *complete* relation. As we define it, completeness additionally requires that any derived witness $x' = T(x)$ is a witness for a valid statement $y'$ (i.e., $(x', y') \in R$); importantly this allows us to argue that simulated proofs are always for *true* statements, which leads to practical instantiations of the scheme. When we instantiate the signature scheme, of course, we need to make sure that the underlying relation meets our completeness requirement. Unfortunately, this is not directly the case for the constructions given in [20], but, as we show, such a difficulty can be overcome by carefully twisting the instantiation of the underlying relations.

*Public-key encryption.* Next, we prove that the PKE scheme by Qin and Liu [49] is secure against bounded tampering. The scheme is based on a variant of the classical Cramer-Shoup paradigm for constructing CCA-secure PKE [16,17]. Specifically, the PKE scheme combines a universal hash-proof system (HPS) together with a one-time lossy filter (OTLF) used to authenticate the ciphertext; the output of a randomness extractor is then used in order to mask the message in a one-time pad fashion. Since the OTLF is unkeyed, the secret key simply consists of the private evaluation key of the HPS, which makes it easier to analyze the security of the PKE scheme in the presence of memory tampering. The bulk of our proof is, indeed, to show that HPS with certain parameters already satisfy bounded tamper resilience.

More in details, every HPS is associated to a set $\mathcal{C}$ of ciphertexts and a subset $\mathcal{V} \subset \mathcal{C}$ of so-called *valid* ciphertexts, together with (the description of) a keyed hash function with domain $\mathcal{C}$. The hash function can be both evaluated privately (using a secret evaluation key) and publicly (on ciphertexts in $\mathcal{V}$, and using a public evaluation key). The main security guarantee is that for any $C \in \mathcal{C} \setminus \mathcal{V}$ the output of the hash function upon input $C$ is unpredictable even given the public evaluation key. In the construction of [49] a ciphertext consists of an element $C \in \mathcal{V}$, from which we derive an hash value $K$ which serves for two purposes:

(i) To extract a random pad via a seeded extractor, used to mask the plaintext; (ii) To authenticate the ciphertext by producing an encoding $\Pi$ of $K$ via the OTLF. The decryption algorithm first derives the value $K$ using the secret evaluation key for the HPS, and then it uses this value to unmask the plaintext provided that the value $\Pi$ can be verified correctly (otherwise decryption results in $\perp$).

In the reduction, the OTLF encoding will be programmed in such a way that, for all ciphertexts asked to the decryption oracle, the encoding is an injective function. This implies that, in order to create a ciphertext with a correct encoding $\Pi$, one has to know the underlying hash value $K$. To prove (standard) CCA security, one argues that all decryption queries with values $C \in \mathcal{V}$ do not reveal any additional information about the secret key, since the corresponding value $K$ could be computed via the public evaluation procedure; as for decryption queries with values $C \in \mathcal{C} \setminus \mathcal{V}$, the corresponding value $K$ is unpredictable, and therefore the decryption oracle will output $\perp$ with overwhelming probability which, again, does not reveal any additional information about the secret key.

The scenario in the case of tampering is more complicated. Consider a decryption oracle instantiated with a tampered secret key $sk' = T(sk)$. A decryption query containing a value $C \in \mathcal{V}$ might now reveal some information about the secret key; however, as we show, this information can be simulated by leaking the public key $pk'$ corresponding to $sk'$. Decryption queries containing values $C \in \mathcal{C} \setminus \mathcal{V}$ are harder to simulate. This is because the soundness property of the HPS only holds for a uniformly chosen evaluation key, while $sk'$, clearly, is not uniform. To overcome this obstacle we distinguish two cases:

- In case the value $T(sk)$ has low entropy, such a value does not reveal too much information on the secret key, and thus, at least intuitively, even if the decryption does not output $\perp$ the resulting plaintext should not decrease the entropy of the secret key by too much;
- In case the value $T(sk)$ has high entropy, we argue that it is safe to use this key within the HPS, i.e. we show that the soundness of the HPS is preserved as long as the secret key hash high entropy (even if it is not uniform).

With the above in mind, the security proof is similar to the ones in [44,49].

*Trading tampering and leakage.* Since our security arguments essentially reduce bounded tampering to bounded leakage (by individuating a short secret-key-dependent hint that allows to simulate polynomially many tampering queries for a given modified key), the theorems we get show a natural tradeoff between the obtained bounds for leakage and tamper resistance.

In particular, our results nicely generalizes previous work, in that we obtain the same bounds as in [20,49] by plugging $\tau = 0$ in our theorem statements.

## 1.2  Related Work

*Bounded leakage.* The signature scheme of Dodis *et al.* [20] generalizes and improves a previous construction by Katz and Vaikuntanathan [39]. Similarly, the PKE construction by Qin and Liu builds upon the seminal work of Naor and Segev [44]; the scheme was further improved in [50].

*Related-key security.* Related-key security was first studied in the context of symmetric encryption [2,3,8,30,43]. With time a number of cryptographic primitives with security against related-key attacks have emerged, including pseudorandom functions [1,4,6,40], hash functions [31], identity-based encryption [7,10], public-key encryption [7,10,42,54], signatures [7,9,10], and more [15,37,51].

All the above works achieve security against an unbounded number of restricted tampering attacks (typically, algebraic relations). Kalai, Kanukurthi, and Sahai [38], instead, show how to achieve security against unrestricted tampering without self-destruct, by assuming a protected mechanism to update the secret key of certain public-key cryptosystems (without modifying the corresponding public key).

*Non-malleable codes.* An alternative approach to achieve tamper-proof security of arbitrary cryptographic primitives against memory tampering is to rely on so-called non-malleable codes. While this solution yields security against an unbounded number of tampering queries, it relies on self-destruct and moreover it requires to further assume that the tampering functions are restricted in granularity (see, e.g., [22,25,41]) and/or computational complexity [5,26,37].

*Tamper-proof computation.* A related line of work (starting with [27,36]), finally, aims at constructing secure compilers protecting against tampering attacks targeting the computation carried out by a cryptographic device (typically in the form of boolean and arithmetic circuits).

## 2 Preliminaries

### 2.1 Notation

*Notation.* For $a, b \in \mathbb{R}$, we let $[a, b] = \{x \in \mathbb{R} : a \leq x \leq b\}$; for $a \in \mathbb{N}$ we let $[a] = \{1, 2, \ldots, a\}$. If $x$ is a string, we denote its length by $|x|$; if $\mathcal{X}$ is a set, $|\mathcal{X}|$ represents the number of elements in $\mathcal{X}$. When $x$ is chosen randomly in $\mathcal{X}$, we write $x \leftarrow_\$ \mathcal{X}$. When $\mathsf{A}$ is an algorithm, we write $y \leftarrow_\$ \mathsf{A}(x)$ to denote a run of $\mathsf{A}$ on input $x$ and output $y$; if $\mathsf{A}$ is randomized, then $y$ is a random variable and $\mathsf{A}(x; r)$ denotes a run of $\mathsf{A}$ on input $x$ and randomness $r$. An algorithm $\mathsf{A}$ is *probabilistic polynomial-time* (PPT) if $\mathsf{A}$ is randomized and for any input $x, r \in \{0, 1\}^*$ the computation of $\mathsf{A}(x; r)$ terminates in at most $poly(|x|)$ steps.

Throughout the paper we let $\kappa \in \mathbb{N}$ denote the security parameter. We say that a function $\nu : \mathbb{N} \to \mathbb{R}$ is negligible in the security parameter $\kappa$ if $\nu(\kappa) = \kappa^{-\omega(1)}$. For two ensembles $\mathcal{X} = \{\mathbf{X}_\kappa\}_{\kappa \in \mathbb{N}}$ and $\mathcal{Y} = \{\mathbf{Y}_\kappa\}_{\kappa \in \mathbb{N}}$, we write $\mathcal{X} \equiv \mathcal{Y}$ if they are identically distributed, $\mathcal{X} \approx_s \mathcal{Y}$ to denote that the corresponding distributions are statistically close, and $\mathcal{X} \approx_c \mathcal{Y}$ to denote that the two ensembles are computationally indistinguishable.

*Languages and relations.* A *decision problem* related to a language $L \subseteq \{0,1\}^*$ requires to determine if a given string $y$ is in $L$ or not. We can associate to any *NP*-language $L$ a polynomial-time recognizable relation $R \subseteq \{0,1\}^* \times \{0,1\}^*$ defining $L$ itself, i.e. $L = \{y : \exists x \text{ s.t. } (x,y) \in R\}$ for $|x| \leqslant poly(|y|)$. The string $y$ is called *theorem*, and the string $x$ is called a *witness* for membership of $y \in L$.

*Random variables.* The min-entropy of a random variable $\mathbf{X}$, defined over a set $\mathcal{X}$, is $\mathbb{H}_\infty(\mathbf{X}) := -\log \max_{x \in \mathcal{X}} \mathbb{P}[\mathbf{X} = x]$, and it measures how $\mathbf{X}$ can be predicted by the best (unbounded) predictor. The average conditional min-entropy of a random variable $\mathbf{X}$ given a random variable $\mathbf{Y}$ and conditioned on an event $E$ is defined as $\widetilde{\mathbb{H}}_\infty(\mathbf{X}|\mathbf{Y}, E) := -\log(\mathbb{E}_{y \leftarrow \mathbf{Y}}[2^{-\mathbb{H}_\infty(\mathbf{X}|\mathbf{Y}=y,E)}])$. We rely on the following basic facts.

**Lemma 1** ([21]). *Let* $\mathbf{X}, \mathbf{Y}$ *and* $\mathbf{Z}$ *be random variables. If* $\mathbf{Y}$ *has at most* $2^\ell$ *possible values, then* $\widetilde{\mathbb{H}}_\infty(\mathbf{X}|\mathbf{Y}, \mathbf{Z}) \geqslant \widetilde{\mathbb{H}}_\infty(\mathbf{X}, \mathbf{Y}|\mathbf{Z}) - \ell \geqslant \widetilde{\mathbb{H}}_\infty(\mathbf{X}|\mathbf{Z}) - \ell$.

**Lemma 2.** *Let* $\mathbf{X}, \mathbf{Y}, \mathbf{Z}$ *be random variables such that* $\mathbf{Y} = f(\mathbf{X}, \mathbf{Z})$ *for an efficiently computable function* $f$. *Then* $\widetilde{\mathbb{H}}_\infty(\mathbf{X}|\mathbf{Y}, \mathbf{Z}, E) \geqslant \widetilde{\mathbb{H}}_\infty(\mathbf{X}|\mathbf{Z}, E) - \beta$, *where the event* $E$ *is defined as* $\{\forall z : \mathbb{H}_\infty(\mathbf{Y}|\mathbf{Z} = z) \leqslant \beta\}$.

*Proof.* Let A be the best predictor for $\mathbf{X}$, given $\mathbf{Y}$ and $\mathbf{Z}$ and conditioned on the event $E$. Consider the predictor A' that upon input $\mathbf{Z}$ first samples an independent copy $\mathbf{X}'$ of the random variable $\mathbf{X}$ and then runs A upon input $f(\mathbf{X}', \mathbf{Z})$. Note that the event $E$ holds for the inputs given to A', therefore the probability that $f(\mathbf{X}', \mathbf{Z}) = f(\mathbf{X}, \mathbf{Z})$ is bounded above by $2^{-\beta}$. This implies the lemma.   □

## 2.2   Public-Key Encryption

A public-key encryption (PKE) scheme is a tuple of algorithms $\mathcal{PKE} = ($Setup, Gen, Enc, Dec$)$ defined as follows. (1) Algorithm Setup takes as input the security parameter and outputs public parameters $pub \in \{0,1\}^*$; all algorithms are implicitly given $pub$ as input. (2) Algorithm Gen takes as input the security parameter and outputs a public/secret key pair $(pk, sk)$; the set of all secret keys is denoted by $\mathcal{SK}$ and the set of all public keys by $\mathcal{PK}$. (3) The randomized algorithm Enc takes as input the public key $pk$, a message $m \in \mathcal{M}$, and randomness $r \in \mathcal{R}$, and outputs a ciphertext $c = $ Enc$(pk, m; r)$; the set of all ciphertexts is denoted by $\mathcal{C}$. (4) The deterministic algorithm Dec takes as input the secret key $sk$ and a ciphertext $c \in \mathcal{C}$, and outputs $m = $ Dec$(sk, c)$ which is either equal to some message $m \in \mathcal{M}$ or to an error symbol $\perp$.

*Correctness.* We say that $\mathcal{PKE}$ satisfies *correctness* if for all $pub \leftarrow_\$ $ Setup$(1^\kappa)$ and $(pk, sk) \leftarrow_\$ $ Gen$(1^\kappa)$ we have that $\mathbb{P}[\text{Dec}(sk, \text{Enc}(pk, m)) = m] = 1$ (where the randomness is taken over the internal coin tosses of algorithm Enc).

$$
\begin{array}{|l|l|l|}
\hline
\end{array}
$$

**Experiment** $\mathbf{Exp}_{\mathcal{PKE},A}^{blt-cca}(\kappa, \ell, \tau)$:

$pub \leftarrow_{\$} \mathsf{Setup}(1^\kappa)$

$(pk, sk) \leftarrow_{\$} \mathsf{Gen}(1^\kappa)$

$b \leftarrow_{\$} \{0,1\};\ \mathcal{Q} \leftarrow \emptyset;\ j \leftarrow 1$

$sk'_0 \leftarrow sk;\ (\forall i \in [\tau])sk'_i \leftarrow \bot;\ c^* \leftarrow \bot$

$(m_0, m_1) \leftarrow A^{\mathsf{Dec}^*(\cdot,\cdot),\mathcal{O}_{sk}^\ell(\cdot),\mathcal{O}_{sk}^\tau(\cdot)}(pk)$

$c^* \leftarrow_{\$} \mathsf{Enc}(pk, m_b)$

$b' \leftarrow A^{\mathsf{Dec}^*(0,\cdot)}(c^*)$

Return

$\quad (b' = b) \wedge (|m_0| = |m_1|) \wedge (c^* \notin \mathcal{Q})$

**Oracle** $\mathsf{Dec}^*(i, c)$:

If $i \notin [0, \tau]$

$\quad$ Return $\bot$

Else if $sk'_i = \bot$

$\quad$ Return $\bot$

Else

$\quad$ If $c^* \neq \bot$

$\quad\quad \mathcal{Q} \leftarrow \mathcal{Q} \cup \{c\}$

$\quad$ Return $\mathsf{Dec}(sk'_i, c)$

**Oracle** $\mathcal{O}_{sk}^\ell(L)$:

Return $L(sk)$

**Oracle** $\mathcal{O}_{sk}^\tau(T)$:

$sk'_j = T(sk)$

$j \leftarrow j + 1$

**Fig. 1.** Experiment defining BLT-IND-CCA security of $\mathcal{PKE}$.

*BLT Security.* We now turn to defining indistinguishability under chosen-ciphertext attacks (IND-CCA) in the bounded leakage and tampering (BLT) setting.

**Definition 1.** *For $\kappa \in \mathbb{N}$, let $\ell = \ell(\kappa)$ and $\tau = \tau(\kappa)$ be parameters. We say that $\mathcal{PKE} = (\mathsf{Setup}, \mathsf{Gen}, \mathsf{Enc}, \mathsf{Dec})$ is $(\tau, \ell)$-BLT-IND-CCA if for all PPT adversaries A there exists a negligible function $\nu : \mathbb{N} \to [0,1]$ such that*

$$\left| \mathbb{P}\left[ \mathbf{Exp}_{\mathcal{PKE},A}^{blt-cca}(\kappa, \ell, \tau) = 1 \right] - \frac{1}{2} \right| \leq \nu(\kappa),$$

*where the experiment $\mathbf{Exp}_{\mathcal{PKE},A}^{blt-cca}(\kappa, \ell, \tau)$ is defined in Fig. 1.*

A few remarks on the definition are in order. In the specification of the BLT-IND-CCA security experiment, oracle $\mathcal{O}_{sk}^\ell$ takes as input (arbitrary polynomial-time computable) functions $L : \mathcal{SK} \to \{0,1\}^*$, and returns $L(sk)$ for a total of at most $\ell$ bits. In a similar fashion, oracle $\mathcal{O}_{sk}^\tau$ takes as input (arbitrary polynomial-time computable) functions $T : \mathcal{SK} \to \mathcal{SK}$, and defines the $i$-th tampered secret key as $sk'_i = T(sk)$; the oracle accepts at most $\tau$ queries. Oracle $\mathsf{Dec}^*$ can be used to decrypt arbitrary ciphertexts $c$ under the $i$-th tampered secret key (or under the original secret key), provided that $c$ is different from the challenge ciphertext.

Notice that A is not allowed to tamper with or leak from the secret key after seeing the challenge ciphertext. As shown in [18] this restriction is necessary already for the case $(\tau, \ell) = (1, 0)$. Finally, we observe that in case $(\tau, \ell) = (0, 0)$ we get, as a special case, the standard notion of IND-CCA security. Similarly, for $\tau = 0$ and $\ell > 0$, we obtain as a special case the notion of "semantic security against a-posteriori chosen-ciphertext $\ell$-key-leakage attacks" from [44].

## 2.3 Signatures

A signature scheme is a tuple of algorithms $\mathcal{SIG} = (\mathsf{Setup}, \mathsf{Gen}, \mathsf{Sign}, \mathsf{Vrfy})$ specified as follows. (1) Algorithm $\mathsf{Setup}$ takes as input the security parameter and outputs public parameters $pub \in \{0,1\}^*$; all algorithms are implicitly given $pub$

Experiment $\mathbf{Exp}_{SIG,A}^{blt\text{-}cma}(\kappa, \ell, \tau)$:

$pub \leftarrow_{\$} \mathsf{Setup}(1^{\kappa})$

$(vk, sk) \leftarrow_{\$} \mathsf{Gen}(1^{\kappa})$

$\mathcal{Q} \leftarrow \emptyset;\ j \leftarrow 1$

$sk_0' \leftarrow sk;\ (\forall i \in [\tau])sk_i' \leftarrow \perp$

$(m^*, \sigma^*) \leftarrow \mathsf{A}^{\mathsf{Sign}^*(\cdot,\cdot), \mathcal{O}_{sk}^{\ell}(\cdot), \mathcal{O}_{sk}^{\tau}(\cdot)}(vk)$

Return

$\qquad (\mathsf{Vrfy}(vk, m^*, \sigma^*) = 1) \wedge (m^* \notin \mathcal{Q})$

Oracle $\mathsf{Sign}^*(i, m)$:

If $i \notin [0, \tau]$

$\qquad$ Return $\perp$

Else if $sk_i' = \perp$

$\qquad$ Return $\perp$

Else

$\qquad \mathcal{Q} \leftarrow \mathcal{Q} \cup \{m\}$

$\qquad$ Return $\mathsf{Sign}(sk_i', m)$

Oracle $\mathcal{O}_{sk}^{\ell}(L)$:

Return $L(sk)$

Oracle $\mathcal{O}_{sk}^{\tau}(T)$:

$sk_j' = T(sk)$

$j \leftarrow j + 1$

**Fig. 2.** Experiment defining BLT-EUF-CMA security of $SIG$.

as input. (2) Algorithm $\mathsf{Gen}$ takes as input the security parameter and outputs a public/secret key pair $(vk, sk)$; the set of all signing keys is denoted by $SK$. (3) The randomized algorithm $\mathsf{Sign}$ takes as input the signing key $sk$, a message $m \in \mathcal{M}$, and randomness $r \in \mathcal{R}$, and outputs a signature $\sigma := \mathsf{Sign}(sk, m; r)$ on $m$. (4) The deterministic algorithm $\mathsf{Vrfy}$ takes as input the verification key $vk$ and a pair $(m, \sigma)$, and outputs a decision bit (indicating whether $(m, \sigma)$ is a valid signature with respect to $vk$).

*Correctness.* We say that $SIG$ satisfies *correctness* if for all messages $m \in \mathcal{M}$ and for all $pub \leftarrow_{\$} \mathsf{Setup}(1^{\kappa})$ and $(vk, sk) \leftarrow \mathsf{Gen}(1^{\kappa})$, algorithm $\mathsf{Vrfy}(vk, m, \mathsf{Sign}(sk, m))$ outputs 1 with all but negligible probability (over the coin tosses of the signing algorithm).

*BLT Security.* We now define what it means for a signature scheme to be existentially unforgeable against chosen-message attacks (EUF-CMA) in the bounded leakage and tampering (BLT) setting.

**Definition 2.** *For $\kappa \in \mathbb{N}$, let $\ell = \ell(\kappa)$ and $\tau = \tau(\kappa)$ be parameters. We say that $SIG = (\mathsf{Setup}, \mathsf{Gen}, \mathsf{Sign}, \mathsf{Vrfy})$ is $(\tau, \ell)$-BLT-EUF-CMA if for all PPT adversaries $\mathsf{A}$ there exists a negligible function $\nu : \mathbb{N} \to [0, 1]$ such that*

$$\mathbb{P}\left[\mathbf{Exp}_{SIG,A}^{blt\text{-}cma}(\kappa, \ell, \tau) = 1\right] \leq \nu(\kappa),$$

*where the experiment $\mathbf{Exp}_{SIG,A}^{blt\text{-}cma}(\kappa, \ell, \tau)$ is defined in Fig. 2.*

The syntax of oracles $\mathcal{O}_{sk}^{\ell}$ and $\mathcal{O}_{sk}^{\tau}$ is the same as before. Oracle $\mathsf{Sign}^*$ can be used to sign arbitrary messages $m$ under the $i$-th tampered signing key $sk_i' = T(sk)$, or under the original signing key $sk$; the goal of the adversary is to forge a signature on a "fresh" message, i.e. a message that was never queried to oracle $\mathsf{Sign}^*$. Note that for $(\tau, \ell) = (0, 0)$ we obtain the standard notion of existential unforgeability under chosen-message attacks. Similarly, for $\tau = 0$ and $\ell > 0$, we obtain the definition of leakage-resilient signatures [39].

# 3   Signatures

In this section we give a generic construction of signature schemes with BLT-EUF-CMA in the standard model. In particular, we show that the construction by Dodis *et al.* [20] is already resilient to bounded leakage and tampering attacks.

## 3.1   The Scheme of Dodis, Haralambiev, Lòpez-Alt, and Wichs

The signature scheme is based on the following ingredients.

*Hard relations.* A leakage-resilient hard relation [20].

**Definition 3.** *A relation $R$ is an $\ell$-leakage-resilient hard relation, with witness space $\mathcal{X}$ and theorem space $\mathcal{Y}$, if the following requirements are met.*

**Samplability:** *There exists a PPT algorithm* SamR *such that for all pairs $(x, y) \leftarrow_{\$} \mathsf{SamR}(1^\kappa)$ we have $(x, y) \in R$, with $x \in \mathcal{X}$ and $y \in \mathcal{Y}$.*

**Verifiability:** *There exists a PPT algorithm that decides if a given pair $(x, y)$ satisfies $(x, y) \in R$.*

**Completeness:** *There exists an efficient deterministic function $\xi$ that given as input any $x \in \mathcal{X}$ returns $y = \xi(x) \in \mathcal{Y}$ such that $(x, y) \in R$.*

**Hardness:** *For all PPT adversaries* A *there exists a negligible function $\nu : \mathbb{N} \to [0, 1]$ such that*

$$\mathbb{P}\left[(x^*, y) \in R: \ (x, y) \leftarrow_{\$} \mathsf{SamR}(1^\kappa); x^* \leftarrow_{\$} \mathsf{A}^{\mathcal{O}_x^\ell(\cdot)}(y)\right] \leq \nu(\kappa),$$

*where the probability is taken over the random coin tosses of* SamR *and* A*, and where oracle $\mathcal{O}_x^\ell(\cdot)$ takes as input efficiently computable functions $L : \mathcal{X} \to \{0, 1\}^*$ and returns $L(x)$ for a total of at most $\ell$ bits.*

*NIZK.* A true-simulation extractable non-interactive zero-knowledge (tSE NIZK) argument system $\mathcal{NIZK} = (\mathsf{I}, \mathsf{P}, \mathsf{V})$ for the relation $R$, supporting labels [20]. Recall that a NIZK argument system supporting labels has the following syntax: (i) Algorithm I takes as input the security parameter $\kappa \in \mathbb{N}$ and generates a common reference string (CRS) $crs \leftarrow_{\$} \mathsf{I}(1^\kappa)$. (ii) Algorithm P takes as input the CRS, a label $\lambda \in \{0, 1\}^*$, and some pair $(x, y) \in R$, and returns a proof $\pi \leftarrow_{\$} \mathsf{P}^\lambda(crs, x, y)$. (iii) Algorithm V takes as input the CRS, a label $\lambda \in \{0, 1\}^*$, and some pair $(x, \pi)$, and returns a decision bit $\mathsf{V}^\lambda(crs, y, \pi)$. Moreover:

**Definition 4.** *We say that $\mathcal{NIZK} = (\mathsf{I}, \mathsf{P}, \mathsf{V})$ is a tSE NIZK for the relation $R$, supporting labels, if the following requirements are met.*

**Correctness:** *For all pairs $(x, y) \in R$ and for all labels $\lambda \in \{0, 1\}^*$ we have that $\mathsf{V}^\lambda(crs, y, \mathsf{P}^\lambda(crs, x, y)) = 1$ with overwhelming probability over the coin tosses of* P, V, *and over the choice of $crs \leftarrow_{\$} \mathsf{I}(1^\kappa)$.*

**Unbounded zero-knowledge:** *There exists a PPT simulator* $S := (S_1, S_2)$ *such that for all PPT adversaries* A *the following quantity is negligible:*[2]

$$\left| \mathbb{P}\left[ b = b' : \begin{array}{l} b \leftarrow_\$ \{0,1\}; (crs, tk) \leftarrow_\$ S_1(1^\kappa); (x, y, \lambda) \leftarrow_\$ A(crs, tk) \\ \pi_0 \leftarrow_\$ P^\lambda(crs, x, y); \pi_1 \leftarrow_\$ S_2^\lambda(tk, y); b' \leftarrow_\$ A(crs, tk, \pi_b) \end{array} \right] - \frac{1}{2} \right|.$$

**True-simulation extractability:** *There exists a PPT extractor* K *such that for all PPT adversaries* A *the following quantity is negligible:*

$$\mathbb{P}\left[ \begin{array}{c} (\lambda^* \notin \mathcal{Q}) \wedge (V^{\lambda^*}(crs, y^*, \pi^*) = 1) \\ \wedge ((x^*, y^*) \notin R) \end{array} : \begin{array}{c} (crs, tk) \leftarrow_\$ S_1(1^\kappa) \\ (y^*, \pi^*, \lambda^*) \leftarrow_\$ A^{\mathcal{O}_{S_2, \tau}(\cdot, \cdot, \cdot)}(crs) \\ x^* \leftarrow_\$ K^{\lambda^*}(tk, y^*, \pi^*) \end{array} \right],$$

*where oracle* $\mathcal{O}_{S_2, \tau}$ *takes as input tuples* $(x_i, y_i, \lambda_i)$ *and returns the same as* $S_2^{\lambda_i}(tk, y_i)$ *as long as* $(x_i, y_i) \in R$ *(and* $\perp$ *otherwise), and* $\mathcal{Q}$ *is the set of all labels* $\lambda_i$ *asked to oracle* $\mathcal{O}_{S_2, \tau}$.

*The signature scheme.* Consider now the following signature scheme $\mathcal{SIG} =$ (Setup, Gen, Sign, Vrfy), based on a relation $R$, and on a non-interactive argument system $\mathcal{NIZK} = (I, P, V)$ for $R$, supporting labels.

- Setup($1^\kappa$): Sample $crs \leftarrow_\$ I(1^\kappa)$ and return $pub := (crs, R)$. (Recall that all algorithms implicitly take $pub$ as input.)
- Gen($1^\kappa$): Run $(x, y) \leftarrow_\$ SamR(1^\kappa)$ and define $vk := y$ and $sk := x$.
- Sign($sk, m$): Compute $\pi \leftarrow_\$ P^m(crs, x, \xi(x))$ and return $\sigma := \pi$; note that the message $m$ is used as a label in the argument system, and that the value $y = \xi(x)$ can be efficiently computed as a function of $x$.
- Vrfy($vk, m, \sigma$): Parse $(vk, \sigma)$ as $vk := y$ and $\sigma := \pi$, and output the same as $V^m(crs, y, \pi)$.

**Theorem 1.** *For* $\kappa \in \mathbb{N}$, *let* $\ell := \ell(\kappa)$, $\ell' := \ell'(\kappa)$, $\tau := \tau(\kappa)$, *and* $n := n(\kappa)$ *be parameters. Assume that* $R$ *is an* $\ell'$-*leakage-resilient hard relation with theorem space* $\mathcal{Y} := \{0,1\}^n$, *and that* $\mathcal{NIZK}$ *is a tSE NIZK for* $R$. *Then the signature scheme* $\mathcal{SIG}$ *described above is* $(\ell, \tau)$-*BLT-EUF-CMA with* $\ell + (\tau + 1) \cdot n \leq \ell'$.

## 3.2   Security Proof

We consider a sequence of mental experiments, starting with the initial game $\mathbf{Exp}_{\mathcal{SIG}, A}^{\text{blt-cma}}(\kappa, \ell, \tau)$ which for simplicity we denote by $\mathbf{G}_0$.

**Game $\mathbf{G}_0$.** This is exactly the game of Definition 2, where the signature scheme $\mathcal{SIG}$ is the scheme described in the previous section. In particular, upon input the $i$-th tampering query $T_i$ the modified secret key $x_i' = T_i(x)$ is computed. Hence, the answer to a query $(i, m)$ to oracle Sign* is computed by parsing $pub = (crs, R)$, computing the statement $y_i' = \xi(x_i')$ corresponding to $x_i'$, and outputting $\sigma := \pi$ where $\pi \leftarrow_\$ P^m(crs, x_i', y_i')$.

---

[2] Strictly speaking we should quantify the definition over all adversaries returning pairs $(x, y) \in R$; alternatively, we can slightly abuse notation and assume that both P and $S_2$ return $\perp$ if that is not the case.

**Game $G_1$.** We change the way algorithm Setup generates the CRS. Namely, instead of sampling $crs \leftarrow\!\!\text{\tiny\$}\; I(1^\kappa)$ we now run $(crs, tk) \leftarrow\!\!\text{\tiny\$}\; S_1(1^\kappa)$ and additionally we replace the proofs output by oracle Sign* by simulated proofs, i.e., $\pi \leftarrow\!\!\text{\tiny\$}\; S_2(tk, y_i')$ where $y_i' = \xi(x_i')$.

**Game $G_2$.** We change the winning condition of the previous game. Namely, the game now outputs one if and only if $\pi^*$ is valid w.r.t. $y$ (as before) and additionally $(x^*, y) \in R$ where the value $x^*$ is computed from the proof $\pi^*$ running the extractor K of the underlying argument system.

We now establish a series of lemmas, showing that the above games are computationally indistinguishable. The first lemma states that $G_0$ and $G_1$ are indistinguishable, down to the unbounded zero-knowledge property of the argument system.

**Lemma 3.** *For all PPT adversaries A there exists a negligible function $\nu_{0,1}$ :* $\mathbb{N} \to [0, 1]$ *such that* $|\mathbb{P}[G_0(\kappa) = 1] - \mathbb{P}[G_1(\kappa) = 1]| \leq \nu_{0,1}(\kappa)$.

*Proof.* We prove a stronger statement, namely that $G_0(\kappa) \approx_c G_1(\kappa)$. By contradiction, assume that there exists a PPT distinguisher $D_{0,1}$ and a polynomial $p_{0,1}(\cdot)$ such that, for infinitely many values of $\kappa \in \mathbb{N}$, we have that $D_{0,1}$ distinguishes between game $G_0$ and game $G_1$ with probability at least $1/p_{0,1}(\kappa)$. Let $q \in poly(\kappa)$ be the number of signature queries asked by $D_{0,1}$. For an index $j \in [q + 1]$ consider the hybrid game $H_j$ that answers the first $j - 1$ queries as in game $G_0$ and all subsequent queries as in game $G_1$. Note that $H_1 \equiv G_1$ and $H_{q+1} \equiv G_0$.

By a standard hybrid argument, we have that there exists an index $j^* \in [q]$ such that $D_{0,1}$ tells apart $H_{j^*}$ and $H_{j^*+1}$ with non-negligible probability $1/q \cdot 1/p_{0,1}(\kappa)$. We build a PPT adversary $A_{0,1}$ that (using distinguisher $D_{0,1}$ and knowledge of $j^* \in [q]$) breaks the non-interactive zero-knowledge property of the argument system. A formal description of $A_{0,1}$ follows.

---

Adversary $A_{0,1}$:
- Receive $(crs, tk)$ from the challenger, where $(crs, tk) \leftarrow\!\!\text{\tiny\$}\; S_1(1^\kappa)$.
- Run $(x, y) \leftarrow\!\!\text{\tiny\$}\; \mathsf{SamR}(1^\kappa)$, set $pub := (crs, R)$, $vk := y$, $x_0' \leftarrow x$, $x_i' \leftarrow \perp$ for all $i \in [\tau]$, and send $(pub, vk)$ to $D_{0,1}$.
- Upon input a leakage query $L$ return $L(x)$ to $D_{0,1}$; upon input a tampering query $T$, set $x_i' = T(x)$.
- Upon input the $j$-th signature query of type $(i, m)$, if $i \notin [0, \tau]$ or $x_i' = \perp$, answer with $\perp$. Otherwise, proceed as follows:
  - If $j \leq j^* - 1$, return $\sigma \leftarrow\!\!\text{\tiny\$}\; P^m(crs, x_i', \xi(x_i'))$ to $D_{0,1}$.
  - Else, if $j = j^*$, forward $(x_i', \xi(x_i'), m)$ to the challenger, receiving back a proof $\pi_b$; return $\sigma := \pi_b$ to $D_{0,1}$.
  - Else, if $j \geq j^* + 1$, forward $\sigma \leftarrow\!\!\text{\tiny\$}\; S_2^m(tk, \xi(x_i'))$ to $D_{0,1}$.
- Output whatever D outputs.

---

For the analysis, note that the only difference between game $H_{j^*}$ and game $H_{j^*+1}$ is on how the $j^*$-th signature query is answered. In particular, in case the hidden bit $b$ in the definition of non-interactive zero-knowledge equals zero, $A_{0,1}$'s simulation produces exactly the same distribution as in $\mathbf{H}_{j^*}$, and otherwise $A_{0,1}$'s simulation produces exactly the same distribution as in $\mathbf{H}_{j^*+1}$. Hence, $A_{0,1}$ breaks the NIZK property with non-negligible advantage $1/q \cdot 1/p_{0,1}(\kappa)$, a contradiction. This concludes the proof.                                                                  □

The second lemma states that $\mathbf{G}_1$ and $\mathbf{G}_2$ are indistinguishable, down to the true-simulation extractability property of the argument system.

**Lemma 4.** *For all PPT adversaries* A *there exists a negligible function* $\nu_{1,2}$ : $\mathbb{N} \to [0,1]$ *such that* $|\mathbb{P}[\mathbf{G}_1(\kappa) = 1] - \mathbb{P}[\mathbf{G}_2(\kappa) = 1]| \leq \nu_{1,2}(\kappa)$.

*Proof.* We prove a stronger statement, namely that $\mathbf{G}_1(\kappa) \approx_c \mathbf{G}_2(\kappa)$. Define the following "bad event" *Bad*, in the probability space of game $\mathbf{G}_1$: The event becomes true if the adversarial forgery $(m^*, \sigma^* := \pi^*)$ is valid (i.e., the proof $\pi^*$ is valid w.r.t. statement $y$ and label $m^*$), but running the extractor $\mathsf{K}(tk, \cdot, \cdot)$ on $(y, \pi^*)$ yields a value $x^*$ such that $(x^*, y) \notin R$.

Notice that $\mathbf{G}_1(\kappa)$ and $\mathbf{G}_2(\kappa)$ are identically distributed conditioning on *Bad* not happening. Hence, by a standard argument, it suffices to bound the probability of provoking event *Bad* by all PPT adversaries A. By contradiction, assume that there exists a PPT adversary $A_{1,2}$ and a polynomial $p_{1,2}(\cdot)$ such that, for infinitely many values of $\kappa \in \mathbb{N}$, we have that $A_{1,2}$ provokes event *Bad* with probability at least $1/p_{1,2}(\kappa)$. We build an adversary $A'$ that (using $A_{1,2}$) breaks true-simulation extractability of the argument system. A formal description of $A'$ follows.

Adversary $A'$:
- Receive *crs* from the challenger, where $(crs, tk) \leftarrow_\$ \mathsf{S}_1(1^\kappa)$.
- Sample $(x, y) \leftarrow_\$ \mathsf{SamR}(1^\kappa)$, set $pub := (crs, R)$, $vk := y$, $x'_0 \leftarrow x$, $x'_i \leftarrow \bot$ (for all $i \in [\tau]$), and forward $(pub, vk)$ to $A_{1,2}$.
- Upon input a leakage query $L$ return $L(x)$ to $A_{1,2}$; upon input a tampering query $T$, set $x'_i = T(x)$.
- Upon input the $j$-th signature query of type $(i, m)$, if $i \notin [0, \tau]$ or $x'_i = \bot$, answer with $\bot$. Otherwise, forward $(x'_i, \xi(x'_i), m)$ to the challenger obtaining a proof $\pi$ as a response, and return $\sigma := \pi$ to $A_{1,2}$.
- Whenever $A_{1,2}$ returns a pair $(m^*, \sigma^*)$, define $\pi^* := \sigma^*$ and output $(y, \pi^*, m^*)$.

For the analysis, we note that $A'$ perfectly simulates signature queries. In fact, by completeness of the underlying relation, the pair $(x'_i, \xi(x'_i))$ is always in the relation $R$, and thus the proof $\pi$ obtained by the reduction is always for a true statement and has exactly the same distribution as in game $\mathbf{G}_1$. As a consequence, $A_{1,2}$ will provoke event *Bad* with probability $1/p_{1,2}(\kappa)$, and thus the pair $(y, \pi^*)$ output by the reduction violates the tSE property of the non-interactive argument with non-negligible probability $1/p_{1,2}(\kappa)$. This finishes the proof.                                                                  □

Finally, we show that the advantage of any PPT adversary in game $\mathbf{G}_2$ must be negligible, otherwise one could violate the hardness of the underlying leakage-resilient relation.

**Lemma 5.** *For all PPT adversaries* A *there exists a negligible function* $\nu_2$ : $\mathbb{N} \to [0,1]$ *such that* $\mathbb{P}\left[\mathbf{G}_2 = 1\right] \leq \nu_2(\kappa)$.

*Proof.* By contradiction, assume there exists a PPT adversary $A_2$ and a polynomial $p_2(\cdot)$ such that, for infinitely many values of $\kappa \in \mathbb{N}$, adversary $A_2$ makes game $\mathbf{G}_2$ output 1 with probability at least $1/p_2(\kappa)$. We construct a PPT adversary $A''$ (using $A_2$) breaking hardness of the leakage-resilient relation $R$. A description of $A''$ follows.

Adversary $A''$:
- Receive $y$ from the challenger, where $(x,y) \leftarrow_\$ \mathsf{SamR}(1^\kappa)$.
- Sample $(crs, tk) \leftarrow_\$ \mathsf{S}_1(1^\kappa)$, set $pub := (crs, R)$, $y_i' \leftarrow \perp$ (for all $i \in [\tau]$), $vk := y$, and forward $(pub, vk)$ to $A_2$.
- Define the leakage function $L_\xi(x) := \xi(x)$ and forward $L_\xi$ to the target leakage oracle $\mathcal{O}_x^\ell$, obtaining a value $y_0'$.
- Upon input a leakage query $L$, forward $L$ to the target leakage oracle $\mathcal{O}_x^\ell$ and return to $A_2$ the answer received from the oracle.
- Upon input the $i$-th tampering query $T$, define the function $L_{T,\xi}(x) := \xi(T(x))$, and forward $L_{T,\xi}$ to the target leakage oracle $\mathcal{O}_x^\ell$; set the value $y_i'$ equal to the answer obtained from the oracle.
- Upon input the $j$-th signature query of type $(i, m)$, if $i \notin [0, \tau]$ or $y_i' = \perp$, answer with $\perp$. Otherwise, run $\pi \leftarrow_\$ \mathsf{S}_2^m(tk, y_i')$ and return $\sigma := \pi$ to $A_2$.
- Whenever $A_{1,2}$ returns a forgery $(m^*, \sigma^*)$, define $\pi^* := \sigma^*$ and output $x^*$ such that $x^* \leftarrow_\$ \mathsf{K}^{m^*}(tk, y, \pi^*)$.

For the analysis, note that $A''$ perfectly simulates signature queries. In fact, for each tampering query $T$ the reduction obtains the statement $y_i'$ corresponding to $x_i' := T(x)$ via a leakage query; given this value a signature for key $x_i'$ is computed by running the zero-knowledge simulator (as defined in $\mathbf{G}_2$). Moreover, the total leakage asked by $A''$ equals $\ell$ (as $A_2$ leaks at most $\ell$ bits from the secret key) plus $n \cdot \tau$ (as for each tampering function $T$ the reduction leaks $n$ bits, and $A_2$ makes at most $\tau$ such queries), plus $n$ bits (as the value $y_0' = \xi(x)$ is needed for simulating signature queries w.r.t. the original secret key), and by assumption $\ell + (\tau + 1) \cdot n \leq \ell'$. Hence, $A''$ breaks the hardness of the leakage-resilient relation with non-negligible probability $1/p_2(\kappa)$. This concludes the proof. $\square$

The proof of the theorem follows by combining the above lemmas.

## 3.3   Concrete Instantiations

We now explain how to instantiate the signature scheme from the previous section using standard complexity assumptions. We need two ingredients: (i) A leakage-resilient hard relation $R$; (ii) A tSE NIZK for the same relation $R$, supporting labels. For the latter component, we rely on the construction due to Dodis *et al.* [20] that allows to obtain a tSE NIZK for arbitrary relations, based on a standard (non-extractable) NIZK for a related relation (see below) and an IND-CCA-secure PKE scheme supporting labels.

Let $\mathcal{PKE} = (\mathsf{Setup}, \mathsf{Gen}, \mathsf{Enc}, \mathsf{Dec})$ be an IND-CCA-secure PKE scheme supporting labels, with message space $\mathcal{X}$. Plugging in the construction from [20] a signature has the form $\sigma := (c, \pi)$, where $c \leftarrow\!\!{}_\$ \mathsf{Enc}^\lambda(pk, x)$ and $\pi$ is a standard NIZK argument for the following derived relation:

$$R^* := \{((y, c, pk, m), (x, r)) : \ (x, y) \in R \wedge c = \mathsf{Enc}^m(pk, x; r)\}. \tag{1}$$

**Diffie-Hellman Assumptions.** In what follows, let $\mathbb{G}$ be a group with prime order $q$ and with generator $g$. Also, let $\mathbb{G}_1$, $\mathbb{G}_2$, $\mathbb{G}_T$ be groups of prime order $q$ and $e : \mathbb{G}_1 \times \mathbb{G}_2 \to \mathbb{G}_T$ be a non-degenerate, efficiently computable, bilinear map.

*Discrete Logarithm.* Let $g \leftarrow\!\!{}_\$ \mathbb{G}$ and $x \leftarrow\!\!{}_\$ \mathbb{Z}_q$. The Discrete Logarithm (DL) assumption holds in $\mathbb{G}$ if it is computationally hard to find $x \in \mathbb{Z}_q$ given $y = g^x \in \mathbb{G}$.

*Decisional Diffie-Hellman.* Let $g_1, g_2 \leftarrow\!\!{}_\$ \mathbb{G}$ and $x_1, x_2, x \leftarrow\!\!{}_\$ \mathbb{Z}_q$. The Decisional Diffie-Hellman (DDH) assumption holds in $\mathbb{G}$ if the following distributions are computationally indistinguishable: $(\mathbb{G}, g_1, g_2, g_1^{x_1}, g_2^{x_2})$ and $(\mathbb{G}, g_1, g_2, g_1^x, g_2^x)$.

*Symmetric External Diffie-Hellman.* The Symmetric External Diffie-Hellman (SXDH) assumption states that the DDH assumption holds in both $\mathbb{G}_1$ and $\mathbb{G}_2$. Such an assumption is not satisfied in case $\mathbb{G}_1 = \mathbb{G}_2$, but it is believed to hold in case there is no efficiently computable mapping between $\mathbb{G}_1$ and $\mathbb{G}_2$ [12,52].

*D-Linear* [35,53]. Let $D \geq 1$ be a constant, and let $g_1, \ldots, g_{D+1} \leftarrow\!\!{}_\$ \mathbb{G}$ and $x_1, \ldots, x_D \leftarrow\!\!{}_\$ \mathbb{Z}_q$. We say that the $D$-linear assumption holds in $\mathbb{G}$ if the following distributions are computationally indistinguishable: $(\mathbb{G}, g_1^{x_1}, \ldots, g_D^{x_D}, g_{D+1}^{x_{D+1}})$ and $(\mathbb{G}, g_1^{x_1}, \ldots, g_D^{x_D}, g_{D+1}^{\sum_{i=1}^D x_i})$. Note that for $D = 1$ we obtain the DDH assumption, and for $D = 2$ we obtain the so-called Linear assumption [53].

**Construction Based on SXDH.** The first instantiation is based on the SXDH assumption, working with asymmetric pairing based groups $(\mathbb{G}_1, \mathbb{G}_2, \mathbb{G}_T)$. The construction below is similar to the one given in [20, Sect. 1.2.2], except that we had to modify the underlying hard relation, in that the one used by Dodis *et al.* does not meet our completeness requirement.[3]

---

[3] In particular, a pair $(x, y) \in R$ is computed by sampling random exponents $r_1, \ldots, r_N \leftarrow\!\!{}_\$ \mathbb{Z}_q$ and outputting $x_i := g^{r_i}$ and $y := \prod_{i=1}^N g_i^{r_i}$, where $g$ is a generator of $\mathbb{G}_2$ and $g_1, \ldots, g_N$ are generators of $\mathbb{G}_1$; thus, by the SXDH assumption, it is hard to compute $y$ given only $x_1, \ldots, x_N$, without knowledge of the randomness $r_1, \ldots, r_N$.

**Hard relation:** Let $N \geq 2$, and $g_1, \ldots, g_N \leftarrow_\$ \mathbb{G}_1$ be generators. The sampling algorithm chooses a random $x := (x_1, \ldots, x_N) \leftarrow_\$ \mathbb{G}_2^N$ and defines $y := \prod_{i=1}^N e(g_i, x_i) \in \mathbb{G}_T$. Notice that the relation satisfies completeness, with mapping function $\xi(\cdot)$ defined by $\xi(x) := \prod_{i=1}^N e(g_i, x_i)$. In the full version [24], we argue that this relation is leakage-resilient under the SXDH assumption.

**Lemma 6.** *Under the SXDH assumption in* $(\mathbb{G}_1, \mathbb{G}_2, \mathbb{G}_T)$, *the above defined relation is an $\ell$-leakage-resilient hard relation for $\ell \leq (N-1) \log q$.*

**PKE:** We use the Cramer-Shoup PKE scheme in $\mathbb{G}_2$ [16], optimized as described in [20]. The public key consists of random generators $(h_1, h_2, h_{3,1}, \ldots, h_{3,N}, h_4, h_5)$ of $\mathbb{G}_2$, and in order to encrypt $x = (x_1, \ldots, x_N) \in \mathbb{G}_2^N$ under label $m \in \{0,1\}^*$ we return a ciphertext:

$$c := (c_1, \ldots, c_{N+3}) = (h_1^r, h_2^r, h_{3,1}^r \cdot x_1, \ldots, h_{3,N}^r \cdot x_N, (h_4 \cdot h_5^t)^r)$$

with $r \leftarrow_\$ \mathbb{Z}_q$, and where $t := H(c_1 || \cdots || c_{N+2} || m)$ is computed using a standard collision-resistant hash function.

**NIZK:** We use the Groth-Sahai proof system [32]. In order to prove that a given pair $x^* := (x, r)$ and $y^* := (y, c, pk, m)$ belongs to the relation of Eq. (1), we first prove that $(x, y) \in R$. This requires to show satisfiability of a one-sided pairing product equation, which can be done with a proof consisting of $2N + 16$ elements in $\mathbb{G}_1$ and $2$ elements in $\mathbb{Z}_q$ (under the SXDH assumption). Next, we prove validity of a ciphertext which requires to show satisfiability of a system of $N + 3$ one-sided multi-exponentiation equations; the latter can be done with a proof consisting of $(N+3) + 2N = 3N + 3$ group elements (under the SXDH assumption).

**Corollary 1.** *Let* $(\mathbb{G}_1, \mathbb{G}_2, \mathbb{G}_T)$ *be asymmetric pairing based groups with prime order $q$. Under the SXDH assumption there exists a signature scheme satisfying BLT-EUFCMA with tampering rate $\rho(\kappa) = O(1/\kappa)$. For $N \geq 2$, the public key consists of a single group element, the secret key consists of $N$ group elements, and a signature consists of $6N + 22$ group elements and $2$ elements in $\mathbb{Z}_q$.*

**Construction Based on DLIN.** The second instantiation is based on the DLIN assumption, working with symmetric pairing based groups $(\mathbb{G}, \mathbb{G}_T)$. The construction below is similar to one of the instantiations given in [20, Sect. 1.2.3], except that we had to modify the underlying hard relation, in that the one used by Dodis *et al.* does not meet our completeness requirement.

**Hard relation:** Let $N \geq 3$, and $g_1, \ldots, g_N, g_1', \ldots, g_N' \leftarrow_\$ \mathbb{G}$ be generators. The sampling algorithm chooses a random $x := (x_1, \ldots, x_N) \leftarrow_\$ \mathbb{G}$ and defines $y_1 := \prod_{i=1}^N e(g_i, x_i) \in \mathbb{G}_T$ and $y_2 := \prod_{i=1}^N e(g_i', x_i)$. Notice that the relation satisfies completeness, with mapping function $\xi(\cdot)$ defined by $\xi(x) := (\prod_{i=1}^N e(g_i, x_i), \prod_{i=1}^N e(g_i', x_i))$. In the full version [24], we argue that this relation is leakage-resilient under the DLIN assumption.

**Lemma 7.** *Under the DLIN assumption in* $(\mathbb{G}, \mathbb{G}_T)$, *the above defined relation is an $\ell$-leakage-resilient hard relation for $\ell \leq (N-2) \log q$.*

**PKE:** We use the Linear Cramer-Shoup PKE scheme in $\mathbb{G}$ [53], optimized as described in [20]. The public key consists of random generators $(h_0, h_1, h_2, h_{3,1}, \ldots, h_{3,N}, h_{4,1}, \ldots, h_{4,N}, h_{5,1}, h_{5,2}, h_{6,1}, h_{6,2})$ of $\mathbb{G}$, and in order to encrypt $x = (x_1, \ldots, x_N) \in \mathbb{G}^N$ under label $m \in \{0,1\}^*$ we return a ciphertext:

$$c := (c_1, \ldots, c_{N+4}) = (h_0^{r_1+r_2}, h_1^{r_1}, h_2^{r_2}, h_{3,1}^{r_1} \cdot h_{4,1}^{r_2} \cdot x_1, \ldots, h_{3,N}^{r_1}$$
$$\cdot h_{4,N}^{r_2} \cdot x_N, (h_{4,1} \cdot h_{5,1}^t)^{r_1} \cdot (h_{4,2} \cdot h_{5,2}^t)^{r_2})$$

with $r_1, r_2 \leftarrow_\$ \mathbb{Z}_q$, and where $t := H(c_1 || \cdots || c_{N+3} || m)$ is computed using a standard collision-resistant hash function.

**NIZK:** We use again the Groth-Sahai proof system. In order to prove that a given pair $x^* := (x, r)$ and $y^* := ((y_1, y_2), c, pk, m)$ belongs to the relation of Eq. (1), we first prove that $(x, (y_1, y_2)) \in R$. This requires to show satisfiability of two one-sided pairing product equations, which can be done with a proof consisting of $3N + 42$ elements in $\mathbb{G}$ and 6 elements in $\mathbb{Z}_q$ (under the DLIN assumption). Next, we prove validity of a ciphertext which requires to show satisfiability of a system of $N+4$ one-sided multi-exponentiation equations; the latter can be done with a proof consisting of $2(N+4)+3N = 5N+8$ group elements (under the DLIN assumption).

**Corollary 2.** *Let $(\mathbb{G}, \mathbb{G}_T)$ be symmetric pairing based groups with prime order $q$. Under the DLIN assumption there exists a signature scheme satisfying BLT-EUFCMA with tampering rate $\rho(\kappa) = O(1/\kappa)$. For $N \geq 3$, the public key consists of two group elements, the secret key consists of $N$ group elements, and a signature consists of $9N + 54$ group elements and 6 elements in $\mathbb{Z}_q$.*

# 4    Public-Key Encryption

We give a construction of an efficient PKE scheme satisfying BLT-IND-CCA security in the standard model. In particular, we prove that the PKE scheme of Qin and Liu [49] is already resilient to bounded leakage and tampering attacks.

## 4.1    The Scheme of Qin and Liu

The encryption scheme is a twist of the well-known Cramer-Shoup paradigm for CCA security [17], and is based on the following ingredients.

*Hash-proof systems.* An $\epsilon$-universal hash-proof system (HPS) $\mathcal{HPS} = (\mathsf{Gen}_{hps}, \mathsf{Pub}, \mathsf{Priv})$. Recall that a HPS has the following syntax: (i) Algorithm $\mathsf{Gen}_{hps}$ takes as input the security parameter, and outputs public parameters $pub := (aux, \mathcal{C}, \mathcal{V}, \mathcal{K}, \mathcal{SK}, \mathcal{PK}, \Lambda_{(\cdot)} : \mathcal{C} \to \mathcal{K}, \mu : \mathcal{SK} \to \mathcal{PK})$ where $aux$ might contain additional structural parameters, and where $\Lambda_{sk}$ is a hash function and, for any $sk \in \mathcal{SK}$, the function $\mu(sk)$ defines the action of $\Lambda_{sk}$ over the subset $\mathcal{V}$ of valid ciphertexts (i.e., $\Lambda_{sk}$ is projective). Moreover the function $\Lambda_{sk}$ is $\epsilon$-almost universal:

**Definition 5.** *A projective hash function $\Lambda_{(\cdot)}$ is $\epsilon$-almost universal, if for all $pk$, $C \in \mathcal{C} \setminus \mathcal{V}$, and all $K \in \mathcal{K}$, it holds that $\mathbb{P}\left[\Lambda_{\mathbf{SK}}(C) = K | \mathbf{PK} = pk, C\right] \leqslant \epsilon$, where $\mathbf{SK}$ is uniform over $\mathcal{SK}$ conditioned on $\mathbf{PK} = \mu(\mathbf{SK})$.*

(ii) Algorithm Pub takes as input a public key $pk = \mu(sk)$, a *valid ciphertext* $C \in \mathcal{V}$, and a witness $w$ for $C \in \mathcal{V}$, and outputs the value $\Lambda_{sk}(C)$. (iii) Algorithm Priv take as input the secret key $sk$ and a ciphertext $C \in \mathcal{C}$, and outputs the value $\Lambda_{sk}(C)$.

**Definition 6.** *A hash-proof system $\mathcal{HPS}$ is $\epsilon$-almost universal if the following holds:*

1. *For all sufficiently large $\kappa \in \mathbb{N}$, and for all possible outcomes of $\mathsf{Gen}_{\mathrm{hps}}(1^{\kappa})$, the underlying projective hash function is $\epsilon(\kappa)$-almost universal.*
2. *The underlying set membership problem is hard. Specifically, for any PPT adversary A the following quantity is negligible:*

$$\mathbf{Adv}^{\mathrm{smp}}_{\mathcal{HPS},\mathsf{A}} := |\mathbb{P}[\mathsf{A}(\mathcal{C}, \mathcal{V}, C_0) = 1 | \, C_0 \leftarrow_{\$} \mathcal{V})]$$
$$- \mathbb{P}[\mathsf{A}(\mathcal{C}, \mathcal{V}, C_1) = 1 | \, C_1 \leftarrow_{\$} \mathcal{C} \setminus \mathcal{V})]|.$$

The lemma below directly follows from the definition of hash-proof system and the notion of min-entropy.

**Lemma 8.** *Let $\Lambda_{(\cdot)}$ be $\epsilon$-almost universal. Then for all $pk$ and $C \in \mathcal{C} \setminus \mathcal{V}$ it holds that $\mathbb{H}_{\infty}(\Lambda_{\mathbf{SK}}(C) | \mathbf{PK} = pk, C) \geqslant -\log \epsilon$ where $\mathbf{SK}$ is uniform over $\mathcal{SK}$ conditioned on $\mathbf{PK} = \mu(\mathbf{SK})$.*

*One-time lossy filters* [49]. A One-Time Lossy Filter (OTLF) $\mathcal{LF} = (\mathsf{Gen}_{\mathrm{lf}}, \mathsf{Eval}, \mathsf{LTag})$ is a family of functions $\mathsf{LF}_{\phi,t}(X)$ indexed by a public key $\phi$ and a tag $t$. Recall that a OTLF has the following syntax: (i) Algorithm $\mathsf{Gen}_{\mathrm{lf}}$ takes as input the security parameter, and outputs a public key $\phi$ and a trapdoor key $\psi$. The public key $\phi$ defines a tag space $\mathcal{T} := \{0,1\}^* \times \mathcal{T}_c$ that contains two disjoint subsets $\mathcal{T}_{\mathrm{inj}}$ and $\mathcal{T}_{\mathrm{loss}}$ and a domain space $\mathcal{D}$. (ii) Algorithm Eval takes as input $\phi$, a tag $t = (t_a, t_c) \in \mathcal{T}$ (where we call $t_a$ the auxiliary tag and $t_c$ the core tag), and $X \in \mathcal{D}$, and outputs $\mathsf{LF}_{\phi,t}(X)$. (iii) Algorithm LTag takes as input $\psi$ and an auxiliary tag $t_a \in \{0,1\}^*$, and outputs a core tag $t_c$ such that $t = (t_a, t_c) \in \mathcal{T}_{\mathrm{loss}}$.

**Definition 7.** *We say that $\mathcal{LF} = (\mathsf{Gen}_{\mathrm{lf}}, \mathsf{Eval}, \mathsf{LTag})$ is an $\ell_{\mathrm{lf}}$-OTLF with domain $\mathcal{D}$ if the following proprieties hold:*

**Lossiness:** *In case the tag $t$ is injective (i.e., $t \in \mathcal{T}_{\mathrm{inj}}$), so is the function $\mathsf{LF}_{\phi,t}(\cdot) := \mathsf{Eval}(\phi, t, \cdot)$. In case $t$ is lossy (i.e., $t \in \mathcal{T}_{\mathrm{loss}}$), then $\mathsf{LF}_{\phi,t}(\cdot)$ has image size at most $2^{\ell_{\mathrm{lf}}}$.*

**Indistinguishability:** *No PPT adversary* A *is able to distinguish lossy tags from random tags, i.e. the following quantity is negligible:*

$$\mathbf{Adv}_{\mathcal{LF},A}^{\mathrm{ind}} := \left| \mathbb{P}\left[A(\phi,(t_a,t_c^0))=1\right] - \mathbb{P}\left[A(\phi,(t_a,t_c^1))=1\right]\right|$$

*where* $(\phi,\psi) \leftarrow_\$ \mathsf{Gen}_{\mathrm{lf}}(1^\kappa), t_a \leftarrow_\$ A(\phi), t_c^0 \leftarrow_\$ \mathcal{T}_c$ *and* $t_c^1 \leftarrow_\$ \mathsf{LTag}(\psi,t_a)$.

**Evasiveness:** *No PPT adversary* A *is able to generate a non-injective tag even given a lossy tag, i.e. the following quantity is negligible:*

$$\mathbf{Adv}_{\mathcal{LF},A}^{\mathrm{evasive}} := \mathbb{P}\left[ \begin{array}{c} (t_a',t_c') \neq (t_a,t_c) \\ (t_a',t_c') \in \mathcal{T} \setminus \mathcal{T}_{\mathrm{inj}} \end{array} : \begin{array}{c} (\phi,\psi) \leftarrow_\$ \mathsf{Gen}_{\mathrm{lf}}(1^\kappa); \\ t_a \leftarrow_\$ A(\phi); t_c \leftarrow_\$ \mathsf{LTag}(\psi,t_a); \\ (t_a',t_c') \leftarrow_\$ A(\phi,(t_a,t_c)) \end{array} \right].$$

*Randomness extractors.* An average-case strong randomness extractor.

**Definition 8.** *An efficient function* $\mathsf{Ext} : \mathcal{X} \times \mathcal{S} \to \mathcal{Y}$ *is an average-case* $(\delta,\epsilon)$-*strong extractor if for all pair of random variables* $(\mathbf{X},\mathbf{Z})$, *where* $\mathbf{X}$ *is defined over a set* $\mathcal{X}$ *and* $\widetilde{\mathbb{H}}_\infty(\mathbf{X}|\mathbf{Z}) \geqslant \delta$, *we have*

$$(\mathbf{Z},\mathbf{S},\mathsf{Ext}(\mathbf{X},\mathbf{S})) \approx_\epsilon (\mathbf{Z},\mathbf{S},\mathbf{U}),$$

*with* $\mathbf{S}$ *uniform over* $\mathcal{S}$ *and* $\mathbf{U}$ *uniform over* $\mathcal{Y}$.

*The encryption scheme.* Consider now the following PKE scheme $\mathcal{PKE} = (\mathsf{Setup}, \mathsf{Gen}, \mathsf{Enc}, \mathsf{Dec})$ with message space $\mathcal{M} := \{0,1\}^m$, based on a HPS $\mathcal{HPS} = (\mathsf{Gen}_{\mathrm{hps}}, \mathsf{Pub}, \mathsf{Priv})$, on a OTLF $\mathcal{LF} = (\mathsf{Gen}_{\mathrm{lf}}, \mathsf{Eval}, \mathsf{LTag})$ with domain $\mathcal{K}$, and on an average-case strong extractor $\mathsf{Ext} : \mathcal{K} \times \{0,1\}^d \to \{0,1\}^m$.

- <u>Setup$(1^\kappa)$</u>: Sample $pub_{\mathrm{hps}} := (aux, \mathcal{C}, \mathcal{V}, \mathcal{K}, \mathcal{SK}, \mathcal{PK}, \Lambda_{(.)}, \mu) \leftarrow_\$ \mathsf{Gen}_{\mathrm{hps}}(1^\kappa)$ and compute $(\phi,\psi) \leftarrow_\$ \mathsf{Gen}_{\mathrm{lf}}(1^\kappa)$. Return $pub := (pub_{\mathrm{hps}}, \phi)$. (Recall that all algorithms implicitly take $pub$ as input.)
- <u>Gen$(1^\kappa)$</u>: Choose a random $sk \leftarrow_\$ \mathcal{SK}$, define $pk = \mu(sk)$, and return $(pk, sk)$.
- <u>Enc$(pk, M)$</u>: Sample $C \leftarrow_\$ \mathcal{V}$ (with witness $w$), $S \leftarrow_\$ \{0,1\}^d$, and a core tag $t_c \leftarrow_\$ \mathcal{T}_c$. Compute $K := \mathsf{Pub}(pk, C, w)$, $\Phi := \mathsf{Ext}(K, S) \oplus M$, and $\Pi := \mathsf{Eval}(\phi,(t_a,t_c),K)$ where $t_a := (C, S, \Phi)$. Output $\hat{C} := (C, S, \Phi, \Pi, t_c)$.
- <u>Dec$(sk, \hat{C})$</u>: Parse $\hat{C} := (C, S, \Phi, \Pi, t_c)$. Compute $\hat{K} := \mathsf{Priv}(sk, C)$ and check if $\mathsf{Eval}(\phi, t, \hat{K}) = \Pi$ where $t := ((C, S, \Phi), t_c)$. If the check fails, reject and output $\perp$; else output $M := \Phi \oplus \mathsf{Ext}(\hat{K}, S)$.

**Theorem 2.** *Let* $\kappa \in \mathbb{N}$ *be the security parameter. Assume that* $\mathcal{HPS}$ *is* $\epsilon$-*almost universal,* $\mathcal{LF}$ *is an* $\ell_{\mathrm{lf}}$-*OTLF with domain* $\mathcal{K}$, *and* $\mathsf{Ext}$ *is an average-case* $(\delta,\epsilon')$-*strong extractor for a negligible function* $\epsilon'$. *Let* $s = s(\kappa)$ *and* $p = p(\kappa)$ *be parameters such that* $s \leqslant \log|\mathcal{SK}|$ *and* $p \geqslant \log|\mathcal{PK}|$ *for any* $\mathcal{SK}, \mathcal{PK}$ *generated by* $\mathsf{Gen}_{\mathrm{hps}}(1^\kappa)$, *and define* $\alpha = -\log\epsilon$ *and* $\beta = s - \alpha$.

*For any* $\delta \leqslant \alpha - \tau(p + \beta + \kappa) - \ell_{\mathrm{lf}} - \ell$ *the PKE scheme* $\mathcal{PKE}$ *described above is* $(\tau,\ell)$-*BLT-IND-CCA with* $\ell + \tau(p + \beta + \kappa) \leqslant \alpha - \ell_{\mathrm{lf}}$.

## 4.2  Security Proof

We consider a sequence of mental experiments, starting with the initial game $\mathbf{Exp}_{\mathcal{PKE},\mathsf{A}}^{\text{blt-cca}}(\kappa,\ell,\tau)$ which for simplicity we denote by $\mathbf{G}_0$.

**Game $\mathbf{G}_0$.** This is exactly the game of Definition 1, where $\mathcal{PKE}$ is the PKE scheme described above. In particular, upon input the $i$-th tampering query $T_i$ the modified secret key $sk_i' = T_i(sk)$ is computed (where $sk$ is the original secret key). Hence, the answer to a query $(i,\hat{C})$ to oracle $\mathsf{Dec}^*$ is computed by parsing $\hat{C} := (C,S,\varPhi,\varPi,t_c)$, computing $\hat{K} := \mathsf{Priv}(sk_i',C)$, and checking $\varPi = \mathsf{Eval}(\phi,((C,S,\varPhi),t_c),\hat{K})$; if the check fails the answer is $\perp$ and otherwise the answer is $M := \varPhi \oplus \mathsf{Ext}(\hat{K},S)$.

**Game $\mathbf{G}_1$.** We change the way the tag $t_c^*$ corresponding to the challenge ciphertext is computed, namely we now let $t_c^* \leftarrow \mathsf{LTag}(\psi,t_a^*)$ (i.e., the tag $t^* = (t_a^*,t_c^*) \in \mathcal{T}_{\text{loss}}$ is now lossy).

**Game $\mathbf{G}_2$.** We add an extra check to the decryption oracle. Namely, upon input a decryption query $(i,(C,S,\varPhi,\varPi,t_c))$ we check whether $t_a := (C,S,\varPhi)$ and $t_c$ satisfy $(t_a,t_c) = (t_a^*,t_c^*)$ (where $t_a^*$ and $t_c^*$ are the auxiliary and core tag corresponding to the challenge ciphertext). If the check succeeds, the oracle returns $\perp$. Notice that $t_a^*$ and $t_c^*$ are initially set to $\perp$, and remain equal to $\perp$ until the challenge ciphertext is generated.

**Game $\mathbf{G}_3$.** We change the way the challenge ciphertext is computed. Namely, we now compute the value $K^*$ as $K^* := \mathsf{Priv}(sk,C^*)$.

**Game $\mathbf{G}_4$.** We change the way the challenge ciphertext is computed. Namely, we now sample $C^*$ as $C^* \leftarrow_{\!\$} \mathcal{C} \setminus \mathcal{V}$.

**Game $\mathbf{G}_5$.** We add an extra check to the decryption oracle; the check is performed only for decryption queries corresponding to tampered secret keys (i.e., $i \geq 1$). At setup, the experiment initializes an additional set $\mathcal{Q}' \leftarrow \emptyset$. Denote by $\mathbf{V}$ the random variable containing all the answers from the decryption and leakage oracles, and define the quantity

$$\gamma_i(\kappa) := \mathbb{H}_\infty(\mathbf{SK}_i'|\mathbf{V} = v, \{\mathbf{SK}_j' = sk_j'\}_{j\in\mathcal{Q}'}, \{\mathbf{PK}_j' = pk_j'\}_{j\in[\tau]\cup\{0\}})$$

where we write $\mathbf{SK}_i'$ for the random variable of the $i$-th tampered secret key and $\mathbf{PK}_i'$ for the random variable of the corresponding public key (by default $pk_i' = \perp$ if $sk_i'$ is undefined and $pk_0' = pk$).
Upon input a decryption query $(i,(C,S,\varPhi,\varPi,t_c))$ such that $i \geq 1$ we proceed exactly as in $\mathbf{G}_4$ but, for all ciphertexts such that $C \in \mathcal{C} \setminus \mathcal{V}$, in case the decryption oracle did not already return $\perp$, we additionally check whether $\gamma_i(\kappa) \leq \beta(\kappa) + \log^2 \kappa$; if that happens, we add the index $i$ to the set $\mathcal{Q}'$ and otherwise we do not modify $\mathcal{Q}'$ and we additionally answer the decryption query with $\perp$.

**Game $\mathbf{G}_6$.** We change the way decryption queries corresponding to the original secret key are answered. Namely, upon input a decryption query $(0,(C,S,\varPhi,\varPi,t_c))$ we proceed as in $\mathbf{G}_5$ but, in case $C \in \mathcal{C} \setminus \mathcal{V}$, we answer the query with $\perp$.

**Game $G_7$.** We change the way the challenge ciphertext is computed. Namely, we now sample $\Phi^* \leftarrow_\$ \{0,1\}^m$. Notice that the challenge ciphertext is now independent of the message being encrypted.

Next, we turn to showing that the above defined games are indistinguishable. In what follows, given a ciphertext $\hat{C} = (C, S, \Phi, \Pi, t_c)$, we say that $\hat{C}$ is *valid* if $C \in \mathcal{V}$ (i.e., if $C$ is a valid ciphertext for the underlying HPS).

**Lemma 9.** *For all PPT adversaries* A *there exists a negligible function* $\nu_{0,1}$ : $\mathbb{N} \rightarrow [0,1]$ *such that* $|\mathbb{P}[\mathbf{G}_0(\kappa) = 1] - \mathbb{P}[\mathbf{G}_1(\kappa) = 1]| \leq \nu_{0,1}(\kappa)$.

*Proof.* We prove a stronger statement, namely that $\mathbf{G}_0(\kappa) \approx_c \mathbf{G}_1(\kappa)$. By contradiction, assume there exists a PPT distinguisher $\mathsf{D}_{0,1}$ and a polynomial $p_{0,1}(\cdot)$ such that, for infinitely many values of $\kappa \in \mathbb{N}$, we have that $\mathsf{D}_{0,1}$ distinguishes between $\mathbf{G}_0$ and $\mathbf{G}_1$ with probability at least $\geq 1/p_{0,1}(\kappa)$. We construct an adversary $\mathsf{A}_{0,1}$ breaking the indistinguishability property of the underlying OTLF $\mathcal{LF}$. At the beginning, adversary $\mathsf{A}_{0,1}$ receives the evaluation key $\phi$ from its own challenger, and simulates the entire experiment $\mathbf{G}_0$ with $\mathsf{D}_{0,1}$ by sampling all other parameters by itself; notice that this can be done because $\mathbf{G}_0$ does not depend on the secret trapdoor $\psi$. Whenever $\mathsf{D}_{0,1}$ outputs $(M_0, M_1)$, adversary $\mathsf{A}_{0,1}$ samples $t_a^*$ as defined in $\mathbf{G}_0$ and returns $t_a^*$ to its own challenger. Upon receiving a value $t_c^*$ from the challenger, $\mathsf{A}_{0,1}$ embeds $t_c^*$ in the challenge ciphertext, and keeps simulating all queries done by $\mathsf{D}_{0,1}$ as before. Finally, $\mathsf{A}_{0,1}$ outputs the same as $\mathsf{D}_{0,1}$.

We observe that $\mathsf{A}_{0,1}$ perfectly simulates the decryption oracle (which is identical in both $\mathbf{G}_0$ and $\mathbf{G}_1$). Moreover, depending on the challenge tag $t_c^*$ being random or lossy, the distribution of the challenge ciphertext produced by $\mathsf{A}_{0,1}$ is identical to that of either $\mathbf{G}_0$ or $\mathbf{G}_1$. Thus, $\mathsf{A}_{0,1}$ retains the same advantage as that of $\mathsf{D}_{0,1}$. This concludes the proof. □

**Lemma 10.** $\mathbf{G}_1 \equiv \mathbf{G}_2$.

*Proof.* Notice that $\mathbf{G}_1$ and $\mathbf{G}_2$ only differ in how decryption queries such that $(t_a, t_c) = (t_a^*, t_c^*)$ are answered. Clearly, such queries are answered identically in the two games for all decryption queries before the generation of the challenge ciphertext. As for decryption queries after the challenge ciphertext has been computed, we distinguish two cases: (i) $\Pi = \Pi^*$, and (ii) $\Pi \neq \Pi^*$. In case (i) we get that $\hat{C} = \hat{C}^*$, and thus both games return $\perp$. In case (ii), note that $\mathbf{G}_1$ checks whether $\Pi = \mathsf{Eval}(\phi, (t_a^*, t_c^*), \mathsf{Priv}(sk_i', C^*))$ and thus it returns $\perp$ whenever $\Pi \neq \Pi^*$. Hence, the two games are identically distributed. □

**Lemma 11.** $\mathbf{G}_2 \equiv \mathbf{G}_3$.

*Proof.* The difference between $\mathbf{G}_2$ and $\mathbf{G}_3$ is only syntactical, as $\mathsf{Priv}(sk, C^*) = K^* = \mathsf{Pub}(pk, C^*, w)$ by correctness of the underlying HPS. □

**Lemma 12.** *For all PPT adversaries* A, *there exists a negligible function* $\nu_{3,4}$ : $\mathbb{N} \rightarrow [0,1]$ *such that* $|\mathbb{P}[\mathbf{G}_3(\kappa) = 1] - \mathbb{P}[\mathbf{G}_4(\kappa) = 1]| \leq \nu_{3,4}(\kappa)$.

*Proof.* We prove a stronger statement, namely that $\mathbf{G}_3(\kappa) \approx_c \mathbf{G}_4(\kappa)$. By contradiction, assume there exists a PPT distinguisher $\mathsf{D}_{3,4}$ and a polynomial $p_{3,4}(\cdot)$ such that, for infinitely many values of $\kappa \in \mathbb{N}$, we have that $\mathsf{D}_{3,4}$ distinguishes between $\mathbf{G}_3$ and $\mathbf{G}_4$ with probability at least $\geq 1/p_{3,4}(\kappa)$. We construct a PPT adversary $\mathsf{A}_{3,4}$ solving the set membership problem of the underlying HPS. $\mathsf{A}_{3,4}$ receives as input $pub_{\mathrm{hps}}$ and a challenge $C^*$ such that either $C^* \leftarrow_{\!s} \mathcal{V}$ or $C^* \leftarrow_{\!s} \mathcal{C} \setminus \mathcal{V}$. Hence, $\mathsf{A}_{3,4}$ perfectly simulates the challenger for $\mathsf{D}_{3,4}$, by sampling all required parameters by itself, and embeds the value $C^*$ in the challenge ciphertext. In case $C^* \leftarrow_{\!s} \mathcal{V}$ we get exactly the same distribution as in $\mathbf{G}_3$, and in case $C^* \leftarrow_{\!s} \mathcal{C} \setminus \mathcal{V}$ we get exactly the same distribution as in $\mathbf{G}_4$. Hence, $\mathsf{A}_{3,4}$ retains the same advantage as that of $\mathsf{D}_{3,4}$. This finishes the proof. $\qquad\square$

For the $j$-th query $(i, \hat{C})$ to the decryption oracle, such that $\hat{C} = (C, S, \Phi, \Pi, t_c)$, we let $Inj_j$ be the event that the corresponding core tag $t_c$ is injective. We also define $Inj := \bigwedge_{j \in [q]} Inj_j$ where $q \in poly(\kappa)$ is the total number of decryption queries asked by the adversary.

**Lemma 13.** *For all PPT adversaries* $\mathsf{A}$ *there exists a negligible function* $\nu_4 :$ $\mathbb{N} \to [0, 1]$ *such that:* $|\mathbb{P}\left[\mathbf{G}_4(\kappa) = 1\right] - \mathbb{P}\left[\mathbf{G}_4(\kappa) = 1 | Inj\right]| \leq \nu_4(\kappa).$

*Proof.* The lemma follows by a simple reduction to the evasiveness property of the OTLF $\mathcal{LF}$. By contradiction, assume there exists a PPT adversary $\mathsf{A}_4$ and a polynomial $p_4(\cdot)$ such that $|\mathbb{P}\left[\mathbf{G}_4(\kappa) = 1\right] - \mathbb{P}\left[\mathbf{G}_4(\kappa) = 1 | Inj\right]| \geq 1/p_4(\kappa)$ for infinitely many values of $\kappa \in \mathbb{N}$. This implies:

$$1/p_4(\kappa) \leq |\mathbb{P}\left[\mathbf{G}_4(\kappa) = 1\right] - \mathbb{P}\left[\mathbf{G}_4(\kappa) = 1 | Inj\right]| \leq \mathbb{P}\left[Inj\right].$$

We build a PPT adversary $\mathsf{B}_4$ with non-negligible advantage in the evasiveness game. The adversary $\mathsf{B}_4$ receives as input a public key $\phi$ for the OTLF and perfectly simulates a run of game $\mathbf{G}_4$ for $\mathsf{A}_4$ by sampling all parameters by itself. After $\mathsf{A}_4$ returns $(M_0, M_1)$, adversary $\mathsf{B}_4$ samples $t_a^*$ as defined in $\mathbf{G}_4$, and forwards $t_a^*$ to its own challenger. Upon receiving $t_c^*$ from the challenger, $\mathsf{B}_4$ embeds $t_c^*$ in the challenge ciphertext for $\mathsf{A}_4$.

Let $\mathcal{Q}$ be the list of decryption queries made by $\mathsf{A}_4$. At the end of the simulation, adversary $\mathcal{B}_4$ picks uniformly at random a ciphertext $\hat{C} = (C, S, \Phi, t_c)$ from the list $\mathcal{Q}$ and outputs the tuple $(t_a := (C, S, \Phi), t_c)$. Clearly, the advantage of $\mathsf{B}_4$ in the evasiveness game is equal to the probability of event $Inj$ happening times the probability of guessing one of the ciphertexts containing a non-injective tag. Let $q(\kappa) \in poly(\kappa)$ be the total number of decryption queries made by $\mathsf{A}_4$. We have obtained,

$$\mathbf{Adv}_{\mathcal{LF}, \mathsf{B}_4}^{\mathrm{evasive}}(\kappa) \geq \mathbb{P}\left[Inj\right]/q(\kappa) \geq 1/q(\kappa) \cdot 1/p_4(\kappa),$$

which is a non-negligible quantity. This concludes the proof. $\qquad\square$

From now on, all of our arguments will be solely information-theoretic, and hence we do not mind if the remaining experiments will no longer be efficient.

**Lemma 14.** *For all (possibly unbounded) adversaries* A *making polynomially many decryption queries, there exists a negligible function* $\nu_{4,5} : \mathbb{N} \to [0,1]$ *such that* $|\mathbb{P}\left[\mathbf{G}_4(\kappa) = 1|Inj\right] - \mathbb{P}\left[\mathbf{G}_5(\kappa) = 1|Inj\right]| \leq \nu_{4,5}(\kappa).$

*Proof.* Recall that $\mathbf{G}_4$ and $\mathbf{G}_5$ differ only in the way decryption queries are handled. In particular, upon input a query $(i, (C, S, \Phi, \Pi, t_c))$ such that $i \geq 1$ and $C \in \mathcal{C} \setminus \mathcal{V}$, the decryption oracle in $\mathbf{G}_5$ checks whether $\gamma_i(\kappa) \leqslant \beta(\kappa) + \log^2 \kappa$. In case that happens, $\mathbf{G}_5$ proceeds identically to $\mathbf{G}_4$ and additionally updates the set $\mathcal{Q}'$ by including the index $i$; otherwise $\mathbf{G}_5$ answers the query with $\perp$. Intuitively, the set $\mathcal{Q}'$ keeps track of the tampered secret keys that did not return $\perp$ upon input an *invalid* ciphertext; the variable $\gamma_i(\kappa)$, instead, measures the conditional min-entropy of the $i$-th tampered secret key conditioned on all values returned by the decryption and leakage oracles, all tampered secret keys within the set $\mathcal{Q}'$, and all public keys corresponding to the tampered secret keys generated so far.

It follows that the distribution of the two games differ only in case the adversary makes a decryption query $(i, (C, S, \Phi, \Pi, t_c))$ such that: (i) $\gamma_i(\kappa) > \beta(\kappa) + \log^2 \kappa$; (ii) $C \in \mathcal{C} \setminus \mathcal{V}$; (iii) $\Pi = \mathsf{Eval}(\phi, (t_a, t_c), \mathsf{Priv}(sk'_i, C))$. Let *Bad* be the event that any (possibly unbounded) adversary makes a decryption query as above. Clearly,

$$|\mathbb{P}\left[\mathbf{G}_4(\kappa) = 1|Inj\right] - \mathbb{P}\left[\mathbf{G}_5(\kappa) = 1|Inj\right]| \leq \mathbb{P}\left[Bad|Inj\right].$$

For all $j \in [q]$, let $Bad_j$ be the event that *Bad* happens for the $j$-th decryption query, which as usual we denote by $(i, (C, S, \Phi, \Pi, t_c))$. Since we are conditioning on *Inj*, we have that there exists a unique value $K$ that is the pre-image of $\Pi$ under function $\mathsf{Eval}(\phi, (t_a, t_c), \cdot)$. Thus, by averaging over all the possible views for the adversary, we obtain:

$$\mathbb{P}\left[Bad_j|Inj\right] = \mathbb{P}\left[\mathsf{Priv}(\mathbf{SK}'_i, C)) = K\right]$$

$$= \sum_{v,pk} \mathbb{P}\left[\mathbf{V} = v, \mathbf{PK} = pk\right] \cdot \mathbb{P}\left[\mathsf{Priv}(\mathbf{SK}'_i, C)) = K|\mathbf{V} = v, \mathbf{PK} = pk\right]$$

$$\leqslant \sum_{v,pk} \mathbb{P}\left[\mathbf{V} = v, \mathbf{PK} = pk\right] \cdot 2^{-\mathbb{H}_\infty(\mathsf{Priv}(\mathbf{SK}'_i, C)|\mathbf{V}=v, \mathbf{PK}=pk)}.$$

Define the set $\mathcal{SK}^*_{K,C} := \{sk : \mathsf{Priv}(sk, C) = K \wedge pk = \mu(sk)\}$. We can write:

$$2^{-\mathbb{H}_\infty(\mathsf{Priv}(\mathbf{SK}'_i, C)|\mathbf{V}=v, \mathbf{PK}=pk)}$$

$$= \max_K \mathbb{P}\left[\mathsf{Priv}(\mathbf{SK}'_i, C) = K|\mathbf{V} = v, \mathbf{PK} = pk\right]$$

$$= \max_K \mathbb{P}\left[\mathbf{SK}'_i \in \mathcal{SK}^*_{K,C}|\mathbf{V} = v, \mathbf{PK} = pk\right]$$

$$\leqslant \max_{K, sk'_i} |\mathcal{SK}^*_{K,C}| \cdot \mathbb{P}\left[\mathbf{SK}'_i = sk'_i|\mathbf{V} = v, \mathbf{PK} = pk\right]$$

$$= \max_K |\mathcal{SK}^*_{K,C}| \cdot 2^{-\mathbb{H}_\infty(\mathbf{SK}'_i|\mathbf{V}=v, \mathbf{PK}=pk)}$$

$$= \max_K \frac{|\mathcal{SK}^*_{K,C}|}{|\mathcal{SK}|} \cdot |\mathcal{SK}| \cdot 2^{-\mathbb{H}_\infty(\mathbf{SK}'_i|\mathbf{V}=v, \mathbf{PK}=pk)}$$

$$\leqslant \epsilon \cdot |\mathcal{SK}| \cdot 2^{-\mathbb{H}_\infty(\mathbf{SK}'_i|\mathbf{V}=v, \mathbf{PK}=pk)} \leqslant \epsilon \cdot |\mathcal{SK}| \cdot 2^{-\beta(\kappa)-\log^2 \kappa} = 2^{-\log^2 \kappa},$$

where in the last line we used the $\epsilon$-almost universality of the underlying HPS, together with the fact that $\gamma_i(\kappa) > \beta(\kappa) + \log^2 \kappa$. Finally, by a union bound over all decryption queries, we obtain that there exists a negligible function $\nu_{4,5} : \mathbb{N} \to [0,1]$ such that $\mathbb{P}[Bad|Inj] \leqslant q \cdot 2^{-\log^2 \kappa} \leq \nu_{4,5}(\kappa)$, which concludes the proof of the lemma. $\qquad\qquad\square$

**Lemma 15.** *For all (possibly unbounded) adversaries* A, *there exists a negligible function* $\nu_{5,6} : \mathbb{N} \to [0,1]$ *such that* $|\mathbb{P}[\mathbf{G}_5(\kappa) = 1|Inj] - \mathbb{P}[\mathbf{G}_6(\kappa) = 1|Inj]| \leq \nu_{5,6}(\kappa)$.

*Proof.* Let *Bad* be the event that the adversary submits a decryption query $(0,(C,S,\varPhi,\varPi,t_c))$ such that: (i) $C \in \mathcal{C} \setminus \mathcal{V}$; (ii) $\varPi = \mathsf{Eval}(\phi, (t_a, t_c), \mathsf{Priv}(sk, C))$. Similarly to the proof of the previous lemma, it suffices to bound the probability of the event *Bad* conditioned on *Inj*. Denote by $(0,(C,S,\varPhi,\varPi,t_c))$ the first decryption query (w.r.t. the original secret key) that triggers event *Bad*. Recall that the view of adversary A in a run of game $\mathbf{G}_5$ consists of its own coin tosses, the public key $pk$, the answers to all queries to the decryption and leakage oracles, and the challenge ciphertext $\hat{C}^*$. In what follows, we write $\mathbf{L}$ for the random variable corresponding to the leakage queries; furthermore, for an index $i \in [\tau]$, we denote with $\mathbf{D}_i$ the random variable corresponding to all decryption queries relative to the $i$-th tampered secret key. Note that we can partition $\mathbf{D}_i$ in two parts: $\mathbf{D}_i^-$ for all decryption queries (w.r.t. the $i$-th tampered secret key) with an invalid ciphertext, and $\mathbf{D}_i^+$ for all decryption queries (w.r.t. the $i$-th tampered secret key) with a valid ciphertext. We also write $\mathbf{W}$ for the random variable corresponding to the overall view in game $\mathbf{G}_5$.

As in the previous lemma, since we are conditioning on event *Inj*, it suffices to analyze the conditional average min-entropy of $\mathsf{Priv}(\mathbf{SK}, C)$ conditioned on the adversarial view.

$$\widetilde{\mathbb{H}}_\infty(\mathsf{Priv}(\mathbf{SK}, C)|\mathbf{W})$$
$$\geqslant \widetilde{\mathbb{H}}_\infty(\mathsf{Priv}(\mathbf{SK}, C)|\mathbf{PK}, \{\mathbf{D}_i\}_{i\in[\tau]}, \mathbf{L}, \hat{\mathbf{C}}^*) \tag{2}$$
$$\geqslant \widetilde{\mathbb{H}}_\infty(\mathsf{Priv}(\mathbf{SK}, C)|\mathbf{PK}, \{\mathbf{D}_i\}_{i\in[\tau]}) - \ell_{\mathrm{lf}} - \ell \tag{3}$$
$$= \widetilde{\mathbb{H}}_\infty(\mathsf{Priv}(\mathbf{SK}, C)|\mathbf{PK}, \{\mathbf{D}_i^+\}_{i\in[\tau]}, \{\mathbf{D}_i^-\}_{i\in\mathcal{Q}'}, \mathcal{Q}') - \ell_{\mathrm{lf}} - \ell \tag{4}$$
$$\geqslant \widetilde{\mathbb{H}}_\infty(\mathsf{Priv}(\mathbf{SK}, C)|\mathbf{PK}, \{\mathbf{D}_i^+\}_{i\in[\tau]}, \{\mathbf{SK}_i'\}_{i\in\mathcal{Q}'}, \mathcal{Q}') - \ell_{\mathrm{lf}} - \ell. \tag{5}$$

Here, Eq. (2) uses the fact that the coin tosses of the adversary are independent of $\mathbf{SK}$, Eq. (3) follows by the chain rule for conditional average min-entropy (cf. Lemma 1), Eq. (4) uses the fact that, by definition of $\mathbf{G}_5$, all decryption queries for keys outside $\mathcal{Q}'$ and with an invalid ciphertext are answered with $\bot$, and Eq. (5) follows by the fact that $\mathbf{D}_i^-$ is a deterministic function of $\mathbf{SK}_i'$.

Let $\mathcal{Q}' = \{i_1, \dots, i_{q'}\}$, as defined in game $\mathbf{G}_5$. Since the fact that $sk_{i_{q'}} \in \mathcal{Q}'$ implies that $\mathbb{H}_\infty(\mathbf{SK}_{i_{q'}}'|\mathbf{W}) \leqslant \beta(\kappa) + \log^2 \kappa$, we can first apply Lemma 2 and then Lemma 1 to obtain

$$\widetilde{\mathbb{H}}_\infty(\mathsf{Priv}(\mathbf{SK}, C)|\mathbf{PK}, \{\mathbf{D}_i^+\}_{i\in[\tau]}, \{\mathbf{SK}_i'\}_{i\in\mathcal{Q}'}, \mathcal{Q}')$$
$$\geqslant \widetilde{\mathbb{H}}_\infty(\mathsf{Priv}(\mathbf{SK}, C)|\mathbf{PK}, \{\mathbf{D}_i^+\}_{i\in[\tau]}, \{\mathbf{SK}_i'\}_{i\in\mathcal{Q}''}, \mathcal{Q}'') - \beta(\kappa) - \log^2 \kappa - \log|\mathcal{Q}'|,$$

where $\mathcal{Q}'' := \mathcal{Q}' \setminus \{i_{q'}\}$. Notice to apply Lemma 2 we need to condition on $sk_{i_{q'}} \in \mathcal{Q}'$, however, such condition holds with probability 1 and by conditioning on a sure event the min-entropy does not change. By iterating the above argument for each key in $\mathcal{Q}'$:

$$\widetilde{\mathbb{H}}_\infty(\mathrm{Priv}(\mathbf{SK}, C)|\mathbf{PK}, \{\mathbf{D}_i^+\}_{i\in[\tau]}, \{\mathbf{SK}_i'\}_{i\in\mathcal{Q}'}, \mathcal{Q}') \tag{6}$$
$$\geq \widetilde{\mathbb{H}}_\infty(\mathrm{Priv}(\mathbf{SK}, C)|\mathbf{PK}, \{\mathbf{D}_i^+\}_{i\in[\tau]}) - \tau \cdot (\beta + \log^2 \kappa + \log \tau),$$

and relying on the fact that the answer to decryption queries for a valid ciphertext and w.r.t. index $j \in [\tau]$ can be computed using the "tampered" projection key $pk_i' = \mu(sk_i')$, we obtain

$$\widetilde{\mathbb{H}}_\infty(\mathrm{Priv}(\mathbf{SK}, C)|\mathbf{PK}, \{\mathbf{D}_i^+\}_{i\in[\tau]}) \geq \widetilde{\mathbb{H}}_\infty(\mathrm{Priv}(\mathbf{SK}, C)|\mathbf{PK}, \{\mathbf{PK}_i'\}_{i\in[\tau]}) \tag{7}$$
$$\geq \alpha - \tau \cdot p,$$

where Eq. (7) follows by Lemmas 1 and 8. Combining together Eqs. (5), (6) and (7), yields:

$$\widetilde{\mathbb{H}}_\infty(\mathrm{Priv}(\mathbf{SK}, C)|\mathbf{W}) \geq \alpha - \tau \cdot (p + \beta + \log^2 \kappa + \log \tau) - \ell_{1f} - \ell.$$

It follows that the decryption oracle in game $\mathbf{G}_5$ does not reject the first invalid ciphertext with probability at most $\epsilon \cdot 2^{\tau(p+\beta+\log^2 \kappa + \log \tau) + \ell_{1f} + \ell}$. A generalization of this argument implies that, for all $j \in [q]$, the probability that the decryption oracle does not reject the $j$-th decryption query of type $(0, \cdot)$ containing an invalid ciphertext is at most $2^{\tau(p+\beta+\log^2 \kappa + \log \tau) + \ell_{1f} + \ell}/(1/\epsilon - q(\kappa))$. Finally, by a union bound over the total number of decryption queries, there exists a negligible function $\nu_{5,6} : \mathbb{N} \to [0, 1]$ such that:

$$\mathbb{P}\left[Bad | Inj\right] \leq \frac{q \cdot 2^{\tau(p+\beta+\log^2 \kappa + \log \tau) + \ell_{1f} + \ell}}{1/\epsilon - q}$$
$$\leq \epsilon \cdot e^{-q\epsilon} \cdot 2^{\tau(p+\beta+\log^2 \kappa + \log \tau) + \ell_{1f} + \ell + \log q}$$
$$\leq 2^{-(\alpha - q\epsilon(\kappa) - \tau(p+\beta+\log^2 \kappa \log \tau) - \ell_{1f} - \ell - \log q)}$$
$$\leq \nu_{5,6}(\kappa).$$

where the last inequality follows by the fact that $\alpha \geq \ell + \ell_{1f} + \tau(p + \beta + \kappa)$ and additionally $\kappa - \log^2 \kappa - \log \tau - \log q/\tau - q\epsilon/\tau \in \omega(\log \kappa)$. $\qquad\square$

**Lemma 16.** *For all (possibly unbounded) adversaries* A, *there exists a negligible function* $\nu_{6,7} : \mathbb{N} \to [0, 1]$ *such that* $|\mathbb{P}\left[\mathbf{G}_6(\kappa) = 1 | Inj\right] - \mathbb{P}\left[\mathbf{G}_7(\kappa) = 1 | Inj\right]| \leq \nu_{6,7}(\kappa)$.

*Proof.* We analyze the conditional average min-entropy of $\mathrm{Priv}(\mathbf{SK}, C^*)$ conditioned on the view of the adversary. By a previous argument, we can write:

$$\widetilde{\mathbb{H}}_\infty(\mathrm{Priv}(\mathbf{SK}, C^*)|\mathbf{W}) \geq \alpha - \tau \cdot (p + \beta + \log^2 \kappa + \log \tau) - \ell_{1f} - \ell,$$

and thus the statement follows by our choice of parameters for the strong average-case extractor. $\qquad\square$

The statement of the theorem now follows by combining the above lemmas together with the fact that in $\mathbf{G}_7$ the challenge ciphertext is independent of the hidden bit $b$, and thus $\mathbb{P}\left[\mathbf{G}_7(\kappa)|Inj\right] = 1/2$ for all (even unbounded) adversaries. This finishes the proof.

### 4.3   Concrete Instantiations

The ratio $\frac{\alpha-\ell-\ell_{\mathrm{lf}}}{p+\beta}$ plays an important role in evaluating the tampering rate of a given instantiation. Ideally, we would like to have an HPS where $\alpha$ is as big as possible while $p$ and $\beta = \alpha - s$ are as small as possible. Below, we give an instantiation based on the Refined Subgroup Indistinguishability (RSI) assumption.

*Instantiation based on RSI.* Let $\xi \in \mathbb{N}$ be a parameter. For security parameter $\kappa \in \mathbb{N}$, let $p$ and $q$ be primes of size respectively $\kappa$ bits and $\xi \cdot \kappa$ bits and define $\bar{p} = 2pq+1$. For this choice of parameters, we have that $\mathbb{Z}_{\bar{p}}^*$ has a unique subgroup of order $N = pq$. Denote by $\mathbb{QR}_{\bar{p}}$ the set of quadratic residues modulo $\bar{p}$; the group $\mathbb{QR}_{\bar{p}}$ can be decomposed as a direct product of $\mathbb{G}_p \times \mathbb{G}_q$ where $\mathbb{G}_p$ and $\mathbb{G}_q$ are cyclic groups of prime order $p$ and $q$ (respectively).

For random $x, y \leftarrow\!\!{}_{\$}\, \mathbb{Z}_{\bar{p}}^*$, one can show that, with overwhelming probability, $g = x^q \bmod \bar{p}$ and $h = y^p \bmod \bar{p}$ are generators of $\mathbb{G}_p$ and $\mathbb{G}_q$ (respectively). Let $pub_{\mathrm{rsi}} := (\mathbb{QR}_{\bar{p}}, \bar{p}, g, h)$. The RSI assumption over $\mathbb{QR}_{\bar{p}}$ states that for all PPT adversary A the following quantity is negligible in the security parameter:

$$\left| \mathbb{P}\left[\mathsf{A}(pub_{\mathrm{rsi}}, g^x \bmod \bar{p}) : x \leftarrow\!\!{}_{\$}\, \mathbb{Z}_{\bar{p}}\right] - \mathbb{P}\left[\mathsf{A}(pub_{\mathrm{rsi}}, y) : y \leftarrow\!\!{}_{\$}\, \mathbb{QR}_{\bar{p}}\right] \right|.$$

The RSI assumption over $\mathbb{QR}_{\bar{p}}$ is conjectured to hold if factoring $N = pq$ is hard [45]. We can derive a HPS as follow. We set $\mathcal{C} := \mathbb{QR}_{\bar{p}}$, $\mathcal{V} := \mathbb{G}_p$, $\mathcal{SK} := \mathbb{Z}_{\bar{p}}$, and $\mathcal{PK} := \mathbb{G}_p$. Given a random secret key $sk \leftarrow\!\!{}_{\$}\, \mathcal{SK}$, the corresponding public key $pk$ is computed as $\mu(sk) := g^{sk} \bmod \bar{p}$. Algorithm Pub, upon input $C := g^w$ (where $w$ is the witness for $C \in \mathcal{V}$) and $pk$ outputs $pk^w \bmod \bar{p}$. Algorithm Priv, upon input $C$ and $sk$, outputs $\varLambda_{sk}(C) := C^{sk} \bmod \bar{p}$. It was shown in [50] that the above construction defines a $1/q$-almost universal HPS based on the RSI assumption. The work of [50] additionally presents a construction of a OTLF achieving $\ell_{\mathrm{lf}} := \log p$ based on the RSI assumption.

Finally, by instantiating the average-case strong extractor using universal hash functions as required by the left-over hash lemma [34] we note that the PKE scheme allows to encrypt messages with bit-length $m = O(\xi\kappa - \tau\kappa - \ell - \kappa)$. We obtain the following result:

**Corollary 3.** *Let $\bar{p}$ be as above. Under the RSI assumption over $\mathbb{QR}_{\bar{p}}$, for any $\xi(\kappa) = \omega(1)$, there exists a PKE scheme satisfying $(\tau, \ell)$-BTL-IND-CCA with tampering rate $\rho(\kappa) = O(1/\kappa - \frac{\ell}{\xi^2\kappa})$. The size of the secret key is $\Omega(\xi\kappa)$, and the PKE scheme allows to encrypt messages with bit-length $m = O(\xi\kappa - \tau\kappa - \ell - \kappa)$.*

# 5    Conclusions and Open Problems

We have shown new constructions of public-key cryptosystems with provable security guarantees against bounded leakage and tampering attacks. The proposed schemes are in the standard model, and can be instantiated efficiently under standard complexity assumptions.

There are several interesting problems left open by our work. First, our constructions only achieve sub-optimal tampering rate $\rho(\kappa) = O(1/\kappa)$, so it would be interesting to find alternative constructions achieving optimal rate in the standard model. Second, it would be interesting to combine related-key attacks with related-randomness attacks [47,48], where the adversary might force a cryptographic scheme to re-use (functions of) its own random coins; a promising idea in this direction is to combine our leakage-to-tamper reduction to so called fully leakage-resilient signatures [14,23], where the adversary can additionally leak on the random coins of the signature algorithm. Third, it remains open how to obtain CCA security for PKE against "after-the-fact" tampering and leakage, where both tampering and leakage can still occur after the challenge ciphertext is generated (in the spirit of [33]). Finally, one could try to come-up with new hash-proof systems meeting the requirements needed for our PKE instantiation under alternative hardness assumptions.

**Acknowledgments.** The authors would like to thank Jesper Buus Nielsen for an interesting conversation regarding the result in Sect. 4. Antonio Faonio was supported by European Research Council Starting Grant 279447.

# References

1. Abdalla, M., Benhamouda, F., Passelègue, A.: An algebraic framework for pseudorandom functions and applications to related-key security. In: Gennaro, R., Robshaw, M. (eds.) CRYPTO 2015. LNCS, vol. 9215, pp. 388–409. Springer, Heidelberg (2015). doi:10.1007/978-3-662-47989-6_19
2. Applebaum, B.: Garbling XOR gates "for free" in the standard model. In: Sahai, A. (ed.) TCC 2013. LNCS, vol. 7785, pp. 162–181. Springer, Heidelberg (2013). doi:10.1007/978-3-642-36594-2_10
3. Applebaum, B., Harnik, D., Ishai, Y.: Semantic security under related-key attacks and applications. In: Innovations in Computer Science, pp. 45–60 (2011)
4. Applebaum, B., Widder, E.: Related-key secure pseudorandom functions: the case of additive attacks. IACR Cryptology ePrint Archive 2014, 478 (2014). http://eprint.iacr.org/2014/478
5. Ball, M., Dachman-Soled, D., Kulkarni, M., Malkin, T.: Non-malleable codes for bounded depth, bounded fan-in circuits. In: Fischlin, M., Coron, J.-S. (eds.) EUROCRYPT 2016. LNCS, vol. 9666, pp. 881–908. Springer, Heidelberg (2016). doi:10.1007/978-3-662-49896-5_31
6. Bellare, M., Cash, D.: Pseudorandom functions and permutations provably secure against related-key attacks. In: Rabin, T. (ed.) CRYPTO 2010. LNCS, vol. 6223, pp. 666–684. Springer, Heidelberg (2010). doi:10.1007/978-3-642-14623-7_36
7. Bellare, M., Cash, D., Miller, R.: Cryptography secure against related-key attacks and tampering. In: Lee, D.H., Wang, X. (eds.) ASIACRYPT 2011. LNCS, vol. 7073, pp. 486–503. Springer, Heidelberg (2011). doi:10.1007/978-3-642-25385-0_26

8. Bellare, M., Kohno, T.: A theoretical treatment of related-key attacks: RKA-PRPs, RKA-PRFs, and applications. In: Biham, E. (ed.) EUROCRYPT 2003. LNCS, vol. 2656, pp. 491–506. Springer, Heidelberg (2003). doi:10.1007/3-540-39200-9_31

9. Bellare, M., Meiklejohn, S., Thomson, S.: Key-versatile signatures and applications: RKA, KDM and joint Enc/Sig. In: Nguyen, P.Q., Oswald, E. (eds.) EUROCRYPT 2014. LNCS, vol. 8441, pp. 496–513. Springer, Heidelberg (2014). doi:10.1007/978-3-642-55220-5_28

10. Bellare, M., Paterson, K.G., Thomson, S.: RKA security beyond the linear barrier: IBE, encryption and signatures. In: Wang, X., Sako, K. (eds.) ASIACRYPT 2012. LNCS, vol. 7658, pp. 331–348. Springer, Heidelberg (2012). doi:10.1007/978-3-642-34961-4_21

11. Biham, E., Shamir, A.: Differential fault analysis of secret key cryptosystems. In: Kaliski, B.S. (ed.) CRYPTO 1997. LNCS, vol. 1294, pp. 513–525. Springer, Heidelberg (1997). doi:10.1007/BFb0052259

12. Boneh, D., Boyen, X., Shacham, H.: Short group signatures. In: Franklin, M. (ed.) CRYPTO 2004. LNCS, vol. 3152, pp. 41–55. Springer, Heidelberg (2004). doi:10.1007/978-3-540-28628-8_3

13. Boneh, D., DeMillo, R.A., Lipton, R.J.: On the importance of checking cryptographic protocols for faults. In: Fumy, W. (ed.) EUROCRYPT 1997. LNCS, vol. 1233, pp. 37–51. Springer, Heidelberg (1997). doi:10.1007/3-540-69053-0_4

14. Boyle, E., Segev, G., Wichs, D.: Fully leakage-resilient signatures. J. Cryptology 26(3), 513–558 (2013)

15. Chen, Y., Qin, B., Zhang, J., Deng, Y., Chow, S.S.M.: Non-malleable functions and their applications. In: Cheng, C.-M., Chung, K.-M., Persiano, G., Yang, B.-Y. (eds.) PKC 2016. LNCS, vol. 9615, pp. 386–416. Springer, Heidelberg (2016). doi:10.1007/978-3-662-49387-8_15

16. Cramer, R., Shoup, V.: A practical public key cryptosystem provably secure against adaptive chosen ciphertext attack. In: Krawczyk, H. (ed.) CRYPTO 1998. LNCS, vol. 1462, pp. 13–25. Springer, Heidelberg (1998). doi:10.1007/BFb0055717

17. Cramer, R., Shoup, V.: Universal hash proofs and a paradigm for adaptive chosen ciphertext secure public-key encryption. In: Knudsen, L.R. (ed.) EUROCRYPT 2002. LNCS, vol. 2332, pp. 45–64. Springer, Heidelberg (2002). doi:10.1007/3-540-46035-7_4

18. Damgård, I., Faust, S., Mukherjee, P., Venturi, D.: Bounded tamper resilience: how to go beyond the algebraic barrier. In: Sako, K., Sarkar, P. (eds.) ASIACRYPT 2013. LNCS, vol. 8270, pp. 140–160. Springer, Heidelberg (2013). doi:10.1007/978-3-642-42045-0_8

19. Damgård, I., Faust, S., Mukherjee, P., Venturi, D.: The chaining lemma and its application. In: Lehmann, A., Wolf, S. (eds.) ICITS 2015. LNCS, vol. 9063, pp. 181–196. Springer, Heidelberg (2015). doi:10.1007/978-3-319-17470-9_11

20. Dodis, Y., Haralambiev, K., López-Alt, A., Wichs, D.: Efficient public-key cryptography in the presence of key leakage. In: Abe, M. (ed.) ASIACRYPT 2010. LNCS, vol. 6477, pp. 613–631. Springer, Heidelberg (2010). doi:10.1007/978-3-642-17373-8_35

21. Dodis, Y., Ostrovsky, R., Reyzin, L., Smith, A.: Fuzzy extractors: how to generate strong keys from biometrics and other noisy data. SIAM J. Comput. 38(1), 97–139 (2008)

22. Dziembowski, S., Pietrzak, K., Wichs, D.: Non-malleable codes. In: Innovations in Computer Science, pp. 434–452 (2010)

23. Faonio, A., Nielsen, J.B., Venturi, D.: Mind your coins: fully leakage-resilient signatures with graceful degradation. In: Halldórsson, M.M., Iwama, K., Kobayashi, N., Speckmann, B. (eds.) ICALP 2015. LNCS, vol. 9134, pp. 456–468. Springer, Heidelberg (2015). doi:10.1007/978-3-662-47672-7_37

24. Faonio, A., Venturi, D.: Efficient public-key cryptography with bounded leakage and tamper resilience. Cryptology ePrint Archive, Report 2016/529 (2016). http://eprint.iacr.org/2016/529

25. Faust, S., Mukherjee, P., Nielsen, J.B., Venturi, D.: Continuous non-malleable codes. In: Lindell, Y. (ed.) TCC 2014. LNCS, vol. 8349, pp. 465–488. Springer, Heidelberg (2014). doi:10.1007/978-3-642-54242-8_20

26. Faust, S., Mukherjee, P., Venturi, D., Wichs, D.: Efficient non-malleable codes and key-derivation for poly-size tampering circuits. In: Nguyen, P.Q., Oswald, E. (eds.) EUROCRYPT 2014. LNCS, vol. 8441, pp. 111–128. Springer, Heidelberg (2014). doi:10.1007/978-3-642-55220-5_7

27. Faust, S., Pietrzak, K., Venturi, D.: Tamper-proof circuits: how to trade leakage for tamper-resilience. In: Aceto, L., Henzinger, M., Sgall, J. (eds.) ICALP 2011. LNCS, vol. 6755, pp. 391–402. Springer, Heidelberg (2011). doi:10.1007/978-3-642-22006-7_33

28. Fiat, A., Shamir, A.: How to prove yourself: practical solutions to identification and signature problems. In: Odlyzko, A.M. (ed.) CRYPTO 1986. LNCS, vol. 263, pp. 186–194. Springer, Heidelberg (1987). doi:10.1007/3-540-47721-7_12

29. Gennaro, R., Lysyanskaya, A., Malkin, T., Micali, S., Rabin, T.: Algorithmic tamper-proof (ATP) security: theoretical foundations for security against hardware tampering. In: Naor, M. (ed.) TCC 2004. LNCS, vol. 2951, pp. 258–277. Springer, Heidelberg (2004). doi:10.1007/978-3-540-24638-1_15

30. Goldenberg, D., Liskov, M.: On related-secret pseudorandomness. In: Micciancio, D. (ed.) TCC 2010. LNCS, vol. 5978, pp. 255–272. Springer, Heidelberg (2010). doi:10.1007/978-3-642-11799-2_16

31. Goyal, V., O'Neill, A., Rao, V.: Correlated-input secure hash functions. In: Ishai, Y. (ed.) TCC 2011. LNCS, vol. 6597, pp. 182–200. Springer, Heidelberg (2011). doi:10.1007/978-3-642-19571-6_12

32. Groth, J., Sahai, A.: Efficient non-interactive proof systems for bilinear groups. In: Smart, N. (ed.) EUROCRYPT 2008. LNCS, vol. 4965, pp. 415–432. Springer, Heidelberg (2008). doi:10.1007/978-3-540-78967-3_24

33. Halevi, S., Lin, H.: After-the-fact leakage in public-key encryption. In: Ishai, Y. (ed.) TCC 2011. LNCS, vol. 6597, pp. 107–124. Springer, Heidelberg (2011). doi:10.1007/978-3-642-19571-6_8

34. Håstad, J., Impagliazzo, R., Levin, L.A., Luby, M.: A pseudorandom generator from any one-way function. SIAM J. Comput. **28**(4), 1364–1396 (1999)

35. Hofheinz, D., Kiltz, E.: Secure hybrid encryption from weakened key encapsulation. In: Menezes, A. (ed.) CRYPTO 2007. LNCS, vol. 4622, pp. 553–571. Springer, Heidelberg (2007). doi:10.1007/978-3-540-74143-5_31

36. Ishai, Y., Prabhakaran, M., Sahai, A., Wagner, D.: Private circuits II: keeping secrets in tamperable circuits. In: Vaudenay, S. (ed.) EUROCRYPT 2006. LNCS, vol. 4004, pp. 308–327. Springer, Heidelberg (2006). doi:10.1007/11761679_19

37. Jafargholi, Z., Wichs, D.: Tamper detection and continuous non-malleable codes. In: Dodis, Y., Nielsen, J.B. (eds.) TCC 2015. LNCS, vol. 9014, pp. 451–480. Springer, Heidelberg (2015). doi:10.1007/978-3-662-46494-6_19

38. Katz, J., Vaikuntanathan, V.: Signature schemes with bounded leakage resilience. In: Matsui, M. (ed.) ASIACRYPT 2009. LNCS, vol. 5912, pp. 703–720. Springer, Heidelberg (2009). doi:10.1007/978-3-642-10366-7_41

39. Katz, J., Vaikuntanathan, V.: Signature schemes with bounded leakage resilience. In: Matsui, M. (ed.) ASIACRYPT 2009. LNCS, vol. 5912, pp. 703–720. Springer, Heidelberg (2009). doi:10.1007/978-3-642-10366-7_41

40. Lewi, K., Montgomery, H., Raghunathan, A.: Improved constructions of PRFs secure against related-key attacks. In: Boureanu, I., Owesarski, P., Vaudenay, S. (eds.) ACNS 2014. LNCS, vol. 8479, pp. 44–61. Springer, Heidelberg (2014). doi:10.1007/978-3-319-07536-5_4

41. Liu, F.-H., Lysyanskaya, A.: Tamper and leakage resilience in the split-state model. In: Safavi-Naini, R., Canetti, R. (eds.) CRYPTO 2012. LNCS, vol. 7417, pp. 517–532. Springer, Heidelberg (2012). doi:10.1007/978-3-642-32009-5_30

42. Lu, X., Li, B., Jia, D.: Related-key security for hybrid encryption. In: Chow, S.S.M., Camenisch, J., Hui, L.C.K., Yiu, S.M. (eds.) ISC 2014. LNCS, vol. 8783, pp. 19–32. Springer, Heidelberg (2014). doi:10.1007/978-3-319-13257-0_2

43. Lucks, S.: Ciphers secure against related-key attacks. In: Roy, B., Meier, W. (eds.) FSE 2004. LNCS, vol. 3017, pp. 359–370. Springer, Heidelberg (2004). doi:10.1007/978-3-540-25937-4_23

44. Naor, M., Segev, G.: Public-key cryptosystems resilient to key leakage. In: Halevi, S. (ed.) CRYPTO 2009. LNCS, vol. 5677, pp. 18–35. Springer, Heidelberg (2009)

45. Nieto, J.M.G., Boyd, C., Dawson, E.: A public key cryptosystem based on a subgroup membership problem. Des. Codes Crypt. 36(3), 301–316 (2005)

46. Otto, M.: Fault Attacks and Countermeasures. Ph.D. thesis, University of Paderborn, Germany (2006)

47. Paterson, K.G., Schuldt, J.C.N., Sibborn, D.L.: Related randomness attacks for public key encryption. In: Krawczyk, H. (ed.) PKC 2014. LNCS, vol. 8383, pp. 465–482. Springer, Heidelberg (2014). doi:10.1007/978-3-642-54631-0_27

48. Paterson, K.G., Schuldt, J.C.N., Sibborn, D.L., Wee, H.: Security against related randomness attacks via reconstructive extractors. In: Groth, J. (ed.) IMACC 2015. LNCS, vol. 9496, pp. 23–40. Springer, Heidelberg (2015). doi:10.1007/978-3-319-27239-9_2

49. Qin, B., Liu, S.: Leakage-resilient chosen-ciphertext secure public-key encryption from hash proof system and one-time lossy filter. In: Sako, K., Sarkar, P. (eds.) ASIACRYPT 2013. LNCS, vol. 8270, pp. 381–400. Springer, Heidelberg (2013). doi:10.1007/978-3-642-42045-0_20

50. Qin, B., Liu, S.: Leakage-flexible CCA-secure public-key encryption: simple construction and free of pairing. In: Krawczyk, H. (ed.) PKC 2014. LNCS, vol. 8383, pp. 19–36. Springer, Heidelberg (2014). doi:10.1007/978-3-642-54631-0_2

51. Qin, B., Liu, S., Yuen, T.H., Deng, R.H., Chen, K.: Continuous non-malleable key derivation and its application to related-key security. In: Katz, J. (ed.) PKC 2015. LNCS, vol. 9020, pp. 557–578. Springer, Heidelberg (2015). doi:10.1007/978-3-662-46447-2_25

52. Scott, M.: Authenticated ID-based key exchange and remote log-in with simple token and PIN number. IACR Cryptology ePrint Archive 2002, 164 (2002). http://eprint.iacr.org/2002/164

53. Shacham, H.: A Cramer-Shoup encryption scheme from the linear assumption and from progressively weaker linear variants. IACR Cryptology ePrint Archive 2007, 74 (2007). http://eprint.iacr.org/2007/074

54. Wee, H.: Public key encryption against related key attacks. In: Fischlin, M., Buchmann, J., Manulis, M. (eds.) PKC 2012. LNCS, vol. 7293, pp. 262–279. Springer, Heidelberg (2012). doi:10.1007/978-3-642-30057-8_16

# Public-Key Cryptosystems Resilient to Continuous Tampering and Leakage of Arbitrary Functions

Eiichiro Fujisaki[✉] and Keita Xagawa

NTT Secure Platform Laboratories, 3-9-11 Midori-cho Musashino-shi,
Tokyo 180-8585, Japan
{fujisaki.eiichiro,xagawa.keita}@lab.ntt.co.jp

**Abstract.** We present the first chosen-ciphertext secure public-key encryption schemes resilient to continuous tampering of arbitrary (efficiently computable) functions. Since it is impossible to realize such a scheme without a self-destruction or key-updating mechanism, our proposals allow for either of them. As in the previous works resilient to this type of tampering attacks, our schemes also tolerate bounded or continuous memory leakage attacks at the same time. Unlike the previous results, our schemes have efficient instantiations, without relying on zero-knowledge proofs. We also prove that there is no secure digital signature scheme resilient to arbitrary tampering functions against a stronger variant of continuous tampering attacks, even if it has a self-destruction mechanism.

**Keywords:** Public-key encryption · Digital signature · Continuous tampering attacks · Bounded or continuous memory leakage

## 1 Introduction

We study the tampering attack security, or equivalently the related-key attack security, of public-key cryptosystems. The tampering attacks allow an adversary to modify the secret of a target cryptographic device and observe the effect of the changes at the output. For instance, the tampering attacks are mounted on the IND-CCA game of a public-key encryption (PKE) scheme, where an adversary may tamper with the secret-key and observe the output of the decryption oracle with the tampered secret.

Theoretical treatment of tampering attack is first considered independently by Gennaro et al. [23] and Bellare and Kohno [6]. The former treated *arbitrary* (efficiently computable) tampering functions, whereas the latter considered a *restricted* class of tampering functions.

Since allowing for all tampering functions is very challenging, Gennaro et al. [23] make a strong compromise that a trusted-third party may publish its verification key (of a secure digital signature scheme) as a part of public parameters where an adversary is not allowed to modify the parameters, and each

© International Association for Cryptologic Research 2016
J.H. Cheon and T. Takagi (Eds.): ASIACRYPT 2016, Part I, LNCS 10031, pp. 908–938, 2016.
DOI: 10.1007/978-3-662-53887-6_33

user may obtain a signature on their *secrets* issued by the trusted-third party. We call this model **the on-line model** (called **the algorithmic tamper-proof security model** in [23]). On the other hand, Bellare and Kohno [6] assume no trusted party. However, its subsequent works [4,5,7,22,28,33,35] allow a trusted party to play a minimum role, where it makes a public parameter, but once it did, it does nothing. An adversary is not allowed to modify the public parameter. We call this model **the common reference string (CRS) model**.

Gennaro et al. [23] suggested that it is *impossible* to realize chosen-ciphertext attack (CCA) secure PKE and digital signature schemes resilient to *all* tampering functions even in the on-line model. Therefore, they allowed a cryptosystem to **self-destruct**, meaning that when detecting tampering, a cryptographic device can erase all internal data, so that an adversary cannot obtain anything more from the device.

Other known ways to bypass the impossibility result are (1) to use **a key-updating mechanism**, i.e., to allow a device to *update* its inner secret with fresh randomness [26], and (2) to allow an adversary to submit a *bounded* number of tampering queries (**the bounded tampering model**) [14].

Tampering is further classified into **persistent** or **non-persistent** (due to [25]). In **persistent tampering attacks**, each tampering is applied to the current version of the secret that has been overwritten by the previous tampering function, i.e., when an adversary queries $(\phi_1, x_1)$ and $(\phi_2, x_2)$ to device $G(s, \cdot)$ in this order, it receives $G(\phi_1(s), x_1)$ and $G(\phi_2(\phi_1(s)), x_2)$, where $\phi_1, \phi_2$ are tampering functions and $x_1, x_2$ are inputs to device $G$. In **non-persistent tampering attacks**, tampering is always applied to the original secret, i.e., an adversary receives $G(\phi_1(s), x_1)$ and $G(\phi_2(s), x_2)$ when submitting the above queries. We insist that for PKE and digital signature schemes without a key-update mechanism, *non-persistent tampering is stronger than persistent tampering*, because an adversary that breaks a cryptosystem in a persistent tampering attack also breaks the same system in a non-persistent tampering attack. It is not clear in a cryptosystem with a key-updating mechanism the similar relation holds.

In this paper we focus on the common reference string (CRS) model (as mentioned above), where we assume a public parameter is generated by a trusted third party and assume that an adversary is not allowed to modify it. This setting is common in many prior works, e.g., [4,5,7,14,22,26,28,33,35].

At CRYPTO 2011, Kalai, Kanukurthi, and Sahai [26] considered **the continual tampering and leakage (CTL) model**, assuming tampering is *persistent*, and PKE and digital signature schemes are allowed to have a key-update algorithm, which updates a secret key with fresh (non-tampered) randomness between periods of tampering and leakage. This security model is considered in the CRS model. The proposed PKE scheme is one-bit-message encryption scheme based on [10] and is only chosen-plaintext attack (CPA) secure. Therefore, in their CTL security model, an adversary is *not* allowed to access the decryption oracle, which means that an adversary cannot observe the effect of tampering at the output of the decryption oracle. Instead, it can observe the effect of tampering at the output of the leakage oracle. We note that this tampering attack is not trivially implied by a leakage attack, because tampered secret $\phi(sk)$ is updated and the adversary can observe a partial information on the updated secret, say $L(\mathsf{Update}(\phi(sk)))$, from the leakage oracle. Their digital signature scheme (with a key-update mechanism) is constructed based on their CTL secure PKE scheme with simulation-sound non-interactive zero-knowledge proofs, which is simply inefficient. They also considered a digital signature scheme without a key-update mechanism in the so-called continuous tampering and bounded leakage (CTBL) model. The digital signature scheme may self-destruct (otherwise, it is impossible to prove the security). They claim that it is secure against persistent tampering attacks in the CTBL model. Remember that, if a digital signature scheme does not have a key-update mechanism, non-persistent tampering is stronger than persistent tampering. We later prove that if a digital signature scheme does not have a key-updating mechanism, it is impossible that it is resilient to continuous *non-persistent* tampering (even if it can self-destruct).

At ASIACRYPT 2013, Damgård, Faust, Mukherjee, and Venturi [14] proposed **the bounded leakage and tampering (BLT) model**. This setting allows **a bounded number** of non-persistent tampering, as well as bounded memory leakage, in the CRS model, where PKE has neither self-destructive nor key-update mechanism. In the BLT model for PKE, in addition to having access to bounded memory leakage oracle, an adversary is allowed to submit *a bounded number* of "pre-challenge"tampering queries $(\phi, \mathsf{CT})$ to the decryption oracle and receive $\mathbf{D}(\phi(sk), \mathsf{CT})$. It may also access the decryption oracle with the original secret-key both in the pre-challenge and post-challenge stages, as in the normal IND-CCA game. They presented a generic construction of IND-CCA BLT secure PKE scheme from an IND-CPA BLT secure PKE scheme with tSE NIZK proofs [15]. An instance of an IND-CPA BLT secure PKE scheme is BHHO PKE scheme [9]. Using the technique of [2], they also consider a variant of the floppy model [2], called **the $\iota$-Floppy model**, where each user has individual secret $y$ different from secret-key $sk$ and is allowed to execute *an invisible key update*, i.e., to update their secret key $sk$ using (non-tampered) secret $y$ with (non-tampered) flesh randomness.

## 1.1   Our Results

We study continuous tampering of arbitrary functions against PKE and digital signature schemes, in the presence of bounded or continuous memory leakage. Due to the impossibility result, we allow PKE and digital signature schemes to have either self-destructive or key-updating mechanism. There is no IND-CCA PKE scheme resilient to post-challenge tampering of arbitrary functions [14]. Indeed, one can break any PKE scheme, by observing the output of the decryption oracle after tampering with the following effciently computable function:

$$\phi(sk) = \begin{cases} sk & \text{if } \mathbf{D}(sk, \mathsf{CT}^*) = m_0, \text{ where } \mathsf{CT}^* \text{ is a challenge ciphertext.} \\ \bot & \text{otherwise.} \end{cases}$$

This attack is unavoidable even with self-destruction, key-updating, and bounded persistent/non-persistent tampering in the on-line model (i.e., in the strongest compromised model). Therefore, we allow tampering queries only in the pre-challenge stage against a PKE scheme.

We present the *first* chosen-ciphertext secure PKE schemes secure against *continuous* (pre-challenge) tampering of *arbitrary* functions. At the same time, our proposals tolerate bounded or continuous memory leakage of arbitrary functions. Interestingly, by putting some parameters in the common reference string and providing a self-destructive mechanism to the decryption algorithm, Qin and Liu's PKE scheme [31] is CTBL-CCA secure, meaning that it is IND-CCA secure resilient to continuous tampering and bounded memory leakage. We also propose the first CTL-CCA secure PKE scheme, meaning that it is IND-CCA secure resilient to continuous tampering and *continual* memory leakage. To the best of our knowledge, this is the *first* IND-CCA secure PKE scheme resilient to *continuous* memory leakage without using zero-knowledge, regardless of tampering.

Our security definitions basically model a *non-persistent* tampering attack, but it is straightforward to modify it to a persistent one. We show that any PKE scheme *without a key-update mechanism* that is CTBL-CCA secure against non-persistent tampering attacks is still CTBL-CCA secure against persistent tampering attacks. So is our CTBL-CCA secure PKE scheme. However, it is not clear that when a PKE scheme has a key-update mechanism, the similar relation holds.

We show that it is impossible to construct a secure digital signature scheme resilient to (continuous) *non-persistent* tampering even if it has a self-destructive mechanism. If a key-update mechanism should run only when tampering is detected, any digital signature scheme with a key-update mechanism is insecure, either.

**Comparison Among Continuous Tampering Models.** Table 1 classifies security models related to our continuous tampering model. Here b-tamp indicates bounded tampering and c-tamp indicates continuous tampering. Similarly, b-leak indicates bounded memory leakage and c-tamp indicates continuous memory leakage. persist indicates persistent tampering and n-persist indicates non-persistent tampering. per./n-per. indicates that the result in this row is effective against both persistent and non-persistent tampering. c-tamp$^-$ indicates the case of KKS signature scheme [26], where an adversary is allowed to submit a *bounded* number of tampering queries within each time period, although the number of tampering queries overall is unbounded. Our result is given in the gray area. Our CTL model imposes a more severe condition in that the scheme is allowed to update secret keys only when it can detect tampering.

## 1.2 Other Related Work

Considering a restricted class of tampering functions, we briefly mention two lines of works.

**Table 1.** Comparison: continuous tampering models and results

| Primitives | Self-dest | Key update | Tampering | Leakage | Security | Notes | Results |
|---|---|---|---|---|---|---|---|
| PKE | w/o | w/o | b-tamp | b-leak | CCA | per./n-per. | DFMV [14] |
| PKE | w/o | w | c-tamp | c-leak | CCA | $\iota$Floppy | DFMV [14] |
| PKE | w | w | b-tamp | - | CCA | post-tamp | Impossible([14]) |
| PKE | w/o | w/o | c-tamp | - | CCA | per./n-per. | Impossible ([23]) |
| PKE | w/o | w | c-tamp | c-leak | CPA | persist | KKS [26] |
| PKE | w | w/o | c-tamp | b-leak | CCA | per./n-per. | This work |
| PKE | w/o | w | c-tamp | c-leak | CCA | n-persist | This work |
| Sig | w/o | w/o | c-tamp | - | CMA | per./n-per. | Impossible ([23]) |
| Sig | w | w/o | c-tamp | b-leak | ? | persist | KKS [26] |
| Sig | w/o | w | c-tamp$^-$ | c-leak | CMA | persist | KKS [26] |
| Sig | w | w/o | c-tamp | - | CMA | n-persist | Impossible (This work) |
| Sig | w/o | w | c-tamp | - | CMA | n-persist | Impossible (This work) |

One research stream derives from Bellare and Kohno's [6], who study tampering (or equivalently related-key) resilient security against specific primitives, such as pseudo-random function (PRF) families, PKE, and identity-based encryption (IBE) schemes. By restricting tampering functions, post-challenge tampering queries can be treated in PKE. Currently, it is known that there is an IBE scheme (and hence, converted to PKE) resilient to polynomial functions [7] (in the CRS model). Qin et al. [33] recently claimed a broader class, but it is not correct [22] (Indeed, there is a counter example [3]). Recently, Fujisaki and Xagawa proposed an IBE scheme resilient to some kind of invertible functions [22]. In the above works, non-persistent tampering is considered, and primitives have neither self-destruction nor key-update mechanism.

The other line of works comes from algebraic manipulation detection (AMD) codes [11,12] and non-malleable codes (NMC) [19], whose codes can detect tampering of a certain class of functions. Dziembowski, Pietrzak, and Wichs [19] presented NMC and its application to tamper-resilient security. In their model, a PKE scheme allows both self-destruction and key-update mechanisms. An adversary accesses target device $G$ with a tampering query $(\phi, x)$ with $\phi \in \Phi$. If the decoding fails, i.e., $\mathsf{Dec}(\phi(\mathsf{Enc}(s)) = \bot$, then $G$ self-destructs. Otherwise, it returns $G(s, x)$ and updates $\mathsf{Enc}(s)$. Faust, Mukherjee, Nielsen, and Ventrui [21] considered *continuous NMC* and apply it to tamper and leakage resilient security (in the split-state model). Recently, Jafargholi and Wichs [25] presented NMCs for a bounded number of any subset of a very broader class of tampering functions. However, since an adversary must choose the subset before seeing the parameters of the codes, this result is not effective against continuous tampering attacks in this paper.

**Independent Work.** Independently of us, Faonio and Venturi [20] has recently showed[1] that the digital signature scheme proposed by Dodis et al. [16] and Qin-Liu PKE scheme [31] are secure in the bounded leakage and tampering (BLT) model [14], where a bounded number of *non-persistent* tampering and bounded memory leakage are allowed in the CRS model. Since we have proved that there is no digital signature scheme resilient to *continuous* non-persistent tampering even if self-destruction is allowed, it is reasonable that the digital signature scheme is proven only secure against bounded tampering. As for the PKE case in which Qin-Liu PKE scheme is proven BLT-CCA secure, the proof analysis is somewhat close to ours, in the sense that it does not use the leakage oracle in a black box way to simulate the effect of tampering (unlike [14]).

## 2  Preliminaries

For $n \in \mathbb{N}$ (the set of natural numbers), $[n]$ denotes the set $\{1, \ldots, n\}$. We let $\mathsf{negl}(\kappa)$ to denote an unspecified function $f(\kappa)$ such that $f(\kappa) = \kappa^{-\omega(1)} = 2^{-\omega(1)\log\kappa}$, saying that such a function is negligible in $\kappa$. We write PPT and DPT algorithms to denote probabilistic polynomial-time and deterministic poly-time algorithms, respectively. For PPT algorithm $A$, we write $y \leftarrow A(x)$ to denote the experiment of running $A$ for given $x$, picking inner coins $r$ uniformly from an appropriate domain, and assigning the result of this experiment to the variable $y$, i.e., $y = A(x; r)$. Let $X = \{X_\kappa\}_{\kappa\in\mathbb{N}}$ and $Y = \{Y_\kappa\}_{\kappa\in\mathbb{N}}$ be probability ensembles such that each $X_\kappa$ and $Y_\kappa$ are random variables ranging over $\{0,1\}^\kappa$. The (statistical) distance between $X_\kappa$ and $Y_\kappa$ is $\mathsf{Dist}(X_\kappa : Y_\kappa) \triangleq \frac{1}{2} \cdot |\Pr_{s\in\{0,1\}^\kappa}[X = s] - \Pr_{s\in\{0,1\}^\kappa}[Y = s]|$. We say that two probability ensembles, $X$ and $Y$, are statistically indistinguishable (in $\kappa$), denoted $X \overset{\mathrm{s}}{\approx} Y$, if $\mathsf{Dist}(X_\kappa : Y_\kappa) = \mathsf{negl}(\kappa)$. In particular, we denote by $X \equiv Y$ to say that $X$ and $Y$ are identical. We say that $X$ and $Y$ are computationally indistinguishable (in $\kappa$), denoted $X \overset{\mathrm{c}}{\approx} Y$, if for every non-uniform PPT $D$ (ranging over $\{0,1\}$), $\{D(1^\kappa, X_\kappa)\}_{\kappa\in\mathbb{N}} \overset{\mathrm{s}}{\approx} \{D(1^\kappa, Y_\kappa)\}_{\kappa\in\mathbb{N}}$.

### 2.1  Entropy and Extractor

The min-entropy of random variable $X$ is defined as $\mathsf{H}_\infty(X) = -\log(\max_x \Pr[X = x])$. We say that a function $\mathsf{Ext} : \{0,1\}^{\ell_s} \times \{0,1\}^n \to \{0,1\}^m$ is an $(k, \epsilon)$-strong extractor if for any random variable $X$ such that $X \in \{0,1\}^n$ and $\mathsf{H}_\infty(X) > k$, it holds that $\mathsf{Dist}((S, \mathsf{Ext}(S, X)), (S, U_m)) \leq \epsilon$, where $S$ is uniform over $\{0,1\}^{\ell_s}$. Let $\mathcal{H} = \{H\}$ be a family of hash functions $H : \{0,1\}^n \to \{0,1\}^m$. $\mathcal{H}$ is called a family of universal hash functions if $\forall x_1, x_2 \in \{0,1\}^n$ with $x_1 \neq x_2$, $\Pr_{H\leftarrow\mathcal{H}}[H(x_1) = H(x_2)] = 2^{-m}$. Then, The Leftover Hash Lemma (LHL) states the following.

---

[1] Their proposal has been submitted to IACR e-Print archive [20] *after* the deadline of ASIACRYPT 2016. So, it is obvious that ours is independent of theirs. We have recently noticed that it will also appear in ASIACRYPT 2016.

**Lemma 1 (Leftover Hash Lemma).** *Assume that the family $\mathcal{H}$ of functions $H : \{0,1\}^n \to \{0,1\}^m$ is a family of universal hash functions. Then for any random variable $X$ such that $X \in \{0,1\}^n$ and $\mathsf{H}_\infty(X) > m$,*

$$\mathsf{Dist}((H, H(X)), (H, U_m)) \leq \frac{1}{2}\sqrt{2^{-(\mathsf{H}_\infty(X)-m)}},$$

*where $H$ is a random variable uniformly chosen over $\mathcal{H}$ and $U_m$ is a random variable uniformly chosen over $\{0,1\}^m$.*

Therefore, $H$ constructs a $(k, 2^{-(k/2+1)})$-strong extractor where $k = \mathsf{H}_\infty(X) - m$.

We use the notion of the average conditional min-entropy defined by Dodis et al. [18] and its "chain rule". Define the average conditional min-entropy of random variable $X$ given random variable $Y$ as

$$\tilde{\mathsf{H}}_\infty(X|Y) \triangleq -\log\left(\mathop{\mathbf{E}}_{y \leftarrow Y}[\max_x \Pr[X = x|Y = y]]\right) = -\log\left(\mathop{\mathbf{E}}_{y \leftarrow Y}[2^{-\mathsf{H}_\infty(X|Y=y)}]\right).$$

**Lemma 2 ("Chain Rule" for Average Min-Entropy [18]).** *When random variable $Z$ takes at most $2^r$ possible values (i.e., $\#\mathsf{Supp}(Z) = 2^r$) and $X, Y$ are random variables, then*

$$\tilde{\mathsf{H}}_\infty(X|(Y, Z)) \geq \tilde{\mathsf{H}}_\infty((X, Y)|Z) - r \geq \tilde{\mathsf{H}}_\infty(X|Z) - r.$$

*In particular,*
$$\tilde{\mathsf{H}}_\infty(X|Z) \geq \mathsf{H}_\infty(X, Z) - r \geq \mathsf{H}_\infty(X) - r.$$

Dodis et al. [18] proved that any strong extractor is an average-case strong extractor for an appropriate setting of the parameters. As a special case, they showed any family of universal hash functions is an average-case strong extractor along with the following generalized version of the leftover hash lemma:

**Lemma 3 (Generalized Leftover Hash Lemma [18]).** *Assume that the family $\mathcal{H}$ of functions $H : \{0,1\}^n \to \{0,1\}^m$ is a family of universal hash functions. Then for any random variables, $X$ and $Z$,*

$$\mathsf{Dist}((H, H(X), Z), (H, U_m, Z)) \leq \frac{1}{2}\sqrt{2^{-(\tilde{\mathsf{H}}_\infty(X|Z)-m)}},$$

*where $H$ is a random variable uniformly chosen over $\mathcal{H}$ and $U_m$ is a random variable uniformly chosen over $\{0,1\}^m$.*

## 2.2    Hash Proof Systems

We recall the notion of the hash proof systems introduced by Cramer and Shoup [13]. Let $\mathcal{C}, \mathcal{K}, \mathcal{SK}$, and $\mathcal{PK}$ be efficiently samplable sets and let $\mathcal{V}$ be a subset in $\mathcal{C}$. Let $\Lambda_{sk} : \mathcal{C} \to \mathcal{K}$ be a hash function indexed by $sk \in \mathcal{SK}$. A hash function family $\Lambda : \mathcal{SK} \times \mathcal{C} \to \mathcal{K}$ is projective if there is a projection $\mu : \mathcal{SK} \to \mathcal{PK}$ such that $\mu(sk) \in \mathcal{PK}$ defines the action of $\Lambda_{sk}$ over subset $\mathcal{V}$.

That is to say, for every $C \in \mathcal{V}$, $K = \Lambda_{sk}(C)$ is uniquely determined by $\mu(sk)$ and $C$. $\Lambda$ is called $\gamma$-entropic [27] if for all $pk \in \mathcal{PK}$, $C \in C\backslash\mathcal{V}$, and all $K \in \mathcal{K}$,

$$\Pr[K = \Lambda_{sk}(C)|(pk, C)] \leq 2^{-\gamma},$$

where the probability is taken over $sk \xleftarrow{\mathsf{U}} \mathcal{SK}$ with $pk = \mu(sk)$. We note that this $\Lambda$ is originally called $2^{-\gamma}$-universal$_1$ in [13]. By definition, we note that $\mathsf{H}_\infty(\Lambda_{sk}(C)|(pk, C)) \geq \gamma$ for all $pk \in \mathcal{PK}$ and $C \in C\backslash\mathcal{V}$.

$\Lambda$ is called $\epsilon$-smooth [13] if $\mathsf{Dist}((pk, C, \Lambda_{sk}(C)), (pk, C, K)) \leq \epsilon$, where $sk \xleftarrow{\mathsf{U}} \mathcal{SK}$, $K \xleftarrow{\mathsf{U}} \mathcal{K}$ and $C \xleftarrow{\mathsf{U}} C\backslash\mathcal{V}$ are chosen *at random* and $pk = \mu(sk)$.

A hash proof system $\mathsf{HPS} = (\mathsf{HPS.param}, \mathsf{HPS.pub}, \mathsf{HPS.priv})$ consists of three algorithms such that $\mathsf{HPS.param}$ takes $1^\kappa$ and outputs an instance of $\mathsf{params} = (\mathsf{group}, \Lambda, C, \mathcal{V}, \mathcal{SK}, \mathcal{PK}, \mu)$, where $\mathsf{group}$ contains some additional structural parameters and $\Lambda$ is a projective hash function family associated with $(C, \mathcal{V}, \mathcal{SK}, \mathcal{PK}, \mu)$ as defined above. The deterministic public evaluation algorithm $\mathsf{HPS.pub}$ takes as input $pk = \mu(sk)$, $C \in \mathcal{V}$ and a witness $w$ such that $C \in \mathcal{V}$ and returns $\Lambda_{sk}(C)$. The deterministic private evaluation algorithm takes $sk \in \mathcal{SK}$ and returns $\Lambda_{sk}(C)$, without taking witness $w$ for $C$ (if it exists). A hash proof system $\mathsf{HPS}$ as above is said to have a hard subset membership problem if two random elements $C \in C$ and $C' \in C\backslash\mathcal{V}$ are computationally indistinguishable, that is, $\{C \,|\, C \xleftarrow{\mathsf{c}} C\}_{\kappa \in \mathbb{N}} \stackrel{\mathsf{c}}{\approx} \{C' \,|\, C' \xleftarrow{\mathsf{U}} C\backslash\mathcal{V}\}_{\kappa \in \mathbb{N}}$.

## 2.3   All-But-One Injective Functions

We recall all-but-one injective functions (ABO) [32], which is a simple variant of all-but-one injective trap-door functions [30].

A collection of $(n, \ell_{\mathsf{lf}})$-all-but-one injective functions with branch collection $\mathcal{B} = \{B_\kappa\}_{\kappa \in \mathbb{N}}$ is given by a tuple of PPT algorithms $\mathsf{ABO} = (\mathsf{ABO.gen}, \mathsf{ABO.eval})$ with the following properties:

- $\mathsf{ABO.gen}$ is a PPT algorithm that takes $1^\kappa$ and any branch $b^* \in B_\kappa$, and outputs a function index $i_{\mathsf{abo}}$ and domain $X$ with $2^n$ elements.
- $\mathsf{ABO.eval}$ is a DPT algorithm that takes $i_{\mathsf{abo}}$, $b$, and $x \in X$, and computes $y = \mathsf{ABO.eval}(i_{\mathsf{abo}}, b, x)$.

We require that $(n, \ell_{\mathsf{lf}})$-all-but-one injective functions given by $\mathsf{ABO}$ satisfies the following properties:

1. For any $b \neq b^* \in B_\kappa$, $\mathsf{ABO.eval}(i_{\mathsf{abo}}, b, \cdot)$ computes an injective function over the domain $X$.
2. The number of elements in the image of $\mathsf{ABO.eval}(i_{\mathsf{abo}}, b^*, \cdot)$ over the domain $X$ is at most $2^{\ell_{\mathsf{lf}}}$.
3. For any $b, b^* \in B_\kappa$, $\{\mathsf{ABO.gen}(1^\kappa, b)\}_{\kappa \in \mathbb{N}} \stackrel{\mathsf{c}}{\approx} \{\mathsf{ABO.gen}(1^\kappa, b^*)\}_{\kappa \in \mathbb{N}}$.

We note that ABO functions can be efficiently constructed under the DDH assumption and the DCR assumption (See Appendix B).

# 3   Continuous Tampering and Bounded Leakage Resilient CCA (CTBL-CCA) Secure Public-Key Encryption

A public-key encryption (PKE) scheme consists of the following four algorithms $\Pi = (\mathsf{Setup}, \mathbf{K}, \mathbf{E}, \mathbf{D})$: The setup algorithm $\mathsf{Setup}$ is a PPT algorithm that takes $1^\kappa$ and outputs public parameter $\rho$. The key-generation algorithm $\mathbf{K}$ is a PPT algorithm that takes $\rho$ and outputs a pair of public and secret keys, $(pk, sk)$. The encryption algorithm $\mathbf{E}$ is a PPT algorithm that takes public parameter $\rho$, public key $pk$ and message $m \in \mathcal{M}$, and produces ciphertext $\mathsf{ct} \leftarrow \mathbf{E}_\rho(pk, m)$; Here $\mathcal{M}$ is uniquely determined by $pk$. The decryption algorithm $\mathbf{D}$ is a DPT algorithm that takes $\rho$, $sk$ and presumable ciphertext $\mathsf{ct}$, and returns message $m = \mathbf{D}_\rho(sk, \mathsf{ct})$. We require for correctness that for every sufficiently large $\kappa \in \mathbb{N}$, it always holds that $\mathbf{D}_\rho(sk, \mathbf{E}_\rho(pk, m)) = m$, for every $\rho \in \mathsf{Setup}(1^\kappa)$, every $(pk, sk)$ generated by $\mathbf{K}(\rho)$, and every $m \in \mathcal{M}$.

We say that PKE $\Pi$ is **self-destructive** *if the decryption algorithm can erase all inner states including $sk$, when receiving an invalid ciphertext $\mathsf{ct}$.* We assume that public parameter $\rho$ is **system-wide**, i.e., *fixed beforehand and independent of all users*, and the only public and secret keys are subject to the tampering attacks. This model is justified in the environment where the common public parameter could be hardwired into the algorithm codes and stored on tamper-proof hardware or distributed via a public channel where tampering is infeasible or could be easily detected.

**CTBL-CCA Security.** For PKE $\Pi$ and an adversary $A = (A_1, A_2)$, we define the experiment $\mathsf{Expt}^{\mathsf{ctbl\text{-}cca}}_{\Pi, A, (\Phi_1, \Phi_2, \lambda)}(\kappa)$ as in Fig. 1. $A$ may adaptively submit (unbounded) polynomially many queries $(\phi, \mathsf{ct})$ to oracle $\mathsf{RKDec}^2$, but $\phi$ should be in $\Phi_i$ appropriately. $A$ may also adaptively submit (unbounded) polynomially many queries $L$ to oracle $\mathsf{Leak}$, before seeing the challenge ciphertext $\mathsf{ct}^*$. The total amount of leakage on $sk$ must be bounded by some $\lambda$ bit length. We note that if $\Pi$ has the **self-destructive** property, $\mathsf{RKDec}$ does not answer any further query, or simply return $\perp$, after it receives an invalid ciphertext such that $\mathbf{D}_\rho(\phi(sk), \mathsf{ct}) = \perp$. We define the advantage of $A$ against $\Pi$ with respects $(\Phi_1, \Phi_2)$ as

$$\mathsf{Adv}^{\mathsf{ctbl\text{-}cca}}_{\Pi, A, (\Phi_1, \Phi_2, \lambda)}(\kappa) \triangleq \big| 2 \Pr[\mathsf{Expt}^{\mathsf{ctbl\text{-}cca}}_{\Pi, A, (\Phi_1, \Phi_2, \lambda)}(\kappa) = 1] - 1 \big|.$$

We say that $\Pi$ is $(\Phi_1, \Phi_2, \lambda)$-CTBL-CCA secure if $\mathsf{Adv}^{\mathsf{ctbl\text{-}cca}}_{\Pi, A, (\Phi_1, \Phi_2, \lambda)}(\kappa) = \mathsf{negl}(\kappa)$ for every PPT $A$.

We say that $\Pi$ is CTBL-CCA secure if it is $(\Phi_{\mathsf{all}}, \{i\eth\}, \lambda)$-CTBL-CCA secure, where $\Phi_{\mathsf{all}}$ is the class of all efficiently computable functions and $i\eth$ denotes the identity function.

*Remark 1.* This security definition models **non-persistent** tampering. However, it is obvious that the persistent tampering version of CTBL-CCA security can be similarly defined.

---

[2] A tampering function is called a related-key derivation (RKD) function in [4,6].

$$\mathsf{Expt}^{\mathsf{ctbl\text{-}cca}}_{\Pi, A, (\Phi_1, \Phi_2, \lambda)}(\kappa):$$
$\rho \leftarrow \mathsf{Setup}(1^{\kappa});$
$(\mathsf{pk}, \mathsf{sk}) \leftarrow \mathbf{K}(\rho);$
$(m_0, m_1, st) \leftarrow A_1^{\mathsf{RKDec}_{\Phi_1}(\cdot, \cdot), \mathsf{Leak}_\lambda(\cdot)}(\rho, \mathsf{pk})$
    such that $|m_0| = |m_1|;$
$\beta^* \leftarrow \{0, 1\};$
$\mathsf{ct}^* \leftarrow \mathbf{E}_\rho(\mathsf{pk}, m_{\beta^*});$
$\beta \leftarrow A_2^{\mathsf{RKDec}_{\Phi_2}(\cdot, \cdot)}(st, \mathsf{ct}^*);$
If $\beta = \beta^*$,
    then return 1; otherwise 0.

$\mathsf{RKDec}_\Phi(\phi, \mathsf{ct}):$
If $\mathsf{ct} = \mathsf{ct}^*$ queried by $A_2$,
    then return $\bot$;
If $\mathbf{D}_\rho(\phi(\mathsf{sk}), \mathsf{ct}) = \bot$,
    then erase $\mathsf{sk}$.
Return $\mathbf{D}_\rho(\phi(\mathsf{sk}), \mathsf{ct})$.

$\mathsf{Leak}_\lambda(L_i):$ $(L_i: i\text{-th query of } A.)$
If $\sum_{j=1}^{i} |L_j(\mathsf{sk})| > \lambda$
    then return $\bot$;
Else return $L_i(\mathsf{sk})$.

**Fig. 1.** The experiment of the CTBL-CCA game.

We now state the following fact.

**Theorem 1.** *Suppose a PKE scheme* $\Pi$ *without a key-update mechanism (as defined in Sect. 5) is CTBL-CCA secure against non-persistent tampering attacks. Then,* $\Pi$ *is also CTBL-CCA secure against persistent tampering attacks.*

*Proof.* For a PKE scheme without a key-update mechanism, persistent tampering queries

$$(\phi_1, \mathsf{ct}_1), (\phi_2, \mathsf{ct}_2), \dots, (\phi_\ell, \mathsf{ct}_\ell)$$

can be simulated non-persistent tampering queries as

$$(\phi_1, \mathsf{ct}_1), (\phi_2 \circ \phi_1, \mathsf{ct}_2), \dots, (\phi_\ell \circ \cdots \circ \phi_1, \mathsf{ct}_\ell).$$

Leakage functions in the persistent tampering attack are also simulated as $L' = L \circ \phi_\ell \cdots \circ \phi_1$, where $\phi_1, \dots, \phi_\ell$ denote all persistent tampering functions submitted before leakage function $L$ is submitted. So, if $\Pi$ is CTBL-CCA secure against non-persistent tampering attacks, then it is CTBL-CCA secure against persistent tampering attacks. ∎

## 4    The CTBL-CCA Secure PKE Scheme

Let $\mathsf{HPS} = (\mathsf{HPS.param}, \mathsf{HPS.pub}, \mathsf{HPS.priv})$ be a hash proof system (described in Sect. 2.2). Let $\mathsf{ABO} = (\mathsf{ABO.gen}, \mathsf{ABO.eval})$ be a collection of all-but-one injective (ABO) functions (described in Sect. 2.3). Let $\mathsf{TCH}$ be a target collision resistant hash family. Let $\mathcal{H} = \{H | H : \{0, 1\}^n \to \{0, 1\}^{\ell_m}\}$ be a family of universal hash functions with $n = |\mathcal{K}|$. Let $\mathsf{OTSig} = (\mathsf{otKGen}, \mathsf{otSign}, \mathsf{otVrfy})$ a strong one-time signature scheme. We assume $\mathsf{vk} = 0 \notin \mathsf{otKGen}$.

At ASIACRYPT 2013, Qin and Liu [31] proposed a new framework for constructing an IND-CCA secure PKE scheme resilient to bounded memory leakage. Assume a PKE scheme based on a hash-proof-system, where an encryption of $m$ is constructed as $CT = (C, H, e)$ where $C \leftarrow \mathcal{V}$ with $w$, $H \leftarrow \mathcal{H}$, and

$e = m \oplus H(\mathsf{HPS.pub}(PK, C, w))$, whereas the decryption is done by computing $m = e \oplus H(\mathsf{HPS.priv}(SK, C))$. Naor and Segev [29] proved that such a PKE scheme is IND-CPA secure resilient to bounded memory leakage. Qin and Liu transformed it to IND-CCA secure one resilient to bounded memory leakage, by using *a one-time lossy filter*. We describe a slight modification of Qin-Liu PKE scheme in Fig. 1. The difference is that (1) our construction divides the original key generation algorithm into the Setup algorithm and the key generation algorithm and puts $\rho$ in the common reference string, and (2) replaces a one-time lossy filter with a combination of a strong one-time signature scheme and an ABO injective function. (Here (2) is not essential. It is just a matter of our preference to use an ABO injective function. Any one-time lossy filter suffices for our purpose.)

We then have the following theorem.

**Theorem 2.** *Let* HPS *be a $\gamma$-entropic hash proof system. Let* ABO *be $(n, \ell_{\mathsf{lf}})$-all-but-one injective function where $n = \log |\mathcal{K}|$. We assume the PKE scheme in Fig. 2 is self-destructive. Then, it is $(\Phi_{\mathsf{all}}, \{\mathsf{id}\}, \lambda)$-CTBL-CCA secure, as long as $\lambda(\kappa) \leq \gamma - \ell_{\mathsf{lf}} - \ell_m - 2\eta - \log(1/\epsilon)$ where $\eta(\kappa) = \omega(\log \kappa)$ and $\epsilon = 2^{-\omega(\log \kappa)}$, and for any PPT adversary A with at most Q queries to* RKDec *oracle,* $\mathsf{Adv}_{\Pi,A,(\Phi_{\mathsf{all}},\{\mathsf{id}\},\lambda)}^{\mathsf{ctbl\text{-}cca}}(\kappa) \leq$

$$2\epsilon_{\mathsf{tcr}} + 2\epsilon_{\mathsf{otsig}} + 4\epsilon_{\mathsf{lossy}} + 4\epsilon_{\mathsf{SD}} + 2^{-\eta+2} + Q \cdot 2^{-(\gamma-\eta-\lambda-\ell_{\mathsf{lf}}-\ell_m-1)} + 2\epsilon,$$

*where $\epsilon_{\mathsf{otsig}}$, $\epsilon_{\mathsf{lossy}}$, and $\epsilon_{\mathsf{SD}}$ denote some negligible functions such that $\mathsf{Adv}_{\mathsf{OTSig},B}^{\mathsf{ot}}(\kappa) \leq \epsilon_{\mathsf{otsig}}$, $\mathsf{Adv}_{\mathsf{ABO},B'}^{\mathsf{lossy}}(\kappa) \leq \epsilon_{\mathsf{lossy}}$, and $\mathsf{Adv}_{\mathsf{HPS},D}^{\mathsf{SD}}(\kappa) \leq \epsilon_{\mathsf{SD}}$ for any PPT adversaries, B, B' and D, respectively.*

**Proof Idea.** Qin-Liu PKE scheme is leakage resilient. So, it is tempting to use the leakage oracle in the black box way to simulate the RKDec oracle (as in [14]). However, the strategy does not work for *continual* tampering, because Qin-Liu PKE scheme is just *bounded* leakage resilient. In addition, even simulating the reply of a single tampering query seems to exceed the leakage bound. So, we need to analyze the exact leakage from tampering.

Let $\mathsf{CT}^* = (C^*, e^*, H^*, \mathsf{vk}^*, \pi^*, \sigma^*)$ be the challenge ciphertext and $b^*$ be the challenge bit. Let $K^* = \Lambda_{SK}(C^*)$ and $e^* = m_{b^*} \oplus H^*(K^*)$. In an early hybrid game of the proof, we set $C^* \notin \mathcal{V}$ and set $\mathsf{T}(\mathsf{vk}^*)$ as a lossy branch, as expected. Since $A(\mathsf{T}(\mathsf{vk}^*), \cdot)$ is lossy now, $SK$ (and hence $K^*$) has large enough entropy after given $\mathsf{CT}^*$. In the pre-challenge stage, we take care of how much entropy on $K^*$ is preserved while answering leakage and tampering queries.

We first observe that when a tampering query $(\phi, \mathsf{CT})$, where $\mathsf{CT} = (C, e, H, \mathsf{vk}, \pi, \sigma)$, is rejected by the decryption oracle, the leaked information on $K^*$ is at most $\log(1/p)$-bit where $p = \Pr[\mathbf{D}(\phi(SK), \mathsf{CT}) = \bot]$. This comes from the following simple lemma.

**Lemma 4.** *For any random variables, X and Z, $\mathsf{H}_\infty(X|Z = z) \geq \mathsf{H}_\infty(X) - \log\left(\frac{1}{\Pr[Z=z]}\right)$.*

| **Set-Up Algorithm** Setup($1^\kappa$): | **Key Generation Algorithm** K($\rho$): |
|---|---|
| params $\leftarrow$ HPS.param($1^\kappa$) | $sk \leftarrow \mathcal{SK}$. |
| where params = | Set $pk := \mu(sk)$. |
| (group, $\Lambda, \mathcal{C}, \mathcal{V}, \mathcal{SK}, \mathcal{PK}, \mu$). | Set $PK := pk$ and $SK := sk$. |
| T $\leftarrow$ TCH where T : $\{0,1\}^* \rightarrow B_\kappa$. | Return $(PK, SK)$ |
| Set $b^* = 0$ as the lossy branch. | |
| $\iota_{abo} \leftarrow$ ABO.gen($1^\kappa, b^*$). | |
| $A(\cdot, \cdot) :=$ ABO.eval($\iota_{abo}, \cdot, \cdot$). | |
| Return $\rho =$ (T, params, $A(\cdot, \cdot)$). | |
| **Encryption Algorithm** $\mathbf{E}_\rho$($PK, m$): | **Decryption Algorithm** $\mathbf{D}_\rho$($SK$, CT): |
| To encrypt a message $m \in \mathbb{G}$, | To decrypt a ciphertext CT, |
| $C \overset{U}{\leftarrow} \mathcal{V}$ with witness $w$. | Parse CT into $(C, e, H, \text{vk}, \pi, \sigma)$. |
| $K =$ HPS.pub($pk, C, w$). | If Vrfy(vk, $(C, e, H, \text{vk}, \pi), \sigma) \neq 1$, |
| (vk, otsk) $\leftarrow$ otKGen($1^\kappa$) | then aborts. |
| $\pi = A(\text{T(vk)}, K)$. $H \leftarrow \mathcal{H}$. | Else $K = \Lambda_{sk}(C)$. |
| $e = m \oplus H(K)$. | If $\pi \neq A(\text{T(vk)}), K)$, |
| $\sigma \leftarrow$ otSign(otsk, $(C, e, \text{vk}, \pi)$). | then aborts. |
| Return CT = $(C, e, H, \text{vk}, \pi, \sigma)$. | Else return $m = e \oplus H(K)$. |

**Fig. 2.** The CTBL-CCA secure PKE scheme based on Qin and Liu's PKE

*Proof.* For any $z \in Z$,

$$-\log\left(\max_x\left(\Pr[X = x | Z = z]\right)\right) = -\log\left(\max_x\left(\frac{\Pr[X = x \wedge Z = z]}{\Pr[Z = z]}\right)\right)$$

$$\geq -\log\left(\max_x\left(\Pr[X = x]\right)\right) - \log\left(\frac{1}{\Pr[Z = z]}\right).$$

∎

By the lemma above, we have

$$\mathsf{H}_\infty(K^* | \mathbf{D}(\phi(SK), \text{CT}) = \bot) \geq \mathsf{H}_\infty(K^*) - \log(1/p). \tag{1}$$

Next, we observe the case that tampering query $(\phi, \text{CT})$ is accepted by the decryption oracle. Since the decryption oracle returns $\mathbf{D}(\phi(SK), \text{CT})$, it would apparently reveal more information on $K^*$ except the fact that CT is a valid ciphertext with respects to $\phi(SK)$[3]. However, it is not true. Indeed, when submitting $(\phi, \text{CT})$, *the adversary has already fixed* $\mathbf{D}(\phi(SK), \text{CT})$. In other word, we have

$$\mathsf{H}_{\mathsf{sh}}\Big(\mathbf{D}(\phi(SK), \text{CT}) \mid (\mathbf{D}(\phi(SK), \text{CT}) \neq \bot), (\phi, \text{CT}), PK\Big) = 0, \tag{2}$$

---

[3] One can always use a "loose" bound such that $\widetilde{\mathsf{H}}_\infty(K^* | \mathbf{D}(\phi(SK), \text{CT})) \geq \mathsf{H}_\infty(K^*) - \lambda$ where $\lambda = \log\Big(\mathbf{D}(\phi(SK), \text{CT})\Big)$. However, the bound is too loose for our purpose.

where $\mathsf{H_{sh}}(X)$ denotes the Shannon entropy of random variable $X$ (i.e., $\mathsf{H_{sh}}(X) := \mathbf{E}_{x \leftarrow X}[\log \frac{1}{\Pr[X=x]}]$). This comes from the fact that $A(\mathsf{T}(\mathsf{vk}), \cdot)$ is injective and $\pi = A(\mathsf{T}(\mathsf{vk}), \Lambda_{\phi(SK)}(C))$ is fixed by $\mathsf{CT}$. Therefore, we have

$$\tilde{\mathsf{H}}_\infty(K^* | \mathbf{D}(\phi(SK), \mathsf{CT}), (\mathbf{D}(\phi(SK), \mathsf{CT}) \neq \bot)) \geq \mathsf{H}_\infty(K^*) - \log(1/p'), \quad (3)$$

where $p' = \Pr[\mathbf{D}(\phi(SK), \mathsf{CT}) \neq \bot]$. Hence, the leaked information on $K^*$ in the "accepted" case is also at most $\log(1/p')$. By definition, $p + p' = 1$.

We note that if the adversary submits a tampering query $(\phi, \mathsf{CT})$ with $p \leq 2^{-\eta} = \mathsf{negl}(\kappa)$ and the unlikely event that $\mathbf{D}(\phi(SK), \mathsf{CT}) = \bot$ really occurs, the leakage on $K^*$ is $\log(1/p) \geq \eta = \omega(\log \kappa)$ bits. The event occurs only with a negligible probability $2^{-\eta}$. We note that if the event occurs with a probability more than $2^{-\eta}$, the leakage on $K^*$ is less than $\eta$ bits. So, we can say that when $\mathbf{D}(\phi(SK), \mathsf{CT}) = \bot$ occurs, the leakage on $K^*$ is bounded by $\eta$-bit except with a negligible probability $2^{-\eta}$. By definition, the event $\mathbf{D}(\phi(SK), \mathsf{CT}) = \bot$ can occur only once. The case with $p' \leq 2^{-\eta} = \mathsf{negl}(\kappa)$ is implied in the next analysis.

Since the decryption algorithm *self-destructs* when rejecting a ciphertext, the adversary's best strategy is to submit a sequence of tampering queries with $p' = \mathsf{non\text{-}negl}$ so that the decryption algorithm can accept as long a prefix of the sequence as possible. Even with this strategy, however, leakage amount on $K^*$ is bounded by $\eta$-bit except with probability $2^{-\eta}$.

We now consider a post-challenge (tampering) query, $(\mathfrak{id}, \mathsf{CT})$, i.e., a normal decryption query, where $\mathsf{CT} = (C, e, H, \mathsf{vk}, \pi, \sigma)$. In the post-challenge stage, we are interested in how to prevent $H^*(K^*)$ from revealing any partial information. Even one bit leakage would possibly break the system. To achieve the goal, we need to reject any invalid ciphertext. The probability relies on the entropy of $K = \Lambda_{SK}(C)$ (where $C \notin \mathcal{V}$). Since the underlying hash proof system is $\gamma$-entropic, we can see that the remaining entropy of $K$ is at least $\gamma - \lambda - \eta - \ell_{\mathsf{lf}} - \ell_m$ (with an overwhelming probability). Here, $\lambda$ is the leakage amount via leakage oracle in the pre-challenge stage, $2^{\ell_{\mathsf{lf}}}$ denotes the number of possible elements of $\pi^*$, where $A(\mathsf{T}(\mathsf{vk}^*), \cdot)$ is lossy, and $\ell_m$ is the bit length of $H^*(K^*)$. Then, the probability that we *cannot* reject an invalid ciphertext is at most $2^{-(\gamma - \lambda - \eta - \ell_{\mathsf{lf}} - \ell_m)}$.

To summarize all the above, (a) just after the pre-challenge stage, the remaining entropy of $K^*$ is at least $\mathsf{H}_\infty(K^*) - \lambda - (\eta + 1)$ with an overwhelming probability. By applying an appropriate universal hash $H^*$, we obtain $H^*(K^*)$ that is statistically close to a true uniform $\ell_m$-bit string. So, $\mathsf{CT}^*$ conceals message $m_{b^*}$ in the statistical sense. (b) In the post-challenge stage, $H^*(K^*)$ reveals no information with an overwhelming probability $1 - Q \cdot 2^{-(\gamma - \lambda - \eta - \ell_{\mathsf{lf}} - \ell_m)}$, where $Q$ is the total number of decryption queries in the post-challenge stage. Like this, the proposal is proven CTBL-CCA secure.

**Proof of Theorem 2.** Here we provide the formal proof of Theorem 2 by using the standard game-hopping strategy. We denote by $S_i$ the event that adversary $A$ wins in **Game** $i$.

- **Game 0**: This game is the original CTBL-CCA game, where $\mathsf{CT}^* = (C^*, e^*, H^*, \mathsf{vk}^*, \pi^*, \sigma^*)$ denotes the challenge ciphertext. By definition, $\Pr[S_0] = \Pr[\beta = \beta^*]$ and $\mathsf{Adv}^{\mathsf{tbl\text{-}cca}}_{\Pi, A, (\Phi_{\mathsf{all}}, \{\mathfrak{id}\}, \lambda)}(\kappa) = |2\Pr[S_0] - 1|$.

- **Game 1**: This game is identical to **Game 0**, except that when we produce the challenge ciphertext $\mathsf{CT}^*$, *we instead computes* $K^* = \mathsf{HPS.priv}(sk, C^*)$. The change is just conceptual and hence, it holds that $\Pr[S_0] = \Pr[S_1]$.
- **Game 2**: This game is identical to **Game 1**, except that $A$ is regarded as a defeat, when it submits tampering query $(\phi, \mathsf{CT})$ such that $\mathsf{T(vk)} = \mathsf{T(vk^*)}$ but $\sigma$ is still a valid signature on $(C, e, H, \mathsf{vk}, \pi)$, where $\mathsf{CT} = (C, e, H, \mathsf{vk}, \pi, \sigma)$ ($\neq \mathsf{CT}^*$). This happens only when $\mathsf{T(vk)} = \mathsf{T(vk^*)}$ with $\mathsf{vk} \neq \mathsf{vk}^*$ or $A$ forges a signature with respects to $\mathsf{vk}^*$. So, we have $\Pr[S_1] - \Pr[S_2] \leq \epsilon_{\mathsf{tcr}} + \epsilon_{\mathsf{otsig}}$.
- **Game 3**: This game is identical to **Game 2**, except that we produce $\rho$ and $\mathsf{CT}^*$ as follows: Before the step 3 in the set-up **Setup**, we run $(\mathsf{vk}^*, \mathsf{otsk}^*) \leftarrow \mathsf{otKGen}(1^\kappa)$ and set $b^* = \mathsf{T(vk^*)}$. Then we do the same things in the subsequent steps. We produce the challenge ciphertext $\mathsf{CT}^*$ similarly in **Game 2** except that we instead use $(\mathsf{vk}^*, \mathsf{otsk}^*)$ generated in the set-up phase. The difference between the probabilities of events, $S_2$ and $S_3$, are close because of indistinguishability between injective and lossy branches. Indeed, we have $\Pr[S_2] - \Pr[S_3] \leq 2\epsilon_{\mathsf{lossy}}$.
- **Game 4**: This game is identical to **Game 3**, except that when producing $\mathsf{CT}^*$, we instead picks up $C^* \xleftarrow{\mathsf{U}} C \backslash \mathcal{K}$. We then have $\Pr[S_3] - \Pr[S_4] \leq 2\epsilon_{\mathsf{SD}}$.
- **Game 5**: This game is identical to the previous game, except that $A$ is regarded as a defeat, when it submits a tampering query $(\phi, \mathsf{CT})$ with $p \leq 2^{-\eta}$ where $p = \Pr[\mathbf{D}(\phi(SK), \mathsf{CT}) = \perp]$ and the (unlikely) event that $\mathbf{D}(\phi(SK), \mathsf{CT}) = \perp$ really occurs. We then have $\Pr[S_4] - \Pr[S_5] \leq 2^{-\eta}$. Without loss of generality, we can assume that $A$ does not make a tampering query with $p > 2^{-\eta}$ in the subsequent games.
- **Game 6**: We say that a sequence of tampering queries made by $A$ is $\eta$-*challenging*, if there is a prefix of the sequence such that the decryption oracle accepts the prefix with probability $\leq 2^{-\eta}$. Let $\mathsf{RDview}$ be a random variable of the transcript between adversary $A$ and oracle $\mathsf{RKDec}$ in the pre-challenge stage and let

$$\mathrm{rdv} = \{(\phi_1, \mathsf{CT}_1, m_1), \ldots, (\phi_{q'}, \mathsf{CT}_{q'}, m_{q'})\} \text{ where } q' \leq Q.$$

be a transcript. If $\mathrm{rdv}$ is $\eta$-challenging, there is the minimum $q_{\min} \leq q'$ such that

$$\Pr[\mathsf{RDview} = \mathrm{rdv}] \leq \Pr\left[\wedge_{i=1}^{q_{\min}} \left(\mathbf{D}(\phi_i(SK), \mathsf{CT}_i) \neq \perp\right)\right] \leq 2^{-\eta}.$$

**Game 6** is identical to the previous game except that $\mathsf{RKDec}$ "self-destructs" at the $(q_{\min} + 1)$-th tampering query of $\eta$-challenging $\mathrm{rdv}$, even if $\mathsf{RKDec}$ accepts the $(q_{\min} + 1)$-th tampering query. (If it rejects an earlier tampering query, it self-destructs at the query.) This experiment is just conceptual and is not required to be executed in a polynomial time. We have $\Pr[S_5] - \Pr[S_6] \leq 2^{-\eta}$, because the prefix is accepted at most $2^{-\eta}$.
- **Game 7**: In this game, for all post-challenge (decryption) query $(\mathsf{id}, \mathsf{CT})$ of $A$, *we return* $\perp$ *if* $C \in C \backslash \mathcal{V}$. This experiment is just conceptual and is not required to be executed in a polynomial time. We evaluate the min-entropy

of $K = \Lambda_{SK}(C)$ derived from the post-challenge tampering query. Let Lview be the random variable of the transcript between adversary $A$ and oracle Leak in the pre-challenge stage. When the first post-challenge decryption query is made, by the "chain rule" of the average-min entropy,

$$\widetilde{H}_\infty(K|(\text{RDview}, \text{Lview}, \pi^*, H^*(K^*))) \geq \widetilde{H}_\infty(K|\text{RDview}) - \lambda - \ell_{\text{lf}} - \ell_m,$$

where $2^{\ell_{\text{lf}}}$ denotes the number of elements in the image of "lossy" function $\pi^* = A(\text{T}(\text{vk}^*), \cdot)$, and $\ell_m$ is the length of $H^*(K^*)$.
By lemma 4, we have

$$H_\infty(K|\text{RDview} = \text{rdv}) \geq H_\infty(K) - \log\left(\frac{1}{\Pr[\text{RDview} = \text{rdv}]}\right) \geq H_\infty(K) - \eta.$$

The second inequality comes from $\Pr[\text{RDview} = \text{rdv}] \geq 2^{-\eta}$, because if rdv is $\eta$-challenging, the adversary cannot make a post-challenge decryption query. Therefore, for $C \in \mathcal{C}\backslash\mathcal{V}$,

$$\widetilde{H}_\infty(K|\text{RDview}) = -\log\left(\mathop{\mathbf{E}}_{\text{rdv}\leftarrow\text{RDview}}[2^{-H_\infty(K|\text{RDview}=\text{rdv})}]\right) \geq \gamma - \eta,$$

because $\Lambda$ is $\gamma$-entropic. Therefore,

$$\widetilde{H}_\infty(K|(\text{RDview}, \text{Lview}, \pi^*, H^*(K^*))) \geq \gamma - \eta - \lambda - \ell_{\text{lf}} - \ell_m.$$

Since $\text{T}(\text{vk}^*) \neq \text{T}(\text{vk})$,

$$\widetilde{H}_\infty(\pi|(\text{RDview}, \text{Lview}, \pi^*, H^*(K^*))) = \widetilde{H}_\infty(K|(\text{RDview}, \text{Lview}, \pi^*, H^*(K^*))),$$

where $\pi = A_{\text{T}(\text{vk}^*)}(\text{T}(\text{vk}), K)$ (injective). This means that RKDec accepts CT with $C \in \mathcal{C}\backslash\mathcal{V}$ only with probability $2^{-(\gamma-\eta-\lambda-\ell_{\text{lf}}-\ell_m)}$. Assuming that $A$ submits $Q$ queries to RKDec in total, the probability that RKDec accepts at least one CT with $C \in \mathcal{C}\backslash\mathcal{V}$ is bounded by $Q \cdot 2^{-(\gamma-\eta-\lambda-\ell_{\text{lf}}-\ell_m)}$. Hence, we have

$$\Pr[S_6] - \Pr[S_7] \leq Q \cdot 2^{-(\gamma-\eta-\lambda-\ell_{\text{lf}}-\ell_m)}.$$

– **Game 8:** This is the last game we make. This game is identical to the previous game except that we replace $H^*(K^*)$ with a uniformly random string from $\{0,1\}^{\ell_m}$. Then it is clear that $\Pr[S_7] = \frac{1}{2}$ because the view of $A$ is independent of $\beta^*$. We now show that the advantages in **Game 7** and **Game 8** are statistically close. Let Reject be the event that $\mathbf{D}(\phi(SK), \text{CT}) = \perp$ in the pre-challenge stage. We note that $\Pr[\text{Reject}] > 2^{-\eta}$, due to **Game 5**. In this game, by definition, all post-challenge queries of "invalid" ciphertexts are rejected. So, the average min-entropy of $K^*$ even after all post-challenge queries are made is equivalent to the average min-entropy of $K^*$ conditioned on the possible events that appear in the pre-challenge stage. That is,

$$\widetilde{H}_\infty(K^*|(\text{RDview}, \text{Reject}, \text{Lview}, \pi^*)) \geq \widetilde{H}_\infty(K^*|\text{RDview}, \text{Reject}) - \lambda - \ell_{\text{lf}}$$
$$\geq \gamma - 2\eta - \lambda - \ell_{\text{lf}}.$$

Remember that $\lambda \leq \gamma - 2\eta - \ell_{\mathsf{lf}} - \ell_m - \log(1/\epsilon)$ and $H^*$ is independent of the view of the post-challenge decryption. By the generalized left-over hash lemma, $H^*(K^*)$ is $\epsilon$-close to the uniform distribution on $\{0,1\}^{\ell_m}$. We then have $\Pr[S_7] - \Pr[S_8] \leq \epsilon$.

By summing up the above inequalities, we have

$$\Pr[S_0] \leq \frac{1}{2} + \epsilon_{\mathsf{tcr}} + \epsilon_{\mathsf{otsig}} + 2\epsilon_{\mathsf{lossy}} + 2\epsilon_{\mathsf{SD}} + 2^{-\eta+1} + Q \cdot 2^{-(\gamma-\eta-\lambda-\ell_{\mathsf{lf}}-\ell_m)} + \epsilon,$$

and conclude the proof of the theorem, with $\mathsf{Adv}^{\mathsf{ctbl\text{-}cca}}_{\Pi,A,(\Phi_{\mathsf{all}},\{i\partial\},\lambda)}(\kappa) = 2\Pr[S_0] - 1.$ ∎

**An Instantiation of CTBL-CCA Secure PKE with $1 - o(1)$ Leakage Rate.** We remark that even if we start with a hash proof system resilient to $1 - o(1)$ leakage rate, we cannot obtain a CTBL-CCA secure PKE scheme with $1 - o(1)$ leakage rate in general. To obtain an optimal leakage rate, we require $\frac{\gamma}{|SK|} = 1 - o(1)$ for a $\gamma$-entropic hash proof system. The cryptosystems of Boneh et al. [9] and Naor-Segev [29] do not satisfy the condition, although they are IND-CPA secure resilient to $1 - o(1)$ leakage rate.

Let $n = pq$ be a composite number of distinct odd primes, $p$ and $q$, and $1 \leq d < p, q$ be a positive integer. It is known that $\mathbb{Z}^{\times}_{n^{d+1}} \cong \mathbb{Z}_{n^d} \times (\mathbb{Z}/n\mathbb{Z})^{\times}$ and any element in $\mathbb{Z}^{\times}_{n^{d+1}}$ is uniquely represented as $(1+n)^{\delta}\gamma^{n^d} \pmod{n^{d+1}}$ for some $\delta \in \mathbb{Z}_{n^d}$ and $\gamma \in (\mathbb{Z}/n\mathbb{Z})^{\times}$. For $\delta \in \mathbb{Z}_{n^d}$, we write $\mathbf{E}^{\mathsf{dj}}(\delta)$ to denote a subset in $\mathbb{Z}^{\times}_{n^{d+1}}$ such that $\mathbf{E}^{\mathsf{dj}}(\delta) = \{(1+n)^{\delta}\gamma^{n^d} \mid \gamma \in (\mathbb{Z}/n\mathbb{Z})^{\times}\}$. It is well known that for any two distinct $\delta, \delta' \in \mathbb{Z}_{n^d}$, it is computationally hard to distinguish a random element in $\mathbf{E}^{\mathsf{dj}}(\delta)$ from a random element in $\mathbf{E}^{\mathsf{dj}}(\delta')$ as long as the decision computational residue (DCR) assumption holds true. Let $\mathcal{C} = \mathbb{Z}^{\times}_{n^{d+1}}$ and $\mathcal{V} = \mathbf{E}^{\mathsf{dj}}(0)$. Let $\mathcal{SK} = \{0,1,\dots,n^{d+1}\} \subset \mathbb{Z}$. Let $g \in \mathcal{V}$ and $\mathcal{PK} = \{\mu(sk) \mid \mu(sk) = g^{sk} \pmod{n^{d+1}}$ where $sk \in \mathcal{SK}\}$ $(= \mathbf{E}^{\mathsf{dj}}(0))$. For $C \in \mathcal{C}$, define $\Lambda_{sk}(C) = C^{sk}$ $\pmod{n^{d+1}}$. Then, $\Lambda : \mathcal{SK} \times \mathcal{C} \to \mathcal{V}$ is projective and $d\log(n)$-entropic and a hash proof system HPS is constructed on $\Lambda$. In addition, $\frac{\text{leakge bound}}{\text{the length of secret-key}} = \frac{d\log(n)-\omega(\log(\kappa))}{(d+1)\log(n)} = 1 - o(1)$.

**Corollary 1.** *By applying the DCR-based hash proof system above and the DCR based instantiation of ABO injective function in Appendix B to the PKE scheme in Fig. 2, it becomes a CTBL-CCA secure PKE scheme with $1 - o(1)$ bounded memory leakage rate under the DCR assumption.*

# 5    Continuous Tampering and Leakage Resilient CCA (CTL-CCA) Secure Public-Key Encryption

We say that PKE has a **key-update** mechanism *if there is a PPT algorithm* Update *that takes* $\rho$ *and* $sk$ *and returns an "updated" secret key* $sk' =$ Update$_{\rho}(sk)$. We assume that the key-updating mechanism Update can be activated only when the decryption algorithm rejects a ciphertext. Therefore, one cannot update his secret key unless the decryption algorithm has detected tampering. We require for $\Pi = (\mathsf{Setup}, \mathsf{Update}, \mathbf{K}, \mathbf{E}, \mathbf{D})$ that for every sufficiently large $\kappa \in \mathbb{N}$ and ever $I \in \mathbb{N}$, it always holds that $\mathbf{D}_\rho(sk_i, \mathbf{E}_\rho(pk, m)) = m$, for every $\rho \in \mathsf{Setup}(1^\kappa)$, every $(pk, sk_0) \in \mathbf{K}(\rho)$, and every $sk_i \in \mathsf{Update}_\rho(sk_{i-1})$ for $i \in [I]$, and every $m \in \mathcal{M}$.

**CTL-CCA Security.** For PKE with a key-update mechanism $\Pi' = (\mathsf{Setup}, \mathsf{Update}, \mathbf{K}, \mathbf{E}, \mathbf{D})$ and an adversary $A = (A_1, A_2)$, we define the experiment $\mathsf{Expt}^{\mathsf{ctl\text{-}cca}}_{\Pi, A, (\Phi_1, \Phi_2, \lambda)}(\kappa)$ as in Fig. 3. $A$ may adaptively submit (unbounded) polynomially many queries $(\phi, ct)$ to oracle RKDec, but it should be $\phi \in \Phi_i$ appropriately. We remark that secret key $sk$ is updated using (non-tampered) flesh randomness only when the decryption algorithm rejects a ciphertext. $A$ may also adaptively submit (unbounded) polynomially many queries $L$ to oracle Leak, before seeing the challenge ciphertext $ct^*$. The total amount of leakage on $sk$ must be bounded by some $\lambda$ bit length within each one period between the key-updating mechanism are activated. We define the advantage of $A$ against $\Pi'$ with respects to $(\Phi_1, \Phi_2)$ as

$$\mathsf{Adv}^{\mathsf{ctl\text{-}cca}}_{\Pi, A, (\Phi_1, \Phi_2, \lambda)}(\kappa) \triangleq |\, 2\Pr[\mathsf{Expt}^{\mathsf{ctl\text{-}cca}}_{\Pi, A, (\Phi_1, \Phi_2, \lambda)}(\kappa) = 1] - 1 \,|.$$

We say that $\Pi$ is $(\Phi_1, \Phi_2, \lambda)$-CTL-CCA secure if $\mathsf{Adv}^{\mathsf{ctl\text{-}cca}}_{\Pi, A, (\Phi_1, \Phi_2, \lambda)}(\kappa) = \mathsf{negl}(\kappa)$ for every PPT $A$.

---

$\mathsf{Expt}^{\mathsf{ctl\text{-}cca}}_{\Pi, A, (\Phi_1, \Phi_2, \lambda)}(\kappa):$
  $\rho \leftarrow \mathsf{Setup}(1^\kappa);$
  $(pk, sk) \leftarrow \mathbf{K}(\rho);$
  LEAKSUM $:= 0;$
  $(m_0, m_1, st) \leftarrow A_1^{\mathsf{RKDec}_{\Phi_1}(\cdot, \cdot), \mathsf{Leak}_\lambda(\cdot)}(\rho, pk)$
    such that $|m_0| = |m_1|;$
  $\beta^* \leftarrow \{0, 1\};$
  $ct^* \leftarrow \mathbf{E}_\rho(pk, m_{\beta^*});$
  $\beta \leftarrow A_2^{\mathsf{RKDec}_{\Phi_2}(\cdot, \cdot)}(st, ct^*);$
  If $\beta = \beta^*,$
    then return 1; otherwise 0.

$\mathsf{RKDec}_\Phi(\phi, ct):$
  If $ct = ct^*$ queried by $A_2,$
    then return $\perp;$
  Return $\mathbf{D}_\rho(\phi(sk), ct).$
  If $\mathbf{D}_\rho(\phi(sk), ct) = \perp,$
    Set $sk \leftarrow \mathsf{Update}_\rho(sk),$
    Set LEAKSUM $:= 0.$
  Else do nothing.

$\mathsf{Leak}_\lambda(L):$
  If LEAKSUM
    $:=$ LEAKSUM $+ |L(sk)| > \lambda.$
    then return $\perp;$
  Else return $L(sk).$

**Fig. 3.** The experiment of the CTL-CCA game.

---

We say that $\Pi$ is simply CTL-CCA secure if it is $(\Phi_{\mathsf{all}}, \{\mathsf{id}\}, \lambda)$-CTL-CCA secure, where $\Phi_{\mathsf{all}}$ denotes the class of all efficiently computable functions and $\mathsf{id}$ denotes the identity function.

*Remark 2.* This security definition models **non-persistent** tampering. However, it is obvious that the persistent tampering version of CTL-CCA security can be similarly defined.

# 6   Random Subspace Lemmas

The following random subspace lemma is provided by Agrawal et al. [2], but we improve the bound using the analysis in Lemma A.1 given by Brakerski et al. [10].

**Lemma 5.** *Let* $2 \leq d < t \leq n$ *and* $\lambda < (d-1)\log(q)$. *Let* $\mathcal{W} \subset \mathbb{F}_q^n$ *be an arbitrary vector subspace in* $\mathbb{F}_q^n$ *of dimension* $t$. *Let* $L : \{0,1\}^* \rightarrow \{0,1\}^\lambda$ *be an arbitrary function. Then, we have*

$$\mathsf{Dist}\bigg( \Big( \mathbf{A}, L(\mathbf{A}v) \Big), \Big( \mathbf{A}, L(u) \Big) \bigg) \leq \sqrt{\frac{2^\lambda}{q^{d-1}}},$$

*where* $\mathbf{A} := (a_1, \ldots, a_d) \leftarrow \mathcal{W}^d$ *(seen as a* $n \times d$ *matrix),* $v \leftarrow \mathbb{F}_q^d$, *and* $u \leftarrow \mathcal{W}$.

If $\mathbf{A} \leftarrow \mathbb{F}_q^{n \times d}$ and $u \leftarrow \mathbb{F}_q$, then it is equivalent to Lemma A.1 given by Brakerski et al. [10]. The proof is given in the full version.

The following is an affine version of Lemma 5.

**Lemma 6.** *Let* $2 \leq d < t \leq n$ *and* $\lambda < (d-1)\log(q)$. *Let* $x \in \mathbb{F}_q^n$ *be an arbitrary vector. Let* $\mathcal{W} \subset \mathbb{F}_q^n$ *be an arbitrary vector subspace in* $\mathbb{F}_q^n$ *of dimension* $t$. *Let* $L : \{0,1\}^* \rightarrow \{0,1\}^\lambda$ *be an arbitrary function. Then, we have*

$$\mathsf{Dist}\bigg( \Big( \mathbf{A}, L(x + \mathbf{A}v) \Big), \Big( \mathbf{A}, L(x + u) \Big) \bigg) \leq \sqrt{\frac{2^\lambda}{q^{d-1}}},$$

*where* $\mathbf{A} := (a_1, \ldots, a_d) \leftarrow \mathcal{W}^d$ *(seen as a* $n \times d$ *matrix),* $v \leftarrow \mathbb{F}_q^d$, *and* $u \leftarrow \mathcal{W}$.

*Proof.* Let $\mathbf{W} \in \mathbb{F}_q^{n \times t}$ be a matrix whose column vectors span $\mathcal{W}$, i.e., $\mathcal{W} = \mathsf{span}(\mathbf{W})$. Now, we have

$$\mathsf{Dist}\bigg( \Big( \mathbf{A}, L(x + \mathbf{A}v) \Big), \Big( \mathbf{A}, L(x + u) \Big) \bigg)$$

$$= \mathsf{Dist}\bigg( \Big( \mathbf{W}\mathbf{R_a}, L(x + \mathbf{W}\mathbf{R_a}v) \Big), \Big( \mathbf{W}r_a, L(x + \mathbf{W}r_u) \Big) \bigg) \quad \text{(where } \mathbf{A} = \mathbf{W}\mathbf{R_a} \; u = \mathbf{W}r_u)$$

$$= \mathsf{Dist}\bigg( \Big( \mathbf{W}\mathbf{R_a}, L'(\mathbf{R_a}v) \Big), \Big( \mathbf{W}\mathbf{R_a}, L'(r_u) \Big) \bigg) \quad \text{(where } L'(y) := L(x + \mathbf{W}y))$$

$$\leq \mathsf{Dist}\bigg( \Big( \mathbf{R_a}, L'(\mathbf{R_a}v) \Big), \Big( \mathbf{R_a}, L'(r_u) \Big) \bigg) \leq \sqrt{\frac{2^\lambda}{q^{d-1}}},$$

where $\mathbf{R_a} \leftarrow \mathbb{F}_q^{t \times d}$, $v \leftarrow \mathbb{F}_q^d$, and $r_u \leftarrow \mathbb{F}_q^t$.

We further provide the following lemma.

**Lemma 7.** *Let* $2 \leq d \leq t' < t \leq n$ *and* $\lambda < (d-1)\log(q)$. *Let* $\mathcal{W} \subset \mathbb{F}_q^n$ *be an arbitrary vector subspace in* $\mathbb{F}_q^n$ *of dimension* $t$. *Let* $L : \{0,1\}^* \rightarrow \{0,1\}^\lambda$ *be an arbitrary function. Then, we have*

$$\mathsf{Dist}\bigg( \Big( \mathbf{A}, L(\mathbf{A}v) \Big), \Big( \mathbf{A}, L(u) \Big) \bigg) \leq \sqrt{\frac{2^\lambda}{q^{d-1}}} + \sqrt{\frac{2^\lambda}{q^{t'-1}}},$$

*where $\mathcal{W}'$ is a random vector subspace in $\mathcal{W}$ of dimension $t'$ (independent of function $L$), $\mathbf{A} := (a_1, \ldots, a_d) \leftarrow \mathcal{W}'^d$ (seen as a $n \times d$ matrix), $v \leftarrow \mathbb{F}_q^d$, and $u \leftarrow \mathcal{W}$.*

*Proof.* Let $\mathbf{W} \in \mathbb{F}_q^{n \times t}$ be a matrix whose column vectors span $\mathcal{W}$, i.e., $\mathcal{W} = \mathrm{span}(\mathbf{W})$. Similarly, let $\mathbf{W}' \in \mathbb{F}_q^{n \times t'}$ be a matrix whose column vectors span $\mathcal{W}'$, i.e., $\mathcal{W}' = \mathrm{span}(\mathbf{W}')$. Then, we have

$$\mathrm{Dist}\Big( \big(\mathbf{A}, L(\mathbf{A}v)\big), \big(\mathbf{A}, L(u)\big) \Big)$$

$$\leq \mathrm{Dist}\Big( \big(\mathbf{A}, L(\mathbf{A}v)\big), \big(\mathbf{A}, L(u')\big) \Big) + \mathrm{Dist}\Big( \big(\mathbf{A}, L(u')\big), \big(\mathbf{A}, L(u)\big) \Big) \quad \text{(where } u' = \mathbf{W}'r'_u\text{)}$$

$$\leq \frac{n}{2}\sqrt{\frac{2^\lambda}{q^{d-1}}} + \mathrm{Dist}\Big( L(u'), L(u) \Big) \quad \text{(where } u = \mathbf{W}r_u\text{)}$$

$$= \sqrt{\frac{2^\lambda}{q^{d-1}}} + \mathrm{Dist}\Big( L'(\mathbf{R}'r'_u), L'(r_u) \Big) \quad \text{(where } \mathbf{W}' = \mathbf{W}\mathbf{R}', L'(y) := L(\mathbf{W}y)\text{)}$$

$$\leq \sqrt{\frac{2^\lambda}{q^{d-1}}} + \sqrt{\frac{2^\lambda}{q^{t'-1}}},$$

where $\mathbf{R}' \leftarrow \mathbb{F}_q^{t \times t'}$, $v \leftarrow \mathbb{F}_q^d$, $r'_u \leftarrow \mathbb{F}_q^{t'}$. and $r_u \leftarrow \mathbb{F}_q^t$. ∎

**Corollary 2.** *Let $2 \leq d \leq t' < t \leq n$ and $\lambda < (d-1)\log(q)$. Let $x \in \mathbb{F}_q^n$ be an arbitrary vector. Let $\mathcal{W} \subset \mathbb{F}_q^n$ be an arbitrary vector subspace in $\mathbb{F}_q^n$ of dimension $t$. Let $L : \{0,1\}^* \to \{0,1\}^\lambda$ be an arbitrary function. Then, we have*

$$\mathrm{Dist}\Big( \big(\mathbf{A}, L(x + \mathbf{A}v)\big), \big(\mathbf{A}, L(x + u)\big) \Big) \leq \sqrt{\frac{2^\lambda}{q^{d-1}}} + \sqrt{\frac{2^\lambda}{q^{t'-1}}},$$

*where $\mathcal{W}'$ is a random vector subspace in $\mathcal{W}$ of dimension $t'$ (independent of function $L$), $\mathbf{A} := (a_1, \ldots, a_d) \leftarrow \mathcal{W}'^d$ (seen as a $n \times d$ matrix), $v \leftarrow \mathbb{F}_q^d$, and $u \leftarrow \mathcal{W}$.*

## 7   The CTL-CCA Secure PKE Scheme

In this section, we present a CTL-CCA-secure PKE scheme. We first provide the intuition behind our construction.

Our starting point is a hash proof system based PKE scheme proposed by Agrawal et al. [2], that is IND-CPA secure resilient to continuous memory leakage in the so-called *Floppy model*, where a decryptor additionally owns secret $\alpha$ to refresh its secret key $sk$ using fresh randomness. The Floppy model assumes secret $\alpha$ is not leaked. The Agrawal et al. scheme is as follows: $pk = (g, g^\alpha, f)$ is a public key and $sk = s$ is the corresponding secret-key such that $f = g^{\langle \alpha, s \rangle}$, where $g$ is a generator of cyclic group $G$ of prime order $q$, $\alpha, s \in (\mathbb{Z}/q\mathbb{Z})^n$. In addition, the decryptor owns $\alpha$ as the key-update key. The encryption of message $m \in G$ under $pk$ is $\mathsf{ct} = (g^c, e) = (g^{r\alpha}, m \cdot f^r)$, while the decryption is computed as $e \cdot (g^{<c,sk>})^{-1}$. The secret key $sk$ is refreshed between each two time periods as $sk := sk + \beta$ where $\beta \leftarrow \ker(\alpha)$ is chosen using secret $\alpha$. Here, $f = g^{<\alpha,s>} = g^{<\alpha,s+\beta>}$, because $<\alpha,\beta> = 0$.

We first convert this scheme to an IND-CPA secure PKE scheme that is resilient to continuous memory leakage in the model of Brakerski et al. [10], where the key-update is executed without additional secret $\alpha$. To do so, we pick up $\ell$ independent vectors, $v_1, \ldots, v_\ell \in \ker(\alpha)$, where $\ell < n - 1 = \dim(\ker(\alpha))$, and publish $\tilde{g}^{\mathbf{V}}$ where $\mathbf{V} = (v_1, \ldots, v_\ell) \in (\mathbb{Z}/q\mathbb{Z})^{n \times \ell}$ is $n \times \ell$ matrix with $v_i$ as $i$-th column. Here we assume asymmetric pairing groups $(e, \mathbb{G}_1, \mathbb{G}_2, \mathbb{G}_T)$ where $g, \tilde{g}$ are generators of $\mathbb{G}_1$ and $\mathbb{G}_2$, respectively. We then set $pk = (g, \tilde{g}, g^\alpha, \tilde{g}^{\mathbf{V}}, Y)$ and $sk = g^s$ such that $Y = e(g, \tilde{g})^{<\alpha,s>}$. Here, the encryption of message $m \in \mathbb{G}_T$ under $pk$ is $\mathsf{ct} = (g^c, e) = (g^{r\alpha}, m \cdot Y^r)$, while the decryption is computed as $e \cdot K^{-1}$, where $K = e(g^c, sk) = e(g, \tilde{g})^{<c,s>}$. The secret key $sk$ is refreshed between each two time periods as $sk := sk \cdot \tilde{g}^\beta$ where $\beta \leftarrow \mathsf{span}(\mathbf{V}) \subset \ker(\alpha)$. We note that random $\tilde{g}^\beta = \tilde{g}^{\mathbf{V}r'}$ can be computed using public $\tilde{g}^{\mathbf{V}}$ with random vector $r' \in \mathbb{F}_q^\ell$. This construction is an IND-CPA secure PKE scheme resilient to continuous memory leakage in the sense of [10] under the extended matrix $d$-linear assumption (on $\mathbb{G}_1$), which is implied by the SXDH assumption. We provide the formal description of the scheme as well as the security proof in Appendix C.

The proposed PKE scheme (as described in Appendix C) is based on a hash proof system where $K = \mathsf{HPS.pub}(Y, g^{r\alpha}, r) = \mathsf{HPS.priv}(g^{r\alpha}, sk) = e(g, \tilde{g})^{<\alpha,s>}$. We then filter the hash key $K$ using the one-time lossy filter technique [31] and finally obtain our CTL-CCA secure construction.

We now describe our full-fledged scheme in Fig. 4.

**Asymmetric Pairing.** Let $\mathsf{GroupG}$ be a PPT algorithm that on input a security parameter $1^\kappa$ outputs a bilinear paring $(\mathbb{G}_1, \mathbb{G}_2, \mathbb{G}_T, e, q, g, \tilde{g})$ such that; $\mathbb{G}_1, \mathbb{G}_2$, and $\mathbb{G}_T$ are cyclic groups of prime order $q$, $g, \tilde{g}$ are generators of $\mathbb{G}_1$ and $\mathbb{G}_2$, respectively, and a map $e : \mathbb{G}_1 \times \mathbb{G}_2 \to \mathbb{G}_T$ satisfies the following properties:

- (Bilinear:) for any $g \in \mathbb{G}_1$, $h \in \mathbb{G}_2$, and any $a, b \in \mathbb{Z}_q$, $e(g^a, h^b) = e(g, h)^{ab}$,
- (Non-degenerate:) $e(g, \tilde{g})$ has order $q$ in $\mathbb{G}_T$, and
- (Efficiently computable:) $e(\cdot, \cdot)$ is efficiently computable.

**Symmetric External Diffie-Hellman (SXDH) Assumption.** The symmetric external DH assumption (SXDH) (on $\mathsf{GroupG}$) is that the DDH problem is hard in both groups, $\mathbb{G}_1$ and $\mathbb{G}_2$. The assumption implies that there is no efficiently computable mapping between $\mathbb{G}_1$ and $\mathbb{G}_2$.

We now present our CTL-CCA secure PKE scheme in Fig. 4.

**Theorem 3.** *The PKE scheme in Fig. 4 is* $(\Phi_{\mathsf{all}}, \{\mathsf{id}\}, \lambda)$-*CTL-CCA secure, as long as* $\lambda(\kappa) < \log(q) - \ell_{\mathsf{lf}} - \ell_m - \eta - \omega(\log \kappa)$ *with* $\eta(\kappa) = \omega(\log \kappa)$, *and for any PPT adversary A with at most Q queries to* $\mathsf{RKDec}$ *oracle,*

---

**Set-Up Algorithm** Setup($1^\kappa$):

$(\mathbb{G}_1,\mathbb{G}_2,\mathbb{G}_T,e,q,g,\tilde{g}) \leftarrow$ GroupG.     $\alpha = (\alpha_1,\ldots,\alpha_n) \leftarrow (\mathbb{Z}/q\mathbb{Z})^n$.

$\mathbf{V} = (v_1,\ldots,v_\ell) \leftarrow \big(\mathrm{Ker}(\alpha)\big)^\ell$, where $\mathbf{V} \in (\mathbb{Z}/q\mathbb{Z})^{n \times \ell}$ and $\ell \leq n-2$.

$g^\alpha := (g_1,\ldots,g_n) = (g^{\alpha_1},\ldots,g^{\alpha_n})$.   $\tilde{g}^\mathbf{V} := (\tilde{g}^{v_1},\ldots,\tilde{g}^{v_\ell})$ where $v_i \in (\mathbb{Z}/q\mathbb{Z})^n$.

T $\leftarrow$ TCH where T $: \{0,1\}^* \to B_\kappa$.   Set $b^* = 0$ as the lossy branch.

$\iota_{\mathrm{abo}} \leftarrow$ ABO.gen($1^\kappa,b^*$).   $A(\cdot,\cdot) :=$ ABO.eval($\iota_{\mathrm{abo}},\cdot,\cdot$).

Return $\rho = (g,\tilde{g},g^\alpha,\tilde{g}^\mathbf{V},\mathsf{T},A(\cdot,\cdot))$.

---

**Key Generation Algorithm** K($\rho$):

$s = (s_1,\ldots,s_n) \leftarrow (\mathbb{Z}/q\mathbb{Z})^n$.

$\tilde{g}^s = (\tilde{g}^{s_1},\ldots,\tilde{g}^{s_n})$.

$Y = e(g^\alpha,\tilde{g}^s) = e(g,\tilde{g})^{\langle \alpha,s \rangle}$.

Set $pk := Y$ and $sk := \tilde{g}^s$.

Return $(pk,sk)$.

**Key Updating Algo.** Update($\rho,sk$):

$r' \leftarrow (\mathbb{Z}/q\mathbb{Z})^\ell$,

Let $sk = \tilde{g}^s$.

Set $sk := sk \cdot \tilde{g}^{\mathbf{V}r'} = \tilde{g}^{s+\mathbf{V}r'}$.

$\big($where $\beta := \mathbf{V}r' \in \mathrm{span}(\mathbf{V}).\big)$

Return $sk$.

---

**Encryption Algorithm** E$_\rho$($pk,m$):

To encrypt a message $m \in \mathbb{G}_T$,

$r \leftarrow \mathbb{Z}/q\mathbb{Z}$.   $K = Y^r$.

$(\mathsf{vk},\mathsf{otsk}) \leftarrow$ otKGen($1^\kappa$).

$\pi = A(\mathsf{T}(\mathsf{vk}),K)$.

$C = (g^\alpha)^r$.   $e = m \cdot K$.

$\sigma \leftarrow$ otSign($\mathsf{otsk},C,e,\mathsf{vk},\pi$)).

Return CT $= (C,e,\mathsf{vk},\pi,\sigma)$.

**Decryption Algorithm** D$_\rho$($sk,$CT):

To decrypt a ciphertext ct,

Parse ct into $(g^c,e,\mathsf{vk},\pi,\sigma)$.

If Vrfy($\mathsf{vk},(g^c,e,\mathsf{vk},\pi),\sigma) \neq 1$,

  then aborts.

Else $K = e(g^c,sk) = e(g,\tilde{g})^{r\langle \alpha,s \rangle}$.

If $\pi \neq A(\mathsf{T}(\mathsf{vk}),K)$,

  then aborts.

Else return $m = e \cdot K^{-1}$.

---

**Fig. 4.** Our CTL-CCA secure PKE scheme

$$\mathsf{Adv}^{\mathrm{ctl\text{-}cca}}_{\Pi,A,(\Phi_{\mathrm{all}},\{\mathrm{id}\},\lambda)}(\kappa) \leq$$

$$2\epsilon_{\mathrm{tcr}} + 2\epsilon_{\mathrm{otsig}} + 4\epsilon_{\mathrm{lossy}} + 4\epsilon_{\mathrm{ex}} + 2^{-\eta+2} + Q \cdot 2^{-(\log(q)-\eta-\lambda-\ell_{\mathrm{lf}}-\ell_m-1)}$$

$$+2Q \cdot \sqrt{\frac{2^\lambda}{q^{\ell-1}}} + 2Q \cdot \sqrt{\frac{2^\lambda}{q^{n-1}}} + \sqrt{\frac{2^\lambda}{q^{n-1}}},$$

$\epsilon_{\mathrm{otsig}}$, $\epsilon_{\mathrm{lossy}}$, and $\epsilon_{\mathrm{ex}}$ denote some negligible functions such that $\mathsf{Adv}^{\mathrm{ot}}_{\mathrm{OTSig},B}(\kappa) \leq \epsilon_{\mathrm{otsig}}$, $\mathsf{Adv}^{\mathrm{lossy}}_{\mathrm{ABO},B'}(\kappa) \leq \epsilon_{\mathrm{lossy}}$, and $\mathsf{Adv}^{\mathrm{ex}}_D(\kappa) \leq \epsilon_{\mathrm{ex}}$ for any PPT adversaries, $B$, $B'$ and $D$, respectively.

Due to the space limitation, the proof is given in the full version.

**An Instantiation of CTL-CCA Secure PKE with $\frac{1}{4}$-o(1) Leakage Rate.** We remark that the underlying hash proof system is $\log(q)$-entropic and we have $|sk| = n\log(q)$. By construction, we require $2 \leq \ell < n-1$. Hence, the best parameter for leakage rate is $n = 4$ and $\ell = 2$, where the resulting CTL-CCA secure PKE scheme has $\frac{1}{4} - o(1)$ leakage rate.

# 8 Impossibility of Non-Persistent Tampering Resilient Signatures

We show that there is no secure digital signature scheme resilient to the non-persistent tampering attacks, if it does not have a key-updating mechanism (See for definition Appendix D). This fact does not contradict [26] (in which they claim a tampering resilient digital signature scheme), because the persistent tampering attack is weaker than the non-persistent attack. To prove our claim, we consider the following adversary. The adversary runs the key-generation algorithm, $\mathsf{Gen}$, and obtains two legitimate pairs of verification and signing keys, $(vk_0, sk_0)$ and $(vk_1, sk_1)$. Then, it sets a set of functions $\{\phi^i_{(sk_0, sk_1)}\}$, such that

$$\phi^i_{(sk_0, sk_1)}(sk) = \begin{cases} sk_0 & \text{if the } i\text{-th bit of } sk \text{ is } 0, \\ sk_1 & \text{otherwise.} \end{cases}$$

For $i = 1, \ldots, |sk|$, the adversary submit $(\phi^i_{(sk_0, sk_1)}, m)$ to the signing oracle and receives $\sigma_i$'s. Then the adversary finds bit $b_i$ such that $\mathsf{Vrfy}(vk_{b_i}, m, \sigma_i) = 1$ for all $i$ and retrieves the entire secret key $sk$. This attack is unavoidable because both $sk_0$ and $sk_1$ are real secret keys and the signing algorithm cannot detect the tampering attack and cannot self-destruct.

If the key-updating algorithm is allowed to run only when a tampering is detected (which is the case of our definition), then there is no secure digital signature scheme resilient to the non-persistent tampering attacks, even if it has both self-destructive and key-updating mechanisms (See for definition Appendix D).

# A    Computational Hardness Assumptions

Let $\mathcal{G}$ be a PPT algorithm that takes security parameter $1^\kappa$ and outputs a triplet $\mathbb{G} = (G, q, g)$ where $G$ is a group of prime order $q$ that is generated by $g \in G$.

**$d$-Linear Assumption.** The $d$-linear assumption [24, 29] (where $d \geq 1$), a generalization of the linear assumption [8], states that there is a PPT algorithm $\mathcal{G}$ such that the following two ensembles are computationally indistinguishable,

$$\left\{ \left( \mathbb{G}, g_1, \ldots, g_d, g_{d+1}, g_1^{r_1}, \ldots, g_d^{r_d}, g_{d+1}^{\sum_{i=1}^{d} r_i} \right) \right\}_{\kappa \in \mathbb{N}}$$

$$\stackrel{c}{\approx} \left\{ \left( \mathbb{G}, g_1, \ldots, g_d, g_{d+1}, g_1^{r_1}, \ldots, g_d^{r_d}, g_{d+1}^{r_{d+1}} \right) \right\}_{\kappa \in \mathbb{N}}$$

where $\mathbb{G} \leftarrow \mathcal{G}(1^\kappa)$, and the elements $g_1, \ldots, g_{d+1} \in G$ and $r_1, \ldots, r_{d+1} \in \mathbb{Z}/q\mathbb{Z}$ are chosen independently and uniformly at random. The DDH assumption (on $\mathcal{G}$) is equivalent to 1-linear assumption (on $\mathcal{G}$) and these assumptions are progressively weaker: For every $d \geq 1$, the $(d+1)$-linear assumption is weaker than the $d$-linear assumption.

**Matrix $d$-Linear Assumption.** We denote by $\mathrm{Rk}_i(\mathbb{F}_q^{m \times n})$ the set of all $m \times n$ matrices over $\mathbb{F}_q$ with rank $i$. The matrix $d$-linear assumption [29] states that there is a PPT algorithm $\mathcal{G}$ such that, for any integers, $m$ and $n$, and for any $d \leq i \leq j \leq \min(m, n)$, the following two ensembles are computationally indistinguishable,

$$\left\{ (\mathbb{G}, g, g^{\mathbf{x}}) \mid \ \mathbb{G} \leftarrow \mathcal{G}(1^\kappa); \ \mathbf{x} \leftarrow \mathrm{Rk}_i(\mathbb{F}_q^{m \times n}) \right\}_{\kappa \in \mathbb{N}}$$

$$\overset{c}{\approx} \left\{ (\mathbb{G}, g, g^{\mathbf{x}}) \mid \ \mathbb{G} \leftarrow \mathcal{G}(1^\kappa); \ \mathbf{x} \leftarrow \mathrm{Rk}_j(\mathbb{F}_q^{m \times n}) \right\}_{\kappa \in \mathbb{N}}.$$

It is known that breaking the matrix $d$-Linear assumption implies breaking the $d$-Linear assumption (on the same $\mathcal{G}$). The following statement holds.

**Lemma 8 ([29]).** *Breaking the matrix $d$-Linear assumption is at least as hard as breaking the $d$-Linear assumption (on the same $\mathcal{G}$).*

**Extended Matrix $d$-Linear Assumption.** We state a stronger version of the matrix $d$-linear assumption, called the extended matrix $d$-linear assumption [2]. For matrix $\mathbf{x} \in \mathbb{F}_q^{n \times m}$, we write $\ker(\mathbf{x})$ to denote the left kernel of $\mathbf{x}$, i.e.,

$$\ker(\mathbf{x}) = \{ v \in \mathbb{F}_q^n \mid v^T \mathbf{x} = 0 \in \mathbb{F}_q^{1 \times m} \}.$$

Here $\ker(\mathbf{x})$ is a subspace in $\mathbb{F}_q^n$ of dimension $(n - \mathrm{rank}(\mathbf{x}))$. The matrix $d$-linear assumption means that it is infeasible to distinguish $g^{\mathbf{x_i}}$ from $g^{\mathbf{x_j}}$, where rank-$i$ matrix $\mathbf{x_i}$ and rank-$j$ matrix $\mathbf{x_i}$ are chosen independently and uniformly for any $d \leq i < j \leq \min(n, m)$. Since $\dim(\ker(\mathbf{x_i})) = n-i$ and $\dim(\ker(\mathbf{x_j})) = n-j$ (with $n - j < n - i$), the matrix $d$-linear assumption does not hold if an adversary additionally receive $n - i$ independent vectors orthogonal to $\mathbf{x}$. However, one cannot yet distinguish them even if $n - j$ independent vectors orthogonal to $\mathbf{x}$ are given, as long as the matrix $d$-linear assumption holds true. The extended matrix $d$-linear assumption [2] states that there is a PPT algorithm $\mathcal{G}$ such that, for any integers, $m$ and $n$, for any $d \leq i \leq j \leq \min(m, n)$, and for any $\ell \leq n - j$, the following two ensembles are computationally indistinguishable,

$$\left\{ (\mathbb{G}, g, g^{\mathbf{x}}, v_1, \ldots, v_\ell) \mid \mathbb{G} \leftarrow \mathcal{G}(1^\kappa); \ \mathbf{x} \leftarrow \mathrm{Rk}_i(\mathbb{F}_q^{m \times n}); v_1, \ldots, v_\ell \leftarrow \ker(\mathbf{x}) \right\}_{\kappa \in \mathbb{N}}$$

$$\overset{c}{\approx} \left\{ (\mathbb{G}, g, g^{\mathbf{x}}, v_1, \ldots, v_\ell) \mid \mathbb{G} \leftarrow \mathcal{G}(1^\kappa); \ \mathbf{x} \leftarrow \mathrm{Rk}_j(\mathbb{F}_q^{m \times n}); v_1, \ldots, v_\ell \leftarrow \ker(\mathbf{x}) \right\}_{\kappa \in \mathbb{N}}.$$

The following statement holds.

**Lemma 9 ([2,10]).** *Breaking the extended matrix $d$-Linear assumption is at least as hard as breaking the $d$-Linear assumption (on the same $\mathcal{G}$).*

The proof is implicitly in [10].

**Decision Computational Residue (DCR) Assumption.** Let $n = pq$ be a composite number of distinct odd primes, $p$ and $q$, and $1 \leq d < p, q$ be a positive integer. We say that the DCR assumption holds if for every PPT $A$, there exists a parameter generation algorithm Gen such that $\mathsf{Adv}_A^{\mathsf{dcr}}(\kappa) =$

$$\Pr[\mathsf{Expt}_A^{\mathsf{dcr}-0}(\kappa) = 1] - \Pr[\mathsf{Expt}_A^{\mathsf{dcr}-1}(\kappa) = 1]$$

is negligible in $\kappa$, where

| $\mathsf{Expt}_A^{\mathsf{dcr}-0}(\kappa)$ : | $\mathsf{Expt}_{d,A}^{\mathsf{dcr}-1}(\kappa)$ : |
|---|---|
| $n \leftarrow \mathsf{Gen}(1^\kappa); R \xleftarrow{\upsilon} \mathbb{Z}_{n^2}^\times$ | $n \leftarrow \mathbb{G}(1^\kappa); R \xleftarrow{\upsilon} \mathbb{Z}_{n^2}^\times$ |
| $c = R^n \bmod n^2$ | $c = (1+n)R^n \bmod n^2$ |
| return $A(n, c)$. | return $A(n, c)$. |

# B   Instantiation of ABO Injective Functions

## B.1   A Matrix Instantiation Based on DDH

Let $\mathcal{G}$ be a PPT algorithm that takes security parameter $1^\kappa$ and outputs a triplet $\mathbb{G} = (G, q, g)$ where $G$ is a group of prime order $q$ that is generated by $g \in G$. Let $\mathcal{B} = \{\mathbb{Z}/q\mathbb{Z}\}$ be a branch collection associated with $\mathbb{G} = (G, q, g)$ generated by $\mathcal{G}$.

- ABO.gen$(1^\kappa, b^*)$ where $b^* \in \mathbb{Z}/q\mathbb{Z}$: Pick up a random column vector $\boldsymbol{u} = (u_i) \in G^\mu$ and a random column vector $\boldsymbol{v} = (v_j) \in G^\mu$. Compute matrix $\mathbf{A} = (A_{i,j}) \in G^{\mu \times \mu}$ as

$$\mathbf{A} = (\boldsymbol{u} \cdot \boldsymbol{v}^T) \boxplus g^{-(b^*)\mathbf{I}_\mu} = \left(u_i v_j g^{-(b^*)\delta_{i,j}}\right) \in G^{\mu \times \mu}$$

where $\boxplus$ denotes the componet-wise product of matrices over $G$, $\mathbf{I}_\mu \in (\mathbb{Z}/q\mathbb{Z})^{\mu \times \mu}$ is the identity matrix and $\delta_{i,j}$ is Kronecker's delta, i.e., $\delta_{i,j} = 1$ if $i = j$ and 0 otherwise. We note that $\mathsf{rank}(\boldsymbol{u} \cdot \boldsymbol{v}^T) = 1$ and, at least with probability $1 - \frac{2\mu}{q}$, $\mathsf{rank}(A) = \mu$. We let $A(b)$ to denote

$$A(b) := A \boxplus g^{bI_\mu} = \left(u_i v_j g^{(b-b^*)\delta_{i,j}}\right) \in G^{\mu \times \mu}.$$

Finally, output $\iota_{\mathsf{abo}} = A(\cdot)$.
- ABO.eval$(\iota_{\mathsf{abo}}, b, x)$: On input matrix $X \in (\mathbb{Z}/q\mathbb{Z})^{\mu \times d}$, output

$$\mathsf{ABO.eval}(\iota_{\mathsf{abo}}, b, x) = A(b) \cdot X \in G^{\mu \times d}.$$

This implementation realizes a collection of $(\mu \cdot d \log(q), (\mu - 1)d \log(q))$-all-but-one injective functions (under the DDH assumption).

## B.2   DCR Based Instantiation

Let $n = pq$ be a composite number of distinct odd primes, $p$ and $q$, and $1 \leq d < p, q$ be a positive integer. It is known that $\mathbb{Z}_{n^{d+1}}^{\times} \cong \mathbb{Z}_{n^d} \times (\mathbb{Z}/n\mathbb{Z})^{\times}$ and any element in $\mathbb{Z}_{n^{d+1}}^{\times}$ is uniquely represented as $(1+n)^{\delta} \gamma^{n^d}$ (mod $n^{d+1}$) for some $\delta \in \mathbb{Z}_{n^d}$ and $\gamma \in (\mathbb{Z}/n\mathbb{Z})^{\times}$. For $\delta \in \mathbb{Z}_{n^d}$, we write $\mathbf{E}^{\mathrm{dj}}(\delta)$ to denote a subset in $\mathbb{Z}_{n^{d+1}}^{\times}$ such that $\mathbf{E}^{\mathrm{dj}}(\delta) = \{(1+n)^{\delta} \gamma^{n^d} \mid \gamma \in (\mathbb{Z}/n\mathbb{Z})^{\times}\}$. It is known that for any two distinct $\delta, \delta' \in \mathbb{Z}_{n^d}$, it is computationally hard to distinguish a random element in $\mathbf{E}^{\mathrm{dj}}(\delta)$ from a random element in $\mathbf{E}^{\mathrm{dj}}(\delta')$ as long as the decision computational residue (DCR) assumption holds true.

- ABO.gen$(1^{\kappa}, b^*)$ where $b^* \in \{0,1\}^{d\kappa}$: Pick up $\kappa/2$-bit distinct odd primes $p, q$ and compute $n = pq$. Then choose $\iota_{\mathsf{abo}} \leftarrow \mathbf{E}^{\mathrm{dj}}(-b^*)$. Output $\iota_{\mathsf{abo}}$.
- ABO.eval$(\iota_{\mathsf{abo}}, b, x)$: On input matrix $x \in \mathbb{Z}_{n^d}$, output

$$\mathsf{ABO.eval}(\iota_{\mathsf{abo}}, b, x) = \left(\iota_{\mathsf{abo}} \cdot (1+n)^b\right)^x (\in \mathbf{E}^{\mathrm{dj}}(b - b^*)^x).$$

This implementation realizes a collection of $(d\log(n), \log((p-1)(q-1)))$-all-but-one injective functions (under the DCR assumption).

# C   The Continuous Leakage Resileint CPA PKE Scheme

We propose an IND-CPA secure PKE scheme resilient to continuous memory leakage, based on Agrawal et al. scheme [2].

- The Key Generation Algorithm: Choose $(\mathbb{G}_1, \mathbb{G}_2, \mathbb{G}_T, e, q, g, \tilde{g}) \leftarrow$ GroupG. Pick up a random column vector $\boldsymbol{\alpha} \leftarrow (\mathbb{Z}/q\mathbb{Z})^n$. Pick up $\ell$ independent column vectors, $\boldsymbol{v_1}, \ldots, \boldsymbol{v_\ell}$, in $(\mathbb{Z}/q\mathbb{Z})^n$ uniformly from $\mathsf{Ker}(\boldsymbol{\alpha})$ where $2 \leq \ell \leq n - 2$. Set $n \times \ell$ matrix $\mathbf{V} = (\boldsymbol{v_1}, \ldots, \boldsymbol{v_\ell})$. Set $g^{\boldsymbol{\alpha}} := (g^{\alpha_1}, \ldots, g^{\alpha_n})^T$. Set $\tilde{g}^{\mathbf{V}} := (\tilde{g}^{\boldsymbol{v_1}}, \ldots, \tilde{g}^{\boldsymbol{v_\ell}})$. Pick up a random column vector $\boldsymbol{s} \leftarrow (\mathbb{Z}/q\mathbb{Z})^n$. Compute $\tilde{g}^{\boldsymbol{s}} = (\tilde{g}^{s_1}, \ldots, \tilde{g}^{s_n})^T$. Compute $Y = e(g^{\boldsymbol{\alpha}}, \tilde{g}^{\boldsymbol{s}}) = e(g, \tilde{g})^{\langle \boldsymbol{\alpha}, \boldsymbol{s} \rangle}$. Set $pk := (g, \tilde{g}, g^{\boldsymbol{\alpha}}, \tilde{g}^{\mathbf{V}}, Y)$ and $sk := \tilde{g}^{\boldsymbol{s}}$. Output $(pk, sk)$.
- The Key Update Algorithm: Take $(pk, sk)$ as input. Choose a random column vector $\boldsymbol{r'} \leftarrow (\mathbb{Z}/q\mathbb{Z})^{\ell}$ and compute $\tilde{g}^{\boldsymbol{\beta}} = \tilde{g}^{\mathbf{V}\boldsymbol{r'}}$. Update $sk := sk \cdot \tilde{g}^{\boldsymbol{\beta}} = \tilde{g}^{\boldsymbol{s} + \boldsymbol{\beta}}$. Note that $\boldsymbol{\beta} \in \mathsf{span}(\mathbf{V}) \subset \ker(\boldsymbol{\alpha})$. Output $sk$.
- The Encryption Algorithm: To encrypt $m \in \mathbb{G}_T$ under $pk$, pick up random $r \leftarrow \mathbb{Z}/q\mathbb{Z}$. Compute $\boldsymbol{C} = g^{r\boldsymbol{\alpha}}$ and $K = Y^r$. Output $\mathsf{CT} = (\boldsymbol{C}, e)$ where $e = m \cdot K$.
- The Decryption algorithm: To decrypt ciphertext $\mathsf{CT} = (g^c, e)$ under $sk$, compute $K = e(g^c, sk) (= e(g, \tilde{g})^{\langle c, s \rangle})$. Output $m = e \cdot K^{-1}$.

We define IND-CPA security of PKE resilient to $\lambda$-continuous memory leakage [10] as $(\emptyset, \emptyset, \lambda)$-CTL-CCA security of PKE.

**Theorem 4.** *The above PKE scheme is $(\emptyset, \emptyset, \lambda)$-CTL-CCA secure, as long as $\lambda(\kappa) < \ell \log(q) - \omega(\log \kappa)$, and for any PPT adversary $A$,*

$$\mathsf{Adv}^{\mathsf{ctl\text{-}cca}}_{\Pi, A, (\emptyset, \emptyset, \lambda)}(\kappa) \leq +4\epsilon_{\mathsf{ex}} + 2Q \cdot \sqrt{\frac{2^\lambda}{q^{\ell-1}}} + 2Q \cdot \sqrt{\frac{2^\lambda}{q^{n-1}}} + \sqrt{\frac{2^\lambda}{q^{n-1}}},$$

*where $Q$ denotes the total number of key-updates in the running time of $A$.*

*Proof.* Here we prove the theorem by using the standard game-hopping strategy. We denote by $S_i$ the event that adversary $A$ wins in **Game** $i$.

- **Game 0**: This game is the original game. We write $\mathsf{CT}^* = (g^{c^*}, e^*)$ where $e^* = m_{b^*} \cdot K^*$ to denote the challenge ciphertext. Let us assume that $Q$ is the maximum number of the key-updates.
  By definition, $\Pr[S_0] = \Pr[b = b^*]$ and $\mathsf{Adv}^{\mathsf{ctl\text{-}cca}}_{\Pi, A, (\emptyset, \emptyset, \lambda)}(\kappa) = |2\Pr[S_0] - 1|$.
- **Game 1**: In this game, we instead produce $\mathsf{CT}^*$ as follows: Compute $K^* = e(g^{c^*}, sk) = e(g, \tilde{g})^{r\langle \boldsymbol{\alpha}, s \rangle}$ and set $e^* = m_{b^*} \cdot K^*$. This change is just conceptual. Then, $\Pr[S_0] = \Pr[S_1]$.
- **Game 2**: This game is identical to **Game 1**, except that we choose $\ell$ independent vectors $\boldsymbol{v}_1, \ldots, \boldsymbol{v}_\ell \leftarrow \ker(\boldsymbol{\alpha}, \boldsymbol{c}^*)$ and set $\mathbf{V} = (\boldsymbol{v}_1, \ldots, \boldsymbol{v}_\ell)$. Since $\boldsymbol{c}^* = r^* \boldsymbol{\alpha}$, $\ker(\boldsymbol{\alpha}, \boldsymbol{c}^*) = \ker(\boldsymbol{\alpha})$. Hence, $\Pr[S_1] = \Pr[S_2]$.
- **Game 3**: This game is identical to **Game 2**, except that when producing $\mathsf{CT}^*$, we instead pick up random vector $\boldsymbol{c}^* \leftarrow \mathbb{F}_q^n$. We note that since $\dim(\ker(\boldsymbol{\alpha}, \boldsymbol{c}^*)) = n - 2 \geq \ell$, we can still choose $\ell$ independent vectors $\boldsymbol{v}_1, \ldots, \boldsymbol{v}_\ell$. The difference between these two games is bounded by the extended matrix $d$-linear assumption.

**Lemma 10.** *Under the extended matrix $d$-linear assumption in Appendix A, we have $\Pr[S_2] - \Pr[S_3] \leq 2\epsilon_{\mathsf{ex}}$.*

*Proof.* Let $\mathbf{x} \in (\mathbb{Z}/q\mathbb{Z})^{n \times 2}$ whose columns are $\boldsymbol{\alpha}$ and $\boldsymbol{c}$, i.e., $\mathbf{x} = (\boldsymbol{\alpha}, \boldsymbol{c})$. Let $\boldsymbol{v}_1, \ldots, \boldsymbol{v}_\ell$ be $\ell$ independent random column vectors chosen via $\boldsymbol{v}_i \leftarrow \ker(\mathbf{x}) = \ker(\boldsymbol{\alpha}, \boldsymbol{c})$ and set $\mathbf{V} = (\boldsymbol{v}_1, \ldots, \boldsymbol{v}_\ell)$. Now given $g^{\mathbf{x}}$ and $\mathbf{V} = (\boldsymbol{v}_1, \ldots, \boldsymbol{v}_\ell)$, we can simulate public and secret keys that the adversary sees during the game, as well as the challenge ciphertext. In the case that $\mathsf{rank}(X) = 1$, we perfectly simulate Game 2. In the case that $\mathsf{rank}(X) = 2$, we perfectly simulate Game 3. Then, we have $\Pr[S_2] - \Pr[S_3] \leq 2\epsilon_{\mathsf{ex}}$. ∎

- **Game 4** is defined as a sequence of $Q + 1$ sub-games denoted by Games, $4.0, \ldots, 4.Q$. For $i = 0, \ldots, Q$, we have
  - **Game 4.$i$**: This game is identical to Game 4.0, except that at the last $i$ key-updates, we instead choose $\boldsymbol{\beta} \leftarrow \ker(\boldsymbol{\alpha})$ and update $sk := sk \cdot \tilde{g}^{\boldsymbol{\beta}}$. We insist that the first $Q - i$ key-updates, $\boldsymbol{\beta}$ is chosen from $\mathsf{span}(\mathbf{V})$, whereas in the last $i$ key-updates, it is chosen from $\ker(\boldsymbol{\alpha})$.
  Game 4.0 is identical to Game 3. The difference between Games, 4.$i$ and 4.$i + 1$, is computationally bounded.

Indeed, by Corollary 2, we have

$$\mathsf{Dist}\Big((\mathbf{V}, L(s + \mathbf{V}r')) : (\mathbf{V}, L(s + \beta))\Big) \leq \sqrt{\frac{2^\lambda}{q^{\ell-1}}} + \sqrt{\frac{2^\lambda}{q^{m-1}}},$$

where $\mathbf{V} \leftarrow \big(\ker(\alpha, c^*)\big)^\ell$, $r' \leftarrow (\mathbb{Z}/q\mathbb{Z})^\ell$, and $\beta \leftarrow \ker(\alpha)$, with $\dim(\ker(\alpha, c^*)) = n - 2$ and $\dim(\ker(\alpha)) = n - 1$. So, we have $\Pr[S_{4.i}] - \Pr[S_{4.i+1}] \leq \sqrt{\frac{2^\lambda}{q^{\ell-1}}} + \sqrt{\frac{2^\lambda}{q^{m-1}}}$, Therefore $\Pr[S_3] - \Pr[S_{4.Q}] \leq Q\sqrt{\frac{2^\lambda}{q^{\ell-1}}} + Q\sqrt{\frac{2^\lambda}{q^{m-1}}}$.

- **Game 5:** This game is identical to **Game**4.$Q$, except that we pick up random $k^* \leftarrow \mathbb{Z}/q\mathbb{Z}$ and compute $K^* = e(g, \tilde{g})^{k^*}$. This $k^*$ is statistically close to $< c^*, s + \beta >$. By Lemma 3,

$$\mathsf{Dist}((c^*, < c^*, s + \beta >, L(s + \beta), \mathsf{view}) : (c^*, k^*, L(s + \beta), \mathsf{view})) \leq \frac{1}{2}2^{-\sqrt{\tilde{\mathsf{H}}_\infty(s+\beta|L(s+\beta),\mathsf{view})}},$$

where view is fixed values containing $\alpha, \mathbf{V}$, and $< \alpha, s >$. Let us repersent $s = s^* + r'\alpha$ such that $s^* \in \ker(\alpha)$ and $r' \in \mathbb{Z}/q\mathbb{Z}$. Since $s^*$ and $\beta$ are only random variables in the above $\tilde{\mathsf{H}}_\infty$, we have

$$\tilde{\mathsf{H}}_\infty(s + \beta|L(s + \beta), \mathsf{view}) = \tilde{\mathsf{H}}_\infty(s^* + \beta|L(s + \beta)) \geq \mathsf{H}_\infty(s^* + \beta) - \lambda = (n - 1)\log(q) - \lambda.$$

Therefore, we have $\Pr[S_{4.Q}] - \Pr[S_5] \leq \frac{1}{2}\sqrt{\frac{2^\lambda}{q^{n-1}}}$. By construction, $\Pr[S_5] = \frac{1}{2}$.

To summarize the above, we have $\Pr[S_0] - \frac{1}{2} =$

$$2\epsilon_{\mathsf{ex}} + Q \cdot \sqrt{\frac{2^\lambda}{q^{\ell-1}}} + Q \cdot \sqrt{\frac{2^\lambda}{q^{n-1}}} + \frac{1}{2}\sqrt{\frac{2^\lambda}{q^{n-1}}}.$$    ∎

# D    Continuos Tampering Secure Signature

A digital signature scheme $\Sigma = (\mathsf{Setup}, \mathsf{KGen}, \mathsf{Sign}, \mathsf{Vrfy})$ consists four algorithms. Setup, the set-up algoritm, takes as input security parameter $1^k$ and outputs public parameter $\rho$. KGen, the key-generation algorithm, takes as input $\rho$ and outputs a pair comprising the verification and signing keys, $(vk, sk)$. Sign, the signing algorithm, takes as input $(\rho, sk)$ and message $m$ and produces signature $\sigma$. Vrfy, the verification algorithm, takes as input verification key $vk$, message $m$ and signature $\sigma$, as well as $\rho$, and outputs a bit. For completeness, it is required that for all $\rho \in \mathsf{Setup}(1^\kappa)$, all $(vk, sk) \in \mathsf{KGen}(\rho)$ and for all $m \in \{0, 1\}^*$, it holds $\mathsf{Vrfy}_\rho(vk, m, \mathsf{Sign}_\rho(sk, m)) = 1$.

We say that digital signature scheme $\Sigma$ is **self-destructive**, if the signing algorithm can erase all inner states including $sk$ and does not work any more, when it can detect tampering. We say that digital signature scheme $\Sigma$ has a **key-updating** mechanism *if there is a PPT algorithm* Update *that takes $\rho$ and sk and returns an "updated" secret key* $sk' = \mathsf{Update}_\rho(sk)$. We assume that the key-updating mechanism Update can be activated only when the signing algorithm detects tampering.

**CTBL-CMA Security.** For digital signature scheme $\Sigma$ and an adversary $A$, we define the experiment $\mathsf{Expt}^{\text{ctbl-cma}}_{\Pi,A,(\Phi,\lambda)}(\kappa)$ as in Fig. 5. We define the advantage of $A$ against $\Pi$ with respects $\Phi$ as

$$\mathsf{Adv}^{\text{ctbl-cma}}_{\Sigma,A,(\Phi,\lambda)}(\kappa) \triangleq \Pr[\mathsf{Expt}^{\text{ctbl-cma}}_{\Sigma,A,(\Phi,\lambda)}(\kappa) = 1].$$

$A$ may adaptively submit (unbounded) polynomially many queries $(\phi, \mathsf{CT})$ to oracle RKSign, but it should be $\phi \in \Phi$. $A$ may also adaptively submit (unbounded) polynomially many queries $L$ to oracle Leak. Finally, $A$ outputs $(m', \sigma')$. We say that $A$ wins if $\mathsf{Vrfy}(\mathsf{vk}, m', \sigma') = 1$ and $m'$ is not asked to RKSign. We note that if Sig has "self-destructive" property, RKSign does not receive any further query from the adversary or simply returns $\perp$. We say that $\Sigma$ is $(\Phi, \lambda)$-CTBL-CMA secure if $\mathsf{Adv}^{\text{tbl-cma}}_{\Sigma,A,(\Phi,\lambda)}(\kappa) = \mathsf{negl}(\kappa)$ for every PPT $A$.

---

$\mathsf{Expt}^{\text{ctbl-cma}}_{\Sigma,A,(\Phi,\lambda)}(\kappa)$:

$\rho \leftarrow \mathsf{Setup}(1^\kappa)$;
$(\mathsf{vk}, \mathsf{sk}) \leftarrow \mathsf{KGen}(\rho)$;
$(m', \sigma') \leftarrow A^{\mathsf{RKSign}_\Phi(\cdot,\cdot),\mathsf{Leak}_\lambda(\cdot)}(\rho, \mathsf{vk})$
If $m' \in \mathsf{List}$ or $\mathsf{Vrfy}_\rho(\mathsf{vk}, m', \sigma') \neq 1$,
    then return 0;
Otherwise 1.

$\mathsf{RKSign}_\Phi(\phi, m)$:
    $\sigma \leftarrow \mathsf{Sign}_\rho(\phi(\mathsf{sk}), m)$;
    If $\sigma = \perp$,
        then erase sk.
    Else return $\sigma$.

---

$\mathsf{Leak}_\lambda(L_i)$: ($L_i$: $i$-th query of $A$.)
    If $\sum_{j=1}^{i} |L_j(\mathsf{sk})| > \lambda$,
        then return $\perp$;
    Else return $L_i(\mathsf{sk})$.

**Fig. 5.** The experiment of the CTBL-CMA game.

**CTL-CMA Security.** For digital signature scheme $\Sigma = (\mathsf{Setup}, \mathsf{KGen}, \mathsf{Update}, \mathsf{Sign}, \mathsf{Vrfy})$ with a key-updating mechanism and an adversary $A$, we define the experiment $\mathsf{Expt}^{\text{ctl-cma}}_{\Sigma,A,(\Phi,\lambda)}(\kappa)$ as in Fig. 6. We define the advantage of $A$ against $\Sigma$ with respects $\Phi$ as

$$\mathsf{Adv}^{\text{ctl-cma}}_{\Sigma,A,(\Phi,\lambda)}(\kappa) \triangleq \Pr[\mathsf{Expt}^{\text{ctl-cma}}_{\Sigma,A,(\Phi,\lambda)}(\kappa) = 1].$$

$A$ may adaptively submit (unbounded) polynomially many queries $(\phi, \mathsf{CT})$ to oracle RKSign, but it should be $\phi \in \Phi$. $A$ may also adaptively submit (unbounded) polynomially many queries $L$ to oracle Leak. Finally, $A$ outputs $(m', \sigma')$. We say that $A$ wins if $\mathsf{Vrfy}(\mathsf{vk}, m', \sigma') = 1$ and $m'$ is not asked to RKSign. We say that $\Sigma$ is $(\Phi, \lambda)$-CTL-CMA secure if $\mathsf{Adv}^{\text{ctl-cma}}_{\Sigma,A,(\Phi,\lambda)}(\kappa) = \mathsf{negl}(\kappa)$ for every PPT $A$.

$$\boxed{\begin{array}{l} \mathsf{Expt}^{\mathsf{ctbl\text{-}cma}}_{\Sigma, A, (\Phi, \lambda)}(\kappa): \\[2ex] \rho \leftarrow \mathsf{Setup}(1^\kappa); \\ (\mathsf{vk}, \mathsf{sk}) \leftarrow \mathsf{KGen}(\rho); \\ (m', \sigma') \leftarrow A^{\mathsf{RKSign}_\Phi(\cdot, \cdot), \mathsf{Leak}_\lambda(\cdot)}(\rho, \mathsf{vk}) \\ \text{If } m' \in \mathsf{List} \text{ or } \mathsf{Vrfy}_\rho(\mathsf{vk}, m', \sigma') \neq 1, \\ \quad \text{then return } 0; \\ \text{Otherwise } 1. \end{array}}$$

$$\boxed{\begin{array}{l} \mathsf{RKSign}_\Phi(\phi, m): \\ \quad \sigma \leftarrow \mathsf{Sign}_\rho(\phi(\mathsf{sk}), m); \\ \quad \text{If } \sigma = \bot, \\ \quad\quad \text{then return } \bot; \\ \quad\quad \mathsf{Set\ sk} \leftarrow \mathsf{Update}_\rho(\mathsf{sk}); \\ \quad\quad \mathsf{Set\ LEAKSUM} := 0; \\ \quad \text{Else return } \sigma. \\ \hline \mathsf{Leak}_\lambda(L): \\ \quad \text{If LEAKSUM} \\ \quad\quad := \mathsf{LEAKSUM} + |L(\mathsf{sk})| > \lambda, \\ \quad\quad \text{then return } \bot; \\ \quad \text{Else return } L(\mathsf{sk}). \end{array}}$$

**Fig. 6.** The experiment of the CTL-CMA game.

# References

1. 51th Annual IEEE Symposium on Foundations of Computer Science, FOCS 2010, IEEE Computer Society (2010)
2. Agrawal, S., Dodis, Y., Vaikuntanathan, V., Wichs, D.: On continual leakage of discrete log representations. In: Sako and Sarkar [36], pp. 401–420
3. Anonymous. A note on the RKA security of continuously non-malleable key-derivation function from PKC 2015. Submitted to PKC 2016
4. Bellare, M., Cash, D.: Pseudorandom functions and permutations provably secure against related-key attacks. In: Rabin, T. (ed.) CRYPTO 2010. LNCS, vol. 6223, pp. 666–684. Springer, Heidelberg (2010). doi:10.1007/978-3-642-14623-7_36
5. Bellare, M., Cash, D., Miller, R.: Cryptography secure against related-key attacks and tampering. In: Lee, D.H., Wang, X. (eds.) ASIACRYPT 2011. LNCS, vol. 7073, pp. 486–503. Springer, Heidelberg (2011). doi:10.1007/978-3-642-25385-0_26
6. Bellare, M., Kohno, T.: A theoretical treatment of related-key attacks: RKA-PRPs, RKA-PRFs, and applications. In: Biham, E. (ed.) EUROCRYPT 2003. LNCS, vol. 2656, pp. 491–506. Springer, Heidelberg (2003). doi:10.1007/3-540-39200-9_31
7. Bellare, M., Paterson, K.G., Thomson, S.: RKA security beyond the linear barrier: IBE, encryption and signatures. In: Wang, X., Sako, K. (eds.) ASIACRYPT 2012. LNCS, vol. 7658, pp. 331–348. Springer, Heidelberg (2012). doi:10.1007/978-3-642-34961-4_21
8. Boneh, D., Boyen, X., Shacham, H.: Short group signatures. In: Franklin, M. (ed.) CRYPTO 2004. LNCS, vol. 3152, pp. 41–55. Springer, Heidelberg (2004). doi:10.1007/978-3-540-28628-8_3
9. Boneh, D., Halevi, S., Hamburg, M., Ostrovsky, R.: Circular-secure encryption from decision Diffie-Hellman. In: Wagner, D. (ed.) CRYPTO 2008. LNCS, vol. 5157, pp. 108–125. Springer, Heidelberg (2008). doi:10.1007/978-3-540-85174-5_7
10. Brakerski, Z., Kalai, Y.T., Katz, J., Vaikuntanathan, V.: Overcoming the hole in the bucket,: Public-key cryptography resilient to continual memory leakage. In: FOCS 2010 [1], pp. 501–510
11. Cramer, R., Dodis, Y., Fehr, S., Padró, C., Wichs, D.: Detection of algebraic manipulation with applications to robust secret sharing and fuzzy extractors. In: Smart, N. (ed.) EUROCRYPT 2008. LNCS, vol. 4965, pp. 471–488. Springer, Heidelberg (2008). doi:10.1007/978-3-540-78967-3_27

12. Cramer, R., Padró, C., Xing, C.: Optimal algebraic manipulation detection codes in the constant-error model. In: Dodis and Nielsen [17], pp. 481–501. http://eprint.iacr.org/2014/116

13. Cramer, R., Shoup, V.: Universal hash proofs and a paradigm for adaptive chosen ciphertext secure public-key encryption. In: Knudsen, L.R. (ed.) EUROCRYPT 2002. LNCS, vol. 2332, pp. 45–64. Springer, Heidelberg (2002). doi:10.1007/3-540-46035-7_4

14. Damgård, I., Faust, S., Mukherjee, P., Venturi, D.: Bounded tamper resilience: how to go beyond the algebraic barrier. In: Sako and Sarkar [36], pp. 140–160. http://eprint.iacr.org/2013/677 and http://eprint.iacr.org/2013/124

15. Dodis, Y., Haralambiev, K., López-Alt, A., Wichs, D.: Cryptography against continuous memory attacks. In: FOCS 2010 [1], pp. 511–520. http://eprint.iacr.org/2010/196

16. Dodis, Y., Haralambiev, K., López-Alt, A., Wichs, D.: Efficient public-key cryptography in the presence of key leakage. In: Abe, M. (ed.) ASIACRYPT 2010. LNCS, vol. 6477, pp. 613–631. Springer, Heidelberg (2010). doi:10.1007/978-3-642-17373-8_35

17. Dodis, Y., Nielsen, J.B. (eds.): TCC 2015. LNCS, vol. 9014. Springer, Heidelberg (2015)

18. Dodis, Y., Ostrovsky, R., Reyzin, L., Smith, A.: Fuzzy extractors: how to generate strong keys from biometrics and other noisy data. SIAM J. Comput. 38(1), 97–139 (2008). Preliminary version in EUROCRYPT 2004

19. Dziembowski, S., Pietrzak, K., Wichs, D.: Non-malleable codes. In: Yao, A.C.C. (ed.) ICS 2010, Beijing, China, Tsinghua University Press, pp. 434–452 (2010). http://eprint.iacr.org/2009/608.D

20. Faonio, A., Venturi, D.: Efficient public-key cryptography with bounded leakage and tamper resilience. IACR Cryptology ePrint Archive 2016, p. 529 (2016)

21. Faust, S., Mukherjee, P., Nielsen, J.B., Venturi, D.: Continuous non-malleable codes. In: Lindell, Y. (ed.) TCC 2014. LNCS, vol. 8349, pp. 465–488. Springer, Heidelberg (2014). doi:10.1007/978-3-642-54242-8_20

22. Fujisaki, E., Xagawa, K.: Efficient RKA-Secure KEM and IBE schemes against invertible functions. In: Lauter, K., Rodríguez-Henríquez, F. (eds.) LATINCRYPT 2015. LNCS, vol. 9230, pp. 3–20. Springer, Heidelberg (2015). doi:10.1007/978-3-319-22174-8_1

23. Gennaro, R., Lysyanskaya, A., Malkin, T., Micali, S., Rabin, T.: Algorithmic tamper-proof (ATP) security: theoretical foundations for security against hardware tampering. In: Naor, M. (ed.) TCC 2004. LNCS, vol. 2951, pp. 258–277. Springer, Heidelberg (2004). doi:10.1007/978-3-540-24638-1_15

24. Hofheinz, D., Kiltz, E.: Secure hybrid encryption from weakened key encapsulation. In: Menezes, A. (ed.) CRYPTO 2007. LNCS, vol. 4622, pp. 553–571. Springer, Heidelberg (2007). doi:10.1007/978-3-540-74143-5_31

25. Jafargholi, Z., and Wichs, D.: Tamper detection and continuous non-malleable codes. In: Dodis and Nielsen [17], pp. 451–480. http://eprint.iacr.org/2014/956

26. Kalai, Y.T., Kanukurthi, B., Sahai, A.: Cryptography with tamperable and leaky memory. In: Rogaway, P. (ed.) CRYPTO 2011. LNCS, vol. 6841, pp. 373–390. Springer, Heidelberg (2011). doi:10.1007/978-3-642-22792-9_21

27. Kiltz, E., Pietrzak, K., Stam, M., Yung, M.: A new randomness extraction paradigm for hybrid encryption. In: Joux, A. (ed.) EUROCRYPT 2009. LNCS, vol. 5479, pp. 590–609. Springer, Heidelberg (2009). doi:10.1007/978-3-642-01001-9_34

28. Liu, F.-H., Lysyanskaya, A.: Tamper and leakage resilience in the split-state model. In: Safavi-Naini, R., Canetti, R. (eds.) CRYPTO 2012. LNCS, vol. 7417, pp. 517–532. Springer, Heidelberg (2012). doi:10.1007/978-3-642-32009-5_30

29. Naor, M., Segev, G.: Public-key cryptosystems resilient to key leakage. In: Halevi, S. (ed.) CRYPTO 2009. LNCS, vol. 5677, pp. 18–35. Springer, Heidelberg (2009). doi:10.1007/978-3-642-03356-8_2

30. Peikert, C., and Waters, B. Lossy trapdoor functions and their applications. In: Ladner, R.E., Dwork, C. (eds.) STOC 2008, pp. 187–196. ACM (2008)

31. Qin, B., Liu, S.: Leakage-resilient chosen-ciphertext secure public-key encryption from hash proof system and one-time lossy filter. In: Sako and Sarkar [36], pp. 381–400

32. Qin, B., Liu, S.: Leakage-flexible CCA-secure public-key encryption: simple construction and free of pairing. In: Krawczyk, H. (ed.) PKC 2014. LNCS, vol. 8383, pp. 19–36. Springer, Heidelberg (2014). doi:10.1007/978-3-642-54631-0_2

33. Qin, B., Liu, S., Yuen, T.H., Deng, R.H., Chen, K.: Continuous non-malleable key derivation and its application to related-key security. In: Katz, J. (ed.) PKC 2015. LNCS, vol. 9020, pp. 557–578. Springer, Heidelberg (2015). doi:10.1007/978-3-662-46447-2_25

34. Qin, B., Liu, S.: Leakage-resilient chosen-ciphertext secure public-key encryption from hash proof system and one-time lossy filter. In: Sako, K., Sarkar, P. (eds.) ASIACRYPT 2013. LNCS, vol. 8270, pp. 381–400. Springer, Heidelberg (2013). doi:10.1007/978-3-642-42045-0_20

35. Wee, H.: Public key encryption against related key attacks. In: Fischlin, M., Buchmann, J., Manulis, M. (eds.) PKC 2012. LNCS, vol. 7293, pp. 262–279. Springer, Heidelberg (2012). doi:10.1007/978-3-642-30057-8_16

# Author Index

Printed in the United States
By Bookmasters